ORGANIC NANOMATERIALS

Synthesis, Characterization, and Device Applications

Edited by

TOMÁS TORRES
Departamento de Química Orgánica
Facultad de Ciencias
Universidad Autónoma de Madrid
Madrid, Spain *and* IMDEA Nanociencia
C/Faraday 9
Ciudad Universitaria de Canto Blanco
Madrid, Spain

GIOVANNI BOTTARI
Departamento de Química Orgánica
Facultad de Ciencias
Universidad Autónoma de Madrid
Madrid, Spain *and* IMDEA Nanociencia
C/Faraday 9
Ciudad Universitaria de Canto Blanco
Madrid, Spain

WILEY

Published by John Wiley & Sons, Inc., Hoboken, New Jersey.
Published simultaneously in Canada.

Library of Congress Cataloging-in-Publication Data

Organic nanomaterials : synthesis, characterization, and device applications / [edited by] Tomas Torres, Giovanni Bottari.
pages cm
Includes index.
ISBN 978-1-118-01601-5 (hardback)
1. Organic compounds–Synthesis. 2. Nanostructured materials. I. Torres, Tomas, editor of compilation. II. Bottari, Giovanni, editor of compilation.
QD262.O64 2013
620.1′17–dc23

2013000335

Printed in the United States of America

10 9 8 7 6 5 4 3 2 1

CONTENTS

PREFACE

In the last decade, much progress has been made in the field of organic nanomaterials. Recent developments in nanoscience and nanotechnology have driven this field forward, thus allowing the preparation of novel materials with controlled morphology and well-defined properties, with clear and exciting technological applications. The new insights into the optoelectronic properties of molecules, together with the recent development of techniques such as scanning probe microscopy, among many others, have pushed chemists to design novel molecular and supramolecular functional architectures. The implications range from the basic molecular self-assembly of complementary organic systems, which constitute an important part of the so-called "bottom-up approach" to exciting new applications of pure organic or hybrid materials, like the ones expected for low dimensional carbon nanostructures, such as fullerenes, nanotubes, and graphenes, or the recent developments in molecular photovoltaics, for example, in nanostructured hybrid materials for energy conversion and storage. The aim of this book entitled *Organic Nanomaterials: Synthesis, Characterization, and Device Applications*, is to present an appropriate and representative coverage of these materials, which constitute one of the most actively pursued fields of science.

This book contains 26 chapters, which have been rationally organized in five main parts. The first part is concerned with introductory and general chapters on nanomaterials and self-assembled nanostructures. Christensen and Tomalia propose a classification strategy for well-defined, soft-matter nanoscale building blocks (Chapter 1), Schenning and González-Rodríguez analyze the role of hydrogen bonding in the nanoscale organization of π-conjugated materials (Chapter 2), and Amabilino reviews some interesting aspects of chiral, organic nanomaterials (Chapter 3). An overview of a class of biochemical nanomaterials is given by Javakhishvili and Hvilsted (Chapter 4), followed by thorough studies of self-assembled porphyrin nanostructures and their potential applications by Shelnutt and Medforth (Chapter 5); finally, electron-transfer functions of nonplanar porphyrins are studied by Fukuzumi and Kojima (Chapter 6).

The second part of the book consists of a series of chapters devoted to carbon nanostructures ranging from fullerenes (including endohedral fullerenes) and carbon nanotubes to graphene, which report on properties, theoretical studies, and applications. The supramolecular aspects of receptors for the molecular recognition of fullerenes are described by Canevet, Pérez, and Martín (Chapter 7), whereas covalent, donor–acceptor ensembles based on phthalocyanines and carbon nanostructures, including graphene, are reviewed by Bottari, Urbani, and Torres (Chapter 8). Breakthroughs in the photophysics of carbon nanotubes are covered by two excellent contributions addressing (a) the photoinduced electron-transfer properties of supramolecular carbon nanotube materials decorated with photoactive sensitizers, outlined by D'Souza, Sandanayaka, and Ito (Chapter 9), and (b) the study of the interactions of porphyrins and phthalocyanines with carbon nanotubes, which is presented by Bartelmeß and Guldi (Chapter 10). The next two chapters are dedicated to endohedral fullerenes, namely to the synthesis of systems encapsulating helium, dihydrogen, and water, by Murata, Murata, and Komatsu (Chapter 11), and to fundamental and applied aspects of endohedral metallofullerenes by Akasaka and co-workers (Chapter 12). In these two interesting chapters, the authors present new insights into the chemistry and properties of endohedral fullerenes. This block of the book mainly devoted to carbon nanostructures is closed with an excellent update on electrochemical characterization and potential applications of carbon materials by Echegoyen and co-workers (Chapter 13), followed by a theoretical approach to solvating insoluble carbon nanostructures by molecular dynamics by Calvaresi and Zerbetto (Chapter 14).

The third group of chapters focuses on different aspects of some inorganic materials, self-assembled monolayers, organic field effect transistors, and molecular self-assembly at solid surfaces. Thus, the topic of inorganic metal capsules with redox-active guests is treated by Macdonell and Cronin (Chapter 15), whereas the following two chapters developed by Huskens and co-workers (Chapter 16) and by Laromaine and Mace (Chapter 17) review the use of stimuli-responsive monolayers and self-assembled monolayers as model biosurfaces, respectively. Finally, the low-dimensionality effects in organic field effect are described by Biscarini and co-workers (Chapter 18), and the block is well-complemented by the growth of organic nanomaterials by molecular self-assembly at solid surfaces, which is developed by Gallego, Otero, and Miranda (Chapter 19).

The fourth part of the book consists of a series of chapters dealing with different areas involving both biological aspects and nanomaterials. In this part, Vélez reports on the interesting area of biofunctionalized surfaces (Chapter 20), whereas Fabbro, Da Ros, and Prato describe carbon nanotube derivatives as anticancer drug delivery systems (Chapter 21), with a special attention to the medical applications of different kinds of carbon nanotube-based nanomaterials. Closing this section, a study on porous nanomaterials for biomedical applications is outlined by De Cola and co-workers (Chapter 22), whereas Foldvari and co-workers report on nanoparticle design for gene therapy (Chapter 23).

The book ends with three comprehensive applied chapters, as examples of the potential use of organic nanostructured materials in nanoscience, which are devoted to sensors and molecular photovoltaics. This part starts with a chapter by Ratera, Tárraga, Molina, and Vecianna which discusses the sensing of Hg(II) ions in water (Chapter 24), and it is followed by two chapters on the main fields of organic solar cells, namely, organic nanomaterials for efficient bulk heterojunctions by Troshin and Sariciftci (Chapter 25), and mesoscopic dye-sensitized solar cells by Nazeeruddin, Ko, and Grätzel (Chapter 26).

Most chapters end with a summarizing conclusion that also serves as an abstract. The combined authors of the chapters give a good representation of the organic nanomaterials, although with different styles as is often the case in multi-author books.

Finally, and most importantly, we are indebted to all the authors for all their efforts in the preparation of their contributions, which we hope the readers will appreciate.

The editors would like to dedicate this book to the memory of our colleague and friend Christian G. Claessens, who recently passed away.

Tomás Torres
Giovanni Bottari

Universidad Autónoma de Madrid, Spain
April 2013

CONTRIBUTORS

Takeshi Akasaka, Life Science Center of Tsukuba Advanced Research Alliance, University of Tsukuba, Tsukuba, Ibaraki 305-8577, Japan; *and* College of Materials Science and Engineering, Huazhong University of Science and Technology, Wuhan, Hubei 430074, China

David B. Amabilino, Institut de Ciència de Materials de Barcelona, Consejo Superior de Investigaciones Científicas, Campus Universitari de Bellaterra, 08193 Cerdanyola del Vallès, Catalonia, Spain

Ildiko Badea, College of Pharmacy and Nutrition, University of Saskatchewan, Saskatoon, Saskatchewan, S7N 5C9, Canada

Juergen Bartelmeß, Department of Chemistry and Pharmacy, Interdisciplinary Center for Molecular Materials (ICMM), Friedrich-Alexander-Universität Erlangen-Nürnberg, 91058 Erlangen, Germany

Mario Barteri, Department of Chemistry, Universitá "La Sapienza", 00195, Rome, Italy

Fabio Biscarini, Università degli Studi di Modena e Reggio Emilia, Dipartimento di Scienze della Vita, via Campi 183, I-41125, Modena Italy and Consiglio Nazionale delle Ricerche (CNR), Istituto per lo Studio dei Materiali Nanostrutturati (ISMN), 40129 Bologna, Italy

Giovanni Bottari, Departamento de Química Orgánica, Facultad de Ciencias, Universidad Autónoma de Madrid, 28049 Madrid, Spain; *and* IMDEA Nanociencia, C/Faraday 9, Ciudad Universitaria de Canto Blanco, E28049 Madrid, Spain

Matteo Calvaresi, Dipartimento di Chimica "G. Ciamician", Università di Bologna, 40126 Bologna, Italy

David Canevet, Laboratoire MOLTECH-Anjou, 49045 ANGERS Cedex 01, France

Stefano Casalini, Consiglio Nazionale delle Ricerche (CNR), Istituto per lo Studio dei Materiali Nanostrutturati (ISMN), 40129 Bologna, Italy

Massimiliano Cavallini, Consiglio Nazionale delle Ricerche (CNR), Istituto per lo Studio dei Materiali Nanostrutturati (ISMN), 40129 Bologna, Italy

Jørn B. Christensen, Department of Chemistry, University of Copenhagen, DK-2100 Copenhagen, Denmark

Tobias Cramer, Consiglio Nazionale delle Ricerche (CNR), Istituto per lo Studio dei Materiali Nanostrutturati (ISMN), 40129 Bologna, Italy

Leroy Cronin, School of Chemistry, University of Glasgow, Glasgow, G12 8QQ, United Kingdom

Tatiana Da Ros, Center of Excellence for Nanostructured Materials (CENMAT), INSTM—Unit of Trieste, Dipartimento di Scienze Chimiche e Farmaceutiche, Università degli Studi di Trieste, 34127 Trieste, Italy

Luisa De Cola, Université de Strasbourg, Institut de Science et d'Ingénierie Supramoléculaires (ISIS), 67083 Strasbourg, France

André Devaux, Department of Chemistry, University of Fribourg, CH- 1700 Fribourg, Switzerland

McDonald Donkuru, College of Pharmacy and Nutrition, University of Saskatchewan, Saskatoon, Saskatchewan, S7N 5C9, Canada

Francis D'Souza, Department of Chemistry, University of North Texas, Denton, TX 76203-5017, United States

Lourdes E. Echegoyen, Department of Chemistry, University of Texas at El Paso, El Paso, TX 79968, United States

Luis Echegoyen, Department of Chemistry, University of Texas at El Paso, El Paso, TX 79968, United States

Mahmoud Elsabahy, School of Pharmacy, University of Waterloo, Waterloo, Ontario, N2L 3G1, Canada. *Current affiliations*: Department of Pharmaceutics, Faculty of Pharmacy, Assiut University, Assiut, Egypt; *and* Laboratory for Synthetic-Biologic Interactions, Department of Chemistry, Texas A&M University, College Station Texas, 77842, United States

Chiara Fabbro, Center of Excellence for Nanostructured Materials (CENMAT), INSTM—Unit of Trieste, Dipartimento di Scienze Chimiche e Farmaceutiche, Università degli Studi di Trieste, 34127 Trieste, Italy. *Current affiliation:* Dipartimento di Scienze Molecolari e Nanosistemi, Università Ca' Foscari di Venezia, 30123 Venezia, Italy

Lai Feng, Department of Physical Science and Technology, School of Energy, Soochow University, Suzhou, Jiangsu 215006, China; *and* Life Science Center of Tsukuba Advanced Research Alliance, University of Tsukuba, Tsukuba, Ibaraki 305-8577, Japan

Marianna Foldvari, Canada Research Chair in Bionanotechnology and Nanomedicine, Waterloo Institute of Nanotechnology, School of Pharmacy, University of Waterloo, Waterloo, Ontario, N2L 3G1, Canada

Shunichi Fukuzumi, Department of Material and Life Science, Graduate School of Engineering, Osaka University and ALCA (JST), 2-1 Yamada-oka, Suita, Osaka 565-0871, Japan; *and* Department of Bioinspired Science, Ewha Womans University, Seoul 120-750, Korea

José M. Gallego, Instituto de Ciencia de Materiales de Madrid, Consejo Superior de Investigaciones Científicas, 28049 Madrid, Spain

Michael Grätzel, Laboratory of Photonics and Interfaces (LPI), Ecole Polytechnique Federale de Lausanne, CH-1015 Lausanne, Switzerland

Dirk M. Guldi, Department of Chemistry and Pharmacy, Interdisciplinary Center for Molecular Materials (ICMM), Friedrich-Alexander-Universität Erlangen-Nürnberg, 91058 Erlangen, Germany

Jurriaan Huskens, Department of Science and Technology, Molecular Nanofabrication Group, MESA$^+$ Institute for Nanotechnology, University of Twente, 7500 AE Enschede, Netherlands

Søren Hvilsted, Danish Polymer Centre, Department of Chemical and Biochemical Engineering, Technical University of Denmark, DK-2800 Kgs. Lyngby, Denmark

Osamu Ito, CarbonPhotoScience Lab., Kita-Nakayama 2-1-6, Izumi-ku, Sendai, 981-3215, Japan

Irakli Javakhishvili, Danish Polymer Centre, Department of Chemical and Biochemical Engineering, Technical University of Denmark, DK-2800 Kgs. Lyngby, Denmark

Pascal Jonkheijm, Department of Science and Technology, Molecular Nanofabrication Group, MESA$^+$ Institute for Nanotechnology, University of Twente, 7500 AE Enschede, Netherlands

Jaejung Ko, Department of New Material Chemistry, Korea University, Jochiwon, Chungnam 339-700, Korea.

Takahiko Kojima, Department of Chemistry, Faculty of Pure and Applied Sciences, University of Tsukuba, 1-1-1 Tennoudai, Tsukuba, Ibaraki 305-8571, Japan

Koichi Komatsu, Institute for Chemical Research, Kyoto University, Uji, Kyoto 611-0011, Japan

Anna Laromaine, Institut de Ciència dels Materials de Barcelona, ICMAB-CSIC, Campus UAB, 08193 Bellaterra, Spain

Francesca Leonardi, Consiglio Nazionale delle Ricerche (CNR), Istituto per lo Studio dei Materiali Nanostrutturati (ISMN), 40129 Bologna, Italy

Fang-Fang Li, Department of Chemistry, University of Texas at El Paso, El Paso, TX 79968, United States

Xing Lu, State Key Laboratory of Material Processing and Die & Mould Technology, College of Materials Science and Engineering, Huazhong University of Science and Technology, Wuhan, Hubei 430074, China; *and* Life Science Center of Tsukuba Advanced Research Alliance, University of Tsukuba, Tsukuba, Ibaraki 305-8577, Japan

Henning Lülf, Université de Strasbourg, Institut de Science et d'Ingénierie Supramoléculaires (ISIS), 67083 Strasbourg, France

Andrew Macdonell, School of Chemistry, University of Glasgow, Glasgow, G12 8QQ, United Kingdom

Charles R. Mace, Diagnostics For All, 840 Memorial Drive, Cambridge, MA 02139, United States

Nazario Martín, Facultad de Ciencias Químicas, Departamento de Química Orgánica, Universidad Complutense de Madrid, 28040 Madrid, Spain; *and* IMDEA Nanociencia, Ciudad Universitaria de Canto Blanco, E28049 Madrid, Spain

Craig J. Medforth, REQUIMTE/Departamento de Química e Bioquímica, Faculdade de Ciências, Universidade do Porto, 4169-007 Porto, Portugal

Rodolfo Miranda, Departamento de Física de la Materia Condensada, Facultad de Ciencias, Universidad Autónoma de Madrid, 28049 Madrid, Spain

Pedro Molina, Departamento de Química Orgánica, Facultad de Química, Universidad de Murcia, Campus de Espinardo, E-30100 Murcia, Spain

Michihisa Murata, Institute for Chemical Research, Kyoto University, Uji, Kyoto 611-0011, Japan

Yasujiro Murata, Institute for Chemical Research, Kyoto University, Uji, Kyoto 611-0011, Japan

Shigeru Nagase, Fukui Institute for Fundamental Chemistry, Kyoto University, Sakyo-ku, Kyoto 606-8103, Japan

Mohammad Khaja Nazeeruddin, Laboratory of Photonics and Interfaces (LPI), Ecole Polytechnique Federale de Lausanne, CH-1015 Lausanne, Switzerland; *and* University, Jochiwon, Chungnam 339-700, Korea

Roberto Otero, Departamento de Física de la Materia Condensada, Facultad de Ciencias, Universidad Autónoma de Madrid, 28049 Madrid, Spain

Emilio M. Pérez, IMDEA Nanociencia, Ciudad Universitaria de Canto Blanco, E28049 Madrid, Spain

Eko Adi Prasetyanto, Université de Strasbourg, Institut de Science et d'Ingénierie Supramoléculaires (ISIS), 67083 Strasbourg, France

Maurizio Prato, Center of Excellence for Nanostructured Materials (CENMAT), INSTM—Unit of Trieste, Dipartimento di Scienze Chimiche e Farmaceutiche, Università degli Studi di Trieste, 34127 Trieste, Italy

Imma Ratera, Department of Molecular Nanoscience and Organic Materials, Institut de Ciència de Materials de Barcelona (ICMAB-CSIC)/CIBER-BBN, Campus de la UAB, 08193 Bellaterra, Spain

David González-Rodríguez, Nanostructured Molecular Systems and Materials Laboratory, Departamento de Química Orgánica, Universidad Autónoma de Madrid, 28049 Madrid, Spain

Atula S. D. Sandanayaka, School of Materials Science, Japan Advanced Institute of Science and Technology (JAIST), Nomi, Ishikawa, 923-1292, Japan

Niyazi Serdar Sariciftci, Linz Institute for Organic Solar Cells (LIOS), Physical Chemistry, Johannes Kepler University of Linz, A-4040 Linz, Austria

Satoru Sato, Life Science Center of Tsukuba Advanced Research Alliance, University of Tsukuba, Tsukuba, Ibaraki 305-8577, Japan

Francesca A. Scaramuzzo, Department of Science and Technology, Molecular Nanofabrication Group, $MESA^+$ Institute for Nanotechnology, University of Twente, 7500 AE Enschede, Netherlands

Albertus P. H. J. Schenning, Laboratory for Functional Organic Materials and Devices, Eindhoven University of Technology, 5600 MB Eindhoven, The Netherlands

John A. Shelnutt, Department of Chemistry, University of Georgia, Athens, GA 30602, United States

Yuta Takano, Life Science Center of Tsukuba Advanced Research Alliance, University of Tsukuba, Tsukuba, Ibaraki 305-8577, Japan

Alberto Tárraga, Departamento de Química Orgánica, Facultad de Química, Universidad de Murcia, Campus de Espinardo, E-30100 Murcia, Spain

Donald A. Tomalia, NanoSynthons LLC, The National Dendrimer and Nanotechnology Center, 1200 N. Fancher Avenue, Mount Pleasant, MI, 48858, United States

Tomás Torres, Departamento de Química Orgánica, Facultad de Ciencias, Universidad Autónoma de Madrid, 28049 Madrid, Spain; *and* IMDEA Nanociencia, C/Faraday 9, Ciudad Universitaria de Canto Blanco, E28049 Madrid, Spain

Pavel A. Troshin, Institute for Problems of Chemical Physics, Russian Academy of Sciences, Chernogolovka, Moscow Region, 142432, Russia

Maxence Urbani, Departamento de Química Orgánica, Facultad de Ciencias, Universidad Autónoma de Madrid, 28049 Madrid, Spain

Jaume Veciana, Department of Molecular Nanoscience and Organic Materials, Institut de Ciència de Materials de Barcelona (ICMAB-CSIC)/CIBER-BBN, Campus de la UAB, 08193 Bellaterra, Spain

Marisela Vélez, Instituto de Catálisis y Petroleoquímica, Consejo Superior de Investigaciones Científicas, Campus de Cantoblanco, 28049 Madrid, Spain

Ronald Verrall, Department of Chemistry, University of Saskatchewan, Saskatoon, Saskatchewan, S7N 5C9, Canada

Adrián Villalta-Cerdas, Department of Chemistry, University of Texas at El Paso, El Paso, TX 79968, United States

Michio Yamada, Department of Chemistry, Tokyo Gakugei University, Koganei, Tokyo 184-8501, Japan

Francesco Zerbetto, Dipartimento di Chimica "G. Ciamician", Università di Bologna, 40126 Bologna, Italy

1

A PROPOSED TAXONOMY AND CLASSIFICATION STRATEGY FOR WELL-DEFINED, SOFT-MATTER NANOSCALE BUILDING BLOCKS

JØRN B. CHRISTENSEN AND DONALD A. TOMALIA

1.1 INTRODUCTION

The field of nanoscience has evolved explosively over the last two decades, generating a vast number of nanoscale structures and objects. These nanoconstructs may be derived from covalently bound molecules in the classical chemical sense or they may be supramolecular, self-assembled structures. That withstanding, in either case "well-defined nanostructures/objects" may be formed from the smaller discrete building blocks (i.e., atoms, small molecules, monomers, polymers, etc.) if certain "bottom-up" assembly processes are involved that ensure structure control of critical nanoscale design parameters (CNDPs) such as (a) size, (b) shape, (c) surface chemistry, (d) flexibility/rigidity, (e) architecture, and (f) composition. These resulting discrete, homogeneous nanostructures/objects constitute an important category of nanomaterials and are distinguished from other heterogeneous nano-assemblies by expressing well-defined interrelationship patterns and quantized stoichiometries with each other [1, 2]. The quantized stoichiometries and interparticle relationships exhibited by these well-defined nano-entities are a consequence of their structure-controlled precursor building blocks and their ability to transfer important critical atomic design parameters (CADPs) or critical molecular design parameters (CMDPs) to the nanoscale level. This structural information is routinely transferred with high integrity to higher complexity in biological systems via certain evolutionary/genealogical aufbau strategies as described in Figure 1.8 and 1.1. Joyce [3, 4] has clearly demonstrated (a) the importance of these genealogical/evolutionary aufbau patterns involving well-defined nanoscale building blocks such as proteins, RNA, and DNA and (b) their ultimate role in the diversification of life.

In the case of abiotic systems, unique structure controlled synthetic strategies and construction rules are usually involved in the formation of these well-defined nanostructures/objects. The importance of understanding these bottom-up construction rules and aufbau patterns cannot be overstated. They will be invaluable for developing classification and taxonomy schemes to define key nano-building blocks (i.e., nano-elements), as well as their hybridization pathways to (i.e., nano-compounds/assemblies) and synthetic evolution to higher complexity in the nanoscale region. In many respects, our present insights are similar to previous historical growth phases in traditional chemistry during the nineteenth century that led to the Mendeleev Periodic Table in 1869. In essence, the patterns and categorization of atomic elements in Mendeleev's periodic table constituted an elegant classification or taxonomy for all the known atomic elements [5, 6]. Mendeleev's seminal atomic element taxonomy [7] has provided powerful insights and critical information useful for *a priori* predictions of elemental physicochemical behavior, as well as chemical reactivity, stoichiometries, assembly patterns, and so on, leading to the formation of traditional small molecular structures and assemblies as illustrated in Figure 1.2. The idea of classifying atomic elements or small molecules according to chosen criteria of similarity features or as a function of their interrelationships patterns/trends is not new. In fact, Swedish botanist Carl von Linnaeus (1707–1778), the "father of taxonomy," initiated such a

Organic Nanomaterials: Synthesis, Characterization, and Device Applications, First Edition. Edited by Tomás Torres and Giovanni Bottari.

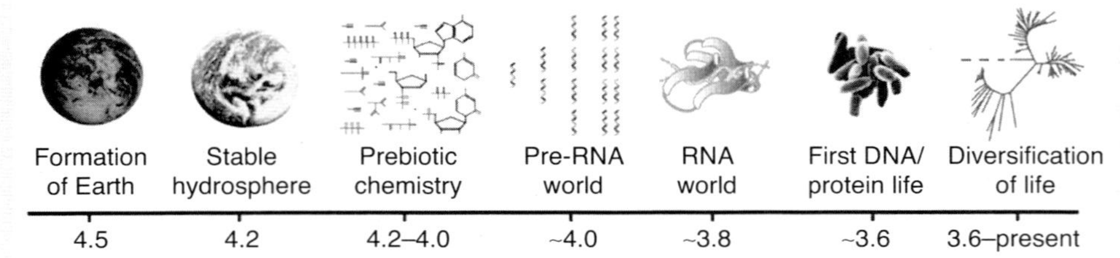

FIGURE 1.1. Timeline of events associated with the early history of life on Earth, with approximate dates in billions of years before the present [4]. Reprinted with permission of Macmillan Publishers Ltd., 2002.

systematic classification for the plant and animal kingdoms over 100 years earlier than Mendeleev, wherein he used a taxonomy scheme to classify biological entities.

These scientific classifications have been described by Mayr and Bock [8] as "the arrangement of well-defined entities" into a hierarchical series of nested classes. A class (i.e., taxon) is defined as "collection of similar entities." Such a taxon consists of collections of entities that share certain similarity features consisting of attributes or traits in common. Similar or related classes at one hierarchical level are combined comprehensively into more exclusive classes at the next lower taxon level to more specifically narrow down the entity description. This process is demonstrated by a Linnaean classification downward from broader taxa to more specific taxa involving diverse entities such as humans, dogs, wolfs, or bacteria as described in Figure 1.3. Variations of the Linnaean taxonomy system have been used to classify a wide range of biological diversity. Many of these classifications are less familiar to chemists and physical scientists; however, they are generally known by biologists and life scientists. Extensive classifications have been reported for complexity at the submicron scale level (e.g., viruses [9,10]) and progress to higher levels of complexity at the micron-scale/macroscale level to include bacteria [11], yeasts [12], fungi [13], plants, and animals as illustrated in Figure 1.3. Presently these taxonomies and classifications are well established for entities

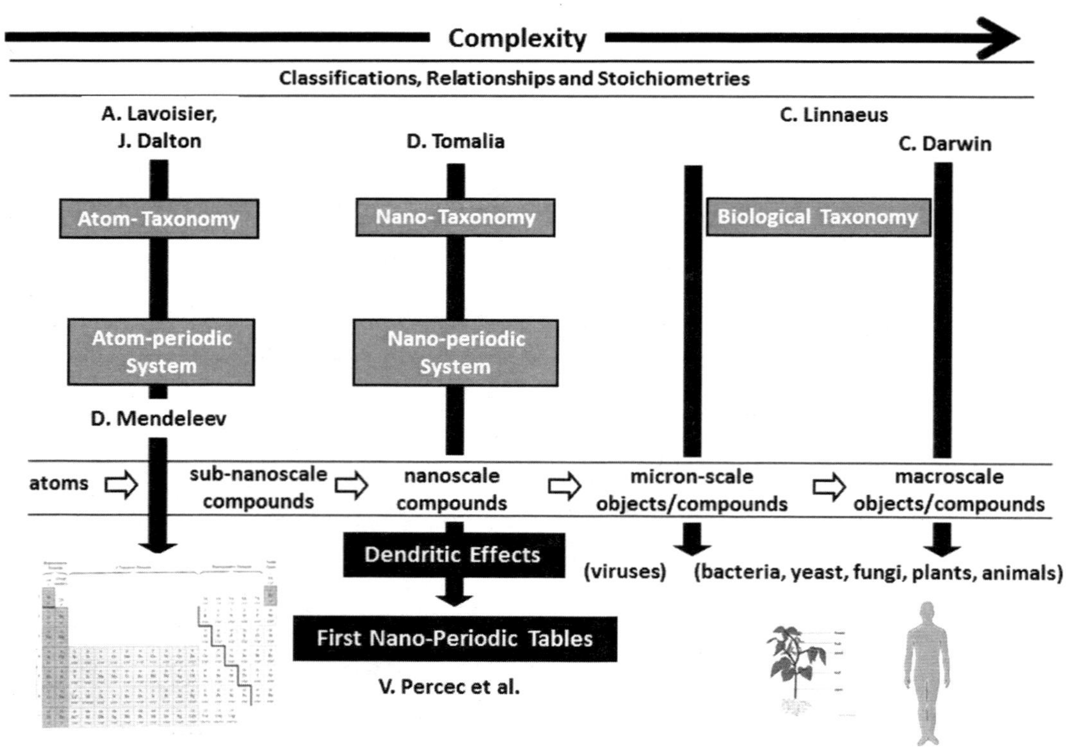

FIGURE 1.2. A comparison of taxonomies at the picoscale nanoscale and micron scale/macroscale as a function of hierarchical complexity.

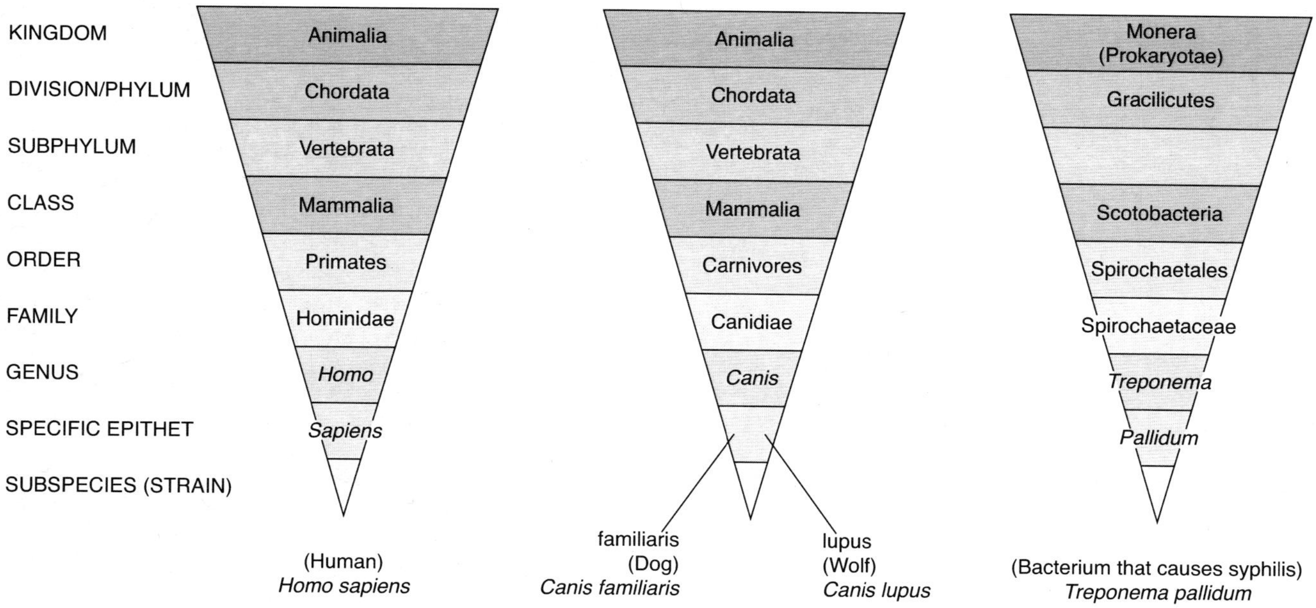

FIGURE 1.3. Linnaean classification of several diversified examples such as a human, a dog, a wolf, and a bacterium [11]. Image reproduced with permission of Copyright John Wiley & Sons.

at the atomic/molecular level as well as higher complexity at the micron/macroscale. However, relatively little has been reported concerning classifications/taxonomies of structures and assemblies at the nanoscale level.

It is widely accepted that well-defined genealogy and evolutionary precursors played a critical role in the development of taxonomies at the micron-scale/macroscale levels [3, 4]. Similarily, it will be of high importance to define analogous taxa for well-defined module/entities at the picoscale, molecular, and nanoscale levels. It is proposed that the CHDPs for these well-defined precursors may very well provide appropriate criteria for evaluating bottom-up construction pathways, roles, and classification schemes for higher-complexity nano-building blocks as described in Figure 1.4. Historically, it is widely recognized that seminal atomic element classifications were critical in the ultimate evolution of an atomic periodic table/system by Mendeleev in 1869 [7]. Based on recent progress reported toward development of a similar nano-periodic system [2] and first examples of nano-periodic tables, a suitable taxonomy strategy appears to be both critical and timely.

In this chapter, we propose further steps toward the development of a universal nano-classification system. This account should be considered a "work in progress" and involves the use of "critical hierarchical design parameters" (CHDPs), namely, (a) sizes, (b) shapes, (c) surface chemistries, (d) flexibility/rigidity, (e) architectures, and (f) elemental compositions as described in Figure 1.4. It is proposed that these CHNPs should be considered as classification criteria (i.e., taxa) and used according to Linnaean principles for classifying various "aufbau precursors" to the 12 hard/soft nano-element categories described later in Section 1.7 (Figure 1.17). It is widely recognized that Linnaean taxonomy principles have been used successfully for classifying evolutionary/genealogical hybrids arising from fundamental entities in biological systems. In a similar manner, at the nanoscale it should be possible to use these CNDP-based taxa as classification criteria for defining future hybridizations of these nano-element categories into anticipated new libraries of nano-compounds and nano-assemblies. Finally, it is proposed that using these CNDPs according to Linnaean classification criteria should be considered as an approach for generating a specific taxonomy-based nomenclature.

1.2 ADAPTATION OF LINNAEAN TAXONOMY PRINCIPLES TO A NEW NANO-CLASSIFICATION SCHEME

Successful Linnaean-like taxonomies have been demonstrated for important soft nano-element categories and nano-compounds and their assemblies. Most noteworthy is the extensive classification of [S-4]-type proteins (Section 1.2.2). Secondly, the development of taxonomies for the combination (i.e., hybridization) of [S-4]-type protein subunits or [S-5]-type viral capsids with [S-6]-type RNA or DNA nano-elements to produce [S-6:S-4] or [S-6:S-5]-type viral nano-compounds, respectively, has been demonstrated as described in Section 1.2.3.

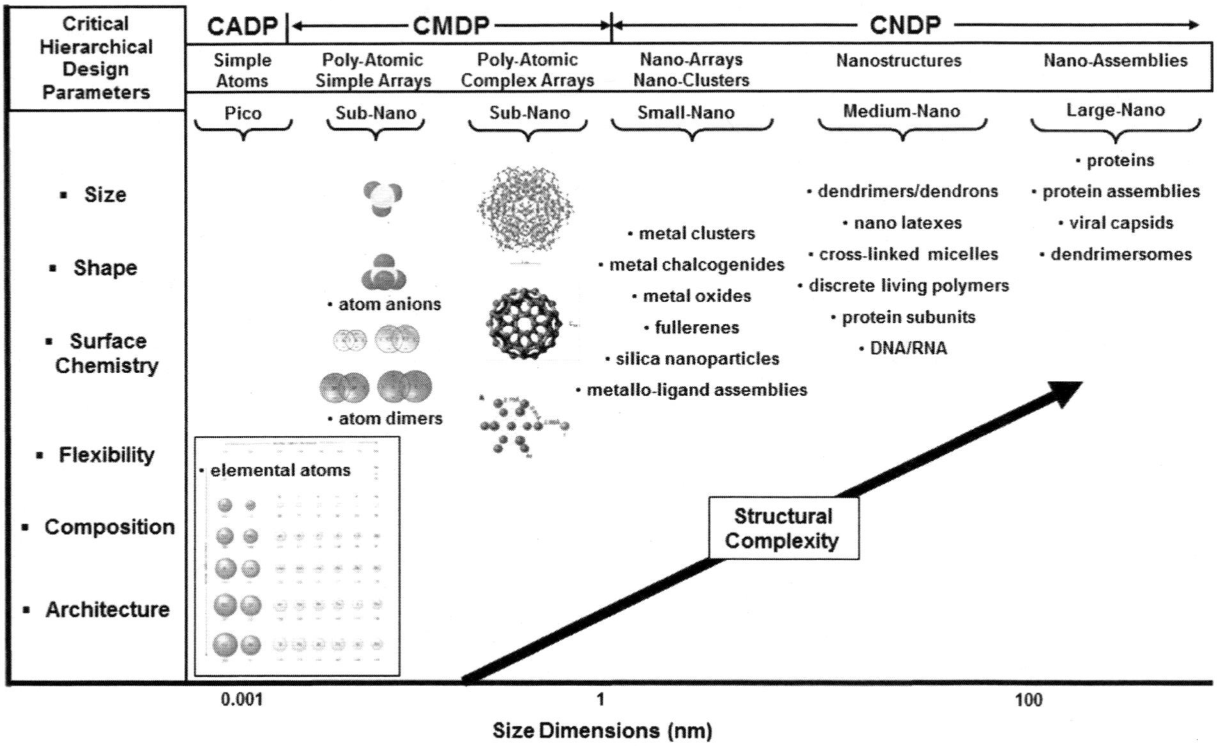

FIGURE 1.4. Critical hierarchical design parameters (CHDPs): (a) size, (b) shape, (c) surface chemistry, (d) flexibility/rigidity, (e) architecture, and (f) elemental composition for various hierarchical structures as a function of dimensions (i.e., atomic-picoscale level (CADP), molecular-subnanoscale (CMDP), and nanometric-nanoscale (CNDP)).

1.2.1 Taxonomy of Biological Structures and Organisms

The term *taxonomy* [Greek *taxis*, arrangement or order, and *nomos*, law or nemein, to distribute or govern] is defined as the science of biological classification. In a broader sense, this taxonomy concept involved the use of a protocol that provides three key features, namely: (a) identification, [*the process of determining that a particular organism belongs to a recognized taxon;* (b) classification [*arrangement of organisms into similar groups or taxa*] and (c) nomenclature [*the branch of taxonomy concerned with the assignment of names to taxonomic groups in agreement with published rules*]. These three parameters have been used routinely for describing classifications and relationships between various biological organism/entities residing in the higher-complexity dimensions of the micron scale–macroscale (i.e., 0.1 μm to meters) [14]. At least eight taxonomic parameters were deemed necessary for defining the higher complexity of micron-scale–macroscale biological systems. They are widely recognized and recalled with the mnemonic "*Do kings play chess on fine grained sand*?" and include the following eight hierarchical parameters: (1) domain, (2) kingdom, (3) phylum, (4) class, (5) order, (6) family, (7) genus, (8) species [15].

A dominant feature that differentiates biological classification (taxonomy) from most other classification systems is *evolution*. The similarity between biological structures or organisms placed in a common taxon is not arbitrary. It is the result of shared descent from their nearest common ancestor. As such, these classifications require an evaluation of evolutionary stages and intrinsic genealogy that precedes each entity classification. In essence, these eight classification parameters not only capture similarities and differences, but also integrate critical biological entity features/properties such as (a) homogeneity, (b) size, (c) morphology/shape, (d) behavior/function, (e) flexibility/rigidity, (f) architecture, and (g) elemental composition as important evolutionary/genealogical selection criteria for these classifications [16]. Such Linnaean-inspired concepts have been studied extensively by essentially every generation of naturalist during the past 200 years, including Charles Darwin. Modified versions of these Linnaean taxonomic criteria and classifications presently serve as the "gold standard" for classifying all biological hierarchy/diversity residing in the micron–meter size range. Such taxonomy concepts have provided a universal and versatile system for classifying all micron-scale–macroscale biological entities (i.e., certain viruses, bacteria, fungi, plants, animals, etc.) and laid the foundation for modern biological taxonomy/classifications and nomenclature. This well-defined taxonomic scheme has served to organize, simplify, and unify wide ranges of diversity and complexity at the micron-scale and macroscale within biology. Most importantly, this well-defined taxonomic system has provided a critical, quantitated protocol for analyzing important patterns, trends, and relationships between classified biological entities. More contemporary modifications have involved the use of molecular level (i.e., DNA/RNA-based) data as a means for classification.

1.2.2 Protein Taxonomies

The nanoscale class of protein structures/assemblies has been proposed as a well-defined soft nano-element category designated as [S-4] based on its quantized CNDPs [2] (see Section 1.7.1). A structural classification of proteins (SCOP) was first pioneered as early as 1995 by Murzin et al. [17]. More recently, a very comprehensive protein taxonomy based on secondary structure was reported [18]. This taxonomy is generated automatically by computer and based solely on secondary protein structure. It takes the form of a cladogram/dendrogram-based "similarity tree" in which proteins with similar secondary structure occupy neighboring leaves. This taxonomy is largely in agreement with SCOP and is a multidimensional classification scheme based on homologous sequences and full three-dimensional structure, as well as information about the chemistry and evolution of the protein. A "similarity tree" based on this taxonomy is as illustrated in Figure 1.5. Much like dendrimer architecture, a similarity tree is heuristically composed of multiple dendron-like domains that are all connected to a central core

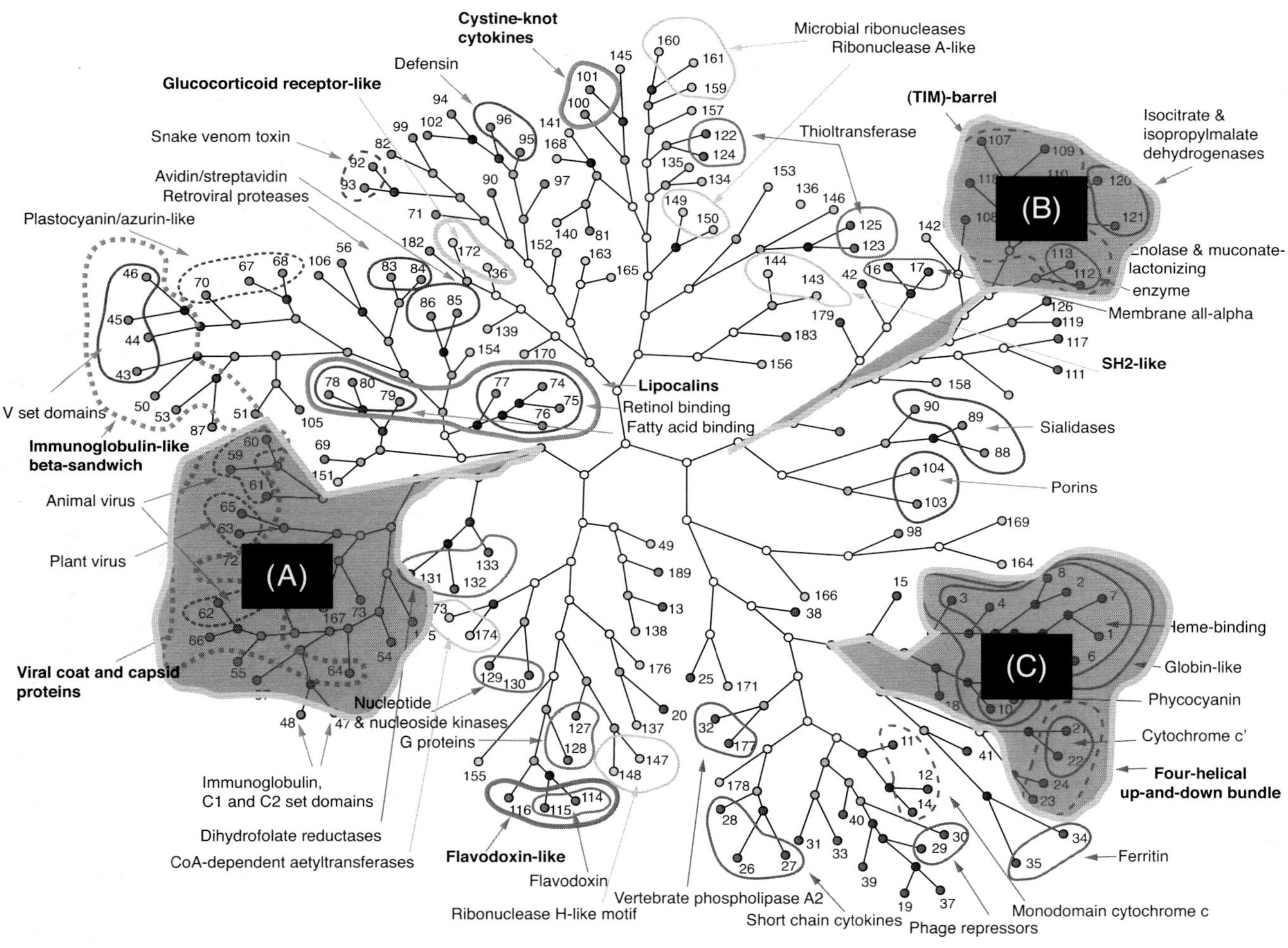

FIGURE 1.5. A similarity tree derived from 183 diverse proteins are examined and classified into various cladograms or similarity domains designated by clusters of protein structures exhibiting similar features or functional properties. For example, domain (**A**) is associated with proteins that self-assemble into viral coats/capsids, domain (**B**) includes a cluster of proteins associated with enzymes and biological catalysts and domain (**C**) identifies specific protein types associated with oxygen transport properties [18]. Reprinted with permission of Macmillan Publishers Ltd., 1999.

of information. These dendron-like domains are referred to as cladograms or dendrograms. The central core contains conserved structural information that feeds into these various cladograms. Based on the hybridization of structural information that is connected to this conserved core, one observes branches of similar yet diverse protein features. As noted in the similarity tree (Figure 1.5), the various cladograms connected to the core serve to cluster similar features into specific domains (i.e., domains (A), (B), and (C)). These domains may be associated with specific properties/functions of interest. Thus, based on secondary protein structure, this taxonomy proved to be a versatile protein classification scheme for evolving differentiating cladograms that associated certain protein types with various biological structural or functional roles. For example, cladogram (A) identifies certain proteins that self-assemble to produce viral coats and capsids as opposed to cladogram (B), which defines clusters of proteins associated with a biological catalysis function (i.e., enzymes, etc.), or cladogram (C), which identifies specific protein types associated with oxygen transport properties (i.e., ferritin, heme-type binding proteins). It is readily apparent that such a taxonomy based on secondary protein structure provides a simple scheme for classifying a large database of complex information. Successful use of such Linnaean taxonomic principles for micronscale/macroscale plants/animals and now nanoscale protein molecules provides optimism for recognizing common features in each of these dimensionally differentiated hierarchical domains. These taxonomic differentiations should be suitable for defining ordered classes and ultimately evolving versatile and useful nomenclatures just as Mendeleev's periodic table of atomic elements led to a classification scheme (i.e., taxonomy) and a universal nomenclature for the elements and their resulting compounds.

1.2.3 Virus Taxonomies

Based on their quantized, well-defined CNDPs, viral capsids have been proposed as an important soft nano-element category designated as [S-5] [2, 19]. As described in Section 1.8.3, they are derived from the stoichiometric self-assembly of category [S-4]-type protein subunits. Viruses, in turn, have been proposed as stoichiometric nano-compounds that are derived from the self-assembly of either [S-4]-type protein subunits around [S-6] RNA or DNA cores (i.e., [S-6]-type nano-elements) or the self-assembly of viral capsids around these RNA/DNA [S-6]-type cores. Meanwhile, an invaluable classification system has been introduced by Baltimore [9] that places viruses into one of seven groups based on an analysis of parameters such as (a) their core nucleic acid type (i.e., DNA or RNA), (b) strandedness (i.e., single strand versus double strand), (c) sense, and (d) method of replication [20] (see Section 1.9.3, Figure 1.31). Such DNA/RNA-driven taxonomy has produced invaluable assistance in the classification and determination of evolutionary termini (i.e., dead ends) confirming extinction or evolutionary gaps raising expectations for biological entities yet to be discovered [21, 22]. On the other hand, a more Linnaean-type taxonomy has been developed for viruses (i.e., the LHT System). This taxonomy is based on a comparison of physico chemical features such as (a) core nucleic acids (i.e., DNA/RNA), (b) symmetry (i.e., helical, icosahedral or complex), (c) presence of envelope components, (d) diameter of capsids, (e) number of capsids, and so on [23] (see Section 1.9.3; Figure 1.30).

1.3 HOW DOES NATURE TRANSFER STRUCTURAL INFORMATION FROM A LOWER HIERARCHICAL LEVEL TO HIGHER COMPLEXITY?

The clustering of similar secondary protein structure as observed in the similarity tree (Figure 1.5), suggests that critical features are conserved and transferred up the hierarchy ladder to produce the the higher complexity observed at the nanoscale in cladograms (A), (B), and (C) above. As such, one must ask: *Are there conserved critical hierarchical design parameters that maintain informational integrity and robustness when transferred to higher complexity*? Anecdotal evidence supports the notion that certain critical design parameters at the atomic, molecular, nanoscale, and microscale level can effectively transfer important structural information to higher hierarchical complexity. Many well-known examples of molecular self-assembly, genetic expression, and evolution clearly demonstrate these principles. The critical role of atoms and molecular level monomers (i.e., alpha-amino acids) is clearly illustrated in Figure 1.6a. As demonstrated in the protein taxonomy (Figure 1.5) it appears that secondary/tertiary structure is far better conserved than specific protein sequence. It is well known that secondary/tertiary protein structures are directly dependent on CNDPs such as (a) size, (b) shape, (c) surface chemistry, and (d) flexibility/rigidity. A clear example is readily illustrated by the conservation of nanoscale protein (i.e., collagen) secondary/tertiary structure throughout its respective hierarchical micron-scale self-assembly steps to produce biological tendons as shown in Figure 1.6b.

In 1917, D'Arcy Thompson (1860–1948) published a seminal biological treatise entitled *On Growth and Form* [16]. This work provided deep insights and answers to the question above. It clearly articulated the importance of preferred and controlled "critical micron-scale/macroscale design parameters" (CMicDPs) such as (a) size, (b) shape, (c) surface function, (d) flexibility/rigidity, (e) architecture, and (f) atomic composition features in the successful evolution of all biological structures and organisms. As early as 1990–1993, it was stated that analogous "critical atomic, molecular and

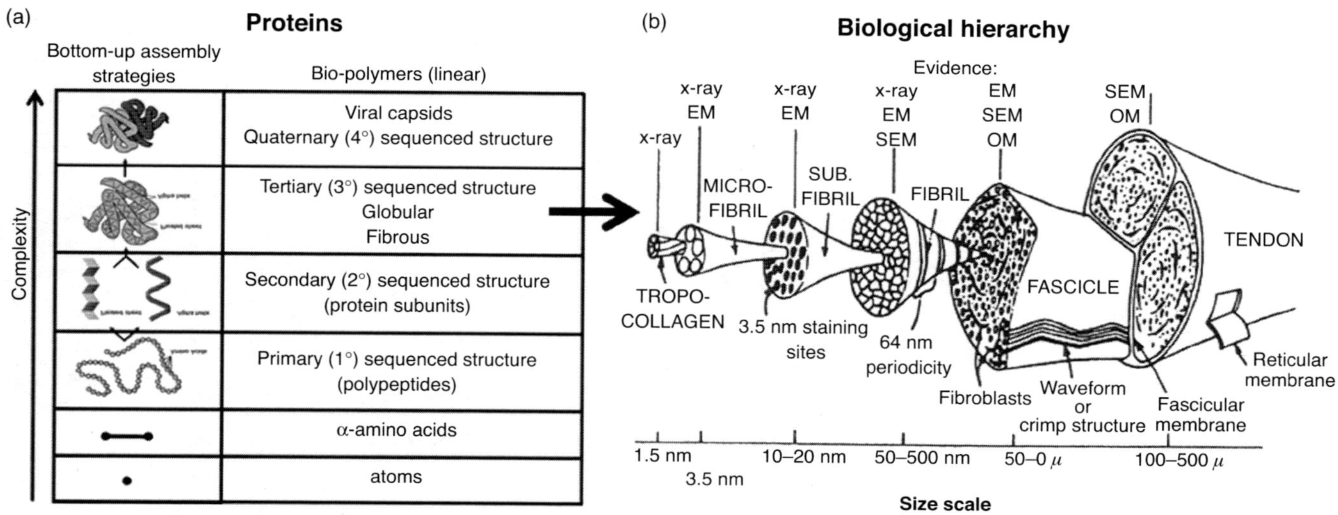

FIGURE 1.6. (**a**) Bottom-up synthesis of proteins with symmetry-breaking events from atoms and monomers leading to self-organization complexity. (**b**) Bottom-up assembly of the fibrous protein, collagen, through six symmetry-breaking levels to produce new emerging properties associated with tendon. Reproduced with permission of Eric Baer.

nanoscale design parameters" were conserved and inextricably associated with the transfer of information in all well-defined structures throughout the atomic to microscale hierarchical continuum [24, 25].

1.4 THE USE OF CLADOGRAMS FOR CLASSIFICATIONS OF WELL-DEFINED BIOLOGICAL (MICRON SCALE/MACROSCALE), ATOMIC (PICOSCALE), AND NANOSCALE BUILDING BLOCKS

1.4.1 Taxonomy of Biological Entities

Cladograms (i.e., dendrograms) as illustrated in Figure 1.5 for nanoscale proteins and in Figure 1.7 for macroscale animals (i.e., vertebrae) have been used traditionally for defining evolutionary (i.e., precursor-type) relationships and pathways leading to clusters of similar entities based on several pervasive criteria. In Figure 1.7**,** the vertebrate classifications (i.e., cladograms) are based on similar evolutionary and genealogical features such as sizes, shapes/morphologies, and functions resulting from hybridization to produce certain well-defined families of organisms. Prior to the advent of DNA sequencing, nearly all cladogram-based biological systematics involved the use of morphology or shape data classifications for defining genealogy and evolutionary pathways. Presently, both molecular and morphological systematics are used extensively for this purpose [11]. Many of these issues have been examined in detail by Zuckerkandl and Pauling [26].

1.4.2 Taxonomy of Atomic Elements

Nearly concurrently, Lavoisier [1743–1794; *Traite Elementaire de Chemie* (1789)], Prout [1754–1844; *Law of Definite Proportions* (1797)], Dalton [(1766–1844; *Law of Multiple Proportions/New System of Chemical Philosophy* (1808)], and others began the classification of less complex hierarchical entities such as the atomic elements. This activity led to Mendeleev's (1834–1907) seminal periodic element proposal wherein he classified and organized known elements according to their periodic properties and chemical behavior. This led to a taxonomic framework (i.e., Mendeleev Periodic Table) that could be adapted to and interpreted according to the earlier Linnaean template for biological structures and organisms. For example, the elements were organized into domains of reactive versus inert types (i.e., unsaturated versus saturated valence shells), vertical elemental groups (i.e., metals, metalloids, and nonmetals; Groups I–VIII), inorganic type elementals versus organic, horizontal periods (i.e., atomic weights), mono-isotope versus poly-isotopic elements, and so on. Rich [27] refers to such a periodic taxonomy as a view of the relationship patterns that exist among the elements. Similarly, one can visualize a crude taxonomy for the atomic elements based on the well-known mnemonic and classical Linnaean taxonomy template used for classifying biological structures/organisms as described in Figures 1.8 and 1.9.

> The elements are different aggregates of the atoms of primordial hydrogen.
>
> —Prout's Hypothesis (1815)

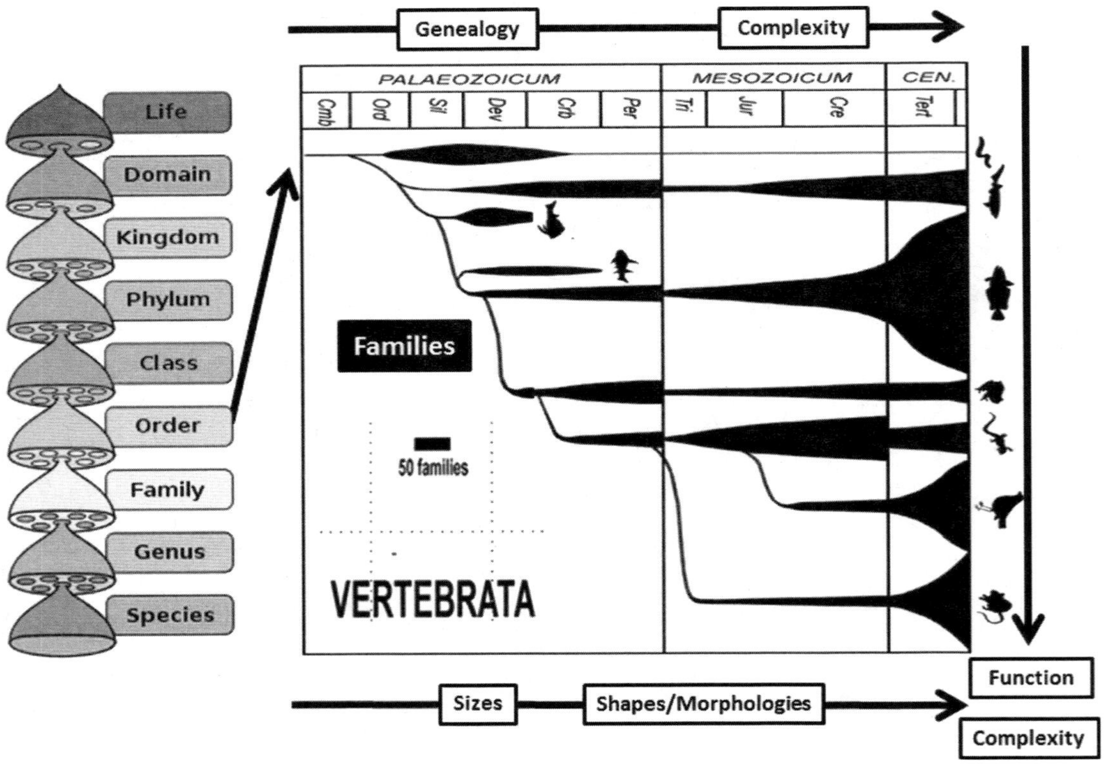

FIGURE 1.7. A cladogramic (i.e., dendrogramic) classification of the Vertebrata order into families using Linnaean taxonomy principles according to sizes, shapes (i.e., morphologies), function, genealogy, and complexity. The darkened dimension bar represents approximately 50 families.

Based on genealogical and evolutionary principles described for the formation of heavier atomic elements from lighter elements [28], one may view Niels Bohr's seminal perspective of the Mendeleev Periodic Table as an evolutionary cladogram for tracing the transformation and evolution of light to heavier atomic elements as shown in Figure 1.8. Based on a Linnaean template, the classification of elements into families based on genealogy or electron shells (i.e., periods) in the context of sizes, shapes, function (i.e., surface chemistry), and complexity is readily apparent.

In general, the Mendeleev Periodic Table has evolved and been viewed as a very articulate taxonomic account of all the known atomic elemental building blocks. Most notable during this periodic table development was the emergence of certain patterns/trends (i.e., vertical elemental groups and horizontal periods) that contained a number of empty positions. These unoccupied elemental positions within these taxonomic patterns suggested the existence of yet undiscovered elements with properties that could be extrapolated from the property patterns/trends of the surrounding elements. Without the benefit of knowing or understanding anything about atomic structure, electronic theory, or quantum mechanics, Mendeleev provided a powerful predictive taxonomic concept that still remains rational and consistent with accepted contemporary understanding of atomic structure and their physicochemical properties.

Furthermore, it is now recognized that the Mendeleev Periodic Table provides a very rich taxonomy for classifying and predicting important CADPs [25] such as (a) size, (b) shape, (c) reactive/inert surface chemistry, (d) flexibility/polarizability and (e) architecture. These periodic trends and property patterns are as described in Figure 1.8 and 1.9. It is obvious, that hybridization of these various co-reactive atomic elements would produce their own unique and differentiated cladograms of small molecule structures and assemblies.

Such an approach might prove to be an interesting template for organizing and classifying various aufbau-type precursors leading to well-defined nano-building blocks (i.e., nano-element categories) as well as their subsequent hybridization into nano-compounds and assemblies. These issues will be discussed later in Section 1.7.1.

1.4.3 In Quest of a Taxonomy for Nonbiological Nanoscale Structures and Assemblies

Hierarchical taxonomies are widely recognized and used routinely for building blocks as small as picoscale atomic elements and well-defined biological structures as well as for

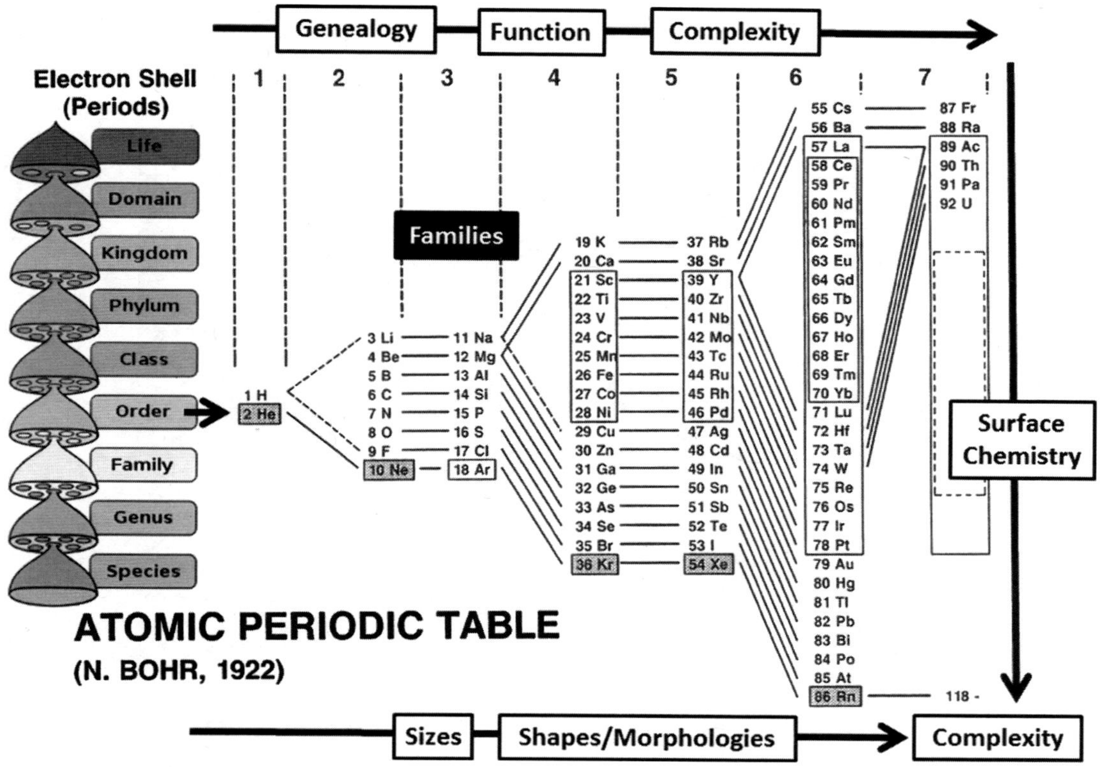

FIGURE 1.8. Linnaean taxonomy categories wherein the "order classification" of the elements is presented in a Bohr-type periodic-type cladogram associated with atomic elemental families that are defined as a function of CADPs that include; vertical periods of electronic shells [(i.e., sizes, orbital shapes, flexibility/rigidity, surface chemistry (reactivity)] and horizontal classes of electronic shells (i.e., sizes, reactivities, increasing masses, etc.).

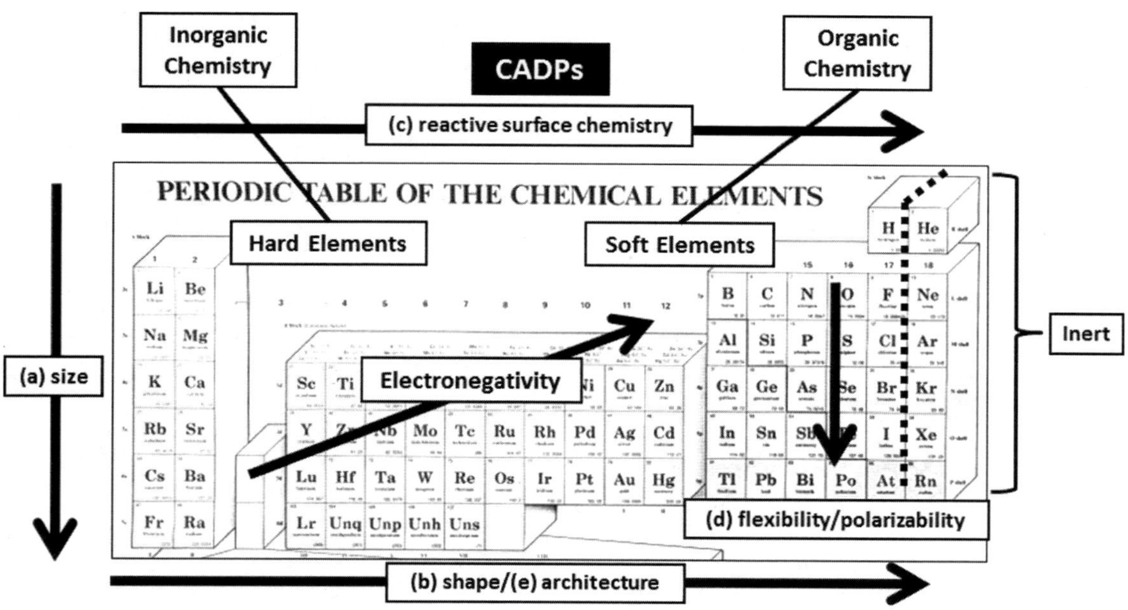

FIGURE 1.9. Structure-controlled critical atomic design parameters (CADPs) for atomic elements classified in a Mendeleev periodic table as a function of (a) size, (b) shape, (c) reactive surface chemistry, (d) flexibility/polarizability, (e) architecture, and (f) elemental composition.

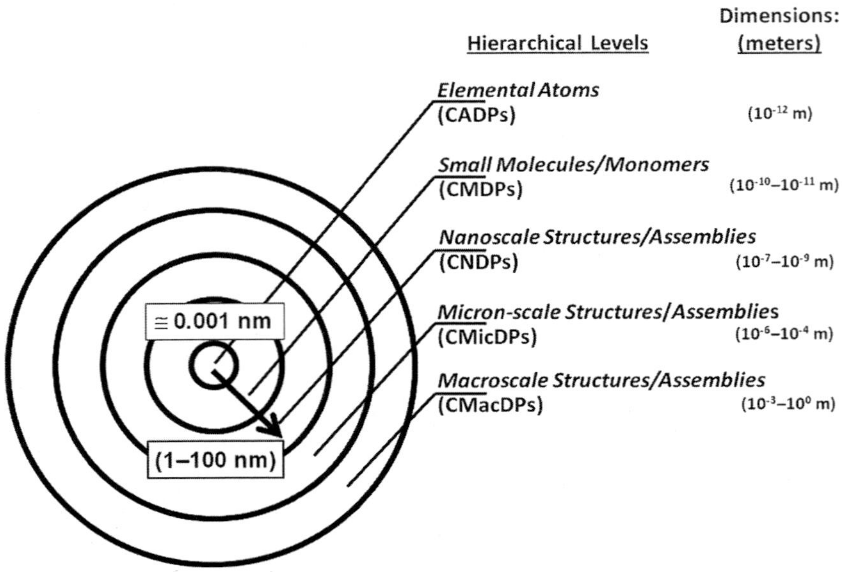

FIGURE 1.10. Concentrically, nested hierarchical levels used to illustrate conservation of critical module design parameters (i.e., sizes, shapes, surface chemistries, flexibilities, and architectures) as a function of their hierarchical levels (i.e., CADP→CMDP→CNDP→CmicDP→CMacDP) associated with their respective dimensions [31].

organisms as large as micron/meter scale. With the exception of biological-type taxonomies offered recently for proteins [29] and for viruses [20] based on messenger DNA/RNA precursors and genealogy, relatively little attention has been focused on nonbiological structures, assemblies, or entities in the 1 to 100-nm dimensional region. As shown in Figure 1.10, this nanoscale region is bracketed by well-demonstrated taxonomies for the atomic elements (i.e., the Mendeleev Periodic Table), as well as by Linnaean-type taxonomies for the micron-scale–macroscale region. As such, it seems plausible to expect a continuum of conserved CHDPs to connect these two hierarchical levels.

It is noteworthy that in spite of enormous activity in the nanotechnology field during the past several decades, only highly specific/limited strategies have been proposed [30]; however, no such universal Linnaean-type taxonomic scheme has been suggested for classifying and unifying well-defined nanostructures, assemblies, or clusters until recently.

1.4.3.1 Taxonomy of Nanostructure/Assemblies

(a) Dendrimers/Dendrons as a Window to a Nanoscale Taxonomy. Recently, we described the use of dendrons/dendrimers as a window to a new nano-periodic system for defining quantized, discrete categories of nano-building blocks (i.e., nano-elements) [31]. They may be thought of as heuristic, core-shell-type atom mimics that share many atom-like combining properties with atoms to yield stoichiometric nano-compounds. Based on more than 12,000 published references in the literature, it has been clearly demonstrated that dendrons/dendrimers are structure controlled nanoscale constructs that exhibit quantized critical nanoscale design parameters (CNDPs) such as (a) size, (b) shape, (d) flexibility/rigidity, and (e) architecture. A variety of dendrimer surface reactions or guest–host assemblies have been shown in the literature to produce stoichiometric nano-compounds including (a) dendrimer–dendrimer, (b) dendrimer–protein, (c) dendrimer–fullerene, and (d) dendrimer–metal nanocluster structures, to mention a few [2]. Meanwhile, as part of this comparison, dendrimers have been shown to exhibit heuristic cladogram properties similar to the atomic elements as illustrated in Figure 1.11, where they are displayed according the Niels Bohr format. Further comparisons can be made by projecting the first three generations of a dendrimer in a Mendeleev-type periodic format [1] as shown in Figure 1.12.

1.5 HEURISTIC MAGIC NUMBER MIMICRY AT THE SUBATOMIC, ATOMIC, AND NANOSCALE LEVELS

Many legendary scientists have made profound historical statements concerning the importance of number theory to the periodicity of matter. The following are just a small sampling:

- John Dalton (1803): "Atoms combine in simple numerical ratios."
- Prout's Hypothesis published anonymously in (1815): "The elements are different aggregates of the atoms of primordial hydrogen."

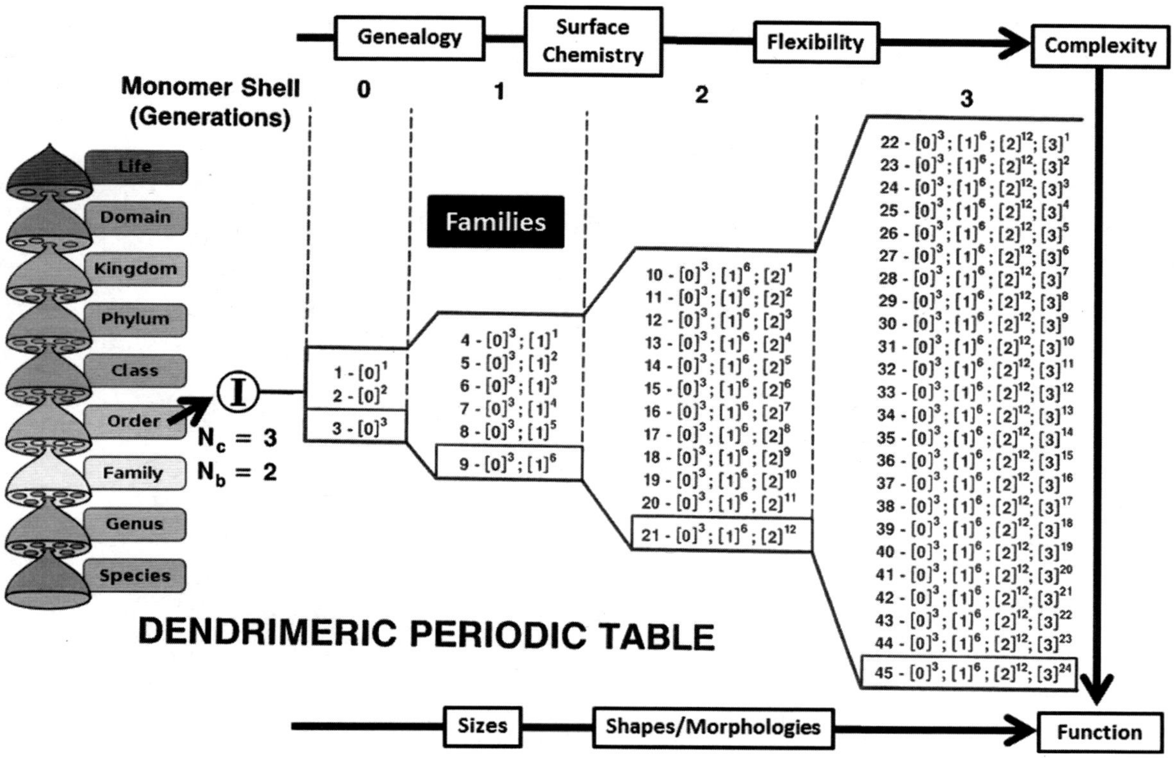

FIGURE 1.11. Linnaean taxonomy categories wherein the "order classification" is associated with a cladogram that defines dendrimer families as a function of core multiplicity ($N_c = 3$) and branch cell multiplicity ($N_b = 2$), as well as CNDPs that include core sizes/shapes, number of monomer shells (generations), flexibility/rigidity (connectivity), and surface chemistry (function).

- Alexander E. B. de Chancourtois (1862): "The properties of the elements are the properties of numbers."

Many of these issues are examined extensively by Boeyens and Levendis [32] and provide a scientifically valid examination of the critical role that "magic numbers" play relative to the periodicity of matter.

A seminal publication by Jena and Castleman [33] describes the routine observation of so-called "magic numbers" associated with clusters that include unique associations and stabilities of atomic particles [34] and, more recently, nanoscale particles such as metal nano-clusters [33, 35–37] and dendrimers [2, 24, 25]. Furthermore, it appears that such magic numbers may also be associated with micron-scale/macroscale objects that are referred to as Fibonacci numbers [38]. As stated by these authors, "it was surprising to observe such similarities between magic numbers in nuclei and atom clusters in spite of the fact that the binding forces in nuclei are very different from those that bind atoms in clusters." That withstanding, it has been found that well-defined numbers of atoms in a cluster of a given size or shape and/or given arrangement/disposition exhibit magic numbers corresponding to various symmetrical, close-packed polygons in two dimensions (2-D) or polyhedrons in three dimensions (3-D) [39]. In this context, magic numbers count the number of spheres in sphere packings of platonic solids and related polyhedral [40]. In essence, magic numbers may be viewed as the result of "atom counting" wherein atoms behave as spheres in well-defined clusters that are structurally controlled as a function of size, shape, surface function, flexibility/rigidity [41], and architecture. These relationships are readily observed in "bravais lattices" wherein cluster geometry depends on crystal structures [39] and precisely defines interrelationship and stoichiometries between participating spheroids (atoms). These "geometrical/mathematical-based magic number" relationships undoubtedly account for their pervasive occurrence at the atomic level, as well as at the nanoscale level when comparing atoms to dendrimers (Figure 1.13) and as well as when comparing atoms to soft clusters (i.e., dendrimers) and hard clusters (i.e., gold nano-clusters) as shown in Figure 1.15.

1.5.1 Heuristic Atom Mimicry of Dendrimers: Nano-Level Core–Shell Analogues of Atoms

A heuristic comparison of the core–shell architectures that are present in dendrimer-based nanoscale modules and

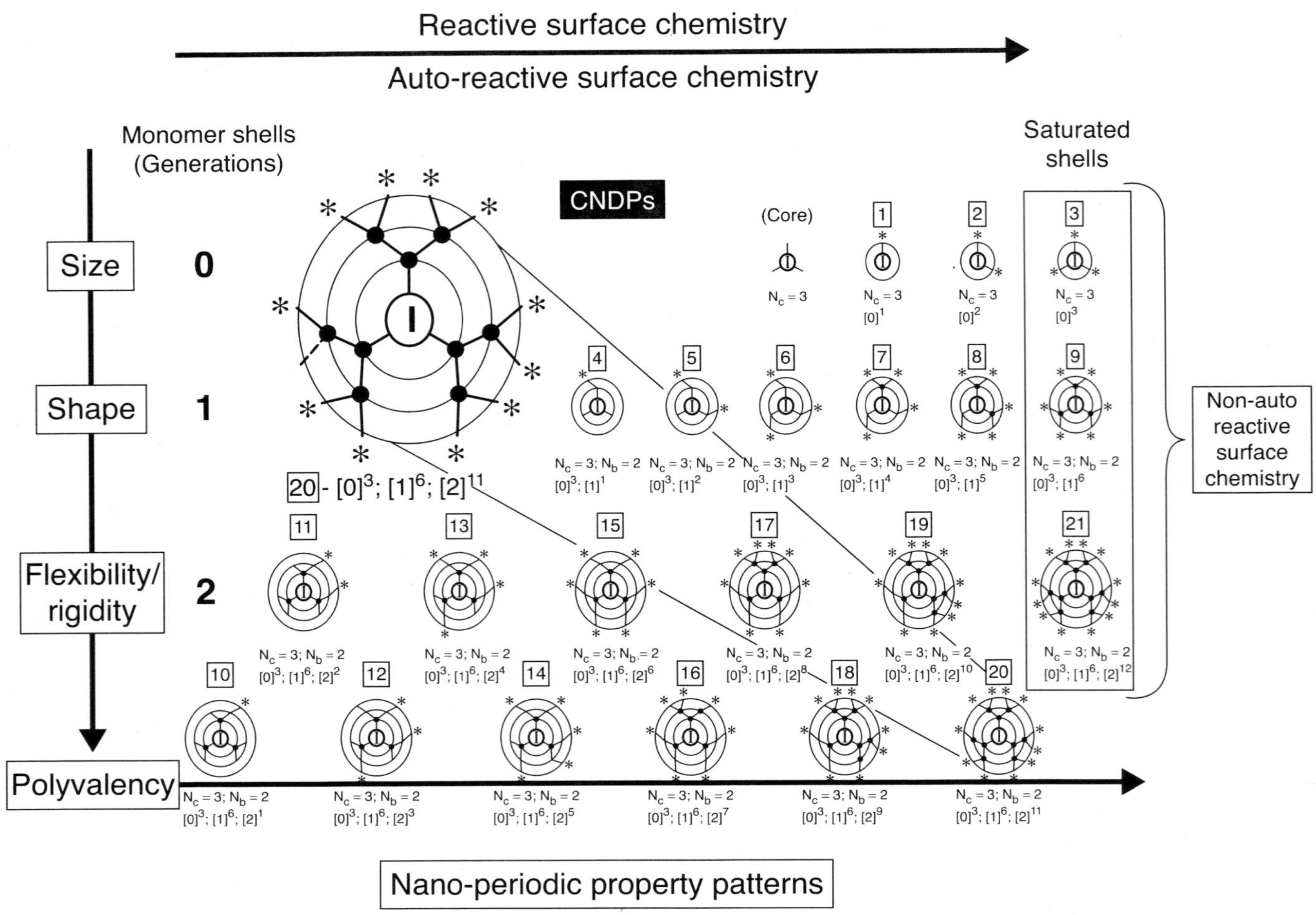

FIGURE 1.12. Structure-controlled critical nanoscale design parameters (CNDPs) for dendrimer-families possessing core multiplicities; $N_c = 3$ and branch cell multiplicities; $N_b = 2$ classified and presented in a Mendeleev-type periodic table as a function of sizes, shapes, flexibility/rigidity and polyvalency.

picoscale atoms was made as early as 1990 [24,42,43]. Furthermore, it appears that a very interesting size continuum exists in the transition from picoscale (atomic) structures to nanoscale dendrimer structures as illustrated in Figure 1.13.

This comparison was used to point out the unique similarities that exist between aufbau components in atoms (i.e., nucleons and electrons) and those that are involved in dendrimer constructions (i.e., cores and branch cell monomers). Remarkable analogies were also noted between recognized, but dimensionally, different parameters shared by both systems such as (a) electron shells versus monomer shells (generations), (b) electron shell versus monomer shell aufbau filling patterns (i.e., mathematically defined), (c) electron shell versus monomer shell saturation levels, (d) atomic weights versus dendrimer molecular weights as a function of shell level and saturation level, and (e) atomic (elemental) reactivity versus dendrimer reactivity as a function of shell saturation level (Figure 1.12). We have referred to these remarkable similarities between picoscale (atomic elements) and nanoscale dendrimers as *atom mimicry*, keeping in mind that picoscale structures are best described by non-Newtonian physics, whereas the dendrimer structures are expected to adhere to and be described by Newtonian physics.

As noted in Figure 1.14, the specific bottom-up building block aufbau leading to dendrimers and higher complexity can be described both mathematically and in the context of CHDPs. This aufbau pathway leading to dendrimers involves a hierarchical transfer sequence of structural information which may be precisely defined mathematically. As such, it appears rational to assume that the CHDPs at each dimensional level are conserved from the picoscale (CADP) → molecular level (CMDP) → nanoscale (CNDP) → micron-scale (CMicDP) levels as illustrated in Figure 1.14.

In summary, one observes the involvement of quantized "magic numbers" of electrons to reach electron shell saturation levels at the picoscale level. Similarly, the

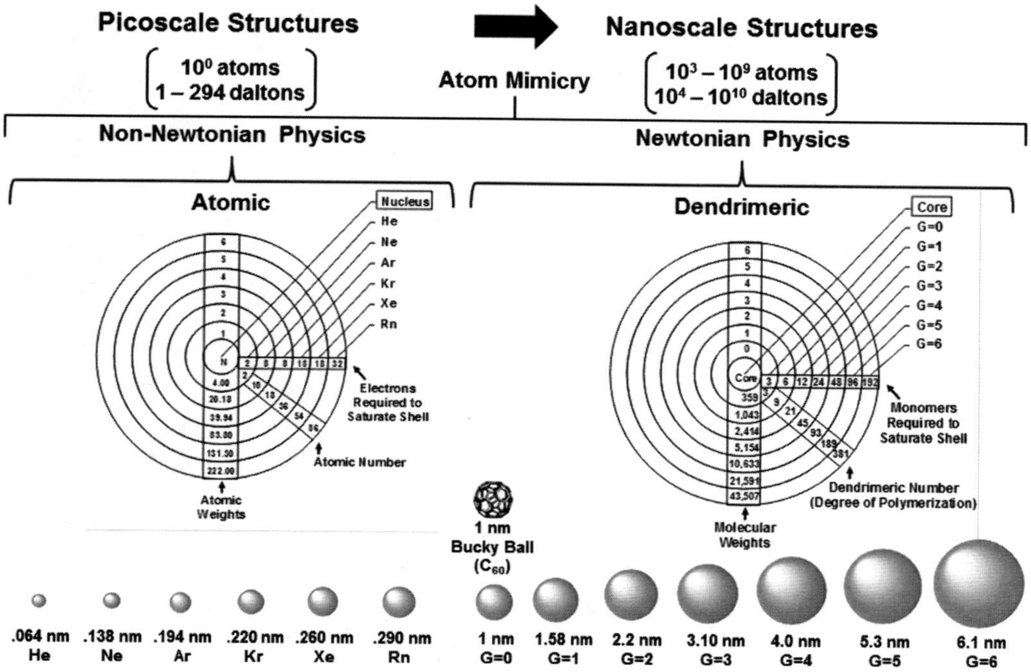

FIGURE 1.13. An example of atom mimicry involving "magic numbers" at the picoscale and nanoscale level, respectively. A comparison of core–shell structures representing picoscale atoms and nanoscale dendrimers, as well as the continuum of sizes that prevails over the two dimensional ranges that are controlled by quantum mechanics and Newtonian physics, respectively [31].

aufbau pathway leading to gold nanoclusters (i.e., hard nano-matter)[44], involves the assembly of discrete numbers of gold atoms to reach shell saturation limits at the nanoscale level. On the other hand, the assembly of discrete numbers of monomer units (i.e., mass) are involved in the case of dendrimers (i.e., soft nano-matter) in order to reach monomer shell saturation limits. In either case, quantized "magic numbers" of metal atoms or monomer units are required to reach a shell saturation limit that mimics atoms and demonstrates "atom mimicry" at the nanoscale level (Figure 1.15).

1.6 ELEMENT CATEGORIES AND THEIR HYBRIDIZATION INTO NANO-COMPOUNDS AND NANO-ASSEMBLIES

1.6.1 A Brief Overview of Nano-classifications (Taxonomies)

One of the first efforts to develop a nanoscale classification (i.e., taxonomy) system evolved from a quest to unify and define nanoscience based on the same "first principles" that underpinned the "central paradigm" for traditional small molecule chemistry [2]. Briefly stated, it was based on a nanomaterials roadmap that focused on well-defined, monodisperse (i.e., >90% monodisperse/homogeneous) nanostructures, nano-assemblies, and exhibited some degree of *atom mimicry*. As illustrated in Figure 1.16, these entities were referred to as nano-element categories. The initial list consisted of 12 nano-element types that were divided into six hard (i.e., inorganic) and six soft (i.e., organic) nanoparticle categories. It was then proposed that based on their *atom mimicry* features, these nano-element categories could be combined and hybridized into combinatorial libraries of nano-compounds/assemblies. Just as is observed with traditional atom/small molecule chemistry, many literature examples have demonstrated analogous periodic behavior at the nanoscale level. These entities exhibited unique nano-periodic property patterns that were dependent on one or more CNDPs. In the case of the [S-1]-type nano-element category (i.e., dendrimers), these CNDP-driven nano-periodic property patterns accounted for essentially all *dendritic effects* reported in the literature [2, 45]. Within the past several years, seminal work by Percec, and co-workers [46] has fulfilled and confirmed these hypotheses by publishing first examples of Mendeleev-like nano-periodic tables. These Percec nano-periodic tables demonstrated that by using CNDPs for [S-1]-type amphiphilic dendrons, it was possible to make *a priori* predictions for their mode of self-assembly with 85–93% accuracy.

The remaining sections of this account will review these issues in more detail and will conclude by connecting these accomplishments to present and future efforts to

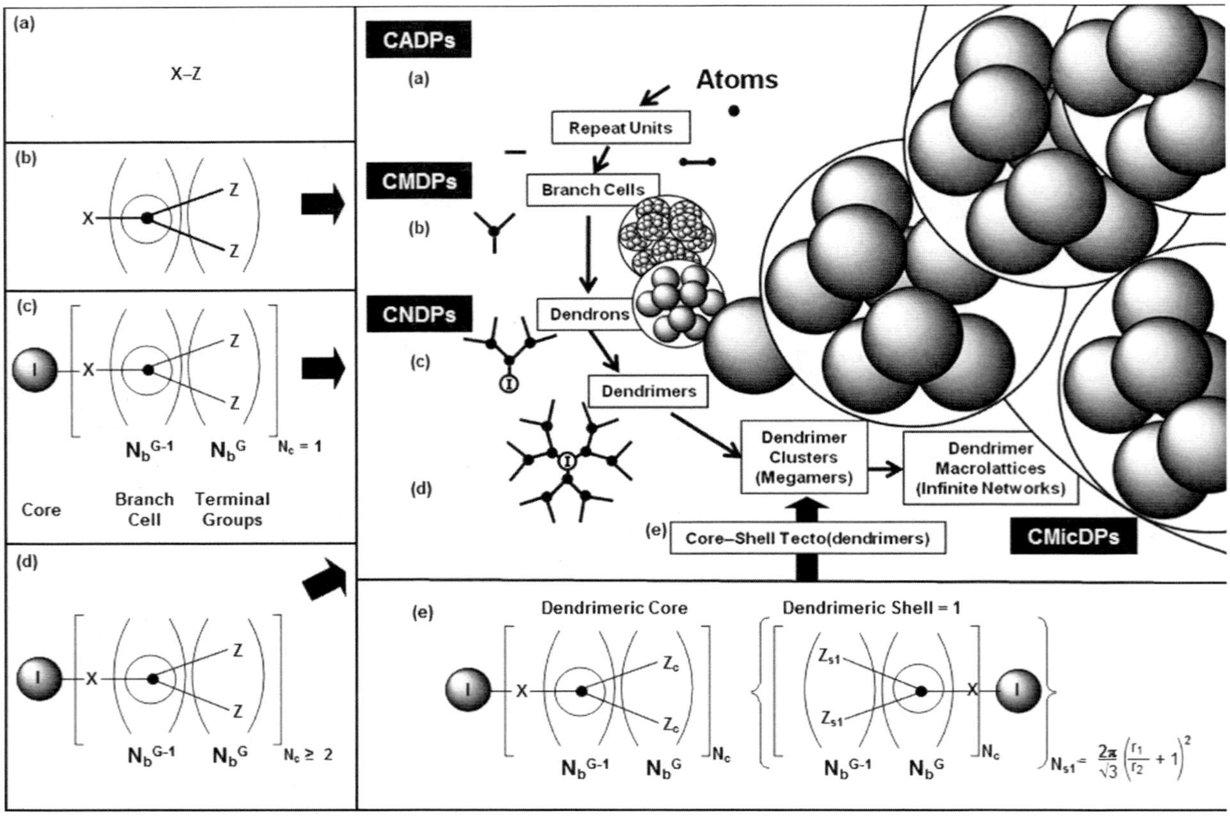

FIGURE 1.14. A mathematically defined, bottom-up aufbau roadmap for constructing and transferring CADP→CMDP's to produce CNDP conserved nanoscale [S-1]-type nano-element category complexity.

develop a comprehensive nano-classification system based on Linnaean-type principles that hopefully might lead to a universal and acceptable approach to a nanomaterials nomenclature.

1.7 A NANO-PERIODIC SYSTEM FOR DEFINING AND UNIFYING NANOSCIENCE

Initial criteria for a nanoscale classification (i.e., taxonomy) system were recently reported by us [2, 31]. Briefly stated, the taxonomy was based on well-defined, monodisperse (i.e., >90%monodisperse/homogeneous) nanostructures, nano-assemblies with dimensions of 1–100 nm, possessing collections of 10^3 to 10^9 atoms and masses of 10^4 to 10^{10} daltons. These well-defined nano-entities were generally obtained by structure-controlled, bottom-up synthetic strategies, as opposed to top-down engineering approaches. Prime examples of such bottom-up aufbau approaches include biological protein or DNA/RNA synthesis, divergent/convergent dendrimer synthesis, and linear polymer synthesis based on Grubbs-type catalysts [47], to mention a few.

Furthermore, it was proposed that these well-defined nano-collections of atoms should possess certain quantized features/properties that manifested some degree of heuristic *atom mimicry*. Atom mimicry was defined as well-defined, nanostructures/assemblies or collections of atoms that exhibited heuristic structural features or quantifiable combining ratios/stoichiometries reminiscent of atoms. These selected candidates were referred to as nano-element categories and divided into types based on (a) conducting, (b) semiconducting, and (c) nonconducting properties. This classification led to two major classes: namely, those reminiscent of traditional inorganic elements (i.e., hard nano-elements) and organic elements (i.e., soft nano-elements). These initial criteria provided broad classifications for dividing well-defined nano-entities into broad taxa or classifications. Using Linnaean-type taxonomy criteria, we now propose further-down selections into more specific taxa (i.e., categories) based on the use of six universal CNDPs. More specifically, these taxa or criteria are related to (a) sizes, (b) shapes, (c) surface chemistries, (d) flexibility/rigidity, (e) architectures, and (f) elemental compositions. These six CHDPs, when structurally controlled, are tangible features that may be quantified and associated with each of the increasingly complex

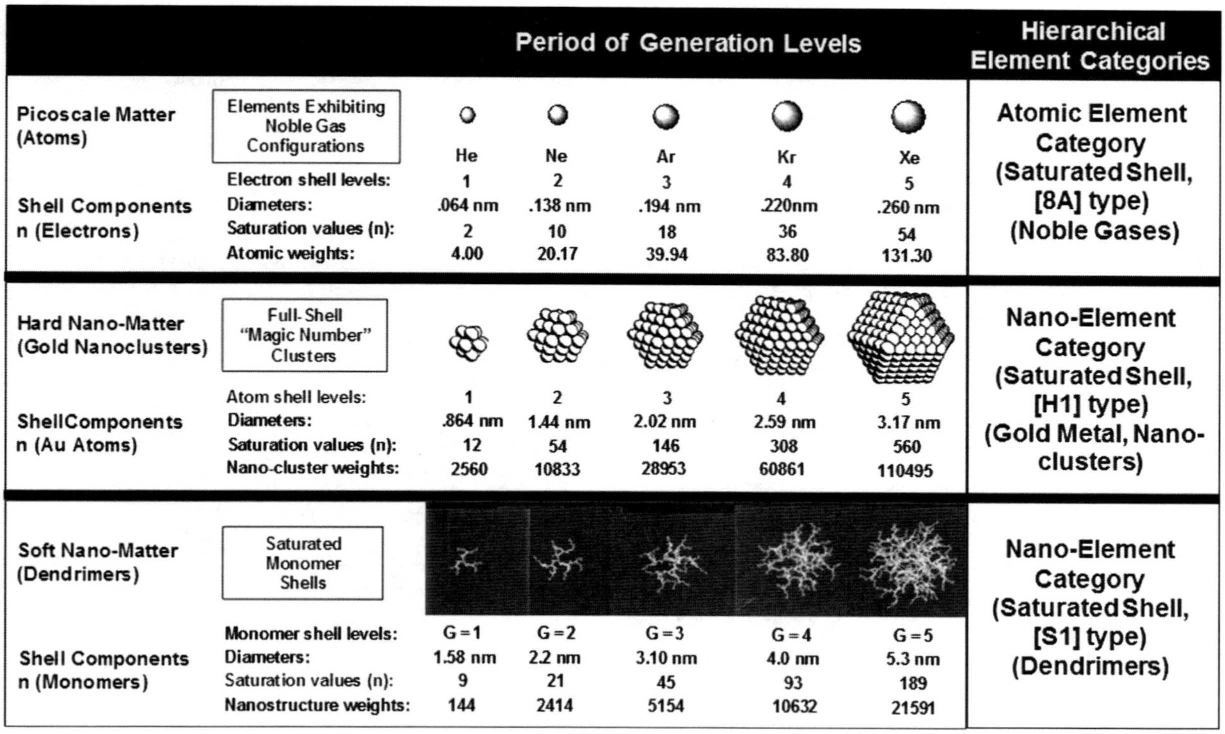

FIGURE 1.15. Comparison of "magic numbers" for atomic picoscale particles, hard nanoparticles, and soft nanoparticle evidence for atom mimicry at the nanoscale level [2]. Center image hard matter reproduced with permission from Elsevier [44].

hierarchical levels as described in Figure 1.2. Furthermore, significant data are accumulating which clearly show that these six taxa (CHDPs) dramatically influence many documented nano-periodic patterns/trends. Quantitation of these six taxa (CNDPs) in the dendrimer field have recently been shown to strongly influence and direct so-called *dendritic effects* [45]. Furthermore, seminal work by Percec and co-workers [46] has shown that these six taxa (i.e., CNDPs) clearly influence the validity of his first examples of Mendeleev-like nano-periodic tables described later in Section 1.8.4 (Figure 1.27).

Using "first principles and step logic" underpinning traditional chemistry, a template and strategy for the evolution of a new "nano-periodic system" and "central dogma" for nanoscience have been advanced [2]. Briefly stated, this concept proposed the following:

1. Creation of a nanomaterials roadmap focused solely on well-defined (i.e., >90% monodisperse) (0-D) and (1-D) nanoscale materials.
2. These well-defined materials were divided into hard and soft nanoparticles, broadly following compositional/architectural criteria for traditional inorganic and organic materials.
3. A preliminary table of *hard and soft nano-element* categories consisting of six hard-matter and six soft-matter particles was proposed, (Figures 1.16 and 1.17). Nano-elemental category selections were based on "atom mimicry" features and the ability to chemically combine or self-assemble like atoms.
4. These hard and soft nano-element categories were shown to produce a wide range of stoichiometric nanostructures by chemical bonding or nonbonding assembly. An abundance of literature examples provides the basis for a combinatorial library of *hard-hard, hard-soft, and soft-soft nano-compounds*, many of which have already been characterized and reported. However, many such predicted constructs remain to be synthesized and characterized [2, 31, 45].
5. Based on the presumed conservation of critical module design parameters (i.e., CADP→CMDP→CNDP) (Figure 1.4), many new emerging *nano-periodic property patterns* have been reported in the literature for both the hard and soft nano-element categories and their compounds [2].

1.7.1 Bottom-Up Synthetic Strategies to Soft Nano-element Categories

Presently, the broad grouping of soft nano-element categories can be divided into two general types: namely, synthetic (i.e., [S-1], [S-2], [S-3]) and biopolymers (i.e., [S-4], [S-5],

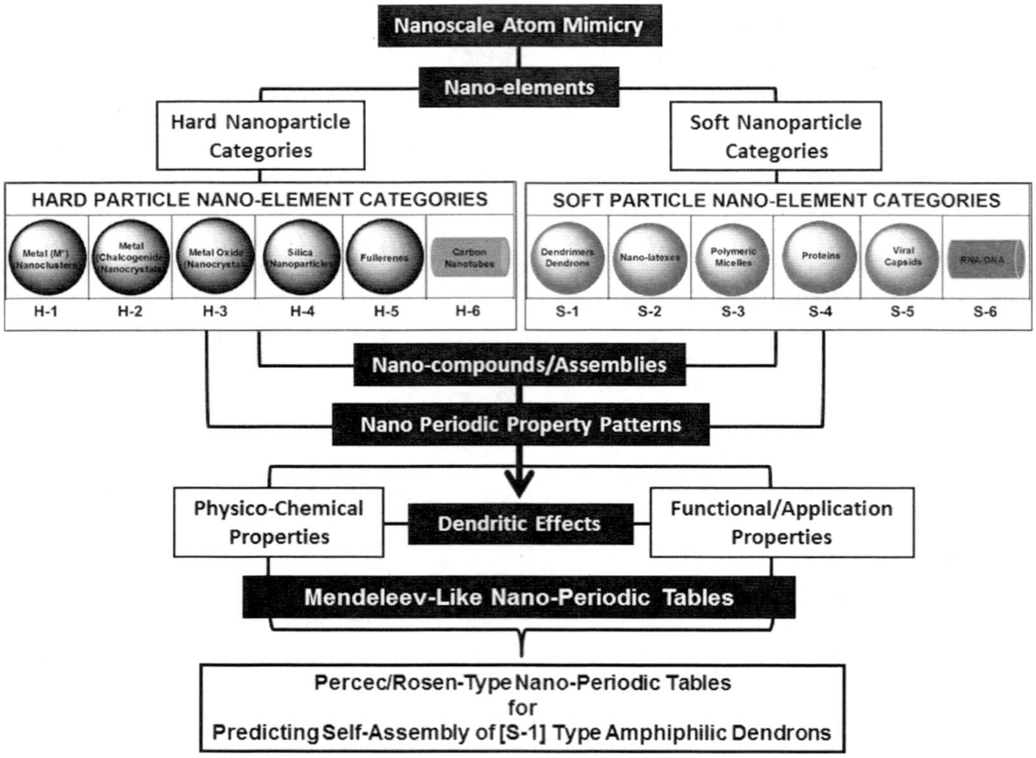

FIGURE 1.16. The first examples of Mendeleev-like nano-periodic tables have recently fulfilled these expected nano property pattern/trend predictions [2, 31]. Percec/Rosen [46] have reported the first three nano-periodic tables for predicting the self-assembly patterns for [S-1] type amphiphilic dendrons with predictive accuracies of 85% to >90% based on knowledge of the primary dendron CNDPs: namely, (a) size, (b) shape, (c) surface/apex chemistry, and (d) flexibility/rigidity [45]. See color insert.

[S-6]) based on the specific aufbau strategies that are used to produce them (Figure 1.18). Several well-known synthesis strategies are shown that have been reported to produce highly structure-controlled aufbau pathways to each category with associated conservation of all CNDPs. It should be noted that specific procedures are generally available to produce these soft element categories in a variety of architectures as noted in Figure 1.18. In this account, we focus mainly on *dendrons and dendrimers* as well as *proteins, viral capsids, and DNA/RNA*—all of which have been proposed as important *soft-matter, nano-element* categories for defining a new nano-periodic sysytem [2]. Since the discovery of dendrimers several decades ago [1, 24, 25, 48–50], significant evidence clearly shows that these nanomaterials react and self-assemble as discrete, quantized nano-units. These well-defined collections of covalently bonded atoms exhibit many features normally associated with traditional elemental atoms (i.e., *atom mimicry*). As such, dendrimers behave much as *soft particle nano-elements* to produce well characterized core-shell (tecto)dendrimers (i.e., *dendrimer-dendrimer-type nano-compounds*). Similarly, combining [S-6]-type RNA/DNA cores with [S-5]-type protein subunits to form nano-compounds such as tobacco mosaic virus as shown in Figure 1.19 and will be described in Section 1.8.3. These results optimistically portend the extension of this concept to a wide variety of other well-defined soft-particle nano-elements, as well as hard-particle nano-elements (i.e., metal oxide nanocrystals, metal chalcogenide nanocrystals, etc.) which have been reported elsewhere [2].

1.8 CHEMICAL BOND FORMATION/VALENCY AND STOICHIOMETRIC BINDING RATIOS WITH DENDRIMERS TO FORM NANO-COMPOUNDS

Using first principles and step logic invoked by Dalton (i.e., *Philosophy for a Chemical System*, 1808 [51] and others, it was possible to experimentally demonstrate that certain quantized nano-modules (i.e., dendrimers, fullerenes, metal nano-clusters, or metal oxide nanocrystals) could be chemically combined or assembled to produce stoichiometric nano-compounds/assemblies possessing well-defined mass combining ratios. Furthermore, these 12 soft and hard nano-elements, designated [S-n] and [H-n], respectively, have been reported to form a wide range of soft-particle and

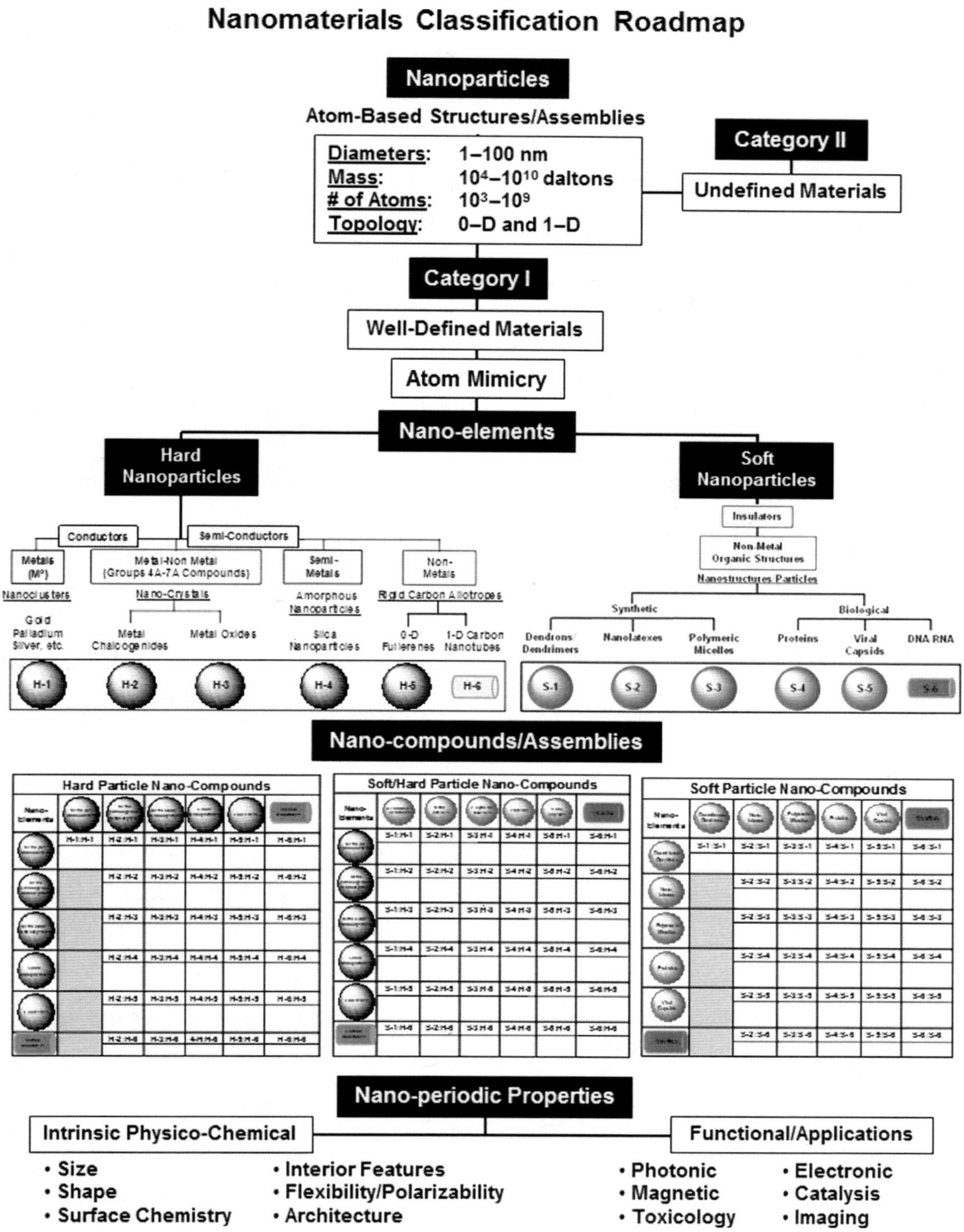

FIGURE 1.17. Concept overview: Using first principles and step logic that led to the "central dogma" for traditional chemistry, the criteria of nanoscale atom mimicry was applied to Category I-type, well-defined nanoparticles. This produced 12 proposed nano-element categories that were classified into six hard particle and six soft-particle nano-element categories. Chemically bonding or assembling these hard and soft nano-elements leads to hard:hard-, soft:hard-, or soft:soft-type nano-compound categories, many of which have been reported in the literature. Based on the discrete, quantized features associated with the proposed nano-elements and their compounds, an abundance of nano-periodic property patterns related to their intrinsic physico-chemical and functional/application properties have been observed and reported in the literature [2].

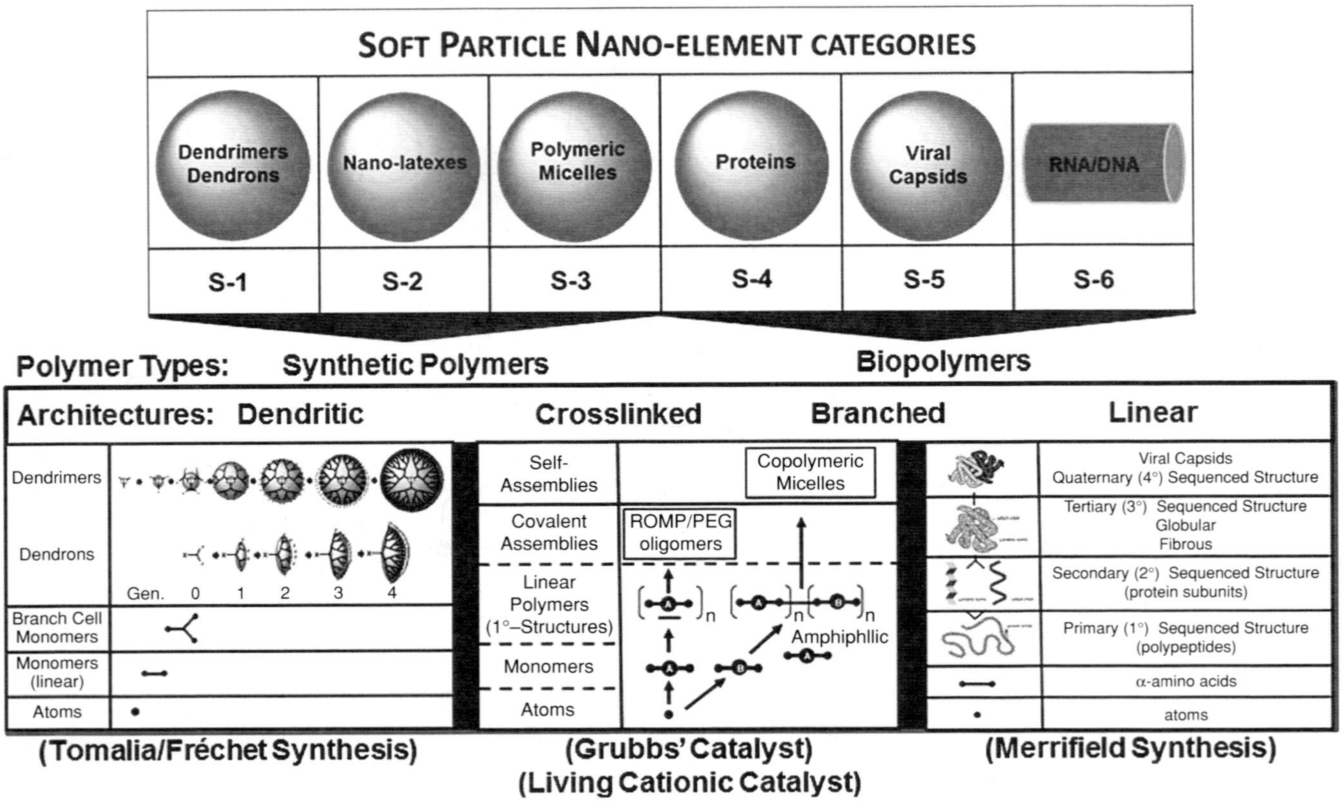

FIGURE 1.18. Initial list of six soft nano-element categories (i.e., [S-1], [S-2], etc.) and associated bottom-up aufbau strategies and precursors. Many of these well-defined precursors or their analogues may exist or organize into nanoscale collections/architectures of atomic elements, monomers, organic/inorganic structures, oligomers, polymers, and so on, that collectively manifest heuristic atom mimicry features. For example, they may collectively exhibit discrete relationships, mass combining ratios, or stoichiometries with other soft or hard nano-element categories. As a component in such a discrete nano-element relationship, these precursors may also collectively exhibit new emerging intrinsic physicochemical trends or functional/application property patterns. *A new expanded nano-element category may have to be considered to accommodate these unique nano-collections of soft and hard matter.*

soft–hard-particle-type nano-compounds. Both the nano-elements and their nano-compounds are widely recognized to exhibit new emerging properties and nano-periodic property patterns [2]. At least three unique core–shell nano-compound category types may be formed by combining appropriate soft and hard nano-elements as illustrated in Figure 1.17. For example, combining [S-1]-type dendrimers with other dendrimers has been shown to produce precise, stoichiometric [S-1:(S-1)$_n$] core–shell-type nano-compounds (Figure 1.19), whereas [H-4]-type fullerenes reacted with [S-1]-type dendrimers to give [S-1:(H-4)$_n$] core-shell type nano-compounds and finally [S-1]-type dendrimers were found to react with [H-1]-type metal nano-clusters to yield stoichiometric [H-1:(S-1)$_n$] core–shell-type nano-compounds. Leading literature references to these nano-compound types and others in the three combinatorial nano-compound libraries (Figure 1.17) are described in greater detail elsewhere [2].

1.8.1 Dendrimer–Dendrimer [S-1:(S-1)$_n$] Core–Shell-Type Nano-compounds

Saturated-shell nano-compounds such as dendrimer–cluster compounds (Figure 1.20) are prepared by a two-step approach that involved, firstly, self-assembly of an excess of carboxylic acid terminated dendrimers (i.e., shell reagent) around a limited amount of amine-terminated dendrimer (i.e., core reagent) in the presence of LiCl. This was followed by covalent amide bond formation between the core and dendrimer shell reagents using a carbodiimide reagent [52–54]. The resulting nano-compounds (i.e., saturated core–shell tecto (dendrimers), referred to as *megamers)* are prime examples of precise poly-dendrimer cluster structures as shown in Figure 1.20.

The stoichiometries of these structures may be predicted mathematically by the Mansfield–Tomalia–Rakesh equation

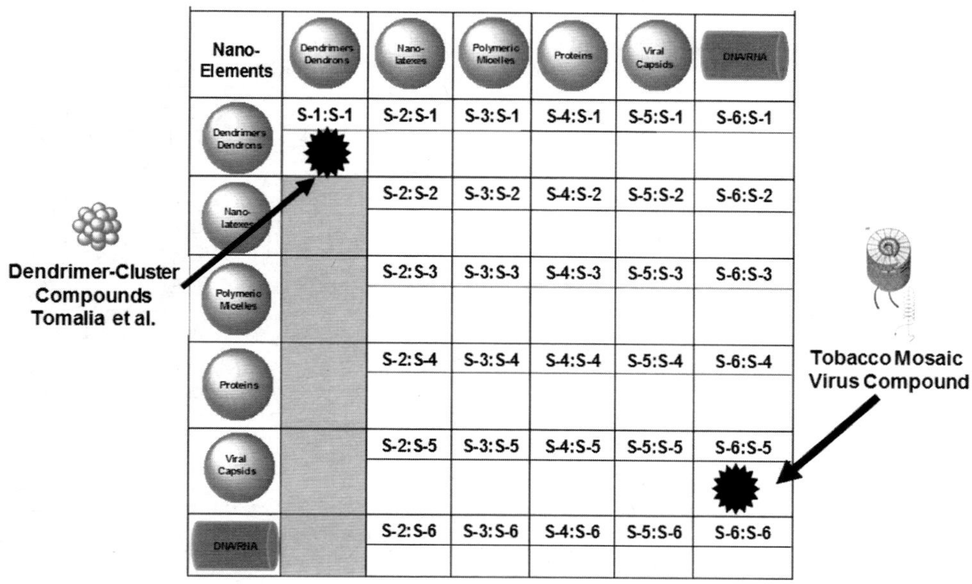

FIGURE 1.19. A soft nano-element combinatorial library indicating possible soft–soft-type nano-compounds. Highlighted dendrimer–cluster compounds [S-1:(S-1)$_n$] and the tobacco mosaic virus [S-6:(S-1)$_{2130}$]-type or [S-6:(S-5)]-type nano-compounds.

Figure 1.21 [1,55] and are dependent upon the relative diameters of dendrimer core reagent relative to dendrimer shell reagent. The predicted stoichiometric limits (N_{max}) are determined by the core–shell spheroid ratios that are shown in the far-right column. These predicted stoichiometries have been unequivocally verified by experimental mass spectrometry, gel electrophoresis, and atomic force field microscopy (AFM) [1, 52, 54, 56]. It is interesting to note that when the core and shell have identical diameters to produce a spheroidal core–shell ratio of one, as is the case for all core–shell, metal nanoclusters, one should expect a core–shell stoichiometry of (N_{max}) = 12 (Figure 1.14). *This is exactly what is predicted according to the* Mansfield–Tomalia–Rakesh equation (Figure 1.21).

Combinatorial libraries of these [S-1:(S-1)$_n$]; core–shell-type nano-compounds can be readily synthesized by simply combining various dendrimer generations according to this protocol. Furthermore, various dendrimer families (i.e., PAMAM, PPI, etc.) may be used to produce a variety of dendrimer cluster–compositional copolymers. As shown in Figure 1.22, using a $G = 7$; PAMAM dendrimer as a core reagent and $G = 5$; PAMAM dendrimers in excess as the shell reagent produces a core-shell nano-compound (i.e., [S-1:(S-1)$_{13}$] with a stoichiometry of 1:13. Many examples

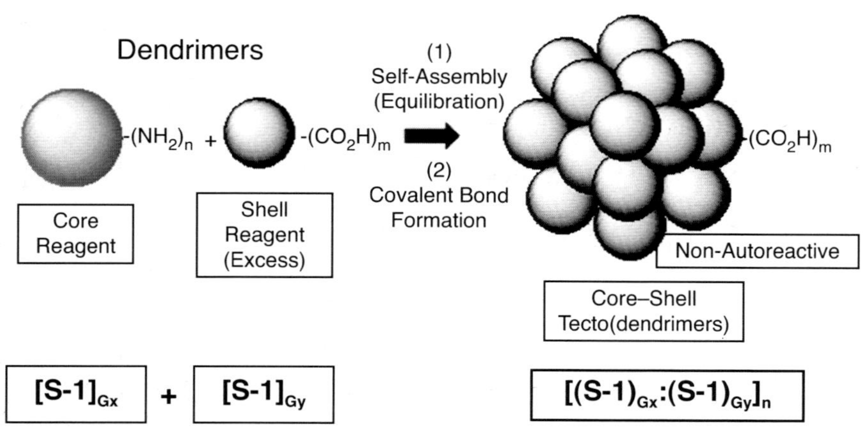

FIGURE 1.20. The saturated-shell-architecture approach to covalent megamer synthesis. All surface dendrimers are carboxylic acid terminated [53].

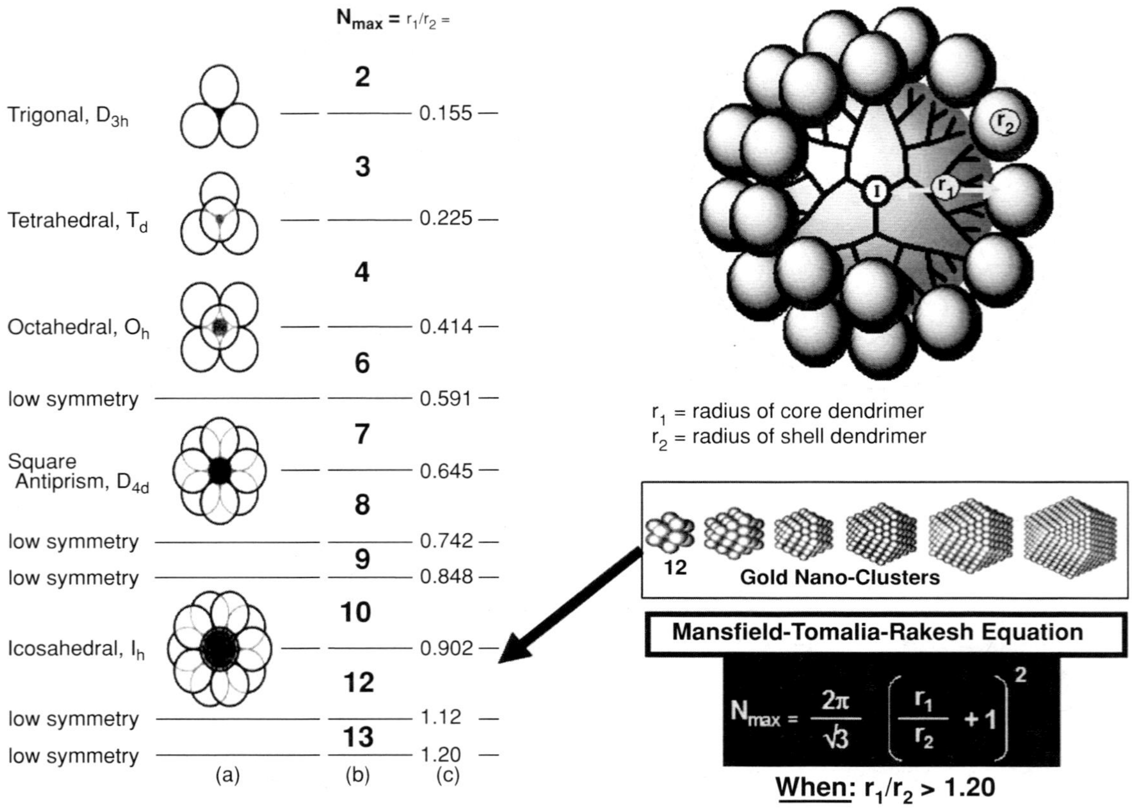

FIGURE 1.21. (**a**) Symmetry properties of core–shell tecto(dendrimer) structures, when $r_1/r_2 < 1.20$. (**b**) Sterically induced stoichiometry (SIS) defined shell capacities ($N_{max.}$), based on the respective core and shell radii, when $r_1/r_2 < 1.20$. (**c**) Mansfield–Tomalia–Rakesh equation for calculating the maximum shell filling value (capacity)($N_{max.}$), when $r_1/r_2 > 1.20$ [1, 31, 55]. It should be noted that when the core spheroid radius equals the shell spheroid radius, as in metal nano-clusters (see Figure 1.15), then the ($N_{max.}$) = 12 as shown above. Metal nano-cluster image reprinted [44] with permission of Elsevier.

of these well–characterized core–shell nano-compounds are described in detail elsewhere [53].

1.8.2 A Quest for Synthetic Mimicry of Biological Quasi-equivalence with [S-1]-Type Amphiphilic Dendrons

As early as 1992, Percec et al. [57] compared the similarity of supramolecular nano-cylinders obtained from his amphiphilic dendrons to the supramolecular assembly of protein subunits to produce the cylindrical viral capsids that surround RNA in tobacco mosaic viruses (TMVs). More recently, Percec [58] reviewed the historical inspiration provided by Klug's seminal Nobel Prize work on the structure of TMV [59, 60]. Percec was able to show unequivocally that dendrons behave much like protein subunits to produce a rich variety of cylindrical and spherical supramolecular dendrimers that exhibit quasi-equivalency much as is noted in many viral capsids. Based on accelerated design strategies involving synthetic amphiphilic dendrons, Percec et al. [61–64] were able to demonstrate the quasi-equivalent mimicry of biological systems by using retrostructural analysis [64] of their periodic and quasi-periodic supramolecule dendrimer assemblies as outlined in Figure 1.23. This remarkable comparison corroborates and documents many dendron libraries and other examples of dendron/dendrimer-based *protein mimicry* [65–67].

1.8.3 Tobacco Mosaic Virus: Compelling Example of a Supramolecular Core–Shell-Type Nano-compound Exhibiting Well-Defined Stoichiometry: Self-Assembly of Protein Subunits [S-4] around a [S-6]; ssRNA Core to Produce [S-6:(S-4)$_{2130}$]

More than three decades ago, important stoichiometric, self-assembly relationships were noted by Klug et al. [59, 60, 68] between the ss-RNA core and the self-assembling protein

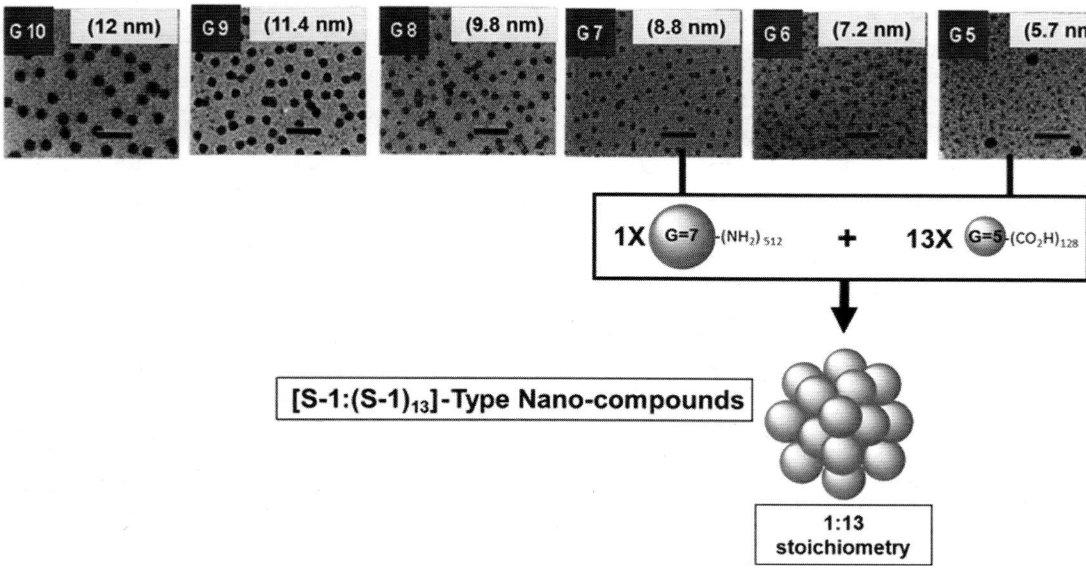

FIGURE 1.22. A size comparison of TEMs for Tomalia-type PAMAM dendrimers; $G = 5$–10. Covalent synthesis of a core–shell tecto (dendrimer) by combining a dendrimer; $G = 7$ core reagent with and excess of dendrimer; $G = 5$ shell reagent to produce the dendrimer cluster with a stoichiometry of 1:13.

subunits in the formation of tobacco mosaic viruses (TMVs). The stoichiometric relationship between the viral core and the viral capsid was carefully documented by x-ray studies. This work rigorously demonstrated that exactly 2130 protein subunits assembled to form a viral capsid shell around a ss-RNA core to produce TMV with diameter = 18 nm, length = 300 nm, and helical symmetry. Elucidation of this self-assembly process together with the unprecedented characterization of this viral assembly by x-ray analysis, garnered the Nobel Prize for A. Klug in 1982. In the context of the systematic, nano-periodic concept [2], this viral construct may be viewed as a supramolecular, stoichiometric core–shell [S-6:$(S\text{-}4)_{2130}$]-type nano-assembly as described in Figure 1.24.

Viral capsids were included in the original list of six soft nano-element categories. These entities consist of a large group of well-defined nano-assemblies/objects that are known to be very well-defined as a function of their CNDPs. They are derived from the self-assembly of protein subunits (i.e., [S-4]-type nano-elements) and fulfill our criteria for their designation as [S-5]-type soft nano-elements. Similarly, RNA and DNA structures (i.e., designated as the [S-6]-type nano-element category) also fulfilled this criteria and were included on this list. As such, it is noteworthy to point out

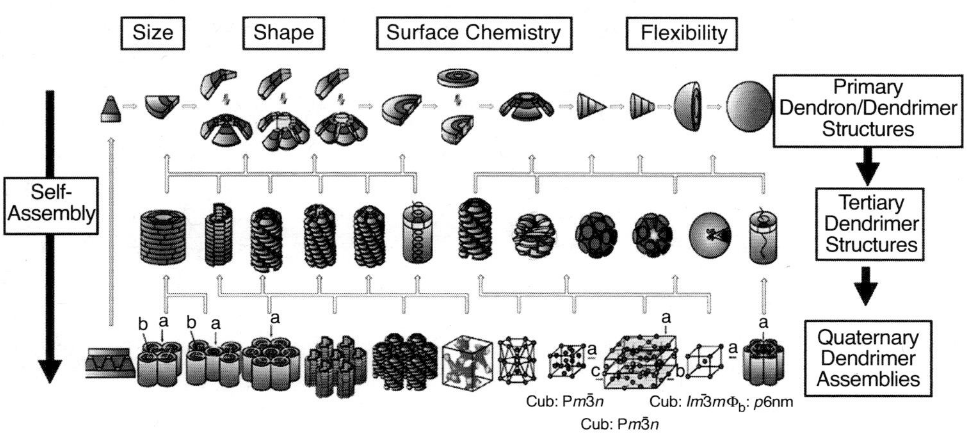

FIGURE 1.23. Dependency of self-assembly patterns leading to tertiary and quaternary dendron assemblies based on primary structure controlled dendron CNDP's such as (a) size, (b) shape, (c) surface/apex chemistry, and (d) flexibility [46]. Copyright 2009 American Chemical Society.

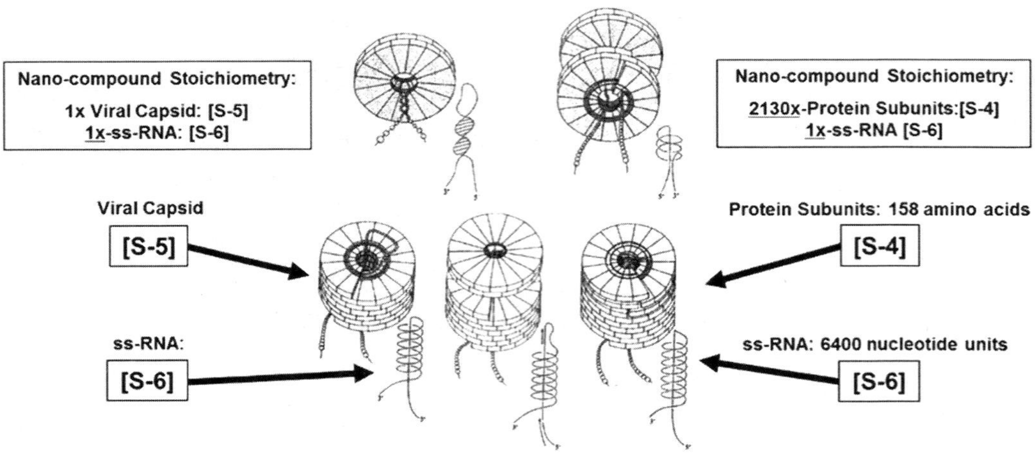

FIGURE 1.24. Tobacco mosaic virus (TMV). An example of a well-defined nano-compound [S-6:$(S\text{-}4)_{2130}$] consisting of an ss-RNA(core):protein subunits (shell) that has nano-dimensions of diameter 18 nm, length 300 nm, and helical symmetry [68]. Adapted with permission from *Scientific American*.

that in the case of TMV, precise stoichiometries are observed for the protein to genome constituents. For example, the protein [S-4] to RNA [S-5] stoichiometry is 2130:1, whereas the viral capsid [S-5] to RNA [S-6] is 1:1.

Similarly, the M13 bacteriophage virus is derived from a single-strand DNA (ssDNA) that is 6407 nucleotides long. This genomic core is encapsulated with ~2700 protein subunits to produce a filament-like cylinder with a diameter of 6.6 nm and length of 880 nm. This viral-type nano-compound exhibits a protein [S-4] to DNA [S-5] stoichiometry of 2700:1. Recent work has shown that these nano-compound building blocks can be self-assembled to produce a wide variety of higher hierarchical organizations that conserve and communicate their CNDPs well into the micron-scale region [69].

It will be interesting to note whether all classified viruses exhibit precise and well-defined stoichiometries between their genomic and protein constituents. *If this is the case, then all viruses may be viewed as nano-compounds in the context of our nano-periodic system.*

Inspired by Klug's work on TMV, the Percec group synthesized and analyzed innumerable libraries of self-assembling amphiphilic dendrons (Figure 1.25) [70,71]. For each library, the dendron primary structures were compared to the tertiary structures of the self-assembled supramolecular dendrimers and the quaternary structure of the crystal lattices. A sampling of these libraries reveals primary dendron structures derived from AB_2; 3,4-, AB_2; 3,5-, AB_3; 3,4,5-dendrons, to mention a few [46]. A typical library for an AB_2; 3,4-disubstituted biphenyl (Bp) dendron family is characterized as a function of dendron CNDPs such as generation (size), surface/apex chemistry, and shape and flexibility as shown in Figure 1.26. These analyses clearly reveal important dendron parameters such as (a) the molecular solid angle (α') of the dendron, (b) morphology (shape) of the supramolecular dendrimer, and (c) aggregation number (μ) (i.e., *supramolecular dendrimer stoichiometry*) varied in a predictive manner to reveal important self-assembly patterns as a function of dendron generation. It should be noted that very precise reproducible stoichiometries were observed for these dendron self-assemblies as evidenced by their discrete aggregation numbers, namely, $[S\text{-}1]_n$ (see Figure 1.26). For example, these library analyses revealed interesting patterns such as: Increasing the generation number causes a change in molecular solid angle (α') and typically a transition from lamellar to columnar and spherical assemblies. Increasing the generation number does not necessarily increase the diameter of the supramolecular dendrimer (D), but generally reduces the aggregation number (i.e., μ) or number of dendrons required to form a supramolecular sphere or the cross section of a supramolecular column. Deviations from these patterns usually indicate the formation of hollow core

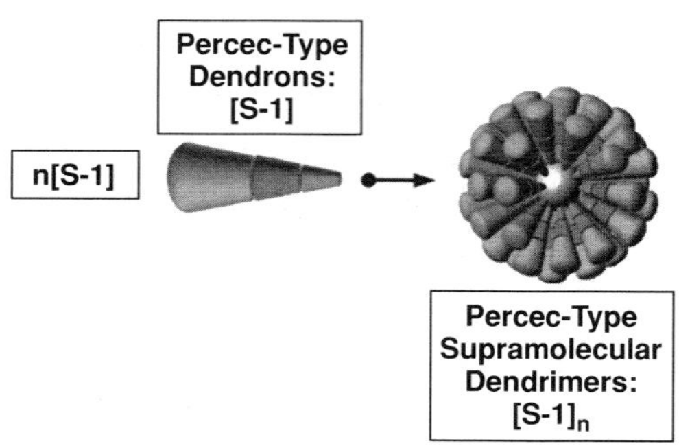

FIGURE 1.25. Amphiphilic dendron self-assembly libraries directed by the critical nanoscale design parameters (CNDPs): (a) size, (b), shape, (c) surface chemistry, and (d) flexibility. Copyright 2009 American Chemical Society [70].

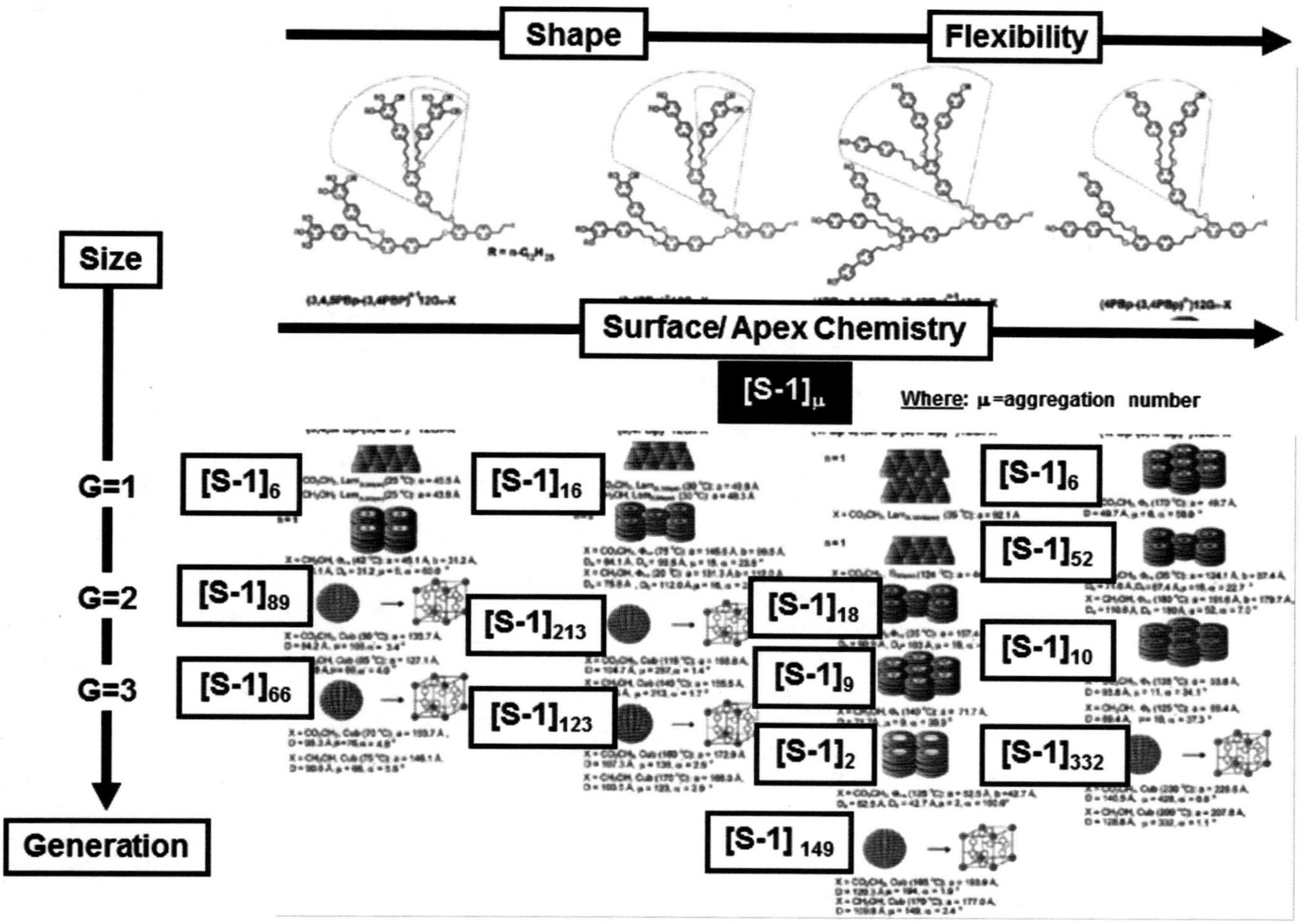

FIGURE 1.26. Structural and retrostructural analysis of supramolecular dendrimers $[S\text{-}1]_{\mu}$ derived from the self-assembly library of AB_2; 3,4-disubstituted-(PBp)-type amphiphilic dendrons; [S-1] [46, 71]. Copyright 2009 American Chemical Society.

supramolecular dendrimers or other novel mechanisms of self-assembly. Generally AB_3; 3,4,5-trisubstitued libraries exhibit more spherical structures when compared to AB_2; 3,4-disubstituted dendron libraries.

Furthermore, it was shown by Percec et al. [46] that simply knowing the four CNDPs—namely, (a) size, (b) shape, (c) surface chemistry, and (d) flexibility of the primary dendron structure—one can predict self-assembly patterns leading to tertiary and quaternary structures with greater than 85–93% accuracy as shown in nano-periodic Table 1.1 (Figure 1.27).

1.8.4 First Nano-periodic Tables for Predicting Amphiphilic Dendron Self-Assembly to Supramolecular Dendrimers Based on the Critical Nanoscale Design Parameters

Like proteins, the primary structures of the amphiphilic dendrons determine their tertiary structure. As such, Percec has compared dozens of his AB_2- and AB_3-derived dendron libraries in an effort to determine trends or "nano-periodic self-assembly patterns" as proposed by others [2]. Percec's seminal comparison produced the first three Mendeleev-like, predictive Nano-periodic Tables I for the self-assembly of aryl ether dendrons, which are described elsewhere [46]. One of these nano-periodic tables is as illustrated in Figure 1.27. These three nano-periodic tables summarize the tertiary and quaternary structures that are formed for similar primary dendron structures, however, using different dendron building blocks. They provide predictive nano-periodic tables that describe general trends in the sequence–structure relationship (i.e., primary→secondary→ tertiary→ quaternary structures). Furthermore, they identify clustered regions where specific structures will be found. The supramolecular dendrimer structures formed may be classified into lamellar, columnar, or spherical morphologies by analogy to β-sheets, helical structures of fibrillar proteins, and the pseudo-spherical structure of globular proteins. In all three

TABLE 1.1. Proposed Linnaean-Type Taxonomy for Classifying Dendrimers

Linnaean Taxa	Similarity Features
• Kingdom	Picoscale, subnanoscale, nanoscale, micron-scale
• Phylum	Synthetic, synthetic-biological, biological
• Class	Linear, cross-linked (bridged), branched, dendritic
• Order	Inorganic, inorganic-organic, organic
• Family	Elemental/connectivity composition (i.e., PAMAM, PPI, PLys)
Genus	(a) Core-type (i.e., NH_3, EDA, DAB, etc.)
	(b) Core/branch cell multiplicities (i.e., $N_c = 3/N_b = 2$)
	(c) Generation level (i.e., $G = 4$)
Species	(a) Terminal/surface chemistry (i.e., electrophilic, nucleophilic, etc.)
	(b) Surface functionality (i.e., amine, hydroxyl, carboxylic, etc.)
	(c) Terminal/surface valency (Z) (i.e., $Z = N_c N_b{}^G = 48$)

Shorthand Nomenclature: *Core:* NH_3; $N_c = 3 \rightarrow N_b = 2$; *dendri*-{poly(amidoamine)-$(NH_2)_{48}$}; ($G = 4$) (PAMAM) dendrimer.

nano-periodic tables, G = 1 dendrons behave similarly and exhibit a high proportion of lamellar and columnar structures including hollow columnar structures.

In the case of the AB_3; *3,4,5–trisubstituted dendron-based* Nano-periodic Table I, most self-assemblies lead to predominately spherical structures at $G = 3$ and entirely spherical forms from $G = 3$–5. The spherical supramolecular dendrimers generally pack to form Cub-type crystal lattices. At $G = 1$, the columnar structures form various ϕ lattices. At $G = 2$, columnar structure pack almost exclusively into the $\O_h$ lattice. These Mendeleev-like nano-periodic tables provide first examples of a systematic format that demonstrate a well-established dependency on conserved CNDPs for *a priori* predictions of [S-1]-type amphiphilic dendron self-assembly behavior.

Very recently, Mirkin and co-workers [72] reported similar nano-periodic property patterns for the assembly of 3-D lattices derived from discrete gold nano-clusters (i.e., [H-1]-type nano-elements) modified with complementary single-strand DNA (i.e., [S-6]-type nano-elements). This work describes first examples of nano-periodic property patterns for hard–soft nano-element combinations. By analogy, to Pauling's historical rules [73] for traditional picoscale atomic element combinations to produce crystal lattices, Mirkin has defined six basic nano-periodic rules that may be used to predict and tune properties for these [H-1:(S-6)$_n$] type 3-D nano-lattices. In conclusion, these remarkable x-ray supported studies reported by Percec and co-workers [46] and Mirkin and co-workers [72] for soft and hard nano-element categories, respectively, provide the first experimental confirmation of the main principles underpinning a proposed nano-periodic system for unifying nanoscience [2,31].

With this as a brief background, we now turn attention in the next section to our original proposal to develop a Linnaean-type taxonomy system for classifying nano-element categories and their subsequent combinatorial hybridizations into nano-compounds and assemblies.

1.9 PROPOSED LINNAEAN-TYPE TAXONOMY FOR SOFT-MATTER-TYPE NANO-ELEMENT CATEGORIES, THEIR COMPOUNDS AND ASSEMBLIES

A broad Linnaean-like template for classification of well-defined nanostructure/objects could involve a simple format such as is described in Table 1.1. For example, classification of an [S-1], ammonia-core, amine-terminated, Tomalia-type poly(amidoamine) (PAMAM) dendrimer might involve the following steps and operations: (a) First, this [S-1]-type nanostructure could be broadly classified according to Linnaean principles, using the taxa shown down to the taxon-specificity of the *order*, namely, a nanoscale, synthetic, dendritic macromolecular structure, (b) the *family* is poly(amidoamine) PAMAM, and the *genus* consists of the core type, core/branch cell multiplicities, and generation level (G); and (c) the *species* can be described as possessing 48-nucleophilic, amine-terminal groups. Meanwhile, a shorthand nomenclature emerges from this classification, which utilizes these taxa and is further described in Figures 1.28–1.29 for various dendrimer families.

1.9.1 A Proposed Dendron/Dendrimer Shorthand Nomenclature

Dendrons and dendrimers have been widely recognized as critical pivotal building blocks in the *dendritic state*. Although several traditional organic nomenclature schemes have been proposed [74,75] to describe and classify the many reported dendron and dendrimer families, these terminologies have not gained wide usage due to their complexity [74,76]. Meanwhile, the evolution of a shorthand nomenclature scheme has become a "works in progress" based on the hybridization of contributions/suggestions extracted from a number of examples in the literature. This shorthand nomenclature involves associating specific dendrimer components

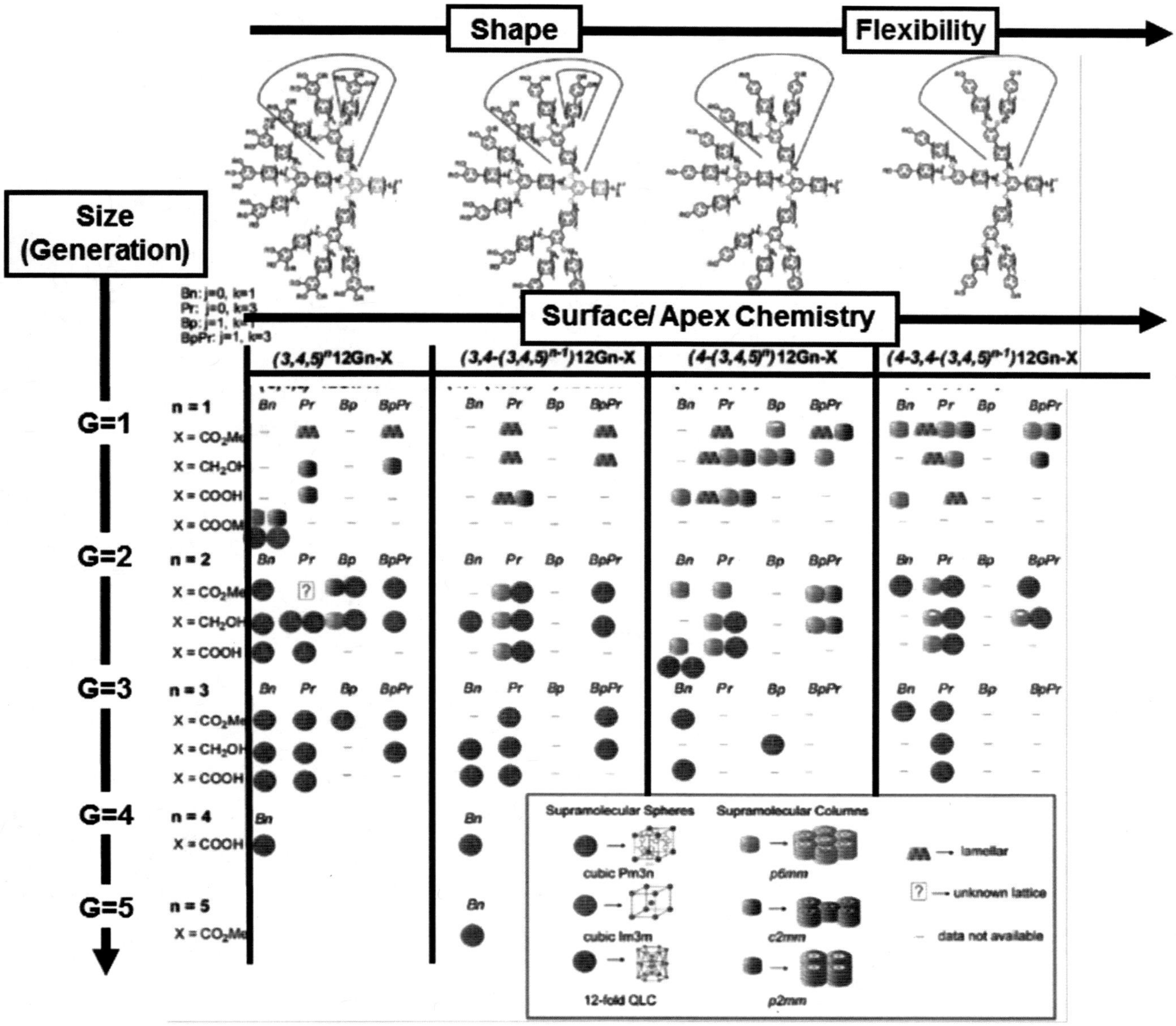

FIGURE 1.27. Nano-Periodic Table I: Primary dendron structures **[S-1]** versus 3-D supramolecular dendrimer structures $[S\text{-}1]_\mu$ for all libraries of AB_3 supramolecular dendrimers. **Bn**, benzyl ether; **Pr**, phenylpropylether; **Bp**, biphenyl-4-methyl ether, **BpPr**, biphenylpropyl ether [46]. Copyright 2009 American Chemical Society.

with the well-known general mathematical expression defining the three major architectural components (i.e., core, interior and terminal groups) (Figure 1.28). These architectural components are in turn described systematically by indicating core → interior branch cell connectivity (i.e., $N_c \rightarrow N_b$ multiplicities), a stoichiometric value for number of terminal groups (Z) and a generation number as described below. Several well-known dendrimer families associated with their scientific originators can be described using this shorthand nomenclature as described with structures (a) and (b)[77] in Figure 1.29.

1.9.2 Classification of [S-1:(S-1)$_n$]-Type Nano-compounds Derived from Dendrimer/Dendron [S-1]-Type Nano-element Categories

The stoichiometry boundaries for combinatorial libraries of dendrimer nano-compounds [S-1:(S-1)$_n$] derived from spheriodal dendrimers based on generational sizes are outlined and defined by the Mansfield–Tomalia–Rakesh equation as described in Figure 1.21. Other variations could involve combinations of different families of dendrimers. For example, combining a PAMAM dendrimer (S-1) with

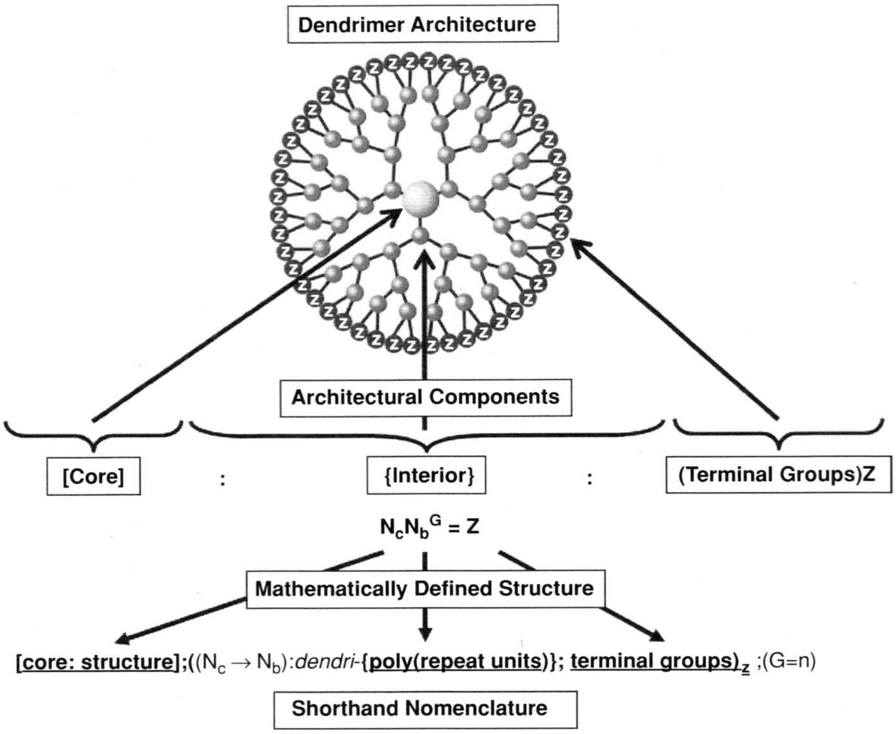

FIGURE 1.28. Association of dendrimer architectural components mathematical expression to define dendrimer structure, where N_c is core multiplicity, N_b is branch cell multiplicity, G represents generation, Z represents teminal groups, and proposed shorthand nomenclature is based on these parameters.

PPI dendrimers (S-1)′ as in Figure 1.29a would produce large series of [(S-1:(S-1)′$_n$] nano-compounds. Of course, many other stoichiometries would be possible by changing the dendrimer shape (i.e., combining spherical with cylindrical shapes). That withstanding, attaching dendrons or dendrimers to the surface of other well-defined spheroidal nanostructures or assemblies such as gold nano-clusters; (i.e., [H-1]-type) would be expected to produce combinatorial series of hard–soft [H-1:(S-1)$_n$]-type nano-compounds. Many of these nano-compounds have been reported in the literature [2, 31].

Using combinations of hard and soft nano-element categories, Mirkin and co-workers [72] recently reported the assembly of gold nano-clusters (i.e., [H-1]-type nano-elements) modified with complementary single-strand DNA (i.e., [S-6]-type nano-elements). These interactive nano-conjugates (i.e., nano-compounds) self-assembled to form libraries of 3-D nano-lattices that mimicked traditional inorganic salt structures. Furthermore, these 3-D, [H-1:(S-6)$_n$], hard–soft-type crystal lattices exhibited crystal structures that unequivocally mimicked elemental atoms and were confirmed by x-ray studies.

As such, it is proposed that systematic cladograms could be developed to classify these possibilities according to similarity principles and features. The development of taxonomies for these various nano-element combinations would be expected to lead to suitable specific nomenclatures as witnessed for other Linnaean-type taxonomies. Nomenclatures have evolved in the case of certain soft nano-element categories, such as [S-4], [S-5], and [S-6] which are associated with the stoichiometric (genomic:capsid) construction of viruses. These issues are described in the next section.

1.9.3 Classification of Nano-compounds (i.e., Viruses) Derived from Proteins [S-4] or Viral Capsids [S-5] and DNA/RNA [S-6]-Type Nano-element Categories

A comprehensive taxonomy based on similarity features for viral capsids and viruses was initiated as early as 1962. This hierarchical classification pioneered by A. Lwoff, R. Horne, and P. Tournier [23], classified viruses according to shared properties with a focus on the following four features:

1. Nature of the nucleic acid core (i.e., RNA/DNA).
2. Symmetry of the capsid.
3. Presence or absence of an envelope.
4. Dimensions of the virion and capsid.

This classification scheme leads to two rather comprehensive cladograms that are divided broadly according to their

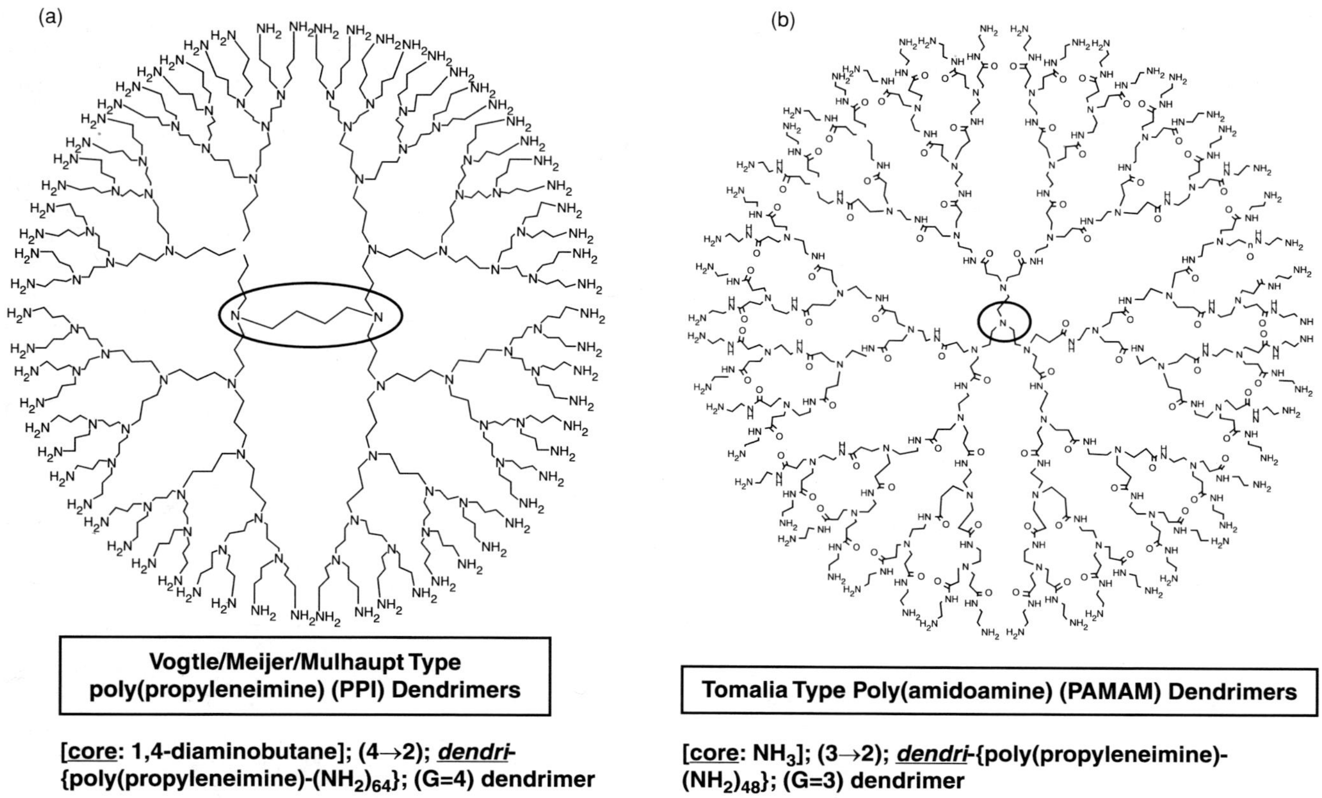

FIGURE 1.29. Various dendrimer families: (**a**) Vögtle/Meijer/Mulhaupt-type poly(propyleneimine) (PPI) dendrimers. (**b**) Tomalia-type poly(amidoamine) (PAMAM) dendrimers. Reproduced with permission from Elsevier [77].

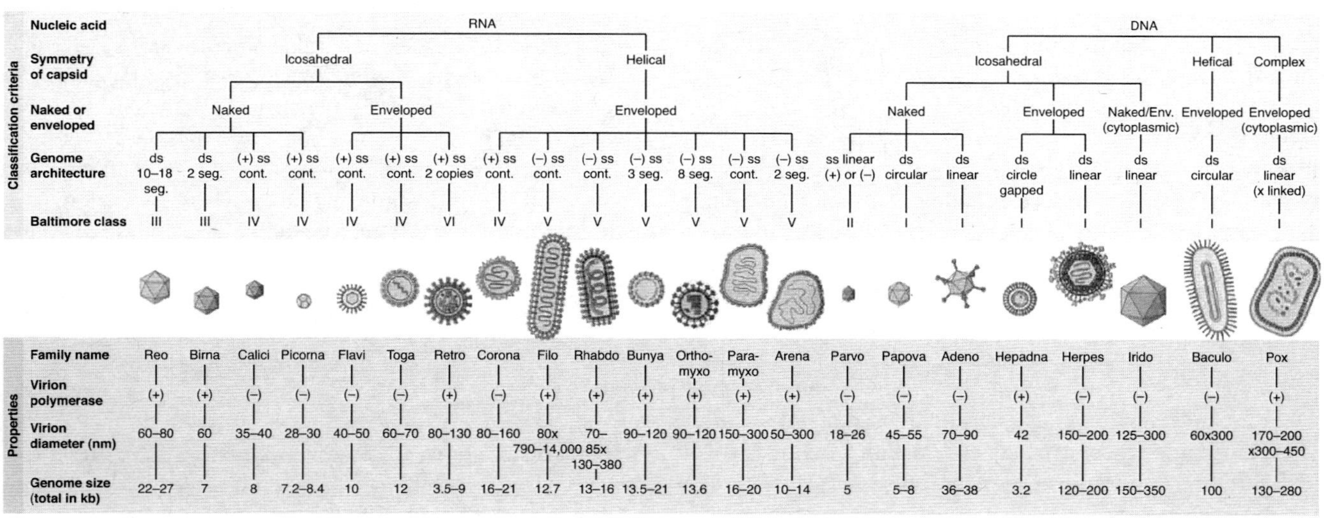

FIGURE 1.30. A virus classification scheme broadly dividing according to RNA/DNA core type, symmetries, envelope features and diameters, and so on. Image courtesy of http://www.nlv.ch/Virologytutorials/graphics/classificationtotal.jpg.

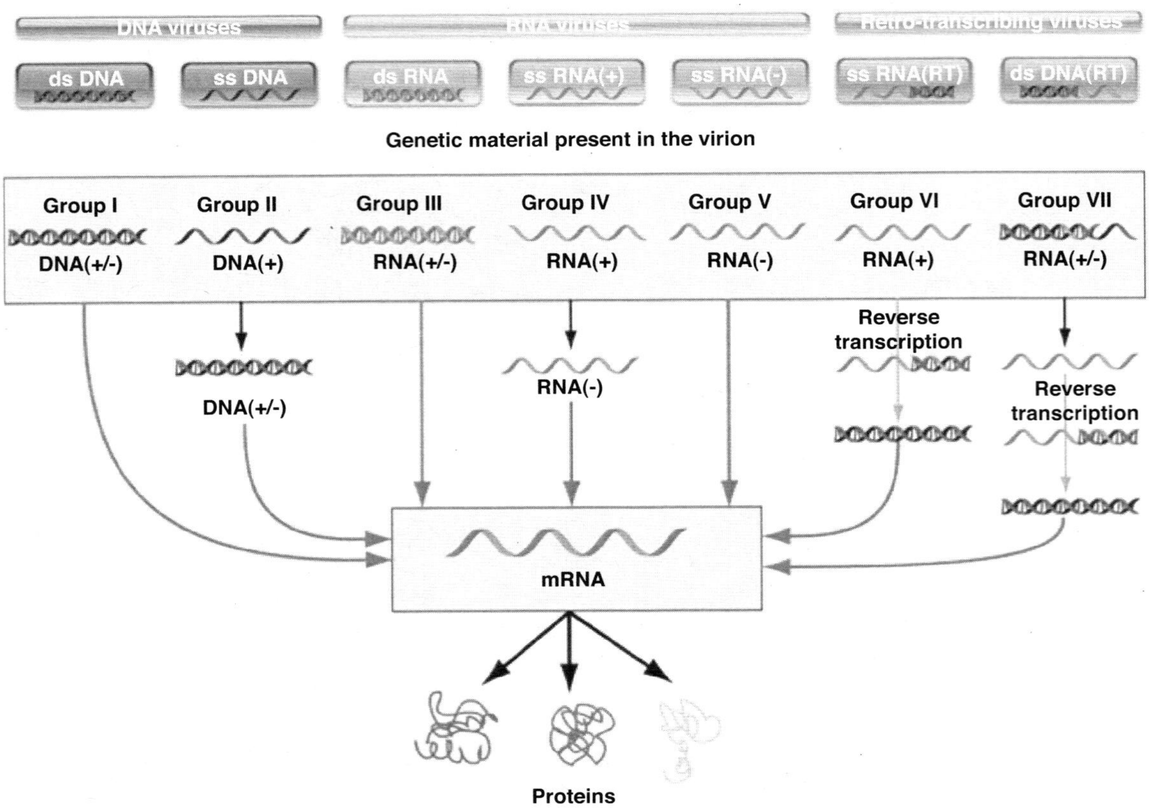

FIGURE 1.31. The Baltimore classification of viruses based on placing all viruses into one of seven groups (I–VII) depending on a combination of their nucleic acid cores (i.e., DNA/RNA), strandedness (i.e., single-stranded or double-stranded), sense, and their method of replication. Courtesy of the SIB Swiss Institute of Bioinformatics.

nucleic acid core types (i.e., RNA or DNA), the resulting symmetries, envelope features, dimensions, and so on, as shown in Figure 1.31 [78].

A second virus classification scheme was introduced by D. Baltimore [9], and it is based on the hypothesis that all viruses must generate positive-strand mRNAs from their genomes in order to produce proteins and replicate themselves. The precise mechanism is different for each virus family, thus leading to seven different strategies for their replication and hence seven different viral families as described in Figure 1.31 [79].

If all viruses could be viewed as stoichiometric nanocompounds derived from well-defined ratios of genomic and protein subunits, it is apparent that suitable cladograms could be developed which would be consistent with both the Lwoff and Baltimore taxonomies described above. Each of these classifications has produced appropriate virus nomenclatures for defining specific viral prototypes.

Presently, viral classifications are focused on the taxon level of ***families*** downward. Members within a virus family are ordered with genomics, the elucidation of evolutionary relationships—that is, analyses of nucleic acid and protein sequence similarities. As such, all virus ***families*** have the suffix viridae (e.g., Caliciviridae, Picornaviridae, Reoviridae, etc.); whereas, ***genera*** have the suffix **-virus**. Within the Picornaviridae there are five genera: *entero*virus, *cardio*virus, *rhino*virus, *aptho*virus and *hepato*virus. Defining actual "species" is the most important but difficult challenge associated with virus nomenclature. There is an element of subjectivity in that task undoubtedly due to subtle mutation-based aberrations in actual practice [20].

1.10 CONCLUSIONS

In this account, we have reviewed the successful use of Linnaean-type taxonomies for micron-scale biological structure/organism classifications, as well as for picoscale, atomic element classifications as evidenced by the Mendeleev Periodic Table. Each of these taxonomies has led to very useful nomenclatures. Each of these taxonomy areas were based on either picoscale or micron-scale aufbau pathways (i.e., evolutionary intermediates) that were organized according to

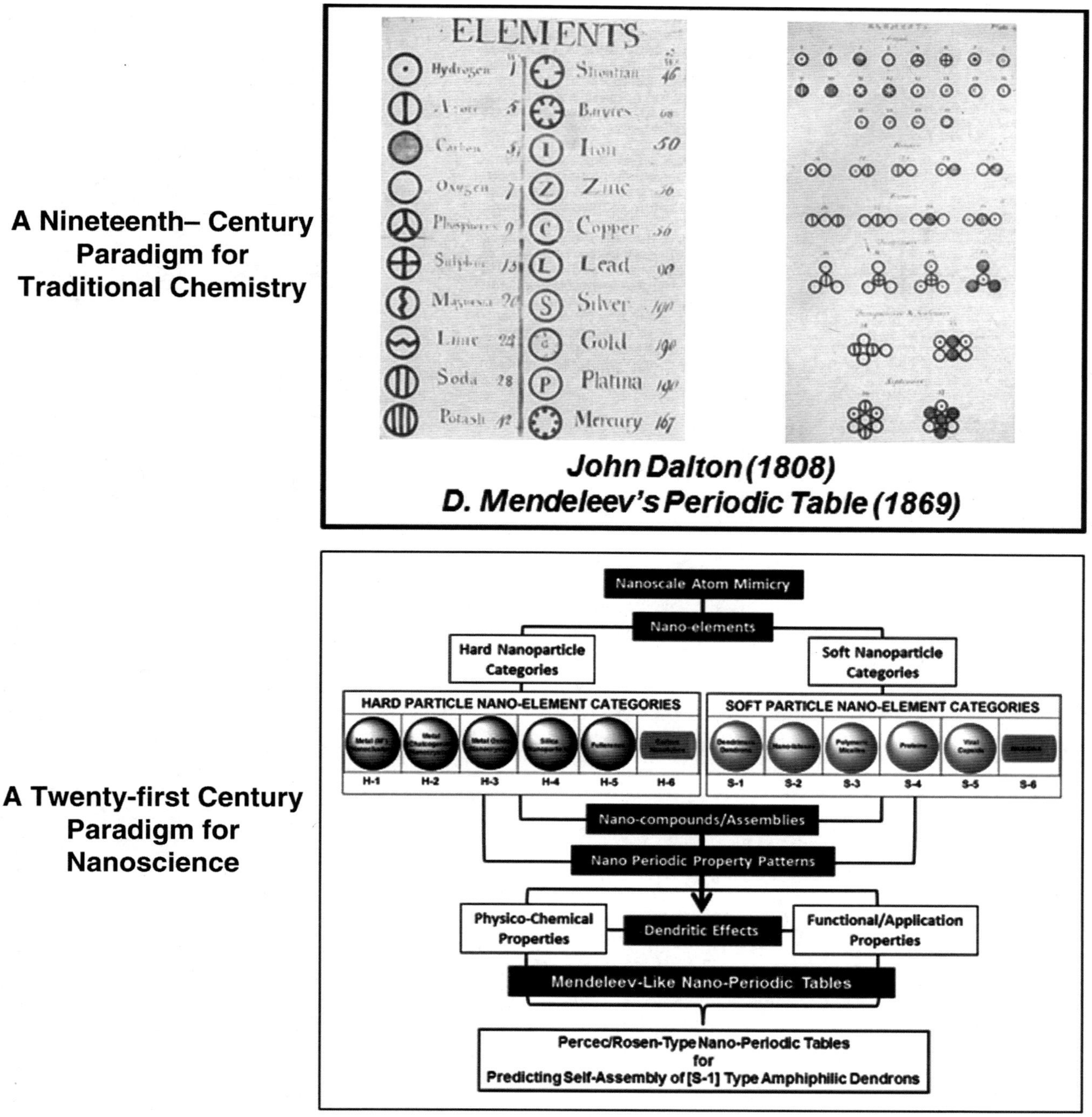

FIGURE 1.32. A brief comparison of nineteenth-century picoscale building blocks [51] (images reproduced with permission from Wiley–VCH Verlag GmbH & Co. KGaA) and twenty-first-century nanoscale building blocks [45].

"features of similarity." This, in turn, led to appropriate cladograms and corresponding "trees of similarity." Mendeleev's Periodic Table of the atomic elements projected in a Niels Bohr format represents such a "tree of similarity for the elements." These Linnaean taxonomy principles have been shown to be useful for classifying three biology-based soft nano-element categories; namely, proteins [S-4], viral capsids [S-5], and DNA/RNA [S-6]-type nano-element categories. These same principles are now proposed as first steps toward the development of a taxonomy and nomenclature for certain well-defined, non-biologically derived, soft nano-element categories and are summarized as follows:

- This is a first step toward the development of a taxonomy and nomenclature for well-defined soft nanomaterials (i.e., [S-1]-, [S-2]-, [S-3]-, [S-4]-, [S-5]-, and [S-6]-type nano-element categories).
- Three of the six soft nano-element categories (i.e., [S-4], [S-5], and [S-6]) enjoy taxonomies/nomenclatures and are biologically based.
- Diverse process strategies that produce the three "nonbiological" soft nano-element categories (i.e., [S-1], [S-2], and [S-3]) involved the construction of specific "macromolecular architectures": namely, (I) linear, (II) cross-linked, and (IV) dendritic architectures—for example, (a) divergent/convergent dendritic amplification (Type IV) dendritic polymers (i.e., [S-1]), (b) micelle-templated polymerization (Type II) cross-linked (bridged polymers) (i.e., [S-2]), and (c) Grubbs-type catalyst (Type I) linear polymers (i.e., [S-3]). As such, initial steps for nonbiological, soft nano-element taxonomies based on "architectural types" should be considered.
- In summary, the eighteenth and nineteenth century witnessed the successful development of Linnaean-based taxonomy systems for classifying vast libraries of micron-scale/macroscale biological structures and organisms. Similarly, during the nineteenth century the chemistry field enjoyed the successful development of a useful Linnaean-like taxonomy system for classifying all of the atomic elements in the form of Mendeleev's Periodic Table [7, 80]. This early classification system ultimately defined and unified traditional chemistry with the emergence of a useful and versatile nomenclature. This present account is an effort to define a new taxonomy and nomenclature for well-defined soft nanomaterials in the context of a nano-periodic system (Figure 1.17, 1.18, and 1.32) derived from these same traditional first principles [2, 19, 31]. It should be considered a "works in progress." However, there is now optimism that continued development and success will lead to a similar Linnaean-type classification system for well-defined soft-matter, nano-element categories and perhaps be extended to hard-matter, nano-element categories with the concurrent emergence of a useful and versatile nanomaterials nomenclature.

ACKNOWLEDGMENTS

We gratefully acknowledge the National Science Foundation for financial support of the CMU-NSF Workshop entitled Periodic Patterns, Relationships and Categories of Well-Defined Nanoscale Building Blocks, NSF Award #0707510, as well as Dr. Mihail Roco, Prof. Virgil Percec, and Dr. Brad Rosen for their valuable discussions. Finally, we wish to express sincere gratitude to Ms. Linda S. Nixon for her perseverance and skill throughout the final manuscript and graphics preparation.

REFERENCES

1. Tomalia, D. A. (2005). Birth of a new macromolecular architecture: Dendrimers as quantized building blocks for nanoscale synthetic polymer chemistry. *Progress in Polymer Science*, **30**, 294–324.
2. Tomalia, D. A. (2009). In quest of a systematic framework for unifying and defining nanoscience. *Journal of Nanoparticle Research*, **11**, 1251–1310.
3. Joyce, G. F. (1992). Directed molecular evolution. *Scientific American*, **267**(6), 90–97.
4. Joyce, G. F. (2002). The antiquity of RNA-based evolution. *Nature*, **418**, 214–221.
5. Mendelejeff, D. (1869). *Zeitschrift fur Chemie*, **12**, 405–406.
6. Mendelejeff, D. (1889). LXIII-The periodic law of the elements. *Journal of the Chemical Society*, **55**, 634–656.
7. Strathern, P. (2000). *Mendeleyev's Dream*. The Berkley Publishing Group, New York.
8. Mayr, E., Bock, W. J. (2002). Classifications and other ordering systems. *Journal Zoological Systematics and Evolutionary Research*, **40**(4), 169–194.
9. Baltimore, D. (1971). Expression of animal virus genomes. *Bacteriological Reviews*, **35**(3), 235–241.
10. Gelderblom, H. R. (1996) Structure and classification of viruses. In: Baron, S., ed. *Medical Microbiology*, 4th edition. The University of Texas Medical Branch, Galveston.
11. Black, J. G. (2008) Microbiology: Principles and explorations. In: *An Introduction to Taxonomy: The Bacteria*, K. Witt, ed., 7th edition, John Wiley & Sons, Hoboken, NJ, Chapter 9.
12. Kurtzman, C. P., Fell, J. W., eds. (2011) *The Yeasts*, 5th edition: *A. Taxonomic Study*. Elsevier, London.
13. Guarro, J., Gene, J., Stchigel, A. M. (1999). Developments in fungal taxonomy. *Clinical Microbiology Reviews*, **12**(3), 454–500.
14. Prescott, L. M., Harley, J. P., Klein, D. A. (1996). *Microbiology*, 3rd edition. Wm. C. Brown Publishers, Dubuque, Ia.
15. Schuh, R. T., Brower, A. V. Z. (2009). *Biological Systematics, Principles and Applications*, 2nd edition. Cornell, Ithaca, NY.
16. Thompson, D. (1987). *On Growth and Form*. Cambridge University Press, London.
17. Murzin, A. G., Brenner, S. E., Hubbard, T., Chothia, C. (1995). Scop: A structural classification of proteins database for the investigation of sequences and structures. *Journal of Molecular Biology*, **247**, 536–540.
18. Przytycka, T., Aurora, R., Rose, G. D. (1999). A protein taxonomy. *Nature Structural Biology*, **6**(7), 672–682.
19. Tomalia, D. A. (2008). Periodic patterns, relationships and categories of well-defined nanoscale building blocks (Http://Www.Nsf.Gov/Crssprgm/Nano/Gc_Charact08_Tomalia_Nsf9_29_08.Pdf). *National Science Foundation Final Workshop Report*, pp. 1–156.

20. Fauquet, C. M., Mayo, M. A., Maniloff, J., Desselberger, U., Ball, L. A., ed. (2005) *Virus Taxonomy: Viiith Report of the International Committee on Taxonomy of Viruses*. Elsevier Academic Press, Amsterdam.
21. Paquin, P., Hedin, M. (2004). *Molecular Ecology*, **13**, 3239–3255.
22. Vertesy, Z., Balint, Z., Kertesz, K., Vigneron, J. P., Lousse, V., Biro, L. P. (2006). Wing scale microstructures and nanostructures in butterflies–natural photonic crystals. *J. Microscopy*, **224**, 108–110.
23. Lwoff, A., Horne, R., Tournier, P. (1962). A system of viruses. *Cold Spring Harbor Symposium on Quantitive Biology*, **27**, 51–55.
24. Tomalia, D. A., Naylor, A. M., Goddard, W. A., III (1990). Starburst dendrimers: Molecular level control of size, shape, surface chemistry, topology and flexibility from atoms to macroscopic matter. *Angewandte Chemie International Edition English*, **29**(2), 138–175.
25. Tomalia, D. A., Durst, H. D. (1993). In: Weber, E.W., ed. *Supramolecular chemistry I—Directed Synthesis and Molecular Recognition. Topics in Current Chemistry*, Vol. 165. Springer Verlag, Berlin, pp. 193–313.
26. Zuckerkandl, E., Pauling, L. (1965). Molecules as documents of evolutionary history. *J. Theoret. Biol.*, **8**, 357–366.
27. Rich, R. L. (1991). A taxonomy of relationships among the elements. *Journal of Chemical Education*, **68**, 828–829.
28. Pullman, B. (1998). *The Atom in the History of Human Thought*. Oxford University Press, New York.
29. Richard, J. S. (1981). The anatomy and taxonomy of protein structure. *Advances in Protein Chemistry*, **34**, 167–339.
30. Gentleman, D. J., Chan, W. C. W. (2009). A systematic nomenclature for codifying engineered nanostructures. *Small*, **5**(4), 426–431.
31. Tomalia, D. A. (2010). Dendrons/dendrimer: Quantized, nanoelement like building blocks for soft-soft and soft-hard nanocompound synthesis. *Soft Matter*, **6**, 456–474.
32. Boeyens, J. C. A., Levendis, D. C. (2010). *Number Theory and the Periodicity of Matter*, Springer, Berlin.
33. Jena, P., Castleman, A. W., Jr. (2006). Clusters: A. bridge across the disciplines of physics and chemistry. *Proceedings of the National Academy of Sciences*, **103**, 10560–10569.
34. Garai, J. (2008). Mathematical formulas describing the sequences of the periodic table. *International Journal of Quantum Chemistry*, **108**, 667–670.
35. Reveles, J. U., Khanna, S. N., Roach, P. J., Castleman, A. W., Jr. (2006). Multiple valence superatoms. *Proceedings of the National Academy of Sciences*, **103**(49), 18405–18410.
36. Qian, M., Reber, A. C., Ugrinov, A., Chaki, N. K., Mandal, S., Saavedra, H. M., Khanna, S. N., Sen, A., Weiss, P. S. (2010). Cluster-assembled materials: Toward nanomaterials with precise control over properties. *ACS Nano*, **4**(1), 235–240.
37. Medel, V. M., Reveles, J. U., Khanna, S. N., Chauhan, V., Sen, P., Castleman, A. W. (2011). Hund's rule in superatoms with transition metal impurities. *Proceedings of the National Academy of Sciences*, **108**, 10062–10066.
38. Vorobiev, N. N. (2002). *Fibonacci Numbers*. Birkhauser Verlag, Basel.
39. Teo, B. K., Sloane, N. J. A. (1985). Magic numbers in polygonal and polyhedral clusters. *Inorganic Chemistry*, **24**, 4545–4558.
40. Teo, B. K., Zhang, H. (2002) Magic numbers in clusters: nucleation and growth sequences, bonding, principles and packing patterns. In: Feldheim, D. L., Foss, C. A., Jr., eds. *Metal Nanoparticles: Synthesis, Characterization and Applications*. Marcel Dekker, New York.
41. Harbola, M. K. (1992). Magic numbers for metallic clusters and the principle of maximum hardness. *Proceedings of the National Academy of Sciences*, **89**, 1036–1039.
42. Tomalia, D. A. (1994). Starburst/cascade dendrimers: Fundamental building blocks for a new nanoscopic chemistry set. *Advances in Materials*, **6**, 529–539.
43. Tomalia, D. A., Frechet, J. M. (2005). Introduction to dendrimers and dendritic polymers. *Progress in Polymer Science*, **30**, 217–219.
44. Schmid, G. (1990). Clusters and colloids: bridges between molecular and condensed material. *Endeavour*, **14**(4), 172–178.
45. Tomalia, D. A. (2012). Dendritic effects: Dependency of dendritic nano-periodic property patterns on critical nanoscale design parameters (CNDPs). *New Journal of Chemistry*, **36**, 264–281.
46. Rosen, B. M., Wilson, D. A., Wilson, C. J., Peterca, M., Won, B. C., Huang, C., Lipski, L. R., Zeng, X., Ungar, G., Heiney, P. A., Percec, V. (2009). Predicting the structure of supramolecular dendrimers via the analysis of libraries of Ab_3 and constitutional isomeric Ab_2 biphenylpropyl ether self-assembling dendrons. *Journal of the American Chemical Society*, **131**, 17500–17521.
47. Fraser, C., Grubbs, R. H. (1995). Synthesis of glycopolymers of controlled molecular weight by ring-opening metathesis polymerization using well-defined functional group tolerant ruthenium carbene catalysts. *Macromolecules*, **28**, 7248–7255.
48. Tomalia, D. A., Baker, H., Dewald, J., Hall, M., Kallos, G., Martin, S., Roeck, J., Ryder, J., Smith, P. (1985). A new class of polymers: starburst dendritic macromolecules. *Polymer Journal (Tokyo)*, **17**, 117–132.
49. Tomalia, D. A., Fréchet, J. M. J. (2002). Discovery of dendrimers and dendritic polymers: A brief historical perspective. *Journal of Polymer Science: Part A: Polymer Chemistry*, **40**, 2719–2728.
50. Tomalia, D. A., Henderson, S. A., Diallo, M. S. (2007) Dendrimers—an enabling synthetic science to controlled organic nanostructures. In: Goddard, W. A., III, Brenner, D. W., Lyshevski, S. E., Iafrate, G. J., eds. *Handbook of Nanoscience, Engineering and Technology*, CRC Press, Boca Raton, FL, pp. 24.1–24.47.
51. Heilbronner, E., Dunitz, J. D. (1993). *Reflections on Symmetry*. VCH Publishers, New York.
52. Tomalia, D. A., Uppuluri, S., Swanson, D. R., Li, J. (2000). Dendrimers as reactive modules for the synthesis of new structure controlled, higher complexity—Megamers. *Pure Appllied Chemistry*, **72**, 2343–2358.

53. Uppuluri, S., Piehler, L. T., Li, J., Swanson, D. R., Hagnauer, G. L., Tomalia, D. A. (2000). Core-Shell Tecto(Dendrimers): I. Synthesis and Characterization of Saturated Shell Models. *Advances in Materials*, **12**(11), 796–800.

54. Betley, T. A., Hessler, J. A., Mecke, A., Banaszak Holl, M. M., Orr, B. G., Uppuluri, S., Tomalia, D. A., Baker, J. R., Jr., (2002). Tapping mode atomic force microscopy investigation of poly(amidoamine) core–shell tecto(dendrimers) using carbon nanoprobes. *Langmuir*, **18**, 3127–3133.

55. Mansfield, M. L., Rakesh, L., Tomalia, D. A. (1996). The random parking of spheres on spheres. *Journal of Chemical Physics*, **105**(8), 3245–3249.

56. Uppuluri, S., Swanson, D. R., Brothers II, H. M., Piehler, L. T., Li, J., Meier, D. J., Hagnauer, G. L., Tomalia, D. A. (1999). Tecto(dendrimer) core–shell molecules: Macromolecular tectonics for the systematic synthesis of larger controlled structure molecules. *Polymer Materials Science and Engineering (ACS)*, **80**, 55–56.

57. Percec, V., Heck, J., Lee, M., Ungar, G., Alvarez-Castillo, A. (1992). Poly{2-vinyloxyethyl 3,4,5-tris[4-(*N*-dodecanyloxy)benzyloxy]benzoate}: A self-assembled supramolecular polymer similar to tobacco mosaic virus. *Journal of Material Chemistry*, **2**, 1033–1039.

58. Percec, V. (2006). Bioinspired supramolecular liquid crystals. *Philosphical Transactions of the Royal Society A*, **364**, 2709–2719.

59. Klug, A. (1983). From macromolecules to biological assemblies (Nobel lecture). *Angewandte Chemie International Edition*, **22**, 565–582.

60. Klug, A. (1999). Tobacco mosiac virus particle structure and the initiation of disassembly. *Philosophical Transactions of the Royal Society of London B*, **354**, 531–535.

61. Percec, V., Johansson, G., Ungar, G., Zhou, J. P. (1996). Fluorophobic effect induces the self-assembly of semifluorinated tapered monodendrons containing crown ethers into supramolecular columnar dendrimers which exhibit a homeotropic hexagonal columnar liquid crystalline phase. *Journal of the American Chemical Society*, **118**(41), 9855–9866.

62. Percec, V., Ahn, C.-H., Cho, W.-D., Jamieson, A. M., Kim, J., Leman, T., Schmidt, M., Gerle, M., Moller, M., Prokhorova, S. A., Sheiko, S. S., Cheng, S. Z. D., Zhang, A., Ungar, G., Yeardley, D. J. P. (1998). Visualizable cylindrical macromolecules with controlled stiffness from backbones containing libraries of self-assembling dendritic side groups. *Journal of the American Chemical Society*, **120**, 8619–8631.

63. Percec, V., Ahn, C.-H., Unger, G., Yeardly, D. J. P., Moller, M. (1998). Controlling polymer shape through the self-assembly of dendritic side-groups. *Nature*, **391**, 161–164.

64. Percec, V., Cho, W.-D., Ungar, G., Yeardley, D. J. P. (2001). Synthesis and structural analysis of two constitutional isomeric libraries of Ab_2-based monodendrons and supramolecular dendrimers. *Journal of the American Chemical Society*, **123**, 1302–1315.

65. Huang, B., Prantil, M. A., Gustafson, T. L., Parquette, J. R. (2003). The effect of global compaction on the local secondary structure of folded dendrimers. *Journal of the American Chemical Society*, **125**, 14518–14530.

66. Tomalia, D. A., Huang, B., Swanson, D. R., Brothers II, H. M., and Klimash, J. W. (2003). Structure control within poly(amidoamine) dendrimers: Size, shape and regio-chemical mimicry of globular proteins. *Tetrahedron*, **59**, 3799–3813.

67. Lockman, J. W., Paul, N. M., Parquette, J. R. (2005). The role of dynamically correlated conformational equilibria in the folding of macromolecular structures. A model for the design of folded dendrimers. *Progress in Polymer Science*, **30**, 423–452.

68. Butler, P. G., Klug, A. (1978). The assembly of a virus. *Scientific American*, **239**(5), 62–69.

69. Chung, W.-J., Oh, J.-W., Kwak, K., Lee, B. Y., Meyer, J., Wang, E., Hexemer, A., Lee, S.-W. (2011). Biomimetic self-templating supramolecular structures. *Nature*, **478**, 364–368.

70. Percec, V., Peterca, M., Dulcey, A. E., Imam, M. R., Hudson, S. D., Nummelin, S., Adelman, P., Heiney, P. A. (2008). Hollow spherical supramolecular dendrimers. *Journal of the American Chemical Society*, **130**, 13079–13094.

71. Rosen, B. M., Wilson, C. J., Wilson, D. A., Peterca, M., Imam, M. R., Percec, V. (2009). Dendron-mediated self-assembly, disassembly, and self-organization of complex systems. *Chemical Reviews*, **109**, 6275–6540.

72. Macfarlane, R. J., Lee, B. Y., Jones, M. R., Harris, N., Schatz, G. C., Mirkin, C. A. (2011). Nanoparticle superlattice engineering with DNA. *Science*, **334**, 204–208.

73. Pauling, L. (1960). *The Nature of the Chemical Bond*, 3rd edition. Cornell University Press, Ithaca, NY.

74. Newkome, G. R., Moorfield, C. N., Vögtle, F. (1996). *Dendritic Molecules*. VCH, Weinheim.

75. Newkome, G. R., Baker, G. R., Young, J. K., Traynham, J. G. (2003). A systematic nomenclature for cascade polymers. *Journal of the Polymer Science, Part A: Polymer Chemistry*, **31**(3), 641–651.

76. Vögtle, F., Richardt, G., Werner, N., Rackstraw, A. J. (2009). *Dendrimer Chemistry: Concepts, Syntheses, Properties, Applications*. Wiley-VCH, Weinheim.

77. Jain, K., Kesharwani, P., Gupta, U., Jain, N. K. (2010). Dendrimer toxicity: Let's meet the challenge. *International Journal of Pharmaceutics*, **394**, 122–142.

78. Buchen-Osmond, C. (2003). The universal virus database Ictvdb. *Computing in Science & Engineering*, **5**(3), 16–25.

79. Gasteiger, E., Gattiker, A., Hoogland, C., Ivanyi, I., Appel, R. D., Bairoch, A. (2003). Expasy: The proteomics server for in-depth protein knowledge and analysis. *Nucleic Acids Research*, **31**, 3784–3788.

80. Scerri, E. R. (2007). *The Periodic Table*. Oxford University Press, New York.

2

ON THE ROLE OF HYDROGEN-BONDING IN THE NANOSCALE ORGANIZATION OF π-CONJUGATED MATERIALS

ALBERTUS P. H. J. SCHENNING AND DAVID GONZÁLEZ-RODRÍGUEZ

2.1 INTRODUCTION

There is little doubt that the development of nanomaterials and nanodevices will become increasingly more important to modern society in the near future. The successful development of nanosized technologies relies on the ability of researchers to efficiently manufacture materials having well-defined supramolecular structures and functions within the nanometer scale. These materials may find application in different areas, from the development of efficient, cheaper and environmentally friendly optoelectronic devices to a whole new generation of catalysts, medicines, or selective membranes.

Despite the fact that traditional optical lithography and etching techniques used for micro- and nanofabrication (the so-called "*Top-Down*" approach) have shown an enormous development in recent years [1], they present several limitations that need to be addressed by other complementary techniques. First of all, they are impractical for structures smaller than 100 nm, since their resolution is limited by the wavelength of the light used and, in general, do not provide a fine control over the internal structure of the material at the nanometer scale. On the other hand, if smaller nanopatterns are required, the use of very expensive equipment and high-energy beams is then only limited to robust inorganic materials. There is a strong emerging need, however, to produce organic (i.e., plastic) materials with perfectly defined supramolecular structure, composition, and function at the nanoscale. The advantages that organic materials can bring to society demands are manifold: they are cheap, easily processable, and biocompatible, and they can be endowed with very diverse functions and applications.

The natural world is a shining example when it comes to the utilization of organic materials. Nature has chosen the unlimited possibilities of organic and metallo-organic molecules to produce a whole diversity of complex, but perfectly defined, nanoobjects having specific functions. The structure of viral capsids, the precise chromophore organization within photosynthetic reaction centers, the integration and disintegration dynamics of the microtubules forming the cell skeleton, or the polymorphism of DNA are just some representative examples. The way the molecules are organized into larger nanostructures in biological systems is by chemical recognition or *self-assembly*, whose importance and complexity represents a continuous inspiration to scientists in their quest for the construction of functional nanostructures through a "*Bottom-Up*" approach. We are entering a scientific period of "*complex matter*" in which the concepts and methodologies of supramolecular chemistry and self-assembly are meeting those of nanotechnology. Self-assembled systems can have unrivaled characteristics and are considered by many scientists as the way forward for the future mass production of nanomaterials, nanodrugs, and nanoelectronics [2].

If such nanoarchitectures are built from molecules with a particular function, one can expect unprecedented properties arising from cooperative interactions between them, which can be of extraordinary impact for several applied

Organic Nanomaterials: Synthesis, Characterization, and Device Applications, First Edition. Edited by Tomás Torres and Giovanni Bottari.

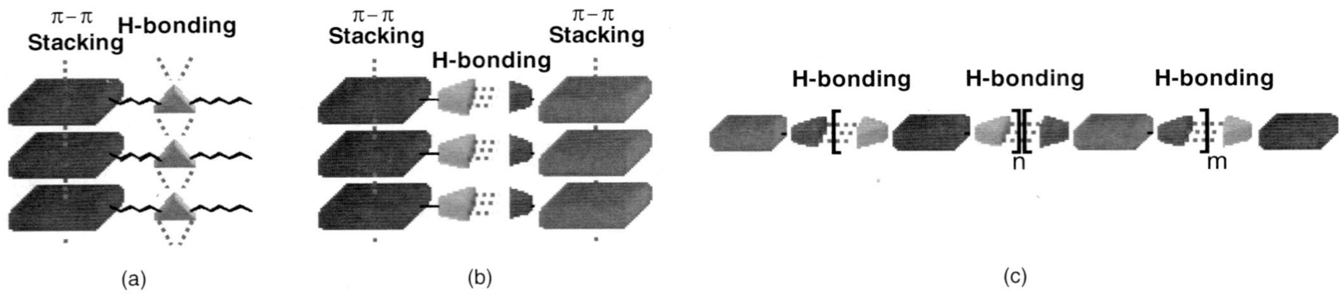

FIGURE 2.1. Models of: (**a**) a stack of π-conjugated molecules stabilized by H-bonding interactions along the polymer axis; (**b**) a stack of π-conjugated molecules where the components are H-bonded in the direction perpendicular to the polymer axis; (**c**) a polymer comprised of π-conjugated molecules where the monomers are associated via H-bonding interactions along the polymer main chain.

fields. π-Conjugated molecules and polymers are a central class of functional molecules that have been extensively used in advanced applications such as sensors and in optoelectronics [3]. For such applications, π-functional materials are required that can form well-organized nanostructures in solution or in the solid state, whose properties can be controlled as a function of the self-assembly process and the molecular structure [4]. Such a control is important for the improved performance and processing of existing materials, as well as for developing new materials with tunable optical and electronic properties. However, controlling the size, shape, and supramolecular structure of self-assembled molecular systems continues to be a major challenge for scientists [2]. For such a goal, one must "program" the molecular building blocks with the appropriate chemical information and be able to modulate the interplay between multiple noncovalent interactions, so that they work in concert to impart stability to the assemblies and, at the same time, define the overall architecture within the nanometer regime [5]. In this context, the use of hydrogen bonds (H-bonds) is regarded as a powerful means to reach a high degree of control over supramolecular architectures, since this noncovalent interaction can be made highly selective and directional [5, 6]. H-bonds are formed when a donor (D) with an available acidic hydrogen atom is interacting with an acceptor (A) carrying available nonbonding electron lone pairs. The strength of this interaction depends mainly on the solvent (a competitor in the formation of H-bonds) and on the chemical nature of the H-bonding donor and acceptor functions, as well as on their number and sequence in a particular molecular fragment [7]. This last parameter has been extensively exploited in order to achieve the high selectivities and association constants typically required to obtain the target nanosized assemblies in solution. In many cases, however, relatively weak H-bonding interactions are used that cooperate with additional noncovalent forces (like π–π stacking or solvophobic interactions) to guide the self-assembly process toward the formation of the desired assembly. In order to accomplish that, a detailed knowledge of the assembly mechanism ruling the supramolecular polymerization is also necessary (i.e., isodesmic or cooperative self-assembly) [8].

The purpose of this chapter is to offer a general vision on the role of H-bonding interactions on the structuring of molecular π-functional materials at the nanoscale. For this chapter, our recent review on hydrogen-bonded supramolecular π-functional materials acted as a starting point [9]. We are focusing now on the most significant recent developments in the field where this versatile noncovalent interaction endows supramolecular systems with a particular property or function. The latter can be very diverse. As will be detailed below, H-bonds have been regularly used to increase the molecule–molecule binding strength, to provide enhanced stability and intermolecular order to the assemblies, and/or to position different semiconducting molecules in a certain arrangement within the nanostructures. The chapter is therefore divided into different sections as a function of the purpose of the H-bonds and the direction in which they are formed with respect to the stacking or polymerization axis. As shown in Figure 2.1, H-bonds can be formed parallel to the stacking axis (Figure 2.1a), perpendicular to the stacking axis (Figure 2.1b), or they can be the main or only non-covalent force that guides monomer association along the supramolecular polymer chain (Figure 2.1c). Particular emphasis will be given to the specific properties and the potential use of the supramolecular architectures obtained.

2.2 H-BONDING ALONG THE STACKING POLYMER AXIS

π–π Stacking between planar aromatic surfaces is a conventional noncovalent interaction that promotes the formation of ordered arrays in crystalline materials and of polymeric fibers in poor solvents [4]. The self-assembly into these stacked wire-like structures, in which semiconducting molecules strongly interact through their π-electron surfaces along the columnar axis, is considered one of the most promising nanoscale architectures in organic optoelectronic devices [10–12]. However, the exact arrangement in which organic semiconducting molecules self-assemble into nanoscopic materials is sometimes difficult to control since aromatic interactions that promote stacking do not have a very specific directionality. The additional use of H-bonding interactions, typically between amide or urea groups, can be interesting if they cooperate with π–π stacking and solvophobic interactions to form ordered arrays along the fiber axis (Figure 2.1a). The main purpose of these H-bonding directors is to increase the kinetic and thermodynamic stability of the nanofibers as well as to enhance the degree of order and suppress relative movements between neighboring π-conjugated cores within the columns. In addition, the presence of amide/urea H-bonds in columnar stacks have often been shown to decrease the stacking intermolecular distance. These groups tend to bend to adapt H-bonding distances (4.6–5.1 Å) to the shorter π–π stacking distances (3.2–3.6 Å), which normally leads to intermolecular twists and to the generation of helical arrays [13–15].

A number of π-functional molecules have been organized into columnar nanostructures with the help of these H-bonding motifs. Recent examples include: π-conjugated oligomers like oligophenylene (OP) [16], oligophenyleneethynylene (OPE) [17], oligophenylenevinylene (OPV) [18–22], oligofluorene (OF) [20, 23–25], or oligothiophene (OT) [20, 26–30]; fused aromatics and discotic molecules such as pyrene [31, 32], naphthalene bisimide (NBI) or perylene bisimide (PBI) [33–40], triphenylenes [41, 42] or azatriphenylenes [43–46], hexabenzocoronene (HBC) [47], phthalocyanine (Pc) [48] or porphyrin (Pp) [49–57]; aromatic amines like phenothiazine or carbazol [58–61]; and electronically rich entities like tetrathiafulvalene (TTF) [62–67], among others. The self-assembled materials thus produced present different unconventional properties that arise from their well-defined organization at the molecular scale. Researchers in this field have been particularly attracted by the possibility of further controlling nano- and mesoscopic order, enhancing certain photophysical features of organic dyes, creating defined pathways for electron and hole migration in π-conjugated materials, or aligning and stabilizing single fibers via cross-linking reactions.

2.2.1 Influence on the nano- and mesoscopic organization

A general consequence of introducing secondary intermolecular H-bonding interactions along the stacks is the formation and/or stabilization of organogels in a variety of solvents. Low-molecular-weight gelators [68] have attracted considerable interest during recent years as supramolecular materials with manifold applications: hydrogels for cell culture [69], templates for mineralization [70] and ion-selective membranes [71], thermo- and mechano-responsive switches and sensors [72], or dye-based organogels with enhanced luminescence properties [73–77].

Yagai et al. [21] studied the self-assembly of OPVs substituted with urea groups in which both the number of these H-bonding motifs (1 (**1**) or 2 (**2**, **3**)) and their relative distance (separated by hexamethylene or dodecamethylene chains) were varied (Figure 2.2a) [21]. All of these compounds were able to gelate nonpolar solvents by formation of tapelike nanostructures, although the OPV having two urea groups separated by the shorter chain clearly showed stronger intermolecular interactions and hence a higher gelation efficiency. However, these authors observed that the same effect has now a disturbing influence on the formation of thermotropic liquid crystalline phases upon gel drying and heating. In this case, the compound with stronger intermolecular interactions displayed a higher transition temperature. A similar conclusion was obtained by Paraschiv et al. [42], who studied the mesophase formation of hexaalkoxytriphenylenes that contained an amide, urea, or thiourea group in one of their alkoxy tails. They showed again that the stronger the H-bonding interactions along the stacks, the more the liquid crystallinity is suppressed. Among the compounds studied, only some thiourea-containing derivatives formed hexagonal columnar mesophases. This observation was explained on the basis of the lower H-bonding ability of this group when compared to urea and amide groups, which introduce a higher rigidity to the triphenylene stacks.

During solvent gelation, a series of hierarchical self-assembly processes can generate quite distinct mesoscopic objects as a function of the solvent employed. The group headed by Ajayaghosh recently demonstrated that OPV oligomers capped with BOC-protected L- (**4**) and D-alanine (**5**) at both edges aggregated into different gel architectures in chloroform and in toluene which, at the same time, produced very distinct mesoscopic structures (Figure 2.2b) [22]. In chloroform, the molecules self-assembled in 1D H-type aggregates that, upon gelation and solvent drying, produced macroporous honeycomb layers. In toluene, in contrast, the OPVs associated into 2D cross-linked tapes that derived in aligned fiber bundles. Similar solvent effects leading to different mesoscopic architectures have also been reported for amphiphilic OT [28] and TTF [63, 67] organogelators.

Despite the relevance of numerous studies and the fast development of characterization techniques, supramolecular chemists are still far of being able to predict precise mesoscopic organizations from a given molecular structure in a given environment [2]. This is mainly because noncovalent interactions are extremely sensitive to variables such as temperature, concentration, or solvent nature [8]. Nonetheless, since self-organization is a hierarchical process that takes place at different length scales, it is clear that having an insight into the primary self-assembling mechanisms that operate at the molecular level is a rational start. Several groups have devoted considerable efforts during recent

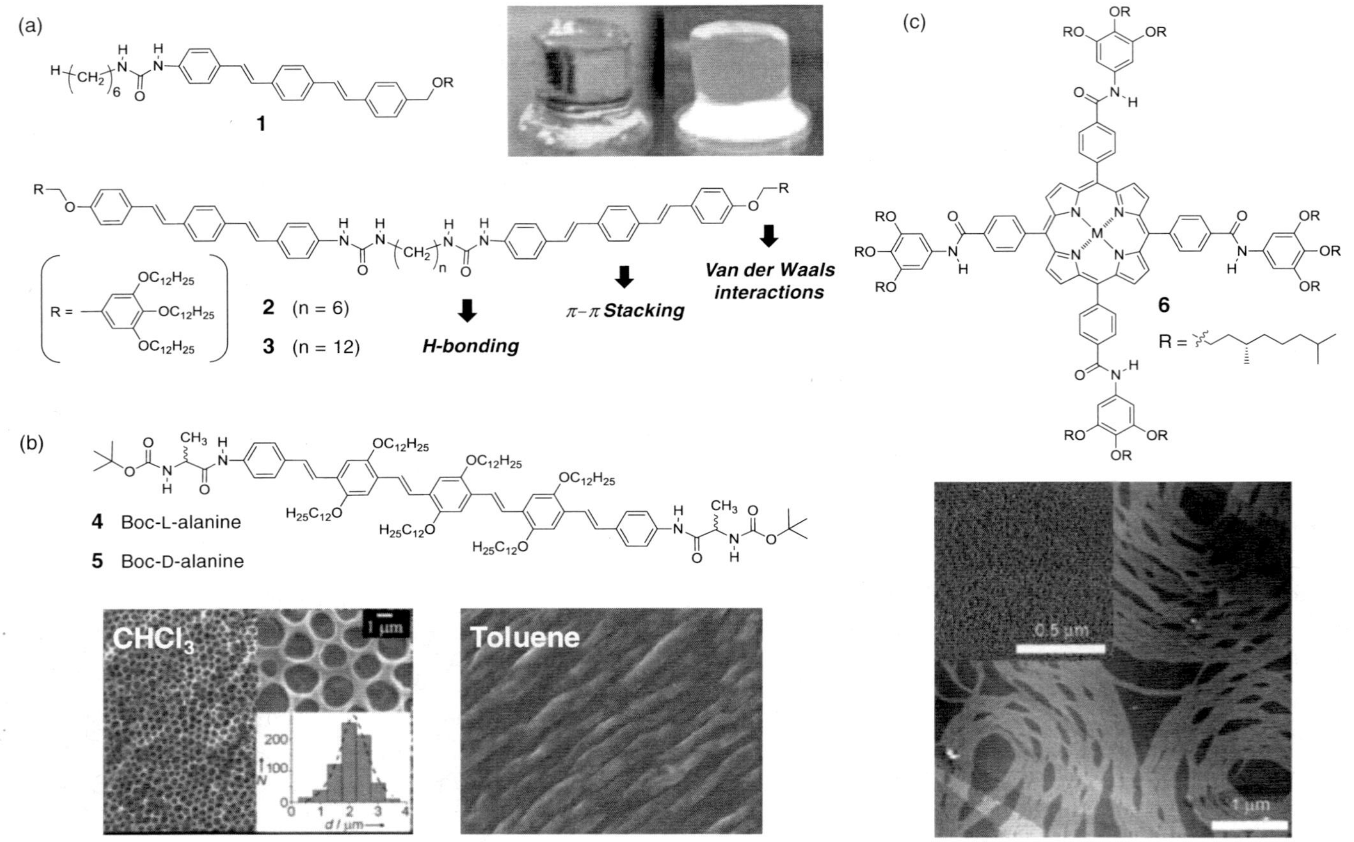

FIGURE 2.2. (**a**) Structure of compounds **1–3**. Photographs of the self-supporting decane gel of **2** deposited on a glass substrate taken under visible light (*left*) and 365 nm UV light (*right*). (**b**) Structure of compounds **4** and **5**. SEM images of (*left*) the honeycomb formed by the gel of **5** after evaporation of chloroform under moisture (inset shows the zoomed images and Lorentzian distribution of the pore size); (*right*) the aligned helical bundles formed by the toluene gel of **4**. (**c**) Structure of porphyrin **6**. AFM height image of **6** deposited from methylcyclohexane on HOPG. **Inset**: AFM height image of **6** deposited from same solution with 500 equivalents of pyridine, showing in this case no fibrilar structures. Adapted from: (a) reference 21; (b) reference 22; (c) reference 54 with the permission of Wiley–VCH.

years to understand the mechanisms that trigger supramolecular polymerizations giving rise to helical assemblies from chiral monomers [78]. The formation of stacked homochiral assemblies has the advantage that the self-assembling process can be additionally monitored by circular dichroism spectroscopy, which is a highly convenient sensitive technique. Many of these monomers have a common chemical structure: a π-conjugated core substituted with lateral chains having amide or urea H-bonding units, which promote columnar aggregation, and a chiral center. In this context, the amide-functionalized porphyrin derivatives **6** studied by Helmich et al. [54, 56, 57] (Figure 2.2c) have proven to be useful monomers to study monomer–aggregate equilibria owing to the intense and characteristic spectroscopic features of these chromophores and the possibility of incorporating different central metals that can interact with additives such as pyridine, ligands-leading to dilution-induced self-assembly and chiral memory effects.

2.2.2 Influence on Photophysical Properties

Amide/urea groups that form H-bonded arrays parallel to the stacking axis have also been employed to control to some extent the relative positioning of chromophores along the columns, thus influencing absorption and emission characteristics. This kind of assembly may bring photophysical features that are not observed in the dissolved monomers and that are mainly related to a given molecular packing [79]. Both the intensity and wavelength of absorption electronic transitions may vary dramatically between the gel and solution states, depending on the polarity of the gelated solvent [58] or the stacking geometry of the molecules. The

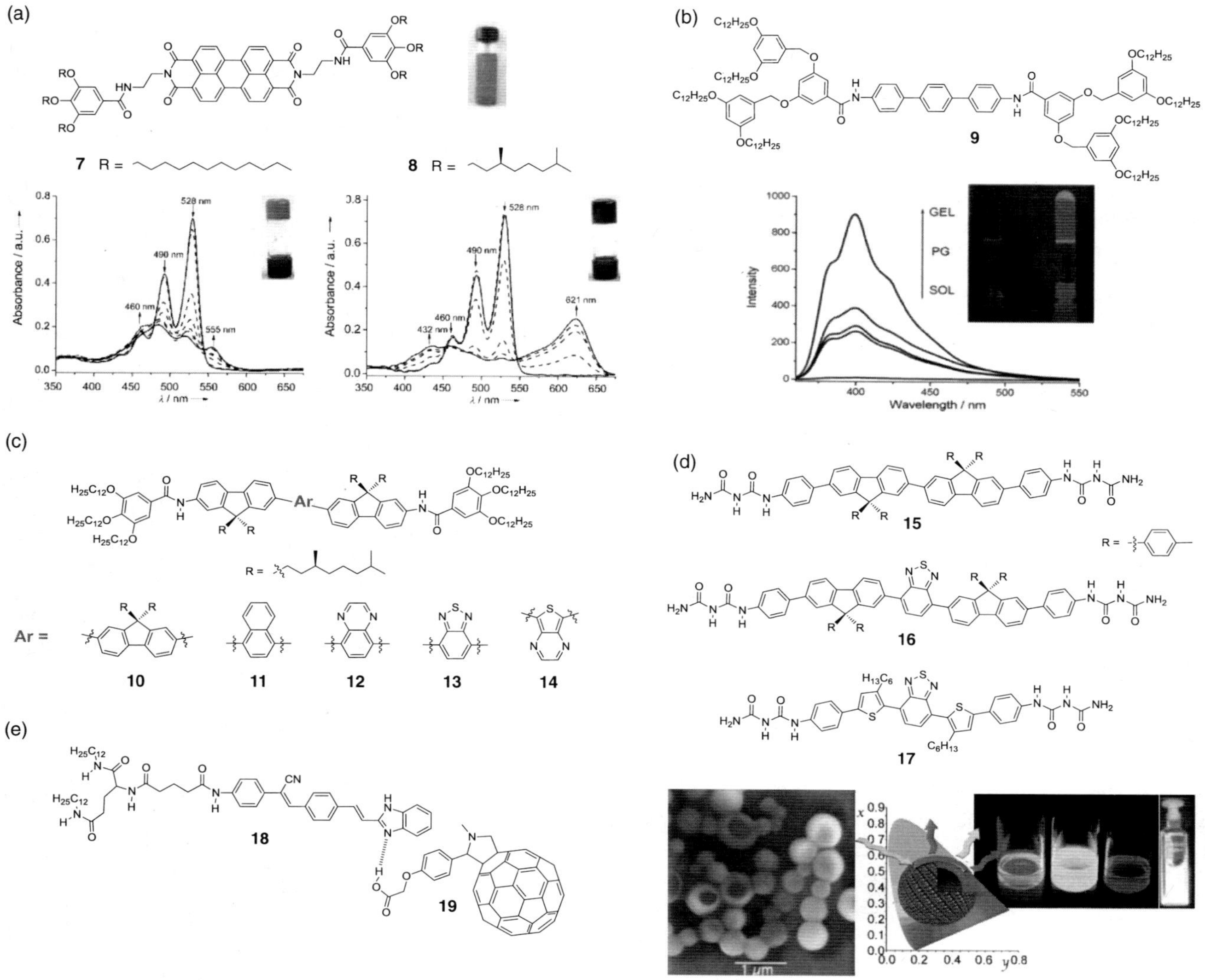

FIGURE 2.3. (**a**) Structure of PBIs **7** and **8**. Solvent-dependent UV–vis absorption spectra of PBIs **7** and **8** at a concentration of 10^{-5} M at 25°C. Arrows indicate the spectral changes upon increasing the methylcyclohexane/$CHCl_3$ ratio. The inserted pictures show the solution and gel colors in each case. (**b**) Structure of terphenylene **9**. Fluorescence spectra of **9** in xylene solution (SOL), partial gels (PG), and gel state at the same concentration (4.2 mM) excited at 350 nm. The inset shows the fluorescence images of the solution (*left*) and gel (*right*) in xylene taken under illumination with 365 nm UV light. (**c**) Structure of fluorene oligomers **10**–**14**. (**d**) Structure of oligomers **15**–**17**. SEM images of the spherical aggregates formed by compound **15** upon dropcasting from dilute anhydrous THF solution ([**15**] = 0.1 mM) onto a SiO_2 substrate. The presence of holes in some of the aggregates is indicative of their hollow nature. (**e**) Structure of the H-bonded complex **18**·**19**. Adapted from: (a) reference 39 with the permission of Wiley-VCH; (b) reference 16 with permission from the American Chemical Society; (c) reference 23; (d) reference 25 with the permission of Wiley-VCH; (e) reference 84. See color insert.

latter effect has been demonstrated by Wurthner and collaborators in a series PBI functional gels in which these chromophores can stack in either parallel (H-type) [34, 35, 39] or tilted (J-type) [36, 37, 39, 80] conformations as a function of the nature of the peripheral H-bonding amide substituents (Figure 2.3a). For example, the stacking of PBI **7**, substituted with long alkyl chains, led to blue-shifted absorption bands, whereas PBI **8** exhibited red-shifted absorption and emission features in the aggregate, as a consequence of the tilted packing imposed by the bulkier side tails [39]. The characteristic absorption features of these aggregated states produced different gel colors ranging from the typical red of PBI dyes to black gels absorbing above 600 nm (Figure 2.3a). H-bond assisted control of H-type *versus* J-type aggregation modes have also been observed in Pp molecules by Shirakawa et al. [49]. Here, the number and position of amide groups around the Pp cores was employed to regulate both the H-/J- stacking geometry, as confirmed by x-ray crystal analyses, and the morphology of the mesoscopic aggregates, which varied from one-dimensional (1D) fibrilar to two-dimensional (2D) sheet-like structures.

As regards to photoluminescence quantum yields, certain molecules that are only slightly fluorescent in solution may exhibit enhanced emission in the self-assembled state. The formation of rigid amide-type H-bonding arrays along fluorophore fibers are very effective in promoting this phenomenon, known as "aggregation-induced enhanced emission" [79, 81], since it is usually originated from J-type aggregate formation and/or planarization and restriction of conformational flexibility of the π-conjugated system in the assembled state. Phenylene oligomers constitute a remarkable example in this context. Chen et al. [16] have reported fluorescence intensities in gels of aromatic *p*-terphenylene molecules with dendritic wedges containing amide groups (**9**; Figure 2.3b) that are almost three orders of magnitude higher than those of the dissolved monomers. Sun and coworkers [17] have also recently studied the gelation and enhanced emission properties of OPE derivatives capped at both ends with two amide groups [17]. Time-resolved emission measurements indicated that the planarization and the restricted conformational flexibility of the gelator aggregates leads to slow nonradiative decays in favor of emissive excited state deactivation pathways. Other fluorophores like PBIs [34, 36], HBCs [47], and star-shaped phenylene-ethynylthiophene molecules substituted with amide/urea groups have also shown enhanced, bathochromically shifted emission bands in the gel state [82].

Within this context, a versatile strategy to vary the emission color of dye-based organogels is by proper election of a combination of fluorophores that are able to mix in H-bonded columnar gel phases. In this kind of material, energy transfer events constitute a fundamental mechanism to harvest and convey the excitation energy between the different dye molecules [83]. Abbel et al. [23,24] reported on related light emitting gels constituted by OF derivatives that were mixed with small amounts of complementary oligomers in which the central aromatic unit was changed (Ar = fluorine (**10**), naphthalene (**11**), quinoxaline (**12**), benzothiadiazole (**13**), or thienopyrazine (**14**); Figure 2.3c), hence giving rise to different absorption and emission wavelengths. All of these compounds were equipped with long alkyl side chains and amide groups at both ends, which promoted the formation of gels having emission colors ranging from blue to red to even white. Very recently, Tseng et al. [25] studied the formation of highly emissive hollow nanospheres of about 200 nm in diameter that were built from aggregated π-conjugated oligomers **15**-**17** (Figure 2.3d). Such building blocks were chosen in view of their strong fluorescence features which, combined, covered most of the visible spectrum (i.e., from 350 to 750 nm). A most remarkable characteristic of the systems prepared by these authors is that, by mixing the appropriate oligomers in the correct ratio, any luminescence color, including white light emission, could be observed from the aggregated objects.

Compared to such energy donor–acceptor mixed systems, there are fewer examples in the literature that deal with electron transfer processes taking place in organized H-bonded π-conjugated gels. Organogelation can, however, create interesting monodimensional ordered arrangements of molecularly interspaced electron donors and acceptors, provided that at least one of them is equipped with suitable H-bonding moieties that promote aggregation. Oligomer **18** (Figure 2.3e) formed fibrilar three-dimensional (3D) network structures in *o*-dichlorobenzene that showed absorption blue-shifts and lower intensity and red-shifted emission features with respect to the molecularly dissolved species [84]. Addition of C_{60} fullerene **19** resulted in a highly efficient fluorescence quenching that was ascribed to a thermodynamically favorable electron transfer process. Interestingly, the addition of pristine C_{60} led to much lower fluorescence quenching. Hence, the authors suggested that the interaction between the carboxylic acid group in **19** with aggregated **18** was found to be crucial in making a good molecular connection leading to efficient charge separation between the electron donor and acceptor.

2.2.3 Hole and Electron Transport

The formation of π–π stacked wire-like structures between semiconducting molecules is also considered as an appealing and promising strategy to facilitate electron and hole transport in organic electronic devices [10–12]. The stabilization of these supramolecular nanostructures by H-bonding interactions along the wire axis can lead to ordered stacks exhibiting high charge-carrier mobilities, which is an essential parameter that affects the performance of organic optoelectronic devices such as field-effect transistors (FETs), light-emitting diodes (LEDs), and photovoltaic cells (PVs). In a seminal work published in 1999, Schoonbeek et al. [26] demonstrated that mono- and bithiophene derivatives like **20** and **21** (Figure 2.4a) could be organized into fibers by the action of pendant urea groups that formed an H-bonded array in the direction of the fiber axis. These thiophene wires displayed relatively high charge mobilities with respect to regular (oligo)thiophenes, which was attributed to a well-defined, closely packed molecular arrangement. The aggregation of the 1D arrays onto different surfaces was next analyzed by the same authors [27]. Whereas oxidic substrates like SiO_2 induce an upright, side-by-side orientation, they tend to lie down flat on graphitic substrates.

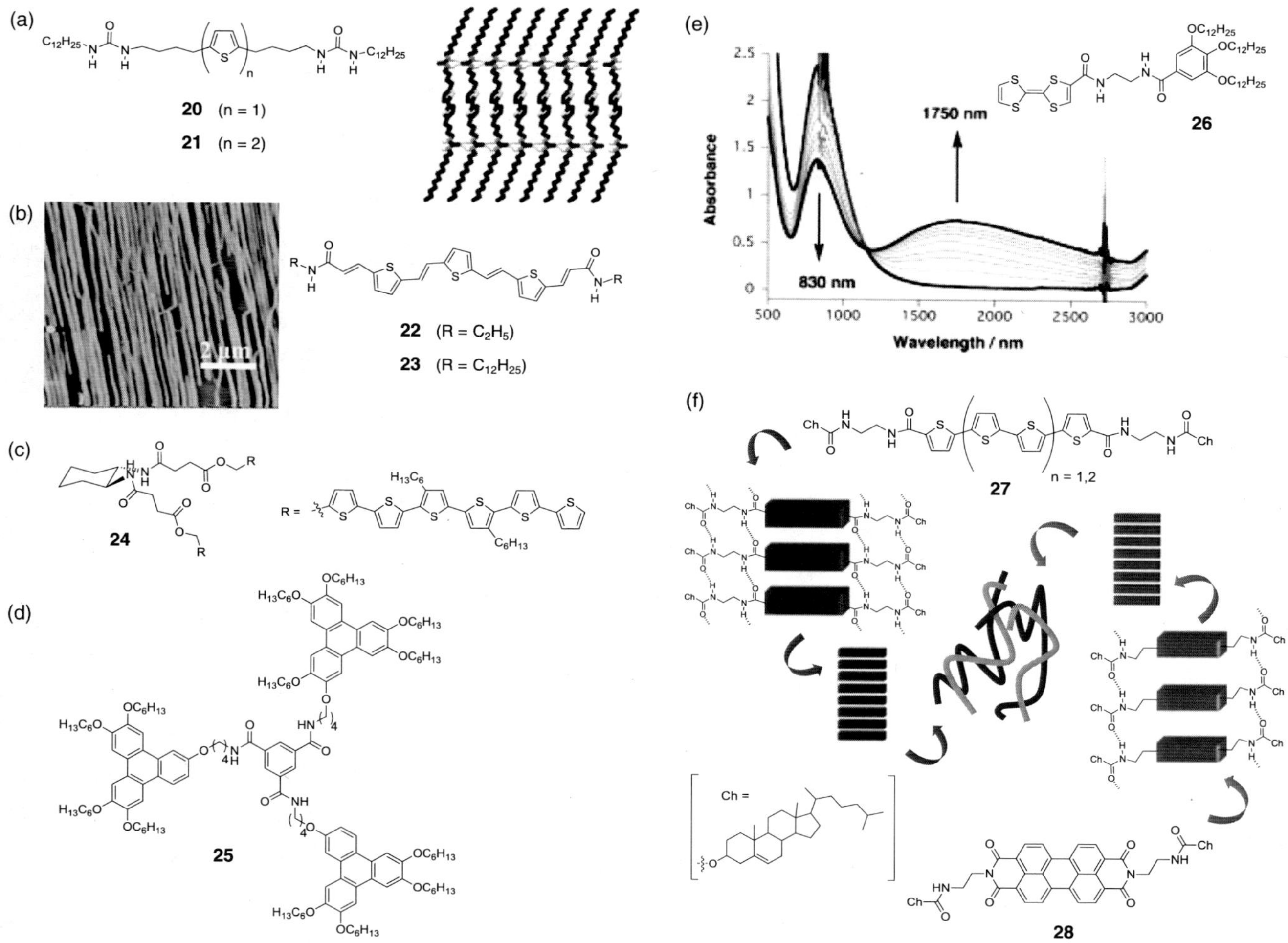

FIGURE 2.4. (**a**) Structure of urea-capped thiophenes **20** and **21** and model of their stacking arrangement. (**b**) Structure of compounds **22** and **23**. *Left*: AFM image of **22** from decane dropcasted on freshly cleaved mica surface ([**22**] = 5 × 10^{-5} M). (**c**) Structure of hairpin-shaped compound **24**. (**d**) Structure of compound **25**. (**e**) Structure of TTF derivative **26**. NIR absorption spectral time-dependent changes of self-assembled **26** prepared from the hexane gel (1.0 g dm^{-3}) in the presence of I_2. The characteristic absorption band at 1750 nm at room temperature after removing the excess I_2 was assigned to the mixed-valence state of the stacked TTF core. (**f**) Self-sorted stacks formed by OT **27** and PBI **28** with entanglement regions that act as heterojunctions. Adapted from: (a) reference 26 with the permission of Wiley–VCH; (b) reference 29 with permission from the American Chemical Society; (c) reference 30; (d) reference 41; (e) reference 85.

More recently, other authors have addressed the relevance of supramolecular order in charge transport along OT stacks. Prasanthkumar et al. [29] studied the effect of gelation on charge-carrier mobilities in oligothienylenevinylene derivatives substituted with amide tails at both ends (**22**, **23**; Figure 2.4b). The samples deposited from $CHCl_3$–dodecane mixtures exhibited epitaxial growth on mica surfaces resulting in aligned supramolecular wires that revealed charge-carrier mobilities 2–3 times higher than less-ordered aggregates drop–casted from $CHCl_3$ solutions. A similar result was obtained by the group of S. I. Stupp when analyzing hole mobilities in hairpin-shaped sexithiophene organogels stabilized by the *trans*-1,2-diamidocyclohexane motif (compound **24**; Figure 2.4c): OFETs fabricated by casting from solvents that promote self-assembly led to higher mobilities [30].

The enhancement of charge carrier mobilities upon formation of ordered columnar stacks have also been observed for several other π-conjugated units. An example is the self-assembly of discotic triphenylene molecules connected to a central 1,3,5 aromatic trisamide unit (**25**; Figure 2.4d) [41]. The presence of this H-bonding unit guides triphenylene stacking into long wires in which the neighboring discs strongly interact through their π-surfaces with a very short twist angle. This organization led to a charge carrier mobility that is about five times higher with respect to the materials obtained with individual triphenylene molecules aggregated solely by π–π stacking interactions. The same aromatic trisamide building block has been employed to organize Pp [53] or OPV [18] trimers into columnar stacks that exhibit a strong tendency to aggregate in solution into long and stable fibers. In the last case it was shown that the topology of the amide groups connecting the central benzene and the OPV chromophore (i.e., -CONH- versus -NHCO-) can determine the helicity and stability of the fibers in solution and the length of the fibrils at a surface [18].

Electroactive materials like TTF that exhibit high electron conductivities in crystalline structures [11] are also relevant candidates for the construction of ordered conducting wires through cooperative π–π stacking and H-bonding interactions. With this idea in mind, several groups have prepared amide- [62–65, 67] or urea-appended [66] TTFs and studied their fibrilar assemblies. Remarkably, when these wires are subjected to annealing and/or doping processes, mixed-valence states are formed, creating materials that display absorbance in the near-infrared [62] (compound **26**; Figure 2.4e) and a considerable conductivity enhancement [64].

The incorporation of amide H-bonding functions at the peripheral chains of metallomacrocycles such as Pcs has been reported (by Kumaran et al. [48] to direct their columnar aggregation onto variety of conducting substrates. These included highly oriented pyrolitic graphite (HOPG) or gold surfaces modified with alkanethiols having amide H-bonding functions, which could serve as nucleation points for stack elongation perpendicular to the substrate. AFM and FT-IR data indicated a layer-by-layer growth of Pc films from dilute solutions and suggested that the Pc planes lie parallel to the solid support. Conductive-tip AFM measurements served to evaluate the conductance across these ordered films, which decreased with the layer height.

Aside from increasing stack stability and intermolecular order, H-bonding directors acting along the fiber axis can in principle be utilized to induce self-sorting processes between two aggregating dyes as a function of the different number and position of H-bonding sites at each dye. Therefore, one could program, for instance, electron-rich and electron-poor π-conjugated molecules with the required information to form self-sorted *p*- and *n*-type nanofibers, respectively. This is a highly appealing picture for organic semiconducting applications, since it may provide separate hole and electron conducting pathways in devices. Sugiyasu et al. [85] studied the formation of fibers from OT and PBI molecules (**27** and **28** in Figure 2.4f) functionalized with chiral cholesterol moieties having four or two H-bonding sites, respectively. This system was designed to assemble into separate *p*- and *n*-type nanofibers based on the fact that the number and position of H-bonding sites at each dye is different. The authors managed to give some evidence on the self-sorting assembly process by means of several spectroscopic techniques. On the other hand, the fluorescence of a cast film of the self-sorted material on an ITO electrode showed significantly quenched fluorescence, indicating photo-induced electron transfer from the OT to the PBI fibers. Furthermore, a photocurrent can be generated upon irradiation of the film. These results were explained by the presence of contact regions between the two fibers that act as heterojunctions for efficient electron transfer and photocurrent generation.

Sakai et al. [86] recently introduced the supramolecular zipper approach to prepare coaxial *p-n* heterojunctions that are grown from gold surfaces. The designed material was composed of rigid *p*-oligophenyl (OP) or *p*-oligophenylethynyl (OPE) rods and pending naphthalenediimide (NDI) molecules bearing either positively charged or negatively charged groups. The combination of different non-covalent interactions like π–π stacking, H-bonding between amide moieties, and ionic interactions resulted in the formation of interdigitated stacks that are vertically aligned with respect to a gold substrate. The controlled layer-by-layer incorporation of multiple NDI derivatives of different electronic character led to efficient light absorption at several wavelengths and, most remarkably, to the generation of a redox gradient. Using this approach, the authors managed nanostructure functional thin films at the molecular level so that the controlled redox gradient along both *n*- and *p*-type materials guided the antiparallel directional flow of electrons and holes to their respective electrodes.

2.2.4 Fiber Alignment and Cross-Linking

Amide or urea H-bonding along the columnar axis lead to fibers that are stable enough to be oriented using different strategies. Hameren et al. [53] employed physical dewetting to construct periodically patterned surfaces covering a range of square millimeters, that is, well into macroscopic length scales. They profited from the columnar self-assembly of porphyrin trimer **29** (Figure 2.5a), which is driven by π–π stacking and H-bonding interactions between the central 1,3,5-trisamide block. The linear patterns consisted of equidistant single stacks of **29** (5 nm wide, up to 1 mm long) and were obtained spontaneously on drop-casting from $CHCl_3$ onto a mica surface by a hierarchical dewetting effect. Similar solvent-vapor annealing processes have been very recently employed to align hexaazatrinaphthylene trisamide-bonded fibers [46]. On the other hand, carbazole-L-isoleucine gelators have been anisotropically aligned in homogeneously oriented smectic mesophases of cyanobiphenyl derivatives [61]. Here, the liquid crystalline matrix served as a template environment for fiber alignment. Fluorescence measurements indicated that the carbazole units were stacked forming a π-conjugated wire consisting of sandwiched excimers, while polarized infrared spectroscopy confirmed the parallel

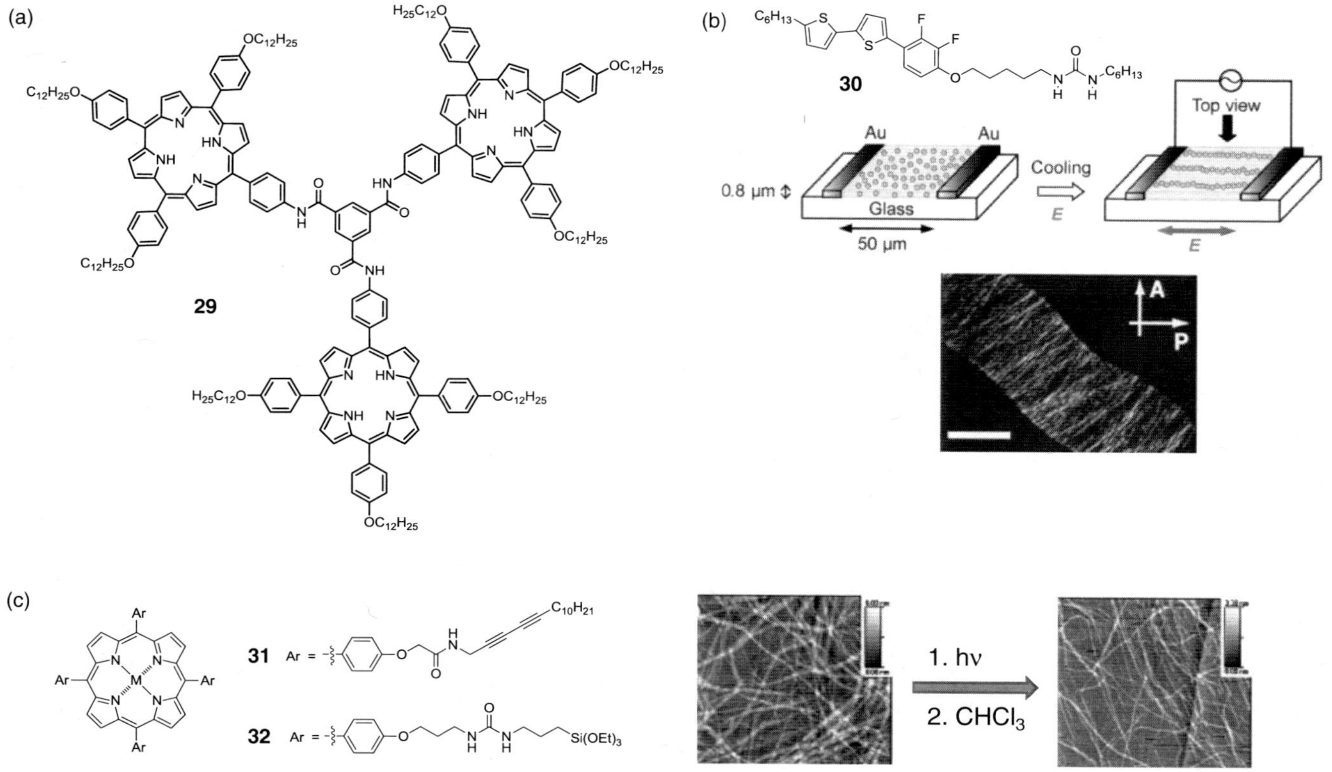

FIGURE 2.5. (**a**) Structure of Pp trimer **29**. (**b**) Structure of compounds **30**. Schematic illustration of the alignment process: the mixture of **30** (50 g L^{-1}) (gray spheres) and dodecylbenzene (light gray shading) was cooled from the isotropic liquid state at 135°C (*left*) while applying the electric field (0.5 V mm^{-1}, 1 kHz) (*right*). *Below*: Polarized optical microscopic images of the aligned fibers in the cell under crossed analyzer (A) and polarizer (P) at room temperature. (**c**) Structure of Pps **31** and **32**, whose aggregates can be stabilized by covalent (photo)polymerization. AFM images of the decalin gel of **31** before (*left*) and after (*right*) UV irradiation and $CHCl_3$ rinsing of the surface. Adapted from: (b) reference 88 with permission from the Royal Society of Chemistry; (c) reference 50 with the permission of the American Chemical Society.

alignment of the C=O stretching band in the H-bonded amide groups with respect to the C≡N stretching band. An additional attribute that can be exploited with these amide/urea H-bonding motifs is that, due to their dipolar nature, they can assist in the alignment of the supramolecular fibers when an electric field is applied [87, 88]. A prerequisite for optimal alignment is that an odd number of these groups must be introduced around the π-conjugated core, or otherwise the dipole moments may cancel out in an antiparallel arrangement. Shoji et al. [88], among others, demonstrated the expediency of this strategy by aligning fibrous aggregates of urea-appended phenylbithiophene **30** (Figure 2.5b). The aligned fibers, which exhibited photoconductive properties, were obtained by applying an electric field between two electrodes while cooling the isotropic solutions of **30** in dodecylbenzene to the gel state. It was also shown that the dielectric anisotropy afforded by the central semifluorinated arene played a key role during the alignment process.

In some cases, when equipped with the appropriate functions, the resulting columnar nanostructures can be covalently fixed once assembled. The formation of directional H-bonds between neighboring stacked molecules may leave (photo)polymerizable groups at the correct distances and angles for postpolymerization under specific conditions. That is the case of diacetylene-substituted hexaazatrinaphthylene [44] or porphyrin (**31**; Figure 2.5c) [50] molecules, which can be photo-cross-linked in the aggregate state under irradiation with UV light. The resulting fibers display a high stability and can be rinsed with solvents to selectively remove unreacted monomers. Covalent immobilization of H-bonded wires has also been performed via in situ sol–gel polycondensation of dyes with covalently linked, peripheral triethoxysilyl groups, as in compound **32** (Figure 2.5c) [31, 51, 52]. The organic/inorganic hybrid superstructures formed in this way enjoy very high thermal and mechanical stabilities, which are a requisite for material applications.

2.3 H-BONDING PERPENDICULAR TO THE STACKING POLYMER AXIS

The previous section dealt with the formation of H-bonding arrays along the stacking axis, which has been demonstrated to be very useful in reinforcing stack stability and directing the internal structure. However, this approach does not fully profit the extraordinary attributes of this selective and directional noncovalent interaction. Some organic fragments display an array of H-bonding donor and acceptor functions that can recognize complementary H-bonding patterns with tunable specificity and strength, which increases with the number of H-bonds formed [5, 7]. When these fragments are incorporated in π-functional molecules, one can promote their homo- or heteroassociation in solution or in the condensed state. Using this approach, there have been numerous examples during the last years of, for instance, π-functional molecules that bind in solution into homo- or heterodimers, trimers, and, in general, linear or cyclic oligomers (quartets, rosettes; *vide infra*) [89, 90]. In this section we will focus on those examples of hierarchically organized functional materials where H-bonding serves to select and define the monomeric building blocks, comprised of one or several different π-conjugated molecules, that subsequently polymerize via π–π stacking in the perpendicular direction with respect to the H-bonding interactions to yield fiber-like materials (Figure 2.1b).

2.3.1 Homo-associated Monomers

One of the most studied systems within this class of supramolecular materials are OPVs **33** (Figure 2.6a), whose self-assembly and properties have been thoroughly investigated during the last 10 years by Meijer, Schenning, and collaborators [91–98]. The derivatives of **33** are functionalized at one terminus with an ureidotriazine moiety, which is known to self-associate into dimers by formation of four complementary H-bonds. Hence, H-bonding assists here in providing homodimers of **33** whose association constant in chloroform, obtained from NMR dilution experiments, is on the order of 2×10^4 L/mol [99]. Apolar solvents like dodecane or methylcyclohexane further promote the polymerization of these dimers by π–π stacking to yield helical fibers of about 100 nm long (Figure 2.6a). The stability of these columnar stacks increases with the oligomer length (i.e., from the OPV trimer to the pentamer) as a result of

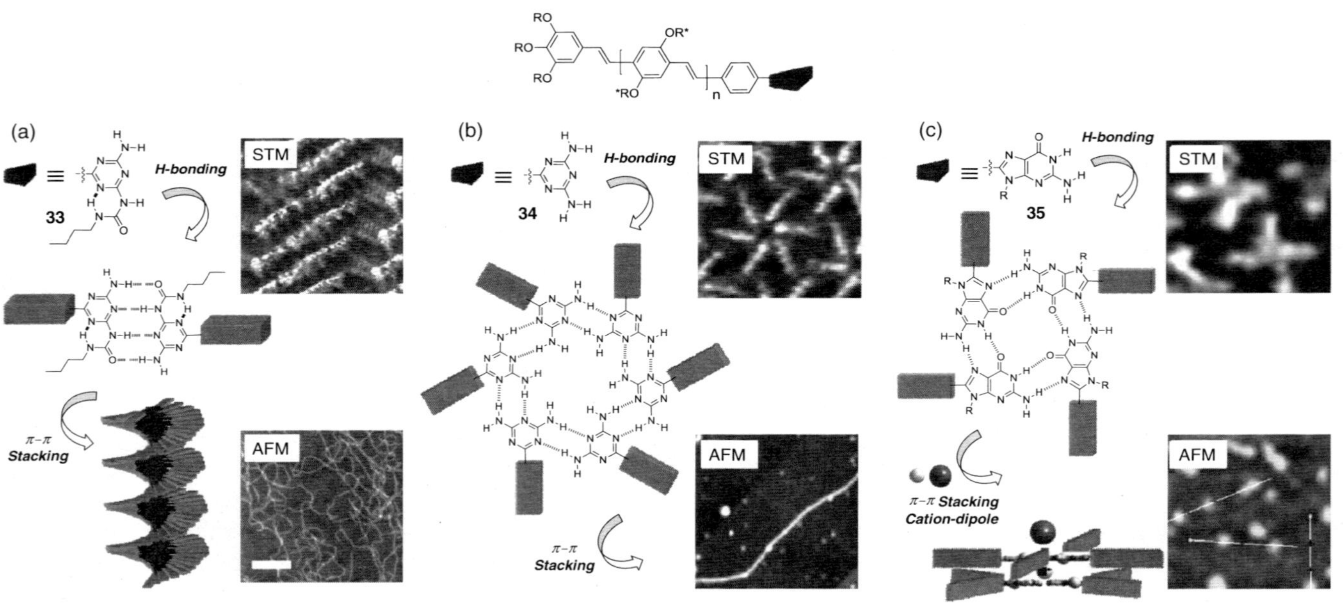

FIGURE 2.6. π–π Stacked aggregates of OPV derivatives **33–35** as a function of the H-bonding unit attached to one end. In each case, the STM images onto HOPG of the H-bonded molecular systems and the AFM images of the aggregates formed are shown on the right. (**a**) Ureidotriazine–OPVs **33** form H-bonded dimers in solution that aggregate by π–π stacking. (**b**) Diaminotriazine–OPVs **34** form H-bonded hexamers in solution that aggregate by π–π stacking. (**c**) Guanine–OPVs **35** form H-bonded tetramers that aggregate in the presence of potassium salts by π–π stacking and cation–dipole interactions to yield octameric nanoparticles. Adapted from: (a) references 94 and 99 with permission from the American Association for the Advancement of Science and the American Chemical Society; (b) reference 110 with the permission of Wiley–VCH; (c) reference 116 with the permission from the American Chemical Society.

the larger size of the homodimer π-conjugated surface [92]. Temperature-dependent spectroscopic measurements clearly indicated that a nucleation-growth mechanism with a high degree of cooperativity operates during fiber formation [94]. When chiral side chains are covalently attached to the OPVs, helical stacks are formed with a preferred handedness that is determined by the nature of the chiral group. Recently it was found that fibers with only one preferred handedness are formed if the assembly process is thermodynamically controlled [98]. Remarkably, *P* or *M* helicity can be also induced either in chiral solvent media [95] or via H-bonding with chiral carboxylic acid guests [96], which bind to the lateral aminotriazine segment that is not involved in dimer formation.

This particular system has served as model for multiple studies conducted in order to weigh the possibilities of using ordered individual fibers in nanosized optoelectronic devices. For instance, these one-dimensional OPV stacks provide a useful and versatile archetype for studying and modeling the mechanism of excitonic migration along π–π stacked molecules [100–107]. Femtosecond-transient absorption measurements revealed fast exciton dynamics that are reminiscent of those found for thin films of poly(phenylenevinylene) [100, 101], which underlines the expediency of the supramolecular approach toward organic materials. When short oligomers having three phenylene units (OPV_3) are mixed within the same stack with a small amount of longer oligomers with four or five phenylene units (OPV_4 or OPV_5), having a lower bandgap, excitation of the short OPVs results in a fast intra-stack energy transfer process to the longer oligomers [102, 103]. An additional relevant conclusion of these studies was that the degree of order within the stacks is a highly determining factor in the dynamics of exciton migration [104–107]. Two types of OPV_3 monomers were employed: (a) monofunctional molecules that form well-defined helical assemblies and (b) bifunctional molecules that aggregate into frustrated stacks that lack any higher helical order. These two types of assemblies were doped with a small amount of the longer OPV that functioned as an energy acceptor. From the analysis of the donor fluorescence quenching and acceptor emission yields, it was concluded that ordered helical stacks, showed much faster exciton mobility than ill-defined aggregates [104]. In fact, only small percentages of energy traps in the well-defined stacks led to almost exclusive acceptor luminescence.

Charge-transport along the columnar stacks was also modeled and investigated in these OPV aggregates in solution by time-resolved microwave conductivity measurements [108]. Despite the ordered nature of the columnar stacks the mobility values withdrawn from these experiments (3×10^{-3} and 9×10^{-3} $cm^2/(V\ s)$ for holes and electrons, respectively) are not very high, and this was ascribed to the relatively large twist angle between adjacent stacked molecules, which is seen as a detrimental factor for optimal orbital overlap and hence for efficient charge migration along the stacks. In addition, in order to assess their conductive properties, the fibers of OPV derivatives such as **33** were transferred to solid supports and deposited over electrodes. High transfer fidelity from solution to a solid substrate, showing negligible fiber dissociation or restructuring, can be achieved provided that the right concentration and suitable, inert substrates (like graphite or silicon oxide) were employed, as demonstrated by a set of AFM studies [92]. The self-assembled molecular wires deposited onto Au–Pd electrodes showed, however, poor electrical conduction. This was not attributed to an inefficient contact between the self-assembled wire and the electrode, but rather to the presence of defects or to an intrinsically poor conductance along the π–π stacks [109].

The extraordinary versatility of H-bonds to connect different molecular units is reflected in that a small change of the chemical structure of the H-bonding motif may lead to a change in the self-recognition mode which, at the same time, is translated in a complete readjustment of the internal chromophore arrangement within the nanostructure. Two relevant examples are given below in relation to the previously described OPV-ureidotriazine system **33**. When the H-bonding unit is modified to diaminotriazine, as in OPV derivatives **34** (Figure 2.6b), these π-conjugated oligomers self-assemble into hexameric rosettes as a result of the formation of pairs of H-bonds between aminotriazine fragments [110]. This organization has been also observed at the 1-phenyloctane/graphite interface by means of STM measurements [110–112]. The planar cyclic hexamers display a large driving force for π–π stacking into tubular chiral objects, as demonstrated in solution by UV–vis, fluorescence, and CD spectroscopy. Furthermore, AFM and small-angle neutron scattering (SANS) studies indicated the formation of fibers with a diameter which matches that of the stacked rosettes (Figure 2.6b).

On the other hand, when the terminal H-bonding motif is changed to guanine (as in OPVs **35**), the formation of tetramers (G-quartets) is observed (Figure 2.6c). In the presence of sodium or potassium salts, these lipophilic H-bonded quartets are able to coordinate the alkaline cations to provide discrete and fairly stable complexes (G-quadruplexes) in organic solvents instead of polymeric fibers [113–115]. In particular, OPV-G derivatives **35** yielded octameric complexes in the presence of potassium salts which displayed negative Cotton effects, bathochromic shifts, and, interestingly, an enhancement of the fluorescence emission intensity [116]. These features are also characteristic of larger organic nanoparticles (>25 nm), habitually produced by reprecipitation methods [117, 118], that show size-dependent emission quantum yields. The OPV-G nanoclusters obtained exhibited an extraordinary stability toward concentration or temperature changes in nonpolar solvents and maintained their structural integrity, without dissociation or further polymerization, when deposited on graphitic or mica substrates [116].

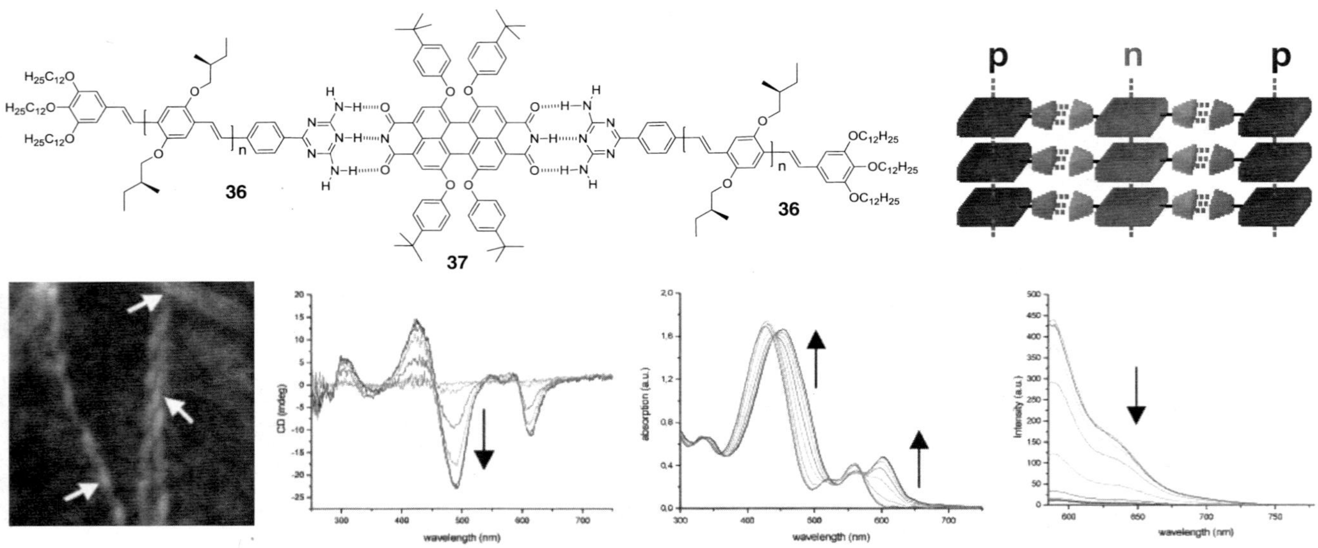

FIGURE 2.7. Structure of the trimer formed by triple H-bond interactions between OPV **36** and PBI **37**. (*Right*) Model of the target *p–n–p* nano-heterojunction. *Below, from left to right*: Tapping mode AFM topographic image of the **36–37–36** complex after spin-coating from methylcyclohexane on a glass/PEDOT:PSS slide. Temperature-dependent CD, UV–vis, and fluorescence spectra of the **36–37–36** triad ($c = 3.7\ 10^{-5}$ mol/L) in methylcyclohexane (arrows indicate the changes upon cooling). Adapted from references 124 and 125 with permission from the American Chemical Society.

The G-quadruplex architecture thus constitutes a very interesting scaffold to which a small number of π-functional molecules can be covalently attached and stacked in a well-defined arrangement to yield small monodisperse nanoobjects. Recent studies by different research groups have also disclosed the singular characteristics of guanine (G) self-assembly to organize different functional molecules, such as Pps [119], pyrenes [120], OTs [121], and paramagnetic radical species like 4-carbonyl-2,2,6,6-tetramethylpiperidine-1-oxyl (TEMPO) [122, 123].

2.3.2 Hetero-associated Monomers

The previous examples demonstrate that H-bonding between self-complementary motifs constitutes a valuable strategy to preorganize a discrete number of well-defined semiconducting oligomers into larger planar π-conjugated structures that then stack into polymeric cylinders. Within this line, an additional step in complexity would be the heteroassociation of π-conjugated molecules with different electronic properties (i.e., dyes absorbing at different wavelengths, electron donor and acceptor molecules, etc.) to construct functional materials.

A complementary motif for this purpose that has been widely utilized in the formation of supramolecular chromophore systems is the heteroassociation of diaminotriazines and imides by triple H-bonding [124–127]. Several groups have studied the coassembly of OPV-diaminotriazine (**36**) and N-unsubstituted PBI (**37**) into *p–n–p* type junctions with the aim that, upon ordered columnar stacking, the self-assembled fibers could be used as models for the antiparallel transport of both electrons and holes (Figure 2.7). The complementary recognition between H-bonding groups led to the connection of two OPVs to both termini of the perylene dye. The resulting planar triad then formed helical columnar stacks of about 7 nm in diameter that further coiled into larger, chiral rod superstructures, as determined by AFM studies (see Figure 2.7). Time-resolved transient absorption measurements confirmed the occurrence of a fast and efficient electron transfer process from the electron donor (OPV) to the acceptor (PBI) within the stacks. However, rather than occurring across the OPV-PBI H-bonds, absorption and CD studies suggested that the charge-transfer process takes place via an intermolecular pathway between J-type stacked donor and acceptor molecules within the columns [126]. This mode of dye ordering along the wires is, on the other hand, not surprising, in view of the thermodynamically favorable π–π stacking interactions between electron-rich and electron-poor π-surfaces.

These and other [128–131] studies carried out with mixtures of conjugated oligomers and PBIs shed light on the importance of the nanoscale organization and relative arrangement of donors and acceptors in the performance of thin films for application in organic field-effect transistors or photovoltaic devices. For instance, Jonkheim et al. [129] managed to successfully prove the existence of two independent pathways for charge transport in ambipolar OFETs built from H-bonded OPV-PBI assemblies. These systems are

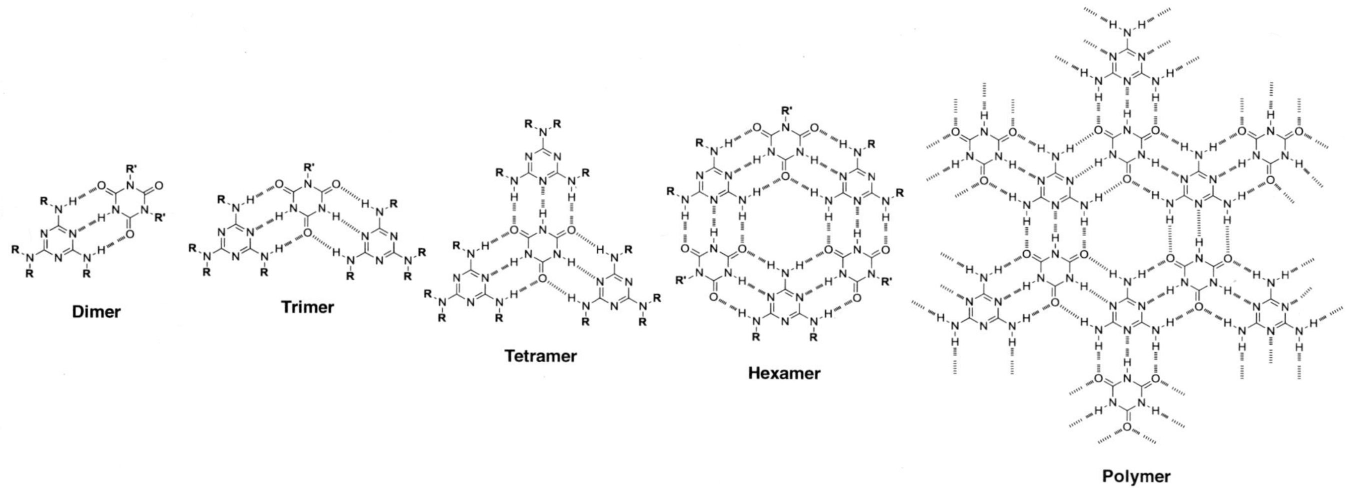

FIGURE 2.8. Structure of different H-bonded oligomers formed from the melamine–cyanurate couple, depending on the number of available triple H-bonding sites in each molecule.

similar to those described above, but now the PBI unit H-binds to a single OPV–diaminotriazine molecule through one of the imide termini. Futhermore, thin films made from alkane solutions of self-assembled OPV–PBI dyads, consisting of uniform rodlike domains, showed superior performances when companed to films built from CH_2Cl_2 solutions where the active components are molecularly dissolved. In sharp contrast, mixtures of OPVs and PBIs that are not connected by H-bonding lack any FET activity, despite the fact that any of these molecules separately show distinct field effects. This was explained on the basis of charge-transfer donor–acceptor π–π complex formation, which leads to an unfavorable supramolecular organization for charge transport. On the other hand, van Herrikhuyzen et al. [128] demonstrated that a 1:1 mixture of N-substituted PBI and OPV–ureidotriazine derivatives like **33** could be orthogonally assembled into separate *n*- and *p*-type stacks in apolar solvents, since the absorption spectra and the melting temperatures of the individual components in the mixture did not change with respect to the separate components in the same experimental conditions. Fluorescence and photoinduced absorption spectroscopy of drop-casted films of these materials revealed that an efficient electron-transfer process takes place between the self-sorted *p*–*n* fibers. However, efficient organic solar cells could not be obtained, presumably due to the nonoptimal parallel orientation of the fibers with respect to the electrodes.

Other H-bonding self-complementary residues have been widely used for the formation of multichromophoric π–π stacked assemblies. Based on the same triple H-bonding association pattern between diaminopyridine and imide motifs, the classical interaction between melamine and barbiturate/cyanurate derivatives can yield very diverse supramolecular structures around which multiple dyes can be assembled. As Figure 2.8 shows, depending on the number of available triple H-bonding sites in each molecule, the melamine–cyanurate association can lead to dimers (one site each), trimers (two sites in one of them and one site in the other), tetramers (three sites in one of them and one site in the other), cyclic hexamers or rosettes (two sites in each residue), or 2D polymeric networks [132] (all three sites available for both molecules) [89, 90].

In recent years, the group of Yagai, among others, has investigated the versatility of this complementary heteroassociation and demonstrated how changes in the structure of the self-assembling monomers can promote the formation of diverse supramolecular architectures [133–138]. For instance, an OPV trimer capped on one end with a melamine unit with one binding site available (**39**; Figure 2.9a) showed distinct self-organization behavior in mixtures with a cyanurate module having 1, 2, or 3 accessible H-bonding sites [133]. As shown in Figure 2.9a, each mixture is supposed to lead to stacked aggregates with 1:1, 1:2, and 1:3 stoichiometric relations between **38** and the corresponding cyanurate. The absorption and emission features of the aggregates changed drastically as a function of the cyanurate-$(\mathbf{38})_{1\text{-}3}$ complex stoichiometry, which was attributed to the different stacking arrangement of the OPVs. On the other hand, a melamine derivative substituted with two azobenzene groups (**40**; Figure 2.9b) was shown to form stable rosettes in the presence of cyanurate derivatives like **40**, but only when the azo-dyes presented an *E* stereochemistry [134]. The planar rosettes formed by combination of melamine **39** and cyanurate **40** can then polymerize to form discotic columns that induce the gelation of apolar solvents like cyclohexane. Irradiation of the gel with UV light promotes the $E \rightarrow Z$ photoisomerization of the azobenzene groups, which results in more sterically demanding species that tend to dissociate [135]. A

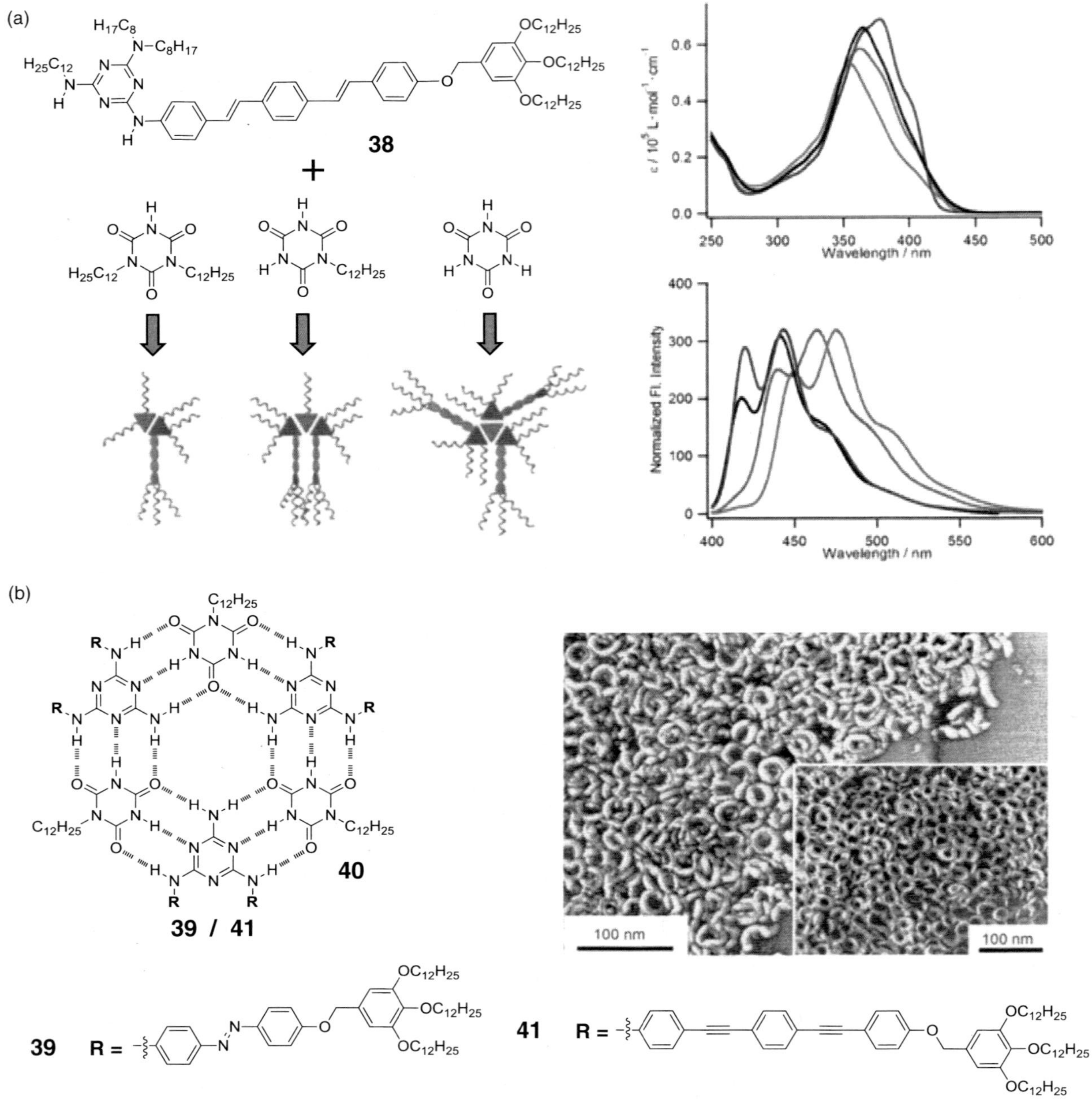

FIGURE 2.9. (**a**) Structure of OPV **38** and schematic representation of its complexation with cyanurates having 1, 2, or 3 available H-bonding sites. *Right*: UV–vis (*top*) and normalized fluorescence spectra (*bottom*) of MCH solutions containing stoichiometric mixtures of **38** ($c = 5 \times 10^{-3}$ M) and the different cyanurates at 10°C. (**b**) Structure of melamine derivatives **39** (substituted with two azobenzene groups) and **41** (substituted with two OPEs). Both compounds, having two H-bonding sites available, form stable rosettes in the presence of 1 equivalent of cyanurate derivatives like **40**, which further aggregate via π–π stacking. In the case of compound **41**, the aggregates formed have a closed toroidal structure, as shown below in the AFM phase image of a sample spin-coated from decane solution on HOPG. Adapted from: (a) reference 133 with permission from the Royal Society of Chemistry; (b) reference 136 with the permission of Wiley–VCH.

related melamine compound substituted with OPE oligomers (**41**; Figure 2.9b) led, outstandingly, to the formation of toroidal objects of nanometer dimensions in the presence of cyanurate **40** [136]. The authors proposed that the cause for this unusual organization was the ~45° tilting of the OPE segments with respect to the H-bonded rosette plane, which produces a curvature in the aggregate that eventually leads to a closed ring superstructure comprised of about 160 monomeric rosettes.

Hoeben et al. [139] also employed this melamine–cyanurate couple to construct trimers constituted by two OPV-diaminotriazine with only one H-bonding site available and one porphyrin–cyanurate conjugate with two H-bonding sites. These ensembles polymerized by π–π stacking to yield helically ordered wires that, interestingly, exhibited efficient energy transfer from the OPV units to the porphyrin dye.

2.4 MAIN-CHAIN H-BONDED π-FUNCTIONAL POLYMERS

We have so far described supramolecular π-functional materials in which H-bonding interactions take place either parallel or perpendicular to the π-stacking axis and were combined with π–π stacking interactions to yield fiber-like polymers. A third division of H-bonded π-conjugated supramolecular materials we have considered comprises those systems in which H-bonding interactions between ditopic self-complementary units takes place along the polymer chain and are the main or only noncovalent force that guides monomer association (Figure 2.1c) [8]. This self-organization mode therefore combines some elements of the previous systems described in Sections 2.2 and 2.3. It resembles on one hand to the supramolecular polymers described in Section 2.2 in that H-bonding interactions contribute to the growth of polymer, but in this case they do not necessarily cooperate with π–π stacking or lead to fibrilar materials. On the other hand, the systems described herein profit from the self-complementary selectivity of some of the H-bonding fragments introduced in Section 2.3 but, in order to construct polymeric systems, ditopic monomers are employed. It should be noted that polymers of a significant molecular weight can only be created in solution providing H-bonding interactions are very strong. Hence, most of the examples described below refer to solid-state systems. All things being equal, this approach, when compared with the most common covalent π-conjugated polymers, holds great promise in terms of producing π-functional materials with a high degree of modularity. In other words, this approach has the potential of fine-tuning the material electronic properties, dynamic features, morphology, and the mechanical and processing characteristics of the polymers formed by proper choice of the nature of the π-conjugated monomers and the H-bonding units. As a matter of fact, as will be shown below, the use of this strategy has been addressed by some authors as a successful means to avoid phase segregation in some organic devices, since the interaction between the self-complementary H-bonding moieties placed at the periphery of the monomeric units can lead to random or alternate supramolecular copolymers. This kind of supramolecular polymerization constitutes therefore an illustration of how to combine the synthetic simplicity and well-defined structure of small molecules with the processability of π-conjugated polymers.

2.4.1 Random (co)Polymers

Supramolecular materials formed by condensation of N-unsubstituted PBIs like **42** represent a simple example of main-chain H-bonded π-conjugated polymers [140–144]. Polymerization is driven by formation of pairs of H-bonds between the antiparallel imide functions. Bulky solubilizing groups are often attached at the perylene bay positions so as to limit stacking interactions and improve the processability of the final self-assembled material. Kaiser et al. observed that this type of PBI compounds (**42**; Figure 2.10a) can be polymerized into aggregates with very pronounced J-type spectral features which, remarkably, can exhibit fluorescence quantum yields near unity [140, 142] or, when substituted with amine groups at the bay positions, absorption maxima in the near infrared (i.e., up to 900 nm) [141]. The supramolecular polymerization mechanism of these N-unsubstituted PBI dyes leading to J-type aggregates has been studied recently in more detail by the group headed by F. Würthner [143]. The presence of the bulky bay substituents not only produced a tilted dye stacking, but also affects the polymerization mechanism which, rather than following an isodesmic model, more general of PBI dyes, was consistent with a cooperative nucleation-growth process. The same group also reported recently on octachloro-PBI molecules (**43**; Figure 2.10b) that crystallized into H-bonded networks [144]. The presence of 8 chlorine atoms at the PBI bay positions resulted in a substantial lowering of the LUMO level and in twisting of the perylene core due to steric interactions. Still, the analysis of the crystal network revealed π–π stacked percolation pathways for charge transport. Indeed, the OTFT devices made from these crystalline materials could operate under air atmosphere exhibiting remarkable electron mobilities and I_{on}/I_{off} ratios at high temperatures.

As described in the previous section, the supramolecular versatility of the imide functions of PBIs for the construction of H-bonded supramolecular polymers should be underlined. In combination with melamine derivatives with two available H-bonding sites, the structural and physical properties of the materials can be modulated [145, 146] and other chromophores or functional groups can be introduced within the polymer backbone [131].

H-bonded supramolecular polymers are also a promising alternative to π-conjugated covalent polymers as far as

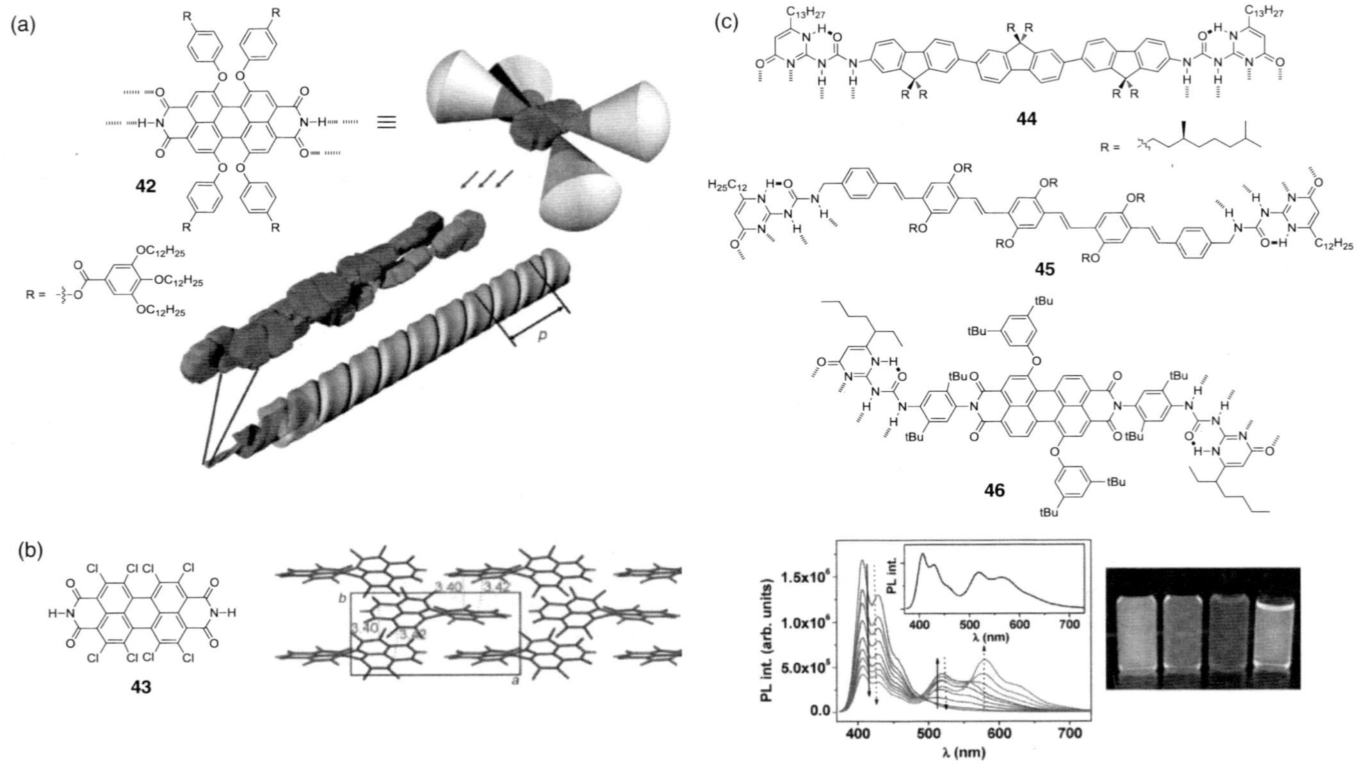

FIGURE 2.10. (**a**) Structure of PBI **42** and schematic representation of its self-assembly process into J-type helical aggregates by H-bonding between the imide groups and π–π stacking between the twisted perylene cores. The gray cones with an apex represent the bay substituents. (**b**) Structure of PBI **43** and packing arrangement in the crystalline state. (**c**) Structure of OF **44**, OPV **45**, and PBI **46** functionalized at the terminal positions by quadruple H-bonding Upy groups. Below on the left, a titration experiment is shown (pure **44**, successive addition of **45**, further addition of **46**). The solid arrows indicate spectral changes upon addition of **45** to **44**, and the dotted arrow represents the change upon addition of **46** to a mixture of **45** and **44**. The inset shows the PL spectrum corresponding to a solution containing **44**:**45**:**46** in a ratio of 59:33:8. On the right, the solutions of pure di-UPy chromophores **44**, **45**, **46** and a white emitting mixture in chloroform under UV irradiation are displayed, in that order. Adapted from: (a) reference 140; (b) reference 144 with permission from Wiley–VCH; (c) reference 150 with permission from the American Chemical Society. See color insert.

light-emitting properties are concerned. The group headed by Schenning and Meijer reported a series of studies focused on the fluorescence emission and energy transfer characteristics of H-bonded polymers of OFs [147–150]. For such purpose, these fluorophores were functionalized at both ends with a 2-ureido-4[1*H*]-pyrimidinone (Upy) residue [7, 151], which dimerizes strongly via four H-bonds with an association constant of 6×10^7 ($CHCl_3$). Hence, the resulting bis-UPy-terminated oligomers, such as **44**, self-assemble into supramolecular polymers [152] that can be end-capped by a variety of functional chain-stoppers (i.e., dyes with a single UPy group) [147]. It was demonstrated that these end-capping fluorophores, namely OPV and PBI dyes, are able to act as energy traps for the excited OF main chain both in solution and in the solid state, and it was shown that H-bonding actually enhances the efficiency of the energy transfer process toward the polymer ends. Femtosecond-resolved photoluminescence spectroscopy was as well employed to study in detail this excitation energy transfer process, which was contrasted with Monte Carlo simulation models [149]. The output of these simulations served as a powerful means to determine the molecular weight distribution in these end-capped supramolecular polymers, because the dynamics of energy migration was found to be dependent on the chain length [148]. The results fitted nicely a Flory distribution, which is based on the assumption of equal reactivity of all functional groups.

Furthermore, the use of three different fluorophores (blue-emitting OF (**44**), green-emitting OPV (**45**), and red-emitting PBI (**46**)), all of which are end-capped with UPy groups at both termini (Figure 2.10c), allowed the preparation of supramolecular polymers where each monomer is randomly distributed along the polymer chain [150]. The resulting material covered a wide region of the absorption spectrum (up to 700 nm; Figure 2.10c). Remarkably, these supramolecular materials showed no phase separation in thin films, and the electroluminescence color could be modulated as a function of each dye monomer content: blue, green, red, or, using appropriate mixing ratios, white emission. This strategy presents utmost interest in view of its high tunability and possibilities: By choosing the right fluorophores, at the right composition, a wide range of emission wavelengths can be obtained in light-emitting diodes [153].

Other H-bonding motifs are also good candidates for the generation of main-chain self-assembled polymers. Barbiturate derivatives are known to form different modes of two-dimensional H-bonded arrays, which were employed recently to organize OPV chromophores into nanosystems that show a reversible transformation between rings and coils [154]. The incorporation of this motif to fullerene C_{60} led to crystalline ribbon-like structures where the C_{60} spheres interact electronically within each H-bonded cable. The group of Bassani examined the anisotropy of such interactions by probing the fluorescence polarization arising from excimer-like states by means of confocal fluorescence microscopy [155]. Compared to 3D crystals of pristine C_{60}, which showed little polarization, the fullerene–barbiturate crystals exhibited very strong polarization of the excimer-like emission. The lower dimensionality of this material was also reflected in electron mobilities in OFET devices that were two orders of magnitude lower than in C_{60} crystals.

Some other moieties contain two self-complementary H-bonding faces in the same covalent structure. In this context, 4,6-diamino-pyrimidin-2(1 *H*)-one is a very interesting example that combines the ADD H-bonding pattern of guanine on one side and the DAA H-bonding pattern of cytosine on the other. The formation of photoresponsive azobenzene supramolecular tapes using this appealing residue has been studied by Yagai et al. [156].

2.4.2 Alternating (co)Polymers

Other than affording the thermodynamic requirements for polymer chain growth, H-bonding units situated at the periphery of π-conjugated cores can also select which monomer is to be incorporated at the termini of the growing chain. Thus, as illustrated in Figure 2.1c and 2.11a, two different monomers substituted at opposite sides with complementary H-bonding fragments will lead to an alternating copolymer.

In this context, the widely used melamine-barbituric acid H-bonding couple has been studied by Huang et al. [157, 158] for the construction of organic photovoltaic devices. Electron-donor OTs symmetrically functionalized at both ends with melamine functions (**47**) were co-deposited with an electron-acceptor fullerene-barbiturate compound (**48**) (Figure 2.11a). The mixture yielded homogeneous films in which the two components formed alternating supramolecular polymers, thus minimizing phase segregation [157]. Photovoltaic devices made of these films led to a 2.5-fold enhancement in light energy to electrical energy conversion.

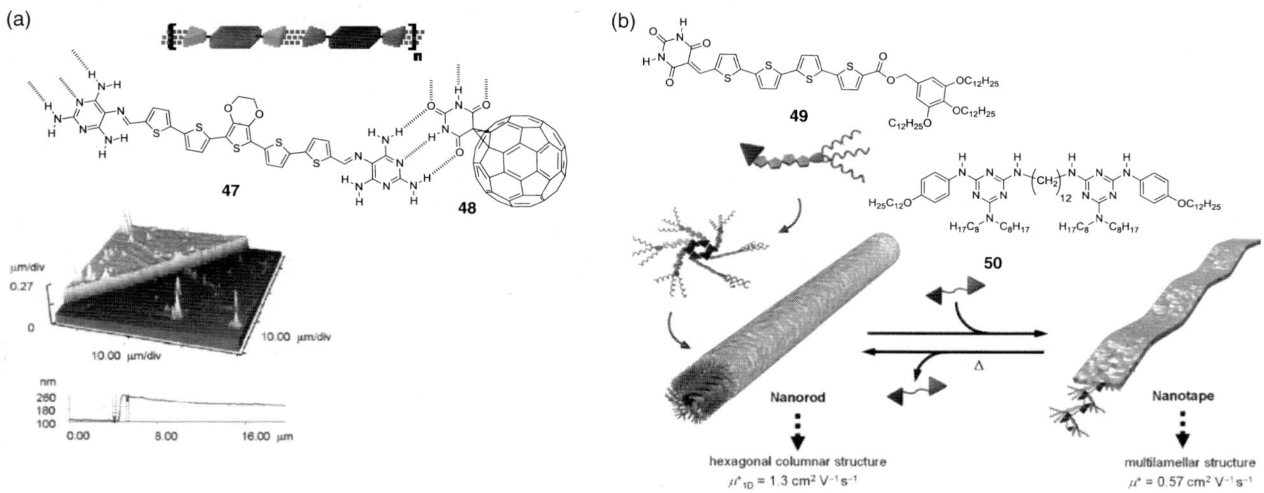

FIGURE 2.11. (**a**) *Top*: Model of an alternating copolymer comprised of π-conjugated molecules where the monomers are associated via H-bonding interactions along the polymer main chain. Structure and association mode of OT-bismelamine **47** and fullerene C_{60} barbiturate derivative **48**. *Below*: AFM image of the film made from equimolar amounts of **47** and **48** showing its thickness and the absence of phase separation. (**b**) Structure of OT **49** and bismelamine **50**. Schematic representation of the interconversion between nanorod and nanotape, each of which exhibit different hole mobilities. Adapted from: (a) reference 157 with permission from the American Chemical Society; (b) reference 165 with permission from Wiley–VCH.

Furthermore, these self-recognizing molecules were organized into molecular-level heterojunctions on gold supports by means of melamine- or barbiturate-grafted alkanethiol self-assembled monolayers [158].

The same melamine–barbituric acid H-bonding motif has also been used by the groups headed by Wurthner and Yagai to generate numerous supramolecular architectures, such as nanoribbons and nanoropes, based on merocyanine [159–162], azobenzene [161], PBI [163, 164], or OT [165] π-conjugated molecules. This last case is a representative example of the strong dependence of a given property on the supramolecular organization of the material which, at the same time, is strongly influenced by the molecular structure and the environmental conditions during the self-assembly process. OT **50**, capped on one end with a barbituric acid residue, spontaneously self-assembled into hexagonally packed nanorods by a combination of H-bonding and π–π stacking interactions, as depicted in Figure 2.11b. In such assemblies, the OT monomers form H-type stacked columns. In the presence of bismelamine derivative **51**, however, a lamellarly packed nanotape architecture was observed by AFM and x-ray diffraction measurements, whereas spectroscopic measurements revealed the disruption of H-type aggregation. The intrinsic charge-carrier mobilities of these two self-organized materials were then evaluated by time-resolved microwave conductivity. Interestingly, hole mobility in the lamellar structure was calculated as $0.57\ cm^2\ V^{-1}\ s^{-1}$, whereas the value of one-dimensional hole mobility in the nanorods was $1.3\ cm^2V^{-1}s^{-1}$ [165]. These values are one order of magnitude higher than the mobility observed for lamellarly organized poly(hexylthiophene)s.

Other structurally similar H-bonding residues can generate alternating self-assembled polymers. The organization of perylene dyes, appended in this case at both ends with thymine or uracil moieties combined with melamine, has been studied in the solid state [166]. Crystal x-ray diffraction results confirmed the formation of long chains constituted by H-bonded arrays that alternate the melamine and the PBI-thymine/uracil chromophores.

2.5 CONCLUSIONS

The purpose of this chapter was to give a general vision on the different roles of H-bonding in producing self-assembled materials comprised of monodisperse π-conjugated molecules. We have focused on three main nanoscale architectures leading to supramolecular polymers, depending on the purpose and direction of the H-bonding arrays. A first case describes supramolecular polymers where H-bonding, typically between amide or urea groups, cooperates with π–π interactions to yield materials in which some properties can be fine-tuned (mesoscale ordering, absorption and emission features, or charge mobilities along the columnar stacks). The last two cases in contrast make use of more sophisticated complementary H-bonding fragments (ureidotriazine, melamine, cyanurate, etc.) which are employed to bring molecules of different characteristics selectively together, thus allowing for the possibility of creating tailored molecular interfaces. We have excluded, nonetheless, numerous examples in which H-bonded polymers or oligomers form 2D monolayers onto diverse surfaces or studies on templated H-bonded polymers, in which a monodisperse covalent polymeric strands guide the assembly of π-conjugated molecules through self-complementary H-bonding interactions. The latter approach has proven extraordinarily useful to define the dimensions of supramolecular nanostructures as well as to control the exact positioning of π-conjugated units with nanometer precision [9].

Despite the presence of numerous challenges ahead, hydrogen-bonded π-conjugated assemblies with custom-made properties are arising as appealing materials for future plastic and nanosized optoelectronic devices. The first optoelectronic devices based on these materials have been constructed demonstrating the proof of principle. H-bonded self-assembly constitutes a convenient modular tool to construct supramolecular π-conjugated materials by simply mixing different suitably "programmed" components, thus avoiding labor-intensive synthesis of new covalent molecules. Polymeric materials obtained in this way often combine the well-defined, uniform electronic properties of monodisperse systems and the processability benefits of polymers. H-bonding interactions produce materials that can be aligned, cross-linked, and, to a certain extent, internally structured at the molecular scale. This supramolecular tool clearly surpasses any other noncovalent interactions in terms of selectivity, directionality, and strength tunability. However, in order to build useful, well-defined architectures, H-bonding must be designed so as to cooperate with other supramolecular forces such as π–π interactions acting in the same or different spatial directions.

ACKNOWLEDGMENTS

The authors would like to acknowledge the many discussions and contributions with all of our former and current colleagues. The research in Eindhoven University of Technology has been supported by the Royal Netherlands Academy of Science (KNAW), the Netherlands Organization for Scientific Research (NWO), and the European Young Investigators Awards (EURYI). DGR would like to acknowledge a Marie Curie Reintegration Grant (SOAFNPCM-230964), a MICINN National Grant (CTQ2011-23659) and an ERC Starting Grant (PROGRAM-NANO- 279548).

REFERENCES

1. Köhler, M.Fritzsche, W. *Nanotechnology: An Introduction to Nanostructuring Techniques*, Wiley–VCH, Weinheim, 2007.

2. Special issue on the status of self-assembly at the beginning of the XXI century: (2002) *Science*, **295**.
3. Hunger, K. *Industrial Dyes*, Wiley-VCH, Weinheim, 2003.
4. Hoeben, F. J. M., Jonkheijm, P., Meijer, E. W., Schenning, A. P. H. J. (2005). About supramolecular assemblies of π-conjugated systems. *Chemical Reviews*, **105**, 1491–1546.
5. Prins, L. J., Reinhoudt, D. N., Timmerman, P. (2001). Non-covalent Synthesis Using Hydrogen Bonding. *Angewandte Chemie International Edition*, **40**, 2382–2426.
6. Steiner, T. (2002). The Hydrogen Bond in the Solid State. *Angewandte Chemie International Edition*, **41**, 48–76.
7. Sijbesma, R. P., Meijer, E. W. (2003) Quadruple hydrogen bonded systems. *Chemical Communications*, 5–16.
8. De Greef, T. F. A., Smulders, M. M. J., Wolffs, M., Schenning, A. P. H. J., Sijbesma, R. P., Meijer, E. W. (2009). Supramolecular polymerization. *Chemical Reviews*, **109**, 5687–5754.
9. González-Rodríguez, D., Schenning, A. P. H. J. (2011). Hydrogen-bonded supramolecular π-functional materials. *Chemistry of Materials*, **23**, 310–325.
10. Schmidt-Mende, L., Fechtenkötter, A., Müllen, K., Moons, E., Friend, R. H., MacKenzie, J. D. (2001). Self-organized discotic liquid crystals for high-efficiency organic photovoltaics. *Science*, **293**, 1119–1122.
11. See Special Issue on Molecular Conductors: (2004). *Chemical Reviews*, **104**.
12. Grozema, F. C., Siebbeles, L. D. A. (2008). Mechanism of charge transport in self-organizing organic materials. *International Reviews in Physical Chemistry*, **27**, 87–138.
13. Nguyen, T. Q., Bushey, M. L., Brus, L. E., Nuckolls, C. (2002). Tuning intermolecular attraction to create polar order and one-dimensional nanostructures on surfaces. *Journal of the American Chemical Society*, **124**, 15051–15054.
14. van Gorp, J. J., Vekemans, J. A. J. M., Meijer, E. W. (2002). *C*3-Symmetrical supramolecular architectures: Fibers and organic gels from discotic trisamides and trisureas. *Journal of the American Chemical Society*, **124**, 14759–14769.
15. Nguyen, T. Q., Martel, R., Avouris, P., Bushey, M. L., Brus, L. E., Nuckolls, C. (2004). Molecular interactions in one-dimensional organic nanostructures. *Journal of the American Chemical Society*, **126**, 5234–5242.
16. Chen, Y., Lv, Y., Han, Y., Zhu, B., Zhang, F., Bo, Z., Liu, C.-Y. (2009). Dendritic effect on supramolecular self-assembly: Organogels with strong fluorescence emission induced by aggregation. *Langmuir*, **25**, 8548–8555.
17. Shen, Y.-T., Li, C.-H., Chang, K.-C., Chin, S.-Y., Lin, H.-A., Liu, Y.-M., Hung, C.-Y., Hsu, H.-F., Sun, S.-S. (2005). Synthesis, optical, and mesomorphic properties of self-assembled organogels featuring phenylethynyl framework with elaborated long- chain pyridine-2,6-dicarboxamides. *Langmuir*, **25**, 8714–8722.
18. van Herrikhuyzen, J., Jonkheim, P., Schenning, A. P. H. J., Meijer, E. W. (2006). The influence of hydrogen bonding and π–π stacking interactions on the self-assembly properties of *C*3-symmetrical oligo(*p*-phenylenevinylene) discs. *Organic & Biomolecular Chemistry*, **4**, 1539–1545.
19. Matmour, R., De Cat, I., George, S. J., Adriaens, W., Leclère, P., Bomans, P. H. H., Sommerdijk, N. A. J. M., Gielen, J. C., Christianen, P. C. M., Heldens, J. T., van Hest, J. C. M., Löwik, D. W. P. M., De Feyter, S., Meijer, E. W., Schenning, A. P. H. J. (2008). Oligo(*p*-phenylenevinylene)-peptide conjugates: Synthesis and self-assembly in solution and at the solid-liquid interface. *Journal of the American Chemical Society*, **130**, 14576–14583.
20. Fang, F.-C., Chu, C.-C., Huang, C.-H., Raffy, G., Del Guerzo, A., Wong, K.-T., Bassani, D. M. (2008). Versatile one-step introduction of multiple hydrogen-bonding sites onto extended π-conjugated systems. *Chemical Communications*, 6369–6371.
21. Yagai, S., Kubota, S., Iwashima, T., Kishikawa, K., Nakanishi, T., Karatsu, T., Kitamura, A. (2008). Supramolecular polymerization and polymorphs of oligo(*p*-phenylene vinylene)-functionalized bis- and monoureas. *Chemistry—A Eropean Journal*, **14**, 5246–5257.
22. Babu, S. S., Mahesh, S., Kartha, K. K., Ajayaghosh, A. (2009). solvent-directed self-assembly of π gelators to hierarchical macroporous structures and aligned fiber bundles. *Chemistry—An Asian Journal*, **4**, 824–829.
23. Abbel, R., van der Weegen, R., Pisula, W., Surin, M., Leclère, P., Lazzaroni, R., Meijer, E. W., Schenning, A. P. H. J. (2009). Multicolour self-assembled fluorene co-oligomers: From molecules to the solid state via white-light-emitting organogels. *Chemistry—A Eropean Journal*, **15**, 9737–9746.
24. Abbel, R., van der Weegen, R., Meijer, E. W., Schenning, A. P. H. J. (2009). Multicolour self-assembled particles of fluorene-based bolaamphiphiles. *Chemical Communications*, 1697–1699.
25. Tseng, K.-P., Fang, F.-C., Shyue, J.-J., Wong, K.-T., Raffy, G., Del Guerzo, A., Bassani, D. M. (2011). Spontaneous generation of highly emissive RGB organic nanospheres. *Angewandte Chemie International Edition*, **50**, 7032–7036.
26. Schoonbeek, F. S., van Esch, J. H., Wegewijs, B., Rep, D., de Hass, M., Klapwijk, T. M., Kellogg, R. M., Feringa, B. L. (1999). Efficient intermolecular charge transport in self-assembled fibers of mono- and bithiophene bisurea compounds. *Angewandte Chemie International Edition*, **38**, 1393–1397.
27. Rep, D. B. A., Roelfsema, R., van Esch, J. H., Schoonbeek, F. S., Kellogg, R. M., Feringa, B. L., Palstra, T. T. M., Klapwijk, T. M. (2000). Self-assembly of low-dimensional arrays of thiophene oligomers from solution on solid substrates. *Advanced Materials*, **12**, 563–566.
28. Brustolin, F., Surin, M., Lemaur, V., Romanazzi, G., Sun, Q., Cornil, J., Lazzaroni, R., Sommerdijk, N. A. J. M., Leclere, P., Meijer, E. W. (2007). The self-assembly of amphiphilic oligothiophenes: Hydrogen bonding and poly(glutamate) complexation. *Bulletin of the Chemical Society of Japan*, **80**, 1703–1715.
29. Prasanthkumar, S., Saeki, A., Seki, S., Ajayaghosh A. (2010). Solution phase epitaxial self-assembly and high charge-carrier mobility nanofibers of semiconducting molecular gelators. *Journal of the American Chemical Society*, **132**, 8866–8867.

30. Tsai, W.-W., Tevis, I. D., Tayi, A. S., Cui, H., Stupp, S. I. (2010). Semiconducting nanowires from hairpin-shaped self-assembling sexithiophenes. *Journal of Physical Chemistry B*, **114**, 14778–14786.
31. Kimura, M., Miki, N., Suzuki, D., Adachi, N., Tatewaki, Y., Shirai, H. (2009). Wrapping of self-organized fluorescent nanofibers with silica wall. *Langmuir*, **25**, 776–780.
32. Zhang, Y., Yang, Z. M., Yuan, F., Gu, H. W., Gao, P., Xu, B. (2004). Molecular recognition remolds the self-assembly of hydrogelators and increases the elasticity of the hydrogel by 106-fold. *Journal of the American Chemical Society*, **126**, 15028–15029.
33. Syamakumari, A., Schenning, A. P. H. J., Meijer, E. W. (2002). Synthesis, optical properties, and aggregation behavior of a triad system based on perylene and oligo (*p* phenylene vinylene) units. *Chemistry—A European Journal*, **8**, 3353–3361.
34. Würthner, F., Hanke, B., Lysetska, M., Lambright, G., Harms, G. S. (2005). Gelation of a highly fluorescent urea-functionalized perylene bisimide dye. *Organic Letters*, **7**, 967–970.
35. Li, X.-Q., Stepanenko, V., Chen, Z., Prins, P., Siebbeles, L. D. A., Würthner, F. (2006). Functional organogels from highly efficient organogelator based on perylene bisimide semiconductor. *Chemical Communications*, 3871–3873.
36. Li, X.-Q., Zhang, X., Ghosh, S., Würthner, F. (2008). Highly fluorescent lyotropic mesophases and organogels based on J-aggregates of core-twisted perylene bisimide dyes. *Chemistry—A European Journal*, **14**, 8074–8078.
37. Würthner, F., Bauer, C., Stepanenko, V., Yagai, S. (2008). A black perylene bisimide super gelator with an unexpected J-type absorption band. *Advanced Materials*, **20**, 1695–1698.
38. Yagai, S., Seki, T., Karatsu, T., Kitamura, A., Würthner, F. (2008). Transformation from H- to J-aggregated perylene bisimide dyes by complexation with cyanurates. *Angewandte Chemie International Edition*, **47**, 3367–3371.
39. Ghosh, S., Li, X.-Q., Stepanenko, V., Würthner, F. (2008). Control of H- and J-type π stacking by peripheral alkyl chains and self-sorting phenomena in perylene bisimide homo- and heteroaggregates. *Chemistry—A European Journal*, **14**, 11343–11357.
40. Tidhar, Y., Weissman, H., Wolf, S. G., Gulino, A., Rybtchinski, B. (2011). Pathway-dependent self-assembly of perylene diimide/peptide conjugates in aqueous medium. *Chemistry—A European Journal*, **17**, 6068–6075.
41. Paraschiv, I., Giesbers, M., van Lagen, B., Grozema, F. C., Abellon, R. D., Siebbeles, L. D. A., Marcelis, A. T. M., Zuilhof, H., Sudhölter, E. J. R. (2006). H-bond-stabilized triphenylene-based columnar discotic liquid crystals. *Chemistry of Materials*, **18**, 968–974.
42. Paraschiv, I., Tomkinson, A., Giesbers, M., Sudhölter, E. J. R., Zuilhof, H., Marcelis, A. T. M. (2007). Amide, urea and thiourea-containing triphenylene derivatives: Influence of H-bonding on mesomorphic properties. *Liquid Crystals*, **34**, 1029–1038.
43. Gearba, R. I., Lehmann, M., Levin, J., Ivanov, D. A., Koch, M. H. J., Barberá, J., Debije, M. G., Piris, J., Geerts, Y. H. (2003). Tailoring discotic mesophases: Columnar order enforced with hydrogen bonds. *Advanced Materials*, **15**, 1614–1618.
44. Yip, H.-L., Zou, J., Ma, H., Tian, Y., Tucker, N. M., Jen, A. K.-Y. (2006). Patterning of robust self-assembled *n*-type hexaazatrinaphthylene-based nanorods and nanowires by microcontact printing. *Journal of the American Chemical Society*, **128**, 13042–13043.
45. Palma, M., Levin, J., Debever, O., Geerts, Y., Lehmann, M., Samori, P. (2008). Self-assembly of hydrogen-bond assisted supramolecular azatriphenylene architectures. *Soft Matter*, **4**, 303–310.
46. Yip, H.-L., Ma, H., Tian, Y., Acton, O., Tucker, N. M., Jen, A. K.-Y. (2011). Solvent-vapor annealing-induced growth, alignment, and patterning of π-conjugated supramolecular nanowires. *Journal of Materials Research*, **26**, 311–321.
47. Dou, X., Pisula, W., Wu, J., Bodwell, G. J., Müllen, K. (2008). Reinforced self-assembly of hexa-*peri*-hexabenzocoronenes by hydrogen bonds: From microscopic aggregates to macroscopic fluorescent organogels. *Chemistry—A European Journal*, **14**, 240–249.
48. Kumaran, N., Veneman, P. A., Minch, B. A., Mudalige, A., Pemberton, J. E., O'Brien, D. F., Armstrong, N. R. (2010). Self-organized thin films of hydrogen-bonded phthalocyanines: Characterization of structure and electrical properties on nanometer length scales. *Chemistry of Materials*, **22**, 2491–2501.
49. Shirakawa, M., Kawano, S-i., Fujita, N., Sada, K., Shinkai, S. (2003). Hydrogen-bond-assisted control of H versus J aggregation mode of porphyrins stacks in an organogel system. *Journal of Organic Chemistry*, **68**, 5037–5044.
50. Shirakawa, M., Fujita, N., Shinkai, S. (2005). A stable single piece of unimolecularly β-stacked porphyrin aggregate in a thixotropic low molecular weight gel: A one-dimensional molecular template for polydiacetylene wiring up to several tens of micrometers in length. *Journal of the American Chemical Society*, **2005**, **127**, 4164–4165.
51. Kishida, T., Fujita, N., Sada, K., Shinkai, S. (2005). Porphyrin gels reinforced by sol–gel reaction via the organogel phase. *Langmuir*, **21**, 9432–9439.
52. Kishida, T., Fujita, N., Sada, K., Shinkai, S. (2005). Sol–gel reaction of porphyrin-based superstructures in the organogel phase: Creation of mechanically reinforced porphyrin hybrids. *Journal of the American Chemical Society*, **2005**, **127**, 7298–7299.
53. van Hameren, R., Schön, P., van Buul, A. M., Hoogboom, J., Lazarenko, S. V., Gerritsen, J. W., Engelkamp, H., Christianen, P. C. M., Heus, H. A., Maan, J. C., Rasing, T., Speller, S., Rowan, A. E., Elemans, J. A. A. W., Nolte, R. J. M. (2006). Macroscopic hierarchical surface patterning of porphyrin trimers via self-assembly and dewetting. *Science*, **314**, 1433–1436.
54. Helmich, F., Lee, C. C., Nieuwenhuizen, M. M. L., Gielen, J. C., Christianen, P. C. M., Larsen, A., Fytas, G., Leclère, P. E. L. G., Schenning, A. P. H. J., Meijer, E. W. (2010). Dilution-induced self-assembly of porphyrin aggregates: A consequence of coupled equilibria. *Angewandte Chemie International Edition*, **49**, 3939–3942.

55. Iavicoli, P., Xu, H., Feldborg, L. N., Linares, M., Paradinas, M., Stafström, S., Ocal, C., Nieto-Ortega, B., Casado, J., López Navarrete, J. T., Lazzaroni, R., De Feyter, S., Amabilino, D. B. (2010). Tuning the supramolecular chirality of one- and two-dimensional aggregates with the number of stereogenic centers in the component porphyrins. *Journal of the American Chemical Society*, **132**, 9350–9362.

56. Helmich, F., Lee, C. C., Schenning, A. P. H. J., Meijer, E. W. (2010). Chiral memory via chiral amplification and selective depolymerization of porphyrin aggregates. *Journal of the American Chemical Society*, **132**, 16753–16755.

57. Helmich, F., Smulders, M. M. J., Lee, C. C., Schenning, A. P. H. J., Meijer, E. W. (2011). Effect of stereogenic centers on the self-sorting, depolymerization, and atropisomerization kinetics of porphyrin-based aggregates. *Journal of the American Chemical Society*, **133**, 12238–12246.

58. Su, L., Bao, C., Lu, R., Chen, Y., Xu, T., Song, D., Tan, C., Shi, T., Zhao, Y. (2006). Synthesis and self-assembly of dichalcone substituted carbazole-based low-molecular mass organogel. *Organic & Biomolecular Chemistry*, **4**, 2591–2594.

59. Yang, X, Lu, R., Xue, P., Li, B., Xu, D., Xu, T. (2008). Carbazole-based organogel as a scaffold to construct energy transfer arrays with controllable fluorescence emission. *Langmuir*, **24**, 13730–13735.

60. Yang, X, Lu, R., Zhou, H., Xue, P., Wang, P., Chen, P., Zhao, Y. (2009). Aggregation-induced blue shift of fluorescence emission due to suppression of TICT in a phenothiazine-based organogel. *Journal of Colloid and Interface Science*, **339**, 527–532.

61. Yabuuchi, K., Tochigi, Y., Mizoshita, N., Hanabusa, K., Kato, T. (2007). Self-assembly of carbazole-containing gelators: Alignment of the chromophore in fibrous aggregates. *Tetrahedron* **63**, 7358–7365.

62. Kitahara, T., Shirakawa, M., Kawano, S-i., Beginn, U., Fujita, N., Shinkai, S. (2005). Creation of a mixed-valence state from one-dimensionally aligned TTF utilizing the self-assembling nature of a low-molecular-weight gel. *Journal of the American Chemical Society*, **127**, 14980–14981.

63. Puigmartí-Luis, J., Minoia, A., Pérez del Pino, A., Ujaque, G., Rovira, C., Lledós, A., Lazzaroni, R., Amabilino, D. B. (2006). Chemical and constitutional influences in the self-assembly of functional supramolecular hydrogen-bonded nanoscopic fibres. *Chemistry—A European Journal*, **12**, 9161–9175.

64. Puigmartí-Luis, J., Laukhin, V., Pérez del Pino, A., Vidal-Gancedo, J., Rovira, C., Laukhina, E., Amabilino, D. B. (2007). Supramolecular conducting nanowires from organogels. *Angewandte Chemie International Edition*, **46**, 238–241.

65. Puigmartí-Luis, J., Pérez del Pino, A., Laukhina, E., Esquena, J., Laukhin, V., Rovira, C., Vidal-Gancedo, J., Kanaras, A. G., Nichols, R. J., Brust, M., Amabilino, D. B. (2008). Shaping supramolecular nanofibers with nanoparticles forming complementary hydrogen bonds. *Angewandte Chemie International Edition*, **47**, 1861–1865.

66. López, J. L., Pérez, E. M., Viruela, P. M., Viruela, R., Ortí, E., Martín, N. (2009). Controlled self-assembly of electron donor nanotubes. *Organic Letters*, **11**, 4524–4527.

67. Puigmartí-Luis, J., Pérez del Pino, A., Laukhin, V., Feldborg, L. N., Rovira, C., Laukhina, E., Amabilino, D. B. (2010). Solvent effect on the morphology and function of novel gel-derived molecular materials. *Journal of Materials Chemistry*, **2010**, **20**, 466–474.

68. Terech, P., Weiss, R. G. (1997). Low molecular mass gelators of organic liquids and the properties of their gels. *Chemical Reviews*, **97**, 3133–3159.

69. Jayawarna, V., Ali, M., Jowitt, T. A., Miller, A. E., Saiaru, A., Gough, J. E., Ulijn, R. V. (2006). Nanostructured hydrogels for three-dimensional cell culture through self-assembly of fluorenylmethoxycarbonyl-dipeptides. *Advanced Materials*, **18**, 611–614.

70. Fujita, N., Asai, M., Yamashita, T., Shinkai, S. (2004). Sol–gel transcription of silica-based hybrid nanostructures using poly(N-vinylpyrrolidone)-coated [60]fullerene, single-walled carbon nanotube and block copolymer templates. *Journal of Materials Chemistry*, **14**, 2106–2114.

71. Beginn, U., Möller, M., *Supramolecular Materials and Technologies*, John Wiley & Sons, Chichester, 1999, pp. 89–176.

72. Beck, J. B., Rowan, S. J. (2003). Multistimuli, multiresponsive metallo-supramolecular polymers. *Journal of the American Chemical Society*, **125**, 13922–13923.

73. Sangeetha, N. M., Maitra U. (2005). Supramolecular gels: Functions and uses. *Chemical Society Reviews*, **34**, 821–836.

74. Ajayaghosh, A., Praveen, V. K. (2007). π-organogels of self-assembled *p*-phenylenevinylenes: Soft materials with distinct size, shape, and functions. *Accounts of Chemical Research*, **40**, 644–656.

75. Ajayaghosh, A., Praveen, V. K., Vijayakumar, C. (2008). Organogels as scaffolds for excitation energy transfer and light harvesting. *Chemical Society Reviews*, **37**, 109–122.

76. Chung, J. W., Yoon, S.-J., Lim, S.-J., An, B.-K., Park, S. Y. (2009). Dual-mode switching in highly fluorescent organogels: Binary logic gates with optical/thermal inputs. *Angewandte Chemie International Edition*, **48**, 7030–7034.

77. Srinivasan, S., Babu, P. A., Mahesh, S., Ajayaghosh, A. (2009). Reversible self-assembly of entrapped fluorescent gelators in polymerized styrene gel matrix: Erasable thermal imaging via recreation of supramolecular architectures. *Journal of the American Chemical Society*, **131**, 15122–15123.

78. Palmans, A. R. A., Meijer, E. W. (2007). Amplification of chirality in dynamic supramolecular aggregates. *Angewandte Chemie International Edition*, **46**, 8948–8968.

79. Babu, S. S., Kartha, K. K., Ajayaghosh, A. (2010). Excited state processes in linear π-system-based organogels. *Journal of Physical Chemistry Letters*, **1**, 3413–3424.

80. Wurthner, F., Kaiser, T. E., Saha-Möller, C. R. (2011). J-aggregates: From serendipitous discovery to supramolecular engineering of functional dye materials. *Angewandte Chemie International Edition*, **50**, 3376–3410.

81. Hong, Y., Lam, J. W. Y., Tang, B. Z. (2009). Aggregation-induced emission: Phenomenon, mechanism and applications. *Chemical Communications*, 4332–4353.

82. Tsou, C.-C., Sun, S.-S. (2006). New fluorescent amide-functionalized phenylethynylthiophene low molecular weight gelator. *Organic Letters*, **8**, 387–390.

83. Ajayaghosh, A., Praveen, V. K., Vijayakumar, C. (2008). Organogels as scaffolds for excitation energy transfer and light harvesting. *Chemical Society Reviews*, **37**, 109–122.

84. Xue, P., Lu, R., Zhao, L, Xu, D., Zhang, X., Li, K., Song, Z., Yang, X., Takafuji, M., Ihara, H. (2010). Hybrid self-assembly of a π gelator and fullerene derivative with photoinduced electron transfer for photocurrent generation. *Langmuir*, **26**, 6669–6675.

85. Sugiyasu, K., Kawano, S.-i., Fujita, N., Shinkai, S. (2008). Self-sorting organogels with *p–n* heterojunction points. *Chemistry of Materials*, **20**, 2863–2865.

86. Sakai, N., Bhosale, R., Emery, D., Mareda, J., Matile, S. (2010). Supramolecular *n/p*-heterojunction photosystems with antiparallel redox gradients in electron- and hole-transporting pathways. *Journal of the American Chemical Society*, **132**, 6923–6925.

87. Sardone, L., Palermo, V., Devaux, E., Credgington, D., de Loos, M., Marletta, G., Cacialli, F., van Esch, J., Samorì, P. (2006). Electric-field-assisted alignment of supramolecular fibers. *Advanced Materials*, **18**, 1276–1280.

88. Shoji, Y., Yoshio, M., Yasuda, T., Funahashi, M., Kato, T. (2010). Alignment of photoconductive self-assembled fibers composed of π-conjugated molecules under electric fields. *Journal of Materials Chemistry*, **20**, 173–179.

89. Yagai, S. (2006). Supramolecular complexes of functional chromophores based on multiple hydrogen-bonding interactions. *Journal of Photochemistry and Photobiology C: Photochemistry Reviews*, **7**, 164–182.

90. Liu, Y., Xiao, S., Li, H., Li, Y., Liu, H., Lu, F., Zhuang, J., Zhu, D. (2004). Self-Assembly and Characterization of A Novel Hydrogen-Bonded Nanostructure. *Journal of Physical Chemistry B*, **108**, 6256–6260.

91. El-ghayoury, A., Schenning, A. P. H. J., van Hal, P. A., van Duren, J. K. J., Janssen, R. A. J., Meijer, E. W. (2001). Supramolecular hydrogen-bonded oligo(*p*-phenylene vinylene) polymers. *Angewandte Chemie International Edition*, **40**, 3660–3663.

92. Jonkheim, P., Hoeben, F. J. M., Kleppinger, R., van Herrikhuyzen, J., Schenning, A. P. H. J., Meijer, E. W. (2003). Transfer of π-conjugated columnar stacks from solution to surfaces. *Journal of the American Chemical Society*, **125**, 15941–15949.

93. Jeukens, C. R. L. P. N., Jonkheijm, P., Wijnen, F. J. P., Gielen, J. C., Christianen, P. C. M., Schenning, A. P. H. J., Meijer, E. W., Maan, J. C. (2005). Polarized emission of individual self-assembled oligo(*p*-phenylenevinylene)-based nanofibers on a solid support. *Journal of the American Chemical Society*, **127**, 8280–8281.

94. Jonkheim, P., van der Schoot, P., Schenning, A. P. H. J., Meijer, E. W. (2006). Probing the solvent-assisted nucleation pathway in chemical self-assembly. *Science*, **313**, 80–83.

95. Wolffs, M., George, S. J., Tomovic, Z., Meskers, S. C. J., Schenning, A. P. H. J., Meijer, E. W. (2007). Macroscopic origin of circular dichroism effects by alignment of self-assembled fibers in solution. *Angewandte Chemie International Edition*, **46**, 8203–8205.

96. George, S. J., Tomovic, Z., Smulders, M. M. J.de Greef, T. F. A., Leclère, P. E. L. G., Meijer, E. W., Schenning, A. P. H. J. (2007). Helicity induction and amplification in an oligo(*p*-phenylenevinylene) assembly through hydrogen-bonded chiral acids. *Angewandte Chemie International Edition*, **46**, 8206–8211.

97. Pisula, W., Tomovic, Z., Wegner, W., Graf, R., Pouderoijen, M. J., Meijer, E. W., Schenning, A. P. H. J. (2008). Liquid crystalline hydrogen bonded oligo(*p*-phenylenevinylene)s. *Journal of Materials Chemistry*, **18**, 2968–2977.

98. Korevaar, P. A., George, S. J., Markvoort, A. J., Smulders, M. M. J., Hilbers, P. A. J., Schenning, A. P. H. J., de Greef, T. F. A., Meijer, E. W. (2012). Pathway complexity in supramolecular polymerization. *Nature*, **481**, 492–496.

99. Gesquière, A., Jonkheijm, P., Hoeben, F. J. M., Schenning, A. P. H. J., De Feyter, S., De Schryver, F. C., Meijer, E. W (2004). 2D-structures of quadruple hydrogen bonded oligo(*p*-phenylenevinylene)s on graphite: Self-assembly behavior and expression of chirality. *Nano Letters*, **4**, 1175–1179.

100. Herz, L. M., Daniel, C., Silva, C., Hoeben, F. J. M., Schenning, A. P. H. J., Meijer, E. W., Friend, R. H., Phillips, R. T. (2003). Fast exciton diffusion in chiral stacks of conjugated *p*-phenylene vinylene oligomers. *Physical Review B*, **68**, 045201-(1–7).

101. Daniel, C., Herz, L. M., Silva, C., Hoeben, F. J. M., Jonkheim, P., Schenning, A. P. H. J., Meijer, E. W. (2003). Exciton bimolecular annihilation dynamics in supramolecular nanostructures of conjugated oligomers. *Physical Review B*, **68**, 235212-(1–9).

102. Hoeben, F. J. M.., Herz, L. M., Daniel, C., Jonkheim, P., Schenning, A. P. H. J., Silva, C., Meskers, S. C. J., Beljonne, D., Phillips, R. T., Friend, R. H., Meijer, E. W. (2004). Efficient energy transfer in mixed columnar stacks of hydrogen-bonded oligo(*p*-phenylenevinylene)s in solution. *Angewandte Chemie International Edition*, **43**, 1976–1979.

103. Beljonne, D., Hennebicq, E., Daniel, C., Herz, L. M., Silva, C., Scholes, G. D., Hoeben, F. J. M., Jonkheim, P., Schenning, A. P. H. J., Meskers, S. C. J., Phillips, R. T., Friend, R. H., Meijer, E. W. (2005). Excitation migration along oligophenylenevinylene-based chiral stacks: Delocalization effects on transport dynamics. *Journal of Physical Chemistry B*, **109**, 10594–10604.

104. Hoeben, F. J. M., Schenning, A. P. H. J., Meijer, E. W. (2005). Energy-transfer efficiency in stacked oligo(*p*-phenylene vinylene)s: Pronounced effects of order. *ChemPhysChem*, **6**, 2337–2342.

105. Daniel, C., Makereel, F., Herz, L. M., Hoeben, F. J. M., Jonkheim, P., Schenning, A. P. H. J., Meijer, E. W., Friend, R. H., Silva, C. (2005). The effects of supramolecular assembly on exciton decay rates in organic semiconductors. *Journal of Chemical Physics*, **123**, 084902-(1–7).

106. Chang, M. H., Hoeben, F. J. M., Jonkheim, P., Schenning, A. P. H. J., Meijer, E. W., Silva, C., Herz, L. M. (2006). Influence of mesoscopic ordering on the photoexcitation transfer dynamics in supramolecular assemblies of oligo-*p*-phenylenevinylene. *Chemical Physics Letters*, **418**, 196–201.

107. Daniel, C., Makereel, F., Herz, L. M., Hoeben, F. J. M., Jonkheim, P., Schenning, A. P. H. J., Meijer, E. W., Silva, C. (2008). Mesoscopic order and the dimensionality of long-range resonance energy transfer in supramolecular semiconductors. *Journal of Chemical Physics*, **129**, 104701-(1–7).

108. Prins, P., Senthilkumar, K., Grozema, F. C., Jonkheijm, P., Schenning, A. P. H. J., Meijer, E. W., Siebbeles, L. D. A. (2005). Charge transport in self-organized ϑ-stacks of *p*-phenylene vinylene oligomers. *Journal of Physical Chemistry B*, **109**, 18267–18274.

109. Durkut, M., Mas-Torrent, M., Hadley, P., Jonkheijm, P., Schenning, A. P. H. J., Meijer, E. W., George, S. J., Ajayaghosh, A. (2006). Electrical transport measurements on self-assembled organic molecular wires. *Journal of Chemical Physics*, **124**, 154704-(1–6).

110. Jonkheijm, P., Miura, A., Zdanowska, M., Hoeben, F. J. M., Jonkheim, P., De Feyter, S., Schenning, A. P. H. J., De Schryver, F. C., Meijer, E. W. (2004). π-conjugated oligo-(*p*-phenylenevinylene) rosettes and their tubular self-assembly. *Angewandte Chemie International Edition*, **43**, 74–78.

111. Miura, A., Jonkheijm, P., De Feyter, S., Schenning, A. P. H. J., Meijer, E. W., De Schryver, F. C. (2005). 2D self-assembly of oligo(*p*-phenylene vinylene) derivatives: From dimers to chiral rosettes. *Small*, **1**, 131–137.

112. Katsonis, N., Xu, H., Haak, R. M., Kudernac, T., Tomović, Z., George, S., Van der Auweraer, M., Schenning, A. P. H. J., Meijer, E. W., Feringa, B. L., De Feyter, S. (2008). Emerging solvent-induced homochirality by the confinement of achiral molecules against a solid surface. *Angewandte Chemie International Edition*, **47**, 4997–5001.

113. Davis, J. T. (2004). G-quartets 40 years later: From 5′-GMP to molecular biology and supramolecular chemistry. *Angewandte Chemie International Edition*, **43**, 668–698.

114. Davis, J. T., Spada, G. P. (2007). Supramolecular architectures generated by self-assembly of guanosine derivatives. *Chemical Society Reviews*, **36**, 296–313.

115. González-Rodríguez, D., van Dongen, J. L. J., Lutz, M., Spek, A. L., Schenning, A. P. H. J., Meijer, E. W. (2009). G-quadruplex self-assembly regulated by Coulombic interactions. *Nature Chemistry*, **1**, 151–155.

116. González-Rodríguez, D., Janssen, P. G. A., Martín-Rapún, R., De Cat, I., De Feyter, S., Schenning, A. P. H. J., Meijer, E. W. (2010). Persistent, well-defined, monodisperse, π-conjugated organic nanoparticles via G-quadruplex self-assembly. *Journal of the American Chemical Society*, **132**, 4710–4719.

117. Xiao, D., Xi, L., Yang, W., Fu, H., Shuai, Z., Fang, Y., Yao, J. (2003). Size-tunable emission from 1,3-diphenyl-5-(2-anthryl)-2-pyrazoline nanoparticles. *Journal of the American Chemical Society*, **125**, 6740–6745.

118. Kim, H.-J., Lee, J., Kim, T.-H., Lee, T. S., Kim, J. (2008). Highly emissive self-assembled organic nanoparticles having dual color capacity for targeted immunofluorescence. *Advanced Materials*, **20**, 1117–1121.

119. Masiero, S., Gottarelli, G., Pieraccini, S. (2009). G-quartets as a self-assembled scaffold for circular porphyrin arrays. *Chemical Communications*, 1995–1996.

120. Martic, S., Liu, X., Wang, S., Wu. G. (2008). Self-assembly of N2-modified guanosine derivatives: Formation of discrete G-Octamers. *Chemistry—A European Journal*, **14**, 1196–1204.

121. Spada, G. P., Lena, S., Masiero, S., Pieraccini, S., Surin, M., Samorì, S. (2008). Guanosine-based hydrogen-bonded scaffolds: Controlling the assembly of oligothiophenes. *Advanced Materials*, **20**, 2433–2438.

122. Graziano, C., Pieraccini, S., Masiero, S., Lucarini, M., Spada, G. P. (2008). A cation-directed switch of intermolecular spin-spin interaction of guanosine derivatives functionalized with open-shell units. *Organic Letters*, **10**, 1739–1742.

123. Neviani, P., Mileo, E., Masiero, S., Pieraccini, S., Lucarini, M., Spada, G. P. (2009). Self-assembled hexadecanitroxide. *Organic Letters*, **11**, 3004–3007.

124. Schenning, A. P. H. J., van Herrikhuyzen, J., Jonkheijm, P., Chen, Z., Würthner, F., Meijer, E.W. (2002). Photoinduced electron transfer in hydrogen-bonded oligo(*p*-phenylene vinylene)-perylene bisimide chiral assemblies. *Journal of the American Chemical Society*, **124**, 10252–10253.

125. Würthner, F., Chen, Z., Hoeben, F. J. M., Osswald, P., You, C-C., Jonkheim, P., van Herrikhuyzen, J., Schenning, A. P. H. J., van der Schoot, P. P. A. M., Meijer, E. W., Beckers, E. H. A., Meskers, S. C. J., Janssen, R. A. J. (2004). Supramolecular *p–n*-heterojunctions by co-self-organization of oligo(*p*-phenylene vinylene) and perylene bisimide dyes. *Journal of the American Chemical Society*, **126**, 10611–10618.

126. Beckers, E. H. A., Chen, Z., Meskers, S. C. J., Jonkheim, P., Schenning, A. P. H. J., Li, X-Q., Osswald, P., Würthner, F., Janssen, R. A. J. (2006). The importance of nanoscopic ordering on the kinetics of photoinduced charge transfer in aggregated ϑ-conjugated hydrogen-bonded donor-acceptor systems. *Journal of Physical Chemistry B*, **110**, 16967–16978.

127. Seitz, W., Jiménez, A. J., Carbonell, E., Grimm, B., Rodríguez-Morgade, M. S., Guldi, D. M., Torres, T. (2010). Synthesis and photophysical properties of a hydrogen-bonded phthalocyanine–perylenediimide assembly. *Chemical Communications*, **46**, 127–129.

128. van Herrikhuyzen, J., Syamakumari, A., Schenning, A. P. H. J., Meijer, E. W. (2004). Synthesis of *n*–type perylene bisimide derivatives and their orthogonal self-assembly with *p*–type oligo(*p*-phenylene vinylene)s. *Journal of the American Chemical Society*, **126**, 10021–10027.

129. Jonkheim, P., Stutzmann, N., Chen, Z., de Leeuw, D. M., Meijer, E. W., Schenning, A. P. H. J., Würthner, F. (2006). Control of ambipolar thin film architectures by Co-self-assembling oligo(*p*-phenylenevinylene)s and perylene bisimides. *Journal of the American Chemical Society*, **128**, 9535–9540.

130. Zhang, J., Hoeben, F. J. M., Pouderoijen, M. J., Schenning, A. P. H. J., Meijer, E. W., De Schryver, F. C., De Feyter, S.

(2006). Hydrogen-bonded oligo(*p*-phenylenevinylene) functionalized with perylene bisimide: Self-assembly and energy transfer. *Chemistry—A Eropean Journal*, **12**, 9046–9055.

131. Yagai, S., Hamamura, S., Wang, H., Stepanenko, V., Seki, T., Unoike, K., Kikkawa, Y., Karatsu, T., Kitamura, A., Würthner, F. (2009). Unconventional hydrogen-bond-directed hierarchical co-assembly between perylene bisimide and azobenzene-functionalized melamine. *Organic & Biomolecular Chemistry*, **7**, 3926–3929.

132. Whitesides, G. M., Simanek, E. E., Mathias, J. P., Seto, C. T., Chin, D. N., Mammen, M., Gordon, D. M. (1995). Noncovalent synthesis: Using physical-organic chemistry to make aggregates. *Accounts of Chemical Research*, **28**, 37–44.

133. Yagai, S., Kubota, S., Unoike, K., Karatsu, T., Kitamura, A. (2008). Cyanurate-guided self-assembly of a melamine-capped oligo(*p*-phenylenevinylene). *Chemical Communications*, 4466–4468.

134. Yagai, S., Nakajima, T., Karatsu, T., Saitow, K.-i., Kitamura, A. (2004). Phototriggered self-assembly of hydrogen-bonded rosette. *Journal of the American Chemical Society*, **125**, 11500–11508.

135. Yagai, S., Nakajima, T., Kishikawa, K., Kohmoto, S., Karatsu, T., Kitamura, A. (2005). Hierarchical organization of photoresponsive hydrogen-bonded rosettes. *Journal of the American Chemical Society*, **126**, 11134–11139.

136. Yagai, S., Mahesh, S., Kikkawa, Y., Unoike, K., Karatsu, T., Kitamura, A., Ajayaghosh, A. (2008). Toroidal nanoobjects from rosette assemblies of melamine-linked oligo(*p*-phenyleneethynylene)s and cyanurates. *Angewandte Chemie International Edition*, **47**, 4691–4694.

137. Yagai, S., Higashi, M., Karatsu, T., Kitamura, A. (2006). Tunable interchromophore electronic interaction of a merocyanine dye in hydrogen-bonded supramolecular assemblies scaffolded by bismelamine receptors. *Chemical Communications*, 1500–1502.

138. Tazawa, T., Yagai, S., Kikkawa, Y., Karatsu, T., Kitamura, A., Ajayaghosh, A. (2010). A complementary guest induced morphology transition in a two-component multiple H-bonding self-assembly. *Chemical Communications*, **46**, 1076–1078.

139. Hoeben, F. J. M., Pouderoijen, M. J., Schenning, A. P. H. J., Meijer, E. W. (2006). Energy transfer in chiral co-assemblies of triple hydrogen-bonded oligo(*p*-phenylene vinylene)s and porphyrin. *Organic & Biomolecular Chemistry*, **4**, 4460–4462.

140. Kaiser, T. E., Wang, H., Stepanenko, V., Würthner, F. (2007). Supramolecular construction of fluorescent J-aggregates based on hydrogen-bonded perylene dyes. *Angewandte Chemie International Edition*, **46**, 5541–5544.

141. Wang, H., Kaiser, T. E., Uemura, S., Würthner, F. (2008). Perylene bisimide J-aggregates with absorption maxima in the NIR. *Chemical Communications*, 1181–1183.

142. Kaiser, T. E., Scheblykin, I. G., Thomsson, D., Würthner, F. (2009). Temperature-dependent exciton dynamics in J-aggregatess: When disorder plays a role. *Journal of Physical Chemistry B*, **113**, 15836–15842.

143. Kaiser, T. E., Stepanenko, V., Würthner, F. (2009). Fluorescent J-aggregates of core-substituted perylene bisimides: Studies on structure–property relationship, nucleation–elongation mechanism, and sergeants-and-soldiers principle. *Journal of the American Chemical Society*, **131**, 6719–6732.

144. Gsänger, M., Oh, J. H., Könemann, M., Höffken, H. W., Krause, A.-M., Bao, Z., Würthner, F. (2010). A crystal-engineered hydrogen-bonded octachloroperylene diimide with a twisted core: An *n*–channel organic semiconductor. *Angewandte Chemie International Edition*, **49**, 740–743.

145. Würthner, F., Thalacker, C., Sauter, A., Schärtl, W., Ibach, W., Hollricher, O. (2000). Hierarchical self-organization of perylene bisimide-melamine assemblies to fluorescent mesoscopic superstructures. *Chemistry—A European Journal*, **6**, 3871–3886.

146. Thalacker, C., Würthner, F. (2002). Chiral perylene bisimide-melamine assemblies: Hydrogen bond-directed growth of helically stacked dyes with chiroptical properties. *Advanced Functional Materials*, **12**, 209–218.

147. Dudek, S. P., Pouderoijen, M., Abbel, R., Schenning, A. P. H. J., Meijer, E. W. (2005). Synthesis and energy-transfer properties of hydrogen-bonded oligofluorenes. *Journal of the American Chemical Society*, **127**, 11763–11768.

148. Schmid, S. A., Abbel, R., Schenning, A. P. H. J., Meijer, E. W., Sijbesma, R. P., Herz, L. M. (2009). Analyzing the molecular weight distribution in supramolecular polymers. *Journal of the American Chemical Society*, **131**, 17696–17704.

149. Schmid, S. A., Abbel, R., Schenning, A. P. H. J., Meijer, E. W., Herz, L. M. (2010). Impact of nuclear lattice relaxation on the excitation energy transfer along a chain of π-conjugated molecules. *Physical Review B*, **81**, 085438-(1–5).

150. Abbel, R., Grenier, C., Pouderoijen, M. J., Stouwdam, J. W., Leclère, P. E. L. G., Sijbesma, R. P., Meijer, E. W., Schenning, A. P. H. J. (2009). White-light emitting hydrogen-bonded supramolecular copolymers based on π-conjugated oligomers. *Journal of the American Chemical Society*, **132**, 833–843.

151. Sijbesma, R. P., Beijer, F. H., Brunsveld, L., Folmer, B. J. B., Hirschberg, J. H. K. K., Lange, R. F. M., Lowe, J. K. L., Meijer, E. W. (1997). Reversible polymers formed from self-complementary monomers using quadruple hydrogen bonding. *Science*, **278**, 1601–1604.

152. Similar Upy-terminated C60 polymers have also been reported. See: Sánchez, L., Rispens, M. T., Hummelen, J. C. (2002). A supramolecular array of fullerenes by quadruple hydrogen bonding. *Angewandte Chemie International Edition*, **41**, 838–840.

153. Vijayakumar, C., Praveen, V. K., Ajayaghosh, A. (2009). RGB emission through controlled donor self-assembly and modulation of excitation energy transfer: A novel strategy to white-light-emitting organogels. *Advanced Materials*, **21**, 2059–2063.

154. Yagai, S., Kubota, S., Saito, H., Unoike, K., Karatsu, T., Kitamura, A., Ajayaghosh, A., Kanesato, M., Kikkawa, Y. (2009). Reversible transformation between rings and coils in

a dynamic hydrogen-bonded self-assembly. *Journal of the American Chemical Society*, **131**, 5408–5410.

155. Chu, C.-C., Raffy, G., Ray, D., Del Guerzo, A., Kauffmann, B., Wantz, G., Hirsch, L., Bassani, D. M. (2010). Self-assembly of supramolecular fullerene ribbons via hydrogen-bonding interactions and their impact on fullerene electronic interactions and charge carrier mobility. *Journal of the American Chemical Society*, **132**, 12717–12723.
156. Yagai, S., Iwashima, T., Kishikawa, K., Nakahara, S., Karatsu, T., Kitamura, A. (2006). Photoresponsive self-assembly and self-organization of hydrogen-bonded supramolecular tapes. *Chemistry—A European Journal*, **12**, 3984–3994.
157. Huang, C.-H., McClenaghan, N. D., Kuhn, A., Hofstraat, J. W., Bassani, D. M. (2005). Enhanced photovoltaic response in hydrogen-bonded all-organic devices. *Organic Letters*, **7**, 3409–3412.
158. Huang, C-H., McClenaghan, N. D., Kuhn, A., Bravic, G., Bassani, D. M. (2006). Hierarchical self-assembly of all-organic photovoltaic devices. *Tetrahedron*, **62**, 2050–2059.
159. Würthner, F., Yao, S., Heise, B., Tschierske, C. (2001). Hydrogen bond directed formation of liquid-crystalline merocyanine dye assemblies. *Chemical Communications*, 2260–2261.
160. Yagai, S., Higashi, M., Karatsu, T., Kitamura, A. (2005). Dye-assisted structural modulation of hydrogen-bonded binary supramolecular polymers. *Chemistry of Materials*, **17**, 4392–4398.
161. Yagai, S., Karatsu, T., Kitamura, A. (2005). Melamine-barbiturate/cyanurate binary organogels possessing rigid azobenzene-tether moiety. *Langmuir*, **21**, 11048–11052.
162. Yagai, S., Kinoshita, T., Higashi, M., Kishikawa, K., Nakanishi, T., Karatsu, T., Kitamura, A. (2007). Diversification of self-organized architectures in supramolecular dye assemblies. *Journal of the American Chemical Society*, **129**, 13277–13287.
163. Yagai, S., Monma, Y., Kawauchi, N., Karatsu, T., Kitamura, A. (2007). Supramolecular nanoribbons and nanoropes generated from hydrogen-bonded supramolecular polymers containing perylene bisimide chromophores. *Organic Letters*, **9**, 1137–1140.
164. Seki, T., Yagai, S., Karatsu, T., Kitamura, A. (2008). Formation of supramolecular polymers and discrete dimers of perylene bisimide dyes based on melamine-cyanurates hydrogen-bonding interactions. *Journal of Organic Chemistry*, **73**, 3328–3335.
165. Yagai, S., Kinoshita, T., Kikkawa, Y., Karatsu, T., Kitamura, A., Honsho, Y., Seki, S. (2009). Interconvertible oligothiophene nanorods and nanotapes with high charge-carrier mobilities. *Chemistry—A European Journal*, **15**, 9320–9324.
166. Hamblin, J., Argent, S. P., Blake, A. J., Wilson, C., Champness, N. R. (2008). Hydrogen-bonding tectons for the construction of bimolecular framework materials. *CrystEngComm*, **10**, 1782–1789.

3

CHIRAL ORGANIC NANOMATERIALS

DAVID B. AMABILINO

3.1 INTRODUCTION

Chirality is unique in that it manifests itself in several ways in natural and synthetic systems over enormous ranges of scale, from atoms to galaxies [1, 2]. The nanometer regime is particularly interesting from various phenomenological viewpoints, and there is a growing interest in the preparation of chiral nanomaterials in general because of the numerous special properties that they can display when compared with more well known macroscopic systems [3]. On the other hand, the origins of chirality in natural systems [4] is surely linked to the occurrence of asymmetry at a nanometer scale, a thought that drives one to consider the different ways in which asymmetry can arise [5].

The material properties related to chiral organic systems are varied, but they all have in common their particular optical behavior. The term "optical activity" (or natural optical activity, to be more precise) when used in regard to chiral systems is most often associated with either (a) circular birefringence—measured in polarimetry experiments and optical rotator dispersion (ORD) because of the different refractive index of the medium to left- and right-handed circularly polarized light, leading to rotation of the plane of linearly polarized light—or (b) circular dichroism (CD)—the differential absorption of left and right circularly polarized light leading to elliptical polarization [6] (Figure 3.1).

So, what are the ways one could make an organic material chiral? There are four main approaches. The most obvious—and most reliable—way for a chemist to introduce chirality into an organic system is by the incorporation of a nonracemic stereogenic element—be it a stereogenic center or some group displaying atropoisomerism—into some molecular fragment in the whole of the sample. This incorporation results not only in the presence of chirality in the unit, but also very often in the transfer of asymmetry into the superstructure [7]. The introduction of the asymmetric center is usually done by stereoselective synthesis from commercially available substituent units, or by the separation of enantiomeric forms of functional units, as is the case of binaphthyl units or helicenes [8, 9], to name but two.

When chirality is induced in a material where the optically active component is a minority, the asymmetry can be transferred to the other chiral or achiral components of the system resulting in a bulk chiral material, as in the case of liquid crystals for example [10]. When a chiral molecule is able to turn a great number of achiral molecules to a conformation where they mimic this inducers chirality, the effect is termed *sergeants and soldiers*, whereby the chiral sergeant makes all the achiral soldiers turn the way it demands [11–14]. On the other hand, in a sample with a nonracemic mixture of enantiomers (i.e., one is in excess) the majority compound can exert a force over the minor component giving rise to superstructural chirality corresponding to the more prevalent isomer. This nonlinear induction of chirality is termed *majority rules* [15–18]. When this effect is operative a scalemic mixture (nonequal mixture of enantiomers) can show the same optical activity as the enantiopure compounds.

Alternatively, spontaneous resolution can occur, whereby two enantiomers condense in separate areas of material [19, 20]. This process occurs rarely in crystalline systems, commonly in monolayer systems (especially on crystalline substrates), and can even take place in the least likely of situations, even in liquid crystals. In most cases, the enantiomers condense in different areas of the bulk sample, but under

Organic Nanomaterials: Synthesis, Characterization, and Device Applications, First Edition. Edited by Tomás Torres and Giovanni Bottari.

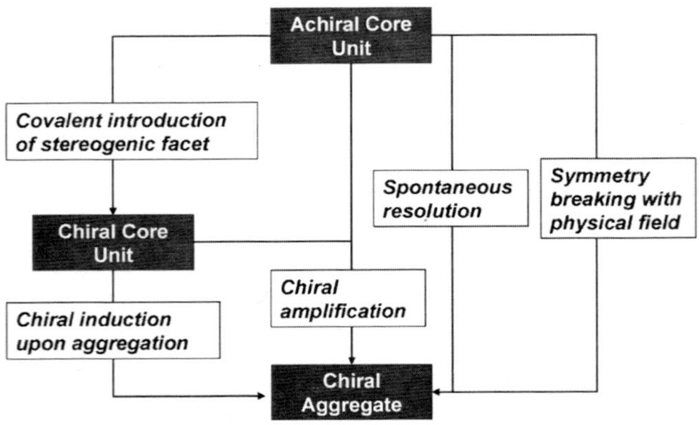

FIGURE 3.1. Routes to chiral nanomaterials, by introduction of stereogenic facet to a core structure giving a chiral core unit which then either aggregates alone, to give the chiral material, or coassembles with an achiral core unit wherein chirality is transferred to the latter. Also depicted are two routes wherein chirality need not be present in the core structure: (1) spontaneous resolution and (2) induction through physical fields such as chiral fluid motion or use of circularly polarized light-induced processes.

certain special conditions, spontaneous deracemization can take place [21], whereby an enantiomeric excess in a condensate appears thanks to racemization in gas or liquid states.

Finally, physical forces can be used to break symmetry. In order to do so, a chiral force must be exerted, such as circularly polarized light [22] or chiral flows that are present in a vortex [23] (Figure 3.1). It is important to note that magnetic fields alone cannot lead to asymmetric induction [24].

Over the length of this chapter, examples showing all of these phenomena will be given. The structure of this admittedly restricted overview will follow a classification based on the dimensionality of the material, rather than the type of compound or application. The reason for this organization of the text is that the methods used for each type of material—molecule, particle, fiber or tube, films, and nanostructured solids—are largely valid across types of molecule and application. All biological materials are chiral, but mention here will be restricted to where the chirality has an obvious role in the properties of the biomaterial. First, the main structural and mechanistic factors determining the asymmetry of organic molecules and their assemblies will be discussed.

3.2 STRUCTURAL AND MECHANISTIC FACTORS IN THE GROWTH OF CHIRAL STRUCTURES

The molecular structure of compounds has an absolutely fundamental role in the way that supramolecular organizations take shape. The balance of hydrophilic and hydrophobic groups and the use of specific supramolecular synthons can change completely the way a functional unit will be arranged. The shape of the molecule is also essential in determining their packing and how chirality is expressed. An example from the macroscopic world, which has analogies in molecules, is that packing compact discs are not the same as packing propellers. In this section we show just some of the influences that can be important.

One of the most reliable effects observed in the transmission of chirality is that whereby changing the position of a stereogenic center with the same absolute configuration along an alkyl chain in a sequential way gives rise to alternating optical activity. This "odd–even" effect is frequently observed in liquid crystals [25–27], self-assembled stacks [28], monolayers [29, 30], and polymers [31, 32]. A good example is the aggregates of asymmetric benzene-1,3,5-tricarboxamides comprising two *n*-octyl and one chiral methylalkyl side chain, which show alpha-helical-type intermolecular hydrogen bonding between neighbouring molecules with the benzene rings atop each other [33]. The position and configuration of the stereogenic center to which a methyl group is attached (at the alpha, beta, or gamma position) in the aliphatic side chains changes the optical activity of the stacks, and also the liquid-crystalline behavior . The self-assembled columns in methylcyclohexane ($\sim 10^{-5}$ M) show Cotton effects in the CD spectra which have a definite odd–even effect. Although the stereogenic center changed its relative position with respect to the core of the columns, the optical activity was independent of the position of the methyl group.

The constitution of aromatic groups also plays a dramatic role in the expression of chirality, through secondary structures. The pyridine substituents that are included in the head group of glutamic acid derivatives plays a determining role in the morphologies displayed by the lipids [34]. The compounds contain the pyridyl nitrogen atom at the 2-, 3-, or 4-position with respect to the lipid pendant group, as shown in Figure 3.2. When the pyridyl nitrogen atom is at the 2-position, an intramolecular hydrogen bond holds the head group in an orientation that favors the formation of lamellae in the thin fibers that are generated upon gelation of DMSO. Both the other isomers form fibers that apparently contain intermolecular hydrogen bonds involving the basic nitrogen atoms. In the case of the 3-pyridyl derivative, twisted tapes are formed, with a right-handed turn only. Finally the 4-pridyl lipid forms helical nanotubes. The type of organization also has an influence on the optical activity of the aggregates, as witnessed by circular dichroism spectroscopy. The work exemplifies the subtleties of molecular structure on the expression of chirality in self-assembled systems.

In the formation of rigid helical polymers, the nature of the spacer in between the chemical functionality which is polymerized and the stereogenic center which induces the chiral conformation in the backbone of the macromolecule proves to be very important [35]. The length, constitution, and conformation of rod-like spacers that link a stereogenic center derived from lactate to an isocyanide group affect the

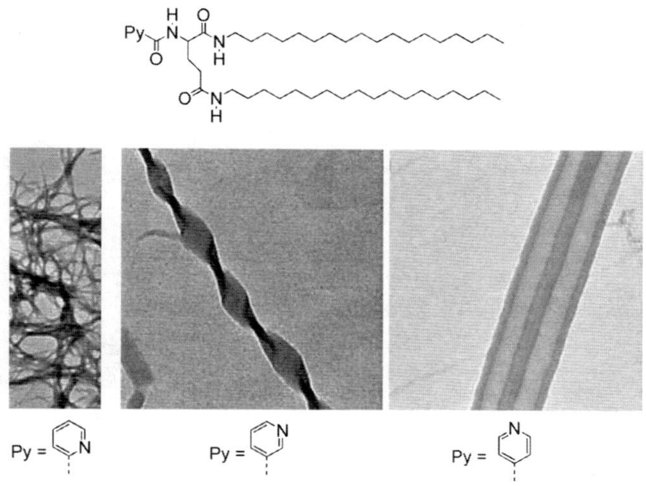

FIGURE 3.2. Pyridyl derivatives of L-glutamic lipids that show three types of chiral self-assembled structure in their gels, and transmission electron microscope images of the different fibers (the width of all the images is 1 μm).

magnitude and sign of induction. Also, the nature of the chiral group (the moieties that are attached to it), the solvent which is used for the polymerization, and even the temperature at which it is performed can influence greatly the chiral induction, in the case of raising the temperature of the reaction even inverting the preferred helicity seen at room temperature [36]. In these cases, the induction of chirality is determined largely kinetically.

Regarding the mechanistic factors in the amplification of chirality in supramolecular systems, one can describe the effects by taking into account two energetic penalties: (a) that related to a reversal of the helicity in a stack—from left- to right-handed and the reverse—and (b) that related to the mismatch of chiralities in a stack—the inclusion of a molecule of opposite handedness to the main one in an aggregate. In supramolecular polymers consisting of benzene-1,3,5-tricarboxamide monomers and in the case of the sergeants-and-soldiers experiments (chiral and achiral monomers), the helix reversal penalty associated with the formation of intermolecular hydrogen bonds was the larger influence, while the mismatch energy is related to steric interactions in the alkyl side chains attached to the stereogenic center [37]. When temperature was increased, the helix reversal penalty remained approximately constant, because the intermolecular hydrogen bonds direct the helicity in the stack and are maintained. On the other hand, the mismatch penalty decreased with increasing temperature, giving opposite effects on the degree of chiral amplification for the sergeants-and-soldiers and the majority-rules experiments. In the former the lowered mismatch penalty decreased chiral amplification, while in the latter a stronger chiral amplification of chirality was seen.

3.3 SINGLE MOLECULE CHIRAL MATERIALS

The ultimate nanomaterial is probably a functional single molecule. In general, and especially historically, the only way to see the single-molecule events is as a dispersion of the solvated and nonaggregated molecules (usually in a liquid) and to examine the ensemble with some spectroscopic technique. However, the advent of scanning probe microscopes and the use of related techniques are allowing true examination of single molecules, at the very least being able to observe them.

Ensemble measurements on chiral polymers reveal their conformations and dynamics. In general, chiral polymers exist in labile conformational states which are in relatively fast equilibrium, in stable chiral conformations in which the dynamics of interchange are slow, or in an intermediate regime where chiral conformations are favored but can interchange upon some stimulus. The latter case can even lead to chiral polymers that contain no stereogenic centers; that is, they exist in atropoisomeric forms [38–40].

The amplification of chirality in polymers lead to important discoveries regarding the phenomena that can occur in nonracemic systems [41], and the passage of chirality along achiral chains has been proven [42]. In particular, the majority rules and sergeants and soldiers experiments were first shown in helical polymers [15, 16]. Polymers with rigid chiral conformations can show sensitive responses to different stimuli and can also show "memory" effects [43]. The use of electrical stimulus to influence the conformation of a polymer chain and its pendant groups has been employed in a poly(isocyanide) bearing tetrathiafulvalene (TTF) groups as redox centers in each monomer [44]. Sequential oxidation of the TTF units results in a dramatic change in the circular dichroism spectrum of the polymer in solution, and each redox state—be it discrete cation radical or dication or a mixtures of neutral and cation radical or cation radical and dication—shows a characteristic optical activity, which can be switched back and forth. The movement of molecular helices has also been observed very recently [45].

Apart from ensemble measurements, the tools of nanoscience have meant that polymers are a particularly interesting case where single-molecule studies can be performed and experiments of this type have revealed intriguing results. For example, noncontact atomic force microscopy showed that single molecules in a chiral polymer could form a unique periodic structure with a pitch of about 20 nm on the surface of a mica substrate [46]. Indeed, atomic force microscopy can reveal extraordinary detail of chiral macromolecular structures [47]. Single-molecule imaging has been used to track the photodegradation reaction of chiral helical polymer after laser irradiation of 405 nm [48]. A phenylacetylene polymer with chiral side chains cast on a mica substrate were imaged using atomic force microscopy, which shows individual macromolecular chains. The kinetics study revealed that the photodegradation of the single polymer chain proceeded stepwise as a quantum phenomenon, and was faster when mobility of the chains was increased. In separate experiments, the motion of this kind of molecule was also investigated by fast-scanning atomic force microscopy. Micro-Brownian motion of the single molecules of the chiral poly(phenylacetylene) on mica under *n*-octylbenzene was revealed [49]. The results revealed that the mean square displacement of a single-polymer chain complied with Einstein's law of Brownian motion.

Photophysical techniques are also powerful for probing the electron and energy transfer processes that take place in single macromolecules bearing chromophores as side groups. The optical behavior of perylene poly(isocyanide)s arising from different structures of the polymer backbone was revealed and rationalized by combining results from single-molecule fluorescence spectroscopy and atomic force microscopy [50]. Short perylene oligomers—which are poorly ordered—fluoresce similarly to the isolated monomers, while long helical perylene macromolecules with a more rigid secondary structure show emission from excimer sites. This feature appears after delocalization of the excitation along the polymer side chains which results from stacking of the perylene units. The photophysical investigation of four poly(isocyanopeptide)s with functional side chains revealed novel energy and charge transfer processes [51]. Two homopolymers containing either platinum(II)porphyrins or perylene-bis(dicarboximide) in the side arms as well as two statistical copolymers with different ratios of porphyrin and perylene-bis(dicarboximide) units were prepared. Proton NMR and circular dichroism spectroscopies confirmed the well-defined chiral secondary structure of the polymer, wherein the peptide hydrogen bonds aid organization in the macromolecules. The presence of the two different chromophores allows charge and/or energy transfer processes to take place. Photoluminescence and transient absorption measurements on nanosecond and picosecond timescales showed that when the porphyrins was irradiated the perylene chromophore was unperceived, while excitation of the perylene chromophores resulted in energy and charge transfer processes with the porphyrin moiety. The additional charge transfer states behave as intermediates that enhance electronic coupling in these multichromophoric systems.

Turning to small molecules, there are a number of systems that reveal chiroptical switching in solution [52]. Perhaps the chiral molecules that are most attractive and applicable in this regard are the molecular rotors based on sterically hindered alkenes [53]. These molecules display unidirectional relative movement of the stators as a result of their inherent chirality [54]. An enormous amount of work has been done on these "molecular machines" or "motors," optimizing the rate of rotation to make it fast and a usable phenomenon, but perhaps one of the nicest examples of their behavior is in

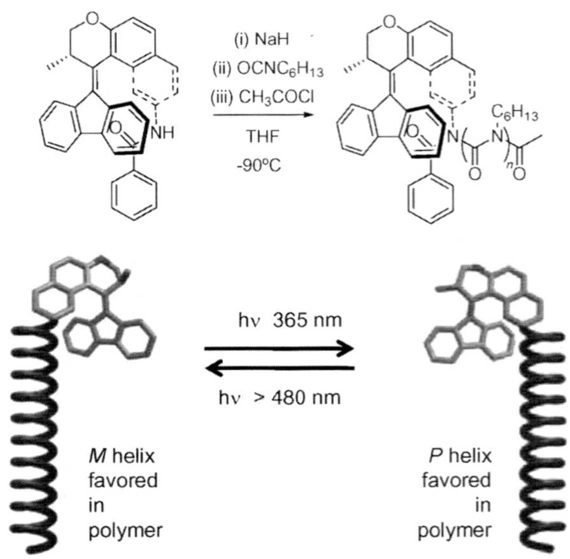

FIGURE 3.3. The synthesis of a macromolecule with poly(isocyanate) chain and a chiral molecular rotor head group which induces a preferred helicity in the helical secondary structure of the polymeric chain that can be switched by light irradiation of the rotor between its diastereomeric isomers.

the passage of the chirality of the hindered alkene to the conformation of an achiral helical polymer (Figure 3.3), using the sergeants-and-soldiers effect, and from there into the helicity of liquid crystals [55]. The diastereomers of the hindered alkene induce opposite handedness in the backbone of poly(hexylisocyanate). Photochemical isomerization of the head group—which acts as a chiroptical switch—between its thermally stable diastereomers leads to *M* or *P* helices in the polymer at the level of the single molecules, which collectively form a lyotropic liquid crystal.

Perhaps the ultimate achievement in chiral organic nanomaterials would be the operation of unimolecular chiroptical devices. Light is perhaps the most obvious way to probe chiral systems, and the observation of the fluorescence and chiroptical signals of single molecules has been achieved [56–58], although it is a practice fraught with complexity [59]. Alternatively, if molecules are immobilized on conducting surfaces, scanning tunneling microscopy can be used to probe the electronic structure of the molecules in contact with the metal. This feat has been achieved for copper(II)phthalocyanine adsorbed on Ag(100), and ab initio electronic structure calculations showed the formation of chiral molecular orbitals in the structurally undistorted molecules [60]. The achiral molecule adsorbs in a chiral orientation on the surface, and the chirality is witnessed exclusively at the electronic level as a result of asymmetric charge transfer between molecules and substrate. The work on this kind of "single molecule" system continues because of this kind of physically intriguing effect.

3.4 CHIRAL ORGANIC NANOPARTICLES

Organic nanoparticles (ONPs) in general [61] are of interest for areas such as sensing [62], optoelectronics, photonics, DNA delivery [63], and molecular recognition [64]. In common with their inorganic cousins, ONPs present characteristics that are distinct from those of either dispersed molecules or the bulk crystal. Intermolecular noncovalent interaction (the so-called many-body interaction) and particle surface defects are very important [65, 66]. Therefore organic quantum dots do have promising characteristics [67].

Chiral organic nanoparticles (CONPs) are particularly attractive for many of the potential applications [68]. Before entering into the properties of these particles, it is interesting to recall the ways in which they might be formed. The preparation of ONPs in general normally takes advantage of apparently straightforward methods, the most common being re-precipitation. This technique involves mixing vigorously a molecularly disperse solution of the organic compound in question in a water-miscible solvent with an aqueous solution that acts as the anti-solvent. The mixing of the two solvents causes rapid entry into a supersaturated solution and leads to the nucleation and growth of ONPs. The size of the particles can be controlled by the parameters regulating growth—for example, the concentration of the molecule in the water-miscible solvent and the stirring rate. Formation of a protective surfactant or a water-soluble polymer layer (e.g., gelatine and poly(vinyl alcohol) around the particles increases their stability. The addition of the protective layers at different aging times is a way of manipulating the size of the ONPs [69]. Evaporation or formation of microemulsions [70] are also popular methods for the formation of ONPs, and more complex methods include vapor-driven self-assembly [71] and laser ablation of organic crystals in a liquid [72].

Regarding the properties of CONPs, perhaps the most striking observation is the size-dependent optical activity of these nanomaterials. The formation of CONPs from the enantiomeric compounds 1,1′-bi-2-naphthol dimethyl ether (BNDE), di-2-naphthylprolinol (DNP), and 2,2′-bis-(*p*-toluenesulfonyloxy)-1,1′-binaphthalene (BTBN) highlight this characteristic [73–75]. The optical activity of the CONPs reflects changes in the absorption spectra, whereby the exciton peaks grow at longer wavelength with increase in particle size. A remarkable feature of the CD spectra of the CONPs of BNDE and DNP is that they are the inverse of those of the dilute monomers. In the case of BTBN, the ratio of the intensities of the first and third Cotton effects grows as the size of the CONPs increases to 60 nm as a result of better excimer formation between chromophores in adjacent molecules. The structural explanation of this effect is the change of dihedral angle in the molecule as the CONPs grow in size. Therefore, exciton chirality can be tuned through the size of the CONPs.

Chirality can be induced into the chromophore of mixed component ONPs. Chiral polymer nanoparticles can induce chirality into hosted achiral molecules [76]. Core–shell-type nanoparticles comprising poly(γ-benzyl-L-glutamate) as a hydrophobic inner core and poly(ethylene oxide) as the hydrophilic shell formed from organic solution of the block copolymer were assembled with an achiral amphiphilic merocyanine, providing a chiral environment to the red-shifted J-aggregates (offset stacking) of the dyes. The CD spectrum of the samples showed a clear Cotton effect arising from the induced chirality in the dyes. On the other hand, the anion exchange of the 4-tolylsulfonate counterions of a tetrapyridinium porphyrin for TRISPHAT or BINPHAT anions (Figure 3.4) using the "ion association synthesis" by sonification of a methanol solution of the components in the presence of poly(vinylpyrrolidone) leads to CONPs with sizes in the range 30–50 nm, practically independently of the counterion used. [77]. The characteristic Soret band of the porphyrins in these nanoparticles shows a significant bathochromic shift when compared with band of the dispersed porphyrin in solution. In addition, this band revealed optical activity in the CD spectrum. The form of the signal was different for the two chiral anions employed, which prompted a theoretical investigation of the conformation of the porphyrin in contact with the counterions. This study

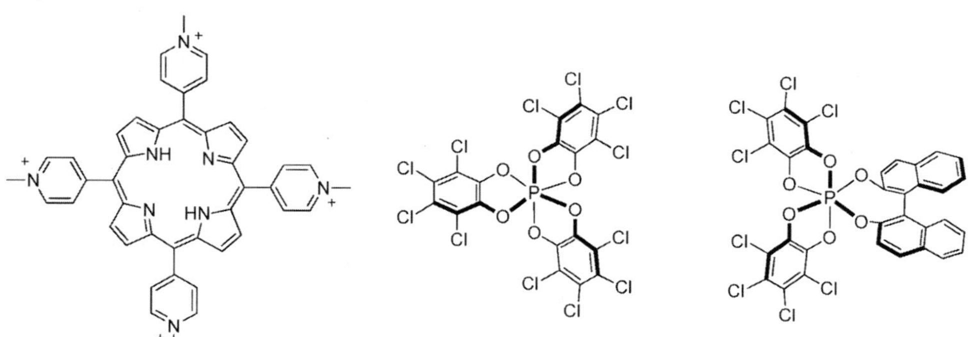

FIGURE 3.4. A cationic porphyrin derivative and the chiral anions with which it forms optically active nanoparticles.

implied that a flattening of the meso-substituents of the porphyrin as well as a ruffling of the porphyrin may be at the origin of these effects, while a coupling between the magnetic and/or electronic transitions of the porphyrin and the chiral anions could influence the optical activity in the visible region.

3.5 CHIRAL FIBERS

Fibrillar aggregates of organic molecules—perhaps the most abundant and easy to observe of nanostructured organic material—comprise a family of nanostructures with intriguing morphologies as well as remarkable properties [78]. DNA and RNA are perhaps archetypal double-helical fibers where the chirality of the sugar backbone is crucial to the structure, and collagen is a naturally occurring substance comprising "nanofibers" each with three intertwined strands [79] and which have numerous roles [80]. In synthetic materials, helical fibers may form by all the routes shown in Figure 3.1, so even achiral molecules can display chiral morphologies, something which does not happen usually in organic structures in natural systems. However, it is important to note that helical morphology is not always observed because chiral molecules often give rise to smooth straight fibers. The following is a personal (and by no means comprehensive) selection of representative examples of organic systems which form chiral fibers.

Aggregates of cyanine dyes have received particular attention because of their interesting properties and excellent optical activity, and they show a rich supramolecular stereochemistry. The preparation of chiral cyanines has a long history [81], but new chiral derivatives continue to show beautiful phenomena. A number of chiral cyanine dyes of different lengths have been synthesized and their optical activity measured and explained on the basis of theory [82]. The assembly of the dyes of this type are aided by π–π stacking, which makes them conducive to the formation of nanoscale fibers. The stacking in the aggregates can follow the fully overlapped π system arrangement (the so-called H-aggregates) or J-aggregates in which the extremes of the π systems overlap.

The rod-like aggregates of chiral bis(merocyanine) derivatives show amplification of chirality in scalemic mixtures [83], which is the majority rules effect. Presence of small enantiomeric excesses resulted in optical activity similar to that of enantiopure material. However, the kinetics of the induction of chirality in this system is complex. The system revealed sigmoidal behavior with relatively large enantiomeric excesses, with the induction period increasing as the excess increases. The measurement of absorption and CD over time revealed no significant difference for short periods after growth initiation, but showed divergence after longer times, an effect interpreted as a nucleation-and-growth process. It was proposed than an autocatalytic role was played when the supramolecular organization of the initial aggregates act as proper templates for elongation into nanorods.

The core substitution of perylene bisimide units leads to molecules that are able to self-assemble in apolar solvents thanks to the supramolecular oligomerization through the imide units (Figure 3.5) [84]. Studies in solution and on

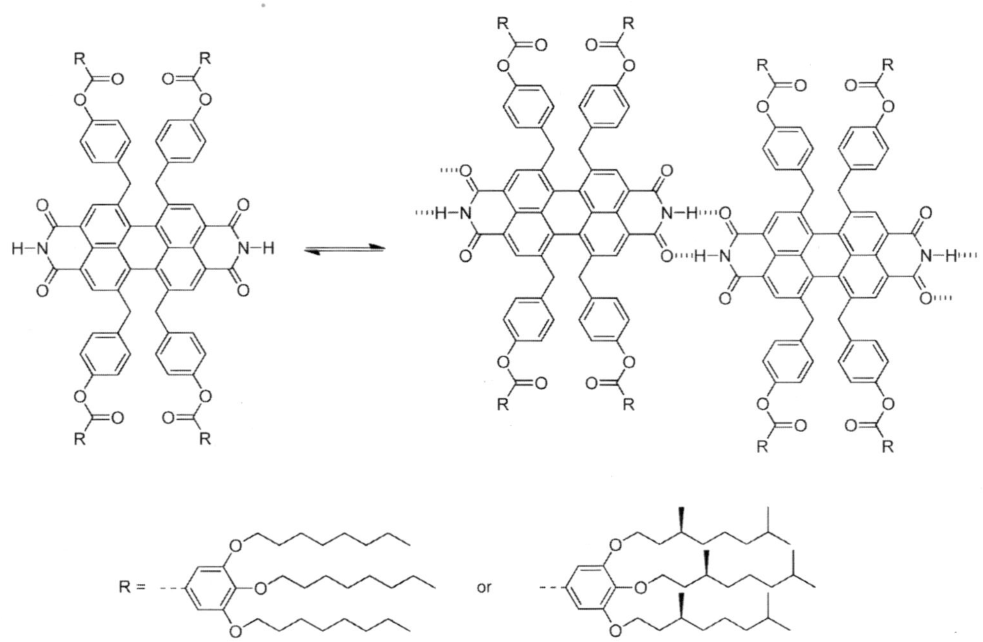

FIGURE 3.5. Core-substituted perylene bisimide molecules that form self-assembled fibers, and whose chiral derivative is able to act as a sergeant over achiral soldier molecules.

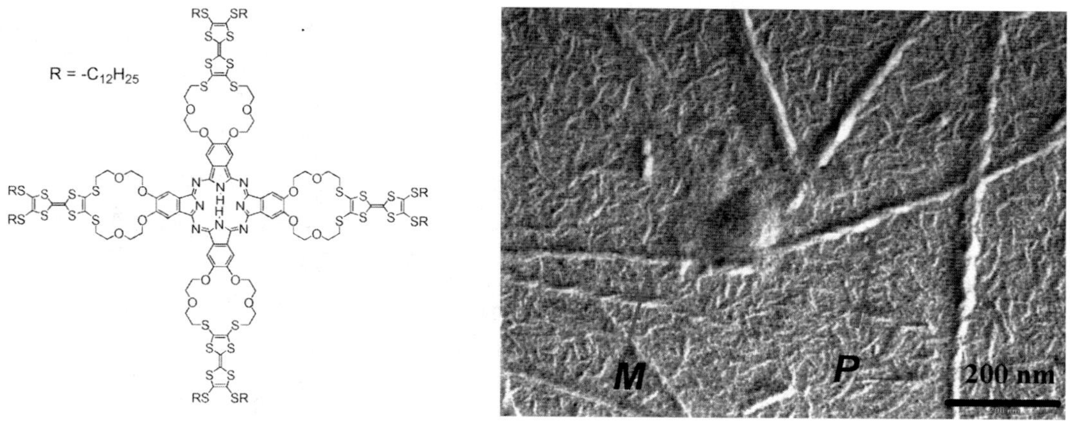

FIGURE 3.6. The achiral TTF-bearing phthalocyanine and a TEM image of the helical fibers it forms.

surfaces by atomic force microscopy show that the compounds self-assemble into extended double string cables, wherein two densely packed hydrogen-bonded supramolecular polymeric chains hold the chromophores in such a way that here is a strong exciton coupling between them. This arrangement makes the J-aggregates highly fluorescent, with quantum yields of almost unity (a remarkable feature in this type of compound). The chiral compound shows a nucleation-growth-type aggregation initiated from the smallest nuclei of two building blocks, rather than the isodesmic mechanism often proposed for the assembly of this kind of molecule. The excitonic coupling displayed by this material is promising for applications in optoelectronic and photovoltaic devices.

Phathalocyanines (Pcs) are a family of compounds with great interest for nanomaterials because of their potential applications in a number of sectors [85]. There are a considerable number of cases of phthalocyanines which form supramolecular chiral fibers [86–88]. One example of an achiral derivative that forms helical tapes from dioxane–chloroform mixtures is the TTF-bearing Pc shown in Figure 3.6 [89]. UV-vis absorption spectroscopy showed that the compound aggregates in $CHCl_3$/MeOH (1:4 v/v), as witnessed by a broadening and blue shift of the major absorbance bands of the Pc. Slow addition of dioxane to a chloroform solution of the compound at 5°C caused gelation of the mixture and transmission electron microscopy (TEM) showed fibers up to several micrometers in length, which is a length consistent with a stack around 100,000 molecules long. Previous studies on crown-ether analogues that lacked the TTF moieties gave similar values [86]. The remarkable aspect of the TEM is the clearly twisted nature of the objects, which are either left- or right-handed. The forces that hold these aggregates together are probably between the TTF moieties and the phthalocyanine aromatic surface that gives rise to the scrolled architecture.

Oligothiophene derivatives show interesting electronic properties—they have been employed in field effect transistors as well as in photovoltaic devices—and for this reason, considerable attention has been paid to their self-assembly. When the oligothiophene core is substituted with chiral oligothyelene-derived chains, optically active aggregates are formed [90]. In solution, oligothiophenes of the type shown in Figure 3.7 form aggregates—as a result of the stacking of the thiophene segments—which can be probed by UV-vis absorption and CD spectroscopies. The experiments show that the stability of the stacks of molecules in *n*-butanol increases with the length of the oligothiophene segment (the heptathiophene derivative aggregates more strongly than the sexithiophene derivative, and so on) as a result of the increased number of π–π interactions. A reversible transition from chiral aggregates into achiral, apparently disordered, aggregates can take place. It is proposed that the end groups

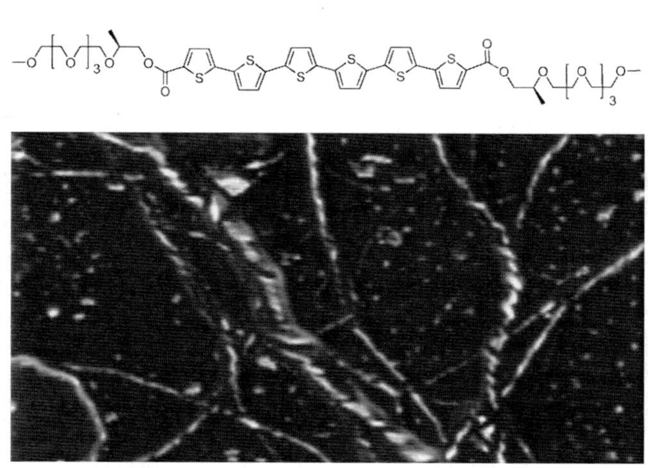

FIGURE 3.7. A chiral sexithiophene derivative and the helical fibrous aggregates it forms (AFM image size 2.5 × 1.25 μm).

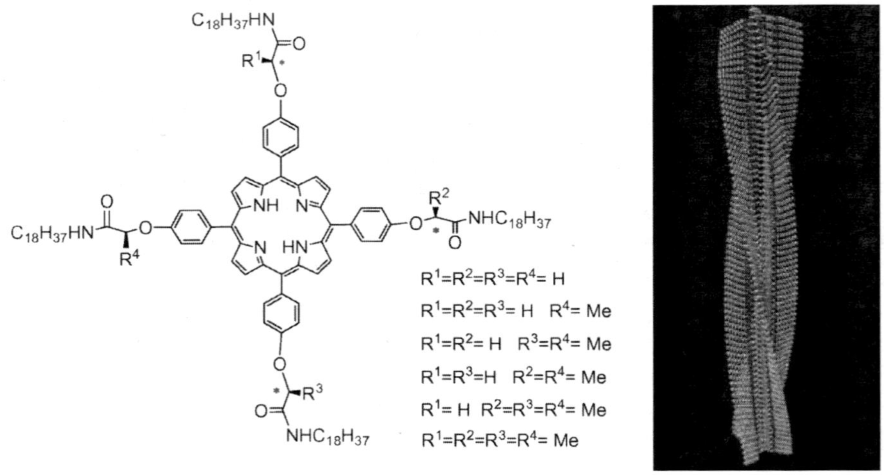

FIGURE 3.8. A series of porphyrin organogelators and a model of the chiral H-aggregates which they form.

are mobile at higher temperatures and no longer induce chiral cc-conformations in the sexithiophene pillars. Amplification of the chirality can be observed when the chiral oligothiophenes are added to achiral analogues. Addition of 30 mol% of "sergeant" sexithiophene derivative to the achiral version is enough to give equivalent optical activity to the entirely chiral sample [91]. The rate of cooling of the mixtures plays an important role in the transfer of chirality: Slow cooling produces the expected optical activity, but fast cooling gives a sample whose CD spectrum is the inverse of that expected. In general, slow cooling leads to the thermodynamically most stable mixed aggregate, while fast cooling gives a kinetic product. The enantiopure compound forms morphologically chiral fibers as a result of hierarchical assembly on silicon surfaces [92]. Only one handedness of fiber is observed in the atomic force microscope images of the samples (Figure 3.7). The assembly process is sensitive to the surface employed in the experiments, as these fibers are not observed either on graphite or on mica, a fact that points to an active role of the surface in propagating the chirality.

The formation of organogels or hydrogels from organic gelators produces nanomaterials which incorporate the liquid medium (organic solvent or water, respectively) or which can be dried to produce xerogels [93–96]. The colloidal structure of the organic material which immobilizes the solvent is usually fibrous and contains aggregates of molecules which one could consider supramolecular polymers (polymers also form gels of the physical and chemical type, that is with covalent bonding reticulating the fiber network) [97]. The forces that hold the molecules together are similar to those in other fibers; and, in general, fibers forming compounds are prone to be gelators (though they are not always). Hydrogen bonding is a particularly common trait in the materials that form gels [98]. The effects of chirality in gelators has been reviewed recently [99].

In principle, one might think that the chirality of a sample of gelator may have an effect on the macroscopic mechanical properties of the gels. In one particular case at least, this does not seem to be a significant factor [100]. On the other hand, it is known that enantiopure gelators have lower critical gel concentrations than their racemic modifications [101]. This feature is very much related to the morphology of the fibers, since while racemic compounds often form flat lamellae, the fibers of enantiopure gelators are often rods or helices.

The number of stereogenic centers in porphyrin-based gelators has a large effect on the critical gel concentration and the morphology of the xerogels [102]. The fibers' primary structure is in the form of H-aggregates held together by hydrogen bonds between the amide moieties in the side chains attached to the porphyrins (Figure 3.8). The compounds with zero or one stereocenters are the best organogelators of the family, while the compound containing four stereogenic centers is a poor gelator of methylcyclohexane. The reason for this effect might be the increased solubility of the compounds with more stereogenic centers.

The idea of forming fibers in gels is attractive because the stacks of π-functional molecules within them can be used for charge transport. The tetrathiafulvalene (TTF) unit is a useful one for this purpose [103], and chirality has been introduced into the materials by β-peptides as a self-assembling component [104]. The aim of this work is to prepare chiral conductors once the TTF units are doped, to give nanomaterials that might show interesting physical properties. In this regard, a non-amphipihilic C_3-symmetrical compound incorporating three TTF forms chiral fibers that are visibly helical in an optical microscope (Figure 3.9), and stems from the chirality in the chiral stacks that they form and which are observed by CD spectroscopy [105]. Although the compound does not form a gel as an achiral derivative does [106], when it is warmed in dioxane and allowed to precipitate, the

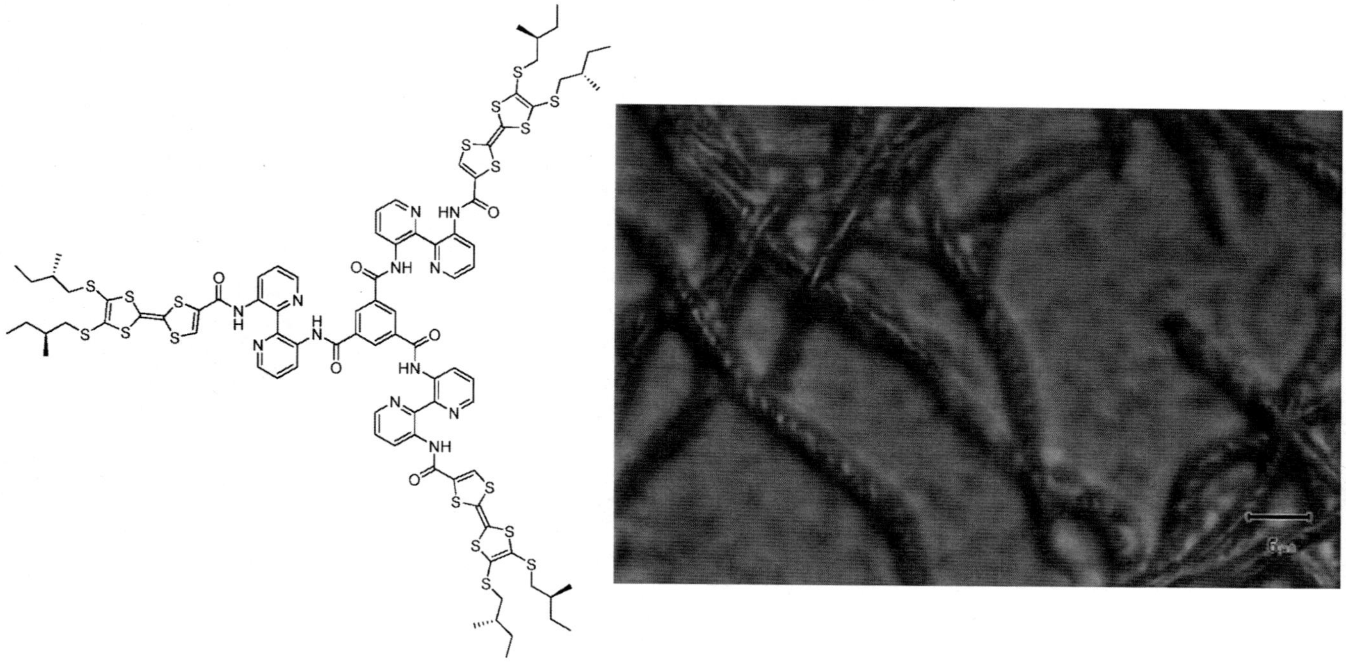

FIGURE 3.9. A non-amphiphilic tris-TTF compound and an optical microscope image of the fibers it forms.

helical fibers self-assemble in a matter of minutes, provided that the solution is sufficiently concentrated. Molecular modeling shows that the molecule forms helical stacks with a chiral arrangement of the TTF units which is in accord with the Cotton effects seen in the CS spectra; furthermore, dynamics on three proximal helices shows how they might intertwine at the second stage of the process, which leads to the macroscopic fibers. It is important to note that the fibers appear well-ordered as they polarize light when observed in the optical microscope between crossed polarizers. The order within stacks is important for transport within stacks of aromatic molecules, as shown for oligophenylenevinylene-based chiral stacks [107].

Another interesting approach to chiral photoactive materials based on organogels was the formation of composite gels with standard low-molecular-weight chiral gelator and an achiral phthalocyanine co-gelator that was subsequently reacted to form cross-links that reinforce the gel [108]. This method is conceptually appealing and awaits further exploitation.

Oligomeric structures of varying types have been self-assembled into chiral fibers, with derivatives of oligopeptides showing particularly rich morphological behavior [109]. Up the molecular length scale, chiral fibers with nanometer-range diameters have also been prepared with polymers. Chiral poly(aniline) nanofibers were prepared in aqueous solution by polymerizing the monomer in concentrated camphor sulfonic acid and using aniline oligomers to accelerate the polymerization reaction that was performed by sequential addition of ammonium persulfate (an oxidant) [110]. An increase in the chirality of PANI nanofibers with increasing aniline–CSA concentrations during the synthesis was observed. The as-prepared PANI nanofiber is doped, and dedoping with NH_4OH (aq) produces a chiral material devoid of stereogenic centers in which the chirality originates only from the conformation of the polymer and/or from the helical packing of the polymer chains induced by the chiral acid in the formation step. The individual nanofibers are between 20 and 40 nm in diameter and several micrometres long. They are mainly twisted and entangled in a network, however. TEM shows some helical single nanofibers within a bundle. An advantage of the conducting polymers is that this kind of nanofiber can also be generated electrochemically [111].

3.6 CHIRAL NANOTUBES

The formation of linear nanostructures with a cavity along their length is an interesting challenge from a number of perspectives. From the supramolecular point of view, the structures formed by chiral lipids are of particular interest [112]. On the other hand, chiral carbon nanotubes have raised tremendous attention in recent years because of their varied electrical properties [113]. Coverage of these very broad areas here would be impossible, but the cited articles will lead the interested reader to the most relevant work. Here recent work on other nanotubes will be discussed briefly.

Non-symmetrically substituted hexabenzocoronene derivatives have been shown to form nanotubular aggregates

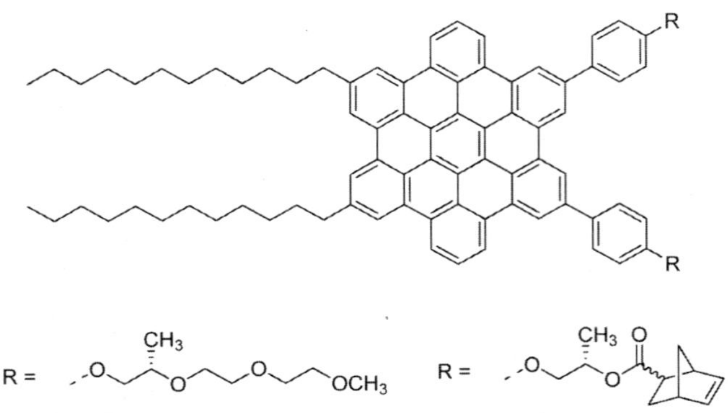

FIGURE 3.10. Hexabenzocoronene derivatives that self-assemble into chiral nanotubes.

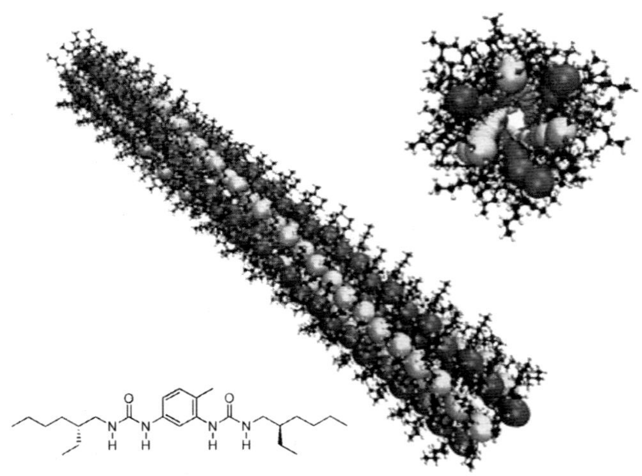

FIGURE 3.11. A chiral bis-urea which self-assembles to give chiral supramolecular nanotubes, whose molecular model is shown with different hydrogen-bonded chains in different colors. See color insert.

[114], and chiral derivatives such as those shown in Figure 3.10 form optically active tubes. TEM micrographs showed that the nanotubes have a high aspect ratio (>1000) and uniform diameters of around 20 nm. The walls are around 3 nm thick, and electron diffraction analysis showed that the aromatic cores are π-stacked with an interplane separation of 3.6 Å. CD spectroscopy shows the helical preference for the chiral derivatives, which display the majority rules phenomenon. Macroscopic fibers of the chiral derivatives are particularly long when compared with their achiral counterparts, and when doped with iodine they reveal a resistivity decrease from 200 MΩcm to 20 Ωcm at 300 K. The doped fibers are semiconductors. The coassembly of chiral and achiral derivatives reveals the sergeants and soldiers effect, and covalent cross-linking also leads to robust conducting chiral materials [115].

Relatively simple bis-ureas have been shown to self-assemble in solution to form nanotubes in which solvent molecules are included (Figure 3.11), and a marked majority-rules effect has been observed in the formation of chiral derivatives [116]. CD spectroscopy showed that chiral side chains help form an asymmetric supramolecular structure in cyclohexane, with strong bisignate Cotton effects observed. On the other hand, in ethanol no signal was observed, apparently because of competition of the compound for hydrogen bonds with the solvent. Interestingly, a chiral solvent—limonene—induced helical nanotubes from a racemic mixture of monomers, which gave the same circular dichroism signature as the enantiopure sample. The helical bias induced by the chiral solvent or by the chiral monomer in majority rules experiments was probed by competition experiments which showed their relative strength: The studies showed that the chirality of the urea is much more dominant than the asymmetrical influence of the solvent in the self-assembled nanotubes. Indeed, solvent rarely influences chiral induction. Be that as it may, the chirality of the solvent does influence the degree of chiral induction in this system, where there is close contact between the assembly and the liquid.

Regarding lipids, chiral lipids derived from glutamic acid self-assemble in absolute alcohol to give an organogels containing long nanotubes with a high aspect ratio [117]. When the enantiomers were mixed in different proportions, the morphology of the assembly could be varied consecutively from nanotubes with a helical seam to nanotwists, wherein the excess enantiomer determined the shape. The bola-amphiphile *N*,*N*-hexadecanedioyl-di-L-glutamic acid forms hydrogels in which the lipid forms nanotubes [118]. Coassembly of the compound with the sodium salt of the azobenzene derivative 4-(phenylazo) benzoic acid led to the formation of a mixed system in which the optical activity could be changed through photoisomerization of the aromatic unit. The isomerization effect on the tubes depended on the coassembly method: When the chromophore was added to preformed nanotubes, the switching was irreversible, but a gel formed with the two components present showed a reversible character.

Molecules based on a naphthalenediimide core with chiral carboxylic acid side-arms (Figure 3.12) have been shown to self-assemble in the solid state in chloroform into noncovalent nanotubes [119, 120]. The amino acid residues have a *syn* geometry with respect to the naphthalenediimide plane in the crystals of the compound, allowing the three S-trityl groups of different molecules to interdigitate. The noncovalent interactions sustaining the tube are hydrogen bonds between the carboxylic groups of adjacent molecules as well as π-stacking interactions between the aromatic cores of the molecule. The supramolecular polymer has a 3_1 helix conformation in the solid, with an internal diameter of approximately 12.4 Å. While the nanotubes are filled with solvent in the absence of guests, they have been shown to act as hosts, accommodating C_{60} in solution [121]. The chirality of

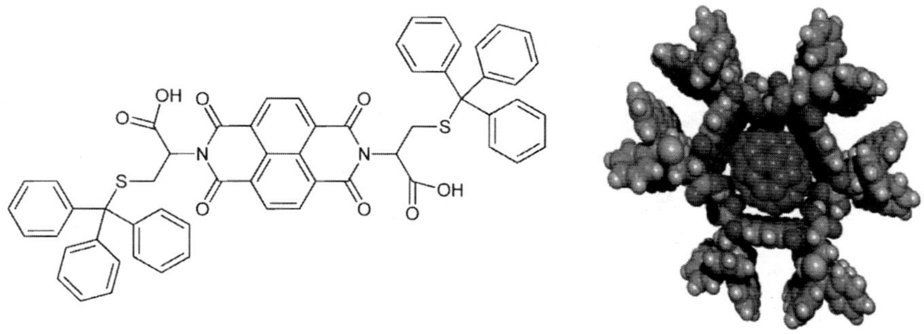

FIGURE 3.12. A chiral naphthalenediimide derivative that forms helical nanotubes capable of incorporating C_{60} (image on the right is a molecular model).

the tube is reflected in the CD spectrum wherein weak but significant signals are observed corresponding to the fullerene. On the other hand, ion pairs and C_{60} have been shown to occupy the cavity of the tubes simultaneously [122], offering the prospect that these chiral cylindrical cavities might be used for multicomponent functional systems.

A quite singular result is the formation of nanotubes at a water–air interface from a molecule with no alkyl chains or hydrophilic groups and therefore does not have an apparent amphiphilic character [123]. The compound, π-conjugated fused pyrazine bearing four trifluoromethyl groups, was spread at the water surface and was transferred onto quartz or other solid surfaces using a lift-off methodology at a determinsed surface pressure. Electron microscopy indicated a tube-like morphology, and circular dichroism spectroscopy of different batches revealed significant optical activity with opposite signs depending on the batch. The hypothesis for the origin of the chirality is in the scrolling up of multilayer sheets of molecules—of approximately 3 nm in thickness—with a slight twisting with respect to the edges of the sheet. The lack of an amphiphilic nature apparently favored the multilayer and scrolling process. Another surprising example of the formation of chiral nanotubes was found in the template-free synthesis of poly(aniline) [124]. When the polymer was prepared by in situ doping polymerization in the presence of (*S*)-(2)-2-pyrrolidone-5-carboxylic acid or its enantiomer as the doping source, SEM and TEM revealed tubes with 80 to 220-nm outer diameter and 50 to 130-nm inner diameter. The tubes show optical activity (as shown in their CD spectra) and conductivity, although the latter is relatively low. The preparation of chiral conducting molecular nanomaterials remains a key challenge.

3.7 CHIRAL MONOLAYERS

There is an increasing interest in the preparation of monolayer systems that contain chiral organic adsorbants, which are connected to the surface either by chemisorption [125] or by physisorption [126]. The research in this area is facilitated by the tools of surface science, such as low-energy electron diffraction and especially scanning tunneling microscopy (STM) which can provide submolecularly resolved images of organic compounds on a variety of conducting surfaces. In both cases (chemisorption and physisorption) the balance of interactions between the adsorbate molecules themselves and between the individual molecules and the surface is determining in the structure of monolayers [127]. On crystalline surfaces particularly, spontaneous resolution is a much more favored process when compared with three-dimensional solid or fluid states or monolayers on water (where no orientational order in the plane of the surface is enforced) [19]. Incorporation of a stereogenic center in the adsorbate molecule gives a thermodynamic preference to one of the two possible orientations with respect to a reference axes on the crystalline surface because a diastereomeric relationship exists, In addition to this effect of handedness of the stereogenic center in the adsorbate, the "footedness" of the molecule—that is, the way in which the groups close to the surface interact strongly with it—plays a critical role [128]. Here only representative examples of functional molecules at surfaces will be given, but the reader is encouraged to see the cited reviews for further details.

The study of the nanostructures of liquid crystals physisorbed to surfaces has proven particularly interesting [129, 130]. Monolayers formed between a bulk liquid crystal and a graphite surface (observed by scanning tunnelling microscopy) are comprised of rows of molecules with a defined tilt angle that was always either clockwise (for *S*) or anticlockwise (for *R*) with respect to the row normal for enantiopure samples. The domains in monolayers of the racemate were either all tilted clockwise or anticlockwise, and the unit cell dimensions were the same as those of the pure enantiomers (Figure 3.13). This result implies spontaneous resolution and formation of a conglomerate on the graphite.

The separation of the enantiomers of solution-deposited self-assembled rosette-type nanostructures on graphite has been observed using atomic force microscopy [131]. The chirality of the assemblies of two achiral components (a quadruple melamine–barbiturate derivative that forms a stacked structure with four hydrogen-bonded hexamers) arises from

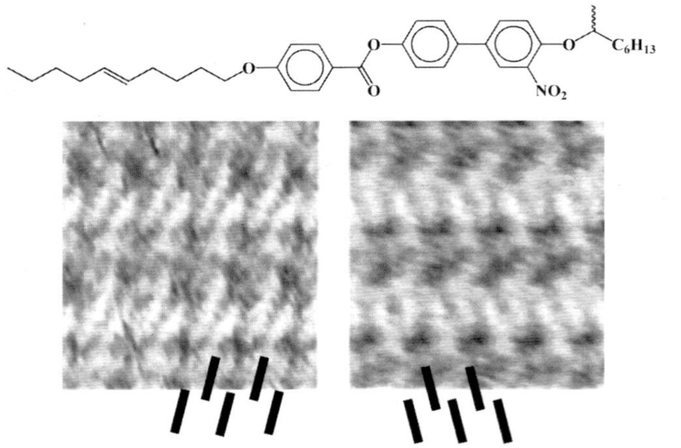

FIGURE 3.13. A chiral liquid crystal and STM images of enantiomorphous domains of the compound resulting from spontaneous resolution. The bars indicate the aromatic cores of the molecule.

the twist between the layers in the superstructure. When chloroform solutions of the racemate were cast onto graphite by slow evaporation of the solvent, the dry films contain rods in which the rosettes are stacked. The domains have oblique unit cells with opposite chiralities. Direct assignment of a *P* or *M* rosette to each domain was not possible, but the evidence suggests strongly that each occupies one of the domains.

Among the many aromatic molecules that have been observed to form chiral structures on surfaces, porphyrins are molecules that are particularly relevant for studies at monolayers, because they may show electronic, magnetic, or catalytic properties at surfaces, and their dimensions usually allow submolecular resolution to be achieved by scanning tunneling microscopy. Thus, the spontaneous resolution of cobalt(II) porphyrins (which can exhibit magnetic and catalytic properties) on a metal surface has been proven by STM [132]. The deformation of the tetraphenylporphyrin structure is noteworthy, and theoretical investigations were essential in the determination of the structure of the layer. The orientation of monolayers of porphyrins physisorbed at the graphite–liquid interface is controlled by incorporating stereogenic centers in the side chains attached to the aromatic core [133]. Again, careful modeling of the experimental results revealed the likely conformations of the molecules at the interface, along with energetically most important factors in the formation of the chiral structures. An interesting possibility for this kind of structure is the switching of the chirality. For now, the switching of orientation in a racemic mixture of chiral conformers has been observed in sandwich-type complexes of phthalocyanines [134].

3.8 CHIRAL FILMS

Thin films of organic materials have a number of applications, and although chiral organic nanomaterials as films are not very abundant, the potential of stereochemistry in determining the arrangement of functional moieties in films is great [135]. In particular, their potential use in photonics is appealing [136]. Polymers and liquid crystals can be particularly interesting for the formation of chiral films. For example, a high degree of circular polarization in the near field of an aperture at the apex of hollow-pyramid probes for scanning optical microscopy was achieved using an annealed thin polymer film containing a chiral polyfluorene derivative [137]. The dichroic properties were measured close to the optical probe. The results showed the feasibility of nano-optics applications exploiting circularly polarized near-field systems.

Supramolecular polymer films also have a number of potential applications [138], and the sensing of chiral systems is an important area [139]. Cyclodextrins are a family of molecules that have a renowned ability to perform stereoselective recognition of small molecules thanks to their chiral cavity. The immobilization of derivatives of cyclodextrins on surfaces is therefore a route to chiral sensing, which has been achieved using a quartz crystal microbalance [140]. The host molecule was attached to the surface of the device by self-assembly of a thiol-derived cyclodextrin onto gold-covered quartz. The transducer showed real-time stereoselective sensing of chiral molecules such as methyl lactate and 2-octanol.

Deposition of the materials is a key determinant in their function, and the surface onto which the materials are cast plays a critical role in the thin film morphology. To take a molecular example, chiral porphyrins with different constitutions form films with varied morphologies when cast onto hydrophobic or hydrophilic surfaces [141]. The porphyrins (with long alkyl tails at the periphery of the conjugated π-electron system), when cast from methanol onto graphite, form round-shape aggregates or larger dewetting type patterns depending on the constitution of the compound. Deposition of the same compounds from chloroform or toluene gives fibril-like structures on graphite, but yield dewetting patterns with well-structured multilayers on mica [142]. Control of the parameters involved in solvent diffusion and evaporation can lead to thin films with interesting structural features. In particular, porphyrin nanowell-array films were prepared by diffusing water into organic solutions of the intrinsically chiral chromophore [143]. It was pointed out that this morphology is interesting for organic photovoltaic devices and biosensors.

3.9 CHIRAL POLYMERS

There are a great number of nanostructured bulk polymers that have interesting materials properties, and a flavor of these characteristics will be given in the following examples. Naturally occurring polymers and their derivatives are particularly appealing for applications because of their availability and low environmental impact. For example, cellulose has been used as the stock material for a very wide range of nanomaterials with a big variety of applications [144]. In the chemical industry, the detection and separation of enantiomers is of utmost importance, and the use of nanoscale materials for use in chiral chromatography is particularly useful [145].

A nice example of the use of nanomaterials in chromatography is the use of materials derived from chitosan (produced from chitin, the structural element in the exoskeleton of crustaceans) are useful as chiral stationary phases in electrochromatography [146]. The material was immobilized onto modified capillaries by copolymerization of the glycidyl methacrylate-modified compound with methacrylamide and bis-acrylamide cross-linkers. Tryptophan enantiomers were separated well by the system, and addition of 80% MeOH into the phosphate buffer allowed the chiral separation of (+/-)-catechin. However, there is room for improvement, as the new stationary phase separated only two groups of tocopherol stereoisomers.

In another area completely, chirality can be induced onto aggregates of achiral polymers which show luminescence, giving rise to materials devoid of stereogenic centers but which display circularly polarized luminescence [147]. The aggregates were formed from a mixture of chloroform (good solvent), alkanol (anti-solvent), and limonene (chiral solvent), whereby the anti-solvent was added to a solution of the polymer in the other two leading to the formation of turbid suspensions of particles. The enantiopurity of the chiral solvent, the size of the aggregates, and the order of addition of the components influence the optical activity.

One of the interesting possibilities for block copolymers is their potential use as templates for other materials (more often than not inorganic) or the formation of composites. The nanostructure of the copolymers determines the structure and properties of the final material. For example, the block copolymer made from styrene and L-lactide is a system that shows a helical phase [148]. Hydrolysis of the lactide fragment leaves chiral channels of nanoscale width in the poly(styrene) which can then be used to template the formation of inorganic materials by sol–gel chemistry. The result is helical hexagonally packed silicon oxide in the polystyrene matrix.

An alternative way to form the helical arrangement in soft polymer materials is to polymerize liquid crystals in their chiral phases. In situ photopolymerization of crosslinking monomers in the isotropic phase of chiral liquid crystals generates optically transparent and isotropic liquid-crystalline composites [149]. In this material, a large electro-optical Kerr constant was determined over a broad temperature range. The authors suggested that the composites could be useful for new flat-panel liquid-crystal displays with high-speed responses, which would not require the rubbing processes used during the habitual fabrication of such devices.

A related approach is that of chiral imprinting in polymers, and a recent and attractive example of this technique is the creation of chiral spaces in the head of diblock copolymer single-chain particles that have a tadpole-like topography [150]. L-Phenylalanine anilide was used as a template during the photopolymerization of a diblock copolymer comprised of poly(*tert*-butyl acrylate) and the random cinnamoyloxyethyl methacrylate and a carboxy-bearing unit derived from it. The polymer exists as a collapsed form of the random polymer block and extended chains of the former block when dissolved in a mixture of chloroform and cyclohexane, hence polymerisation of the cinnamoyloxyethyl residues with the chiral template present generates a chiral environment. Removal of the template releases the spaces that are then capable of chiral recognition. The approach shows the sophistication in the control of nanostructure which can be achieved in polymeric systems, and it bodes well for their future application.

3.10 CHIRAL NANOPOROUS SOLIDS

Materials containing pores of nanometer dimensions are receiving a great deal of interest for a number of reasons, and chiral frameworks are particularly relevant [151]. The applications of these materials range from catalysis to environment-responsive systems. Yet, in their vast majority, these types of material are not purely organic, but rely on inorganic or coordination chemistry. There are certain important examples though.

Certain crystalline polymers contain nanomet scale pores, which are interesting for sensing applications [152]. Syndiotactic polystyrene (s-PS) is a good example of a polymorphic macromolecule whose structure can be varied through crystallization conditions. The nanoporous low packing density ($\rho \approx 0.98$ g cm^{-3}) helical crystalline phases can be formed when guests are removed from the solvent-crystallised sample by extraction [153]. In this way, the polymer can be used as a sensor for chiral guest materials, as has been proven for carvone employing CD as the detection method [152]. Polymeric porous materials are therefore of interest for sensing chirality, as well as for separation and sequestering. There are a wide variety of porous polymers that await exploitation in this regard.

Top-down and bottom-up techniques often come together for the preparation of polymer nanosystems, and a recent example of a porous material that exemplifies this approach is the formation of artificial nanochannels for chiral sensing [154]. The conical nanochannels (640 nm wide at the rim and 11 nm wide at the base) were prepared by chemical etching of a polyethylene terephthalate (PET) membrane. Carboxyl groups were present on the nanochannel surface in the polymer after etching, which allowed coupling of mono-6-amino-β-cyclodextrin. The functionalized nanochannel exhibited selective chiral recognition of L-His, which was shown by the changes in the ionic current flowing through the nanochannel. No significant changes in the ionic current were found when the modified channel was exposed to solutions of D-His or other aromatic amino acids. This result paves the way for the use of similarly imaginative approaches for sensing, separation, and even more exotic properties in porous materials.

3.11 CONCLUDING REMARKS

The preparation of chiral organic nanomaterials is achieved with increasing control in the present day, by the combination of both bottom-up and top-down approaches. Relatively cheap sources of asymmetry can be incorporated into functional units through stereoselective synthesis to generate chiral molecules whose assembly can be influenced using a variety of conditions. There remain great opportunities regarding the application of these materials, and the development of systems that can be used for optical properties or chiral separations are particularly promising. The use of lithography and imprinting techniques hold special promise.

Theory is of great importance in the interpretation of many phenomena related to chiral nanomaterials, ranging from the structure to the optical activity [155]. One expects that the increasing accuracy of molecular modeling will make it an indispensable tool in the prediction and interpretation of properties of chiral nanomaterials. It will also be of tremendous aid in the prediction of structural features that encourage the hierarchical transfer of asymmetry from the molecular scale upward.

The opportunities that exist for the use of the special characteristics given by chiral systems means that study of the properties of chiral organic nanomaterials is especially important. The multidisciplinary efforts aimed at uncovering synergies between optical and electronic properties and chiral nanostructure are perceived to be particularly relevant.

ACKNOWLEDGMENTS

The author thanks all the people who contributed to the work he has published in the area of chiral nanomaterials, and he is grateful to the funding bodies who supported the research, especially the MICINN (Spain, CTQ2010-16339), the Generalitat de Catalunya (2009 SGR 158), and the European Community's Seventh Framework Programme under grant agreement n. NMP4-SL-2008-214340, project RESOLVE. A special thank you to Mathieu Linares for the modeling illustration in Figure 3.11.

REFERENCES

1. Wagnière, G. H. (2007). *On Chirality and the Universal Asymmetry*. Verlag Helvetica Chimica Acta, Wiley–VCH, Zurich.
2. Kondepudi, D. K. (2004). In Palyi, G., Zucchi, C., Caglioti, L., eds. *Progress in Biological Chirality*. Elsevier, Amsterdam, pp. 159–171.
3. Amabilino, D. B., ed. (2009). *Chirality at the Nanoscale*, Wiley–VCH, Zurich.
4. Guijarro, A., Yus, M. (2009). *The Origin of Chirality in the Molecules of Life*. RSC Publishing, London.
5. Avalos, M., Babiano, R., Cintas, P., Jimenez, J. L., Palacios, J. C. (2010). Homochirality and chemical evolution: New vistas and reflections on recent models. *Tetrahedron: Asymmetry*, **21**, 1030–1040.
6. Barron, L. D. (2004). *Molecular Light Scattering and Optical Activity*, 2nd edition, Cambridge University Press, New York.
7. Palmans, A. R. A, Meijer, E. W. (2007). Amplification of chirality in dynamic supramolecular aggregates. *Angewandte Chemie International Edition*, **46**, 8948–8968.
8. Miyano, S., Okada, S., Hotta, H., Takeda, M., Kabuto, C., Hashimoto, H. (1989). Optical resolution of 2′-methoxy-1,1′-binaphthyl-2-carboxylic acid, and application to chiral derivatizing agent and application to chiral derivatizing agent for HPLC separation of enaniomeric alcohols and amines. *Bulletin of the Chemical Society of Japan*, **62**, 1528–1533.
9. Martin, R. H., Marchant, M. J. (1974). Resolution and optical properties α_{max}, ORD and CD of hepta-helicene, octa-helicene and nonahelicene. *Tetrahedron*, **30**, 343–345.
10. Pieraccini, S., Masiero, S., Ferrarini, A., Spada, G. P. (2011). Chirality transfer across length-scales in nematic liquid crystals: Fundamentals and applications. *Chemical Society Reviews*, **40** 258–271.
11. Green, M. M., Reidy, M. P., Johnson, R. J., Darling, G., O'Leary, D. J., Willson, G. (1989). Macromolecular stereochemistry—The out-of-proportion influence of optically-active co-monomers on the conformational characteristics of polyisocyanates—The sergeants and soldiers experiment. *Journal of the American Chemical Society*, **111**, 6452–6454.
12. Gu, H., Nakamura, Y., Sato, T., Teramoto, A., Green, M. M., Jha, S. K., Andreola, C., Reidy, M. P. (1998). Optical rotation of random copolyisocyanates of chiral and achiral monomers: Sergeant and soldier copolymers. *Macromolecules* **31**, 6362–6368.
13. de Jong, J. J. D, Tiemersma-Wegman, T. D., van Esch, J. H., Feringa, B. L. (2005). Dynamic chiral selection and amplification using photoresponsive organogelators. *Journal of the American Chemical Society*, **127**, 13804–13805.
14. Mikami, K., Daikuhara, H., Inagaki, Y., Yokoyama, A., Yokozawa, T. (2011). Chiral amplification based on sergeants and soldiers effect in helically folded poly(naphthalene carboxamide). *Macromolecules*, **44**, 3185–3188.
15. Green, M. M., Peterson, N. C., Sato, T., Teramoto, A., Cook, R., Lifson, S. (1995). A helical polymer with a cooperative response to chiral information. *Science*, **268**, 1860–1866.
16. Green, M. M., Garetz, B. A., Munoz, B., Chang, H., Hoke, S., Cooks, R. G. (1995). Majority rules in the copolymerization of mirror-image isomers. *Journal of the American Chemical Society*, **117**, 4181–4182.

17. van Gestel, J., Palmans, A. R. A, Titulaer, B., Vekemans, J. A. J. M., Meijer, E. W. (2005). "Majority-rules" operative in chiral columnar stacks of C-3-symmetrical molecules. *Journal of the American Chemical Society*, **127**, 5490–5494.

18. Mateos-Timoneda, M. A., Crego-Calama, M., Reinhoudt, D. N. (2006). Amplification of chirality in hydrogen-bonded tetrarosette helices. *Chemistry—A European Journal*, **12**, 2630–2638.

19. Pérez-García, L., Amabilino, D. B. (2002). Spontaneous Resolution under Supramolecular Control. *Chemical Society Reviews*, **31**, 342–356.

20. Pérez-García, L., Amabilino, D. B. (2007). Spontaneous resolution, whence and whither: From enantiomorphic solids to chiral liquid crystals, monolayers and macro- and supra-molecular polymers and assemblies. *Chemical Society Reviews*, **36**, 941–967.

21. Avalos, M., Babiano, R., Cintas, P., Jiménez, J. L., Palacios, J. C. (2004). Symmetry breaking by spontaneous crystallization is it the most plausible source of terrestrial handedness we have long been looking for? A reappraisal. *Origins of Life and Evolution of the Biosphere*, **34**, 391–405.

22. Inoue, Y. (1992). Asymmetric photochemical reactions in solution. *Chemical Reviews*, **92**, 741–770.

23. Rubires, R., Farrera, J. A., Ribo, J. M. (2001). Stirring effects on the spontaneous formation of chirality in the homoassociation of diprotonated meso-tetraphenylsulfonato porphyrins. *Chemistry—A European Journal*, **7**, 436–446.

24. Avalos, M., Babiano, R., Cintas, P., Jimenez, J. L., Palacios, J. C., Barron, L. D. (1998). Absolute asymmetric synthesis under physical fields: Facts and fictions. *Chemical Reviews*, **98**, 2391–2404.

25. Bualek, S., Patumtevapibal, S., Siripitayananon, J. (1981). Odd–even effect in the helical twisting power of chiral molecules in induced cholesteric mesophases. *Chemical Physics Letters*, **79**, 389–391.

26. Yoshizawa, A., Matsuzawa, K., Nishiyama, I. (1995). Coupling between chirality and odd–even effect of twin materials in smectic liquid-crystalline phases. *Journal of Materials Chemistry*, **5**, 2131–2137.

27. Wang, J., Li, C. Y., Jin, S., Weng, X., Van Horn, R. M., Graham, M. J., Zhang, W. B., Jeong, K. U., Harris, F. W., Lotz, B., Cheng, S. Z. D (2010). Helical crystal assemblies in nonracemic chiral liquid crystalline polymers: Where chemistry and physics meet. *Industrial & Engineering Chemistry Research*, **49**, 11936–11947.

28. Henze, O., Feast, W. J., Gardebien, F., Jonkheijm, P., Lazzaroni, R., Leclere, P., Meijer, E. W., Schenning, A. P. H. J. (2006). Chiral amphiphilic self-assembled alpha,alpha′-linked quinque-, sexi-, and septithiophenes: Synthesis, stability and odd–even effects. *Journal of the American Chemical Society*, **128**, 5923–5929.

29. Yablon, D. G., Wintgens, D., Flynn, G. W. (2002). Odd/even effect in self-assembly of chiral molecules at the liquid–solid interface: An STM investigation of coadsorbate control of self-assembly. *Journal of Physical Chemistry B*, **106**, 5470–5475.

30. Tao, F., Goswami, J., Bernasek, S. L. (2006). Self-assembly and odd–even effects of cis-unsaturated carboxylic acids on highly oriented pyrolytic graphite. *Journal of Physical Chemistry B*, **110**, 4199–4206.

31. Amabilino, D. B., Ramos, E., Serrano, J.-L., Sierra, T., Veciana, J. (1998). Long-range chiral induction in chemical systems with helical organization—promesogenic monomers in the formation of poly(isocyanides) and in the organization of liquid crystals. *Journal of the American Chemical Society*, **120**, 9126–9134.

32. Zhi, J. G., Zhu, Z. G., Liu, A. H., Cui, J. X., Wan, X. H., Zhou, Q. F. (2008). Odd-even effect in free radical polymerization of optically active 2,5-bis[(4′-alkoxycarbonyl)-phenyl]styrene. *Macromolecules*, **41**, 1594–1597.

33. Stals, P. J. M, Smulders, M. M. J, Martin-Rapun, R., Palmans, A. R. A, Meijer, E. W. (2009). Asymmetrically substituted benzene-1,3,5-tricarboxamides: Self-assembly and odd-even effects in the solid state and in dilute solution. *Chemistry—A European Journal*, **15**, 2071–2080.

34. Duan, P., Zhu, X., Liu, M. (2011). Isomeric effect in the self-assembly of pyridine-containing L-glutamic lipid: Substituent position controlled morphology and supramolecular chirality. *Chemical Communications*, **47**, 5569–5571.

35. Amabilino, D. B., Ramos, E., Serrano, J. L., Sierra, T., Veciana, J. (2005). Chiral teleinduction in the polymerization of isocyanides. *Polymer*, **46**, 1507–1521.

36. Kajitani, T., Okoshi, K., Yashima, E. (2008). Helix-sense-controlled polymerization of optically active phenyl isocyanides. *Macromolecules*, **41**, 1601–1611.

37. Smulders, M. M. J, Filot, I. A. W, Leenders, J. M. A, van der Schoot, P., Palmans, A. R. A, Schenning, A.P.H.J., Meijer, E. W. (2010). Tuning the extent of chiral amplification by temperature in a dynamic supramolecular polymer. *Journal of the American Chemical Society*, **132**, 611–619.

38. Farina, M. (1987). The stereochemistry of linear macromolecules. *Topics in Stereochemistry*, **17**, 1–111.

39. Okamoto, Y., Yashima, E. (1990). Asymmetric polymerisation of methacrylates. *Progress in Polymer Science*, **15**, 263–298.

40. Nolte, R. J. M (1994). Helical poly(isocyanides). *Chemical Society Reviews*, **23**, 11–19.

41. Green, M. M., Jha, S. K. (1997). The road to chiral amplification in polymers originated in Italy. *Chirality*, **9**, 424–427.

42. Ousaka, N., Inai, Y. (2009). Transfer of noncovalent chiral information along an optically inactive helical peptide chain: Allosteric control of asymmetry of the C-terminal site by external molecule that binds to the N-terminal site. *Journal of Organic Chemistry*, **74**, 1429–1439.

43. Hase, Y., Nagai, K., Iida, H., Maeda, K., Ochi, N., Sawabe, K., Sakajiri, K., Okoshi, K., Yashima, E. (2009). Mechanism of Helix Induction in Poly(4-carboxyphenyl isocyanide) with Chiral Amines and Memory of the Macromolecular Helicity and Its Helical Structures *Journal of the American Chemical Society*, **131**, 10719–10732.

44. Gomar-Nadal, E., Veciana, J., Rovira, C., Amabilino, D. B. (2005). Chiral teleinduction in the formation of a

macromolecular multistate chiroptical redox switch. *Advanced Materials*, **17**, 2095–2098.

45. Ohta, E., Sato, H., Ando, S., Kosaka, A., Fukushima, T., Hashizume, D., Yamasaki, M., Hasegawa, K., Muraoka, A., Ushiyama, H., Yamashita, K., Aida, T. (2011). Redox-responsive molecular helices with highly condensed pi-clouds. *Nature Chemistry*, **3**, 68–73.
46. Shinohara, K. I., Kitami, T., Nakamae, K. (2004). Direct measurement of structural diversity in single molecules of a chiral helical pi-conjugated polymer. *Journal of Polymer Science Part A—Polymer Chemistry*, **42**, 3930–3935.
47. Kumaki, J., Sakurai, S., Yashima, E. (2009). Visualization of synthetic helical polymers by high-resolution atomic force microscopy. *Chemical Society Reviews*, **38**, 737–746.
48. Shinohara, K. I., Kodera, N., Oohashi, T. (2010). Single-molecule imaging of photodegradation reaction in a chiral helical pi-conjugated polymer chain. *Journal of Polymer Science Part A—Polymer Chemistry*, **48**, 4103–4107.
49. Shinohara, K., Kodera, N., Ando, T. (2009). Single-molecule imaging of a micro-brownian motion of a chiral helical pi-conjugated polymer as a molecular spring driven by thermal fluctuations. *Chemistry Letters*, **38**, 690–691.
50. Hernando, J., de Witte, P. A. J, van Dijk, E. M. H. P., Korterik, J., Nolte, R. J. M, Rowan, A. E., Garcia-Parajo, M. F., van Hulst, N. F. (2004). Investigation of perylene photonic wires by combined single-molecule fluorescence and atomic force microscopy. *Angewandte Chemie International Edition*, **43**, 4045–4049.
51. Huang, Y. S., Yang, X. D., Schwartz, E., Lu, L. P., Albert-Seifried, S., Finlayson, C. E., Koepf, M., Kitto. H. J., Ulgut, B., Otten, M. B. J, Cornelissen, J. J. L. M., Nolte, R. J. M, Rowan, A. E., Friend, R. H. (2011). Sequential energy and electron transfer in polyisocyanopeptide-based multichromophoric arrays. *Journal of Physical Chemistry B*, **115**, 1590–1600.
52. Canary, J. W. (2009). Redox-triggered chiroptical molecular switches. *Chemical Society Reviews*, **38**, 747–756.
53. Feringa, B. L. (2007). The art of building small: From molecular switches to molecular motors. *Journal of Organic Chemistry*, **72**, 6635–6652.
54. Vicario, J., Walko, M., Meetsma, A., Feringa, B. L. (2006). Fine tuning of the rotary motion by structural modification in light-driven unidirectional molecular motors. *Journal of the American Chemical Society*, **128**, 5127–5135.
55. Pijper, D., Jongejan, M. G. M, Meetsmia, A., Feringa, B. L. (2008). Light-controlled supramolecular helicity of a liquid crystalline phase using a helical polymer functionalized with a single chiroptical molecular switch. *Journal of the American Chemical Society*, **130**, 4541–4552.
56. Hassey, R., Swain, E. J., Hammer, N. I., Venkataraman, D., Barnes, M. D. (2006). Probing the chiroptical response of a single molecule. *Science*, **314**, 1437–1439.
57. Hassey, R., McCarthy, K. D., Basak, E. S. D, Venkataraman, D., Barnes, M. D. (2008). Single-molecule chiroptical spectroscopy: Fluorescence excitation of individual helicene molecules in polymer-supported thin-films. *Chirality*, **20**, 1039–1046.
58. Hassey-Paradise, R., Cyphersmith, A., Tilley, A. M., Mortsolf, T., Basak, D., Venkataraman, D., Barnes, M. D. (2010). Dissymmetries in fluorescence excitation and emission from single chiral molecules. *Chirality*, **21**, E265–E276.
59. Tang, Y. Q., Cook, T. A., Cohen, A. E. (2009). Limits on fluorescence detected circular dichroism of single helicene molecules. *Journal of Physical Chemistry A*, **113**, 6213–6216.
60. Mugarza, A., Lorente, N., Ordejon, P., Krull, C., Stepanow, S., Bocquet, M. L., Fraxedas, J., Ceballos, G., Gambardella, P. (2010). Orbital specific chirality and homochiral self-assembly of achiral molecules induced by charge transfer and spontaneous symmetry breaking. *Physical Review Letters*, **105**, 115702.
61. Masuhara, H., Nakanishi, H., Sasaki, K. (2003). *Single Organic Nanoparticles*, Springer Verlag, New York.
62. An, B. K., Kwon, S. K., Jung, S. D., Park, S. Y. (2002). Enhanced emission and its switching in fluorescent organic nanoparticles. *Journal of the American Chemical Society*, **124**, 14410–14415.
63. Kommareddy, S., Amiji, M. (2007). Poly(ethylene glycol)-modified thiolated gelatin nanoparticles for glutathione-responsive intracellular DNA delivery. *Nanomedicine—Nanotechnology Biology and Medicine*, **3**, 32–42.
64. Chung, T. W., Cho, K. Y., Nah, J. W., Akaike, T., Cho, C. S. (2002). Chiral recognition of bilirubin by polymeric nanoparticles. *Langmuir*, **18**, 6462–6464.
65. Fu, H. B., Yao, J. N. (2001). Size effects on the optical properties of organic nanoparticles. *Journal of the American Chemical Society*, **123**, 1434–1439.
66. Fu, H., Loo, B. H., Xiao, D., Xie, R., Ji, X., Yao, J., Zhang, B., Zhang, L. (2002). Multiple emissions from 1,3-diphenyl-5-pyrenyl-2-pyrazoline nanoparticles: Evolution from molecular to nanoscale to bulk materials. *Angewandte Chemie-International Edition*, **41**, 962–965.
67. Arikan, M., Pepino, R. T., Ingvarsson, S., Shelykh, I. (2009). Optical spectra of organic quantum dots in the strong coupling regime. *Superlattices and Microstructures*, **47**, 139–144.
68. Gautier, C., Bürgi, T. (2009). In: Amabilino, D. B., ed. *Chirality at the Nanoscale*. Wiley-VCH, Zurich.
69. Xie, R., Xiao, D., Fu, H., Ji, X., Yang, W., Yao, J. (2001). Effect of PVA on the growth and the optical properties of perylene nanocrystals. *New Journal of Chemistry*, **25**, 1362–1364.
70. Debuigne, F., Jeunieau, L., Wiame, M., Nagy, J. B. (2000). Synthesis of organic nanoparticles in different W/O microemulsions. *Langmuir*, **16**, 7605–7611.
71. An, B.-K., Kwon, S.-K. Park, S. Y. (2007). Photopatterned arrays of fluorescent organic nanoparticles. *Angewandte Chemie International Edition*, **46**, 1978–1982.
72. Tamaki, Y., Asahi, T., Masuhara, H. (2002). Nanoparticle formation of vanadyl phthalocyanine by laser ablation of its crystalline powder in a poor solvent. *Journal of Physical Chemistry A*, **106**, 2135–2139.
73. Xiao, D., Yang,W., Yao, J., Xi, L., Yang, X., Shuai, Z. (2004). Size-dependent exciton chirality in (R)-(+)-1,1′-bi-2-naphthol dimethyl ether nanoparticles. *Journal of the American Chemical Society*, **126**, 15439–15444.

74. Xi, L., Fu, H., Yang, W., Yao, J. (2005). CD inversion and fluorescence enhancement in organic nanoparticles of (R)-di-2-naphthylprolinol. *Chemical Communications*, 492–494.
75. Zhang, Y., Peng, A., Wang, J., Yang, W., Yao, J. (2006). Size-tunable exciton chirality and fluorescence emission in (R)-(-)-2,2′-bis-(*p*-toluenesulfonyloxy)-1,1′-binaphthalene nanoparticles. *Journal of Photochemistry and Photobioly A Chemistry*, **181**, 94–98.
76. Chung, T. W., Nah, J. W., Akaike, T., Parka, Y.-H., Cho, C.-S. (2000). Induced circular dichroism of merocyanine dye in the self-assembled polymeric nanoparticles. *Polymer*, **41**, 6415–6419.
77. Yao, H., Sasahara, H., Kimura, K. (2011). Organic porphyrin nanoparticles with induced optical activity: Ion-based synthesis from achiral chromophore and chiral counterions. *Chemistry of Materials*, **23**, 913–922.
78. Brizard, A., Oda, R., Huc, I. (2005). Chirality effects in self-assembled fibrillar networks. *Topics in Current Chemistry*, **256**, 167–218.
79. Shoulders, M. D., Raines, R. T. (2009). Collagen structure and stability. *Annual Review of Biochemistry*, **78**, 929–958.
80. Franchi, M., Trire, A., Quaranta, M., Orsini, E., Ottani, V. (2007). Collagen structure of tendon relates to function. *The Scientific World Journal*, **7**, 404–420.
81. Reichardt, C. (1995). Chiral polymethine dyes: A remarkable but forgotten π system. *Journal of Physical Organic Chemistry*, **8**, 761–763.
82. Eggers, L.Kolster, K.Buss, V. (1997). Helically twisted chiral cyanine dyes: Influence of chromophore length on observed and calculated rotator strengths. *Chirality*, **9**, 243–249.
83. Lohr, A., Wurthner, F. (2008) Evolution of homochiral helical dye assemblies: Involvement of autocatalysis in the "Majority-Rules" effect. *Angewandte Chemie International Edition*, **47**, 1232–1236.
84. Kaiser, T. E., Stepanenko, V., Wurthner, F. (2009). Fluorescent J-aggregates of core-substituted perylene bisimides: Studies on structure–property relationship, nucleation–elongation mechanism, and sergeants-and-soldiers principle. *Journal of the American Chemical Society*, **131**, 6719–6732.
85. de la Torre, G., Claessens, C. G., Torres, T. (2007). Phthalocyanines: Old dyes, new materials. Putting color in nanotechnology. *Chemical Communications*, 2000–2015.
86. Engelkamp, H., Middelbeek, S., Nolte, R. J. M (1999). Self-assembly of disk-shaped molecules to coiled-coil aggregates with tunable helicity. *Science*, **284**,785–788.
87. Fox, J. M., Katz, T. J., Van Elshocht, S., Verbiest, T., Kauranen, M., Persoons, A., Thongpanchang, T., Krauss, T., Brus, L. (1999). Synthesis, self-assembly, and nonlinear optical properties of conjugated helical metal phthalocyanine derivatives. *Journal of the American Chemical Society*, **121**, 3453–3459.
88. Elemans, J. A. A. W., Rowan, A. E., Nolte, R. J. M. (2003). Mastering molecular matter. Supramolecular architectures by hierarchical self-assembly. *Journal of Materials Chemistry*, **13**, 2661–2670.
89. Sly, J., Kasák, P., Gomar-Nadal, E., Rovira, C., Górriz, L., Thordarson, P., Amabilino, D. B., Rowan, A. E., Nolte, R. J. M (2005). Chiral molecular tapes from novel tetra(thiafulvalene-crown-ether)-substituted phthalocyanine building blocks. *Chemical Communications*, 1255–1257.
90. Leclère, Ph., Surin, M., Viville, P., Lazzaroni, R., Kilbinger, A. F. M, Henze, O., Feast, W. J., Cavallini, M., Biscarini, F., Schenning, A. P. H. J., Meijer, E. W. (2004). About oligothiophene self-assembly: From aggregation in solution to solid-state nanostructures. *Chemistry of Materials*, **16**, 4452–4466.
91. Brunsveld, L., Schenning, A. P. H. J., Broeren, M. A. C., Janssen, H. M., Vekemans, J. A. J. M., Meijer, E. W. (2000). Chiral amplification in columns of self-assembled N,N′,N″-tris((*S*)-3,7-dimethyloctyl)benzene-1,3,5-tricarboxamide in dilute solution. *Chemistry Letters*, 292–293.
92. Leclère, Ph., Surin, M., Henze, O., Jonkheijm, P., Biscarini, F., Cavallini, M., Feast, W. J., Kilbinger, A. F. M., Lazzaroni, R., Meijer, E. W., Schenning, A. P. H. J. (2004). Surface-controlled self-assembly of chiral sexithiophenes. *Journal of Materials Chemistry*, **14**, 1959–1963.
93. Sangeetha, N. M., Maitra, U. (2005). Supramolecular gels: Functions and uses. *Chemical Society Reviews*, **34**, 821–836.
94. Ajayaghosh, A., Praveen, V. K. (2007). Pi-organogels of self-assembled *p*-phenylenevinylenes: Soft materials with distinct size, shape, and functions. *Accounts of Chemical Research*, **40**, 644–656.
95. Hirst, A. R., Escuder, B., Miravet, J. F., Smith, D. K. (2008). High-tech applications of self-assembling supramolecular nanostructured gel-phase materials: From regenerative medicine to electronic devices. *Angewandte Chemie International Edition*, **47**, 8002–8018.
96. Wang, Y. J., Tang, L. M., Yu, J. (2009). Supramolecular hydrogel based on low-molecular-weight gelators: From structure to function. *Progress in Chemistry*, **21**, 1312–1324.
97. Suzuki, M., Hanabusa, K. (2010). Polymer organogelators that make supramolecular organogels through physical cross-linking and self-assembly. *Chemical Society Reviews*, **39**, 455–463.
98. Gonzalez-Rodriguez, D., Schenning, A. P. H. J. (2011). Hydrogen-bonded Supramolecular pi-Functional Materials. *Chemistry of Materials*, **23**, 310–325.
99. Smith, D. K. (2009). Lost in translation? Chirality effects in the self-assembly of nanostructured gel-phase materials. *Chemical Society Reviews*, **38**, 684–694.
100. Terech, P., Pasquier, D., Bordas, V., Rossat, C. (2000). Rheological properties and structural correlations in molecular organogels. *Langmuir*, **16**, 4485–4494.
101. Kim, J. U., Schollmeyer, D., Brehmer, M., Zentel, R. (2011). Simple chiral urea gelators, (*R*)- and (*S*)-2-heptylurea: Their gelling ability enhanced by chirality. *Journal of Colloid and Interface Science*, **357**, 428–433.
102. Iavicoli, P., Xu, H., Feldborg, L. N., Linares, M., Paradinas, M., Stafström, S., Ocal, C., Nieto-Ortega, B., Casado, J., López Navarrete, J. T., Lazzaroni, R., De Feyter, S., Amabilino, D. B. (2010). Tuning the supramolecular chirality of one- and two-dimensional aggregates with the number of stereogenic centers in the component porphyrins. *Journal of the American Chemical Society*, **132**, 9350–9362.

103. Yang, X. Y., Zhang, D. Q., Zhang, G. X., Zhu, D. B. (2011). Tetrathiafulvalene (TTF)-based gelators: Stimuli responsive gels and conducting nanostructures. *Science China—Chemistry*, **54**, 596–602.

104. Torres, E., Puigmartí-Luis, J., Pérez del Pino, Á., Ortuño, R. M., Amabilino, D. B. (2010). Use of unnatural β-peptides as a self-assembling component in functional organic fibers. *Organic & Biomolecular Chemistry*, **8**, 1661–1665.

105. Danila, I., Riobé, F., Piron, F., Puigmartí-Luis, J., Wallis, J. D., Linares, M., Ågren, H., Beljonne, D., Amabilino, D. B., Avarvari, N. (2011). Hierarchical Chiral expression from the nano- to meso-scale in synthetic supramolecular helical fibers of a non-amphiphilic C_3-symmetrical π-functional molecule. *Journal of the American Chemical Society*, **133**, 8344–8353.

106. Danila, I., Riobé, F., Puigmartí-Luis, J., Pérez del Pino, Á., Wallis, J. D., Amabilino, D. B., Avarvari, N. (2009). Supramolecular electroactive organogel and conducting nanofibers with C3-symmetrical architectures. *Journal of Materials Chemistry*, **19**, 4495–4504.

107. Beljonne, D., Hennebicq, E., Daniel, C., Herz, L. M., Silva, C., Scholes, G. D., Hoeben, F. J. M, Jonkheijm, P., Schenning, A.P.H.J., Meskers, S. C. J, Phillips, R. T., Friend, R. H., Meijer, E. W. (2005). Excitation migration along oligophenylenevinylene-based chiral stacks: Delocalization effects on transport dynamics. *Journal of Physical Chemistry B*, **109**, 10594–10604.

108. Diaz, D. D., Cid, J. J., Vazquez, P., Torres, T. (2008). Strength enhancement of nanostructured organogels through inclusion of phthalocyanine-containing complementary organogelator structures and in situ cross-linking by click chemistry. *Chemistry—A European Journal*, **14**, 9261–9273.

109. Palmer, L. C., Stupp, S. I. (2008). Molecular self-assembly into one-dimensional nanostructures. *Accounts of Chemical Research*, **41**, 1674–1684.

110. Li, W. G., Wang, H. L. (2004). Oligomer-assisted synthesis of chiral polyaniline nanofibers. *Journal of the American Chemical Society*, **126**, 2278–2279.

111. Weng, S. H., Lin, Z. H., Chen, L. X., Zhou, J. Z. (2010). Electrochemical synthesis and optical properties of helical polyaniline nanofibers. *Electrochimica Acta*, **55**, 2727–2733.

112. Spector, M. S., Price, R. R., Schnur, J. M. (1999). Chiral lipid tubules. *Advanced Materials*, **11**, 337–340.

113. Zhang, Y. N., Zheng, L. X. (2010). Towards chirality-pure carbon nanotubes. *Nanoscale*, **2**, 1919–1929.

114. Yamamoto, T., Fukushima, T., Aida, T. (2008). Self-assembled nanotubes and nanocoils from pi-conjugated building blocks. *Advances in Polymer Science*, **220**, 1–27.

115. Yamamoto, T., Fukushima, T., Kosaka, A., Jin, W., Yamamoto, Y., Ishii, N., Aida, T. (2008). Conductive one-handed nanocoils by coassembly of hexabenzocoronenes: Control of morphology and helical chirality. *Angewandte Chemie International Edition*, **47**, 1672–1675.

116. Isare, B., Linares, M., Zargarian, L., Fermandjian, S., Miura, M., Motohashi, S., Vanthuyne, N., Lazzaroni, R., Bouteiller, L. (2010). Chirality in dynamic supramolecular nanotubes induced by a chiral solvent. *Chemistry—A European Journal*, **16**, 173–177.

117. Zhu, X. F., Li, Y. G., Duan, P. F., Liu, M. H. (2010). Self-assembled ultralong chiral nanotubes and tuning of their chirality through the mixing of enantiomeric components. *Chemistry—A European Journal*, **16**, 8034–8040.

118. Cao, H., Jiang, J., Zhu, X. F., Duan, P. F., Liu, M. H. (2011). Hierarchical co-assembly of chiral lipid nanotubes with an azobenzene derivative: Optical and chiroptical switching. *Soft Matter*, **7**, 4654–4660.

119. Pantos, G. D., Pengo, P., Sanders, J. K. M (2007). Hydrogen-bonded helical organic nanotubes. *Angewandte Chemie International Edition*, **46**, 194–197.

120. Anderson, T. W., Sanders, J. K. M, Pantos, G. D. (2010). The sergeants-and-soldiers effect: Chiral amplification in naphthalenediimide nanotubes. *Organic & Biomolecular Chemistry*, **8**, 4274–4280.

121. Pantos, G. D., Wietor, J. L., Sanders, J. K. M (2007). Filling helical nanotubes with C60. *Angewandte Chemie International Edition*, **46**, 2238–2240.

122. Tamanini, E., Pantos, G. D., Sanders, J. K. M (2010). Ion pairs and C60: Simultaneous guests in supramolecular nanotubes. *Chemistry—A European Journal*, **16**, 81–84.

123. Yao, P. P., Wang, H. F., Chen, P. L., Zhan, X. W., Kuang, X., Zhu, D. B., Liu, M. H. (2009). hierarchical assembly of an achiral pi-conjugated molecule into a chiral nanotube through the air/water interface. *Langmuir*, **25**, 6633–6636.

124. Yang, Y. S., Wan, M. X. (2002). Chiral nanotubes of polyaniline synthesized by a template-free method. *Journal of Materials Chemistry*. **12**, 897–901.

125. Raval, R. (2009). Chiral expression from molecular assemblies at metal surfaces: Insights from surface science techniques. *Chemical Society Reviews*, **38**, 707–721.

126. Elemans, J. A. A. W., De Cat, I., Xu, H., De Feyter, S. (2009). Two-dimensional chirality at liquid–solid interfaces *Chemical Society Reviews*, **38**, 722–736.

127. Amabilino, D. B., De Feyter, S., Lazzaroni, R., Gomar-Nadal, E., Veciana, J., Rovira, C., Abdel-Mottaleb, M. M., Mamdouh, W., Iavicoli, P., Psychogyiopoulou, K., Linares, M., Minoia, A., Xu, H., Puigmartí-Luis, J. (2008). Monolayer self-assembly at liquid–solid interfaces: Chirality and electronic properties of molecules at surfaces. *Journal of Physics. Condensed Matter*, **20**, 184003.

128. Mark, A. G., Forster, M., Raval, R. (2011). Recognition and ordering at surfaces: The importance of handedness and footedness. *ChemPhysChem*, **12**, 1474–1480.

129. Stevens, F., Dyer, D. J., Walba, D. M. (1996). Direct observation of enantiomorphous monolayer crystals from enantiomers by scanning tunneling microscopy. *Angewandte Chemie International Edition*, **35**, 900–901.

130. Walba, D. M., Stevens, F., Clark, N. A., Parks, D. C. (1996). Detecting molecular chirality by scanning tunneling microscopy. *Accounts of Chemical Research*, **29**, 591–597.

131. Schönherr, H., Crego-Calama, M., Vancso, G. J., Reinhoudt, D. N. (2004). Spontaneous resolution of racemic hydrogen-bonded nanoassemblies on graphite revealed by atomic force microscopy. *Advanced Materials*, **16**, 1416–1420.

132. Donovan, P., Robin, A., Dyer, M. S., Persson, M., Raval, R. (2010). Unexpected deformations induced by surface

interaction and chiral self-assembly of Co(II)-tetraphenylporphyrin (Co-TPP) adsorbed on Cu(110): A combined STM and periodic DFT study. *Chemistry—A European Journal*, **16**, 11641–11652.

133. Linares, M., Iavicoli, P., Psychogyiopoulou, K., Beljonne, D., De Feyter, S., Lazzaroni, R., Amabilino, D. B. (2008). Chiral expression at the solid–liquid interface: A joint experimental and theoretical study of the self-assembly of chiral porphyrins on graphite. *Langmuir*, **24**, 9566–9574.

134. Lei, S. B., Deng, K., Yang, Y. L., Zeng, Q. D., Wang, C., Jiang, J. Z. (2008). Electric driven molecular switching of asymmetric tris(phthalocyaninato) lutetium triple-decker complex at the liquid/solid interface. *Nano Letters*, **8**, 1836–1843.

135. O'Neill, M., Kelly, S. M. (2011). Ordered materials for organic electronics and photonics. *Advanced Materials*, **23**, 566–584.

136. Kopp, V. I., Zhang, Z. Q., Genack, A. Z. (2003). Lasing in chiral photonic structures. *Progress in Quantum Electronics*, **27**, 369–416.

137. Savoini, M., Biagioni, P., Lakhwani, G., Meskers, S. C. J, Duo, L., Finazzi, M. (2009). Near-field circular polarization probed by chiral polyfluorene. *Optics Letters*, **34**, 3571–3573.

138. Tsukruk, V. V. (1997). Assembly of supramolecular polymers in ultrathin films. *Progress in Polymer Science*, **22**, 247–311.

139. Ariga, K., Richards, G. J., Ishihara, S., Izawa, H., Hill, J. P. (2010). Intelligent chiral sensing based on supramolecular and interfacial concepts. *Sensors*, **10**, 6796–6820.

140. Xu, C., Ng, S. C., Chan, H. S. O (2008). Self-assembly of perfunctionalized β-cyclodextrins on a quartz crystal microbalance for real-time chiral recognition. *Langmuir*, **24**, 9118–9124.

141. Iavicoli, P., Simón-Sorbed, M., Amabilino, D. B. (2009). Surface aggregate morphology of chiral porphyrins as a function of constitution and amphiphilic nature. *New Journal of Chemistry*, **33**, 358–365.

142. Iavicoli, P., Linares, M., Pérez del Pino, A., Lazzaroni, R., Amabilino, D. B. (2008). Self-assembly of a chiral porphyrin at surfaces. *Superlattices and Microstructures*, **44**, 556–562.

143. Peng, X. B., Huang, Y. Y., Gao, C., Peng, J. B., Komatsu, N., Osuka, A., Cao, Y. (2010). Macroscopic films of porphyrin nanowell-arrays via solvent diffusion-induced self-assembly. *Journal of Physical Chemistry C*, **114**, 18449–18454.

144. Klemm, D., Kramer, F., Moritz, S., Lindstrom, T., Ankerfors, M., Gray, D., Dorris, A. (2011). Nanocelluloses: A new family of nature-based materials. *Angewandte Chemie International Edition*, **50**, 5438–5466.

145. Sancho, R., Minguillón, C. (2009). The chromatographic separation of enantiomers through nanoscale design. *Chemical Society Reviews*, **38**, 797–805.

146. Chen, J. L., Hsieh, K. H. (2011). Nanochitosan crosslinked with polyacrylamide as the chiral stationary phase for open-tubular capillary electrochromatography. *Electrophoresis*, **32**, 398–407.

147. Kawagoe, Y., Fujiki, M., Nakano, Y. (2010). Limonene magic: Noncovalent molecular chirality transfer leading to ambidextrous circularly polarised luminescent pi-conjugated polymers. *New Journal of Chemistry*, **34**, 637–647.

148. Tseng, W.-H., Chen, C.-K., Chiang, Y.-W., Ho, R.-M., Akasaka, S., Hasegawa, H. (2009). Helical nanocomposites from chiral block copolymer templates. *Journal of the American Chemical Society*, **131**, 1356–1357.

149. Haseba, Y., Kikuchi, H., Nagamura, T., Kajiyama, T. (2005). Large electro-optic Kerr effect in nanostructured chiral liquid-crystal composites over a wide temperature range. *Advanced Materials*, **17**, 2311–2315.

150. Njikang, G., Liu, G.J., Hong, L.Z. (2011). chiral imprinting of diblock copolymer single-chain particles. *Langmuir*, **27**, 7176–7184.

151. Bradshaw, D., Claridge, J. B., Cussen, E. J., Prior, T. J., Rosseinsky, M. J. (2005). Design, chirality, and flexibility in nanoporous molecule-based materials. *Accounts of Chemical Research*, **38**, 273–282.

152. Pilla, P., Cusano, A., Cutolo, A., Giordano, M., Mensitieri, G., Rizzo, P., Sanguigno, L., Venditto, V., Guerra, G. (2009). Molecular sensing by nanoporous crystalline polymers. *Sensors*, **9**, 9816–9857.

153. Petraccone, V., de Ballesteros, O. R., Tarallo, O., Rizzo, P., Guerra, G. (2008). Nanoporous polymer crystals with cavities and channels. *Chemistry of Materials*, **20**, 3663–3668.

154. Han, C.P., Hou, X., Zhang, H.C., Guo, W., Li, H.B., Jiang, L. (2011). Enantioselective recognition in biomimetic single artificial nanochannels. *Journal of the American Chemical Society*, **133**, 7644–7647.

155. Mukhopadhyay, P., Wipf, P., Beratan, D. N. (2009). Optical signatures of molecular dissymmetry: Combining theory with experiments to address stereochemical puzzles. *Accounts of Chemical Research*, **42**, 809–819.

4

BIOCHEMICAL NANOMATERIALS BASED ON POLY(ε-CAPROLACTONE)

Irakli Javakhishvili and Søren Hvilsted

4.1 INTRODUCTION

Owing to its excellent biocompatibility, tunable degradation kinetics, high permeability to drug molecules, and full excretion from an organism upon completion of a treatment, poly(ε-caprolactone) (PCL) has enjoyed extensive application in the field of surgical sutures, fixation devices, tissue engineering, and controlled drug delivery and release systems [1]. PCL is approved by the US Food and Drug Agency (FDA). PCL exhibits splendid rheological and viscoelastic properties that facilitate manufacturing of a vast range of implants and devices. PCL nanospheres encapsulating active therapeutic agents have been prepared, as well as used in ocular and oral drug delivery [1]. In addition, PCL can be exploited as a versatile and resourceful multifunctional building block for construction of micellar nanoparticulate drug delivery vehicles. Block copolymer micelles—spherical, nanosized supramolecular assemblies of amphiphilic copolymers with core-shell type architecture—afford enhanced bioavailability, diminished toxic effects, and increased permeability through the physiological barriers compared to the free drugs [2]. The core of the pharmaceutical polymeric micelle should display a high loading capacity and a controlled drug-release profile. Furthermore, good compatibility between the compartmentalized drug and the polymer chains that form the micellar core is necessary. The micelle corona should afford effective steric stabilization, along with reactive functional groups for the modification of the micellar surface. The corona dictates micelle hydrophilicity, charge, and valency. These properties define: pharmacokinetics, biocompatibility, circulation time, biodistribution, surface adsorption, interaction with biosurfaces, and targetability [3]. Poly(ethylene glycol) (PEG) has been vastly employed as the hydrophilic block because of its low cost, low toxicity, effectual steric stabilization, and protection and also because it has been approved by regulatory agencies. The hydrophobic block is often comprised of units of propylene oxide, aspartic acid, β-benzoyl-L-aspartate, γ-benzyl-L-glutamate, D,L-lactic acid, and ε-caprolactone (ε-CL) [2]. Thus, PEG-*b*-PCL micelles have been detected in various organelles of a cell as a result of nonspecific uptake that considerably enhanced internalization of the compartmentalized model agent compared to the free agent [4].

However, polymeric micelles tend to disintegrate upon dilution as the polymer concentration falls below the critical micelle concentration (cmc). Therefore, it would be advantageous to stabilize the micelle core by trapping and arresting the radiating chains and thus creating a unimolecular container with rigid structure [5].

This is dedicated to introduction of the synthetic layout for preparation of functional biomaterials based on PCL intended for drug delivery applications. Issues such as biocompatibility and robustness of the aggregates—ways of stabilization of the core–shell type morphologies—will be addressed. Furthermore, the cascade of high-fidelity reactions leading to the library of heterobifunctional PCL building blocks will be discussed.

4.2 LIVING POLYMERIZATION OF ε-CAPROLACTONE

Polymeric drug delivery devices often consist of aliphatic polyesters, polyethers, polyamides, polyanhydrides, and

Organic Nanomaterials: Synthesis, Characterization, and Device Applications, First Edition. Edited by Tomás Torres and Giovanni Bottari.

polyurethanes. Polyesters intended for the drug and gene delivery or tissue engineering applications must display controlled biodegradation rate, bioadherence, and suitable glass transition temperature and crystallinity [6, 7]. Ring-opening polymerization (ROP) allows synthesis of the polymers with predetermined molecular weight characteristics under mild reaction conditions from lactones of various ring sizes and with or without different pendant functional groups. Furthermore, plethora of functional initiators permits incorporation of specific end functionalities [6, 7]. Thus, ROP facilitates construction of the polyesters with tailor-made properties. Cyclic monomers that have been polymerized via ROP include ε-CL, δ-valerolactone (δ-VL), β-butyrolactone, lactide, and glycolide [7].

ROP encompasses mechanisms involving anionic, cationic, and coordinative initiating species [6]. While ionic species engage in adverse inter- and intramolecular transesterification reactions responsible for decreasing the molecular weight and broadening the molecular weight distribution (MWD), the organometallic compounds that incorporate metals possessing *d*-orbitals of favorable energy establish good control over the polymerization reaction. Provided that the reaction conditions are appropriately adjusted, ROP of lactones and lactides is a living/controlled process resulting in the polymer of predetermined molecular weight and narrow MWD. ROP involves two major mechanistic routes defined by the organometallics used: (a) monomer activation by catalyst complexation with the carbonyl group of the monomer followed by nucleophilic attack by any nucleophile present in the polymerization mixture and (b) coordination–insertion, that is, metal alkoxide's coordination with the carbonyl group of the monomer followed by the rupture of the acyl–oxygen bond of the monomer and insertion into the metal–alkoxide bond [6].

The following paragraphs elucidate advantages and limitations of the metal-based catalytic systems in the ROP of ε-CL and thus justify the catalyst of choice for our investigations.

Aluminum- and tin-based systems are the ones most frequently used in ROP of ε-CL [8]. Aluminum is not as active a catalyst as many other metals employed in the ROP of lactones, but it offers good control over the polymerization reaction [8]. One of the most scrutinized catalytic systems for the ROP of ε-CL is aluminum (III) isopropoxide, which affords excellent control while suppressing intramolecular transesterification and backbiting reactions [8]. The major shortcoming of using $Al(O^iPr)_3$ is the forced presence of the isopropoxide chain end. However, it can be remedied by employing triethyl aluminum as a catalyst and a functional alcohol as an initiator [7]. In general, aluminum alkoxides (Et_2AlOR, $EtAl(OR)_2$, and $Al(OR)_3$) provide high selectivity, good control over the molecular weight and chain end functionalities of polyesters. α-Terminus of the polyester chain is capped by RO—the alkoxy group of the initiator—while the hydrolysis of the propagating chains affords ω-hydroxyl end group. However, different nucleophiles can be employed to terminate the chains with desirable functionality. High degree of control over the polymerization reaction allows preparation of various macromolecular architectures including comb, star, graft, and hyperbranched (co)polyesters [6]. Numerous aluminum Schiff base complexes have also been utilized in the ROP of ε-CL [8].

Stannous (II) ethylhexanoate (tin octoate, $Sn(Oct)_2$) is the most frequent catalyst of choice for the ROP of ε-CL. The advantages that tin octoate offers are commercial availability, ease of handling, efficiency, and solubility in numerous organic solvents. In combination with an alcohol, tin octoate allows synthesis of the polymer in a controlled manner [8]. The polymerization mechanism consists of (a) transformation of $Sn(Oct)_2$ into tin alkoxide upon reaction with an alcohol and a (b) subsequent coordination–insertion step. Hence, the molecular weight of the polymer can be tuned by adjusting the monomer-to-alcohol molar ratio [6]. Thus, together with the alcohol initiators with embedded functionalities, this simple catalytic system affords excellent control over the molecular weight and ensures near-quantitative integration of end groups [7]. Furthermore, it has been approved by FDA as a food additive [6]. The disadvantage of tin octoate is the necessity of applying elevated temperatures to the reaction mixture provoking inter- and intramolecular esterification reactions that broaden the MWD [8]. Other tin-based catalytic complexes such as tin triflate and dibutyltin dimethoxide, have also been employed [8]. An extensive survey of the metal catalysts for the polymerization of ε-CL is presented in a couple of recent reviews [8, 9].

Besides polymerizations mediated by the metal-based catalytic complexes, enzymatic ROP (eROP) has also been exercised for preparation PCL. Since it is crucial for biomedical applications to avoid toxic metallic contaminants, the eROP seems to be an attractive alternative. Lipase-catalyzed eROP of ε-CL has been investigated thoroughly. The degree of control in eROP is inferior to that attained in chemical ROP, with polydispersity index (PDI) being higher than 2 [6]. However, the chemoselective enzymatic catalyst affords incorporation of various terminal functional groups without protection and deprotection chemistries [8].

"All-organic" initiating systems include *n*-BuOH/$HCl{\cdot}Et_2O$ combination—cationic catalyst for the controlled ROP of ε-CL and δ-VL [6, 10]. The nucleophilic catalysts such as phosphines, tertiary amines, *N*-imidazolium carbenes, and thiazaolium carbenes have been used in the ROP of lactide initiated by alcohols. The mechanism involves nucleophilic attack of the catalyst on the carbonyl group of the monomer, resulting in the formation of an active intermediate that can react with an alcohol [6]. Aza-compounds, phosphazene bases, and various organic acids can also serve as catalysts in ROP of ε-CL [8].

Derivatization of aliphatic polyesters is more demanding and painstaking than that of nondegradable polymers. The reaction conditions that result in rupture of the ester bonds may lead to premature polymer decomposition. Therefore, substantial investigations have been conducted for developing the ways for preparation of functional polyesters. Incorporation of the functional groups along the polyester backbone may be achieved through polymerization of the monomers containing functional groups in α- or γ-position, or by attaching functionalities in the α-position of the carbonyl of the preformed chain [6].

ε-CL substituted by acrylate [11], protected hydroxyl group [12], bromide [13], and PEG [14] has been prepared, and then homo- and copolymerized by ROP. ε-CL with a pendant acrylate group is a difunctional monomer that can be polymerized by both ROP and controlled radical polymerization (CRP) in a living/controlled way [11]. PCL with protected hydroxyl groups along the chain is an excellent macromolecular scaffold for initiation of ROP of various monomers. Making use of orthogonal protecting groups allows selective deprotection and grafting from the PCL macroinitiator [12]. Introducing these functional groups along the polymer chain allows myriads of chemical transformations resulting in the polyesters with desirable properties. For instance, brominated homo- and copolymers of ε-CL have been quaternized by pyridine forming degradable polycations that form nanoparticles with a strong interaction with plasmid DNA [15]. Incorporation of pendant hydroxyl groups leads to the polyesters with amphiphilic nature. The amphiphilic PCL-*g*-PEG with a comb-like structure comprising biodegradable hydrophobic backbone has been prepared by the ROP of ε-CL bearing a PEG substituent [16]. The major downside of using substituted lactones is the necessity of protecting those functional groups that may interfere with metal alkoxides during the ROP. Also, the deprotection protocol must be sufficiently mild in order to retain the polyester backbone intact [6].

4.3 COPOLYMERS WITH POLY(ε-CAPROLACTONE)

The previous section provided some insight about the possibilities of constructing PCL with the same or different functional groups at the chain ends via ROP. It has been explicitly stated and supported by overwhelming amount of experimental data that length and functionality of the PCL blocks can be adjusted with great precision. This allows preparation of various polymeric architectures featuring PCL building units by combination of mechanistically distinct polymerization techniques. A heterobifunctional initiator concept for combining CRP technique—nitroxide-mediated polymerization (NMP) or atom transfer radical polymerization (ATRP)—with the ROP was originally introduced by Hawker et al. [17]. Block copolymers of ε-CL and styrene (St), and ε-CL and methyl methacrylate (MMA) have been prepared by employing hydroxy-substituted alkoxyamine or 2,2,2-tribromoethanol as dual initiators. There was no limiting factor defining the sequence of the polymerizations. Jakubowski et al. [18] applied the dual initiator strategy utilizing 2-hydroxyethyl 2-bromoisobutyrate (HEBI) to initiate ATRP of *N*,*N*-dimethylamino-2-ethyl methacrylate (DMAEMA) on one hand and initiate ROP of ε-CL on the other. In this case, the sequence of the polymerization reactions had dramatic importance: PDMAEMA block had to be synthesized after the PCL block, since the ROP of ε-CL could not be carried out in the presence of the PDMAEMA due to the complexation of the amine groups with $Sn(Oct)_2$ that served as the catalyst for ROP. Another example of linking two different polymerization types is the introduction of 2-bromopropionyl moiety at the PEG termini [19] to obtain a difunctional macroinitiator (MI) for ATRP of St and thus provide a convenient method to produce well-defined amphiphilic block copolymers. These MIs have since been extensively used for preparation of complex multiblock polymer materials [20, 21].

A large number of amphiphilic block, star and graft copolymers where PCL serves as the hydrophobic component has been prepared. The initiating strategy or moiety and ω-terminal function determine the resulting architecture. While MIs normally result in block copolymers, well-defined multifunctional initiators on a relatively small core molecule give rise to star polymers. Finally, graft or comb-shaped polymers will result if many initiating sites co-reside on a parent polymer backbone.

4.3.1 Block Copolymers

A variety of block copolymers based on PCL with potential in biomedical application have been developed. Table 4.1 summarizes various examples and provides the most important features of the specific reaction sequence including the achieved morphology and intended biomedical function if that is provided.

The block copolymer dextran-S-S-PCL is prepared by coupling PCL-SH with orthopyridyl disulfide end-capped dextran. Doxorubicin-loaded micelles assembled from these diblock copolymers exhibit rapid internalization and drug release inside cells, which is ascribed to the disulfide bridge rupture induced by glutathione present in the intracellular compartments [22]. In another approach, ROP of ε-CL catalyzed by $Sn(Oct)_2$ is performed using amino-terminated poly(Z-L-lysine) as the MI. After deblocking of the amine functional groups, poly(L-lysine)-*b*-PCL forms core–shell type and vesicular morphologies, depending on the ratio of hydrophilic and hydrophobic blocks [23]. The pH- and temperature-responsive properties of PDMAEMA have been exploited in combination with valuable features offered by PCL. PDMAEMA can be absorbed by endocytosis and

TABLE 4.1. Amphiphilic Block Copolymers Incorporating PCL

Hydrophilic Component	Polymerization Details[a]	Resulting Morphology	Intended Functions	Reference
Dextran	Coupling polymer blocks	Micelles	Doxorubicin loading and glutathione triggered release	22
Poly(L-lysine)	Poly(L-lysine) MI is employed in ROP of ε-CL	Core–shell type, vesicular		23
PDMAEMA	ATRP from the PCL MI	Micelles of varying size	Nonviral DNA vector	24
P(NIPAAm-*co*-DMAAm)	The copolymer serves as MI for ROP of ε-CL	Micelles with LCST ~40°C	Doxorubicin loading	25
PVP	PVP-OH is employed as MI	Spherical micelles in the range of 30–80 nm		26
PVP	MI strategy based on PVP-OH	Spherical micelles with an average diameter of 105 nm	Pyrene loading	27
PHPMAm or PVP	Free radical polymerization in the presence of HS-PCL-SH	Micelles in the range of 30–200 nm	Compartmentalization of doxorubicin and amphotericin B	28
mPEG, PDMAEMA	mPEG-OH is employed in the ROP, which is then followed by ATRP	Spherical micelles and vesicles of about 100 nm; The size depends on pH	Naproxen loading	29
PGAEMA	ATRP from Br-PCL-Br	Aggregates of 120–150 nm in aqueous solutions		30
PVBG	ROP by 1-pyrenemethanol, ATRP of the protected monomer, deblocking	Spherical micelles with the diameter ranging from 20 to 90 nm; hollow nanospheres after removal of PCL core		31
PMPC	ROP followed by ATRP	Vesicles in the range of 40–500 nm with 20 nm thick membranes	Silicified vesicles: organic/inorganic nanostructures	32
mPEG, PAA	ROP initiated by mPEG-OH followed by ATRP	Vesicles with hydrodynamic diameter from about 180 nm to 240 nm	BSA adsorption as well as entrapment	33
mPEG, PPEEA	ROP	Spherical micelles	siRNA binding and delivery into cellular cytoplasm for gene silencing	34
PG2MA	ROP employing HEBI, ATRP of protected monomer	Micelles – 45–88 nm, vesicles – 120 nm		35
P(mPEGMA-*co*-GAMA)	ROP catalyzed by $Al(O^iPr)_3$, ATRP of the protected monomer	Spherical micelles with the diameter of 150 ± 30 nm	"Cluster effect": multivalent recognition between proteins and carbohydrate residues	36
PEG, P2VP	Anionic polymerizations, ROP, biotin ligation	Spherical and flower-like micelles (34 and 27 nm)	pH-responsive drug delivery device with targeting vectors	37
PGMA	eROP catalyzed by Novozyme 435, ATRP	Spherical aggregates		38
P4VP	Dual initiator: ROP followed by NMP	Spherical and rod-like nanostructures	Transport of $AuCl_4^-$ from aqueous to organic phase	39
PAEMA	ROP followed by ATRP	Vesicles with diameter of about 190 ± 49 nm and membrane thickness of about 12.4 ± 0.8 nm	Reduction of $AuCl_4^-$ on the vesicle surface; silicification. Potential in drug delivery	40

[a]Since ε-CL is normally polymerized by ROP, this is not mentioned, and only in special cases a remark is added.

can be utilized as a nonviral DNA vector. The following synthetic protocol was exercised: ROP of ε-CL initiated by $Al(O^iPr)_3$, end-functionalization with 2-bromoisobutyryl bromide (B*i*BB) to obtain the ATRP initiating sites, and subsequent CuBr/1,1,4,7,10,10-hexamethyltriethylenetetramine (HMTETA)-catalyzed ATRP of DMAEMA. The size of the micelles depends on the copolymer concentration, pH, and the relative amount of PCL [24]. Free radical copolymerization of *N*-isopropylacrylamide (NIPAAm) and *N,N*-dimethylacrylamide (DMAAm) in the presence of the chain transfer agent 2-mercaptoethanol provided hydroxyl end-capped P(NIPAAm-*co*-DMAAm). After fractionation, the copolymers with the PDI <1.80 have been obtained [25]. These MIs were then chain-extended by the $Sn(Oct)_2$-catalyzed ROP of ε-CL in xylene at 140°C. The polymeric micelle obtained from this block copolymer displayed the lower critical solution temperature (LCST) at 40°C. The micelle was successfully loaded with the anticancer drug doxorubicin. However, no enhancement in the doxorubicin release at 42.5°C from that at 37°C was detected. Poly(vinyl-2-pyrrolidone) (PVP) is a hydrophilic, biocompatible polymer, which exhibits low cytotoxicity, shields DNA from extracellular nucleases, and imparts greater stability to peptides in vivo. Hydroxyl end-capped PVP has been prepared by free radical polymerization using 2-mercaptoethanol as a chain transfer agent [26]. It was then reacted with $AlEt_3$ to obtain the MI for the ROP of ε-CL. The ROP was performed in CH_2Cl_2 at 40°C. Diblock copolymers with narrow MWD (PDI = 1.15–1.35) have been obtained. Micelles have been prepared by a diafiltration method. The diameters of these self-assembled structures have been assessed by dynamic light scattering (DLS). Depending on the ratio of hydrophobic and hydrophilic blocks, the size of the micelles varied between about 30 and 80 nm. Spherical aggregates have been visualized by transmission electron microscopy (TEM). The size distribution of the micelles obtained from TEM was larger compared to that obtained from DLS due to the micelle aggregation upon drying [26].

Vinyl-2-pyrrolidone has also been polymerized by cobalt-mediated radical polymerization employing Co(II) acetylacetonate as catalyst [27]. The ω-end group of the PVP was modified to incorporate a hydroxyl functional group by making use of the free radical initiator bearing a hydroxyl function. PVP-OH was then exploited as a MI in the ROP of ε-CL catalyzed by $Sn(Oct)_2$. The polymerization was performed in toluene at 110°C. Size exclusion chromatography (SEC) confirmed successful formation of the diblock copolymer and high initiation efficiency of the PVP-OH MI. The cmc has been estimated by monitoring the pyrene incorporation. Spherical micelles have been observed in TEM investigations. Average diameter of 105 nm has been determined by the DLS experiments [27].

Symmetrical or ABC-type triblock copolymers with central PCL blocks have also been prepared by different strategies. The bioresorbable polymeric chain transfer agent HS-PCL-HS has been synthesized by derivatizing HO-PCL-OH with 3,3′-dithiobis(propionic acid), and then reducing the disulfide bonds by 1,4-dithiothreitol [28]. α,ω-PCL-dithiol turned out to be stable with no disulfide bridges forming due to the air oxidation. Free radical polymerization of *N*-(2-hydroxypropyl)methacrylamide (HPMAm) in the presence of HS-PCL-SH resulted in the amphiphilic triblock copolymer PHPMAm-*b*-PCL-*b*-PHPMAm. The PHPMAm block is hydrophilic, biocompatible, and nonimmunogenic. In a similar fashion, vinyl-2-pyrrolidone was polymerized to form PVP-*b*-PCL-*b*-PVP. The PVP shell serves as both lyoprotectant and cryoprotectant. Micelles obtained from these triblock copolymers displayed the capacity to entrap anticancer drugs such as doxorubicin and amphotericin B [28].

Monomethoxy PEG (mPEG) was used as the MI for the conventional ROP of ε-CL in toluene at 120°C [29]. The terminal hydroxyl group of the block copolymer mPEG-*b*-PCL was converted into an ATRP initiating site by reacting with B*i*BB. The third block was then formed by ATRP of DMAEMA mediated by the CuBr/*N,N,N′,N″,N″*-pentamethyldiethylenetriamine (PMDETA) catalytic complex in DMF at 50°C. Narrow MWD (PDI = 1.20–1.28) have been achieved. A fluorescence probe method was exercised to determine the critical aggregation concentration of the block copolymers. TEM investigations revealed spherical micelles and vesicles with an average size of about 100 nm, depending on the composition of the triblock copolymers. Effect of the pH on the micelle size has been elucidated by DLS: Increase of the pH resulted in increase of the particle size. The micelles have been loaded with an anti-inflammatory drug naproxen, and the drug release profiles have been examined [29].

Thorough investigation of PEG-*b*-PCL self-assembled nanostructures revealed that the average block proportions determine the dominant morphology of the block copolymer aggregates in aqueous solutions. Giant and flexible worm micelles can be prepared. It was demonstrated that "chain-end cleavage" of the PCL block triggers worm-to-sphere transition [41].

These few examples comprise just a tiny fraction of the vast pool of amphiphilic block copolymers with PCL assuming the functions of a hydrophobic constituent. One can easily perceive that depending on the composition and weight fractions of individual blocks, these amphiphiles can assemble into various nanoscopic structures sizes of which can be manipulated by external stimuli. Thus, virtually innumerable combinations can be realized. It is remarkable that the overwhelming majority of these aggregates are intended to function as drug delivery vehicles.

4.3.2 Star Copolymers

The star configuration is attractive for many applications since generally it is possible to prepare high-molecular-weight

TABLE 4.2. Amphiphilic Star Polymers Based on PCL

Hydrophilic Component	Polymerization Details[a]	Resulting Morphology	Intended Functions	Reference
PEG	Per-6-(*tert*-butyldimethylsilyl)-β-cyclodextrin multifunctional initiator, ROP, CuAAC	Spherical and worm-like aggregates	Conjugation of ibuprofen	42
PmPEGMA	Pentaerythritol and dipenta-erythritol multivalent initiators, ROP, ATRP	Star block copolymer unimers	Compartmentalization of dyes and commercially available drugs	43
PAA	Boltron HBP H40 multifunctional initiator, ROP, ATRP		Encapsulation of fragrances	44
PAA	Hydroxypropylcellulose (HPC) multifunctional initiator, ROP, ATRP, cross-linking		Encapsulation of pyrene in water	45
PAA	Per-6-(*tert*-butyldimethylsilyl)-β-cyclodextrin, ROP, ATRP	Spherical micelles with the diameter from 10 to 50 nm, wormlike aggregates		46
mPEG, PEEP	mPEG-(N_3)OH, CuAAC, ROP of EEP			47
PPEGA or NIPAAm	RAFT, ROP, CuAAC			48
mPEG	mPEG serves as the macroinitiator for ROP, chain end transformation, Michael addition to hy-PEI		Gene delivery	49
PGAMA	Tetrafunctional initiator, ROP, ATRP	Micelles in water		50
PGAEMA	ROP from PAMAM-$(OH)_{12}$, ATRP	Bilayered vesicles (~ 100 nm), spherical micelles (~ 150 nm), large compound aggregates (~305 nm)	Selective binding to Concavalin A	51
mPEG	ROP of ε-CL, γ-Et_3SiO-ε-CL, and α-A-ε-CL. Cross-linking affords cyclic copolyester. Deprotection and mPEG grafting—sun-shaped polymer	Spherical micelles		52
PDMAEMA	Pentafunctional initiator, ROP, RAFT	Spherical micelles with pH-responsive and thermoresponsive shell		53
PGAMA	ROP from a tetrafunctional sugar moiety with a blocked primary alcohol; capping, deblocking and ATRP	Aggregates with the average size below 80 nm		54

[a] Since ε-CL is normally polymerized by ROP, this is not mentioned and only in special cases a remark is added.

polymers with relatively low viscosity as compared to their linear counterparts. Furthermore, amphiphilic star polymers are characterized by a core–shell-type structure, abundance of functionalities, and tunable surface properties (Table 4.2).

Per-6-(*tert*-butyldimethylsilyl)-β-cyclodextrin has been employed as the multifunctional initiator in the ROP of ε-CL catalyzed by $Sn(Oct)_2$ to generate 14-arm PCL star polymer. The terminal hydroxyl groups of PCL arms are conjugated with the model drug ibuprofen. Cu (I)-mediated azide-alkyne 1,3-cycloaddition (CuAAC) is utilized to link 7 PEG chains to the cyclodextrin core. These miktoarm star copolymers assemble into spherical as well as worm-like aggregates in water [42].

Pentaerythritol and dipentaerythritol were used as the initiators for preparation of the PCL block resulting in 4- and 6-arm star-shaped structures, respectively. The derivatization of the hydroxyl end groups of the PCL arms by B*i*BB is followed by CuBr–PMDETA-mediated ATRP of poly(ethylene glycol) methyl ether methacrylate (mPEGMA). The star block copolymers exist as unimers in water. Compartmentalization of dyes and commercially available drugs was demonstrated [43].

In another development, Boltron HBP H40 hyperbranched polyester served as the multifunctional initiator in ROP of ε-CL [44]. The hydroxyl end groups of the obtained star-shaped polymer are converted to the activated halides by reacting with B*i*BB. ATRP of *tert*-butyl acrylate (*t*BA) is catalyzed by CuBr–2,2′-bipyridyl complex. Removal of *tert*-butyl protecting groups leads to amphiphilic star block copolymer with a poly(acrylic acid) (PAA) corona capable of encapsulating small hydrophobic and volatile molecules such as fragrances [44]. A similar approach was taken when HPC was employed as the multifunctional initiating scaffold [45]. The resulting HPC-*g*-PCL-*b*-PAA was cross-linked with a diamine. The star copolymers had the capacity to encapsulate pyrene in water.

ROP of ε-CL has been conducted in bulk at 125°C employing per-6-(*tert*-butyldimethylsilyl)-β-cyclodextrin and $Sn(Oct)_2$ as the initiator and catalyst, respectively. Relatively narrow MWD (PDI $\leq$ 1.26) has been obtained. The terminal hydroxyl groups of the PCL chains have then been blocked by reacting them with acetyl chloride. Desilylation of the seven primary hydroxyl groups of the β-cyclodextrin moiety has been carried out under mild conditions ($BF_3{\cdot}(C_2H_5)_2O$). Afterwards, they have been converted into the ATRP initiating sites by treating them with B*i*BB. This MI has been utilized in CuBr–PMDETA-mediated ATRP of *t*BA. P*t*BA arms of various degree of polymerization (DP) have been grafted from the MI. Selective removal of the *tert*-butyl groups afforded the amphiphilic miktoarm star copolymer bearing 14 PCL and 7 PAA arms. The cmc values of these amphiphiles were fairly low (0.54–2.34 $mg{\cdot}L^{-1}$). Depending on the weight fraction of the PAA block, spherical micelles with the number average mean diameters from 10 to 50 nm were observed; worm-like aggregates have also been identified by DLS and TEM [46].

An amphiphilic ABC 3-miktoarm star terpolymer was prepared by combination of ROP of ε-CL and "click" chemistry [47]. Here, an alkyne-terminated PCL was "clicked" in the CuAAC reaction with the PEG chain containing both azide and hydroxyl functionalities in the same terminus, which had emerged from the original epoxy end group. The secondary alcohol at the junction point was finally utilized as the initiating site in ROP of 2-ethoxy-2-oxo-1,3,2-dioxaphospholane (EEP) [47].

Another new "click" chemistry approach leading to an AB_2-type miktoarm star polymer employed appropriately functionalized polymers prepared by reversible addition-fragmentation chain transfer (RAFT) polymerization and ROP [48]. An azide-functional RAFT agent was utilized in CRP of poly(ethylene glycol) acrylate (PEGA) and NIPAAm to yield azide end-capped polymers. 2,3-Dihydroxypropyl prop-2-ynylcarbamate was synthesized and used as a difunctional initiator carrying a terminal alkyne function in the ROP of ε-CL mediated by $Sn(Oct)_2$ (PDI $<$1.30). Thus, the azide end-capped polymers were coupled with the alkyne-functional PCL via the "click" reaction catalyzed by CuBr-PMDETA complex. The star polymers had low PDI ($\sim$1.3) and were expected to have potential applications as drug delivery biomaterials [48].

The amphiphilic star copolymers that have been briefly touched upon in this section exhibit interesting solution properties: They may exist as unimers or may assemble into a multitude of shapes. Usually, the purpose is selective or even targeted drug delivery since these structures can accommodate various small organic molecules on one hand and may carry targeting motifs on the surface on the other. Moreover, star polymers offer improved stability compared to micellar structures since they are characterized by covalently linked arms, and do not depend on thermodynamic parameters. Provided that they offer similar beneficial properties such as encapsulation, shielding, smuggling, and controlled release of drugs like conventional micelles, stars are a step forward owing to their sturdy constitution eliminating premature rupture of the device in vivo.

4.3.3 Graft Copolymers

The graft copolymers are yet another topological variety which exploits naturally occurring or synthetic functional polymers, and in combination with PCL offers an assortment of amphiphiles capable of self-assembly into potentially useful nanostructures. Table 4.3 lists a few examples of graft copolymers and some of their applications.

ROP of α-propargyl-ε-CL together with ε-CL has been followed by conjugation of the α-azido-PDMAEMA chains to the polyester backbone by CuAAC. Water-soluble copolymers were prepared by quaternization of the amino groups with methyl iodide. The graft copolymers assembled in core–shell-type structures. These micelles can be loaded with a hydrophobic drug clofazimine [55]. Also, the biodegradable and biocompatible polysaccharide—chitosan—with mucoadhesive properties has been exploited. Thus, grafting PCL onto chitosan may result in amphiphilic copolymers with the capacity to assemble into a micelle structure with a positively charged hydrophilic shell and hydrophobic inner compartment and therefore may have potential in drug delivery. On the other hand, the positively charged surface may facilitate cell adhesion and thus may find application in tissue engineering. The ω-hydroxyl functional group of a PCL chain initiated by dodecanol was modified by reacting it with succinic anhydride; and the ensuing esterification with *N*-hydroxysuccinimide afforded the PCL chain with terminal activated carboxylic group [56]. These activated polyester chains were grafted onto

TABLE 4.3. Amphiphilic Graft Copolymers Featuring PCL Block

Hydrophilic Component	Polymerization Details[a]	Resulting Morphology	Intended Functions	Reference
PDMAEMA	RO*co*P of α-propargyl-ε-CL and ε-CL, CuAAC	Micelles	Clofazimine loading	55
Chitosan	Activated PCL grafting	Nanoparticles of 100–215 nm in water		56
PAEMA	ROP initiated by HEMA, RAFT, galactosyl and porphyrin ligation	Spherical micelles with the diameter of around 50 nm	Doxorubicin loading, selective binding, fluorescence enhancement	57
mPEG	P(ε-CL-*co*-(2-oxepane-1,5-dione) is grafted with ω-aminooxy-mPEG and *p*-methoxy-O-benzylhydroxylamine	Spherical aggregates of 116 ± 35-nm diameter, cylindrical aggregates of 12 ± 2-nm diameter and 190 ± 37-nm length		58
mPEG	HEMA-initiated ROP, RAFT of PCL MM and DMAPAAm with mPEG-based macro-RAFT agent	Spherical micelles with the diameter ranging from 97 to 110 nm	DNA condensing, doxorubicin loading	59
mPEG	ROP of ε-CL followed by ROP of a mixture of cyclic carbonates some of which carried a pendant mPEG	Spherical micelles with an average diameter of 93 nm	Potential in drug delivery	60

[a]Since ε-CL is normally polymerized by ROP, this is not mentioned and only in special cases a remark is added.

6-*O*-triphenylmethyl-chitosan by making use of the free amine functionalities of the latter. Various grafting densities have been attained. Afterwards, the triphenylmethyl protecting groups were removed garnishing the amphiphilic chitosan-*g*-PCL macromolecules. These macromolecules formed nanoparticles in water with the diameters ranging from about 100 to 215 nm. Zeta-potential measurements suggested that these nanoparticles exhibited cationic surfaces in the aqueous milieu at pH 6.5 [56].

To conclude with, graft copolymers incorporating PCL block, alongside block and star copolymers, provide a means to build quite sophisticated biomaterials by making use of ever-advancing arsenal of chemical reactions. Again, materials for drug delivery applications are predominant. This stems from the biocompatibility and biodegradability of hydrophobic as well as hydrophilic components, along with the ability of these assemblies to provide a wonderful nanoenvironment as a shelter for therapeutic agents.

4.4 HETEROBIFUNCTIONAL PCL-DERIVED NANOMATERIALS

In the following section, we shall elucidate how synthetic strategy and precise control of the size and functionality of the PCL building block provide a means of designing various advanced biomaterials. The basic approach relies on preparation of the well-defined heterobifunctional PCL of low molecular weight through the incorporation of specific chemical functionalities which on one hand are embedded in the employed initiator (F1) and on the other hand are integrated by quantitative derivatization (F2) of the terminal hydroxyl function as illustrated in Scheme 4.1.

The quantitative esterification reactions (F2) introduce temporarily protected thiol functional group, *n*-alkene, and methacrylic moiety. It should be noted at this point that both ends of the PCL chain can be efficiently utilized in the "click" coupling by employing CuAAC and thiol-ene reactions. We have previously shown that many different functionalities like mono- and dicarboxylic acids [61, 62], sulfonates and amines [63] with azide tags can be "clicked" onto alkyne-substituted polystyrene. Additionally, in more recent developments, alkyne-substituted fluorescent molecules, a 12-crown-4 ether, PEG ($M_n = 5000$ Da), a fluoropropyl ester, and biotin have been "clicked" onto the azide-substituted surface of the conducting polythiophene [64, 65]. Thus, the alkyne and azide tags at the PCL terminus provide a plethora of possibilties for ligating numerous functionalities and structural fragments at the chain end. The versatility of the heterobifunctional PCL will be elaborated below through three very different examples from our own research.

4.4.1 Gold Nanoparticles with an Amphiphilic block Copolymer Corona

The first example illustrates the preparation of gold nanoparticles (GNPs) coated with an amphiphilic block copolymer.

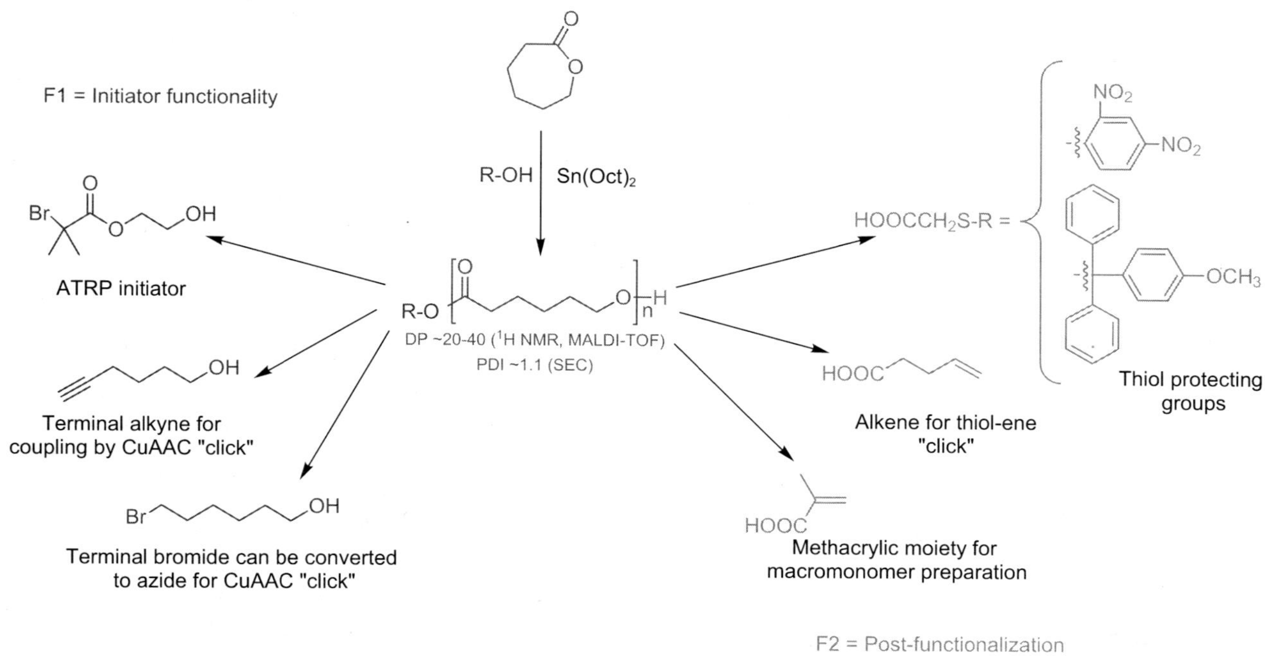

SCHEME 4.1. Different possibilities of synthesizing heterobifunctional PCL building blocks by use of functional initiators (F1) and post-functionalization (F2).

These hydrophilic core–shell-type structures are intended for bladder cancer therapy. GNPs are envisioned to be superior to polymeric micelles as candidates in the construction of drug delivery devices. The encapsulation and controlled release of drugs, as well as the adequate masking from reticuloendothelial system (RES), remain as a challenge in the case of polymeric micelles [66]. On the contrary, monolayer protected GNPs offer improved stability, low toxicity, and versatility of surface functionalities in addition to small size. These features render them unrecognizable by RES. By tailoring the surface properties of the GNPs, the system can provide site-specificity and extended circulation time [66, 67]. For instance, antibody-conjugated GNPs provide high contrast for noninvasive imaging of targeted cancer tissues [68]. Provided that a suitable polymeric shell has been constructed, monolayer protected GNPs may serve as excellent reservoirs for hydrophobic drugs. Thus, several factors have been considered in the design process. The final nanoparticles should be composed of biocompatible, biodegradable, and bioresorbable components. Moreover, they should be water-soluble if intended to be administered directly in the bladder. At the same time, these nanoparticles must contain a hydrophobic layer that facilitates loading of lipophilic anticancer drugs (e.g., paclitaxel and doxorubicin). Taking into consideration these requirements, PCL is the obvious choice for building a hydrophobic inner shell. If that then could be linked to a mucus adhesive polymer such as PAA, which would ultimately form the outer layer, the particles after adhesion would warrant a long residence time allowing the loaded drug to diffuse to the cancer cells in contrast to the drugs administered orally. In the latter case the drugs are quickly excreted from the bladder the first time it is emptied. Finally, the gold core could have potential in both sensing and thermal ablative therapy [69] (by exploiting surface plasmon resonance properties of the gold clusters). The hydrophobic layer should be covalently attached to the surface of GNPs to obtain a robust drug delivery platform that can endure severe dilution in the bladder. Tethering of the polymeric ligands on the gold surface can be realized through the gold–thiol interactions. Thus, terminal of the PCL block should incorporate thiol functionality. The selected synthetic strategy [70] is outlined in Scheme 4.2.

The utilized difunctional initiator, HEBI, was first employed in the optimized ROP of ε-CL to result in a heterobifunctional PCL with a verified DP of 30 and a PDI of 1.09. Validation of the heterobifunctionality and determination of molecular weights were achieved by extensive use of ^{1}H NMR spectroscopy and SEC in combination. At this stage, temporarily protected thiol functionality was introduced by esterification of the PCL hydroxyl terminal with α-(4-monomethoxytritylthio)acetic acid relying on the acid-labile 4-monomethoxytrityl group, which has been successfully employed in peptide synthesis as a sulfhydryl protection of mercapto acids [71]. The resulting MI could then be utilized for ATRP of *t*BA. Although the ultimate target is a PAA hydrophilic block, the water soluble acrylic acid cannot be polymerized directly. As a consequence the *t*BA precursor monomer was utilized. The polymerization catalyzed

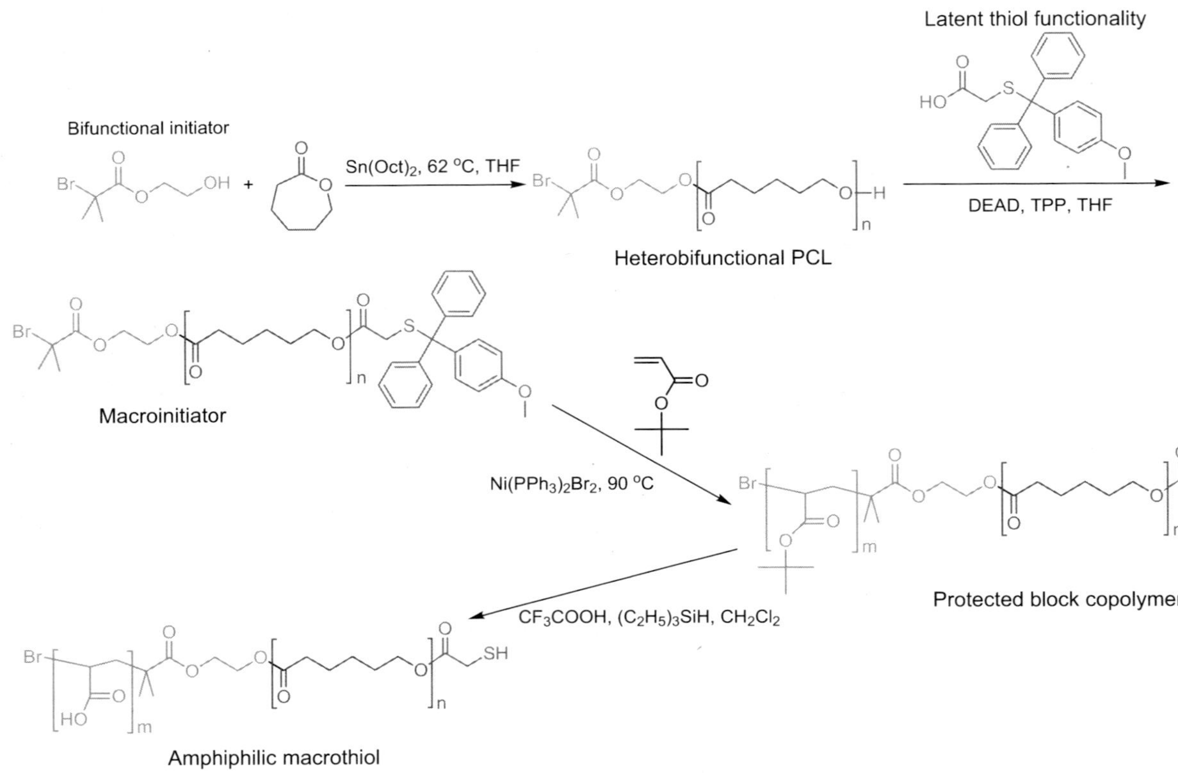

SCHEME 4.2. Synthetic sequence for preparation of amphiphilic block copolymer precursor for the water soluble GNPs intended for bladder cancer therapy.

by $Ni(PPh_3)_2Br_2$ was very slow, however, the resulting protected block copolymers had the targeted relatively narrow MWD [70]. It should be emphasized that copper-catalyzed ATRP of *t*BA in combination with a 2,4-dinitrophenyl protecting group, as originally promoted by Hedrick and co-workers [72], did not lead to well-defined block copolymers. Poor control over the ATRP may be attributed to the loss of the catalyst activity due to possible ligation of copper with the 2,4-dinitrophenyl group or with the polyester backbone. The double, concurrent deprotection of the protected block copolymer could then be performed in one step by use of triflouroacetic acid (TFA) to yield the thiol terminated amphiphilic block copolymer. The GNPs were then synthesized according to a modified procedure [73] consisting of the reduction of gold (III) chloride trihydrate in the presence of the thiol end-capped amphiphilic diblock copolymer. By using thiol-deficient conditions (threefold excess of the gold salt to the thiol-terminated block copolymer), well-separated, aggregation-free GNPs with a mean diameter of 9.0 ± 3.1 nm were obtained as demonstrated by a combination of UV–vis spectroscopy and TEM; an example of the latter is shown in Figure 4.1A. TEM affords visualization of the gold core solely. The polymeric shell cannot be pictured due to the minute difference in contrast between the shell and the relatively thick carbon film which serves as the support of the TEM sample. However, we may hypothesize that in aqueous solution the corona consists of (a) an extended, hydrated layer of PAA and (b) a collapsed, dehydrated layer of PCL. In fact, the diameter of the nanoparticles obtained from TEM does not correlate with that extrapolated from UV–vis spectrum [74]. One can only surmise that the absorption peak has experienced the hypsochromic shift observed by Tenhu and co-workers [75, 76]. The gold content of these nanoparticles on the order of 16 wt% could easily be determined by use of thermogravimetric analysis (TGA) performed in nitrogen atmosphere.

The anticipated function of a drug-loaded GNP is sketched in Figure 4.1B: The GNP's adhesion to the bladder's mucous membrane imparts to the anticancer drug an extended residence time in the bladder, and therefore possibly sufficient time to diffuse to the cancer cells. Thus, the PCL layer may be loaded with hydrophobic anticancer therapeutic agents, while the PAA layer may facilitate anchoring of the nanoparticle to mucous membranes resulting in site-specific delivery. Carboxylic groups of PAA may be employed for the ligation of biologically active molecules to boost the cellular uptake. The gold core ensures structural stability, and it may serve as a marker in localization studies.

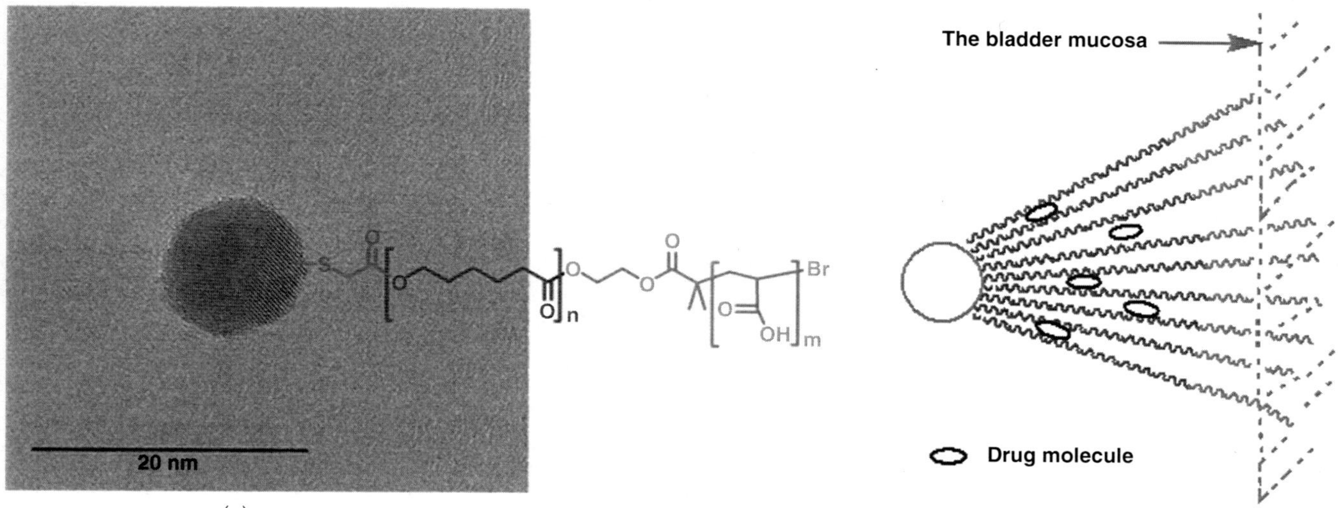

(a) (b)

FIGURE 4.1. (**A**) TEM image of one GNP decorated with one amphiphilic PCL-*b*-PAA chain. (**B**) A small fraction of the GNP corona (loaded with drug molecules) adhering to bladder mucosa. See color insert.

4.4.2 Amphiphilic Rodcoil L-Lysine Dendrons

The second example demonstrates a facile synthesis of a linear-dendritic cholesteryl-PCL-*b*-L-lysine-G_2 by thiol–ene and azide–alkyne "click" reactions [77]. Here the inspiration stems from the intriging possibilities of creating self-assembled polymeric micelles by the proper combination of constituent building blocks assisted by the recently developed powerful synthetic tools. In particular, amphiphilic linear-dendritic block copolymers have gained significant importance due to their peculiar properties that drive self-assembly into various shapes, and they have the capacity to form complex supramolecular architectures [78]. For instance, Würsch et al. [79] have designed linear-dendritic amphiphiles that resemble naturally occuring phospholipids. The hydrophilic dendrons based on 2,2-bis(hydroxymethyl)propionic acid (A) were combined with the linear segment of the hydrophobic PCL (B) through the ROP of ε-CL initiated by the dendrons possessing multiple functionalities. A symmetric linear-dendritic triblock copolymer (ABA) was also prepared via the Mitsunobu coupling of the linear-dendritic block copolymer (AB) with the acid-functionalized dendrons. Li et al. [80] developed a route towards poly(L-lactide)-*b*-dendritic poly(L-lysine) based on the metal free ROP of the L-lactide initiated by the hydroxyl end-capped BOC-protected poly(L-lysine) dendrons. It was pointed out that poly(L-lysine) could play a role as a functional vector for DNA transfection. Klok et al. [81] reported on the synthesis of a novel class of amphiphilic linear-dendritic block copolymers, which they termed rodcoil dendrons. The preparation of these intricate structures involved ROP of L-lactide initiated by cholesterol, and substitution of the hydroxyl terminus of this cholesteryl-oligo(L-lactic acid) with an L-lysine dendron. Liquid crystalline nature of the cholesteryl moiety could induce self-assembly of the rodcoil dendron, and the L-lysine dendritic wedge could influence the development of the nanostructure. Both cholesterol and L-lysine could interact with the cell membrane and thus enhance cell adhesion. [81]. With these aims in mind, our strategy became the extensive exploitation of the heterobifunctional PCL tailored for effective orthogonal "click" couplings with the suitable bioactive moieties. The synthetic strategy is outlined in Scheme 4.3.

The ROP of ε-CL was succesfully initiated by 5-hexyn-1-ol, and the resulting PCL was esterified with 4-pentenoic acid in a Mitsunobu protocol [77]. The α-hexynyl-ω-pentenoyl-PCL had a DP of 20 as determined by ^{1}H NMR spectroscopy and verified by MALDI-TOF mass spectrometry. Both techniques also substantiated a near to 100% end-group fidelity, whereas the PDI was calculated to be 1.11 by SEC. The L-lysine second-generation (G_2) dendron with an azide group in the focal position was conveniently prepared from the corresponding bromide after reaction with sodium azide. The bromide dendron was first obtained by a simple, newly introduced DCC/4-DMAP coupling of 3-bromo-1,2-propanediol with the BOC-protected L-lysine. A CuAAC reaction between the α-hexynyl terminal of the PCL chain and the azide-functional dendritic wedge produced the linear-dendritic block copolymer PCL-*b*-L-lysine-G_2. The extent of the reaction was near-quantitative. It should be pointed out that the simple but efficient "click" approach taken here supersedes numerous tedious protecting-deprotecting steps utilized previously [81] when coupling dendrons to linear polymers. The final step—attachment of the cholesteryl moiety—is particularly simple because a commercially

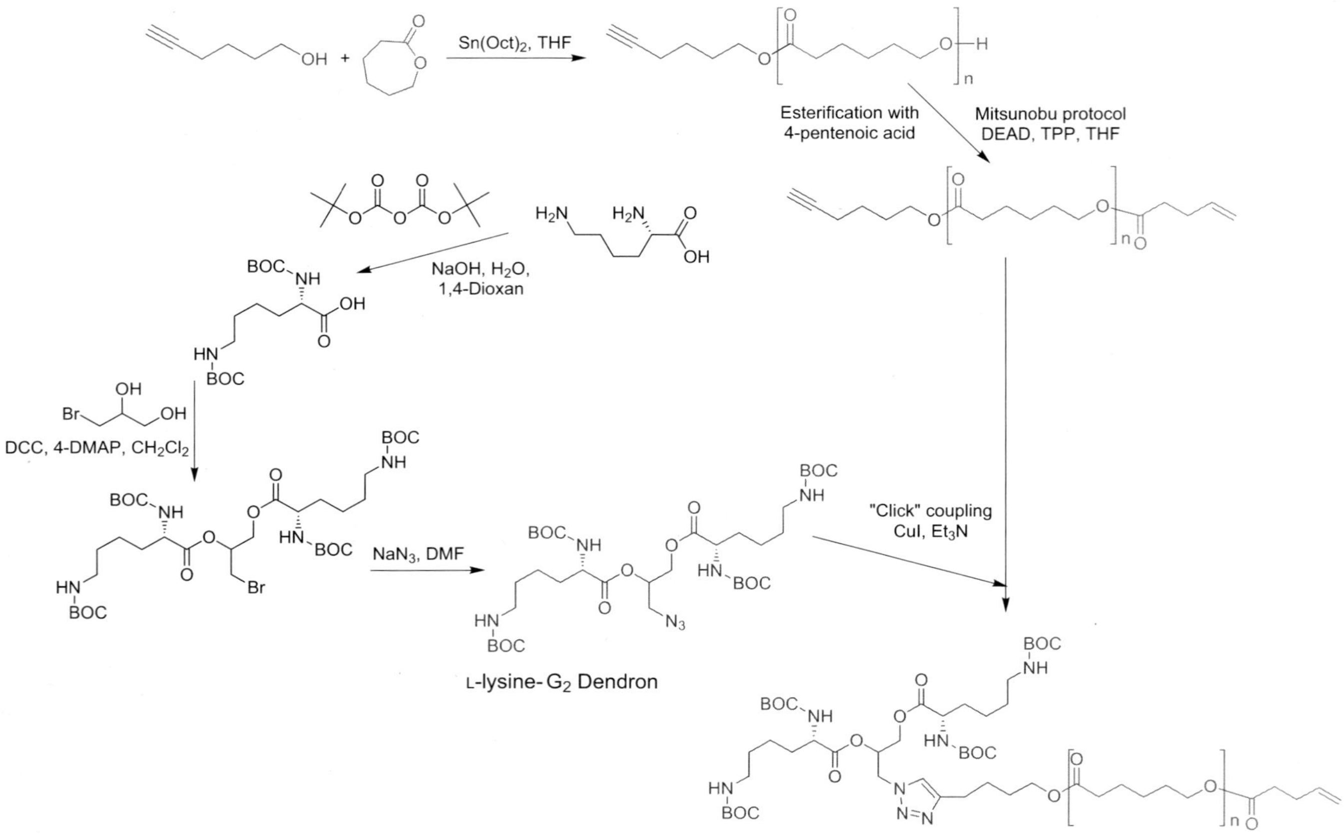

SCHEME 4.3. Preparation of α-hexynyl-ω-pentenoyl-PCL, L-lysine-G_2 dendron, and the CuAAC coupling between the L-lysine-G_2 dendron and PCL.

available thiocholesterol could be employed in a UV-initiated thiol–ene "click" reaction as illustrated in Scheme 4.4.

Also in this case, a very high coupling effieciency is achieved. Meticulous monitoring of the extent of both "click" couplings by ^{1}H NMR experiments confirmed highly efficient linking of these architectural elements (Figure 4.2): The resonance signal attributed to the acetylenic proton (d) disappears, whereas the resonance peak assigned to the methylene group next to the alkyne function (c) exhibits downfield shift.

In addition, the peaks attributable to the dendritic wedge have been identified. The signals originating from the alkene moiety are detected at 5.00 ppm and 5.80 ppm, proving the preservation of this functional group. Furthermore, the integrals of the peaks assigned to the terminal alkene (b, 5.80 ppm) and CH_2NHBOC (e, 2.93–3.22 ppm) fragment of the dendron are in good agreement with the theoretical values. These ^{1}H NMR data unequivocally point to the total exhaustion of the terminal alkene groups as well: Resonances a and b have shifted upfield. The presence of the cholesteryl moiety is also evident (h). MALDI-TOF spectrometry revealed that only the weight increases expected from the addition of the respective bioactive units were observed [77]. This was also reflected in the product PDI obtained by SEC, which in both cases were basically unaltered. The ultimate deprotection is effieciently carried out by a standard TFA treatment as shown in Scheme 4.4. In the specific example scrutinized here, the relatively few synthetic steps necessary to produce such a complex functional polymer must be emphasized. Furthermore, since the couplings of the very different molecular constituents are based on successful orthogonal "click" chemistry, the modularity of the concept should be highlighted. Moreover, both biomimetic and biofunctional components could in principle be exchanged for other molecular fragments, provided that the necessary "click" tags are maintained for the respective "click" reactions, and thus the strategy paves the way for the design of other intriguing biofunctional materials.

4.4.3 Miktoarm Core Cross-Linked Nanoparticles with Biologically Active Moieties on the Surface

A robust but still flexible synthetic strategy for design of nanomaterials with biological activity was the aim and target of the final development based on the heterobifunctional PCL. It was mentioned in the opening sections that micellar drug delivery devices suffer from the intrinsic instability upon

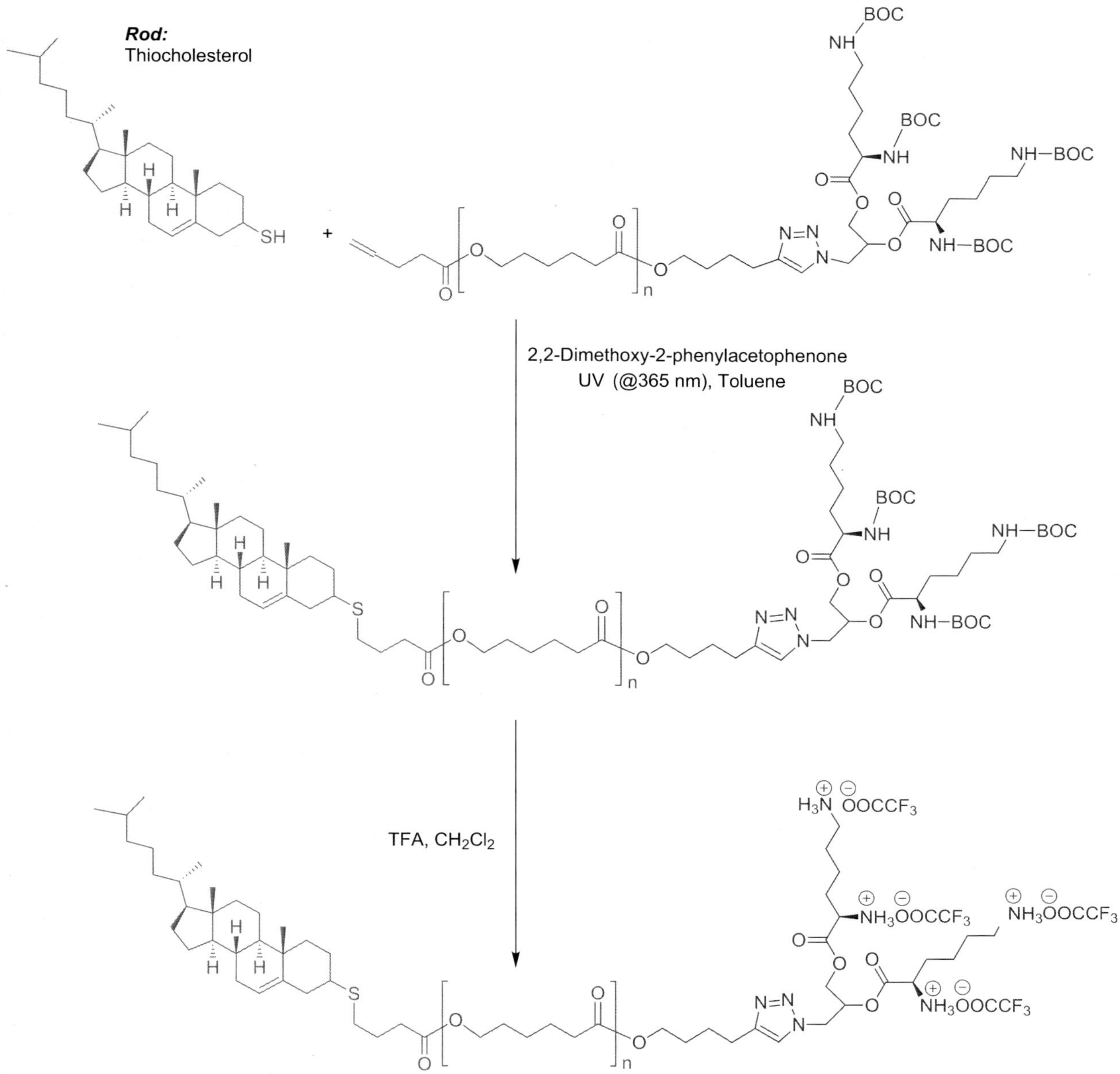

SCHEME 4.4. Combining thiocholesterol with PCL-*b*-L-lysine-G_2 by thiol–ene "click" chemistry and deprotection of cholesteryl-PCL-*b*-L-lysine-G_2.

dilution. The robustness of the system could be attained by covalent cross-linking of functional macromonomers (MM) in combination with a difunctional cross-linker by CRP (e.g., ATRP), resulting in core cross-linked star (CCS) polymers. These nano-sized structures may serve as unimolecular containers for transporting desired payload. Flexibility in the design could be secured by effective coupling possibilities, like "click" chemistry, of biologically interesting moieties. Examples of the first strategy are well known—for example, based on polystyrene (PS) and divinyl benzene [82, 83]. In a similar manner, surface functionalized CCS polymers based on P*t*BA with epoxy, cyano, hydroxyl or amine surface groups have been prepared [84]. Well-defined, globular miktoarm CCS block copolymers with mixed poly(butyl acrylate) and PEG arms have furthermore been synthesized by stepwise addition of the cross-linker and the initiator to the MMs [85]. Also miktoarm CCS copolymers with biocompatible PCL and PS arms are known [86–88]. Hawker and co-workers [89] have reported synthesis of CCS polymers with amine and hydroxyl surface functional groups employing functional initiators for NMP. Linear dendritic MMs carrying Fréchet-type or polyester dendrons have been utilized

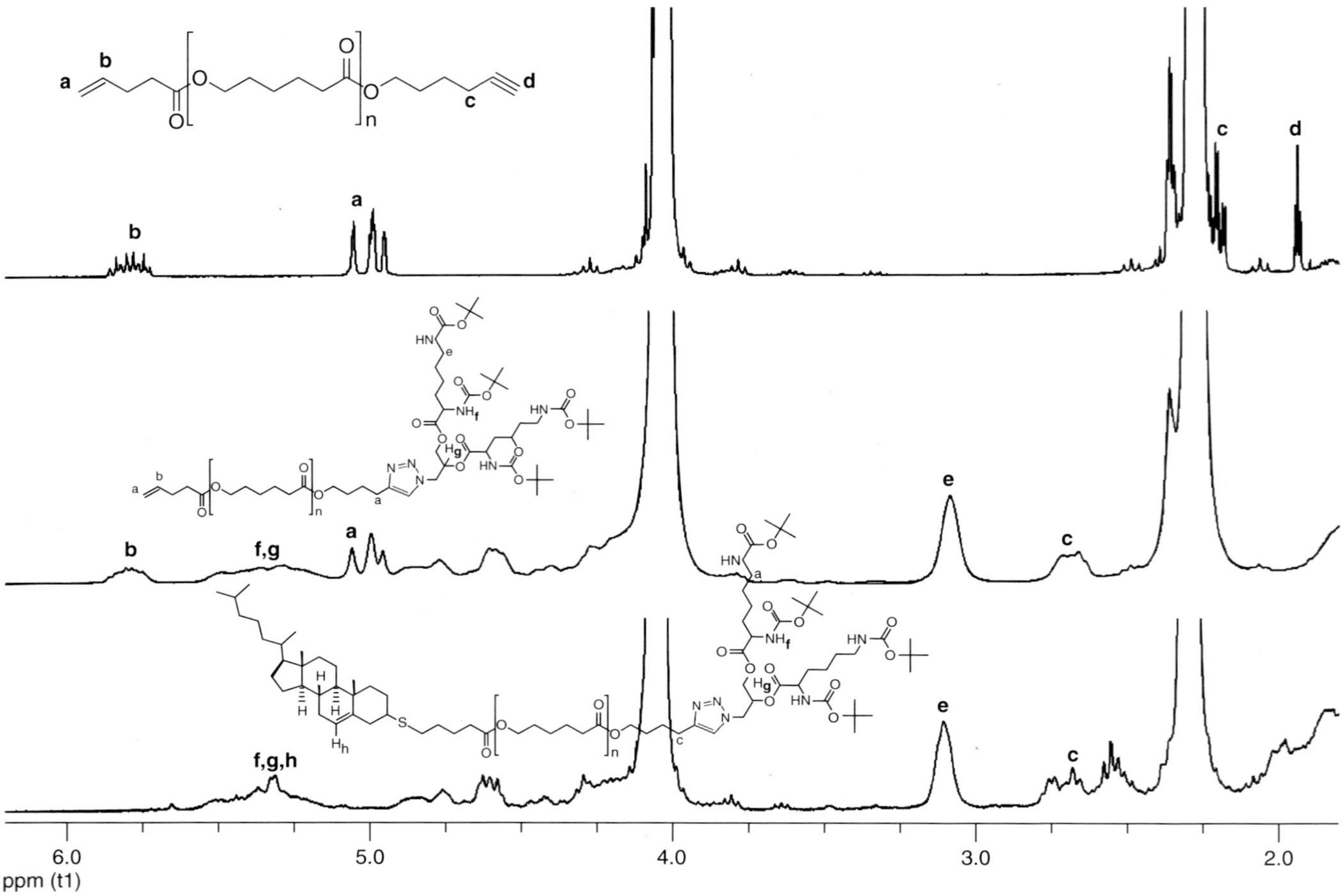

FIGURE 4.2. Overlay of ^{1}H NMR spectra of the heterobifunctional PCL and the products of consecutive "click" reactions.

for preparation of CCS polymers with dendron-decorated peripheries [90]. In our approach, however, we focused on the design of methacrylic MMs with α-azide groups for introduction of potential biological activity [91]. Scheme 4.5 displays the preparation of the functional MMs.

Initially, the bromo-functional PCL is prepared by our optimized ROP protocol of ε-CL initiated by 6-bromo-1-hexanol. The bromo-terminated PCL with a DP = 29 and a PDI = 1.09 is then esterified with methacrylic acid to furnish a MM. This route was chosen despite the existence of paths for preparation of azide-functional PCL based on nucleophilic substitution of the bromo end group of Br-PCL-OH by either sodium azide [92] or ROP of ε-CL employing azide-containing initiator [93]. However, in our view there is a potential risk of azide reduction by triphenyl phosphine (TPP) in the subsequent Misunobu coupling of N_3-PCL-OH and methacrylic acid. Thus, the azide functionality is then introduced by converting the bromo end group of the methacrylic MM by reaction with sodium azide to produce MM1. Although all the analyses results imply that the product is pure, it should be stressed that the N_3-PCL-methacrylate (MM1) suffers limited shelf-life at room temperature: Apparently the azide group and the methacrylic double bond undergo uncatalyzed cycloaddition [91]. At this stage a hormone-synthetic polymer conjugate is prepared by a Cu(I)-catalyzed cycloaddition between the MM1 and 17α-ethynylestradiol resulting in a hormone-bearing methacrylate, MM2. The ^{1}H NMR spectrum in Figure 4.3 displays all the characteristic resonances.

The resonance signals attributed to the methacrylic acid residue (**h**, **i**, and **j**) confirmed the preservation of the vinyl double bond. The triazole proton (**k**) appeared at 7.41 ppm, and the signals originating from the estradiol moiety were detected as well. In a similar manner, ethynylferrocene was used to produce another methacrylate-based macromonomer, MM3. It should be noted that the ferrocene is not assigned any biological activity; instead it was anticipated to equip the CCS polymers with inherent contrast properties in TEM experiments. The ^{1}H NMR spectroscopy corroborated formation of the target structure (Figure 4.3). A linear-dendritic MM containing an L-lysine dendritic wedge is designed as outlined in Scheme 4.6. Here, an α-alkynyl-PCL with a

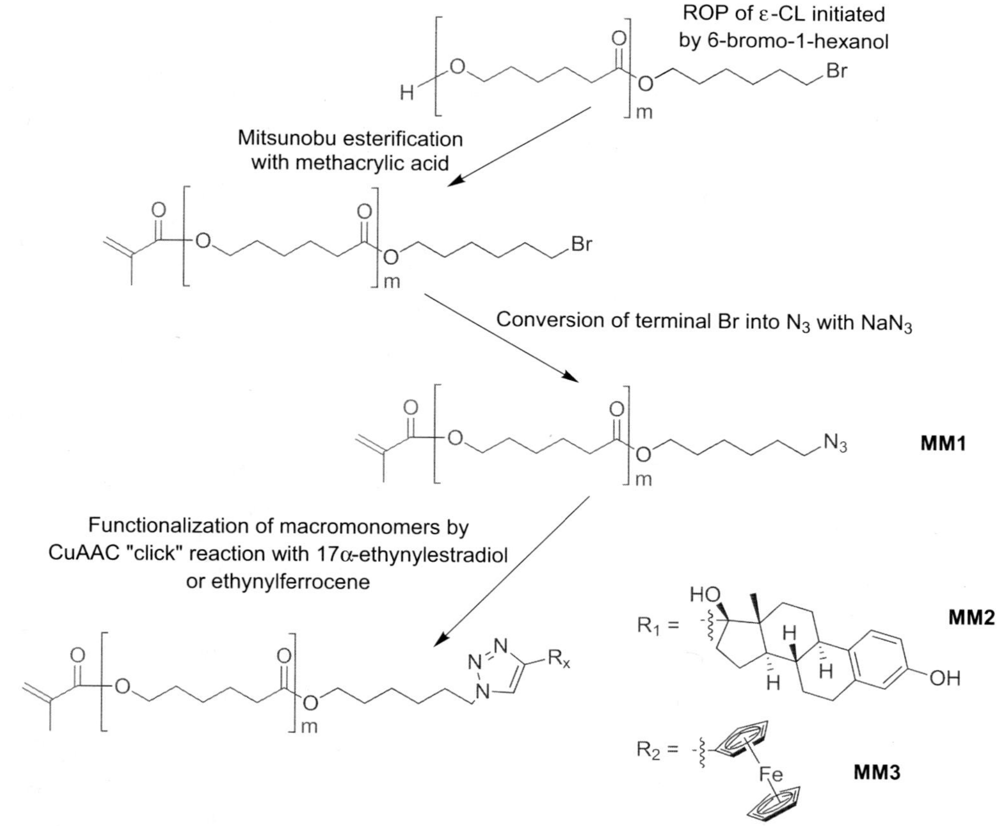

SCHEME 4.5. Preparation of functional macromonomers: MM1, MM2, MM3.

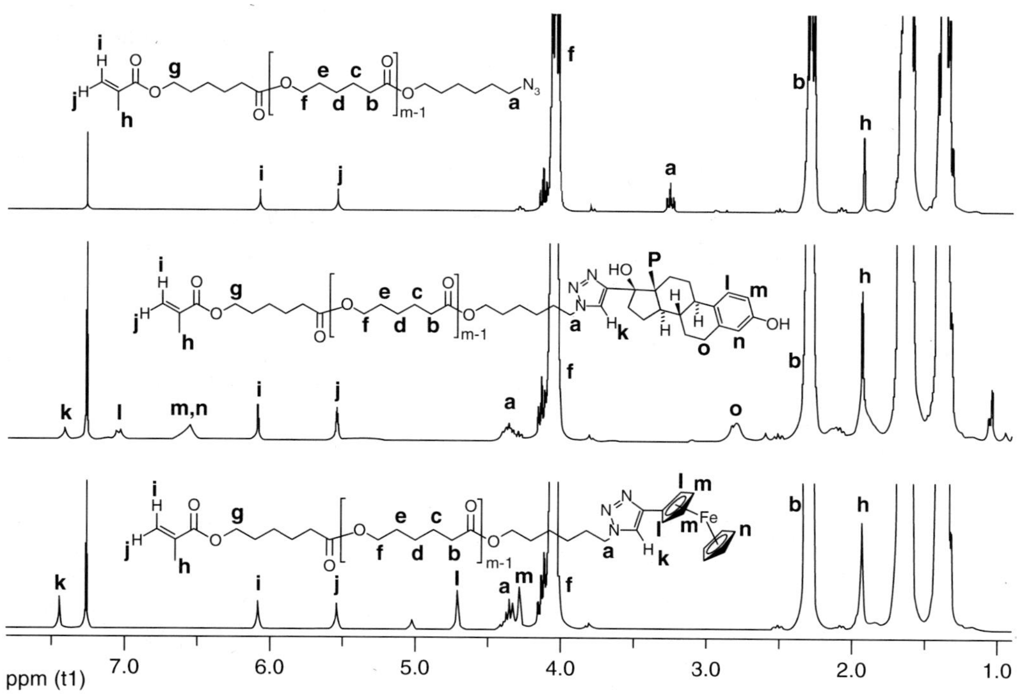

FIGURE 4.3. Overlay of ^{1}H NMR spectra of the MM1, MM2, and MM3.

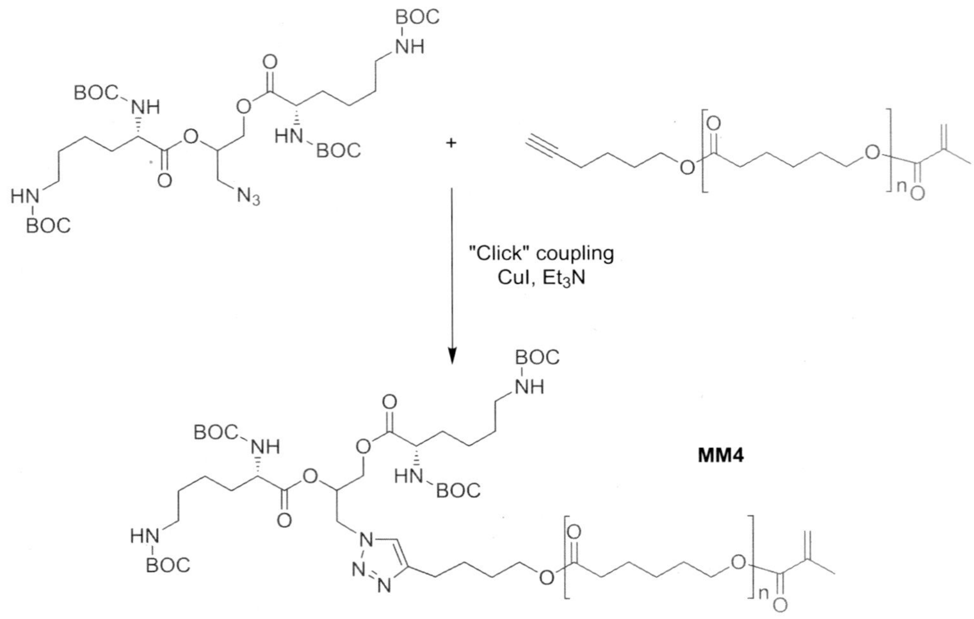

SCHEME 4.6. Preparation of the linear-dendritic macromonomer MM4.

DP = 21 and a DPI = 1.09 is initially prepared according to the previously developed procedure as shown in Scheme 4.3. It is then esterified with methacylic acid to produce a MM that is "clicked" in the CuAAC reaction with the azide-functionalized L-lysine-G_2 dendron (Scheme 4.3), and the linear-dendritic MM—PCL-*b*-(di-BOC-L-lysine)-G_2 methacrylate (MM4)—is obtained. In the CCS polymer formation with both difunctional cross-linker and MMs, the number of experimental variables are high, and the polymerization conditions need careful balancing in order to avoid gelation, star–star coupling taking into account the process being radical in nature with the coexistence of several growing chains, resonable yield, and so on. The ATRP reactions employing ethyl α-bromoisobutyrate (EB*i*B) as initiator and CuBr–HMTETA catalytic complex in toluene at 50°C, as depicted in Schemes 4.7 and 4.8, furnish the best nanoparticles.

The CCS1 formation results in a large M_w value ($\sim 1.6 \times 10^6$ Da, determined with a light scattering detector); however, PDI is fairly narrow (1.43). The unprecedented large M_w is attributed to star–star coupling througout ATRP, and the aforementioned azide–methacrylate cycloaddition reaction at the later stage. Even though a certain amount of peripheral azide functions are preserved (confirmed by ^{1}H NMR experiments) for further derivatization, the unpredictable nature of the reaction renders CCS1 less attractive, and the synthetic layout may not be considered as the optimal one. A copolymerization employing MM1 and MM4 resulted in CCS2 wth $M_w \sim 2.0 \times 10^6$ Da (PDI = 1.55). The structural investigation by ^{1}H NMR spectroscopy revealed the presence of both arms. However, the azide–methacrylate cycloaddition during the cross-linking process as well as after isolation of the star product led to star–star coupling and consumption of an undefined amount of the azide functions. The remaining azide groups on the surface of CCS2 were "clicked" with ethynylferrocene in the conventional CuAAC reaction as depicted in Scheme 4.7, resulting in CCS3. Since uncatalyzed cycloaddition between azide and methacrylic moieties is inevitable, and undermines reaction control and orthogonality concept, an alternative, truly modular approach is then adopted. It relies on introduction of the desired functionality or structural unit by CuAAC "click" reaction before the cross-linking step, thus tying up the azide termini. It must be noted that this pathway for preparation of MMs makes the design less flexible since the CCS polymer's surface chemistry is dictated by the employed MM.

The derivatized MM2, MM3, and MM4 are activated in the CCS copolymer manufacturing process resulting in CCS4 and CCS5 that contain both peripheral BOC-protected L-lysine dendrons and estradiol or ferrocene moieties as shown in Scheme 4.8.

SEC analysis of the obtained products revealed no unreacted MMs and $M_w = 2.65 \times 10^5$ Da and PDI = 1.29 (for CCS4). Fairly low PDI suggests that the ATRP is under control, and star–star couplings are most likely suppressed. Similar results are obtained for CCS5. ^{1}H NMR spectroscopy data (Figure 4.4) unambiguously point toward the incorporation of both MMs into the miktoarm CSS copolymer.

The resonance signals arising from the PCL backbone and the nanogel core overlap. The peaks related to the methyl

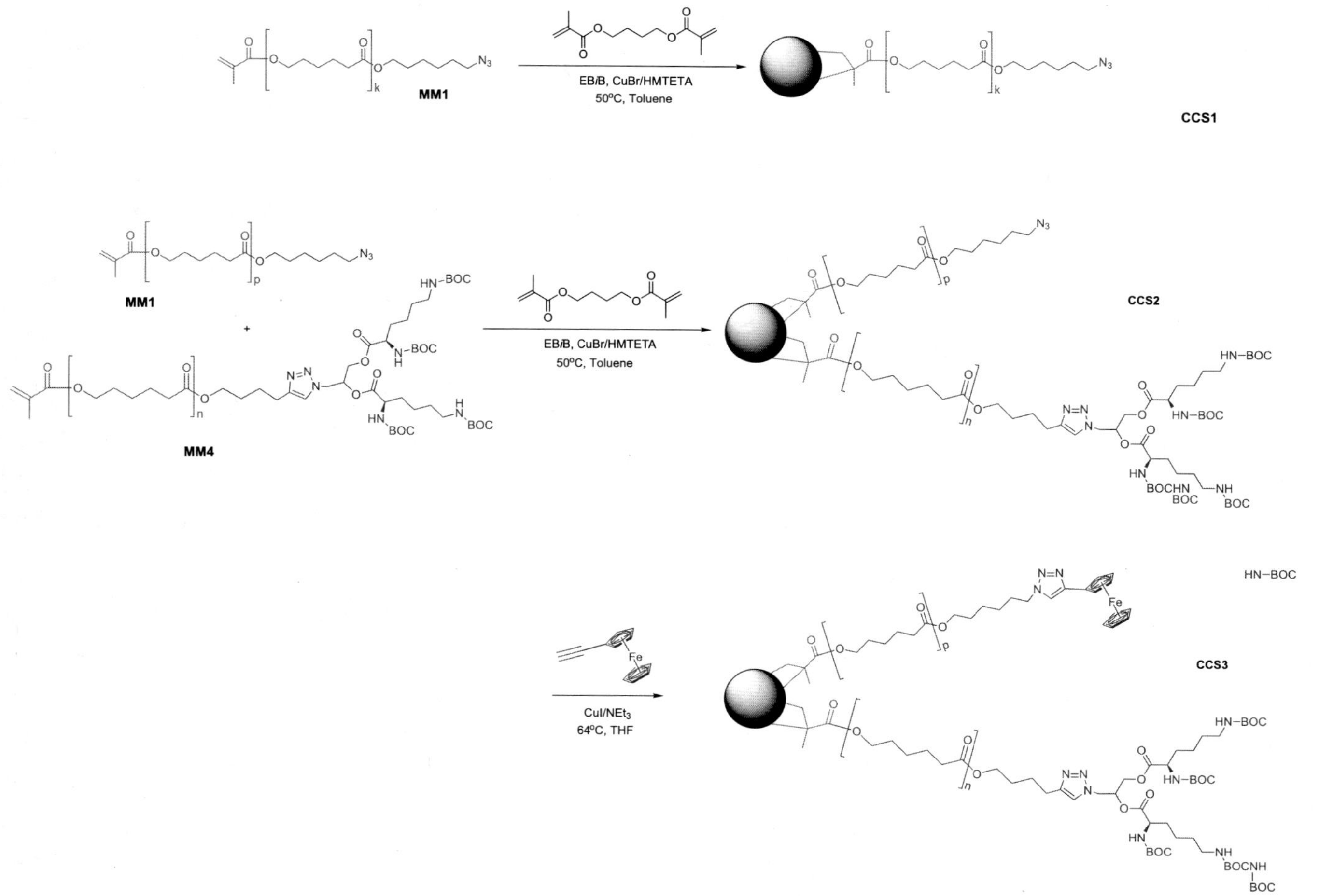

SCHEME 4.7. Preparation of miktoarm CCS copolymers with subsequent post-functionalization of star by "click" on arms with azide terminals.

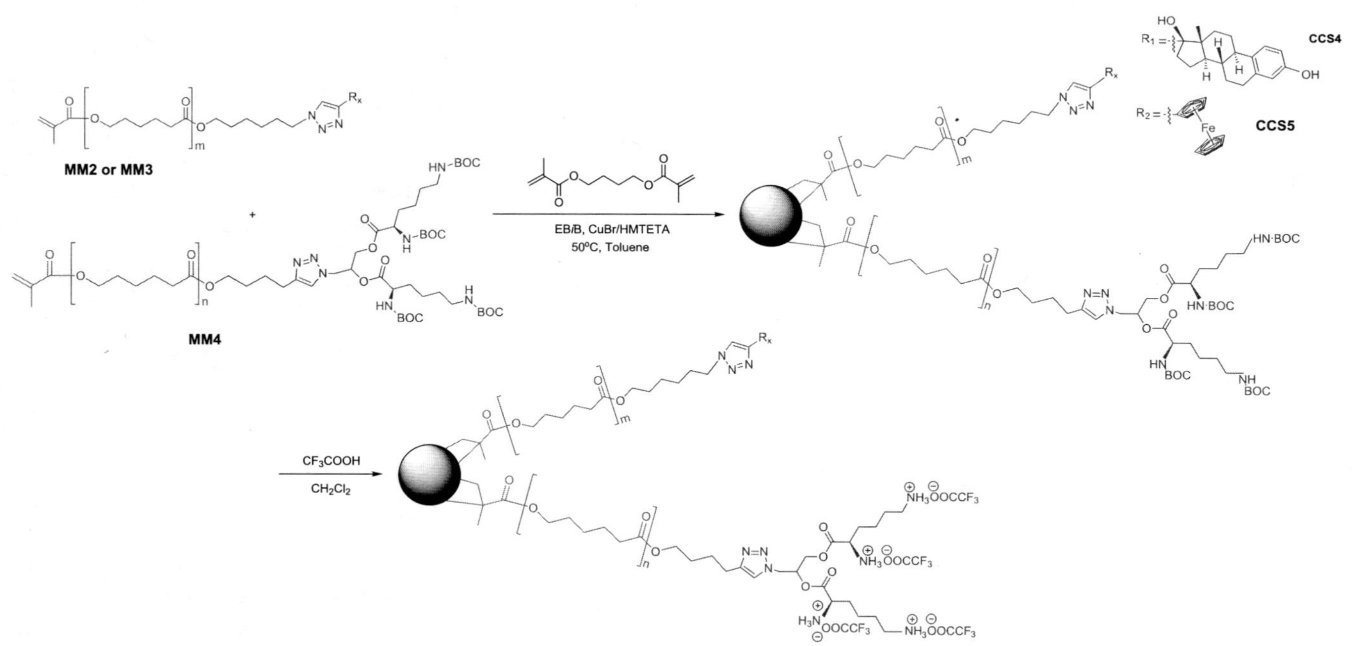

SCHEME 4.8. Preparation of miktoarm CCS copolymers by copolymerization of functional macromonomers.

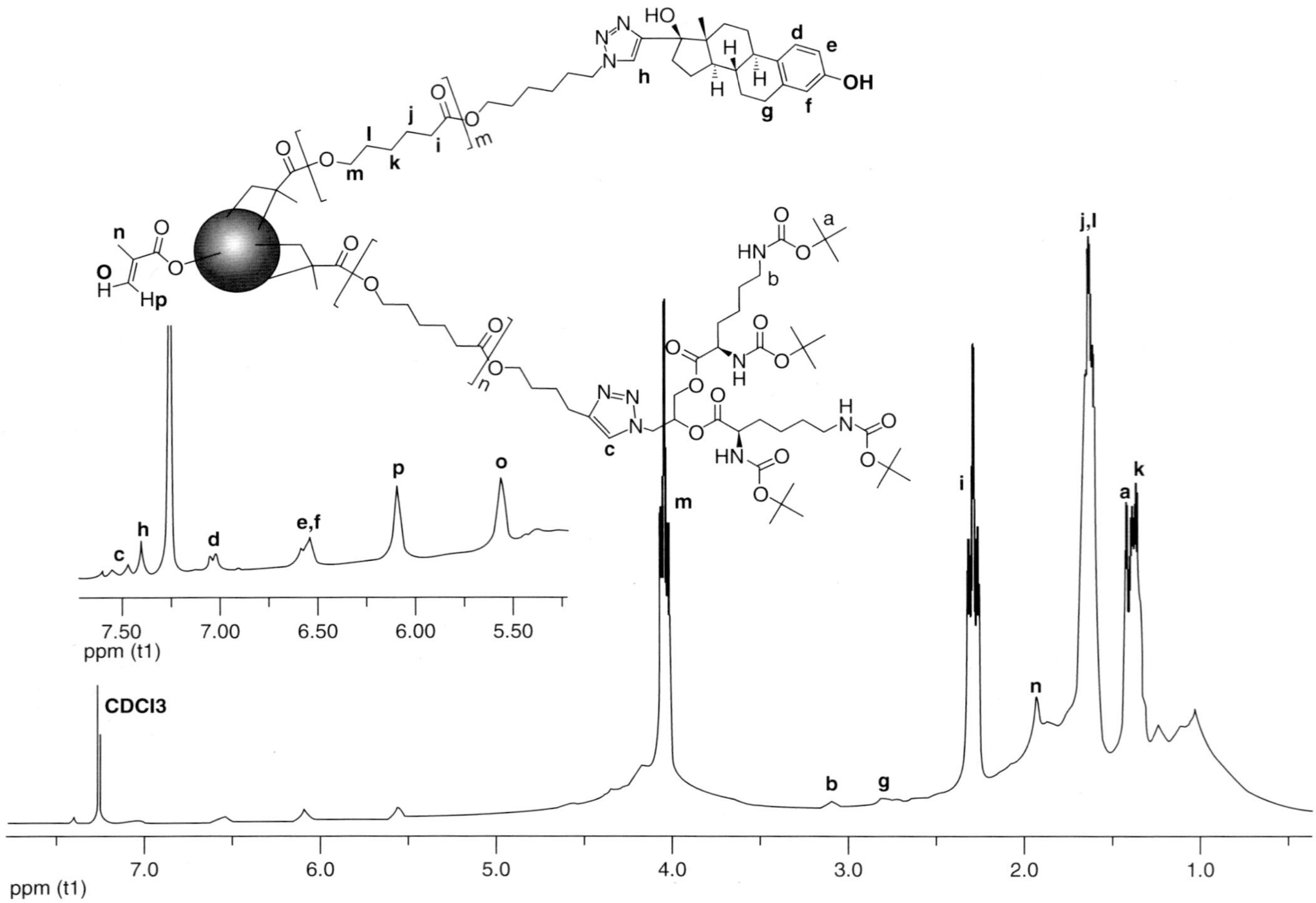

FIGURE 4.4. ^{1}H NMR spectrum of the BOC-protected L-lysine dendron and estradiol containing CCS4.

group of the BOC protection (**a**), the methylene group adjacent to the BOC-protected amino group (**b**), and the triazole protons (**c**) of the MM4 can be identified. The signals, originating from the estradiol moiety and the triazole proton of the MM2, have been assigned as well. The resonances **o** and **p** are attributed to the methacrylic acid residues of the core: The segmental mobility of the core has not been frozen completely. However, the intensities of the resonances from both estradiol and L-lysine are too weak to be employed in estimation of the ratio between the different arms. Therefore, TGA has been exploited for this purpose. By taking advantage of the fact that the decomposition of the BOC protecting groups as well as the core of the CCS copolymers occur at relatively well-defined temperature intervals, it was possible to estimate that the CCS4 consists of, on average, 24 PCL-*b*-di-BOC-L-lysine-G_2 and 27 PCL-estradiol arms whereas the CCS5 is comprised of 15 PCL-*b*-di-BOC-L-lysine-G_2 and 12 PCL-ferrocene arms.

Furthermore, the miktoarm CCS5 has been visualized by TEM (Figure 4.5). Clusters of the CCS5 were observed without additional staining. Spherical nanoparticles in the range of 10–20 nm can be identified. The clustering of the nanoparticles may be the result of poor repulsive interactions among the coronas of the CCS copolymers. Low contrast and aggregation do not allow accurate evaluation of the size and distribution of these nanoparticles.

Thus, the synthesis of miktoarm CCS copolymers consisting of linear-dendritic and linear PCL arms has been accomplished. The cornerstone for preparation of these CCS polymers is the library of well-defined MMs, which in turn is based on heterobifunctional PCLs. These heterotelechelic building blocks are characterized with low PDI and high degree of functionality. Elaboration of the structure and composition of the intermediate and final products proves the robustness, flexibility, efficiency, and high degree of control offered by the proposed synthetic sequences. The miktoarm CCS4 after removal of BOC protecting groups possesses (a) a hydrophilic, cationically charged exterior decorated with targeting ligands and (b) a hydrophobic, bioerodible core, and it may have potential in targeted drug delivery.

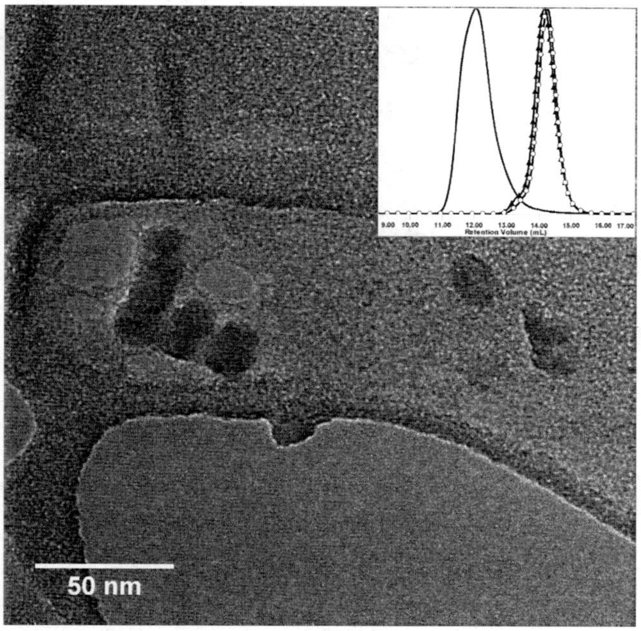

FIGURE 4.5. TEM image of the miktoarm CCS5. The image has been acquired without additional staining. **Inset**: Normalized SEC RI detector signals of MM3 (▲), MM4 (○), and CCS5 (—).

4.5 CONCLUSIONS AND OUTLOOK

The long-established biocompatible and bioresorbable character of PCL in combination with the multitude of highly efficient reactions leading to the library of functional PCLs as well as PCL-based block copolymers have, over the years, provided novel PCL-derived biofunctional materials with ever-increasing complexity. While earlier developments were based simply on controllable degradation rates, the architectures of later design are characterized with (a) multiple functionalities and (b) tunable block length and copolymer composition. The potential of these materials in drug delivery is yet to be explored. With the advance of very precise and efficacious chemical tools, greater degree of control and complexity in the design of PCL-based nano-sized biomaterials is anticipated.

ACRONYMS AND ABBREVIATIONS

α-A-ε-CL	1-(2-oxooxepan-3-yl)ethyl acrylate
ATRP	atom transfer radical polymerization
b	block
B*i*BB	2-bromoisobutyryl bromide
BOC	*tert*-butyloxycarbonyl
BSA	bovine serum albumin
CCS	core-crosslinked star
ε-CL	ε-caprolactone
cmc	critical micelle concentration
CRP	controlled radical polymerization
CuAAC	Cu (I)-mediated azide-alkyne 1,3-cycloaddition
Da	dalton
DCC	*N,N′*-dicyclohexylcarbodiimide
DEAD	diethyl azodicarboxylate
DLS	dynamic light scattering
DMAAm	*N,N*-dimethylacrylamide
DMAEMA	*N,N*-dimethylamino-2-ethyl methacrylate
4-DMAP	4-(dimethylamino)pyridine
DMAPAAm	*N*-[3-(dimethylamino)propyl] methacrylamide
DMF	*N,N*-dimethylformamide
DNA	deoxyribonucleic acid
DP	degree of polymerization
EB*i*B	ethyl α-bromoisobutyrate
EEP	2-ethoxy-2-oxo-1,3,2-dioxaphospholane
eROP	enzymatic ring-opening polymerization
FDA	US Food and Drug Agency
g	graft
G_2	second-generation dendron
GAMA	6-*O*-methacryloyl-D-galactopyranose
GNP	gold nanoparticle
HEBI	2-hydroxyethyl 2-bromoisobutyrate
HEMA	2-hydroxyethyl methacrylate
HMTETA	1,1,4,7,10,10-hexamethyltriethylenetetramine
HPC	hydroxypropylcellulose
HPMA	*N*-(2-hydroxypropyl)methacrylamide
Hy-PEI	hyperbranched polyethylenimine
LCST	lower critical solution temperature
MALDI-TOF	matrix-assisted laser desorption and ionization time of flight
MI	macroinitiator
MM	macromonomer
MMA	methyl methacrylate
M_n	number average molecular weight
mPEG	monomethoxy poly(ethylene glycol)
mPEGMA	poly(ethylene glycol) methyl ether methacrylate
M_w	weight average molecular weight
MWD	molecular weight distribution
n	normal
NIPAAm	*N*-isopropylacrylamide
NMP	nitroxide-mediated polymerization
NMR	nuclear magnetic resonance
PAA	poly(acrylic acid)
PAEMA	poly(2-aminoethyl methacrylate hydrochloride)
PAMAM	poly(amido amine)
PCL	poly(ε-caprolactone)
PDI	polydispersity index
PDMAEMA	poly(*N,N*-dimethylamino-2-ethyl methacrylate)

PEEP	poly(2-ethoxy-2-oxo-1,3,2-dioxaphospholane)
PEG	poly(ethylene glycol)
PEGA	poly(ethylene glycol) acrylate
PGAEMA	poly(D-gluconamidoethyl methacrylate)
PGAMA	poly(6-*O*-methacryloyl-D-galactopyranose)
PGMA	poly(glycidyl methacrylate)
PG2MA	poly(glycerol monomethacrylate)
PHPMAm	poly(*N*-(2-hydroxypropyl) methacrylamide)
PMDETA	*N,N,N′,N′,N″*-pentamethyldiethylenetriamine
PMPC	poly(2-(methacryloyloxy)ethyl phosphorylcholine)
PmPEGMA	poly(poly(ethylene glycol) methyl ether methacrylate)
P(mPEGMA-*co*-GAMA)	poly(poly(ethylene glycol) methyl ether methacrylate-*co*-6-*O*-methacryloyl-D-galactopyranose)
P(NIPAAM-*co*-DMAAm)	poly(*N*-isopropylacrylamide-*co-N,N*-dimethylacrylamide)
PPEEA	poly(2-aminoethyl ethylene phosphate)
PPEGA	poly(poly(ethylene glycol) acrylate)
PS	polystyrene
P*t*BA	poly(*tert*-butyl acrylate)
PVBG	poly(6-*O*-(4-vinylbenzyl)-D-galactose)
PVP	poly(vinyl-2-pyrrolidone)
P2VP	poly(2-vinylpyridine)
P4VP	poly(4-vinylpyridine)
RAFT	reversible addition-fragmentation chain transfer
RES	reticuloendothelial system
RI	refractive index
ROP	ring-opening polymerization
SEC	size exclusion chromatography
siRNA	small interfering ribonucleic acid
$Sn(Oct)_2$	tin octoate
St	styrene
*t*BA	*tert*-butyl acrylate
TEM	transmission electron microscopy
TFA	trifluoroacetic acid
TGA	thermogravimetric analysis
THF	tetrahydrofuran
TPP	triphenylphosphine
UV	ultraviolet
δ-VL	δ-valerolactone

REFERENCES

1. Woodruff, M. A., Hutmacher, D. W. (2010). The return of a forgotten polymer—Polycaprolactone in the 21st century. *Progress in Polymer Science*, **35**, 1217–1256.
2. Torchilin, V. P. (2007). Micellar nanocarriers: Pharmaceutical perspectives. *Pharmaceutical Research*, **24**, 1–16.
3. Torchilin, V. P. (2004). Targeted polymeric micelles for delivery of poorly soluble drugs. *Cellular and Molecular Life Sciences*, **61**, 2549–2559.
4. Savic, R., Luo, L., Eisenberg, A., Maysinger, D. (2003). Micellar nanocontainers distribute to defined cytoplasmic organelles. *Science*, **300**, 615–618.
5. Wang, F., Bronich, T. K., Kabanov, A. V., Rauh, R. D., Roovers, J. (2005). Synthesis and evaluation of a star amphiphilic block copolymer from poly(ε-caprolactone) and poly(ethylene glycol) as a potential drug delivery carrier. *Bioconjugate Chemistry*, **16**, 397–405.
6. Jérôme, C., Lecomte, P. (2008). Recent advances in the synthesis of aliphatic polyesters by ring-opening polymerisation. *Advanced Drug Delivery Reviews*, **60**, 1056–1076.
7. Dove, A. P. (2008). Controlled ring-opening Polymerisation of cyclic esters: Polymer blocks in self-assembled nanostructures. *Chemical Communications*, 6446–6470.
8. Labet, M., Thielemans, W. (2009). Synthesis of polycaprolactone: a review. *Chemical Society Reviews*, **38**, 3484–3504.
9. Arbaoui, A., Redshaw, C. (2010). Metal catalysts for ε-caprolactone polymerisation. *Polymer Chemistry*, **1**, 801–826.
10. Shibasaki, Y., Sanada, H., Yokoi, M., Sanda, F., Endo, T. (2000). Activated monomer cationic polymerization of lactones and the application to well-defined block copolymer synthesis with seven-membered cyclic carbonate. *Macromolecules*, **33**, 4316–4320.
11. Mecerreyes, D., Humes, J., Miller, R. D., Hedrick, J. L., Detrembleur, C., Lecomte, P., Jérôme, R., San Roman, J. (2000). First example of an unsymmetrical difunctional monomer polymerizable by two living/controlled methods. *Macromolecular Rapid Communications*, **21**, 779–784.
12. Stassin, F., Halleux, O., Dubois, P., Detrembleur, C., Lecomte, P., Jérôme, R. (2000). Ring opening copolymerization of ε-caprolactone, γ-(triethylsilyloxy)-ε-caprolactone and γ-ethylene ketal-ε-caprolactone: A route to hetero-graft copolyesters. *Macromolecular Symposia*, **153**, 27–39.
13. Detrembleur, C., Mazza, M., Halleux, O., Lecomte, P., Mecerreyes, D., Hedrick, J. L., Jérôme, R. (2000). Ring-opening polymerization of γ-bromo-ε-caprolactone: A novel route to functionalized aliphatic polyesters. *Macromolecules*, **33**, 14–18.
14. Rieger, J., Bernaerts, K. V., Du Prez, F. E., Jérôme, R., Jérôme, C. (2004). Lactone end-capped poly(ethylene oxide) as a new building block for biomaterials. *Macromolecules*, **37**, 9738–9745.
15. Vroman, B., Mazza, M., Fernandez, M. R., Jérôme, R., Préat, V. (2007). Copolymers of ε-caprolactone and quaternized ε-caprolactone as gene carriers. *Journal of Controlled Release*, **118**, 136–144.
16. Rieger, J., Van Butsele, K., Lecomte, P., Detrembleur, C., Jérôme, R., Jérôme, C. (2005). Versatile functionalization and grafting of poly(ε-caprolactone) by Michael-type addition. *Chemical Communications*, 274–276.
17. Hawker, C. J., Hedrick, J. L., Malmström, E. E., Trollsås, M., Mecerreyes, D., Moineau, G., Dubois, P., Jérôme, R. (1998).

Dual living free radical and ring opening polymerizations from a double-headed initiator. *Macromolecules*, **31**, 213–219.

18. Jakubowski, W., Lutz, J.-F., Slomkowski, S., Matyjaszewski, K. (2005). Block and random copolymers as surfactants for dispersion polymerization. I. Synthesis via atom transfer radical polymerization and ring-opening polymerization. *Journal of Polymer Science, Part A: Polymer Chemistry*, **43**, 1498–1510.
19. Jankova, K., Chen, X., Kops, J., Batsberg W. (1998). Synthesis of amphiphilic PS-*b*-PEG-*b*-PS by atom transfer radical polymerization. *Macromolecules*, **31**, 538–541.
20. Matysjaszewski, K., Braunecker, W. A., (2007). Radical polymerization. In: Matysjaszewski, K., Gnanou, Y., Leibler, L. eds. *Macromolecular Engineering: Precise Synthesis, Materials Properties, Applications*. Wiley-VCH Verlag GmbH & Co. KGaA, Weinheim, pp. 161–215.
21. Hansen, N. M. L., Jankova, K., Hvilsted, S. (2007). Fluoropolymer materials and architectures prepared by controlled radical polymerizations. *European Polymer Journal*, **43**, 255–293.
22. Sun, H., Guo, B., Li, X., Cheng, R., Meng, F., Liu, H., Zhong, Z. (2010). Shell-sheddable micelles based on dextran-SS-poly(ε-caprolactone) diblock copolymer for efficient intracellular release of doxorubicin. *Biomacromolecules*, **11**, 848–854.
23. Motala-Timol, S., Jhurry, D., Zhou, J., Bhaw-Luximon, A., Mohun, G., Ritter, H. (2008). Amphiphilic poly(L-lysine-*b*-caprolactone) block copolymers: Synthesis, characterization, and solution properties. *Macromolecules*, **41**, 5571–5576.
24. Bougard, F., Jeusette, M., Mespouille, L., Dubois, P., Lazzaroni, R. (2007). Synthesis and supramolecular organization of amphiphilic diblock copolymers combining poly(*N*,*N*-dimethylamino-2-ethyl methacrylate) and poly(ε-caprolactone). *Langmuir*, **23**, 2339–2345.
25. Nakayama, M., Okano, T., Miyazaki, T., Kohori, F., Sakai, K., Yokoyama, M. (2006). Molecular design of biodegradable polymeric micelles for temperature-responsive drug release. *Journal of Controlled Release*, **115**, 46–56.
26. Chung, T. W., Cho, K. Y., Lee, H.-C., Nah, J. W., Yeo, J. H., Akaike, T., Cho, C. S. (2004). Novel micelle-forming block copolymer composed of poly(ε-caprolactone) and poly(vinyl pyrrolidone). *Polymer*, **45**, 1591–1597.
27. Jeon, H. J., You, Y. C., Youk, J. H. (2009). Synthesis and characterization of amphiphilic poly(*N*-vinyl pyrrolidone)-*b*-poly(ε-caprolactone) copolymers by a combination of cobalt-mediated radical polymerization and ring-opening polymerization. *Journal of Polymer Science: Part A: Polymer Chemistry*, **47**, 3078–3085.
28. Lele, B. S., Leroux, J. C. (2002). Synthesis and micellar characterization of novel amphiphilic A–B–A triblock copolymers of *N*-(2-hydroxypropyl)methacrylamide or *N*-vinyl-2-pyrrolidone with poly(ε-caprolactone). *Macromolecules*, **35**, 6714–6723.
29. Zhang, W., He, J., Liu, Z., Ni, P., Zhu. X. (2010). Biocompatible and pH-responsive triblock copolymer mPEG-*b*-PCL-*b*-PDMAEMA: Synthesis, self-assembly, and application. *Journal of Polymer Science: Part A: Polymer Chemistry*, **48**, 1079–1091.
30. Narain, R., Armes, S. P. (2003). Synthesis and aqueous solution properties of novel sugar methacrylate-based homopolymers and block copolymers. *Biomacromolecules*, **4**, 1746–1758.
31. Lu, F. Z., Meng, J. Q. Du, F. S. Li, Z. C., Zhang, B. Y. (2005). Pyrene end-labeled diblock glycopolymers: Synthesis and aggregation. *Macromolecular Chemistry and Physics*, **206**, 513–520.
32. Du, J. Z., Armes, S. P. (2009). Preparation of biocompatible zwitterionic block copolymer vesicles by direct dissolution in water and subsequent silicification within their membranes. *Langmuir*, **25**, 9564–9570.
33. Wittemann, A., Azzam, T., Eisenberg, A. (2007). Biocompatible polymer vesicles from biamphiphilic triblock copolymers and their interaction with bovine serum albumin. *Langmuir*, **23**, 2224–2230.
34. Sun, T. M., Du, J. Z., Yan, L. F., Mao, H. Q., Wang, J. (2008) Self-assembled biodegradable micellar nanoparticles of amphiphilic and cationic block copolymer for siRNA delivery. *Biomaterials*, **29**, 4348–4355.
35. Giacomelli, C., Borsali, R. (2008). ATRP of silylated glycerol monomethacrylate in organic medium for convenient synthesis of amphiphilic copolymers. *Macromolecular Rapid Communications*, **29**, 573–579.
36. Suriano, F., Coulembier, O., Degée, P., Dubois, P. (2008). Carbohydrate-based amphiphilic diblock copolymers: Synthesis, characterization, and aqueous properties. *Journal of Polymer Science: Part A: Polymer Chemistry*, **46**, 3662–3672.
37. Van Butsele, K., Cajot, S., Van Vlierberghe, S., Dubruel, P., Passirani, C., Benoit, J.-P., Jérôme, R., Jérôme, C. (2009). pH-Responsive flower-type micelles formed by a biotinylated poly(2-vinylpyridine)-block-poly(ethylene oxide)-block-poly(ε-caprolactone) triblock copolymer. *Advanced Functional Materials*, **19**, 1416–1425.
38. Sha, K., Li, D., Li, Y., Liu, X., Wang, S., Guan, J., Wang, J. (2007). Synthesis, characterization, and micellization of an epoxy-based amphiphilic diblock copolymer of ε-caprolactone and glycidyl methacrylate by enzymatic ring-opening polymerization and atom transfer radical polymerization. *Journal of Polymer Science: Part A: Polymer Chemistry*, **45**, 5037–5049.
39. Lu, C.-H., Huang, C.-F., Kuo, S.-W., Chang, F.-C. (2009). Synthesis and characterization of poly(ε-caprolactone-*b*-4-vinylpyridine): Initiation, polymerization, solution morphology, and gold metalation. *Macromolecules*, **42**, 1067–1078.
40. Du, J., Armes, S. P. (2008). Preparation of primary amine-based block copolymer vesicles by direct dissolution in water and subsequent stabilization by sol–gel chemistry. *Langmuir*, **24**, 13710–13716.
41. Geng, Y., Discher, D. E. (2005). Hydrolytic degradation of poly(ethylene oxide)-*block*-polycaprolactone worm micelles. *Journal of the American Chemical Society*, **127**, 12780–12781.
42. Gou, P.-F., Zhu, W.-P., Shen, Z.-Q. (2010). Synthesis, self-assembly, and drug-loading capacity of well-defined cyclodextrin-centered drug-conjugated amphiphilic $A_{14}B_7$ miktoarm star copolymers based on poly(ε-caprolactone) and poly(ethylene glycol). *Biomacromolecules*, **11**, 934–943.

43. Schramm, O. G., Meier, M. A. R., Hoogenboom, R., van Erp, H. P., Gohy, J.-F., Schubert, U. S. (2009). Polymeric nanocontainers with high loading capacity of hydrophobic drugs. *Soft Matter*, **5**, 1662–1667.

44. Ternat, C., Kreutzer, G., Plummer, C. J. G., Nguyen, T. Q., Herrmann, A., Ouali, L., Sommer, H., Fieber, W., Velazco, M. I., Klok, H.-A., Månson, J.-A. E. (2007). Amphiphilic multi-arm star-block copolymers for encapsulation of fragrance molecules. *Macromolecular Chemistry and Physics*, **208**, 131–145.

45. Östmark, E., Nyström, D., Malmström, E. (2008). Unimolecular nanocontainers prepared by ROP and subsequent ATRP from hydroxypropylcellulose. *Macromolecules*, **41**, 4405–4415.

46. Gou, P.-F., Zhu, W.-P., Xu, N., Shen, Z.-Q. (2010). Synthesis and self-assembly of well-defined cyclodextrin-centered amphiphilic $A_{14}B_7$ multimiktoarm star copolymers based on poly(ε-capro-lactone) and poly(acrylic acid). *Journal of Polymer Science: Part A: Polymer Chemistry*, **48**, 2961–2974.

47. Yuan, Y. Y., Wang, Y. C., Du, J. Z., Wang, J. (2008). Synthesis of amphiphilic ABC 3-miktoarm star terpolymer by combination of ring-opening polymerization and "click" chemistry. *Macromolecules*, **41**, 8620–8625.

48. Vora, A., Singh, K., Webster, D. C. (2009). A new approach to 3-miktoarm star polymers using a combination of reversible addition-fragmentation chain transfer (RAFT) and ring opening polymerization (ROP) via "click" chemistry. *Polymer*, **50**, 2768–2774.

49. Liu, Y., Nguyen, J., Steele, T., Merkel, O., Kissel, T. (2009). A new synthesis method and degradation of hyperbranched polyethylenimine grafted polycaprolactone block mono-methoxyl poly(ethylene glycol) copolymers (hy-PEI-g-PCL-b-mPEG) as potential DNA delivery vectors. *Polymer*, **50**, 3895–3904.

50. Chen, Y. M., Wulff, G. (2002). ABA and star amphiphilic block copolymers composed of polymethacrylate bearing a galactose fragment and poly(ε-caprolactone). *Macromolecular Rapid Communications*, **23**, 59–63.

51. Dai, X. H., Zhang, H. D., Dong, C. M. (2009). Fabrication, biomolecular binding, in vitro drug release behavior of sugar-installed nanoparticles from star poly(ε-caprolactone)/glycopolymer biohybrid with a dendrimer core. *Polymer*, **50**, 4626–4634.

52. Li, H., Jérôme, R., Lecomte, P. (2008). Amphiphilic sun-shaped polymers by grafting macrocyclic copolyesters with PEO. *Macromolecules*, **41**, 650–654.

53. Li, J., Ren, J., Cao, Y., Yuan, W. (2010). Synthesis of biodegradable pentaarmed star-block copolymers via an asymmetric BIS-TRIS core by combination of ROP and RAFT: From star architectures to double responsive micelles. *Polymer*, **51**, 1301–1310.

54. Suriano, F., Coulembier, O., Dubois, P. (2010). Synthesis of amphiphilic A_3B mikto-arm copolymers from a sugar core: Combination of hydrophobic PCL and hydrophilic glycopolymers for biocompatible nanovector preparation. *Journal of Polymer Science: Part A: Polymer Chemistry*, **48**, 3271–3280.

55. Darcos, V., Habnouni, A. E., Nottelet, B., Ghzaoui, A. E., Coudane, J. (2010). Well-defined PCL-*graft*-PDMAEMA prepared by ring-opening polymerisation and click chemistry. *Polymer Chemistry*, **1**, 280–282.

56. Guan, X., Quan, D., Shuai, X., Liao, K., Mai K. (2007). Chitosan-*graft*-poly(ε-caprolactone)s: An optimized chemical approach leading to a controllable structure and enhanced properties. *Journal of Polymer Science: Part A: Polymer Chemistry*, **45**, 2556–2568.

57. Wu, D. Q., Li, Z. Y., Li, C., Fan, J. J., Lu, B., Chang, C., Cheng, S. X., Zhang, X. Z., Zhuo, R. X. (2010). Porphyrin and galactosyl conjugated micelles for targeting photodynamic therapy. *Pharmaceutical Research*, **27**, 187–199.

58. Iha, R. K., Van Horn, B. A., Wooley, K. L. (2010). Complex, degradable polyester materials via ketoxime ether-based functionalization: amphiphilic, multifunctional graft copolymers and their resulting solution-state aggregates. *Journal of Polymer Science: Part A: Polymer Chemistry*, **48**, 3553–3563.

59. Zhu, J. L., Cheng, H., Jin, Y., Cheng, S. X., Zhang, X. Z., Zhuo, R. X. (2008) Novel polycationic micelles for drug delivery and gene transfer. *Journal of Materials Chemistry*, **18**, 4433–4441.

60. Zhang, X., Chen, F., Zhong, Z., Zhuo, R. (2010). Amphiphilic block-graft copolymers with a degradable backbone and polyethylene glycol pendant chains prepared via ring-opening polymerization of a macromonomer. *Macromolecular Rapid Communications*, **31**, 2155–2159.

61. Thomsen, A. D., Malmström, E., Hvilsted, S. (2006). Novel polymers with high carboxylic acid loading. *Journal of Polymer Science: Part A: Polymer Chemistry*, **44**, 6360–6377.

62. Daugaard, A. E., Hvilsted, S. (2008). The influence of pendant carboxylic acid loading on surfaces of statistical poly[(4-hydroxystyrene)-*co*-styrene]s. *Macromolecular Rapid Communication*, **29**, 1119–1125.

63. Dimitrov, I., Jankova, K., Hvilsted, S. (2010). Synthesis of polystyrene-based random copolymers with balanced number of basic or acidic functional groups. *Journal of Polymer Science: Part A: Polymer Chemistry*, **48**, 2044–2052.

64. Daugaard, A. E., Hvilsted, S., Hansen, T. S., Larsen, N. B. (2008). Conductive polymer functionalization by click chemistry. *Macromolecules*, **41**, 4321–4327.

65. Hansen, T. S., Daugaard, A. E., Hvilsted, S., Larsen, N. B. (2009). Spatially selective functionalization of conducting polymers by "electroclick" chemistry. *Advanced Materials*, **21**, 1–4.

66. Gibson, J. D.; Khanal, B. P.; Zubarev, E. R. (2007). Paclitaxel-functionalized gold nanoparticles. *Journal of the American Chemical Society*, **129**, 11653–11661.

67. Hong, R., Han, G., Fernández, J. M., Kim, B.-J., Forbes, N. S., Rotello, V. M. (2006). Glutathione-mediated delivery and release using monolayer protected nanoparticle carriers. *Journal of the American Chemical Society*, **128**, 1078–1079.

68. Agarwal, A., Huang, S. W., O'Donnell, M., Day, K. C., Day, M., Kotov, N., Ashkenazi, S. (2007). *Journal of Applied Physics*, **102**, 064701.

69. Hirsch, L. R., Stafford, R. J., Bankson, J. A., Sershen, S. R., Rivera, B., Price, R. E., Hazle, J. D., Halas, N. J., West, J.

L. (2003). Nanoshell-mediated near-infrared thermal therapy of tumors under magnetic resonance guidance. *Proceedings of the National Academy of Sciences of the United States of America*, **100**, 13549–13554.

70. Javakhishvili, I., Hvilsted, S. (2009). Gold nanoparticles with thiol-derivatized amphiphilic poly(ε-caprolactone)-*b*-poly(acrylic acid). *Biomacromolecules*, **10**, 74–81.

71. Mourtas, S., Gatos, D., Kalaitzi, V., Katakalou, C., Barlos, K. (2001). S-4-Methoxytrityl mercapto acids: Synthesis and application. *Tetrahedron Letters*, **42**, 6965–6967.

72. Carrot, G., Hilborn, J. G., Trollsås, M., Hedrick, J. L. (1999). Two general methods for the synthesis of thiol-functional polycaprolactones. *Macromolecules*, **32**, 5264–5269.

73. Azzam, T., Eisenberg, A. (2007). Monolayer-protected gold nanoparticles by the self-assembly of micellar poly(ethylene oxide)-*b*-poly(ε-caprolactone) block copolymer. *Langmuir*, **23**, 2126–2132.

74. Hussain, I., Graham, S., Wang, Z., Tan, B., Sherrington, D. C., Rannard, S. P., Cooper, A. I., Brust, M. (2005). Size-controlled synthesis of near-monodisperse gold nanoparticles in the 1–4 nm range using polymeric stabilizers. *Journal of the American Chemical Society*, **127**, 16398–16399.

75. Raula, J., Shan, J., Nuopponen, M., Niskanen, A., Jiang, H., Kauppinen, E., Tenhu, H. (2003). Synthesis of gold nanoparticles grafted with a thermoresponsive polymer by surface-induced reversible-addition-fragmentation chain-transfer polymerization. *Langmuir*, **19**, 3499–3504.

76. Shan, J., Chen, J., Nuopponen, M., Viitala, T., Jiang, H., Peltonen, J., Kauppinen, E., Tenhu, H. (2006). Optical properties of thermally responsive amphiphilic gold nanoparticles protected with polymers. *Langmuir*, **22**, 794–801.

77. Javakhishvili, I., Tanner, S., Binder, W. H., Hvilsted, S. (2010). Facile synthesis of linear-dendritic cholesteryl-poly(ε-caprolactone)-*b*-(L-lysine)$_{G2}$ by thiol–ene and azide–alkyne "click" reactions. *Polymer Chemistry*, **1**, 506–513.

78. Gitsov, I. (2008). Hybrid linear dendritic macromolecules: From synthesis to applications. *Journal of Polymer Science: Part A: Polymer Chemistry*, **46**, 5295–5314.

79. Würsch, A., Möller, M., Glauser, T., Lim, L. S., Voytek, S. B., Hedrick, J. L., Frank, C. W., Hilborn, J. G. (2001). Dendritic-linear A*x*B*x* block copolymers prepared via controlled ring-opening polymerization of lactones from orthogonally protected multifunctional initiators. *Macromolecules*, **34**, 6601–6615.

80. Li, Y., Li, Q., Li, F., Zhang, H., Jia, L., Yu, J., Fang, Q., Cao, A. (2006). Amphiphilic poly(L-lactide)-*b*-dendritic poly(L-lysine)s synthesized with a metal-free catalyst and new dendron initiators: Chemical preparation and characterization. *Biomacromolecules*, **7**, 224–231.

81. Klok, H.-A., Hwang, J. J., Hartgerink, J. D., Stupp, S. I. (2002). Self-assembling biomaterials: L-lysine-dendron-substituted cholesteryl-(L-lactic acid)$_n$. *Macromolecules*, **35**, 6101–6111.

82. Xia, J., Zhang, X., Matysjaszewski, K. (1999). Synthesis of star-shaped polystyrene by atom transfer radical polymerization using an "arm first" approach. *Macromolecules*, **32**, 4482–4484.

83. Gao, H., Matysjaszewski, K. (2009). Synthesis of functional polymers with controlled architecture by CRP of monomers in the presence of cross-linkers: From stars to gels. *Progress in Polymer Science*, **34**, 317–350.

84. Zhang, X., Xia, J., Matysjaszewski, K. (2000). End-functional poly(*tert*-butyl acrylate) star polymers by controlled radical polymerization. *Macromolecules*, **33**, 2340–2345.

85. Gao, H., Matysjaszewski, K. (2008). Synthesis of low-polydispersity miktoarm star copolymers via a simple "Arm-First" method: Macromonomers as arm precursors. *Macromolecules*, **41**, 4250–4257.

86. Du, J., Chen, Y. (2004). Star polymer, PCL-PS heteroarm star polymer by ATRP, and core-carboxylated PS star polymer thereof. *Macromolecules*, **37**, 3588–3594.

87. Wiltshire, J. T., Qiao, G. G. (2006). Selectively degradable core cross-linked star polymers. *Macromolecules*, **39**, 9018–9027.

88. Wiltshire, J. T., Qiao, G. G. (2008). Synthesis of core cross-linked star polymers with adjustable coronal properties. *Macromolecules*, **41**, 623–631.

89. Bosnan, A. W., Vestberg, R., Heumann, A., Fréchet, J. M. J., Hawker, C. J. (2003). A modular approach toward functionalized three-dimensional macromolecules: From synthetic concepts to practical applications. *Journal of the American Chemical Society*, **125**, 715–728.

90. Connal, L. A., Vestberg, R., Hawker, C. J., Qiao, G. G. (2007). Synthesis of dendron functionalized core cross-linked star polymers. *Macromolecules*, **40**, 7855–7863.

91. Javakhishvili, I., Hvilsted, S. (2010). Miktoarm core-crosslinked star copolymers with biologically active moieties on peripheries. *Polymer Chemistry*, **1**, 1650–1661.

92. Degée, P., Dubois, P., Jérôme, R., Teyssié, P. (1992). Macromolecular engineering of polylactones and polylactides. 9. Synthesis, characterization, and application of ω-primary amine poly(ε-caprolactone). *Macromolecules*, **25**, 4242–4248.

93. Hoskins, J. N., Grayson, S. M. (2009). Synthesis and degradation behavior of cyclic poly(ε-caprolactone). *Macromolecules*, **42**, 6406–6413.

5

SELF-ASSEMBLED PORPHYRIN NANOSTRUCTURES AND THEIR POTENTIAL APPLICATIONS

John A. Shelnutt and Craig J. Medforth

5.1 INTRODUCTION

The field of organic nanomaterials covers a wide range of research on nanoscale materials composed of organic precursors, and the aim of the research is to produce tailored nanostructures and functional properties [1–3]. These materials include dendrimers, supermolecules, artificially designed proteins, polymers, and a wide variety of other self-assembled nanostructures and films. Organic nanomaterials are used as functional subunits in the form of thin films, fibers, wires, and light antennae in proposed and realized nanoscale devices. The design of nanoscale architectures composed of functional organic components is a major goal in the development of new functional biopolymers, ultrasensitive sensors for biomolecules and chemicals, separations materials and membranes, optoelectronics, photonics, and solar energy and other devices that will be indispensable for a sustainable economy based on renewable energy. Solar energy, which provides the only renewable energy source of sufficient magnitude to meet projected global energy requirements [4], stands to gain from the development of organic nanomaterials because there are many current technologies for collecting and transforming solar energy that can benefit from their use. In particular, photovoltaics, dye-sensitized solar cells, artificial photosynthesis, and fuel cells may benefit from advancements in organic nanomaterials. Consequently, a considerable amount of research and development is devoted to the pursuit of improved solar energy systems based on organic nanomaterials [1–4].

The favorable light-harvesting and charge-transport functions of porphyrins (and the closely related phthalocyanines) point to a class of highly attractive organic molecules for producing functional solar nanomaterials as related tetrapyrrole pigments play key roles in biological photosynthesis [5, 6]. As a consequence, these molecules have garnered a great deal of attention for a range of solar technologies [7–14], partly due to their chemical stability and the ease with which their molecular structure can be synthetically manipulated to provide new functions and structures.

Recently, our group and others have found that porphyrins self-organize as nanoscale and microscale structures that mimic related biological structures of photosynthetic organisms (e.g., chlorosomal rods of green bacteria). These nanostructures form by self-assembly and other synthetic methods [15–17]. These synthetic porphyrin-based structures not only are of interest for solar energy technologies, but also have potential application as optoelectronic and photonic materials for nano- and microscale electronic devices.

In this chapter, we describe the recent developments in the synthesis of isolable self-assembled porphyrin nanostructures [18, 19], describe the early stages of the characterization of their electronic, optical, and photocatalytic properties, and speculate on their potential utilization in nanoengineered systems for the production of fuels and power from solar energy. Section 5.2 summarizes the several methods of synthesis of self-assembled porphyrin nanostructures and describes studies of the nano- and microscale morphological

Organic Nanomaterials: Synthesis, Characterization, and Device Applications, First Edition. Edited by Tomás Torres and Giovanni Bottari.

features of the porphyrin nanomaterials (primarily by electron microscopy). Section 5.3 summarizes the optical, electronic, optoelectronic, and photocatalytic properties of the porphyrin nanostructures, focusing finally on their remarkable ability to grow nanoscale metal structures onto their outer surface to form metal nanocomposites. This is made possible by the reductive photocatalytic ability of certain porphyrin metal complexes incorporated into the nanostructures to reduce metal ions from solution with the aid of a soluble sacrificial electron donor and visible light. This self-metallization process can be used to add catalytic sites (e.g., platinum nanocatalysts for hydrogen production) and metal conductors (e.g., for electrically connecting the porphyrin nanostructures to other structures). Section 5.4 focuses on the utilization of porphyrin nanostructures, and particularly a new type of binary solid formed by the self-assembly of ionic porphyrins [18–20], in systems for solar energy generation, storage, and utilization. The binary ionic solids are formed from oppositely charged porphyrins that can independently be chosen based on their desired functional properties without significantly altering the basic ionic crystalline structure of the solid. Cooperative interactions and collective properties of these new nanomaterials can lead to entirely new or enhanced functionality not possessed by the constituent porphyrin anions and cations alone. The potential uses of these highly tunable porphyrin nanostructures, called *cooperative binary ionic* (CBI) solids [18–20], and their metal nanocomposites in possible solar devices are described [21]. Finally, we conclude with speculations on the future directions of research on self-assembled porphyrin nanostructures and potential new uses of CBI nanomaterials in Section 5.5.

5.2 SYNTHESIS AND STRUCTURE

5.2.1 Overview

Before discussing the self-assembled porphyrin nanostructures, we provide an overview of the types of materials included in this review. This chapter is focused on self-assembled nanostructures prepared in solution with synthetic porphyrins; we do not address similar structures that may arise from the self-assembly of naturally occurring porphyrin derivatives such as chlorophyll and bacteriochlorophyll. The porphyrin nanostructures are free-standing, can be isolated as discrete units, and at least one dimension or feature of the structure is less than 100 nm. Porphyrins are the dominant and typically the only molecular entity in the nanostructure. This limitation excludes many porphyrin-based materials that are also of interest in optoelectronics and energy research, such as porphyrin layers on surfaces, porphyrin nanoparticles, vapor-deposited films, and so on. The reader is directed to other recent reviews that include some of these topics [7, 11, 22, 23].

It has long been known that porphyrins self-assemble to form oligomeric structures and aggregates, although the lack of control over the size, dispersity, and morphology of these materials is problematic for many potential applications. Beginning in the last decade, reports began to appear describing the synthesis of isolable porphyrin nanostructures with well-defined shapes and sizes [18, 24–32]. The following subsections review the syntheses and structures of some of these new organic nanomaterials.

5.2.2 Synthesis by Reprecipitation

Reprecipitation is probably the simplest method for producing nanostructures containing only a single type of porphyrin, and many porphyrin nanostructures have been prepared in this way. In 2007, Wang et al. [29] described the synthesis of square porphyrin nanosheets from SnPyTriPP (**1**) by injection of a 1.0 mM solution of the porphyrin into vigorously stirred water at room temperature. SEM images of the product (Figure 5.1) show porphyrin sheets with edge lengths of 0.3–1.0 μm. AFM studies revealed that the sheets have uniform thicknesses of only 7–12 nm, with some of the nanosheets having a triangular sheet growing atop and oriented along the diagonal of the square sheet. XRD data indicate that the sheets are crystalline. A layered structure with a hydrogen-bonding motif (Sn—O···H—O—H···N_{py}) between layers was observed in the x-ray crystal structure of the related porphyrin Sn^{IV}T(4-Py)P [33], prompting us to suggest a similar structure for the SnPyTriPP nanosheets. Subsequently, we obtained the x-ray structure of a single

FIGURE 5.1. SEM image of nanosheets prepared by the reprecipitation of SnPyTriPP.

crystal of SnPyTriPP showing similar layers of porphyrins that are linked via a hydrogen-bonded network of water molecules [34].

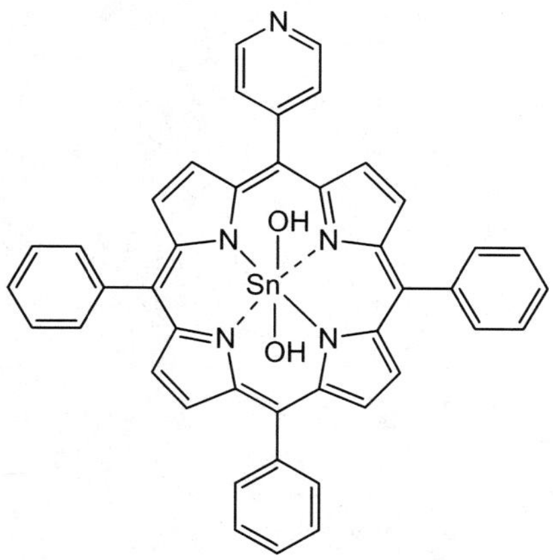

1

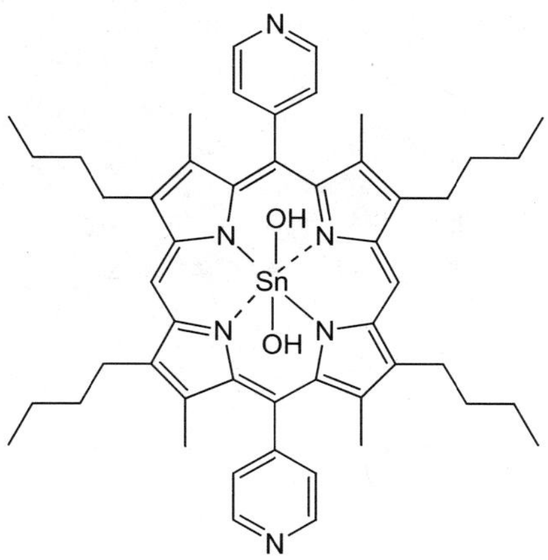

2

A large number of studies of the reprecipitation reactions of tin porphyrins to produce nanocrystals have now been published by Lee et al. [25–27]. For example, reprecipitation of SnDiPyP (**2**) gives nanorods, with the length of the nanorods being controlled by the addition of the triblock copolymer Pluronic F127 (Figure 5.2) [25]. Lee et al. have also reported the effects of the bulky substituent groups and hydrophobicity on the reprecipitation of the tin porphyrin complexes **3a–c** (R = methyl, *n*-butyl, or *n*-octyl) [26] and the structures formed by **2**, **3** (R = *n*-butyl), and **4** for a fixed set of reaction conditions [27].

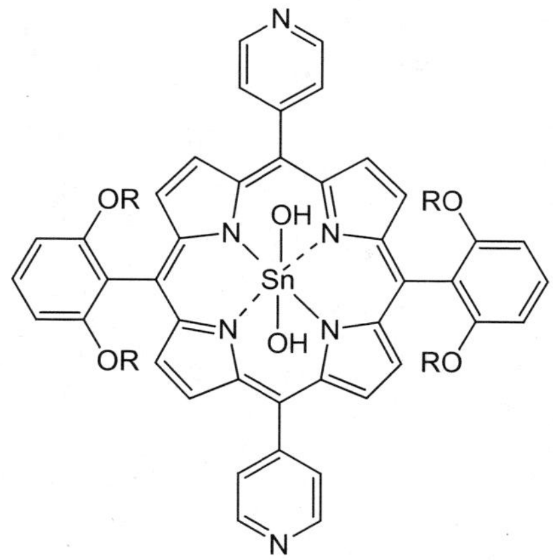

3a-c

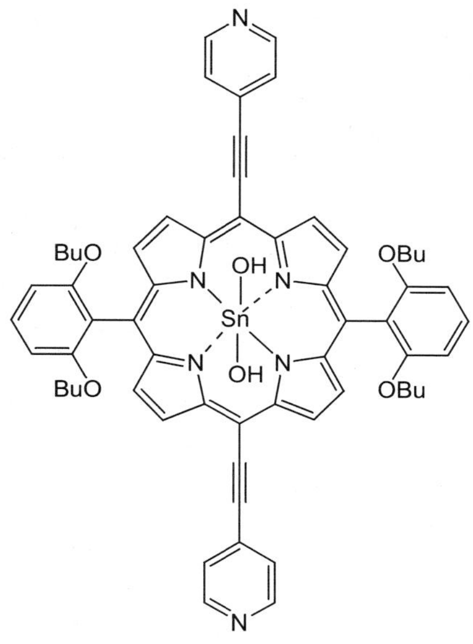

4

5.2.3 Self-Assembly by Coordination Polymerization

Coordination polymerization provides another useful method for synthesizing porphyrin nanostructures. Although not entirely unrelated to the reprecipitation method, coordination polymerization is treated separately herein. Porphyrin nanostructures have been prepared by two types of coordination polymerization reactions (Figure 5.3). In type A (Figure 5.3a), a single metalloporphyrin species containing both a coordinating metal (e.g., Zn(II)) and one or more ligating peripheral substituent groups (e.g., pyridyl groups) is polymerized. A type A example is the polymerization of ZnT(4-Py)P, where coordination occurs between the core metal ions

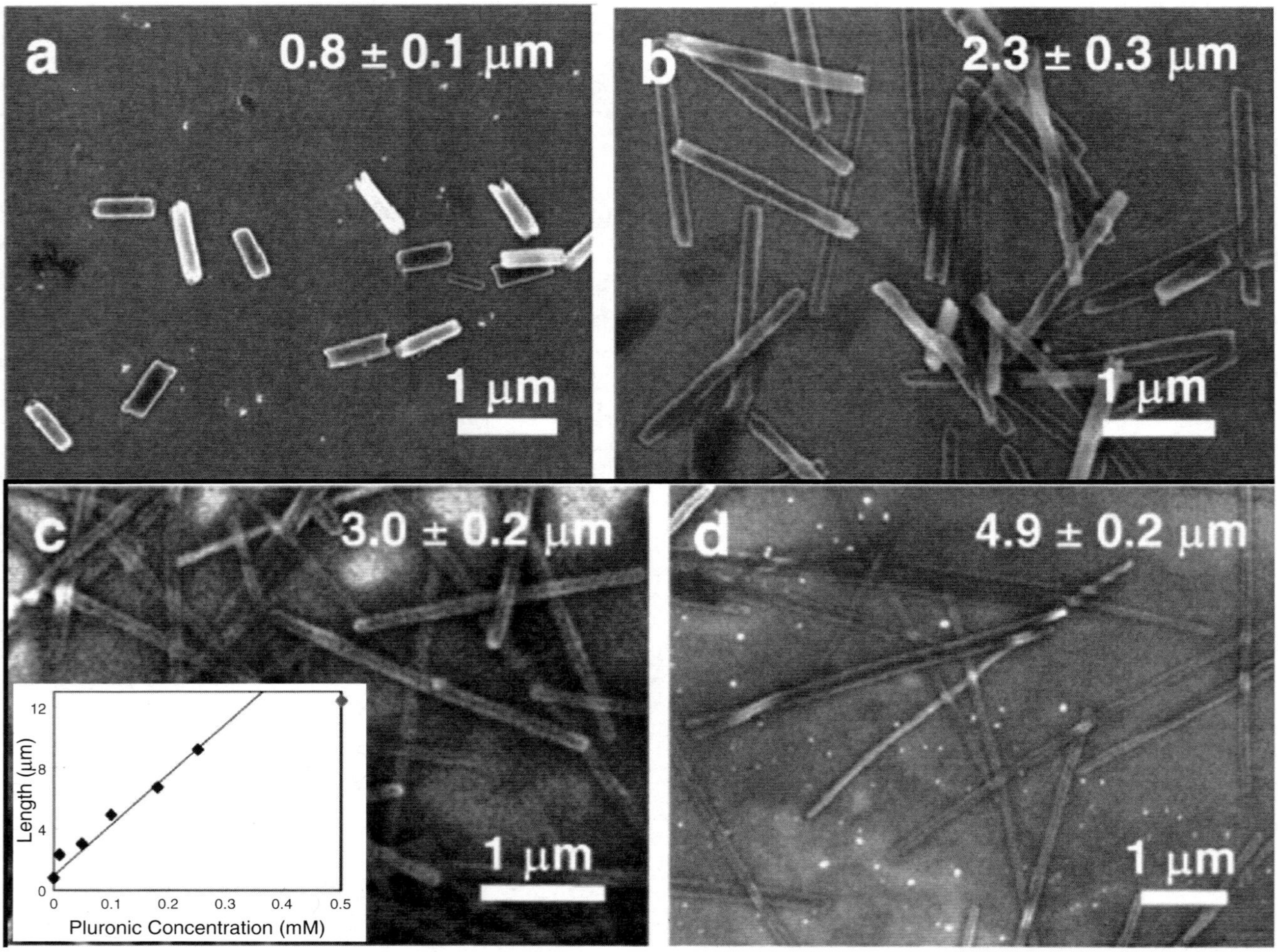

FIGURE 5.2. SEM images of porphyrin nanorods obtained by reprecipitation of SnDiPyP. The length of the nanorods varies with the concentration of Pluronic F127 surfactant (**inset**), but their width is approximately constant. Adapted from reference 25.

and the pyridyl groups giving a polymer [24]. In type B coordination polymers (Figure 5.3b), porphyrin molecules with ligating peripheral substituents (e.g., T(4-Py)P) are joined together by an added metal complex (e.g., H_2PtCl_6) [30], with the metal complex providing the linkers between the porphyrin substituent groups to produce the polymer. Numerous x-ray crystal structures of porphyrin coordination polymers are available in the literature [35–39].

The first example of a porphyrin nanostructure formed by coordination polymerization to be reported was obtained from a type A reaction by Hu et al. [24]. A 0.25 mM solution of ZnT(4-Py)P in DMF was added to a continuously stirred 0.4 mM solution of CTAB in water at room temperature. Polymerization produced hollow hexagonal nanoprisms that were 95 $\pm$ 3 nm wide and 529 $\pm$ 12 nm long (Figure 5.4). Changing the surfactant concentration alters the length of the nanoprisms leaving the width unaltered. The surfactant may serve as a template for nanostructure growth. A comparison of XRD data with single-crystal x-ray data [40] suggested a six-coordinate zinc atom with each porphyrin linked to four neighboring porphyrins (shown as an idealized structure in Figure 5.3a).

A porphyrin nanostructure formed by type B coordination polymerization has been reported by Wang et al. [30]. Reaction of metal complexes of MT(4-Py)P (M = Sn^{IV}, Co^{III}, and Fe^{III}) with chloroplatinic acid (H_2PtCl_6) gives nanospheres. With Sn^{IV}T(4-Py)P, the nanospheres (Figure 5.5a) have an average diameter of 161.1 $\pm$ 12.8 nm. XRD results show that the structures are amorphous. EDX analysis indicated a formula of $((Sn^{IV}T(4\text{-}Py)P)(Pt^{II}Cl_2)_2)_n$ which is illustrated in the idealized structure shown in Figure 5.3b. Nanospheres were also prepared with other porphyrins including Co^{III}T(4-Py)P (Figure 5.5b) and Fe^{III}T(4-Py)P (not shown). The nanospheres were very robust, retaining their structure upon

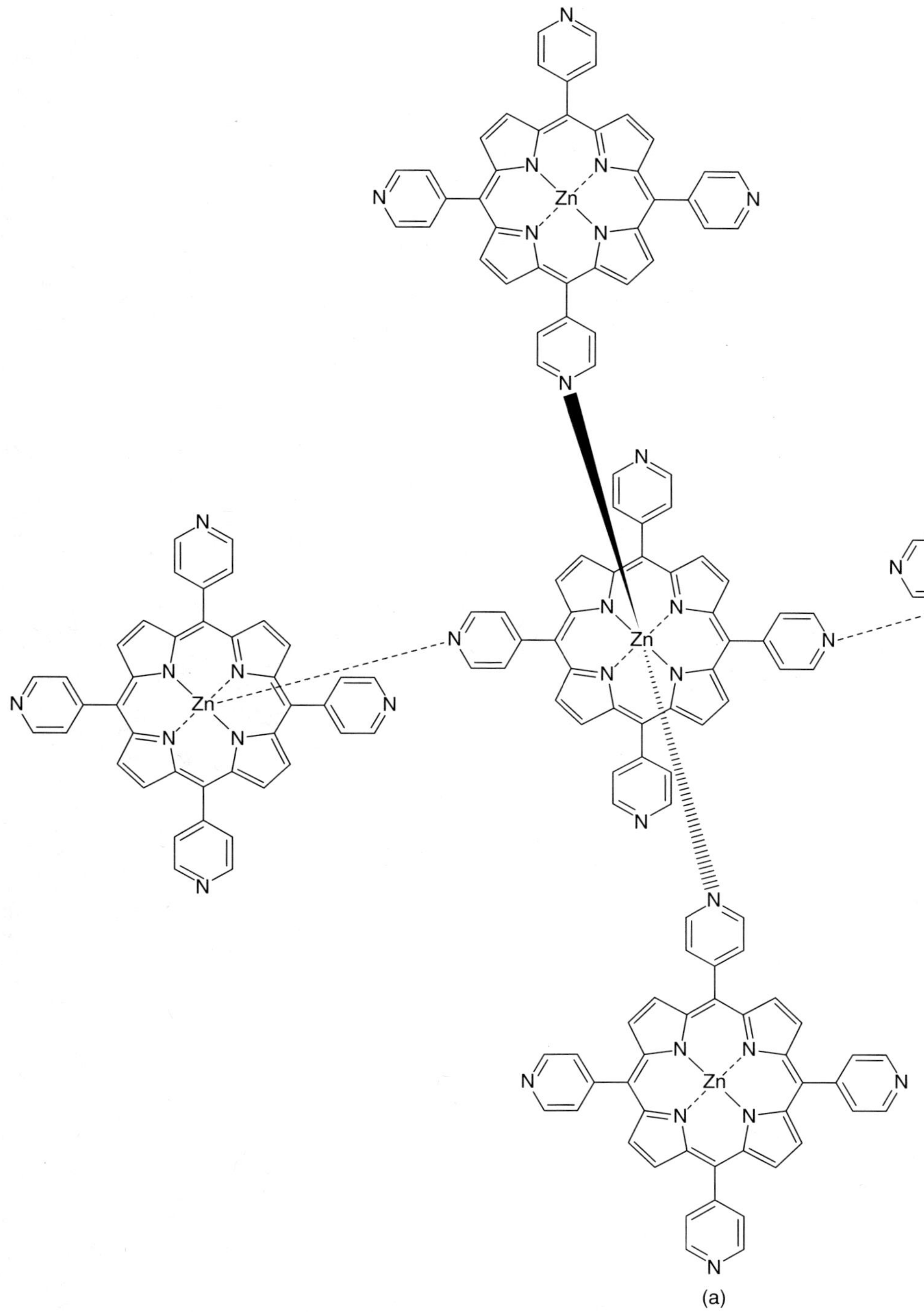

FIGURE 5.3. Representations of (**a**) type A (self-coordination) coordination polymers and (**b**) type B (added metal-induced complexation) coordination polymers. Adapted from reference 19. (*Continued*)

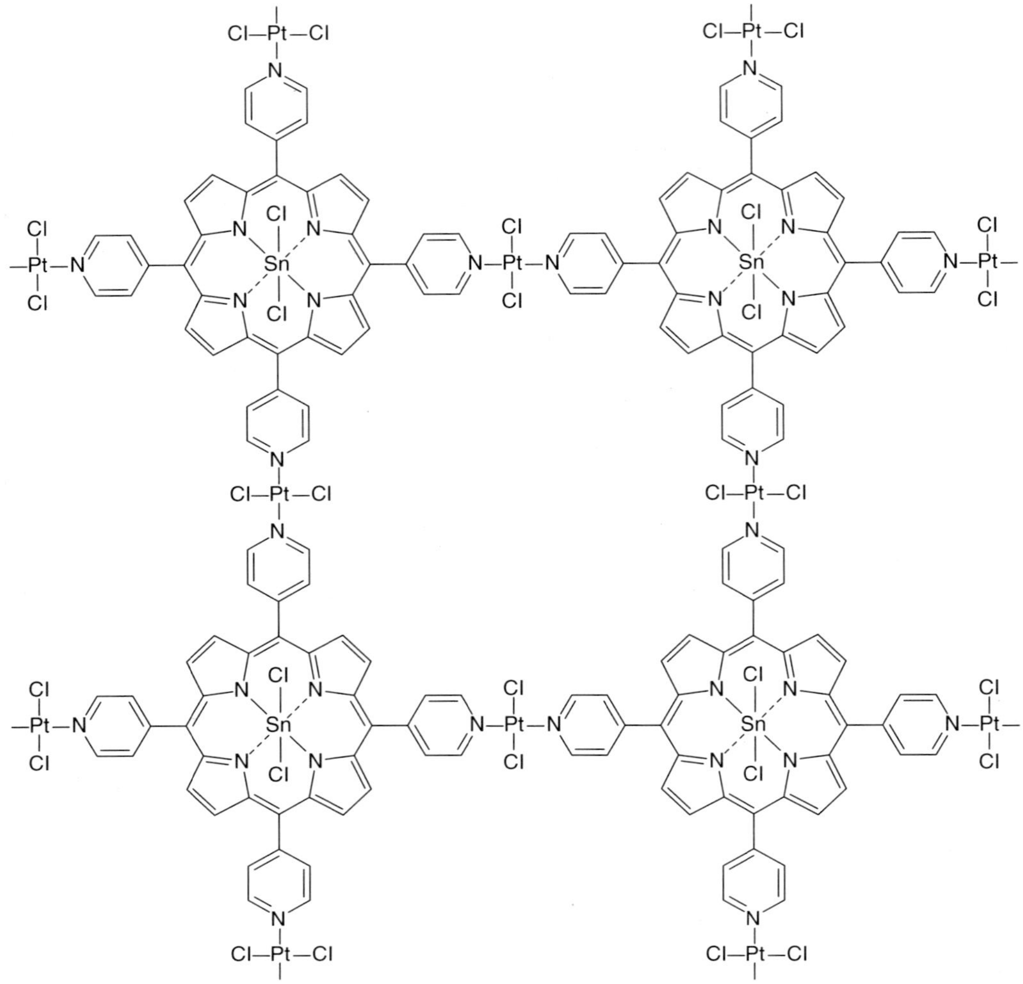

(b)

FIGURE 5.3. (*Continued*)

drying. The nanospheres are insoluble in water (pH range 2–12) and organic solvents (ethanol, dichloromethane); pyridine is the only solvent found that dissolves the nanospheres.

5.2.4 Synthesis by Ionic Self-Assembly

One of the most promising new methods for preparing porphyrin nanostructures is ionic self-assembly [41]. Organic structures formed by ionic self-assembly conventionally involve the reaction between an organic molecule and an oppositely charged ion or polyelectrolyte, with the goal of producing an ionic solid in which the organic species provides useful functionality. For porphyrins, examples of this type of solid are the crystalline solids formed by a water-soluble porphyrin anion or cation with small counterions (e.g., $CF_3SO_3^-$, ClO_4^-, Na^+, etc.) [42–44]. More interesting are binary organic solids composed of

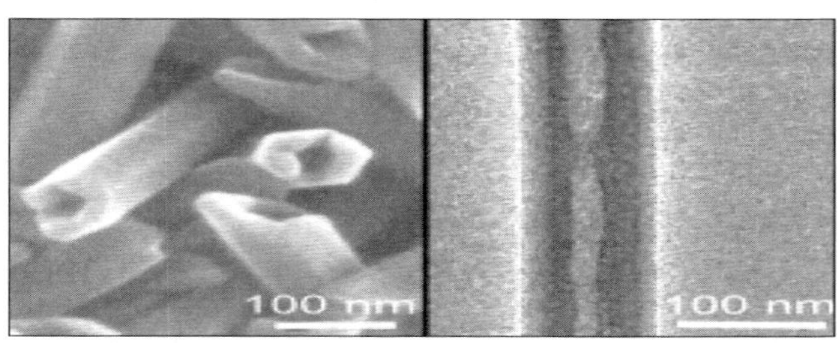

FIGURE 5.4. SEM image (**left**) and TEM image (**right**) of hollow hexagonal nanoprisms prepared by coordination polymerization of ZnT(4-Py)P. The nanostructures are composed of a type A self-coordination polymer. Adapted from reference 24.

two oppositely charged porphyrins for which both types of porphyrin (polyelectrolytes) may provide distinct and possibly complementary functionality. This approach for producing porphyrin nanostructures was first reported by Wang et al. [32] in 2004 using the extensively studied water-soluble porphyrin TPPS (**5**) and T(4-Py)P (as the tin complex, **6**). When allowed to self-assemble at pH 2 (0.01 M HCl), the species present in solution are H_4TPPS^{2-} (**7**) and a mixture of $Sn^{IV}T(4\text{-Py})P(OH)_2$ (net charge 4+) and $Sn^{IV}T(4\text{-Py})P(OH)(H_2O)$ (net charge 5+) (**8**).

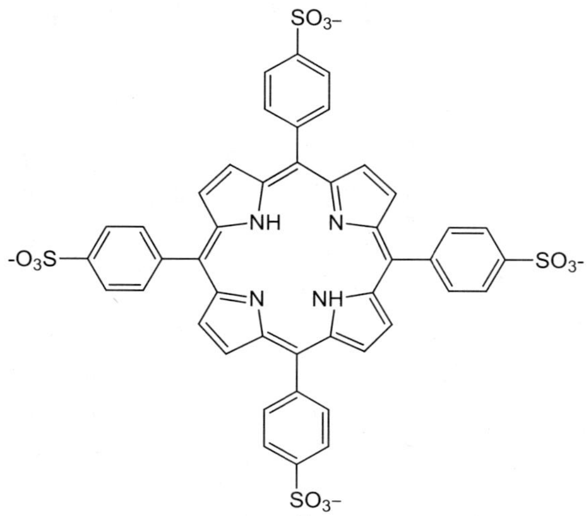

5

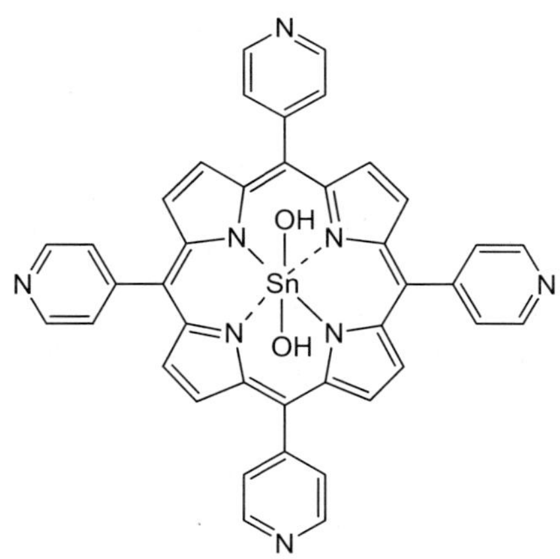

6

Remarkably, the ionic self-assembly of these porphyrins produces a greenish precipitate that consists almost entirely of nanotubes (Figure 5.6). The nanotubes are typically 30–70 nm wide and can be micrometers in length. The walls of the tubes vary but are typically 20 nm thick. The nanotubes are hollow as can be seen in the high-magnification TEM image of a SnT(4-Py)P–H_4TPPS nanotube caught in an end on orientation (Figure 5.6, inset).

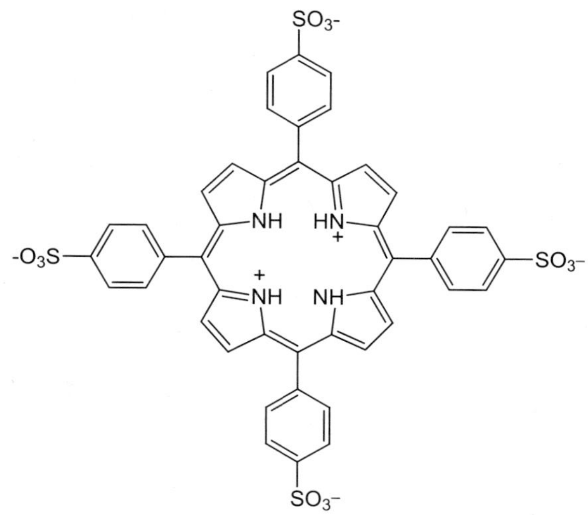

7

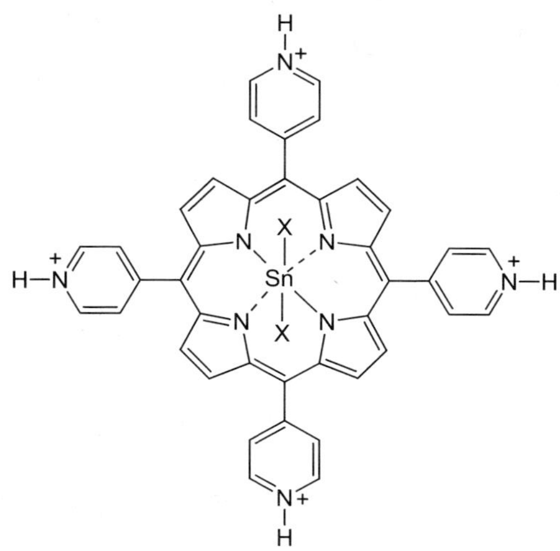

8

Analysis of the composition of the nanotubes shows that the two porphyrins are present in a ratio of approximately 2.0–2.5 TPPS to 1.0 SnT(4-Py)P [32]. This is consistent with a binary ionic solid [41] formed by the two porphyrins without additional counterions, as the observed composition balances the net molecular charges of −2 and +4/+5 for the porphyrins in the solid. Indeed, EDX spectra detected no counterions in the nanotubes, even when heavy counterions were employed (e.g., the reactions were carried out in 0.01 M HI). The solubility of this ionic solid is very low, resulting in a high yield of the nanotubes.

The inset TEM image in Figure 5.6 also reveals a multilamellar cylindrical structure, suggesting a mechanism by which the binary porphyrin tubes might form. Previous studies show that H_4TPPS^{2-} self-aggregates are nanotubes with walls only one layer of porphyrins thick [45–47]. In the case

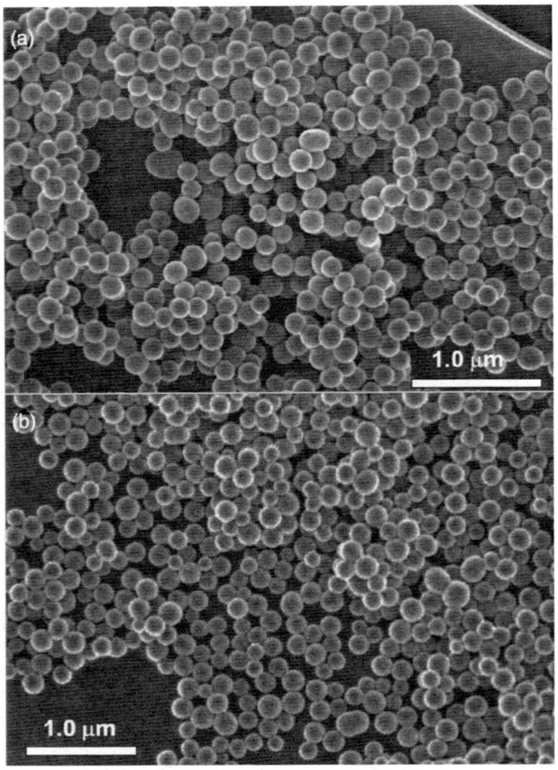

FIGURE 5.5. SEM image of coordination polymer nanospheres prepared from $Sn^{IV}T(4\text{-}Py)P$ and H_2PtCl_6 (**a**), and from $Co^{III}T(4\text{-}Py)P$ and H_2PtCl_6 (**b**). The nanospheres (**b**) formed in the presence of 1.7 and 12.3 mM HCl have size distributions of 67.3 ± 17.7 and 175.6 ± 19.7 nm. Adapted from reference 19.

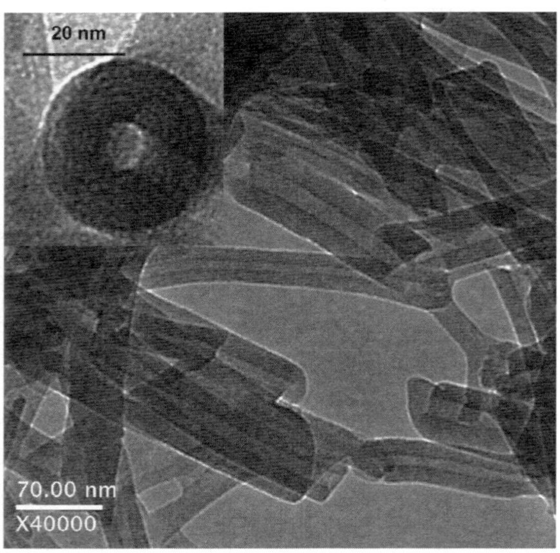

FIGURE 5.6. TEM images of porphyrin nanotubes obtained by the self-assembly of TPPS and $Sn^{IV}T(4\text{-}Py)P$ at pH 2. **Insert**: High magnification TEM image of a porphyrin nanotube caught in the vertical orientation showing the multilamellar structure. Adapted from reference 19.

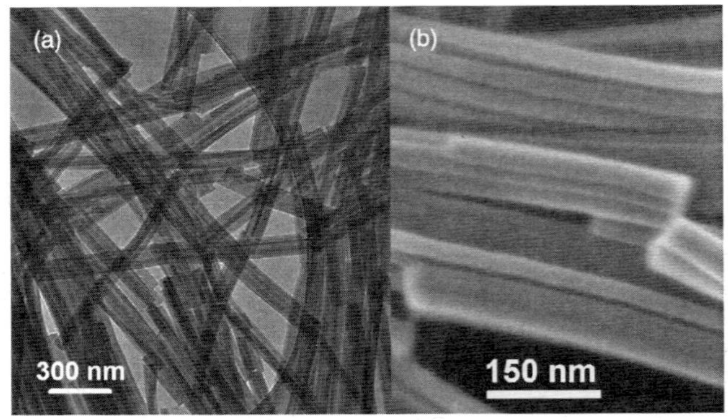

FIGURE 5.7. TEM (**a**) and SEM (**b**) images of porphyrin nanofiber bundles obtained by ionic self-assembly of TPPS and $Sb^{V}OTPP$. Adapted from reference 19.

of the binary ionic nanotubes, the H_4TPPS^{2-} nanotubes may nucleate the addition of successive layers of SnT(4-Py)P and H_4TPPS^{2-} to give the multilamellar structure shown in the inset in Figure 5.6. The internal diameters of the single-layer H_4TPPS^{2-} nanotubes (17.9 nm) [46] and the binary nanotubes are certainly consistent with this model.

Other ionic self-assembly experiments using porphyrin anions and cations have produced a range of novel porphyrin nanomaterials. In 2006, Wang et al. [28] reported that the ionic self-assembly of free base TPPS (H_2TPPS^{4-}, **5**) with oxo-antimony(V) TPP ($Sb^{V}OTPP$, **9**) gives nanofiber bundles (Figure 5.7). The reaction was carried out in a phase transfer reaction with TPPS in water and $Sb^{V}OTPP$ in dichloromethane, producing strands of pink material in the dichloromethane layer. Elemental analysis and optical absorption studies indicated that the material was composed of H_2TPPS^{4-} and $Sb^{V}OTPP^{+}$ in a 1:4 ratio. This is again consistent with the formation of the binary ionic solid without the presence of other small anions or cations. TEM and SEM images of the pink material show a high yield of porphyrin nanofiber bundles that are 70–140 nm wide and 1–2 microns long.

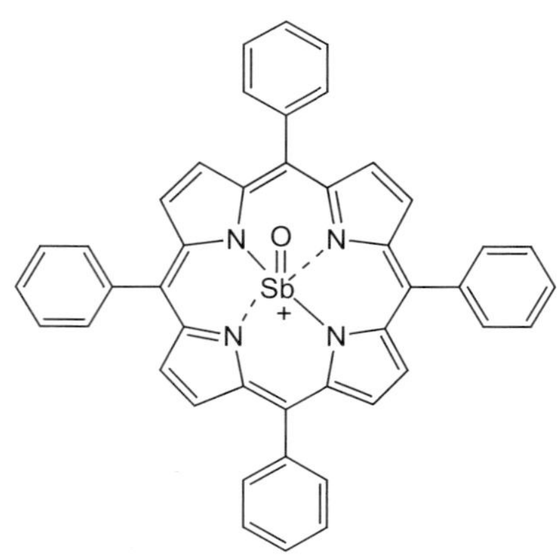

9

For the foregoing binary ionic porphyrin nanostructures, one of the porphyrins is present as a minor component (20–30 mol%). Technologically more interesting structures might be expected when the two porphyrins contribute equally to the solid (as is typical in organic charge-transfer solids for example). Recently, Martin et al. [20] have described porphyrin structures of this type obtained by ionic self-assembly of ZnT(N-EtOH-4-Py)P^{4+} (**10**) and $SnTPPS^{4-}$ (**11**) in equal proportion. These porphyrins produce the remarkable "four-leaf clover"-like structures shown in Figure 5.8. Although these structures are microscale, they possess nanoscale features as shown in the inset of Figure 5.8.

FIGURE 5.8. SEM images of "four-leaf clovers" obtained by ionic self-assembly of ZnT(N-EtOH-4-Py)P and SnTPPS. (**inset**) High-magnification image of a clover from an edge-on orientation. Adapted from reference 20.

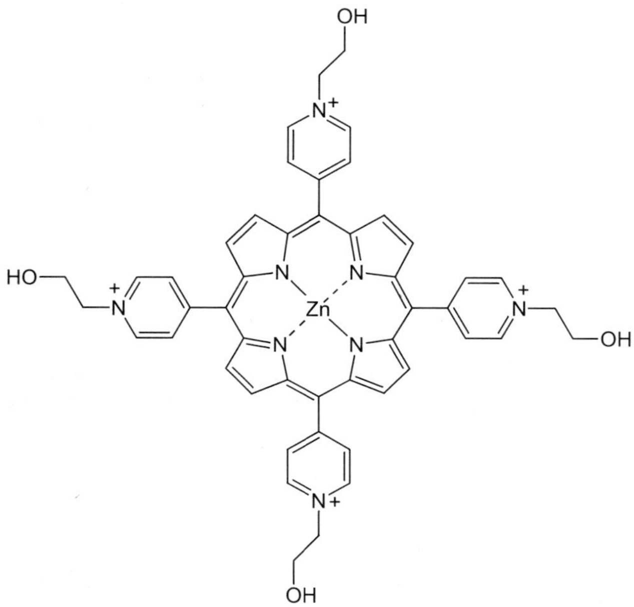

10

Martin et al. [20] and Tian et al. [21] have shown that similar fourfold symmetric microstructures (SEM images b in Figure 5.9) were formed at room temperature by the different combinations of Zn and Sn (Sn–Zn, Zn–Sn, Sn–Sn, and Zn–Zn) in these two porphyrin ligands—that is, all four binary combinations of the porphyrin cations and anions ZnT(N-EtOH-4-Py)P^{4+} (**10**), $SnTPPS^{4-}$ (**11**), SnT(N-EtOH-4-Py)P^{4+} (**12**), and $ZnTPPS^{4-}$ (**13**). The clover-like structures for all of these metal combinations showed basically the same XRD patterns for the materials synthesized at room temperature, demonstrating that they all share a common crystal structure.

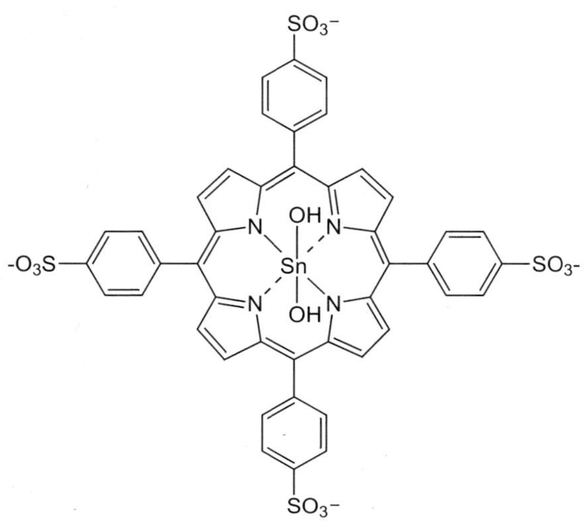

11

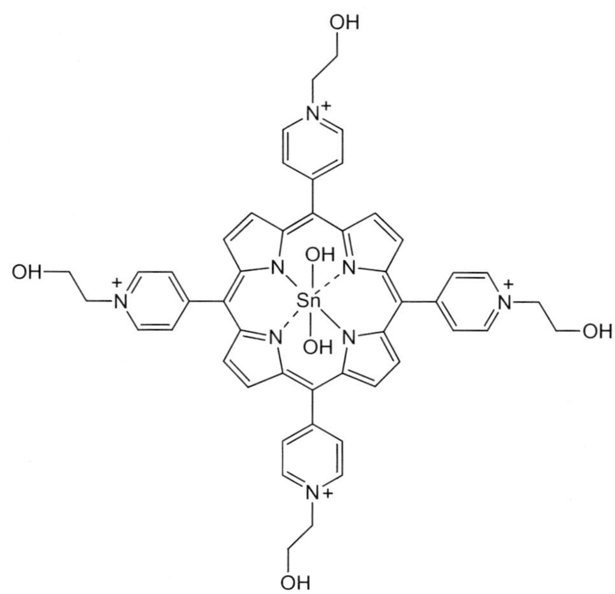

12

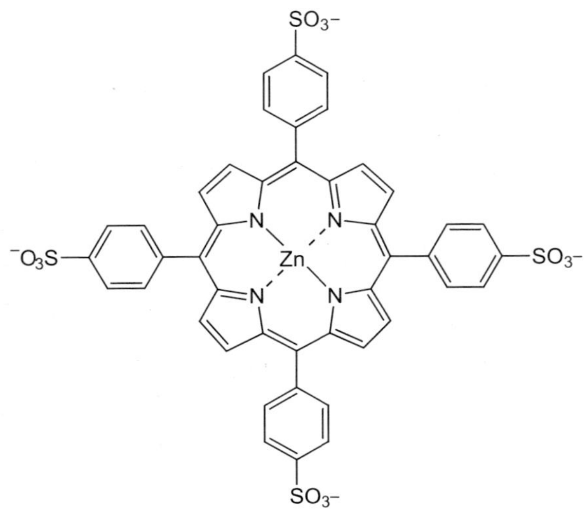

13

More recently, Tian et al. [21] demonstrated that each combination of Zn and Sn in $TPPS^{4-}$ and T(N-EtOH-4-Py)P^{4+} produces a distinct family of structures in which the morphology of the clovers changes as the growth temperature is varied. As shown in Figure 5.9a–d, these structural families are morphologically related as would be expected since they share the same basic crystal structure for room temperature growth. Moreover, XRD studies of high- and low-temperature structures within a family (the Zn–Sn clovers) confirm that they also share the same crystal structure at different temperatures, although the reflections are narrower for the high temperature structure. The latter indicates that higher crystallinity or larger crystallites are present in the high temperature microstructures. These clover-like structures self-assemble within minutes, suggesting that they form by diffusion-limited crystallization [20]. This is consistent with the strong influence of ionic strength [20] and particularly temperature [20, 21] on the morphology because these factors influence diffusion rates and solubility. Within limits, the metals contained in the porphyrin rings can be varied while keeping the same crystal structure. Studies have shown related clover-like structures for Fe(III), Mn(III), and Co(III) complexes as well as those of Zn and Sn, although not all possible metal combinations have been examined for these other metals. This finding suggests that the interactions between the ionic substituents on the porphyrin ring, along with weaker hydrophobic and H-bonding interactions, determine the observed structures at least for these metals and the synthesis conditions examined.

In contrast with altering the metals in the porphyrin core, changing the peripheral ionic substituents leads to a different crystal structure. For example, SnT(N-Me-4-Py)P^{4+} (**14**), which merely changes the pyridyl nitrogen substituents of SnT(N-EtOH-4-Py)P^{4+} from 2-hydroxyethyl groups to methyl groups, combines with $ZnTPPS^{4-}$ to give nanosheets (Figure 5.10) under some synthesis conditions instead of the clover-like Zn–Sn structures (Figure 5.9).

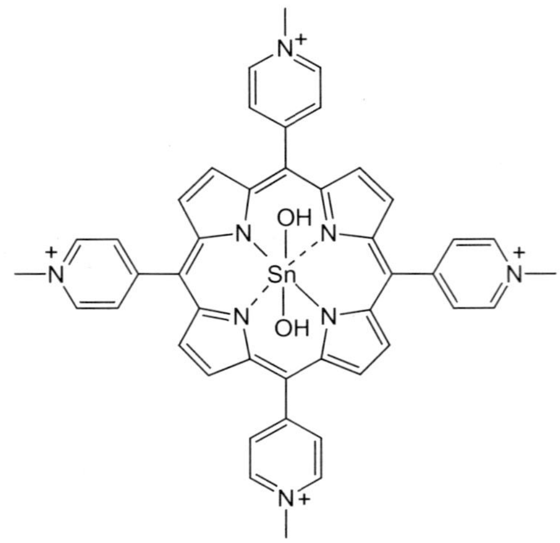

14

The Zn–Sn nanosheets form crystals large enough for single-crystal x-ray structure determination using synchrotron radiation, and the crystal structure of this first cooperative binary ionic solid has been reported [48].

5.3 OPTICAL, ELECTRONIC, AND PHOTOCATALYTIC PROPERTIES

5.3.1 UV-Visible Absorption and Emission Spectra

Optical spectroscopy has been used to investigate the light absorption and emission properties of suspensions of the porphyrin nanostructures. The spectra provide information about the intermolecular interactions and collective properties of the porphyrins within the nanostructures. For example, Figure 5.11 shows the UV-visible absorption spectrum of the H_4TPPS–SnT(4-Py)P nanotubes as a suspension at pH 2, together with the absorption spectra of the corresponding monomeric porphyrins in solution at the same pH [32]. The spectrum of the nanotubes is very different from the constituent porphyrin spectra. The optical spectrum of the nanotubes shows weak Soret bands at the wavelengths of the monomeric porphyrins as well as intense bands at 496 and 714 nm that are usually associated with J-aggregation. Suspensions of the nanotubes are a brownish green color; but when illuminated by a bright visible light source, they exhibit intense resonant light scattering, giving them a brilliant green color. Part of the intensity of the J-aggregate bands in the extinction spectrum (Figure 5.11) actually comes from

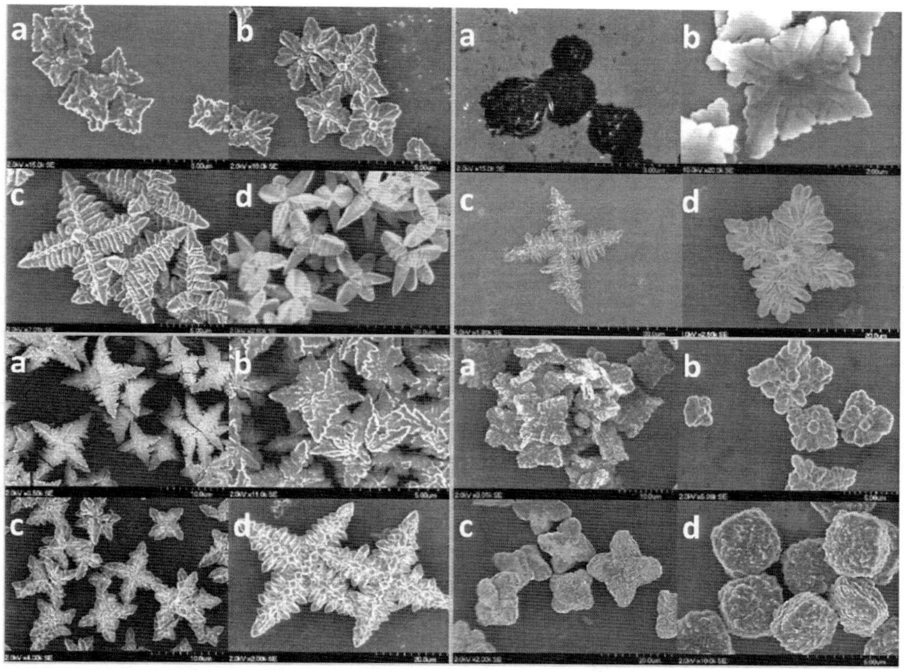

FIGURE 5.9. SEM images of clovers formed from Zn and Sn metal complexes of TPPS and T(*N*-EtOH-4-Py)P at different growth temperatures (**a**, 10°C; **b**, 20°C; **c**, 60°C; **d**, 80°C) where the combination of metals is Sn–Zn (*left top*), Zn–Sn (*right top*), Zn–Zn (*left bottom*) or Sn–Sn (*right bottom*). Adapted from reference 21.

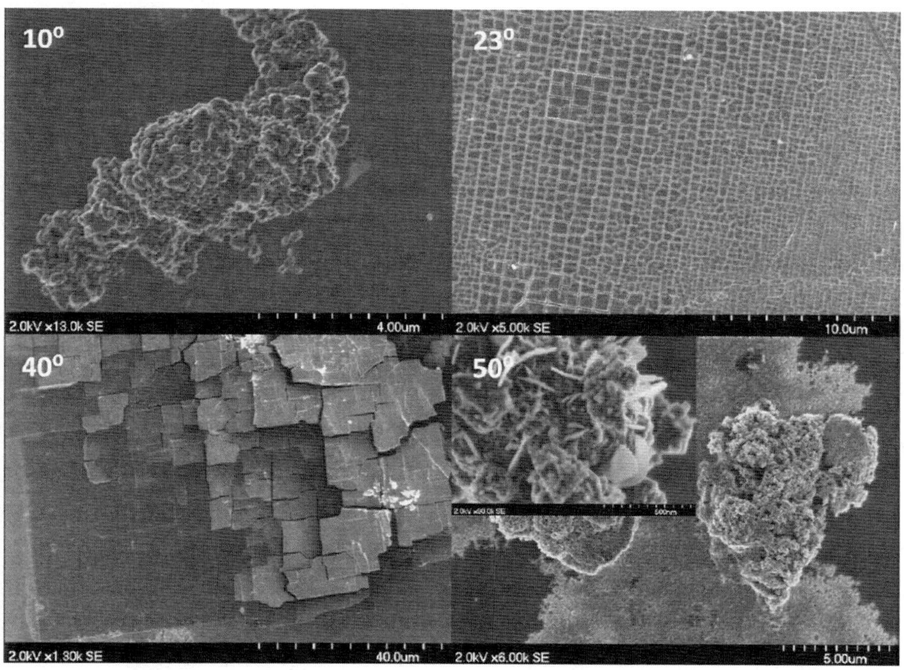

FIGURE 5.10. SEM images of CBI nanosheets formed from $ZnTPPS^{4-}$ and $SnT(N\text{-Me-4-Py})P^{4+}$ at 10°C, 23°C, 40°C, and 50°C growth temperatures.

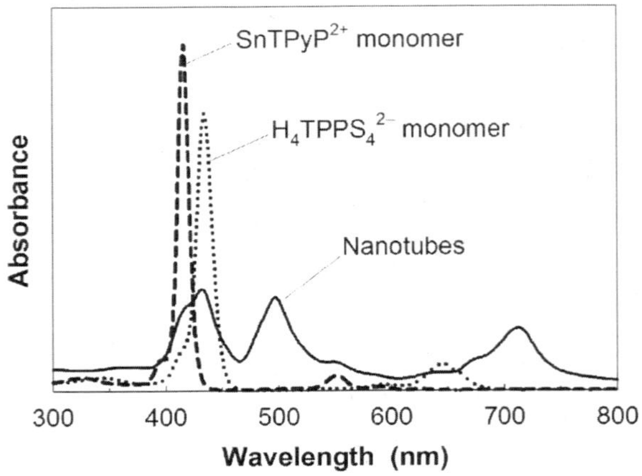

FIGURE 5.11. UV–Visible absorption spectrum of TPPS–SnT(4-Py)P nanotubes at pH 2. The absorption spectra of the monomeric porphyrins at the same pH are shown for comparison. Adapted from reference 19.

resonant light scattering rather than absorption of light with the resonant light scattering coming mainly from the green absorption band at 496 nm. J-aggregate bands arise from exchange coupling of Soret transitions of adjacent molecules. Interactions between the transition dipoles lead to a band of states with the lowest energy state receiving the most of the intensity, so the J-aggregate Soret bands are red-shifted compared to the Soret bands of monomeric species. The porphyrin molecules in the nanotubes must therefore stack as in J-aggregates, resulting in the coupling of their electronic transition dipoles. The presence of J-aggregate bands also implies that excitons are delocalized over multiple molecules. Essentially, the coupled porphyrin molecules act as a single very large molecule giving the intense light scattering. J-aggregate bands have also been observed in self-aggregates of H_4TPPS^{2-} and are related to the offset stacking arrangement of the porphyrin rings [49–51].

Recent resonance Raman studies shed further light on the arrangement and electronic interaction of the H_4TPPS^{2-} and $Sn^{IV}T$(4-Py)P molecules in the nanotubes [52]. Laser excitation in the region of the J-aggregate bands (496.5 or 501.7 nm) enhances Raman lines for only H_4TPPS^{2-} whereas excitation in the monomer region (406.7 or 413.1 nm) enhances the spectrum of both porphyrins, although the lines of $Sn^{IV}T$(4-Py)P predominate. This indicates that the J-aggregates within the nanotubes are formed from H_4TPPS^{2-} only and that $Sn^{IV}T$(4-Py)P molecules do not take part in the J-aggregate structure. This is in agreement with the proposed [32] lamellar structure of the nanotubes composed of single layers of slip-stacked H_4TPPS^{2-} molecules alternating with more sparse layers of the Sn porphyrin molecules. The Raman data support this model by suggesting that the H_4TPPS^{2-} J-aggregated stacks are segregated from the sparse $Sn^{IV}T$(4-Py)P layers.

Note that the monomer-like Soret bands of the nanotubes are not the result of dissociation of the nanotubes in solution, because they are also seen in spectra of the nanotube precipitate in the solid state. The emission spectra of the nanotubes display negligible intensity indicating quenching of the strong fluorescence of the constituent monomeric Sn and diacid porphyrins.

Exciton delocalization does not always occur to the same extent in porphyrin nanostructures. Other examples of porphyrin nanostructures with less red-shifted bands in their UV–visible spectra are the TPPS–Sb^{V}OTPP nanofiber bundles (Figure 5.7) [28], the SnT(4-Py)P–$PtCl_2$ coordination polymer nanospheres (Figure 5.5a) [30], the SnPyTriPP nanosheets (Figure 5.1) [29], and the ZnTPPS–SnT(N-EtOH-4-Py)P nanosheets [48]. In each case, the absorption bands observed more closely resemble those of the monomeric porphyrins, although the bands of the nanostructures are much broader and red-shifted. This likely indicates the formation of aggregate structures with weaker electronic coupling and less electron delocalization than is seen for the nanotubes. In contrast with the nanotubes, the nanofiber bundles, nanospheres, and nanosheets have stronger emission spectra suggesting weaker intermolecular interaction. The emission spectra nonetheless differ from those of the monomers, either by changes in relative intensities or by the presence of additional emission bands [28, 29].

The Zn–Sn porphyrin clover-like nanostructures exhibit J-aggregate bands (Figure 5.12), but at present the exact nature of these bands is undetermined [21]. The J-aggregate bands of the clovers associated with the Soret transition

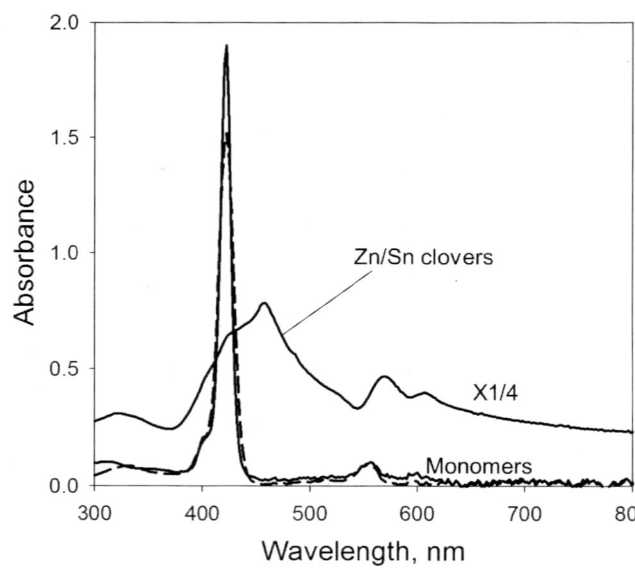

FIGURE 5.12. UV–visible spectrum of Zn/Sn clovers (upper solid line) and the constituent porphyrins ZnTPPS (solid line) and SnT(N-EtOH-4-Py)P (dashed line). The J-aggregate Soret band is near 460 nm.

are not as red-shifted as for the nanotubes and they are broader; thus their shape may indicate multiple bands. Reliable UV-visible spectra of the clovers from other metal combinations in the porphyrin have not been obtained because their large size results in almost complete absorption of the incident light and intense scattering at all wavelengths.

5.3.2 Electronic and Optoelectronic Properties

The conductivity and photoconductivity of some of the self-assembled porphyrin nanostructures have been investigated. Smith and co-workers [45, 53, 54] have investigated the photoconductivity of the porphyrin nanorods and nanotapes formed by self-assembly of $H_4TPPS_4^{2-}$ at low pH. The nanorods are insulating in the dark, but become photoconductive upon exposure to 488-nm light [53]. The photoconductivity grows over hundreds of seconds and decays slowly when the light is turned off. More recently, Riley et al. [55] investigated the photoconductivity and persistence of photoconductivity of $H_4TPPS_4^{2-}$ nanorods and bundles of nanorods. The photoconductivity grows by a factor of 15 over several days. The effects of O_2, Xe, and gating on photoconductivity were examined, and the conductivity changes are attributed to light-induced and gate-voltage induced adsorption and desorption of O_2. In addition, Yeats et al. [54] showed that meso-tetrakis-mono-phenyl-tri-sulfonatophenylporphyrin ($TPPS_3$) formed nanotapes on surfaces. The nanotapes are photoconductive in much the same way as the $H_4TPPS_4^{2-}$ nanorods but exhibit lower quantum efficiency.

For the four-leaf clover-like structures that were synthesized using Zn and Sn complexes of $TPPS^{4-}$ and T(*N*-EtOH-4-Py)P^{4+} (Figure 5.9) [20, 21], the metals in the porphyrin were chosen to enable the porphyrin macrocycle to function as an electron donor (Zn^{II}) or as an electron acceptor (Sn^{IV}). The donor–acceptor ability is determined from the ring reduction potentials, which are −1.61 V for ZnOEP versus −0.90 V for $Sn(OH)_2$OEP, in DMSO solvent and TBAP supporting electrolyte with an SCE reference electrode [56]. Care must be taken in using the redox potentials of the octaethylporphyrin metal derivatives to estimate the energy levels for other porphyrin ligands, however, because the electron-donating/withdrawing peripheral substituent groups also influence the actual redox potentials. For the four-leaf clover-like structures that were synthesized using $ZnTPPS^{4-}$ and SnT(N-EtOH-4-Py)P^{4+} (see Zn–Sn clovers in Figure 5.9 (top right panel)) [21], the electron-withdrawing pyridinium substituents act to make the Sn porphyrin an even better acceptor, increasing the separation of the donor and acceptor energy levels. Thus the CBI solid that is formed becomes a weak charge-transfer solid that could be conductive or photoconductive.

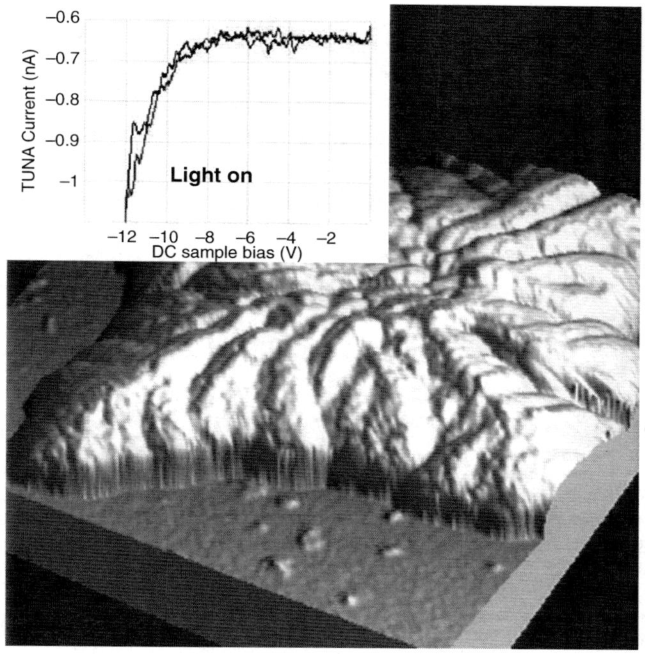

FIGURE 5.13. AFM studies of a ZnTPPS/SnT(N-EtOH-4-Py) clover. amd the TUNA current–voltage curve for conduction through the clover from the p-doped conductive Si substrate upon illumination with cool white light. Adapted from reference 20. See color insert.

Conducting AFM studies of individual Zn–Sn clovers (Figure 5.13) reveal significant photoconductivity between the conducting Si substrate and the conducting tip at a point on the top surface of the clover (i.e., through the clover). This can be seen from the current–voltage curve in Figure 5.13. When the bias of the applied potential is switched, no photocurrent is observed up to 12 V and the low residual current is constant in the dark for either bias condition.

The photoconductivity of the Zn–Sn clovers can be rationalized by the electron-transfer processes expected to occur in the donor–acceptor CBI solid. The excited state of the Zn porphyrin can reduce the Sn porphyrin to the radical anion, leaving a hole on the Zn porphyrin. Electron transfer separates the hole and electron and the two may become uncorrelated, leading to free charge carriers and a photocurrent. The reduction of the Sn porphyrin acceptor is even more energetically favorable if the Sn porphyrin is also photoexcited. Besides these electron-transfer and charge-separation processes that depend on the redox potentials (energy levels), the peripheral ionic substituents may electrostatically influence other factors controlling electron transfer and the stability of the charge-separated species that may be formed. For example, for the Zn–Sn clovers the electron-accepting Sn porphyrin core is surrounded by positively charged substituents. The four positive charges may stabilize the electron on the Sn porphyrin. Similarly, the four negative charges at the periphery of the Zn porphyrin may stabilize the holes. For the Zn–Sn clovers, the disposition of the ionic substituents is thus favorable to electron transfer and may stabilize the charge-separated Zn porphyrin π-radical cations (holes) and Sn porphyrin π-radical anions (electrons). This electrostatic stabilization of the charge-separated holes and electrons may also be favorable for the formation of free charge carriers.

When the metals are reversed (i.e., the Sn–Zn clovers), the driving force for charge separation is expected to be decreased because the Zn porphyrin has the electron-withdrawing pyridinium substituents and becomes a worse donor, closing the energy gap between donor and acceptor. In addition, the positively charged hole resides on the porphyrin ring with the positively charged substituents and the transferred electron resides on the porphyrin with the negatively charged substituents, a situation that may be energetically unfavorable. Indeed, the Sn–Zn clovers, unlike the Zn–Sn clovers, are not photoconductive [20].

These findings suggest that clovers and other binary ionic nanomaterials possess unique properties that might be advantageous for optoelectronics and for solar energy applications in particular. Most importantly, the interaction between the functional porphyrin cores in these nanomaterials can be tuned by varying metals and the peripheral ionic substituents. This raises the exciting possibility of designing nanomaterials with specific structures and functions to optimize processes such as exciton mobility, charge separation, conductivity, and catalytic activity.

5.3.3 Photocatalysis and Self-Metallization to Form Nanocomposites

Tin, antimony, and zinc porphyrins are known to function as photocatalysts in homogeneous solution [57–60]. The first step in the photocatalytic cycle is the absorption of a visible or UV photon by the porphyrin and the formation of a long-lived excited state, usually the triplet π–π* state, P*. This photoexcited species can be rapidly reduced by an electron donor or oxidized by an electron acceptor. For a reductive photocatalytic cycle such as that mediated by a Sn porphyrin (SnP), an electron donor such as ascorbic acid, EDTA, or triethanolamine reduces the SnP to form the long-lived radical anion $SnP^{-\bullet}$. (The SnP^*–$SnP^{-\bullet}$ reduction potential is approximately +1.1 V versus SCE for SnOEP, calculated by adding the triplet-state energy to the potential for SnP–$SnP^{-\bullet}$.) In homogeneous solution under alkaline conditions, the lifetime of $SnP^{-\bullet}$ is approximately 10 s when triethanolamine is the electron donor. $SnP^{-\bullet}$ is a strong reductant (SnP/$SnP^{-\bullet}$, −0.66 V versus SCE) that is capable of efficiently reducing a variety of metal ions (e.g., Au^{I}, Pt^{II}) to the zero valent metal. As an example, the photocatalytic reduction of Au^{I} by a Sn porphyrin in the presence of a sacrificial electron donor (ED) is described by Eqs. (5.1)–(5.3):

$$SnP + h\nu \rightarrow SnP^* \quad (5.1)$$

$$SnP^* + ED \rightarrow SnP^{-\bullet} + ED_{ox} \quad (5.2)$$

$$SnP^{-\bullet} + Au^+ \rightarrow SnP + Au^0 \quad (5.3)$$

Zn porphyrins may carry out reductive chemistry in a similar manner through an oxidative photocatalytic cycle in which the excited state ZnP* initially reduces a metal ion or electron relay molecule forming the Zn porphyrin radical cation. The radical cation then gets reduced back to the neutral Zn porphyrin by the electron donor. Either a reductive photocycle or oxidative photocycle could potentially reduce metal ions to the zero valent metal, but the Sn porphyrin photocycle is especially efficient at reductive metallization reactions because of the long lifetime of the radical anion.

Many of the porphyrin nanostructures have been prepared using Sn porphyrins in the hope that these nanostructures will have the capacity for reductive photocatalysis in a manner similar to the Sn porphyrins in solution. A number of studies have now confirmed the photocatalytic ability of nanostructures containing the Sn (or Sb) porphyrins [28–31]. The first example of the self-metallization of a porphyrin nanostructure was reported by Wang et al. [31] for the porphyrin nanotubes shown in Figure 5.6. Gold(I) thiourea, gold(I) thiosulfate, and K_2PtCl_4 solutions were used as the sources of metal ions and ascorbic acid provided the sacrificial electron donor. TEM images of some of the metalized nanostructures obtained using the gold(I) complexes are shown in Figures 14b and the inset image.

Reduction of gold(I) thiourea complex results in metal growth occurring exclusively inside the nanotube, along with the formation of a metal ball once the metal growth process is no longer confined by the nanotube (Figure 5.14b). The nanotube can be dissolved away while leaving the gold rod and ball intact (Figure 5.14c). Intriguingly, changing the complex to Au(I) thiosulfate alters the location where metal growth occurs, with most of the metal growth now occurring on the outside of the nanotube (see Figure 5.14, inset). The high specificity of metal growth may be related to the interactions between the charges on the gold complexes and the surface charges of the nanotube.

Growth of platinum on the surface of the nanotubes using K_2PtCl_4 as the metal source produces varying patterns of metal growth that depend upon the concentration of the platinum ions and the irradiation time. Photocatalytic platinization sensitized by the nanostructures is more complex than that for gold because platinum becomes catalytic for the oxidation of ascorbic acid and reduction of Pt complex in the presence of Pt nanoparticles of a large enough size [61–71]. This autocatalytic reaction results in a rapid dark reaction

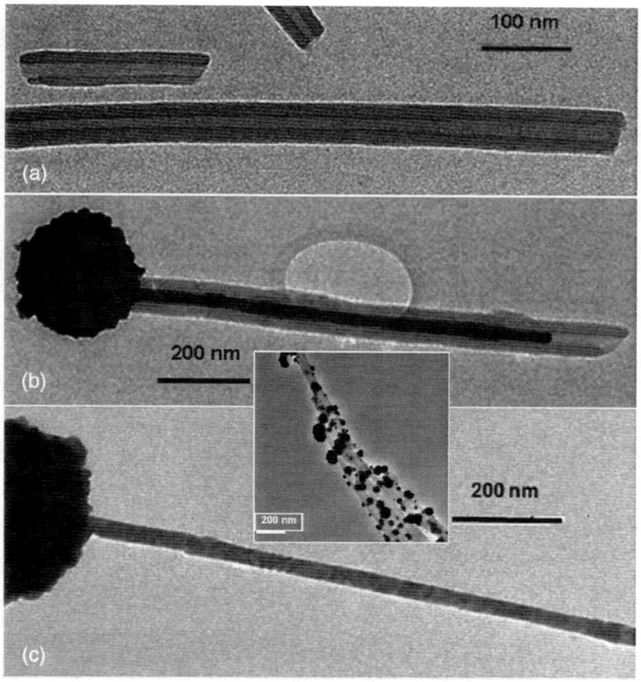

FIGURE 5.14. TEM images of TPPS/$Sn^{IV}T(4\text{-}Py)P$ nanotubes (**a**), a gilded nanotube obtained using Au(I) thiourea as the metal source (**b**), and the gold wire and ball remaining after the nanotube has been dissolved (**c**). **Inset**: Gold nanoparticles on nanotubes obtained using anionic Au(I) thiosulfate as the metal complex. Adapted from reference 19.

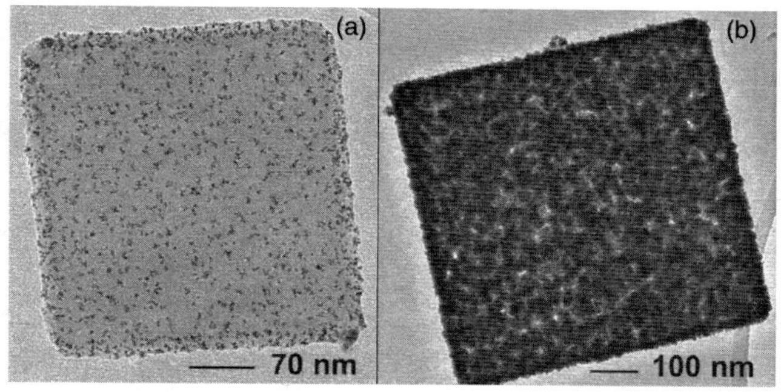

FIGURE 5.15. (**a**) TEM images porphyrin of SnPyTriPP nanosheet decorated with photodeposited Pt nanoparticles. (**b**) Nanosheet encased within a thin porous mat of autocatalytically grown Pt dendrites. Adapted from reference 19.

that may grow particles as well as dendritic metal structures. For the porphyrin nanotubes, a low concentration of Pt (0.1 mM) and a short irradiation time (15 min) gives platinum growth mainly on the exterior of the tube (not shown) [31]. A 10-fold increase in the platinum concentration gives metal growth on the exterior and interior of the nanotube. Finally, when both the platinum concentration and the growth time are increased (1 mM and 35 min), the outside of the tube is covered in large platinum dendrites and the core of the tube is filled with dendritic platinum. The ability of K_2PtCl_4 to grow platinum metal on both the inside and outside of the tube, in contrast with the selectivity seen for the gold complexes, may be related to the fact that aged K_2PtCl_4 solution forms an equilibrium mixture of 42% $Pt(H_2O)_2Cl_2$, 53% $Pt(H_2O)Cl_3^-$, and 5% $PtCl_4^{2-}$ [72], a large fraction of which is in a neutral form and not significantly influenced by charges on the surface of the nanotubes.

Since this first report on the nanotubes, several other metallization reactions of porphyrin nanostructures have been described. These include the metallization of the porphyrin nanofiber bundles made by self-assembly of H_2TPPS and Sb^VOTPP [28]. Metals, including platinum, gold, silver, and a platinum–gold alloy, were grown on the nanofiber bundles as nanoparticles and nanodendrites. The square porphyrin nanosheets prepared by reprecipitation of SnPyTriPP could also be platinized [29]. To demonstrate control over Pt loading on the sheets, they were prepared with a low loading of platinum nanoparticles (Figure 5.15a) and with a heavier loading that formed a porous mat surrounding the sheet like a pizza box (Figure 5.15b).

Porphyrin nanospheres prepared by the coordination polymerization of $Sn^{IV}T(4\text{-}Py)P$ with chloroplatinic acid were converted into Pt nanocomposites (not shown) using a range of metallization methods (*vide infra*) [30] including by photocatalytic reduction of a Pt salt. Some of the Pt ions within the nanospheres (see Figure 5.3b) could also be reduced chemically by stirring the nanospheres in 0.1 M $NaBH_4$ for 3 hours or by hydrothermal treatment in an autoclave at 150°C for 41 hours.

The first column of Figure 5.16 shows the unplatinized room temperature porphyrin clovers (Figure 5.9, images b) for all four pairs of the Zn and Sn porphyrins selected from **10**, **12** and **11**, **13**. Platinization using ascorbic acid as the electron donor occurs for all four types of clovers, although the density of Pt nanoparticles and the size of the nanoparticles vary somewhat for clovers with different combinations of porphyrins (Figure 5.16, middle column) [21].

The detailed mechanism by which the CBI porphyrin nanostructures reduce metal complex and deposit metal nanoparticles on their surface is currently unknown; that is, the relative contribution of the Zn and Sn porphyrins to the reduction has not been determined. At a minimum, light must be adsorbed by the porphyrins at the surface which then react in the excited state to ultimately reduce Pt complex. Porphyrin molecules near the surface can also contribute to the self-metallization reaction by absorbing photons and transferring energy to the surface porphyrins by exciton migration. For those structures that are conductive or photoconductive, electron or hole transfer to surface porphyrins in contact with the reactant solution may also contribute to self-metallization. The relative contributions of these processes

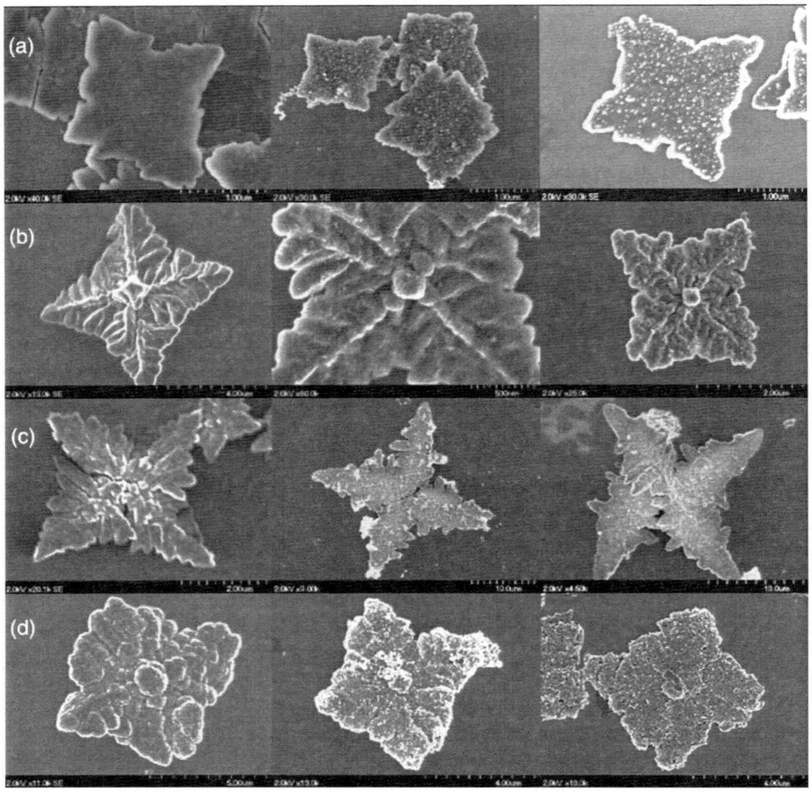

FIGURE 5.16. SEM images of the structures prepared at room temperature for all four combinations of Zn(II) and Sn(IV) in the two porphyrins (*left*), the corresponding unwashed platinized clovers (*middle*), and the washed platinized clovers after two weeks of continuous hydrogen generation (*right*): Zn/Sn (**a**), Sn/Zn (**b**), Zn/Zn (**c**), and Sn/Sn (**d**). Adapted from reference 21.

are only beginning to be investigated. Understanding the relationship between the structure of the CBI solids and their photoactivity will be crucial for optimum use of these new materials in various optoelectronic and solar energy technologies.

5.4 APPLICATIONS OF PORPHYRIN NANOSTRUCTURES AND NANOCOMPOSITES TO THE GENERATION, STORAGE, AND UTILIZATION OF SOLAR ENERGY

5.4.1 Solar Hydrogen Production

Solar water splitting has long been viewed as the "Holy Grail" of renewable energy research [73], and nanoscale feature sizes are essential to improving the efficiency of such solar fuel-producing systems [4]. For example, nanoscale light-harvesting structures are needed because the exciton migration distances are typically less than 20 nm in organic materials requiring that charge-separation sites be located within this limit. High absorptivities are necessary so that nanoscale structures can efficiently harvest light. Moreover, in systems that harness sunlight to generate fuel, the high surface areas provided by nanoscale materials are necessary to increase surface photocatalytic reaction rates.

Previous studies have shown that adding a Pt colloid to a solution of tin porphyrin in the presence of an electron donor leads to the production of hydrogen gas [57,58]. In a reaction similar to that resulting in self-metallization [Eqs. (5.1)–(5.3)], electrons from the Sn porphyrin π-anion radical are used to reduce protons to hydrogen at a catalytic Pt metal nanoparticle according to Eq. (5.4) [replacing Eq. (5.3) in the reductive photocycle):

$$2\text{SnP}^{-\bullet} + 2\text{H}^{+} \xrightarrow{\text{Pt}} 2\text{SnP} + \text{H}_2 \qquad (5.4)$$

The porphyrin nanocomposites with platinum nanoparticles (e.g., the platinized clovers shown in Figure 5.16) contain the necessary components for photocatalytic production of hydrogen via a similar reaction mechanism. In the binary structures, the Zn porphyrin component may also photocatalyze hydrogen evolution by an oxidative photocycle. Significantly, the porphyrins and platinum nanoparticles that are close together in the nanocomposite could more directly interact to improve efficiency and decrease degradation in comparison with what would occur with homogeneous porphyrin solutions containing colloidal catalysts. That is, the close proximity of the charge-separation location to the Pt catalyst might improve efficiency, and bimolecular porphyrin radical anion reactions in solution that are known to lead to porphyrin degradation might be avoided in the solid state. As an example, preliminary studies have shown that hydrogen gas is indeed produced when platinized nanotubes are irradiated in the presence of ascorbic acid [31]. Although this system was found to be unstable, it suggests that the tin porphyrins inside the platinized nanotubes can carry out reductive photocatalysis using the platinum particles on the tubes as catalytic proton reduction sites. It is unknown whether the TPPS molecules in the nanotubes also play a role as light-harvesting subunits.

Our ultimate goal is to construct a nanoscale photosynthetic system that incorporates the other nanoscale components required to produce a fully functional solar water splitting nanodevice. The ability to organize and direct the bottom-up arrangement of components of a photosynthetic system is a central scientific challenge in artificial photosynthesis [12]. The photocatalytic properties of the porphyrin nanostructures provide a pathway to comprehensive nanocomposites incorporating both water oxidation and proton reduction photocatalysts. There are many possible architectures for such devices based on the self-assembled nanostructures described in the preceding sections. For example, hydrogen-generating platinized porphyrin nanostructures (such as the platinized porphyrin nanotubes) coupled through a nanoscale electrical conductor to inorganic semiconductor-based water-oxidation catalysts (such as nanoparticles or a film) could carry out the two halves of the water-splitting reaction. The (Z-scheme) energetics of this type of nanodevice are illustrated in Figure 5.17.

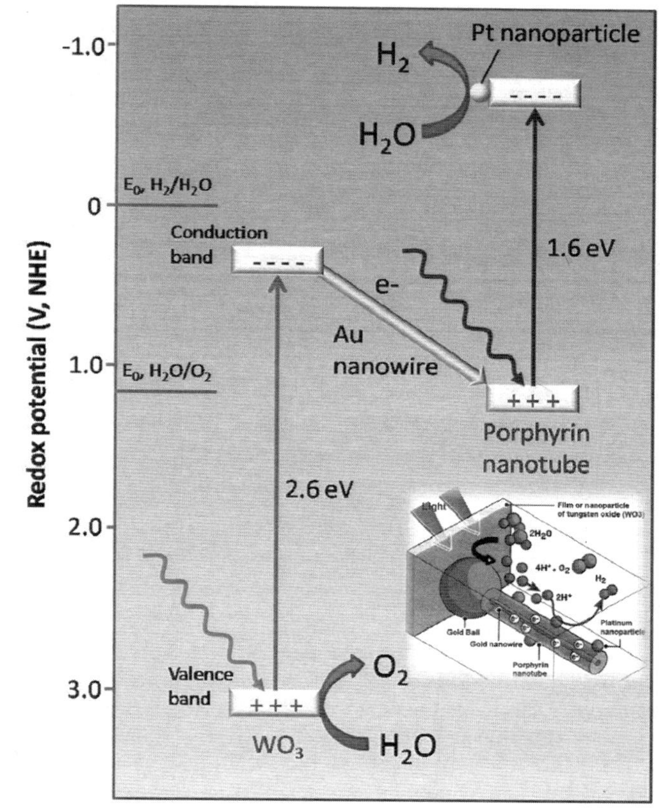

FIGURE 5.17. Energetics of a proposed water-splitting nanodevice. Excited-state redox potentials are calculated based on ground-state potentials for SnOEP [87] and the triplet ground-state energy gap; potentials for water are for pH 0. Adapted from reference 19.

In a nanodevice like that envisioned in Figure 5.17, absorption of photons by the tungsta nanoparticle promotes electrons into the conduction band, which then flow to and along the gold interconnect to the porphyrin nanostructure. At the interface with the porphyrin nanostructure, the electrons are available to reduce the photoexcited Sn porphyrin molecules producing radical anions (or to reduce Zn porphyrin radical cations back to the neutral porphyrin). Exciton transport to the porphyrin nanostructure–gold conductor interface would make the entire porphyrin nanostructure able to harvest and convert light energy. If the porphyrin nanostructure is also conductive, the highly reducing electrons (radical anions) may propagate to the catalytic platinum nanoparticles on the surface where they could reduce protons to hydrogen. Note that the gold conductor may also function as a spacer to protect the organic nanostructure from degradation by strongly oxidizing holes in the tungsten valence band. These holes at the surface of the tungsten particle can oxidize water to oxygen. The construction of this type of nanodevice utilizes the photocatalytic properties of the porphyrin nanostructure to grow the catalytic nanoparticles and conductive metal components.

Generally, there are two uses for porphyrin nanostructures in artificial photosynthesis systems as illustrated in Figure 5.18. The first uses the porphyrin as the semiconductor structure shown on the right side of Figure 5.18a. A tungsta particle might serve as the water-oxidation semiconductor shown on the left side of Figure 5.18a. The gold conductor of the envisioned nanodevice replaces the conventional redox couple (e.g., Fe^{2+}/Fe^{3+}) shown in Figure 5.18a. The semiconductors carry out both the light-harvesting and charge-separation processes familiar from biological photosynthesis.

The second system (Figure 5.18b) uses the porphyrin nanostructure purely as a light-harvesting antenna that collects the light energy and transmits it to a reaction center (labeled Receptor in Figure 5.18b) where charge separation occurs. This system closely mimics the light-harvesting proteins of biological photosynthesis. The best analogue in

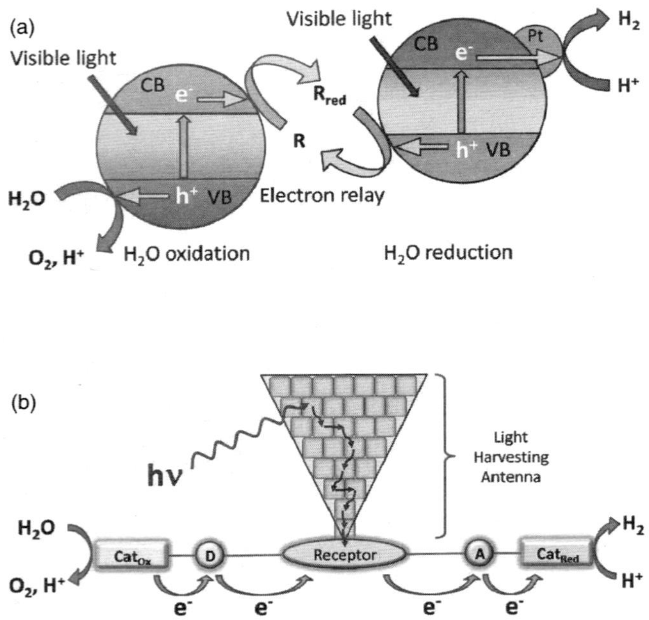

FIGURE 5.18. A two-semiconductor artificial photosynthesis system for light-driven water splitting (**a**) requires the nanotube to transport electrons and excitons. The biomimetic approach uses the porphyrin nanostructure as a light-harvesting component (**b**) and requires additional components (e.g., energy receptor and relay molecules, A = electron acceptor, D = electron donor) to split water. Adapted from reference 19. See color insert.

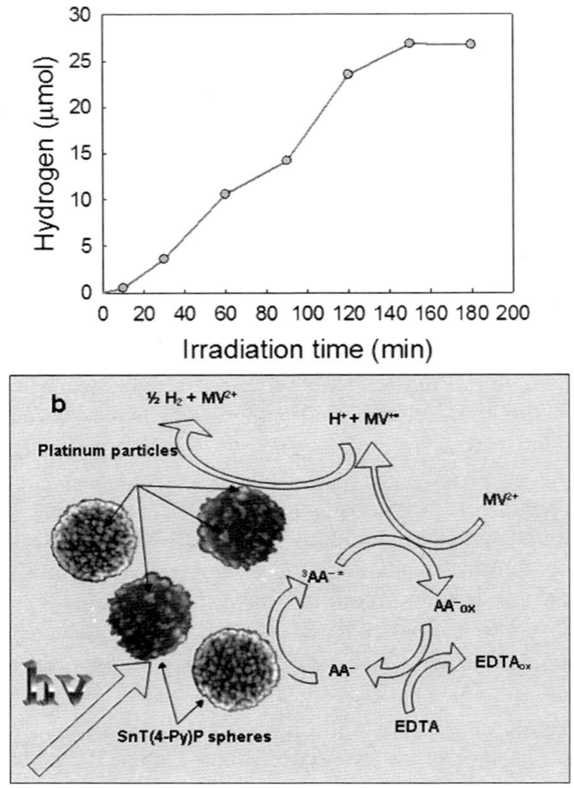

FIGURE 5.19. (**a**) Total hydrogen produced versus irradiation time for the platinized SnT(4-Py)P nanospheres and (**b**) illustration of the hydrogen-production photocycle for the platinized SnT(4-Py)P nanospheres. Adapted from reference 19.

nature is the light-harvesting structure in green bacteria—the chlorosomal nanorods. The chlorosomal rods are composed of self-assembled bacteriochlorophyll molecules (a porphyrin derivative) and involve no protein component, unlike the light-harvesting proteins of plants. In this system design (Figure 5.18b), the porphyrin structure only carries out the light-harvesting and energy-transfer functions, and, consequently, additional components are necessary for charge separation (Receptor) and electron transport (electron donor D, electron acceptor A). Hydrogen-generating systems of both types prepared using porphyrin nanostructures are described in the next two subsections.

5.4.2 Hydrogen Production Using Porphyrin Nanostructures as Light-Harvesting Arrays

The coordination-polymer nanospheres shown in Figure 5.5a function as the light-harvesting components of a system for production of hydrogen gas (Figure 5.19a) [30]. Platinization of the nanospheres gives a nanocomposite that is the basis of the photosynthetic system shown in Figure 5.19b. Anthracene-9-carboxylic acid (AA^-) is added to a suspension of the nanospheres to serve as an excitation-energy transfer molecule (Receptor in Figure 5.18b). Excited-state energy is transferred from the nanospheres to a low-lying triplet state of AA^- to produce $^3AA^{-*}$. This species transfers an electron to methyl viologen (MV^{2+}) to produce reduced methyl viologen (MV^+). MV^+ is the soluble electron acceptor (A in Figure 5.18b) that transfers electrons to Pt nanoparticles residing on the surface of the coordination polymer nanospheres that catalyze the reduction of protons to hydrogen. The oxidized carrier (AA^-_{ox}) is then regenerated by the sacrificial electron donor EDTA (D in Figure 5.18b).

A plot of hydrogen production versus irradiation time for this photosynthetic system is shown in Figure 5.19a. Total hydrogen production increases steadily for about 2 hours; recharging the system with EDTA results in the resumption of hydrogen production, and recharging can be repeated at least several times. Although the tin porphyrins of the nanospheres might have been able to directly deliver electrons to platinum for H_2 generation, control experiments show that no hydrogen is produced without the other components (AA^-, MV^{2+}). This indicates that the nanospheres provide a light-harvesting role in this case rather than a reductive photocatalytic function arising from the tin porphyrin. Moreover, the excitation energy must be derived from the porphyrins in the nanospheres because AA^- does not absorb visible light [74]. The platinized nanospheres demonstrate the feasibility of using porphyrin nanostructures as simple light-harvesters in artificial photosynthesis systems for the production of hydrogen. Thus, the porphyrin nanostructures complement the existing body of work on using ordered porphyrin arrays as artificial light-harvesting antennae [8, 11].

5.4.3 Hydrogen Production Using Porphyrin Nanostructures as Organic Semiconductors

Porphyrin nanostructures may also function as hydrogen-producing organic semiconductors as illustrated in the right side of Figure 5.18a. Especially promising in this regard are the cooperative binary ionic solids (e.g., Figures 5.8–5.10), which function as organic semiconductors as demonstrated by the photoconductivity of the Zn–Sn clovers [20]. Figure 5.20 gives the hydrogen-evolution results obtained for the platinized CBI clovers shown in Figure 5.16 (middle panels) using visible light (0.1 W cm^{-1}) and triethanolamine as the electron donor at pH 3 [21]. In the absence of the electron relay (MV^{2+}), small amounts of H_2 are detected for the donor–acceptor Zn–Sn clover nanocomposite, presumably as a result of direct electron transfer from the photoexcited

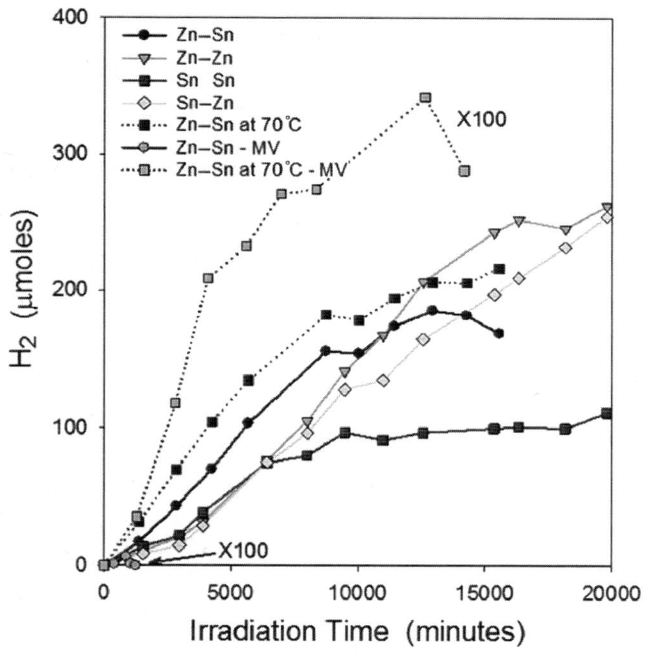

FIGURE 5.20. Total H_2 generated by the platinized porphyrin clovers (Zn–Sn, Zn–Zn, Sn–Sn, Sn–Zn) versus irradiation time with visible light from a tungsten lamp for clovers grown at room temperature (solid lines) and grown at 70°C for the Zn–Sn clovers (dotted lines). Zn–Sn clovers with MV (black circles) and without MV (dark gray circles) are expanded by a factor of 100. Adapted from reference 21.

clovers to the Pt nanoparticles on the surface. Not much hydrogen is produced because the charge-carrier diffusion length is typically less than 20 nm in organic semiconductors. The short diffusion length means that most of the porphyrin solid is too far from the Pt nanoparticles for electrons to be transported to them. This is the first report of porphyrin nanostructures photosensitizing hydrogen production without the aid of an electron relay.

Of course when the electron relay is added to the suspension of platinized clovers, the whole surface of the porphyrin nanostructure becomes active in the photocatalytic cycle and electrons are transported by $MV^{+\cdot}$ from surface reaction sites to the Pt nanocatalyst through the aqueous phase. With the relay present, the rate of H_2 production increases by a factor of 100 and the reaction system becomes much more stable. In fact, the rate of hydrogen production for some of the nanocomposites is sustained for two weeks without signs of significant degradation in the rate of H_2 generation or the appearance of the nanostructures (Figure 5.16, right column).

5.4.4 Novel Electrocatalysts for Fuel Cells

Besides the possible uses of porphyrin nanocomposites in the production of hydrogen using sunlight, the reductive photocatalytic ability of porphyrins and porphyrin nanostructures can also be harnessed to produce novel electrocatalysts for the *utilization* of solar fuels. Hydrogen fuel cells, also known as polymer electrolyte membrane (PEM) fuel cells [75], are one of the most promising technologies to replace the internal combustion engine in a "green" economy based on the use of hydrogen as a transportation fuel. Platinum is employed as an electrocatalyst at both the anode and cathode of a PEM fuel cell. Due to the cost and scarcity of platinum, an important area of fuel cell research is aimed at increasing the efficiency of platinum usage by nanostructuring the metal to increase active surface area and improve the performance and durability of the electrocatalyst [76].

A novel strategy for producing nanostructured platinum using the photocatalytic properties of tin porphyrins and porphyrin nanostructures has been developed in the last decade [65–70]. The photocatalytic nanostructuring of platinum is described here in part as broader background to the methods used in the synthesis of the metal nanocomposites described in Section 5.3.3. There is also the possibility of utilizing cobalt porphyrin–platinum nanocomposites as hybrid fuel cell electrocatalysts. Cobalt porphyrin-modified electrodes have been investigated for catalyzing the oxygen reduction reaction, suggesting that a Co porphyrin–Pt hybrid membrane electrode might be of interest for fuel cells. The preparation of nanocomposites with Co porphyrin-containing nanostructures and other metals. Such as Ni are also of interest.

Using isolated tin porphyrins, the photocatalytic nanostructuring method involves three phenomena combined in a new way. First, the discovery [66] of dendritic growth of platinum allowed the production of many nanoscale structures, including three-dimensional globular dendrites [66,68], two-dimensional sheet-like dendritic growth [61,62,64–66], and nanowires [63]. The second discovery was that soft molecular assemblies (e.g., surfactant structures like micelles, bicelles, liposomes) could template and control the dendritic growth allowing elaborate metal structures (e.g., spheres, cages, foam-like structures) to form from the dendritic platinum [61–70]. The third discovery was that the tin porphyrin photocatalyst could be used to localize the nucleation and metal growth and thereby control size, size distribution, and location at which the metal grows [61–70].

We have already described the growth of gold nanowires inside photoactive porphyrin nanotubes where growth is initiated and sustained by the photocatalytic reaction that is mediated by the porphyrin nanotubes. Similarly, the formation of the CBI clover nanocomposites with gold and platinum has been described in Section 5.3.3. In this case, the photocatalyst is the porphyrin clover and the template is the surface of the clovers. The principal difference between the work described in this section and in Section 5.3.3 is thus the nature of the photocatalyst and template—that is, an isolated porphyrin and soft or hard template versus a porphyrin nanostructure that also acts as a template.

A high-magnification TEM image showing the structure of the dendritic platinum sheets that grow within a surfactant bilayer of a bicelle is illustrated in Figure 5.21a [61,62]. The bright area is the Pt dendritic sheet that has branches that are typically 2–5 nm wide and 2–3 nm thick with spacing between the branches of 1–2 nm. Confined growth within the surfactant bilayer of the bicelle gives the two-dimensional nanosheet morphology.

The effect on the dendritic growth of platinum metal in other surfactant assemblies containing porphyrin photocatalyst molecules is seen in Figure 5.21b and in Figure 5.22. Figure 5.21b shows the uniform foam-like spheres of dendritic platinum nanosheet resulting from metal growth in the bilayers of aggregated unilamellar liposomes that contain a low concentration of tin porphyrins. Only 10 seconds of irradiation with white light initiates Pt growth centers at the beginning of the reaction leaving constant autocatalytic growth times for the dendritic Pt foam before the Pt complex is exhausted. From the initial nanoparticle seed, a dendritic sheet grows within the liposomal aggregates in an isotropic manner until the Pt complex is exhausted, leaving the spherical nanostructures consisting of highly convoluted Pt nanosheet. The Pt nanosheet is essentially identical to that shown in high magnification in Figure 5.21a. This high-surface-area foam-like structure shows sintering/ripening resistance and other properties that may be important in catalytic and electrocatalytic applications [64].

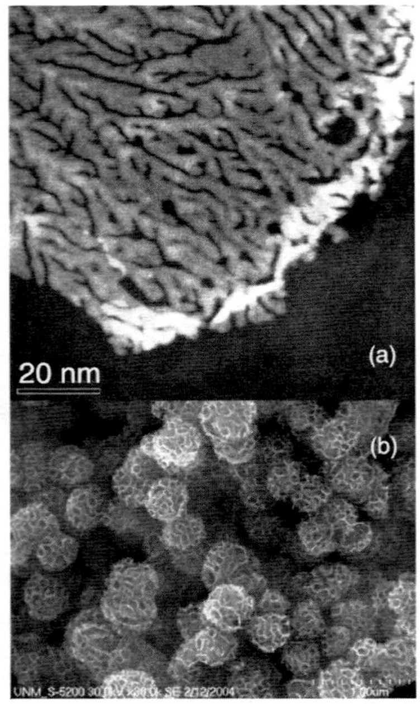

FIGURE 5.21. SEM images of the platinum dendritic nanosheets (**a**) and the foam-like spheres of convoluted dendritic Pt nanosheets (**b**) grown within the bilayers of aggregated DSPC-cholesterol liposomes.

Similarly, in the presence of SDS micelles containing photocatalytic tin porphyrins, three-dimensional dendrites of uniform size and size distribution result from growth in visible light with aged K_2PtCl_4 complex and ascorbic acid (AA) (Figure 5.22a) [66]. In the absence of the tin porphyrin photocatalyst or light, the platinum nanostructures have a wide distribution of sizes (Figure 5.22b). In the dark reaction, slow chemical reduction of Pt ions by AA produces Pt nanoparticle growth centers (seeds) at very low concentrations and at varying times by the simplified redox reaction shown in Eq. 5:

$$Pt^{2+} + 2AA \rightarrow Pt^0 + 2AA_{ox} \tag{5.5}$$

When these seeds reach a certain size (~500 atoms) [71], the Pt nanoparticles become autocatalytic for the platinum reduction reaction and platinum growth becomes rapid. The slow, continuous formation of seed nanoparticles over the entire reaction time leads to a wide range of autocatalytic growth time periods, producing a broad size distribution. When a catalytic amount of tin porphyrin photocatalyst (SnOEP) is incorporated into the SDS micelles and the reaction mixture is irradiated with visible light for several minutes, the product has a much narrower size distribution as well as a smaller average size (Figure 5.22a) [66,68]. The narrow product distribution results from the tin porphyrin complex serving as a reductive photocatalyst that produces a large number of platinum seeds at the start of the reaction. These seeds then grow

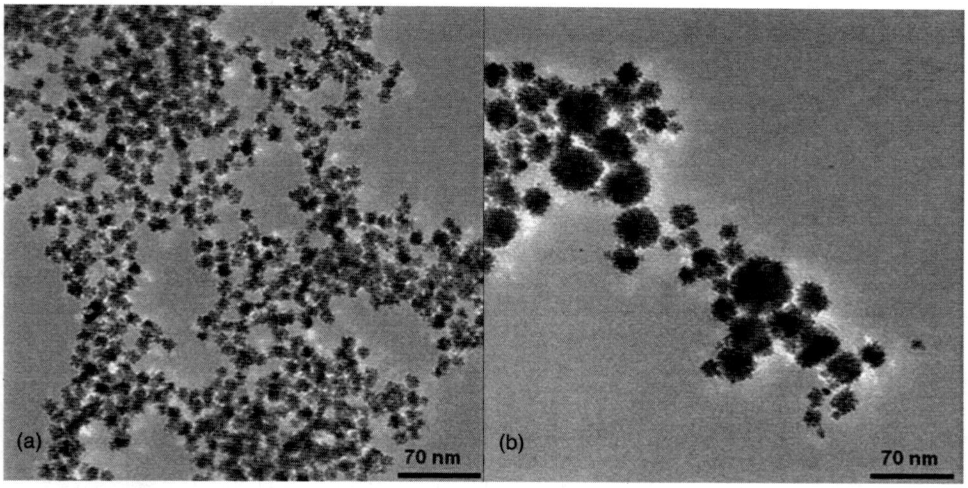

FIGURE 5.22. TEM images of platinum nanostructures produced in the presence of SDS with (**a**) or without (**b**) a tin porphyrin photocatalyst and white light. Adapted from reference 19.

for nearly equal times via the autocatalytic growth mechanism, producing the observed uniform-sized product. The dendrites are single crystals in spite of their complex morphology. In the case of growth in micelles [66,68], the growth is isotropic producing globular dendrites (Figure 5.22), but growth in bilayer structures such as liposomes, vesicles, and bicelles forms the approximately 2-nm (6 atoms) thick dendritic sheets shown in Figure 5.21.

Figure 5.23 shows platinum nanocages that are produced in liposomes [67] and platinum shells grown on Sn lipoporphyrin-modified benzene nanodroplets in aqueous nanoemulsions [70]. Related platinum nanostructures can be grown in other lipid assemblies [62, 63, 65, 67]. The platinum structures shown in Figure 5.23 are size-controlled by porphyrin-mediated photocatalytic seed particle growth, light intensity, and initial Pt salt concentration.

5.5 FUTURE DIRECTIONS AND CONCLUSIONS

5.5.1 Overview

The investigation of porphyrin and phthalocyanine nanomaterials is in its infancy, but already there appear to be several technological areas that may derive benefit from their robustness, optical and electronic properties, nanoscale size, and wide range of morphologies. Especially promising are materials formed from the new cooperative binary ionic solids, which offer tunable properties, crystal structures, and morphologies. The CBI nanomaterials offer not only the functional properties of their constituent porphyrins, but also the possibility of emergent properties that arise because of collective and cooperative behaviors of the porphyrins in the solid state. In the following subsections, we discuss several additional technological areas in which porphyrin nanomaterials may be useful.

5.5.2 Carbon Dioxide Reduction

Besides the production of hydrogen using sunlight as described in the preceding section, another goal in the renewable energy field is the reduction of carbon dioxide to precursors for carbon-based liquid fuels. Self-assembled porphyrin nanostructures offer new opportunities to realize this vision. For example, numerous studies have reported that cobalt porphyrins catalyze the electrochemical reduction of CO_2 to CO [9, 77, 78]. The carbon monoxide produced provides a feedstock for the production of carbon-based fuels through syngas reactions. Studies in our laboratories [79] have verified that gas diffusion electrodes dip-coated with cobalt porphyrins are promising electrodes for the electrocatalytic reduction of CO_2 to CO under aqueous conditions. The reduction potentials for CO_2 reduction using aqueous electrolyte are considerably lower than for nonaqueous systems (approximately -1.0 V versus -2.0 V), and negligible amounts of side-products (e.g., formate) are produced. In addition, quantum chemical studies have recently suggested a mechanism by which the electrochemical reduction of carbon dioxide is mediated by the Co porphyrin electrocatalyst [80, 81]. The calculations suggest that the CO_2 reduction reaction proceeds via a Co(I) porphyrin complex as illustrated in Figure 5.24a.

There are several ways in which nanostructures containing cobalt porphyrins might be used to improve the efficiency and rates of reduction of CO_2. The simplest approach is to use nanostructures containing cobalt porphyrins as electrode coatings to see if this improves the efficiency of CO_2 reduction or the durability of the electrode. Conventional dip coating gives micrometer-sized crystals of the Co porphyrin

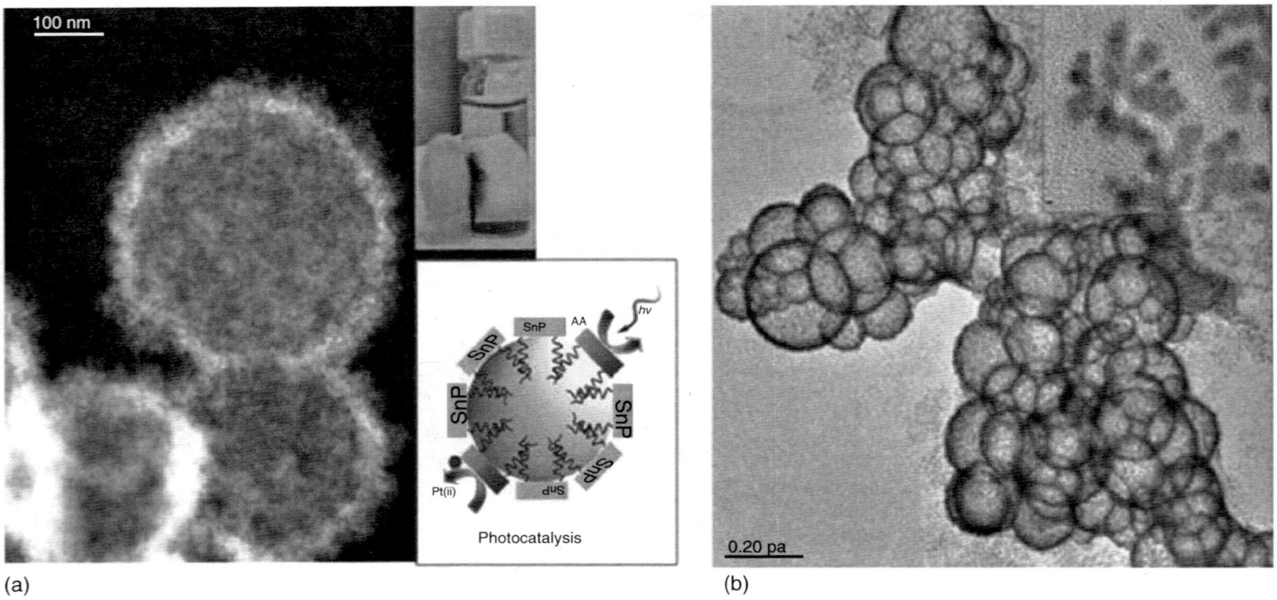

FIGURE 5.23. (**a**) Porphyrin photocatalysis controls the thickness of the spherical Pt shells grown on benzene nanodroplets stabilized by a tin lipoporphyrin (dark-field TEM image). (**b**) Many tin porphyrins in a single liposome generate small dendritic sheets that link together to form spherical cage nanostructures. **Insets**: (**a**) Illustration of the tin lipoporphyrin-stabilized benzene droplets in water and the attraction of the platinum nanospheres containing magnetite nanoparticles to a magnetic stirring bar and (**b**) high magnification TEM image of the small dendritic sheets that join to form the nanocages. Adapted from reference 19.

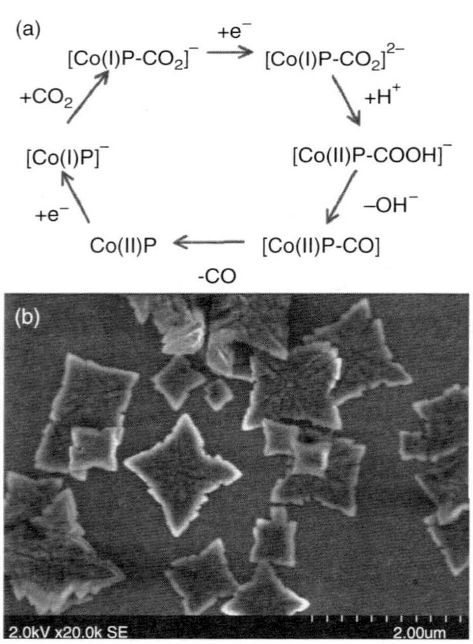

FIGURE 5.24. (**a**) Reaction mechanism for CO_2 reduction by cobalt porphyrin modified electrodes determined from quantum chemical calculations. (**b**) SEM image of a microscale clover-like structure with nanoscale features composed of ZnTPPS and Co^{III}T(N-EtOH-4-Py)P. Adapted from references 80 and 19.

on the carbon electrode. Application of porphyrin nanostructures might lead to better performance and materials utilization.

A more ambitious approach involves coating the electrode with a nanomaterial containing a second porphyrin that can photoelectrochemically assist the CO_2 reduction reaction. For example, a CBI nanostructure containing cobalt and zinc porphyrins might use the photoexcited zinc porphyrin to reduce the cobalt porphyrin to the active Co(I) state [80, 81]. The function of the electrode would then be to reduce the oxidized Zn porphyrin back to the resting neutral species at a much more positive electrode potential than is required for dark CO_2 reduction. For such binary porphyrin nanomaterials, it may also be possible that the electrode would not be necessary. The photoexcited Zn porphyrin could reduce the Co porphyrin to the active Co(I) species, and the Zn porphyrin cation might be reduced back to the resting neutral state chemically by a redox couple that also drives O_2 evolution.

Figure 5.24b shows an SEM image of CBI structures composed of equal proportions of Zn and Co porphyrins. A tin porphyrin clover could also be used in a similar fashion except that reduction of the excited Sn porphyrin to the radical anion by the electrode would occur first, followed by reduction of the Co porphyrin by the Sn porphyrin radical anion. A Zn–Co or Sn–Co porphyrin CBI nanomaterial for

CO_2 reduction could potentially operate purely photocatalytically (without an electrode), but would still need to be coupled to the O_2 generating half reaction through a redox couple. A potential problem with this approach might result from the ability of Co porphyrins to quench the excited state of the Zn or Sn porphyrin and dissipate the energy by nonradiative decay. Other CBI structures can be imagined that might solve this potential problem by isolating the energy-transfer and electron-transfer processes.

5.5.3 Photovoltaics and Dye-Sensitized Solar Cells

Major contributors to the cost of silicon-based photovoltaics are the pure crystalline photoactive materials and the need for complex processing technologies, both of which are energy intensive. An alternative is to use semiconducting organic materials that are inexpensive to synthesize via low-temperature and high-throughput processes. However, organic solar cells suffer from problems such as low solar absorptivities, short exciton diffusion lengths, low charge mobilities, and the resulting low solar efficiencies. To overcome these problems, the interface between the electron-donor layer and the electron-acceptor layer has been nanostructured by the creation of either a bulk heterojunction or an ordered interdigitated or bicontinuous heterojunction. The design of the heterojunction is ultimately a trade-off between maximizing the absorption of light and minimizing recombination of the photogenerated charge carriers. An example of the design of a novel bulk heterojunction is the use of a blend of different phthalocyanine chromophores to maximize light absorption [82].

The CBI porphyrin nanostructures described here might be useful in the construction of new types of heterojunctions for photovoltaic cells, especially given the possibilities for tuning the structure and electronic properties of the materials. The absorptivities of porphyrins and phthalocyanines are much higher than the light-absorbing polymers typically used in organic solar cells, potentially allowing a thinner active layer. Porphyrin-based nanomaterials could be simply used in bulk heterojunctions, but might also be aligned for the production of ordered heterojunctions. Specifically, the cooperative binary ionic solids (Figures 5.8–5.10) seem well-suited to the production of active heterojunctions of solar cells. CBI nanomaterials such as the clovers might have segregated electron donor (hole transport) and acceptor (electron transport) columns that are interdigitated at the (2-nm) molecular level. As illustrated in Figure 5.25, an active CBI layer—for example, consisting of a thin layer of the porphyrin clover-like material between electron-conducting (BCP) and hole-conducting (PEDOT) polymers—could provide several properties desired for solar cells. These include the high solar absorptivity and high surface areas between the donor and acceptor stacks for efficient charge separation. The CBI active layer may also provide high enough electron and hole mobilities to be compatible with the nanoscale charge-collection distances needed for high solar efficiencies.

Dye-sensitized solar cells (DSSCs) are another area where porphyrin nanostructures and arrays may provide opportunities for improved performance and durability [10, 83, 84]. In DSSCs, the dye is typically adsorbed onto the surface of a wide band-gap semiconductor (e.g., TiO_2) at the light-active electrode in order to collect visible light and inject electrons into the conduction band of the semiconductor from the excited state of the dye molecules. The oxidized dye molecule is then reduced by a redox couple (e.g., I^-/I_3^-), which is regenerated by electrons at the other electrode after completing the external electrical circuit. Porphyrins have been used as dyes in DSSCs usually as a monolayer on the semiconductor surface, although some researchers have investigated the use of covalently bound porphyrin arrays [84]. Porphyrin arrays are used to improve the absorptivity, allowing the nanostructured semiconductor layer to be thinner and thus more efficient.

Self-assembled porphyrin nanostructures might be used in a similar way, providing an even larger collector array. Specifically, the porphyrin nanostructure would serve as a very large light-harvesting array much like the chlorosomal nanorods of bacteriochlorophyll in green bacteria, which are the most efficient biological light-harvesting systems known [85]. However, the porphyrin nanostructure would function best in DSSCs if it were also conducting so that the whole structure could interact with the redox couple. Whether porphyrin nanostructures can efficiently transport energy and electrons and can be effectively and stably attached to the surface of semiconductor materials like TiO_2 remains to be determined, but this approach provides unexplored opportunities for increasing DSSC performance.

5.5.4 Conclusions

Studies of the self-assembly of porphyrins in solution have led to the synthesis of discrete nanostructures including nanotubes, hollow hexagonal nanoprisms, nanofiber bundles, high-aspect ratio square nanosheets, and more elaborate morphologies. Methods used to prepare these porphyrin nanomaterials include ionic self-assembly, reprecipitation, and coordination polymerization. The nanostructures can be isolated intact, and many have been imaged using electron microscopy (SEM, TEM) and AFM. Some of the materials have also been investigated using spectroscopic techniques such as UV-visible absorption, fluorescence, and resonance Raman spectroscopies. The self-assembled nanostructures display a range of useful optical and electronic properties, including resonant light scattering indicative of exciton delocalization within the structures.

Many of the porphyrin nanostructure are photocatalytically active and have the ability to self-metalize when irradiated by visible light in the presence of a solution of metal

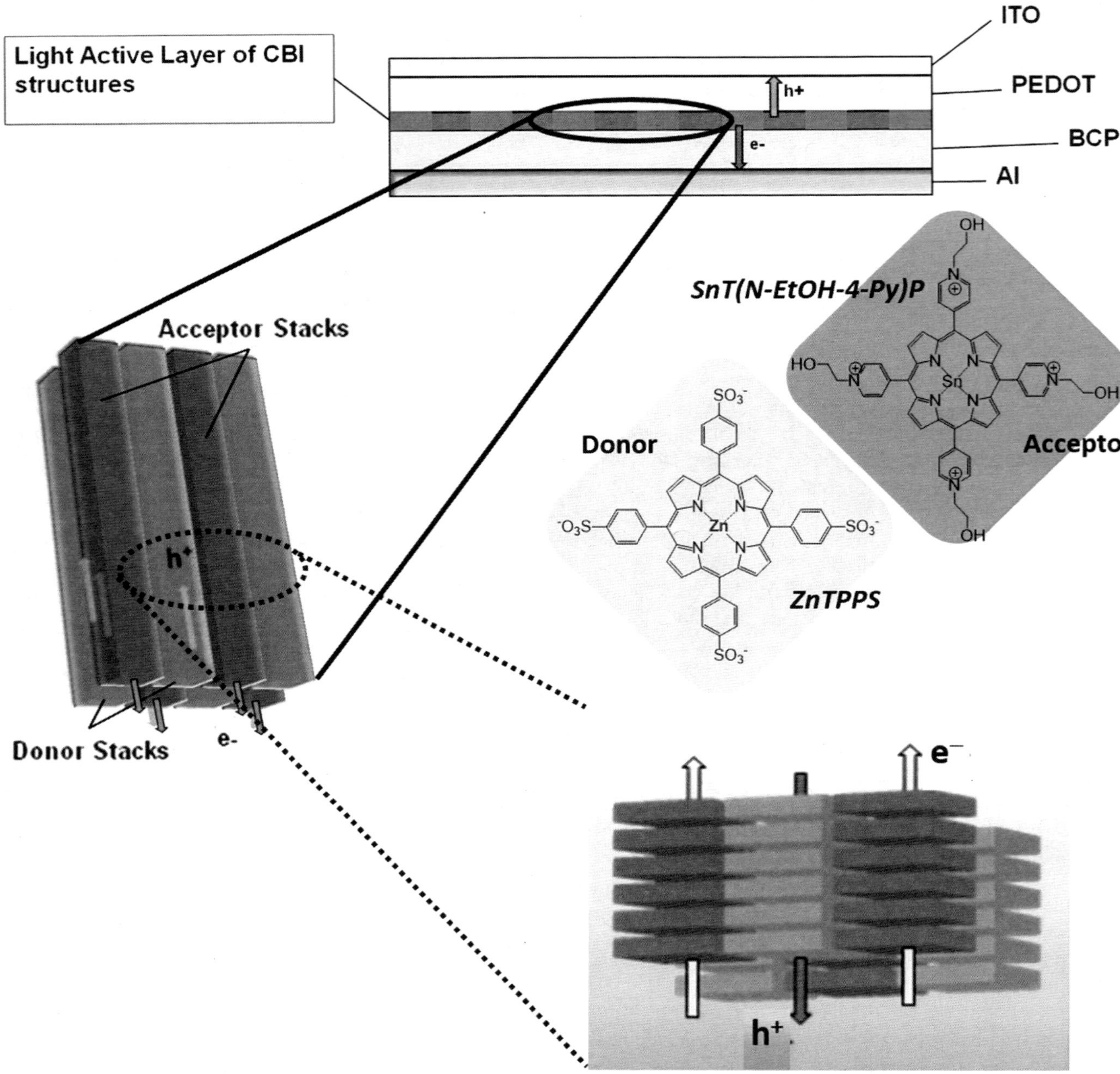

FIGURE 5.25. Hypothetical heterojunction solar cell nanoengineered from electron donor–acceptor cooperative binary ionic solids. Adapted from reference 19.

ions and a sacrificial electron donor. This allows the production of remarkable metal nanocomposites with multiple functionalities. For example, platinum nanoparticles grown on the surface of the nanostructures can be used as proton reduction catalysts for the production of hydrogen. Studies of porphyrin nanostructure–platinum nanocomposites with sacrificial electron donors have demonstrated light-driven hydrogen production via reductive photocatalysis or light-harvesting mechanisms. Other metal structures can also be photocatalytically grown and might be used as electrical connectors for coupling to other functional units such as water oxidation catalysts. This approach could potentially allow the production of solar-powered water-splitting devices and nanodevices.

The nascent field of self-assembled porphyrin nanostructures and their metal nanocomposites provides an intriguing option for the construction of materials for energy-related areas such as solar fuels production and electrical generation by organic photovoltaics and dye-sensitized solar cells. Other applications such as catalysis, sensors [86], nanotagging, and optoelectronics need to be explored as well. Many basic and engineering questions still remain regarding these new organic nanomaterials and their usefulness in such applications. On a basic research level, comparatively little is known about the internal structure of the porphyrin nanomaterials and the arrangement of and interactions between the porphyrin subunits. Comparative crystal structure studies would be helpful in understanding the range of solids that can be made and how structure relates to function. In addition, the current understanding of properties such as excitation energy transfer, charge separation, and exciton migration is limited. Having access to information about the relationship between the molecular properties of the porphyrin subunits and the structure and function of the porphyrin nanostructure would be very helpful, especially in light of the considerable latitude for tuning the functional subunits and porphyrin nanomaterials for applications in solar energy and other fields.

ACRONYMS

AA	ascorbic acid
AFM	atomic force microscopy
BCP	bathocuproine
CTAB	cetyl trimethylammonium bromide
DMF	dimethylformamide
DSPC	1,2-Distearoyl-sn-glycero-3-phosphocholine
DSSC	dye-sensitized solar cells
EDTA	ethylenediaminetetraacetic acid
EDX	energy dispersive x-ray spectroscopy
PEDOT	poly(3,4-ethylenedioxythiophene)
PEM	polymer electrolyte membrane
SCE	saturated calomel electrode
SDS	sodium dodecyl sulfate
SEM	scanning electron microscopy
TEM	transmission electron microscopy
XRD	x-ray diffraction

ACKNOWLEDGMENTS

This work was supported by the US Department of Energy (DE-FG02-02ER15369). Sandia is a multiprogram laboratory operated by Sandia Corporation, a Lockheed Martin Company, for the United States Department of Energy's National Nuclear Security Administration under contract DE-AC04-94AL85000. CJM and JAS would like to thank the many people who have contributed to the work described in this chapter: Zhongchun Wang (Soladigm, Inc.), Kathleen Martin (Sandia National Laboratories), John Jacobsen (UC Davis), Frank van Swol (Sandia National Laboratories), James Miller (Sandia National Laboratories), Neil Schore (University of California at Davis), Yujiang Song (Dalian Institute of Physical Chemistry), Robert M. Garcia (Sandia National Laboratories), Kuangchiu Ho (University of New Mexico), Weishi Wang (University of Central Florida), Ricardo Franco (Universidade de Nova Lisboa), and Zhiyong Li (Hewlett-Packard Laboratories, Palo Alto). C.J.M. is currently the recipient of a Marie Curie Fellowship from the Fundao Fundação para a Ciência e a Tecnologia, Portugal and the Marie Curie Action Cofund.

REFERENCES

1. Pramod, P., Thomas, K. G., George, M. V. (2009). Organic nanomaterials: morphological control for charge stabilization and charge transport. *Chemistry—An Asian Journal*, **4**, 806.
2. Zhang, C., Zhao, Y. S., Yao, J. (2011). Organic composite nanomaterials: energy transfers and tunable luminescent behaviors. *New Journal of Chemistry*, **35**, 973.
3. Grimsdale, A. C., Müllen, K. (2005). The chemistry of organic nanomaterials. *Angewandte Chemie International Edition*, **44**, 5592.
4. Lewis, N. S., Crabtree, G. (2005). *Basic Research Needs for Solar Energy Utilization*, Department of Energy, Washington, DC.
5. Blankenship, R. E. (2002). *Molecular Mechanisms of Photosynthesis*. Blackwell Science, London.
6. Magnuson, A., Anderlund, M., Johansson, O., LindBlad, P., Lomoth, R., Styring, S., Sundström, V., Hammarström, L. (2009). Biomimetic and microbial approaches to solar fuel generation. *Accounts of Chemical Research*, **42**, 1899.
7. Hammarström, L., Hammes-Schiffer, S. (2009). Artificial photosynthesis and solar fuels. *Accounts of Chemical Research*, **42**, 1859.
8. Gust, D., Moore, T. A., Moore, A. L. (2009). Solar fuels via artificial photosynthesis. *Accounts of Chemical Research*, **42**, 1890.
9. Morris, A. J., Meyer, G. J., Fujita, E. (2009). Molecular approaches to the photocatalytic reduction of carbon dioxide for solar fuels. *Accounts of Chemical Research*, **42**, 1983.
10. Youngblood, W. J., Lee, S.-H. A., Maeda, K., Mallouk, T. E. (2009). Visible light water splitting using dye-sensitized oxide semiconductors. *Accounts of Chemical Research*, **42**, 1966.
11. Aratani, N., Kim, D., Osuka, A. (2009). Discrete cyclic porphyrin arrays as artificial light-harvesting antenna. *Accounts of Chemical Research*, **42**, 1922.
12. Wasielewski, M. R. (2009). Self-assembly strategies for integrating light harvesting and charge separation in artificial photosynthetic systems. *Accounts of Chemical Research*, **42**, 1910.
13. Sun, S.-S., Sariciftci, N. S., eds. (2005). *Organic Photovoltaics: Mechanisms, Materials, and Devices*. CRC Press, Boca Raton, FL.

14. Collings, A. F., Critchley, C., eds. (2005). *Artificial Photosynthesis: From Basic Biology to Industrial Application*. Wiley–VCH, Weinheim.

15. Hasobe, T. (2010). Supramolecular nanoarchitectures for light energy conversion. *Physical Chemistry Chemical Physics*, **12**, 44.

16. Ariga, K., Hill, J. P., Lee, M. V., Vinu, A., Charvet, R., Acharya, S. (2008). Challenges and breakthroughs in recent research on self-assembly. *Science and Technology Advanced Materials*, **9**, 014109.

17. Elemans, J., Van Hameren, R., Nolte, R. J. M., Rowan, A. E. (2006). Molecular materials by self-assembly of porphyrins, phthalocyanines, and perylenes. *Advanced Materials*, **18**, 1251.

18. Medforth, C. J., Wang, Z., Martin, K. E., Song, Y., Jacobsen, J. L., Shelnutt, J. A. (2009). Self-assembled porphyrin nanostructures. *Chemical Commununications*, 7261.

19. Medforth, C. J., Shelnutt, J. A. (2011). Self-assembled porphyrin nanostructures. In: Kadish, K. M., Smith, K. M., Guilard, R., eds. *Handbook of Porphyrin Science*, Vol. 11. World Scientific Publishing, Hackensack, NJ, pp. 181.

20. Martin, K. E., Wang, Z., Busani, T., Garcia, R. M., Chen, Z., Jiang, Y., Song, Y., Jacobsen, J. L., Vu, T. T., Schore, N. E., Swartzentruber, B. S., Medforth, C. J., Shelnutt, J. A. (2010). Donor–acceptor biomorphs from the ionic self-assembly of porphyrins. *Journal of the American Chemical Society*, **132**, 8194.

21. Tian, Y., Martin, K. E., Shelnutt, J. Y.-T., Evans, L., Busani, T., Miller, J. E., Medforth, C. J., Shelnutt, J. A. (2011). Morphological families of self-assembled porphyrin structures and their photosensitization of hydrogen generation. *Chemical Communications*, **47**, 6069.

22. Drain, C. M., Bazzan, G., Milic, T., Vinodu, M., Goeltz, J. C. (2005). Formation and applications of stable 10 nm to 500 nm supramolecular porphyrinic materials. *Israel Journal of Chemistry*, **45**, 255.

23. Drain, C. M., Varotto, A., Radivojevic, I. (2009). Self-organized porphyrinic materials. *Chemical Reviews*, **109**, 1630.

24. Hu, J. S., Guo, Y. G., Liang, H. P., Wan, L. J., Jiang, L. (2005). Three-dimensional self-organization of supramolecular self-assembled porphyrin hollow hexagonal nanoprisms. *Journal of the American Chemical Society*, **127**, 17090.

25. Lee, S. J., Hupp, J. T., Nguyen, S. T. (2008). Growth of narrowly dispersed porphyrin nanowires and their hierarchical assembly into macroscopic columns. *Journal of the American Chemical Society*, **130**, 9632.

26. Lee, S. J., Jensen, R. A., Malliakas, C. D., Kanatzidis, M. G., Hupp, J. T., Nguyen, S. T. (2008). Effect of secondary substituent on the physical properties, crystal structures, and nanoparticle morphologies of (porphyrin)Sn(OH)2:diversity enabled via synthetic manipulation. *Journal of Materials Chemistry*, **18**, 3640.

27. Lee, S. J., Malliakas, C. D., Kanatzidis, M. G., Hupp, J. T., Nguyen, S. T. (2008). Amphiphilic porphyrin nanocrystals: Morphology tuning and hierarchical assembly. *Advanced Materials*, **20**, 3543.

28. Wang, Z., Ho, K. J., Medforth, C. J., Shelnutt, J. A. (2006). Porphyrin nanofiber bundles from phase-transfer ionic self-assembly and their photocatalytic self-metallization. *Advanced Materials*, **18**, 2557.

29. Wang, Z., Li, Z., Medforth, C. J., Shelnutt, J. A. (2007). Self-assembly and self-metallization of porphyrin nanosheets. *Journal of the American Chemical Society*, **129**, 2440.

30. Wang, Z., Lybarger, L. E., Wang, W., Medforth, C. J., Miller, J. E., Shelnutt, J. A. (2008). Monodisperse porphyrin nanospheres synthesized by coordination polymerization. *Nanotechnology*, **19**, 395604.

31. Wang, Z., Medforth, C. J., Shelnutt, J. A. (2004). Self-metallization of photocatalytic porphyrin nanotubes. *Journal of the American Chemical Society*, **126**, 16720.

32. Wang, Z., Medforth, C. J., Shelnutt, J. A. (2004). Porphyrin nanotubes by ionic self-assembly. *Journal of the American Chemical Society*, **126**, 15954.

33. Jo, H. J., Jung, S. H., Kim, H.-J. (2004). Synthesis and hydrogen-bonded supramolecular assembly of *trans*-dihydroxotin(IV) tetrapyridylporphyrin complexes *Bulletin of the Korean Chemical Society*, **25**, 1869.

34. Beavers, C. M., Jacobsen, J., Schore, N. E., Medforth, C. J., Shelnutt, J. A., Wang, Z. SnPyTriPP crystal structure, unpublished results.

35. Carlucci, L., Ciani, G., Proserpio, D. M., Porta, F. (2003). Open network architectures from the self-assembly of $AgNO_3$ and 5,10,15,20-tetra(4-pyridyl)porphyrin (H_2tpyp) building blocks: The exceptional self-penetrating topology of the 3D network of $[Ag_8(Zn^{II}tpyp)7(H_2O)_2](NO_3)_8$. *Angewandte Chemie, International Education*, **42**, 317.

36. Ohmura, T., Usuki, A., Fukumori, K., Ohta, T., Ito, M., Tatsumi, K. (2006). New porphyrin-based metal-organic framework with high porosity: 2-D infinite 22.2-Å, square-grid coordination network. *Inorganic Chemistry*, **45**, 7988.

37. Kumar, R., Goldberg, I. (1998). Supramolecular assembly of heterogeneous multiporphyrin arrays—structures of [{Zn-II(tpp)}(2)(tpyp)] and the coordination polymer [{[Mn-III(tpp)](2)(tpyp)(ClO_4)(2)}(infinity)]. *Angewandte Chemie International Edition*, **37**, 3027.

38. Pan, L., Huang, X., Phan, H. L., Emge, T. J., Li, J., Wang, X. (2004). 1-D infinite array of metalloporphyrin cages. *Inorganic Chemistry*, **43**, 6878.

39. Hagrman, D., Hagrman, P. J., Zubieta, J. (1999). Solid-state coordination chemistry: the self-assembly of microporous organic-inorganic hybrid frameworks constructed from tetrapyridylporphyrin and bimetallic oxide chains or oxide clusters. *Angewandte Chemie International Edition*, **38**, 3165.

40. Krupitsky, H., Stein, Z., Goldberg, I., Strouse, C. E. (1994). Crystal structure of ZnT(4-Py)P. *Journal Inclusion Phenomena*, **18**, 177.

41. Faul, C. F. J., Antonietti, M. (2003). Ionic self-assembly: Facile synthesis of supramolecular materials. *Advanced Materials*, **15**, 673.

42. La, T., Richards, R. A., Lu, R. S., Bau, R., Miskelly, G. M. (1995). Solution chemistry and crystal structure of nickel tetrakis(2,3,5,6-tetrafluoro-*N,N,N*-trimethyl-4-aniliniumyl)porphyrin trifluoromethanesulfonate ($NiTF_4TMAP(CF_3SO_3)_4$). *Inorganic Chemistry*, **34**, 5632.
43. Ivanca, M. A., Lappin, G., Scheidt, W. R. (1990). Water-soluble ferric porphyrinates: Solution and solid-state species. *Inorganic Chemistry*, **30**, 711.
44. Kanemitsu, H., Harada, R., Ogo, S. (2010). A water-soluble iridium(III) porphyrin. *Chemical Communications*, **46**, 3083.
45. Schwab, A. D., Smith, D. E., Rich, C. S., Young, E. R., Smith, W. F., de Paula, J. C. (2003). Porphyrin nanorods. *Journal of Physical Chemistry B*, **107**, 11339.
46. Vlaming, S. M., Augulis, R., Stuart, M. C. A., Knoester, J., van Loosdrecht, P. H. M. (2009). Excitonic spectra and the microscopic structure of self-assembled porphyrin nanotubes. *Journal of Physical Chemistry B*, **113**, 2273.
47. Friesen, B. A., Nishida, K. R. A., McHale, J. L., Mazur, U. (2009). New nanoscale insights into the internal structure of tetrakis(4-sulfonatophenyl) porphyrin nanorods. *Journal of Physical Chemistry C*, **113**, 1709.
48. Tian, Y., Beavers, C. M., Busani, T., Martin, K. E., Jacobsen, J. L. Mercado, B. Q., Swartzentruber, B. S., van Swol, F., Medforth, C. J., Shelnutt, J. A. (2012). Binary Ionic Porphyrin Nanosheets: Electronic and Light-harvesting Properties Regulated by Crystal Structure. *Nanoscale*, 4, 1695.
49. Koti, A. S. R., Taneja, J., Periasamy, N. (2003). Control of coherence length and aggregate size in the J-aggregate of porphyrin. *Chemical Physics Letters*, **375**, 171.
50. Micali, N., Romeo, A., Lauceri, R., Purrello, R., Mallamace, F., Scolaro, L. M. (2000). Fractal structures in homo- and heteroaggregated water soluble porphyrins. *Journal of Physical Chemistry B*, **104**, 9416.
51. Lauceri, R., Gurrieri, S., Bellacchio, E., Contino, A., Monsu'scolaro, L., Romeo, A., Toscano, A., Purrello, R. (2000). J-type aggregates of the anionic meso-tetrakis(4-sulfonatophenyl)porphine induced by hindered cationic porphyrins. *Supramolecular Chemistry*, **12**, 193.
52. Franco, R., Jacobsen, J. L., Wang, H., Wang, Z., Istvan, K., Schore, N. E., Medforth, C. J., Shelnutt, J. A. (2010). Internal organization and electronic structure of porphyrin nanotubes revealed by resonance raman spectroscopy. *Physical Chemistry Chemical Physics*, **12**, 4072.
53. Schwab, A. D., Smith, D. E., Bond-Watts, B., Johnston, D. E., Hone, J., Johnson, A. T., de Paula, J. C., Smith, W. F. (2004). Photoconductivity of self-assembled porphyrin nanorods. *Nano Letters*, **4**, 1261.
54. Yeats, A. L., Schwab, A. D., Massare, B., Johnston, D. E., Johnson, A. T., de Paula, J. C., Smith, W. F. (2008). Photoconductivity of self-assembled nanotapes made from meso-tri(4-sulfonatophenyl)monophenylporphine. *Journal of Physical Chemistry C*, **112**, 2170.
55. Riley, C. K., Muller, E. A., Feldman, B. E., Cross, C. M., Van Aken, K. L., Johnston, D. E., Lu, Y., Johnson, A. T., de Paula, J. C., Smith, W. F. (2010). Effects of O_2, Xe, and gating on the photoconductivity and persistent photoconductivity of porphyrin nanorods. *Journal of Physical Chemistry C*, **114**, 19227.
56. Kadish, K. M., Royal, G., Van Caemelbecke, E., Gueletti, L. (2000). Metalloporphyrins in Nonaqueous Media: Database of Redox Potentials. In: Kadish, K. M., Smith, K. M., Guilard, R., eds. *The Porphyrin Handbook*, Vol. 9, Academic Press, San Diego, CA, p. 1.
57. Kruger, W., Fuhrhop, J.-H. (1982). Evolution of hydrogen from photochemically reduced water-soluble tin(IV)-porphyrins. *Angewandte Chemie International Edition*, **21**, 131.
58. Song, X.-Z., Jia, S.-L., Miura, M., Ma, J.-G., Shelnutt, J. A. (1998). Electron transfer photosensitized by a tin lipoporphyrin in solution, micelles, and at water-organic solvent interfaces. *Journal of Photochemistry and Photobiology*, **113**, 233.
59. Shelnutt, J. A. (1983). Photoreduction of methylviologen sensitized by dihydroxytin(IV) uroporphyrin. *Journal of the American Chemical Society*, **105**, 7179.
60. Kalyanasundarum, K., Shelnutt, J. A., Grätzel, M. (1988). Sensitization and photoredox reactions of Zn(II)- and Sb(V)(O)(Cl)-uroporphyrins in aqueous media. *Inorganic Chemistry*, **27**, 2820.
61. Garcia, R. M., Song, Y., Dorin, R. M., Wang, H., Moreno, A. M., Jiang, Y.-B., Tian, Y., Qiu, Y., Medforth, C. J., Coker, E. N., van Swol, F., Miller, J. E., Shelnutt, J. A. (2010). Templated growth of platinum nanowheels using the inhomogeneous reaction environment of bicelles. *Physical Chemistry Chemical Physics*, **13**, 4846.
62. Song, Y., Dorin, R. M., Garcia, R. M., Jiang, Y.-B., Wang, H., Li, P., Qiu, Y., van Swol, F., Miller, J. E., Shelnutt, J. A. (2008). Synthesis of platinum nanowheels using a bicellar template. *Journal of the American Chemical Society*, **130**, 12602.
63. Song, Y., Garcia, R. M., Dorin, R. M., Wang, H., Qiu, Y., Coker, E. N., Miller, J. E., Shelnutt, J. A. (2007). Synthesis of platinum nanowire networks using a soft template. *Nano Letters*, **7**, 3650.
64. Song, Y., Hickner, M. A., Challa, S. R., Dorin, R. M., Garcia, R. M., Wang, H., Jiang, Y.-B., Li, P., Qiu, Y., van Swol, F., Medforth, C. J., Miller, J. E., Nwoga, T., Kawahara, K., Li, W., Shelnutt, J. A. (2009). Evolution of dendritic platinum nanosheets into ripening-resistant holey sheets. *Nano Letters*, **9**, 1534.
65. Song, Y., Steen, W. A., Peña, D., Jiang, Y.-B., Medforth, C. J., Huo, Q., Pincus, J. L., Qiu, Y., Sasaki, D. Y., Miller, J. E., Shelnutt, J. A. (2006). Foam-like nanostructures created from dendritic platinum sheets on liposomes. *Chemical Materials*, **18**, 2335.
66. Song, Y., Yang, Y., Medforth, C. J., Pereira, E., Singh, A. K., Xu, H., Jiang, Y., Brinker, C. J., van Swol, F., Shelnutt, J. A. (2004). Controlled synthesis of 2-D and 3-D dendritic platinum nanostructures. *Journal of the American Chemical Society*, **126**, 635.
67. Song, Y. J., Garcia, R. M., Dorin, R. M., Wang, H. R., Qiu, Y., Shelnutt, J. A. (2006). Synthesis of platinum nanocages by using liposomes containing photocatalyst molecules. *Angewandte Chemie Intlernational Edition*, **45**, 8126.

68. Song, Y. J., Jiang, Y. B., Wang, H. R., Pena, D. A., Qiu, Y., Miller, J. E., Shelnutt, J. A. (2006). Platinum nanodendrites. *Nanotechnology*, **17**, 1300.

69. Wang, H., Song, Y., Wang, Z., Medforth, C. J., Miller, J. E., Evans, L., Li, P., Shelnutt, J. A. (2008). Silica–metal core–shells and metal shells synthesized by porphyrin-assisted photocatalysis. *Chemistry of Materials*, **20**, 7434.

70. Wang, H., Song, Y. J., Medforth, C. J., Shelnutt, J. A. (2006). Interfacial synthesis of dendritic platinum nanoshells templated on benzene nanodroplets stabilized in water by a photocatalytic lipoporphyrin. *Journal of the American Chemical Society*, **128**, 9284.

71. Greenbaum, E. (1988). Interfacial photoreactions at the photosynthetic membrane interface: An upper limit for the number of platinum atoms required to form a hydrogen-evolving platinum metal catalyst. *Journal of Physical Chemistry*, **92**, 4571.

72. Ciacchi, L. C., Pompe, W., De Vita, A. (2001). Initial nucleation of platinum clusters after reduction of K_2PtCl_4 in aqueous solution: A first principles study. *Journal of the American Chemical Society*, **123**, 7371–7380.

73. Bard, A. J., Fox, M. A. (1995). Artificial photosynthesis—solar splitting of water to hydrogen and oxygen. *Accounts of Chemical Research*, **28**, 141.

74. Johansen, O., Mau, A. W.-H., Sasse, W. H. F. (1983). Highly efficient photoreduction of water using energy transfer. *Chemical Physics Letters*, **94**.

75. Barbir, F. (2005). *PEM Fuel Cells: Theory and Practice*. Academic Press, Burlington, MA, p. 456.

76. Zhong, C.-J., Luo, J., Fang, B., Warjala, B. N., Njoki, P. N., Loukrakpam, R., Yin, J. (2010). Nanostructured catalysts in fuel cells. *Nanotechnology*, **21**, 062001.

77. Furuya, N., Koide, S. (1991). Electroreduction of carbon dioxide by metal phthalocyanines. *Electrochimica Acta*, **36**, 1309.

78. Furuya, N., Matsui, K. (1989). Electroreduction of carbon dioxide on gas-diffusion electrodes modified by metal phthalocyanines. *Journal of Electroanalytical Chemistry*, **271**, 181.

79. Ryba, G., Shelnutt, J. A., Prairie, M. R., Assink, R. A. (1997). *Solar Photocatalytic Conversion of CO_2 to Methanol*. National Technical Information Service, Oak Ridge, TN, p. 28.

80. Leung, K., Nielsen, I. M. B., Sai, N., Medforth, C. J., Shelnutt, J. A. (2010). Cobalt-porphyrin catalyzed electrochemical reduction of carbon dioxide in water. 2. Mechanism from first principles. *Journal of Physical Chemistry A*, **37**, 10174.

81. Nielsen, I. M. B., Leung, K. (2010). Cobalt-porphyrin catalyzed electrochemical reduction of carbon dioxide in water. 1. A density functional study of intermediates. *Journal of Physical Chemistry, A* **114**, 10166.

82. Varotto, A., Nam, C.-Y., Radivojevic, I., Tome, J. P. C., Cavaleiro, J. A. S., T., B. C., Drain, C. M. (2010). Phthalocyanine blends improve bulk heterojunction solar cells. *Journal of the American Chemical Society*, **132**, 2552.

83. Imahori, H., Umeyama, T., Ito, S. (2009). Large π-aromatic molecules as potential sensitizers for highly efficient dye-sensitized solar cells. *Accounts of Chemical Research*, **42**, 1809.

84. Campbell, W. M., Burrell, A. K., Officer, D. L., Jolley, K. W. (2004). Porphyrins as light harvesters in the dye-sensitised TiO_2 solar cell. *Coordination Chemistry Reviews*, **248**, 1363.

85. Ganapathy, S., Oostergetel, G. T., Wawrzyniak, P. K., Beus, M., Gomez Maqueo Chew, A., Buda, F., Boekema, E. J., Bryant, D. A., Holzwarth, A. R., de Groot, H. J. M. (2009). Alternating *syn–anti* bacteriochlorophylls form concentric helical nanotubes in chlorosomes. *Proceedings of the National Academy of Science USA*, **106**, 8525.

86. Dini, F., Martinelli, E., Pomarico, G., Paolesse, R., Monti, D., Filippini, D., D'Amico, A., Lundstrom, I., Di Natale, C. (2009). Chemical sensitivity of self-assembled porphyrin nanoaggregates. *Nanotechnology*, **20**, 055502.

87. Davis, D. G. (1978). Electrochemistry of porphyrins. In: Dolphin, D., ed. *The Porphyrins*, Vol. V. Academic Press, New York.

6

NANOSTRUCTURES AND ELECTRON-TRANSFER FUNCTIONS OF NONPLANAR PORPHYRINS

SHUNICHI FUKUZUMI AND TAKAHIKO KOJIMA

6.1 INTRODUCTION

Porphyrins are naturally occurring and hetero-macrocyclic compounds containing four pyrrole rings with extended π-conjugation. Because porphyrins have four nitrogens and the nitrogen lone pairs converge to the central point of the porphyrin, they can act as a tetradentate chelating ligand when a metal ion binds [1]. In the metal complexes of porphyrins (metalloporphyrins), axial ligands play important roles in controlling the conformation of the metal complexes and characteristics of metal center [1]. When axial coordination is utilized as linkage among metal centers in metalloporphyrins, the regulation of strength of the axial coordination is an important factor to stabilize porphyrin-based molecular assemblies [2–4]. Metalloporphyrins have been used as functional molecules both in the natural [5–15] and artificial systems [16–22] by virtue of its rich photochemical and redox properties. With respect to redox properties, their biological functions involve O_2 transport (hemoglobin) [5], O_2 storage (myoglobin) [8, 23], electron transfer (cytochromes) [9, 10], oxidation reaction, and other enzymatic reactions (cytochrome P450 [12, 24–26] and peroxidases [13, 14], catalases [15], etc.). Such variety of functionalities of heme proteins and heme enzymes are derived from axial coordinated amino acid residues, distal amino acid residues around the heme, and distortion of porphyrin ring due to interactions between apoprotein and the heme [27]. With regard to photochemical properties, chlorophyll, which is a porphyrin analogue, plays important roles in photon capture and energy transport, as well as in conversion of light energy to chemical energy in photosynthetic systems as light-harvesting antenna complexes (LHC) and the reaction center (RC) in nature [28–33]. A chlorophyll dimer (the special pair) in the photosynthetic reaction center attains photon energy transferred from light-harvesting antenna [28–30], and then the photoexcited special pair releases an electron to adjacent chlorophyll to give rise to a charge-separated (CS) state between the special pair and the quinone B [31–33]. A large number of porphyrin derivatives have been synthesized to mimic the light-harvesting and charge separation processes in the photosynthesis [19–22, 34–36]. In most cases, synthesized porphyrins are planar molecules due to the π-conjugation including all atoms in macrocycles with sp^2-hybridized orbitals. Due to conformational requirements, however, four types of nonplanar deformation of the porphyrin ring structure have been realized; "saddle," "ruffle," "wave," and "dome," as shown in Figure 6.1 [37, 38]. Such a curved π-surface can be utilized to construct novel self-assembled three-dimensional structures that would not be realized by using planar π-systems. In addition, distortion of the macrocycle π-plane is expected to change the acid–base and redox properties of porphyrins.

In this chapter, we describe recent development regarding unique nanostructures and functions of nonplanar porphyrins in comparison with those of planar porphyrins. The self-assembly of nonplanar porphyrins can be exploited to produce well-defined discrete porphyrin nanostructures. The facile synthesis and robust nature of these nonplanar porphyrin-based materials allow access to a novel class of nanomaterials with potential applications in important areas such as catalysis, sensors, and solar energy conversion. Thus, utilization of nonplanar porphyrins and their metal complexes may be regarded as a useful strategy to pave a way to a new category of organic nanomaterials composed of porphyrin compounds.

Organic Nanomaterials: Synthesis, Characterization, and Device Applications, First Edition. Edited by Tomás Torres and Giovanni Bottari.

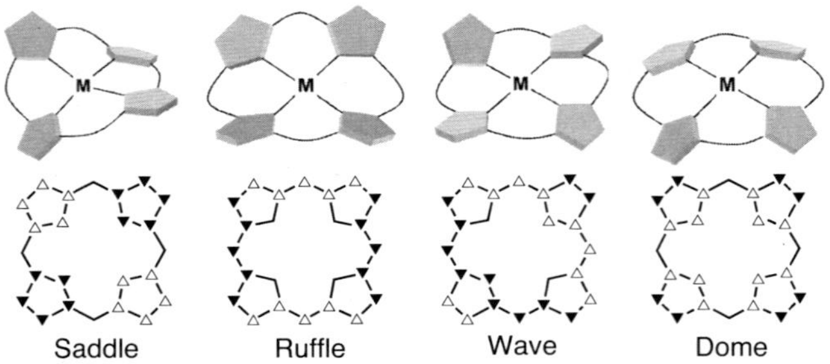

FIGURE 6.1. The four symmetrical nonplanar distortions for porphyrin compounds (above). Open triangles (Δ) correspond to atoms above the mean plane of the 24-atom porphyrin core, and filled reverse triangles (▼) correspond to atoms below the plane; atoms not represented by triangles are in the plane.

6.2 INTERMOLECULAR PHOTOINDUCED ELECTRON TRANSFER OF NONPLANAR PORPHYRINS

Dodecaphenylporphyrin (H_2DPP), which contains a phenyl ring at each β-pyrrole and *meso* position of the porphyrin macrocycle, adopts a nonplanar saddle distortion due to steric repulsion among the ring substituents [38–41]. Such a nonplanar distortion resulted in double protonation of H_2DPP with an acid to yield the porphyrin dication (H_4DPP^{2+}), which acts as an electron acceptor in contrast to unprotonated porphyrins, which usually act as electron donors rather than acceptors [42, 43]. The photoinduced intermolecular electron-transfer kinetics for the reduction of a series of nonplanar porphyrins, namely, hydrochloride salts of saddle-distorted dodecaphenylporphyrin ($[H_4DPP]Cl_2$), tetrakis(2, 4,6-trimethylphenyl)porphyrin ($[H_4TMP]Cl_2$), tetraphenylp orphyrin ($[H_4TPP]Cl_2$), and octaphenylporphyrin ($[H_4OPP]Cl_2$), has been investigated in comparison with that of a planar porphyrin, (acetonitrile-κN)(tetrakis(pentafluorophenyl) porphyrinato)zinc(II) ($[Zn(F_{20}TPP)(MeCN)]$), by laser flash photolysis measurements [44]. The driving-force dependence of logarithm of the rate constant (log k_{et}) for intermolecular electron transfer from a series of electron donors to the triplet excited states of porphyrins ($^3\{[H_4DPP]Cl_2\}^*$, $^3[Zn(F_{20}TPP)(MeCN)]^*$, $^3\{[H_4OPP]Cl_2\}^*$, $^3\{[H_4TMP]Cl_2\}^*$, and $^3\{[H_4TPP]Cl_2\}^*$) is shown in Figure 6.2. The dependence of k_{et} on the driving force of electron transfer ($-\Delta G_{et}$) is well-fitted by the Marcus equation for intermolecular outer-sphere electron transfer [Eq. (6.1)]:

$$k_{et} = k_{12} Z \exp[-(\lambda/4)(1 + \Delta G_{et}/\lambda)^2/k_B T]/ (k_{12} + Z \exp[-(\lambda/4)(1 + \Delta G_{et}/\lambda)^2/k_B T]) \quad (6.1)$$

where k_{12} is the diffusion rate constant in MeCN that is taken as 2.0×10^{10} M^{-1} s^{-1}, $Z[= (k_B T/h)(k_{12}/k_{21})]$ is the collision frequency that is taken as 1×10^{11} M^{-1} s^{-1}, k_B is the Boltzmann constant, h is the Planck constant, and λ is the reorganization energy of electron transfer [44, 45].

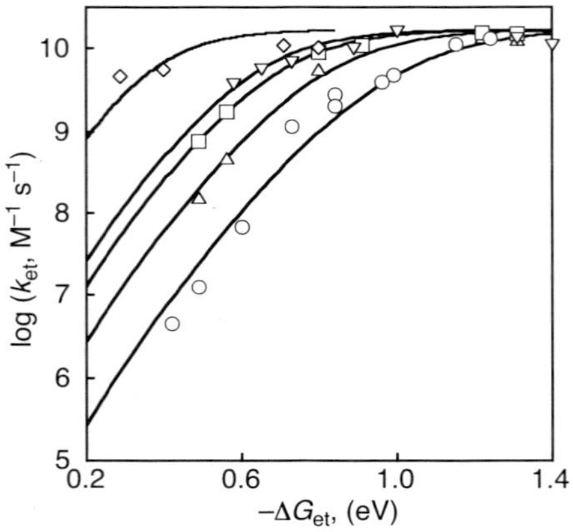

FIGURE 6.2. Driving force dependence of log k_{et} for electron transfer from a series of electron donors to $^3\{[H_4DPP]Cl_2\}^*$ (○) and $^3[Zn(F_{20}TPP)(MeCN)]^*$ (◇) in MeCN at 298 K and $^3\{[H_4OPP]Cl_2\}^*$ (Δ) and $^3\{[H_4TPP]Cl_2\}^*$ (□) and $^3\{[H_4TMP]Cl_2\}^*$ (▽) in MeCN containing ~30 equivalents of HCl at 298 K, and the fit of the curve based on the Marcus theory of electron transfer [Eq. (6.4)] is shown by the solid line with λ = 1.69, 1.45, 1.29, 1.21, and 0.84 eV.

Diprotonated porphyrin dications exhibit saddle distortion due to steric repulsion of N–H protons [46–52], as shown in Figure 6.3 [44], where the calculated structures of $[H_4DPP]Cl_2$, $[H_4OPP]Cl_2$, $[H_4TPP]Cl_2$, and $[H_4TMP]Cl_2$ agree with the reported crystal structures to exhibit nonplanar saddle distortion, whereas the porphyrin ring of $[Zn(F_{20}TPP)(MeCN)]$ is planar [53]. The DFT-optimized structures of $[Zn(F_{20}TPP)(MeCN)]$ and one-electron reduced species of $[H_4DPP]Cl_2$, $[H_4OPP]Cl_2$, $[H_4TPP]Cl_2$, $[H_4TMP]Cl_2$, and $[Zn(F_{20}TPP)(MeCN)]$ are also shown in Figure 6.3 [44].

The degree of the distortion in the doubly protonated porphyrins is evaluated by root-mean-square out-of-plane displacement (ΔRMS), which is defined as [Eq. (6.2)] [46]:

$$\Delta RMS = \sqrt{\frac{1}{24}\sum_{i=1}^{24} \delta_i^2} \quad (6.2)$$

where δ_i is the orthogonal displacement of atom i in the macrocycle from the mean plane.

The reorganization energy of electron transfer generally consists of solvent reorganization energy (λ_s) and intramolecular reorganization energy (λ_i): $\lambda = \lambda_s + \lambda_i$ [45]. The λ values for the electron-transfer reduction of saddle-distorted porphyrin dications are significantly larger as compared with the reorganization energy for the electron-transfer reduction of planar $[Zn(F_{20}TPP)(MeCN)]$, increasing linearly with increasing distortion of the porphyrin ring (ΔRMS) (Figure 6.4). Such a linear correlation between λ and ΔRMS

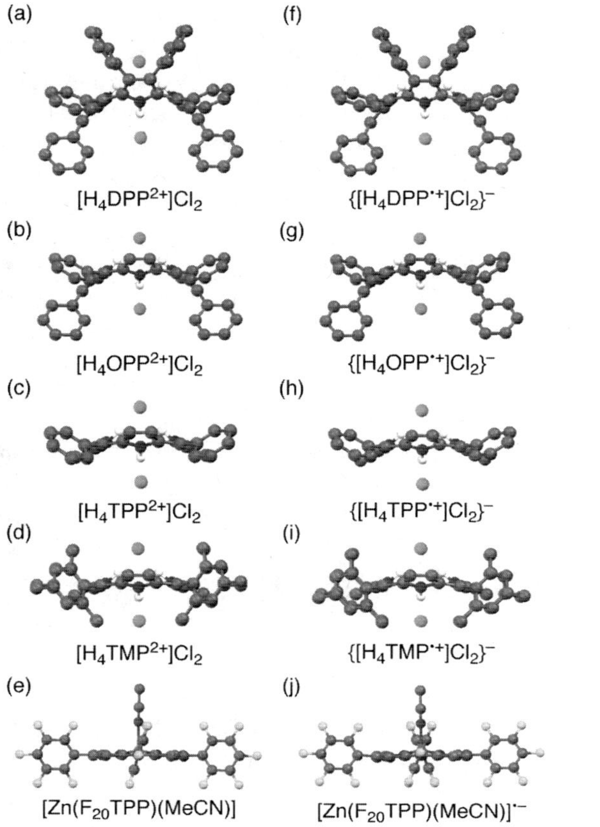

FIGURE 6.3. Optimized structures of (**a**) $[H_4DPP^{2+}]Cl_2$, (**b**) $[H_4OPP^{2+}]Cl_2$, (**c**) $[H_4TPP^{2+}]Cl_2$, (**d**) $[H_4TMP^{2+}]Cl_2$, (**e**) $[Zn(F_{20}TPP)(MeCN)]$, (**f**) $\{[H_4DPP^{\bullet+}]Cl_2\}^-$, (**g**) $\{[H_4OPP^{\bullet+}]Cl_2\}^-$, (**h**) $\{[H_4TPP^{\bullet+}]Cl_2\}^-$, (**i**) $\{[H_4TMP^{\bullet+}]Cl_2\}^-$, and (**j**) $[Zn(F_{20}TPP)(MeCN)]^{\bullet-}$ obtained by DFT calculations at the B3LYP/3-21G level of theory. Hydrogen atoms of phenyl groups are omitted for clarity.

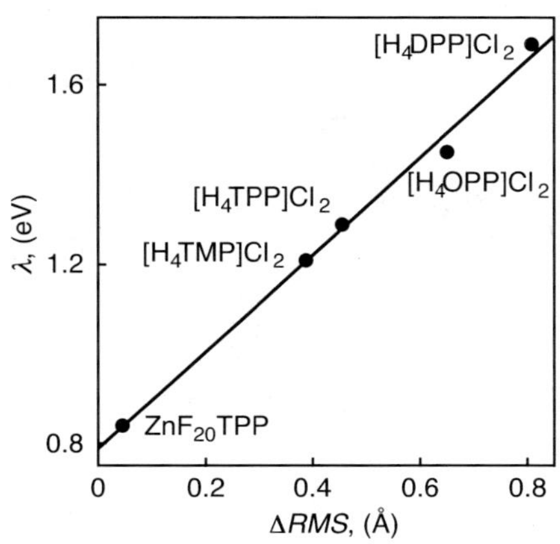

FIGURE 6.4. A relationship between λ values in photoinduced electron-transfer reduction of diprotonated porphyrin dications and $[Zn(F_{20}TPP)]$ in MeCN and root-mean-square out-of-plane displacements (ΔRMS).

suggests that the λ values of saddle-distorted porphyrin dications largely depend on intramolecular reorganization energy (λ_i), which is determined by the resultant structural change of the entity, because the solvent term of reorganization energy may remain the same [44]. The differences in ΔRMS values of porphyrin dications before and after electron transfer are approximately the same, whereas the difference in the distance of hydrogen bonding among the pyrrole N–H protons and the chloride anions ($N\text{–}H\cdots Cl^-$) become larger, with increase of ΔRMS values. Thus, the elongation of the hydrogen bond may be the main factor for the increase of λ_i. Similar observation was reported for the electron-transfer oxidation of metalloporphyrins, which exhibit large reorganization energies due to the large changes in coordination bond lengths around metal centers before and after electron transfer [54]. Such a correlation between the molecular distortion and reorganization energy was also observed in the type 1 copper site of M98Q amicyanin [55]. The weakening of hydrogen bonding among pyrrole N–H protons and chloride anions gives rise to the larger distortion as represented by the larger ΔRMS in the one-electron reduced species to exemplify the larger λ_i values [46].

The λ values of nonplanar saddle-distorted metalloporphyrins ([Zn(DPP)] and [Al(DPP)(PhCOO)]) were also determined in the intermolecular photoinduced electron-transfer oxidation with a series of electron acceptors, being 0.27 eV larger than those of planar metalloporphyrins without and with an axial ligand ([Zn(TPP)] and [Al(TPP)(PhCOO)], respectively) when they are compared at the same driving force of photoinduced electron transfer [56]. Thus, the conformational distortion of the porphyrin ring in DPP^{2-} results in a significant increase in the bond reorganization energy of photoinduced intermolecular electron-transfer oxidation and reduction of metalloporphyrins.

6.3 PHOTOINDUCED ELECTRON TRANSFER IN SUPRAMOLECULAR COMPLEXES OF NONPLANAR PORPHYRINS

6.3.1 Hydrogen-Bond Complexes

Beside the intermolecular photoinduced electron transfer of nonplanar porphyrins (Figure 6.2), intrasupramolecular photoinduced electron transfer of a saddle-distorted H_4DPP^{2+} was investigated in a series of hydrogen-bonded supramolecular complexes composed of H_4DPP^{2+} and ferrocene derivatives bearing a carboxylate group [57]. The x-ray crystal structure of a ferrocenecarboxylic acid (FcCOOH) salt of H_2DPP ($H_4DPP(FcCOO)_2$) is shown in Figure 6.5, where two ferrocenecarboxylate ($FcCOO^-$) anions are hydrogen-bonded with H_4DPP^{2+} asymmetrically [57]. On one side, the saddle-distorted H_4DPP^{2+} forms two-point hydrogen bonding among two of the N–H protons of pyrroles and two of

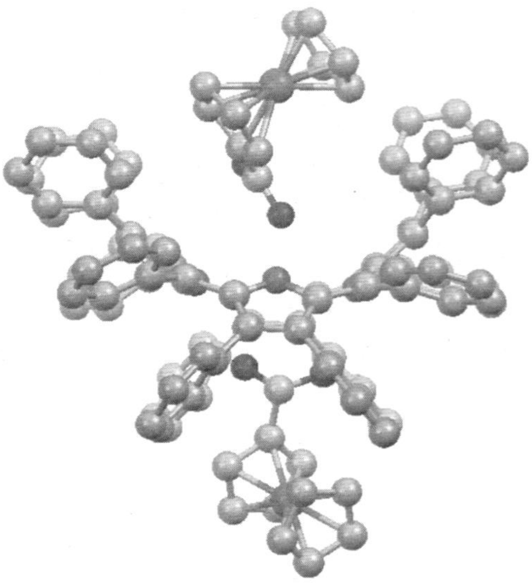

FIGURE 6.5. Crystal structure of $H_4DPP(FcCOO)_2$.

the oxygen atoms in the carboxylate of $FcCOO^-$ with the interatomic (N···O) distance of 2.726 (5) and 2.587 (4) Å. On the other side, one of the oxygen atoms of the carboxylate group in $FcCOO^-$ is hydrogen-bonded with two of the N–H protons of pyrroles. Both $FcCOO^-$ anions were placed in a cleft formed by the two phenyl groups attached to a pyrrole tilted up toward the ferrocene parts, forming CH/π interactions with the phenyl groups.

The supramolecular complex ($H_4DPP(FcCOO)_2$) in Figure 6.5 is also formed in solution, and the binding constants obtained by spectroscopic titration indicate the strong binding (10^8–10^{10} M^{-2}) even in a polar and coordinating solvent, benzonitrile (PhCN) [57]. ^1H NMR signals of the ferrocene moiety exhibited upfield shifts due to shielding by the ring current of the porphyrin ring in $H_4DPP(FcCOO)_2$, and the largest upfield shift was observed for the proton adjacent to the carboxylate, indicating formation of hydrogen bonding between the pyrrole moieties and the carboxylate group of $FcCOO^-$ in $CDCl_3$ [57].

The highest occupied molecular orbital (HOMO) and the lowest unoccupied molecular orbital (LUMO) of $H_4DPP(FcCOO)_2$ were estimated by DFT calculations using the B3LYP/6-31G(d) basis set on the basis of the crystal structure as shown in Figure 6.6, which indicates that the HOMO and the LUMO are localized on the $FcCOO^-$ moiety and the H_4DPP^{2+} moiety, respectively. Thus, photoinduced electron transfer is expected to occur from the $FcCOO^-$ moiety to the singlet excited state of H_4DPP^{2+} upon photoexcitation of the diprotonated porphyrin.

The occurrence of efficient electron transfer from $FcCOO^-$ to the singlet excited state of H_4DPP^{2+} ($^1[H_4DPP^{2+}]^*$) was confirmed by the transient absorption bands at 550 and 1030 nm due to $^1[H_4DPP^{2+}]^*$ observed upon femtosecond laser excitation at 430 nm of a deaerated PhCN solution of $H_4DPP(FcCOO)_2$ [57]. The rate constants of photoinduced intrasupramolecular electron transfer (k_{ET}) and back electron transfer (k_{BET}) were found to be 5.0×10^{11} s^{-1} and 6.1×10^{10} s^{-1}, respectively [58]. The driving force dependence of the rate constants of intrasupramolecular photoinduced electron transfer from a series of hydrogen-bonded electron donors with a carboxyl group to $^1[H_4DPP^{2+}]^*$ (k_{ET}) and back electron transfer to afford the corresponding ground states (k_{BET}) is shown in Figure 6.7,

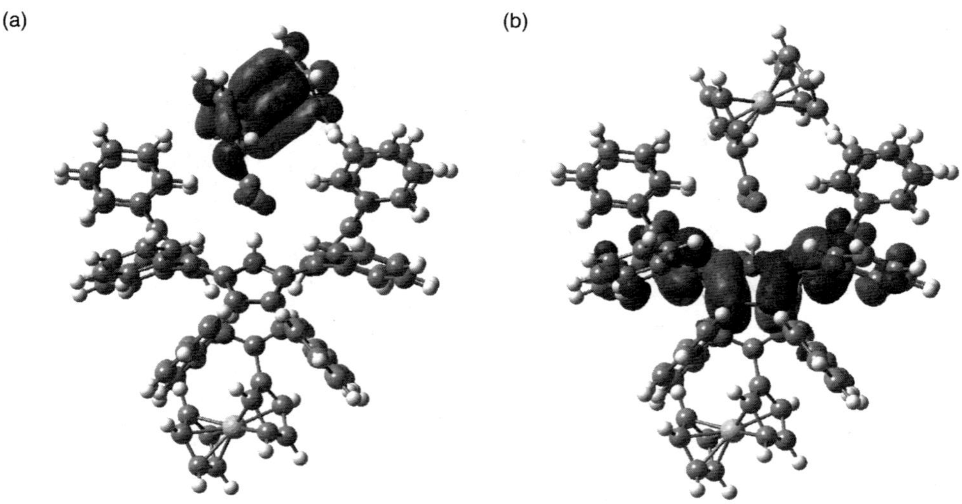

FIGURE 6.6. (**a**) HOMO and (**b**) LUMO orbitals of $H_4DPP(FcCOO)_2$ calculated by the DFT method at the B3LYP/6-31G(d) level.

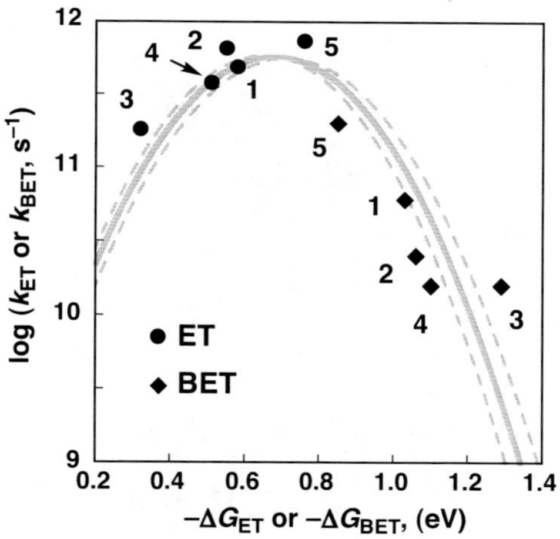

FIGURE 6.7. Driving force dependence of log k_{ET} (•) or k_{BET} (♦) for intrasupramolecular photoinduced electron transfer and back electron transfer in supramolecular complexes of H_4DPP^{2+} with hydrogen-bonded electron donors [1, FcCOOH; 2, FcAcOH (ferroceneacetic acid); 3, BrFcCOOH (1-bromo-1′-carboxyferrocene); 4, Fc(COOMe)(COOH) (1-methoxycarbonyl-1′-carboxyferrocene); 5, TTFCOOH (2-carboxytetrathiafulvalene); 6, FcPhCOOH (4-carboxyphenylferrocene); 7, FcbphCOOH (4-methoxycarbonyl-4′-ferrocenylbiphenyl))] in PhCN at room temperature. The fitting to the Marcus theory of the electron transfer [Eq. (6.8)] is shown by the solid line with $\lambda = 0.68$ eV and $V = 43$ cm^{-1} and the dotted lines with $\lambda = 0.65$ and 0.71 eV and $V = 43$ cm^{-1}, respectively.

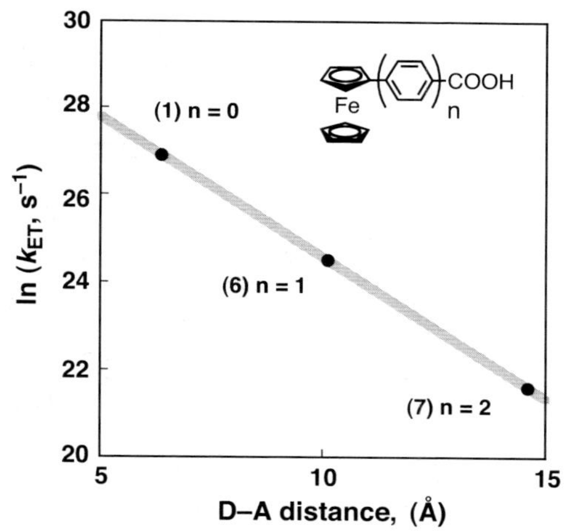

FIGURE 6.8. Distance dependence of ln k_{ET} for intrasupramolecular electron transfer from electron donors ((**1**) $FcCOO^-$, (**6**) $FcPhCOO^-$, and (**7**) $FcbphCOO^-$) to $^1[H_4DPP^{2+}]^*$, where the distance between an electron donor and an electron acceptor is defined as that between the Fe atom of the ferrocene moiety and the center of a mean plane of the porphyrin ring.

which is well-analyzed by the Marcus equation of nonadiabatic intramolecular electron transfer [Eq. (6.3)] [58],

$$k_{ET} = \left(\frac{4\pi^3}{h^2\lambda k_B T}\right)^{1/2} V^2 \exp\left[-\frac{(\Delta G_{ET} + \lambda)^2}{4\lambda k_B T}\right] \quad (6.3)$$

where V is the electronic coupling matrix element, h is the Planck constant, T is the absolute temperature, ΔG_{ET} is the free energy change of electron transfer, and λ is the reorganization energy of electron transfer. The curve fit of the results for a variety of electron donors by using [Eq. (6.3)] (Figure 6.7) afforded a small reorganization energy ($\lambda = 0.68 \pm 0.03$ eV) and a large electronic coupling matrix element ($V = 43 \pm 7$ cm^{-1}).

It should be noted that the λ value (0.68 ± 0.03 eV) of intrasupramolecular electron transfer is significantly smaller than that of intermolecular photoinduced electron transfer (1.69 eV for $[H_4DPP]^{2+}$) in Figure 6.2, probably because the elongation of the hydrogen bond observed in intermolecular electron transfer (*vide supra*) is reduced in the supramolecular complex. The λ value of the supramolecular complex is comparable to those of covalently linked donor–acceptor systems consisting of electrically neutral porphyrins as electron donors (0.41–0.66 eV) [59–66]. The large electronic coupling matrix element may result from the strong interaction between H_4DPP^{2+} and electron-donor molecules with a carboxylate group because the HOMO of $FcCOO^-$ extends to the carboxylate moiety (Figure 6.6a) [57].

The distance dependence of intrasupramolecular photoinduced electron transfer and back electron transfer was also examined by changing of the number of phenylene linker between the ferrocene moiety and the carboxyl group of ferrocenecarboxylic acid derivatives (see the inset of Figure 6.8). The distance (r) between electron donor and acceptor is defined as that between the Fe atom of ferrocene and the center of a mean plane of the porphyrin ring of H_4DPP^{2+}. The distance dependence of the rate constant of intrasupramolecular electron transfer (k_{ET}) is given by Eq. (6.4):

$$k_{ET} = k_0 \exp(-\beta r) \quad (6.4)$$

where k_0 is the rate constant of adiabatic intrasupramolecular electron transfer, r is the donor–acceptor center-to-center distance, and β is the decay coefficient factor (damping factor), which depends primarily on the nature of the bridging molecule [67]. From the slope of log k_{ET} versus r in Figure 6.8, the β value was determined to be 0.64 Å^{-1}. This value for the hydrogen-bonded supramolecular complex is comparable to those of covalently phenylene-bridged multiporphyrin systems. The distance dependence of the rate constant of back electron transfer (k_{BET}) was also determined to be 0.53 Å^{-1}. Thus, intrasupramolecular photoinduced electron transfer and back electron transfer in supramolecular complexes of saddle-distorted H_4DPP^{2+} with hydrogen-bonded

(a) (b)

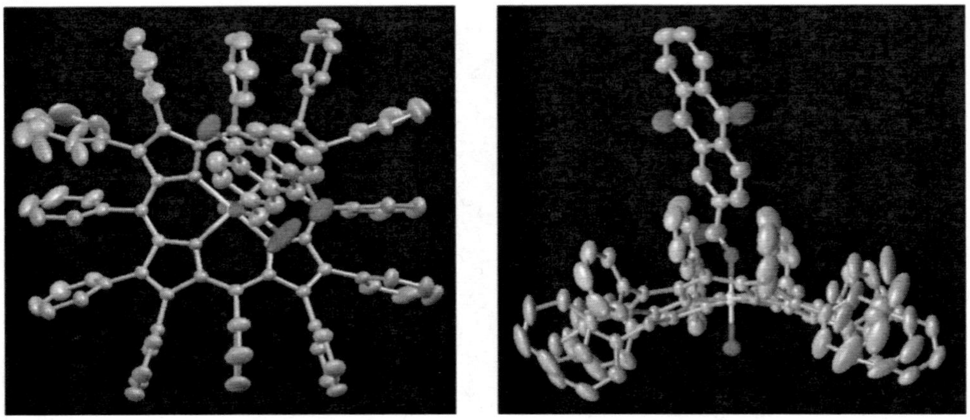

FIGURE 6.9. Crystal structure of [Al(DPP)(AqCO*O*)(H_2O)]•PhCN from different directions. (**a**) Top view. (**b**) Side view.

electron donors are as efficient as covalently bonded electron-donor systems [62, 68].

6.3.2 Coordination Complexes of $[Al(DPP)]^+$ versus $[Al(TPP)]^+$

Intracomplex photoinduced electron transfer of coordination complexes of both planar $[Al(TPP)]^+$ and nonplanar $[Al(DPP)]^+$ with anthraquinone-2-carboxylate ($AqCOO^-$) that can act as an electron acceptor was also examined to clarify the effect of distortion of the porphyrin ring on the electron transfer [56]. The strong Lewis acidity of the Al^{3+} ion enables the formation of strong axial binding of $AqCOO^-$. The difference in the binding geometries between [Al(DPP)(AqCO*O*)] and [Al(TPP)(AqCO*O*)] is shown by the crystal structures in Figures 6.9 and 6.10, respectively. In the saddle-distorted [Al(DPP)(AqCO*O*)] complex, the $AqCOO^-$ moiety is nearly perpendicular to the porphyrin ring (Figure 6.9). There is no intracomplex π–π interaction between the saddle-distorted porphyrin ligand and the axial $AqCOO^-$ ligand, and the $AqCOO^-$ moiety's range of motion is limited by the steric hindrance of the phenyl groups of DPP^{2-}.

In contrast to the saddle-distorted [Al(DPP)(AqCO*O*)] complex (Figure 6.9), the strong axial coordination of $AqCOO^-$ to the Al^{3+} ion of planar $Al(TPP)^+$ and intracomplex π–π interactions are found between the porphyrin ligand and the $AqCOO^-$ ligand as shown in Figure 6.10. The intracomplex π–π interactions are also observed between the centroid of the phenyl ring with the carboxyl group of the $AqCOO^-$ ligand and the *meso* carbon just below the centroid in the porphyrin ring in the face-to-face fashion, as well as between the centroid of the quinone moiety of the $AqCOO^-$ ligand and the *ortho* carbon of the *meso*-phenyl group in the edge-to-face fashion.

In both cases of planar [Al(TPP)(AqCO*O*)] and nonplanar [Al(DPP)(AqCO*O*)], efficient electron transfer from the singlet excited state of Al porphyrins to the anthraquinone moiety occurs followed by rapid back electron transfer to

(a) (b)

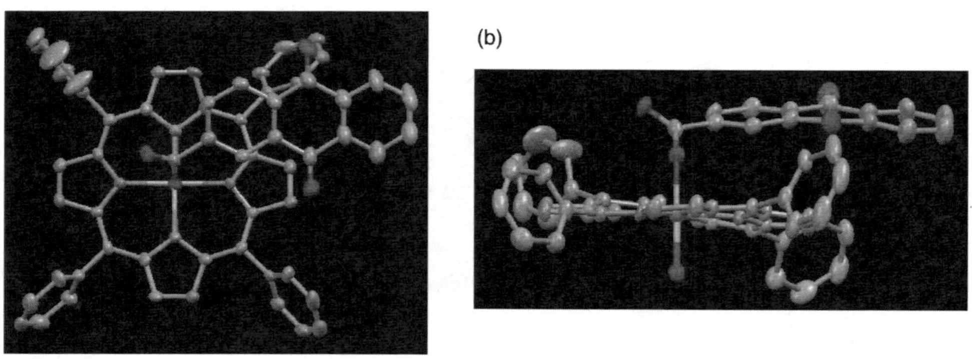

FIGURE 6.10. Crystal structure of [Al(TPP)(AqCO*O*)(H_2O)]•PhCN from different directions. (**a**) Top view. (**b**) Side view.

produce the triplet excited state [56]. The rate constant of electron transfer from the $AqCOO^-$ moiety to the singlet excited state of Al(DPP) ($k_{ET} = 2.5 \times 10^{10}\ s^{-1}$) is larger than that of Al(TPP) ($k_{ET} = 1.3 \times 10^{10}\ s^{-1}$) [56]. The faster photoinduced electron transfer in [Al(DPP)(AqCOO)] than that in [Al(TPP)(AqCOO)] results from the easier oxidation of Al(DPP) ($E_{ox} = 0.55$ V vs. SCE) than that of Al(TPP) ($E_{ox} = 0.74$ V vs. SCE) due to the distortion of the porphyrin ring [56]. The λ values of [Al(TPP)(AqCOO)] and [Al(DPP)(AqCOO)] were determined to be 0.86 and 0.98 eV, respectively [56]. As is the case in intermolecular electron transfer (*vide supra*), the λ value of saddle-distorted porphyrin is larger than that of planar porphyrin. The close proximity between the Al porphyrins and the $AqCOO^-$ moiety in both cases (Figures 6.9 and 6.10) makes the intracomplex photoinduced electron transfer and back electron transfer adiabatic, when the electron-transfer rate is solely determined by the driving force and reorganization energy of electron transfer but not by the geometric difference between the donor and acceptor moieties [56], because the rate constant of adiabatic electron transfer does not depend on structures and electronic coupling between electron donors and acceptors [58].

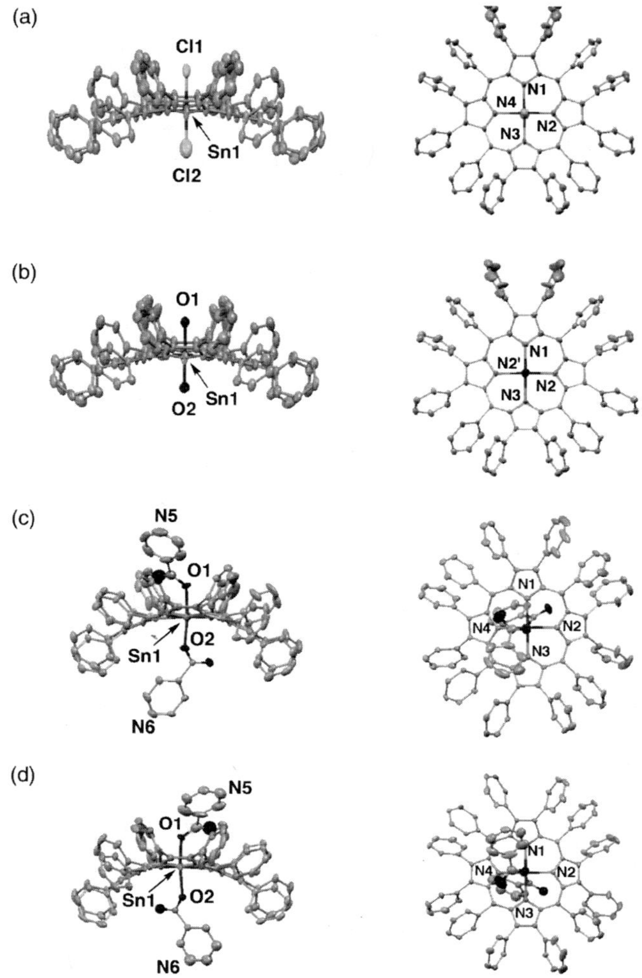

FIGURE 6.11. Side views (*left*) and top views (*right*) of crystal structures of Sn(DPP) complexes: (**a**) $[Sn(DPP)(Cl)_2]$ (**1**). (**b**) $[Sn(DPP)(OH)_2]$ (**2**). (**c**) $[Sn(DPP)(4\text{-}PyCOO)_2]$ (**3**). (**d**) $[Sn(DPP)(3\text{-}PyCOO)_2]$ (**4**). Thermal ellipsoids are drawn at the 50% probability level.

6.3.3 Coordination Complexes of $Sn(DPP)^{2+}$

The Sn^{4+} ion is a stronger Lewis acid than the Al^{3+} ion, and thus $Sn(DPP)^{2+}$ can form more stable coordination bond than $Al(DPP)^+$ for the construction of axial coordination complexes of Sn(DPP) with electron acceptors with carboxylate groups [69]. As an electron acceptor, μ_3-oxo-centered and carboxylato-bridged trinuclear ruthenium clusters, $[Ru_3(\mu_3\text{-}O)(OC(O)R)_6L_3]^{n+}$, have been utilized because of small reorganization energies of electron transfer with small structural change upon the electron transfer [70–73].

X-ray crystal structures of Sn(DPP) complexes, $[Sn(DPP)(Cl)_2]$ (**1**), $[Sn(DPP)(OH)_2]$ (**2**), $[Sn(DPP)(4\text{-}PyCOO)_2]$ (**3**), and $[Sn(DPP)(3\text{-}PyCOO)_2]$ (**4**) (3-PyCOOH: 3-pyridine carboxylic acid, 4-PyCOOH: 4-pyridine carboxylic acid) are shown in Figure 6.11. All the Sn(DPP) complexes exhibit saddle-distorted structures to a similar extent. The Sn–O (carboxylato) bond lengths in **3** and **4** (2.077(2) Å for Sn1–O1 and 2.087(2) Å for Sn1–O2 in **3**, 2.077(2) Å for Sn1–O1 and 2.089(2) Å for Sn1–O2 in **4**) are significantly shorter than those of planar $[Sn(TPyP)(4\text{-}PyCOO)_2]$ (Sn-O: 2.087(3) and 2.091(4) Å for $[Sn(TPyP)(4\text{-}PyCOO)_2]$ (*trans*-diisonicotinate[*meso*-tetrakis(4-pyridyl)porphyrinato]tin(IV)) [74]. Thus, the distortion of porphyrin plane strengthens the axial coordination, because saddle distortion enhances Lewis acidity of a metal center due to poor overlap between orbitals ($d_{x2\text{-}y2}$, p_x, and p_y) of the metal center and the lone pairs of pyrrole nitrogens [75].

The Sn(DPP) complexes with trinuclear Ru(III) clusters (**9–12** in Figure 6.12) were prepared by the reaction of acetato-bridged or benzoato-bridged clusters ($[Ru_3(\mu_3\text{-}O)(OAc)_6(Py)_2(MeOH)](PF_6)$ (**5**), $[Ru_3(\mu_3\text{-}O)(OBz)_6(H_2O)_3](PF_6)$ (**6**), and $[Ru_3(\mu_3\text{-}O)\ (OBz)_6(Py)_2(CO)]$ (**7**) and $[Ru_3(\mu_3\text{-}O)(OBz)_6(Py)_2(MeOH)](PF_6)$ (**8**)) with excess amount of **3** or **4** in CH_2Cl_2/acetone mixed solution [69]. The Ru-cluster moieties in the Sn(DPP) complexes showed one reversible reduction wave due to $Ru^{III}{}_3/Ru^{II}Ru^{III}{}_2$ redox couples (0.03 V for **9**, 0.11 V for **10**, −0.13 V for **11**, and −0.11 V for **12**) and one reversible oxidation wave ascribed to $Ru^{III}{}_3/Ru^{III}{}_2Ru^{IV}$ redox couples (1.06 V for **9**, 1.06 V for **10**, 0.90 V for **11**, and 0.93 V for **12**) [69]. The Sn(DPP) moieties exhibited reversible reduction waves (–0.91 V for **9**, −0.90 V for **10**, −0.89 V for **11**, and −0.91 V for **12**) due to $Sn(DPP)/Sn(DPP)^{\bullet -}$ and reversible oxidation waves (1.06 V for **9**, 1.06 V for **10**, 1.04 V for **11**, and 1.04 V for **12**) due to $Sn(DPP)/Sn(DPP)^{\bullet +}$, which were overlapped with the

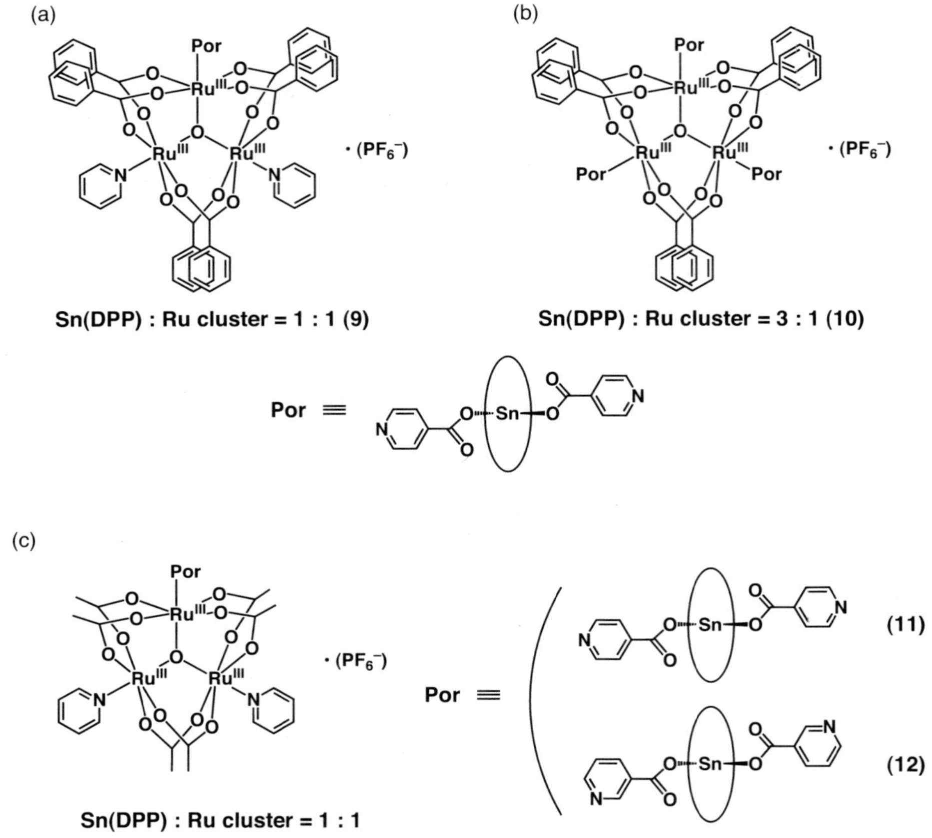

FIGURE 6.12. Structures of conglomerates of (**a**) **9**, (**b**) **10**, and (**c**) **11** and **12**.

oxidation waves of the Ru cluster moieties in the case of **9** and **10** [69]. The reversible redox couples of **9–12** indicate that the Sn(DPP) complexes with trinuclear Ru(III) clusters are stable in solution and in the course of redox reactions.

Femtosecond laser flash photolysis measurements of **9–12** revealed that efficient electron transfer from the singlet excited state of the Sn(DPP) moiety to the trinuclear Ru clusters occurs efficiently to produce the electron-transfer state ($Sn(DPP)^{\bullet+}$–$Ru^{II}Ru^{III}{}_2$) in the Sn(DPP) complexes with trinuclear Ru(III) clusters [69]. The rate constants of photoinduced electron transfer (k_{ET}) and back electron transfer to the ground state (k_{BET}) were determined from the rise and decay of the transient absorption due to the electron-transfer state [69]. The driving-force dependence of log k_{ET} and log k_{BET} is shown in Figure 6.13. The fitting of the data to the Marcus equation [Eq. (6.3)] afforded the reorganization energy of $\lambda = 0.58$ eV and the electronic coupling matrix element $V = 28.2$ cm^{-1} as shown as the solid line in Figure 6.13. This reorganization energy of electron transfer is relatively small in comparison with other related dyads giving rise to charge-separated states or electron-transfer states [61,62,76,77]. As a consequence, both photoinduced electron transfer and back

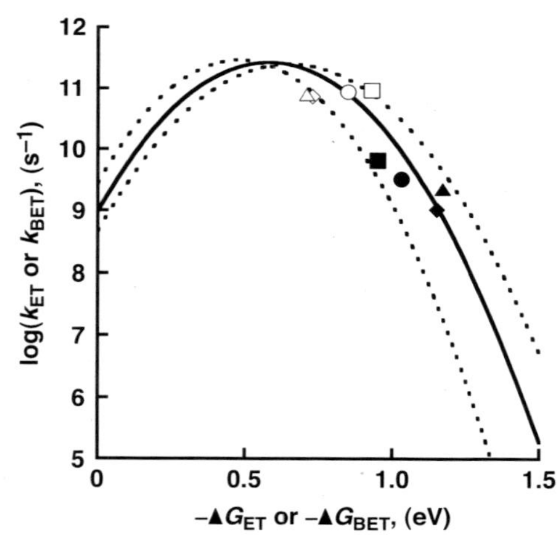

FIGURE 6.13. Driving-force dependence of log k_{ET} (closed shapes) or k_{BET} (open shapes) for intramolecular electron transfer of conglomerates in PhCN at 298 K; **9** (circle), **10** (square), **11** (triangle), and **12** (rhombus). The fit of the curve based on the Marcus theory of electron transfer [Eq. (6.3)] is shown by the solid line with $\lambda = 0.58$ eV and $V = 28.2$ cm^{-1} and by the dotted lines with $\lambda = 0.50$ and 0.66 eV and $V = 28.2$ cm^{-1}, respectively.

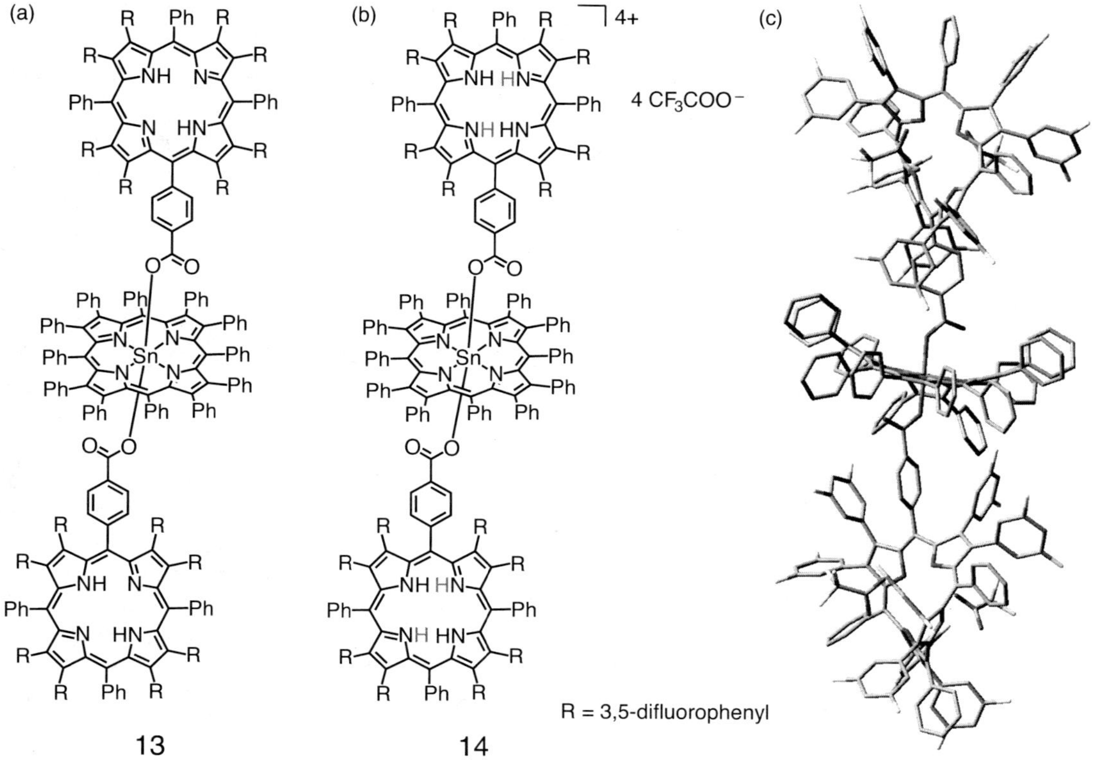

FIGURE 6.14. Porphyrin triads (**a**) **13** and (**b**) **14**. (**c**) The optimized structure of **14** calculated by the AM1 method.

electron transfer occur in the Marcus inverted region, where the rate of electron transfer is slowed down with increasing the driving force.

As described above, the $Sn(DPP)^{2+}$ unit forms stable complexes with trinuclear Ru^{III} clusters by axial coordination in which $Sn(DPP)^{2+}$ acts as a good electron donor [69]. On the other hand, diprotonated H_2DPP dication (H_4DPP^{2+}) has been used as a good electron acceptor (*vide supra*) [44, 57]. Thus, the combination of $Sn(DPP)^{2+}$ with H_4DPP^{2+} carboxylate derivatives using two axial coordination bonds of carboxylates to the Sn(IV) center afford a robust electron donor–acceptor triad composed of only porphyrin derivatives [78]. The reaction of $Sn(DPP)(OH)_2$ (**2**) with $H_2F_{16}DPPCOOH$ (2,3,7,8,12,13,17,18-octakis(3,5-difluorophenyl)-5-(4-carboxyphenyl)-10,15,20-triphenylporphyrin) afforded a porphyrin triad, $Sn(DPP)(H_2F_{16}DPPCOO)_2$ (**13**) in which the $Sn(DPP)^{2+}$ unit is linked with the two $H_2F_{16}DPPCOO^-$ units by strong coordination bonds (Figure 6.14a) [78]. The $H_2F_{16}DPPCOO^-$ unit of $Sn(DPP)(H_2F_{16}DPPCOO)_2$ was diprotonated by the reaction with trifluoroacetic acid (CF_3COOH) to afford a robust electron acceptor–donor–acceptor porphyrin triad, $Sn(DPP)\{(H_4F_{16}DPPCOO)(CF_3COO)_2\}_2$ (**14**) in which the $Sn(DPP)^{2+}$ unit and the $H_4F_{16}DPP^{2+}COO^-$ ($H_4F_{16}DPPCOO^+$) unit act as an electron donor and an acceptor, respectively (Figure 6.14b).

Femtosecond laser flash photolysis measurements of **13** in PhCN revealed that efficient energy transfer occurs from the singlet excited state of the $Sn(DPP)^{2+}$ unit to the $H_2F_{16}DPPCOO^-$ unit to generate singlet excited state of $H_2F_{16}DPPCOO^-$. In contrast to the case of **13**, the transient absorption spectra of **14** that contains the diprotonated form ($H_4F_{16}DPPCOO^+$) clearly indicated occurrence of fast electron transfer from the singlet excited state of the $Sn(DPP)^{2+}$ unit to the $H_4F_{16}DPPCOO^+$ unit [78]. The resulting singlet electron-transfer state composed of $Sn(DPP)^{\bullet+}$ and $[H_4F_{16}DPPCOO]^{\bullet}$ decays to the ground state with the rate constant of 1.4×10^{10} s^{-1} in competition with generation of the triplet electron-transfer state, which was detected by the nanosecond transient absorption spectroscopy [78]. The lifetime of the triplet electron-transfer state was determined to be 50 μs, which is much longer than that of the singlet electron-transfer state (71 ps) because of the spin-forbidden character of the back electron-transfer process [78]. The existence of the long-lived triplet electron-transfer state was confirmed by electron transfer from the one-electron-reduced diprotonated porphyrin moiety in the triplet electron-transfer state to an

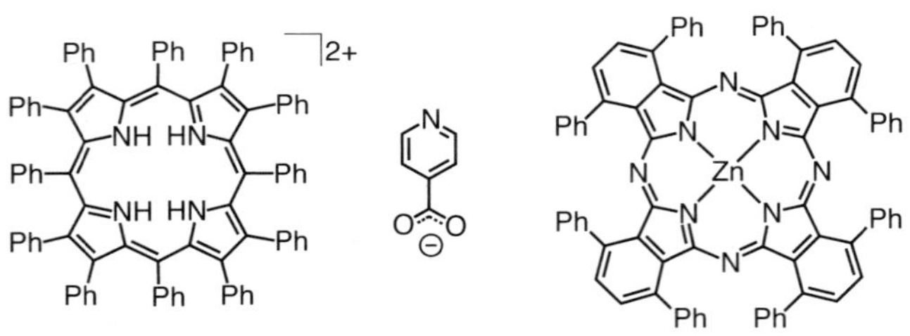

FIGURE 6.15. Components employed for a supramolecular complex $(H_4DPP)\{Zn(OPPc)(4\text{-}PyCOO)\}_2$.

electron acceptor (*p*-chloranil) to produce *p*-chloranil radical anion [78].

6.4 SUPRAMOLECULAR CONGLOMERATE COMPOSED OF SADDLE-DISTORTED ZINC(II)-PHTHALOCYANINE AND H_4DPP^{2+}

Hydrogen bonding utilized in $H_4DPP(FcCOO)_2$ and axial coordination in Al(III)- and Sn(IV)-porphyrins with electron acceptors with carboxylate groups (*vide supra*) are combined to construct a discrete supramolecular conglomerate composed of H_4DPP^{2+} and two saddle-distorted Zn(II)-octaphenylphthalocyanine (Zn(OPPc) in Figure 6.15 [79]. Zn(OPPc) and H_4DPP^{2+} are linked by 4-pyridine carboxylate (4-$PyCOO^-$) as an axial ligand of the Zn(OPPc) units and a hydrogen-bonded counter anion of H_4DPP^{2+}. Because porphyrins exhibit strong Soret bands around 400–450 nm, whereas phthalocyanines show strong Q bands around 700–800 nm, the combination of those two systems can cover nearly the whole range of the visible region and can be a useful strategy for development of photofunctional materials for efficient light-energy conversion [80–83].

The supramolecular conglomerate $(H_4DPP)\{Zn(OPPc)(4\text{-}PyCOO)\}_2$ was obtained by the reaction of $H_4DPP(4\text{-}PyCOO)_2$ with Zn(OPPc) in toluene [79]. The crystal structure of the conglomerate is shown in Figure 6.16, where the conglomerate contains one H_4DPP^{2+} ring and two Zn(OPPc) complexes linked by two 4-$PyCOO^-$ ions [79]. The bridging 4-$PyCOO^-$ ion coordinates to the Zn^{II} center of Zn(OPPc) through the pyridine nitrogen atom, whereas its carboxylate group forms two-point intermolecular hydrogen bonding with two pyrrole protons of the H_4DPP^{2+} unit. The Zn^{II} ions in the Zn(OPPc) units are located toward the pyridine ligands 0.438(2) Å (Zn1) and 0.462(3) Å (Zn2) above the least-square planes composed of the 24 atoms of each OPPc ligand. The crystal packing of $(H_4DPP)\{Zn(OPPc)(4\text{-}PyCOO)\}_2$ exhibited a one-dimensional nanowire structure into the crystallographic *b* axis (Figure 6.16a) by intermolecular π–π interactions of the peripheral phenyl groups of the OPPc ligand. The nanowires are extended into the direction of the crystallographic *ac* plane (Figure 6.16c) also with intermolecular π–π interactions. This conglomerate holds its architecture in $CDCl_3$ as evidenced by ^1H diffusion ordered spectroscopy [79].

Femtosecond laser photoexcitation of the conglomerate clearly gives rise to generation of the intrasupramolecular electron-transfer state involving radical cations of both one-electron-reduced $[H_4DPP]^{2+}$ and one-electron-oxidized Zn(OPPc) via the singlet excited states of both the $[H_4DPP]^{2+}$ and the Zn(OPPc) moieties [79]. First, the singlet excited state of the $[H_4DPP]^{2+}$ unit (1.66 eV) undergoes fast intrasupramolecular energy transfer to produce the singlet excited state of the Zn(OPPc) moiety (1.52 eV). This was followed by intrasupramolecular electron transfer from

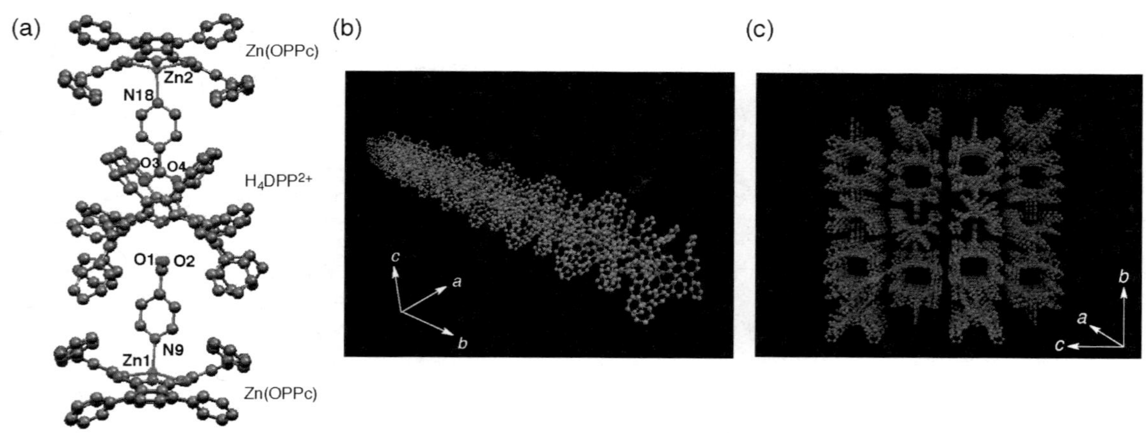

FIGURE 6.16. (**a**) Crystal structure of $(H_4DPP)\{Zn(OPPc)(4\text{-}PyCOO)\}_2$. Hydrogen atoms are omitted for clarity. (**b**) One-dimensional nanowire structure of $(H_4DPP)\{Zn(OPPc)(4\text{-}PyCOO)\}_2$ in the crystal directed to the crystallographic *b* axis. (**c**) Crystal packing of $(H_4DPP)\{Zn(OPPc)(4\text{-}PyCOO)\}_2$. Solvent molecules of crystallization are omitted for clarity. See color insert.

$^1[Zn(OPPc)]^*$ to H_4DPP^{2+} to afford the electron-transfer state, which has the lifetime of 667 ps, which is much longer than the lifetime (85 ps) of the electron-transfer state in a covalently linked ZnPc–SnPor dyad system [84].

6.5 NONPLANAR PORPHYRIN NANOCHANNELS

Porphyrins have been extensively utilized as building blocks to prepare zeolite-like porous nanomaterials by the self-assembly of metalloporphyrins [85–90]. To date, a variety of porphyrin nanostructures have been reported, including nanorods that exhibit photoconductivity [91–93], for the surface fabrication of electrodes toward the developments of photovoltaic cells [94–99]. Shelnutt and co-workers have developed a porphyrin nanotube that can form platinum nanoparticles on the surface to catalyze photochemical H_2 evolution [100, 101]. There are many crystal structures of porphyrins that contain pores filled with generally disordered solvates [102–104]. However, the construction of multicomponent supramolecular assemblies of porphyrins to provide novel structures and integrated functionalities, which cannot be realized by each building block alone, has been a challenge.

The use of a curved π-surface of saddle-distorted H_2DPP described above provides a promising way to construct unique supramolecular architecture such as a porphyrin nanochannel (PNC) as shown in Figure 6.17, which can encapsulate redox active species. Indeed, H_2DPP has been utilized as a ligand for a Mo^V center to access a porphyrin nanotube that contains unstable and unprecedented tetranuclear Mo^{VI}-oxo clusters in its inner space [105]. Supramolecules that are based on the self-assembly of saddle-distorted H_2DPP derivatives owing to intermolecular interactions can give isolated inner spaces for the protection of the encapsulated guest molecules from external

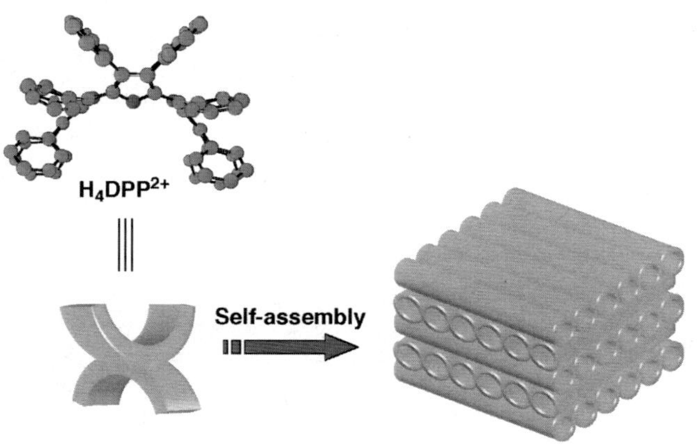

FIGURE 6.17. Formation of porphyrin nanochannel (PNC) derived from self-assembly of saddle-distorted H_4DPP^{2+}.

lethal attack [105–108]. This structural motif can contribute toward the stabilization unstable and metastable species in the cavity. Based on this strategy, novel functionalities of porphyrin-based multicomponent materials at a molecular level have been obtained in a well-organized manner as single crystals as described below.

Co-crystallization of $[H_4DPP^{2+}]Cl_2{\cdot}4CH_3CN$ and an excess amount of hydroquinone (H_2Q) in $CHCl_3/CH_3CN$ afforded a cationic nanochannel in which H_2Q molecules are encapsulated as shown in Figure 6.18 [109]. Two saddle-distorted porphyrin dications create a ring having 1-nm diameter with intermolecular π–π interactions between peripheral phenyl groups with 3.46-Å separation. The porphyrin rings are connected together by weak π–π interactions with distances ranging from 3.56 to 3.71 Å to form the channels. The channels run below and above the individual porphyrin mean plane with an angle of 71° in the directions of [1 1 0] and [1 −1 0] in the crystal as shown in the crystal packing in Figure 6.18b. The chloride anion forms hydrogen bonding with the pyrrole N–H protons with a distance of 3.22 Å. The hydroquinone molecule is included due to intermolecular π–π interactions with H_4DPP^{2+} and hydrogen bonding between the hydroxyl groups of hydroquinone with the nitrogen atoms of the acetonitrile molecules in the skeleton units in the distance of 2.85 Å [109, 110]. This hydrogen bonding fixes the configuration of the hydroxyl groups to be the *trans* orientation. Tetrafluorohydroquinone (H_2QF_4) and tetrachlorohydroquinone (H_2QCl_4) can also be encapsulated in the PNC architecture [110].

When benzoquinone derivatives, which are two-electron oxidized species of the corresponding hydroquinones, were employed as guest molecules, no encapsulation was observed under the same conditions as the case of H_2Q [110]. Thus, PNC recognizes the electronic character of guest molecules, rather than specific functional groups or shapes in the molecules. This selection rule may result from the cationic character of H_4DPP^{2+} that favors electron-rich substrates owing to electrostatic and charge-transfer donor–acceptor interactions. When tetrabromohydroquinone (H_2QBr_4) was employed as a guest molecule, however, the sterically bulkier H_2QBr_4 was not encapsulated in the PNC. Thus, steric factors are also important for encapsulation of guest molecules into the PNC.

The porphyrin nanochannels including electron-donating molecules undergo photoinduced electron transfer from the guest donor molecules to the singlet state of the H_4DPP^{2+} to produce radical cations of the guest molecules and $H_4DPP^{\bullet+}$ upon photoirradiation as detected by EPR and laser flash photolysis measurements [110]. EPR spectra of hydroquinone radical cations were observed under photo-irradiation ($\lambda >$ 340 nm) of PNC–hydroquinone crystals at room temperature [110]. The produced hydroquinone radical cations were stable because photoinduced electron transfer from hydroquinones to the singlet excited state of H_4DPP^{2+} affords

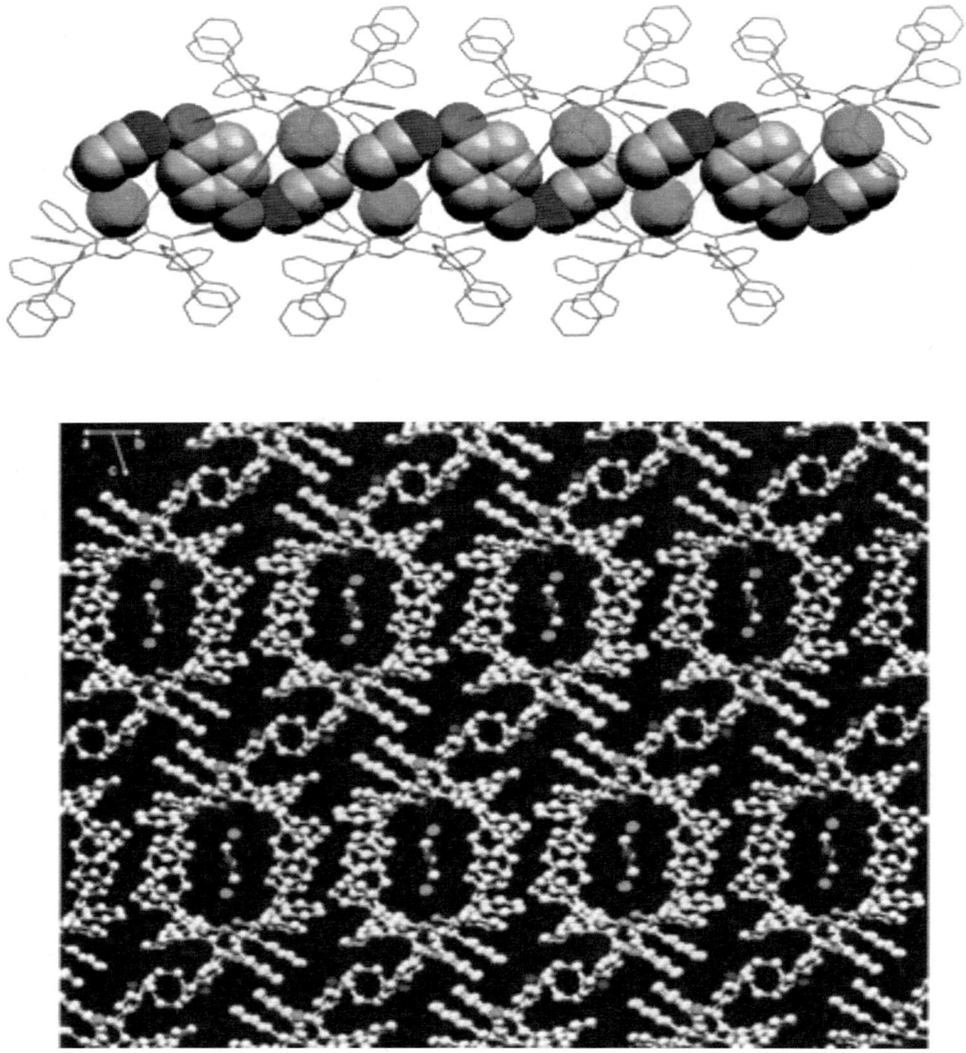

FIGURE 6.18. Crystal structure of PNC-H_2Q. The H_4DPP^{2+} unit is described with a wireframe.

hydroquinone radical cations fixed and protected by the nanochannel and $H_4DPP^{\bullet+}$, the latter of which disproportionates to yield H_4DPP^{2+} and H_4DPP to suppress the back electron transfer [110]. The analysis of the observed hyperfine splitting of the hydroquinone radical cations revealed that the configuration of both OH groups of H_2Q are fixed in the trans orientation in the PNC guest-inclusion site as seen in the crystal structure in Figure 6.18 [110]. Femtosecond laser flash photolysis of a KBr pellet of PNC-H_2Q revealed the occurrence of fast electron transfer from H_2Q to the singlet excited state $^1(H_4DPP^{2+})^*$. The rate constant of the photoinduced electron transfer was determined to be $2.1 \times 10^{11}\ s^{-1}$ at 298 K [110]. The porphyrin radical cation ($H_4DPP^{\bullet+}$) produced in the photoinduced electron transfer decays obeying second-order kinetics [110]. The bimolecular decay involves disproportionation of $H_4DPP^{\bullet+}$ and back electron transfer from $H_4DPP^{\bullet+}$ to $H_2Q^{\bullet+}$ to produce the triplet excited state, $^3(H_4DPP^{2+})^*$, because the triplet excited energy of H_4DPP^{2+} is lower than the electron-transfer state [110]. The disproportionation of $H_4DPP^{\bullet+}$ may proceed rapidly through electron hopping between adjacent $H_4DPP^{\bullet+}$ and H_4DPP^{2+} in the PNC–H_2Q crystals, which results in the formation of $H_2Q^{\bullet+}$ that remains the stable product as observed by EPR spectra (*vide supra*).

6.6 PHOTOCONDUCTIVITY OF PORPHYRIN NANOCHANNELS

Efficient photoinduced electron transfer in electron donor–PNC crystals described above has been applied to develop photoconductive materials (*vide infra*). Organic molecular semiconductors showing photoconductivity are important components to develop well-organized photofunctional

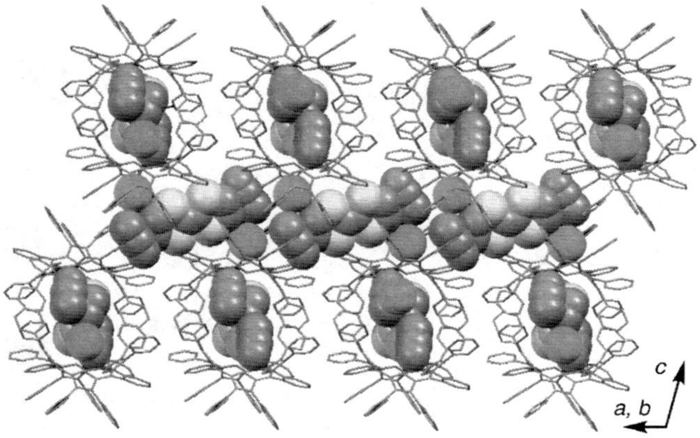

FIGURE 6.19. The crystal structure of PNC-TTF.

materials in nanometer scale [111–113]. In the solid-state structures of organic photoconductors, weak noncovalent interactions including π–π interactions play indispensable roles in terms of molecular ordering and construction of conduction pathways [112]. Thus, electron donor–PNC crystals provide an excellent opportunity to examine the conduction pathway.

PNC including stronger electron-donating guest molecules than hydroquinone derivatives such as TTF was also prepared and the crystal structure of the porphyrin nanochannel including TTF (PNC-TTF) exhibits a column structure along with the crystallographic *c* axis by virtue of intermolecular π–π interactions among peripheral phenyl groups attached at the β-positions of pyrroles of $[H_4DPP]Cl_2$ as shown in Figure 6.19 [114].

Parallel to the crystallographic *ab* plane, weaker inter-column π–π interactions are recognized. The driving force of the inclusion of TTF into PNC is π–π interaction with H_4DPP^{2+} via the sulfur of TTF and the nitrogen of the pyrrole moiety in $[H_4DPP]Cl_2$. Such π–π interaction plays an important role in efficient electron transfer from TTF to the singlet excited state of $[H_4DPP]Cl_2$. The occurrence of photoinduced electron transfer was confirmed by solid-state femtosecond laser flash photolysis and EPR measurements [110, 114].

A single crystal of PNC–TTF exhibited photoconductivity upon photoirradiation at 633 nm with a He–Ne laser. Photocurrent measurements were made by a two-terminal method using gold paste and gold wire 50 μm ϕ) as shown in Figure 6.20a [114]. The photocurrent showed direction-dependence toward the crystallographic *c* axis. This indicates that the intermolecular π–π interaction is the main conduction pathway. When the electric voltage was applied along with the crystallographic *c* axis, the photocurrent reached 26 nA (Figure 6.20b) [114]. On the other hand, when the electric voltage was applied along the direction perpendicular to the crystallographic *c* axis, the maximum of photocurrent was 2.9 nA, which was only ~1/10 compared with that observed in the direction of the *c* axis (Figure 6.20b) [114]. As shown in Figure 6.19, the intermolecular π–π interaction among H_4DPP^{2+} units was found mainly along with the crystallographic *c* axis for the β-phenyl groups. This result clearly indicates that the main conduction path is the intracolumn π–π interactions among the β-phenyl groups to facilitate electron hopping [114].

6.7 SUMMARY AND CONCLUSIONS

As described above, utilization of nonplanar porphyrins and their metal complexes provides a unique strategy to develop a new category of molecular and supramolecular assemblies based on porphyrin compounds. In particular, a non-planar saddle distortion of dodecaphenylporphyrin (H_2DPP) resulted in double protonation of H_2DPP with an acid to yield the porphyrin dication (H_4DPP^{2+}), which can act as an electron acceptor in sharp contrast to unprotonated porphyrins, which usually act as electron donors. The saddle-distorted H_4DPP^{2+} can form a variety of hydrogen-bonded

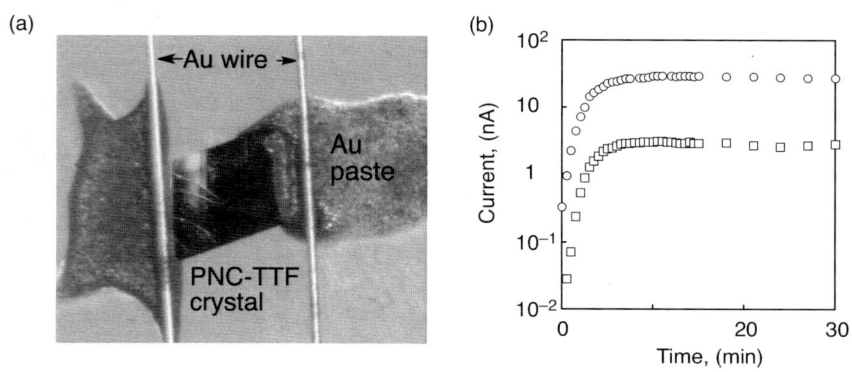

FIGURE 6.20. (**a**) A sample manipulation for photoconductivity measurements of a single crystal of PNC–TTF. (**b**) Direction dependence of photocurrent under AM 1.5 illumination at room temperature: parallel to the crystallographic *c* axis (circle) and perpendicular to the *c* axis (square).

supramolecular complexes with electron donors bearing a carboxylate group, which exhibit efficient photoinduced electron transfer. In metalloporphyrins, the saddle distortion of the porphyrin plane also strengthens the axial coordination, affording stable complexes of electron donor–acceptor ensembles based on axial coordination to the metal center, in which efficient photoinduced electron transfer occurs to produce the electron-transfer state. Axial coordination and hydrogen bonding are combined to construct a discrete supramolecular conglomerate composed of H_4DPP^{2+} and two saddle-distorted Zn(II)-octaphenylphthalocyanine (Zn(OPPc), which also exhibits efficient photoinduced electron transfer. The use of a curved π-surface of saddle-distorted H_4DPP^{2+} also provides a promising way to construct unique supramolecular architecture such as a porphyrin nanochannel (PNC) in which electron donors can be encapsulated to exhibit photoconductivity along the direction of intermolecular π–π interaction among the H_4DPP^{2+} units in the crystals. Thus, the unique characteristics of nonplanar porphyrins will pave a way to construction of multicomponent photofunctional conglomerates based on self-assembled porphyrin nanoarchitectures.

ACKNOWLEDGMENTS

The authors gratefully acknowledge the contributions of their collaborators and co-workers mentioned in the cited references, as well as support by a Grant-in-Aid (Nos. 20108010 and 237500141), a Global COE program, "the Global Education and Research Center for Bio-Environmental Chemistry," from the Ministry of Education, Culture, Sports, Science and Technology, Japan and KOSEF/MEST through WCU project (R31-2008-000-10010-0), Korea.

REFERENCES

1. Kadish, K. M., Smith, K. M., Guilard, R. eds. (2000). *Porphyrin Handbook*, Vols. 1–8. Academic Press, San Diego, CA.
2. Shoji, O., Okada, S., Satake, A., Kobuke, Y. (2005). *Journal of the American Chemical Society*, **127**, 2201–2210.
3. Hajjaj, F., Yoon, Z. S., Park, J., Satake, A., Kim, D., Kobuke, Y. (2006). *Journal of the American Chemical Society*, **128**, 4612–4623.
4. Fukuzumi, S., Honda, T., Ohkubo, K., Kojima, T. (2009). *Dalton Transactions*, 3880–3889.
5. Ho, C. (1992). *Advances in Protein Chemistry*, **43**, 153–312.
6. (a) Paoli, M., Nagai, K. (2006). In: Messerschmidt, A., Huber, R., Wieghardt, K., Poulos, T., eds. *Handbook of Metalloproteins*, John Wiley & Sons, Chichester, UK, pp. 16–30.
7. Dickerson, R. E., Geiss, I. (1983). *Hemoglobin: Structure, Function, Evolution, and Pathology*. Benjamin/Cummings, Menlo Park, CA. (b) In: Ho, C., Lukin, J. A., eds. (2000). *Encyclopedia of Life Science*; Nature Publishing Group, London.
8. Flögel, U., Merx, M. W., Gödecke, A., Decking, U. K. M., Schrader, J. (2001). *Proceedings of the National Academy of Sciences USA*, **98**, 735–740.
9. Simonneaux, G., Bondon, A. (2005). *Chemical Reviews*, **105**, 2627–2646.
10. Bertini, I., Cavallaro, G., Rosato, A. (2006). *Chemical Reviews*, **106**, 90–115.
11. Sono, M., Roach, M. P., Coulter, E. D., Dawson, J. H. (1993). *Chemical Reviews*, **96**, 2841–2888.
12. Denisov, I. G., Makris, T. M., Sligar, S. G., Schlichting, I. (2005). *Chemical Reviews*, **105**, 2253–2278.
13. Dunford, H. B. (2002). *Coordination Chemical Reviews*, **233–234**, 311–318.
14. Rouzer, C. A., Marnett, L. J. (2003). *Chemical Reviews*, **103**, 2239–2304.
15. Loewen, P. C., Carpena, X., Rovira, C., Ivancich, A., Perez-Luque, R., Haas, R., Odenbreit, S., Nicholls, P., Fita, I. (2004). *Biochemistry*, **43**, 3089–3103.
16. Collman, J. P., Boulotav, R., Sunderland, C. J., Fu, L. (2004). *Chemical Reviews*, **104**, 561–588.
17. Hayashi, T., Hisaeda, Y. (2002). *Accounts of Chemical Research*, **35**, 35–43.
18. Watanabe Y., Nakajima, H., Ueno, T. (2007). *Accounts of Chemical Research*, **40**, 554–562.
19. Gust, D., Moore, T. A., Moore, A. L. (2009). *Accounts of Chemical Research*, **42**, 1890–1898.
20. Fukuzumi, S. (2003). *Organic and Biomolecular Chemistry*, **1**, 609–620.
21. Fukuzumi, S. (2008). *Physical Chemistry Chemical Physics*, **10**, 2283–2297.
22. Fukuzumi, S., Kojima, T. (2008). *Journal of Materials Chemistry*, **18**, 1427–1439.
23. Springer, B. A., Sligar, S. G., Olson, J. S., Phillips, G. N. Jr. (1994). *Chemical Reviews*, **94**, 699–714.
24. Sono, M., Roach, M. P., Coulter, E. D., Dawson, J. H. (1996). *Chemical Reviews*, **96**, 2841–2888.
25. Williams, P. A., Cosme, J., Ward, A., Angove, H. C., Vinkovic, D. M., Jhoti, H. (2003). *Nature*, **424**, 464–468.
26. Meunier, B., de Visser, S. P., Shaik, S. (2004). *Chemical Reviews*, **104**, 3947–3980.
27. Perutz, M. F. (1970). *Nature*, **228**, 726–734.
28. McDermott, G., Prince, S. M., Freer, A. A., Hawthornthwaite–Lauless, A. M., Papiz, M. Z., Cogdell, R. J., Isaacs, N. W. (1995). *Nature*, **374**, 517–521.
29. Koepke, J., Hu, X., Muenke, C., Schulten, K., Michel, H. (1996). *Structure*, **4**, 581–597.
30. McLuskey, K., Prince, S. M., Cogdell, R. J., Isaacs, N. W. (2001). *Biochemistry*, **40**, 8783–8789.
31. Zouni, A., Witt, H.-T., Kern, J., Fromme, P., Krauß, N., Saenger, W., Orth, P. (2001). *Nature*, **409**, 739–743.
32. Jordan, P., Fromme, P., Witt, H. T., Klukas, O., Saenger, W., Krauß, N. (2001). *Nature*, **411**, 909–917.

33. Melkozernov, A. N., Barber, J., Blankenship, R. E. (2006). *Biochemistry*, **45**, 331–345.
34. Wasielewski, M. R. (2009). *Accounts of Chemical Research*, **42**, 1910–1921.
35. Fukuzumi, S., Guldi, D. M. (2001). In: Balzani, V., ed. *Electron Transfer in Chemistry*, Vol. 2. Wiley–VCH, New York, pp. 270–337.
36. Bottari, G., de la Torre, G., Guldi, D. M., Torres, T. (2010). *Chemical Reviews*, **110**, 6768–6816.
37. Jentzen, W., Simpson, M. C., Hobbs, J. D., Song, X., Ema, T., Nelson, N. Y., Medforth, C. J., Smith, K. M., Veyrat, M., Mazzanti, M., Ramasseul, R., Marchon, J.–C., Takeuchi, T., Goddard, W. A. III, Shelnutt, J. A. (1995). *Journal of the American Chemical Society*, **117**, 11085–11097.
38. Nurco, D. J., Medforth, C. J., Forsyth, T. P., Olmstead, M. M., Smith, K. M. (1996). *Journal of the American Chemical Society*, **118**, 10918–10919.
39. Shelnutt, J. A., Medforth, C. J., Berber, M. D., Barkigia, K. M., Smith, K. M. (1991). *Journal of the American Chemical Society*, **113**, 4077–4087.
40. Medforth, C. J., Senge, M. O., Smith, K. M., Sparks, L. D., Shelnutt, J. A. (1992). *Journal of the American Chemical Society*, **114**, 9859–9869.
41. Medforth, C. J., Hobbs, J. D., Rodriguez, M. R., Abraham, R. J., Smith, K, M., Shelnutt, J. A. (1995). *Inorganic Chemistry*, **34**, 1333–1341.
42. Fukuzumi, S., Imahori, H. (2001). In: Balzani, V., ed. *Electron Transfer in Chemistry*, Vol. 2. Wiley–VCH, Weinheim, pp. 927–975.
43. Kadish, K. M., Van-Caemelbecke, E., Royal, G. (2000). In: Kadish, K. M., Smith, K. M., Guilard, R., eds. *The Porphyrin Handbook*, Vol. 8. Academic Press, San Diego, pp. 1–114.
44. Nakanishi T., Ohkubo, K., Kojima, T., Fukuzumi, S. (2009). *Journal of the American Chemical Society*, **131**, 577–584.
45. Marcus, R. A. (1964). *Annual Review of Physical Chemistry*, **15**, 155–196.
46. Jentzen, W., Turowska-Tyrk, I., Scheidt, W. R., Shelnutt, J. A. (1996). *Inorganic Chemistry*, **35**, 3559–3567.
47. Stone, A., Fleischer, B. (1968). *Journal of the American Chemical Society*, **90**, 2735–2748.
48. Fleischer, E. B. (1970). *Accounts of Chemical Research*, **3**, 105–112.
49. Cheng, B., Munro, O. Q., Marques, H. M., Scheidt, W. R. (1997). *Journal of the American Chemical Society*, **119**, 10732–10742.
50. Juillard, S., Ferrand, Y., Simonneaux, G., Toupet, L. (2005). *Tetrahedron*, **61**, 3489–3495.
51. Rayati, S., Zakavi, S., Ghaemi, A., Carroll P. J. (2008). *Tetrahedron Letters*, **49**, 664–667.
52. Kojima, T., Nakanishi, T., Harada, R., Ohkubo, K., Yamauchi, S., Fukuzumi, S. (2007). *Chemistry—A European Journal*, **11**, 8714–8725.
53. Birnbaum, E. R., Hodge, J. A., Grinstaff, M. W., Schaefer, W. P., Henling, L., Labinger, J. A., Bercaw, J. E., Gray, H. B. (1995). *Inorganic Chemistry*, **34**, 3625–3632.
54. Fukuzumi, S., Nakanishi, I., Tanaka, K., Suenobu, T., Tabard, A., Guilard, R., Van Caemelbecke, E., Kadish, K. M. (1999). *Journal of the American Chemical Society*, **121**, 785–790.
55. Ma, J. K., Mathews, S., Davidson, V. L. (2007). *Biochemistry*, **46**, 8561–8568.
56. Kanematsu, M., Naumov, P., Kojima, T., Fukuzumi, S. (2011). *Chemistry—A European Journal*, **17**, 12372–12384.
57. Honda, T., Nakanishi, T., Ohkubo, K., Kojima, T., Fukuzumi, S. (2010). *Journal of the American Chemical Society*, **132**, 10155–10163.
58. Marcus, R. A., Sutin, N. (1985). *Biochimica Biophysica Acta*, **811**, 265–322.
59. Fukuzumi, S., Ohkubo, K., Imahori, H., Shao, J., Ou, Z., Zheng, G., Chen, Y., Pandey, R. K., Fujitsuka, M., Ito, O., Kadish, K. M. (2001). *Journal of the American Chemical Society*, **123**, 10676–10683.
60. Hutchison, J. A., Sintic, P. J., Brotherhood, P. R., Scholes, C., Blake, I. M., Ghiggino, K. P., Crossley, M. J. (2009). *Journal of Physical Chemistry C*, **113**, 11796–11804.
61. Imahori, H., Tamaki, K., Guldi, D. M., Luo, C., Fujitsuka, M., Ito, O., Sakata, Y., Fukuzumi, S. (2001). *Journal of the American Chemical Society*, **123**, 2607–2617.
62. Imahori, H., Guldi, D. M., Tamaki, K., Yoshida, Y., Luo, C., Sakata, Y., Fukuzumi, S. (2001). *Journal of the American Chemical Society*, **123**, 6617–6628.
63. Imahori, H., Sekiguchi, Y., Kashiwagi, Y., Sato, T., Araki, Y., Ito, O., Yamada, H., Fukuzumi, S. (2004). *Chemistry—A European Journal*, **10**, 3184–3196.
64. Ohkubo, K., Kotani, H., Shao, J., Ou, Z., Kadish, K. M., Li, G., Pandey, R. K., Fujitsuka, M., Ito, O., Imahori, H., Fukuzumi, S. (2004). *Angewandte Chemie International Edition*, **43**, 853–856.
65. Ohkubo, K., Imahori, H., Shao, J., Ou, Z., Kadish, K. M., Chen, Y., Zheng, G., Pandey, R. K., Fujitsuka, M., Ito, O., Fukuzumi, S. (2002). *Journal of Physical Chemistry A*, **106**, 10991–10998.
66. Okamoto, K., Mori, Y., Yamada, H., Imahori, H., Fukuzumi, S. (2004). *Chemistry—A European Journal*, **10**, 474–483.
67. Closs, G. L., Miller, J. R. (1988). *Science*, **240**, 440–447.
68. Pettersson, K., Wiberg, J., Ljungdahl, T., Mårtensson, J., Albinsson, B. (2006). *Journal of Physical Chemistry A*, **110**, 319–326.
69. Kojima, T., Hanabusa, K., Ohkubo, K., Shiro, M., Fukuzumi, S. (2010). *Chemistry—A European Journal*, **16**, 3646–3655.
70. Baumann, J. A., Salmon, D. J., Wilson, S. T., Meyer, T. J., Hatfield, W. E. (1978). *Inorganic Chemistry*, **17**, 3342–3350.
71. Abe, M., Sasaki, Y., Yamaguchi, T., Ito, T. (1992). *Bulletin of the Chemical Society of Japan*, **65**, 1585–1590.
72. Ito, T., Hamaguchi, T., Nagino, H., Yamaguchi, T., Kido, H., Zavarine, I. S., Richmond, T., Washington, J., Kubiak, C. P. (1999). *Journal of the American Chemical Society*, **121**, 4625–4632.
73. Abe, M., Michi, T., Sato, A., Kondo, T., Zhou, W., Ye, S., Uosaki, K., Sasaki, Y. (2003). *Angewandte Chemie International Edition*, **42**, 2912–2915.

74. Kim, H., Jo, H. J., Kim, J., Kim, S., Kim, D., Kim, K. (2005). *Crystal Engineering Communications*, **7**, 417–420.

75. Kojima, T., Nakanishi, T., Honda, T., Harada, R., Shiro, M., Fukuzumi, S. (2009). *European Journal of Inorganic Chemistry*, 727–734.

76. Martín-Gomis, L., Ohkubo, K., Fernández-Lázaro, F., Fukuzumi, S., Sastre-Santos, A. (2008). *Journal of Physical Chemistry C*, **112**, 17694–17701.

77. D'Souza, F., Chitta, R., Ohkubo, K., Tasior, M., Subbaiyan, N. K., Zandler, M. E., Rogacki, M. K., Gryko, D. T., Fukuzumi, S. (2008). *Journal of the American Chemical Society*, **130**, 14263–14272.

78. Honda, T., Nakanishi, T., Ohkubo, K., Kojima, T., Fukuzumi, S. (2010). *Journal of Physical Chemistry C*, **114**, 14290–14299.

79. Kojima, T., Honda, T., Ohkubo, K., Shiro, M., Kusukawa, T., Fukuda, T., Kobayashi, N., Fukuzumi, S. (2008). *Angewandte Chemie International Edition*, **47**, 6712–6716.

80. Kadish, K. M., Smith, K. M., Guilard, R., eds. (2003). *The Porphyrin Handbook*, Academic Press, San Diego, CA.

81. Rio, Y., Rodríguez-Morgade, M. S., Torres, T. (2008). *Organic Biomolecular Chemistry*, **6**, 1877–1894.

82. Mack, J., Kobayashi, N. (2011). *Chemical Reviews*, **111**, 281–321.

83. Bottari, G., de la Torre, G., Guldi, D. M., Torres. T. (2010). *Chemical Reviews*, **110**, 6768–6816.

84. Fortage, J., Göransson, E., Blart, E., Becker, H.-C., Hammarström, L., Odobel, F. (2007). *Chemical Communications*, 4629–4631.

85. Bryn, M. P., Curtis, C. J., Hsiou, Y., Khan, S. I., Sawin, P. A., Tendick, S. K., Terzis, A., Strouse, C. E. (1993). *Journal of the American Chemical Society*, **115**, 9480–9497.

86. Bryn, M. P., Curtis, C. J., Goldberg, I., Hsiou, Y., Khan, S. I., Sawin, P. A., Tendick, S. K., Strouse, C. E. (1991). *Journal of the American Chemical Society*, **113**, 6549–6557.

87. Bryn, M. P., Strouse, C. E. (1991). *Journal of the American Chemical Society*, **113**, 2501–2508.

88. Bryn, M. P., Curtis, C. J., Khan, S. I., Sawin, P. A., Tsurumi, R., Strouse, C. E. (1990). *Journal of the American Chemical Society*, **112**, 1865–1874.

89. Diskin-Posner, Y., Patra, G. K., Goldberg, I. (2001). *European Journal of Inorganic Chemistry*, 2515–2523.

90. Diskin-Posner, Y., Dahal, S., Goldberg, I. (2000). *Angewandte Chemie International Edition*, **39**, 1288–1292.

91. Schwab, A. D., Smith, D. E., Rich, C. S., Young, E. R., Smith, W. F., de Paula, J. C. (2003). *Journal of Physical Chemistry B*, **107**, 11339–11345.

92. Schwab, A. D., Smith, D. E., Bond-Watts, B., Johnston, D. E., Hone, J., Johnson, A. T., de Paula, J. C., Smith, W. F. (2004). *Nano Letters*, **4**, 1261–1265.

93. Drain, C. M. (2002). *Proceedings of the National Academy of Science USA*, **99**, 5178–5182.

94. Hasobe, T., Kamat, P. V., Troiani, V., Solladié, N., Ahn, W. T. K., Kim, S. K., Kim, D., Kongkanand, A., Kuwabata, S., Fukuzumi, S. (2005). *Journal of Physical Chemistry B*, **109**, 19–23.

95. Hasobe, T., Hattori, S., Kamat, P. V., Fukuzumi, S. (2006). *Tetrahedron*, **62**, 1937–1946.

96. Hasobe, T., Imahori, H., Kamat, P. V., Ahn, T. K., Kim, D., Hanada, T., Hirakawa, T., Fukuzumi, S. (2005). *Journal of the American Chemical Society*, **127**, 1216–1228.

97. Hasobe, T., Fukuzumi, S., Kamat, P. V. (2005). *Journal of the American Chemical Society*, **127**, 11884–11885.

98. Hasobe, T., Fukuzumi, S., Hattori, S., Kamat, P. V. (2007). *Chemistry—Asian Journal*, **2**, 265–272.

99. Hasobe, T., Saito, K., Kamat, P. V., Troiani, V., Qiu, H., Solladié, N., Ahn, T. K., Kim, K. S., Kim, S. K., Kim, D., D'Souza, F., Fukuzumi, S. (2007). *Journal of Material Chemistry*, **17**, 4160–4170.

100. Wang, Z., Medforth, C., Shelnutt, J. A. (2004). *Journal of the American Chemical Society*, **126**, 16720–16721.

101. Tian, Y., Martin, K. E., Shelnutt, J. Y.-T., Evans, L., Busani, T., Miller, J. E., Medforth, C. J., Shelnutt, J. A. (2011). *Chemical Communications*, **47**, 6069–6071.

102. Li, L.-L., Yang, C.-J., Chen, W.-H., Lin, H.-J. (2003). *Angewandte Chemie International Edition*, **42**, 1505–1508.

103. Kosal, M. E., Chou, J.-H., Wilson, S. R., Suslick, K. S. (2002). *Nature Materials*, **1**, 118–121.

104. Suslick, K. S., Bhyrappa, P., Chou, J.-H., Kosal, M. E., Nakagaki, S., Smithenry, D. W., Wilson, S. R. (2005). *Accounts of Chemical Research*, **38**, 283–291.

105. Harada, R., Matsuda, Y., Okawa, H., Kojima, T. (2004). *Angewandte Chemie International Edition*, **43**, 1825–1828.

106. Yokoyama, A., Kojima, T., Ohkubo, K., Fukuzumi, S. (2010). *Inorganic Chemistry*, **49**, 11190–11198.

107. Kojima, T., Harada, R., Nakanishi, T., Kaneko, K., Fukuzumi, S. (2007). *Chemistry of Materials*, **19**, 51–58.

108. Yokoyama, A., Kojima, T., Fukuzumi, S. (2011). *Dalton Transactions*, **40**, 6445–6450.

109. Harada, R., Kojima, T. (2005). *Chemical Communications*, 716–718.

110. Kojima, T., Nakanishi, T., Harada, R., Ohkubo, K., Yamauchi, S., Fukuzumi, S. (2007). *Chemistry—A European Journal*, **13**, 8714–8725.

111. Adam, D., Schuhmacher, P., Simmerer, J., Haussling, L., Siemensmeyer, K., Etzbach, K. H., Ringsdorf, H., Haarer, D. (1994). *Nature*, **371**, 141–143.

112. Weiss, D. S., Abkowitz, M. (2010). *Chemical Reviews*, **110**, 479–526.

113. Inabe, T., Tajima, H. (2004). *Chemical Reviews*, **104**, 5503–5534.

114. Nakanishi, T., Kojima, T., Ohkubo, K., Hasobe, T., Nakayama, K.-i., Fukuzumi, S. (2008). *Chemistry of Materials*, **20**, 7492–7500.

7

TWEEZERS AND MACROCYCLES FOR THE MOLECULAR RECOGNITION OF FULLERENES

David Canevet, Emilio M. Pérez, and Nazario Martín

7.1 INTRODUCTION

Fullerenes have been in the spotlight of chemical research right from their discovery in 1985 [1], earning Kroto, Curl, and Smalley one of the most immediate Nobel prizes in chemical sciences, in 1996. Most of the research efforts in fullerene chemistry have been focused in their ability to work as electron acceptors [2]. As a consequence, fullerene derivatives are nowadays the benchmark *n*-type materials for the construction of organic photovoltaic devices [3]. Developments in the supramolecular chemistry of fullerenes are expected to aid their application in such devices by means of facilitating their purification from fullerene soot and enabling the design of self-assembling fullerene materials. In fact, the search for molecular receptors for fullerenes was initiated just seven years after the discovery of C_{60}, and immediately after it became available in sufficient quantities by contact-arc vaporization of graphite [4].

In 1992, the group headed by Diederich and Ringsdorf [5] reported the first purposely designed hosts for fullerenes. Their receptors were based on a macrocyclic azacrown ether in which the nitrogen groups were alkylated or acylated with aromatic groups, further substituted with long alkane chains, forming a lipophilic "cup." Simultaneously, Wennerström and co-workers [6] had reported the association of C_{60} by two units of γ-cyclodextrin, permitting the solubilization of the fullerene in water. At approximately the same time, the groups headed by Raston [7a] and Shinkai [7b], working independently, reported the selective purification of C_{60} from fullerene soot by its selective association with *p-tert*-butylcalix[8]arene [7].

From these early works, it became immediately apparent that fullerenes are very peculiar guest molecules from the point of view of molecular recognition. Their unusual shape (approximately spherical, maximizing the surface-to-volume ratio) and chemical nature (unpolarized polyenes) restrict the types of noncovalent forces one can utilize for their association to dispersion forces (π–π and van der Waals) [8]. Thus, the simplest approach for the design of hosts for fullerenes is the construction of a nonpolar cavity of the appropriate size to fit the fullerene guest. This strategy takes advantage of the intrinsically low solubility of fullerenes, and the recognition event is based fundamentally on van der Waals and solvophobic interactions.

A complementary strategy involves the identification of organic molecules capable of establishing positive noncovalent interactions with the fullerenes, habitually large aromatic molecules with electron donor character which can interact with the fullerenes through π–π stacking and charge-transfer interactions. Once the recognizing motif is pinpointed, if a single unit does not suffice to build an efficient receptor, a suitable spacer is designed so as to hold two or more units at the right distance to recognize the fullerene in question. Within this strategy, perhaps the most popular design has been the connection of two units of the recognizing fragment to form what are often termed "tweezers" for fullerene [8c, 9]. Tweezers-like receptors are usually synthetically affordable, and their open structures are sufficiently flexible to adapt to the guest.

On the other hand, the vast majority of the most efficient hosts for fullerenes reported to date feature macrocyclic architectures. The extraordinary binding abilities of

Organic Nanomaterials: Synthesis, Characterization, and Device Applications, First Edition. Edited by Tomás Torres and Giovanni Bottari.

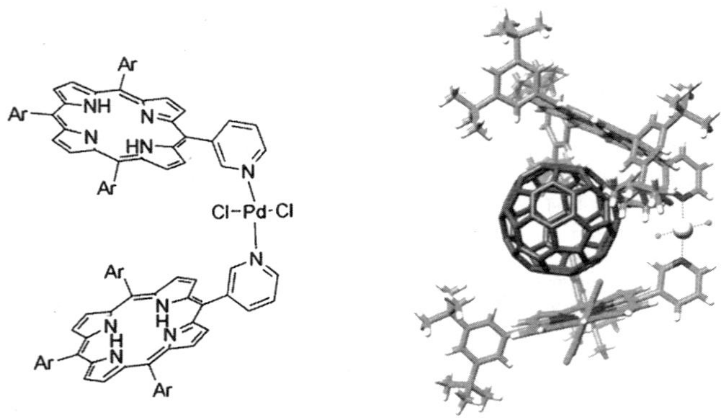

1 **1·C_{60}**

FIGURE 7.1. Chemical structure of the "jaws" receptor reported by Boyd and Reed, and x-ray crystal structure of its complex with C_{60}. Reproduced from reference 8c, with permission of the IUPAC.

such receptors are based on a combination of both of the strategies mentioned above: They typically show a cavity of the right size and shape to accommodate the fullerene guest, which is decorated with specific recognition units, maximizing the interactions between host and guest. In contrast to tweezers-like designs, macrocyclic structures are often more demanding from the point of view of synthesis, and the increase in preorganization sometimes results in very poor association constants due to a mismatch between the sizes of host and guest [8a].

In this chapter, we will offer an overview of some of the most accomplished examples of tweezers and macrocycles used in the molecular recognition of fullerenes [10].

7.2 PORPHYRIN-BASED MOLECULAR TWEEZERS AND MACROCYCLES

Porphyrins are well known to interact favorably with fullerenes both in solution and in the solid state, which has been exploited in the design of a wide variety of receptors for fullerenes [11]. Several tweezers featuring porphyrins as binding motifs have also been reported, spearheaded by the work of Boyd and Reed on their "jaws" receptors. The first example of this series was reported in 2000; the structure of the receptor and its complex with C_{60} are depicted in Figure 7.1 [12].

In receptor **1** the porphyrin units are linked through coordination on a palladium metal center. The use of Pd results in partial metalation of the porphyrin units. Binding of C_{60} and C_{70} to receptor **1** was unambiguously demonstrated by a variety of techniques, including FAB and MALDI mass spectrometry, ^{1}H and ^{13}C NMR. Remarkably, they succeeded in obtaining single crystals of the **1**·C_{60} complex. In the crystal, the closest Pd to fullerene carbon distance is 2.856 Å, indicating very tight binding of the guest. The binding constant of this receptor toward C_{60} was estimated to be log $K_a = 3.7$ in toluene-d_8 at room temperature. Following their preliminary communication, Boyd and Reed reported very similar receptors, in which the coordination link was substituted with covalent spacers, namely amides of both the isophthalic and terephthalic acids. These receptors were proven to be worse binders for fullerenes than the receptors connected through a Pd center [13].

A particularly interesting example of molecular tweezers for fullerene was reported by Shinkai and co-workers in 2002 [14]. Their receptor (**2** in Figure 7.2) can be switched from a state in which it cannot bind fullerene ("off") to one in which it can ("on") through the coordination of two pyridine addends to either the porphyrins or an external Pd center. When C_{60} is added to a solution of **2** in its "off" state, the authors observed no changes in its UV–vis spectrum. On the contrary, when the same experiment is carried out with **2** in its "on" state, the absorption corresponding to the Soret band of the porphyrins decreases in intensity and is bathochromically shifted, an indication of the association of C_{60}. Indeed, through UV–vis titrations, it was estimated that **2** "on" binds [60]fullerene with a respectable stability constant of log $K_a = 3.7$ in toluene:CH_2Cl_2 (50:1).

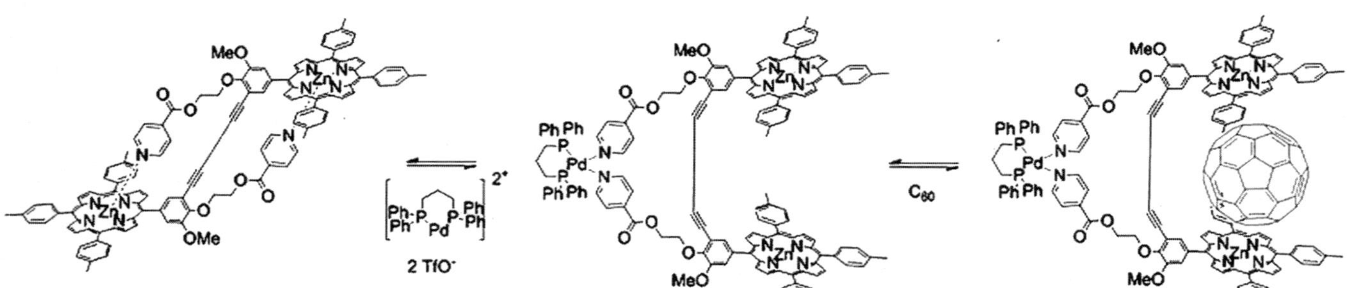

2 "off" **2 "on"** **2·C_{60}**

FIGURE 7.2. Structure of the switchable receptor synthesized by Shinkai and co-workers; also presented is scheme showing the receptor switching from the "off" to the "on" state, along with its binding to C_{60}.

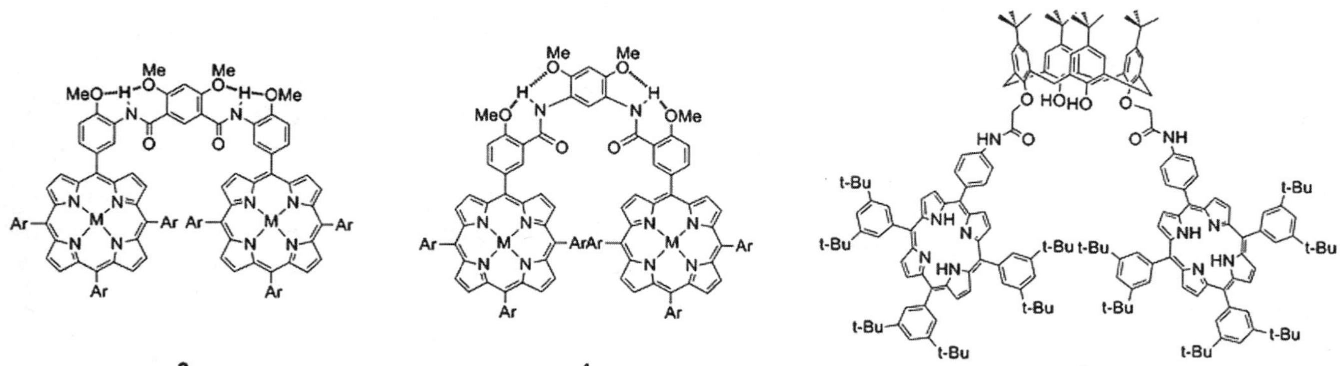

3 4 5

FIGURE 7.3. Chemical structure of receptors **3** and **4** reported by Li; and structure of receptor **5** that showed the highest affinity for C_{60} and C_{70} among those described by Armaroli, Boyd and Reed.

More recently, the group of Li made use of secondary H-bonding interactions within the spacer to introduce a higher degree of preorganization into the tweezers-like design (Figure 7.3) [15]. As a consequence of this, both porphyrin units are held in a *syn* conformation, enhancing their ability to associate fullerenes. The authors estimated the binding constants of these receptors toward C_{60} and C_{70} through UV–vis titrations. A significant increase of over one order of magnitude in the binding constants was observed for both **3** and **4** when compared to receptor **1**. For example, **3** binds C_{60} with a binding constant of log $K_a = 5.0$ in toluene at room temperature, whereas **1** shows a binding constant of log $K_a = 3.7$ under the same experimental conditions.

Comparing receptors **3** and **4**, it is immediately apparent that **3** is a stronger binder, associating C_{60} approximately four times better than **4**. Although the authors of the original manuscript do not comment on this, it beautifully illustrates how even extremely small structural changes, like the change in connectivity of the amide groups between **3** and **4**, can result in relatively high changes in the binding constants.

The collaboration between the groups headed by Armaroli, Boyd, and Reed has produced a family of porphyrin tweezers for fullerene in which the covalent linkers are calix[4]arenes [16]. The structure of the most efficient binder for C_{60} and C_{70} among the six receptors synthesized is shown in Figure 7.3. Receptor **5** showed binding constants of log $K_a = 5.4$ and log $K_a = 6.4$ in toluene at room temperature toward C_{60} and C_{70}, respectively.

In 2008, the group of Morin described a picturesque example of bisporphyrin tweezers for fullerenes. In this case, the two porphyrin units are not covalently but instead mechanically linked, making use of a [3]rotaxane structure [17] in which the porphyrins decorate the macrocycles (Figure 7.4) [18]. The increase in conformational freedom provided by the mechanical bond as compared to the examples where rigid spacers are utilized results in a significant decrease in the binding constants. For example, **6** binds C_{60} in toluene at room temperature with log $K_a = 3.7$, close to two orders of magnitude smaller than **5**. Nevertheless, this example illustrates the possibility of utilizing different kinds of spacers, including unconventional ones.

We will conclude this section dedicated to porphyrin tweezers for fullerenes with the work on tripodal porphyrin receptors reported by the groups headed by Pascu and Sanders [19]. At first sight their receptors do not strictly fit in the category of molecular tweezers, since they feature three recognizing motifs each, but as we shall see, their binding mode is tweezers-like. The structures of the tripodal hosts **7** and **8** are shown in Figure 7.5.

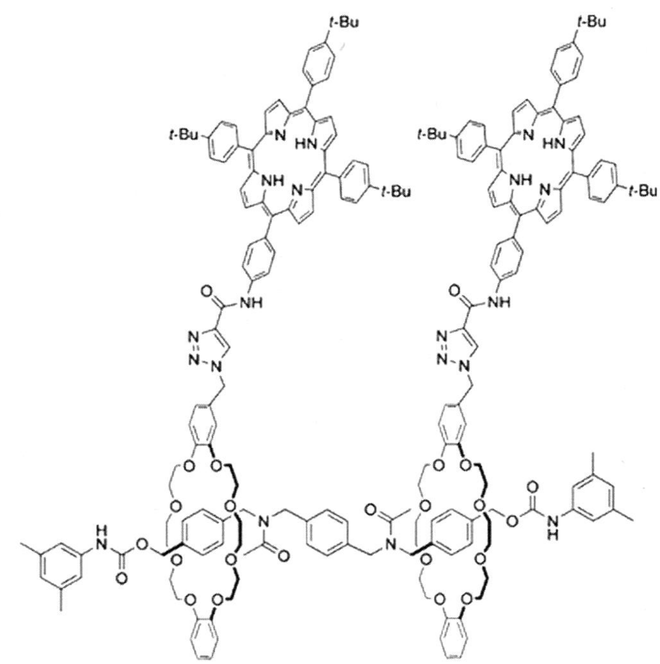

6

FIGURE 7.4. Structure of the [3]rotaxane **6** shown in the co-conformation in which it can bind C_{60}. Reproduced from reference 8c, with permission of the IUPAC.

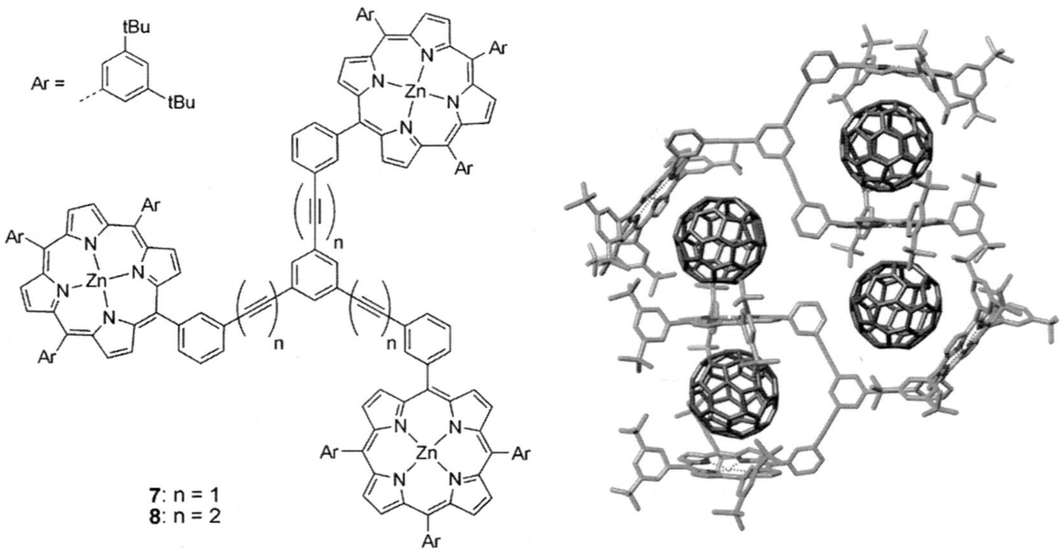

FIGURE 7.5. Chemical structure of the tripodal receptors studied by Pascu and Sanders and solid-state structure of the **8**·C_{60} co-crystal.

Given their chemical structure, in principle several binding modes toward C_{60} can be expected for **7** and **8**, including one in which all three porphyrin units are wrapped around a single C_{60} molecule. In d_8-toluene solution, the authors studied the binding event through ^{1}H NMR titrations and Job's plots. For host **8** a 1:1 binding stoichiometry and a log $K_a = 3.4$ toward C_{60} at room temperature were determined. The case of receptor **7** is more complicated: During the titration of **7** against C_{60}, dissimilar binding isotherms and stoichiometries were suggested depending on the ^{1}H NMR signal under scrutiny. The signals which were shifted upfield suggested a 1:1 binding event, with a log K_a between 2.5 and 3, whereas the analysis of the downfield shifted protons indicated a 2:1 C_{60}:**7** stoichiometry. To explain these data, the authors carried out careful qualitative analysis of the NMR data, to conclude that, in solution, **8** binds C_{60} by wrapping it with all three porphyrin units simultaneously, adopting a bowl-like conformation. The smaller receptor, **7**, cannot adopt such conformation; thus it binds C_{60} in a tweezers-like fashion, leaving one porphyrin unbound and explaining the difference in binding constants.

This picture changes when moving from solution to the solid phase. When **8**·C_{60} co-crystals were studied under x-ray diffraction, a dimeric structure was found in which two units of **8**, each binding one fullerene in a pincer-like fashion, are linked through association of two more units of C_{60}, bound to the external face of the tweezers binding site of one of the molecules of **8** and to the third porphyrin moiety of the other (Figure 7.5). This change in binding mode is explained on account of the decrease in the entropic cost of an intermolecular binding event such as that shown in Figure 7.7 when moving from solution to the solid state. In the case of host **7**, a similar tweezers-like disposition with the third porphyrin unit linking two units of the receptor was found, but in this case the guest of the tweezers cavity was not C_{60} but toluene instead, the solvent of crystallization.

In 1999, Aida, Saigo, and co-workers synthesized a cyclic dimer of a zinc porphyrin, **9** (Figure 7.6), and studied its ability to associate C_{60} [20]. Compound **9** was actually the best host for fullerene described at the time, with log $K_a = 5.8$ in benzene at room temperature. This finding opened an avenue for the development of a whole family of cyclic metalloporphyrin dimers with outstanding properties. A comprehensive account of such systems up to 2007 has already been published [21], so in the following we will focus on the most recent developments only.

Tashiro, Aida, and co-workers showed that a single iridium porphyrin was capable of interacting with [60]fullerene forming a 1:1 complex with a binding constant of log $K_a = 5.3$ in benzene. In accordance with such a high binding constant for a single porphyrin, the corresponding dimeric macrocycle, **10** (Figure 7.6), shows a log $K_a = 8.1$ toward C_{60} in 1,2-dichlorobenzene (*o*-DCB) at room temperature, which up until now constitutes a world record in complex stability [22]. More importantly, structural analysis both in the solid state, through x-ray diffraction, and in solution, through NMR, showed that the corresponding porphyrin dimer **10** forms coordination bonds with fullerenes in a η^2 fashion at very low temperatures, distorting the guest in an ellipsoidal shape. Regarding porphyrin-based receptors, frontiers between supramolecular and covalent chemistry have clearly been reached with iridium (III) porphyrins.

Such a powerful family of receptors is particularly promising for applications like the enantioselective recognition of

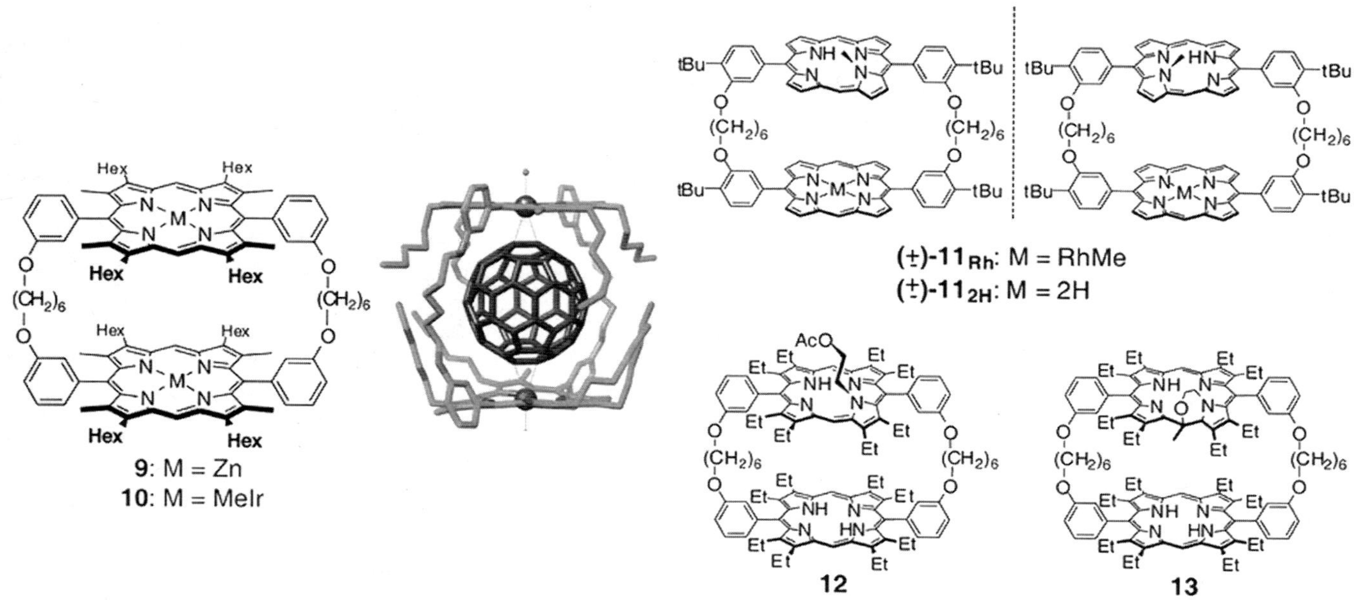

FIGURE 7.6. (**Left**) Structures of compounds **9** and **10**. (**Middle**) The x-ray structure of C_{60}·**10** complex. (**Right**) Structures of chiral derivatives **11–13**.

higher fullerenes. In this context, insertion of asymmetry is a requisite. Regarding cyclic porphyrin hosts, this was performed in an elegant manner in compounds **11–13** (Figure 7.6). Instead of inserting chiral centers at the periphery of the hosts, desymmetrization was achieved on a porphyrin ring that directly interacts with the electronic cloud of the fullerene host. Studying the interactions of these hosts with the smallest chiral fullerene, C_{76}, Aida and co-workers [23] were able to show that compound $\mathbf{11_{Rh}}$ is suitable for sensing chirality in a racemic mixture but also for quantifying the enantiomeric excess via NMR experiments. Though chiral and strongly interacting with C_{76}, host $\mathbf{11_{Rh}}$ did not display any enantioselectivity. Luckily, this feature is an asset to quantify the enantiomeric excess in chiral fullerene mixtures since binding constants for both enantiomers have to be equal for this purpose.

A second generation of chiral hosts, **12** and **13** (Figure 7.6), was described very recently [24]. These receptors present three major differences with **9** and **10**: (i) Pyrrole rings are substituted with ethyl groups in the β and β′ positions, (ii) porphyrin units are not metalated, and (iii) porphyrin units are more distorted. Such differences are expected to help chiral discrimination since metalation increases the affinity of porphyrin rings for fullerenes excessively and thus decreases the selectivity, and a more distorted structure is meant to be more discriminating toward both enantiomers. With this in mind, macrocycle **13**, one of the most distorted phlorin derivatives ever described, was expected to be the most suitable for enantioselective extractions. However, its distortion was also responsible for a decrease in the binding ability toward C_{76} that prevented such an extraction. A first step was finally achieved in 2010 with derivative **12** which allows the obtention of enriched fractions in (+) or (−)-C_{76} with a 7% ee after a single extraction [24].

A new trend in the field of fullerene hosting consists in increasing the number of interacting porphyrins within the macrocycles. The first example was recently reported by Anderson and collaborators [25] with compound **14** (Figure 7.7). As calculated from UV–vis and fluorescence titrations, the resulting rigid structure presents remarkable advantages in terms of fullerene recognition. In particular, **14** shows extremely high binding constants toward C_{60} (log $K_a = 6.2$), C_{70} (log $K_a = 8.2$), and C_{86} (log $K_a > 9$), all in toluene at room temperature. Moreover, the remarkable difference between the stabilities of the associates of **14** with C_{60} and with the higher fullerenes also endorses **14** with a high degree of selectivity, as nicely illustrated with competition experiments followed by mass spectrometry.

To date, the receptor displaying the highest number of porphyrin units was described by Osuka and co-workers [26] with the so-called "nanobarrel" **15**, which features four linked porphyrin units (Figure 7.7). In this work, the authors report the elegant synthesis of this host and the solid-state structures of both **15** and C_{60}·**15**. These reveal that **15** displays concave porphyrin units that allow a stronger interaction with the fullerene guest. Despite the expected synergic effect of the four porphyrin units, the calculated binding constant (log $K_a = 5.7$ in toluene) is not as large as could have been anticipated. This is presumably due to the use of nickel porphyrins. Indeed, previous studies showed a decrease in the binding ability of one order of magnitude when switching from zinc to nickel in macrocyclic porphyrin dimers [27]. The extreme

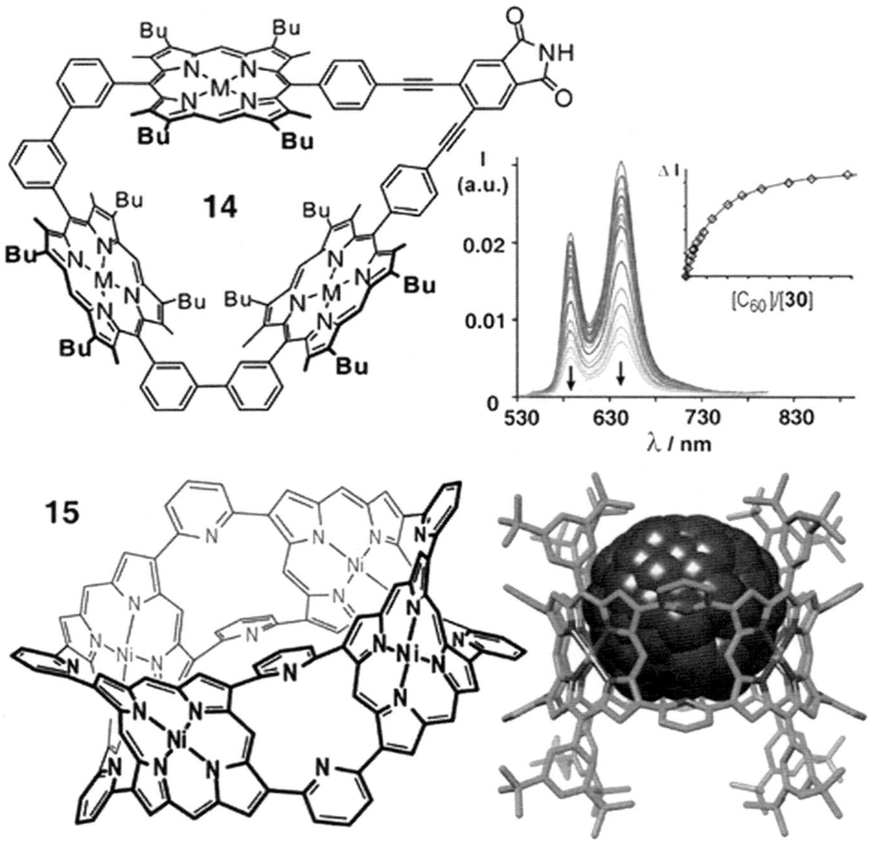

FIGURE 7.7. (**Top**) Receptor **14** displaying three porphyrin units (*left*); evolution of the fluorescence emission of **14** (9.3×10^{-8} M) upon addition of C_{60} in toluene (*right*), reproduced with permission from reference 42. (**Bottom**) Nanobarrel **15** (*left*) and the x-ray crystal structure of its complex with C_{60} (*right*). Copyright Wiley–VCH Verlag GmbH & Co. KGaA. Reproduced with permission from reference 8a.

rigidity of host **15** could also be detrimental, since a certain degree of flexibility may be required to optimize the porphyrin-C_{60} distances [20].

The attractive porphyrin–fullerene interaction has also been explored in dynamic combinatorial chemistry (DCC) [28]. First attempts were performed by Sanders and collaborators [29] with the precursor **16** (Scheme 7.1). This dithiol derivative can be efficiently dimerized in the presence of bipyridine or 1,4-diazabicyclo[2.2.2]octane (DABCO) to afford the cyclic dimer **17**. Given the strong affinity of fullerenes for this host, the influence of C_{60} on the dynamic equilibrium leading to **17** was also investigated. Unfortunately, the radicals R-S$^{\bullet}$ generated during the experiment react with the fullerene template, ruling out the possibility of using dynamic disulfide chemistry in the presence of a fullerene template.

Later, this templating approach could be successfully followed by Langford and coworkers [30]. The strategy lies on the utilization of a template fullerene guest that preorganizes porphyrin subunits before the cyclization reaction. In this manner, the final equilibrium can be displaced toward the best fullerene host (Scheme 7.1).

For instance, the metathesis reaction of porphyrin derivative **18** in the absence of fullerene template mainly leads to the formation of the macrocyclic dimer **19** ($y = 50\%$) and affords the cyclic trimer **20** as a side product ($y = 9\%$). On the other hand, the same reaction carried out in the presence of C_{60} or C_{70} results in the trimer **20** as the major product ($y = 59\%$) while dimer **19** becomes the side product ($y = 4\%$). Performing the reaction with either fullerene does not significantly modify the final state, which is not surprising since **20** is sufficiently large and flexible to accommodate both C_{60} and C_{70}. In principle, using higher fullerenes would be of particular interest to extend this concept and prepare molecules capable of extracting them selectively.

Increasing the binding ability of fullerene receptors is also interesting for the preparation of functional materials although examples along these directions remain scarce. To the best of our knowledge, the majority of the work concerning this issue has been reported by the team headed by Fumito

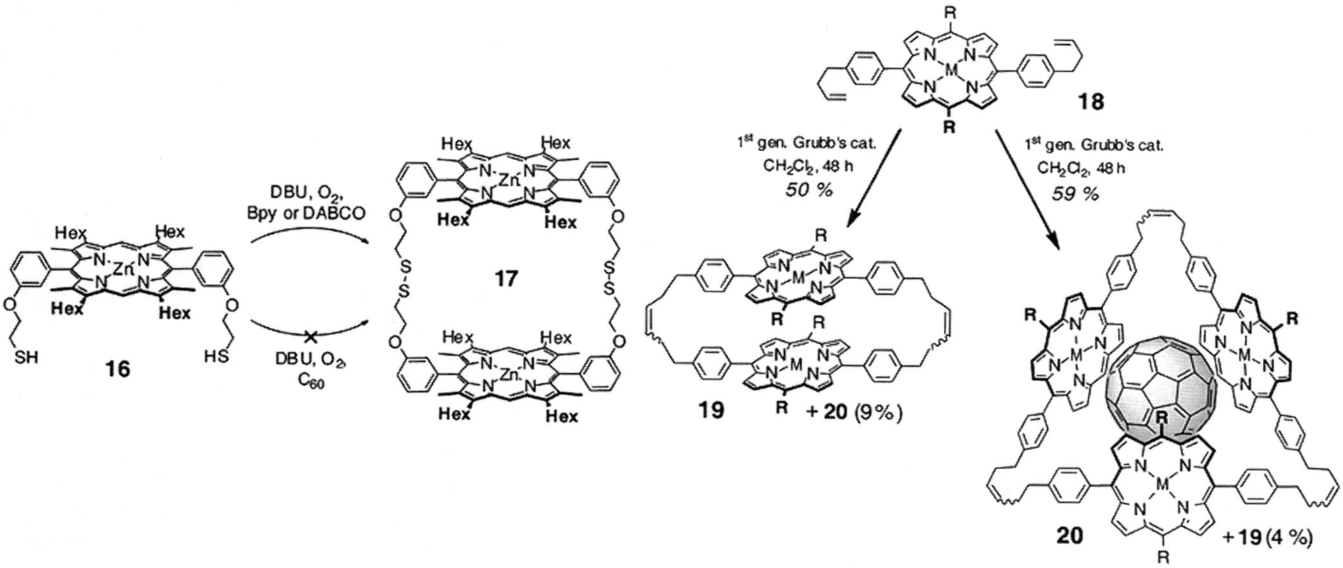

SCHEME 7.1. Attempted template synthesis leading to **17** and templated synthesis of macrocycle **20**.

Tani [31]. The corresponding studies lie on the self-assembly of macrocyclic dimers **21** and **22** (Figure 7.8).

Thanks to the pyridyl rings substituting porphyrin units and to the rigidity of these molecules, an original network of C–H· · ·N hydrogen bonds between the pyrrole β-hydrogens and the pyridyl nitrogens and π–π interactions between adjacent pyridyl rings is formed, allowing the growth of supramolecular nanotubes. In the case of **22**, this arrangement is maintained when C_{60} is co-crystallized with the host [31c]. The linear organization of the guests in this material encouraged authors to thoroughly study the conductivity of the corresponding materials [31b]. These measurements notably demonstrated that such an organization does give rise to an anisotropic and high conductivity (σ = 0.72 $cm^2\ V^{-1}\ s^{-1}$) along the fullerene axis. A very different situation arises when co-crystallizing C_{60} and **21** [31a]. Indeed, nanotubes are no longer observed in the solid state, and fullerenes are organized in a zig-zag fashion. This structural modification consequently results in a decreased conductivity (0.13–0.16 $cm^2\ V^{-1}\ s^{-1}$) and a lower anisotropy. Also noteworthy are the results obtained with C_{60}·**22** concerning organic photovoltaics, since authors were able to prepare a photoelectrochemical cell capable of converting solar energy with a 0.33% efficiency.

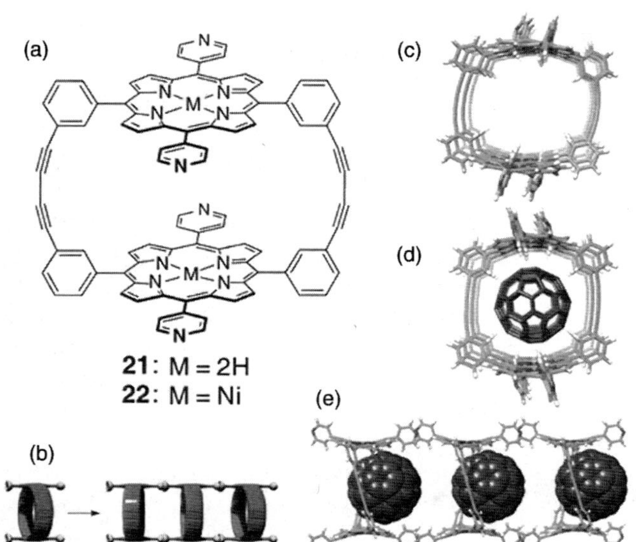

FIGURE 7.8. (**a**) Compounds **21** and **22**, (**b**) their self-assembly, (**c**) x-ray crystal structures of **22**, and (**d**, **e**) C_{60}·**22**. Copyright Wiley–VCH Verlag GmbH & Co. KGaA. Reproduced with permission from reference 8a. See color insert.

7.3 FULLY ORGANIC MOLECULAR TWEEZERS AND MACROCYCLES

The work carried out with porphyrins has by and large dominated the design of molecular receptors for fullerenes [32]. Nevertheless, our group [8c, 33] and others [34] have recently been investigating other recognition motifs. In this second part we will discuss the application of the molecular tweezers and macrocycle designs to those novel recognition fragments.

Corannulene, a large conjugated compound consisting of five benzene rings fused into a central five-member ring, was first synthesized by flash vacuum pyrolysis in 1966 [35]. Following this, several related examples of bowl-shaped

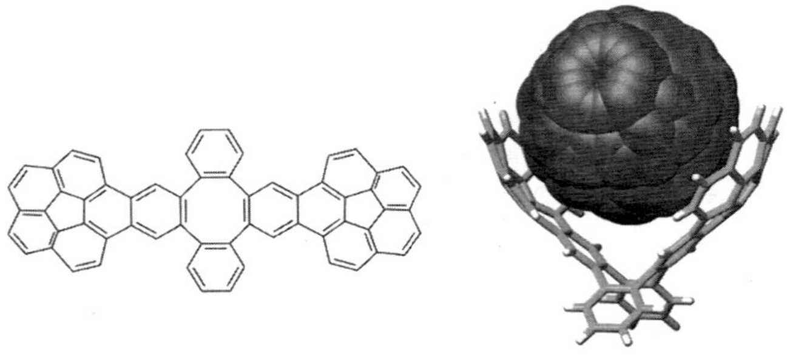

23 **23**·C_{60}

FIGURE 7.9. Chemical structure of corannulene-based molecular tweezers **23**, and solid-state structure of the **23**·C_{60} complex. Please note that the fullerene unit is disordered in the crystal. Reproduced from reference 8c, with permission of the IUPAC.

conjugated molecules, which in many cases can be seen as fullerene fragments, have been synthesized. Various attempts at constructing molecular receptors for fullerenes which incorporate one of these bowl-shaped molecules as recognizing elements have been reported [34]; but, as could be expected considering their size—nearly identical to that of fullerene and thus too small to associate it—chemical derivatization is necessary to observe binding.

The most successful example of receptor for fullerene based on corannulene is precisely a molecular tweezers, the "buckycatcher" [36]. In buckycatcher **23**, two units of corannulene are linked through a rigid polycyclic structure (Figure 7.9). Host **23** forms stable complexes with C_{60} (log $K_a = 3.9$, toluene-d_8, room temperature), and x-ray diffraction studies of mixtures of **23** and C_{60} allowed the determination of the solid-state structure of the complex (Figure 7.9).

Recently, we realized that the shape complementarity between the concave aromatic face of 2-[9-(1,3-dithiol-2-ylidene)anthracen-10(9*H*)-ylidene]-1,3-dithiole (exTTF, **24**, Figure 7.10a) and the exterior of fullerenes should lead to large and positive noncovalent interactions. In fact, DFT calculations predict positive binding energies of up to 7.00 kcal mol^{-1} between a single unit of exTTF and C_{60} in the gas phase. However, we have not observed conclusive experimental evidence of association in either UV–vis or NMR titrations.

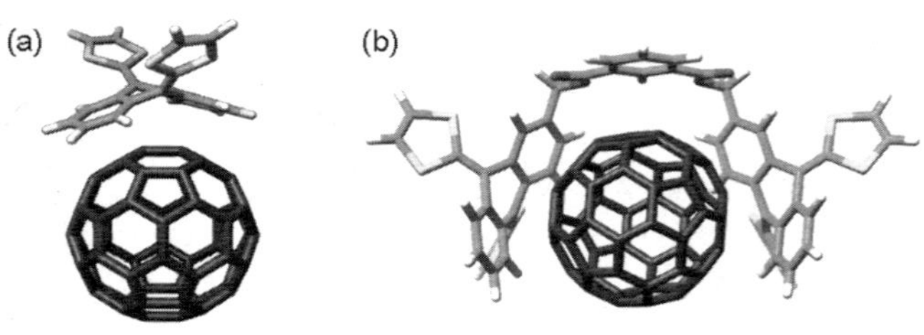

24·C_{60} **25**·C_{60}

FIGURE 7.10. (**a**) Molecular model showing the shape complementarity between **24** and C_{60}. (**b**) Energy-minimized structure of the **25**·C_{60} complex, as predicted by calculations at the BH&H/6-31G** level. See color insert.

In light of these results and the precedents described above, we designed receptor **25** (Figure 7.10b) in which two exTTF units serve as recognizing fragments, and an isophthalate diester acts as a spacer [37]. Receptor **25** was designed as a proof-of-principle of the validity of exTTF as a building block for fullerene receptors, so easy synthetic access was a priority. Host **25** was obtained in excellent yields from readily available exTTF methyl alcohol and commercially available isophthaloyl dichloride. Upon addition of fullerene to a solution of **25**, we observed changes indicative of association in the UV–vis spectrum of the receptor. Namely, the absorption band characteristic of exTTF, centered at $\lambda = 434$ nm, decreases in intensity with increasing concentration of C_{60}. Besides this, after subtraction of the absorption of fullerene, we observe the concomitant appearance of a charge-transfer band at $\lambda = 482$ nm. Nonlinear regression of these spectral changes allowed us to estimate a binding constant of log $K_a = 3.5$ in chlorobenzene at room temperature. Binding was also observed in the gas phase through MALDI-TOF mass spectrometry and in ^{1}H NMR titrations in toluene-d_8. Considering the modest degree of preorganization of receptor **25**, the considerable stability of the **25**·C_{60} complex demonstrates that exTTF is indeed a very valid fragment as a recognizing motif for fullerene. This is most probably due to the unique combination of shape and electronic complementarity between exTTF (concave, electron-rich) and fullerene (convex, electron-poor). Indeed, in later work, we have also demonstrated that this is also the case for other π-extended TTF derivatives [38].

Since exTTF is known to undergo photoinduced electron transfer (PET) when covalently connected to fullerene [3d], we decided to investigate whether intermolecular PET processes were also possible in the **25**·C_{60} complex. To do so, we synthesized an analogue of **25** featuring a terephthalic ester spacer, and we carried out photophysical studies in solution in collaboration with the group of Dirk M. Guldi [21]. We were glad to observe that intracomplex PET does occur from the exTTF units of both receptors to the bound C_{60}. The charge-separated states presented short lifetimes in the range of 3.5–12.7 ps, since the formation of the complexes implies partial orbital overlap between the electroactive units, which facilitates both charge separation and charge recombination processes.

The fairly high binding constant of receptor **25** toward C_{60} got us interested in the role that the concave shape of the recognizing unit **24** played in the overall stabilization of the complex. Shortly before our investigations, the group headed by Kawase had termed the increase in noncovalent interactions between curved aromatic hosts and guests "concave-convex interactions" and suggested that these might play a distinct contribution to the stabilization of the complexes. In order

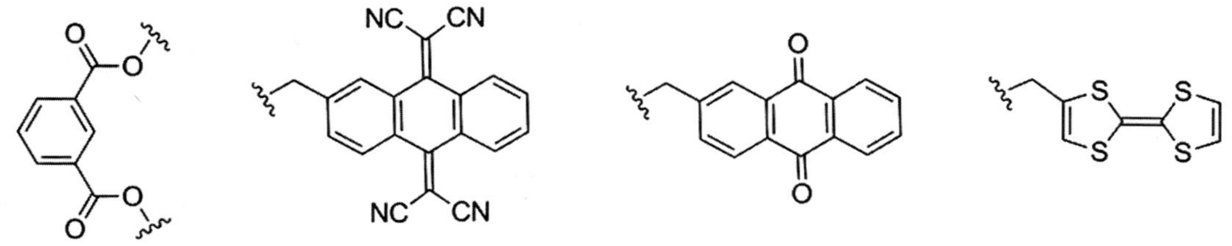

26 27 28

FIGURE 7.11. Chemical structures of receptors **26–28**.

to shed light on whether these concave–convex interactions contributed to stabilize our complexes, and to what extent, we designed and synthesized a collection of structurally related receptors **26–28** (Figure 7.11) in which 11,11,12,12-tetracyanoanthraquinodimethane (TCAQ), anthraquinone, and tetrathiafulvalene (TTF) were utilized as recognizing units. These hosts, together with **25**, provided a full collection of receptors in which the size, the shape, and the electronic character of the recognizing motifs were selectively modified. The binding constants of receptors **25–28** toward C_{60} were investigated through ^{1}H NMR titrations in $CDCl_3$ [39]. Receptor **25** is electron-rich and incorporates five aromatic rings (two per recognizing unit plus the isophthalic spacer) and a large concave surface. As a consequence, **25** is the strongest binder for C_{60}, with a log $K_a = 3.5$. In tweezers **26**, TCAQ is the recognizing element. Thus, it portrays identical number of aromatic rings and surface available for recognition as **25** with close to identical curvature, but is electron-poor. The variation in electronic nature has the outcome of a drop-off in log K_a to 3.2. A similar decrease in the association constant is observed when moving from **26** to **27**. In this case, the surface available for van der Waals interactions is similar to that of **25** and **26**, but **27** lacks both the concave–convex and the electronic complementarity. Host **27** shows a binding constant of log $K_a = 2.9$. Lastly, no sign of association with C_{60} was observed in the case of tweezers **28**, in which the electron-rich, small, and nonaromatic TTF served as recognizing unit.

Comparison of the binding constants of **25** and **26** toward C_{60} suggests a noticeable contribution of coulombic interactions. However, the fact that **28** does not show any sign of complexation toward C_{60} implies that it is not quantitatively comparable to those of π–π and van der Waals forces, since all tweezers featuring large recognizing motifs are capable of associating C_{60} under our experimental conditions. The contribution of concave–convex interactions is illustrated by the cases of receptors **26** and **27**. In spite of the more electron-poor character of **26** when compared to **27**, its binding constant toward C_{60} is larger, which can only be justified by the concave shape of the TCAQ recognizing units.

Calixarenes and CTVs were proven to be efficient receptors toward fullerenes early on. The possibility of associating two or more of these units in order to improve the affinity and/or the selectivity of the receptor for fullerene guests was tested by Matsubara et al. [40] with the CTV unit and Wang and Gutsche [41] with calixarene-based derivatives. In the former case, the authors described the recognition properties of CTV-based hosts **29–31** (Figure 7.12). Surprisingly, the authors did not report important modifications in terms of binding ability. For instance, derivative **2** displays just a light improvement in affinity toward C_{60} compared to CTV **29** (log $K_a = 4.3$ and log $K_a = 4.0$ in benzene at 298 K,

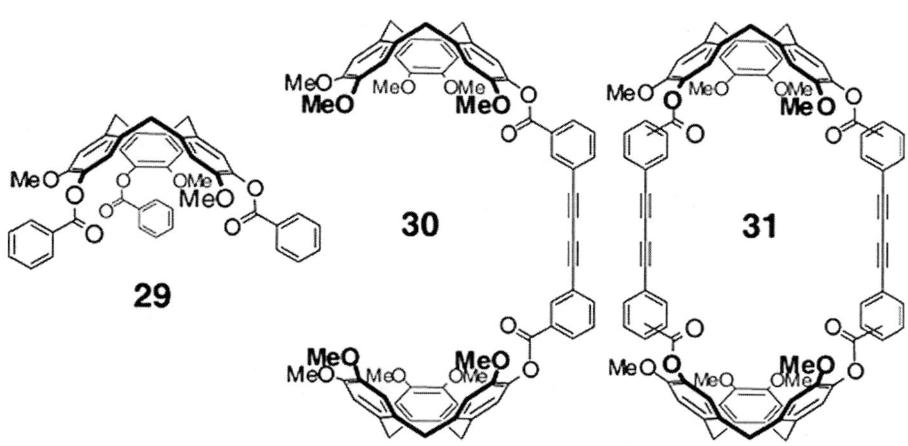

FIGURE 7.12. Structures of compounds **29–31**.

respectively) [40b, 40c]. The insertion of a second linker in *m*-**31** does not lead to a significant increase in the binding constant either, with log $K_a = 4.3$ under identical experimental conditions [40a]. This presumably results from the long and rigid 1,4-diphenylbuta-1,3-diyne spacer, which prevents efficient synergic binding by both CTV moieties. The fact that cage *p*-**31** binds C_{60} worse than **29** (log $K_a = 3.8$ in the same conditions) also supports this assumption. In this sense, the reduction of the butadiyne spacers to their alkyl counterparts would have probably been relevant, as reported later with macrocyclic porphyrin dimers [20]. In the case of Wang and Gutsche, results were even worse since their calix[4]- and -[6]arene-based cages did not show any affinity toward C_{60}, which can be once again ascribed to a lack of flexibility of the linkers [41]. However, one should keep in mind that rigidity does not necessarily constitute a problem: Provided that the size and shape of the receptor match the dimensions of the guest, and that the latter can access the cavity of the host, rigidity can be considered as an advantage. Indeed, a nonflexible host displays a high degree of preorganization, which typically results in the formation of more stable complexes.

This is nicely illustrated with the cyclo-*p*-phenylene acetylene (CPPA) family described by Kawase et al. [34,42]. The very rigid cyclo-*p*-hexaphenyleneacetylene proved to be an efficient host for [60]- and [70]fullerenes with binding constants of log $K_a = 4.2$ and 4.3 in benzene at room temperature, as measured by UV–vis titrations [42c, 42d]. Fluorescence measurements were also performed and showed a decrease in the emission intensity when adding aliquots of C_{60}. The corresponding Stern–Volmer constant proved to be as high as log $K_{sv} = 4.8$. Comparatively, the Stern–Volmer constant with C_{70} was higher (log $K_{sv} = 5.1$), in agreement with the slightly higher affinity of cyclo-*p*-hexaphenyleneacetylene toward C_{70}.

Regarding cyclic-*p*-aryleneacetylenes and their binding ability toward fullerenes, a significant improvement could be achieved thanks to rational modifications of the host structure [42e]. Indeed, the authors increased the aromatic surface of the host by replacing some of the phenyl units by naphthyl ones. In this manner, the dispersion forces responsible for the recognition phenomenon are strengthened, which allows receptor **32a** to bind C_{60} with an association constant as high as log $K_a = 5.0$ in benzene at room temperature. Nicely, the supramolecular associations **32a**·**32b** proved to be an efficient receptor toward C_{60} affording the so-called "onion-type" carbon nanostructure C_{60}·**32a**·**32b** represented in Figure 7.13 [42a].

Very recently, the team headed by Pasini described several chiral receptors displaying two or three axially chiral *(R)*-binapthyl moieties linked through rigid spacers (Figure 7.14) [43]. Their host–guest properties are strongly dependent on their molecular structure. For example, while compound **36** binds [60]fullerene with log $K_a = 3.5$ in toluene at 298 K,

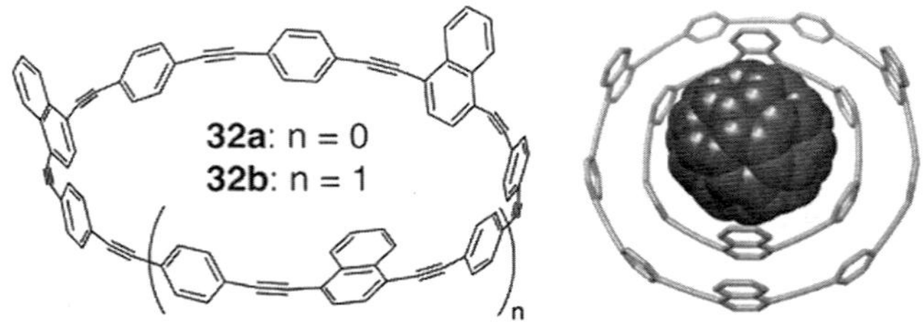

FIGURE 7.13. (**Left**) Compounds **32a** and **32b** and (**Right**) a representation of their onion-like supramolecular nanostructure with C_{60}.

35 does not have any detectable affinity toward C_{60} under the same conditions. The fact that macrocycles **37b** and **37c** (i.e., *meta* and *para* substituted analogues) form complexes of different stoichiometry with C_{60} is also a nice illustration of a small structural variation inducing a significant difference on the binding properties. Indeed, **37c** binds [60]fullerene in a 1:1 fashion while **37b**·C_{60} exists as a mixture of complexes of different stoichiometries. Using Hill's equation, the authors deduced an apparent association constant for each binding event of log $K_{app} = 3.1$ in toluene at 298 K and a Hill coefficient of 2.9, which the authors interpret as meaning that 2.9 molecules of **37b** are interacting with the fullerene guest on average [44]. Beyond the simple recognition properties of these hosts, their most interesting feature certainly lies on their inherent chirality. Besides, the circular dichroism study performed on **37**·C_{60} shows a Cotton effect in absorption bands peculiar to C_{60}. Such a chiral induction offers promising perspectives for the chiral resolution of higher fullerenes but also for developing new methods to functionalize these carbon nanostructures enantioselectively [45].

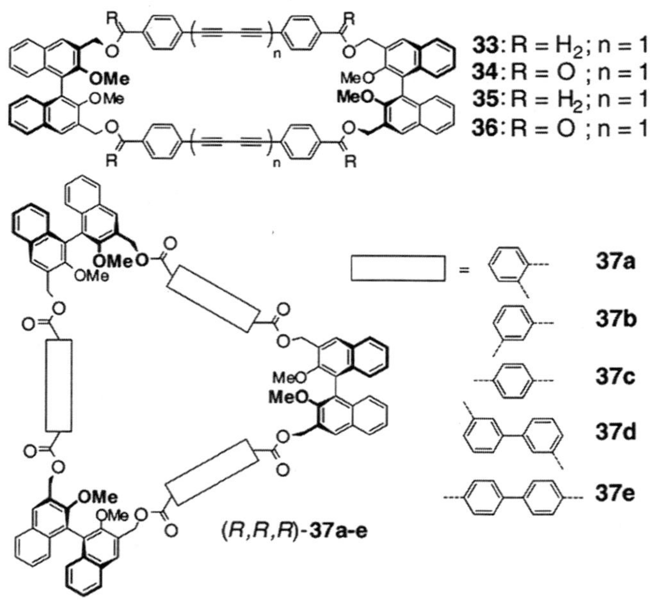

FIGURE 7.14. Pasini's binaphthalene-based receptors **33**–**37**.

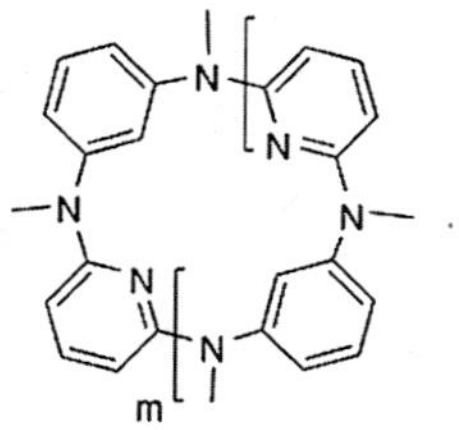

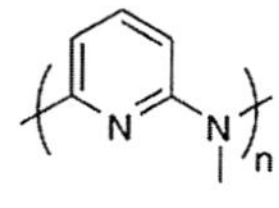

38a: m = 1
38b: m = 3

39a: n = 5
39b: n = 6
39c: n = 7
39d: n = 8
39e: n = 9
39f : n = 10
39f : n = 11

FIGURE 7.15. Macrocycles **38** and **39** involving pyridine rings.

Nitrogen-containing heterocycles, such as pyridine and pyrrole, have recently been used for the preparation of fullerene hosts. The interaction between the host and the C_n guest is generally the result of dispersion forces. As a consequence, the presence of heteroatoms within the aromatic moieties is likely to modify the electronic distribution and the polarizability of the corresponding receptors and, thus, their binding properties. In this context, the team of Mei-Xiang Wang reported important contributions with two families of compounds, namely azacalix[m]arene[n]pyridines and azacalix[n]pyridines [46]. The preparation of derivatives **38a** and **38b** (Figure 7.15) is remarkably straightforward since they are both obtained in three steps starting from 1,3-diaminobenzene and 2,6-dibromopyridine. Unlike **38b**, the addition of **38a** to a C_{60} solution does not lead to any color change, suggesting that no interaction takes place with the smaller macrocycle **38a**, which was confirmed by UV–vis and fluorescence titrations. These studies demonstrate that **38b** forms 1:1 complexes with both C_{60} and C_{70} and authors report binding constants of log K_{SV} = 4.8 and 5.1 in toluene at 298 K, respectively, although neglecting the effect of possible dynamic quenching. Later, the same team reported the synthesis of a new series of macrocycles constructed with pyridine rings only (**39a–g** in Figure 7.15) [47]. The recognition properties of these hosts toward fullerenes were studied by fluorescence spectroscopy, taking only into account static quenching. Surprisingly, this exhaustive work does not show significant differences between receptors **39a–g** in terms of fullerene recognition, even though their diameters (at least in the solid state) are extremely different. The same assessment could be set out for two fullerene receptors, described by the same team, endowed with one and two calix[1]arene[4]pyridine units [48]. In this particular case, the authors measured a lower binding constant for the host made up of two recognition units.

The 1,8-naphthypyridine moiety is another example of a nitrogen-containing aromatic unit recently utilized to prepare **40**, a receptor also displaying two triptycene units (Figure 7.16) [49]. The recognition phenomenon was confirmed by means of fluorescence spectroscopy, which showed that **40** binds both fullerenes in a 1:1 stoichiometry, with log K_{sv} = 4.9 in toluene at room temperature.

As far as triptycene derivatives are concerned, we wish to underline the role of this fragment in the recognition process through the example of compound **41** [50]. The structure of the latter molecule can be compared with the calix[6]arene molecule. However, while calix[6]arene binds fullerenes C_{60} and C_{70} in a 2:1 stoichiometry, **41** forms 1:1 inclusion complexes with both C_{60} and C_{70}. This particular feature is due to the rigidity of the triptycene moieties that prevents the existence of the double-cone conformation observed in the 2:1 inclusion complexes between calix[6]arene and fullerenes. Thanks to the high degree of preorganization supplied by this unit, which is likely to increase both the affinity and the selectivity of receptors, we believe that triptycene-based receptors are appealing hosts for future developments in the supramolecular chemistry of fullerenes.

To date, pyrrole-based macrocyclic receptors for fullerenes remain scarce. The first example was described by Sessler, Jeppesen, and collaborators [51] with derivative **42**, which consists in the fusion of four tetrathiafulvalene units to a calix[4]pyrrole scaffold (Figure 7.17). Compound **42** mainly exists in the 1,3-alternate conformation in solution and interacts with C_{60} only very weakly. Yet, thanks to the affinity of calix[4]pyrroles for chloride anion, it is possible to switch this molecule to its cone conformation, which affords the deep and electron-rich cavity **42**·Cl^-.

The authors demonstrated that, in dichloromethane, C_{60} is surrounded by two units of **42**·Cl^- and that the formation of the 2:1 complex is governed by two sequential equilibria, which lead to C_{60}·(**42**·Cl^-) (log K_1 = 3.4) and

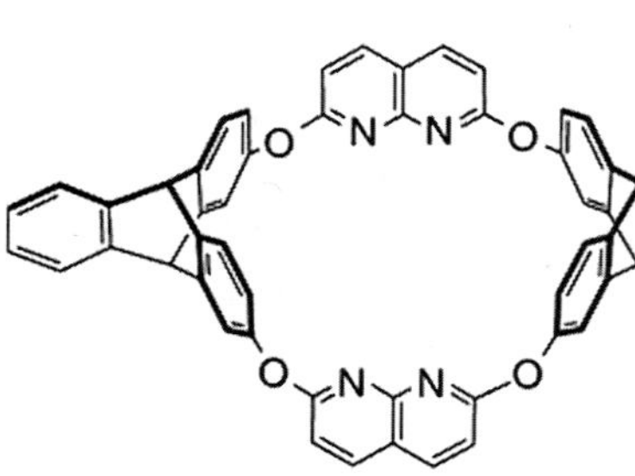

40

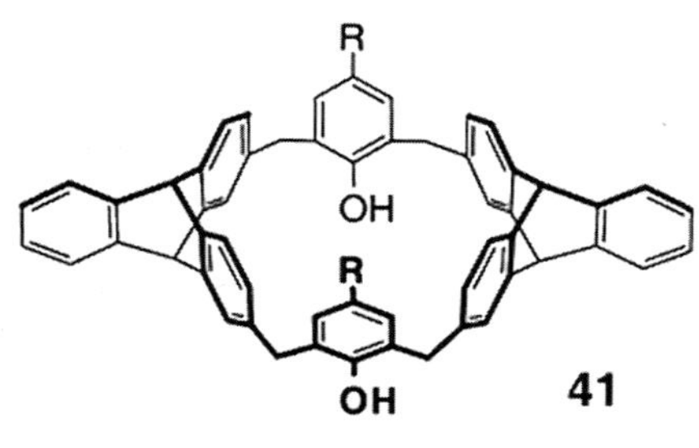

41

FIGURE 7.16. Rigid macrocycles based on triptycene units proposed by Chen and co-workers.

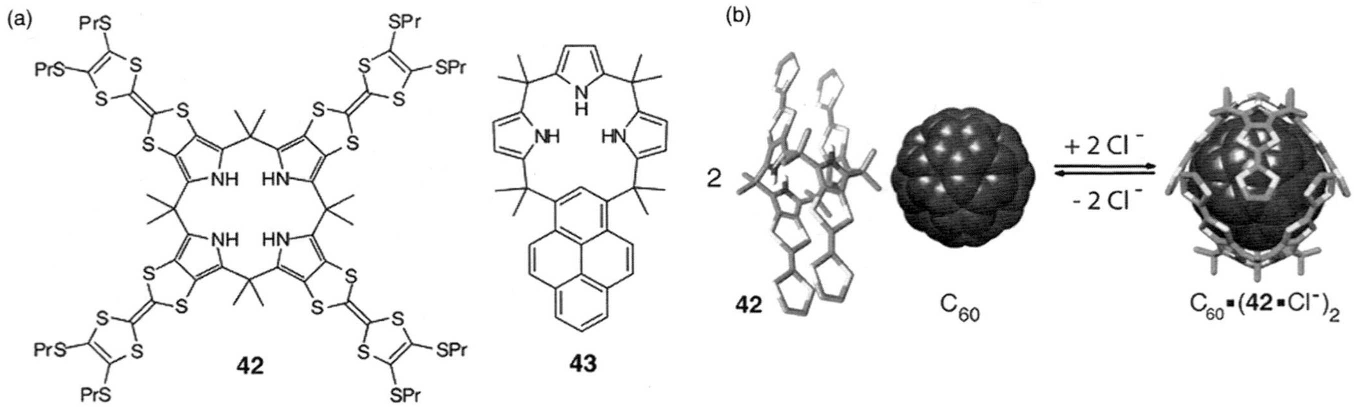

FIGURE 7.17. (**a**) Calix[4]pyrrole **42** and receptor **43**. (**b**) Schematic representation of the fullerene encapsulation and release upon complexation and precipitation of chloride anions.

C_{60}·(**42**·Cl^-)$_2$ (log $K_2 = 4.1$), respectively. Further achievements were reported later by taking advantage of the features of compound **42** [52]. For example, it was found possible to control the encapsulation and the release of the fullerene guest, thanks to complexation/precipitation sequences (Figure 7.17). Starting from a mixture of **42** and C_{60}, the addition of tetrabutylammonium chloride provokes the inclusion of the guest within C_{60}·(**42**·Cl^-)$_2$, while the subsequent addition of sodium tetraphenylborate precipitates chloride anions and regenerates the free fullerene and **42**, in its 1,3-alternate conformation.

In the latter example, the recognition phenomenon is undoubtedly the result of both the shape of the receptor, which surrounds the convex fullerene guest, and the electron-rich nature of the host, which contains four tetrathiafulvalene units. Indeed, electronic complementarity is also relevant when aiming at designing such receptors, since fullerenes are well-known electron acceptors. Calix[4]pyrrole does not intrinsically interact with C_{60}, whether with or without chloride anion. Consequently, whether the pyrrole rings were participating in the recognition process or not was not clear, until recently. Macrocycle **43**, initially designed as a fluoride receptor, is made up of three pyrrole and one pyrene units and forms stable C_{60}·(**43**·F^-) and C_{70}·(**43**·F^-) 1:1 complexes in a toluene-d_8:CD_3CN (95:5) mixture [29]. Considering that neither calix[4]pyrrole nor pyrene alone are able to generate stable complexes with fullerenes, example **43** definitely confirms that pyrrole rings contribute to the recognition of fullerenes.

Considering the success of the exTTF-based tweezers described earlier [21,37], we anticipated that adding preorganization through the formation of macrocyclic hosts would produce even more efficient hosts for fullerenes. According to this, macrocycles **44–46**, in which we varied both the aromatic and the alkenyl linkers between the exTTF units, were synthesized (Figure 7.18). This proved to be a worthwhile strategy since macrocycle **44b** is one of the best fully organic receptors ever reported, with a binding constant of log $K_a = 6.5$ in chlorobenzene, a very good solvent for fullerenes, and up to log $K_a = 7.5$ in benzonitrile, a less competitive solvent [33].

We have thoroughly investigated the recognition properties of a whole family of exTTF-based macrocycles **44–46** by systematically modifying the spacer (*p*-phenylene, *m*-phenylene and 2,6-naphthylene) and the alkenyl chain length ($n = 1, 2, 3$) [34]. This study allowed us to discover some remarkably efficient macrocyclic hosts for fullerenes; for instance, **46c** binds C_{70} with log $K_a = 6.1$ in chlorobenzene at room temperature. Moreover, we also showed how small structural variations lead to important changes in the binding ability. For example, **44c**, binds C_{60} with log $K_a = 3.5$, a binding constant three orders of magnitude smaller than its closely related congener **44b**. Beyond measuring very different binding constants along this series of macrocycles, we also found changes in the stoichiometry of the complexes. For example, **44a** and **45a** bind [60]- and [70]fullerene in a 1:1 and a 2:1 fashion, respectively. Yet, a different situation arises with **46a** which forms 1:1 complexes with both fullerenes, thanks to a slightly larger diameter.

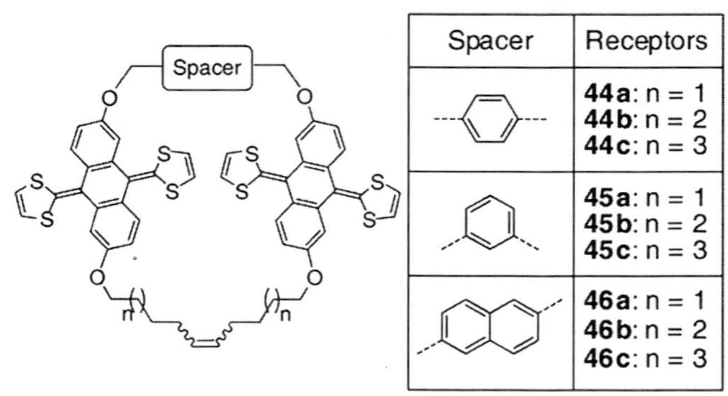

FIGURE 7.18. exTTF-based receptors **44–46**.

It is well known that fullerenes interact with crown-ether derivatives since the pioneering work reported by Mukherjee and co-workers [53]. Taking advantage of this feature, our group recently described an excellent host for [60]- and [70]fullerenes which involves two crown-ether units around a central exTTF [54]. In this manner, the synergistic n–π and π–π interactions of crown-ethers and exTTF allow this derivative to bind C_{60} and C_{70} with a micromolar affinity in chlorobenzene.

As far as crown-like macrocycles are concerned, we would finally like to stress an important breakthrough reported by Akasaka and co-workers [55], which deals with endohedral fullerenes. In this context, the authors described the selective precipitation of La@C_{82} and La_2@C_{80} in the presence of the azacrown derivative 1,4,7,10,13,16-hexaazacyclooctadecane, making easier the separation by HPLC techniques. In this sense, Akasaka's findings are of utmost importance for a more global utilization of endohedral fullerenes.

7.4 CONCLUSIONS AND OUTLOOK

In this chapter, we have reviewed two of the most frequent strategies for the synthesis of fullerene hosts: the constructions of tweezers and macrocycles. Tweezers-like receptors tend to show relatively modest binding constants, in the range of log $K_a = 2$–4. On the positive side, their synthesis is often much more straightforward than that of more elaborate macrocyclic hosts. Thus, the molecular tweezers design is particularly useful in the search for new recognition motifs, where synthetic accessibility is a prerequisite. Comparatively, macrocyclic hosts show the largest binding constants toward fullerenes, reaching values of log $K_a > 6$ for C_{60}. However, the increase in preorganization requires that the structure of the macrocycle is finely tuned to match their host, since even very small structural changes can lead to a dramatic decrease in the binding constant.

The construction of molecular hosts for fullerenes is expected to have a direct impact on their application in organic photovoltaic devices, both through the development of simplified purification procedures and by providing the tools necessary to understand their self-assembly into organized materials at the nanoscale. Considering the advantages and disadvantages of each design, we anticipate that both tweezers and macrocycles will continue to play a prominent role in advancing towards these goals.

REFERENCES AND NOTES

1. Kroto, H. W., Heath, J. R., O'Brien, S. C., Curl, R. F., Smalley, R. E. (1985). C_{60}: Buckminsterfullerene. *Nature*, **318**, 162–163.
2. Martín, N. (2006). New challenges in fullerene chemistry. *Chemical Communications*, 2093–2104.
3. (a) Thompson, B. C., Fréchet, J. M. J. (2008). Polymer-fullerene composite solar cells. *Angewandte Chemie International Edition*, **47**, 58–77. (b) Delgado, J. L., Bouit, P.-A., Filippone, S., Herranz, M. A., Martín, N. (2010). Organic photovoltaics: A chemical approach. *Chemical Communications*, **46**, 4853–4865. (c) Guldi, D. M., Illescas, B. M., Atienza, C. M., Wielopolski, M., Martín, N. (2009). Fullerene for organic electronics. *Chemical Society Reviews*, **38**, 1587–1597. (d) Martín, N., Sánchez, L., Herranz, M. A., Illescas, B., Guldi, D. M. (2007). Electronic communication in tetrathiafulvalene (TTF)/C_{60} systems: Toward molecular solar energy conversion materials? *Accounts of Chemical Research*, **40**, 1015–1024.
4. (a) Krätschmer, W., Lamb, L. D., Fostiropoulos, K., Huffman, D. R. (1990). Solid C_{60}: A new form of carbon. *Nature*, **347**, 354–358. (b) Haufler, R. E., Conceiçao, J., Chibante, L. P. F., Chai, Y., Byrne, N. E., Flanagan, S., Haley, M. M., O'Brien, S. C., Pan, C., et al. (1990). Efficient production of C_{60} (buckminsterfullerene), $C_{60}H_{36}$, and the solvated buckide ion. *Journal of Physical Chemistry*, **94**, 8634–8636.
5. Diederich, F., Effing, J., Jonas, U., Jullien, L., Plesnivy, T., Ringsdorf, H., Thilgen, C., Weinstein, D. (1992). C_{60} and C_{70} in cages? Mono- and multilayers of azacrown compounds and fullerenes. *Angewandte Chemie International Edition English*, **31**, 1599–1602.
6. Andersson, T., Nilsson, K., Sundahl, M., Westman, G., Wennerström, O. (1992). C_{60} embedded in γ-cyclodextrin: a water-soluble fullerene. *Journal of the Chemical Society, Chemical Communications*, 604–606.
7. (a) Atwood, J. L., Koutsantonis, G. A., Raston, C. L. (1994). Purification of C_{60} and C_{70} by selective complexation with calixarenes. *Nature*, **368**, 229–231. (b) Suzuki, T., Nakashima, K., Shinkai, S. (1994). Very convenient and efficient purification method for fullerene (C_{60}) with 5,11,17,23,35,41,47-octa-*tert*-butylcalix[8]arene-49,50,51,52,53,54,55,56-octol. *Chemistry Letters*, 699–702.
8. (a) Canevet, D., Pérez, E. M.Martín, N. (2011). Wraparound hosts for fullerenes: tailored macrocycles and cages. *Angewandte Chemie International Edition*, **50**, 9248–9259. (b) Pérez, E. M. (2011). Energy, supramolecular chemistry, fullerenes, and the sky. *Pure and Applied Chemistry*, **83**, 201–211. (c) Pérez, E. M., Martín, N. (2010). Molecular tweezers for fullerenes. *Pure and Applied Chemistry*, **82**, 523–533. (d) Pérez, E. M., Martín, N. (2008). Curves ahead: Molecular receptors for fullerenes based on concave–convex complementarity. *Chemical Society Reviews*, **37**, 1512–1519.
9. Harmata, M. (2004). Chiral molecular tweezers. *Accounts of Chemical Research*, **37**, 862–873.
10. We find it is advisable to express binding constants as logarithms, particularly for large association constants. For the sake of comparison, throughout the chapter we will express all binding constants as log K_a.
11. (a) Olmstead, M. M., Costa, D. A., Maitra, K., Noll, B. C., Phillips, S. L., Van Calcar, P. M., Balch, A. L. (1999). Interaction of curved and flat molecular surfaces. The structures of crystalline compounds composed of fullerene (C_{60}, $C_{60}O$,

C_{70}, and $C_{120}O$) and metal octaethylporphyrin units. *Journal of the American Chemical Society*, **121**, 7090–7097. (b) Boyd, P. D. W., Reed, C. A. (2005). Fullerene–porphyrin constructs. *Accounts of Chemical Research*, **38**, 235–242.

12. Sun, D., Tham, F. S., Reed, C. A., Chaker, L., Burgess, M., Boyd, P. D. W. (2000). Porphyrin–fullerene host–guest chemistry. *Journal of the American Chemical Society*, **122**, 10704–10705.
13. Sun, D., Tham, F. S., Reed, C. A., Chaker, L., Boyd, P. D. W. (2002). Supramolecular fullerene–porphyrin chemistry. Fullerene complexation by metalated "jaws porphyrin" hosts. *Journal of the American Chemical Society*, **124**, 6604–6612.
14. Ayabe, M., Ikeda, A., Shinkai, S., Sakamoto, S., Yamaguchi, K. (2002). A novel 60]fullerene receptor with a Pd(II)-switched bisporphyrin cleft. *Chemical Communications*, 1032–1033.
15. Wu, Z.-Q., Shao, X.-B., Li, C., Hou, J.-L., Wang, K., Jiang, X.-K., Li, Z.-T. (2005). Hydrogen-bonding-driven preorganized zinc porphyrin receptors for efficient complexation of C_{60}, C_{70}, and C_{60} derivatives. *Journal of the American Chemical Society*, **127**, 17460–17468.
16. Hosseini, A., Taylor, S., Accorsi, G., Armaroli, N., Reed, C. A., Boyd, P. D. W. (2006). Calix[4]arene-linked bisporphyrin hosts for fullerenes: Binding strength, solvation effects, and porphyrin-fullerene charge transfer bands. *Journal of the American Chemical Society*, **128**, 15903–15913.
17. Kay, E. R., Leigh, D. A., Zerbetto, F. (2007). Synthetic molecular motors and mechanical machines. *Angewandte Chemie International Edition*, **46**, 72–191.
18. Marois, J.-S., Cantin, K., Desmarais, A., Morin, J.-F. (2008). [3]Rotaxane-porphyrin conjugate as a novel supramolecular host for fullerenes. *Organic Letters*, **10**, 33–36.
19. Tong, L. H., Wietor, J.-L., Clegg, W., Raithby, P. R., Pascu, S. I., Sanders, J. K. M. (2008). Supramolecular assemblies of tripodal porphyrin hosts and C_{60}. *Chemistry—A European Journal*, **14**, 3035–3044.
20. Tashiro, K., Aida, T., Zheng, J.-Y., Kinbara, K., Saigo, K., Sakamoto, S., Yamaguchi, K. (1999). A cyclic dimer of metalloporphyrin forms a highly stable inclusion complex with C_{60}. *Journal of the American Chemical Society*, **121**, 9477–9478.
21. Gayathri, S. S., Wielopolski, M., Pérez, E. M., Fernández, G., Sánchez, L., Viruela, R., Ortí, E., Guldi, D. M., Martín, N. (2009). Discrete supramolecular donor–acceptor complexes. *Angewandte Chemie International Edition*, **48**, 815–819.
22. Yanagisawa, M., Tashiro, K., Yamasaki, M., Aida, T. (2007). Hosting fullerenes by dynamic bond formation with an iridium porphyrin cyclic dimer: A "chemical friction" for rotary guest motions. *Journal of the American Chemical Society*, **129**, 11912–11913.
23. Shoji, Y., Tashiro, K., Aida, T. (2006). Sensing of chiral fullerenes by a cyclic host with an asymmetrically distorted π-electronic component. *Journal of the American Chemical Society*, **128**, 10690–10691.
24. Shoji, Y., Tashiro, K., Aida, T. (2010). One-pot enantioselective extraction of chiral fullerene C_{76} using a cyclic host carrying an asymmetrically distorted, highly π-basic porphyrin module. *Journal of the American Chemical Society*, **132**, 5928–5929.
25. Gil-Ramírez, G., Karlen, S. D., Shundo, A., Porfyrakis, K., Ito, Y., Briggs, G. A. D., Morton, J. J. L., Anderson, H. L. (2010). A cyclic porphyrin trimer as a receptor for fullerenes. *Organic Letters*, **12**, 3544–3547.
26. Song, J., Aratani, N., Shinokubo, H., Osuka, A. (2010). A porphyrin nanobarrel that encapsulates C_{60}. *Journal of the American Chemical Society*, **132**, 16356–16357.
27. Zheng, J.-Y., Tashiro, K., Hirabayashi, Y., Kinbara, K., Saigo, K., Aida, T., Sakamoto, S., Yamaguchi, K. (2001). Cyclic dimers of metalloporphyrins as tunable hosts for fullerenes: A remarkable effect of rhodium(III). *Angewandte Chemie International Edition*, **40**, 1857–1861.
28. Corbett, P. T., Leclaire, J., Vial, L., West, K. R., Wietor, J.-L., Sanders, J. K. M., Otto, S. (2006). Dynamic combinatorial chemistry. *Chemical Reviews*, **106**, 3652–3711.
29. Kieran, A. L., Pascu, S. I., Jarrosson, T., Sanders, J. K. M. (2005). Inclusion of C_{60} into an adjustable porphyrin dimer generated by dynamic disulfide chemistry. *Chemical Communications*, 1276–1278.
30. Mulholland, A. R., Woodward, C. P., Langford, S. J. (2011). Fullerene-templated synthesis of a cyclic porphyrin trimer using olefin metathesis. *Chemical Communications*, **47**, 1494–1496.
31. (a) Nobukuni, H., Shimazaki, Y., Uno, H., Naruta, Y., Ohkubo, K., Kojima, T., Fukuzumi, S., Seki, S., Sakai, H., Hasobe, T., Tani, F. (2010). Supramolecular structures and photoelectronic properties of the inclusion complex of a cyclic free-base porphyrin dimer and C_{60}. *Chemistry—A European Journal*, **16**, 11611–11623. (b) Nobukuni, H., Tani, F., Shimazaki, Y., Naruta, Y., Ohkubo, K., Nakanishi, T., Kojima, T., Fukuzumi, S., Seki, S. (2009). Anisotropic high electron mobility and photodynamics of a self-assembled porphyrin nanotube including C_{60} molecules. *Journal of Physical Chemistry C*, **113**, 19694–19699. (c) Nobukuni, H., Shimazaki, Y., Tani, F., Naruta, Y. (2007). A nanotube of cyclic porphyrin dimers connected by nonclassical hydrogen bonds and its inclusion of C_{60} in a linear arrangement. *Angewandte Chemie International Edition*, **46**, 8975–8978.
32. Tashiro, K., Aida, T. (2007). Metalloporphyrin hosts for supramolecular chemistry of fullerenes. *Chemical Society Reviews*, **36**, 189–197.
33. Pérez, E. M., Illescas, B. M., Herranz, M. A., Martín, N. (2009). Supramolecular chemistry of π-extended analogues of TTF and carbon nanostructures. *New Journal of Chemistry*, **33**, 228–234.
34. Kawase, T., Kurata, H. (2006). Ball-, bowl-, and belt-shaped conjugated systems and their complexing abilities: Exploration of the concave–convex π–π interaction. *Chemical Reviews*, **106**, 5250–5273.
35. Barth, W. E., Lawton, R. G. (1966). Dibenzo[*ghi,mno*]fluo ranthene. *Journal of the American Chemical Society*, **88**, 380–381.
36. Sygula, A., Fronczek, F. R., Sygula, R., Rabideau, P. W., Olmstead, M. M. (2007). A double concave hydrocarbon buckycatcher. *Journal of the American Chemical Society*, **129**, 3842–3843.

37. Pérez, E. M., Sánchez, L., Fernández, G., Martín, N. (2006). exTTF as a building block for fullerene receptors. Unexpected solvent-dependent positive homotropic cooperativity. *Journal of the American Chemical Society*, **128**, 7172–7173.

38. Pérez, E. M., Sierra, M., Sánchez, L., Torres, M. R., Viruela, R., Viruela, P. M., Ortí, E., Martín, N. (2007). Concave tetrathiafulvalene-type donors as supramolecular partners for fullerenes. *Angewandte Chemie International Edition*, **46**, 1847–1851.

39. Pérez, E. M., Capodilupo, A. L., Fernández, G., Sánchez, L., Viruela, P. M., Viruela, R., Ortí, E., Bietti, M., Martín, N. (2008). Weighting non-covalent forces in the molecular recognition of C_{60}. Relevance of concave–convex complementarity. *Chemical Communications*, 4567–4569.

40. (a) Matsubara, H., Oguri, S.-Y., Asano, K., Yamamoto, K. (1999). Syntheses of novel cyclotriveratrylenophane capsules and their supramolecular complexes of fullerenes. *Chemistry Letters*, 431–432. (b) Matsubara, H., Shimura, T., Hasegawa, A., Semba, M., Asano, K., Yamamoto, K. (1998). Syntheses of novel fullerene tweezers and their supramolecular inclusion complex of C_{60}. *Chemistry Letters*, 1099–1100. (c) Matsubara, H., Hasegawa, A., Shiwaku, K., Asano, K., Uno, M., Takahashi, S., Yamamoto, K. (1998). Supramolecular inclusion complexes of fullerenes using cyclotriveratrylene derivatives with aromatic pendants. *Chemistry Letters*, 923–924.

41. Wang, J., Gutsche, C. D. (1998). Calixarenes. 48. Complexation of fullerenes with bis-calix[*n*]arenes synthesized by tandem Claisen rearrangement. *Journal of the American Chemical Society*, **120**, 12226–12231.

42. (a) Kawase, T., Tanaka, K., Shiono, N., Seirai, Y., Oda, M. (2004). Onion-type complexation based on carbon nanorings and a buckminsterfullerene. *Angewandte Chemie International Edition*, **43**, 1722–1724. (b) Kawase, T., Fujiwara, N., Tsutumi, M., Oda, M., Maeda, Y., Wakahara, T., Akasaka, T. (2004). Supramolecular dynamics of cyclic [6]paraphenyleneacetylene complexes with [60]- and [70]fullerene derivatives: Electronic and structural effects on complexation. *Angewandte Chemie International Edition*, **43**, 5060–5062. (c) Kawase, T., Tanaka, K., Fujiwara, N., Darabi, H. R., Oda, M. (2003). Complexation of a carbon nanoring with fullerenes. *Angewandte Chemie International Edition*, **42**, 1624–1628. (d) Kawase, T., Seirai, Y., Darabi, H. R., Oda, M., Sarakai, Y., Tashiro, K. (2003). All-hydrocarbon inclusion complexes of carbon nanorings: Cyclic [6]- and [8]paraphenyleneacetylenes. *Angewandte Chemie International Edition*, **42**, 1621–1624. (e) Kawase, T., Tanaka, K., Seirai, Y., Shiono, N., Oda, M. (2003). Complexation of carbon nanorings with fullerenes: Supramolecular dynamics and structural tuning for a fullerene sensor. *Angewandte Chemie International Edition*, **42**, 5597–5600.

43. (a) Coluccini, C., Dondi, D., Caricato, M., Taglietti, A., Boiocchi, M., Pasini, D. (2010). Structurally-variable, rigid and optically-active D_2 and D_3 macrocycles possessing recognition properties towards C_{60}. *Organic and Biomolecular Chemistry*, **8**, 1640–1649. (b) Caricato, M., Coluccini, C., Dondi, D., Vander Griend, D. A., Pasini, D. (2010). Nesting complexation of C_{60} with large, rigid D_2 symmetrical macrocycles. *Organic and Biomolecular Chemistry*, **8**, 3272–3280.

44. Although it is often considered a direct indication of the number of available binding sites, the Hill coefficient is best thought of as an interaction coefficient reflecting the extent of cooperativity, with a maximum value equal to the number of binding sites.

45. Filippone, S., Maroto, E. E., Martín-Domenech, A., Suárez, M., Martín, N. (2009). An efficient approach to chiral fullerene derivatives by catalytic enantioselective 1,3-dipolar cycloadditions. *Nature Chemistry*, **1**, 578–582.

46. Wang, M.-X., Zhang, X.-H., Zheng, Q.-Y. (2004). Synthesis, structure, and [60]fullerene complexation properties of azacalix[*m*]arene[*n*]pyridines. *Angewandte Chemie International Edition*, **43**, 838–842.

47. (a) Gong, H.-Y., Zhang, X.-H., Wang, D.-X., Ma, H.-W., Zheng, Q.-Y., Wang, M.-X. (2006). Methylazacalixpyridines: Remarkable bridging nitrogen-tuned conformations and cavities with unique recognition properties. *Chemistry—A European Journal*, **12**, 9262–9275. (b) Liu, S.-Q., Wang, D.-X., Zheng, Q.-Y., Wang, M.-X. (2007). Synthesis and structure of nitrogen bridged calix[5]- and -[10]-pyridines and their complexation with fullerenes. *Chemical Communications* 3856–3858. (c) Zhang, E.-X., Wang, D.-X., Zheng, Q.-Y., Wang, M.-X. (2008). Synthesis of large macrocyclic azacalix[n]pyridines (n = 6–9) and their complexation with fullerenes C_{60} and C_{70}. *Organic Letters*, **10**, 2565–2568.

48. Wu, J.-C., Wang, D.-X., Huang, Z.-T., Wang, M.-X. (2010). Synthesis of diverse N,O-bridged calix[1]arene[4]pyridine-C_{60} dyads and triads and formation of intramolecular self-inclusion complexes. *Journal of Organic Chemistry*, **75**, 8604–8614.

49. Hu, S.-Z., Chen, C.-F. (2010). Triptycene-derived oxacalixarene with expanded cavity: synthesis, structure and its complexation with fullerenes C_{60} and C_{70}. *Chemical Communications*, **46**, 4199–4201.

50. Tian, X.-H., Chen, C.-F. (2010). Triptycene-derived calix[6]arenes: synthesis, ftructures, and their complexation with fullerenes C_{60} and C_{70}. *Chemistry—A European Journal*, **16**, 8072–8079.

51. Nielsen, K. A., Cho, W.-S., Sarova, G. H., Petersen, B. M., Bond, A. D., Becher, J., Jensen, F., Guldi, D. M., Sessler, J. L., Jeppesen, J. O. (2006). Supramolecular receptor design: Anion-triggered binding of C_{60}. *Angewandte Chemie International Edition*, **45**, 6848–6853.

52. (a) Nielsen, K. A., Martín-Gomis, L., Sarova, G. H., Sanguinet, L., Gross, D. E., Fernández-Lázaro, F., Stein, P. C., Levillain, E., Sessler, J. L., Guldi, D. M., Sastre-Santos, A., Jeppesen, J. O. (2008). Binding studies of tetrathiafulvalene-calix[4]pyrroles with electron-deficient guests. *Tetrahedron*, **64**, 8449–8463. (b) Nielsen, K. A., Sarova, G. H., Martin-Gomis, L., Fernández-Lázaro, F., Stein, P. C., Sanguinet, L., Levillain, E., Sessler, J. L., Guldi, D. M., Sastre-Santos, A., Jeppesen, J. O. (2008). Chloride anion controlled molecular "switching". Binding of 2,5,7-trinitro-9-dicyanomethylenefluorene-C_{60} by tetrathiafulvalene calix[4]pyrrole and photophysical generation of two different

charge-separated states. *Journal of the American Chemical Society*, **130**, 460–462.

53. (a) Saha, A., Nayak, S. K., Chottopadhyay, S., Mukherjee, A. K. (2003). Spectrophotometric study of complexation of dicyclohexano-24-crown-8 with [60]- and [70]fullerenes and other acceptors. *Journal of Physical Chemistry B*, **107**, 11889–11892. (b) Bhattacharya, S., Sharma, A., Nayak, S. K., Chattopadhyay, S., Mukherjee, A. K. (2003). NMR study of complexation of crown ethers with [60]- and [70]fullerenes. *Journal of Physical Chemistry B*, **107**, 4213–4217.

54. Grimm, B., Santos, J., Illescas, B. M., Muñoz, A., Guldi, D. M., Martín, N. (2010). A new exTTF-crown ether platform to associate fullerenes: Cooperative n–π and π–π effects. *Journal of the American Chemical Society*, **132**, 17387–17389.

55. Tsuchiya, T., Sato, K., Kurihara, H., Wakahara, T., Nakahodo, T., Maeda, Y., Akasaka, T., Ohkubo, K., Fukuzumi, S., Kato, T., Mizorogi, N., Kobayashi, K., Nagase, S. (2006). Host–guest complexation of endohedral metallofullerene with azacrown ether and its application. *Journal of the American Chemical Society*, **128**, 6699–6703.

8

COVALENT, DONOR–ACCEPTOR ENSEMBLES BASED ON PHTHALOCYANINES AND CARBON NANOSTRUCTURES

Giovanni Bottari, Maxence Urbani, and Tomás Torres

In memory of Christian G. Claessens

8.1 INTRODUCTION

Phthalocyanines (Pcs) are thermally and chemically stable macrocycles that possess interesting physicochemical properties such as a rich redox chemistry and an intense absorption in the red/near-infrared (NIR) region of the solar spectrum, with extinction coefficients as high as 200,000 M^{-1} cm^{-1} and high fluorescence quantum yields. Moreover, the planar and aromatic structure of these macrocycles allows for their organization into stacks through the occurrence of π–π supramolecular interactions [1, 2]. These unique features render Pcs ideal molecular components for the preparation of donor–acceptor (D–A) ensembles. In such D–A systems, the Pc role is twofold: (i) It behaves as an antenna, since it absorbs very efficiently a substantial region of the solar spectrum [3, 4] and (ii) once photoexcited, it acts as an electron donor for the acceptor moiety. These characteristics make these macrocycles promising building blocks for their incorporation in photovoltaic and artificial photosynthetic systems.

In this context, a wide range of covalent and supramolecular Pc-based D–A systems have been prepared and studied incorporating electroactive acceptor units of diverse nature and redox character such as fullerenes, carbon nanotubes (CNTs), graphene, perylenediimide, anthraquinone, ferrocene (Fc), ruthenium bypyridine complexes, flavin, or porphyrins [5–7]. Among the acceptor moieties used, members of the carbon nanostructures' family such as C_{60} fullerene, CNTs, and graphene hold a privileged position due to their unique physicochemical properties. C_{60} fullerene possesses excellent electron acceptor properties, which can give rise, when implemented in D–A systems, to the formation of stable radical ion pair species [8–14]. Similarly to C_{60} fullerene, CNTs also present a high affinity for electrons which, once accepted, can be transported along their one-dimensional (1-D) tubular structure [15–19]. For these reasons, both, fullerenes and CNTs are widely used as acceptor materials in organic and hybrid solar cells. More recently, graphene, one of the latest entries in the family of carbon nanostructures and one of the "rising stars" in the field of nanotechnology [20], has generated a tremendous interest due to its extraordinary physical and mechanical properties that render it an outstanding material for electronics, material science, and photoconversion systems [21].

In this chapter, we review some of the most representative examples of covalent D–A systems based on Pcs, fullerene, CNTs, and graphene where the Pc plays the dual role of antenna/electron donor unit, while the carbon nanostructure is the electron acceptor moiety. Photostimulation of these covalent D–A systems lead, in most of the cases, to the formation of long-lived, charge-separated (CS) states which can span from ps up to ms, both in solution and/or in the solid state. In such systems, the nature of the spacer connecting the donor and the acceptor units, which ultimately control the distance, orientation, and electronic communication between these active moieties, has also been varied, and the effect of these changes on the CS dynamics has been investigated. Some of these covalent ensembles have also been incorporated, with some success, as active components in photovoltaic devices. Furthermore, the planar and π-conjugated surface of Pcs opens up the possibility to control, by using

Organic Nanomaterials: Synthesis, Characterization, and Device Applications, First Edition. Edited by Tomás Torres and Giovanni Bottari.

supramolecular interactions, the nanoscopic organization of these D–A ensembles over large length scales, with the ultimate goal of extending or even improving some of the chemical and physical properties presented by these systems in the organized phases with respect to the molecularly dispersed species.

8.2 DONOR–ACCEPTOR, COVALENTLY LINKED PHTHALOCYANINE–FULLERENE SYSTEMS

Among the acceptor units employed for the preparation of covalent, Pc-based systems, C_{60} fullerene enjoys a privileged position due, mainly, to its excellent electron affinity, which renders this spherical molecule a perfect molecular component for the construction of photo- and electroactive systems. The extraordinary electron acceptor properties of C_{60} fullerene [8–13], coupled with its small reorganization energy (λ) and its ability to promote ultrafast charge separation together with very slow charge recombination features, have prompted the incorporation of this carbon nanomaterial in a large number of D–A covalent systems based on Pcs, where photoinduced electron transfer (PET) and solar energy conversion applications are sought.

The first report on a molecular system comprising both a Pc and a C_{60} fullerene moiety covalently linked (i.e., a Pc–C_{60} dyad) appeared in 1997 by Hanack, Hirsch, and coworkers. Pc–C_{60} dyad **1** was prepared through a [4 + 2] Diels–Alder cycloaddition reaction between Ni(II)Pc **2** bearing two terminal double bonds (i.e., diene) and C_{60} fullerene (i.e., dienophile) (Scheme 8.1) [22, 23].

Electrochemistry measurements carried out on **1** showed five reversible reduction peaks, which were assigned by spectroelectrochemistry to be either fullerene- or Pc-based. However, these measurements showed that the reduction potentials of both the Pc and the fullerene moieties in **1** did not change significantly compared to those of the respective parent compounds (i.e., Pc **2** and a C_{60} fullerene monoadduct lacking the Pc moiety), thus ruling out any ground-state electronic communication between the two electroactive subunits in **1**.

The third-order nonlinear optical properties of a Cu(II)Pc–C_{60} Diels–Alder adduct analogue of **1** have also been investigated, giving rise to high second-order hyperpolarizability values [24]. Moreover, it was found that nanoparticle dispersions of this Cu(II)Pc–C_{60} dyad exhibit enhanced optical limiting performances [25] and nonlinear absorption enhancement [26].

Since that first Pc–C_{60} dyad, a large number of covalently linked Pc–C_{60} systems (i.e., dyads, triads, tetrads, etc.) have been prepared and the physicochemical properties (i.e., redox, photophysical, liquid crystalline, etc.) of these conjugates studied with the aim of shedding light on the structure–properties relationships of these ensembles.

Among these Pc–C_{60} dyads, compounds **3a–c**, consisting in a metallated or free-base Pc macrocycle directly linked to a fulleropyrrolidine acceptor moiety, probably represents the structurally simpler one. Dyads **3a–c** were prepared by 1,3-dipolar cycloaddition reaction of azomethine ylides, generated in situ from formyl Pcs **4a–c** and *N*-methylglycine, to C_{60} fullerene (also known as the Prato–Maggini reaction), obtaining Pc–C_{60} dyads **3a–c** in reasonable yields (i.e., 40% (**3a**), 43% (**3b**), 41% (**3c**)) (Scheme 8.2). Such reaction, which results in the formation of a pyrrolidine ring, represents one of the most effective synthetic strategies used nowadays for the functionalization of fullerenes, as well as CNTs and more recently graphene.

UV–vis studies on dyads **3a–c** revealed a small ground-state electronic communication between the two spatially close, active units (i.e., Pc and C_{60}). Electrochemical investigation showed that the Pc-based oxidation and reduction potentials were positively shifted in dyads **3a**,**b** in comparison to the redox potentials of some reference compounds (i.e., Pcs **4a**,**b** and a fullerene derivative lacking the Pc moiety). Conversely, the C_{60}-based reduction potentials for **3a**,**b** were negatively shifted with respect to those of a fullerene derivative reference compound, thus suggesting some degree of ground-state, intra- and/or intermolecular interactions between the electron-donating Pc unit to the electron accepting fullerene moiety. A detailed analysis of the photophysical properties (i.e., steady-state and transient absorption and

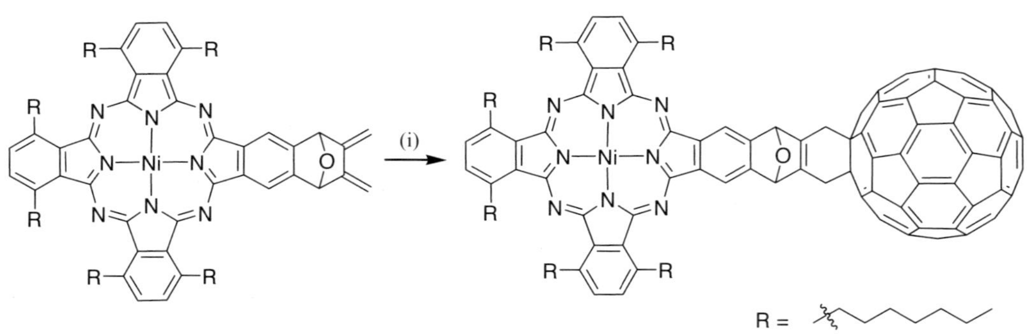

2

1

SCHEME 8.1. Synthesis of Pc–C_{60} fullerene dyad **1**. (i) C_{60} fullerene, toluene, reflux.

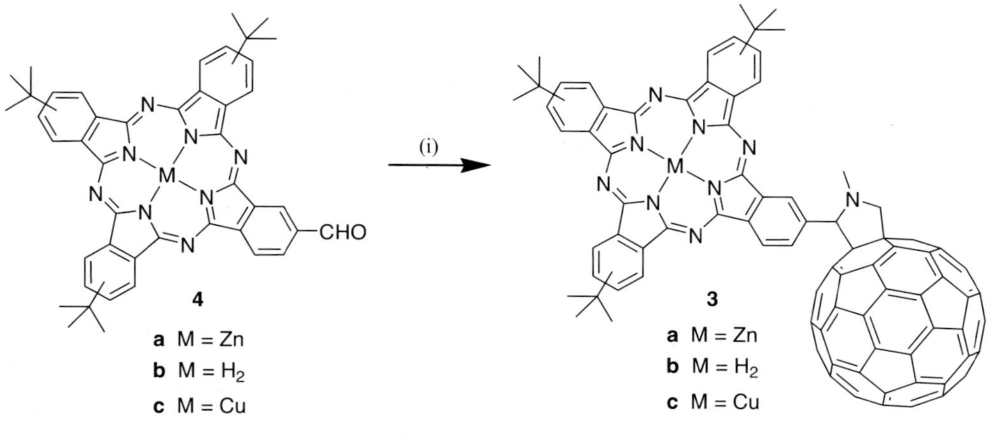

SCHEME 8.2. General reaction conditions for the preparation of Pc–C_{60} dyads **3a–c** from aldehyde-substituted Pcs **4**. (i) C_{60} fullerene, *N*-methylglycine, toluene, reflux.

time-resolved fluorescence measurements) of dyads **3a–c** in solution was also carried out, confirming the occurrence of PET processes [27,28].

The photovoltaic properties of Pc–C_{60} dyad **3a** as photoactive component in organic solar cells have also been investigated [29,30]. The incorporation of covalently linked Pc–C_{60} dyads in solar cells could be envisioned as a valuable possibility to avoid macroscopic phase segregation between the donor and the acceptor units, an important problem in organic photovoltaic devices, thus ensuring efficient charge separation along the active layer of the photovoltaic device. Photophysical characterization of spin-coated films of dyad **3a** showed lifetimes of about 0.2 ms, which is several orders of magnitude longer than that reported for the same dyad in solution, as a result of the stabilization of the charge transfer state in the solid. Solar cells using thin films of Pc–C_{60} **3a** as the active material were fabricated by spin-coating the dyad from a toluene solution onto a poly(3,4-ethylenedioxythiophene) poly(styrenesulfonate) (PEDOT:PSS) modified indium tin oxide (ITO) electrode, although the photovoltaic performances of the resulting device resulted to be quite low. Following the same motivation, devices composed of a solution-processed blend of poly((2-methoxy-5-((3,7-dimethyloctyl)oxy)-*p*-phenylene) vinylene) (MDMO-PPV) and dyad **3a** were also prepared with the aim of combining the absorption of the Zn(II)Pc moiety at around 400 and 700 nm, with the absorption of MDMO-PPV around 500 nm, which was also expected to contribute to the photocurrent [31]. In this case, an energy transfer process was found to take place from MDMO-PPV to the Zn(II)Pc component of the dyad. The open circuit voltage (V_{OC}) of the solution-processed device was 0.5 V, which was significantly larger than the 0.32 V value obtained with devices using only Pc–C_{60} **3a**. However, the short-circuit current (I_{sc}) in the mixture was lower, thereby indicating charge transport problems within the device.

A series of Pc–C_{60} dyads bearing different spacers (i.e., single (**5**), double (**6**), or triple (**7**) bonds) between the Pc and C_{60} fullerene moieties have also been prepared with the aim of studying how the nature of the spacer between the donor and the acceptor units influences the stabilization of the photogenerated radical ion pair states (Scheme 8.3) [32]. The preparation of dyads **5–7** involves the use of a common Pc starting material **8**, which has also been extensively used in order to prepare several other D–A Pc-based systems. Several factors have contributed to the extensive use of Pc **8** for the preparation of D–A Pc-based systems such as the possibility to prepare it in reasonable yields (> 40%) and its easy chromatographic purification. Moreover, the *tert*-butyl groups that decorate the macrocycle periphery in **8** confer to this molecule high solubility, and they help to reduce the typical strong self-aggregation tendency of the Pcs, which often represents a problem when dealing with Pcs reaction mixtures that need to be purified. Besides the above-mentioned points, the most important feature of Pc **8** is, probably, the presence of an iodine atom directly attached to the Pc framework, which allows the functionalization of this macrocycle with, virtually, any functional group through the use of metal-catalyzed reactions.

In order to prepare the aforementioned Pc–C_{60} dyads, the aldehyde-containing Pcs **9–11** were prepared by metal-catalyzed reaction on Pc **8** (in the case of Pc **11**, this Pc was obtained upon oxidation of Pc **12** containing a propargylic alcohol). The three formyl-containing Pcs **9–11** were finally subjected to a 1,3-dipolar cycloaddition reaction with C_{60} fullerene and *N*-methylglycine in refluxing toluene to afford conjugates **5–7** in 20%, 23%, and 21% yields, respectively. Detailed photophysical investigations of compounds **5–7** revealed, in all cases, the occurrence of an intramolecular electron transfer dynamic that evolves from the photoexcited Pc macrocycle to the electron-accepting C_{60} moiety. Moreover, it was observed in **5–7** that the nature of the linker (i.e.,

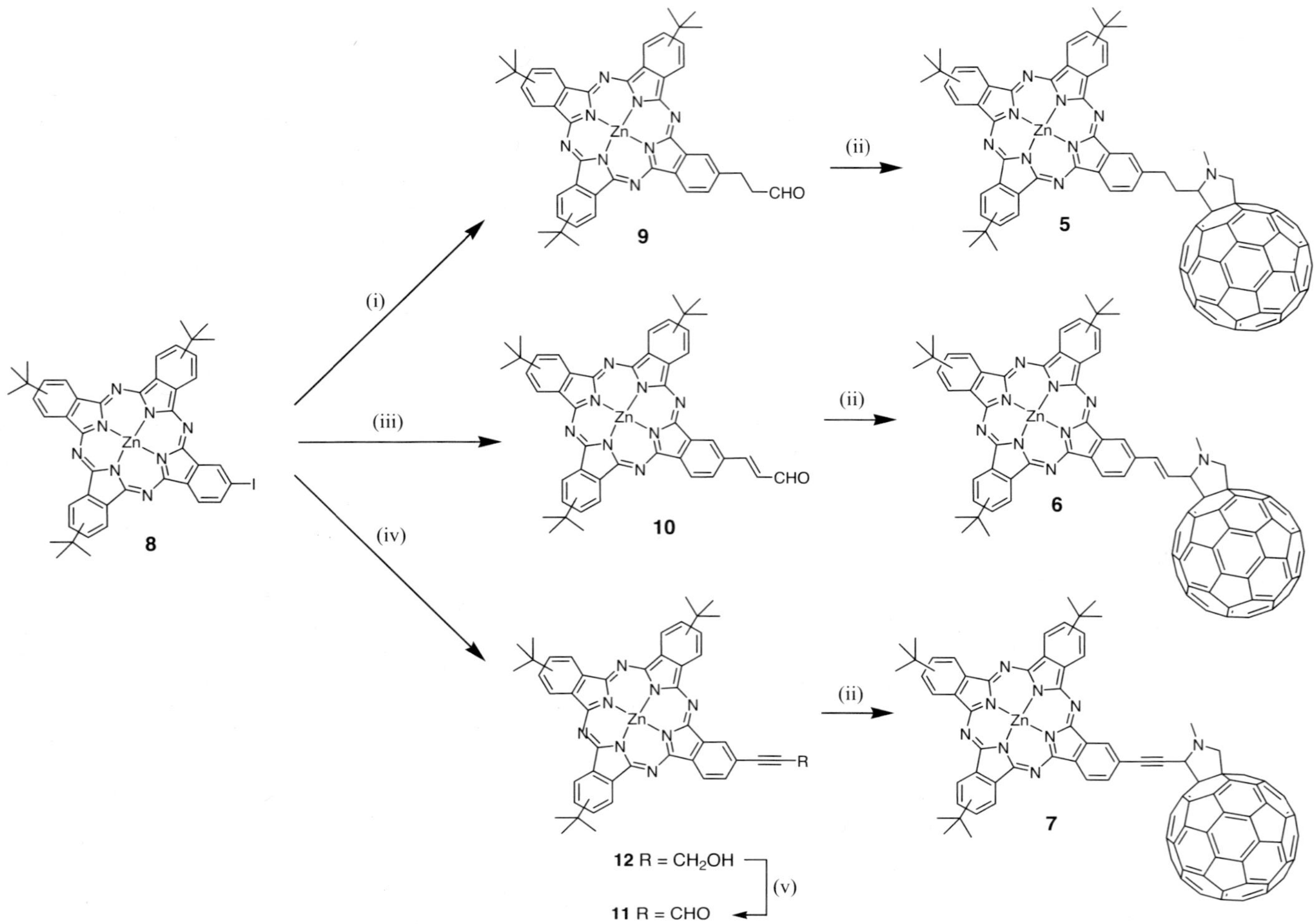

SCHEME 8.3. Synthesis of Zn(II)Pc–C_{60} conjugates **5–7**. (i) Allylic alcohol, $Pd(OAc)_2$, $NaHCO_3$, Bu_4NCl, DMF, 30°C. (ii) C_{60} fullerene, *N*-methylglycine, toluene, reflux. (iii) Acrolein, $Pd(OAc)_2$, $NaHCO_3$, Bu_4NCl, DMF, 20°C. (iv) Propargyl alcohol, $Pd(PPh_3)_2Cl_2$, CuI, THF, NEt_3, RT. v) pyridinium chlorochromate, CH_2Cl_2, RT.

conjugation and flexibility) has an influence on the lifetime of the photogenerated radical ion pair states as a consequence of the differently hybridized carbon atoms, with the longest lifetimes observed in the case of dyad **7** (i.e., 1200 ps in anisole), followed by dyad **6** (i.e., 979 ps in anisole) and dyad **5** (i.e., 755 ps in anisole).

A structurally rigid Pc–C_{60} dyad (**13**) in which the two active components are separated by a [2.2]*para*-cyclophane unit has also been prepared in order to study the effect of the *para*-cyclophane, which is regarded as a pseudoconjugated spacer, on the communication between the Pc and C_{60} (Figure 8.1) [33].

It was observed that the use of such bridge in **13** exerts an appreciable stabilization of the radical ion pair lifetime relative to **5–7** (i.e., 2600 ps in anisole). This fact is likely due to the structural and electronic features of the *para*-cyclophane spacer which (i) weakens the electronic coupling between Zn(II)Pc and C_{60} and (ii) keeps the donor and the acceptor at a larger distance with respect to dyads **3a** or **5–7**.

Pc–C_{60} dyads **14** are constituted by a fullerene moiety covalently connected via a flexible, nonconjugated spacer to a Pc bearing either *tert*-butyl (i.e., **14a**) or trifluoroethoxy (i.e., **14b**) groups at its peripheral positions (Figure 8.2) [34]. Both dyads were obtained starting from a common fullerene derivative bearing a terminal phthalonitrile moiety that was then reacted with 3-*tert*-butyl phthalonitrile or a persubstituted, trifluoroethoxy-coated phthalonitrile, leading to Pc–C_{60} dyads **14a** and **14b**, respectively. Electrochemical and spectroscopic studies on both dyads revealed that whereas in the case of ensemble **14a**, the occurrence of an efficient intramolecular PET process was detected, dyad **14b** did not show any sign of electronic communication between Pc and C_{60}. This result is presumably due to the strong

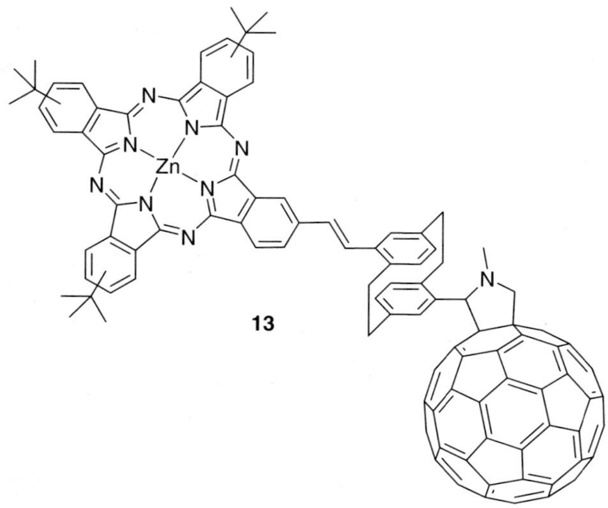

FIGURE 8.1. Structure of Pc–C_{60} conjugate **13**.

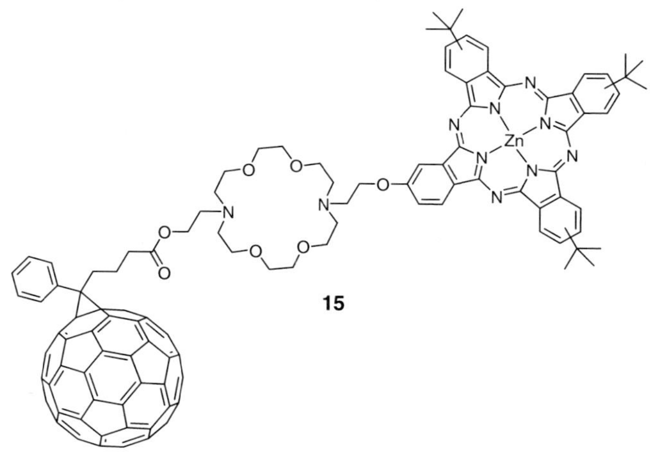

FIGURE 8.3. Molecular structure of crown ether-containing Pc–C_{60} conjugate **15**.

electron-withdrawing nature of the numerous trifluoroethoxy groups in **14b**, which confer an acceptor character to the Pc macrocycle equivalent to that of C_{60} fullerene.

A flexible azacrown macrocycle has also been used as linker between a Pc and C_{60} fullerene in **15** with the aim of studying the effect of possible ion-induced conformational changes on the electronic communication between the active units (Figure 8.3) [35].

UV–vis studies on such conjugate showed no electronic communication between C_{60} and the Pc macrocycle. Moreover, absorption, fluorescence, and electrochemical measurements on dyad **15** revealed that the ground- and excited-state properties of this dyad were insensitive to both the nature of the solvent or the presence of alkali (i.e., Na^+ or K^+) cations.

Pc–C_{60} dyads (**16–18**) have also been prepared in which the fullerene moiety has been doubly functionalized using

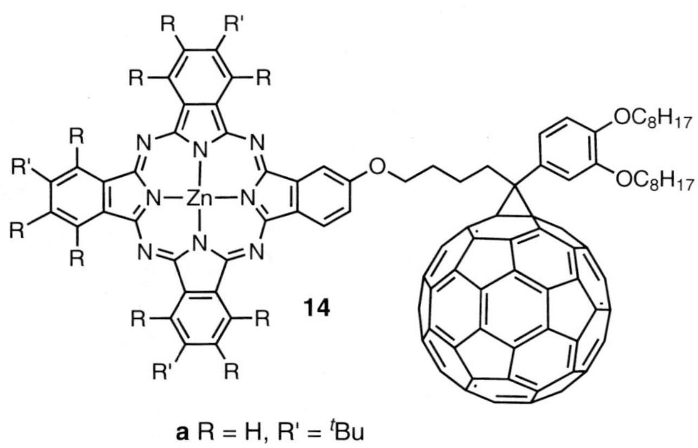

FIGURE 8.2. Molecular structure of Pc–C_{60} dyads **14a**,**b**.

a Bingel–Hirsch reaction (Figure 8.4) [36]. Similarly to the Prato–Maggini reaction, the Bingel–Hirsch reaction is a reaction widely used in fullerene chemistry in order to functionalize these spherical carbon nanostructures using, in general, a malonate derivative as starting material in the presence of a base, leading to a fullerene cyclopropane. The photophysical properties of dyads **16–18** have been carried out in order to elucidate to which extent the geometrical restrictions imposed by the double addition on the fullerene scaffold, which set the Pc and the C_{60} fullerene moieties in spatial proximity, would influence the occurrence of PET processes. Examination of the electron-transfer energetics revealed that these Pc–C_{60} dyads exist mainly in an extended conformation as opposed to a face-to-face orientation common for structurally similar, double-bridged porphyrin–fullerene dyads.

Moreover, evidences of a PET involving an intramolecular exciplex intermediate was observed for Pc–C_{60} dyads **16** and **18** in which excitation of the Pc moiety in both dyads results in a rapid PET from the macrocycle to C_{60} via an exciplex state in both polar and nonpolar solvents [37]. The $Pc^{\bullet+}$–$C_{60}{}^{\bullet-}$ CS state then decays directly to the ground state in 30–70 ps in a polar solvent, whereas in nonpolar solvents, roughly 20% of the dyads undergo transition from the CS state to the Pc triplet state $^3Pc^*$–C_{60} before relaxation to the ground state. In such systems, the formation of the CS state was also confirmed by electron spin resonance measurements at low temperature in both polar and nonpolar solvents. Dyad **16** was also organized in Langmuir–Blodgett (LB) films, and its photophysical properties were studied [38]. These measurements showed the formation of a long-lived (i.e., μs), intermolecular CS state that was attributed to the organization of the dyad within the film.

Pc–fullerene conjugates have also been prepared in which changes in the nature of the acceptor moiety (i.e., endohedral

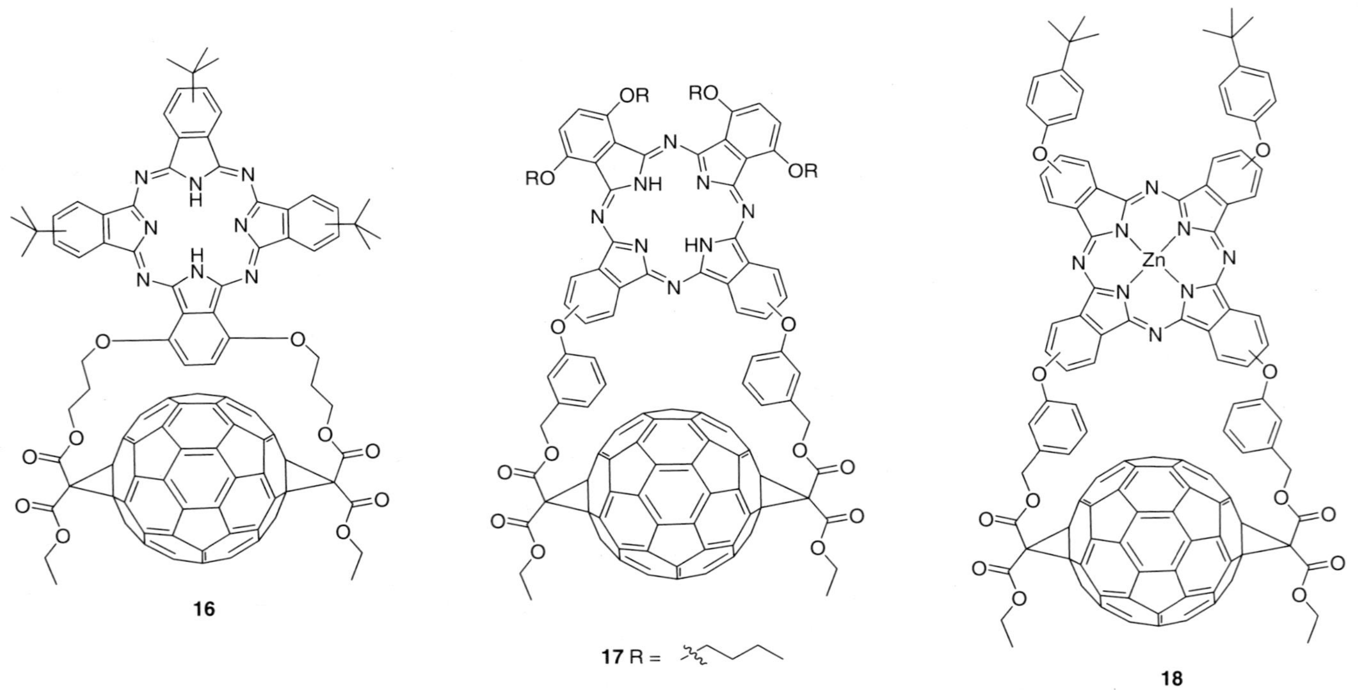

FIGURE 8.4. Molecular structures of the three Pc–C_{60} dyads **16–18**, each of them having a doubly functionalized C_{60} moiety.

metallofullerenes (EMFs) instead of C_{60} fullerene) or the donor unit (i.e., double-decker Pcs instead of planar Pcs) have been carried out.

The synthesis of Pc-based D–A dyads **19a**,**b** and **20** in which a Pc macrocycle has covalently connected to a trimetallic nitride templated EMF (TNT-EMF) has recently been reported [39]. TNT-EMFs, which represent one of the latest entry within the family of fullerene compounds, are constituted by a carbon cage encapsulating a trimetallic nitride cluster. These metallofullerenes, isolated for the first time in 1999 [40], present several advantages with respect to C_{60} fullerene, such as higher absorption coefficients in the visible region of the electromagnetic spectrum and lower energy gaps between the highest occupied molecular orbital (HOMO) and the lowest unoccupied molecular orbital (LUMO), while showing an electron-accepting ability similar to C_{60}.

$M_3N@C_{80}$-based (M = Y, Sc) dyads **19a**,**b** and **20** containing a Pc macrocycle (as well as some other acceptor units such as Fc or extended-tetrathiafulvalene) have been prepared via Prato–Maggini (**19a**,**b**) or Bingel–Hirsch (**20**) reactions (Figure 8.5).

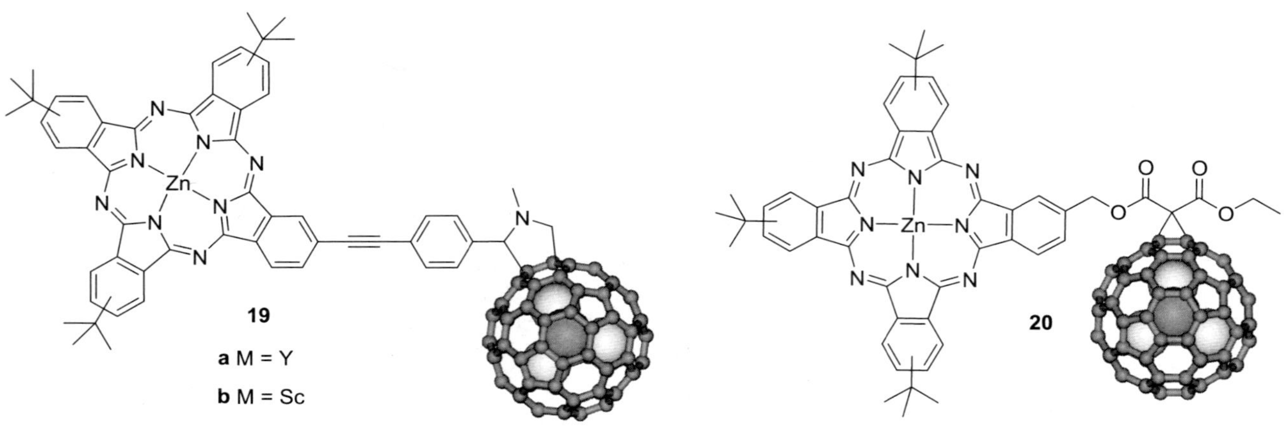

FIGURE 8.5. Molecular structures of $M_3N@C_{80}$-based (M = Y, Sc) Pc–endohedral fullerene dyads **19a**,**b** and **20**.

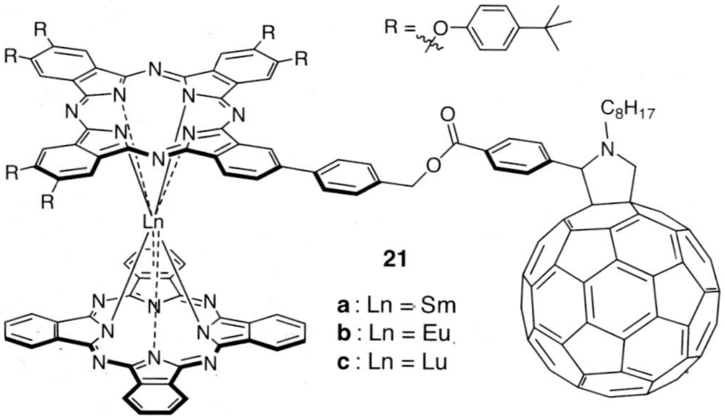

FIGURE 8.6. Molecular structures of double-decker lanthanide(III) bis(phthalocyaninato)–C_{60} dyads **21a-c**.

However, the synthesis of these Pc–EMF dyads was found to be more difficult and lower-yielding than their C_{60}-based counterparts, due to the inherent lower chemical reactivity of the EMFs with respect to C_{60} fullerene. Moreover, the poor chemical stability of the systems obtained precluded an accurate electrochemical and photophysical characterization of these ensembles.

Double-decker Pcs have also been employed, as an alternative to planar Pcs, for the preparation of C_{60}-containing, D–A systems. Sandwich complexes **21a–c** were prepared and characterized by using different techniques such as electrochemistry, UV–vis, and ^{1}H NMR spectroscopies (Figure 8.6) [41]. In the case of the ^{1}H NMR characterization, addition of hydrazine hydrate to solutions of the complexes in [D$_7$]DMF was necessary in order to convert the free-radical, double-decker species into the respective protonated counterparts. Transient absorption experiments on these dyads revealed the formation of CS states with lifetimes of ~3 ns in toluene, but only when irradiating at 387 nm, a wavelength at which both the Pc lanthanide and C_{60} are excited. On the other hand, the exclusive excitation of the Pc unit at 665 nm leads to the formation of a short-lived, singlet excited state that is unable to populate the electron transfer CS state. With the aim of (i) improving the light-harvesting ability of covalently linked D–A Pc–C_{60} ensembles and (ii) increasing the lifetime of the photogenerated radical ion pair, molecular systems containing multiple Pc donor and/or C_{60} acceptor units such as **22–24** have been synthesized (Figure 8.7) [42–44]. However, detailed photophysical studies on such systems showed that the presence of more than one donor and/or acceptor moiety does not necessarily lead to an increase in the photogenerated CS lifetime (i.e., 2.2 ns in anisole for **24**, 1 ns in benzonitrile for **22**, and 180 ps in benzonitrile for **23**). In the case of triad **24**, its photovoltaic performances were also tested, showing very low short-circuit current (I_{sc}) and open-circuit voltage (V_{oc}) values.

The formation of silicon–oxygen bonds has been also used as a means to obtain Pc-based ensembles containing multiple C_{60} moieties, which have been placed at the axial position of a Si(IV)Pc macrocycle. In this context, Si(IV)Pc–$(C_{60})_2$ triads such as **24** [45] and **25** [46] bearing two axial fullerene substituents have been prepared (Figure 8.8). The synthesis of both ensembles involves the reaction of a Si(IV)Cl_2Pc macrocycle with a C_{60} fullerene derivative bearing an alcohol (in the case of **24**) or an acid (in the case of **25**) group. In both triads, the presence of two bulky C_{60}-functionalized ligands at their axial positions avoids the typical aggregation of Pcs by hampering the Pc–Pc intermolecular interactions.

In the case of compound **24**, electrochemical studies revealed that the first reduction of the Pc moiety was anodically shifted with respect to that of a reference Pc macrocycle lacking the C_{60} fullerene axial ligand. This observation was rationalized in terms of the strong electron-withdrawing effect of the two axial fullerene substituents, which led to the stabilization of the first reduced state of the Pc. However,

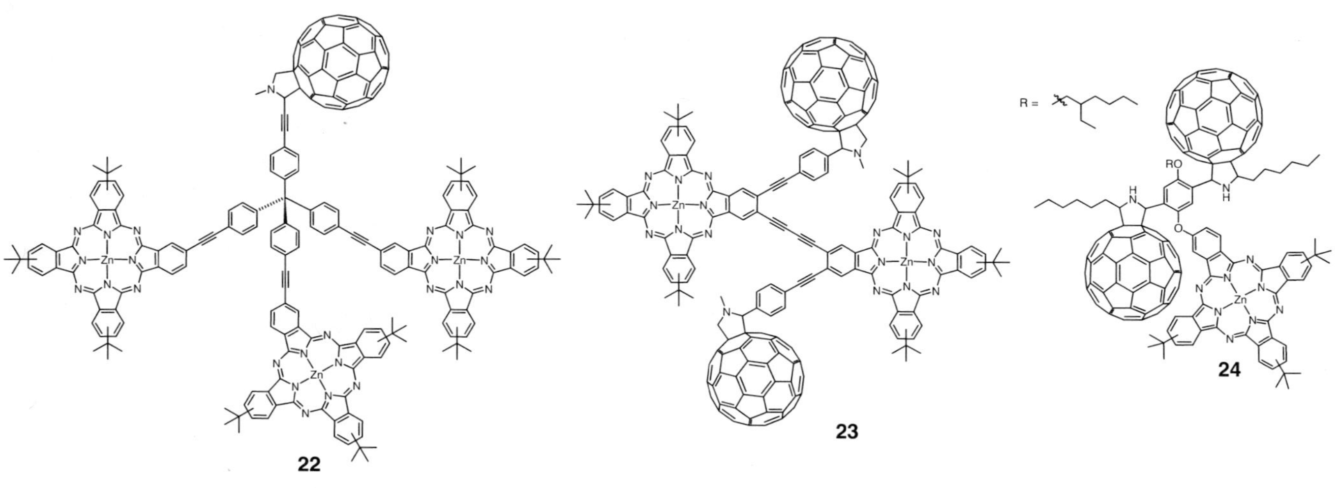

FIGURE 8.7. Molecular structures of Pc$_3$–C_{60} tetrad **22**, Pc$_2$–$(C_{60})_2$ tetrad **23** and Pc–$(C_{60})_2$ triad **24**.

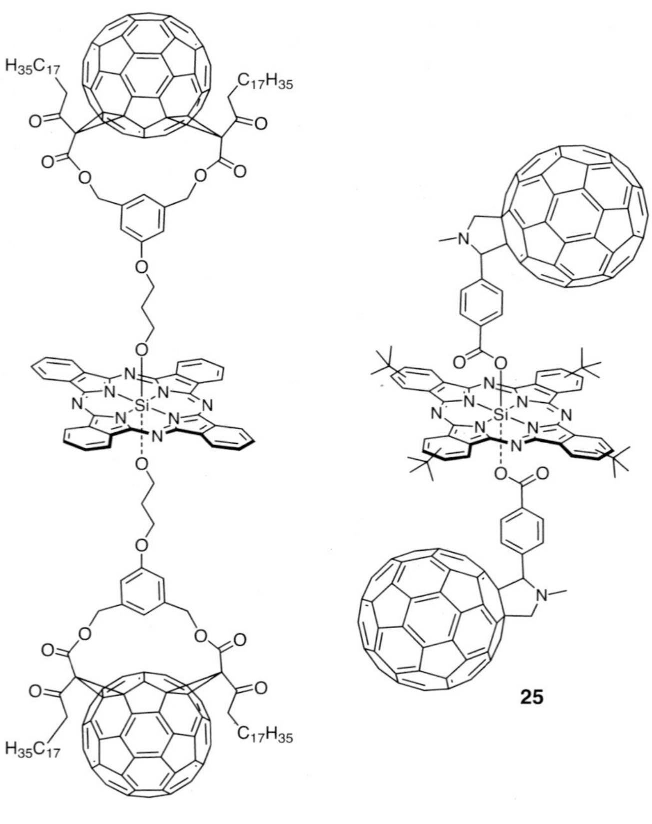

24

25

FIGURE 8.8. Molecular structures of axially substituted Si(IV)Pc–$(C_{60})_2$ triads **24** and **25**.

the first oxidation of the Pc moiety in **24**, as well as the first reduction potential of C_{60}, did not change much with respect to some reference compounds. Similarly to triad **24**, also in the case of Si(IV)Pc–$(C_{60})_2$ **25**, no ground-state interactions were detected by UV–vis and electrochemical studies. However, in the case of **25**, photoexcitation of the triad results in formation of the CS state by PET from the singlet excited state of the Si(IV)Pc moiety to C_{60}, with a lifetime of the CS state of 5 ns in benzonitrile at 298 K.

A Si(IV)Pc-based pentad **26** comprising two different electron-accepting entities, namely C_{60} and naphthalene diimide (NDI), axially coordinated to a central Si(IV)Pc macrocycle has been constructed (Figure 8.9) [47]. NDI was chosen as an electron-acceptor moiety because of its low reduction potential and the location of its singlet excited state, which is higher in energy than those of Si(IV)Pc and C_{60}, thus providing a large driving force for the CS process to occur. Photoirradiation of **26** at 380 nm leads to the exclusive formation of the NDI singlet-excited-state species (at this wavelength the extinction coefficients of both Pc and C_{60} are quite low), which evolves into a Si(IV)$Pc^{\bullet+}(NDI)_2^{\bullet-}(C_{60})_2$ radical ion pair state that outlasts that seen in a related Si(IV)Pc–NDI_2 system lacking the two C_{60} moieties (i.e., 1000 versus 250 ps). This stabilization of the CS state could be explained considering that electron shifts from the $NDI^{\bullet-}$ species to the adjoining C_{60} moiety to generate Si(IV)$Pc^{\bullet+}(NDI)_2(C_{60})_2^{\bullet-}$ can occur, as demonstrated by the fullerene radical anion fingerprint observed at around 1000 nm. These results demonstrate that, from a photophysical standpoint, the presence of the two adjacent, and suitably positioned, electron-accepting moieties can be beneficial, in some cases, for the stabilization of a photogenerated CS state.

The strategy of preparing systems containing several Pc and/or C_{60} units has reached a high level of complexity with the preparation of dendritic and polymeric species.

The axial coordination of fullerene-rich dendrons terminated with an hydroxyl functionality to a Si(IV)Pc has been used in order to obtain a series of Pc–C_{60} dendritic ensembles such as **27** (Figure 8.10) [48]. In such systems, it was observed that the increase in the dendron generation from 2 up to 8 (**27**) was accompanied by a stabilization of the photoinduced CS state, probably due to a more efficient migration of the photogenerated electron charges among the C_{60} subunits.

On the other hand, random Pc–C_{60} copolymer systems (**28**) containing several electroactive Pc and C_{60} pendant

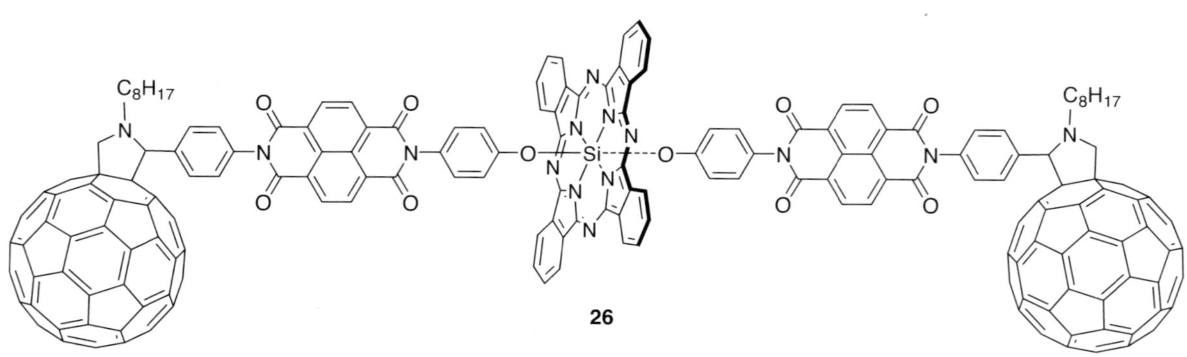

26

FIGURE 8.9. Molecular structure of Si(IV)Pc **26** bearing 4,5,8-naphthalenenediimide (NDI)–C_{60} moieties.

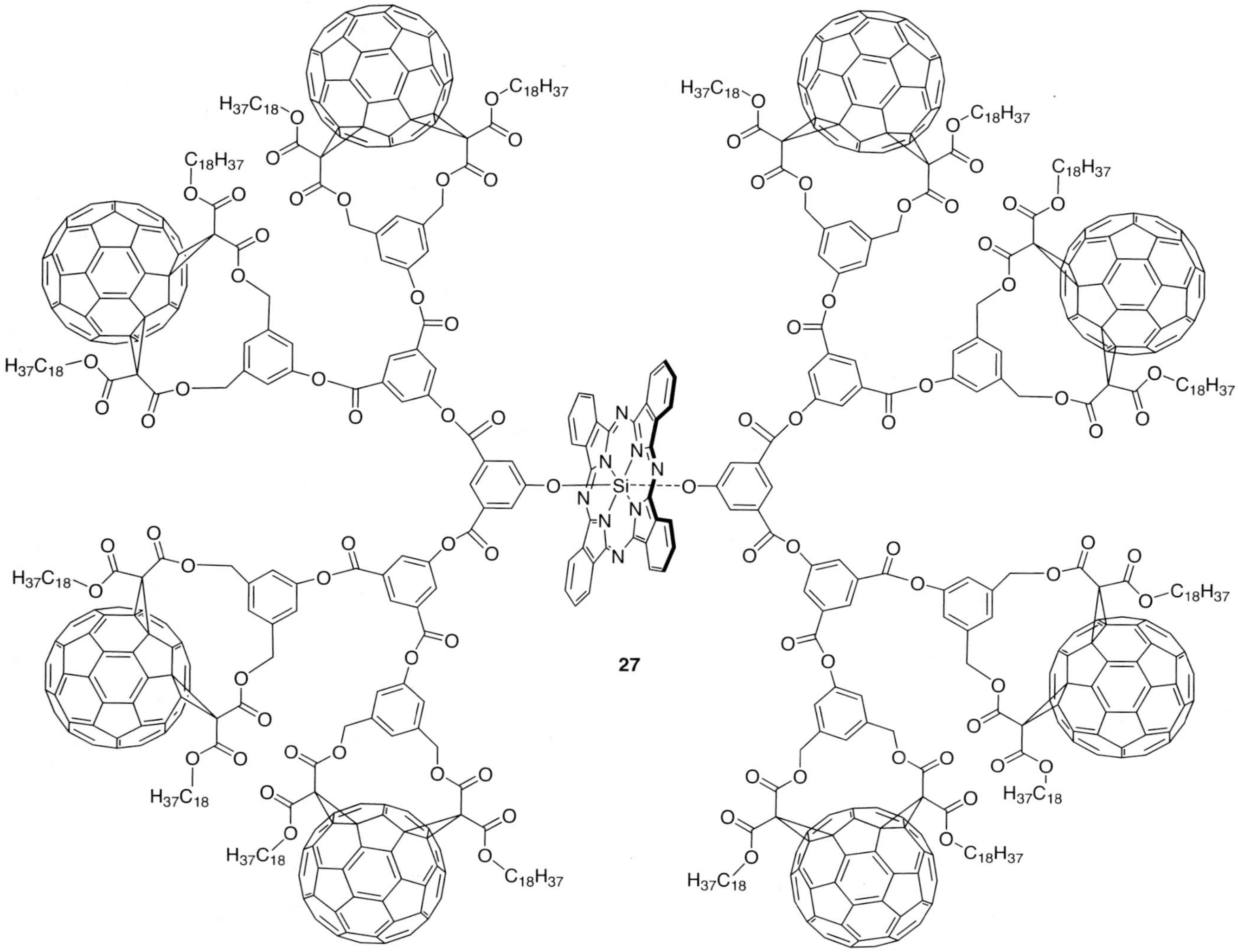

27

FIGURE 8.10. Molecular structure of Pc **27** comprising a Si(IV)Pc axially substituted with two fullerodendrimer units.

units have also been prepared by ring-opening metathesis polymerization of a Pc and C_{60} derivatives, both containing a polymerizable norbornene moiety, in the presence of a Grubbs' catalyst (Scheme 8.4) [49]. Fluorescence studies on these polymeric systems showed a quenching of the Pc fluorescence which was strongly dependent on the relative Pc–C_{60} content in the polymer, being larger for those polymers having a higher fullerene-to-Pc ratio (i.e., **28a**). Interestingly, for these systems, photoinduced CS lifetimes of the order of microseconds were obtained, suggesting a stabilizing effect of the photogenerated charges when going from discrete Pc–fullerene molecular ensembles to polymeric Pc–fullerene frameworks. Photovoltaic measurements on such Pc–C_{60} polymers showed that the covalent backbone of the polymers maintains the Pc and fullerene moieties in close proximity, which facilitates the occurrence of an efficient CS, whereas it hampers the transport of charges to the electrodes, which turned out to be very poor, with the overall power conversion efficiency (PCE) of the resulting devices being low.

8.3 PHTHALOCYANINE–C_{60} COVALENT SYSTEMS PRESENTING LONG-RANGE ORDER

Although several reports on the preparation of covalent Pc–C_{60} systems have appeared during the last decade, scarcer are the examples in which these covalent architectures have been organized over large length scales using supramolecular interactions. In this context, the organization of Pc–C_{60} ensembles, and more generally of any D–A system, is highly desirable since it could lead to the emergence of interesting chemical and/or physical properties not showed

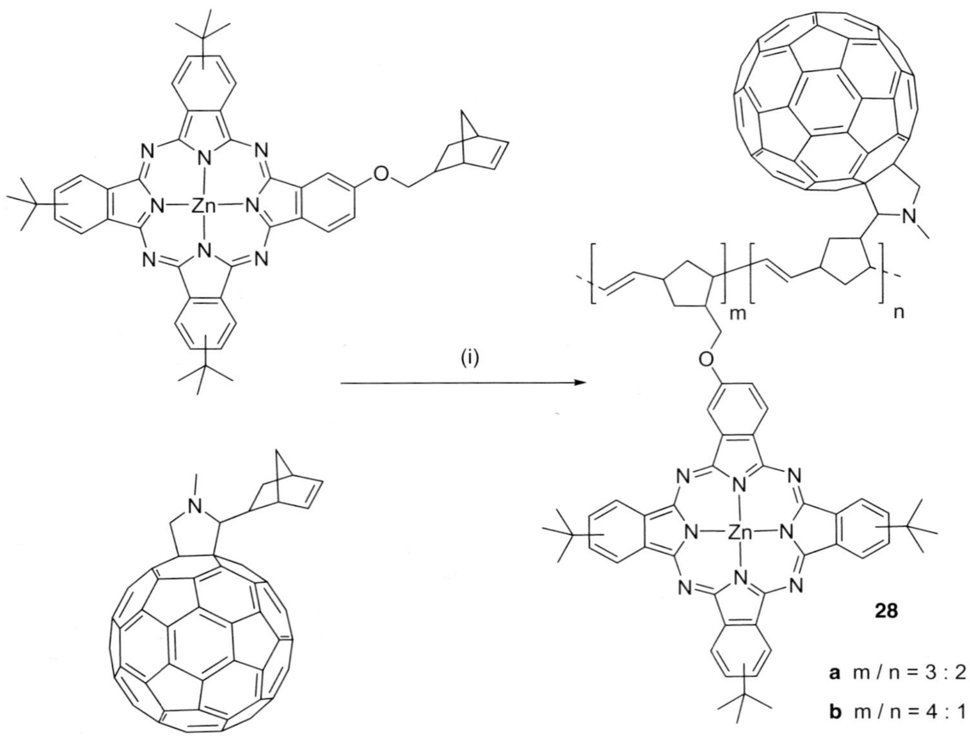

SCHEME 8.4. Synthesis of Pc–C_{60} polymers **28a,b**. (i) Grubbs catalyst [1,3-dimesityl-4,5-dihydroimidazol-2-ylidene) (tricyclohexylphosphine)$Cl_2RuCH(C_6H_5)$] (2%), toluene, rt.

by their molecularly dispersed counterparts, an aspect particularly appealing especially for applications of these systems in molecular photovoltaics and field effect transistor, where the order of both the acceptor and the donor components in the solid state is a key issue to consider in order to obtain high carrier mobilities.

Within the "toolbox" of supramolecular interactions that have been used to self-assemble covalently linked Pc–C_{60} systems, the combination of π–π stacking and liquid-crystalline interactions is particularly attractive because, in appropriately substituted Pc macrocycles, it allows generation of highly ordered, columnar ensembles [50]. In this context, few reports have appeared on the preparation of fully mesogenic Pc–C_{60} dyads [51–53].

In the former report, a series of mesogenic Pc–C_{60} dyads (**28c–f**) was prepared by an esterification reaction between unsymmetrically substituted Pcs bearing a terminal alcohol group and a fullerene derivative bearing a terminal acid moiety (Figure 8.11) [51]. The thermotropic properties of these dyads were studied by polarized optical microscopy (POM) and differential scanning calorimetry (DSC) revealing the formation of liquid-crystalline mesophases only in the case of Pc–C_{60} ensembles **28e,f**. This result suggests that, for such systems, a long spacer is needed in order to allow the bulky C_{60} moiety to be accommodated in the columnar mesophase formed by the stacked Pc macrocycles.

Similarly, a mesogenic Pc–C_{60} dyad (**29**) has been prepared consisting in a hexadodecyl-substituted Zn(II)Pc covalently connected through a flexible spacer to a C_{60} fullerene via a Bingel–Hirsch cyclopropanation reaction (Figure 8.11) [52]. Also in this case, POM and DSC studies on this dyad revealed liquid-crystalline behavior of this ensemble between 80°C and 180°C. Complementary XRD studies showed that Pc–C_{60} dyad **29** adopts a rectangular symmetry within the columnar mesophase (i.e., Col_r), each rectangular unit having a column at its center and four others at its corners.

More recently, a liquid-crystalline, Zn(II)Pc–C_{60} dyad (**30**) containing six 4-dodecyloxyphenoxy groups around the Pc macrocycle and a short, semiflexible bridge between the Pc and the fullerene units has been reported which is able to self-assemble forming segregated, D–A columns (Figure 8.11) [53]. Heated films of Zn(II)Pc–C_{60} dyad **30** showed remarkably high electron (μ_e) (i.e., 0.11 $cm^2\ V^{-1}\ s^{-1}$) and hole (μ_h) (i.e., 0.26 $cm^2\ V^{-1}\ s^{-1}$) mobilities, which are among the highest values reported for organic materials with D–A heterojunction. Moreover, it was observed that the μ_e and μ_h values increased 7 and 26 times after a heating-cooling treatment with respect to pristine films of dyad **30**, suggesting that the heating of the sample allows the Zn(II)Pc–C_{60} molecules to adopt a more regular inter- and intracolumn arrangement within the liquid-crystalline mesophase, leading

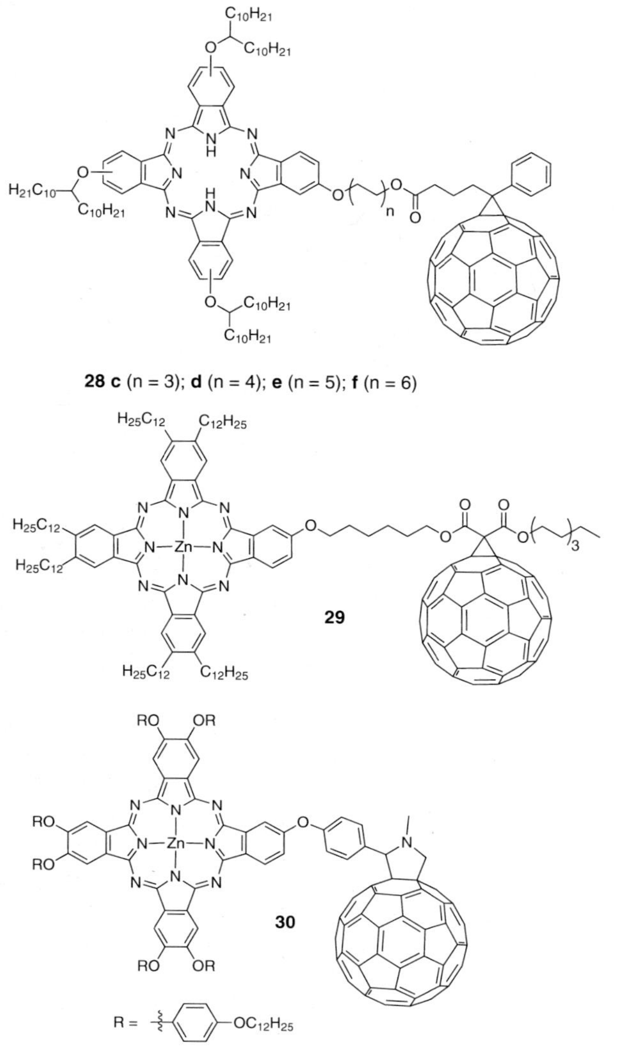

FIGURE 8.11. Molecular structures of mesogenic Pc–C_{60} dyads **28c-f–30**.

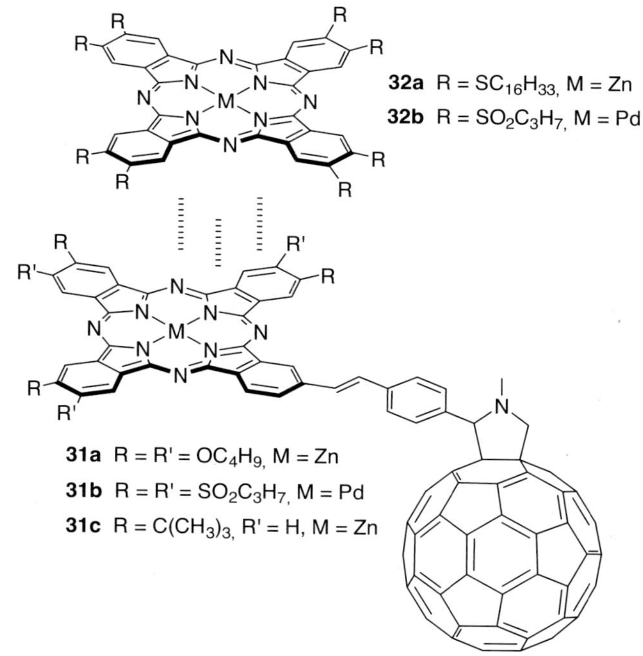

FIGURE 8.12. Molecular structures of Pc–C_{60} dyads **31** and symmetric Pc **32**.

to an improvement of the ambipolar charge-transport properties of the films.

In 2008, a report on an indirect and easy way to incorporate a series of photoactive Pc–C_{60} dyads into a liquid-crystalline architecture was reported consisting in blending a nonmesogenic Pc–C_{60} dyad (**31a–c**) with a mesogenic, symmetrically-substituted Zn(II)Pc (**32a**) (Figure 8.12) [54]. The use of blends, in which a mesogen is able to induce mesomorphism on a nonmesogenic Pc-based functional material, represents an interesting strategy for the incorporation of photoactive D–A Pc–C_{60} systems in liquid crystalline architectures. Such approach, in fact, would help to mitigate the synthetic problems related to the preparation and isolation of mesogenic, unsymmetrically substituted Pc compounds.

The formation of long-range, ordered Pc–C_{60} nanoaggregates obtained combining π–π stacking and hydrophilic–hydrophobic noncovalent interactions has also been reported using an amphiphilic Pc–C_{60} dyad (**33**, Figure 8.13a) [55]. UV–vis and light-scattering studies demonstrated that such an amphiphilic system aggregates when dispersed in water. Insights into the morphology of these aggregates formed by dyad salt **33** in water were gathered by transmission electron microscopy (TEM) studies which revealed the formation of uniform, micrometer-long nanorods (Figure 8.13b). Steady-state and transient absorption studies demonstrated that the self-organization ability of the amphiphilic ensemble **33** in water has also a profound influence on the photophysical properties of these 1-D nano-objects. For such a system, an impressive stabilization of more than six orders of magnitude was observed for the CS lifetime of self-assembled dyad **33** (i.e., 1.4 ms) with respect to a structurally related Pc–C_{60} dyad analogous to **33**, which lacks the hydrophilic ammonium unit and thus, which is not able to form nanotubules (i.e., ~3 ns).

A similar stabilization of the photoinduced CS state was observed for a discrete D–A supramolecular triad constituted by an electron-deficient, symmetrically substituted Pd(II)Pc (**32b**) and an electron-rich Zn(II)Pc–C_{60} dyad (**31a**) bearing a *para*-phenylenevinylene spacer, which is one of the most proficient building blocks for the design of molecular wires with β values as low as 0.01 $Å^{-1}$ [56], separating the Pc and the C_{60} units (Figure 8.12) [57].

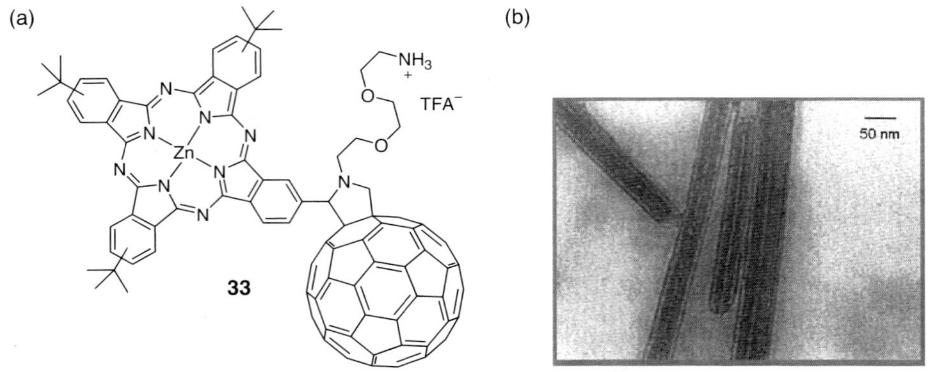

FIGURE 8.13. **(a)** Molecular structure of Pc–C_{60} dyad **33**. **(b)** TEM image of the nanotubules formed by dyad **33** in water. The image in **(b)** is reprinted with permission from reference 55. Copyright 2005, American Chemical Society.

The formation in solution of the D–A, supramolecular complex **31a/32b** was inferred by the changes in the absorption and fluorescence spectra of Pc–C_{60} dyad **31a** during the titration with symmetric Pc **32b**, whereas the Job plot method was used to ascertain the 1:1 stoichiometry of the **31a/32b** supramolecular triad. A numerical analysis of the titration data revealed an association constant (K_a) for the supramolecular complex **31a/32b** of $\sim 5 \times 10^5$ M^{-1} in $CHCl_3$ as a result of strong D–A π–π stacking interactions between the electron-deficient, alkylsulfonyl-substituted Pc **32b** and the electron-rich, alkoxy-substituted Pc **31a**. Transient absorption studies on supramolecular complex **31a/32b** gave a radical ion pair lifetime of 475 ns in THF, a value considerably higher than that observed for Zn(II)Pc–C_{60} **31a** alone (i.e., 130 ns), thus demonstrating, as in the previous case, that self-organization can provide an efficient, and simple, strategy to stabilize the photogenerated radical ion pair state in D–A, Pc–C_{60} ensembles.

Supramolecular interactions have also been used to promote the organization of a structurally rigid, covalently linked Pc–C_{60} conjugate (**34**) on solid surfaces (Figure 8.14a) [58]. Dyad **34** is able, when dropcasted from a toluene solution, to self-organize on highly ordered pyrolytic graphite (HOPG) and graphite oxide, giving rise to the formation of micrometer-long fibers and films as revealed by AFM experiments (Figure 8.14b). Conductive AFM (C-AFM) measurements were carried out on such supramolecular fibers and films in order to address the electrical properties of these nanostructures obtaining electrical conductivity values as high as 30 μA for bias voltages ranging from 0.30 to 0.55 V. Control experiments revealed that the high electrical conductivity values observed for this solid-supported, self-assembled Pc–C_{60} conjugate were strongly related to the supramolecular order of the dyad within the nanostructures.

It has been shown that the same Pc–C_{60} dyad **34** was also able to self-organize by means of noncovalent interactions on the outer wall of SWCNTs grown by chemical vapor deposition on a silica surface [59]. A possible explanation for the formation of such ensemble could be the strong affinity of dyad **34** for graphitic surfaces (*vide supra*) coupled to the poor affinity of this conjugate for hydrophilic surfaces such as silica which lead to the long-range organization of

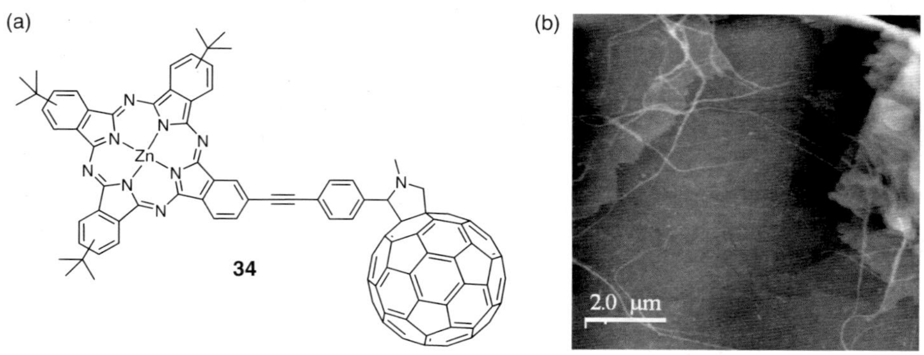

FIGURE 8.14. **(a)** Molecular structure of Pc–C_{60} conjugate **34**. **(b)** AFM topographic image of dyad **34** drop casted on HOPG. The image in **(b)** is reprinted with permission from reference 58. Copyright 2008, Wiley-VCH. See color insert.

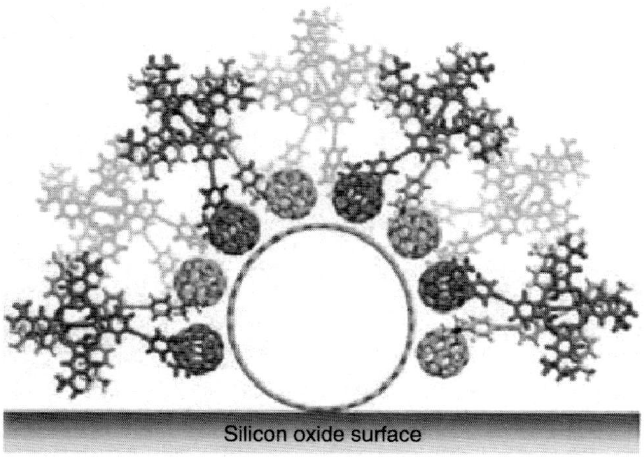

FIGURE 8.15. Frontal view of the proposed supramolecular organization of Pc–C_{60} dyad **34** on a 2.5-nm SWCNT on a silicon oxide surface. The image is reprinted with permission from reference 59. Copyright 2010, Royal Society of Chemistry. See color insert.

34 on the curved, 1-D, graphite-like surface of the SWCNTs (Figure 8.15).

8.4 COVALENTLY LINKED PHTHALOCYANINE–CARBON NANOTUBE ENSEMBLES

Similarly to fullerenes, carbon nanotubes (CNTs)[60–62] have also gained a prominent role as key materials in the field of nanotechnology [63], due to extraordinary physical properties such as electrical, optical limiting (OL), linear and nonlinear optical (NLO) properties, as well as thermal and mechanical properties, which have prompted their application in various technological areas, including electronics, sensing [64, 65], energy conversion [17, 66–70], fuel cells, gas storage, or biological functions (drug carrier, photodynamic therapy) [71, 72].

Even though multiwalled carbon nanotubes (MWCNTs) were the first type of tubular carbon nanotubes discovered [73], the leading carbon nanotubular structures in terms of scientific relevance are SWCNTs, which can be formally considered as a graphene sheet that has been rolled up on itself to form a seamless nanocylinder. SWCNTs are formed as a mixture of tubes with different diameters and helicities, and can possess either a metallic or a semiconducting electronic character. The sp^2 character of the SWCNTs' carbon atoms, together with the tubular structures of these carbon nanoform, gives rise to continuous electronic states in the SWCNTs' conduction band band, making these tubes suitable, for instance, for collecting electrons from the excited state of an organic conjugated molecule. In turn, these electrons might be transported under nearly ballistic conditions along the tubular 1-D, SWCNTs' axis. In this context, the functionalization of CNTs with molecules having photoactive properties offers the possibility for the development of novel hybrid materials with improved optoelectronic properties and great promises in converting solar energy into electricity [18].

To date, two general approaches for the functionalization of SWCNTs have been reported: (a) the covalent attachment of molecules to the open edges or sidewalls of SWCNTs and (b) the noncovalent interactions of aromatic molecules or macromolecules to the outer nanotube walls. In principle, the noncovalent functionalization is particularly attractive because, following this approach, the electronic structure of the nanotubes remains essentially unaffected. However, the stability of the ensembles resulting from the covalent addition of functional molecules to the nanotube is much higher, which is desirable in terms of preparation and reproducibility of possible devices. Moreover, the covalent approach allows some control of the degree of functionalization, difficult to achieve in the case of supramolecular ensembles, thus allowing to modulate the properties of the resulting hybrid material. However, some aspects of the covalent functionalization of SWCNTs are not satisfactorily controlled yet. First, the position of the attached molecules along the tube is difficult to determine. Indeed, the reactivity at the tips of the tube is higher, as a consequence of the larger pyramidalization of the carbon atoms, but some reactivity is also expected along the sidewalls as a consequence of the π-orbital misalignment of the bonds at a certain angle to the tube axis. Second, SWCNTs are formed as a mixture of tubes with different diameters, helicities and electronic character (metallic or semiconducting), being, in general, the metallic SWCNTs more reactive in addition reactions than semiconducting ones, and in both cases the reactivity is inversely proportional to the diameter of the tubes. This selectivity has been utilized for the separation of carbon nanotubes, with the aim of obtaining SWCNTs with a single electronic character, either metallic or semiconducting [74].

Up to date, several systems based on the covalent and noncovalent attachment of electron donor molecules such as Fc [75, 76] tetrathiafulvalene [77, 78], porphyrins (Pors)[79–89] or Pcs to CNTs have been prepared and studied, showing the possibility to generate long-lived, photogenerated CS states, which, in some cases, have been used to generate photocurrent in photoelectrochemical cells for solar energy conversion [90–93]. In this context, the excellent optical and electronic properties of Pcs make these macrocycles ideal molecular partners for CNTs toward the preparation of novel functional material.

Nowadays, the preferred method of covalent side- and end-walled functionalization of SWCNTs are conventional, solvent-based, chemical techniques, even though some recent works have reported the use of new, nonconventional techniques in solvent-free, mild conditions for both purification [94, 95] and functionalization of CNTs, such as ball milling [96] or microwave-assisted methods [97, 98]. These new methodologies present many advantages compared to the conventional methods but, despite their potential interest, have been scarcely used so far.

The most conventional procedure for the covalent CNTs' functionalization involves the treatment of the CNT material with a large excess of a strong oxidizing acid, usually mixtures of concentrated sulfuric and nitric acid, at reflux over a long period of time. This oxidizing treatment opens the SWCNTs, giving rise to short, uncapped CNTs (tubes with an average length of 100–300 nm) bearing oxygen-containing groups, such as carboxylates, at the end of the tubes and at defective sites of the CNTs' sidewalls. Therefore, the carboxyl groups can be readily derivatized to the corresponding, highly reactive acid chlorides by treatment with $SOCl_2$ and subsequently coupled to amines or alcohols.

Using this oxidative route, covalent Pc–SWCNT ensembles **35** were prepared through an amidation reaction between acyl chloride-functionalized SWCNTs and unsymmetrically substituted Zn(II)Pcs peripherally substituted with a free amino group and various solubilizing groups (*tert*-butyl (**35a**) or dioctylaminocarbonylmethoxy (**35b**)) which helped in giving good solubility to the final Pc–SWCNT hybrid ensembles (Figure 8.16) [99].

The obtained Pc–SWCNT hybrids **35** could be characterized by UV–vis and IR spectroscopies, as well as TEM, the latter technique showing that the nanotubes are no longer aggregated into large bundles. Although the Pcs units bear several solubilizing groups, the solubility of the resulting hybrid materials proved to be too low in most of the organic solvents thus preventing their photophysical characterization. The same amidation coupling was used to prepare a Pc–SWNT system by reacting an acyl chloride-functionalized SWCNTs with an octaamino-substituted, double-decker erbium (III) bisphtalocyaninato complex (Eu(III)Pc_2) **36** (Figure 8.17) [100].

UV–vis absorption spectrum of D–A ensemble **36** suggests strong intramolecular interaction between the Pc units and SWCNTs, also confirmed by the Raman spectrum, which should facilitate the photoinduced CT.

The strategy used to obtain covalent Pc–CNT assemblies involving an esterification or amidation reaction between the appropriately functionalized CNT and Pc entities (i.e., acid chloride-functionalized MWCNTs and alcohol- or amine-derivatized Pcs) has been slightly modified in order to obtain Pc–MWCNT hybrid ensemble **37a** (Figure 8.18) [101].

The preparation of **37a** involved the synthesis of phthalonitrile-bonded MWCNTs in a two-step reaction starting from carboxylic acid-functionalized MWCNT (which were prepared using the same oxidative treatment used for SWCNTs) which were esterified using hexanediol followed by an *ipso*-substitution of the resulting hydroxyl-terminated MWCNTs with 4-nitrophthalonitrile. Finally, the Pcs were formed by the reaction of the phthalonitrile-functionalized MWCNTs with 4-lauryloxy phthalonitrile in the presence of a copper(II) salt, to obtain the final Cu(II)Pc–MWCNT hybrid material **37a**, which showed good solubility in a variety of organic solvents. TGA led to an estimation of a 10%

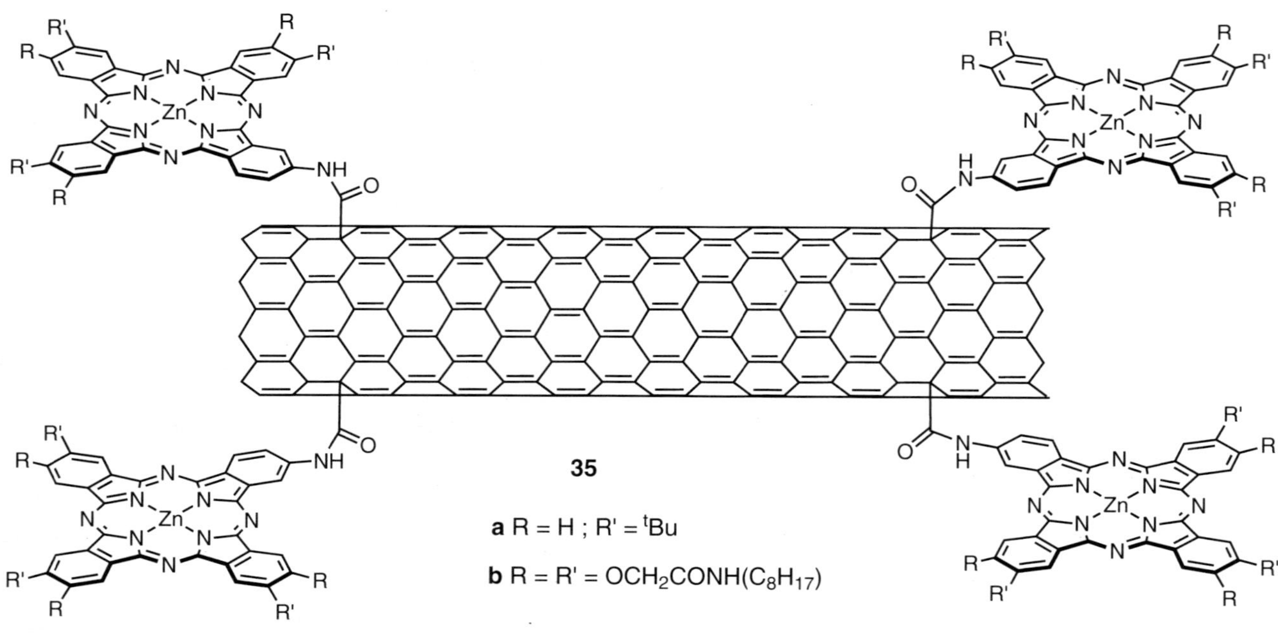

FIGURE 8.16. Zn(II)Pc–SWCNT hybrids **35** obtained by an amidation reaction between an acyl chloride-functionalized SWCNT and unsymmetrically substituted Zn(II)Pc-NH_2.

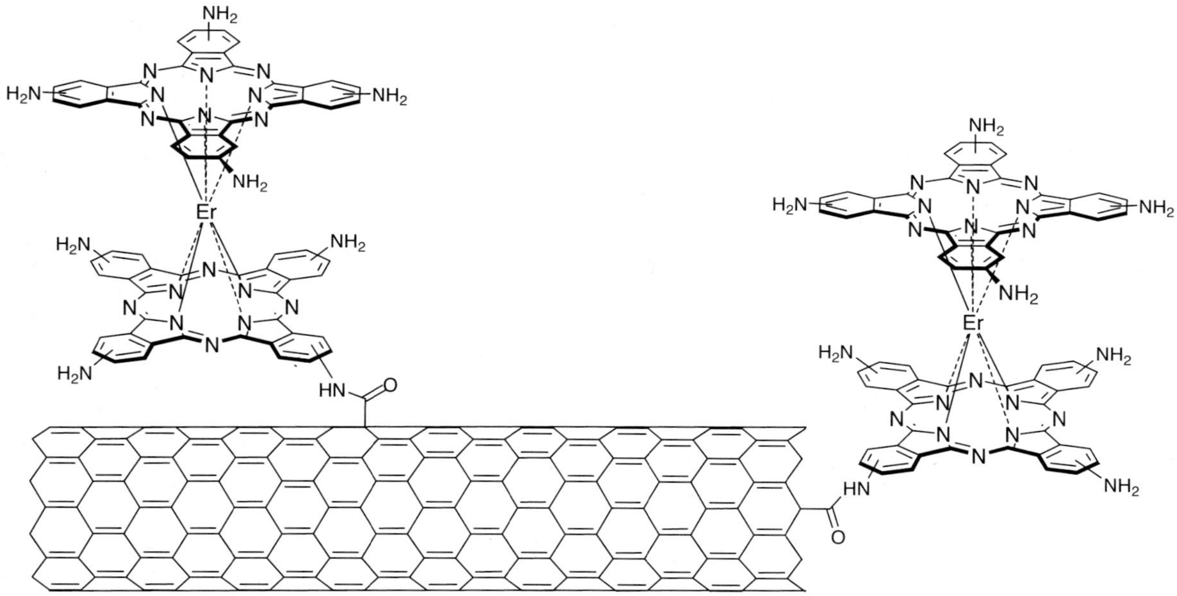

36

FIGURE 8.17. SWCNT–Eu(III)Pc_2 hybrid **36** obtained by amidation reaction between an acyl chloride-functionalized SWCNT and an octaamino-substituted, double-decker erbium (III) bisphtalocyaninato complex Eu(III)Pc_2.

weight content of Cu(II)Pc within the hybrid material. The Cu(II)Pc–MWCNT hybrid material showed better photoconductivity than pristine Cu(II)Pc and the Cu(II)Pc–MWCNT blended composite.

A covalent hybrid material **37b** based on MWCNTs and Pcs have also been prepared by direct amidation reaction between carboxylic acid-derivatized MWCNTs and a tetraamino-substituted Mn(II)Pc (Figure 8.19) [102]. Photophysical studies carried out on **37b** showed the occurrence of an intramolecular photoinduced CT process from the light-harvesting Pcs to the MWCNTs.

More recently, the same synthetic strategy was used to prepare H_2Pc–MWCNT hybrid **37c**, which showed enhanced optical limiting properties (Figure 8.19) [103]. Photophysical studies on **37c** revealed a substantial decrease of the Pc fluorescence intensity compared to the amino-functionalized H_2Pc precursor, thus suggesting a quenching of the Pc singlet excited state by the covalently linked MWCNT.

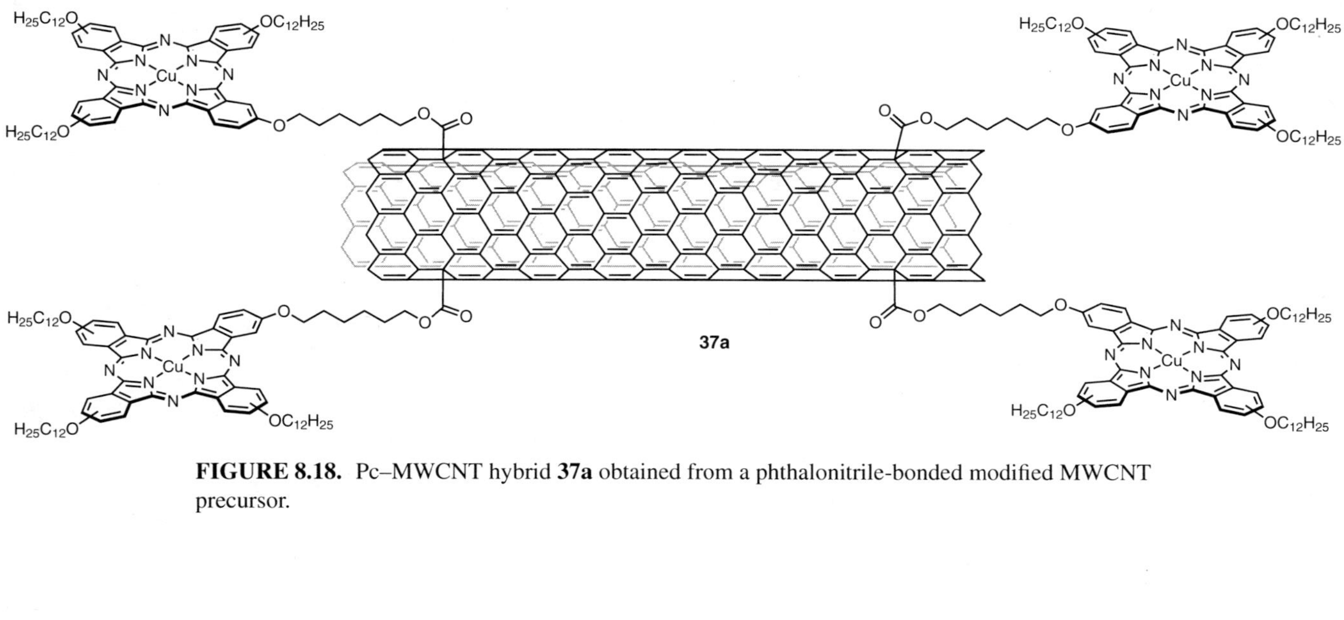

37a

FIGURE 8.18. Pc–MWCNT hybrid **37a** obtained from a phthalonitrile-bonded modified MWCNT precursor.

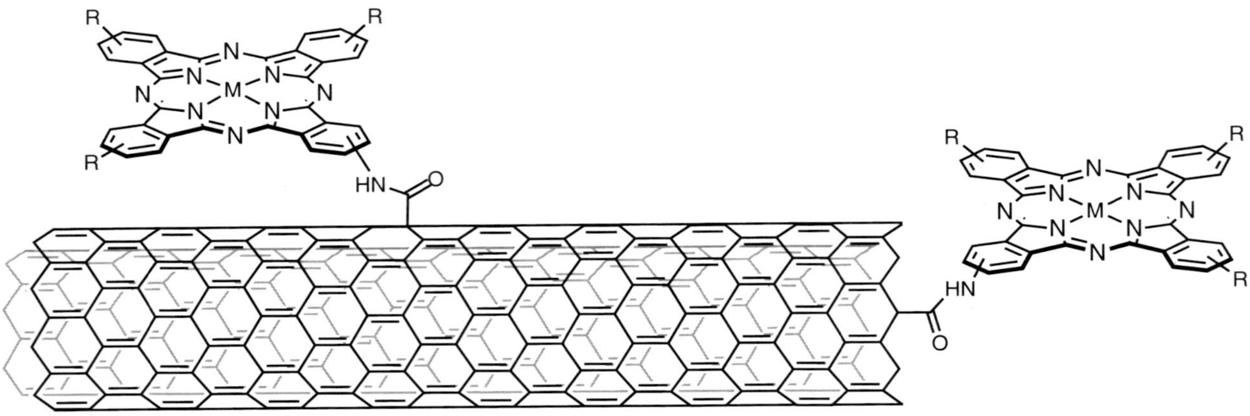

37

b : R = NH_2 ; M = Mn

c : R = ^{t}Bu ; M = H_2

FIGURE 8.19. Pc–MWCNT hybrids **37b,c** obtained by amidation reaction between carboxylic acid-functionalized MWCNTs and amino-functionalized Pcs.

The use of acid chloride-functionalized CNTs as a starting reactive species remains the most used and widespread method for the preparation of covalent, CNT-based hybrid materials. However, an important drawback of this methodology is that only amidation or esterification reactions can be used, thus narrowing the possibility to use a wide range of functional groups. Moreover, this synthetic strategy gives rise to functionalization of CNTs mostly at their endwall, thus leading to a non-uniform distribution of the functional groups along the nanotube.

In this context, other synthetic methods have been pursued for the preparation of covalent Pc–CNT ensembles. Since the sp^2 carbon atoms constituting the nanotubes' sidewall (which represent the large majority of the CNT's carbon atoms) are chemically less reactive than those at the endwall, a more uniform functionalization of the nanotube would require, in principle, the use of more reactive species such as aryl radicals, aryl and diazonium cations, nitrenes, carbenes, or 1,3-dipoles.

Recently, amino-functionalized SWCNTs (SWCNT-NH_2) have been prepared in a single-step reaction, involving a diazonium salt intermediate as a reactive species [104]. The amino functionality present on such CNTs allows to carry out a variety of reactions for further functionalization of CNTs with a wide range of building blocks. This approach has been used for the preparation of Pc–SWCNT **38** by amino coupling between a modified SWCNT-NH_2 and a Zn(II)Pc–COOH derivative using *N,N′*-dicyclohexylcarbodiimide as an activating coupling agent (Figure 8.20) [105]. Raman spectroscopy studies on **38** evidenced that the functionalization of the nanotube occurred preferentially at the sidewall, as suggested by the significant intensity of the D bands in the Raman spectrum of the hybrid material. The TGA analyses on both **38** and its amino-derivatized SWCNT precursor revealed that, whereas the SWCNT precursor is highly functionalized (i.e., 1 amine group per 37 carbon atoms), the ratio dropped to 1 amine group per 1430 carbon atoms in the case of the Pc–SWCNT hybrid. Following the same synthetic strategy, covalently linked Co(II)Pc–SWCNT complex **39** was prepared by reacting amino-derivatized SWCNT with a tetracarboxylic acid-functionalized Pc and its potential use in electrocatalysis studied (Figure 8.20) [106].

The Prato–Maggini reaction, certainly one of the most successful functionalization strategies used in fullerene chemistry, has also been successfully employed for the covalent functionalization of CNTs. This approach has proven to be a powerful method to enhance the CNTs processability [107, 108] since it usually leads to CNT-based materials with good solubility in organic media. This strategy has been used in order to prepare ensemble **40** (Figure 8.21).

Pc–SWCNT hybrid **40** was prepared by using two different synthetic strategies. The first one involved the initial functionalization of the nanotube by a 1,3-dipolar cycloaddition reaction with an azomethyne ylide generated in situ by reaction between *N*-octylglycine and 4-carboxybenzaldehyde. Subsequently, the obtained carboxylic acid-functionalized SWCNT was coupled with a hydroxymethyl-substituted Zn(II)Pc leading to the final Pc–SWCNT hybrid material **40**. The other approach consisted in a direct 1,3-dipolar cycloaddition reaction of an

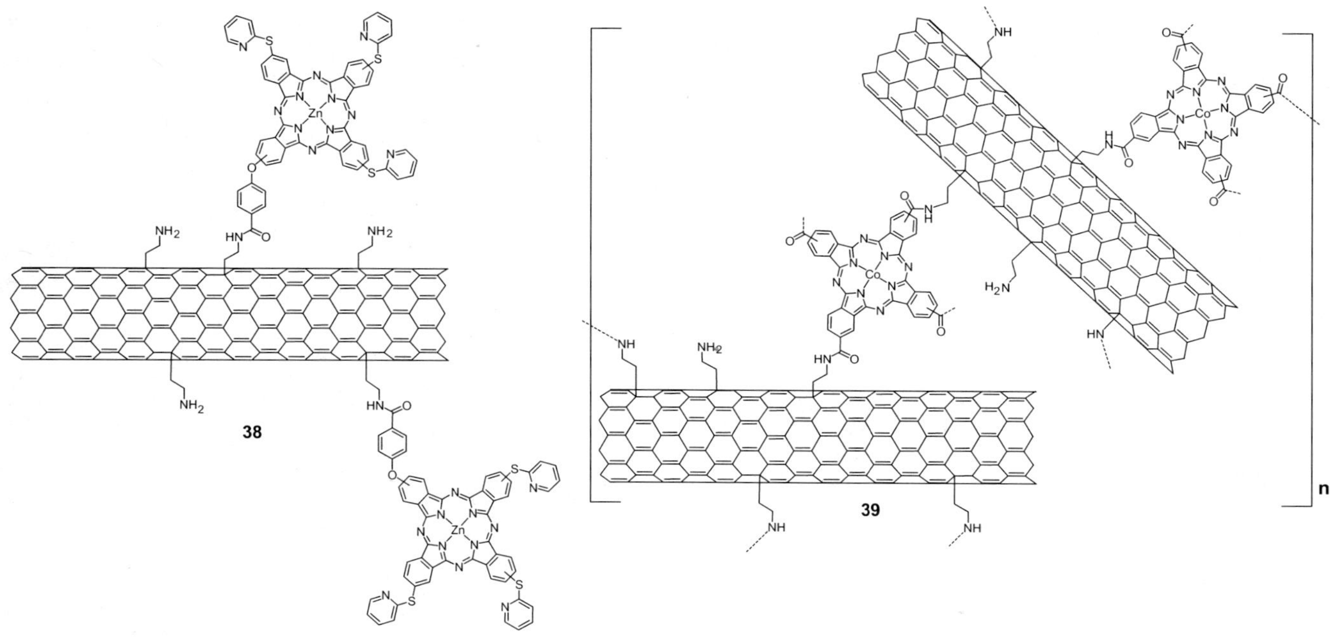

FIGURE 8.20. Pc–SWCNT hybrids **38** and **39** obtained by an amidation reaction between SWCNT-NH_2 and mono (**38**) or tetra (**39**) carboxylic acid Pcs.

azomethyne ylide species generated in situ from the reaction between an aldehyde-functionalized Zn(II)Pc derivative and *N*-octylglycine on the pristine SWCNT material. TGA studies on the Pc–SWCNT hybrid materials obtained by the two different routes showed, in the case of the product resulting from the stepwise strategy, a higher Pc content in the Pc–SWCNT hybrid material (1 Pc unit per 300 carbon atoms) than for the direct synthesis (1 Pc unit per 500 carbon atoms). These results could be rationalized considering that in the former synthetic route a large excess of the cheap and

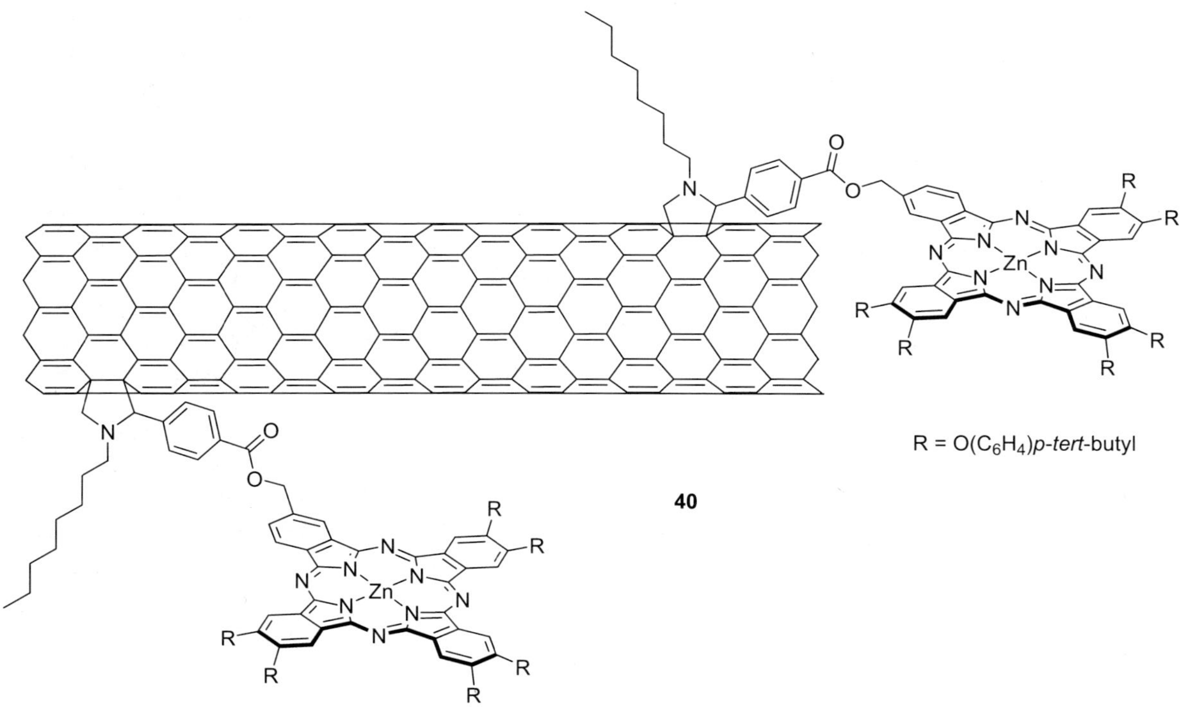

FIGURE 8.21. Pc–SWCNT ensemble **40** obtained by the Prato–Maggini reaction.

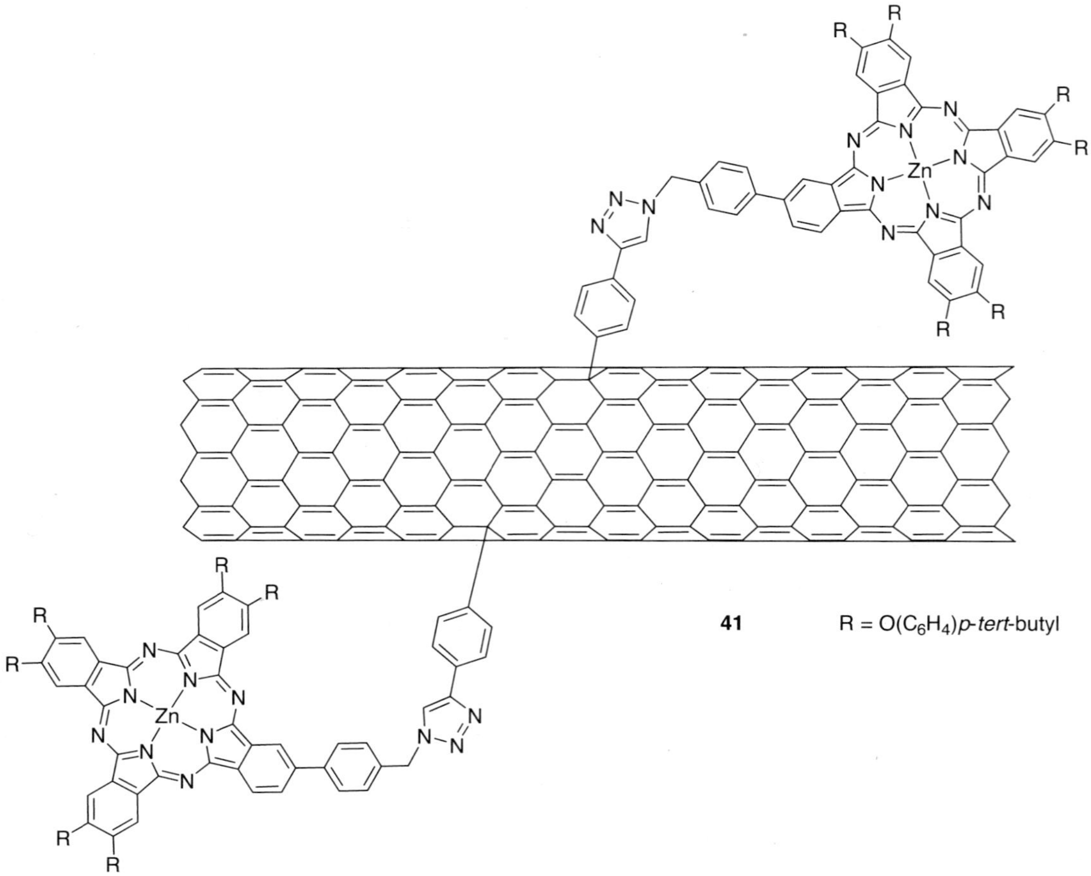

FIGURE 8.22. Pc–SWCNT hybrid **41** obtained by Huisgen 1,3-dipolar cycloaddition.

commercially available *N*-octylglycine and 4-carboxybenzaldehyde reactants was employed in the first step of the reaction, which is also the one that sets the maximum number of potential functional units (i.e., carboxylic acid) introduced on the CNT surface, probably resulting in a higher sidewall functionalization of the nanotube. However, it has to be noted that an important drawback of this stepwise route lies in the difficulty to control the complete post-functionalization of the carboxylic acid groups in the SWCNT-COOH material (in the case of **40** by an esterification reaction with a hydroxyl-substituted Pc). To overcome this problem, "click chemistry" reactions have started to be used for the preparation of CNT-based materials since, besides being clean, versatile, modular, and tolerant to a wide variety of functional groups, they often involve simple work-up and purification procedures leading to high yields. Among the "click chemistry" reactions, the Huisgen cycloaddition, a 1,3-dipolar cycloaddition between azide and acetylene derivatives catalyzed by a Cu(I) salt, certainly represents nowadays one of the most effective and used reactions. Such reaction has been used in order to prepare a highly functionalized Pc–SWCNT material (**41**) (Figure 8.22) [109].

In a first step, SWCNT was converted into a phenylacetylene-functionalized SWCNT treating the pristine SWCNT material with 4-(trimethylsilyl)ethynyl)aniline following a procedure developed by Tour and co-workers for the addition of aromatic radicals to the CNT sidewalls [110]. Such ethynyl-terminated SWCNTs were then reacted with azide-functionalized Zn(II)Pcs leading to the formation of the Pc–SWCNT hybrid **41**. TGA studies on the ethinyl-functionalized SWCNT material evidenced the presence of about 1 phenylacetylene unit group per 165 carbon atoms. This value fully fits with the estimation of the number of Pc units introduced after the click reaction. This result highlights the success of this approach in order to have a high degree of SWCNTs' functionalization. Photophysical studies on Pc–SWCNT hybrid **41** evidenced the occurrence of a photoinduced electron transfer process from the Zn(II)Pc to the carbon nanotube upon photoexcitation of the Pc moieties.

8.5 PHTHALOCYANINE–GRAPHENE ENSEMBLES

In the last few years graphene, a single-atom-thick two-dimensional carbon material and one of the latest entry within the family of carbon nanostructures, has gained increasing interest due to its striking mechanical, optical, and electrical features, making it a potentially interesting partner for Pcs.

So far two different synthetic strategies have been employed for the preparation of Pc–graphene ensembles. One approach is based on the noncovalent functionalization of reduced graphene oxide (rGO) and unsubstituted Pcs [111, 112], or between exfoliated graphene and poly-*para*-phenylenevinylene oligomers containing pendant Pc moieties [113]. In the other approach, graphene oxide (GO) is usually employed as starting material and then covalently functionalized with Pcs [114].

Recently, the first example of a covalent Pc–graphene ensemble (**42**) obtained by covalent linkage of Pcs to non-modified graphene obtained by liquid-phase exfoliation of graphite has been reported (Figure 8.23).

A few-layer graphene dispersion was obtained by gentle sonication of graphite flakes in *N*-methyl pyrrolidone. Covalent functionalization of the resulting graphene-based material with Pcs was achieved in two steps consisting in a 1,3-dipolar cycloaddition reaction among exfoliated graphite, an excess of *N*-methylglycine, and 4-formylbenzoic acid, affording a graphene-based material containing pending phenylcarboxylic acid units, followed by an esterification reaction of such graphene material with an alcohol-terminated, metal-free Pc in the presence of condensation agents, yielding Pc-graphene nanoconjugate **42**. The resulting nanomaterial **42** was characterized by a number of analytical techniques such as TGA, Fourier transform infrared spectroscopy (FT-IR), AFM, TEM, and Raman experiments, as well as steady-state and time-resolved spectroscopic techniques.

TGA revealed that the number of Pcs attached to the basal plane of graphene in conjugate **42** is approximately 1 per 1600 carbon atoms, being the low degree of functionalization beneficial, since it mostly preserves the electronic features of graphene. Additional information about the structural and electronic features of **42** were gathered from Raman experiments on dropcasted DMF dispersions of nanoconjugate **42**. Such studies supported the few-layer graphene nature of the hybrid material as well as the presence of sp^3 carbon atoms at the basal plane of graphene.

Steady-state absorption experiments on **42** further supported the covalent functionalization of the few-layer graphene material, as revealed by the presence of the Pc Soret-band transitions, superimposed to the features corresponding to the graphene absorptions and light scattering.

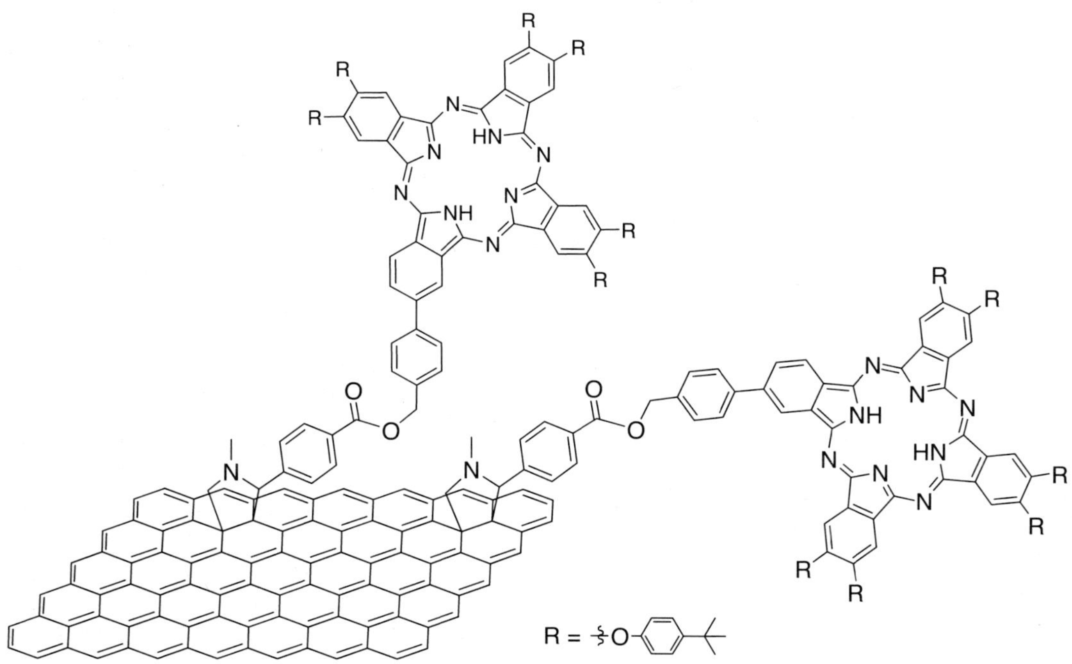

42

FIGURE 8.23. Pc–graphene nanoconjugate **42**.

On the other hand, steady-state fluorescence measurements revealed a substantial quenching of the Pc emission, thus suggesting some degree of electronic communication between the Pc and graphene. Transient absorption measurements were finally used to shed light onto the nature of such communications (electron versus energy transfer) revealing, upon excitation of the ensemble at 387 nm, the rapid formation and deactivation of the Pc singlet excited-state characteristics. Simultaneously with the Pc singlet excited-state decay, the formation of a new transient species evolving at 515 and 850 nm was observed which was related to the one-electron oxidized Pc species accompanied by a broad maximum at 1100 nm due to the new conduction band electrons injected from the photoexcited Pc into graphene. Multiwavelength analysis afforded lifetimes of 3.3 ± 0.5 and 270 ± 10 ps in DMF, attributable to a fast charge separation and a slow charge recombination process, respectively.

8.6 CONCLUSIONS AND OUTLOOK

During the past few decades, significant efforts have been directed toward the preparation and study of synthetic model compounds trying to mimic the multifaceted functions of natural photosynthetic systems, ranging from light harvesting to CT and photoconversion. In this context, D–A materials comprising Pcs as donor moieties, along with carbon nanostructures such as fullerenes, SWCNTs, or graphene as acceptor materials, are among the most interesting and investigated systems.

Pcs are compounds that present, besides a rich redox chemistry, an intense optical absorption in the red/NIR of the solar spectrum, thus representing perfect light-harvesting systems and ideal components for the construction of artificial photosynthetic systems. On the other hand, carbon nanostructures such as C_{60} fullerene, SWCNTs, or graphene possess unique electron-accepting features that render them perfect molecular partners for photo- and electroactive systems such as Pcs.

In this connection, a large number of molecular ensembles based on the combination of Pcs with these carbon nanostructures have been prepared, leading to innovative materials and composites with unique optoelectronic properties. Particularly, photoinduced electron-transfer processes occur in the large majority of these covalent, Pc–carbon nanostructure ensembles leading to the formation of long-lived CS lifetimes as demonstrated by photophysical investigations.

Although important advances have been made toward the preparation and study of covalent, Pc–carbon nanostructure systems, more work is needed in order to finely predict and control the physicochemical properties of the resulting ensembles, both in solution and in condensed phases. In this sense, future challenges for this field include not only a better control of the photophysical features of these D–A systems, but also, and probably more importantly, the implementation of these architectures into efficient, photovoltaic devices. In this context, the search for new D–A systems with appropriate HOMO–LUMO levels for maximizing charges injection into the electrodes will be beneficial. At the same time, the possibility to control/promote the long-range order of these D–A systems over large length scales should also be strongly pursued, since the order of both the acceptor and the donor active components within the photovoltaic devices is a key issue for achieving high charges mobilities, an important point especially for the development of efficient organic solar cells. The challenges and opportunities that rest on these systems are clearly enormous and should attract the efforts of many researchers in this important area.

ACKNOWLEDGMENTS

We would like to thank our colleagues and co-workers whose names appear in the references of this chapter for contributions to the area highlighted here. Financial support from MICINN and MEC, Spain (CTQ2011-24187/BQU, CONSOLIDER INGENIO 2010, CSD2007-00010 on Molecular Nanoscience, and PLE2009-0070), and the Comunidad de Madrid (MADRISOLAR-2, S2009/PPQ/1533) is acknowledged.

REFERENCES

1. Elemans, J. A. A. W., van Hameren, R., Nolte, R. J. M., Rowan, A. E. (2006). Molecular materials by self-assembly of porphyrins, phthalocyanines, and perylenes. *Advanced Materials*, **18**, 1251–1266.
2. Martinez-Diaz, V. M., Bottari, G. (2009). Supramolecular organization of phthalocyanines: From solution to surface. *Journal of Porphyrins and Phthalocyanines*, **13**, 471–480.
3. Rio, Y., Rodriguez-Morgade, M. S., Torres, T. (2008). Modulating the electronic properties of porphyrinoids: A voyage from the violet to the infrared regions of the electromagnetic spectrum. *Organic and Biomolecular Chemistry*, **6**, 1877–1894.
4. Martinez-Diaz, M. V., de la Torre, G., Torres, T. (2010). Lighting porphyrins and phthalocyanines for molecular photovoltaics. *Chemical Communications*, **46**, 7090–7108.
5. Gonzalez-Rodriguez, D., Bottari, G. (2009). Phthalocyanines, subphthalocyanines and porphyrins for energy and electron transfer applications. *Journal of Porphyrins and Phthalocyanines*, **13**, 624–636.
6. D'Souza, F., Ito, O. (2009). Supramolecular donor-acceptor hybrids of porphyrins/phthalocyanines with fullerenes/carbon nanotubes: Electron transfer, sensing, switching, and catalytic applications. *Chemical Communications*, 4913–4928.

7. Bottari, G., de la Torre, G., Guldi, D. M., Torres, T. (2010). Covalent and noncovalent phthalocyanine–carbon nanostructure systems: Synthesis, photoinduced electron transfer, and application to molecular photovoltaics. *Chemical Reviews*, **110**, 6768–6816.
8. Martin, N., Sanchez, L., Illescas, B., Perez, I. (1998). C_{60}-based electroactive organofullerenes. *Chemical Reviews*, **98**, 2527–2547.
9. Guldi, D. M., Prato, M. (2000). Excited-state properties of C_{60} fullerene derivatives. *Accounts of Chemical Research*, **33**, 695–703.
10. Fukuzumi, S., Guldi, D. M. (2001). *Electron Transfer in Chemistry*, Vol. 2, Balzani, V., ed. Wiley-VCH, New York, p 270.
11. Langa, F., Nierengarten, J.-F., eds. (2007). Nanoscience and Nanotechnology Series. *Fullerenes: Principles and Applications*, The Royal Society of Chemistry, Cambridge, U.K.
12. Delgado, J. L., Herranz, M. A., Martin, N. (2008). The nanoforms of carbon. *Journal of Materials Chemistry*, **18**, 1417–1426.
13. Guldi, D. M., Illescas, B. M., Atienza, C. M., Wielopolski, M., Martin, N. (2009). Fullerene for organic electronics. *Chemical Society Reviews*, **38**, 1587–1597.
14. Imahori, H., Umeyama, T., Kurotobi, K., Takano, Y. (2012). Self-assembling porphyrins and phthalocyanines for photoinduced charge separation and charge transport. *Chemical Communications*, **48**, 4032–4045.
15. Reich, S., Thomsen, C., Maultzsh, J., eds. (2004). *Carbon Nanotubes: Basic Concepts and Physical Properties*, VCH, Weinheim, Germany.
16. Tasis, D., Tagmatarchis, N., Bianco, A., Prato, M. (2006). Chemistry of carbon nanotubes. *Chemical Reviews*, **106**, 1105–1136.
17. Sgobba, V., Guldi, D. M. (2008). Carbon nanotubes as integrative materials for organic photovoltaic devices. *Journal of Materials Chemistry*, **18** 153–157.
18. Sgobba, V., Guldi, D. M. (2009). Carbon nanotubes-electronic/electrochemical properties and application for nanoelectronics and photonics. *Chemical Society Reviews*, **38**, 165–184.
19. Akasaka, T., Wudl, F., Nagase, S., eds. (2010). Chemistry of Nanocarbons, Wiley.
20. Geim, A. K., Novoselov, K. S. (2007). The rise of graphene. *Nature Materials*, **6**, 183–191.
21. Luo, B., Liu, S., Zhi, L. (2012). Chemical approaches toward graphene-based nanomaterials and their applications in energy-related areas. *Small*, **8**, 630–646.
22. Linssen, T. G., Duerr, K., Hanack, M., Hirsch, A. (1995). A green fullerene: synthesis and electrochemistry of a Diels–Alder adduct of [60]fullerene with a phthalocyanine. *Journal of the Chemical Society, Chemical Communications*, 103–104.
23. Durr, K., Fiedler, S., Linssen, T., Hirsch, A., Hanack, M. (1997). Synthesis and electrochemistry of Diels-Alder adducts of [60]fullerene with a phthalocyanine and a hemiporphyrazine. *Chemische Berichte*, **130**, 1375–1378.
24. Huang, W., Wang, S., Liang, R., Gong, Q., Qiu, W., Liu, Y., Zhu, D. (2000). Ultrafast third-order non-linear optical response of Diels–Alder adduct of fullerene C_{60} with a metallophthalocyanine. *Chemical Physics Letters*, **324**, 354–358.
25. Zhu, P., Wang, P., Qiu, W., Liu, Y., Ye, C., Fang, G., Song, Y. (2001). Optical limiting properties of phthalocyanine–fullerene derivatives. *Applied Physics Letters*, **78**, 1319–1321.
26. Tian, Z., He, C., Liu, C., Yang, W., Yao, J., Nie, Y., Gong, Q., Liu, Y. (2005). Optical limiting property of nanoparticles from a copper phthalocyanine–fullerene dyad. *Materials Chemistry and Physics*, **94**, 444–448.
27. Guldi, D. M., Gouloumis, A., Vazquez, P., Torres, T. (2002). Charge-transfer states in strongly coupled phthalocyanine fullerene ensembles. *Chemical Communications*, 2056–2057.
28. Guldi, D. M., Zilbermann, I., Gouloumis, A., Vazquez, P., Torres, T. (2004). Metallophthalocyanines: Versatile electron-donating building blocks for fullerene dyads. *Journal of Physical Chemistry B*, **108**, 18485–18494.
29. Loi, M. A., Denk, P., Hoppe, H., Neugebauer, H., Winder, C., Meissner, D., Brabec, C., Sariciftci, N. S., Gouloumis, A., Vazquez, P., Torres, T. (2003). Long-lived photoinduced charge separation for solar cell applications in phthalocyanine-fulleropyrrolidine dyad thin films. *Journal of Materials Chemistry*, **13**, 700–704.
30. Loi, M. A., Denk, P., Hoppe, H., Neugebauer, H., Meissner, D., Winder, C., Brabec, C. J., Sariciftci, N. S., Gouloumis, A., Vazquez, P., Torres, T. (2003). A fulleropyrrolidine–phthalocyanine dyad for photovoltaic applications. *Synthetic Metals*, **137**, 1491–1492.
31. Neugebauer, H., Loi, M. A., Winder, C., Sariciftci, N. S., Cerullo, G., Gouloumis, A., Vazquez, P., Torres, T. (2004). Photophysics and photovoltaic device properties of phthalocyanine–fullerene dyad:conjugated polymer mixtures. *Solar Energy Materials and Solar Cells*, **83**, 201–209.
32. Quintiliani, M., Kahnt, A., Wolfle, T., Hieringer, W., Vazquez, P., Gorling, A., Guldi, D. M., Torres, T. (2008). Synthesis and photoinduced electron-transfer properties of phthalocyanine–[60]fullerene conjugates. *Chemistry-A European Journal*, **14**, 3765–3775.
33. Kahnt, A., Guldi, D. M., de la Escosura, A., Martinez-Diaz, M. V., Torres, T. (2008). [2.2]paracyclophane: A pseudoconjugated spacer for long-lived electron transfer in phthalocyanine–C_{60} dyads. *Journal of Materials Chemistry*, **18**, 77–82.
34. Sukeguchi, D., Yoshiyama, H., Shibata, N., Nakamura, S., Toru, T., Hayashi, Y., Soga, T. (2009). Synthesis and spectroscopic investigation of trifluoroethoxy-coated phthalocyanine linked with fullerene. *Journal of Fluorine Chemistry*, **130**, 361–364.
35. Sastre, A., Gouloumis, A., Vazquez, P., Torres, T., Doan, V., Schwartz, B. J., Wudl, F., Echegoyen, L., Rivera, J. (1999). Phthalocyanine-azacrown-fullerene multicomponent system: Synthesis, photoinduced processes, and electrochemistry. *Organic Letters*, **1**, 1807–1810.
36. Isosomppi, M., Tkachenko, N. V., Efimov, A., Vahasalo, H., Jukola, J., Vainiotalo, P., Lemmetyinen, H. (2006).

Photoinduced electron transfer of double-bridged phthalocyanine–fullerene dyads. *Chemical Physics Letters*, **430**, 36–40.

37. Niemi, M., Tkachenko, N. V., Efimov, A., Lehtivuori, H., Ohkubo, K., Fukuzumi, S., Lemmetyinen, H. (2008). Exciplex mediated photoinduced electron transfer reactions of phthalocyanine–fullerene dyads. *Journal of Physical Chemistry A*, **112**, 6884–6892.
38. Lehtivuori, H., Kumpulainen, T., Efimov, A., Lemmetyinen, H., Kira, A., Imahori, H., Tkachenko, N. V. (2008). Photoinduced electron transfer in Langmuir–Blodgett monolayers of double-linked phthalocyanine–fullerene dyads. *Journal of Physical Chemistry C*, **112**, 9896–9902.
39. Pinzon, J. R., Cardona, C. M., Herranz, M. A., Plonska-Brzezinska, M. E., Palkar, A., Athans, A. J., Martin, N., Rodriguez-Fortea, A., Poblet, J. M., Bottari, G., Torres, T., Gayathri, S. S., Guldi, D. M., Echegoyen, L. (2009). Metal nitride cluster fullerene $M_3N@C_{80}$ ($M = Y, Sc$) based dyads: Synthesis, and electrochemical, theoretical and photophysical studies. *Chemistry—A European Journal*, **15**, 864–877.
40. Stevenson, S., Rice, G., Glass, T., Harlch, K., Cromer, F., Jordan, M. R., Craft, J., Hadju, E., Bible, R., Olmstead, M. M., Maltra, K., Fisher, A. J., Balch, A. L., Dorn, H. C. (1999). Small-bandgap endohedral metallofullerenes in high yield and purity. *Nature*, **401**, 55–57.
41. Ballesteros, B., de la Torre, G., Shearer, A., Hausmann, A., Herranz, M. A., Guldi, D. M., Torres, T. (2010). Lanthanide(III) bis(phthalocyaninato)–[60]fullerene dyads: Synthesis, characterization, and photophysical properties. *Chemistry—A European Journal*, **16**, 114–125.
42. Gouloumis, A., de la Escosura, A., Vazquez, P., Torres, T., Kahnt, A., Guldi, D. M., Neugebauer, H., Winder, C., Drees, M., Sariciftci, N. S. (2006). Photoinduced electron transfer in a new bis(C_{60})-phthalocyanine triad. *Organic Letters*, **8**, 5187–5190.
43. Quintiliani, M., Kahnt, A., Vazquez, P., Guldi, D. M., Torres, T. (2008). Implementing a tripodal relay station in a phthalocyanine–[60] fullerene conjugate. *Journal of Materials Chemistry*, **18**, 1542–1546.
44. Kahnt, A., Quintiliani, M., Vazquez, P., Guldi, D. M., Torres, T. (2008). A bis(C_{60})-bis(phthalocyanine) nanoconjugate: Synthesis and photoinduced charge transfer. *ChemSusChem*, **1**, 97–102.
45. Kim, K. N., Choi, C. S., Kay, K.-Y. (2005). A novel phthalocyanine with two axial fullerene substituents. *Tetrahedron Letters*, **46**, 6791–6795.
46. Martin-Gomis, L., Ohkubo, K., Fernandez-Lazaro, F., Fukuzumi, S., Sastre-Santos, A. (2007). Synthesis and photophysical studies of a new nonaggregated C_{60}-silicon phthalocyanine–C_{60} triad. *Organic Letters*, **9**, 3441–3444.
47. El-Khouly, Mohamed E., Kim, Jung H., Kay, K.-Y., Choi, Chan S., Ito, O., Fukuzumi, S. (2009). Synthesis and photoinduced intramolecular processes of light-harvesting silicon phthalocyanine-naphthalenediimide–fullerene connected systems. *Chemistry—A European Journal*, **15**, 5301–5310.
48. El-Khouly, M. E., Kang, E. S., Kay, K.-Y., Choi, C. S., Aaraki, Y., Ito, O. (2007). Silicon-phthalocyanine-cored fullerene dendrimers: Synthesis and prolonged charge-separated states with dendrimer generations. *Chemistry-A European Journal*, **13**, 2854–2863.
49. de la Escosura, A., Martinez-Diaz, M. V., Torres, T., Grubbs, R. H., Guldi, D. M., Neugebauer, H., Winder, C., Drees, M., Sariciftci, N. S. (2006). New donor–acceptor materials based on random polynorbornenes bearing pendant phthalocyanine and fullerene units. *Chemistry—An Asian Journal*, **1**, 148–154.
50. Simon, J., Bassoul, P. (1993). In: Leznoff, C. C., Lever, A. B. P., eds. *Phthalocyanines*: *Properties and Applications*, Vol. 2. VCH, Weinheim, Germany, Chapter 6.
51. Geerts, Y. H., Debever, O., Amato, C., Sergeyev, S. (2009). Synthesis of mesogenic phthalocyanine–C_{60} donor–acceptor dyads designed for molecular heterojunction photovoltaic devices. *Beilstein Journal of Organic Chemistry*, **5**, No. 49.
52. Ince, M., Martinez-Diaz, M. V., Barbera, J., Torres, T. (2011). Liquid crystalline phthalocyanine–fullerene dyads. *Journal of Materials Chemistry*, **21**, 1531–1536.
53. Hayashi, H., Nihashi, W., Umeyama, T., Matano, Y., Seki, S., Shimizu, Y., Imahori, H. (2011). Segregated donor–acceptor columns in liquid crystals that exhibit highly efficient ambipolar charge transport. *Journal of the American Chemical Society*, **133**, 10736–10739.
54. de la Escosura, A., Martinez-Diaz, M. V., Barbera, J., Torres, T. (2008). Self-organization of phthalocyanine-[60]fullerene dyads in liquid crystals. *Journal of Organic Chemistry*, **73**, 1475–1480.
55. Guldi, D. M., Gouloumis, A., Vazquez, P., Torres, T., Georgakilas, V., Prato, M. (2005). Nanoscale organization of a phthalocyanine–fullerene system: Remarkable stabilization of charges in photoactive 1-D nanotubules. *Journal of the American Chemical Society*, **127**, 5811–5813.
56. Giacalone, F., Segura, J. L., Martin, N., Guldi, D. M. (2004). Exceptionally small attenuation factors in molecular wires. *Journal of the American Chemical Society*, **126**, 5340–5341.
57. de la Escosura, A., Martinez-Diaz, M. V., Guldi, D. M., Torres, T. (2006). Stabilization of charge-separated states in phthalocyanine–fullerene ensembles through supramolecular donor–acceptor interactions. *Journal of the American Chemical Society*, **128**, 4112–4118.
58. Bottari, G., Olea, D., Gomez-Navarro, C., Zamora, F., Gomez-Herrero, J., Torres, T. (2008). Highly conductive supramolecular nanostructures of a covalently linked phthalocyanine–C_{60} fullerene conjugate. *Angewandte Chemie—International Edition*, **47**, 2026–2031.
59. Bottari, G., Olea, D., Lopez, V., Gomez-Navarro, C., Zamora, F., Gomez-Herrero, J., Torres, T. (2010). Ordering phthalocyanine–C_{60} fullerene conjugates on individual carbon nanotubes. *Chemical Communications*, **46**, 4692–4694.
60. Harris, P. (2001). *Carbon Nanotubes and Related Structures: New Materials for the Twenty-First Century*, Cambridge University Press, Cambridge, U.K.
61. Dresselhaus, M. S., Dresselhaus, G., Avouris, P. (2001). *Carbon Nanotubes: Synthesis, Structure, Properties and Applications*. Springer-Verlag: Berlin.
62. Reich, S., Thomsen, C., Maultzsh, J. (2004). *Carbon Nanotubes: Basic Concepts and Physical Properties*, VCH: Weinheim, Germany.

63. Poole, C. P., Owens, F. J. (2003). *Introduction to Nanotechnology*. Wiley-Interscience, Weinheim, Germany.

64. Lee, M., Im, J., Lee, B. Y., Myung, S., Kang, J., Huang, L., Kwon, Y. K., Hong, S. (2006). Linker-free directed assembly of high-performance integrated devices based on nanotubes and nanowires. *Nature Nanotechnology*, **1**, 66–71.

65. Borghetti, J., Derycke, V., Lenfant, S., Chenevier, P., Filoramo, A., Goffman, M., Vuillaume, D., Bourgoin, J. P. (2006). Optoelectronic switch and memory devices based on polymer-functionalized carbon nanotube transistors. *Advanced Materials*, **18**, 2535–2540.

66. Ago, H., Petritsch, K., Shaffer, M. S. P., Windle, A. H., Friend, R. H. (1999). Composites of carbon nanotubes and conjugated polymers for photovoltaic devices. *Advanced Materials*, **11**, 1281–1285.

67. Barazzouk, S., Hotchandani, S., Vinodgopal, K., Kamat, P. V. (2004). Single-wall carbon nanotube films for photocurrent generation. A prompt response to visible-light irradiation. *Journal of Physical Chemistry B*, **108**, 17015–17018.

68. Chaudhary, S., Lu, H., Müller, A. M., Bardeen, C. J., Ozkan, M. (2007). Hierarchical placement and associated optoelectronic impact of carbon nanotubes in polymer–fullerene solar cells. *Nano Letters*, **7**, 1973–1979.

69. Kongkanand, A., Martínez Domínguez, R., Kamat, P. V. (2007). Single wall carbon nanotube scaffolds for photoelectrochemical solar cells. Capture and transport of photogenerated electrons. *Nano Letters*, **7**, 676–680.

70. Sandanayaka, A. S. D., Ito, O., Zhang, M., Ajima, K., Iijima, S., Yudasaka, M., Murakami, T., Tsuchida K. (2009). Photoinduced electron transfer in zinc phthalocyanine loaded on single-walled carbon nanohorns in aqueous solution. *Advanced Materials*, **21**, 4366–4371.

71. Kam, N. W. S., O'Connell, M., Wisdom, J. A., Dai, H. (2005). Carbon nanotubes as multifunctional biological transporters and near-infrared agents for selective cancer cell destruction. *Proceedings of the National Academy of Sciences of the United States of America*, **102**, 11600–11605.

72. Zhang, M., Murakami, T., Ajima, K., Tsuchida, K., Sandanayaka, A. S. D., Ito, O., Iijima, S., Yudasaka, M. (2008). Fabrication of ZnPc/protein nanohorns for double photodynamic and hyperthermic cancer phototherapy. *Science*, **105**, 39, 14773–14778

73. Iijima, S. (1991). Helical microtubules of graphitic carbon. *Nature*, **354**, 56–58.

74. Qiu, H., Maeda, Y., Akasaka, T. (2009). Facile and scalable route for highly efficient enrichment of semiconducting single-walled carbon nanotubes. *Journal of the American Chemical Society*, **131**, 16529–16533.

75. Guldi, D. M., Marcaccio, M., Paolucci, D., Paolucci, F., Tagmatarchis, N., Tasis, D., Vazquez, E., Prato, M. (2003). Single-wall carbon nanotube–ferrocene nanohybrids: Observing intramolecular electron transfer in functionalized SWNTs. *Angewandte Chemie-International Edition*, **42**, 4206–4209.

76. Yang, X., Lu, Y., Ma, Y., Li, Y., Du, F., Chen, Y. (2006). Noncovalent nanohybrid of ferrocene with single-walled carbon nanotubes and its enhanced electrochemical property. *Chemical Physics Letters*, **420**, 416–420.

77. Herranz, M. A., Martin, N., Campidelli, S., Prato, M., Brehm, G., Guldi, D. M. (2006). Control over electron transfer in tetrathiafulvalene-modified single-walled carbon nanotubes. *Angewandte Chemie-International Edition*, **45**, 4478–4482.

78. Herranz, M. A., Ehli, C., Campidelli, S., Gutierrez, M., Hug, G. L., Ohkubo, K., Fukuzumi, S., Prato, M., Martin, N., Guldi, D. M. (2008). Spectroscopic characterization of photolytically generated radical ion pairs in single-wall carbon nanotubes bearing surface-immobilized tetrathiafulvalenes. *Journal of the American Chemical Society*, **130**, 66–73.

79. Cioffi, C., Campidelli, S., Sooambar, C., Marcaccio, M., Marcolongo, G., Meneghetti, M., Paolucci, D., Paolucci, F., Ehli, C., Aminur Rahman, G. M., Sgobba, V., Guldi, D. M., Prato M. (2007). Synthesis, characterization, and photoinduced electron transfer in functionalized single wall carbon nanohorns. *Journal of the American Chemical Society*, **129**, 3938–3945.

80. Murakami, H., Nomura, T., Nakashima, N. (2003). Noncovalent porphyrin-functionalized single-walled carbon nanotubes in solution and the formation of porphyrin–nanotube nanocomposites. *Chemical Physics Letters*, **378**, 481–485.

81. Guldi, D. M., Rahman, G. M. A., Jux, N., Tagmatarchis, N., Prato, M. (2004). Integrating single-wall carbon nanotubes into donor-acceptor nanohybrids. *Angewandte Chemie-International Edition*, **43**, 5526–5530.

82. Baskaran, D., Mays, J. W., Zhang, X. P., Bratcher, M. S. (2005). Carbon nanotubes with covalently linked porphyrin antennae: Photoinduced electron transfer. *Journal of the American Chemical Society*, **127**, 6916–6917.

83. Guldi, D. M., Taieb, H., Rahman, G. M. A., Tagmatarchis, N., Prato, M. (2005). Novel photoactive single-walled carbon nanotube–porphyrin polymer wraps: Efficient and long-lived intracomplex charge separation. *Advanced Materials*, **17**, 871–875.

84. Tanaka, H., Yajima, T., Matsumoto, T., Otsuka, Y., Ogawa, T. (2006). Porphyrin molecular nanodevices wired using single-walled carbon nanotubes. *Advanced Materials*, **18**, 1411–1415.

85. Guldi, D. M., Rahman, G. M. A., Qin, S. H., Tchoul, M., Ford, W. T., Marcaccio, M., Paolucci, D., Paolucci, F., Campidelli, S., Prato, M. (2006). Versatile coordination chemistry towards multifunctional carbon nanotube nanohybrids. *Chemistry—A European Journal*, **12**, 2152–2161.

86. Chitta, R., Sandanayaka, A. S. D., Schumacher, A. L., D'Souza, L., Araki, Y., Ito, O., D'Souza, F. (2007). Donor–acceptor nanohybrids of zinc naphthalocyanine or zinc porphyrin noncovalently linked to single-wall carbon nanotubes for photoinduced electron transfer. *Journal of Physical Chemistry C*, **111**, 6947–6955.

87. Ehli, C., Rahman, G. M. A., Jux, N., Balbinot, D., Guldi, D. M., Paolucci, F., Marcaccio, M., Paolucci, D., Melle-Franco, M., Zerbetto, F., Campidelli, S., Prato, M. (2006). Interactions in single wall carbon nanotubes/pyrene/porphyrin nanohybrids. *Journal of the American Chemical Society*, **128**, 11222–11231.

88. Alvaro, M., Atienzar, P., De la Cruz, P., Delgado, J. L., Troiani, V., Garcia, H., Langa, F., Palkar, A., Echegoyen, L. (2006).

Synthesis, photochemistry, and electrochemistry of single-wall carbon nanotubes with pendent pyridyl groups and of their metal complexes with zinc porphyrin. Comparison with pyridyl-bearing fullerenes. *Journal of the American Chemical Society*, **128**, 6626–6635.

89. Yu, J., Mathew, S., Flavel, B. S., Johnston, M. R., Shapter, J. G. (2008). Ruthenium porphyrin functionalized single-walled carbon nanotube arrays. A step toward light harvesting antenna and multibit information storage. *Journal of the American Chemical Society*, **130**, 8788–8796.
90. Guldi, D. M., Rahman, G. M. A., Prato, M., Jux, N., Qin, S., Ford, W. (2005). Single-wall carbon nanotubes as integrative building blocks for solar-energy conversion. *Angewandte Chemie-International Edition*, **44**, 2015–2018.
91. Sgobba, V., Rahman, G. M. A., Guldi, D. M., Jux, N., Campidelli, S., Prato, M. (2006). Supramolecular assemblies of different carbon nanotubes for photoconversion processes. *Advanced Materials*, **18**, 2264–2269.
92. Hasobe, T., Fukuzumi, S., Kamat, P. V. (2006). Organized assemblies of single wall carbon nanotubes and porphyrin for photochemical solar cells: Charge injection from excited porphyrin into single-walled carbon nanotubes. *Journal of Physical Chemistry B*, **110**, 25477–25484.
93. D'Souza, F., Sandanayaka, A. S. D., Ito, O. (2010). SWNT-based supramolecular nanoarchitectures with photosensitizing donor and acceptor molecules. *Journal of Physical Chemistry Letters*, **1**, 2586–2593.
94. Lin, W., Moon, K.-S., Zhang, S., Ding, Y., Shang, J., Chen, M., Wong, C.-P. (2010). Microwave makes carbon nanotubes less defective. *ACS Nano*, **4**, 1716–1722.
95. Vázquez, E., Georgakilas, V., Prato, M. (2002). Microwave-assisted purification of HIPCO carbon nanotubes . *Chemical Communications*, 2308–2309.
96. Rubio, N., Fabbro, C., Herrero, M. A., de la Hoz, A., Meneghetti, M., Fierro, J. L. G., Prato, M., Vázquez, E. (2011). Ball-milling modification of single-walled carbon nanotubes: Purification, cutting, and functionalization. *Small*, **7**, 665–674.
97. Vázquez, E., Prato, M. (2009). Carbon nanotubes and microwaves: Interactions, responses, and applications. *ACS Nano*, **3**, 3819–3824.
98. Brunetti, F. G., Antonia Herrero, M.de M. Munõz, J., Díaz-Ortiz, A., Alfonsi, J., Meneghetti, M., Prato, M., Vázquez E. (2008). Microwave-induced multiple functionalization of carbon nanotubes. *Journal of the American Chemical Society*, **130**, 8094–8100.
99. de la Torre, G., Blau, W., Torres, T. (2003). A survey on the functionalization of single-walled nanotubes. The chemical attachment of phthalocyanine moieties. *Nanotechnology*, **14**, 765–771.
100. Xu, H.-B., Chen, H.-Z., Shi, M.-M., Bai, R., Wang, M. (2005). A novel donor-acceptor heterojunction from single-walled carbon nanotubes functionalized by erbium bisphthalocyanine. *Materials Chemistry and Physics*, **94**, 342–346.
101. Yang, Z., Pu, H., Yuan, J., Wan, D., Liu, Y. (2008). Phthalocyanines–MWCNT hybrid materials: Fabrication, aggregation and photoconductivity properties improvement. *Chemical Physics Letters*, **465**, 73–77.
102. Yang, Z.-L., Chen, H.-Z., Cao, L., Li, H.-Y., Wang, M. (2004). Synthesis and photoconductivity study of carbon nanotube bonded by tetrasubstituted amino manganese phthalocyanine. *Materials Science and Engineering: B*, **B106**, 73–78.
103. He, N., Chen, Y., Bai, J., Wang, J., Blau, W. J., Zhu, J. (2009). Preparation and optical limiting properties of multiwalled carbon nanotubes with π-conjugated metal-free phthalocyanine moieties. *Journal of Physical Chemistry B*, **113**, 13029–13035.
104. Ellison, M. D., Gasda, P. J. (2008). Functionalization of single-walled carbon nanotubes with 1,4-benzenediamine using a diazonium reaction. *Journal of Physical Chemistry C*, **112**, 738–740.
105. Wadzanai, C., Nyokong T. (2010). Characterization of amine-functionalized single-walled carbon nanotube-low symmetry phthalocyanine conjugates. *Carbon*, **48**, 2831–2838.
106. Mugadza, T., Nyokong T. (2010). Covalent linking of ethylene amine functionalized single-walled carbon nanotubes to cobalt (II) tetracarboxyl-phthalocyanines for use in electrocatalysis. *Synthetic Metals*, **160**, 2089–2098.
107. Georgakilas, V., Kordatos, K., Prato, M., Guldi, D. M., Holzinger, M., Hirsch A. (2002). Organic functionalization of carbon nanotubes. *Journal of the American Chemical Society*, **124**, 5, 760–761.
108. Georgakilas, V., Bourlinos, A., Gournis, D., Tsoufis, T., Trapalis, C., Mateo-Alonso, A., Prato, M. (2008). Multipurpose organically modified carbon nanotubes: From functionalization to nanotube composites. *Journal of the American Chemical Society*, **130**, 8733–8740.
109. Campidelli, S., Ballesteros, B., Filoramo, A., Diaz Diaz, D., de la Torre, G., Torres, T., Rahman, G. M. A., Ehli, C., Kiessling, D., Werner, F., Sgobba, V., Guldi, D. M., Cioffi, C., Prato, M., Bourgoin, J.-P. (2008). Facile decoration of functionalized single-wall carbon nanotubes with phthalocyanines via "click chemistry". *Journal of the American Chemical Society*, **130**, 11503–11509.
110. Bahr, J. L., Tour, J. M. (2001). Highly functionalized carbon nanotubes using in situ generated diazonium compounds. *Chemistry of Materials*, **13**, 3823–3824.
111. Chunder, A., Pal, T., Khondaker, S. I., Zhai, L. (2010). Reduced graphene oxide/copper phthalocyanine composite and its optoelectrical properties. *Journal of Physical Chemistry B*, **114**, 15129–15135.
112. Zhang, X.-F., Xi, Q. (2011). A graphene sheet as an efficient electron acceptor and conductor for photoinduced charge separation. *Carbon*, **49**, 3842–3850.
113. Malig, J., Jux, N., Kiessling, D., Cid, J.-J., Vázquez, P., Torres, T., Guldi, D. M. (2011). Towards tunable graphene/phthalocyanine–PPV hybrid systems. *Angewandte Chemie International Edition*, **50**, 3561–3565.
114. Zhu, J., Li, Y., Chen, Y., Wang, J., Zhang, B., Zhang, J., Blau, W. J. (2011). Graphene oxide covalently functionalized with zinc phthalocyanine for broadband optical limiting. *Carbon*, **49**, 1900–1905.

9

PHOTOINDUCED ELECTRON TRANSFER OF SUPRAMOLECULAR CARBON NANOTUBE MATERIALS DECORATED WITH PHOTOACTIVE SENSITIZERS

FRANCIS D'SOUZA, ATULA S. D. SANDANAYAKA, AND OSAMU ITO

9.1 INTRODUCTION

Recently, supramolecular nanocarbon architectures have received considerable attention due to their outstanding electronic properties suitable for optoelectronic and light-induced applications [1–8]. This is especially true for chemically functionalized single-walled carbon nanotubes (SWCNTs) with photoactive molecules, recognized to be useful materials for photocatalytic and light-energy harvesting applications [9–20]. Covalent functionalization is structurally the simpler method, although the synthetic procedures are intricate due to the poor solubility of SWCNT in solvents used for syntheses. Furthermore, covalent functionalization converts the sp^2 carbons of some double bonds of SWCNT to sp^3 carbons, thus making them less applicable for light-induced applications [11]. On the other hand, noncovalent self-assembly methods retain completely the π-networks of SWCNT, which is a key feature required to aggressively exploit their photo and redox applications [10, 20].

Among the non-covalent functionalization approaches, the simplest method is the direct π–π stacking of aromatic sensitizers such as porphyrins (MP) and phthalocyanines (MPc) onto the SWCNT surface [11]. However, such direct π–π interactions promote close contact between the entities, significantly altering their electronic structure and corresponding individual characteristic features. Usage of spacer-connected sensitizer molecules can avoid such direct interaction, which would also extend the lifetime of the charge-separated states generated upon light illumination [21]. For this purpose, appropriate linkages are necessary to connect the sensitizers to SWCNTs. In the case of noncovalent strategy, the π-electron aromatic compounds such as pyrene, to which the desired groups are appended, can be employed as adsorbent onto the SWCNTs surface via π–π stacking [18–20]. When covalent bonds are used to connect the photosensitizers to the pyrene unit, a "double-decker" architecture could be visualized. However, when additional intermolecular interactions such as metal–ligand coordination, ion-paring, inclusion complexation, and hydrogen bonding are employed to connect the sensitizer via appropriately functionalized pyrene receptors, a "triple-decker" architecture could be visualized [12]. Both two- and three-component architectures are highly viable and versatile for building photoactive supramolecules for various applications as shown in Scheme 9.1 [20]. Dendrimer-possessing photosensitizers can also be used to construct SWCNT–photosensitizer nanohybrids, since the dendrons intertwine with SWCNT and probably prevent direct interaction between the photosensitizer and SWCNT.

This chapter documents the recent reports on (a) supramolecular construction of nanoarchitectures using photosensitizing electron-donor and electron-acceptor molecules with mainly diameter-sorted SWCNT as one of the components and (b) key findings in the areas of photoinduced charge separation (CS), photocatalysis, and photoelectrochemistry from our laboratories.

Organic Nanomaterials: Synthesis, Characterization, and Device Applications, First Edition. Edited by Tomás Torres and Giovanni Bottari.

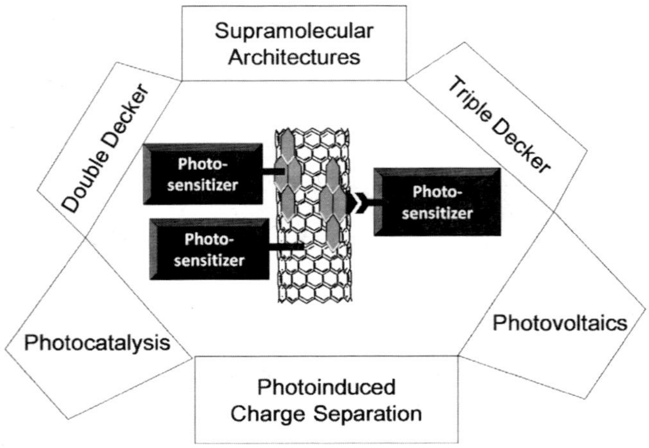

SCHEME 9.1. Supramolecular SWCNT–photosensitizer via multicomponent architecture and their different light-induced applications.

9.2 MODULATING ELECTRON TRANSFER PATH IN DIAMETER-SORTED SWCNTs

An energy level diagram depicting the possible photoinduced processes for ZnP and SWCNT is illustrated in Figure 9.1 (right side). For ZnP–SWCNT, the narrow band gap of SWCNT is positioned inside the wider HOMO–LUMO gap of ZnP, which is quite different from the usual organic donor–acceptor molecular pairs [22]. Therefore, two types of photoinduced CS processes could be envisioned as shown by curved arrows, namely, electron transfer (ET) from the ZnP-LUMO to the conduction level of SWCNT, generating $ZnP^{\bullet+}$–$SWCNT^{\bullet-}$ ion pair (upper arrow), and ET from the valance band of SWCNT to the half-filled HOMO level of the excited $^1ZnP^*$ generating $ZnP^{\bullet-}$–$SWCNT^{\bullet+}$ ion pair (lower arrow) [23]. In the case of SWCNT–C_{60}, upon excitation of the fullerene entity of the nanohybrids, an electron transfer from the valence band of SWCNT to the HOMO level of $^1C_{60}{}^*$ occurs to result in $C_{60}{}^{\bullet-}$–$SWCNT^{\bullet+}$ ion pair as illustrated in Figure 9.1 (left-side arrow) [24, 25].

When diameter-sorted SWCNTs are used, ZnP/SWCNT (*n,m*) nanohybrids with lower conduction-band level can be easily reduced by the photoexcited ZnP (upper arrow), whereas SWCNT(*n,m*) with higher valence-band level can easily be oxidized by photoexcited C_{60} and also ZnP (lower arrows) [24].

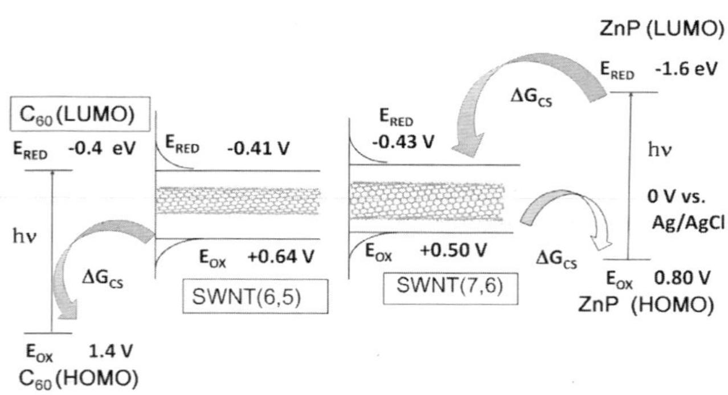

FIGURE 9.1. Schematic energy level diagrams for photo-induced charge separation processes of SWCNT(*n,m*) hybrids with ZnP (*right side*) and C_{60} (*left side*). The arrows represent the charge-separation paths. Modified from references 22–24.

9.3 COVALENTLY LINKED ARCHITECTURES

Among the available methods for covalent functionalization of SWCNT, Prato's method of ylide-adduct formation, analogous to fulleropyrrolidine synthesis [26], is a widely used approach. As sensitizer, metalloporphyrin (MP) derivatives are often used due to their close structural resemblance to chlorophyll pigments in natural photosynthesis [27, 28]. The length of the spacer unit between the MP and SWCNT is also known to play an important role for optimum performance. Longer and flexible spacer units allow direct contact between MP and SWCNT, thus permitting stronger π–π interactions. Recently, Arai et al. [29] reported a covalently bonded ZnP–(sp)–SWCNT with short spacer (sp) unit as shown in Figure 9.2a. The ZnP–(sp)–SWCNT hybrids are soluble in DMF and reveal the absorption spectral features of both ZnP at 420 and 550–600 nm, along with SWCNT in the near-IR region. Steady-state fluorescence and the time profiles (Figure 9.2b) revealed efficient quenching, which could be attributed to either the ET process or the energy-transfer (EnT) process. Interestingly, the transient absorption spectra of the nanohybrids revealed bands at 650 nm corresponding to the formation of $ZnP^{\bullet+}$ and in the near-IR region attributable to the formation of $SWCNT^{\bullet-}$ (Figure 9.2c (left)), confirming the occurrence of the CS process via $^1ZnP^*$. Hence, the fluorescence quenching is ascribed to the CS process via $^1ZnP^*$ whose rate is evaluated to be $k^S{}_{CS} = 4.2 \times 10^9\ s^{-1}$ (Figure 9.2b). The absorption intensities of $ZnP^{\bullet+}$–(sp)–$SWCNT^{\bullet-}$ rise quickly, supporting the fast CS process via $^1ZnP^*$, and then decay slowly giving a charge recombination (CR) rate, k_{CR} of $4.0 \times 10^6\ s^{-1}$. The reciprocal of k_{CR} provided an estimate of the lifetime of the radical ion pair, τ_{RIP} being about 250 ns (Figure 9.2c(left)).

Addition of benzylviologen dication (BV^{2+}) as a second electron acceptor and benzyl-dihydronicotinamide (BNAH) as a sacrificial electron donor to the hybrid solution resulted in the disappearance of radical ion bands of $ZnP^{\bullet+}$–(sp)–$SWCNT^{\bullet-}$, concurrently producing an absorption band characteristic of $BV^{\bullet+}$ at 622 nm (Figure 9.2c (right)). These observations indicate that the electron of $SWCNT^{\bullet-}$ mediates to BV^{2+}, producing $BV^{\bullet+}$ [30]. From the time profile, $BV^{\bullet+}$ is found to persist over a microsecond time scale.

By the steady-state absorption measurements during light illumination of ZnP–(sp)–SWCNT in the presence of BV^{2+} and BNAH, accumulation of $BV^{\bullet+}$ (60% maximal yield) is

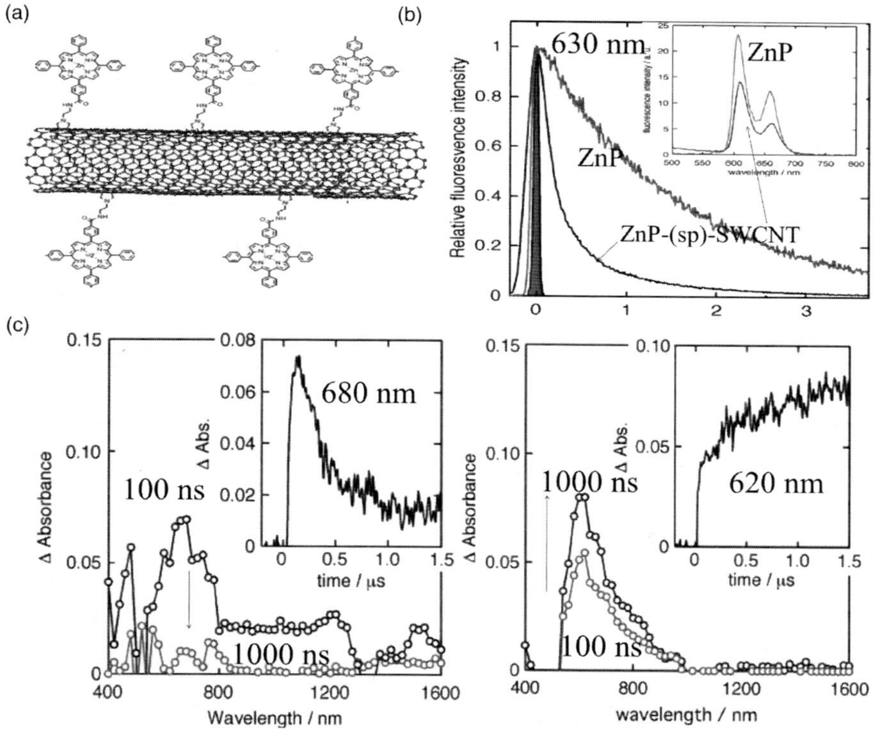

FIGURE 9.2. (**a**) Proposed structure of ZnP–(sp)–SWCNT hybrid. (**b**) Fluorescence time profiles and steady-state fluorescence spectra of ZnP–(sp)–SWCNT. (**c**) (*Left*) Transient spectra of ZnP–(sp)–SWCNT with 532-nm excitation. (*Inset*: time profile at 680 nm.) (*Right*) Visible light excitation of ZnP–(sp)–SWCNT in the presence of BV^{2+} and BNAH. (*Inset*: time profile at 620 nm.) Modified from reference 29.

observed [29]. In a series of control experiments, the absence of either the donor or the acceptor entities decreased the yields considerably (Figure 9.3a), supporting the occurrence of a catalytic ET reaction from BNAH to BV^{2+} producing $BV^{\bullet+}$ and $BNAH^{\bullet+}$. Since BV^{2+} and BNAH are not photoexcited with the visible light, the photoinduced CS state, $ZnP^{\bullet+}$–(sp)–$SWCNT^{\bullet-}$, mediates the electron of $SWCNT^{\bullet-}$ to BV^{2+} producing $BV^{\bullet+}$, whereas $ZnP^{\bullet+}$ transfers the hole to BNAH producing $BNAH^{\bullet+}$, which irreversibly dissociates to benzyl-nicotinamide cation (BNA^+); that is, BNAH acts as a sacrificial hole transfer reagent to $ZnP^{\bullet+}$ (Figure 9.3a). These observations support a photocatalytic action of ZnP–(sp)–SWCNT to generate $BV^{\bullet+}$.

The excess electron on $SWCNT^{\bullet-}$ produced by the visible light illumination of ZnP–(sp)–SWCNT can also be utilized for H_2 evolution in the presence of an appropriate metal catalyst. For example, photosensitized H_2 generation is observed for ZnP–(sp)–SWCNT hybrids in the presence of excess of colloidal Pt protected with poly-vinylpyrrolidone (Pt-PVP) and BNAH in ethanol–aqueous buffer solution under visible light (>500 nm) irradiation. The H_2 yield is pH-dependent as shown in Figure 9.3b; high H_2 yield is observed in acidic and neutral solutions, whereas low H_2 yield is seen in alkali solution. This implies that $ZnP^{\bullet+}$–(sp)–$SWCNT^{\bullet-}$ supplies excess electrons to Pt catalysis, from where H^+ accepts an electron to form H_2 (scheme in Figure 9.3b). Added BNAH traps the hole of $ZnP^{\bullet+}$ to prevent CR with $SWCNT^{\bullet-}$, resulting in an increase in H_2 evolution [29].

When a solar cell was built using an FTO–SnO_2 electrode modified by ZnP–(sp)–SWCNT coating, the photon-to-current responses are observed by the Xe light (>500 nm) illumination. The incident photon-to-current efficiency (IPCE) spectrum is shown in Figure 9.4a, in which three peaks appeared at 420, 560, and 610 nm, resembling the absorption peaks of the ZnP moiety of ZnP–(sp)–SWCNT, suggesting that the photocurrent is derived from the light absorption of the ZnP unit of the ZnP–(sp)–SWCNT according to the scheme in Figure 9.4b. The maximal IPCE value is 7% at 420 nm, which is almost the same level as the reported values of other similar porphyrin–SWCNT supramolecular electrodes [31,32].

9.4 DOUBLE-DECKER ARCHITECTURES VIA π–π STACKING AND COVALENT BONDING

Among the noncovalent functionalization approaches, the simplest method is to utilize glue molecules such as

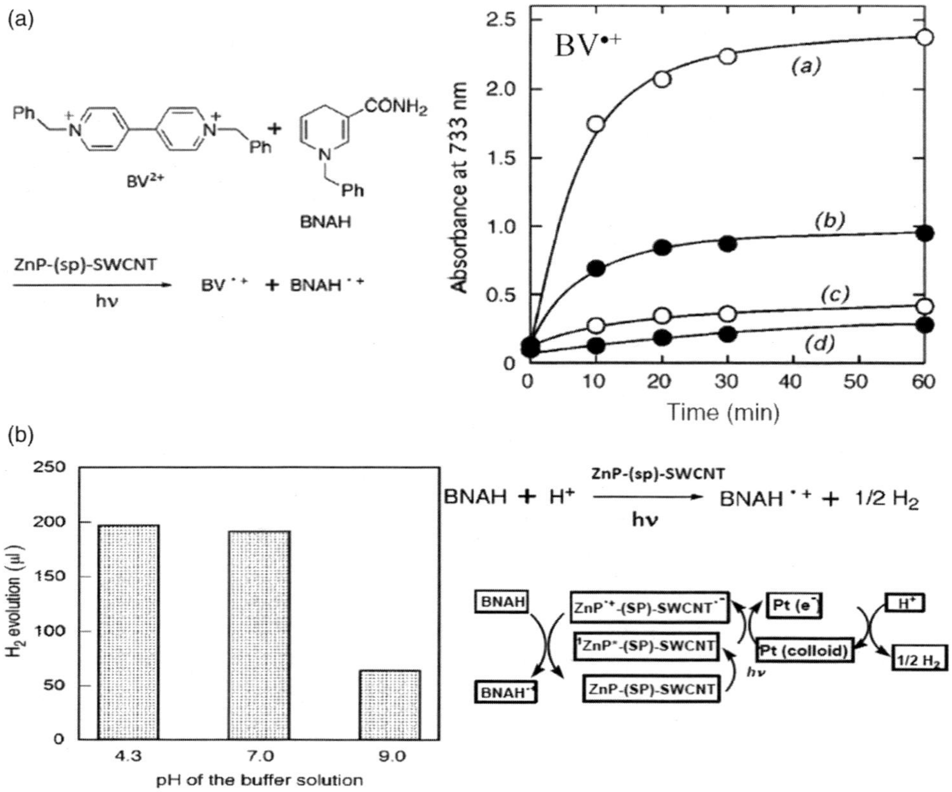

FIGURE 9.3. (**a**) Photocatalytic electron pooling resulting in the accumulation of $BV^{\bullet+}$ during light irradiation of the ZnP moiety of the ZnP–(sp)–SWCNT hybrids with the continuous Xe light (>500 nm) in the presence of BV^{2+} and BNAH in deaerated DMF. (**a**) ZnP–(sp)–SWCNT, (**b**) SWCNT, (**c**) ZnTPP, and (**d**) BV^{2+} and BNAH. (**b**) Photocatalytic H_2 evolution in ethanol–aqueous solution with the Xe light (>500 nm, 60 min) illumination of ZnP–(sp)–SWCNT in the presence of BNAH and colloidal Pt-PVP at RT. H_2-evolution mechanism. Modified from reference 29.

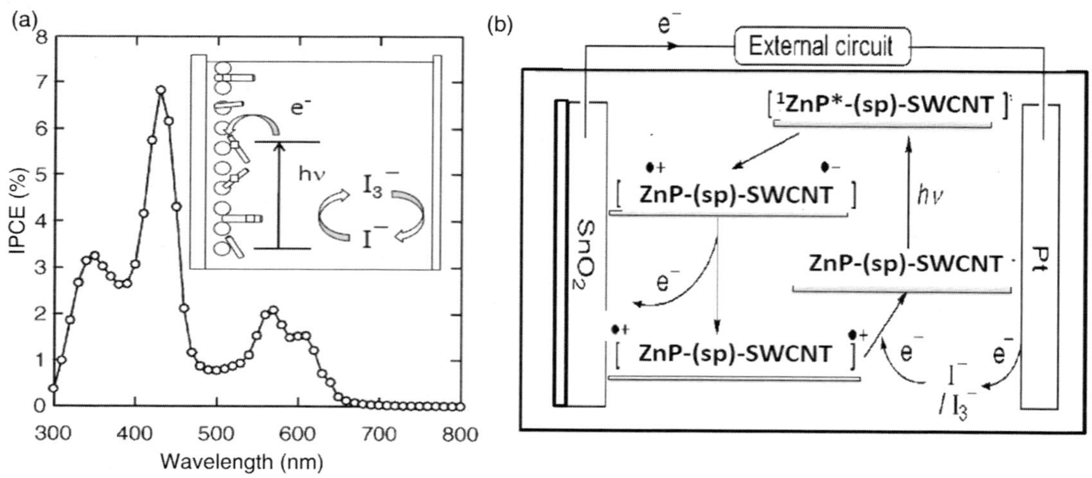

FIGURE 9.4. (**a**) The incident photon-to-photocurrent efficiency (IPCE) versus wavelength of the illumination light of the FTO/SnO_2 electrode coated with ZnP–(sp)–SWCNT. (*Inset*): Photovoltaic cell in CH_3CN. (**b**) Proposed mechanism for electron flow in the solar cell. Modified from reference 29.

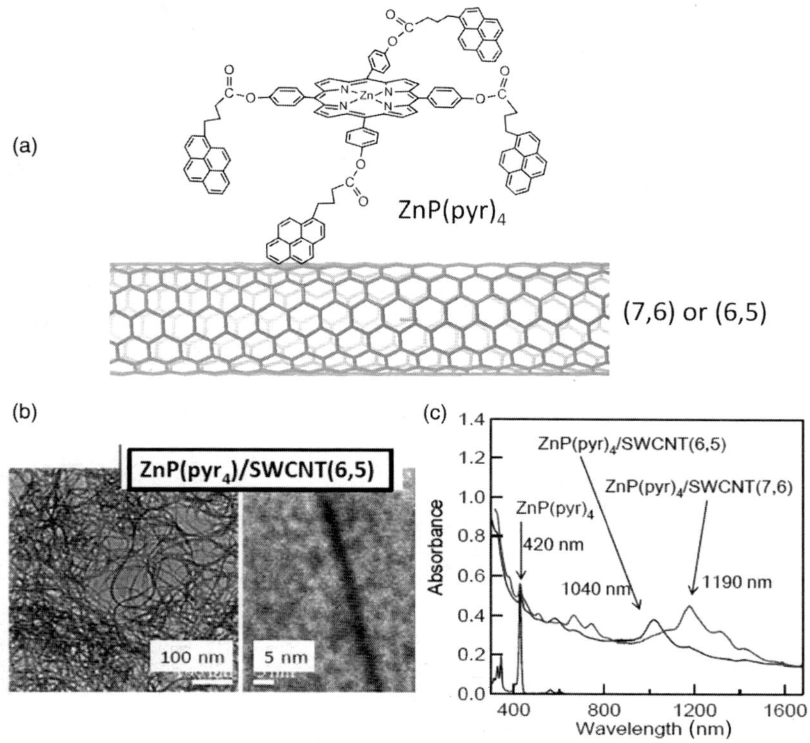

FIGURE 9.5. (**a**) Structure of $ZnP(pyr)_4$ used to form $ZnP(pyr)_4$–SWCNT(*n,m*) nanohybrids in which the pyrene entities adsorb directly onto the surface of SWCNT, perhaps leaving an appropriate space between ZnP and SWCNT(*n,m*). (**b**) TEM images of the indicated nanohybrids at different magnification scales. (**c**) Absorption spectra of the nanohybrids in DMF. Modified from reference 22.

π-electron aromatic compounds, to which the desired photosensitizers are appended to adhere to the surface of SWCNT [20]. As indicated in Figure 9.1, porphyrin and C_{60} compounds are good candidates as the photoinduced CS counterparts with respect to SWCNT.

9.4.1 Porphyrins and Phthalocyanines as Photosensitizers

Figure 9.5a shows the three-component molecular systems of ZnP functionalized with four pyrene (pyr) entities ($ZnP(pyr)_4$), in which ZnP is a typical visible-light sensitizer and pyr entities are π-stacking agents to SWCNT [22]. By employing the recently available diameter-sorted semiconducting SWCNT(*n,m*) [33], $ZnP(pyr)_4$–SWCNT(*n,m*) hybrids are constructed by treating the individual components in DMF (dimethylformamide). The obtained homogeneous solution was stable and transparent for several days. TEM images of the dried $ZnP(pyr)_4$–SWCNT(*n,m*) samples are shown in Figure 9.5b, in which the treatment with $ZnP(pyr)_4$ unraveled the tangled SWCNT(*n,m*) in DMF; an expanded image also shows that each SWCNT(*n,m*) looks blurry and thicker, probably due to the attachments of the $ZnP(pyr)_4$ and its aggregates. Figure 9.5c shows optical absorption spectra of the nanohybrids in DMF; the sharp peaks in the 600 to 1400 nm region are attributed to SWCNT(*n,m*) and the peak at 420 nm is due to the Soret band of the ZnP moiety.

Photophysical properties are monitored by the fluorescence measurements of the ZnP moiety in the 600 to 700-nm region. As shown in Figure 9.6a, the ZnP-fluorescence intensity quenching is observed in the presence of SWCNT(*n,m*), supporting appreciable interaction between the $^1ZnP^*$ and SWCNT surface. From the accelerated ZnP-fluorescence decays upon attachment of SWCNT(*n,m*) as shown in inset of Figure 9.6a, the dynamic events such as ET and EnT via $^1ZnP^*$ can be considered as quenching mechanism for such supramolecular systems. The energies of the radical ion pairs (RIPs) are evaluated to be 1.21 eV for $ZnP(pyr)_4^{\bullet+}$–SWCNT$(6,5)^{\bullet-}$ and 1.23 eV for $ZnP(pyr)_4^{\bullet+}$–SWCNT$(7,6)^{\bullet-}$ from the difference between E_{OX}(ZnP) and E_{RED}(SWCNT) (see Figure 9.1). Thus, the free-energy changes for the CS process (ΔG_{CS}) via pyr$-^1ZnP^*$ (1.90 eV) are both negative ($\Delta G_{CS} = -0.69$ eV for $ZnP(pyr)_4$-SWCNT(6,5) and $\Delta G_{CS} = -0.67$ eV for $ZnP(pyr)_4$-SWCNT(7,6)), predicting exothermic CS processes.

Evidence for the CS process between $^1ZnP^*$ and SWCNT(*n,m*) is obtained by constructing a photocatalytic cycle to extract the electron and accumulate it as a reduced chemical product of hexylviologen dication (HV^{2+}) in the

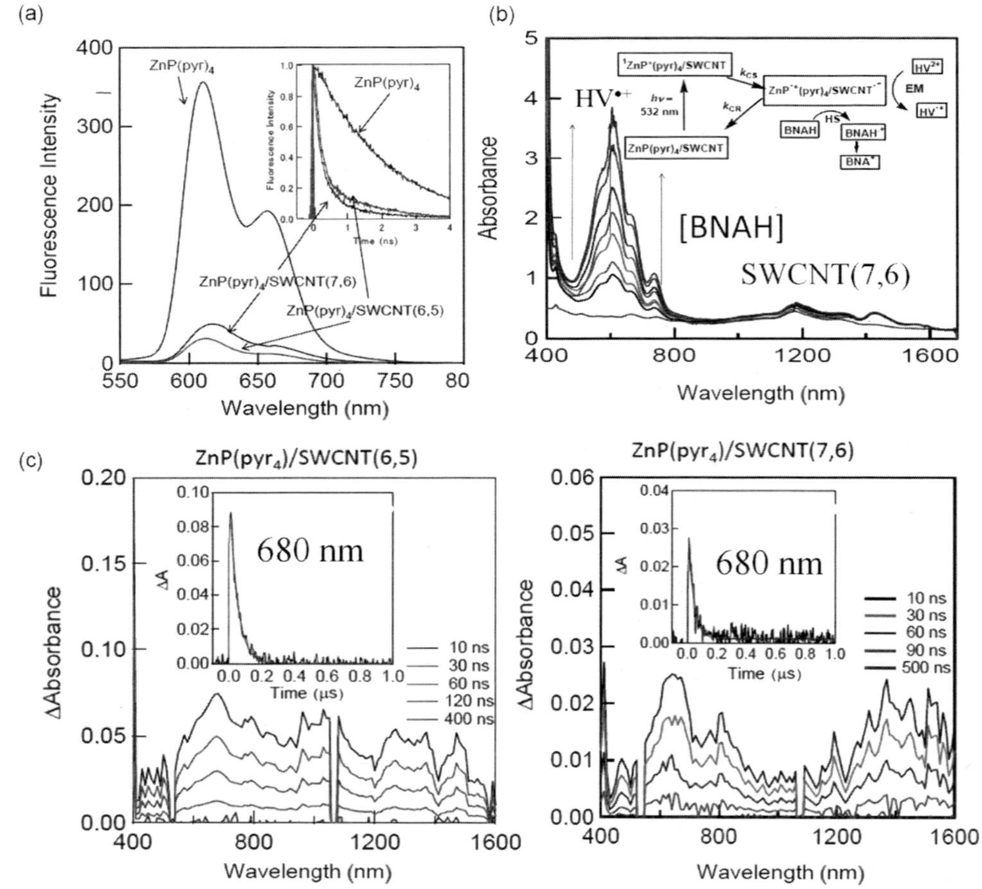

FIGURE 9.6. (**a**) Steady-state fluorescence spectra and the decay time profiles of $ZnP(pyr)_4$ in the presence of SWCNT(6,5) and SWCNT(7,6) in DMF. (**b**) Accumulation of the $HV^{\bullet+}$ absorption band with light irradiation of $ZnP(pyr)_4$ of $ZnP(pyr)_4$–SWCNT(7,6) in the presence of HV^{2+} (0.5 mM) and BNAH (1 mM step) with 532-nm laser light (5 shots). *Inset*: Proposed scheme (EM, electron migration; HS, hole shift). (**c**) Nanosecond transient absorption spectra of $ZnP^{\bullet+}(pyr)_4$–$SWCNT^{\bullet-}$ and the time profile of $ZnP^{\bullet+}$ at 680 nm in Ar-saturated DMF. Modified from reference 22.

presence of BNAH by the visible light irradiation of the ZnP moiety in $ZnP(pyr)_4$–SWCNT(*n,m*). Figure 9.6b shows the accumulation of $HV^{\bullet+}$-absorption band at 620 nm with repeated 532-nm laser light irradiation. These observations support that the electron of $SWCNT^{\bullet-}$ mediates to HV^{2+}, generating $HV^{\bullet+}$ (scheme in inset of Figure 9.6b, in which BNAH acts as sacrificial electron donor to $ZnP^{\bullet+}$. The photocatalytic conversion from HV^{2+} to $HV^{\bullet+}$ was 60% for SWCNT(7,6) and 40% for SWCNT(6,5) [22].

To determine the actual quenching process, the transient absorption measurements are performed as shown in Figure 9.6c. The main absorption at 660 nm is attributed to $ZnP^{\bullet+}$, while the absorptions in the near-IR region are due to $SWCNT^{\bullet-}$, supporting the generation of $ZnP^{\bullet+}(pyr)_4$–$SWCNT^{\bullet-}$, which persists for about 50 ns as seen in inset of Figure 9.6c. By careful examination of the spectra in the near-IR region, the $SWCNT(6,5)^{\bullet-}$ shows the bands at 1000, 1250, 1350, and 1450 nm, whereas $SWCNT(7,6)^{\bullet-}$ shows bands at 1200, 1350, and 1550 nm bands, which could be used as a new evidence of the diameter-enriched $SWCNT^{\bullet-}$. The fast rise of $ZnP^{\bullet+}$ at 680 nm occurs within the nanosecond laser pulse width (6 ns), suggesting that the fluorescence lifetimes mainly correspond to the CS process; therefore, a $k^S_{CS} = 5.1 \times 10^9\ s^{-1}$ is evaluated for $ZnP(pyr)_4$–SWCNT(7,6), which is slightly larger than $k^S_{CS} = 3.4 \times 10^9\ s^{-1}$ evaluated for $ZnP(pyr)_4$–SWCNT(6,5). The absorption bands of $ZnP^{\bullet+}(pyr)_4$–$SWCNT^{\bullet-}$ decay within 100 ns, giving $k_{CR} = 2.7 \times 10^7\ s^{-1}$, $\tau_{RIP} = 40$ ns for SWCNT(7,6) and $k_{CR} = 2.1 \times 10^7\ s^{-1}$, $\tau_{RIP} = 50$ ns for SWCNT(6,5) (Figure 9.6).

Photoelectrochemical cells are also constructed with $ZnP(pyr)_4$–SWCNT(*n,m*) on the ITO/SnO_2 electrodes in the presence of KI/I_2 redox mediator. Although the IPCE values are not high, the $ZnP(pyr)_4$–SWCNT(7,6) electrode shows higher efficiency than that of the $ZnP(pyr)_4$–SWCNT(6,5) electrode. This trend relates to slightly easily reducible

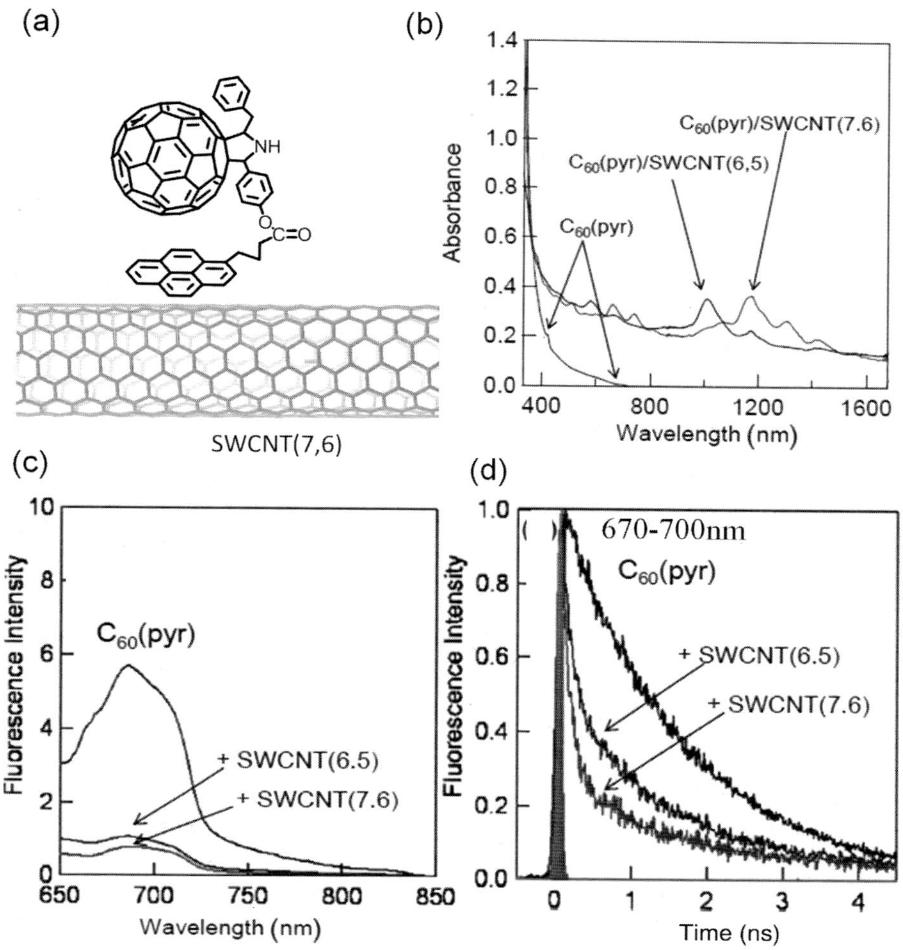

FIGURE 9.7. (**a**) SWCNT–pyrC_{60} assembly via π–π interactions. (**b**) Absorption spectra. (**c**) Steady-state fluorescence spectra. (**d**) Fluorescence time profiles. All the spectra of the SWCNT–pyrC_{60} nanohybrids were recorded in DMF. Modified from reference 25.

SWCNT(7,6) (−0.41 V versus Ag/AgCl) compared to that of less reducible SWCNT(6,5) (−0.43 V versus Ag/AgCl) [22].

9.4.2 Fullerene as Photosensitizer

The self-assembly protocols have also been applied to investigate the electron-donor ability of SWCNTs by constructing hybrids with strong electron acceptors such as fullerene, C_{60}. Here, the pyr moiety covalently linked with fullerene (pyrC_{60}) is used to π–π stack the diameter sorted semiconducting SWCNTs, thereby forming nanohybrids (see Figure 9.7a) [25]. Here, it is very interesting to investigate the donor ability of SWCNT, when photosensitizing electron acceptor, C_{60}, is employed as a counterpart of the donor–acceptor hybrid (see Figure 9.1).

The optical absorption spectra of the nanohybrids (Figure 9.7b) illustrate the characteristic absorption bands of fullerene in the shorter region than 700 nm with a shoulder at 430 nm and the SWCNTs covering the visible and near-IR region; the pyrene absorptions are overlapped in the shorter wavelength region (<350 nm). Considerable red shifts of the SWCNT peaks are observed due to π-stacking with pyrene units directly and with C_{60} units indirectly. From the shifts of the absorption peaks at the longest wavelength of SWCNT(*n,m*), the band-gap variations of SWCNT(*n,m*) after the nano-hybridizations can be evaluated. Band gaps of 1.23 eV for SWCNT(6,5)–pyrC_{60}, and 1.05 eV for SWCNT(7,6)–pyrC_{60} are evaluated using this approach.

The steady-state fluorescence spectra of SWCNT(*n,m*)–pyrC_{60} observed by predominate excitation of the C_{60} moiety (Figure 9.7c) demonstrate that the pyrC_{60}-fluorescence peak appears around 690 nm and that the fluorescence intensities are quenched over 80% upon formation of the nanohybrids with SWCNT(*n,m*). The energies of the RIPs can be evaluated to be 1.01 eV for SWCNT(6,5)$^{•+}$–pyr$C_{60}^{•-}$ and 0.87 eV for SWCNT(7,6)$^{•+}$–pyr$C_{60}^{•-}$ from the difference between E_{OX}(SWCNT) and E_{RED}(C_{60}) as shown in Figure 9.1

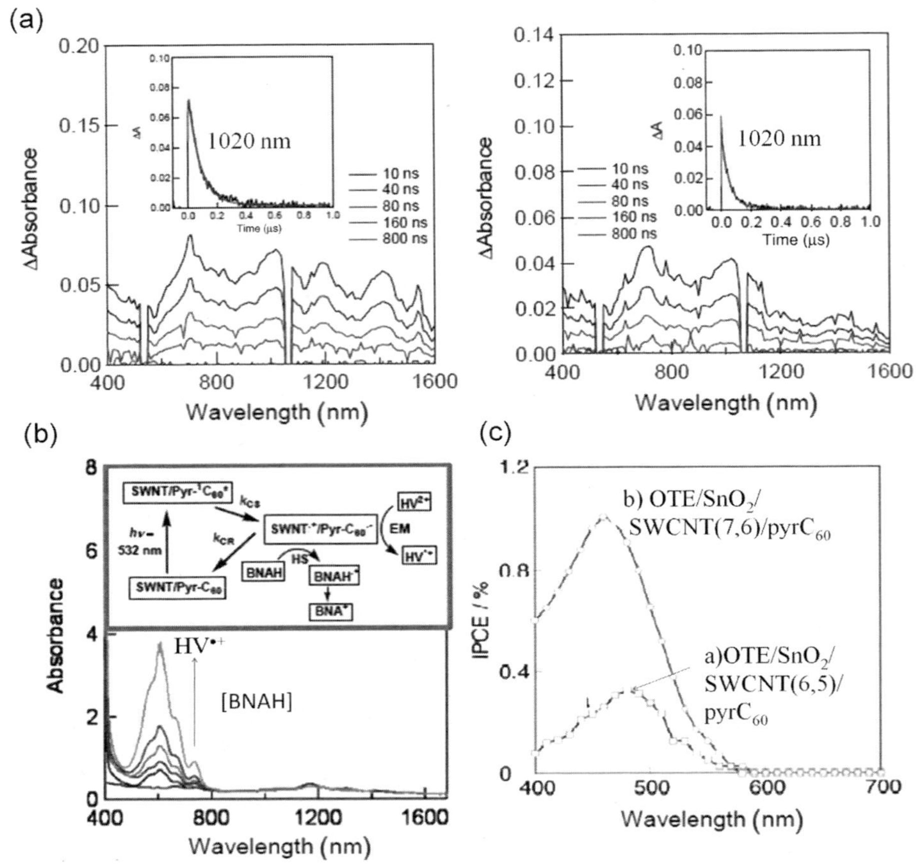

FIGURE 9.8. (**a**) Nanosecond transient absorption spectra of SWCNT(*n,m*)–pyrC_{60} in DMF ((left) SWCNT(6,5)-pyrC_{60} and (right) SWCNT(7,6)-pyrC_{60}). *Inset*: Time-profiles at 1020 nm; λ_{ex} = 532 nm. (**b**) Steady-state absorption spectra of SWCNT(7,6)–pyrC_{60} measured after five laser shots at 532 nm in the presence of HV^{2+} (0.5 mM) and BNAH 0–5 mM with 1 mM step in Ar-saturated DMF. (**c**) IPCE curves recorded for the FTO–SnO_2–SWCNT(*n,m*)–pyrC_{60} electrode in the presence of I^-/I_3^- mediator in acetonitrile. Modified from reference 25.

(left-side). Thus, the ΔG_{CS} values via $^1C_{60}^*$ (1.80 eV) are both negative ($\Delta G_{CS} = -0.79$ eV for SWCNT(6,5)–pyrC_{60} and $\Delta G_{CS} = -0.93$ eV for SWCNT(7,6)–pyrC_{60}), predicting exothermic CS process. The k_{CS} values are calculated by the difference of the fluorescence lifetimes to be 4.75×10^9 s^{-1} for SWCNT(6,5)–pyrC_{60} and 6.09×10^9 s^{-1} for SWCNT(7,6)–pyrC_{60} (Figure 9.7d). This trend in the k_{CS} values is along the exothermicity of the CS process via $^1C_{60}^*$.

The transient absorption measurements are performed as shown in Figure 9.8a. The main absorption at 1000 nm is attributed to $C_{60}^{\bullet-}$, while the absorption at 700 nm can be ascribed to $^3C_{60}^*$ due to noninteracting pyrC_{60} with SWCNT. In the near-IR region, the broad bands are attributed to SWCNT$^{\bullet+}$, supporting the generation of SWCNT$^{\bullet+}$–pyr$C_{60}^{\bullet-}$, which persists for about 200 ns as seen in the inset of Figure 9.8a. With a careful look at the spectra in the near-IR region, the SWCNT(6,5)$^{\bullet+}$ reveals bands at 1200, 1350, and 1550 nm, whereas the SWCNT(7,6)$^{\bullet+}$ show weaker bands at 1200- and 1400-nm regions. The absorption bands of SWCNT(7,6)$^{\bullet+}$ decay within 100 ns, giving $k_{CR} = 1.8 \times 10^7$ s^{-1} ($\tau_{RIP} = 60$ ns), whereas $k_{CR} = 1.2 \times 10^7$ s^{-1} ($\tau_{RIP} = 80$ ns) for SWCNT(6,5)$^{\bullet+}$. The ratio of k_{CS}/k_{CR} is ~360, suggesting considerable charge stabilization in these nanohybrids.

Electron mediation experiments are performed by adding HV^{2+} and BNAH to the SWNT(*n,m*)–pyrC_{60} solution. By the absorption spectral measurements after the repeated laser-light irradiation (532 nm) of SWCNT(7,6)–pyrC_{60}, accumulation of $HV^{\bullet+}$ is observed by its 620-nm absorption band as shown in Figure 9.8b, in which the $HV^{\bullet+}$ increases with the BNAH concentration, supporting the role of BNAH as a sacrificial HS reagent with changing to BNA^+. These observations support that the 532-nm light irradiation induces the CS process within SWCNT–pyrC_{60}, generating SWCNT$^{\bullet+}$–pyr$C_{60}^{\bullet-}$, from which the electron mediates to HV^{2+} generating $HV^{\bullet+}$, while the hole shifts to BNAH as shown in the inset scheme of Figure 9.8b. The maximal conversion of $HV^{\bullet+}$ is evaluated to be 70% for SWCNT(7,6)–pyrC_{60},

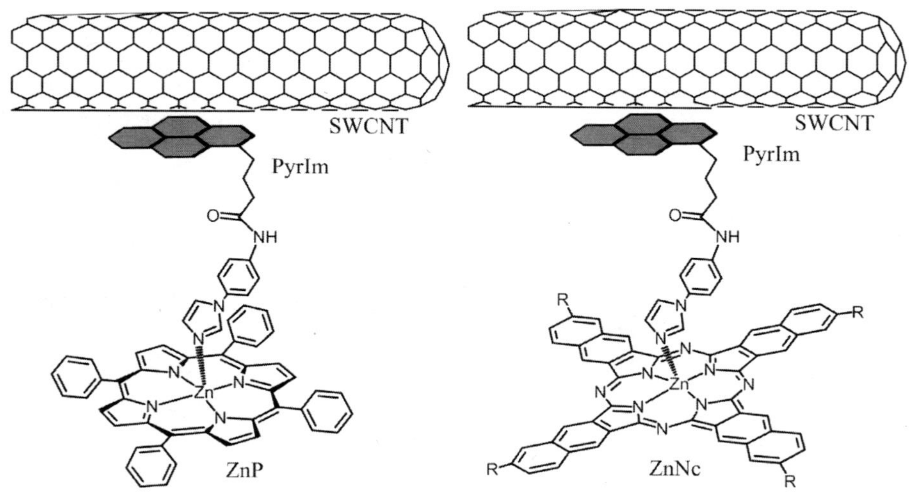

FIGURE 9.9. Strategy adopted for the construction of SWCNT–pyrIm→ZnP and SWCNT–pyrIm→ZnNc donor–acceptor hybrids via metal–ligand axial coordination method. Modified from reference 35.

which is higher than that for SWCNT(6,5)–pyrC_{60} (46%). The SWCNT(7,6)–pyrC_{60} nanohybrids with a narrower band gap and easily oxidizable SWCNT(7,6) provided higher efficiencies of photocatalytic process involving electron pooling experiments [25].

Furthermore, photoelectrochemical measurements are performed for SWCNTs–pyrC_{60} casted on OTE/SnO_2 electrode. The IPCE spectra are measured as shown in Figure 9.8c, in which the increasing tendency of the IPCE from 600 nm to the maxima at 450-480 nm is similar to the absorption behavior of pyrC_{60}, whereas the decrease from the maxima to 400 nm may be due to the absorption of SWCNT(*n,m*) interrupting the C_{60} excitation. The IPCE of SWCNT(7,6)–pyrC_{60} reaches 1% as a maximal value at 460 nm, which is about 3 times larger than that of SWCNT(6,5)–pyrC_{60}. This order is the same as for the $HV^{\bullet+}$-accumulation conversions [25].

As a short summary, electron-donor ability of semiconducting SWCNTs toward the photoexcited C_{60} is conclusively demonstrated. The electron-donor ability of SWCNT(*n,m*) depends on the energy level of the conduction band. Furthermore, the experimentally determined kinetic parameters of CS and subsequent CR processes are helpful in rationally explaining the efficiencies of the electron pooling and electrochemical photocurrent for SWCNT(7,6)–pyrC_{60} nanohybrids.

9.5 TRIPLE-DECKER ARCHITECTURES VIA π–π STACKING AND COORDINATION BOND FORMATION

Metal–ligand coordination has been one of the widely used supramolecular approaches to construct donor–acceptor systems with coordinatively unsaturated metallomacrocycle sensitizers, especially for building fullerene–porphyrin and fullerene phthalocyanines dyads [34]. This strategy has been extended toward building donor–acceptor hybrids bearing SWCNT using a coordinating ligand imidazole appended pyrene, pyr-Im, as a connector to ZnP or ZnPc. As examples, SWCNT–ZnP and SWCNT–zinc naphthalocyanine (ZnNc) nanohybrids are shown in Figure 9.9, in which the pyrene entity of the connector pyr-Im molecule π-stacks to SWCNT, leaving the Im group to coordinate to central Zn ion of ZnP or ZnNc [35].

In the fluorescence measurements, the ZnNc-fluorescence intensity is quenched and the fluorescence lifetime becomes short upon formation of SWCNT–pyrIm→ZnNc as shown in Figure 9.10a. In the transient absorption spectra, the peak at 980 nm is a diagnostic evidence of the $ZnNc^{\bullet+}$ (Figure 9.10b) and the 1400-nm band is assigned to $SWCNT^{\bullet-}$, supporting the generation of $SWCNT^{\bullet-}$–pyrIm→$ZnNc^{\bullet+}$ as CS product. Steady-state absorption spectra of SWCNT–pyrIm→ZnNc shown in Figure 9.10c reveal the characteristic absorption band at 780 nm due to ZnNc and SWCNT absorption bands in the near-IR region.

The $HV^{\bullet+}$ accumulation during the visible light illumination is shown in the absorption spectra (Figure 9.10d), in which the catalytic action of ZnNc is also confirmed by invariance of its absorption intensity at 780 nm during $HV^{\bullet+}$ accumulation under visible light illumination [35]. The $HV^{\bullet+}$ conversions are 93% for SWCNT–pyrIm→ZnNc and 83% for SWCNT–pyrIm→ZnP in DMF.

9.6 TRIPLE-DECKER ARCHITECTURES VIA π–π STACKING AND ION-PAIR INTERACTIONS

Ion-paring construction of the nanohybrids can also be applied for SWCNTs using the tetrapyrrole macrocycles

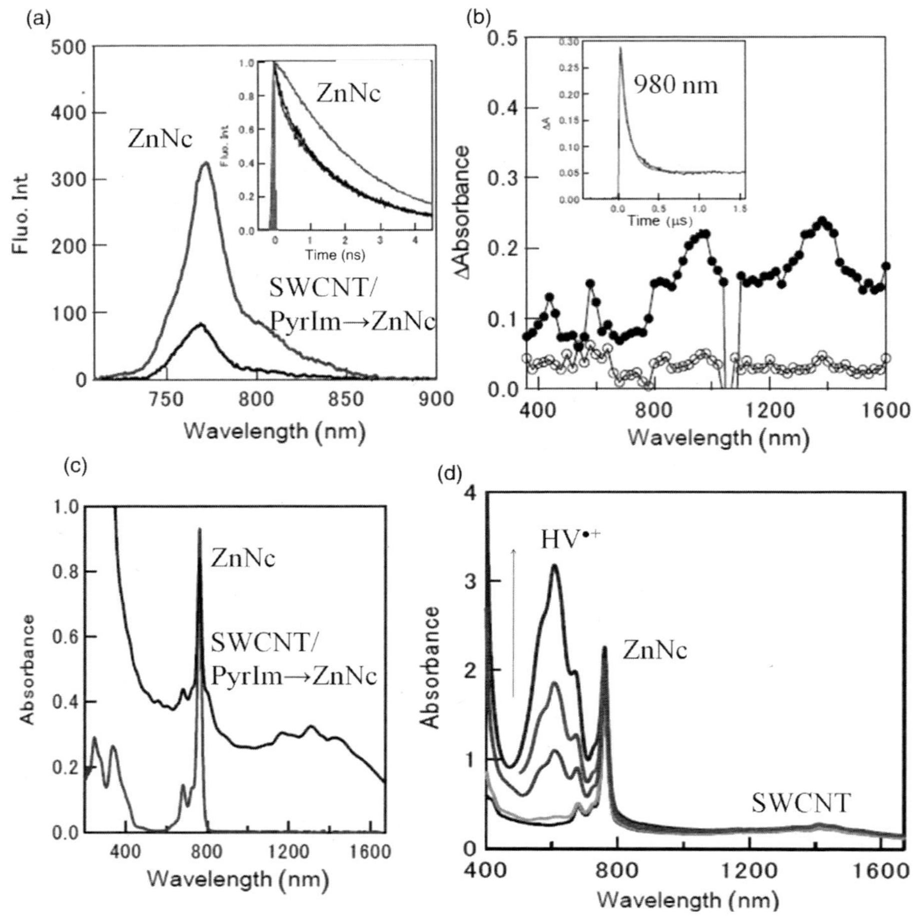

FIGURE 9.10. Spectral data for SWCNT–pyrIm→ZnNc in DMF. (**a**) Fluorescence spectra and time profiles. (**b**) Transient absorption spectra and time profile. (**c**) Absorption spectra. (**d**) The $HV^{\bullet+}$ accumulation during the visible light illumination of the ZnNc part in a SWCNT–pyrIm→ZnNc hybrid. Modified from reference 35.

bearing peripheral charges as photosensitizers and the pyrene derivatives with opposite charges [23, 36]. Two examples are shown in Figure 9.11a; in the first case, negatively charged ($TPPS^-$)M (tetrakis(4-sulfonatophenyl)porphyrin) is ion-paired with positively charged $pyrNH_3^+$ that is π-stacked to SWCNT. In the second case, positively charged ($TMPyP^+$)M (tetrakis(4-*N*-methylpyridy)porphyrin), is ion-paired with negatively charged $pyrCOO^-$, in which the pyr moiety is π-stacked to SWCNT. Formations of SWCNT(*n,m*)–$pyrNH_3^+$–($TPPS^-$)M and SWCNT(*n,m*)–$pyrCOO^-$–($TMPyP^+$)M are confirmed by TEM imaging and spectroscopic measurements. Occurrence of photoinduced electron transfer via ($TPPS^-$)M and($TMPyP^+$)M is confirmed by the transient absorption and fluorescence measurements [23, 36]. The $HV^{\bullet+}$ accumulation under visible light illumination of the ($TPPS^-$)Zn part is also observed for these nanohybrids. The formation of $SWCNT^{\bullet-}$–$pyrNH_3^+$–($TPPS^-$)$Zn^{\bullet+}$ ion-pair species via $^1(TPPS^-)Zn^*$ is envisioned, since ($TPPS^-$)Zn has electron-rich porphyrin ring. As for SWCNT–$pyrCOO^-$/($TMPyP^+$)M, the electron-deficient ($TMPyP^+$)M is capable of accepting an electron from SWCNT, generating $SWCNT^{\bullet+}$–$pyrNH_3^+$–($TMPyP^+$)$Zn^{\bullet-}$ ion-pair species (see the energy diagrams shown in Figure 9.1). The $SWCNT^{\bullet+}$–$pyrNH_3^+$–($TPPS^-$)$M^{\bullet-}$ is also able to accumulate $HV^{\bullet+}$ in a photocatalytic cycle by electron mediation from ($TPPS^-$)$M^{\bullet-}$ to HV^{2+}.

Photoelectrochemical studies using FTO–SnO_2 electrodes modified with these donor–acceptor nanohybrids unanimously demonstrated the ability of these nanohybrids to harvest light energy into electricity. Importantly, the photocurrent generation follows the trend observed for CS; that is, IPCE of a maximum of 8% is achieved for photocells with SWCNT(7,6) and ($TPPS^-$)Zn as sensitizer, while the conversion efficiencies are smaller for the photo cells derived from SWCNT(6,5) (see Figure 9.11b). These results suggest that it is possible to achieve higher light energy conversion efficiencies by the selection of the appropriate SWCNTs with right band gap on combination of appropriate porphyrins [23].

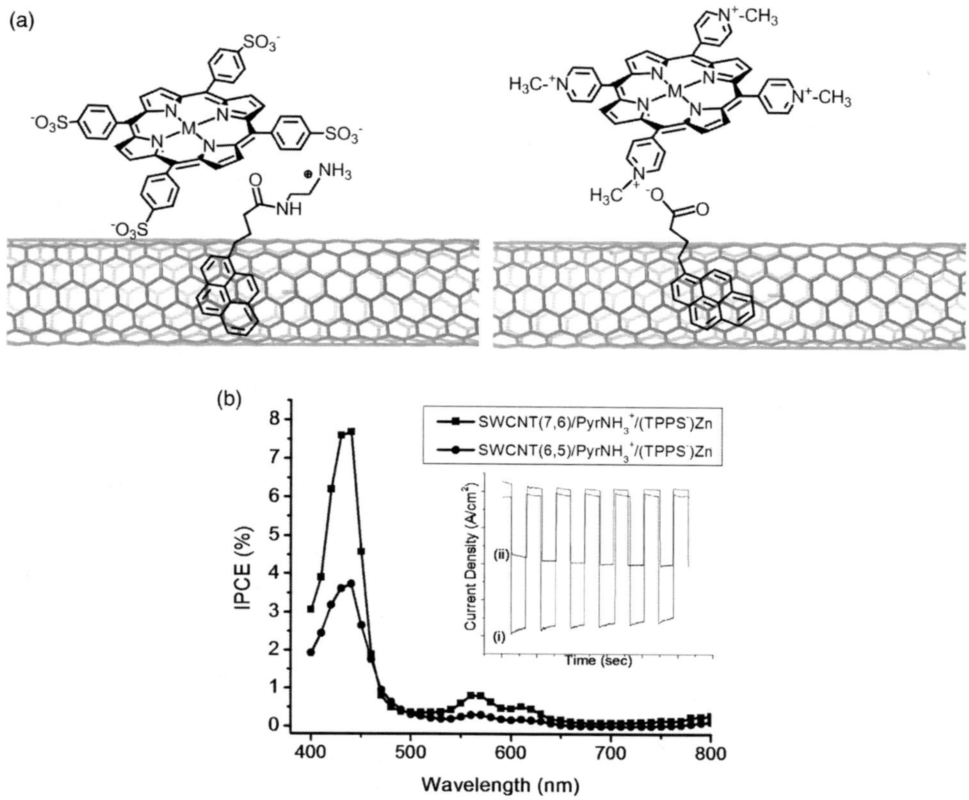

FIGURE 9.11. (**a**) Proposed structures of the SWCNT(n,m)–PyrNH$_3^+$–(TPPS$^-$)M and SWCNT(n,m)–PyrCOO$^-$–(TMPyP$^+$)M donor–acceptor nanohybrids. (**b**) IPCE curves of SWCNT(n,m)–PyrNH$_3^+$–(TPPS$^-$)Zn in the presence of I$^-$–I$_3^-$ mediator in acetonitrile. *Inset*: Light on–off current cycle. Modified from reference 23.

9.7 TRIPLE-DECKER ARCHITECTURES VIA π–π STACKING AND CROWN ETHER INCLUSION COMPLEX FORMATION

Crown ether inclusion complexation has been widely adopted for the construction of donor–acceptor conjugates using crown ether-appended porphyrins with alkyl ammonium cation functionalized fullerenes [37–41]. This methodology can also be extended to SWCNT "triple-decker" architectures using alkyl ammonium functionalized pyrene as an intermediate glue.

9.7.1 Porphyrin/Phthalocyanine-SWCNT Triple-Decker Architectures via π–π Stacking and Crown Ether–Alkyl Ammonium Cation Binding

Cation-dipole interactions have been utilized to assemble the SWCNT–donor hybrids as shown in Figure 9.12 [42, 43], in which ZnP or ZnPc used as donors are functionalized to bear one or four benzo-18-crown-6 (crown) entities as ion–dipole interacting sites, while π-stacking pyrene is functionalized with an alkyl ammonium ion (PyrNH$_3^+$). Cation–crown ether inclusion complexation and pyrene–SWCNT π–π interactions result in SWCNT–pyrNH$_3^+$:(crown)$_n$ZnP (n = 1 or 4) or SWCNT–PyrNH$_3^+$:(crown)$_n$ZnPc nanohybrids which are stable and soluble in DMF.

After charactering the SWCNT(n,m)/pyrNH$_3^+$:(crown)$_n$MP conjugates using TEM and optical techniques, electron transfer via photoexcited ZnP and ZnPc is investigated using time-resolved emission and transient absorption techniques. For example, from the fluorescence quenching of ZnP and ZnPc upon complexing with SWCNT via inclusion complexation, the k^S_{CS} values on the order of (3–5) × 10^9 s^{-1} are evaluated for SWCNT(n,m)–PyrNH$_3^+$:(crown)$_n$MP in DMF [43].

Evidence for CS in the newly assembled conjugates is obtained from the transient absorption spectral studies by the observation of characteristic peak signals of the radical cation and anion species. The nanosecond transient absorption spectra of SWCNT(n,m)–pyrNH$_3^+$:**2** observed by the 532-nm laser light, which predominantly excites the ZnP moiety, are shown in Figures 9.13a and 9.13b, in which the transient absorption peak corresponding to the formation of the ZnP$^{•+}$(2$^{•+}$) is evident around 680 nm. The corresponding counterpart, SWCNT$^{•-}$, shows its characteristic bands

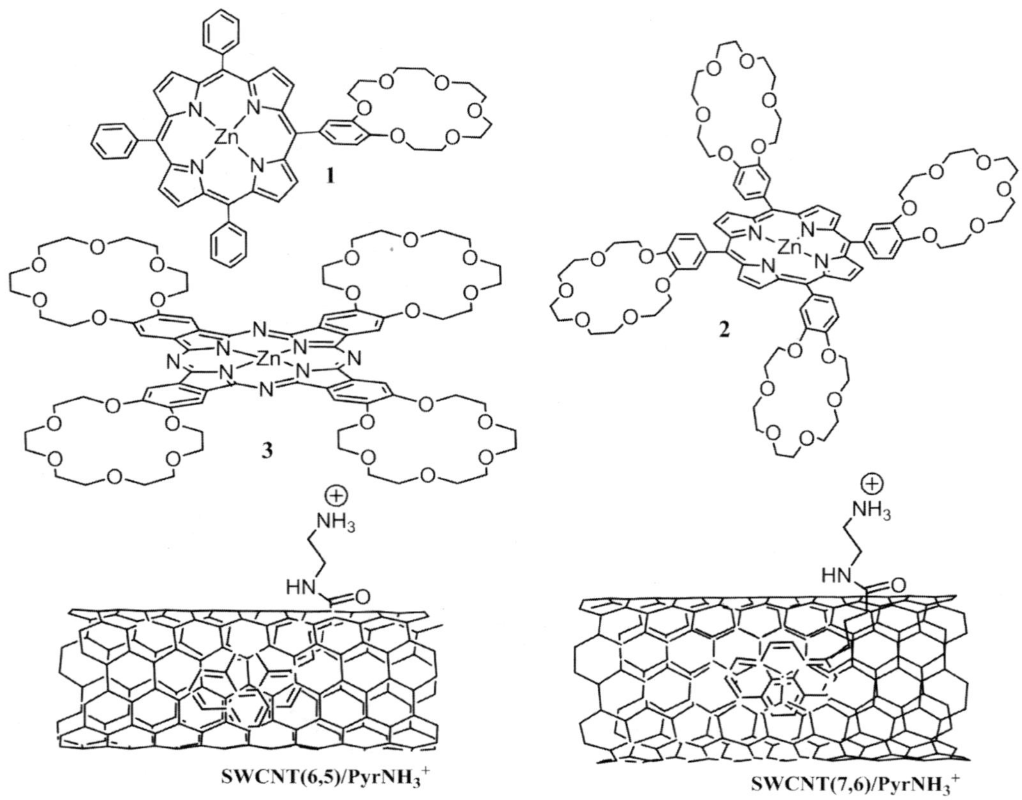

FIGURE 9.12. Molecular systems used for ZnP–SWCNT and ZnPc–SWCNT "triple-decker" architectures via π–π stacking and crown ether–alkyl ammonium cation binding. Modified from references 42 and 43.

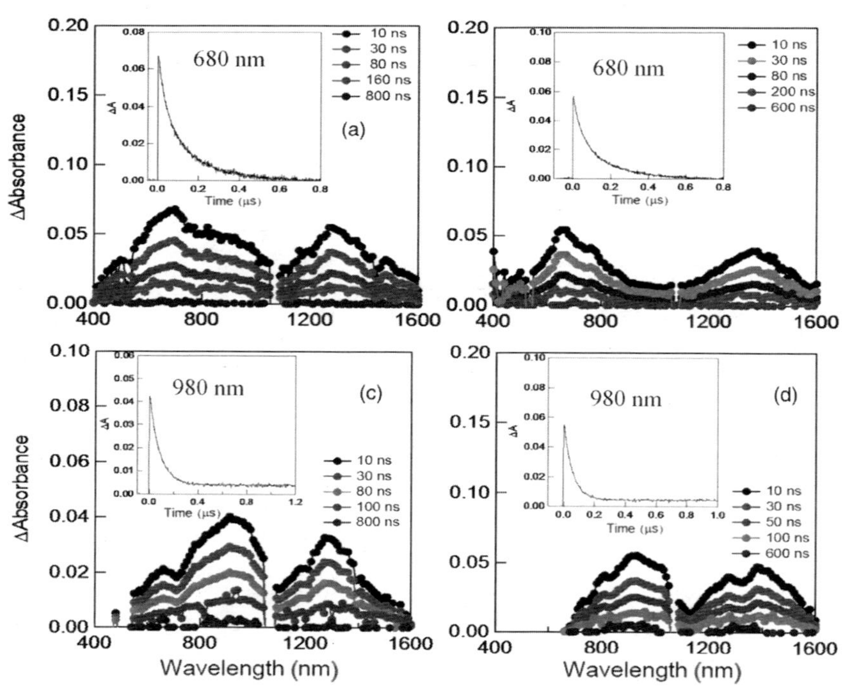

FIGURE 9.13. Nanosecond transient absorption spectra in DMF using 532-nm laser light. (**a**) SWCNT(6,5)–pyrNH$_3^+$:**2**, (**b**) SWCNT(7,6)–pyrNH$_3^+$:**2**, (**c**) SWCNT(6,5)–pyrNH$_3^+$:**3**, and (**d**) SWCNT(7,6)–pyrNH$_3^+$:**3**, recorded at different time intervals. The insets show decay profile of the MP π-cation radical band at 680 nm for **2** and at 980 nm for **3**. Modified from reference 43.

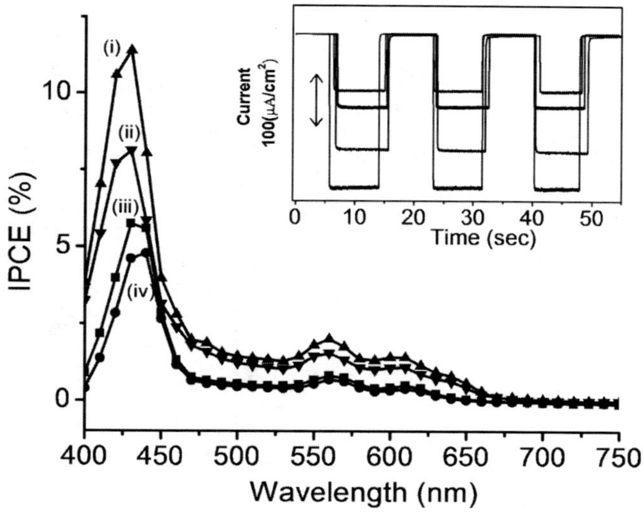

FIGURE 9.14. IPCE of (i) FTO–SnO_2–SWCNT(7,6)–$PyrNH_3^+$:**1**, (ii) FTO–SnO_2–SWCNT(6,5)–$PyrNH_3^+$:**1**, (iii) FTO–SnO_2–SWCNT(7,6)–$PyrNH_3^+$:**2**, and (iv) FTO–SnO_2–SWCNT(6,5)–$PyrNH_3^+$:**2** in the presence of I^-–I_3^- mediator in acetonitrile. *Inset*: Light on–off cycle. Modified from reference 43.

around 1400 nm in the NIR region as a broad peak with some fine structures. In the transient spectral features for the SWCNT(*n,m*)–$PyrNH_3^+$:**3**, $ZnPc^{\bullet+}$ ($3^{\bullet+}$) appear around 980 nm, providing convincing proof of electron transfer in these nanohybrids, combining with the 1400-nm band of $[SWCNT(n,m)]^{\bullet-}$. The k_{CR} is estimated to be (1.0–1.6) $\times 10^7$ s^{-1} by monitoring the decay of these bands [43].

Photoelectrochemical studies using FTO/SnO_2 electrodes modified with these donor–acceptor conjugates unanimously demonstrated the ability of the SWCNT(7,6)–$pyrNH_3^+$:MP conjugates to harvest light energy into electricity. The IPCE spectral trace is similar to the absorption spectra of ZnP as shown in Figure 9.14. The figure inset shows on-off light switching, in which the reproducibility of the current indicates higher the stability of the systems. The IPCE maxima located around 420 nm follow the trend observed for charge separation; that is, an IPCE =12% is achieved for FTO–SnO_2–SWCNT(7,6)–$pyrNH_3^+$:**1**. For tetra-crown porphyrin **2**, a slightly lower IPCE value is observed. The higher charge-separation efficiency is established for SWCNT(7,6) with narrower band gap compared with the thin SWCNT(6,5) with wider band gap. In the case of ZnPc, the IPCE maxima are observed at 670 nm, similar to its absorption band, although the maximal values are about 1% [43].

9.7.2 Fullerene–SWCNT Hybrids via π-Stacking and Cation–Crown Binding

In another example, SWCNT–$pyrNH_3^+$ is self-assembled to benzo-18-crown-6 functionalized C_{60} via cation–dipole interactions to form the donor–acceptor hybrids of the type shown in Figure 9.15a [44]. These are characterized by the TEM and UV–visible–near-IR methods. The TEM image of SWCNT in Figure 9.15b reveals thickening of the tube upon treatment with $pyrNH_3^+$:$CrownC_{60}$.

Transient absorption spectra recorded for these nanohybrids showed the 1400-nm band of $SWCNT^{\bullet+}$ and the 1000-nm band of $C_{60}^{\bullet-}$, supporting the formation of $SWCNT^{\bullet+}$–$pyrNH_3^+$:$CrownC_{60}^{\bullet-}$ (Figure 9.16a). The band at 700 nm is characteristic of $^3C_{60}^*$, suggesting the presence of loosely bound $CrownC_{60}$. Indeed, the decay of $^3C_{60}^*$ is slower than those of $SWCNT^{\bullet+}$ (1400 nm) and $CrownC_{60}^{\bullet-}$. From the decay curve, k_{CR} is estimated to be 1×10^7 s^{-1} in DMF. From the C_{60}-fluorescence quenching, k_{CS} is estimated to be 3.5×10^9 s^{-1} in DMF. Thus, a value of $k_{CS}/k_{CR} = 350$ is obtained, suggesting considerable charge stabilization in SWCNT–$pyrNH_3^+$:$CrownC_{60}$ in DMF [44]. In a photocatalytic experiment, addition of HV^{2+} and BNAH to the

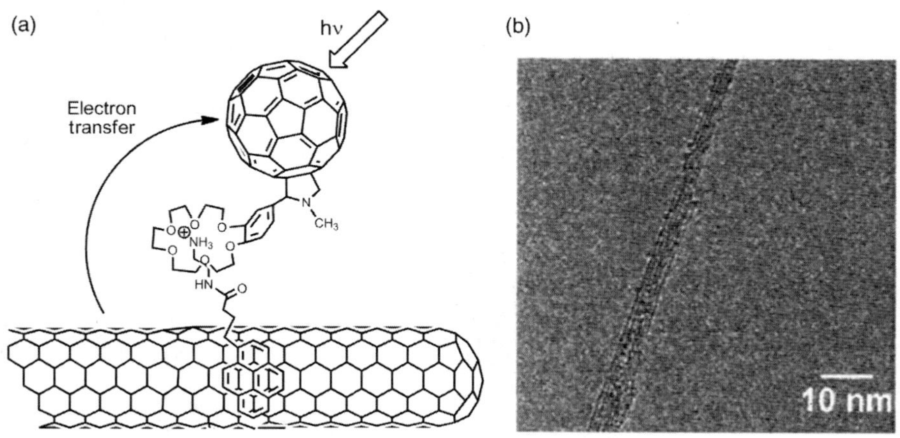

FIGURE 9.15. (**a**) Molecular structure and strategy adopted for SWCNT–$pyrNH_3^+$:$CrownC_{60}$ hybrid formation. (**b**) TEM image of a single SWCNT–$pyrNH_3^+$:$CrownC_{60}$ (dry sample). Modified from reference 44.

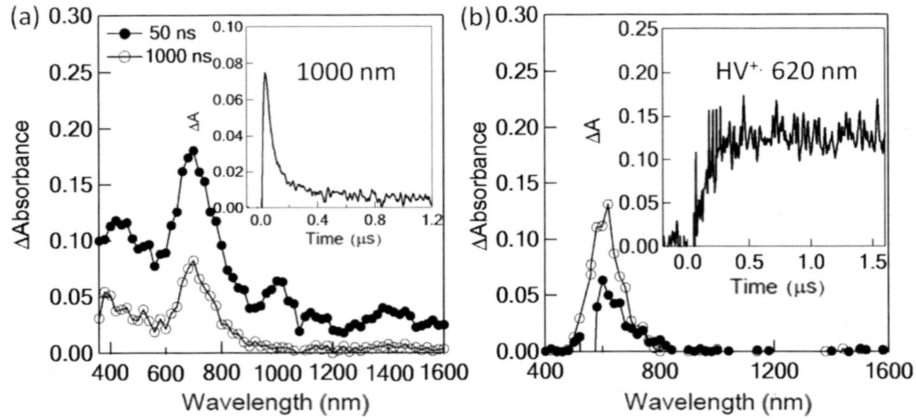

FIGURE 9.16. (**a**) Transient absorption spectra of SWCNT–pyrNH_3^+:CrownC_{60} hybrids in DMF with 532-nm laser light in DMF. (**b**) Spectral features upon addition of HV^{2+} and BNAH to the nanohybrids solution producing $HV^{\bullet+}$ in DMF. Modified from reference 44.

nanohybrid solution resulted in the disappearance of all of the transient absorption bands of $SWCNT^{\bullet+}$–pyrNH_3^+:Crown$C_{60}^{\bullet-}$. Instead, a new band at 620 nm corresponding to the formation of $HV^{\bullet+}$ appeared, suggesting occurrence of the electron mediation process from the Crown$C_{60}^{\bullet-}$ to HV^{2+} generating $HV^{\bullet+}$ (Figure 9.16b). Indeed, the yield of $HV^{\bullet+}$ accumulation increased by the repeated irradiation of the 532-nm laser light in DMF, giving a maximum conversion efficiency of 40%. The possible mechanism of this photocatalytic process is similar to the scheme in Figure 9.6(b).

9.8 DENDRIMER ARCHITECTURE

As a simple case, dendrimers containing C_{60} groups can surround SWCNTs, resulting in the homogeneous dispersion of nanohybrids in solution. An example of C_{60} dendrimer C_{60}(G*n*-COO$^-$), $n = 1.5$, is shown in Figure 9.17a [45]. Since C_{60} dendrimer is water-soluble, SWCNT–C_{60} dendrimer gave the homogeneous aqueous solution. The dry sample showed TEM image of a jaggy SWCNT, supporting the formation of SWCNT–C_{60} dendrimer nanohybrids shown in Figure 9.17b (upper right). Absorption spectra showed the visible bands due to C_{60} moiety and near-IR band of SWCNT (Figure 9.17b (lower right)).

From the C_{60}-fluorescence quenching with SWCNT (Figure 9.18 (a)), the k_{CS} values are estimated to be 3.9×10^9 s^{-1} for $n = 1.5$ and 3.6×10^9 s^{-1} for $n = 2.5$, although they contain intramolecular contribution to the CS process from carboxylate to C_{60} in aqueous solution [45]. Transient spectra showed the characteristic 1000 nm of fullerene anion radical and near-IR bands due to

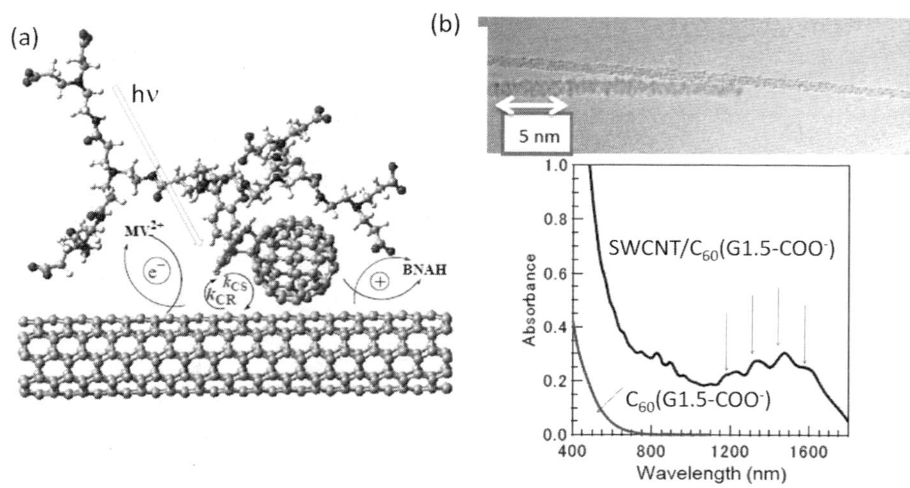

FIGURE 9.17. (**a**) C_{60} dendrimer, C_{60}(G*n*-COO$^-$); $n = 1.5$ with SWCNT ($n = 1.5$). (**b**) TEM image of SWCNT–C_{60}(G*n*-COO$^-$) (*upper right*) and the absorption spectra of SWCNT–C_{60}(G*n*-COO$^-$) (*lower right*). Modified from reference 45.

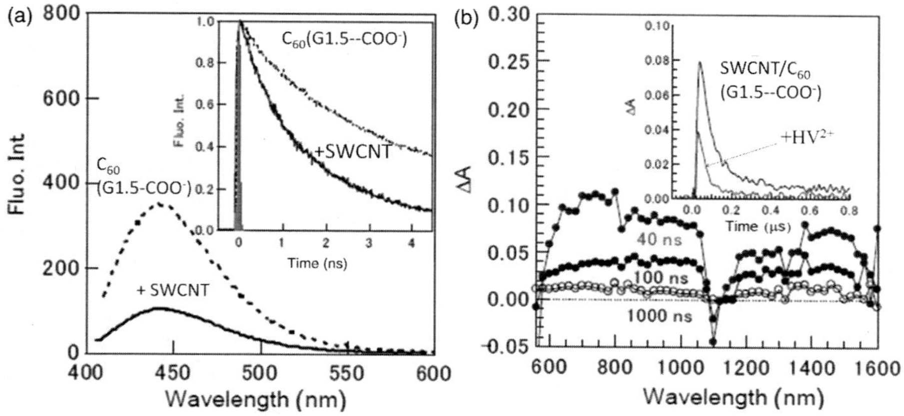

FIGURE 9.18. (**a**) Fluorescence spectra and time profile of SWCNT–C_{60}(G1.5-COO$^-$). (**b**) Transient absorption spectra of SWCNT–C_{60}(G1.5–COO$^-$) (excitation at 532 nm) in water. Modified from reference 45.

SWCNT$^{\bullet+}$–$C_{60}{}^{\bullet-}$(Gn-COO$^-$). The k_{CR} values are found to be about 1×10^7 s^{-1} for dendrimers ($n = 1.5$ and $n = 2.5$), revealing some charge stabilization. Photocatalytic reduction of MV^{2+} is also observed by electron mediation from $C_{60}{}^{\bullet-}$ to MV^{2+} as seen in rapid decay in inset of Figure 9.18 (b) [45].

9.9 SUMMARY

In this chapter, we have shown that the self-assembled supramolecular approaches developed recently in our laboratories as simple and versatile methods to construct donor–acceptor hybrids involving single-wall carbon nanotubes (SWCNT) and photosensitizers to study photophysical and photochemical process. Charge stabilization to some extent is observed in all of the studied donor–acceptor hybrids, revealing the advantages of the self-assembly methods that avoid direct contact between SWCNT and photosensitizers entities. Among the diameter-sorted nanotubes, the overall performances of SWCNT(7,6) derived photocells and photocatalytic schemes are better compared to SWCNT(6,5) due to favorable redox states. Practical applications related to photocatalysis for accumulation of photoreduced products via electron mediation through the nanohybrids, along with photoelectrochemistry, to directly convert light energy into electricity, are demonstrated. While the efficiency of photoinduced electron transfer and photocatalysis seems to be quite efficient, the efficiencies of light energy conversion are moderate. The highest value observed is 12% IPCE for ZnP–SWCNT nanohybrids assembled via π-stacking and ion-pairing methods, clearly showing the need for further research to improve light energy conversion efficiency. These nanohybrid systems are also utilized in catalytic reduction of H^+ to H_2 generation. Definitively, better schemes to improve the efficiency of this process are needed. Utilization of diameter-sorted nanotubes has shed light to gather structure–activity aspects; however, higher numbers of diameter-sorted tubes are required for the subsequent studies and applications. Much more could be anticipated in the coming years on supramolecular donor–acceptor nanohybrids formed by diameter-sorted SWCNTs.

ACKNOWLEDGMENTS

The authors are thankful to our collaborators at the Tohoku University (Sendai, Japan), JAIST (Nomi, Japan)), and Wichita State University (Wichita, KS, USA), as well as to Drs. K. Miyazawa and T. Wakahara at NIMS. This work was financially supported by the National Science Foundation (Grant Nos. CHE-0804015 and CHE-1110942, and EPS-0903806 to FD) and matching support from the State of Kansas through the Kansas Technology Enterprise Corporation.

REFERENCES

1. Harris, P. J. F. (2001). *Carbon Nanotubes and Related Structures: New Materials for the Twenty-First Century*. Cambridge University Press, Cambridge.
2. Dresselhaus, M. S., Dresselhaus, G., Avouris, P., eds. (2001). *Carbon Nanotubes: Synthesis, Structure and Applications*. Springer Publishing, New York.
3. Reich, S., Thomsen, C., Maultzsch, J. (2004). *Carbon Nanotubes: Basic Concepts and Physical Properties*. Wiley–VCH, Weinheim.
4. Roth, S., Carroll, D. (2004). *One-Dimensional Metals: Conjugated Polymers, Organic Crystals, Carbon Nanotubes*. Wiley–VCH, Weinheim.
5. Meyyappan, M. (2006). *Carbon Nanotubes, Science and Application*. Wiley-VCH, Weinheim.

6. Poole, C. P., Owens, F. J., eds. (2003). *Introduction to Nanotechnology*, Wiley-Interscience, Weinheim.
7. Wolf, E. L., ed. (2004). *Nanophysics and Nanotechnology: An Introduction to Modern Concepts in Nanoscience*. John Wiley & Sons, New York.
8. Saito, R., Dresselhaus, G., Dresselhaus, M. S. (1998). *Physical Properties of Carbon Nanotubes*. Imperial College Press, London.
9. Sgobba, V., Rahman, G. M. A., Guldi, D. M. In: Basiuk, V. A., ed. (2006). *Carbon Nanotubes in Electron Donor–Acceptor Nanocomposites, Chemistry of Carbon Nanotubes*. American Scientific Publishers, Stevenson Ranch, CA.
10. Sgobba, V., Rahman, G. M. A., Ehli, C., Guldi, D. M. (2007). Covalent and non-covalent approaches toward multifunctional carbon nanotubes materials. In: Langa, F., Nierengarten, J. F., eds. *Fullerenes—Principles and Applications*. Royal Society of Chemistry, Cambridge.
11. Sgobba, V., Guldi, D. M. (2009). Carbon nanotube-electronic/electrochemical properties and application for nanoelectronics and photonics. *Chemical Society Reviews*, **38**, 165–184.
12. D'Souza, F., Ito, O. (2010). Tetrapyrrole–nanocarbon hybrids: Self-assembly and photoinduced electron transfer. In: Kadish, K. M., Guilard, R., Smith, K. M., eds. *Handbook of Porphyrin Science*, Vol. 1. World Science Publishers, Stevenson Ranch, CA, Chapter 4, pp. 307–437.
13. Bottari, G., de la Torre, G., Guldi, D. M., Torres, T. (2010). Covalent and noncovalent phthalocyanine–carbon nanostructure systems: Synthesis, photoinduced electron transfer, and application to molecular photovoltaics. *Chemical Reviews*, **110**, 6768–6816.
14. Karousis, N., Tagmatarchis, N., Tasis, D. (2010). Current progress on the chemical modification of carbon nanotubes. *Chemical Reviews*, **110**, 5366–5397.
15. Hasobe, T. (2010). Supramolecular nanoarchitectures for light energy conversion. *Physical Chemistry Chemical Physics*, **12**, 44–57.
16. Fukuzumi, S., Kojima, T. (2008). Photofunctional nanomaterials composed of multiporphyrins and carbon-based π-electron acceptors. *Journal of Materials Chemistry*, **18**, 1427–1439.
17. Delgado, J. L., Herranz, M. Á., Martín, N. (2008). The nano forms of carbon. *Journal of Materials Chemistry*, **18**, 1417–1426.
18. D'Souza F., Ito, O. (2009). Supramolecular donor–acceptor hybrids of porphyrins/phthalocyanines with fullerenes/carbon nanotubes: Electron transfer, sensing, switching, and catalytic applications. *Chemical Communications (Feature Article)*, **45**, 4913–4928.
19. Sandanayaka, A. S. D., Ito, O. (2009). Photoinduced electron transfer in supramolecules composed of porphyrin/phthalocyanine and nanocarbon materials. *Journal of Porphyrins and Phthalocyanines*, **13**, 1017–1033.
20. D'Souza, F., Sandanayaka, A. S. D., Ito, O. (2010). SWNT-based supramolecular nanoarchitectures with photosensitizing donor and acceptor molecules. *Journal of Physical Chemistry Letters (Perspective)*, **1**, 2586–2593.
21. Chen, R. J., Zhang, Y., Wang, D., Dai, H. (2001). Noncovalent sidewall functionalization of single-walled carbon nanotubes from pretein immobilization. *Journal of American Chemical Society*, **123**, 3838–3839.
22. Maligaspe, E., Sandanayaka, A. S. D., Hasobe, T., Ito O., D'Souza F. (2010). Sensitive efficiency of photovoltaic cell to band gaps of semi-conductive SWNTs with the supramolecularly attached zinc porphyrin bearing pyrene glues. *Journal of the American Chemical Society*, **132**, 8158–8164.
23. Das, S. K., Subbaiyan, N. K., D'Souza, F., Sandanayaka, A. S. D., Hasobe, T., Ito, O. (2011). Photoinduced processes of the supramolecularly functionalized semi-conductive SWCNTs with porphyrins via ion-pairing interactions. *Energy and Environment Chemistry*, **4**, 707–716.
24. Tanaka, Y., Hirana, Y., Niidome, Y., Kato, K., Saito, S., Nakashima, N. (2009). Experimentally determined redox potentials of individual *(n,m)* single-walled carbon nanotubes. *Angewandte Chemie International Edition*, **48**, 7655–7659.
25. Sandanayaka, A. S. D., Maligaspe, E., Hasobe, T., Ito, O., D'Souza, F. (2010). Diameter dependent electron transfer in supramolecular nanohybrids of (6,5)- or (7,6)-enriched semiconducting SWNT as donors and fullerene as acceptor. *Chemical Communications*, **46**, 8749–8751.
26. Maggini, M., Scorrano, G., Prato, M. (1993). Addition of azomethine ylides to C_{60}: Synthesis, characterization, and functionalization of fullerene pyrrolidines. *Journal of the American Chemical Society*, **115**, 9798–9799.
27. El-Khouly, M. E., Ito, O., Smith, P. M., D'Souza, F. (2004). Intermolecular and supramolecular photoinduced electron transfer processes of fullerene–porphyrin/phthalocyanine systems. *Journal of Photochemistry and Photobiology, C*. **5**, 79–104.
28. Kadish, K. M., Smith, K. M., Guilard, R. eds. (2000). *The Porphyrin Handbook*. Vol. 1–10. Academic Press, San Diego.
29. Arai, T., Nobukuni, S., Sandanayaka, A. D. S., Ito, O. (2009). Zinc porphyrin covalently bound to the side wall of single-walled carbon nanotubes via flexible bonds: Photoinduced electron transfer in polar solvent. *Journal of Physical Chemistry*, **113**, 14493–14499.
30. Fukuzumi, S., Suenobu, T., Patz, M., Hirasaka, T., Itoh, S., Fujitsuka, M., Ito, O. (1998). Selective one-electron and two-electron reduction of C_{60} with NADH dimer analogues via photoinduced electron transfer. *Journal of American Chemical Society*, **120**, 8060–8068.
31. Guldi, D. M., Rahman, G. M. A., Qin, S. H., Tchoul, M., Ford, W. T., Marcaccio, M, Paolucci. D., Paolucci, F., Campidelli, S., Prato, M. (2006). Versatile coordination chemistry towards multifunctional carbon nanotube nanohybrids. *Chemistry–A European Journal*, **12**, 2152–2161.
32. Boul, P. J., Cho, D. G., Rahman, G. M. A., Marquez, M., Ou, Z., Kadish, K. M., Guldi, D. M., Sessler, J. L. (2007). Sapphyrin-nanotube assemblies. *Journal of American Chemical Society*, **129**, 5683–5687.
33. Bachili, S. M., Balzani, L., Herrara, J. E., Pompeo, F., Resasco, D. E., Weisman, R. B. (2003). Narrow *(n,m)* distribution of single walled carbon nanotubes grown using a solid supported

catalyst. *Journal of American Chemical Society*, **125**, 11187–11197.

34. D'Souza, F., Ito, O. (2005). Photoinduced electron transfer in supramolecular systems of fullerenes functionalized with ligands capable of bindings to zinc porphyrins and zinc phthalocyanines. *Coordination Chemical Reviews*, **249**, 1410–1422.
35. Chitta, R., Sandanayaka, A. S. D., Shumacher, A. L., Araki, Y., Ito, O., D'Souza, F. (2007). Donor–acceptor nanohybrids of zinc naphthalocyanine or zinc porphyrin non-covalently linked to single-wall carbon nanotubes for photoinduced electron transfer. *Journal of Physical Chemistry C*, **111**, 6947–6955.
36. Sandanayaka, A. S. D., Chitta, R., Subbaiyan, N. K., D'Souza, L., Ito, O., D'Souza, F. (2009). Photoinduced charge-separation in ion-paired porphyrin-single wall carbon nanotube donor–acceptor hybrids. *Journal of Physical Chemistry C*, **113**,13425–13432.
37. D'Souza, F., Maligaspe, E., Sandanayaka, A. S. D., Subbaiyan, N. K., Karr, P. A., Hasobe, T., Ito, O. (2010). Photoinduced charge-separation in supramolecular phthalocyanine–multifullerene conjugate assembly by crown ether-alkyl amine cation interaction. *Journal of Physical Chemistry C*, **114**, 10951–10595.
38. Maligaspe, E., Tkachenko, N. V., Subbaiyan, N. K., Chitta, R., Zandler, M.E,. Lemmetyinen, H., D'Souza, F. (2009). Photosynthetic antenna-reaction center mimicry: Sequential energy- and electron-transfer in a self-assembled supramolecular triad composed of boron dipyrrin, zinc porphyrin and fullerene. *Journal of Physical Chemistry A*, **113**, 8478–8489.
39. D'Souza, F., Maligaspe, E., Ohkubo, K., Zandler, M. E., Subbaiyan, N. K., Fukuzumi, S. (2009). Photosynthetic reaction center mimicry: Low-reorganization energy driven charge stabilization in self-assembled co-facial zinc phthalocyanine dimer–fullerene conjugate. *Journal of American Chemical Society*, **131**, 8787–8797.
40. D'Souza, F., Chitta, R., Gadde, S., Zandler, M. E., McCarty, A. L., Karr, P.A, Sandanayaka, A. S. D., Araki, Y., Ito, O. (2006). Design, syntheses, and studies of supramolecular porphyrin–fullerene conjugates using bis 18-crown-6 appended porphyrins and pyridine and/or alkyl ammonium functionalized fullerenes. *Journal of Physical Chemistry B*, **110**, 5905–5913.
41. Maligaspe, E., D'Souza, F. (2010). NOR and AND logic gates based on supramolecular porphyrin–fullerene conjugates. *Organic Letters*, **12**, 624–627.
42. D'Souza F., Chitta, R., Sandanayaka, A. S. D., Subbaiyan, N. K., D'Souza, L, Araki, Y., Ito, O. (2007). Self-assembled single-walled carbon nanotube:zinc-porphyrin hybrids through ammonium ion-crown ether interaction: Construction and electron transfer. *Chemistry—A European Journal*, **13**, 8277–8284.
43. Sandanayaka, A.S.D., Subbaiyan, N.K., Chitta, R., Maligaspe, E., Hasobe, T., Ito, O., D'Souza, F. (2011). Diameter sorted SWCNT–porphyrin and SWCNT–phthalocyanine conjugates self-assembled for light energy harvesting. *Chem Phys Chem*, **12**, 2266–2274.
44. D'Souza, F., Chitta, R., Sandanayaka, A. S. D., Subbaiyan, N. K., D'Souza, L., Araki, Y., Ito, O. (2007). Supramolecular carbon nanotube–fullerene donor–acceptor hybrids for photoinduced electron transfer. *Journal of American Chemical Society*, **129**, 15865–15871.
45. Sandanayaka, A. S. D., Takaguchi, Y., Sako, Y., Tamura, M., Ito, O. (2010). Photoinduced electron transfer of single walled carbon nanotubes surrounded by fullerodendrimers in aqueous media. *Advanced Science Letters*, **3**, 353–357.

10

INTERFACING PORPHYRINS/PHTHALOCYANINES WITH CARBON NANOTUBES

JUERGEN BARTELMEß AND DIRK M. GULDI

10.1 INTRODUCTION

The bacterial photosynthetic reaction center provides meaningful incentives for the optimization of charge separation processes in artificial model systems—*nanometer scale structures* [1–6]. Common to all these systems is a relay of short-range energy/electron transfer reactions, evolving among chlorophyll and quinone moieties [7]. Among many key parameters that govern electron transfer reactions, the reorganization energy imposes probably the most far reaching impact. The reorganization energy (λ) is the energy required to structurally reorganize the donor, the acceptor, and their solvation spheres upon electron transfer [8, 9]. For example, the primary electron transfer processes of photosynthesis are characterized by an extremely small reorganization energy (0.2 eV), attained by the transmembrane protein environment [10,11]. This aspect is central for achieving the ultrafast charge separation and retarding the energy wasting charge recombination, which is highly exergonic ($-\Delta G_{CR}{}^{\circ} = 1.2$ eV) [12].

Nanometer scale structures—fullerenes and carbon nanotubes (CNTs)—are the focus of considerable interest because they can be used to test fundamental ideas about the roles of dimensionality and confinement in materials of greatly reduced size (see Figure 10.1) [13–16].

One group of active *nanometer scale structures* that we consider is the carbon nanotubes (CNTs) [17–20]. Of the wide range of nanostructures available, CNT, in general, and single-wall carbon nanotubes (SWNTs), in particular, stand out as unique materials. In fact, CNTs have emerged as a new class of materials with exceptionally high tensile strength [21–25], the highest thermal conductivity known [26–32], and extraordinary field emission properties [33–37].

SWNTs are one-dimensional nanostructures formed by hexagon networks of carbon atoms, rolled up to create seamless cylinders, along a chiral vector. While their diameters are typically in the range of nanometers, individually CNT reach lengths of up to 4 cm, leading to high aspect ratios. Notably, the conductivity of CNTs varies between semiconducting and metallic and depends, in large part, on the chiral angle of the tubes and their diameter. They readily accept electrons, which can then be transported under nearly ideal conditions along the axis.

On the other hand, multiple concentric graphene cylinders—multiwall carbon nanotubes (MWNTs)-exhibit metallic or semiconducting properties, which depend solely on their outermost shell [38, 39]. On account of the large number of concentric cylindrical graphitic tubes present in MWNT, they are considered even more suitable in electron donor–acceptor ensembles than SWNTs. It is notable that, in line with this purely structural assumption, appreciable differences were seen as far as the stabilization of charge separation is concerned [40,41].

Due to the unique ability of SWNTs to form both metallic and semiconducting species, they represent appealing candidates for key building blocks in the construction of photovoltaic devices [42–51]. A major setback is, however, that carbon nanotubes are subject to strong, mutual interactions through van der Waals forces and, therefore, yield agglomerates of intimately intertwined long tubes. These cohesive forces account for the poor SWNT dispersibility in most common solvents: They hinder, in large part, their chemical manipulation and, more importantly, the realization of the

Organic Nanomaterials: Synthesis, Characterization, and Device Applications, First Edition. Edited by Tomás Torres and Giovanni Bottari.

FIGURE 10.1. Carbon nanostructures. See color insert.

full potential of SWNTs in practical applications. To transfer, their outstanding properties from the nanoscale to the macroscale, the chemical and physical modification of the SWNT and MWNT surface are essential steps [52–58].

All the mentioned nanometer scale structures—fullerenes and CNTs—are natural electron acceptors. Proof of this concept is obtained mathematically even by qualitative molecular orbital theory. The argument starts from isolated C_2 fragments that are brought together starting from infinite distance. Each fragment has a π orbital and a π^* orbital. As they are brought together to form, for example, a CNT, the two degenerate sets of π and π^* orbitals mix prevalently between themselves and spread in energy. The low-lying end of the π^* orbitals is very stable and readily accept electrons [59–61]

Porphyrinoid and especially metalloporphyrinoid systems—with their rich and extensive absorptions throughout the visible region of the solar spectrum—hold particularly great promise as integrative building blocks with increased absorptive cross sections [62–65]. Over the course of recent years, they emanate as light-harvesting building blocks in the construction of molecular architectures. Their high electronic excitation energy, typically exceeding 2.0 eV, powers a strongly exergonic electron transfer and intercedes hereby the conversion between light and chemical/electrical energy. Another important feature of porphyrins is their highly delocalized π-electron systems. Upon an uptake or release of electrons, such delocalization results in minimal structural change upon electron transfer. Rich redox properties render porphyrins and porphyrin analogous as essential components in important biological electron transport systems including photosynthesis and respiration.

One of the major challenges that still lies ahead is to regulate the forces, which will ultimately dictate size and shape in relation to function of the resulting nanometer scale structures. Can molecular tailoring of fullerenes and CNTs contribute to the induction of new assemblies? To address such issues, we have probed in recent years novel electron donor–acceptor ensembles, in which nanometer scale structures constitute the acceptor moiety and biomimetic organization principles—*hydrogen bonding, complementary electrostatics, π–π stacking, and metal coordination*—thereby ensuring the hierarchical integration of multiple components into well-ordered arrays. These spontaneous organization principles permit engineering of novel functional electron donor–acceptor (nano)conjugates and (nano)hybrids and, simultaneously, achieve predetermined architectures of controlled sizes and outer-shell structures, with high directionality and selectivity. These nanometer scale structures have been successfully utilized to convert photolytically generated radical ion pairs into electrical or chemical energy by constructing integrated artificial photosynthetic assemblies on modified indium–tin-oxide electrodes.

An intriguing concept involves using the surface of semiconducting and metallic CNT as templates to integrate porphyrins or phthalocyanines that serve as visible light-harvesting chromophores.

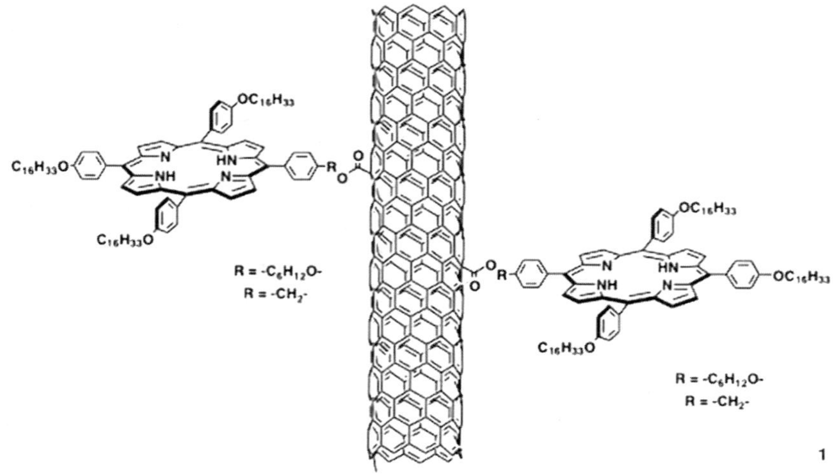

1

10.2 RESULTS AND DISCUSSIONS

A first example reports on the efficient covalent tethering of SWNTs with porphyrins through the esterification of SWNT-bound carboxylic acid groups (**1**) [66]. Based on the initially established approach, the surface-bound carboxylic acid groups were produced by acid treatment. In particular, the work started with two porphyrin derivatives containing terminal hydroxyl groups [5-(*p*-hydroxyhexyloxyphenyl)-10,15,20-tris(*p*-hexadecyloxyphenyl)-21*H*,23*H*-porphine] and [5-(*p*-hydroxymethylphenyl)-10,15,20-tris(*p*-hexadecy loxyphenyl)-21*H*,23*H*-porphine] [67]. FTIR and thermal analysis indicated that amide or ester bonds were formed to bridge between the excited state electron donor and the SWNT. In the corresponding electron donor–acceptor nanoconjugates (*i.e.*, SWNT-H_2P) the photoexcited porphyrins deactivate, through a transduction of excited-state energy. Interestingly, the rates and efficiencies of the excited-state transfer depend on the length of the tether that links the porphyrins with the SWNT. Thus, the materials with the shorter tethers showed the least fluorescence quenching.

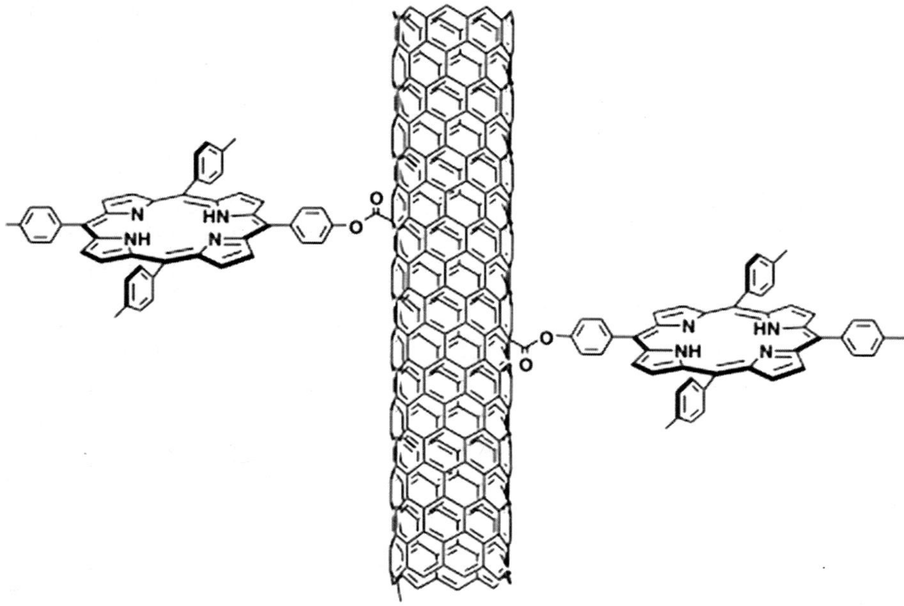

2

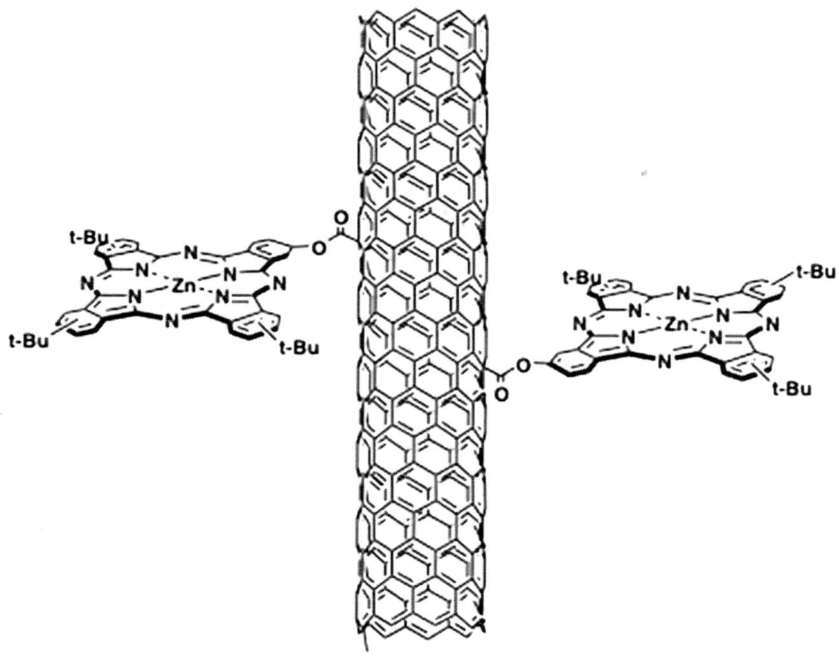

3

Extending the scope of the aforementioned work, the carboxylic functionalities proved to be convenient for linking 5-*p*-hydroxyphenyl-10-15-20-tritolylporphyrin to either SWNTs or MWNTs (**2**) [66, 68]. According to thermogravimetric analyses, the amount of grafted porphyrins was estimated to range from 8% to 22%, depending on the initially employed porphyrin concentrations.

Unsymmetrically substituted zinc aminophthalocyanines were linked to SWNT by a reaction of the terminal carboxylic acid groups of shortened chemically etched SWNT (**3**) [69]. However, the resulting materials are nearly insoluble in common organic solvents. The solubility/dispersability of the SWNT-ZnPc conjugates is, however, an indispensable task for a ready manipulation and a feasible solution phase processing them into devices.

Reasonable alternatives involve a straightforward *Prato* reaction with *N*-octylglycine and a formyl-containing ZnPc or a stepwise approach that involves a former Prato cycloaddition to the double bonds of SWNT using *p*-formyl benzoic acid followed by esterification of the derivatized nanotubes with an appropriate ZnPc. The occurrence of electron transfer from photoexcited ZnPc to SWNT in **4** [70] is observed in transient absorption experiments, which reflect the absorption of the one-electron oxidized ZnPc cation and the concomitant bleaching of the van Hove singularities typical from SWNT. Charge separation (*i.e.*, 2.0×10^{10} s^{-1}) and charge recombination (*i.e.*, $7 \pm 0.5 \times 10^{-5}$ s) dynamics reveal a notable stabilization of the radical ion pair product in dimethylformamide.

We described the functionalization of SWNT with 4-(2-trimethylsilyl)ethynylaniline and the subsequent attachment of ZnPc derivative using the reliable *Huisgen* 1,3-dipolar cycloaddition. The motivation of this study was the preparation of an SWNT platform, which allows the facile fabrication of more complex functional nanometer-scale structures, such a SWNT–ZnPc **5** [71]. **5** was fully characterized by a combination of analytical techniques such as Raman, absorption and emission spectroscopy, atomic force and scanning electron microscopy, and thermogravimetric analysis. **5** was also investigated with a series of steady-state and time-resolved spectroscopy experiments, and a photoinduced communication between the two photoactive components was identified. Such beneficial features lead to monochromatic internal photoconversion efficiencies of 17.3%, when **5** was tested as photoactive material in an ITO photoanode.

Quite different is the approach that involves SWNT functionalization by placing pyridyl isoxazolino functionalities along the sidewalls of short SWNTs. The synthesis is based on the 1,3-dipolar cycloaddition of a nitrile oxide onto SWNTs. The resulting SWNT-py form complexes with ZnP. Formation of the SWNT-py/ZnP complex **6** [72] was firmly established by a detailed electrochemical study. However, photochemical excitation of the complex between SWNT-py/ZnP does not lead to electron transfer with the generation of charge separated states. Fluorescence and laser flash studies indicate that the main process is energy transfer from the singlet excited-state of ZnP to SWNT-py.

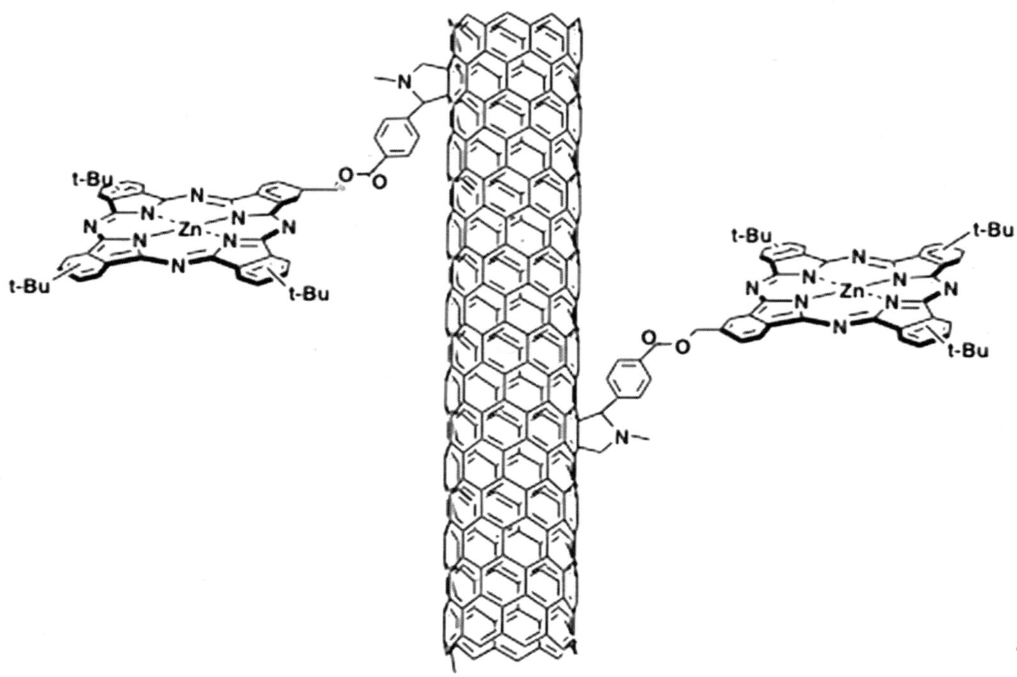

4

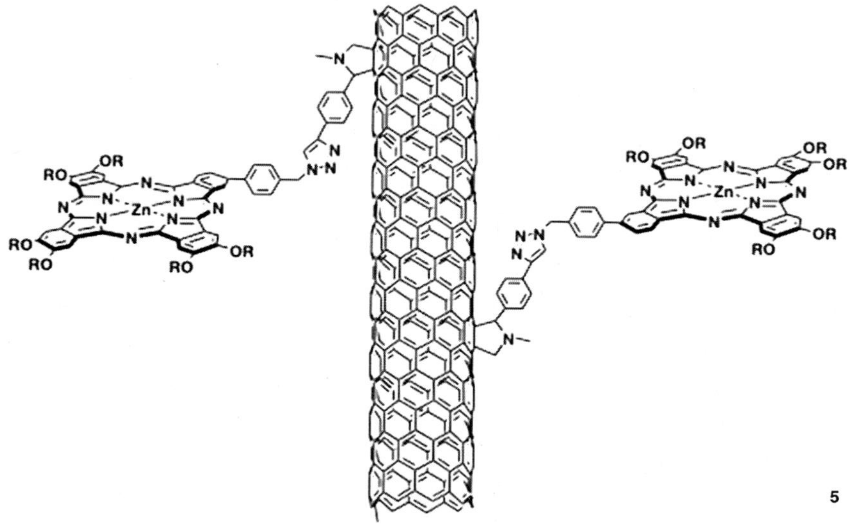

5

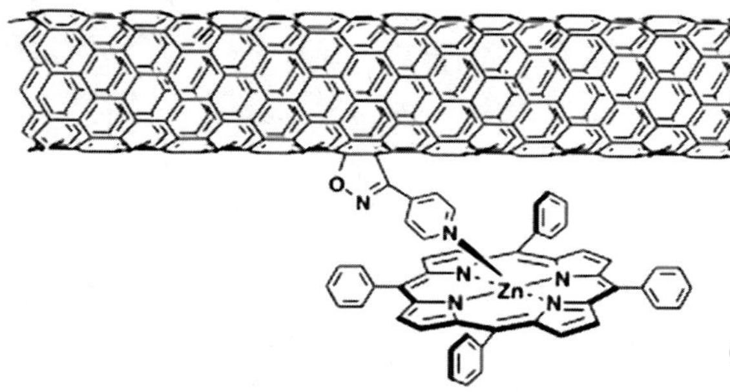

6

SWNT functionalization is also the basis when applying Suzuki coupling reactions (**7**) [73]. This type of coupling reactions represents an efficient method for introducing ZnP onto the SWNT sidewalls. Despite all perceptions, covalently functionalized SWNTs were found to serve as efficient quenchers even after their perfectly conjugated sidewall structure has been disrupted.

An important consideration when associating CNT with electron donor building blocks is to preserve—as much as possible—the unique electronic structures of the CNTs. A versatile approach involves grafting SWNT with polymers such as poly(sodium 4-styrenesulfonate) (SWNT-PSS^{n-} **8** [45, 66, 74]) with a SWNT-to-PSS^{n-} ratio of 55/45. In turn,

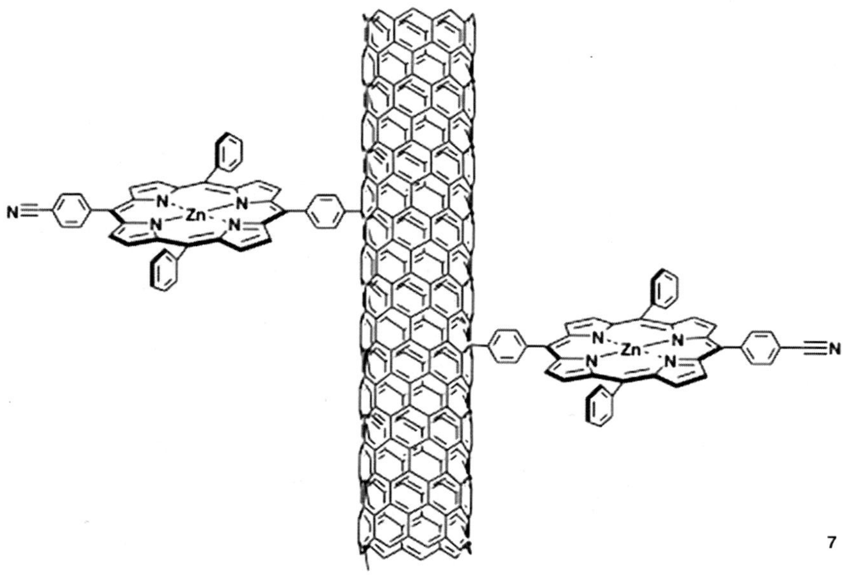

7

highly stable water-dispersable SWNTs were prepared with relative ease. The attached PSS functionalities also assist in exfoliating individual SWNT-PSS^{n-} from the larger bundles and stabilizing them. AFM and TEM analysis corroborated the presence of SWNTs with lengths reaching several micrometers and diameters around 1.2 nm. A Coulomb complex formation was achieved with SWNT-PSS^{n-} and an octapyridinium H_2P salt. Several spectroscopic techniques like absorption, fluorescence, and TEM were used to monitor the complex formation between SWNT-PSS^{n-} and octapyridinium H_2P salt yielding SWNT-PSS^{n-}-electrostatic-H_2P. Importantly, photoexcitation of the octapyridinium H_2P chromophore in the newly formed nanohybrid structure results in efficient intrahybrid charge separation (0.3 ns), which leads, subsequently, to radical ion pair formation. The charge separation is governed by a large thermodynamic driving force of 0.81 eV. The newly formed radical ion pair exhibits a remarkably long lifetime of 14 μs under anaerobic conditions, which constitutes one of the longest reported for any CNT ensemble found so far.

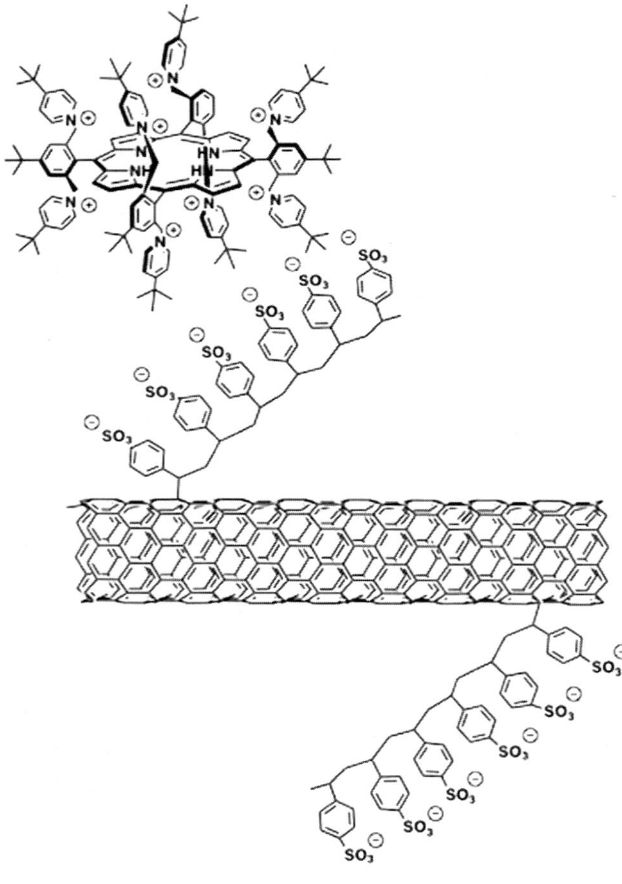

8

Similarly, dispersable SWNTs—grafted with poly(4-vinylpyridine) (PVP)—were assayed in coordination tests with ZnP. Grafting of the polymer to the SWNT sidewalls was supported by Raman and near-IR spectra, and the composition of the resulting SWNT–PVP was estimated as 61/39. Kinetic and spectroscopic evidence corroborates the successful formation of SWNT-PVP/ZnP nanohybrids (**9**) in solutions [75]. Within this SWNT-PVP/ZnP nanohybrid, static electron transfer quenching 2.0 ($\pm$ 0.1) $\times$ 10^9 s^{-1} converts the photoexcited ZnP chromophore into a microsecond-lived radical ion pair state—that is, one-electron oxidized ZnP and reduced SWNT.

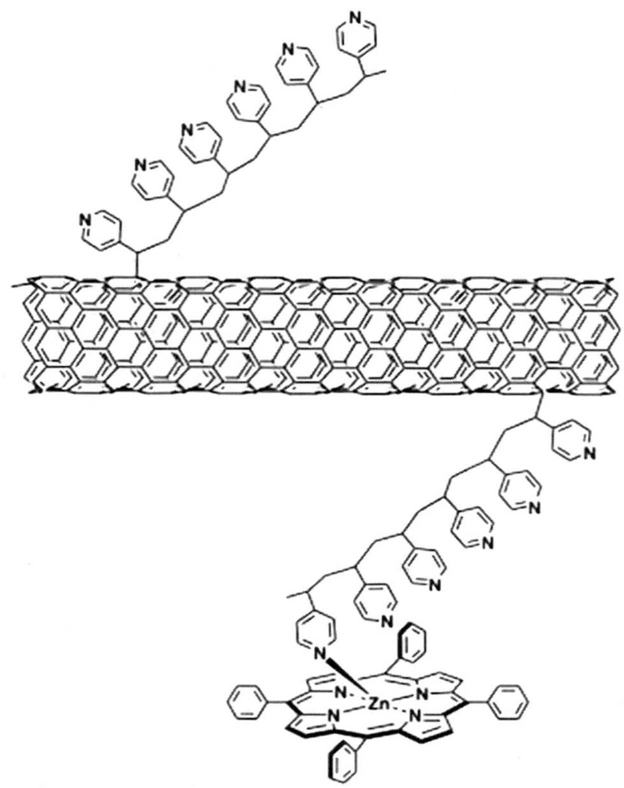

9

Reacting poly((vinylbenzyl)trimethylammonium chloride) (PVBTA^{n+}) through free-radical polymerization of the monomer in the presence of SWNT yielded polymer-grafted SWNT. A donor–acceptor nanohybrid (**10**) [76, 77] have been afterward prepared using electrostatic/van der Waals interactions between covalent SWNT-PVBTA^{n+} and 5,15-bis-[2′,6′-bis-{2″,2″-bis-(carboxy)-ethyl}-methyl-4′-*tert*-butyl-pheny]-10,20-bis-(4′-*tert*-butylphenyl) porphyrin octasodium (H_2P^{8-} and/or ZnP^{8-}).

A series of SWNT that are functionalized with PAMAM dendrimers (**11**) [78] were described. Importantly, the dendrimers are linked directly to the SWNT surface using a divergent methodology. This approach allows increasing the number of functional groups on the nanotubes without provoking significant damages to the conjugated π-system

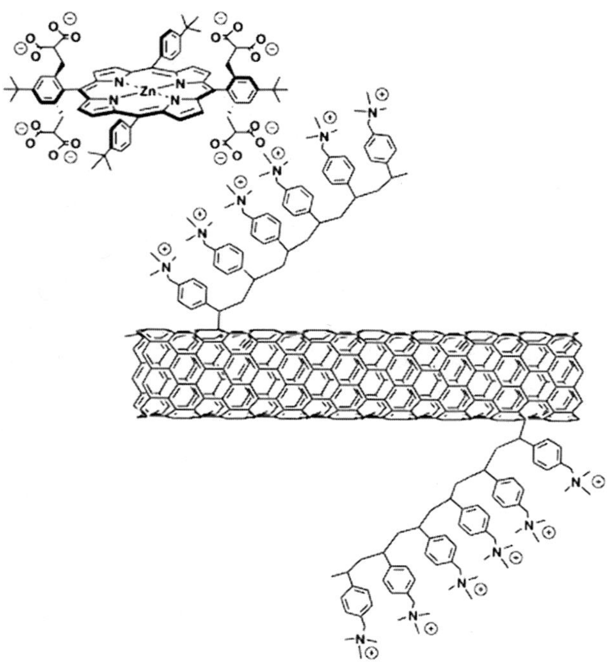

10

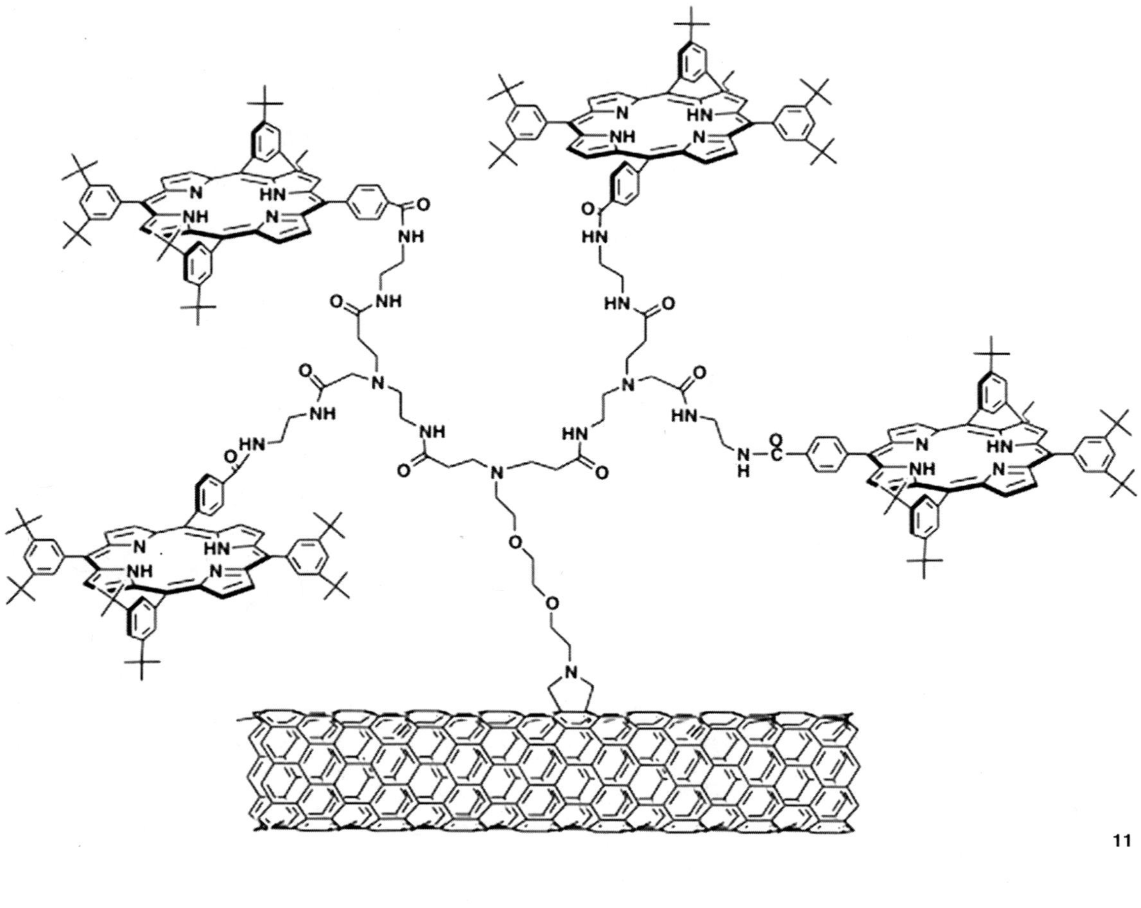

11

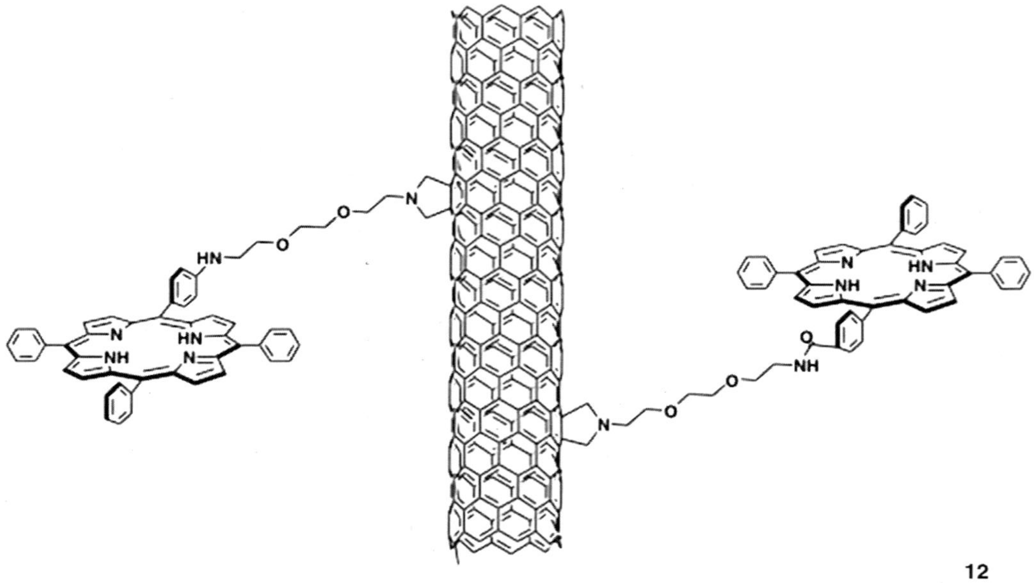

12

of SWNT and comes as a reasonable alternative to **12** [79] where the lack of sufficient H_2P absorption hampers further photovoltaic applications. **11** shows that several tetraphenylporphyrin moieties can be linked to the periphery of the dendrimers. The photophysical properties of the resulting nanoconjugates (*i.e.*, SWNT-$(H_2P)_n$) have been investigated with a series of steady-state and time-resolved spectroscopy. The fluorescence kinetics provide evidence for two transient decays, one very short-lived (*i.e.*, 0.04 ± 0.01 ns) and one long-lived (*i.e.*, 8.6 ± 1.2 ns). A possible explanation is that some porphyrin units do not interact with the nanotubes, thus exhibiting a fluorescence lifetime similar to that of the free porphyrin. Complementary transient absorption measurements not only corroborate the fast decay of the photoexcited H_2P, but also confirm that intraconjugate charge separation evolves from the excited porphyrin to the SWNT.

As a leading example, our success in the facile supramolecular association of pristine SWNT with linearly polymerized porphyrin polymers toward versatile electron donor–acceptor nanohybrids should be considered. The target SWNT nanohybrids, which are dispersable in organic media, were realized through the use of soluble and redox-inert poly(methylmethacrylate) (PMMA) bearing surface-immobilized porphyrins (*i.e.*, H_2P-polymer). Conclusive evidence for H_2P-polymer/SWNT interactions came from absorption spectroscopy: The fingerprints of SWNT and H_2P-polymer are discernable throughout the UV, visible, and near-infrared part of the spectrum. A similar conclusion was also derived from TEM and AFM. AFM images illustrate, for example, the debundling of individual SWNT. An additional feature of H_2P-polymer/SWNT (**13**) [66, 76, 80] is an intrahybrid charge separation, which has been shown to last for 2.1 ± 0.1 μs.

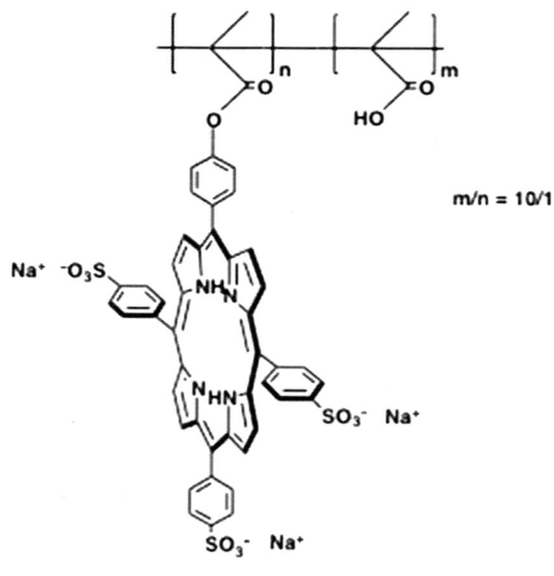

13

SWNT/$PVBTA^{n+}$ are water-suspendable SWNT, in which $PVBTA^{n+}$ are noncovalently wrapped around them—related to **10** [76, 77]. Versatile electron donor–acceptor nanohybrids have been prepared using electrostatic/van der Waals interactions between SWNT/$PVBTA^{n+}$ nanohybrids and porphyrins (H_2P^{8-} and/or ZnP^{8-}). Several spectroscopic, microscopic, transient, and photoelectrochemical measurements were employed to characterize the resulting supramolecular complexes. The photoexcitation of the nanohybrids afforded long-lived radical ion pairs with lifetimes as long as 2.2 μs.

interaction between SWNT and H_2P was evaluated by UV–vis and fluorescence spectra and by chemical removal of H_2P from SWNTs. In the electron donor–acceptor nanohybrids, the Soret and Q-bands of H_2P moieties were significantly broadened and their fluorescence was almost completely quenched. Apparent extinction coefficients at the Soret bands were decreased to ~20%. Successful complexation with, for example, the ZnP-polymer was manifested in a 127-nm bathochromic shift of the Q-band absorption. Additional evidence for interactions among ZnP-polymer, ZnP-trimer, ZnP, and SWNTs was obtained from fluorescence spectra, where the ZnP fluorescence was significantly quenched. The fluorescence quenching has been ascribed to energy transfer between the photoexcited porphyrin and SWNT.

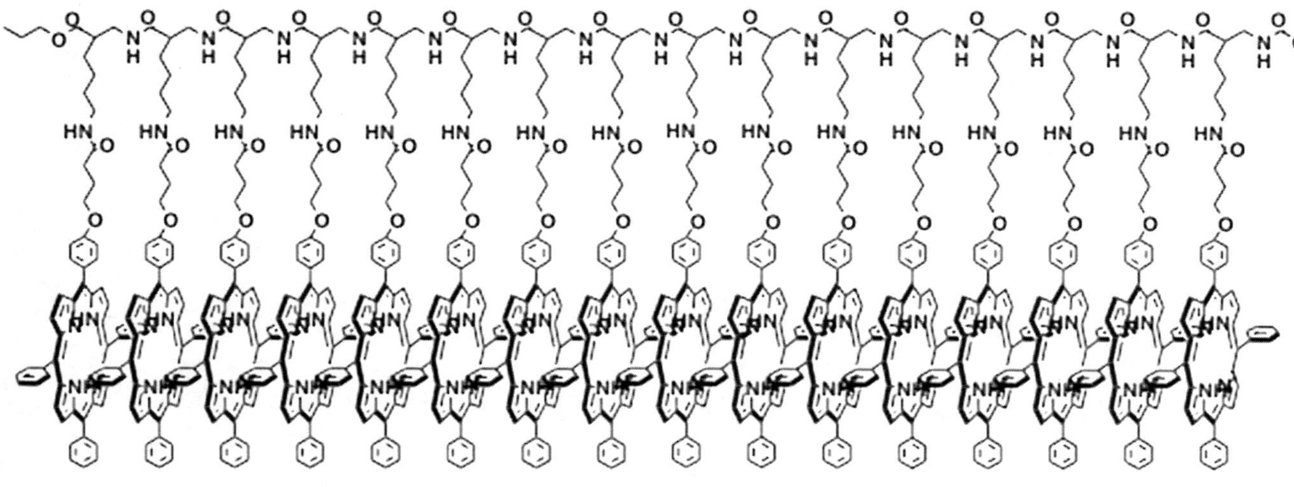

14

Strategies for SWNT separation are critical to develop nanotubes as useful nanoscale building blocks. Herein, diameter-selective dispersion of SWNT has been accomplished through noncovalent complexation with a flexible porphyrinic polypeptides bearing 16 porphyrin units in DMF. Supramolecular formation occurs through wrapping of the peptide backbone and π–π interactions between porphyrins and SWNT to extract the large-diameter nanotubes (~1.3 nm) as revealed by UV-vis–near-IR and Raman spectroscopy as well as high-resolution transmission electron micropscopy. The photoexcitation of this electron donor–acceptor hybrid (**14**) [81] affords long-lived charge-separated states.

Following basically a similar strategy, SWNT were found to strongly interact with a highly soluble, conjugated ZnP-polymer (**15**), [82] a novel triply fused ZnP-trimer (**16**) [83], and just ZnP (**17**) [84, 85]; a rare earth phthalocyanine (*i.e.*, $HErPc_2$) [87]. Stable H_2P/SWNT composites were also realized by condensating tetraformylporphyrins (**18**) and diaminopyrenes on SWNT [87]. The degree of

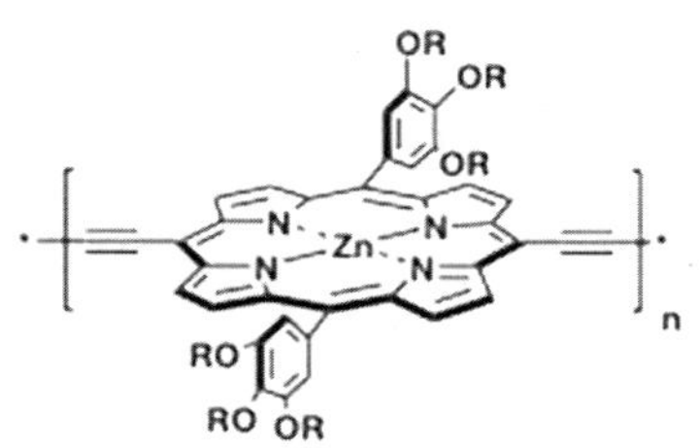

$R = (CH_2)_{15}CH_3$

15

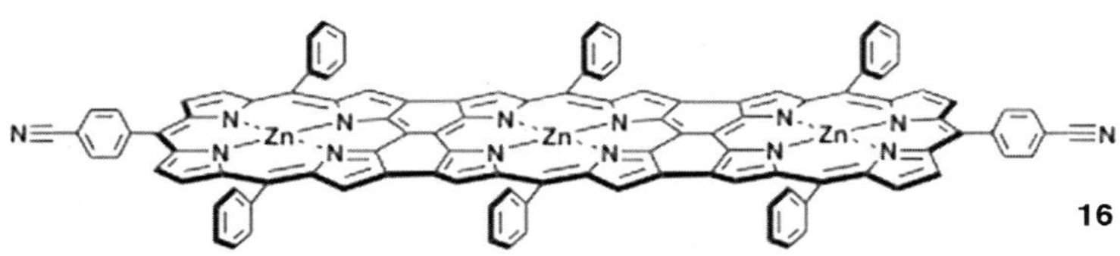

16

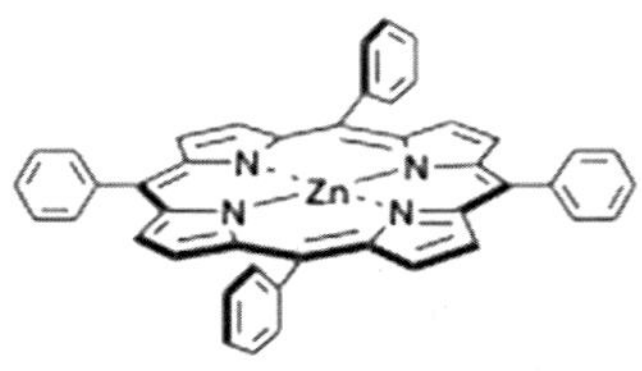

17

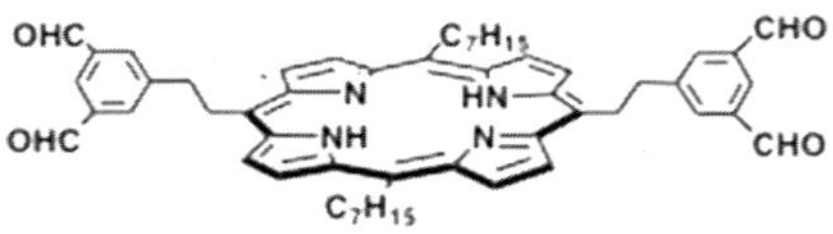

18

Supramolecular structures consisting of dendritic porphyrins (**19**) and SWNT have been prepared and characterized as an efficient donor/acceptor system [88]. Noncovalent interactions enable the pair system to produce suitable electron transfer through a process occurring from the dendritic porphyrin core to the graphenic wall of carbon nanotubes. Important is structure/architecture correlation based on photoelectrical investigations.

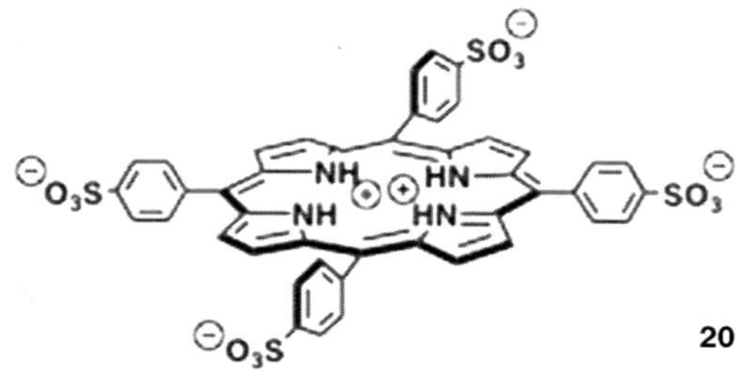

20

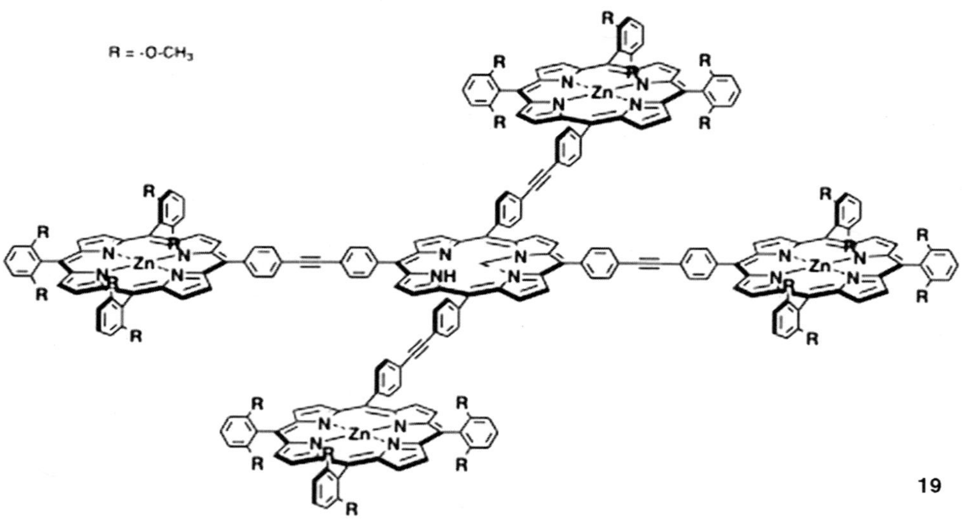

19

Generally, SWNT interactions with H_2P turned out to be appreciably stronger than what is typically seen when employing ZnP. AFM images of, for example, SWNT/H_2P revealed smaller bundle sizes than those recorded for SWNT/ZnP with even some individual specimen (*i.e.*, diameters of 1.5 ± 0.2 nm). Overall, SWNT/H_2P interactions opens the way to suspensions that are more stable than those of SWNT/ZnP.

The trend of stability was further corroborated with a water-soluble H_2P derivative, namely, [*meso*-(tetrakis-4-sulfonatophenyl) porphine dihydrochloride] (**20**) [89]. In these SWNT suspensions, which were shown to be stable for several weeks, eminent interactions protect H_2P against protonations to the corresponding diacid. Moreover, these strongly interacting hybrids have been successfully aligned onto hydrophilic polydimethylsiloxane surfaces by flowing SWNT solutions along a desired direction and then transferred to silicon substrates by stamping.

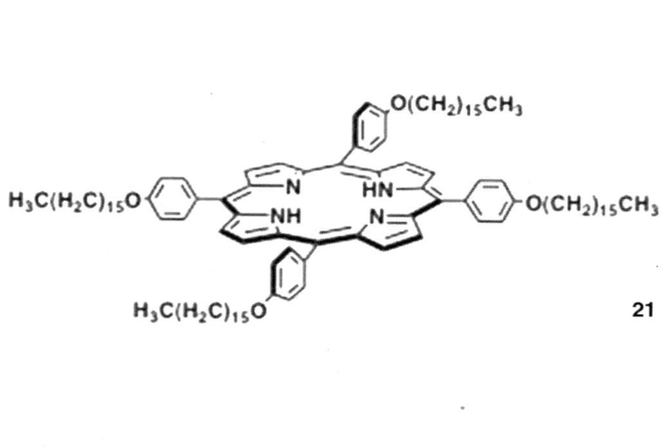

21

Quite interesting is the following work: SWNTs were effectively worked up with H_2P [5,10,15, 20-tetrakis (hexadecyloxyphenyl)-21*H*,23*H*-porphine] (**21**) [90]. Porphyrin functionalized SWNT were separated from non-functionalized SWNT by dissolving in THF and, following this, separated from H_2P by treatment with acetic acid and vigorous centrifugation. After heating the SWNT fractions to 800°C in a nitrogen atmosphere, spectroscopic analysis showed that the insoluble SWNTs are enriched in metallic SWNTs, while the other, formerly porphyrin functionalized SWNT fraction is enriched in semiconducting SWNTs. We must assume selective interactions between H_2P and semiconducting SWNT as the inception to a successful separation of metallic and semiconducting nanotubes.

Solubilization/dispersion of SWNTs was also accomplished with protoporphyrin IX (**22**) [91,92] and tetraphenylporphyrin iron(III) chloride (**23**) [92]. Transmission electron and atomic force microscopies, as well as UV–vis–near-IR spectroscopy, revealed that the protoporphyrins IX individually dissolved SWNT in polar solvents under mild conditions of sonication using a bath-type sonicator, followed by centrifugation at 1000*g*. The absorption and fluorescence spectral measurements of the octaalkylporphyrins in dimethylformamide containing 1 vol% tetrahydrofuran with various amounts of SWNT showed that the absorption maxima of the octaalkylporphyrins decreased with an increase in the concentration of SWNT without wavelength shift, and the fluorescence of the porphyrins was quenched. These spectral behaviors have been used to extract information about the SWNT/porphyrin interactions.

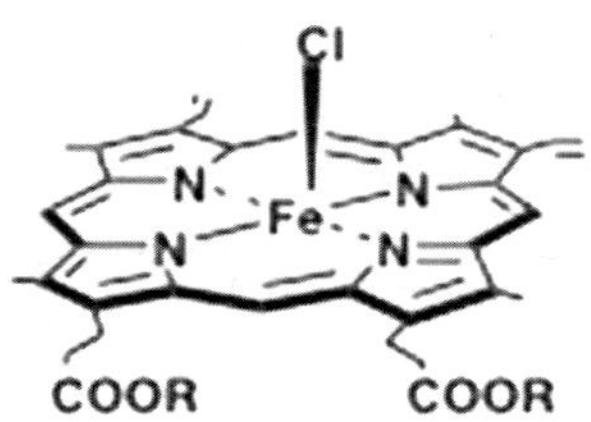

22

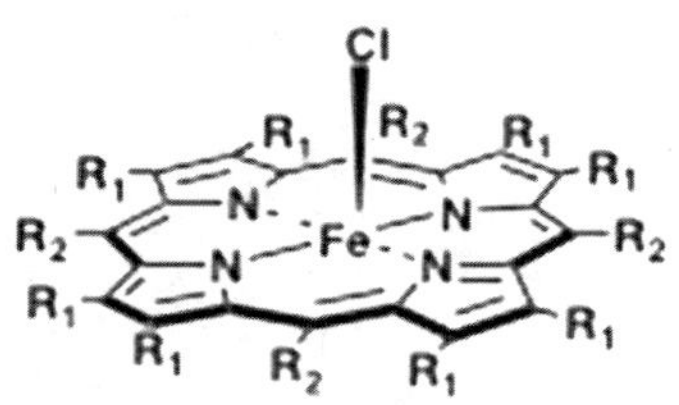

23

SWNT rings have been synthesized via noncovalent hybridization of carbon nanotubes and porphyrins (**24**) [93]. Optical absorption spectroscopy, scanning electron microscopy, and transmission electron microscopy results showed that the formation of the SWNT rings is strongly affected by the molecular architecture of the porphyrins, the hybrid concentration, the SWNT-to-porphyrin mass ratio, and the substrate on which the hybrid films were coated. The electron donation features of the substituents—para-positions of the phenyl rings in 5,10,15,20-tetraphenyl-21*H*,23*H*-porphine—and the structural regularity of the substituent groups facilitate the formation of J-aggregates.

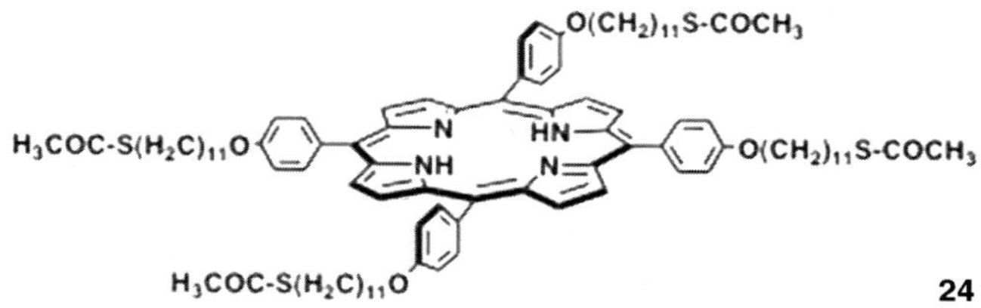

24

Ionic pyrene derivatives [*i.e.*, 1-(trimethylammonium acetyl) pyrene (pyrene^{+}) (**25**)] were used to solubilize

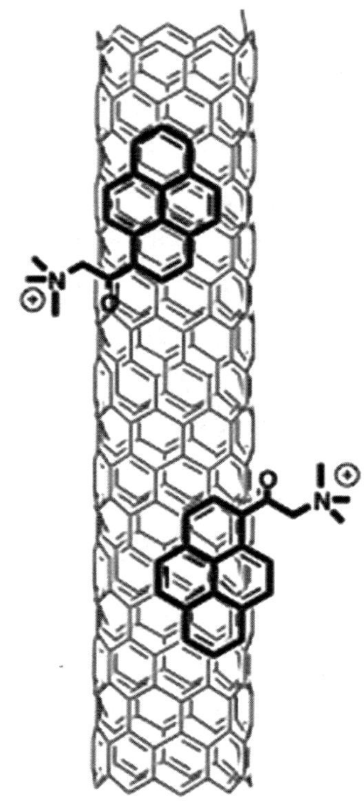

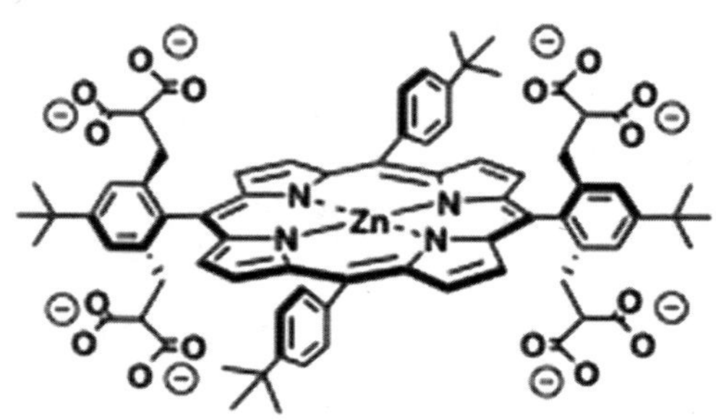

25

SWNTs through π–π interactions with solubilities as high as 0.2 mg/mL and a stability that reaches several months under ambient conditions without showing any apparent precipitations. In terms of electronic interactions between the two π-systems the most important data stem from steady-state voltammetry with SWNT/pyrene$^+$: large potential shifts of 0.4 V with respect to free pyrene$^+$. Now that the surface of SWNT is covered with positively charged ionic head groups, van der Waals and electrostatic interactions are utilized to complex oppositely charged electron donors. Water-soluble porphyrins 5,15-bis-[2′,6′-bis-{2″,2″-bis-(carboxy)-ethyl}-methyl-4′-*tert*-butyl-pheny]-10,20-bis-(4′-*tert*-butylphenyl) porphyrin octasodium (H_2P^{8-}) salt and the related zinc complex (ZnP^{8-}) were selected as ideal candidates for the development of SWNT-based photoactive systems. In the composite systems, fluorescence and transient absorption studies showed rapid intrahybrid electron transfer (0.2 $\pm$ 0.05 ns). The driving force for photoinduced electron transfer reactions is appreciably large with values around 1.0 eV. The differential absorption changes of the SWNT/pyrene$^+$/ZnP^{8-} or SWNT/pyrene$^+$/H_2P^{8-} assemblies are governed by broad absorptions in the visible between 600 and 800 nm because of ZnP^{8-}- or H_2P^{8-}-centered redox products [40, 66, 76, 77, 94–97].

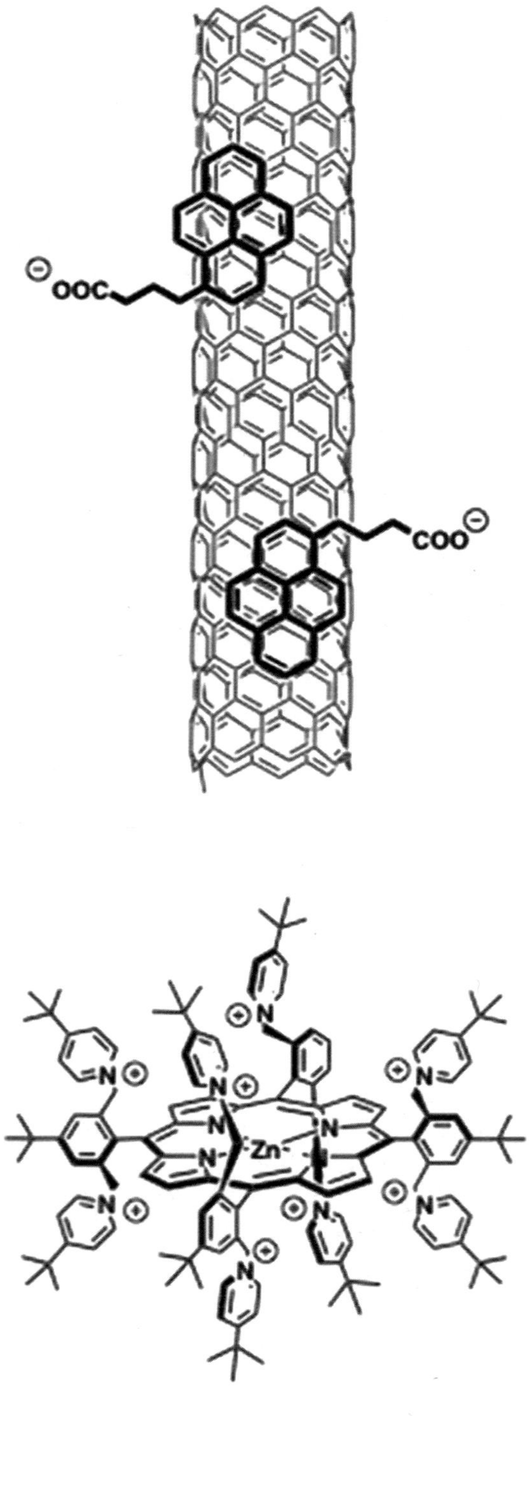

26

Multiple concentric cylindrical graphitic tubes (MWNTs) are more suitable for achieving charge transfer and charge transport [40]. MWNTs interact similarly to SWNT with pyrene$^+$ and produce stable nanocomposites. It is interesting that a better delocalization of electrons in MWNTs helps to significantly enhance the stability of the radical ion pair (5.8 $\pm$ 0.2 μ) relative to the analogous SWNT (0.4 $\pm$ 0.05 μ). Percolation of the charge inside the concentric wires in MWNTs decelerates the decay dynamics that are associated with the charge recombination.

Applying similar π–π interactions, SWNT were integrated with a series of negatively charged pyrene derivatives (pyrene$^-$), such as 1-pyreneacetic acid, 1-pyrenecarboxylic acid, 1-pyrenebutyric acid, 8-hydroxy-1,3,6-pyrenetrisulfonic acid, and a series of water-soluble, positively charged porphyrins into functional nanohybrids (**26**) through a combination of associative van der Waals and electrostatic interactions. Solubilities of SWNT/pyrene$^-$ nanocomposites were around 2.0 mg/mL. The resulting nanohybrids were photoexcited with visible light and were found to behave similar to the previously described systems [45, 76, 96, 97].

Photoinduced electron transfer in donor–acceptor self-assembled SWNT/zinc naphthalocyanine (ZnNc) or ZnP nanohybrids has been demonstrated. The nanohybrids were constructed by solubilizing SWNT by π–π stacking of pyrene functionalized imidazole moieties. Through the use of the imidazole ligand of the soluble donor ZnNc and ZnP entities were axially coordinated to yield SWNT/ZnNc and SWNT/ZnP nanohybrids **27** and **28** [98]. Steady-state and time-resolved emission studies revealed efficient fluorescence quenching of the donor entities. Nanosecond transient absorption spectra revealed that the photoexcitation of the ZnNc or ZnP moiety resulted in the one-electron oxidation of the donor unit with a simultaneous one-electron reduction of SWNT.

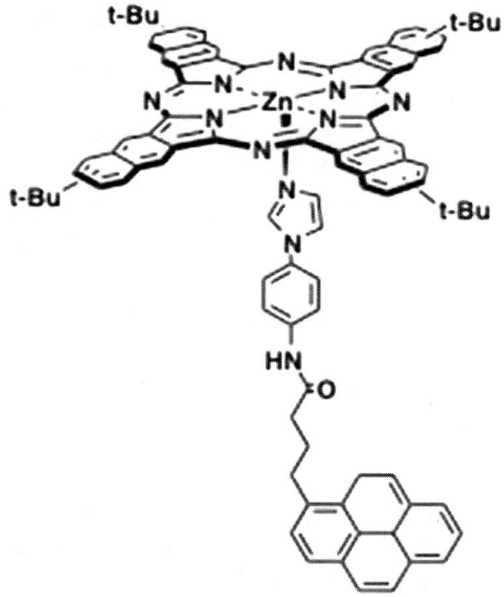

27

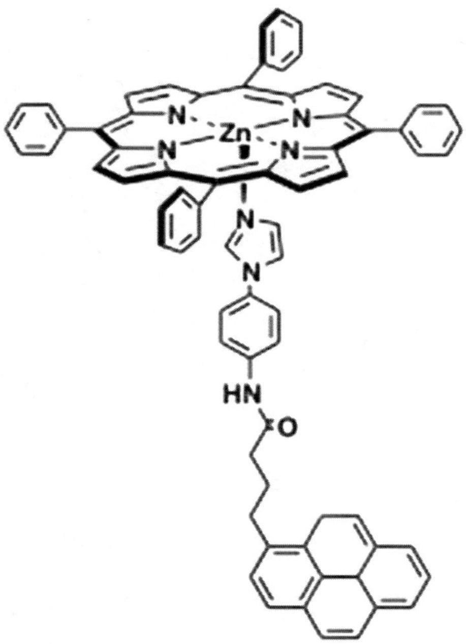

28

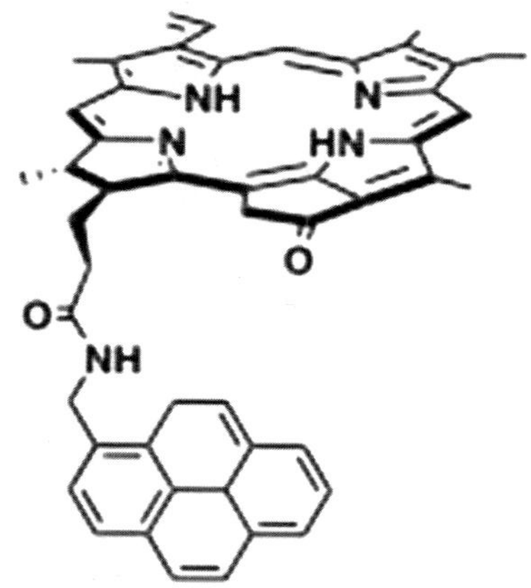

29

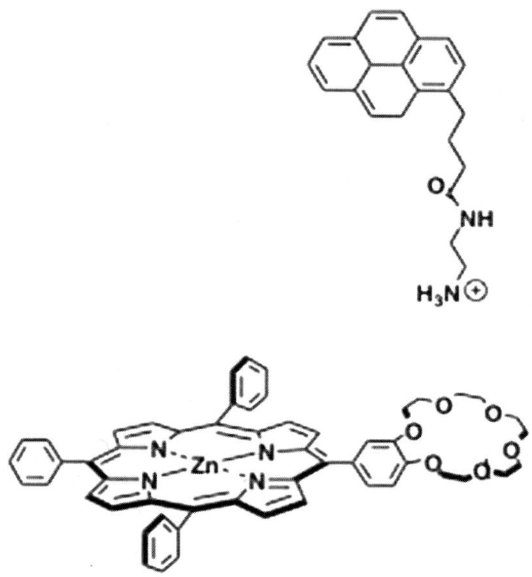

30

Similarly, pyrene-mediated noncovalent attachment of a chlorophyll derivative, *pyro*-pheophorbide *a* (**29**), to a soluble single-wall carbon nanotube has been pursued [99].

An ammonium ion–crown ether interaction has been successfully used to construct SWNT/ZnP (**30**) [100]. The [18]crown-6 to alkyl ammonium ion binding strategy resulted in SWNT/ZnP nanohybrids that are stable and soluble in DMF. Both steady-state and time-resolved emission studies revealed efficient quenching of the singlet excited state of ZnP, and free-energy calculations suggested that electron transfer quenching occurred. Nanosecond transient absorption spectral results supported the charge-separation quenching process. A stabilization of charges that reaches into the time domain of 100 μs demonstrates that a hydrogen-bonding motif is a successful self-assembly method to build electron donor–acceptor nanohybrids, which are useful for light-energy harvesting and photovoltaic applications.

The concept of pyrene as "glue" was further used in a recent study, where ZnP was functionalized with four pendent pyrene entities forming ZnP(pyr)$_4$ **31** [101], leading to stable SWNT/ZnP hybrids due to strong π–π interactions. TEM and Raman studies revealed the successful exfoliation of individual SWNTs from SWNT bundles. Free-energy calculations showed the possibility of a charge transfer from the photoexcited porphyrin to SWNT, which could be verified by nanosecond transient absorption experiments. Interestingly, in solar cell prototypes, (7,6) SWNTs, owing to a narrower band gap, showed a three times higher maximum

IPCE% (~1%) than (6,5) SWNT with a larger band gap. This is the first report on photoinduced electron transfer in self-assembled SWNT/ZnP hybrids having (7,6)- and (6,5)-enriched carbon nanotubes.

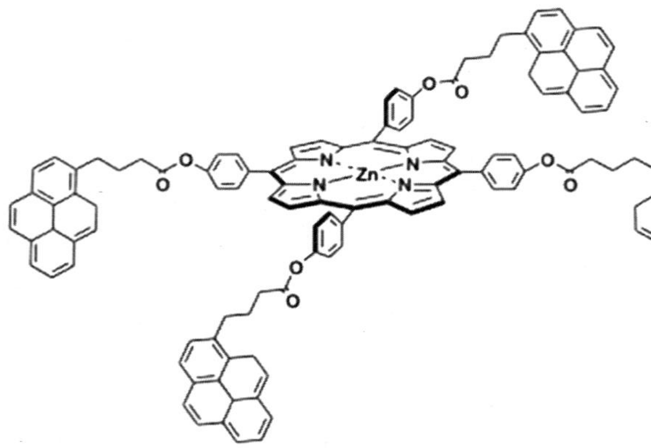

31

In our own recent studies, we succeeded in preparing stable and highly concentrated SWNT hybrid suspensions in a solvent mixture of 25% THF and 75% DMF in a simple one-step procedure. Centrifugation was possible to purify the obtained SWNT suspensions [102]. Photosensitization was accomplished by either ZnPc–pyrene (**32**) or H_2Pc–pyrene (**33**). Absorption, fluorescence, and time-resolved transient absorption studies verified mutual interactions between ZnPc or H_2Pc and SWNT. The well debundling of SWNT was corroborated by NIR-fluorescence spectroscopy and AFM. Raman measurements, on the other hand, revealed that SWNT remain undamaged during the process. Solar cell prototypes constructed with these SWNT suspensions showed excellent performance for SWNT/H_2Pc-pyrene with IPCE% as high as 21%, with an applied bias of 0.1 V, which is a promising value for this kind of SWNT solar cells.

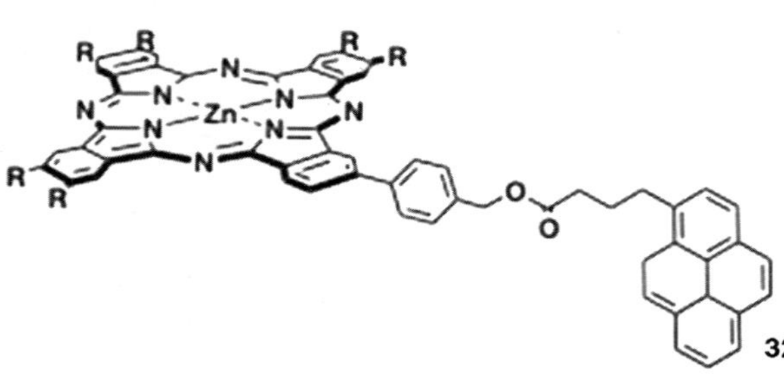

32

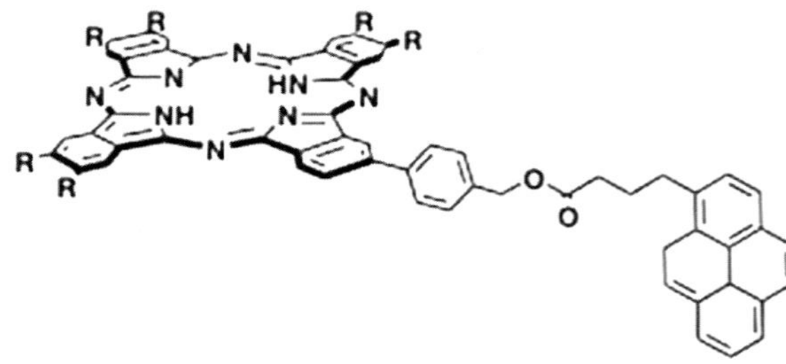

33

The development of the aforementioned one-step process for the preparation of stable SWNT hybrid suspension motivated us to probe other molecules, namely ZnP-ZnPc (**34**) and H_2P-ZnPc (**35**) in the solvent mixture of 25% THF and 75% DMF [103]. **34** as well as **35** show panchromic light harvesting, and thus they are promising building blocks for solar energy conversion devices. We were able to prepare stable suspensions of SWNT/ZnP-ZnPc and SWNT/H_2P-ZnPc and could verify ground- and excited-state interaction by absorption and fluorescence spectroscopy. Femtosecond transient absorption spectroscopy showed first an energy transfer from photoexcited ZnP or H_2P to ZnPc, followed by electron transfer to SWNT, forming a charge-separated state with a lifetime of ~100 ps.

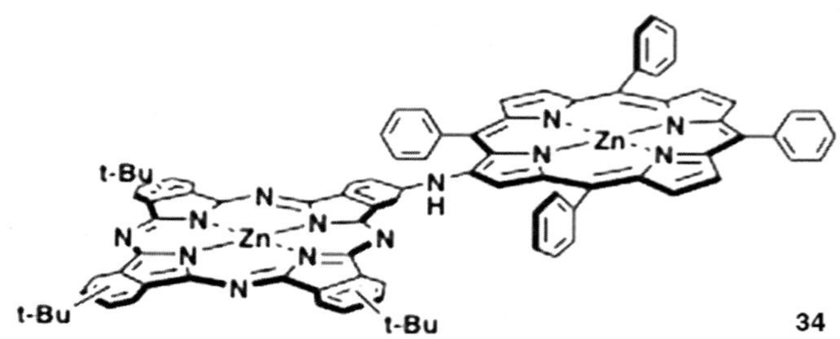

34

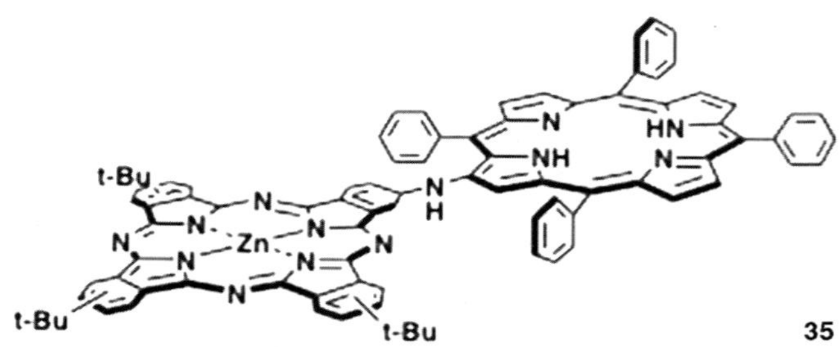

35

Further studies of noncovalently functionalized SWNT with ZnPc were carried out by using a series of ZnPc-*para*(phenylene vinylene) oligomers (ZnPc-PPV) with a varied number of repetition units (**36**), an extended hexyl spacer between ZnPc and PPV (**37**), and a –CN functionalized PPV oligomer (**38**) [104, 105]. The oligomer backbone of **38** is, due to the functionalization with CN groups, electron poor, namely *n-type* character, while SWNT are *p-type*. This leads to the observation that **38** readily forms a high concentrated SWNT hybrid suspension in THF, while the other oligomers failed in this context. The suspensions were extraordinary stable and could be purified by centrifugation. Absorption, fluorescence, Raman, and AFM studies verified the presence of well-debundled, undamaged, and highly functionalized SWNT hybrids, while femtosecond transient absorption measurements showed the formation of a short-lived charge-separated state upon photoexcitation which decays with 1.9×10^9 s^{-1}, three orders of magnitude slower than the charge separation occurs [104]. The other oligomers were utilized to functionalize SWNT as well. Interestingly, ZnPc–PPV oligomer **37**, with the extended spacer, was able to form highly stable SWNT suspensions, without reaching the performance of **38**. With a more elaborate process, SWNTs could be suspended with all the oligomers (**36–38**). Studies that were concerned with the interactions between ZnPc-PPV and SWNT have demonstrated that the structure of the ZnPc-PPV has by far a larger influence on the performance of the ZnPc-PPV oligomers to suspend SWNTs than does the number of repetition units. On the contrary, the number of repetition units has an influence on the surface coverage of the SWNT by ZnPc-PPV: The shorter the oligomer, the more surface is covered [105].

Another recent study investigates the interactions between water-soluble H_2Pc and SWNT in H_2O and D_2O, respectively. Hydrophilicity was introduced by the functionalization of the Pc with dendritic wedges of first-to-third generations with hydroxymethyl functions, yielding H_2Pc **39–41** [106]. Following the fundamental characterization of H_2Pc **39–41**, their ground- and excited-state interactions with SWNT were probed by absorption, fluorescence, and Raman spectroscopy as well as by AFM. Femtosecond transient absorption studies of SWNT/**42** in D_2O verified charge separation upon photoexcitation, with a lifetime of this newly generated state of about 250 ps.

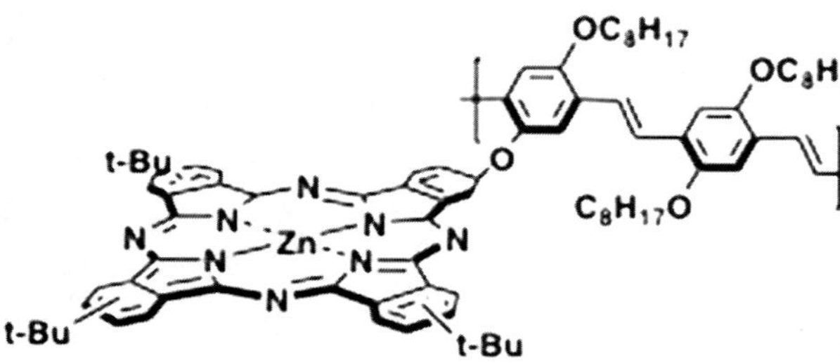

36

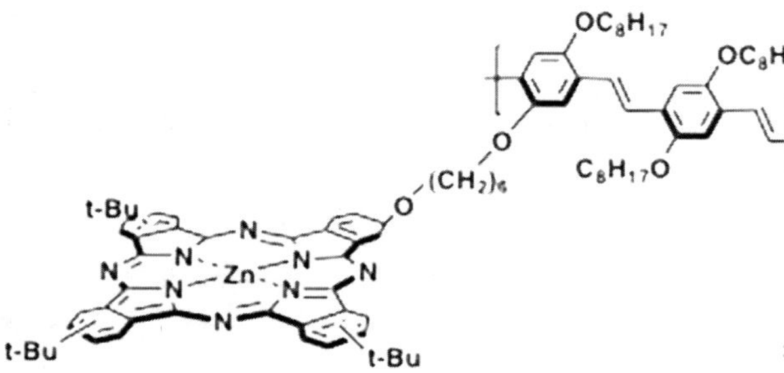

37

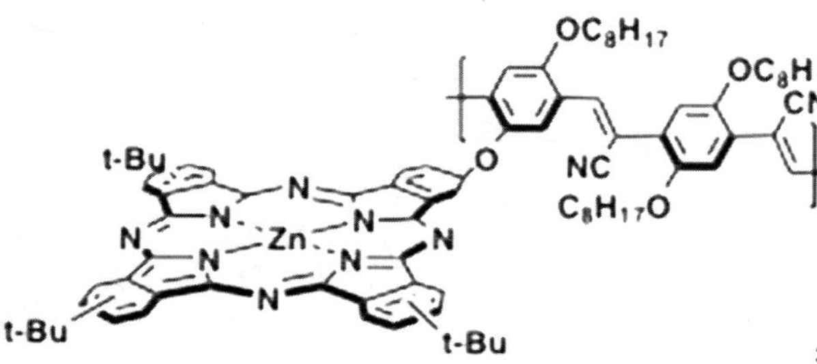

38

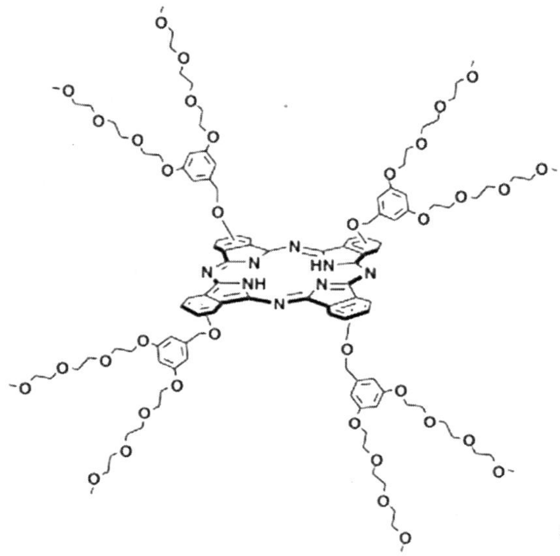

39

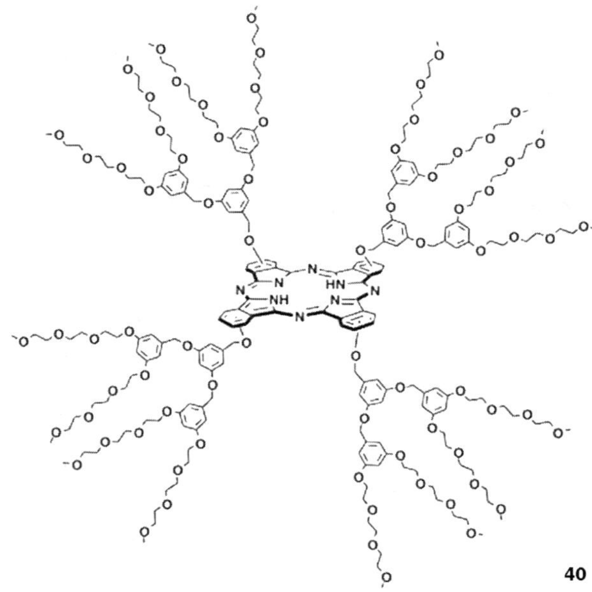

40

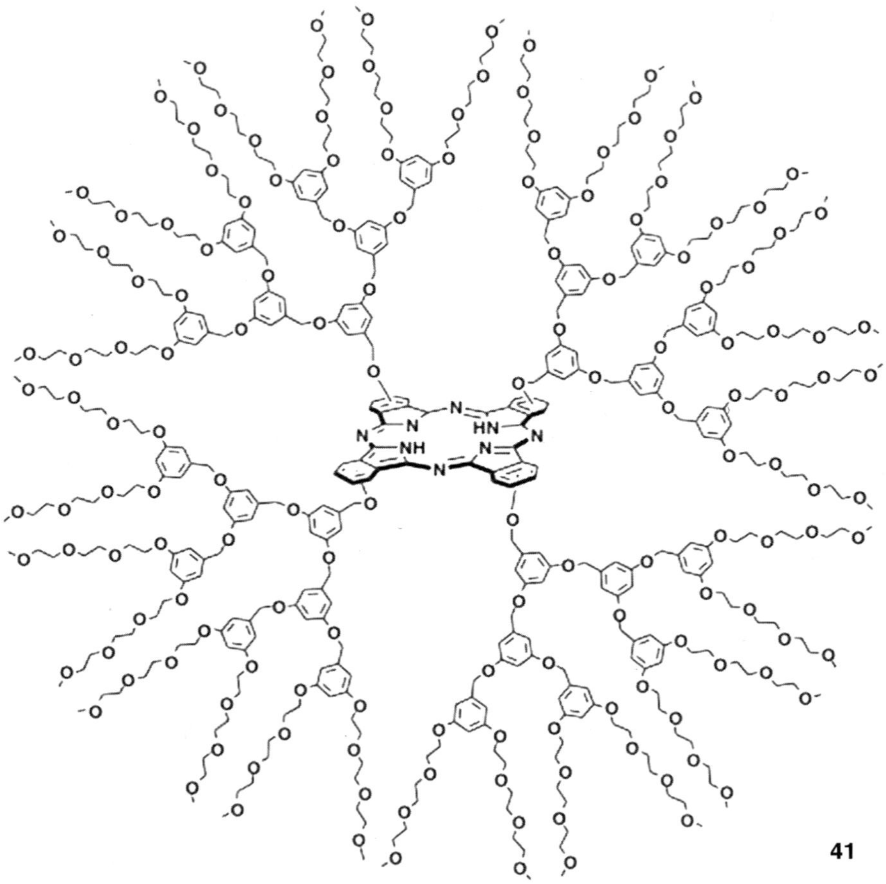

41

10.3 OUTLOOK

It is hoped that artificial photosynthetic systems will one day be able to harness some of the energy in sunlight for applications such as solar energy, optoelectronics and molecular machines. These systems must be able to harvest the light, separate the electrons and "holes" formed when the light is absorbed, and then transport these charges to catalytic sites where the oxidation of water and reduction of carbon dioxide will occur. Notable progress has been made in all of these areas, but researchers have still not developed components that are both efficient and robust; moreover, these different components have not been integrated into a working system. An issue that is often overlooked is self-regulation and repair. Consider, for example, photosynthesis in green plants; this process is optimized for the conversion of solar energy into chemical potential at ambient light levels, but it is also able to adapt to changing light conditions, such as sudden increases in brightness that would, in the absence of a protection mechanism, lead to the production of significant quantities of toxic free radicals in the cells.

REFERENCES

1. Deisenhofer, J., Epp, O., Miki, K., Huber, R., Michel, H. (1985). Structure of the protein subunits in the photosynthetic reaction centre of *Rhodopseudomonas viridis* at 3Å resolution. *Nature*, **318**, 618–624.
2. Collings, A. F., Critchley, C. (2005). *Artificial Photosynthesis: From Basic Biology to Industrial Application*. Wiley-VCH. Weinheim.
3. Balzani, V. (2001). *Electron Transfer in Chemistry*. Wiley-VHC, Weinheim.
4. Guldi, D. M. (2002). Fullerene–porphyrin architectures; photosynthetic antenna and reaction center models. *Chemical Society Reviews*, **31**, 22–36.
5. Imahori, H., Mori, Y., Matano, Y. (2003). Nanostructured artificial photosynthesis. *Journal of the Photochemistry and Photobiology, C*, **4**, 51–83.
6. Gust, D., Moore, T. A., Moore, A. L. (1993). Molecular mimicry of photosynthetic energy and electron transfer. *Accounts of Chemical Research*, **26**, 198–205.

7. Johnson, D. G., Niemczyk, M. P., Minsek, D. W., Wiederrecht, G. P., Svec, W. A., Gaines, G. L., Wasielewski, M. R. (1993). Photochemical electron transfer in chlorophyll–porphyrin–quinone triads: The role of the porphyrin-bridging molecule. *Journal of the American Chemical Society*, **115**, 5692–5701.
8. Guldi, D. M. (2003). Molecular porphyrin–fullerene architectures. *Pure and Applied Chemistry*, **75**, 1069–1075.
9. Guldi, D. M. (2003). Small reorganization energy of fullerenes in charge separation processes. *Spectrum*, **16**, 8–11.
10. Lakhno, V. D. (2005). Dynamical theory of primary processes of charge separation in the photosynthetic reaction center. *Journal of Biological Physics*, **31**, 145–159.
11. Marchi, M., Gehlen, J. N., Chandler, D., Newton, M. (1993). Diabatic surfaces and the pathway for primary electron transfer in a photosynthetic reaction center. *Journal of the American Chemical Society*, **115**, 4178–4190.
12. Guldi, D. M., Fukuzumi, S. (2002). *Developments in Fullerene Science*, **4**, 237–240.
13. Fritzsche, W., Köhler, M. (2007). *Nanotechnology: An Introduction to Nanostructuring Techniques*. Wiley-VCH, Weinheim, p. 332.
14. Rao, C. N. R., Müller, A., Cheetham, A. K. (2004). *The Chemistry of Nanomaterials: Synthesis, Properties and Applications*, Wiley-VCH, Weinheim, p. 761.
15. Wolf, E. L. (2004). *Nanophysics and Nanotechnology: An Introduction to Modern Concepts in Nanoscience*. Wiley-VCH, Weinheim, p. 308.
16. Cao, G. (2004). *Nanostructures and Nanomaterials: Synthesis, Properties & Applications*, Imperial College Press, London, p. 448.
17. Reich, S., Thomsen, C., Maultzsch, J. (2004). *Carbon Nanotubes: Basic Concepts and Physical Properties*. Wiley-VCH, Weinheim, p. 224.
18. Dresselhaus, M. S., Dresselhaus, G., Avouris, P. (2001). *Carbon Nanotubes: Synthesis, Structure, Properties and Applications*. Springer, Berlin.
19. Harris, P. J. F. (2002). *Carbon Nanotubes and Related Structures: New Materials for the Twenty-First Century*. Cambridge University Press, Cambridge, p. 294.
20. Roth, S., Carroll, D. (2004). *One-Dimensional Metals: Conjugated Polymers, Organic Crystals, Carbon Nanotubes*, Wiley-VCH. Weinheim, p. 264.
21. Yu, M.-F., Files, B. S., Arepalli, S., Ruoff, R. S. (2000). Tensile loading of ropes of single wall carbon nanotubes and their mechanical properties. *Physics Review Letters*, **84**, 5552–5555.
22. Iijima, S., Brabec, C., Maiti, A., Bernholc, J. (1996). Structural flexibility of carbon nanotubes. *Journal of Chemical Physics*, **104**, 2089–2092.
23. Treacy, M. M. J., Ebbesen, T. W., Gibson, J. M. (1996). Exceptionally high young's modulus observed for individual carbon nanotubes. *Nature*, **381**, 678–680.
24. Falvo, M. R., Clary, G. J., Taylor, R. M., Chi, V., Brooks, F. P., Washburn, S., Superfine, R. (1997). Bending and buckling of carbon nanotubes under large strain. *Nature*, **389**, 582–584.
25. Calvert, P. (1999). Nanotube composites: A recipe for strength. *Nature*, **399**, 210–211.
26. Huxtable, S. T., Cahill, D. G., Shenogin, S., Xue, L., Ozisik, R., Barone, P., Usrey, M., Strano, M. S., Siddons, G., Shim, M., Keblinski, P. (2003). Interfacial heat flow in carbon nanotube suspensions. *Nature Materials*, **2**, 731–734.
27. Zhang, H. L., Li, J. F., Yao, K. F., Chen, L. D. (2005). Spark plasma sintering and thermal conductivity of carbon nanotube bulk materials. *Journal of Applied Physics*, **97**, 114310.
28. Mingo, N., Broido, D. A. (2005). Length dependence of carbon nanotube thermal conductivity and the "problem of long waves". *Nano Letters*, **5**, 1221–1125.
29. Sun, Y., Miyasato, T., Kirimoto, K., Kusunoki, M. (2005). Metallic–nonmetallic transition on the conductivity temperature dependence of multiwall carbon nanotubes. *Applied Physics Letters*, **86**, 223108.
30. Kim, P., Shi, L., Majumdar, A., McEuen, P. L. (2001). Thermal transport measurements of individual multiwalled nanotubes. *Physics Review Letters*, **87**, 215502.
31. Hone, J., Batlogg, B., Benes, Z., Johnson, A. T., Fischer, J. E. (2000). Quantized phonon spectrum of single-wall carbon nanotubes. *Science*, **289**, 1730–1733.
32. Pop, E., Mann, D., Wang, Q., Goodson, K., Dai, H. (2006). Thermal conductance of an individual single-wall carbon nanotube above room temperature. *Nano Letters*, **6**, 96–100.
33. Misewich, J. A., Martel, R., Avouris, P., Tsang, J. C., Heinze, S., Tersoff, J. (2003). Electrically induced optical emission from a carbon nanotube FET. *Science*, **300**, 783–786.
34. de Jonge, N., Allioux, M., Doytcheva, M., Kaiser, M., Teo, K. B. K., Lacerda, R. G., Milne, W. I. (2004). Characterization of the field emission properties of individual thin carbon nanotubes. *Applied Physics Letters*, **85**, 1607.
35. Jo, S. H., Wang, D. Z., Huang, J. Y., Li, W. Z., Kempa, K., Ren, Z. F. (2004). Field emission of carbon nanotubes grown on carbon cloth. *Applied Physics Letters*, **85**, 810.
36. Yoon, S. M., Chae, J., Suh, J. S. (2004). Comparison of the field emissions between highly ordered carbon nanotubes with closed and open tips. *Applied Physics Letters*, **84**, 825.
37. Buldum, A., Lu, J. P. (2003). Electron field emission properties of closed carbon nanotubes. *Physics Review Letters*, **91**, 236801.
38. Liang, X. L., Peng, L. M., Chen, Q., Che, R. C., Xia, Y., Xue, Z. Q., Wu, Q. D. (2003). High-field response and direct identification of metallic and semiconducting multiwalled carbon nanotubes. *Physics Reviews B*, **68**, 073403.
39. Collins, P. G., Arnold, M. S., Avouris, P. (2001). Engineering carbon nanotubes and nanotube circuits using electrical breakdown. *Science*, **292**, 706–709.
40. Guldi, D. M., Rahman, G. M. A., Jux, N., Balbinot, D., Tagmatarchis, N., Prato, M. (2005). Multiwalled carbon nanotubes in donor–acceptor nanohybrids—Towards long-lived electron transfer products. *Chemical Communications*, 2038–2040.

41. Feng, W., Fujii, A., Ozaki, M., Yoshino, K. (2005). Perylene derivative sensitized multi-walled carbon nanotube thin film. *Carbon*, **43**, 2501–2507.

42. Rahman, G. M. A., Guldi, D. M., Cagnoli, R., Mucci, A., Schenetti, L., Vaccari, L., Prato, M. (2005). Combining single wall carbon nanotubes and photoactive polymers for photoconversion. *Journal of the American Chemical Society*, **127**, 10051–10057.

43. Landi, B. J., Castro, S. L., Ruf, H. J., Evans, C. M., Bailey, S. G., Raffaelle, R. P. (2005). CdSe quantum dot-single wall carbon nanotube complexes for polymeric solar cells. *Solar Energy Materials and Solar Cells*, **87**, 733–746.

44. Nakayama, K.-I., Asakura, Y., Yokoyama, M. (2004). Photovoltaic device using composite films of polymer and carbon nanotube cut by acid treatment. *Molecular Crystals and Liquid Crystals*, **424**, 217–224.

45. Guldi, D. M., Rahman, G. M. A., Prato, M., Jux, N., Qin, S., Ford, W. T. (2005). Single wall carbon nanotubes as integrative building blocks for solar energy conversion. *Angewandte Chemie International Edition*, **44**, 2015–2018.

46. Rud, J., Lovell, L., Senn, J., Qiao, Q., McLeskey, J. (2005). Water soluble polymer/carbon nanotube bulk heterojunction solar cells. *Journal of Materials Science*, **40**, 1455–1458.

47. Jang, S.-R., Vittal, R., Kim, K.-J. (2004). Incorporation of functionalized single-wall carbon nanotubes in dye-sensitized TiO_2 solar cells. *Langmuir*, **20**, 9807–9810.

48. Kymakis, E., Amaratunga, G. A. J. (2003). Photovoltaic cells based on dye-sensitisation of single-wall carbon nanotubes in a polymer matrix. *Solar Energy Materials and Solar Cells*, **80**, 465–472.

49. Jung, K.-H., Hong, J. S., Vittal, R., Kim, K.-J. (2002). Enhanced photocurrent of dye-sensitized solar cells by modification of TiO_2 with carbon nanotubes. *Chemistry Letters*, 864–865.

50. Kymakis, E., Amaratunga, G. A. J. (2002). Single-wall carbon nanotube/conjugated polymer photovoltaic devices. *Applied Physics Letters*, **80**, 112.

51. Landi, B. J., Raffaelle, R. P., Castro, S. L., Bailey, S. G. (2005). Single-wall carbon nanotube–polymer solar cells. *Progress in Photovoltaics*, **13**, 165–172.

52. Tasis, D., Tagmatarchis, N., Georgakilas, V., Prato, M. (2003). Soluble carbon nanotubes. *Chemistry—A European Journal*, **9**, 4000–4008.

53. Dyke, C. A., Tour, J. M. (2004). Overcoming the insolubility of carbon nanotubes through high degrees of sidewall functionalization. *Chemistry—A European Journal*, **10**, 812–817.

54. Ikeda, A., Hayashi, K., Konishi, T., Kikuchi, J.-I. (2004). Solubilization and debundling of purified single-walled carbon nanotubes using solubilizing agents in an aqueous solution by high-speed vibration milling technique. *Chemical Communications*, 1334–1335.

55. Shen, K., Curran, S., Xu, H., Rogelj, S., Jiang, Y., Dewald, J., Pietrass, T. (2005). Single-walled carbon nanotube purification, pelletization, and surfactant-assisted dispersion: a combined tem and resonant micro-raman spectroscopy study. *Journal of Physical Chemistry B*, **109**, 4455–4463.

56. Grossiord, N., Regev, O., Loos, J., Meuldijk, J., Koning, C. E. (2005). Time-dependent study of the exfoliation process of carbon nanotubes in aqueous dispersions by using UV–visible spectroscopy. *Analytical Chemistry*, **77**, 5135–5139.

57. Melle-Franco, M., Marcaccio, M., Paolucci, D., Paolucci, F., Georgakilas, V., Guldi, D. M., Prato, M., Zerbetto, F. (2004). Cyclic voltammetry and bulk electronic properties of soluble carbon nanotubes. *Journal of the American Chemical Society*, **126**, 1646–1647.

58. Guldi, D. M., Holzinger, M., Hirsch, A., Georgakilas, V., Prato, M. (2003). First comparative emission assay of single-wall carbon nanotubes—Solutions and dispersions. *Chemical Communications*, 1130–1131.

59. Li, J., Zhang, Y., Zhang, M. (2002). The electronic structure and its theoretical simulation of carbon nanotube with finited length. Part I: The frontier orbitals and its properties of short armchair nanotubes. *Chemical Physics Letters*, **364**, 328–337.

60. Yoshida, M., Aihara, J.-I. (1999). Validity of the weighted HOMO–LUMO energy separation as an index of kinetic stability for fullerenes with up to 120 carbon atoms. *Physical-Chemistry and Chemical Physics*, **1**, 227–230.

61. Aihara, J.-I., Kanno, H. (2004). General features of the polyene references graph-theoretically defined for fullerenes. *Chemical Physics Letters*, **398**, 440–444.

62. Kalyanasundaram, K. (1997), *Photochemistry of Polypyridine and Porphyrin Complexes*. Academic Press. London, p. 642.

63. Gouterman, M., Rentzepis, P. M., Straub, K. D. (1987). *Porphyrins: Excited States and Dynamics, ACS Symposium Series*. *American Chemical Society*, p. 371.

64. Dolphin, D. (1978). *The Porphyrins*, Academic Press, New York, p. 546.

65. Kadish, K. M., Smith, K. M., Guilard, R. (2003). *The Porphyrin Handbook*, Academic Press. New York, p. 3310.

66. Guldi, D. M., Rahman, G. M. A., Sgobba, V., Ehli, C. (2006). Multifunctional molecular carbon materials—From fullerenes to carbon nanotubes. *Chemical Society Reviews*, **35**, 471–487.

67. Li, H., Martin, R. B., Harruff, B. A., Carino, R. A., Allard, L. F., Sun, Y.-P. (2004). Single-walled carbon nanotubes tethered with porphyrins: synthesis and photophysical properties. *Advanced Materials*, **16**, 896–900.

68. Baskaran, D., Mays, J. W., Zhang, X. P., Bratcher, M. S. (2005). Carbon nanotubes with covalently linked porphyrin antennae: photoinduced electron transfer. *Journal of the American Chemical Society*, **127**, 6916–6917.

69. de la Torre, G., Blau, W., Torres, T. (2003). A survey on the functionalization of single-walled nanotubes. The chemical attachment of phthalocyanine moieties. *Nanotechnology*, **14**, 765–771.

70. Ballesteros, B., de la Torre, G., Ehli, C., Rahman, G. M. A., Agullo-Rueda, F., Guldi, D. M., Torres, T. (2007). Single-wall carbon nanotubes bearing covalently linked phthalocyanines—photoinduced electron transfer. *Journal of the American Chemical Society*, **129**, 5061–5068.

71. Campidelli, S., Ballesteros, B., Filoramo, A., Díaz, D. D., de la Torre, G., Torres, T., Rahman, G. M. A., Ehli, C., Kiessling,

D., Werner, F., Sgobba, V., Guldi, D. M., Cioffi, C., Prato, M., Bourgoin, J.-P. (2008). Facile decoration of functionalized single-wall carbon nanotubes with phthalocyanines via "click chemistry". *Journal of the American Chemical Society*, **130**, 11503–11509.

72. Alvaro, M., Atienzar, P., de la Cruz, P., Delgado, J. L., Troiani, V., Garcia, H., Langa, F., Palkar, A., Echegoyen, L. (2006). Synthesis, photochemistry, and electrochemistry of single-wall carbon nanotubes with pendent pyridyl groups and of their metal complexes with zinc porphyrin. comparison with pyridyl-bearing fullerenes. *Journal of the American Chemical Society*, **128**, 6626–6635.

73. Guo, Z., Du, F., Ren, D., Chen, Y., Zheng, J., Liu, Z., Tian, J. (2006). Covalently porphyrin-functionalized single-walled carbon nanotubes: A novel photoactive and optical limiting donor–acceptor nanohybrid. *Journal of Materials Chemistry*, **16**, 3021–3030.

74. Guldi, D. M., Rahman, G. M. A., Ramey, J., Marcaccio, M., Paolucci, D., Paolucci, F., Qin, S., Ford, W. T., Balbinot, D., Jux, N., Tagmatarchis, N., Prato, M. (2004). Donor–acceptor nanoensembles of soluble carbon nanotubes. *Chemical Communications*, 2034–2035.

75. Guldi, D. M., Rahman, G. M. A., Qin, S., Tchoul, M., Ford, W. T., Marcaccio, M., Paolucci, D., Paolucci, F., Campidelli, S., Prato, M. (2006). Versatile coordination chemistry towards multifunctional carbon nanotube nanohybrids. *Chemistry—A European Journal*, **12**, 2152–2161.

76. Guldi, D. M., Rahman, G. M. A., Zerbetto, F., Prato, M. (2005). Carbon nanotubes in electron donor–acceptor nanocomposites. *Accounts of Chemical Research*, **38**, 871–878.

77. Rahman, G. M. A., Troeger, A., Sgobba, V., Guldi, D. M., Jux, N., Tchoul, M. N., Ford, W. T., Mateo-Alonso, A., Prato, M. (2008). Improving photocurrent generation: supramolecularly and covalently functionalized single-wall carbon nanotubes–polymer/porphyrin donor–acceptor nanohybrids. *Chemistry—A European Journal*, **14**, 8837–8846.

78. Campidelli, S., Sooambar, C., Lozano Diz, E., Ehli, C., Guldi, D. M., Prato, M. (2006). Dendrimer-functionalized single-wall carbon nanotubes: synthesis, characterization, and photoinduced electron transfer. *Journal of the American Chemical Society*, **128**, 12544–12552.

79. Ehli, C., Campidelli, S., Brunetti, F. G., Prato, M., Guldi, D. M. (2007). Single-wall carbon nanotube porphyrin nanoconjugates. *Journal of Porphyrins and Phthalocyanines*, **11**, 442–447.

80. Guldi, D. M., Taieb, H., Rahman, G. M. A., Tagmatarchis, N., Prato, M. (2005). Novel photoactive single-walled carbon nanotube–porphyrin polymer wraps: efficient and long-lived intracomplex charge separation. *Advanced Materials*, **17**, 871–875.

81. Saito, K., Troiani, V., Qiu, H., Solladie, N., Sakata, T., Mori, H., Ohama, M., Fukuzumi, S. (2007). Nondestructive formation of supramolecular nanohybrids of single-walled carbon nanotubes with flexible porphyrinic polypeptides. *Journal of Physical Chemistry C*, **111**, 1194–1199.

82. Cheng, F., Adronov, A. (2006). Noncovalent functionalization and solubilization of carbon nanotubes by using a conjugated zn–porphyrin polymer. *Chemistry—A European Journal*, **12**, 5053–5059.

83. Cheng, F., Zhang, S., Adronov, A., Echegoyen, L., Diederich, F. (2006). Triply fused Zn^{II}-porphyrin oligomers: synthesis, properties, and supramolecular interactions with single-walled carbon nanotubes (SWNTs). *Chemistry—A European Journal*, **12**, 6062–6070.

84. Rahman, G. M. A., Guldi, D. M., Campidelli, S., Prato, M. (2006). Electronically interacting single wall carbon nanotube–porphyrin nanohybrids. *Journal of Materials Chemistry*, **16**, 62–65.

85. Ni Mhuircheartaigh, É. M., Giordani, S., Blau, W. J. (2006). Linear and nonlinear optical characterization of a tetraphenylporphyrin–carbon nanotube composite system. *Journal of Physical Chemistry B*, **110**, 23136–23141.

86. Cao, L., Chen, H. Z., Zhou, H. B., Zhu, L., Sun, J. Z., Zhang, X. B., Xu, J. M., Wang, M. (2003). Carbon-nanotube-templated assembly of rare-earth phthalocyanine nanowires. *Advanced Materials*, **15**, 909–913.

87. Satake, A., Miyajima, Y., Kobuke, Y. (2005). Porphyrin–carbon nanotube composites formed by noncovalent polymer wrapping. *Chemistry Materials*, **17**, 716–724.

88. Valentini, L., Trentini, M., Mengoni, F., Alongi, J., Armentano, I., Ricco, L., Mariani, A., Kenny, J. M. (2007). Synthesis and photoelectrical properties of carbon nanotube–dendritic porphyrin light harvesting molecule systems. *Diamond and Related Materials*, **16**, 658–663.

89. Chen, J., Collier, C. P. (2005). Noncovalent functionalization of single-walled carbon nanotubes with water-soluble porphyrins. *Journal of Physical Chemistry B*, **109**, 7605–7609.

90. Li, H., Zhou, B., Lin, Y., Gu, L., Wang, W., Fernando, K. A. S., Kumar, S., Allard, L. F., Sun, Y.-P. (2004). Selective interactions of porphyrins with semiconducting single-walled carbon nanotubes. *Journal of the American Chemical Society*, **126**, 1014–1015.

91. Turdean, G. L., Popescu, I. C., Curulli, A., Palleschi, G. (2006). Iron(III) protoporphyrin IX—Single-wall carbon nanotubes modified electrodes for hydrogen peroxide and nitrite detection. *Electrochimica Acta*, **51**, 6435–6441.

92. Murakami, H., Nakamura, G., Nomura, T., Miyamoto, T., Nakashima, N. (2007). Noncovalent porphyrin-functionalized single-walled carbon nanotubes: Solubilization and spectral behaviors. *J. Porphyrin Phthalocyanine*, **11**, 418–427.

93. Geng, J., Ko, Y. K., Youn, S. C., Kim, Y.-H., Kim, S. A., Jung, D.-H., Jung, H.-T. (2008). Synthesis of swnt rings by noncovalent hybridization of porphyrins and single-walled carbon nanotubes. *Journal of Physical Chemistry C*, **112**, 12264–12271.

94. Ehli, C., Rahman, G. M. A., Jux, N., Balbinot, D., Guldi, D. M., Paolucci, F., Marcaccio, M., Paolucci, D., Melle-Franco, M., Zerbetto, F., Campidelli, S., Prato, M. (2006). Interactions in single wall carbon nanotubes/pyrene/porphyrin nanohybrids. *Journal of the American Chemical Society*, **128**, 11222–11231.

95. Tagmatarchis, N., Prato, M., Guldi, D. M. (2005). Soluble carbon nanotube ensembles for light-induced electron transfer interactions. *Physica E*, **29**, 546–550.

96. Guldi, D. M., Rahman, G. M. A., Jux, N., Tagmatarchis, N., Prato, M. (2004). Integrating single-wall carbon nanotubes into donor–acceptor nanohybrids. *Angewandte Chemie International Edition*, **43**, 5526–5530.

97. Guldi, D. M., Rahman, G. M. A., Jux, N., Balbinot, D., Hartnagel, U., Tagmatarchis, N., Prato, M. (2005). Functional single-wall carbon nanotube nanohybrids associating SWNTs with water-soluble enzyme model systems. *Journal of the American Chemical Society*, **127**, 9830–9838.

98. Chitta, R., Sandanayaka, A. S. D., Schumacher, A. L., D'Souza, L., Araki, Y., Ito, O., D'Souza, F. (2007). Donor–acceptor nanohybrids of zinc naphthalocyanine or zinc porphyrin noncovalently linked to single-wall carbon nanotubes for photoinduced electron transfer. *Journal of Physical Chemistry C*, **111**, 6947–6955.

99. Kavakka, J. S., Heikkinen, S., Kilpeläinen, I., Mattila, M., Lipsanen, H., Helaja, J. (2007). Noncovalent attachment of pyro-pheophorbide a to a carbon nanotube. *Chemical Communications*, 519–521.

100. D'Souza, F., Chitta, R., Sandanayaka, A. S. D., Subbaiyan, N. K., D'Souza, L., Araki, Y., Ito, O. (2007). Self-assembled single-walled carbon nanotube:zinc–porphyrin hybrids through ammonium ion–crown ether interaction: construction and electron transfer. *Chemistry—A European Journal*, **13**, 8277–8284.

101. Maligaspe, E., Sandanayaka, A. S. D., Hasobe, T., Ito, O., D'Souza, F. (2010). Sensitive efficiency of photoinduced electron transfer to band gaps of semiconductive single-walled carbon nanotubes with supramolecularly attached zinc porphyrin bearing pyrene glues. *Journal of the American Chemical Society*, **132**, 8158–8164.

102. Bartelmess, J., Ballesteros, B., de la Torre, G., Kiessling, D., Campidelli, S., Prato, M., Torres, T., Guldi, D. M. (2010). Phthalocyanine–pyrene conjugates: a powerful approach toward carbon nanotube solar cells. *Journal of the American Chemical Society*, **132**, 16202–16211.

103. Bartelmess, J., Soares, A. R. M., Martínez-Díaz, M. V., Neves, M. G. P. M. S., Tomé, A. C., Cavaleiro, J. A. S., Torres, T., Guldi, D. M. (2011). Panchromatic light harvesting in single wall carbon nanotube hybrids. immobilization of porphyrin–phthalocyanine conjugates. *Chemical Communications*, **47**, 3490–3492.

104. Bartelmess, J., Ehli, C., Cid, J.-J., García-Iglesias, M., Vázquez, P., Torres, T., Guldi, D. M. (2011). Tuning and optimizing the intrinsic interactions between phthalocyanine-based ppv oligomers and single-wall carbon nanotubes toward *n*-type/*p*-type. *Chem. Sci.*, **2**, 652–660.

105. Bartelmess, J., Ehli, C., Cid, J.-J., García-Iglesias, M., Vázquez, P., Torres, T., Guldi, D. M. (2011). Screening interactions of zinc phthalocyanine–PPV oligomers with single wall carbon nanotubes—A comparative study. *Journal of Materials Chemistry*, **21**, 8014–8020.

106. Hahn, U., Engmann, S., Oelsner, C., Ehli, C., Guldi, D. M., Torres, T. (2010). Immobilizing water-soluble dendritic electron donors and electron acceptors—phthalocyanines and perylenediimides—onto single wall carbon nanotubes. *Journal of the American Chemical Society*, **132**, 6392–6401.

11

ORGANIC SYNTHESIS OF ENDOHEDRAL FULLERENES ENCAPSULATING HELIUM, DIHYDROGEN, AND WATER

Michihisa Murata, Yasujiro Murata, and Koichi Komatsu

11.1 INTRODUCTION

Endohedral fullerenes, the fullerenes encapsulating some small atom(s) or molecule(s) [1, 2], are not simply the subject of academic curiosity but have turned out to have important applicability. They have potential utility in various fields such as molecular electronics [3], magnetic resonance imaging as a contrast agent [4], and nuclear magnetic resonance (NMR) analysis [5, 6]. However, their production is based on nonefficient and hard-to-control physical methods such as co-vaporization of carbon and metal atoms [1, 2] and high-pressure/high-temperature treatment with noble gases [5–7]. Under such circumstances, it is highly desirable to "synthesize" these species by means of organic chemistry. In fact, such an organic synthetic method termed "molecular surgery" was proposed by Rubin and has been pursued by his group and others [8–11].

This method consists of four steps: (1) creation of an opening on the fullerene cage, (2) enlargement of the opening, (3) insertion of some guest species through the enlarged opening, and (4) closure of the opening without losing any of the guest species. Thus, starting from fullerene C_{60}, various sizes of opening were made on the cage [12–17], and small atoms or molecules such as He [17–19], H_2 [15–18, 20], H_2O [13, 21, 22], CO [22, 23], NH_3 [24], N_2 [23], and CH_4 [25] have been inserted into the open-cage fullerenes (steps 1 to 3). However, there has been no attempt reported for the shrinkage or closing of the once created large opening on the cage until our work has been completed (step 4) [26, 27].

In this chapter we present the first successful achievement of the molecular surgery involving complete steps 1 through 4, to give novel endohedral fullerenes containing helium, dihydrogen, or water. In contrast to the endohedral metallofullerenes, where the electronic properties of carbon cages are significantly altered by electron transfer from the encapsulated metals, such change has not been observed at all in these complexes encapsulating the neutral species (He, H_2, H_2O). The very minute interaction between C_{60} and the guest species is supported by the theoretical calculations, which show that highest occupied molecular orbital (HOMO) and lowest unoccupied molecular orbital (LUMO) energy levels of the C_{60} cage are almost unchanged regardless of the kind of the entities inside (Table 11.1). Hence the endohedral fullerenes prepared by the molecular surgery method can provide unique opportunities to elucidate intrinsic nature of the atoms and molecules isolated from the external environment.

It is also important to point out that the size of the inner cavity of C_{60} is limited to roughly 4 Å in inner diameter. As summarized in Table 11.1 and Figure 11.1, the encapsulation energies, calculated for the hypothetical equilibrium $X + C_{60} \rightleftarrows X@C_{60}$, are negative (i.e., attractive) for even relatively large molecules such as methane and ammonia, while that for carbon dioxide is shown to be unfavorable due to its bulkiness (5.5 Å along the longer axis). However, the molecular surgery method possesses a potential to produce a variety of unprecedented endohedral fullerenes encapsulating molecules such as carbon monoxide, dioxygen, dinitrogen, methane, ammonia, or formaldehyde.

Organic Nanomaterials: Synthesis, Characterization, and Device Applications, First Edition. Edited by Tomás Torres and Giovanni Bottari.

TABLE 11.1. Electronic Properties of Various Endohedral Fullerenes and Encapsulation Energies[a] Calculated for the Hypothetical Equilibrium: ($X + C_{60} \rightleftarrows X@C_{60}$)

Guest (X)	MW	HOMO (eV)	LUMO (eV)	Dipole Moment (Debye)	ΔE^b (kcal mol^{-1})	ΔE^c (kcal mol^{-1})
None	720	−7.25	−2.70	0		
He	724	−7.24	−2.70	0	−2.8	−2.4
H_2	722	−7.24	−2.70	0	−9.0	−8.2
CO	748	−7.25	−2.72	0.03	−17.0	−11.9
H_2O	738	−7.22	−2.71	0.43	−19.4	−13.5
O_2	752	−7.24	−2.71	0	−18.1	−13.7
N_2	748	−7.24	−2.72	0.02	−18.0	−14.1
CH_4	736	−7.25	−2.71	0	−16.8	−14.2
NH_3	737	−7.23	−2.71	0.38	−22.0	−16.8
H_2CO	750	−7.23	−2.73	0.60	−15.6	−9.8
CO_2	764	−7.19	−2.81	0	2.9	10.1

[a]All calculations were carried out at the M06-2X/6-31G(d,p) level without any symmetry assumptions.
[b]Encapsulation energies without basis set superposition error (BSSE) correction.
[c]Encapsulation energies with BSSE correction.

11.2 HOW WE STARTED THE RESEARCH—REACTIONS OF C_{60} WITH POLYAZA-AROMATICS

Previously, we were interested in the reactions of C_{60} with diaza- [28], triaza- [29], and tetraaza-aromatic compounds [30,31]. In particular, the *solid-state* reaction of C_{60} with the diaza-aromatic phthalazine (**1**) and with tetraaza-aromatic compound **2**, using what we call the high-speed vibration milling (HSVM) method, afforded dimeric fullerenes **3** and **4**. The reactions are considered to proceed through the first [4 + 2] reaction followed by denitrogenation forming a diene unit, along with the second [4 + 2] reaction as shown in Scheme 11.1. On the other hand, exactly the same reactants (i.e., C_{60} and phthalazine) under liquid-phase, thermal reaction conditions in 1-chloronaphthalene, afforded a benzo-derivative of open-cage fullerene **5**, instead of the dimeric fullerene. The reaction mechanism is considered as shown in Scheme 11.2. Exactly in the same manner, the reaction of C_{60} with 4,6-dimethyl-1,2,3-triazine gave a similar open-cage fullerene derivative as a somewhat unstable compound [29]. This finding of the facile reaction to open the fullerene cage has led us to explore the more effective and suitable reaction for the first step of "molecular surgery."

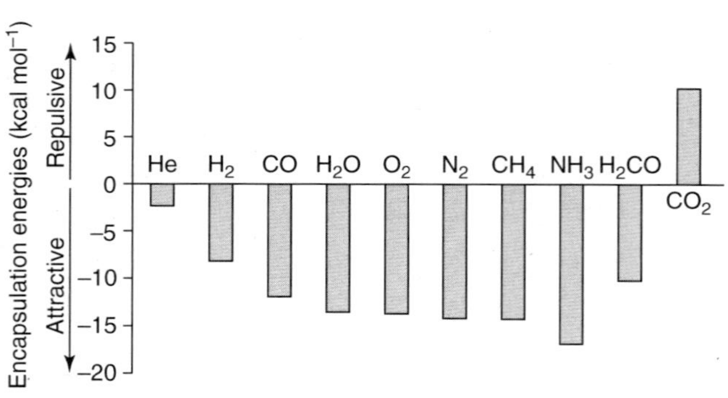

FIGURE 11.1. Encapsulation energies of C_{60} for various guest species calculated at the M06-2X/6-31G(d,p) level with BSSE corrections after the optimization.

11.3 ENDOHEDRAL C_{60} ENCAPSULATING DIHYDROGEN, $H_2@C_{60}$

11.3.1 Synthesis

After several trials, we found that the reaction of C_{60} with 5,6-diphenyl-3-(2-pyridyl)-1,2,4-triazine (**6**) can efficiently give the fullerene derivative **7** with an eight-membered-ring opening in 85% yield based on consumed C_{60} (59% conversion). The mechanism is supposed to be the same as depicted in Scheme 11.2. The x-ray structure of **7** indicated that C=C double bonds on the rim of the opening are considerably distorted (twist angle, 39.0° and 38.8°) and should be highly reactive. Thus, singlet oxygen, generated in situ by photo-irradiation of **7**, selectively reacted with one of these double bonds to cleave it, resulting in enlargement of the opening to give diketone **8** with a 12-membered-ring opening in 60% yield (Scheme 11.3) [14]. For further enlargement of the opening, compound **8**, which turned out to be a strong π-electron accepter, was activated by addition of a typical π-electron donor, tetrakis(dimethylamino)ethylene (TDAE), and was reacted with elemental sulfur to afford compound **9** having a 13-membered-ring opening as a single product in 77% yield (Scheme 11.3, reaction (a)). The x-ray crystallography of **9** confirmed its structure with the size of the opening being 5.64 Å for a longer axis and 3.75 Å for a shorter axis [14]. A selenium atom can also be inserted to give **10** by using CH_3SNa as an activator (Scheme 11.3, reaction (b)) [16].

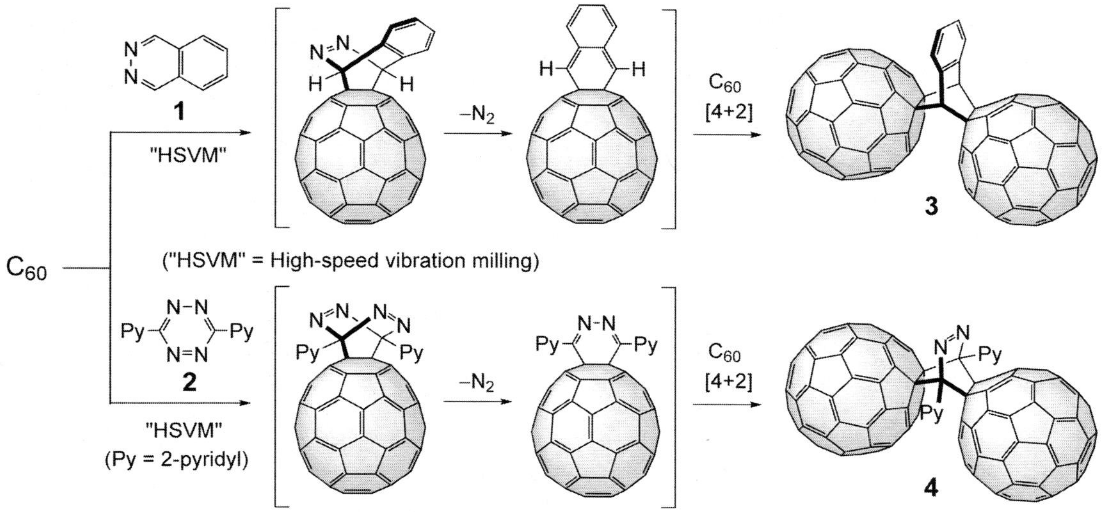

SCHEME 11.1.

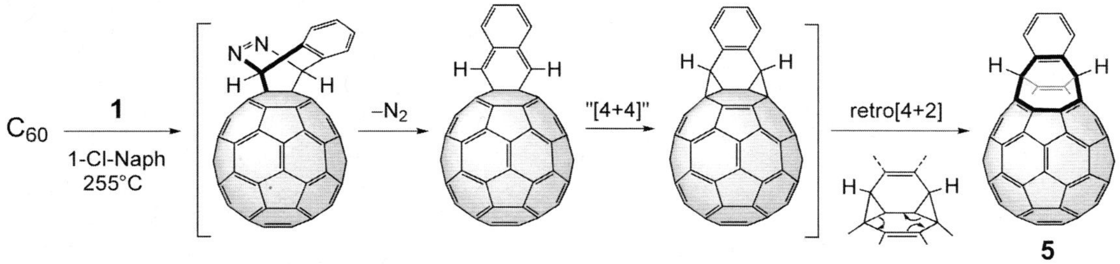

SCHEME 11.2.

The 100% incorporation of dihydrogen inside the cage was achieved by treating a powder of **9** placed in an autoclave with high-pressure dihydrogen gas (800 atm) at 200°C for 8 hours [20]. A sharp ^{1}H NMR signal of the encapsulated dihydrogen was observed at δ −7.25 ppm in ODCB-d_4. The synchrotron x-ray diffraction analysis with maximum entropy method clearly revealed that dihydrogen is present at the center of the inner cavity of **9** [32]. On the other hand, the ejection of dihydrogen from H_2@**9** takes place at the temperature range, around 160–190°C, following first-order kinetics with an activation energy (E_a) of 34.2 kcal mol^{-1}.

As the first step of the restoration of the C_{60} cage, removal of the sulfur atom out of the rim of the opening of H_2@**9** was conducted as shown in Scheme 11.4 (steps a and b): the sulfide moiety was oxidized to sulfoxide and subsequently removed as an SO unit, thus generating molecule H_2@**8** [26, 27]. Then, the McMurry reaction worked efficiently, leading to the formation of H_2@**7** in 88% yield (Scheme 11.4, step c). At each of these steps, dihydrogen was completely retained inside the cage. The final step to remove all the remaining organic addends on the fullerene cage was accomplished by simply heating a powder of H_2@**7** in a glass tube under vacuum at 340°C for 2 hours. The crude product was dissolved in carbon disulfide and passed through a silica-gel column to give a purple solution containing desired $H_2@C_{60}$ (contaminated by 9% empty C_{60}) in 67% yield [26, 27].

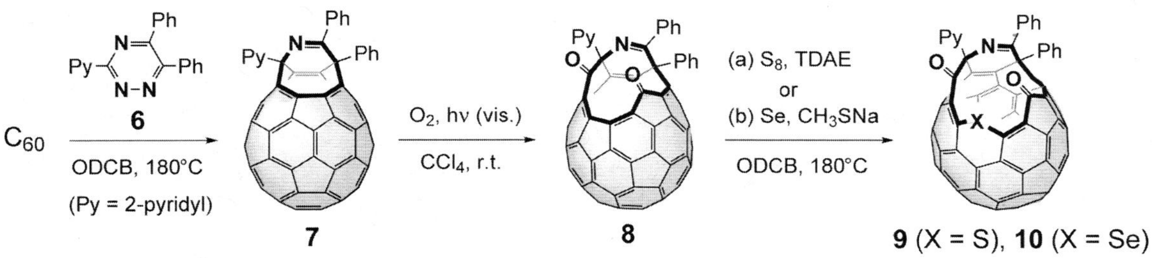

SCHEME 11.3.

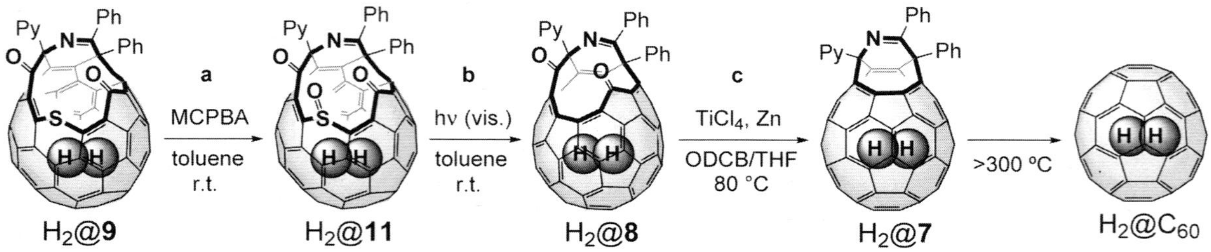

SCHEME 11.4.

Thus, $H_2@C_{60}$ was synthesized in a total yield of 22% from $H_2@$**9**, which can be obtained in 40% yield from consumed C_{60} [14, 20].

11.3.2 Properties

Complete separation of $H_2@C_{60}$ from empty C_{60} was achieved by recycling HPLC on semipreparative Cosmosil Buckyprep columns (two directly connected columns; 250 mm length, 10 mm inner diameter; mobile phase, toluene; flow rate, 4 mL min^{-1}): $H_2@C_{60}$ was obtained as a pure material after 20 recycles (total retention time, 399 minutes; the retention time for empty C_{60}, 395 minutes). As indicated by the theoretical calculations, there exists the attractive interaction between C_{60} and dihydrogen ($\Delta E = -8.2$ kcal mol^{-1}, Table 11.1). The very weak van der Waals interaction must have contributed to this separation. The ^{1}H NMR showed a signal of dihydrogen at δ −1.44 ppm in ODCB-d_4, which is 5.98 ppm upfield shifted from that of dissolved free dihydrogen. This value is comparable to the upfield shift reported for a ^{3}He NMR signal of $^3He@C_{60}$ [5, 6]. The ^{13}C NMR spectrum of $H_2@C_{60}$ exhibited a signal at δ 142.84 ppm in ODCB-d_4, which is slightly shifted to downfield by 0.08 ppm relative to that of C_{60}.

The electrochemical behavior of $H_2@C_{60}$ investigated by cyclic voltammetry (CV) was quite similar to that of C_{60}. Three stepwise reduction waves (−1.13, −1.54, −1.99 V versus Fc/Fc$^+$, in ODCB) and one oxidation peak (at +1.62 V in 1,1,2,2-tetrachloroethane) were recorded [26]. When the measurement was conducted in toluene–acetonitrile (5:1) at −10°C under vacuum [33], the fourth to sixth reduction waves appeared (−2.39, −2.95 V, −3.5 V), which were shifted slightly and gradually to more negative potentials (by 0.04, 0.07, and 0.15 V, respectively) than C_{60}, due to a weak repulsive interaction between the dihydrogen and negatively multicharged C_{60} cage [27].

11.3.3 Derivatives

The NMR chemical shift of an encapsulated atom inside fullerene cages is very sensitive to the change in the magnetic field inside the cage, as extensively investigated using $^3He@C_{60}$ by the use of ^{3}He NMR spectroscopy. Although $^3He@C_{60}$ is only available at the occupation level of 0.1–1% by using rather harsh conditions (650°C and 3000 atm of ^{3}He gas) [5–7, 34], $H_2@C_{60}$ at occupation level of 100% is now prepared by the organic synthetic approach as described above. In addition, it is significantly convenient that ^{1}H NMR spectroscopy can be used instead of ^{3}He NMR. Thus, in order to examine the NMR chemical shift of dihydrogen, some representative derivatives of $H_2@C_{60}$ were synthesized (Scheme 11.5) [35–37]. The solid-state mechanochemical dimerization of $H_2@C_{60}$ (occupation level of 91%) using the HSVM (high speed vibration milling) technique under the same conditions as reported previously [38, 39] afforded the dumbbell-shaped dimer, $(H_2@C_{60})_2$ [26, 27], similarly to the reaction of empty C_{60}. The inside dihydrogen does not affect the reactivity in this reaction. The chemical shift for dihydrogen of $(H_2@C_{60})_2$ is δ −4.04 ppm, which is 8.58 ppm upfield shifted from that of free dihydrogen similarly to the case for $^3He@C_{120}$ [39] (8.81 ppm upfield shift from the signal of free ^{3}He). The chemical shifts for dihydrogen of $H_2@$**12**, $H_2@$**13**, and $H_2@$**14** are δ −3.27, −4.30, and −4.64 ppm, respectively, in ODCB-d_4 [27]. Since the chemical shifts sensitively change according to the difference in the structures of the organic addends, the encapsulated

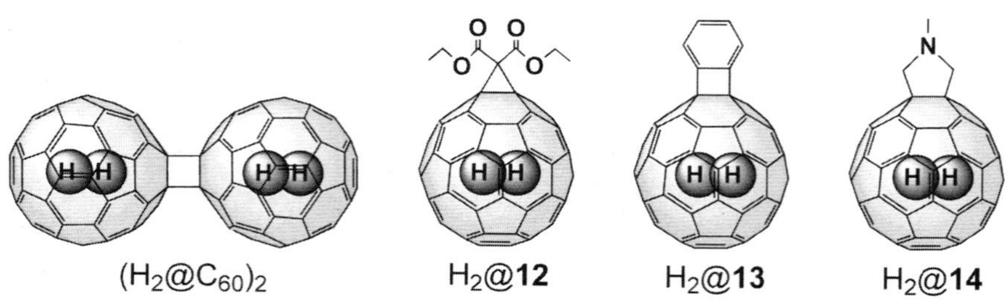

SCHEME 11.5.

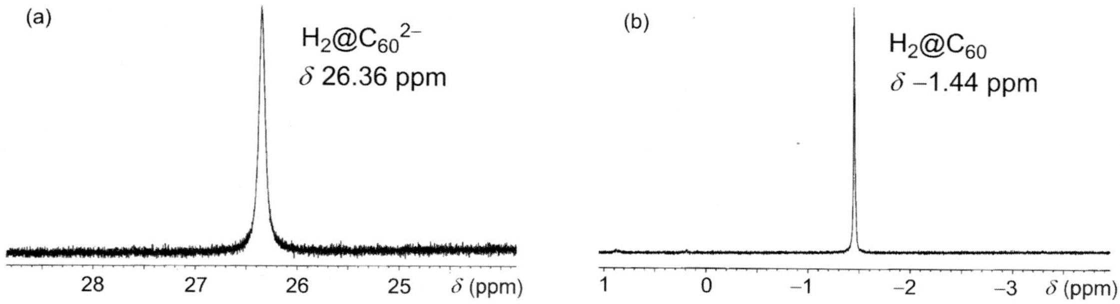

FIGURE 11.2. ^1H NMR spectra (300 MHz) of (**a**) $H_2@C_{60}^{2-}$ in CD_3CN and (**b**) $H_2@C_{60}$ in ODCB-d_4.

dihydrogen can also be used as a good probe to investigate the chemical reactions at the exterior of the cage, just as a ^3He atom inside C_{60} has been used for this purpose [40–42].

11.3.4 Utilization of the Encapsulated Dihydrogen as an NMR Probe

The NMR chemical shift for dihydrogen of $H_2@C_{60}$ is sensitive to the magnetic field inside the fullerene cage [27]. Thus, the chemical shift can be also used as a tool to disclose the spherical aromaticity of the fullerene π-systems. The spherical aromaticity of hexaanion of C_{60} and C_{70} has already been studied based on the chemical shift of encapsulated ^3He [43]. Among the possible anionic states, a dianion of C_{60} is particularly important in synthetic chemistry for the introduction of two organic groups on the cage [44,45]. Although the "$2(N+1)^2$ rule" [46], describing the spherical aromaticity of I_h-symmetrical fullerenes, suggests that the 62-π-electron system should not have high aromaticity, little had been known about the aromaticity of C_{60}^{2-}.

To clarify this issue, the dianion of $H_2@C_{60}$ was generated by treating with CH_3SNa in CD_3CN [47,48]. The resulting dark red solution showed vis/near-IR absorptions at λ_{max} = 830 and 944 nm and a broad ^{13}C NMR signal around 183 ppm, indicating the generation of $H_2@C_{60}^{2-}$ [49–51]. The ^1H NMR signal for dihydrogen of $H_2@C_{60}^{2-}$ appeared at surprisingly low field such as δ 26.36 ppm (Figure 11.2). This signal shifted to downfield by 27.8 ppm relative to that of neutral $H_2@C_{60}$ (δ −1.44 ppm in ODCB-d_4) [26, 27]. This result demonstrates that the spherical aromaticity of C_{60} drastically decreases upon two-electron reduction. The NICS (nucleus independent chemical shifts) [52] calculations (B3LYP/6-31G*) for all the hexagons (diamagnetic in the neutral state) and pentagons (paramagnetic in the neutral state) of C_{60}^{2-} suggested that the ring currents of all hexagons become *paramagnetic* while those of all pentagons become *diamagnetic* upon two-electron reduction. Because there exist more hexagons than pentagons in the C_{60} cage, the antiaromatic character of hexagons overwhelms the aromatic character of pentagons, resulting in a strong deshielding inside the cage.

11.4 ENDOHEDRAL C_{70} ENCAPSULATING DIHYDROGEN, $H_2@C_{70}$ AND $(H_2)_2@C_{70}$

The size of the inner cavity of C_{70} is estimated to be approximately 5 Å along the long axis and 4 Å along the short axis, which is larger than that of C_{60} (~4 Å in inner diameter). Therefore, it is expected to be possible to insert more than one small molecule through a chemically created opening on the surface of C_{70}.

In the manner similar to the method for synthesizing $H_2@C_{60}$, a thermal reaction of C_{70} with 3,6-di(2-pyridyl)pyridazine (**15**) was conducted to give molecule **16** in 73% yield, based on consumed C_{70} (55% conversion) (Scheme 11.6, step a). Then, the reaction of **16** with singlet oxygen afforded oxidized product **17**, by one double-bond

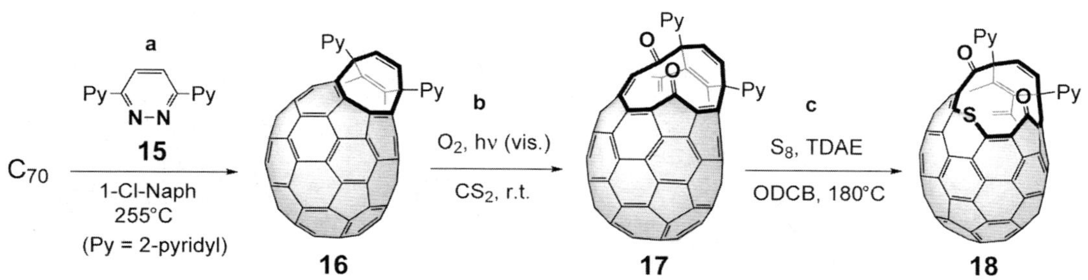

SCHEME 11.6.

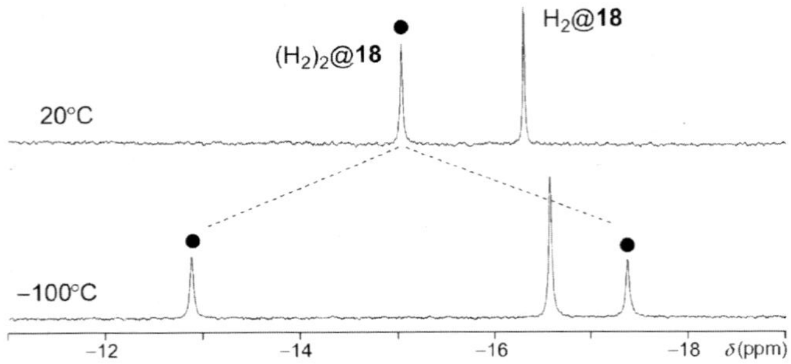

FIGURE 11.3. ^{1}H NMR spectra (400 MHz, CS_2-CD_2Cl_2 (4:1)) of the mixture of $(H_2)_2$@**18** and H_2@**18** (1:2) taken at 20°C and −100°C.

cleavage, in 60% yield, based on consumed **16** (81% conversion) (Scheme 11.6, step b). The opening of **17** was enlarged by the insertion of a sulfur atom to give **18** with a 13-membered-ring opening in 94% yield (Scheme 11.6, step c) [53].

The insertion of dihydrogen to **18** was carried out by applying 890 atm of dihydrogen gas at 230°C for 8 hours. The encapsulation of dihydrogen was clearly demonstrated by appearance of a new intense signal of dihydrogen in H_2@**18** in the ^{1}H NMR spectrum at unusually high field, δ −16.51 ppm in ODCB-d_4. Noteworthy is that a small signal was also observed at δ −15.22 ppm corresponding to $(H_2)_2$@**18**. The integrated relative intensity was determined to be 1.94 H and 0.13 H, respectively, by comparison with the intensity of a pyridyl proton signal at δ 8.68 ppm (1.00 H).

The appearance of two molecules of dihydrogen of $(H_2)_2$@**18** as a singlet is ascribed to the rapid exchange of each position of two molecules of dihydrogen, giving the time-averaged NMR signal. Thus, $(H_2)_2$@**18** was enriched by the recycling HPLC (Cosmosil Buckyprep column, toluene) and the low-temperature NMR experiment was conducted (Figure 11.3). The ^{1}H NMR spectrum at 20°C exhibited a slightly broadened signal of $(H_2)_2$@**18** (δ −15.04 ppm) along with a sharp signal of H_2@**18** (δ −16.30 ppm). The difference in the line shape indicates that the motion of the incorporated two molecules of dihydrogen is already slightly restricted. The existence of two molecules of dihydrogen inside **18** was indisputably confirmed by the appearance of new two signals at −100°C. The line-shape analysis showed that the two molecules of dihydrogen exchange their positions at the rate of only 50 times per second at −100°C, while the rate increases up to 4.9×10^5 times per second at 20°C. The Arrhenius plot gave the preexponential factor (A) and the activation energy (E_a) as $10^{11.7}$ and 8.0 kcal mol^{-1}, respectively. The activation parameters were also determined to be $\Delta G^\ddagger$ (25°C) = 9.4 kcal mol^{-1}, $\Delta H^\ddagger$ = 7.4 kcal mol^{-1}, and $\Delta S^\ddagger$ = −7 cal K^{-1} mol^{-1}.

The closure of the opening was carried out by applying the same reactions as those employed for the synthesis of $H_2@C_{60}$. As shown in Scheme 11.7, the mixture of $(H_2)_2$@**18** and H_2@**18** (3:97) was oxidized with MCPBA to give the sulfoxides, which were subjected to photoelimination of the SO unit to afford compounds $(H_2)_n$@**17** (n = 2 (3%), 1 (97%)) in 57% yield. Then, the McMurry reaction gave compounds $(H_2)_n$@**16** in 61% yield. Finally, thermolysis of $(H_2)_n$@**16** provided endohedral fullerene $(H_2)_n@C_{70}$ (n = 2 (3%), 1 (97%)) (contaminated by 10% empty C_{70}) in 56% yield as a dark brown powder [54].

The ^{1}H NMR spectrum of the product in ODCB-d_4 exhibited a small signal for $(H_2)_2@C_{70}$, along with an intense signal for $H_2@C_{70}$ at δ −23.80 and −23.97 ppm, respectively, with an integrated ratio being 6:97. In the ^{13}C NMR spectrum in ODCB-d_4, five signals appeared at δ 150.95, 148.39, 147.71, 145.72, and 131.24 ppm for $H_2@C_{70}$. All of the signals were slightly shifted to downfield as compared to those of C_{70} (δ 150.90, 148.36, 147.69, 145.67, and 131.17 ppm) in the range $\Delta\delta$ 0.02–0.07 ppm. The $\Delta\delta$ values are smaller than that between $H_2@C_{60}$ and C_{60} ($\Delta\delta$ 0.08 ppm) [26, 27]. This

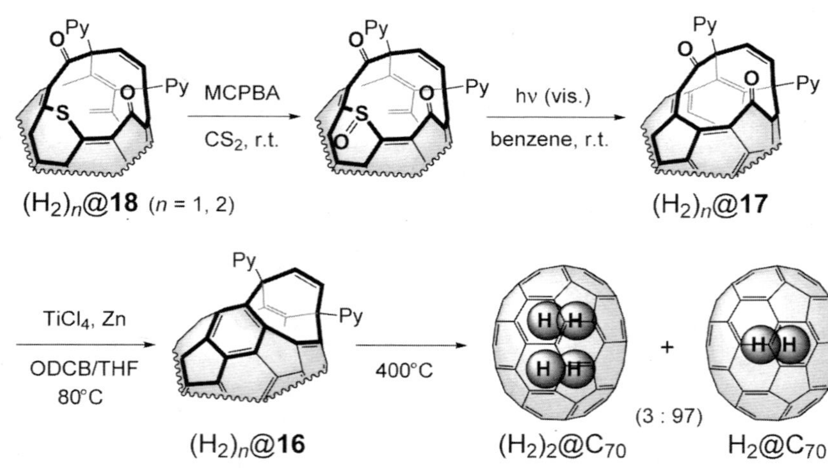

SCHEME 11.7.

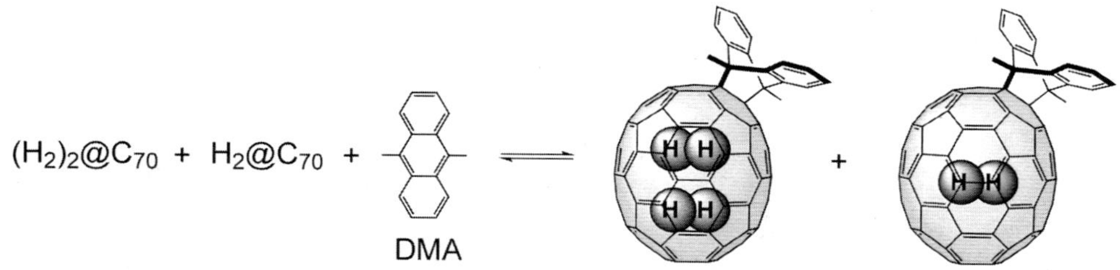

(H_2)$_2$@**19** H_2@**19**

SCHEME 11.8.

indicates that the van der Waals interaction between inner dihydrogen and outer C_{70} is quite minute, even as compared to that of H_2@C_{60}, reflecting the larger inner space.

Despite the minute interaction, theoretical calculations (MPWB1K/6-31G**) [55] with basis set superposition error (BSSE) correction indicated that the encapsulation of two molecules of dihydrogen in C_{70} is exothermic by −9.3 kcal mol^{-1}, which is higher than that of H_2@C_{70} (−6.9 kcal mol^{-1}), indicating that more interaction is present between the two molecules of dihydrogen and the C_{70} cage. Therefore, there is a possibility that the difference in chemical reactivity of the outer C_{70} cage becomes appreciable when one or two molecule(s) of dihydrogen exist(s) inside the cage. To clarify this, the Diels–Alder reaction of (H_2)$_2$@C_{70} and H_2@C_{70} with 9,10-dimethylanthracene (DMA) [56, 57] was investigated (Scheme 11.8). Thus, a solution of a mixture of (H_2)$_2$@C_{70}, H_2@C_{70}, and C_{70} (molar ratio, 2:70:28) and DMA in ODCB-d_4 was prepared [42]. The NMR spectrum exhibited new signals for the encapsulated dihydrogen of adducts (H_2)$_2$@**19** and H_2@**19** at δ −21.80 and −22.22 ppm, respectively, in addition to the signals of unreacted (H_2)$_2$@C_{70} and H_2@C_{70}. The equilibrium constants K_2 and K_1 for the addition of DMA to (H_2)$_2$@C_{70} and to H_2@C_{70}, respectively, are summarized in Table 11.2. As shown, the K_2 value is smaller than the K_1 value by more than 15% at each temperature, demonstrating the "apparently" decreased reactivity of (H_2)$_2$@C_{70} toward DMA. The van't Hoff plot of ln K_2 or ln K_1 versus T^{-1} gave excellent linear fits and provided ΔH_2 and ΔH_1 as −13.4 and −13.8 kcal mol^{-1}, respectively, and ΔS_2 and ΔS_1 as −32.9 and −33.7 cal mol^{-1} K^{-1}, respectively. Thus, the two molecules of dihydrogen encapsulated inside C_{70} were found to affect both ΔH and ΔS in this reaction although to a very minute degree [54].

TABLE 11.2. Equilibrium Constants K_2 and K_1 for Addition of DMA to (H_2)$_2$@C_{70} and H_2@C_{70} in ODCB-d_4 at 30°C, 40°C, and 50°C

	T (°C)	30	40	50
(H_2)$_2$@C_{70}	K_2 (M^{-1}):	296	143	74.7
H_2@C_{70}	K_1 (M^{-1}):	364	177	88.4

11.5 ENDOHEDRAL FULLERENES ENCAPSULATING HELIUM, He@C_{60} AND He@C_{70}

Next, we challenged to synthesize endohedral C_{60} encapsulating helium. As described above, the attractive interaction between helium and C_{60} is quite small (ΔE = −2.4 kcal mol^{-1}, Table 11.1), indicating that the encapsulation of helium should require high pressure, even as compared to the conditions used in the encapsulation of dihydrogen to **9** (i.e., 800 atm).

The synthesis of He@C_{60} was conducted by applying the synthetic procedure developed for H_2@C_{60}, though we selected sulfoxide derivative **11** instead of sulfide **9** so that we could carry out the size reduction of the opening by removing the SO unit immediately after the insertion of helium [58] in order to prevent escape of once trapped helium inside the cage. Thus, powdery **11** was heated at 115°C under pressurized ^{4}He gas (1230 atm) for 1 h (Scheme 11.9). The resulting powder was immediately dissolved in benzene and photoirradiated to afford He@**8** at the He occupation level of 30%.

As for a hypothetical equilibrium—for example, He + C_{60} ⇄ He@C_{60}—an encapsulation energy (ΔE = −1.99 kcal mol^{-1}) and an equilibrium constants (K_{eq} = 6.41 × 10^{-4} atm^{-1} at 300 K, 1.71 × 10^{-4} atm^{-1} at 500 K) have been theoretically evaluated [59]. According to these data, the occupation level of helium inside C_{60} in the hypothetical equilibrium under 1230 atm can be calculated as 44% at 300 K and 17% at 500 K. Since the interaction between helium and the C_{60} cage is minute, these estimations may be applied to the equilibrium for He@**11**. Thus, the experimental data for the incorporation of helium in **11** at the occupation level of 30% at 388 K (115°C) appears to indicate that the equilibrium state has already been reached. The lower occupation level of helium than that of dihydrogen [20] can be mainly ascribed to the smaller encapsulation energy of helium (ΔE = −2.4 kcal mol^{-1}) than that of dihydrogen (ΔE = −8.2 kcal mol^{-1}) as has been shown in Figure 11.1 and Table 11.1.

Then, the opening of He@**8** (He@**8**/**8** = 3/7) was transformed to an eight-membered ring by the McMurry reaction affording He@**7** (He@**7**/**7** = 3/7). Thermal treatment of powdery He@**7** gave He@C_{60} in 76% isolated yield at an occupation level of 30% [58]. The occupation level was increased up

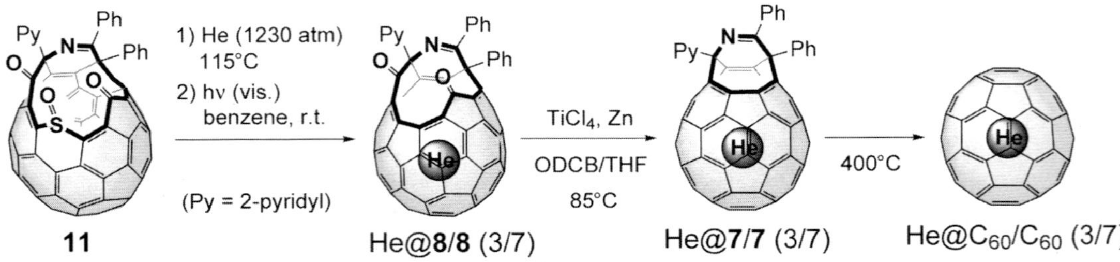

SCHEME 11.9.

to 60% by the use of recycling HPLC (Cosmosil Buckyprep column, toluene). The ^{13}C NMR spectrum exhibited a signal for He@C_{60} at $\delta = 142.80$ ppm, which was very slightly shifted to downfield by $\Delta\delta = 0.02$ ppm relative to that of empty C_{60} ($\delta = 142.78$ ppm). This $\Delta\delta$ value is apparently smaller than that observed for H_2@C_{60} ($\Delta\delta = 0.08$ ppm) [26, 27], Ar@C_{60} ($\Delta\delta = 0.17$ ppm) [60], Kr@C_{60} ($\Delta\delta = 0.39$ ppm) [61], and Xe@C_{60} ($\Delta\delta = 0.95$ ppm) [62], again indicating the quite minute electronic interaction between the C_{60} π-system and the encapsulated helium.

He@C_{70} (occupation level of 30%) was also synthesized by applying the same procedure described above. However, regrettably we could not observe any evidence for the formation of He_2@C_{70} under these conditions, in contrast to the formation of $(H_2)_2$@C_{70} under the similar conditions [53, 54].

11.6 SPIN CHEMISTRY

Spin chemistry is a field that is concerned with the nuclear and electronic spin control of the rates on fundamental chemical and physical steps in thermal and photochemical reactions. Since macroscopic synthesis of H_2@C_{60} was achieved in 2005 [26, 27], research works on spin chemistry of H_2@C_{60} have been extensively conducted by Professor N. J. Turro and co-workers [63]. As is well known, the molecule of H_2 exists as two distinct species—that is, allotropes of elemental hydrogen. One allotrope, termed as *para*-H_2 (pH_2), is a lower-energy species possessing antiparallel nuclear spins. The other allotrope, termed as *ortho*-H_2 (oH_2), is a higher-energy species possessing parallel nuclear spins. A profound difference between them is that pH_2 is NMR silent whereas oH_2 is NMR active. From the viewpoint of spin chemistry, H_2@C_{60} and H_2@C_{70} are very interesting molecular systems in which the H_2 molecule is completely isolated by the carbon cage from the outside world surrounding it.

Various interesting studies on spin chemistry have been conducted using H_2@C_{60}, these include, for example, (i) systematic comparison of the nuclear spin-lattice relaxation of H_2 and H_2@C_{60} in various solvents [64,65], (ii) conversion of pH_2 to oH_2 and of oH_2 to pH_2 trapped inside C_{60} by spin catalysis [66], (iii) communication of the encapsulated H_2 inside C_{60} with an electronically excited dioxygen, $^1O_2$, in the outside world [67], and (iv) controlling the rate of nuclear spin interconversion of the encapsulated H_2 by covalently attaching the paramagnetic spin catalyst to the C_{60} cage [68].

One additional example employing C_{70} is as follows. Although fullerenes C_{60} and C_{70} are rather similar in their structures, electrochemical properties, and chemical reactivity, there is a large difference in their lifetimes of the electronic triplet state which is generated via intersystem crossing of the excited singlet state; 135 μs for C_{60} and 11.8 ms for C_{70}, respectively [2]. The equilibrium between free pH_2 and oH_2 takes months or longer to achieve in the absence of a paramagnetic spin catalyst. It is quite intriguing to see whether the excited triplet state of C_{60} or C_{70} works as a spin catalyst toward the conversion between pH_2 and oH_2 [69]. Thus, a mixture of H_2@C_{60} and H_2@C_{70} was dissolved together in toluene-d_8, and the resulting mixture was irradiated with visible light at 77 K. The ^1H NMR analysis of the mixture using HD@C_{60} and HD@C_{70} as internal standards for comparison of the peak intensities before and after irradiation showed that the oH_2 inside C_{70} was readily depleted after irradiation, whereas no change was observed for H_2@C_{60}. This result shows that the photoinduced spin conversion is effective only for H_2@C_{70} with 100-fold longer lifetime of the triplet than the triplet of H_2@C_{60}. As shown in Scheme 11.10, the triplet H_2@C_{70} generated by photo-irradiation has enough time to work as a spin catalyst to cause the conversion of the oH_2 to the pH_2 encapsulated in C_{70}. On the other hand, thermally induced spin conversion with liquid O_2 as a spin catalyst was found to be effective for both H_2@C_{60} and H_2@C_{70} [69].

11.7 SYNTHESIS AND PROPERTIES OF H_2O@C_{60}

Needless to say, water is very important for material, environments, and human life. It exists as oligomeric form in gas phase, and as polymeric form in both liquid and solid phases by the strong intermolecular hydrogen bonding [70]. Owing to the hydrogen bonding, water has characteristic properties such as high boiling and melting points and high dielectric constant, and it acts as both acid and base in bulk. However, a single molecule of H_2O without any hydrogen bonding to other organic molecule or coordination to metal is quite rare

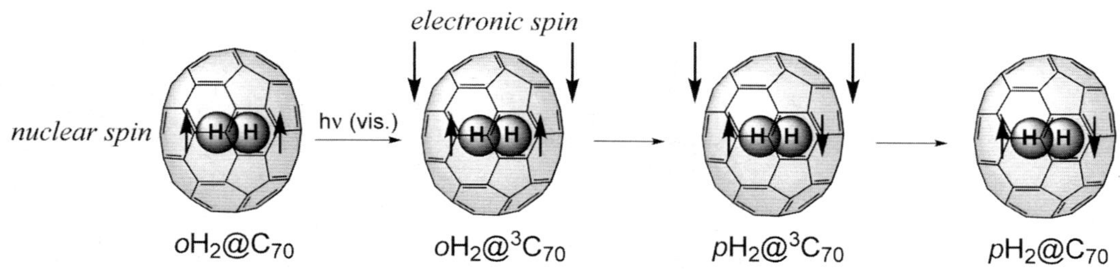

SCHEME 11.10.

[13, 21, 22, 71, 72]. If we can encapsulate H_2O in a fullerene cage, this would give us a good opportunity to study the intrinsic properties of a H_2O molecule as a completely isolated single molecule.

In order to apply the molecular surgery method to encapsulation of a larger molecule than H_2, it is necessary to make fullerenes having an opening large enough for such a molecule to pass through [13, 21, 22]. However, it would be more difficult to close such a large opening than a small one. If the size of an opening on fullerenes can be controlled dynamically—that is, a small opening changes into a larger one in situ under high-temperature conditions and regenerates itself again after insertion of a molecule through it—then wide varieties of endohedral fullerenes would be synthesized. Here, we describe the macroscopic synthesis of $H_2O@C_{60}$ based on such a concept of dynamic control of the opening size [73].

The synthesis of new open-cage C_{60} derivatives is shown in Scheme 11.11. Applying our previous synthetic route [14, 53], diketones **22a** and **22b** were synthesized in 29% and 41% yields by the reaction of C_{60} with pyridazine derivatives **20a** and **20b** followed by photochemical reaction of intermediate **21**. Then, the reaction of **22** with 2.3

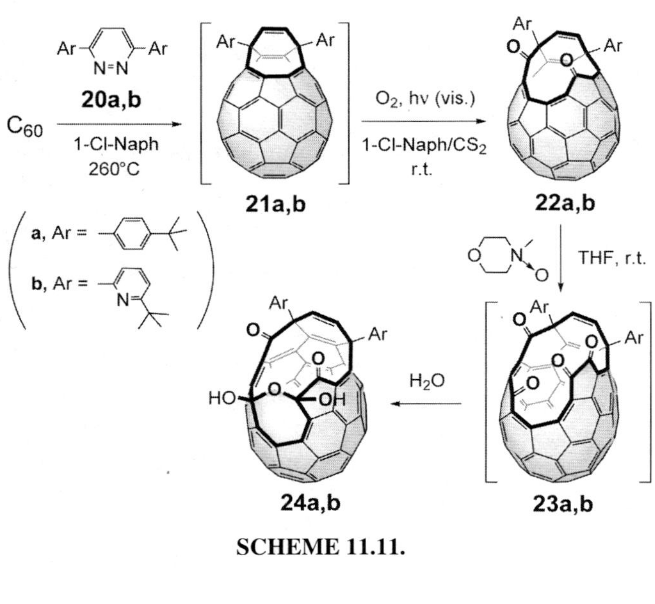

SCHEME 11.11.

equivalents of *N*-methylmorpholine *N*-oxide in wet THF at room temperature afforded **24a** and **24b** in good yields.

The structure of **24a** was determined unambiguously by the single crystal x-ray analysis, showing that the opening is made of a 13-membered ring containing two hemiacetal carbons in addition to two carbonyl carbons. When **24a** was heated in refluxing toluene for 30 min, it was completely transformed to a new compound, most probably **23a**, as examined by HPLC analysis (Buckyprep, toluene). The size of the opening of tetraketone **23** is apparently larger than that of hemiacetal **24**, implying that insertion of an H_2O molecule into **23** could be now possible.

Actually, quantitative encapsulation of an H_2O molecule in **24b** was achieved when **24b** was heated in wet toluene under the high pressure of 9000 atm at 120°C for 36 h as shown in Scheme 11.12. The high-pressure is critical because the encapsulation at ambient pressure was only 8%. The formation of $H_2O@$**24b** is explained by the dynamic control of the opening size as mentioned above: that is, elimination of water from **24b** generates **23b** into which insertion of an H_2O molecule takes place through the 16-membered-ring opening generated in situ, followed by addition of water at the opening edge to regenerate **24b** structure.

Next, rather simple method for closing the opening of $H_2O@$**24b** was developed compared with that of $H_2@$**9** [26,27]. The reaction of $H_2O@$**24b** with an excessive amount of $P(Oi\text{-}Pr)_3$ in refluxing toluene gave $H_2O@$**21b** in 50% isolated yield without the loss of the encapsulated H_2O molecule. This reaction would have involved elimination of water from $H_2O@$**24b** to generate $H_2O@$**23b** followed by the two successive carbonyl couplings on $H_2O@$**23b** and then on $H_2O@$**22b**, to give $H_2O@$**21b**. Then, $H_2O@$**21b** was mixed with neutral Al_2O_3 in a mortar, and the resulting mixture was heated at 360°C for 1 h under vacuum to give $H_2O@C_{60}$ as a brown powder in 29% isolated yield after purification with silica gel chromatography.

The direct evidence for existence of the encapsulated H_2O molecule inside C_{60} was obtained from the single crystal x-ray analysis of the molecular complex $H_2O@C_{60}\cdot$(nickel (II) octaethylporphyrin)$_2$ at 100 K as shown in Figure 11.4. It was observed that the oxygen atom of the encapsulated H_2O molecule is located just at the center of the C_{60} cage. This is very similar to the structure of $Kr@C_{60}$ [74] and $H_2@C_{60}$

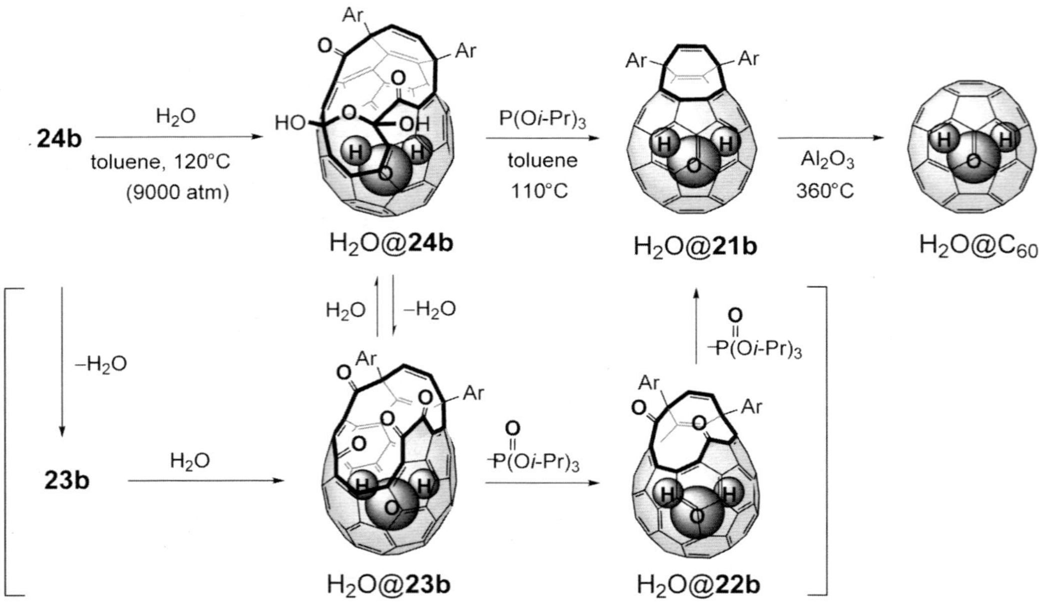

SCHEME 11.12.

[75] and is different from that of $Li^{+}@C_{60}\cdot SbCl_6^{-}$ with the encapsulated Li^{+} ion located at off-center position of C_{60} [76]. The position of the hydrogen atom of the encapsulated H_2O molecule was refined experimentally, and the OH bonds were found to be directed toward the Ni atoms through the C_{60} cage, demonstrating that the motion of the water is stopped inside the C_{60} cage under these conditions.

The physical properties of the outer C_{60} cage are not affected by the encapsulated water molecule as judged from the UV–vis and IR spectra. In the ^{13}C NMR spectra, the signal corresponding to $H_2O@C_{60}$ is observed at 142.89 ppm as a sharp singlet, indicating that the encapsulated water molecule rotates freely at ambient temperature. The difference in ^{13}C NMR chemical shift between $H_2O@C_{60}$ (142.89 ppm in ODCB-d_4) and empty C_{60} is 0.11 ppm, which is larger than that for $H_2@C_{60}$ ($\Delta\delta = 0.08$) [26,27] and smaller than that for $Ar@C_{60}$ ($\Delta\delta = 0.17$) [60], reflecting that the van der Waals size of H_2O is in between H_2 and Ar. Upon cyclic voltammetry in ODCB, an irreversible oxidation peak and four quasi-reversible reduction waves were observed at the almost same potential of emplty C_{60} (+1.32, −1.08, −1.46, −1.91, and −2.38 V versus Fc/Fc^{+} for $H_2O@C_{60}$; +1.32, −1.08, −1.47, −1.92, and −2.39 V for empty C_{60}). This result indicates that the single molecule of H_2O is electrochemically stable under the hydrophobic environment inside C_{60}.

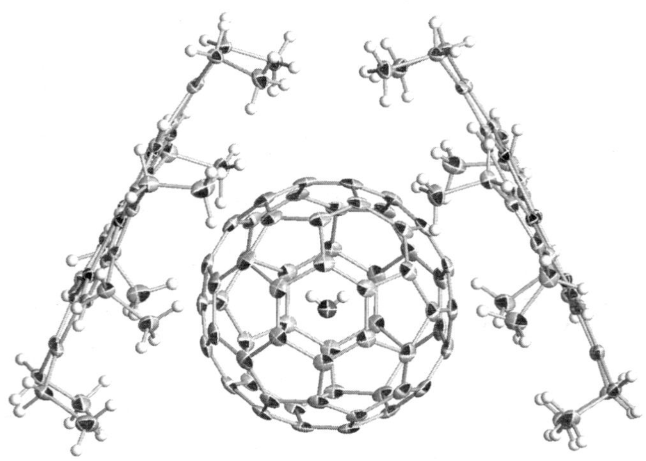

FIGURE 11.4. X-ray structure of the molecular complex $H_2O@C_{60}\cdot$(nickel(II) octaethylporphyrin)$_2$ at the 50% probability level.

11.8 APPLICATION OF OPEN-CAGE FULLERENES TO ORGANIC SOLAR CELLS

We have described our recent achievements in the molecular surgical synthesis of endohedral fullerenes. The open-cage fullerenes, which are important synthetic intermediates for the endohedral fullerenes, are of interest also in materials science because the electronic properties of the fullerene cage can be controlled by functional group(s) at the opening. In this final section, we wish to demonstrate the application of some open-cage fullerenes as n-type semiconductor materials in bulk-heterojunction (BHJ) solar cells [77].

Since an active layer of BHJ solar cells is prepared by the wet process, compounds **25** and **26**, which were functionalized by highly solubilizing two *n*-octyl groups (Scheme 11.13), were prepared by the same reaction as depicted in Scheme 11.3 above. On the other hand, in order

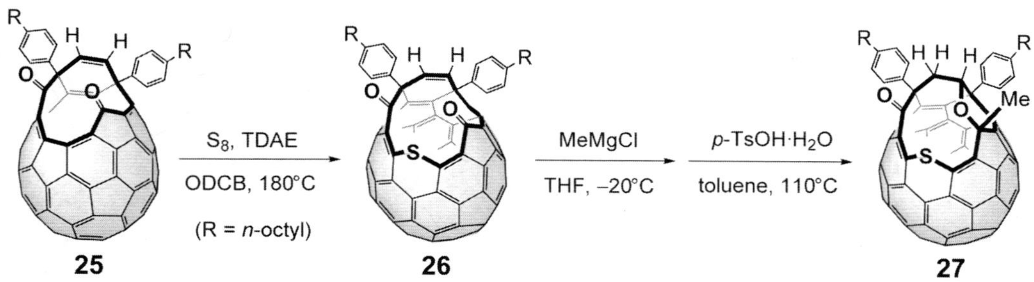

SCHEME 11.13.

to raise the LUMO energy level, a carbonyl group of **26** was modified via addition of methyl Grignard reagent and subsequent transannular cyclization under acidic conditions to afford **27** (Scheme 11.13) [78]. The reduction potentials of **25–27**, C_{60}, and the benchmark acceptor PCBM ((6,6)-phenyl C_{61}-butyric acid methyl ester [79]) determined by cyclic voltammetry are summarized in Table 11.3 together with calculated LUMO levels. All molecules **25–27** exhibited pseudo-reversible first to third reduction waves in the same manner as those for C_{60} and PCBM. It is worth noting that the first reduction potentials ($E_{red}{}^1$) of **26** and **27** markedly shifted to a negative side upon the simple modification of **25**; that is, the sulfur insertion to **25** caused the negative shift by $\Delta E_{red}{}^1 = 0.12$ V, and the modification of the carbonyl group resulted in the further shift by $\Delta E_{red}{}^1 = 0.11$ V, as supported by the results of DFT calculations. Thus, the present approach provides a useful way to control the LUMO energy level of the C_{60} cage. Since an open-circuit voltage (V_{oc}) of the BHJ solar cells is closely associated with the energy difference between acceptor's LUMO and donor's HOMO [80, 81], the higher LUMO energy level of **27** than that of PCBM is expected to contribute to improve the V_{oc} value, when these are incorporated in the BHJ solar cells.

The BHJ solar cells were fabricated using **25–27** as the acceptor materials with a combination of the representative donor polymer, poly(3-hexylthiophene) (P3HT), by basically following the procedure optimized for PCBM:P3HT-based device [82]. The devices with a layered structure of glass/ITO/PEDOT:PSS/**25-27**:P3HT/TiO_x/Al were prepared. The device performances for the BHJ solar cells under AM 1.5 G conditions of 100 mW cm^{-2} are summarized in Table 11.4. As expected, **25**:P3HT-based device marked a low V_{oc} value (0.49 V), reflecting the low-lying LUMO energy level of **25**. The value was clearly improved to $V_{oc} = 0.62$ V for a **26**:P3HT-based device and further to $V_{oc} = 0.74$ V for a **27**:P3HT-based device. As shown in Figure 11.5, the linear regression indicated a clear correlation ($r^2 = 1.0$) between the LUMO energy levels of **25–27** and the V_{oc} values, although the point for PCBM:P3HT appears to deviate downward. It is worth noting that the V_{oc} of **27**:P3HT-based device was superior to that of PCBM:P3HT device ($V_{oc} = 0.60$ V) by $\Delta V_{oc} = 140$ mV. This ΔV_{oc} is much larger than the value obtained by the introduction of electron-donating groups on the phenyl ring of PCBM ($\Delta V_{oc} = ca.$ 30 mV) [83]. These data demonstrated that the open-cage fullerene derivatives are potential *n*-type materials with tunable LUMO levels for the BHJ solar cells.

On the other hand, the short-circuit current densities (J_{sc}) of **26**:P3HT and **27**:P3HT-based devices ($J_{sc} = 6.49$ and 6.57 mA cm^{-2}, respectively) slightly decreased with reference to that of PCBM:P3HT-based device ($J_{sc} = 7.76$ mA cm^{-2}), while the fill factors (FF) were almost identical. One possible reason would be that the presence of two *n*-octyl chains prevented the effective intermolecular π–π interaction between the C_{60} cages and hence decreased the charge mobility [84].

TABLE 11.3. Electrochemical Reduction Potentials[a] of C_{60}, 25–27, and PCBM and Calculated LUMO Levels[b]

Compound	$E_{red}{}^1$ (V)	$E_{red}{}^2$ (V)	$E_{red}{}^3$ (V)	LUMO (eV)
C_{60}	−1.09	−1.49	−1.95	−3.23
25	−1.00	−1.40	−2.03	−3.32
26	−1.12	−1.52	−2.01	−3.18
27	−1.23	−1.62	−2.14	−3.00
PCBM	−1.17	−1.57	−2.09	−3.04

[a] Values for $1/2(E_{pa} + E_{pc})$ in V versus Fc/Fc$^+$; in ODCB with 0.1 M $Bu_4N^+BF_4^-$ as supporting electrolyte; a Pt wire as a counter electrode and a glassy carbon as a working electrode; scan rate, 20 mV s^{-1}.
[b] Calculated at the B3LYP/6-31G* level of theory with replacement of the *n*-octyl groups by methyl groups.

TABLE 11.4. Characteristics of Bulk-Heterojunction Solar Cells[a]

Active Layer	V_{oc} (V)	J_{sc} (mA cm^{-2})	FF	η_e (%)
25:P3HT	0.49	5.20	0.53	1.35
26:P3HT	0.62	6.49	0.62	2.49
27:P3HT	0.74	6.57	0.64	3.11
PCBM:P3HT	0.60	7.76	0.67	3.12

[a]Open-circuit voltage (V_{oc}), short-circuit current density (J_{sc}), fill factor (FF), and PCE (power conversion efficiency) (η_e) with **25–27** and PCBM as the acceptor materials and P3HT as the donor material. The active area of the device is 5.2 mm^2.

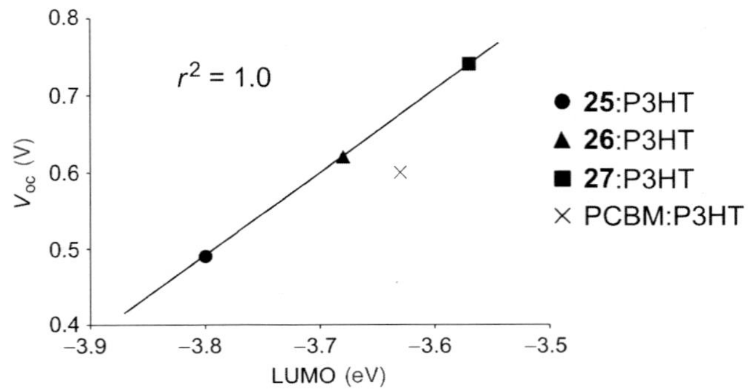

FIGURE 11.5. The open-circuit voltages (V_{oc}) of BHJ solar cells using **25**–**27** and PCBM as the acceptor materials and P3HT as the donor material, plotted against LUMO levels of the acceptor materials estimated based on the first reduction potentials ($E_{red}{}^1$, V versus Fc/Fc^+) on CV with the equation; LUMO level = $-(E_{red}{}^1 + 4.8)$. Linear regression was carried out with the data for **25**–**27**:P3HT.

As a result, the device using **27** provided the best PCE (power conversion efficiency) ($\eta_e = 3.11\%$), which is comparable to that of the benchmark PCBM:P3HT-based device ($\eta_e = 3.12\%$).

11.9 OUTLOOK

As described above, organic synthetic routes applying the molecular surgery turned out to be powerful methods to synthesize endohedral fullerenes encapsulating a helium atom, hydrogen molecule(s), and its isotope(s), and the methodology was further extended to the synthesis of $H_2O@C_{60}$ in which a molecule with a permanent dipole was encapsulated. It is expected that the dipole moment affects the physical properties of outer C_{60} cage such as phase transition temperature in the solid state and charge mobility in the thin film when electric potential is applied. Furthermore, we will be able to develop novel spin chemistry of the H_2O molecule.

It should be noted that all these compounds cannot be synthesized by the physical method generally used to obtain endohedral metalofullerenes. For further developments, rational design and practical synthesis of novel open-cage fullerenes are still required. They should have a larger opening that can be closed after insertion of guest molecule(s). Now application of high pressure in a gas phase as well as in a liquid phase toward open-cage fullerenes is found to be effective for the insertion of the guest molecule(s). As the guest molecules, CO, O_2, NH_3, or CH_4 might be envisioned, among which O_2 would be the most interesting from the viewpoint of its behavior as a ground-state triplet molecule. We wish that a variety of novel endohedral fullerenes encapsulating small molecule(s) will be realized in the near future. We believe that these novel materials will open new research areas in basic science as well as application fields as functional materials.

REFERENCES

1. Akasaka, T., Nagase, S. (2002). *Endofullerenes: A New Family of Carbon Clusters*. Kluwer Academic Publishers, Dordrecht, The Netherlands.
2. Kadish, K. M., Ruoff, R. S. (2000). *Fullerenes: Chemistry, Physics and Technology*. John Wiley and Sons, Inc. New York.
3. Tsuchiya, T., Kumashiro, R., Tanigaki, K., Matsunaga, Y., Ishitsuka, M. O., Wakahara, T., Maeda, Y., Takano, Y., Aoyagi, M., Akasaka, T., Liu, M. T. H., Kato, T., Suenaga, K., Jeong, J. S., Iijima, S., Kimura, F., Kimura, T., Nagase, S. (2008). Nanorods of endohedral metallofullerene derivative. *Journal of the American Chemical Society*, **130**, 450–451.
4. Shu, C.-Y., Wang, C.-R., Zhang, J.-F., Gibson, H. W., Dorn, H. C., Corwin, F. D., Fatouros, P. P., Dennis, T. J. S. (2008). Organophosphonate functionalized $Gd@C_{82}$ as a magnetic resonance imaging contrast agent. *Chemistry of Materials*, **20**, 2106–2109.
5. Saunders, M., Cross, R. J., Jiménez-Vázquez, H. A., Shimshi, R., Khong, A. (1996). Noble gas atoms inside fullerenes. *Science*, **271**, 1693–1697.
6. Saunders, M., Jiménez-Vázquez, H. A., Cross, R. J., Mroczkowski, S., Freedberg, D. I., Anet, F. A. L. (1994). Probing the interior of fullerenes by ^3He NMR spectroscopy of endohedral $^3He@C_{60}$ and $^3He@C_{70}$. *Nature*, **367**, 256–258.
7. Saunders, M., Jiménez-Vázquez, H. A., Cross, R. J., Mroczkowski, S., Gross, M. L., Giblin, D. E., Poreda, R. J. (1994). Incorporation of helium, neon, argon, krypton, and xenon into fullerenes using high pressure. *Journal of the American Chemical Society*, **116**, 2193–2194.
8. Rubin, Y. (1999). Ring opening reactions of fullerenes: Designed approaches to endohedral metal complexes. *Topics in Current Chemistry*, **199**, 67–91.
9. Murata, M., Murata, Y., Komatsu, K. (2008). Surgery of fullerenes. *Chemical Communications*, 6083–6094.
10. Murata, M., Murata, Y., Komatsu, K., Molecular surgery toward organic synthesis of endohedral fullerenes. *Chemistry of Nanocarbons*. Wiley-Blackwell, Oxford, 2010, pp. 215–237.
11. Vougioukalakis, G. C., Roubelakis, M. M., Orfanopoulos, M. (2010). Open-cage fullerenes: Towards the construction of nanosized molecular containers. *Chemical Society Reviews*, **39**, 817–844.
12. Schick, G., Jarrosson, T., Rubin, Y. (1999). Formation of an effective opening within the fullerene core of C_{60} by an unusual reaction sequence. *Angewandte Chemie International Edition*, **38**, 2360–2363.
13. Iwamatsu, S.-i., Uozaki, T., Kobayashi, K., Re, S., Nagase, S., Murata, S. (2004). A bowl-shaped fullerene encapsulates a water into the cage. *Journal of the American Chemical Society*, **126**, 2668–2669.
14. Murata, Y., Murata, M., Komatsu, K. (2003). Synthesis, structure, and properties of novel open-cage fullerenes having heteroatom(s) on the rim of the orifice. *Chemistry—A European Journal*, **9**, 1600–1609.
15. Iwamatsu, S.-i., Murata, S., Andoh, Y., Minoura, M., Kobayashi, K., Mizorogi, N., Nagase, S. (2005). Open-cage

fullerene derivatives suitable for the encapsulation of a hydrogen molecule. *Journal of Organic Chemistry*, **70**, 4820–4825.

16. Chuang, S.-C., Murata, Y., Murata, M., Mori, S., Maeda, S., Tanabe, F., Komatsu, K. (2007). Fine tuning of the orifice size of an open-cage fullerene by placing selenium in the rim: Insertion/release of molecular hydrogen. *Chemical Communications*, 1278–1280.
17. Chuang, S.-C., Murata, Y., Murata, M., Komatsu, K. (2007). The outside knows the difference inside: Trapping helium by immediate reduction of the orifice size of an open-cage fullerene and the effect of encapsulated helium and hydrogen upon the NMR of a proton directly attached to the outside. *Chemical Communications*, 1751–1753.
18. Rubin, Y., Jarrosson, T., Wang, G.-W., Bartberger, M. D., Houk, K. N., Schick, G., Saunders, M., Cross, R. J. (2001). Insertion of helium and molecular hydrogen through the orifice of an open fullerene. *Angewandte Chemie International Edition*, **40**, 1543–1546.
19. Stanisky, C. M., Cross, R. J., Saunders, M., Murata, M., Murata, Y., Komatsu, K. (2005). Helium entry and escape through a chemically opened window in a fullerene. *Journal of the American Chemical Society*, **127**, 299–302.
20. Murata, Y., Murata, M., Komatsu, K. (2003). 100% Encapsulation of a hydrogen molecule into an open-cage fullerene derivative and gas-phase generation of $H_2@C_{60}$. *Journal of the American Chemical Society*, **125**, 7152–7153.
21. Xiao, Z., Yao, J., Yang, D., Wang, F., Huang, S., Gan, L., Jia, Z., Jiang, Z., Yang, X., Zheng, B., Yuan, G., Zhang, S., Wang, Z. (2007). Synthesis of [59]fullerenones through peroxide-mediated stepwise cleavage of fullerene skeleton bonds and X-ray structures of their water-encapsulated open-cage complexes. *Journal of the American Chemical Society*, **129**, 16149–16162.
22. Iwamatsu, S.-i., Stanisky, C. M., Cross, R. J., Saunders, M., Mizorogi, N., Nagase, S., Murata, S. (2006). Carbon monoxide inside an open-cage fullerene. *Angewandte Chemie International Edition*, **45**, 5337–5340.
23. Stanisky, C. M., Cross, R. J., Saunders, M. (2009). Putting atoms and molecules into chemically opened fullerenes. *Journal of the American Chemical Society*, **131**, 3392–3395.
24. Whitener, K. E., Jr., Frunzi, M., Iwamatsu, S.-i., Murata, S., Cross, R. J., Saunders, M. (2008). Putting ammonia into a chemically opened fullerene. *Journal of the American Chemical Society*, **130**, 13996–13999.
25. Whitener, K. E., Jr., Cross, R. J., Saunders, M., Iwamatsu, S.-i., Murata, S., Mizorogi, N., Nagase, S. (2009). Methane in an open-cage [60]fullerene. *Journal of the American Chemical Society*, **131**, 6338–6339.
26. Komatsu, K., Murata, M., Murata, Y. (2005). Encapsulation of molecular hydrogen in fullerene C_{60} by organic synthesis. *Science*, **307**, 238–240.
27. Murata, M., Murata, Y., Komatsu, K. (2006). Synthesis and properties of endohedral C_{60} encapsulating molecular hydrogen. *Journal of the American Chemical Society*, **128**, 8024–8033.
28. Murata, Y., Kato, N., Komatsu, K. (2001). The reaction of fullerene C_{60} with phthalazine: The mechanochemical solid-state reaction yielding a new C_{60} dimer versus the liquid-phase reaction affording an open-cage fullerene. *Journal of Organic Chemistry*, **66**, 7235–7239.
29. Murata, Y., Murata, M., Komatsu, K. (2001). The reaction of fullerene C_{60} with 4,6-dimethyl-1,2,3-triazine: Formation of an open-cage fullerene derivative. *Journal of Organic Chemistry*, **66**, 8187–8191.
30. Murata, Y., Suzuki, M., Komatsu, K. (2001). Synthesis and electrochemical properties of novel dimeric fullerenes incorporated in a 2,3-diazabicyclo[2.2.2]oct-2-ene framework. *Chemical Communications*, 2338–2339.
31. Murata, Y., Suzuki, M., Rubin, Y., Komatsu, K. (2003). Structure of the hydration product of the C_{60}-di(2-pyridyl)-1,2,4,5-tetrazine adduct. *Bulletin of the Chemical Society of Japan*, **76**, 1669–1672.
32. Sawa, H., Wakabayashi, Y., Murata, Y., Murata, M., Komatsu, K. (2005). Floating single hydrogen molecule in an open-cage fullerene. *Angewandte Chemie International Edition*, **44**, 1981–1983.
33. Xie, Q., Pérez-Cordero, E., Echegoyen, L. (1992). Electrochemical detection of $C_{60}{}^{6-}$ and $C_{70}{}^{6-}$: Enhanced stability of fullerides in solution. *Journal of the American Chemical Society*, **114**, 3978–3980.
34. Cross, R. J., Khong, A., Saunders, M. (2003). Using cyanide to put noble gases inside C_{60}. *Journal of Organic Chemistry*, **68**, 8281–8283.
35. Camps, X., Hirsch, A. (1997). Efficient cyclopropanation of C_{60} starting from malonates. *Journal of the Chemical Society, Perkin Transactions 1*, 1595–1596.
36. Hoke II, S.H., Molstad, J., Dilettato, D., Jay, M. J., Carlson, D., Kahr, B., Cooks, R. G. (1992). Reaction of fullerenes and benzyne. *Journal of Organic Chemistry*, **68**, 5069–5071.
37. Maggini, M., Scorrano, G.; Prato, M. (1993). Addition of azomethine ylides to C_{60}: Synthesis, characterization, and functionalization of fullerene pyrrolidines. *Journal of the American Chemical Society*, **115**, 9798–9799.
38. Wang, G.-W., Komatsu, K., Murata, Y., Shiro, M. (1997). Synthesis and X-ray structure of dumb-bell-shaped C_{120}. *Nature*, **387**, 583–586.
39. Komatsu, K., Wang, G.-W., Murata, Y., Tanaka T., Fujiwara, K. (1998). Mechanochemical synthesis and characterization of the fullerene dimer C_{120}. *Journal of Organic Chemistry*, **63**, 9358–9366.
40. Saunders, M., Jiménez-Vázquez, H. A., Bangerter, B. W., Cross, R. J. (1994). ^{3}He NMR: A powerful new tool for following fullerene chemistry. *Journal of the American Chemical Society*, **116**, 3621–3622.
41. Rüttimann, M., Haldimann, R. F., Isaacs, L., Diederich, F., Khong, A., Jiménez-Vázquez, H. A., Cross, R. J., Saunders, M. (1997). π-Electron ring-current effects in multiple adducts of $^3He@C_{60}$ and $^3He@C_{60}$: A ^{3}He NMR study. *Chemistry—A European Journal*, **3**, 1071–1076.
42. Wang, G.-W., Saunders, M., Cross, R. J. (2001). Reversible Diels-Alder addition to fullerenes: A study of equilibria using

^{3}He NMR spectroscopy. *Journal of the American Chemical Society*, **123**, 256–259.

43. Shabtai, E., Weitz, A., Haddon, R. C., Hoffman, R. E., Rabinovitz, M., Khong, A., Cross, R. J., Saunders, M., Cheng, P.-C., Scott, L. T. (1998). ^{3}He NMR of He@$C_{60}{}^{6-}$ and He@$C_{70}{}^{6-}$. New records for the most shielded and the most deshielded ^{3}He inside a fullerene. *Journal of the American Chemical Society*, **120**, 6389–6393.
44. Subramanian, R., Kadish, K. M., Vijayashree, M. N., Gao, X., Jones, M. T. (1996). Chemical generation of $C_{60}{}^{2-}$ and electron transfer mechanism for the reactions with alkyl bromides. *The Journal of Physical Chemistry*, **100**, 16327–16335.
45. Fukuzumi, S., Suenobu, T., Hirasaka, T., Arakawa, R., Kadish, K. M. (1998). Formation of C_{60} adducts with two different alkyl groups via combination of electron transfer and S_N2 reactions. *Journal of the American Chemical Society*, **120**, 9220–9227.
46. Hirsch, A., Chen, Z., Jiao, H. (2000). Spherical aromaticity in I_h symmetrical fullerenes: The $2(N + 1)^2$ rule. *Angewandte Chemie International Edition*, **39**, 3915–3917.
47. Allard, E., Delaunay, J., Cheng, F., Cousseau, J., Ordfflna, J., Javier, G. (2001). Novel C_{60}-based building blocks derived from $C_{60}{}^{2-}$ anion. *Organic Letters*, **3**, 3503–3506.
48. Murata, M., Ochi, Y., Tanabe, F., Komatsu, K., Murata, Y. (2008). Internal magnetic fields of dianions of fullerene C_{60} and its cage-opened derivatives studied with encapsulated H_2 as an NMR probe. *Angewandte Chemie International Edition*, **47**, 2039–2041.
49. Trulove, P. C., Carlin, R. T., Eaton, G. R., Eaton, S. S. (1995). Determination of the singlet-triplet energy separation for $C_{60}{}^{2-}$ in DMSO by electron paramagnetic resonance. *Journal of the American Chemical Society*, **117**, 6265–6272.
50. Subramanian, R., Boulas, P., Vijayashree, M. N., D'Souza, F., Jones, M. T., Kadish, K. M. (1994). A facile and selective method for the solution-phase generation of $C_{60}{}^{-}$ and $C_{60}{}^{2-}$. *Journal of the Chemical Society, Chemical Communications*, 1847–1848 (1994).
51. Boyd, P. D. W., Bhyrappa, P., Paul, P., Stinchcombe, J., Bolskar, R. D., Sun, Y., Reed, C. A. The $C_{60}{}^{2-}$ fulleride ion. *Journal of the American Chemical Society*, **117**, 2907–2914.
52. Schleyer, P. v. R., Maerker, C., Dransfeld, A., Jiao, H., Hommes, N. J. R. v. E. (1996). Nucleus-independent chemical shifts: A simple and efficient aromaticity probe. *Journal of the American Chemical Society*, **118**, 6317–6318.
53. Murata, Y., Maeda, S., Murata, M., Komatsu, K. (2008). Encapsulation and dynamic behavior of two H_2 molecules in an open-cage C_{70}. *Journal of the American Chemical Society*, **130**, 6702–6703.
54. Murata, M., Maeda, S., Morinaka, Y., Murata, Y., Komatsu, K. (2008). Synthesis and reaction of fullerene C_{70} encapsulating two molecules of H_2. *Journal of the American Chemical Society*, **130**, 15800–15801.
55. Slanina, Z., Pulay, P., Nagase, S. (2006). H_2, Ne, and N_2 energies of encapsulation into C_{60} evaluated with the MPWB1K functional. *Journal of Chemical Theory and Computation*, **2**, 782–785.
56. Lamparth, I., Maichle-Mossmer, C., Hirsch, A. (1995). Reversible template-directed activation of equatorial double bonds of the fullerene framework: Regioselective direct synthesis, crystal structure, and aromatic properties of T_h-$C_{66}(COOEt)_{12}$. *Angewandte Chemie International Edition in English*, **34**, 1607–1609.
57. Frunzi, M., Cross, R. J., Saunders, M. (2007). Effect of xenon on fullerene reactions. *Journal of the American Chemical Society*, **129**, 13343–13346.
58. Morinaka, Y., Tanabe, F., Murata, M., Murata, Y., Komatsu, K. (2010). Rational synthesis, enrichment, and ^{13}C NMR spectra of endohedral C_{60} and C_{70} encapsulating a helium atom. *Chemical Communications*, **46**, 4532–4534.
59. Patchkovskii, S., Thiel, W. (1997). Equilibrium yield for helium incorporation into buckminsterfullerene: Quantum-chemical evaluation. *The Journal of Chemical Physics*, **106**, 1796–1799.
60. Takeda, A., Yokoyama, Y., Ito, S., Miyazaki, T., Shimotani, H., Yakigaya, K., Kakiuchi, T., Sawa, H., Takagi, H., Kitazawa, K., Dragoe, N. (2006). Superconductivity of doped Ar@C_{60}. *Chemical Communications*, **912**.
61. Yamamoto, K., Saunders, M., Khong, A., Cross, R. J., Grayson, M., Gross, M. L., Benedetto, A. F., Weisman, R. B. (1999). Isolation and spectral properties of Kr@C_{60}, a stable van der Waals molecule. *Journal of the American Chemical Society*, **121**, 1591.
62. Syamala, M. S., Cross, R. J., Saunders, M. (2002). ^{129}Xe NMR spectrum of xenon inside C_{60}. *Journal of the American Chemical Society*, **124**, 6216–6219.
63. Turro, N. J., Chen, J. Y.-C., Sartori, E., Ruzzi, M., Marti, A., Lawler, R., Jockusch, S., López-Gejo, J., Komatsu, K., Murata, Y. (2010). The spin chemistry and magnetic resonance of H_2@$C_{60.}$ From the Pauli principle to trapping a long lived nuclear excited spin state inside a buckyball. *Accounts of Chemical Research*, **43**, 335–345.
64. Sartori, E., Ruzzi, M., Turro, N. J., Decatur, J. D., Doetschman, D. C., Lawler, R. G., Buchachenko, A. L., Murata, Y., Komatsu, K. (2006). Nuclear relaxation of H_2 and H_2@C_{60} in organic solvents. *Journal of the American Chemical Society*, **128**, 14752–14753.
65. Sartori, E., Ruzzi, M., Turro, N. J., Komatsu, K., Murata, Y., Lawler, R. G., Buchachenko, A. L. (2008). Paramagnet enhanced nuclear relaxation of H_2 in organic solvents and in H_2@C_{60}. *Journal of the American Chemical Society*, **130**, 2221–2225.
66. Turro, N. J., Marti, A. A., Chen, J. Y.-C., Jockusch, S., Lawler, R. G., Ruzzi, M., Sartori, E., Chuang, S.-C., Komatsu, K., Murata, Y. (2008). Demonstration of a chemical transformation inside a fullerene. The reversible conversion of the allotropes of H_2@C_{60}. *Journal of the American Chemical Society*, **130**, 10506–10507.
67. López-Gejo, J., Marti, A. A., Ruzzi, M., Jockusch, S., Komatsu, K., Tanabe, F., Murata, Y., Turro, N. J. (2007). Can H_2 inside C_{60} communicate with the outside world? *Journal of the American Chemical Society*, **129**, 14554–14555.
68. Li, Y., Lei, X., Jockusch, S., Chen, J. Y.-C., Frunzi, Johnson, J. A., Lawler, R. G., Murata, Y., Komatsu, K., Turro, N. J.

(2010). A magnetic switch for spin-catalyzed interconversion of nuclear spin isomers. *Journal of the American Chemical Society*, **132**, 4042–4043.

69. Frunzi, M., Jockusch, S., Chen, J.Y.-C., Calderon, R. M. K., Lei, X., Murata, Y., Komatsu, K., Guldi, D. M., Lawler, R. G., Turro, N. J. (2011). A photochemical on–off switch for tuning the equilibrium mixture of H_2 nuclear spen isomers as a function of temperature. *Journal of the American Chemical Society*, **133**, 14232–14235.
70. Ludwig, R. (2001). Water: From clusters to the bulk. *Angewandte Chemie International Edition*, **40**, 1808–1827.
71. Zhang, Q., Pankewitz, T., Liu, S., Klopper, W., Gan, L. (2010). Switchable open-cage fullerene for water encapsulation. *Angewandte Chemie International Edition*, **49**, 9935–9938.
72. Noh, T. H., Heo, E., Park, K. H., Jung, O.-S. (2011). Motion of an isolated water molecule within a flexible coordination cage: Structural properties and catalytic effects of ionic palladium(II) complexes. *Journal of the American Chemical Society*, **133**, 1236–1239.
73. Kurotobi, K., Murata, Y. (2011). A single molecule of water encapsulated in fullerene C_{60}. *Science*, **333**, 613–616.
74. Lee, H. M., Olmstead, M. M., Suetsuna, T., Shimotani, H., Dragoe, N., Cross, R. J., Kitazawa, K., Balch, A. L. (2002). Crystallographic characterization of Kr@C_{60} in (0.09Kr@C_{60}/0.91C_{60})·{Ni^{II}(OEP)}·$2C_6H_6$. *Chemical Communications*, **46**, 1352–1353.
75. Kohama, Y., Rachi, T., Jing, J., Li, Z., Tang, J., Kumashiro, R., Izumisawa, S., Kawaji, H., Atake, T., Sawa, H., Murata, Y., Komatsu, K., Tanigaki, K. (2009). Rotational sublevels of an ortho-hydrogen molecule encapsulated in an isotropic C_{60} cage. *Physical Review Letters*, **103**, 073001-1–073001-4.
76. Aoyagi, S., Nishibori, E., Sawa, H., Sugimoto, K., Takata, M., Miyata, Y., Kitaura, R., Shinohara, H., Okada, H., Sakai, T., Ono, Y., Kawachi, K., Yokoo, K., Ono, S., Omote, K., Kasama, Y., Ishikawa, S., Komuro, T., Tobita, H. (2010). A layered ionic crystal of polar Li@C_{60} superatoms. *Nature Chemistry*, **2**, 678–683.
77. Murata, M., Morinaka, Y., Murata, Y., Yoshikawa, O., Sagawa, T., Yoshikawa, S. (2011). Modification of the σ-framework of [60]fullerene for bulk-heterojunction solar cells. *Chemical Communications*, **47**, 7335–7337.
78. Murata, M., Morinaka, Y., Kurotobi, K., Komatsu, K., Murata, Y. (2010). Reaction of cage-opened fullerene derivative with grignard reagents and subsequent transannular cyclization. *Chemistry Letters*, **39**, 298–299.
79. Hummelen, J. C., Knight, B. W., LePeq, F., Wudl, F. (1995). Preparation and characterization of fulleroid and methanofullerene derivatives. *Journal of Organic Chemistry*, **60**, 532–538.
80. Vandewal, K., Tvingstedt, K., Gadisa, A., Inganäs, O., Manca, J. V. (2009). On the origin of the open-circuit voltage of polymer–fullerene solar cells. *Nature Materials*, **8**, 904–909.
81. Brabec, C. J., Cravino, A., Meissner, D., Sariciftci, N. S., Fromherz, T., Rispens, M. T., Sanchez, L., Hummelen, J. C. (2001). Origin of the open circuit voltage of plastic solar cells. *Advanced Functional Materials*, **11**, 374–380.
82. Hayakawa, A., Yoshikawa, O., Fujieda, T., Uehara, K., Yoshikawa, S. (2007). High performance polythiophene/fullerene bulk-heterojunction solar cell with a TiO_x hole blocking layer. *Applied Physics Letters*, **90**, 163517-1–163517-3.
83. Kooistra, F. B., Knol, J., Kastenberg, F., Popescu, L. M., Verhees, W. J. H., Kroon, J. M., Hummelen, J. C. (2007). Increasing the open circuit voltage of bulk-heterojunction solar cells by raising the LUMO level of the acceptor. *Organic Letters*, **9**, 551–554.
84. Piliego, C., Holcombe, T. W., Douglas, J. D., Woo, C. H., Beaujuge, P. M., Fréchet, J. M. J. (2010). Synthetic control of structural order in *N*-alkylthieno[3,4-c]pyrrole-4,6-dione-based polymers for efficient solar cells. *Journal of the American Chemical Society*, **132**, 7595–7597.

12

FUNDAMENTAL AND APPLIED ASPECTS OF ENDOHEDRAL METALLOFULLERENES AS PROMISING CARBON NANOMATERIALS

Michio Yamada, Xing Lu, Lai Feng, Satoru Sato, Yuta Takano, Shigeru Nagase, and Takeshi Akasaka

12.1 INTRODUCTION

Since the first proposal in 1985 immediately after the discovery of the prominent stability of C_{60} [1], encapsulation of metal atom(s) inside hollow fullerene cages (endohedral metallofullerenes, EMFs) has been explored vigorously to produce novel forms of fullerene-based nanomaterials. Smalley and co-workers [2] prepared carbonaceous soot using laser vaporization of graphite rods impregnated with appropriate metal oxides, revealing that the soot contained lanthanum-encaged fullerenes, which can be subsequently extracted with toluene. Today, such lanthanum EMFs are described as La@C_n, in which the symbol @ denotes a La atom located inside a C_n fullerene cage. Spurred by this success, great efforts have been made for widespread production, characterization, and application of various EMFs [3]. This chapter summarizes recent advances in the chemistry of EMFs in terms of fundamental and applied aspects as promising candidates for future fruitful applications.

12.2 SYNTHESIS, SEPARATION, AND PURIFICATION OF EMFs

In fact, EMFs can be synthesized in several ways, similar to the production of empty fullerenes involving generation of carbon-rich vapor or plasma in a helium or argon atmosphere. Among these, the standard direct-current (DC) arc-discharge method using a Krätschmer–Huffman generator has been applied routinely to prepare EMFs in macroscopic quantities [4]. Soot containing EMFs are grown in the course of diffusion of carbon vapor followed by cooling, during which time the metal atoms interacting with carbon atoms are entrapped inside carbon clusters to form EMFs. It is most likely that formation of EMFs is a thermodynamically controlled process. The collected soot can be heated to reflux in organic solvents such as 1,2,4-trichlorobenzene (TCB) or *N*,*N*-dimethylformamide (DMF) to obtain extracts that contain a mixture of soluble EMFs together with empty fullerenes.

The basic approach to separate and purify EMFs is the use of high-performance liquid chromatography (HPLC). To date, various HPLC columns such as Buckyclutcher, PBB, PYE, Buckyprep, and Buckyprep-M [5] have been developed and applied for the effective separation of empty fullerenes and EMFs. However, because of their similar structures, a tedious and time-consuming multistage HPLC procedure is typically required for isolation of each EMF isomer. To solve this problem, alternative methods for purification of EMFs have been proposed, involving electrochemical reduction [6], chemical reduction [7], and sublimation followed by chemical oxidation [8]. These methods are based on their unique redox properties, which differ greatly from those of empty fullerenes. For instance, selective electrochemical reduction followed by filtration was performed to separate lanthanide EMFs from extracts based on lower reduction

Organic Nanomaterials: Synthesis, Characterization, and Device Applications, First Edition. Edited by Tomás Torres and Giovanni Bottari.

potentials than those of empty fullerenes. Anions of EMFs are highly soluble in polar solvents, but they are insoluble in nonpolar solvents. Therefore, neutral empty fullerenes were excluded via utilization of their different solubility. Supramolecular complexation with azacrown ether instead of bulk electrolysis was also adopted to reduce lanthanum EMFs to their anions [9]. Dorn and co-workers [10] demonstrated the purification of trimetallic nitride EMFs by taking advantage of their extraordinary kinetic chemical stability with respect to the other fullerenes in Diels–Alder reactions with a cyclopentadiene-functionalized resin. In a related work, Stevenson et al. [11] reported the use of cyclopentadienyl and amino functionalized silica to bind contaminant fullerenes selectively, leading to purification of trimetallic nitride EMFs.

12.3 STRUCTURE ELUCIDATION OF EMFs

As for lanthanide EMFs, $M@C_{82}$ (M = encaged metal atoms) is most abundantly extracted as a representative monometallic EMF. Because of the difficulty in performing single-crystal x-ray analysis as a result of their low crystallinity and their round structures, ^{13}C NMR spectroscopy remains the most useful tool for the determination of their cage structure and symmetry. However, the paramagnetic nature of $M@C_{82}$ prevents NMR spectral analysis because the transfer of three electrons from the metal atom to C_{82} cage yields the electronic structure of $M^{3+}C_{82}{}^{3-}$, which is paramagnetic. Akasaka and co-workers accomplished determination of cage symmetries in the paramagnetic EMFs, $M@C_{82}$ (M = La [12], Ce [13], Pr [14], and Y [15]) using ^{13}C NMR measurements of their anionic forms $[M@C_{82}]^-$, prepared by bulk-controlled potential electrolysis (Scheme 12.1). For instance, the ^{13}C NMR spectrum of $[La@C_{82}]^-$ exhibited 11 distinct lines with full intensity and 7 lines with half intensity, verifying that the major isomer of $La@C_{82}$ has C_{2v} symmetry (therefore, the molecule can be expressed more precisely as $La@C_{2v}$-C_{82}), which agrees with the theoretical study showing that the C_{2v} isomer is energetically most stable for accommodation of the La atom. Subsequently, mapping of the bond connectivity and definitive assignment of the ^{13}C NMR lines was accomplished in $[M@C_{2v}$-$C_{82}]^-$ (M = La and Ce) using a two-dimensional incredible natural abundance double quantum transfer experiment (INADEQUATE) NMR [16, 17].

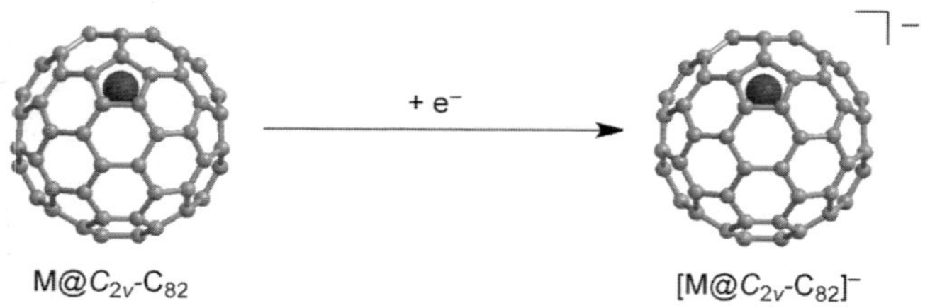

SCHEME 12.1. Electrochemical reduction of $M@C_{2v}$-C_{82}.

The remaining question for $M@C_{2v}$-C_{82} was where the encaged metal atom is located. Regarding $Ce@C_{2v}$-C_{82} anion, ^{13}C NMR spectral analysis of the paramagnetic effects of an f electron on the Ce atom provided information related to the metal atom position inside C_{2v}-C_{82} [17]. Paramagnetic interaction of the f electron spin with ^{13}C nucleus spins on the cage carbons consists of Fermi contact and pseudocontact interactions, each of which contributes to chemical shifts in the NMR spectrum. In theory, the former depends on $1/T$ (T = absolute temperature) and the latter depends on $1/T^2$ [18]. The actual ^{13}C NMR signals of $Ce@C_{2v}$-C_{82} exhibited not $1/T$ dependence but $1/T^2$ dependence. The results show that pseudocontact interaction produces a more dominant contribution than Fermi contact interaction because no meaningful connection exists between the metal atom and the cage carbons. Fermi contact interaction is isotropic, whereas pseudocontact interaction is anisotropic. Therefore, the metal atom position in $Ce@C_{2v}$-C_{82} anion was provided by linear line fitting of the ^{13}C NMR chemical shifts as a function of $1/T^2$. The paramagnetic NMR spectral analysis revealed the distance between Ce and the center of the hexagonal ring along the C_2 axis as 2.1–2.8 Å, which shows good agreement with DFT calculations.

The paramagnetic NMR spectral analysis was also applied for dimetallic EMFs such as $Ce_2@D_2(10611)$-C_{72} [19], $Ce_2@D_{3h}(5)$-C_{78} [20], $Ce_2@I_h$-C_{80} [21], and $Ce_2@D_{5h}$-C_{80} [22] to reveal the behavior of the metal atoms in the fullerene cages. A peculiar case was evident with the latter two molecules: $Ce_2@I_h$-C_{80} and $Ce_2@D_{5h}$-C_{80}. The ^{13}C NMR spectrum of $Ce_2@I_h$-C_{80} exhibited two carbon signals with an integral ratio of 3:1, showing that the two Ce atoms are equivalent and not fixed at a specific position. This is an analogy with the dynamic random motion of two La atoms in $La_2@I_h$-C_{80}, which was confirmed by variable-temperature (VT) ^{139}La NMR spectral analysis [23]. The circuit of two La^{3+} cations produces a magnetic field at the La positions. This is reflected in the nuclear magnetic relaxation rate and therefore the line width. Normally, the interaction between nuclear spins and the induced magnetic field (spin–rotation relaxation) does not contribute greatly to the relaxation time in solution because neighbors dominantly quench the molecular rotation. However, the rotation of the La atoms can be preserved because of the fullerene cage protection, so that the spin–rotation interaction has a drastic effect on the relaxation process, leading to increased line width with increasing temperature. Therefore, the dynamic motion of La atoms in fullerenes is visible using ^{139}La NMR measurements. In contrast, the dynamic motion of Ce atoms in fullerenes is visible by paramagnetic ^{13}C NMR spectral analysis. In fact, the observed paramagnetic effects attributable to the Ce atoms in $Ce_2@I_h$-C_{80} were very small compared to those for

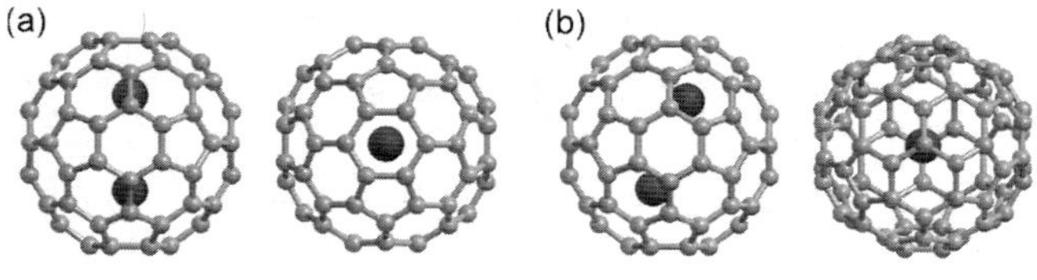

FIGURE 12.1. Two orthogonal views of (**a**) D_{2h} and (**b**) D_{3d} configrations of $M_2@I_h$-C_{80}.

Ce@C_{2v}-C_{82} [17, 21]. The observation was rationalized by the three-dimensional random motion of the Ce atoms, which engenders circulation of the f electrons, thereby reducing the paramagnetic effects. Regarding the second isomer, $Ce_2@D_{5h}$-C_{80}, two full-intensity lines and four half-intensity lines were observed in the ^{13}C NMR spectrum [22]. The observed ^{13}C NMR pattern can be attributed unambiguously to the D_{5h}-C_{80} cage that obeys the isolated-pentagon rule (IPR). In addition, results of the paramagnetic ^{13}C NMR spectral analysis coupled with theoretical calculations revealed that the Ce atoms circulate two-dimensionally along a band of 10 contiguous hexagons in the D_{5h}–C_{80} cage.

Details of the trajectory of metal atoms in $M_2@I_h$-C_{80} have been investigated using computational calculations. Shimotani et al. [24] reported that the D_{3d} configuration is the global minimum in total energy, being more stable by 1.9 kcal/mol than the D_{2h} configuration (Figure 12.1). Frequency analysis showed one imaginary frequency for the D_{2h} configuration, which indicates that it is not a local minimum but a saddle point. The potential energy surface calculations displayed that La ions travel between 10 equivalent D_{3d} positions through D_{2h} positions and consequently form a pentagonal dodecahedral trajectory. These results differ greatly from the theoretical study by Nagase and co-workers [25]. Later, Hao and co-workers [26] conducted geometry optimization and vibrational frequency analyses for $La_2@I_h$-C_{80}, which revealed that the most stable configuration for $La_2@I_h$-C_{80} is that of D_{2h} symmetry, which was merely regarded as a saddle point by Shimotani et al. Most notably, Hao and co-workers draw the same conclusions as those inferred by Shimotani et al. when using nonrelativistic basis sets of the same size within the DFT frame. That is, the D_{3d} configuration is more stable in energy than the D_{2h} configuration, although the energy difference is only 0.025 kcal/mol. The calculations imply that the relativistic effects definitely play an important role for $La_2@I_h$-C_{80}. Based on the calculations, a new La_2 pair motion scenario is proposed in which two La ions, passing through saddle point C_{2h}, travel between equivalent D_{2h} configurations. The trajectory of this La_2 motion forms a pentagonal dodecahedron. Muthukumar et al. [27] calculated the structural and electronic properties of $Ce_2@I_h$–C_{80} using DFT and found that $Ce_2@I_h$-C_{80} has a D_{3d} symmetric ground-state structure and that Ce binds on top of one carbon atom in a six-membered ring. This binding site also differs from that of La in $La_2@I_h$-C_{80}, which is D_{2h} symmetric. Furthermore, they reported that the binding site minimizes the Ce(f)–Ce(f) overlap in favor of Ce–C bonding, thereby producing the D_{3d} configuration.

To elucidate the metal atoms' positions and movements, the nature of the metal–cage bonding in EMFs should be elucidated. EPR [28] and ultraviolet photoelectron spectroscopy (UPS) [29] studies of $M@C_{82}$ have revealed that EMFs can be treated as zwitterions and can be described as $M^{3+}@C_{82}{}^{3-}$. This ionic model is useful to explain carbon cage isomerism and the metal atoms' behavior in $M@C_{2v}$-C_{82} and $M_2@I_h$-C_{80}. Kobayashi and Nagase [30] reported that the $C_2(3)$ cage is the lowest energy isomer of the neutral C_{82} among the nine IPR isomers, while the C_{2v} cage is the most stable for $C_{82}{}^{2-}$ and $C_{82}{}^{3-}$. The results agree well with the fact that La@C_{2v}-C_{82} is regarded as the most stable in terms of thermodynamics [31]. Such is also the case for $La_2@I_h$-C_{80}. The I_h-$C_{80}{}^{6-}$ was calculated as the most stable among the seven IPR isomers of $C_{80}{}^{6-}$, although the most stable isomer of neutral C_{80} is D_{2h} symmetric [32]. The electrostatic potential of C_{2v}–$C_{82}{}^{3-}$ displayed that all the contour lines are highly negative inside the cage. In particular, the bottom of the electrostatic potential valley was found at an off-center position, at which a cationic metal atom can be highly stabilized. This position is in good agreement with the metal atom position in Ce@C_{2v}–C_{82} obtained using paramagnetic NMR spectral analysis of its anion [17]. It is particularly interesting that the electrostatic potentials of I_h-$C_{80}{}^{6-}$ showed almost concentric circles with no clear minima, suggesting that specific positions are not especially stable. Therefore, it is reasonable to consider the existence of three-dimensional random rotation of the metal atoms in $M_2@I_h$–C_{80}.

Kobayashi and Nagase [33] also conducted a topological analysis of the electron density distribution $\rho(r)$ and its Laplacian $\nabla^2\rho(r)$ for $Sc_2@C_{84}$ and Ca@C_{72} according to the "atoms in molecules (AIM)" theory developed by Bader [33,34]. The AIM analysis showed that the values of electron density at bond critical points (BCPs) were very small and the Laplacian of the density at BCP was positive, suggesting that the metal atoms interact with specific cage carbons and that metal–carbon bonds are highly ionic in both cases. Additionally, they explained that the carbon–carbon bonds are strongly covalent with π character, even after receiving electrons from Sc atoms.

Popov [35] performed systematic computational investigations combined with vibrational spectroscopy for trimetallic nitride EMFs, and he reported that the bonding in EMFs has a rather complex nature, with both ionic and covalent contributions being equally important.

12.4 ELECTRONIC PROPERTIES

Electronic properties of EMFs differ greatly from those of empty fullerenes because endohedral metal atom doping

TABLE 12.1. Reactivities Toward Disilirane and Redox Potentials in V versus Fc/Fc$^+$

	Reactivity			
Fullerene	$h\nu$	Heat (80°C)	$^{ox}E_1$	$^{red}E_1$
La@C_{2v}-C_{82}	Yes	Yes	+0.07	−0.42
Ce@C_{2v}-C_{82}	Yes	Yes	+0.08	−0.41
Pr@C_{2v}-C_{82}	Yes	Yes	+0.07	−0.39
Gd@C_{2v}-C_{82}	Yes	Yes	+0.09	−0.39
Y@C_{2v}-C_{82}	Yes	Yes	+0.10	−0.37
Ce_2@D_2(10611)-C_{72}	Yes	No	+0.24	−0.68
La_2@D_{3h}(5)-C_{78}	Yes	Yes	+0.26	−0.40
Ce_2@D_{3h}(5)-C_{78}	Yes	Yes	+0.25	−0.52
La_2@I_h-C_{80}	Yes	Yes	+0.56	−0.31
Ce_2@$I_h C_{80}$	Yes	Yes	+0.57	−0.39
Ce_2@D_{5h}-C_{80}	Yes	Yes	+0.20	−0.40
Sc_3C_2@I_h-C_{80}	Yes	Yes	−0.03	−0.50
Sc_2C_2@C_{3v}(8)-C_{82}	Yes	No	+0.47	−0.94
Sc_3N@I_h-C_{80}	Yes	No	+0.62	−1.22
C_{60}	Yes	No	+1.21	−1.12
C_{70}	Yes	No	+1.19	−1.09

"Yes" signifies that a 1:1 adduct of fullerene with disilirane was formed, and "no" signifies that the formation was not detected.

involves electron transfer from the metal atoms to the carbon cage to stabilize them energetically. In this context, electrochemical properties of EMFs have been widely investigated using cyclic voltammetry (CV) and differential pulse voltammetry (DPV). Results revealed that EMFs have much smaller HOMO–LUMO gaps than those of C_{60} [36,37]. This finding indicates that EMFs can act as a stronger electron donor as well as a stronger electron acceptor than C_{60} does. It is noteworthy that the electronic properties of Ce@C_{2v}-C_{82} [14], Pr@C_{2v}-C_{82} [38], and Gd@C_{2v}-C_{82} [39] closely resemble those of La@C_{2v}-C_{82} [40], although Ce, Pr, and Gd atoms have 4f electrons, suggesting that the presence of 4f electrons plays no important role in the electrochemical events in EMFs (see Table 12.1 in the next section). Similarly, redox potentials of La_2@I_h-C_{80} [41] and Ce_2@I_h-C_{80} [21] resemble each other. The first reduction of M@C_{2v}-C_{82} is facile because the SOMO is filled up to produce the closed-shell electronic structure. It is noteworthy that the first reduction potentials of La_2@I_h-C_{80} and Ce_2@I_h-C_{80} are smaller than those of La@C_{2v}-C_{82} and Ce@C_{2v}-C_{82} despite their closed-shell electronic structures. The DFT calculations showed that SOMO of M@C_{2v}-C_{82} are distributed to the cage whereas the LUMO of M_2@I_h-C_{80} are localized on the two M^{3+} cations. The difference in the molecular orbitals between M@C_{2v}-C_{82} and M_2@I_h-C_{80} might explain their peculiar electronic nature. Although the first reduction potentials of La_2@D_{5h}-C_{80} and Ce_2@D_{5h}-C_{80} are respectively similar to those of La_2@I_h-C_{80} and Ce_2@I_h-C_{80}, the first oxidation potentials of the D_{5h} isomers are higher than those of the I_h isomers. This observation suggests that the molecular symmetry plays a dominant role in electrochemical properties.

12.5 CHEMICAL REACTIVITY

In general, monometallic and dimetallic EMFs are more reactive than C_{60} because of their small HOMO–LUMO gaps. Furthermore, the molecular symmetries of EMFs are lower than C_{60}. Therefore, the reaction of EMFs often produces multiple adducts and many structural isomers. In this context, regioselective derivatization is desired for EMFs to introduce functions. Akasaka et al. [42] reported thermal bis-silylation of La@C_{2v}-C_{82} as the first report on chemical derivatization of EMFs using 1,2-disilacyclopropane (disilirane). The reactivity of La@C_{2v}-C_{82} toward disilirane contrasted strikingly with that of C_{60}, which requires photoirradiation. Later, bis-silylation was conducted to other EMFs such as La_2@I_h-C_{80} [43], Ce_2@I_h-C_{80} [21], and Ce_2@D_{3h}(5)-C_{78} [20]. The bis-silylation of M@C_{2v}-C_{82} did not take place in a regioselective manner. However, M_2@I_h-C_{80} (M = La, Ce) and Ce_2@D_{3h}(5)-C_{78} reacted with disilirane to form single isomers of the monoadducts under thermal conditions. It is noteworthy that no multiple additions occurred in the reaction of EMFs with disilirane, although excess disilirane was loaded. The bis-silylation is considered to involve electron transfer from disilirane to EMFs at the ground states. Therefore, strong electron affinity of EMFs is necessary for the thermal reaction. Once addition of disilirane takes place, the EMFs become electron-rich because of the electron-donating character of the silyl substituent. Therefore, they lose the electron affinity to suppress the reactivity. However, reaction of Sc_3N@I_h-C_{80} with disilirane proceeded not thermally but photochemically [44]. The lower reactivity of Sc_3N@I_h-C_{80} toward disilirane than that of M_2@I_h-C_{80} (M = La and Ce) is ascribed to the difference in the energy levels and spatial distribution of the LUMOs between Sc_3N@I_h-C_{80} and La_2@I_h-C_{80}. Electrochemical studies revealed that the first reduction potential of Sc_3N@I_h-C_{80} is much higher than that of La_2@I_h-C_{80}. Similarly, Sc_2C_2@C_{3v}(8)-C_{82} do not react thermally with disilirane but only photochemically because of its high first reduction potential [45]. This trend is perfectly comparable to the trend of the first reduction potentials of the EMFs. In this context, reactivity of EMFs with disilirane can serve as a chemical probe to evaluate the electron-accepting character of EMFs (Scheme 12.2 and Table 12.1).

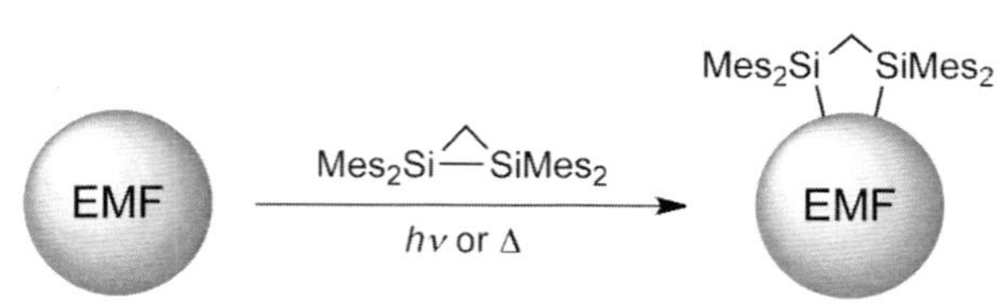

SCHEME 12.2. Reaction of EMFs with disilirane. Mes stands for 2,4,6-trimethylphenyl.

Chemical derivatization is also effective for preparation of single crystals that are suitable for x-ray crystallography. Addition of electrophilic adamantylidene (Ad) carbene and x-ray analysis of the adducts have been achieved for La@C_{2v}-C_{82} [46], Ce@C_{2v}-C_{82} [47], Y@C_{2v}-C_{82} [48], Gd@C_{2v}-C_{82} [49], La_2@D_2(10611)-C_{72} [50], La_2@D_{3h}(5)-C_{78} [51], La_2@I_h-C_{80} [52], Sc_2C_2@C_{2v}(5)-C_{80} [53], Sc_3C_2@I_h-C_{80} [54], and Sc_2C_2@C_{3v}(8)-C_{82} [55] to elucidate their cage structures and positions of the encapsulated species. Interestingly, structural analyses of the major isomers of La_2@D_2(10611)-C_{72}(Ad) revealed that Ad carbene selectively attacks the carbons adjacent to the [5,5]-junctions, instead of the [5,5]-carbons themselves, proving that the fused-pentagon bonds are stabilized by the internal La atoms via charge transfer and are therefore less reactive than the adjacent ones [50a]. The x-ray structure of a bis-adduct of La_2@D_2(10611)-C_{72}(Ad)$_2$ reveals that the second Ad group tends to attack the unsubstituted fused-pentagon region by a combined effect of the high curvature and the steric hindrance of the Ad group [50b].

Recent development in chemical functionalization of EMFs involves 1,3-dipolar cycloaddition [56], Bingel–Hirsch reaction [57], Diels–Alder reaction [58], benzyne addition [59], radical reaction [60], and carbosilylation [61]. These investigations have shed light on the distinct chemical reactivities of EMFs.

The pyramidalization of trigonal carbon atoms in fullerenes is related to the local strain, thereby playing an important role in the reactivities of fullerenes. Haddon [62] proposed π-orbital axis vector (POAV) indices to facilitate quantitative analysis of the chemical reactivities. Planar trigonal carbon atoms, such as those in benzene, have POAV angles of 0.0°, whereas the pyramidalized carbon atoms in C_{60} have a POAV value of 11.6°. The POAV values of carbon atoms in the optimized structure of La@C_{2v}-C_{82} are 7.36°–11.20°. In this context, carbon atoms having large POAV values are presumed to be highly reactive. Regarding polar reactions, charge densities and distributions of frontier orbitals are also of importance in terms of reactivity. Regioselectivity in Ad carbene addition [46] and the Bingel–Hirsch reaction [57a–c] of La@C_{2v}-C_{82} is rationalized based on the charge densities. Because of the electron transfer from the La atom to the cage, two carbon atoms nearest to the La atom are negatively charged. In addition, the carbon atoms have large POAV values of 11.20° and 11.28°. Therefore, electrophilic Ad carbene attacks the carbon atoms to afford preferential formation of the monoadduct isomer. However, some positively charged carbon atoms have large POAV values opposite the position of the La atom. In fact, nucleophiles selectively attack one of them to afford the corresponding monoadducts and bis-adducts in the Bingel–Hirsch reaction [57a–c]. In the case of radical coupling reactions, SOMO spin densities should be considered together with POAV values to rationalize the reactivities. Radical coupling reactions of La@C_{2v}-C_{82} occurred at carbon atoms having large SOMO spin densities and POAV values to give EPR-silent monoadducts [60a]. Diels–Alder reaction of La@C_{2v}-C_{82} with cyclopentadiene proceeded smoothly to afford a single product [58a]. Subsequently, single-crystal x-ray crystallographic analysis of La@C_{2v}–C_{82}–1,2,3,4,5-pentamethylcyclopentadiene (Cp*) derivative was conducted, revealing that the addition position is the same as that in the Bingel–Hirsch reaction [58b]. In addition, La@C_{2v}-C_{82}(Cp*) crystallizes to form a dimmer as found in the crystal structure of a Bingel bisadduct La@C_{2v}-C_{82}[CH(COOEt)$_2$]$_2$, owing to its open-shell electronic structure (Figure 12.2). It is particularly interesting that the *beta*-LUMO and *alpha*-HOMO of La@C_{2v}-C_{82} have no large orbital coefficients on the carbon atoms, implying that other orbitals play a main role in the reaction. In contrast, thermal reactions such as 1,3-dipolar cycloaddition [56], bis-silylation [20, 21, 43, 44], and carbosilylation [61] provide thermodynamically stable isomers.

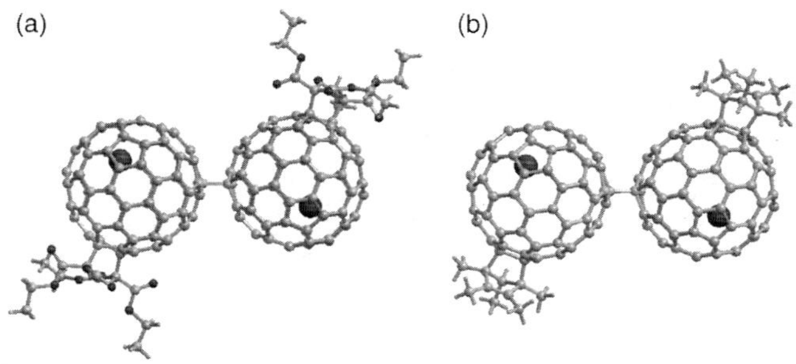

FIGURE 12.2. X-ray structures of dimeric (**a**) La@C_{2v}-C_{82}[CH(COOEt)$_2$]$_2$ and (**b**) La@C_{2v}-C_{82}(Cp*).

12.6 CONTROL OF DYNAMIC MOTION OF METAL ATOMS IN FULLERENE CAGES

The positions and movements of the metal atoms are important determinants of the chemical and physical properties of EMFs. In this sense, control of the dynamic motion of metal atoms in fullerene cages would help in designing functional devices for molecular electronics. Akasaka and co-workers [63] found that chemical functionalization of EMFs can regulate not only the electronic properties but also the metal positions and movements of the encaged metal atoms (Figure 12.3). Single-crystal x-ray crystallographic analysis and paramagnetic ^{13}C NMR spectral analysis of the Ce_2@I_h-C_{80}–disilirane monoadduct revealed that the two Ce atoms are localized at two positions directed toward the hexagonal ring along the equatorial plane of the fullerene sphere [21]. The results clearly demonstrate that bis-silylation can regulate the three-dimensional random motion of the encapsulated metal atoms to a restricted manner. Later, VT-^{139}La NMR and x-ray

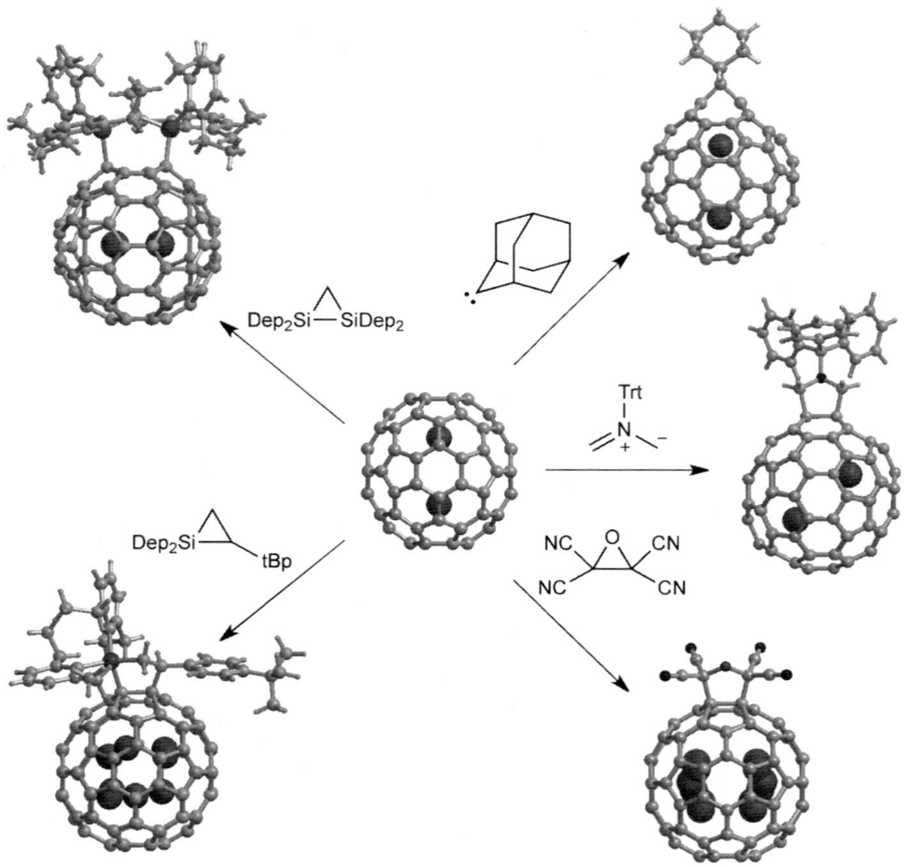

FIGURE 12.3. Chemical functionalization of $M_2@I_h$-C_{80} and x-ray structures of the derivatives. Dep, 2,6-diethylphenyl; tBp, 4-*tert*-butylphenyl; Trt, triphenylmethyl.

analyses of the $La_2@I_h$-C_{80}–disilirane monoadduct revealed that the metal atoms are allowed to hop two-dimensionally along the equator of the bis-silylated cage [43]. This hopping motion shows good agreement with the computational studies that preceded the experimental findings [64].

The 1,3-dipolar cycloaddition of $M_2@I_h$-C_{80} (M = La [56a,b] and Ce [56b]) with *N*-trityl azomethine ylide afforded two regioisomers, among which the structures of [6,6]-fulleropyrrolidines were elucidated unambiguously using VT-^{139}La NMR and x-ray analyses. Results revealed that the metal atoms are fixed at slantwise positions on the mirror plane in the [6,6]-fulleropyrrolidines. The metal position is also supported by the paramagnetic ^{1}H NMR spectral analysis. In contrast, the analysis indicated that the metal atoms in the [5,6]-fulleropyrrolidine are collinear with the pyrrolidine ring. The difference in the metal atom positions is explainable by the electrostatic potential maps inside the pyrrolidino cages in terms of the ionic model. This observation suggests that the metal positions can be controlled by the addition positions on the cage.

Ad carbene addition of $M_2@I_h$-C_{80} (M = La and Ce) produced single monoadducts having [6,6]-open structures, which were formed by cleavage of the [6,6]-bond junction [52]. The two metal atoms are found to be collinear with the spiro carbon in the x-ray structure. The paramagnetic NMR spectral analysis exhibited that the metal atoms do not move around but instead stand still in the NMR timescale. It is noteworthy that the La–La distance is highly elongated to 4.031 Å by the carbene addition. This remarkable elongation results from expansion of the inner space of the cage caused by the bond cleavage, which reduces electrostatic repulsion among the metal cations.

Recently, thermal carbosilylation of $La_2@I_h$-C_{80} with silacyclopropane (silirane) has been reported [61]. The carbosilylation took place at the [5,6]-junction bond to achieve two diastereomers of the monoadducts. The ^{139}La NMR spectral analyses of the adducts disclosed that the encaged La atoms in the carbosilylated C_{80} do not stand still but instead have a certain dynamic movement. It is particularly interesting that the x-ray crystallographic analysis of one diastereomer of the carbosilylated derivatives showed disorder of the La atoms at eight sites with site occupation factors (SOFs) of 0.097–0.639 in the cage interior. The presence of eight La sites reveals small energy differences among various orientations of the La atoms, which allow the La atoms to move around inside the cage. Additionally, it is interesting that all the La sites are positioned along a 10 contiguous hexagons inside the cage. Therefore, it is speculated that the circuit along the band of 10 contiguous hexagons is the predominant trajectory of the movement of La atoms.

Cycloaddition of $La_2@I_h$-C_{80} with tetracyanoethylene oxide (TCNEO) also occurred regioselectively to form the [5,6]-closed derivative [65]. X-ray analysis of the TCNEO adduct showed that disorder involving the La atom locations is apparent in the cage interior, as revealed in the x-ray structure of the carbosilylated derivative. In all, 12 La atoms have SOFs of 0.046–0.378 at 90 K. Observation of the disordered La sites suggests that the La atoms have a large amplitude motion inside a fullerene cage. The existence of the dynamic movement was also supported by VT-^{139}La NMR spectral analysis.

12.7 ELECTRONIC MODULATION OF EMFs BY EXOHEDRAL CHEMICAL FUNCTIONALIZATION

Recent electrochemical investigation has shown that electronic affinity of EMFs decreases upon introduction of carbon substituents, as found for C_{60}. The trend is caused not only by the decrease of the π-electron system of the cage surface but also by the inductive effects of the carbon substituents. When the reaction involves a C–C bond cleavage, the cathodic shift of the first reduction potential of the derivative is small because the derivatized cage can retain its π-electron system [52]. Meanwhile, the reduction potentials are shifted cathodically to a great degree by bis-silylation. The effect can be rationalized by the strong electron-donating character of the silyl substituents because of the β-effect. In addition, the electronic modulation takes place step-by-step according to the number of silicon atoms introduced on the cage. In fact, the first reduction potentials of the bis-silylated $La_2@I_h$-C_{80} derivative and two diastereomers of the carbosilylated $La_2@I_h$-C_{80} are, respectively, –0.70, and –0.50 to –0.53 V, [61] whereas that of $La_2@I_h$-C_{80} is –0.31 V versus the ferrocene/ferrocenium couple.

In contrast, addition of TCNEO group is effective for enhancing the electron-accepting character [65]. This electrochemical study revealed that the first reduction potential of $La_2@I_h$-C_{80}–TCNEO adduct was shifted anodically by 100 mV to reach –0.21 V, which is the lowest value among the reported EMFs and their derivatives to date. These results demonstrate clearly that the electronic properties of EMFs can be modulated by chemical functionalization of the outer surface.

12.8 MISSING EMFs

Understanding the structures and properties of fullerenes and EMFs has largely relied on their solubility in common solvents. For those species which are soluble, such as C_{60} and $La@C_{82}$, and which can be handled with solution-processed laboratory techniques, investigation has been particularly intense. However, many fullerene species, especially monometallic EMFs with cage sizes of C_{72}–C_{80}, are not extractable. They remain in the extraction residue. For that reason, such molecules are called missing fullerenes [2b]. Great efforts have been exerted during the past two decades to obtain these missing fullerenes.

Although pyridine is effective to extract the insoluble $M@C_{60}$ through a mechanism that remains unclear, pure samples have never been obtained [66]. In 1998, Diener and Alford [67] reported an electrochemical reduction method, which showed astonishingly high efficiency toward such insoluble species as $Gd@C_{60}$, $Gd@C_{74}$, and C_{74}. Several years later, Bolskar and Alford proposed a chemical oxidation method to extract insoluble species such as $Tm@C_{60}$ and $Tm@C_{70}$, as well as $Gd@C_{2n}$ ($2n = 60, 72$–80) [8]. However, these reduction or oxidation methods invariably relied on soot sublimates as starting materials, thereby rendering the results poorly reproducible. Moreover, these methods afforded only mixtures of a certain class of EMFs and/or fullerenes. Their separation appeared impossible because of their ionic states and strong interactions with counterions in solution.

Recently, we developed a chemical derivatization method to achieve insoluble EMFs in the form of their dichlorophenyl derivatives. Successful results were obtained for $La@C_{2n}$ ($2n = 72, 74, 80, 82$) [68]. By accident, we observed that, during the process using TCB to extract EMFs, some insoluble EMFs could be functionalized by trace dichlorophenyl radicals generated by refluxing TCB. Consequently, the derivatives became soluble and extractable. However, considering the fact that no dichlorophenyl derivative of the most abundant (but soluble) $La@C_{82}$ was detected, generation of the dichlorophenyl radicals and their addition to these insoluble species might involve a heterogeneous catalytic process: $La@C_{82}$ is soluble in TCB and is therefore exempted from derivatization [69].

The x-ray results of the derivatives of these missing metallofullerenes—that is, $La@C_{2n}(C_6H_3Cl_2)$ ($2n$ = 72, 74, 80, 82)—are presented in Figure 12.4. In fact, $La@C_2(10612)$-C_{72} has a non-IPR cage with one pair of fused pentagons. The La atom interacts strongly with the fused-pentagon carbons, causing the dichlorophenyl group

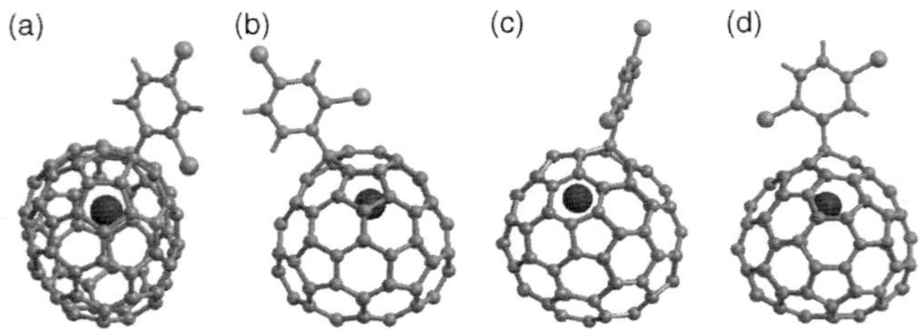

FIGURE 12.4. X-ray structures of (**a**) La@C_2(10612)-$C_{72}C_6H_3Cl_2$, (**b**) La@D_{3h}-$C_{74}C_6H_3Cl_2$, (**c**) La@C_{2v}(3)-$C_{80}C_6H_3Cl_2$, and (**d**) La@C_{3v}(7)-$C_{82}C_6H_3Cl_2$.

to be singly bonded to an adjacent carbon of the [5,5]-junction [68a]. Actually, La@D_{3h}-C_{74}, La@C_{2v}(3)-C_{80}, and La@C_{3v}(7)-C_{82} all have IPR cages [68b–d]. As a common phenomenon, the addend is singly bonded to a cage carbon that is very close to the internal metal. Calculations show that the cage carbons of addition all have high radical characters, as evidenced by both high POAV and high SOMO spin density. Such high radical characters of these molecules are responsible for their high reactivity toward radicals.

12.9 METAL CARBIDE EMFs: STRUCTURES AND CHEMISTRY

The fullerene interior can host not only pure metals but also metal clusters. Among cluster fullerenes, EMFs with a metal carbide moiety inside are very interesting. In 2001, Sc_2C_2@D_{2d}(23)-C_{84} was identified as the first carbide cluster EMF [70]. Subsequently, many carbide EMFs have been reported, including several examples that were assigned initially as conventional EMFs. Furthermore, Sc_2C_{84} was assigned as Sc_2@D_{2d}(23)-C_{84} using the synchrotron powder XRD/MEM/Rietveld method [71]. However, our recent NMR and single-crystallographic results have revealed that it should be Sc_2C_2@C_{3v}(8)-C_{82} [72, 73]. Similarly, Sc_3C_{82}, which had been assigned incorrectly as Sc_3@C_{82}, was confirmed as Sc_3C_2@I_h-C_{80} using x-ray results of its carbene adduct [54]. More interestingly, the five atoms of the Sc_3C_2 moiety were revealed as coplanar.

Sc-containing carbide clusters inside other small fullerenes were also found. For example, Sc_2C_2@C_{2v}(6073)-C_{68} reportedly has a different non-IPR cage from that of Sc_3N@D_3(6140)-C_{68} because of its different number of transferred electrons and the orientation of the clusters as well [74]. To date, carbide clusters with Sc, Y, and Er have been found frequently in EMFs. One example containing a large cluster was recently reported for Gd_2C_2@D_3(85)-C_{92} [75].

Very recently, we identified a new carbide EMF Sc_2C_2@C_{2v}(5)-C_{80}. Formation of the cage is found to be templated by the internal cluster via the dual effect of both electron transfer and topology. More interestingly, results show that the dynamic motion of the internal Sc_2C_2 cluster is strongly dependent on temperature. For that reason, this EMF is useful as a molecular thermometer [53].

12.10 SYNTHESIS AND PHOTOPHYSICS OF EMF-BASED DYADS

Mimicking natural photosynthetic systems has been a hot topic in recent years. Empty fullerenes such as C_{60}, C_{70}, and C_{84} have been widely used as important molecular building blocks because of their rich redox chemistry and low reorganization energy in photoinduced electron transfer (PET) events [76, 77]. Generally, these fullerenes are good electron acceptors (A) when conjugated with electron donors (D) such as some dye molecules and polymers. Consequently, great scientific effort has been devoted to various fullerene-based D–A nanostructure systems, ranging from the smallest D–A unit (a dyad) to optoelectronic devices (a solar cell) for practical applications. Very recently, some interest has emerged in utilizing endohedral fullerenes in this field because of their unique electronic structures and even richer redox chemistry.

The first D–A system comprising both an EMF and an electron donor moiety was reported by Echegoyen and co-workers in 2008 [78, 79]. They prepared M_3N@I_h-C_{80}-ferrocene (Fc) (M = Sc, Y) dyads (**1a**, **1b**) via a Prato reaction (Figure 12.5), in which the Fc moiety was closely linked to the 2-position of that pyrrolidine ring. These [3 + 2] cycloadditions occur at either a [5,6] or a [6,6]-bond of the I_h-C_{80} cage, depending on the size of endohedral M_3N cluster. Particularly, dyad **1a** was characterized as a [5,6]-adduct with acceptable stability within two months, whereas dyad **1b** was a [6,6]-adduct that facilely underwent decomposition under light exposure and high temperature. Substitution of the Fc by a phthalocyanine (Pc) moiety gave rise to dyads **2a** and **2b**, which are either poorly yielded or decomposed during isolation. Thus, it appears that the 2-substituted pyrrolidine ring of M_3N@I_h-C_{80} has poor stability when bearing a donor moiety. Two other isomeric dyads (**3**, **4**) of Sc_3N@I_h-C_{80}-triphenylamine (TPA) were demonstrated in further works [80]. Noteworthy is the substitution position of TPA moiety, which is on either the *N*-position or 2-position of the [5,6]-pyrrolidine ring, resulting in the formation of respective dyads **3** and **4**. Importantly, **3** exhibits higher thermal stability with respect to **4**. This result sheds light on the relation between the structures and the stabilities of these pyrrolidine adducts, which might benefit the future design for the construction of stable M_3N@I_h-C_{80}-based dyads.

However, M_2@I_h-C_{80} (M = La, Ce) has been viewed as a counterpart of trimetallic nitride template (TNT) fullerenes M_3N@I_h-C_{80} (M = Sc, Y, Lu) because of their same fullerene cage and theoretically proposed isoelectronic structures. However, the chemical reactivity of M_2@I_h-C_{80} varies

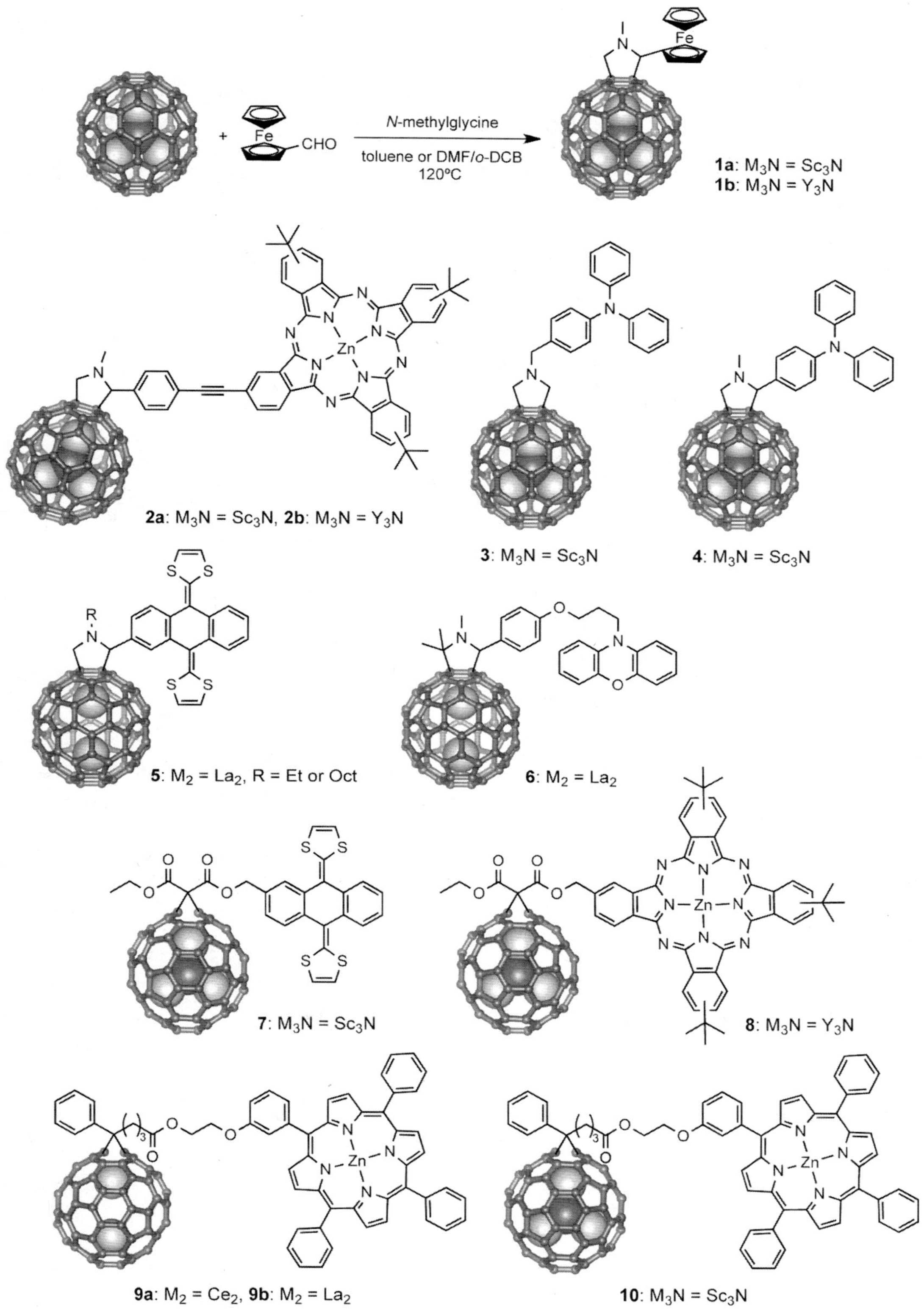

FIGURE 12.5. Schematic structures of dyads **1–10**.

greatly from that of TNT fullerenes because of their different endohedral clusters. More recently, our group synthesized dyads **5** and **6** using $La_2@I_h$-C_{80} as an electron acceptor and π-extended tetrathiafulvalene (ex-TTF) or phenoxazine as an electron donor [81, 82]. These dyads, which were all characterized as [5,6]-pyrrolidine adducts, exhibited good stability under ambient conditions.

The CV and DPV measurements of **1**, **3–5** in ODCB showed that their voltammograms do not correspond exactly to the superposition of those for [5,6]-pyrrolidine adducts and pristine donor molecules. For **1**, the oxidation potential of Fc moiety is positively shifted by 150 mV relative to that of pristine Fc, [78] whereas in the case of **3–5**, the oxidation potentials of TPA and ex-TTF moieties are all shifted to more negative potentials by 60–190 mV [80, 81], as compared with the pristine ones, indicating remarkable electronic communications between the two active units in the ground states.

The photophysical properties of **1** and **3–5** were investigated to elucidate the extent to which the EMF moiety would promote or influence the PET event occurrence. During the charge separation process, the radical ion pairs of **1**, **3–5** were formed within a period ranging from a few picoseconds to tens of picoseconds according to the transient absorption measurements, which is either faster or comparable to those of C_{60} analogues. On the other hand, in the charge recombination process, radical ion pairs of **1**, **3**, and **4** were characterized respectively as having longer lifetimes than those of C_{60} analogues. A 'stabilizing' effect of $Sc_3N@I_h$-C_{80} moiety was therefore proposed. Regarding $La_2@I_h$-C_{80}-based dyad **5**, the lifetime (3.7 ns in toluene) of that well-defined radical ion pair $(La_2@I_h\text{-}C_{80})^{\bullet-}$-(ex-TTF)$^{\bullet+}$ is much shorter than that of the C_{60} analogue (21 ns in toluene) [83]. Herein, the $La_2@I_h$-C_{80} moiety otherwise exhibits some 'destabilizing' effects. Considering all these reported examples, $Sc_3N@I_h$-C_{80} and $La_2@I_h$-C_{80} moieties might indeed promote the charge separation process, but afford different contributions to the electron recombination process. Nevertheless, from the viewpoint of chemistry, this difference shall be confirmed by comparing similarly constructed dyads that possess not only the same linker but also the same donor moiety. Consequently, the discrimination effect induced by either topologies or donor moieties would be ruled out, and the effects of endohedral cluster can be clarified.

The Bingel–Hirsch reaction is another typical methodology that has been developed for various functionalized fullerenes. However, in the Bingel–Hirsch reactions of EMFs, the presence of reagents (i.e., DBU, CBr_4, and I_2) that are commonly used in C_{60} reactions often engenders unwanted reductions or oxidations of EMFs. Thus, this methodology is under limited consideration in preparing functionalized EMFs. The efforts to synthesize dyads **7** and **8** via Bingel–Hirsch reactions unfortunately failed because of unexpected decompositions during isolation or purification procedure, which might be promoted by the proximity of the endohedral fullerenes [79].

In our group, another synthetic strategy was used for the preparation of EMF–zinc porphyrin (ZnP) dyads. This synthesis in fact involves a [1 + 2]-cycloaddition on either $M_2@I_h$-C_{80} (M = La, Ce) or $Sc_3N@I_h$-C_{80} in the presence of in-situ-generated diazo compound, respectively giving rise to the formation of $M_2@I_h$-C_{80}-ZnP and $Sc_3N@I_h$-C_{80}-ZnP dyads (**9**, **10**) [84, 85]. Taking advantage of their [6,6]-fulleroid structures, all these dyads exhibit high stabilities under ambient conditions. Also, important is the comparison between these similarly constructed dyads. In the ground state, in spite of the long and flexible linker, an unusually strong electronic interaction between the two active units of **9** was demonstrated by spectroscopic and electrochemical studies, whereas for **10**, such interaction is weaker. More impressive is the key difference between the photoexcited **9** and **10**. In detail, **9** undergoes a fast charge separation process upon photoexcitation and yields a radical ion pair, whose nature, namely $(M_2@I_h\text{-}C_{80})^{\bullet-}$–$(ZnP)^{\bullet+}$) versus $(M_2@I_h\text{-}C_{80})^{\bullet+}$–$(ZnP)^{\bullet-}$), varies with solvent polarity. Herein, the different strength in electronic coupling, which involves $[Ce_2]^{6+}$/ZnP versus $[I_h\text{-}C_{80}]^{6-}$/ZnP, was proposed to affect the electron-transfer kinetics and the electron-transfer mechanism. However, **10** affords the same $(Sc_3N@I_h\text{-}C_{80})^{-\bullet}$–$(ZnP)^{\bullet+}$) radical ion pair regardless of the solvent.

12.11 EMFs AS ACTIVE COMPONENTS IN ORGANIC SOLAR CELLS

Organic solar cells have rapidly developed in recent years with the aim of achieving a low-cost, lightweight, flexible, and environmentally friendly solar plane. A typical bulk-heterojunction (BHJ) solar cell consists of a glass substrate coated with a transparent electrode, an active layer, and a metallic top electrode. Particularly, the active layer functions as the light-absorbing unit, thereby generating an exciton through promotion of an electron from the HOMO to the LUMO level. In the next step, the exciton dissociates to generate charge carriers that move toward the corresponding electrodes. In this sense, the active layer is a crucial solar cell component. To date, one of the most successful constructions of the active layer is the combination of poly(*p*-phenylenevinylene)- or polythiophene-type polymers with PCBM, a soluble derivative of C_{60}, giving rise to devices with power conversion efficiencies (PCE) greater than 4% [86].

More recently, a derivative of TNT fullerene, [6,6]-phenyl-$Lu_3N@I_h$-C_{81}-butyric hexyl ester ($Lu_3N@I_h$-C_{80}-PCBH, **11**, see Figure 12.6), has been involved in the active layer with the commonly used donor polymer poly(3-hexyl)thiophene (P3HT) [87, 88]. The performance of the

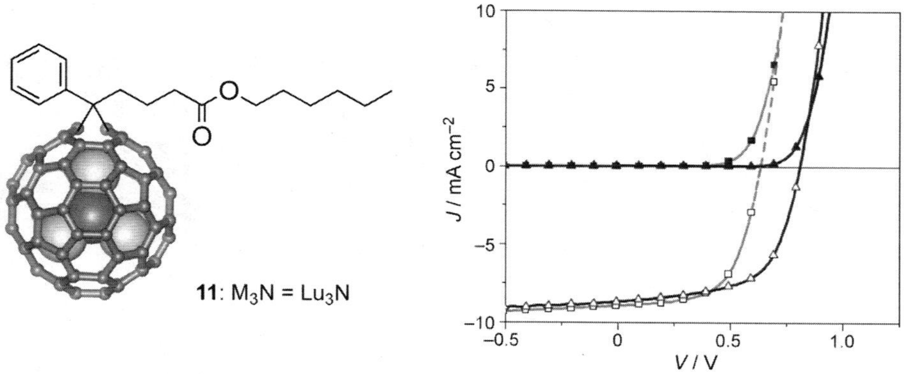

FIGURE 12.6. $Lu_3N@I_h$-C_{80}-PCBH (**11**) and current–voltage characteristics of P3HT/$Lu_3N@I_h$-C_{80}-PCBH (black) and P3HT/C_{60}-PCBM (gray). Solid triangles represent the dark *J–V* curves. Open triangles represent the devices' *J–V* curve under simulated solar illumination. Reprinted from reference 87 with permission from Wiley–VCH.

corresponding photovoltaic devices was investigated by optimizing the blend ratio of two active components and the thermal annealing process. Consequently, the most efficient electron transport is achieved at 50 wt% $Lu_3N@I_h$-C_{80}-PCBH loading. Additionally, results show that a combination of pre-production and post-production annealing processes can improve the surface morphology and the electrode contact. Finally, the optimized P3HT/$Lu_3N@I_h$-C_{80}-PCBH devices with PCE of 4.2% and open-circuit voltage (V_{oc}) of 0.89 V were achieved under AM1.5G solar simulation conditions. These values are all higher than those of P3HT/C_{60}-PCBM reference devices, indicating the advantage of using $Lu_3N@I_h$-C_{80}-PCBH as electron acceptor material. The essence of such improvement can be attributed to the higher LUMO energy of $Lu_3N@I_h$-C_{80} derivatives, which is closer to the LUMO level of the polymer donor P3HT than that of C_{60}. In other words, minimizing the LUMO energy offset of the donor–acceptor system enhanced the solar cell performance through larger V_{oc}.

12.12 CARRIER TRANSPORT PROPERTIES OF EMFs

In view of organic electronics, carrier transport properties of EMFs are the key criteria for applications. It is considered that orderly alignment of EMFs is important to gain high carrier transport properties. In this context, chemical functionalization of EMFs enhances crystallinity and allows formation of well-arrayed one-dimensional (1D) nanostructures. Slow evaporation of a CS_2 solution containing La@C_{2v}-C_{82}(Ad) allowed the formation of needle-like precipitates (nanorods) [89]. The field-effect transistor (FET) properties of the nanorods were measured using SiO_2/doped-Si substrates and Ag electrodes. The observation shows that the actions were *p*-type. On the other hand, whiskers of other empty fullerenes are well known to show *n*-type semiconductivity [90]. The observed *p*-type action would be explained by the work function of Ag and the HOMO level of La@C_{2v}-C_{82}(Ad).

To evaluate intrinsic charge-carrier mobilities (μ), the flash-photolysis time–resolved microwave conductivity (TRMC) measurement has served as an electrodeless measurement tool in conjugated organic materials [91]. The TRMC measurements of a single-crystal and nanorods of La@C_{2v}-C_{82}(Ad) respectively exhibited anisotropic and extremely high electron mobility of μ >10 cm^2 V^{-1} s^{-1} along the *c* axis and long axis direction (Figure 12.7) [92]. For comparison, TRMC measurements were performed for dropcast La@C_{2v}-C_{82}(Ad). The obtained μ value of 7 × 10^{-2} cm^2 V^{-1} s^{-1} for the dropcast La@C_{2v}-C_{82}(Ad) is much smaller than those for the single-crystal nanorods of La@C_{2v}-C_{82}(Ad), which indicates that the orderly aligned nanostructures of La@C_{2v}-C_{82}(Ad) is indeed important for the high performance. The crystal structure for La@C_{2v}-C_{82}(Ad) is presented in Figure 12.8a. In the crystal structure, the shortest intermolecular distance between the La@C_{2v}-C_{82}(Ad) units in the *c*-axis direction is 2.60 Å. The LUMO of La@C_{2v}-C_{82}(Ad) has a large distribution on the nearest carbon atoms (Figure 12.8b) and contributes to the high electron mobility. Furthermore, the single-crystal La@C_{2v}-C_{82}(Ad) exhibits a considerable dark current in the *I–V* trace measurement, suggesting that the single crystal functions as an organic conductor without carrier generation.

The co-crystal La@C_{2v}-C_{82}·Ni^{II}(octaethylporphyrin) (OEP) exhibits anisotropic and high electron mobility as 0.9 cm^2 V^{-1} s^{-1} in the *c*-axis direction. As inferred from the crystal structure, the shortest intermolecular distance between La@C_{2v}-C_{82} is 3.02 Å, as shown in Figure 12.9. In terms of the interaction of the molecular orbital, this short intermolecular distance for La@C_{2v}-C_{82} causes strong orbital interaction, which also engenders highly anisotropic

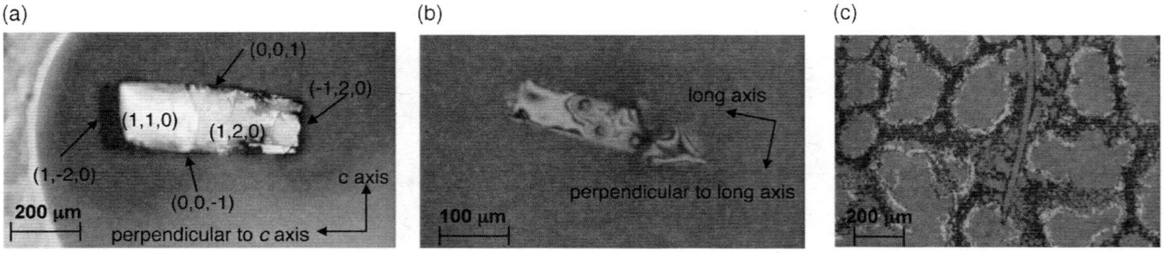

FIGURE 12.7. Photographs of (**a**) single-crystal La@C_{2v}-C_{82}(Ad). (**b**) nanorods, and (**c**) dropcast La@C_{2v}-C_{82}(Ad).

conductivity transients. This electron mobility of La@C_{2v}-C_{82}·Ni^{II}(OEP) is much higher than that of single-crystal Ni^{II}(OEP) ($\mu = 0.07$ cm^2 V^{-1} s^{-1}), suggesting that the observed high electron mobility of La@C_{2v}-C_{82}·Ni^{II}(OEP) was derived from La@C_{2v}-C_{82}. Consequently, thoroughly ordered molecular arrangement with La@C_{2v}-C_{82} exhibits a high electron transport property.

The band structures of La@C_{2v}-C_{82}(Ad) and La@C_{2v}-C_{82}·Ni^{II}(OEP) were calculated using DMol^3 code as shown in Figure 12.10 [93]. For single-crystal La@C_{2v}-C_{82}(Ad), the energy difference between the valence–band top and the conduction-band bottom (band gap) is only 0.005 eV, based on our calculations. At room temperature, the single-crystal La@C_{2v}-C_{82}(Ad) has numerous valence electrons, which have sufficient thermal energy, excited across the small band gap to the conduction band. Consequently, the single-crystal La@C_{2v}-C_{82}(Ad) can be regarded as a semi-metal. Meanwhile, La@C_{2v}-C_{82}·Ni^{II}(OEP) crystal possesses metallic behavior because some bands cross the Fermi level along the $\Gamma > Y$ and $C > Z$ directions. The calculations demonstrate that the single-crystal La@C_{2v}-C_{82}(Ad) and La@C_{2v}-C_{82}·Ni^{II}(OEP) are good media for electron transportation. The effective mass of the electron of the conduction band bottom and hole of valence band top were also calculated. For the single-crystal La@C_{2v}-C_{82}(Ad), the calculated effective masses 0.97 and 0.91 m_0 (where m_0 signifies the mass of free electron) respectively suggest nearly free electron behavior. However, for single-crystal La@C_{2v}-C_{82}·Ni^{II}(OEP), along the $\Gamma > Z$ direction, which corresponds to the c axis of the crystal, the respective effective masses of the electron and hole are 9.89 m_0 and 11.03 m_0. The larger effective masses for single-crystal La@C_{2v}-C_{82}·Ni^{II}(OEP) compared to those in the single-crystal La@C_{2v}-C_{82}(Ad) can explain its lower electron mobility.

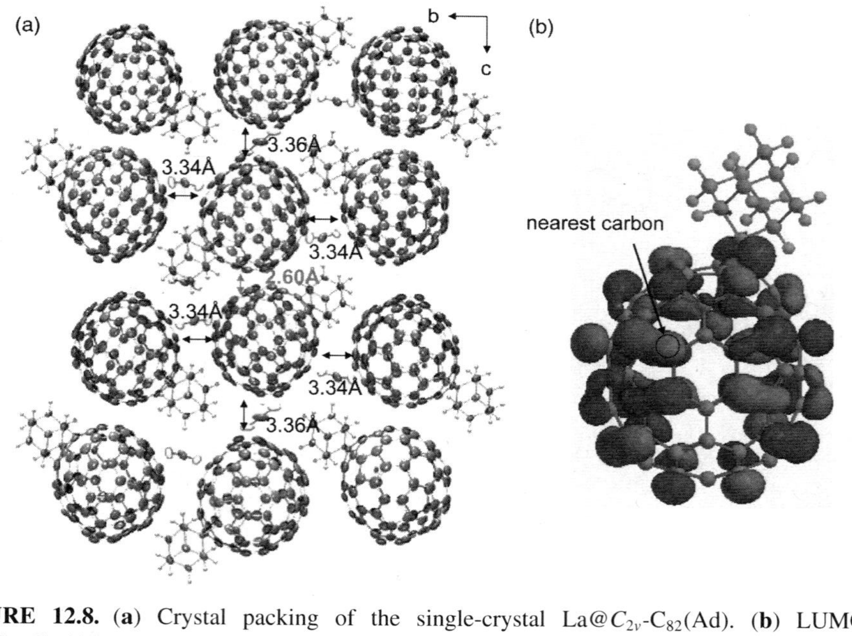

FIGURE 12.8. (**a**) Crystal packing of the single-crystal La@C_{2v}-C_{82}(Ad). (**b**) LUMO of La@C_{2v}-C_{82}(Ad).

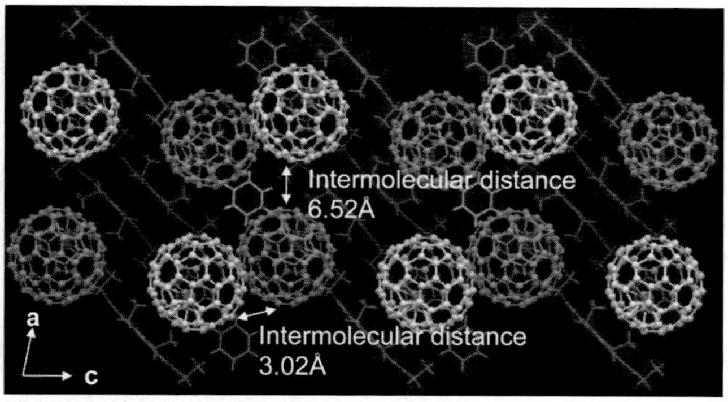

FIGURE 12.9. Crystal packing of La@C_{2v}-C_{82}·Ni^{II}(OEP)·1.5-benzene. The purple and orange colors denote porphyrin and benzene molecules, respectively. See color insert.

12.13 CONCLUSION

During the last few years, several new approaches for separating and purifying EMFs were reported, which enable us to pursue the fundamental and applied aspects of EMFs. Molecular structures of EMFs bearing new cage isomers and new encapsulates have been elucidated unequivocally using not only NMR spectroscopy but also single-crystal x-ray crystallography. In this respect, positions and dynamic behaviors of metal atoms inside fullerene cages are visible using paramagnetic ^{13}C NMR spectral analysis for cerium EMFs and VT-^{139}La NMR spectral analysis for lanthanum EMFs. Chemical derivatization is useful to prepare single crystals suitable for x-ray analysis. DFT computational studies at improved levels of theory, along with their comparison to experimental data, enrich investigations into the structural and chemical properties of EMFs.

Recent studies also demonstrated that functionalized EMFs possess fascinating and promising properties. For instance, metal atoms' movements in M_2@I_h-C_{80} can be regulated by exohedral chemical derivatization. In terms of applied aspects, various EMF-based D–A dyads have been synthesized and the unique photophysical properties were investigated. Photovoltaic devices using derivatized EMFs were also prepared and results revealed the advantage of using Lu_3N@I_h-C_{80}-PCBH as electron acceptor material. In addition, TRMC measurements showed that the single crystal of La@C_{2v}-C_{82}(Ad) possesses a high electron mobility and semi-metallic property. Further experimental and theoretical studies on EMFs and functionalized EMFs will deepen understanding on such novel nature of EMFs.

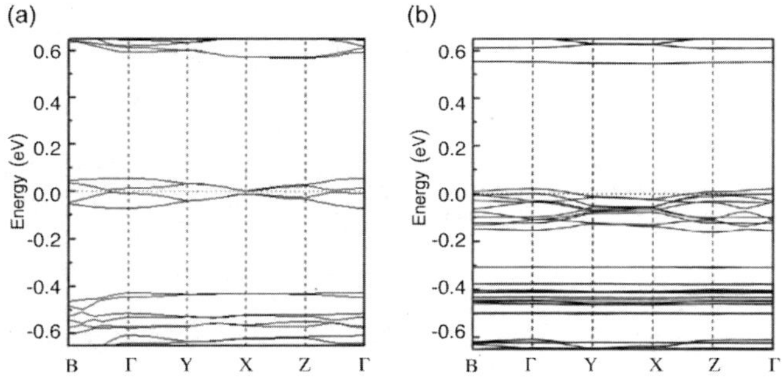

FIGURE 12.10. Band structures of (**a**) La@C_{2v}-C_{82}(Ad) and (**b**) La@C_{2v}-C_{82}·Ni^{II}(OEP).

REFERENCES

1. Kroto, H. W., Heath, J. R., O'Brien, S. C., Curl, R. F., Smalley, R. E. (1985). C_{60}: Buckminstefullerene. *Nature*, **318**, 162–163.
2. (a) Heath, J. R., O'Brien, S. C., Zhang, Q., Liu, Y., Curl, F. R., Tittle, F. K., Smalley, R. E. (1985). Lanthanum complexes of spheroidal carbon shells. *Journal of the American Chemical Society*, **107**, 7779–7780. (b) Chai, Y., Guo, T., Jin, C., Haufler, R. E., Chibante, L. P. F., Fure, J., Wang, L., Alford, J. M., Smalley, R. E. (1991). Fullerenes with metals inside. *The Journal of Physical Chemistry*, **95**, 7564–7568.
3. (a) Akasaka, T., Nagase, S., eds. (2000). *Endofullerenes: A New Family of Carbon Clusters*. Kluwer, Dordrecht, The Netherlands. (b) Akasaka, T., Wudl, F., Nagase, S., eds. (2010). *Chemistry of Nanocarbons*. Wiley-Blackwell, London. (c) Chaur, M. N., Melin, F., Ortiz, A. L., Echegoyen, L. (2009). Chemical, electrochemical, and structural properties of endohedral metallofullerenes. *Angewandte Chemie International Edition*, **48**, 7514–7538. (d) Dunsch, L., Yang, S. (2007). Metal nitride cluster fullerenes: Their current state and future prospects. *Small*, **3**, 1298–1320.
4. Krätschmer, W., Lamb, L. D., Fostiropoulos, K., Huffman, D. R. (1990). Solid C_{60}: A new form of carbon. *Nature*, **347**, 354–358.
5. Buckyclutcher column was provided by Regis Chemical. PBB, PYE, Buckyprep, and Buckyprep-M columns are available from Nacalai Tesque.
6. Tsuchiya, T., Wakahara, T., Shirakura, S., Maeda, Y., Akasaka, T., Kobayashi, K., Nagase, S., Kato, T., Kadish, K. M. (2004). Reduction of endohedral metallofullerenes: A convenient method for isolation. *Chemistry of Materials*, **16**, 4343–4346.
7. (a) Sun, B., Gu, Z. (2002). Solvent-dependent anion studies on enrichment of metallofullerene. *Chemistry Letters*, 1164–1165. (b) Tsuchiya, T., Wakahara, T., Lian, Y., Maeda, Y., Akasaka, T., Kato, T., Mizorogi, N., Nagase, S. (2006). Selective extraction and purification of endohedral metallofullerene from carbon soot. *The Journal of Physical Chemistry B*, **110**, 22517–22520.
8. Bolskar, R. D., Alford, J. M. (2003). Chemical oxidation of endohedral metallofullerenes: Identification and separation of distinct classes. *Chemical Communications*, 1292–1293.
9. Tsuchiya, T., Sato, K., Kurihara, H., Wakahara, T., Nakahodo, T., Maeda, Y., Akasaka, T., Ohkubo, K., Fukuzumi, S., Kato, T., Mizorogi, N., Kobayashi, K., Nagase, S. (2006). Host–guest complexation of endohedral metallofullerene with azacrown ether and its application. *Journal of the American Chemical Society*, **128**, 6699–6703.

10. Ge, Z., Duchamp, J. C., Cai, T., Gibson, H. W., Dorn, H. C. (2005). Purification of endohedral trimetallic nitride fullerenes in a single, facile step. *Journal of the American Chemical Society*, **127**, 16292–16298.
11. Stevenson, S., Thompson, M. C., Coumbe, H. L., Mackey, M. A., Coumbe, C. E., Phillips, J. P. (2006). Chemically adjusting plasma temperature, energy and reactivity (CAPTEAR) method using NO_x and combustion for selective synthesis of $Sc_3N@C_{80}$ metallic nitride fullerenes. *Journal of the American Chemical Society*, **128**, 8829–8835.
12. Akasaka, T., Wakahara, T., Nagase, S., Kobayashi, K., Waelchli, M., Yamamoto, K., Kondo, M., Shirakura, S., Okubo, S., Maeda, Y., Kato, T., Kako, M., Nakadaira, Y., Nagahata, R., Gao, X., Caemelbecke, E. V., Kadish, K. M. (2000). $La@C_{82}$ anion. An unusually stable metallofullerene. *Journal of the American Chemical Society*, **122**, 9316–9317.
13. Wakahara, T., Okubo, S., Kondo, M., Maeda, Y., Akasaka, T., Waelchli, M., Kako, M., Kobayashi, K., Nagase, S., Kato, T., Yamamoto, K., Gao, X., Caemelbecke, E. V., Kadish, K. M. (2002). Ionization and structural determination of the major isomer of $Pr@C_{82}$. *Chemical Physics Letters*, **360**, 235–239.
14. Wakahara, T., Kobayashi, J., Yamada, M., Maeda, Y., Tsuchiya, T., Okamura, M., Akasaka, T., Waelchli, M., Kobayashi, K., Nagase, S., Kato, T., Kako, M., Yamamoto, K., Kadish, K. M. (2004). Characterization of $Ce@C_{82}$ and its anion. *Journal of the American Chemical Society*, **126**, 4883–4887.
15. Feng, L., Wakahara, T., Tsuchiya, T., Maeda, Y., Lian, Y., Akasaka, T., Mizorogi, N., Kobayashi, K., Nagase, S., Kadish, K. M. (2005). Structural characterization of $Y@C_{82}$. *Chemical Physics Letters*, **405**, 274–277.
16. Tsuchiya, T., Wakahara, T., Maeda, Y., Akasaka, T., Waelchli, M., Kato, T., Mizorogi, N., Kobayashi, K., Nagase, S. (2005). 2D NMR characterization of the $La@C_{82}$ anion. *Angewandte Chemie International Edition*, **44**, 3282–3285.
17. Yamada, M., Wakahara, T., Lian, Y., Tsuchiya, T., Akasaka, T., Waelchli, M., Mizorogi, N., Nagase, S., Kadish, K. M. (2006). Analysis of lanthanide-induced NMR shifts of the $Ce@C_{82}$ anion. *Journal of the American Chemical Society*, **128**, 1400–1401.
18. Bleaney, B. (1972). Nuclear magnetic resonance shifts in solution due to lanthanide ions. *Journal of Magnetic Resonance*, **8**, 91–100.
19. Yamada, M., Wakahara, T., Tsuchiya, T., Maeda, Y., Akasaka, T., Mizorogi, N., Nagase, S. (2008). Spectroscopic and theoretical study of endohedral dimetallofullerene having a non-IPR fullerene cage: $Ce_2@C_{72}$. *The Journal of Physical Chemistry A*, **112**, 7627–7631.
20. Yamada, M., Wakahara, T., Tsuchiya, T., Maeda, Y., Kako, M., Akasaka, T., Yoza, K., Horn, E., Mizorogi, N., Nagase, S. (2008). Location of the metal atoms in $Ce_2@C_{78}$ and its bis-silylated derivative. *Chemical Communications*, 558–560.
21. Yamada, M., Nakahodo, T., Wakahara, T., Tsuchiya, T., Maeda, Y., Akasaka, T., Kako, M., Yoza, K., Horn, E., Mizorogi, N., Kobayashi, K., Nagase, S. (2005). Positional control of encapsulated atoms inside a fullerene cage by exohedral addition. *Journal of the American Chemical Society*, **127**, 14570–14571.
22. Yamada, M., Mizorogi, N., Tsuchiya, T., Akasaka, T., Nagase, S. (2009). Synthesis and characterization of the D_{5h} isomer of endohedral dimetallofullerene $Ce_2@C_{80}$: Two-dimensional circulation of encapsulated metal atoms inside a fullerene cage. *Chemistry-A European Journal*, **15**, 9486–9493.
23. Akasaka, T., Nagase, S., Kobayashi, K. (1997). ^{13}C and ^{139}La NMR studies of $La_2@C_{80.}$ First evidence for circular motion of metal atoms in endohedral dimetallofullerenes. *Angewandte Chemie International Edition in English*, **36**, 1643–1645.
24. Shimotani, H., Ito, T., Iwasa, Y., Taninaka, A., Shinohara, H., Nishibori, E., Takata, M., Sakata, M. (2004). Quantum chemical study on the configurations of encapsulated metal ions and the molecular vibration modes in endohedral dimetallofullerene $La_2@C_{80}$. *Journal of the American Chemical Society*, **126**, 364–369.
25. (a) Kobayashi, K., Nagase, S., Akasaka, T. (1995). A theoretical study of C_{80} and $La_2@C_{80}$. *Chemical Physics Letters*, **245**, 230–236. (b) Kobayashi, K., Nagase, S. (1996). Structures and electronic states of endohedral dimetallofullerenes: $M_2@C_{80}$ ($M = Sc$, Y, La, Ce, Pr, Eu, Gd, Yb and Lu). *Chemical Physics Letters*, **262**, 227–232.
26. Zhang, J., Hao, C., Li, S., Mi, W., Jin, P. (2007). Which configuration is more stable for $La_2@C_{80}$, D_{3d} or D_{2h}? Recomputation with ZORA methods with ADF. *The Journal of Physical Chemistry C*, **111**, 7862–7867.
27. Muthukumar, K., Larsson, A. (2008). Explanation of the different preferential binding sites for Ce and La in $M_2@C_{80}$ ($M = Ce$, La)—A density functional theory prediction. *Journal of Materials Chemistry*, **18**, 3347–3351.
28. Johnson, R. D., de Vries, M. S., Salem, J., Bethune, D. S., Yannoni, C. S. (1992). Electron paramagnetic resonance studies of lanthanum-containing C_{82}. *Nature*, **355**, 239–240.
29. Hino, S., Takahashi, H., Iwasaki, K., Matsumoto, K., Miyazaki, T., Hasegawa, S., Kikuchi, K., Achiba, Y. (1993). Electronic structure of metallofullerene LaC_{82}: Electron transfer from lanthanum to C_{82}. *Physical Review Letters*, **71**, 4261–4263.
30. Kobayashi, K., Nagase, S. (1998). Structures and electronic states of $M@C_{82}$ ($M = Sc$, Y, La and lanthanides). *Chemical Physics Letters*, **282**, 325–329.
31. Slanina, Z., Kobayashi, K., Nagase, S. (2004). Computed temperature development of the relative stabilities of $La@C_{82}$ isomers. *Chemical Physics Letters*, **388**, 74–78.
32. Kobayashi, K., Nagase, S., Akasaka, T. (1996). Endohedral dimetallofullerenes $Sc_2@C_{84}$ and $La_2@C_{80.}$ Are the metal atoms still inside the fullerene cages? *Chemical Physics Letters*, **261**, 502–506.
33. Kobayashi, K., Nagase, S. (1999). Bonding features in endohedral metallofullerenes. Topological analysis of the electron density distribution. *Chemical Physics Letters*, **302**, 312–316.
34. Bader, R. F. W. (1990). *Atoms in Molecules—A Quantum Theory*, Oxford University Press, Oxford.
35. Popov, A. A. (2008). Metal-cage bonding, molecular structures and vibrational spectra of endohedral fullerenes: Bridging experiment and theory. *Journal of Computational and Theoretical Nanoscience*, **6**, 1–26.

36. Yamada, M., Tsuchiya, T., Akasaka, T., Nagase, S. (2010). In-depth understanding of *p*-electron systems: New vistas in fullerene endohedrals. *Pure and Applied Chemistry*, **82**, 757–767.

37. Yamada, M., Akasaka, T., Nagase, S. (2010). In: D'Souza, F., Kadish, K. M., eds. *Fundamentals and Applications of Carbon Nano Materials*. World Scientific Publishing Co., Singapore, Chapter 5.

38. Akasaka, T., Okubo, S., Kondo, M., Maeda, Y., Wakahara, T., Kato, T., Suzuki, T., Yamamoto, K., Kobayashi, K., Nagase, S. (2000). Isolation and characterization of two $Pr@C_{82}$ isomers. *Chemical Physics Letters*, **319**, 153–156.

39. Suzuki, T., Kikuchi, K., Oguri, F., Nakao, Y., Suzuki, S., Achiba, Y., Yamamoto, K., Funasaka, H., Akasaka, T. (1996). Electrochemical properties of fullerenolanthanides, *Tetrahedron*, **52**, 4973–4982.

40. Suzuki, T., Maruyama, Y., Kato, T., Kikuchi, K., Nakao, Y., Suzuki, S., Achiba, Y., Yamamoto, K., Funasaka, H., Takahashi, T. (1995). Electrochemical properties of fullerenolanthanides. *Synthetic Metals*, **70**, 1443–1446.

41. Suzuki, T., Maruyama, Y., Kato, T., Kikuchi, K., Nakao, Y., Achiba, Y., Kobayashi, K., Nagase, S. (1995). Electrochemistry and ab initio study of the dimetallofullerene $La_2@C_{80}$. *Angewandte Chemie International Edition in English*, **34**, 1094–1096.

42. Akasaka, T., Kato, T., Kobayashi, K., Nagase, S., Yamamoto, K., Funasaka, H., Takahashi, T. (1995). Exohedral adducts of $La@C_{82}$. *Nature*, **374**, 600–601.

43. Wakahara, T., Yamada, M., Takahashi, S., Nakahodo, T., Tsuchiya, T., Maeda, Y., Akasaka, T., Kako, M., Yoza, K., Horn, E., Mizorogi, N., Nagase, S. (2007). Two-dimensional hopping motion of encapsulated La atoms in silylated $La_2@C_{80}$. *Chemical Communications*, 2680–2682.

44. (a) Iiduka, Y., Ikenaga, O., Sakuraba, A., Wakahara, T., Tsuchiya, T., Maeda, Y., Nakahodo, T., Akasaka, T., Kako, M., Mizorogi, N., Nagase, S. (2005). Chemical reactivity of $Sc_3N@C_{80}$ and $La_2@C_{80}$. *Journal of the American Chemical Society*, **127**, 9956–9957. (b) Wakahara, T., Iiduka, Y., Ikenaga, O., Nakahodo, T., Sakuraba, A., Tsuchiya, T., Maeda, Y., Kako, M., Akasaka, T., Yoza, K., Horn, E., Mizorogi, N., Nagase, S. (2006). Characterization of the bis-silylated endofullerene $Sc_3N@C_{80}$. *Journal of the American Chemical Society*, **128**, 9919–9925.

45. Akasaka, T., Nagase, S., Kobayashi, K., Suzuki, T., Kato, T., Kikuchi, K., Achiba, Y., Yamamoto, K., Funasaka, H., Takahashi, T. (1995). Synthesis of the first adducts of the dimetallofullerenes $La_2@C_{80}$ and $Sc_2@C_{84}$ by addition of a disilirane. *Angewandte Chemie International Edition in English*, **34**, 2139–2141.

46. Maeda, Y., Matsunaga, Y., Wakahara, T., Takahashi, S., Tsuchiya, T., Ishitsuka, M. O., Hasegawa, T., Akasaka, T., Liu, M. T. H., Kokura, K., Horn, E., Yoza, K., Kato, T., Okubo, S., Kobayashi, K., Nagase, S., Yamamoto, K. (2004). Isolation and characterization of a carbene derivative of $La@C_{82}$. *Journal of the American Chemical Society*, **126**, 6858–6859.

47. Takano, Y., Aoyagi, M., Yamada, M., Nikawa, H., Slanina, Z., Mizorogi, N., Ishitsuka, M. O., Tsuchiya, T., Maeda, Y., Akasaka, T., Kato, T., Nagase, S. (2009). Anisotropic magnetic behavior of anionic $Ce@C_{82}$ carbene adducts. *Journal of the American Chemical Society*, **131**, 9340–9346.

48. Lu, X., Nikawa, H., Feng, L., Tsuchiya, T., Maeda, Y., Akasaka, T., Mizorogi, N., Slanina, Z., Nagase, S. (2009). Location of the yttrium atom in $Y@C_{82}$ and its influence on the reactivity of cage carbons. *Journal of the American Chemical Society*, **131**, 12066–12067.

49. Akasaka, T., Kono, T., Takematsu, Y., Nikawa, H., Nakahodo, T., Wakahara, T., Ishitsuka, M. O., Tsuchiya, T., Maeda, Y., Liu, M. T. H., Yoza, K., Kato, T., Yamamoto, K., Mizorogi, N., Slanina, Z., Nagase, S. (2008). Does $Gd@C_{82}$ have an anomalous endohedral structure? Synthesis and single crystal X-ray structure of the carbene adduct. *Journal of the American Chemical Society*, **130**, 12840–12841.

50. (a) Lu, X., Nikawa, H., Nakahodo, T., Tsuchiya, T., Ishitsuka, M. O., Maeda, Y., Akasaka, T., Toki, M., Sawa, H., Slanina, Z., Mizorogi, N., Nagase, S. (2008). Chemical understanding of a non-IPR metallofullerene: Stabilization of encaged metals on fused-pentagon bonds in $La_2@C_{72}$. *Journal of the American Chemical Society*, **130**, 9129–9136. (b) Lu, X., Nikawa, H., Tsuchiya, T., Maeda, Y., Ishitsuka, M. O., Akasaka, T., Toki, M., Sawa, H., Slanina, Z., Mizorogi, N., Nagase, S. (2008). Bis-carbene adducts of non-IPR $La_2@C_{72}$: Localization of high reactivity around fused pentagons and electrochemical properties. *Angewandte Chemie International Edition*, **47**, 8642–8645.

51. Nikawa, H., Kikuchi, T., Wakahara, T., Nakahodo, T., Tsuchiya, T., Rahman, G. M. A., Akasaka, T., Maeda, Y., Yoza, K., Horn, E., Yamamoto, K., Mizorogi, N., Nagase, S. (2005). Missing metallofullerene $La@C_{74}$. *Journal of the American Chemical Society*, **127**, 9684–9685.

52. Yamada, M., Someya, C., Wakahara, T., Tsuchiya, T., Maeda, Y., Akasaka, T., Yoza, K., Horn, E., Liu, M. T. H., Mizorogi, N., Nagase, S. (2008). Metal atoms collinear with the spiro carbon of 6,6-open adducts, $M_2@C_{80}(Ad)$ (M = La and Ce, Ad = adamantylidene). *Journal of the American Chemical Society*, **130**, 1171–1176.

53. Kurihara, H., Lu, X., Iiduka, Y., Mizorogi, N., Slanina, Z., Tsuchiya, T., Akasaka, T., Nagase, S. (2011). $Sc_2C_2@C_{80}$ rather than $Sc_2@C_{82}$: Templated formation of unexpected $C_{2v}(5)$-C_{80} and temperature-dependent dynamic motion of internal Sc_2C_2 cluster. *Journal of the American Chemical Society*, **133**, 2382–2385.

54. Iiduka, Y., Wakahara, T., Nakahodo, T., Tsuchiya, T., Sakuraba, A., Maeda, Y., Akasaka, T., Yoza, K., Horn, E., Kato, T., Liu, M. T. H., Mizorogi, N., Kobayashi, K., Nagase, S. (2005). Structural determination of metallofullerene Sc_3C_{82} revisited: A surprising finding. *Journal of the American Chemical Society*, **127**, 12500–12501.

55. Iiduka, Y., Wakahara, T., Nakajima, K., Nakahodo, T., Tsuchiya, T., Maeda, Y., Akasaka, T., Yoza, K., Liu, M. T. H., Mizorogi, N., Nagase, S. (2007). Experimental and theoretical studies of the scandium carbide endohedral metallofullerene

$Sc_2C_2@C_{82}$ and its carbene derivative. *Angewandte Chemie International Edition*, **46**, 5562–5564.

56. (a) Yamada, M., Wakahara, T., Nakahodo, T., Tsuchiya, T., Maeda, Y., Akasaka, T., Yoza, K., Horn, E., Mizorogi, N., Nagase, S. (2006). Synthesis and structural characterization of endohedral pyrrolidinodimetallofullerene: $La_2@C_{80}(CH_2)_2NTrt$. *Journal of the American Chemical Society*, **128**, 1402–1403. (b) Yamada, M., Okamura, M., Sato, S., Someya, C. I., Mizorogi, N., Tsuchiya, T., Akasaka, T., Kato, T., Nagase, S. (2009). Two regioisomers of endohedral pyrrolidinodimetallofullerenes $M_2@I_h\text{-}C_{80}(CH_2)_2NTrt$ (M = La, Ce; Trt = trityl): Control of metal atom positions by addition positions. *Chemistry—A European Journal*, **15**, 10533–10542. (c) Cardona, C. M., Kitaygorodskiy, A., Ortiz, A., Herranz, M. A., Echegoyen, L. (2005). The first fulleropyrrolidine derivative of Sc3N@C80: Pronounced chemical shift differences of the geminal protons on the pyrrolidine ring. *The Journal of Organic Chemistry*, **70**, 5092–5097. (d) Cai, T., Ge, Z., Iezzi, E. B., Glass, T. E., Harich, K., Gibson, H. W., Dorn, H. C. (2005). Synthesis and characterization of the first trimetallic nitride templated pyrrolidino endohedral metallofullerenes. *Chemical Communications*, 3594–3596. (e) Cardona, C. M., Kitaygorodskiy, A., Echegoyen, L. (2005). Trimetallic nitride endohedral metallofullerenes: Reactivity dictated by the encapsulated metal cluster. *Journal of the American Chemical Society*, **127**, 10448–10453. (f) Echegoyen, L., Chancellor, C. J., Cardona, C. M., Elliott, B., Rivera, J., Olmstead, M. M., Balch, A. L. (2006). X-ray crystallographic and EPR spectroscopic characterization of a pyrrolidine adduct of $Y_3N@C_{80}$. *Chemical Communications*, 2653–2655. (g) Cardona, C. M., Elliott, B., Echegoyen, L. (2006). Unexpected chemical and electrochemical properties of $M_3N@C_{80}$ (M = Sc, Y, Er). *Journal of the American Chemical Society*, **128**, 6480–6485. (h) Cai, T., Slebodnick, C., Xu, L., Harich, K., Glass, T. E., Chancellor, C., Fettinger, J. C., Olmstead, M. M., Balch, A. L., Gibson, H. W., Dorn, H. C. (2006). A pirouette on a metallofullerene sphere: Interconversion of isomers of *N*-tritylpyrrolidino I_h $Sc_3N@C_{80}$. *Journal of the American Chemical Society*, **128**, 6486–6492. (i) Chen, N., Zhang, E.-Y., Tan, K., Wang, C.-R., Lu, X. (2007). Size effect of encaged clusters on the exohedral chemistry of endohedral fullerenes: A case study on the pyrrolidino reaction of $Sc_xGd_{3-x}N@C_{80}$ ($x = 0$–3). *Organic Letters*, **9**, 2011–2013. (j) Cai, T., Xu, L., Gibson, H. W., Dorn, H. C., Chancellor, C. J., Olmstead, M. M., Balch, A. L. (2007). $Sc_3N@C_{78}$: Encapsulated cluster regiocontrol of adduct docking on an ellipsoidal metallofullerene sphere. *Journal of the American Chemical Society*, **129**, 10795–10800. (k) Wang, T., Wu, J., Xu, W., Xiang, J., Lu, X., Li, B., Jiang, L., Shu, C., Wang, C. (2010). Spin divergence induced by exohedral modification: ESR study of $Sc_3C_2@C_{80}$ fulleropyrrolidine. *Angewandte Chemie International Edition*, **49**, 1786–1789.

57. (a) Feng, L., Nakahodo, T., Wakahara, T., Tsuchiya, T., Maeda, Y., Akasaka, T., Kato, T., Horn, E., Yoza, K., Mizorogi, N., Nagase, S. (2005). A singly bonded derivative of endohedral metallofullerene: $La@C_{82}CBr(COOC_2H_5)_2$. *Journal of the American Chemical Society*, **127**, 17136–17137. (b) Feng, L., Wakahara, T., Nakahodo, T., Tsuchiya, T., Piao, Q., Maeda, Y., Lian, Y., Akasaka, T., Horn, E., Yoza, K., Kato, T., Mizorogi, N., Nagase, S. (2006). The Bingel monoadducts of $La@C_{82}$: Synthesis, characterization, and electrochemistry. *Chemistry—A European Journal*, **12**, 5578–5586. (c) Feng, L., Tsuchiya, T., Wakahara, T., Nakahodo, T., Piao, Q., Maeda, Y., Akasaka, T., Kato, T., Yoza, K., Horn, E., Mizorogi, N., Nagase, S. (2006). Synthesis and characterization of a bisadduct of $La@C_{82}$. *Journal of the American Chemical Society*, **128**, 5990–5991. (d) Lukoyanova, O., Cardona, C. M., Rivera, J., Lugo-Morales, L. Z., Chancellor, C. J., Olmstead, M. M., Rodriguez-Fortea, A., Poblet, J. M., Balch, A. L., Echegoyen, L. (2007). "Open rather than closed" malonate methano-fullerene derivatives. The formation of methanofulleroid adducts of $Y_3N@C_{80}$. *Journal of the American Chemical Society*, **129**, 10423–10430. (e) Cai, T., Xu, L., Shu, C., Champion, H. A., Reid, J. E., Anklin, C., Anderson, M. R., Gibson, H. W., Dorn, H. C. (2008). Selective formation of a symmetric $Sc_3N@C_{78}$ bisadduct: Adduct docking controlled by an internal trimetallic nitride cluster. *Journal of the American Chemical Society*, **130**, 2136–2137.

58. (a) Maeda, Y., Miyashita, J., Hasegawa, T., Wakahara, T., Tsuchiya, T., Nakahodo, T., Akasaka, T., Mizorogi, N., Kobayashi, K., Nagase, S., Kato, T., Ban, N., Nakajima, H., Watanabe, Y. (2005). Reversible and regioselective reaction of $La@C_{82}$ with cyclopentadiene. *Journal of the American Chemical Society*, **127**, 12190–12191. (b) Maeda, Y., Sato, S., Inada, K., Nikawa, H., Yamada, M., Mizorogi, N., Hasegawa, T., Tsuchiya, T., Akasaka, T., Kato, T., Slanina, Z., Nagase, S. (2010). Regioselective exohedral functionalization of $La@C_{82}$ and its 1,2,3,4,5-pentamethylcyclopentadiene and adamantylidene adducts. *Chemistry—A European Journal*, **16**, 2193–2197. (c) Iezzi, E. B., Duchamp, J. C., Harich, K., Glass, T. E., Lee, H. M., Olmstead, M. M., Balch, A. L., Dorn, H. C. (2002). A symmetric derivative of the trimetallic nitride endohedral metallofullerene, $Sc_3N@C_{80}$. *Journal of the American Chemical Society*, **124**, 524–525. (d) Lee, H. M., Olmstead, M. M., Iezzi, E., Duchamp, J. C., Dorn, H. C., Balch, A. L. (2002). Crystallographic characterization and structural analysis of the first organic functionalization product of the endohedral fullerene $Sc_3N@C_{80}$. *Journal of the American Chemical Society*, **124**, 3494–3495.

59. (a) Lu, X., Nikawa, H., Tsuchiya, T., Akasaka, T., Toki, M., Sawa, H., Mizorogi, N., Nagase, S. (2010). Nitrated benzyne derivatives of $La@C_{82}$: Addition of NO_2 and its positional directing effect on the subsequent addition of benzynes. *Angewandte Chemie International Edition*, **49**, 594–597. (b) Li, F. F., Pinzon, J. R., Mercado, B. Q., Olmstead, M. M., Balch, A. L., Echegoyen, L. (2011). [2 + 2] Cycloaddition reaction to $Sc_3N@I_h\text{-}C_{80}$. The formation of very stable [5,6]- and [6,6]-adducts. *Journal of the American Chemical Society*, **133**, 1563–1571.

60. (a) Takano, Y., Yomogida, A., Nikawa, H., Yamada, M., Wakahara, T., Tsuchiya, T., Ishitsuka, M. O., Maeda, Y., Akasaka, T., Kato, T., Slanina, Z., Mizorogi, N., Nagase, S. (2008). Radical coupling reaction of paramagnetic endohedral metallofullerene $La@C_{82}$. *Journal of the American Chemical Society*, **130**, 16224–16230. (b) Shu, C., Slebodnick, C., Xu, L., Champion, H., Fuhrer, T., Cai, T., Reid, J. E., Fu, W., Harich, K., Dorn, H. C., Gibson, H. W. (2008). Highly regioselective derivatization of trimetallic nitride templated endohedral

metallofullerene via a facile photochemical reaction. *Journal of the American Chemical Society*, **130**, 17755–17760.

61. Yamada, M., Minowa, M., Sato, S., Kako, M., Slanina, Z., Mizorogi, N., Tsuchiya, T., Maeda, Y., Nagase, S., Akasaka, T. (2010). Thermal carbosilylation of endohedral dimetallofullerene $La_2@I_h$-C_{80} with silirane. *Journal of the American Chemical Society*, **132**, 17953–17960.
62. (a) Haddon, R. C. (1993). Chemistry of the Fullerenes: The manifestation of strain in a class of continuous aromatic molecules. *Science*, **261**, 1545–1550. (b) Haddon, R. C. (1997). C_{60}: Sphere or polyhedron? *Journal of the American Chemical Society*, **119**, 1797–1798.
63. Yamada, M., Akasaka, T., Nagase, S. (2010). Endohedral metal atoms in pristine and functionalized fullerene cages. *Accounts of Chemical Research*, **43**, 92–102.
64. Kobayashi, K., Nagase, S., Maeda, Y., Wakahara, T., Akasaka, T. (2003). $La_2@C_{80}$: Is the circular motion of two La atoms controllable by exohedral addition? *Chemical Physics Letters*, **374**, 562–566.
65. Yamada, M., Minowa, M., Sato, S., Slanina, Z., Tsuchiya, T., Maeda, Y., Nagase, S., Akasaka, T. (2011). Regioselective cycloaddition of La2@Ih-C80 with tetracyanoethylene oxide: Formation of an endohedral dimetallofullerene adduct featuring enhanced electron-accepting character. *Journal of the American Chemical Society*, **133**, 3796–3799.
66. Kubozono, Y., Maeda, H., Takabayashi, Y., Hiraoka, K., Nakai, T., Kashino, S., Emura, S., Ukita, S., Sogabe, T. (1996). Extractions of $Y@C_{60}$, $Ba@C_{60}$, $La@C_{60}$, $Ce@C_{60}$, $Pr@C_{60}$, $Nd@C_{60}$, and $Gd@C_{60}$ with aniline. *Journal of the American Chemical Society*, **118**, 6998–6999.
67. Diener, M. D., Alford, J. M. (1998). Isolation and properties of small-bandgap fullerenes. *Nature*, **393**, 668–671.
68. (a) Wakahara, T., Nikawa, H., Kikuchi, T., Nakahodo, T., Rahman, G. M. A., Tsuchiya, T., Maeda, Y., Akasaka, T., Yoza, K., Horn, E., Yamamoto, K., Mizorogi, N., Slanina, Z., Nagase, S. (2006). $La@C_{72}$ having a non-IPR carbon cage. *Journal of the American Chemical Society*, **128**, 14228–14229. (b) Nikawa, H., Kikuchi, T., Wakahara, T., Nakahodo, T., Tsuchiya, T., Rahman, G. M. A., Akasaka, T., Maeda, Y., Yoza, K., Horn, E., Yamamoto, K., Mizorogi, N., Nagase, S. (2005). Missing metallofullerene $La@C_{74}$. *Journal of the American Chemical Society*, **127**, 9684–9685. (c) Nikawa, H., Yamada, T., Cao, B., Mizorogi, N., Slanina, Z., Tsuchiya, T., Akasaka, T., Nagase, S. (2009). Missing metallofullerene with C_{80} cage. *Journal of the American Chemical Society*, **131**, 10950–10954. (d) Akasaka, T., Lu, X., Kuga, H., Nikawa, H., Mizorogi, N., Slanina, Z., Tsuchiya, T., Yoza, K., Nagase, S. (2010). Dichlorophenyl derivatives of $La@C_{3v}(7)$-C_{82}: Endohedral metal induced localization of pyramidalization and spin on a triple-hexagon junction. *Angewandte Chemie International Edition*, **49**, 9715–9719.
69. Lu, X., Akasaka, T., Nagase, S. (2011). Chemistry of endohedral metallofullerenes: The role of metals. *Chemical Communications*, **47**, 5942–5957.
70. Wang, C.-R., Kai, T., Tomiyama, T., Yoshida, T., Kobayashi, Y., Nishibori, E., Takata, M., Sakata, M., Shinohara, H. (2001). A scandium carbide endohedral metallofullerene: $(Sc_2C_2)@C_{84}$. *Angewandte Chemie International Edition*, **40**, 397–399.
71. Takata, M., Nishibori, E., Umeda, B., Sakata, M., Yamamoto, E., Shinohara, H. (1997). Structure of endohedral dimetallofullerene $Sc_2@C_{84}$. *Physical Review Letters*, **78**, 3330–3333.
72. Iiduka, Y., Wakahara, T., Nakajima, K., Tsuchiya, T., Nakahodo, T., Maeda, Y., Akasaka, T., Mizorogi, N., Nagase, S. (2006). ^{13}C NMR spectroscopic study of scandium dimetallofullerene, $Sc_2@C_{84}$ vs. $Sc_2C_2@C_{82}$. *Chemical Communications*, 2057–2059.
73. Iiduka, Y., Wakahara, T., Nakajima, K., Nakahodo, T., Tsuchiya, T., Maeda, Y., Akasaka, T., Yoza, K., Liu, M. T. H., Mizorogi, N., Nagase, S. Experimental and theoretical studies of the scandium carbide endohedral metallofullerene $Sc_2C_2@C_{82}$ and its carbene derivative. *Angewandte Chemie International Edition*, **46**, 5562–5564.
74. Shi, Z.-Q., Wu, X., Wang, C.-R., Lu, X., Shinohara, H. (2006). Isolation and characterization of $Sc_2C_2@C_{68}$: A metal-carbide endofullerene with a non-IPR carbon cage. *Angewandte Chemie International Edition*, **45**, 2107–2111.
75. Yang, H., Lu, C., Liu, Z., Jin, H., Che, Y., Olmstead, M. M., Balch, A. L. (2008). Detection of a family of gadolinium-containing endohedral fullerenes and the isolation and crystallographic characterization of one member as a metal-carbide encapsulated inside a large fullerene cage. *Journal of the American Chemical Society*, **130**, 17296–17300.
76. Haddon, R. C., Brus, L. E., Raghavachari, K. (1986). Electronic structure and bonding in icosahedral C_{60}. *Chemical Physics Letters*, **125**, 459–464.
77. (a) Fukuzumi, S., Nakanishi, I., Suenobu, T., Kadish, K. M. (1999). Electron-transfer properties of C_{60} and *tert*-butyl-C_{60} radical. *Journal of the American Chemical Society*, **121**, 3468–3474. (b) Fukuzumi, S., Ohkubo, K., Imahori, H., Guldi, D. M. (2003). Driving force dependence of intermolecular electron-transfer reactions of fullerenes. *Chemistry—A European Journal*, **9**, 1585–1593. (c) Imahori, H., El-Khouly, M. E., Fujitsuka, M., Ito, O., Sakata, Y., Fukuzumi, S. (2001). Solvent dependence of charge separation and charge recombination rates in porphyrin-fullerene dyad. *The Journal of Physical Chemistry A*, **105**, 325–332.
78. Pinzon, J. R., Plonska-Brzezinska, M. E., Cardona, C. M., Athans, A. J., Gayathri, S. S., Guldi, D. M., Herranz, M. A., Marin, N., Torres, T., Echegoyen, L. (2008). $Sc_3N@C_{80}$-ferrocene electron-donor/acceptor conjugates as promising materials for photovoltaic applications. *Angewandte Chemie International Edition*, **47**, 4173–4176.
79. Pinzon, J. R., Cardona, C. M., Herranz, M. A., Plonska-Brzezinska M. E., Palkar, A., Athans, A. J., Martin, N., Rodriguez-Fortea, A., Poblet, J. M., Bottari, G., Torres, T., Gayathri, S. S., Guldi, D. M., Echegoyen, L. (2009). Metal nitride cluster fullerene $M_3N@C_{80}$ ($M = Y, Sc$) based dyads: Synthesis, and electrochemical, theoretical and photophysical studies. *Chemistry—A European Journal*, **15**, 864–877.
80. Pinzon, J. R., Gasca, D. C., Sankaranarayanan, S. G., Bottari, G., Torres, T., Guldi, G. M., Echegoyen, L. (2009). Photoinduced charge transfer and electrochemical properties of

triphenylamine I_h-$Sc_3N@C_{80}$ donor–acceptor conjugates. *Journal of the American Chemical Society*, **131**, 7727–7734.

81. Takano, Y., Herranz, M. A., Martin, N., Shankara, G. R., Guldi, G. M., Tsuchiya, T., Nagase, S., Akasaka, T. (2010). Donor–acceptor conjugates of lanthanum endohedral metallofullerene and π-extended tetrathiafulvalene. *Journal of the American Chemical Society*, **132**, 8048–8055.
82. Tsuchiya, T. et al., unpublished result.
83. Diaz, M. C., Herranz, M. A., Illescas, B. M., Martin, N. (2003). Probing charge separation in structurally different C_{60}/exTTF ensembles. *The Journal of Organic Chemistry*, **68**, 7711–7721.
84. Guldi, G. M., Feng, L., Shankara, G. R., Nikawa, H., Yamada, M., Mizorogi, N., Tsuchiya, T., Akasaka, T., Nagase, S., Herranz, M. A., Martin, N. (2010). A molecular $Ce_2@I_h$-C_{80} switch—Unprecedented oxidative pathway in photoinduced charge transfer reactivity. *Journal of the American Chemical Society*, **132**, 9078–9086.
85. Feng, L., Radhakrishnan, S. G., Mizorogi, N., Slanina, Z., Nikawa, H., Tsuchiya, T., Akasaka, T., Nagase, S., Martin, N., Guldi, N. (2011). Synthesis and charge-transfer chemistry of $La_2@I_h$-C_{80}/$Sc_3N@I_h$-C_{80}–zinc porphyrin conjugates: Impact of endohedral cluster. *Journal of the American Chemical Society*, **133**, 7608–7618.
86. (a) Grätzel, M. (2009). Recent advances in sensitized mesoscopic solar cells. *Accounts of Chemical Research*, **42**, 1788–1798. (b) Delgado, J. L., Bouit, P.-A., Filippone, S., Herranz, M. A., Martin, N. (2010). Organic photovoltaics: a chemical approach. *Chemical Communications*, **46**, 4853–4865.
87. Ross, R. B., Cardona, C. M., Guldi, G. M., Sankaranarayanan, S. G., Reese, M. O., Kopidakis, N., Peet, J., Walker, B., Bazan, G. C., Van Keuren, E., Holloway, B. C., Drees, M. (2009). Endohedral fullerenes for organic photovoltaic devices. *Nature Materials*, **8**, 208–212.
88. Ross, R. B., Cardona, C. M., Swain, F. B., Guldi, G. M., Sankaranarayanan, S. G., Van Keuren, E., Holloway, B. C., Drees, M. (2009). Tuning conversion efficiency in metallo endohedral fullerene-based organic photovoltaic devices. *Advanced Functional Materials*, **19**, 2332–2337.
89. Tsuchiya, T., Kumashiro, R., Tanigaki, K., Matsunaga, Y., Ishitsuka, M. O., Wakahara, T., Maeda, Y., Takano, Y., Aoyagi, M., Akasaka, T., Liu, M. T. H., Kato, T., Suenaga, K., Jeong, J. S., Iijima, S., Kimura, F., Kimura, T., Nagase, S. (2008). Nanorods of endohedral metallofullerene derivative. *Journal of the American Chemical Society*, **130**, 450–451.
90. Ogawa, K., Kato, T., Ikegami, A., Tsuji, H., Aoki, N., Ochiai, Y., Bird, J. P. (2006). Electrical properties of field-effect transistors based on C_{60} nanowhiskers. *Applied Physics Letters*, **88**, 112109.
91. Seki, S., Yoshida, Y., Tagawa, S., Asai, K., Ishigure, K., Furukawa, K., Fujiki, M., Matsumoto, N. (1999). Effects of structural defects on hole drift mobility in arylsubstituted polysilanes. *Philosophical Magazine B*, **79**, 1631–1645. (b) Grozema, F. C., Siebbeles, L. D. A., Warman, J. M., Seki, S., Tagawa, S., Scherf, U. (2002). Hole conduction along molecular wires: s-bonded silicon versus p-bond-conjugated carbon. *Advanced Materials*, **14**, 228–231. (c) Saeki, A., Seki, S., Sunagawa, T., Ushida, K., Tagawa, S. (2006). Charge-carrier dynamics in polythiophene films studied by in-situ measurement of flash-photolysis time-resolved microwave conductivity (FP-TRMC) and transient optical spectroscopy (TOS). *Philosophical Magazine*, **86**, 1261–1276. (d) Yamamoto, Y., Fukushima, T., Jin, W., Kosaka, A., Hara, T., Namura, T., Saeki, A., Seki, S., Tagawa, S., Aida, T. (2006). A glass hook allows fishing of hexa-*peri*-hexabenzocoronene graphitic nanotubes: Fabrication of a macroscopic fiber with anisotropic electrical conduction. *Advanced Materials*, **18**, 1297–1300. (e) Saeki, A., Seki, S., Koizumi, Y., Sunagawa, T., Ushida, K., Tagawa, S. (2005). Increase in the mobility of photogenerated positive charge carriers in polythiophene. *The Journal of Physical Chemistry B*, **109**, 10015–10019. (f) Saeki, A., Seki, S., Takenobu, T., Iwasa, Y., Tagawa, S. (2008). Mobility and dynamics of charge carriers in rubrene single crystals studied by flash-photolysis microwave conductivity and optical spectroscopy. *Advanced Materials*, **20**, 920–923.
92. Sato, S., Seki, S., Honsho, Y., Wang, L., Nikawa, H., Luo, G., Lu, J., Haranaka, M., Tsuchiya, T., Nagase, S., Akasaka, T. (2011). Semi-metallic single-component crystal of soluble $La@C_{82}$ derivative with high electron mobility. *Journal of the American Chemical Society*, **133**, 2766–2771.
93. (a) Delley, B. (1990). An all-electron numerical method for solving the local density functional for polyatomic molecules. *Journal of Chemical Physics*, **92**, 508–517. (b) Delley, B. (2000). From molecules to solids with the $DMol^3$ approach. *Journal of Chemical Physics*, **113**, 7756–7764.

13

AN UPDATE ON ELECTROCHEMICAL CHARACTERIZATION AND POTENTIAL APPLICATIONS OF CARBON MATERIALS

FANG-FANG LI, ADRIÁN VILLALTA-CERDAS, LOURDES E. ECHEGOYEN, AND LUIS ECHEGOYEN

13.1 INTRODUCTION

The vast spectrum of unique properties and structures of carbon nanomaterials have literally revolutionized the world of materials research. In the last 25 years, the discovery of a large variety of allotropic forms of carbon has moved the field of carbon-based chemistry into new horizons, demonstrating that elemental carbon is intrinsically more complex than perhaps any other element. From theoretical calculations to serendipitous discoveries of new structures, the emerging applications of carbon in nanotechnology, electronics, and optics was simply unimaginable two decades ago.

Fullerenes and carbon nanotubes (CNTs) represent two rapidly growing families of the many allotropic forms of carbon [1–3]. The recent isolation of graphene (single layers of graphite) has added a striking dimensionality to the potential applications of this element [4]. In fullerene chemistry, researchers have studied how functionalization of the various sizes and structures of the carbon cages leads to specific changes in electronic properties (e.g., redox potentials and HOMO–LUMO gaps). These studies have led to revolutionary designs of materials with properly tuned properties for specific applications (e.g., dyads, molecular transistors, sensors). The potential use of fullerenes with encapsulated inorganic moieties in materials science is also being explored due to their physicochemical and organic–inorganic hybrid electronic properties [5–9]. Similarly, electrochemical studies of CNTs and CNT-modified electrodes have generated much interest as they promise to be useful for applications in electronic nanodevices, sensors, and energy storage [6, 10–12].

In this chapter, we will focus exclusively on the most recent electrochemical studies and advances toward promising applications of fullerenes and some of their closest relatives: endohedral metallofullerenes (EMFs) and carbon nano-onions (CNOs). A thorough discussion of carbon nanotubes and graphene is beyond the scope of this chapter. A good number of reviews have appeared recently, which are excellent to bring the reader up to date in the field [13–16].

13.2 PRISTINE FULLERENES

The industrial production of fullerenes consists of continuous hydrocarbon burning to create a sooting flame near 2000 Kelvin [17–23]. After the discovery of C_{60}, many other larger, empty-cage fullerenes, the higher fullerenes were discovered and isolated from the soot mixture [1, 22–26]. C_{60} and C_{70} are the most abundant and consequently the subject of most research, but many different cage sizes (defined by number of carbon atoms) and shapes (defined by isomer symmetry) can be extracted from the soot, adding new levels of complexity to an already complex mixture of materials [1, 27–29]. Extensive theoretical studies have been performed to predict and categorize the formation of the most probable sizes and shapes of empty fullerenes [29–32], but their isolation and characterization continues to be a challenge. To date, empty fullerene cages from 60 to 90 carbon atoms

Organic Nanomaterials: Synthesis, Characterization, and Device Applications, First Edition. Edited by Tomás Torres and Giovanni Bottari.

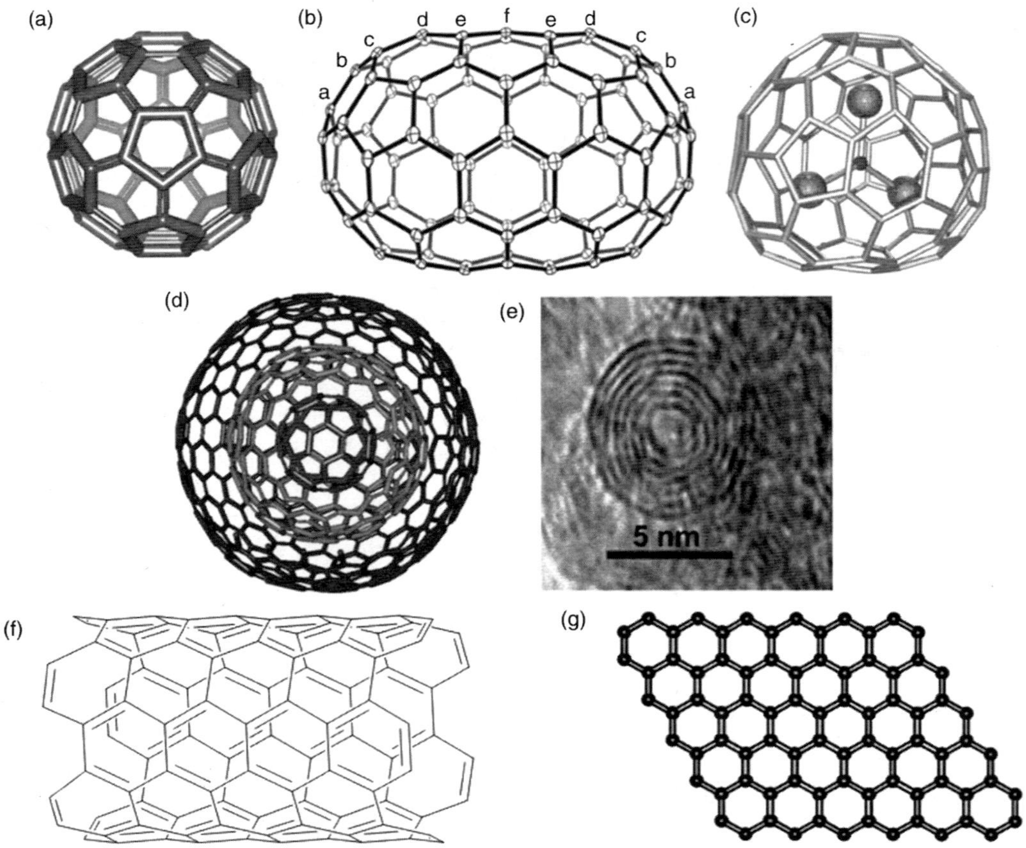

FIGURE 13.1. General examples of carbon materials: (**a**) I_h–C_{60}. (**b**) $D_{5h}(1)$–C_{90}. Reprinted with permission from reference 35. Copyright © 2010 Wiley–VCH Verlag GmbH & Co. KGaA, Weinheim. (**c**) Endohedral fullerene crystal structure of $Gd_3N@C_2(22010)C_{78}$. Reprinted with permission from reference 36. Copyright © American Chemical Society. (**d**) Three-shell fullerene (or nano-onion) I_h–$C_{60}@C_{240}@_{540}$. (**e**) High-resolution transmission electron microscopy (HRTEM) of a seven-shell carbon nano onion. Reprinted with permission from reference 37. Copyright © 2007, Wiley–VCH Verlag GmbH & Co. KGaA, Weinheim. (**f**) Single-wall carbon nanotube. (**g**) Graphene.

have been isolated [33–35] and fully characterized. The most recently crystallized higher, empty-cage series of fullerene isomers were $C_1(30)$–C_{90}, $C_1(32)$–C_{90}, and $D_{5h}(1)$–C_{90} (Figure 13.1b) reported by Balch and co-workers [34, 35]. The first two do not display the prototypical high symmetry of icosahedral C_{60}; however, $D_{5h}(1)$–C_{90} is not only highly symmetric but displays a tubular shape [34].

13.2.1 Electronic Properties and Electrochemistry of Pristine Fullerenes

C_{60} has been vastly explored due to its relative high abundance and singular isomeric form. Its electrochemical properties have been thoroughly studied under a wide variety of solvents, supporting electrolytes and temperatures [8], as these factors have a significant effect on the redox potential values. The most outstanding features of the electrochemistry and electronic properties of C_{60} are reviewed below as an introduction to discussions on the properties of C_{60} derivatives and other carbon allotropes.

C_{60} has eight potentially available oxidation states, which make it a highly attractive organic molecule for further modification and tuning of its redox chemistry. It has a closed-shell electronic configuration with a fivefold degenerate HOMO and an energetically low-lying and triply degenerate LUMO (Figure 13.2), which allows it to accept up to six electrons in solution [38–41]. Each one of the reduction steps is electrochemically reversible with a relatively constant separation between reduction steps of 450 ± 50 mV. With each addition of one electron, the C_{60} cage does not suffer any structural modification from its quasi-spherical architecture [42–45]. As confirmed by x-Ray crystallography, no Jahn–Teller distortion is observed [45], a property that is crucial for the long-lived charge separated states required for photovoltaic applications [46]. C_{60} is not easily oxidizable [47, 48]. The half-wave potential for the first oxidation in

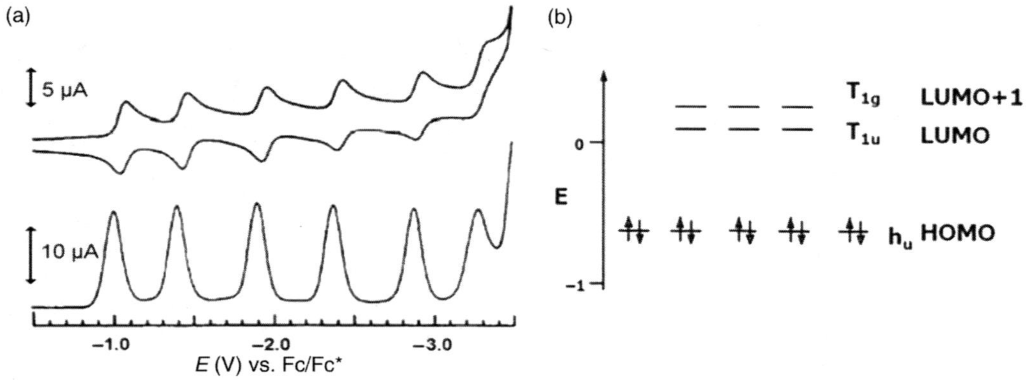

FIGURE 13.2. (**a**) Cathodic electrochemistry of C_{60}. Cyclic voltammetry (*top*) and differential pulse voltammetry (*bottom*) in toluene/acetilonitride 5:1 v/v at −10°C under vacuum. (**b**) Schematic representation of the electronic configuration of HOMO–LUMO for pristine C_{60}. Reprinted with permission from reference 41. Copyright © 1992, American Chemical Society.

1,1,2,2-tetrachloroethane shows a chemically reversible, electrochemically quasi-reversible, one-electron process at +1.26 V versus Fc/Fc^+ (see Table 13.1) [49].

The electrochemistry of the larger C_{70} cage also shows six one-electron, electrochemically reversible reduction steps. However, its electronic configuration is determined by its D_{5h} cage symmetry. It displays a doubly degenerate LUMO capable of accepting four electrons and a slightly more energetic LUMO+1 capable of accepting an additional two electrons [41, 50]. Two oxidation steps have been reported for C_{70} in 1,1,2,2-tetrachloroethane, consisting of an electrochemically quasi-reverible first oxidation at +1.20 V and a chemically irreversible second oxidation at +1.75 V (see Table 13.1) [49].

Cage size and shape play a very significant role in the electronic properties and hence in the relative stability of empty cage fullerene structures. An increment in the number of carbon atoms results in distinctive changes in symmetry and an increase in the number of possible isomers [8]. In general, from the electrochemistry standpoint, as the size of the cage increases the reduction and oxidation steps become easier (Table 13.1). This implies that the electron

TABLE 13.1. Half-Wave Reduction/Oxidation Potentials (in V versus Fc/Fc^+) of Different Size and Shape Empty Cage Fullerenes [8, 52–54][a]

Fullerene	Solvent	Temp. (°C)	E_2^{ox} (V)	E_1^{ox} (V)	E_1^{red} (V)	E_2^{red} (V)	E_3^{red} (V)	E_4^{red} (V)	E_5^{red} (V)	E_6^{red} (V)	$(E_1^{ox} - E_1^{red})$ (V)
I_h–C_{60}	PhMe/MeCN	−10			−0.98	−1.37	−1.87	−2.35	−2.85	−3.26	
	PhMe/DMF	−60			−0.82	−1.26	−1.82	−2.33	−2.89	−3.34	
	Liq. NH_3	−70			−1.04	−1.56	−2.00	−2.37	−2.43	−3.03	
	o-DCB	n/r		1.32	−1.13	−1.50	−1.94	−2.41			2.45
	TCE	r.t.		1.26	−1.06						2.32
D_{5h}–C_{70}	PhMe/MeCN	−10			−0.97	−1.34	−1.78	−2.21	−2.70	−3.70	
	o-DCB	n/r		1.21	−1.10	−1.46	−1.86	−2.27			2.31
	TCE	r.t.	1.75	1.20	1.02						2.22
D_2–C_{76}	*o*-DCB	n/r		0.73	−1.00	−1.30	−1.76	−2.15			1.73
	TCE	r.t.	1.30	0.81	−0.83	−1.12				1.64	1.64
C_{2v}–C_{78}	PhMe/MeCN	−15		0.90	−0.72	−1.08	−1.79	−2.18	−2.45	−2.73	1.62
D_3–C_{78}	DCM	r.t.		0.74	−0.64	−0.94	−1.7	−2.05			1.38
C_2(a)–C_{82}	*o*-DCB	—		0.72	−0.69	−1.04	−1.58	−1.94			1.41
C_s(14)–C_{84}	*o*-DCB	—		0.86	−0.78	−1.08	−1.6	−1.98			1.64
C_2(11)–C_{84}	*o*-DCB	—		0.85	−0.58	−0.87	−1.63	−1.97			1.43
D_2(22)–C_{84}	*o*-DCB	—		0.93	−0.79	−1.1	−1.4	−1.75			1.72
D_{2d}(23)–C_{84}	*o*-DCB	—		0.95	−0.77	−1.07	−1.36	−1.7			1.72

[a]The supporting electrolyte was $TBAPF_6$ in all cases, except for I_h-C_{60} in liquid NH_3 which required KI.

affinities of the higher fullerenes are lower in energy than those of smaller cages. The same principle applies for their ionization energies. It has also been shown that cage symmetry influences the redox potentials, which can be rationalized based on the different degenerate orbitals and electronic configurations. For a given symmetry, increasing the number of carbons causes an anodic shift of the reduction potentials due to the presence of more *p*-orbital coupling in the sp^2 spherical network and the larger charge separation–delocalization [51]. On the other hand, isomers of the same number of cages with carbons may display different electronic properties due to changes in the degeneracy of the LUMO and LUMO+1 (Table 13.1) [8, 52–54]. The variation in electrochemical potential has been successfully used as a technique to characterize different cage isomers with the same number of carbons [53, 54].

The redox chemistry of higher fullerenes is clearly as rich as that of C_{60} and C_{70}, making the exploration of their properties truly fascinating. Their low abundance and complex mixture of isomers have so far made them impractical for materials science applications, but fast progress in preparation, separation, and purification methods continues. As noted above, the relative stability of fullerene structures depends on their individual electronic configuration. For a given cage size, only those isomers with energetically stable electronic configurations are observed [29]. Many isomers presumed stable based on theoretical calculations and the isolated pentagon rule (IPR) have been found not to form due to their unfavorable electronic configuration [29]. Interestingly, very stable endohedral metallofullerenes (EMFs) have been prepared featuring cage symmetries that were never isolated before in their empty form because they are energetically unstable [8]. This phenomenon is discussed in more depth in Section 13.3. Importantly, no exceptions to the IPR have ever been observed for empty fullerene cages.

The breadth of studies on empty, pristine (nonfunctionalized) fullerenes and their electrochemical properties is vast and beyond the scope of this chapter. Supplementary reading is available to those wishing to learn more about this fascinating group of carbon materials. We recommend several reviews and book chapters that have recently appeared in the literature [5, 6, 8, 29, 55].

Advances in the understanding of the electronic structure of fullerene and other carbon nanomaterials have been achieved by combining the study of their electronically charged states, via electrochemical charging or direct contact with redox active species (chemical doping), with in situ spectroscopic measurements such as UV–vis–NIR, Raman, or electron spin resonance (ESR) in solution. Spectroelectrochemistry uses electrochemical charging, which provides several advantages over chemical doping because it controls the redox environment and allows for a wider selection of counter ions as electrolytes in solution [6, 55]. For a more fundamental understanding of this technique and its application to study carbon nanomaterials we suggest reviewing the work of Kavan and co-workers [6, 55, 56]. Here we present recent examples in which the implementation of spectroelectrochemistry provides structural and electronic information not accessible by other means.

To understand the ESR spectra of fullerenes, it is important to consider the electronic configuration of the HOMO and the LUMO. As previously discussed, C_{60} has a triply degenerate LUMO and a closed-shell configuration with a fivefold degenerate HOMO. Therefore, the pristine material is diamagnetic and ESR silent. However, the anion/cation species are paramagnetic and ESR active as extensively reported [40, 55, 57]. In 2002 Dunsch and co-workers reported the in situ spectroelectrochemistry of C_{60}, showing the correlated change in ESR–vis–NIR spectra to the first two reduction potentials of C_{60} [57]. For cations of C_{60}, the in situ ESR spectra are difficult to obtain due to the high ionization energy and the high reactivity of the positively charged species [57].

In 2007, an in situ study of $C_2(3)$–C_{82} was reported, where the first reduction and oxidation steps were followed by ESR and vis–NIR spectroscopy (Figure 13.3) [58]. It was found that the formation of the anion could be easily and clearly detected by ESR, which gave rise to a sharp signal with a line width (ΔB_{pp}) of 0.15 G and a g factor of 2.00009 without the interference of signals from reducing agents. Also, the generation of new optical bands in the vis–NIR range could be clearly assigned to the monoanion formation due to the absence of other active species in solution. This study also reported the ESR spectrum of the stable cation of $C_2(3)$–C_{82} at room temperature in non–acidic medium, conditions that were never before suitable to obtain ESR spectra of empty fullerene cationic species [58]. Further studies of the same material demonstrated a redox potential dependence of the ESR–vis–NIR spectra up to the fourth reduction step [59]. The decrease in the intensity of the ESR signals as the dianion and tetraanion formed confirmed the diamagnetic character of these two species (Figure 13.4a). The ESR spectrum obtained during reduction at the potential of the third reduction step contained two signals assigned to the trianion and the monoanion. The trianion signal disappeared upon reoxidation. Notably, the small signal for the monoanion remained present due to the synproportionation reaction of the dianion with the neutral species still present. Subtracting the monoanion signal gave rise to a clear ESR spectrum of the trianion, characterized by a sharp line ($\Delta B_{pp} = 0.17$ G) and a g factor of 2.00020, clearly distinct from the monoanion signal. (Figure 13.4b) [59].

Spectroelectrochemistry has also been used to accurately detect the presence of other paramagnetic isomer impurities in samples of higher fullerenes. In 2008 Dunsch and coworkers [60] reported the in situ ESR measurement of the first reduction step of a $D_{2d}(23)$–C_{84} sample. It was found that even when the cyclic voltammetry and ^{13}C-NMR did

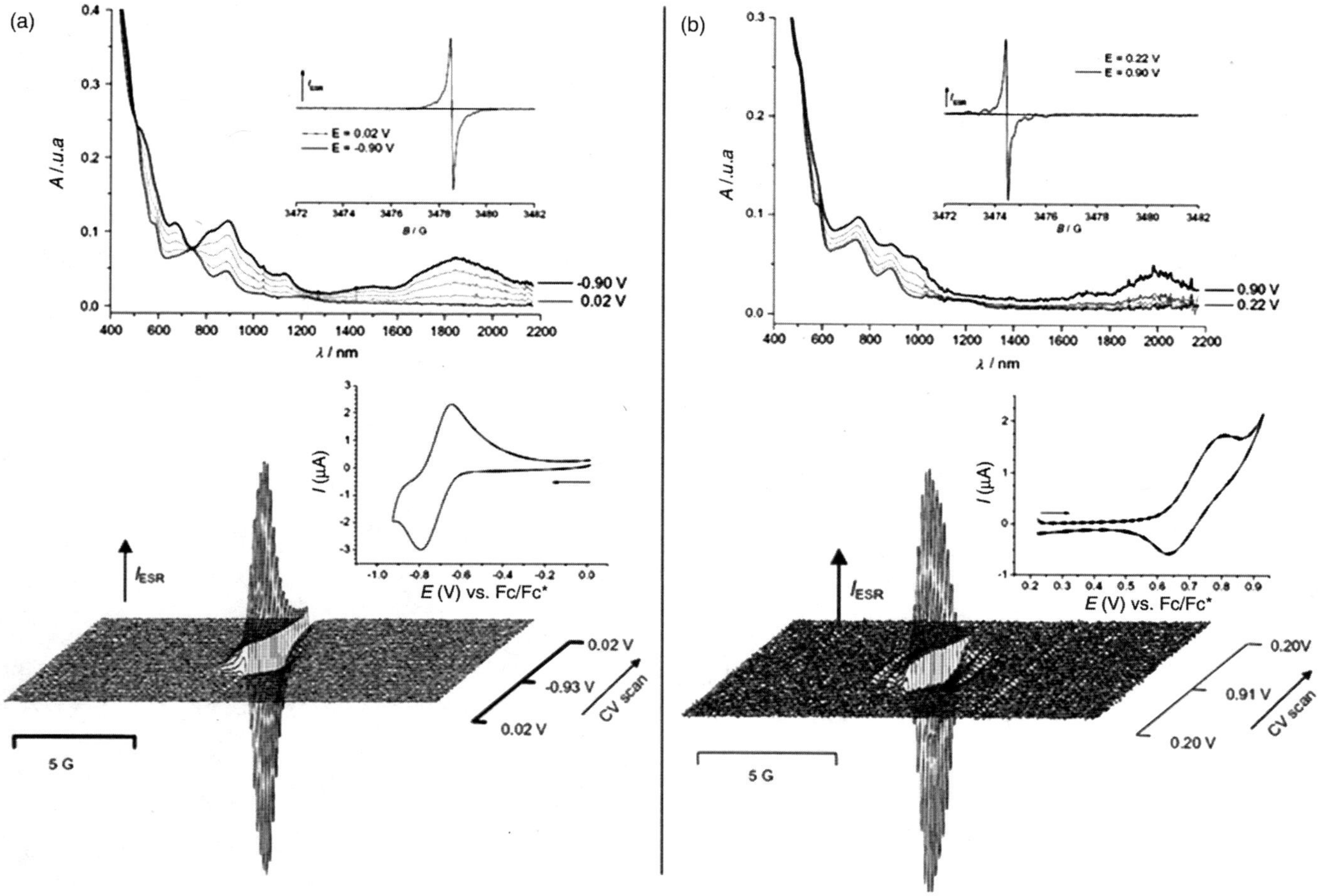

FIGURE 13.3. ESR–vis–NIR spectra observed by in situ spectroelectrochemistry of $C_2(3)$–C_{82}. (**a**) First reduction step. (**b**) First oxidation step. Reprinted with permission from reference 58. Copyright © 2007, Elsevier.

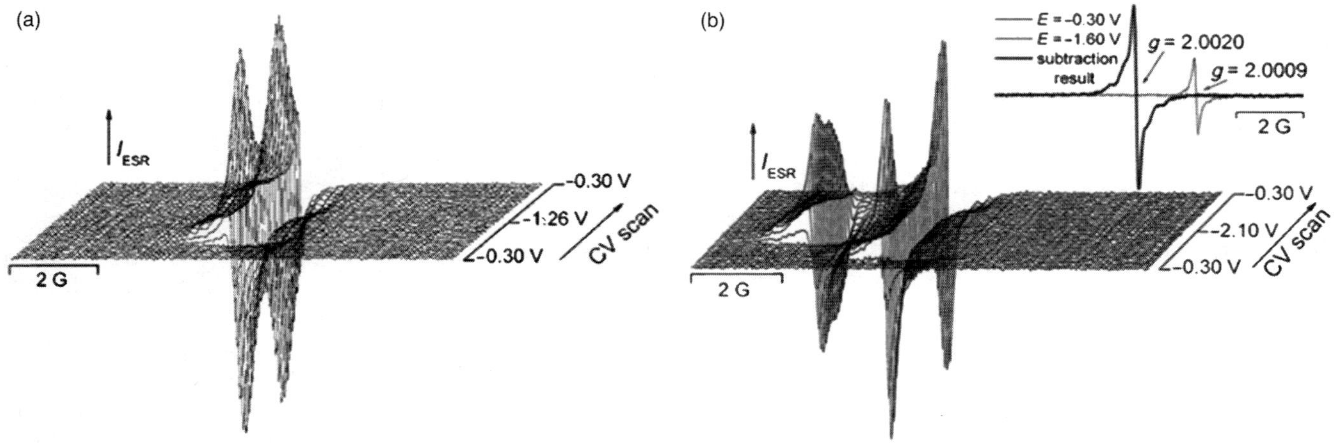

FIGURE 13.4. Potential dependence of the ESR spectrum of $C_2(3)$–C_{82} in potential region of (**a**) first and second reduction steps and (**b**) first to fourth reduction steps. Reprinted with permission from reference 59. Copyright © 2008, Wiley–VCH Verlag GmbH & Co. KGaA, Weinheim.

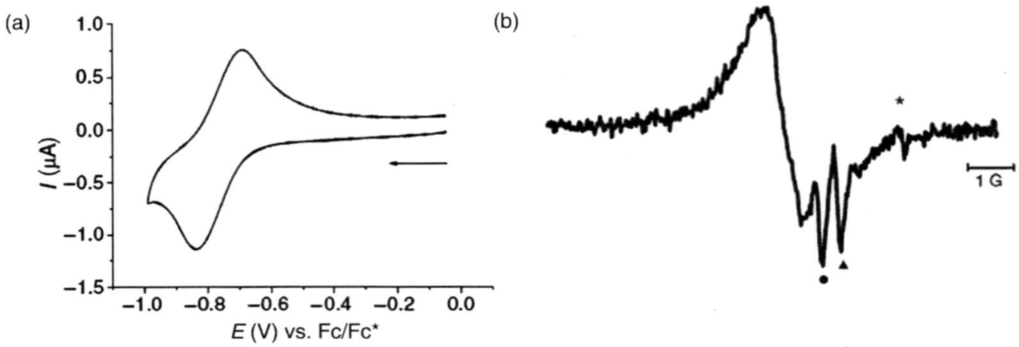

FIGURE 13.5. (**a**) Cyclic voltammogram of the first reduction step of $D_{2d}(23)$–C_{84}. (**b**) ESR spectrum observed during reduction of $D_{2d}(23)$–C_{84} at −0.95 V versus Fc/Fc^+; • corresponds to $C_2(3)$-C_{82}, ▲ corresponds to $C_s(16)$-C_{84}, and ∗ is an unidentified impurity. Reprinted with permission from reference 60. Copyright © 2008, Elsevier.

not allow a conclusive assignment of the presence of specific fullerene impurities, the in situ ESR spectrum of the first reduction clearly provided a method to detect the presence of three impurities and identify two of them as $C_2(3)$–C_{82} and $C_s(16)$–C_{84} (Figure 13.5). Comparison of in situ ESR spectra of the monoanion of C_{84} isomers provided clear evidence that as the cage symmetry decreases, the line width also decreases [54].

13.3 ENDOHEDRAL FULLERENES

Endohedral fullerenes represent the next frontier for exploration within the fullerene family. They consist of the guest, typically composed of single atoms, small molecules, or inorganic clusters, surrounded by a carbon cage (Figure 13.1c). A general description of this encapsulation follows the formula $X@C_{2n}$, where X describes the composition of the inner core and C_{2n} represents the even number of carbons that constitute the cage. Combinations of guest and cage have given rise to extensive theoretical and experimental research on the electronic interactions between them and the effect that these interactions have on cage reactivity, on electrochemical behavior, and, consequently, on the potential applications of these materials [6, 36, 61, 62]. The synthesis of endohedral fullerenes involves several different methods, depending on the desired core to be encapsulated [5]. One of the unanswered questions that is of fundamental importance is, What drives endohedral fullerene formation? Although it appears as if there is some degree of randomness, increasing evidence is pointing to preferential formation of endohedral structures based on electronic exchange between guest and cage. For example, endohedral fullerenes with encapsulated inorganic clusters, as metallic, bimetallic, trimetallic nitrides, and metal sulfides [63–65], among others [5, 66–72], feature a nonrandom electron transfer from the cluster to the cage that gives rise to unique stabilization of the whole structure.

13.3.1 Electronic Properties and Electrochemistry of Endohedral Fullerenes

When these materials were first prepared, the appearance of energetically unfavorable cages when metals were encapsulated suggested the existence of an interaction between metal/cluster and the cage that conferred electronic stability to these materials. In fact, in the case of trimetallic nitride EMFs ($M_3N@C_{2n}$), it has been found that formation involves the donation of six electrons from the HOMO of the cluster (M_3N) to the LUMO of the cage: $M_3N^{6+}@C_{2n}{}^{6-}$ [73–75]. This electron transfer changes the overall electron count of the cage and consequently its electronic configuration and reactivity. The electrochemical behavior of these materials perfectly reflects this phenomenon. Since the HOMO is located on the encapsulated inorganic moiety, oxidation occurs at a lower potential than that of the empty cage analogue. To illustrate this, note that the first oxidation of the empty cage $C_2(a)$–C_{82} is reported to occur at +0.72 V (Table 13.1) whereas $La@C_{82}$, $Pr@C_{82}$, $Ce@C_{82}$, $Ga@C_{82}$, and $Y@C_{82}$ show the first oxidation step in the range of +0.10 to +0.07 V and the second oxidation step at +1.08 to +1.07 V (Table 13.2). Although a direct comparison cannot give an exact assessment because of differences in cage symmetry, the fact remains that encapsulation of these metals inside C_{82} decreases the oxidation potential by 600 mV or more relative to the closest empty cage analogue. For this reason, the use of EMFs in optoelectronics as moderate electron donor materials is now envisioned.

The electron transfer from metal to cage upon formation of EMF's also decreases the electron affinity of the material, causing the reduction steps to occur at more negative potentials than empty cages of similar size (Table 13.2) [8, 9, 76]. Overall, the change in electronic structure of EMFs, as observed through their redox chemistry, gives rise to a narrowing of the HOMO–LUMO gaps, an important property for their incorporation in a variety of electronic-based applications [6, 73]. As a result of all these findings, a new avenue

TABLE 13.2. Redox Potential (in V versus Fc/Fc^+) of EMF Materials [8, 61, 65, 73][a]

Fullerene	Solvent	Temp. (°C)	E_2^{ox} (V)	E_1^{ox} (V)	E_1^{red} (V)	E_2^{red} (V)	E_3^{red} (V)	E_4^{red} (V)	E_5^{red} (V)	E_6^{red} (V)	$(E_1^{ox} - E_1^{red})$ (V)
$La@(C_{2v})–C_{82}$	*o*-DCB	r.t.	1.07	0.07	−0.42	−1.37	−1.53	−2.26	−2.46		0.49
$Pr@(C_{2v})–C_{82}$	*o*-DCB	r.t.	1.08	0.07	−0.39	−1.35	−1.46	−2.21			0.46
$Ce@(C_{2v})–C_{82}$	*o*-DCB	r.t.	1.08	0.08	−0.41	−1.41	−1.53	−1.79	−2.25	−2.5	0.49
$Gd@(C_{2v})–C_{82}$	*o*-DCB	r.t.	1.08	0.09	−0.39	−1.38	−2.22				0.48
$Y@(C_{2v})–C_{82}$	*o*-DCB	r.t.	1.07	0.10	−0.37	−1.34	−2.22	−2.47			0.47
$La_2@(D_2)–C_{72}$	*o*-DCB	r.t.	0.75	0.24	−0.68	−1.92					0.92
$Ce_2@(D_2)–C_{72}$	*o*-DCB	r.t.	0.82	0.18	−0.81	−1.86					0.99
$La_3N@C_{88}$	*o*-DCB	r.t.		0.21	−1.36	−1.67					1.57
$Ce_3N@C_{88}$	*o*-DCB	r.t.		0.08	−1.30	−1.57					1.38
$Sc_3N@(I_h)–C_{80}$	*o*-DCB	r.t.		0.59	−1.29	−1.56					1.88
$Sc_3N@(D_{5h})–C_{80}$	*o*-DCB	r.t.		0.34	−1.33						1.67
$Lu_3N@(I_h)–C_{80}$	*o*-DCB	r.t.		0.64	−1.40						2.04
$Lu_3N@(D_{5h})–C_{80}$	*o*-DCB	r.t.		0.45	−1.41						1.86
$Y_3N@(I_h)–C_{80}$	*o*-DCB	r.t.		0.64	−1.44	−1.83	−2.38				2.08
$Sc_2C_2@C_{82}$	Pyridine	r.t.		0.16	−0.95	−1.38					1.11
$Sc_3C_2@(I_h)C_{80}$	*o*-DCB	r.t.		−0.06	−0.50	−1.64	−1.82				0.44
$Sc_2(\mu_2\text{-}S)@C_s(6)C_{82}$	*o*-DCB	r.t.	0.65	0.39	−0.98	−1.12	−1.73				1.23
$Sc_2(\mu_2\text{-}O)@C_s(6)C_{82}$	*o*-DCB	r.t.	0.72	0.35	−0.96	−1.28	−1.74				1.18

[a]The supporting electrolyte was $TBAPF_6$ in all cases, except for $Sc_2C_2@C_{82}$ and $Sc_3C_2@(I_h)C_{80}$ in TBAP.

of research has opened up. In recent years, totally unpredicted cage symmetries are being prepared with different clusters encapsulated inside [5, 6, 8]. Most interestingly, combinations of unexpected inorganic cores and unusual carbon cages have been prepared, whose individual components cannot be independently isolated. A remarkable case of this phenomenon was reported by Beavers et al. [36] with the crystal structure of $Gd_3N@C_2(22010)C_{78}$ (Figure 13.1c). This material consists of a C_2-symmetric carbon cage that was never isolated before and that was completely unexpected due to the presence of two pairs of fused pentagons on the carbon structure, a double violation of the IPR [24, 29, 77, 78]. The isolation and identification of this material has not only challenged previous assumptions because it suggests that IPR may not generally apply to endohedral structures, but it has opened new avenues of exploration within the fullerene field.

As with pristine fullerenes, ESR spectroelectrochemistry has brought new light into the understanding of the electronic structure of EMFs. In 2007 Dunsch and co-workers [79, 80] reported the ESR spectra of the radical cation and anion of another non-IPR EMF, $Sc_3N@D_3(6140)–C_{68}$, generated by electrochemical oxidation and reduction, respectively (Figure 13.6). Backed by density functional theory (DFT) calculations, the ESR spectra of these two species revealed that, although there is some spin delocalization between the cluster and the cage, the contribution of the internal Sc_3N to the spin density is only minor and that most of the unpaired spin is delocalized on the fullerene cage. The ESR patterns of both cation and anion radicals show the 22 lines and intensity distribution that are characteristic of three equivalent Sc nuclei. Corresponding g factors of 2.0010 and 2.0023 were obtained. The strong participation of the carbon cage was revealed by the hyperfine splitting values of 1.28 G and 1.75 G, respectively for the cation and anion, when compared to the corresponding g value for the radical anion of $Sc_3N@I_h–C_{80}$ previously determined [80]. Indeed, the ESR spectrum of the latter also displays the 22 characteristic lines of three equivalent Sc nuclei, but its hyperfine splitting of 55.6 G and g factor of 1.9984 indicates that, as opposed to the C_{68} analogue, its spin density is localized almost exclusively in the cluster. This assertion was confirmed by Popov et al. [81] using DFT calculations.

In 2010 Vargova et al. [82] also reported the in situ ESR spectroelectrochemistry of $Y@C_{82}$. In this study the attenuation of the ESR intensity of the pristine paramagnetic material during the electrochemical scanning in the potential range of the first and second cathodic steps confirmed, in a precise and controlled way, the formation of the anion ($Y@C_{82}{}^{1-}$) and trianion ($Y@C_{82}{}^{3-}$) species. This was explained on the basis of the expected diamagnetic character (ESR silent) of these reduced states. It was also found that between the first and second reduction steps, from mono- to trianion states, there is no detectable paramagnetic intermediate species such as dianion formed by the symproportionation reaction of $Y@C_{82}{}^{1-}$ and $Y@C_{82}{}^{3-}$. This could only be detected by in situ ESR in the timescale of the cyclic voltammetry measurement [82].

13.3.2 Functionalized C_{60} and Endohedral Fullerenes

Chemical functionalization of pristine fullerenes has produced a great library of materials with carefully designed

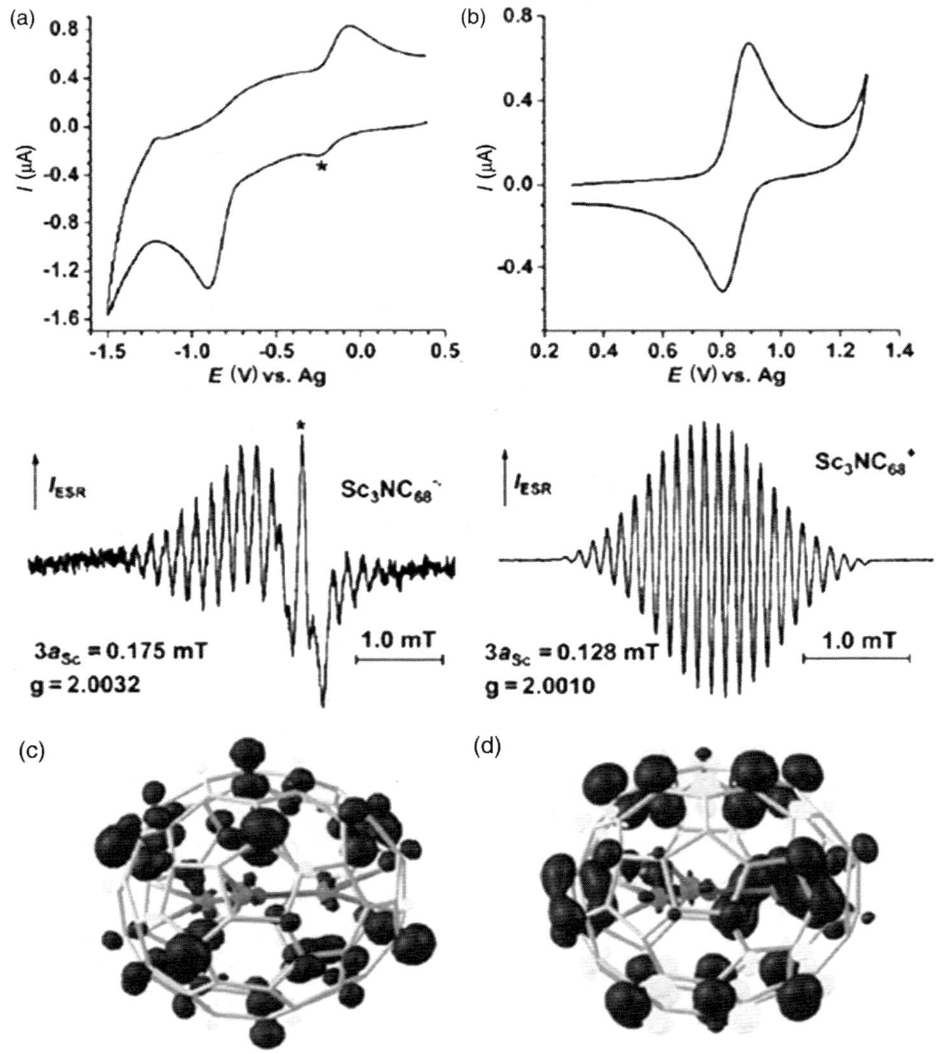

FIGURE 13.6. (**a**) In situ cyclic voltammograms and representative ESR spectra of (**a**) $[Sc_3N@C_{68}]^{\bullet -}$ and (**b**) $[Sc_3N@C_{68}]^{\bullet +}$ radical ions observed in the course of first reduction and first oxidation peaks, respectively (asterisk denotes a narrow impurity signal at $g = 2.0005$). Spin density distribution in (**c**) $[Sc_3N@C_{68}]^{\bullet -}$ and (**d**)) $[Sc_3N@C_{68}]^{\bullet +}$. Reprinted with permission from reference 80. Copyright © 2008, American Chemical Society.

physicochemical properties such as solubility, electron transport ability, thermal stability and photostability, redox chemistry, and electronic configuration [83–87]. New electrochemical interactions can be accomplished by coupling the electronic properties of the fullerene and addend (in single and multiple additions) [88]. The change in redox chemistry can be attributed to the overall change in the electronic configuration of the cage, resulting from changes in the sp^2 to sp^3 hybridization of the bonded carbons and the loss of symmetry caused by derivatization [83–87].

Similarly, functionalizing the cage of endohedral fullerenes adds a new dimension to the control of electronic and other properties for their many potential applications. Since 2002, derivatization of endohedral fullerenes has been focused on chemical processes, specifically on clycloaddition reactions [89–110].

The next few sections are dedicated to evidence of the electrochemical tuning that can be achieved via functionalization through selected examples that have been targeted for optoelectronic applications.

13.4 ELECTROSYNTHESIS OF C_{60} AND C_{70} FULLERENE DERIVATIVES

Anionic C_{60} is an electron donor and behaves as a nucleophile [7, 111]. Protocols that use fullerene anions as a

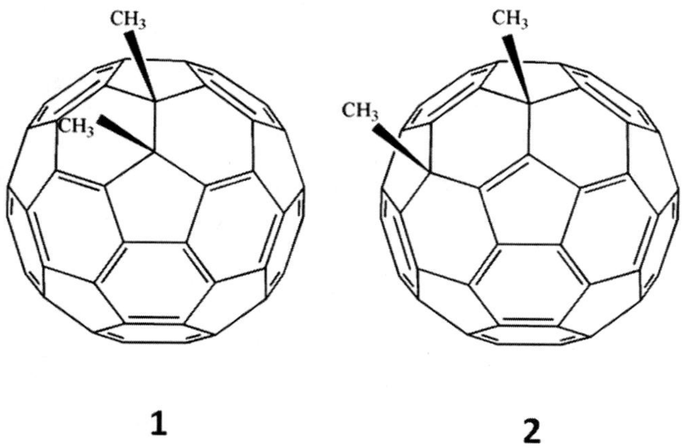

FIGURE 13.7. Structures of the first two electrosynthesized derivatives of C_{60}.

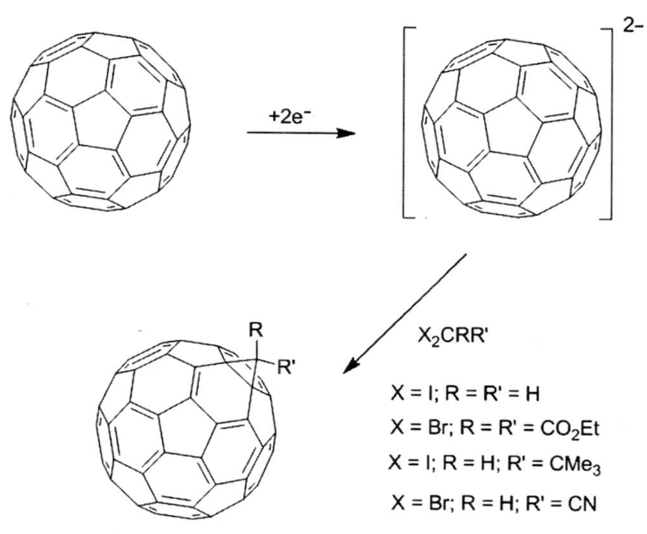

FIGURE 13.8. Reactions of C_{60}^{2-} with a variety of organic halides.

reactive building block to prepare fullerene derivatives which are difficult to prepare in other ways have been well established [112–124]. The synthetic chemistry of anionic C_{60} has been shown to be an important complement to that of neutral C_{60}, since fullerene anions can be easily generated by either chemical [114, 115, 125] or electrochemical reductive methods [112, 113, 116–118, 124, 126, 127]. For chemical reduction of fullerenes with alkali metals, control over the electrons transferred is difficult and consequently there is little control over the number of addends. Electrochemical reduction of fullerenes, on the other hand, is more selective because the formation of the desired anions—for example, C_{60}^{-}, C_{60}^{2-} and C_{60}^{3-}—can be readily achieved by controlling the potential, E, and the charge transferred, q, to the fullerene cages. Electrochemical reduction is thus easier than chemical reduction for functionalization of fullerenes when fullerene anions are used as building blocks.

The first functionalization of a fullerene using anions generated under controlled potential bulk electrolysis (CPE) conditions was reported by Kadish and co-workers [112] in 1993. Compounds **1** and **2**, 1,2- and 1,4-$(CH_3)_2C_{60}$ (Figure 13.7), were selectively synthesized via the formation of C_{60}^{2-} by controlled potential electrolysis (CPE) and addition of excess methyl iodide. This novel method for preparing derivatized fullerenes provided an alternative tool to synthesize fullerene derivatives.

Subsequently, the selective synthesis of [6,6] methanofullerenes using the reactions of C_{60}^{2-} generated by electrolysis with a variety of organic halides (Figure 13.8) was explored by Echegoyen and co-workers [113] in 1996. The same procedure reported by Kadish and co-workers was used, but a different solvent (MeCN-toluene) and supporting electrolyte ($TBAPF_6$) were used to dissolve the C_{60} sample. A potential of −0.7 V versus $Ag/AgNO_3$ was applied to generate the dark-red C_{60}^{2-} dianion solution, followed by the addition of excess electrophile. The reaction was kept running without any applied potential until the rest potential did not change, indicating that the reaction was complete. The products were characterized by NMR and UV–vis. In that study, monoanionic C_{60} was also prepared by CPE setting the potential at −0.22 V versus $Ag/AgNO_3$. The addition of organic halide resulted in no nucleophilic reaction of C_{60}^{-}, indicating that the monoanion is not reactive toward the electrophile and proving that the C_{60} monoanion is not as reactive as the dianion.

The reaction of C_{60} anions with alkyl and arylhalides was detected by cyclic voltammetry in 1996 by Dunsch and co-workers [126]. New waves besides those of C_{60} were observed as a result of new products formed by the reactions. The study of C_{60} in the presence of iodomethane was conducted first. When the direction of the potential scan was reversed after the second reduction potential of C_{60}, an additional reduction wave at −0.7 V versus Ag/AgCl appeared, suggesting that C_{60}^{2-} had reacted with iodomethane. The additional wave did not appear when the applied potential was reversed at a value lower than the second reduction potential of C_{60}, indicating that the reaction occurs only when the dianionic fullerene is present. This result confirmed the conclusion reached by the previous report from Echegoyen and co-workers [113] described above. In that same study, Dunsch and co-workers also detected the formation of fullerene adducts in the presence of iodobenzene and 1-iodonaphthalene. However, these reactions occurred only when the applied potential was high enough to form C_{60}^{3-} rather than C_{60}^{2-}. As opposed to iodomethane, iodobenzene and 1-iodonaphthalene form adducts only with trianionic C_{60}. Interestingly, aryl dihalides such as 1,2-diiodobenzene were found to react, however, with C_{60}^{2-} and not with C_{60}^{3-}.

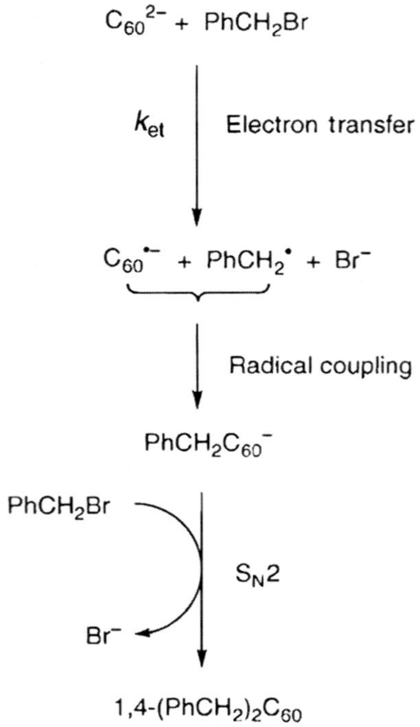

FIGURE 13.9. Reaction mechanism of C_{60}^{2-}. Reprinted from reference 114 with permission. Copyright © 1996, American Chemical Society.

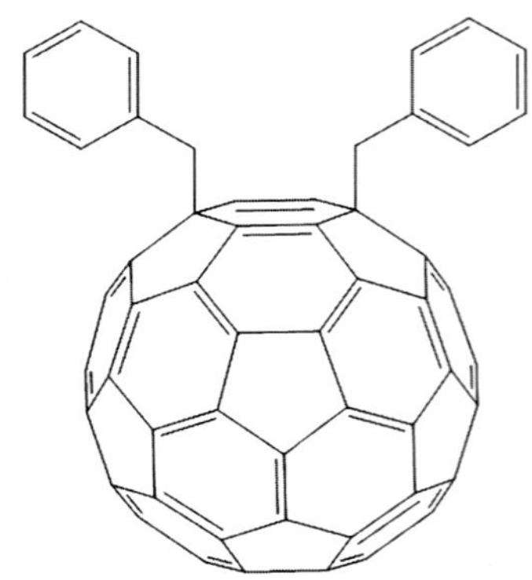

3

FIGURE 13.10. Molecular structure of 1,4-$(PhCH_2)_2C_{60}$ (**3**).

The reaction mechanism between chemically and electrochemically generated C_{60} dianions remained unclear until Kadish and Fukuzumi reported the reaction between C_{60}^{2-} and alkyl bromides ($PhCH_2Br$). Kinetic studies and theoretical calculations for these reactions provided valuable information about the mechanism of nucleophilic reactions of fullerene dianions (Figure 13.9) [114]. In this proposed mechanism, the initial electron transfer from C_{60}^{2-} to benzyl bromide ($PhCH_2Br$) generates the $C_{60}^{\bullet-}/PhCH_2^{\bullet}$ radical pair, which is produced by cleavage of the C–Br bond as a result of the electron transfer reduction of $PhCH_2Br$. A radical coupling occurs rapidly between the radicals of $C_{60}^{\bullet-}$ and $PhCH_2^{\bullet}$, giving rise to the monoadduct anion $PhCH_2C_{60}^{-}$, which in turn undergoes the S_N2 reaction with another $PhCH_2Br$ molecule to yield the bisadduct of 1,4-$(PhCH_2)_2C_{60}$ (**3**) (see Figure 13.10).

The establishment of the reaction mechanism for nucleophilic reactions of fullerene dianions led to further studies of the reactions of fullerene anions with various electrophilies. In 1998, Kadish et al. [116] synthesized di- and tetraaddition derivatives of C_{60} following a similar procedure. C_{60}^{2-} was generated by exhaustive electrolysis under a nitrogen atmosphere. The potentiostat was then switched off, and an excess amount of RBr (R = $C_6H_5CH_2$, 2-$BrC_6H_4CH_2$, 3-$BrC_6H_4CH_2$ and 4-$BrC_6H_4CH_2$) was added to the resulting C_{60}^{2-} anion solution. The yield of each C_{60} derivative was about 50%, and 5% of unreacted C_{60} was recovered. The method used for the synthesis of bisadducts was also applied to produce the tetraadduct. $(C_6H_5CH_2)_2C_{60}$ (**3**) was used as the starting material instead of C_{60}. The dianion of $(C_6H_5CH_2)_2C_{60}$ was generated by bulk electrolysis and excess benzyl bromide ($C_6H_5CH_2Br$) was added. The product, $(C_6H_5CH_2)_4C_{60}$, was isolated by HPLC and characterized with UV–vis and ^{13}C NMR spectrum. No crystal structures were reported for these compounds. The mechanism proposed for the synthesis of $(C_6H_5CH_2)_2C_{60}$ from C_{60}^{2-} by Kadish and Fukuzumi also fits the reaction of $[(C_6H_5CH_2)_2C_{60}]^{2-}$ with $C_6H_5CH_2Br$. The 1998 study also reported the redox potentials of all bisadducts and tetraadducts, which are summarized in Table 13.3. The bisadduct $(C_6H_5CH_2)_2C_{60}$ displays three reversible one-electron reductions at $E_{1/2} = -0.051$, -0.99, and -1.50 V versus SCE. The addition of two benzyl groups on the C_{60} cage causes the reduction potentials to shift to more negative values than those of pristine C_{60}. The reduction potentials of the three Br-substituted benzyl C_{60} derivatives are also listed in Table 13.3. Given that Br is electron-withdrawing, the compounds containing Br were assumed to exhibit less negative potentials than those without Br. However, as shown in Table 13.3, the presence of Br on derivatives **4**, **5**, and **6** (see Figure 13.11) did not significantly change the reduction potentials with respect to those of $(C_6H_5CH_2)_2C_{60}$. Derivatives of C_{60} can also accept multiple electrons, but they are generally less electron-deficient than C_{60} due to cleavage of the π-electron conjugation of C_{60}, as evidenced by the cathodic shift of the reduction potentials. The tetraadduct **7** also displays reversible reductive electrochemical behavior. The half-wave potential for each reduction is -0.68, -1.11, and -1.88 V versus SCE, respectively, in 0.1 M PhCN-TBAP. The reduction potentials of $(C_6H_5CH_2)_4C_{60}$ are cathodically shifted by 170, 120, and 380 mV, respectively, compared with those observed for the

TABLE 13.3. Reduction Potentials of C_{60} and C_{70} and Their Derivatives (in V versus SCE)[a]

Compounds	Solvent	Temp. (°C)	E_1^{ox} (V)	E_1^{red} (V)	E_2^{red} (V)	E_3^{red} (V)	E_4^{red} (V)
C_{60}	PhCN	r.t.		−0.42	−0.84	−1.33	
$(C_6H_5CH_2)_2C_{60}$	PhCN	r.t.		−0.51	−0.99	−1.50	
$(2\text{-}BrC_6H_4CH_2)C_{60}$,	PhCN	r.t.		−0.51	−0.96	−1.48	
$(3\text{-}BrC_6H_4CH_2)C_{60}$	PhCN	r.t.		−0.47	−0.99	−1.48	
$(4\text{-}BrC_6H_4CH_2)C_{60}$	PhCN	r.t.		−0.48	−1.00	−1.48	
$(C_6H_5CH_2)_4C_{60}$	PhCN	r.t.		−0.68	−1.11	−1.88	
1,4,10,24-$(C_6H_5CH_2)_4C_{60}$	PhCN	r.t.		−0.69	−1.12	−1.88	
1,2,4,15-$(C_6H_5CH_2)_4C_{60}$	PhCN	r.t.		−0.74	−1.15	−1.85	
1,2-$(PhCH_2)_2C_{60}$	PhCN	r.t.		−0.62	−1.04	−1.57	
1,2-$H(PhCH_2)C_{60}$	PhCN	r.t.		−0.54	−0.94	−1.50	
$PhCH_2C_{60}–C_{60}PhCH_2$	PhCN	r.t.	−0.06	−0.50	−0.94	−1.36	
[6,6] cyclic phenylimidate	DMF	r.t.		−0.30			
1,2-benzal-3-*N*-4-*O*-cyclic phenylimidate C_{60}	DMF	r.t.		−0.38			
1,4-dibenzyl-2,3-cyclic phenylimidate C_{60}	DMF	r.t.		−0.43	−0.88		−1.86
C_{70}	PhCN	r.t.		−0.42	−0.83	−1.27	−1.70
$(PhCH_2)_2C_{70}$ major component	PhCN	r.t.		−0.37	−0.84	−1.37	−1.84
$(PhCH_2)_2C_{70}$ minor component	PhCN	r.t.		−0.52	−0.94	−1.37	−1.84

[a]The supporting electrolyte was TBAP.

bisadduct $(C_6H_5CH_2)_2C_{60}$. Clearly, the number of addends on C_{60} can have a significant effect on the electrochemical properties of these organofullerenes. Note that there are three possible isomers for the C_{60} bisadduct: 1,2-[6,6] closed, 1,2-[5,6] open, and 1,4-addition. As for the C_{60} tetraadduct, numerous isomers are possible. However, no discussion on isomer identity or separation was provided in this report.

In 2000, Kadish et al. [118] synthesized two isomers of $(C_6H_5CH_2)_4C_{60}$ by CPE, using the same procedure described above. The structures of both isomers were characterized unambiguously by x-ray single-crystal diffraction. One was found to be a 1,4,10,24-$(C_6H_5CH_2)_4C_{60}$ (isomer **7a**), which is the same product reported in 1998 [116], and the other was the 1,2,4,15-$(C_6H_5CH_2)_4C_{60}$ (isomer **7b**) as shown in Figure 13.12. The electrochemical data for both isomers are listed in Table 13.3. Cyclic voltammograms in 0.1M PhCN-TBAP show that both isomers undergo three reversible one-electron reductions within the potential range of the solvent. Despite having the same number of addends, the $E_{1/2}$ for each reduction is different for each isomer due to the different positions of the benzyl group on the C_{60} cage. For example, the first reduction potential of isomer **7a** is less negative than the one of isomer **7b**, indicating that the first reduction of **7a** is easier than that for **7b**. However, not all reductions follow the same trend, the third reduction of **7a** is more difficult than that of **7b**. Therefore, not only the number but also the addition position of the addends affects the electrochemical behavior of C_{60} derivatives, which is in contrast

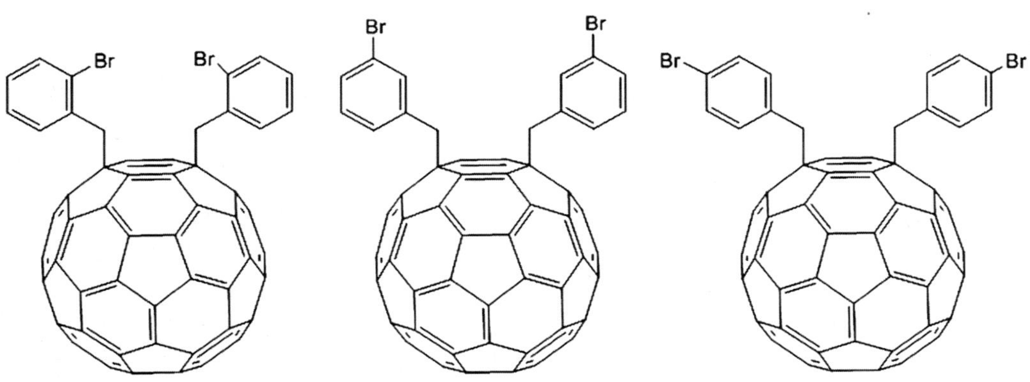

4 5 6

FIGURE 13.11. Molecular structures of $(2\text{-}BrC_6H_4CH_2)C_{60}$ (**4**), $(3\text{-}BrC_6H_4CH_2)C_{60}$ (**5**), and $(4\text{-}BrC_6H_4CH_2)C_{60}$ (**6**).

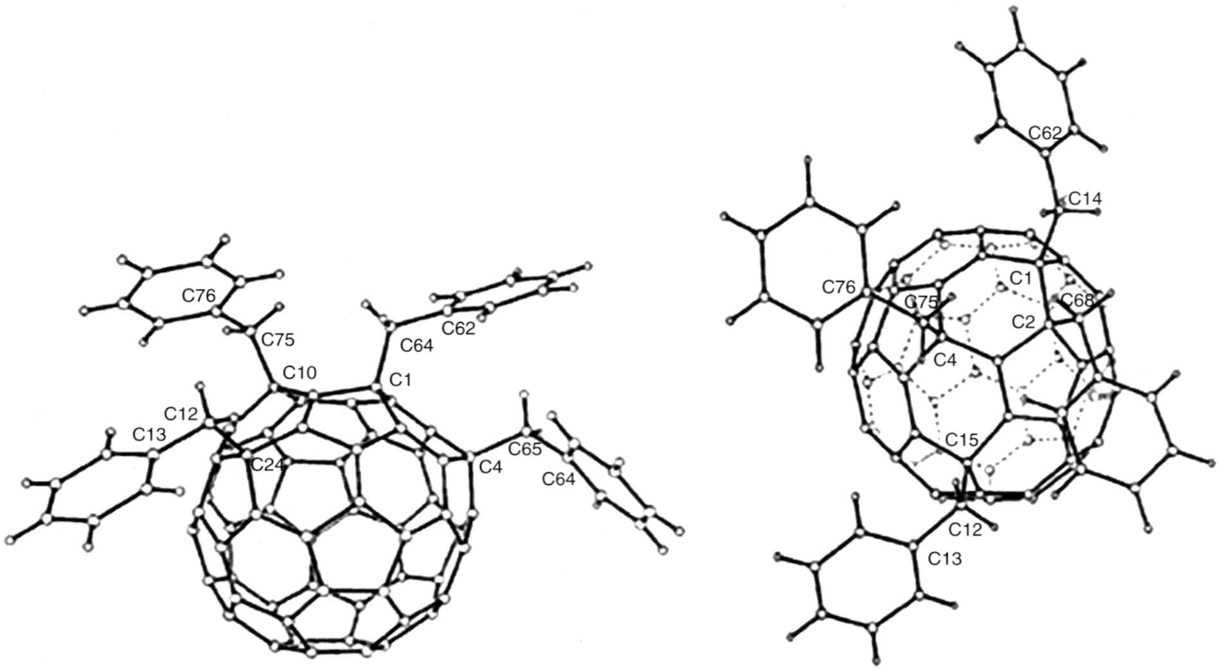

7a 7b

FIGURE 13.12. Crystal structures of 1,4,10,24-$(C_6H_5CH_2)_4C_{60}$ (isomer **7a**) and 1,2,4,15-$(C_6H_5CH_2)_4C_{60}$ (isomer **7b**). Reprinted from reference 118 with permission. Copyright © 2000, American Chemical Society.

with the results obtained for bisadducts of methanofullerenes, for which their $E_{1/2}$ values are independent of the position of the addends [128].

In 2007 a reinvestigation of the reaction between $C_{60}{}^{2-}$ and benzyl bromide in benzonitrile containing 0.1 M tetra-*n*-butylammonium perchlorate (TBAP) showed that there are more reaction products than previously reported [124]. C_{60} was subjected to CPE at −1.10 V versus SCE in freshly distilled PhCN containing 0.1 M TBAP under a nitrogen atmosphere. Excess benzyl bromide was added to the resulting anion solution after the potentiostat was switched off. The reaction was allowed to proceed for 40 min with stirring. The crude product was purified by HPLC using a semipreparative silica gel column using a mixture of cyclohexane/toluene (70:30 v/v) as the eluent. The four products were identified as methanofullerene $C_{61}HPh$ (**8**), 1,2-dihydrofullerene $HPhCH_2C_{60}$ (**9**), 1,4-$(PhCH_2)_2C_{60}$ (**3**), and 1,2-$(PhCH_2)_2C_{60}$ (**10**), shown in Figure 13.13. The previously reported 1,2-$(PhCH_2)_2C_{60}$ was actually identified as the [6,6]-methanofullerene, $C_{61}HPh$, as determined by x-ray single-crystal diffraction. A new product assigned as 1,2-$(PhCH_2)_2C_{60}$ was obtained for the first time. It exhibits three well-defined reversible one-electron reductions with $E_{1/2}$ values of −0.62, −1.04, and −1.57 V versus SCE. The reduction potentials of 1,2-$(PhCH_2)_2C_{60}$ are cathodically shifted by 110, 50, and 70 mV, respectively, with respect to the reduction potentials of 1,4-$(PhCH_2)_2C_{60}$. This was the first time that such a large difference in potentials had been observed between the 1,2- and 1,4-isomers of a C_{60} bisadduct. The result suggests that the 1,2-adduct is less electron-deficient than the 1,4-adduct when there is no substituent effect.

Most reactions of electrochemically generated dianionic C_{60} with organic halides have been performed in benzonitrile solution using the same procedure described above. The mechanism proposed by Fukuzumi and Kadish seems to apply for these reactions as well. However, studies on the

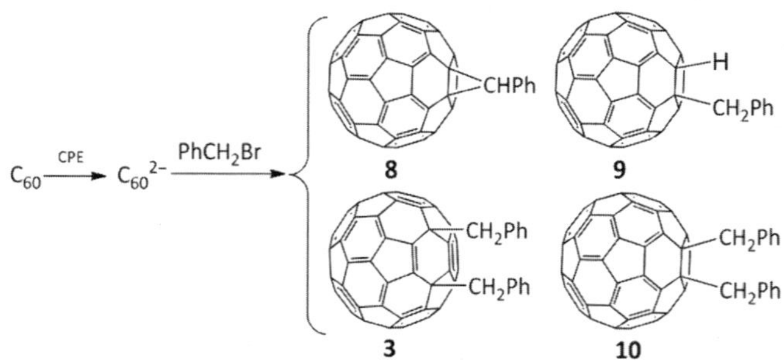

FIGURE 13.13. Reaction of C_{60} dianion with benzyl bromide.

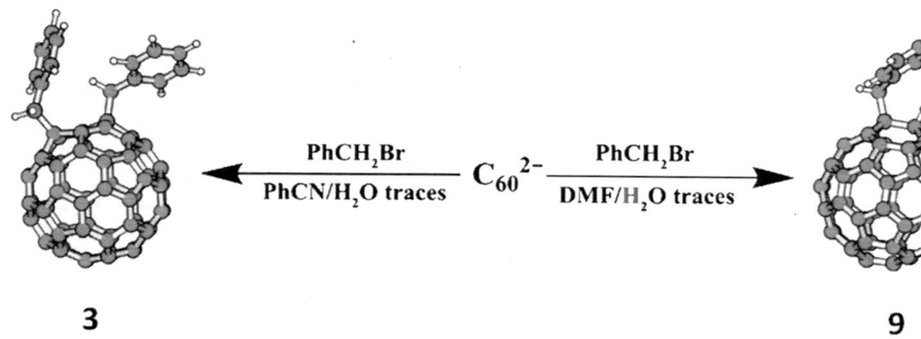

FIGURE 13.14. Reactions of $C_{60}{}^{2-}$ with benzyl bromide in PhCN and DMF. Reprinted from reference 129 with permission. Copyright © 2010, American Chemical Society.

reaction of electrochemically generated $C_{60}{}^{2-}$ in solvents other than benzonitrile are limited. Recently, Gao and coworkers [129] reported that the reaction of $C_{60}{}^{2-}$ with the same organic halides described above but carried out in N,N-dimethylformamide (DMF) can give rise to different products. 1,2-Dihydro-[60]fullerenes (1,2-HRC_{60}) were obtained as the major products instead of the 1,4-R_2C_{60} adducts obtained in benzonitrile (see Figure 13.14) [129, 130]. The procedure in DMF was similar to that in benzonitrile. Interestingly, three fractions were isolated by HLPC with 1,2-$H(PhCH_2)C_{60}$ as the major product, indicating the existence of a solvent effect on the reactivity of $C_{60}{}^{2-}$. The structure of 1,2-$H(PhCH_2)C_{60}$ was solved by x-ray single-crystal diffraction. The formation of 1,2-dihydro-[60]fullerenes in DMF is quite unusual according to the previously proposed mechanism. Therefore the origin of the hydrogen atom in 1,2-$H(PhCH_2)C_{60}$ was investigated using $PhCD_2Br$ as the reagent and DMF-d_7 as the solvent. The results indicated that neither $PhCH_2Br$ nor DMF are the source of the hydrogen atom. Further investigation indicated that the DMF solvent effect on the reactivity of $RC_{60}{}^{-}$ could be explained in terms of traces of water present in the solvent. H_2O in the DMF solvation shell facilitates the formation of hydrogen bonding with $PhCH_2C_{60}{}^{-}$ in a manner similar to that of other carbanions [130], with the subsequent proton transfer from water to the C_{60} sphere. A decrease in the nucleophilicity of $PhCH_2C_{60}{}^{-}$ results in the formation of 1,2-$H(PhCH_2)C_{60}$ as a major product. Water is not miscible with benzonitrile, so the proton transfer from H_2O to $C_{60}{}^{2-}$ is not efficient, which accounts for the formation of 1,4-$(PhCH_2)C_{60}$ as the major product when using benzonitrile as solvent.

The electrochemical properties of 1,2-$H(PhCH_2)C_{60}$ (**9**) were studied using cyclic voltammetry in 0.1 M PhCN-TBAP. The voltammogram shows three reversible one-electron redox processes with $E_{1/2}$ at −0.54, −0.94, and −1.50 V versus SCE (Figure 13.15a). The anionic species

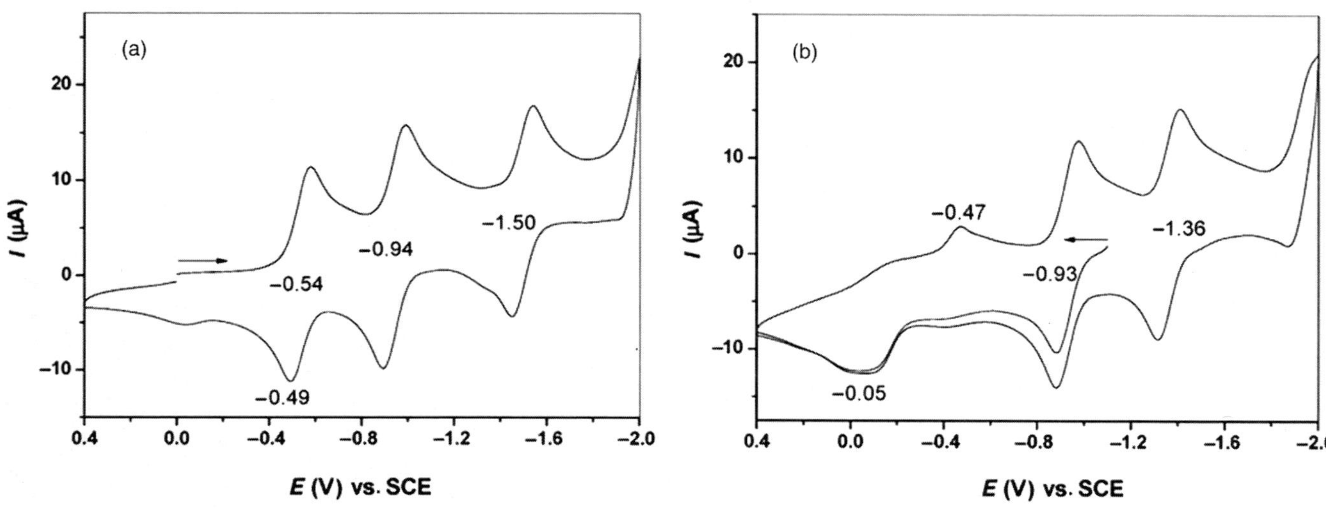

FIGURE 13.15. Cyclic voltammograms of 1,2-$H(PhCH_2)C_{60}$: (**a**) in neutral form, and (**b**) after transferring two electrons to 1,2-$H(PhCH_2)C_{60}$. The measurements were carried out in 0.1 M PhCN–TBAP at a scan rate of 0.1 V/s. Reprinted from reference 131 with permission. Copyright © 2011, American Chemical Society.

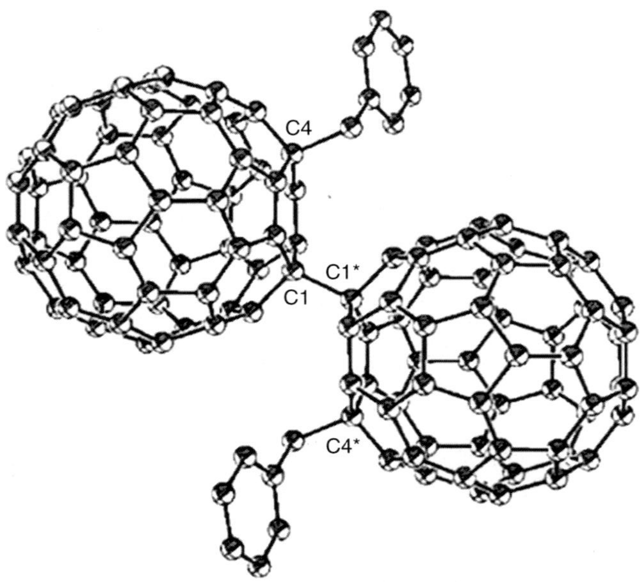

11

FIGURE 13.16. ORTEP diagram of the meso form of the singly bonded $PhCH_2C_{60}$–$C_{60}PhCH_2$ dimer. Hydrogen atoms and solvent molecules are omitted for clarity. Reprinted from reference 131 with permission. Copyright © 2011, American Chemical Society.

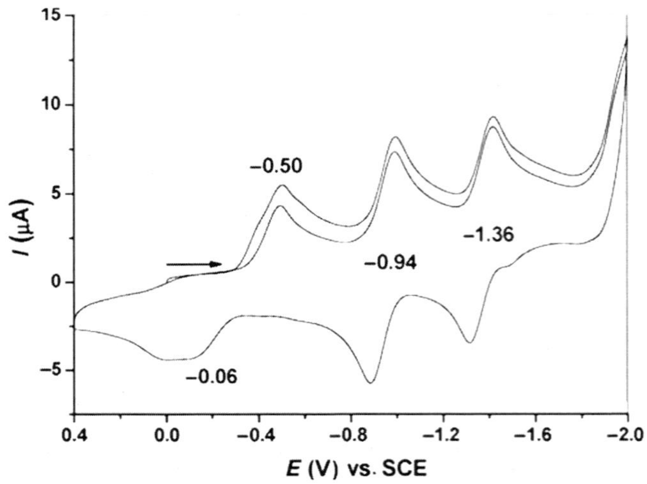

FIGURE 13.17. Cyclic voltammogram of the singly bonded $PhCH_2C_{60}$–$C_{60}PhCH_2$ dimmer in 0.1 M PhCN–TBAP with a scan rate of 0.1 V/s. Reprinted from reference 131 with permission. Copyright © 2011, American Chemical Society.

are stable on the cyclic voltammetric time scale. However, when reduced by CPE at −1.10 V versus SCE, which is ~150 mV more negative than the $E_{1/2}$ of the second reduction, the cyclic voltammogram of the resulting anionic solution appeared completely different from that of the original 1,2-H($PhCH_2$)C_{60}, indicating that a chemical reaction had occurred (Figure 13.15b). The solution was reoxidized at 0 V versus SCE. The final product was identified as the singly bonded $PhCH_2C_{60}$–$C_{60}PhCH_2$ dimer, which consists of meso and racemic regioisomers, inseparable by HPLC. The crystal structure of the meso regioisomer (**11**) was obtained and is shown in Figure 13.16 [131].

The cyclic voltammogram of this C_{60} dimer (**11**) is shown in Figure 13.17. The relatively broad first reduction wave results from the different reduction potentials of the meso and racemic dimers and a chemical reaction that occurs during this reduction process. The second and third waves correspond to two redox processes associated with the dissociated dimer and the formation of anionic monomeric $PhCH_2C_{60}^-$. Accordingly, the redox couples at −0.94 and −1.36 V can be assigned to $PhCH_2C_{60}^-$/$PhCH_2C_{60}^{2-\bullet}$ and $PhCH_2C_{60}^{2-\bullet}$/$PhCH_2C_{60}^{3-}$, respectively. The oxidation wave at −0.06 V is assumed to result from the oxidation of $PhCH_2C_{60}^-$ to $PhCH_2C_{60}^\bullet$ radical, which then undergoes a radical recombination process to reform the singly bonded $PhCH_2C_{60}$–$C_{60}PhCH_2$ dimer. The similarity between the cyclic voltammograms of the dimer (**11**) and dianonic 1,2-H($PhCH_2$)C_{60} (**9**) demonstrates that dimerization occurs after two electrons are transferred to 1,2-H($PhCH_2$)C_{60}. Voltammetric titration of 1,2-H($PhCH_2$)C_{60} with TBAOH provided an effective way to elucidate the mechanism. The fullerenyl hydrogen was homolytically cleaved from the C_{60} sphere as a result of the two-electron reduction. This electrochemically induced dehydrogenation was observed early in 1993 with $C_{60}H_2$ [132].

The protocols for using dianionic C_{60} generated by CPE as a building block for fullerene functionalizations are now well established. However, studies of the reactivity of trianionic C_{60} are limited in comparison. In 1991, Dubois et al. [40] reported the spectroelectrochemical study of C_{60} and C_{70} anions. For the first time, C_{60}^{3-} and C_{70}^{3-} produced by electrolysis were observed to be unstable. They displayed irreversible behavior in CH_2Cl_2 Subsequent electrochemical and vis–NIR spectroelectrochemical studies by Eaton and coworkers [133] confirmed previously reported observations about the instability of C_{60}^{3-}. Until 2000, no product analyses of these reactions had been reported. In 2000, Beulen and Echegoyen [134] produced C_{60}^{3-} by controlled potential electrolysis with an applied potential ~150 mV more negative than the third reduction potential of C_{60}. A chemical reaction clearly occurred during electrolysis, and after reoxidation of the resulting solution, methanofullerenes of the type $C_{60} > (CH_2)_n$ in ~75% yield were obtained. These results confirmed that trianionic C_{60} is not stable and has the ability to react with the solvent. Detailed studies on the reactivities of C_{60}^{3-} have been recently carried out by Gao et al. [135].

In 2008, Gao and co-workers [135] found an unexpected reactivity for C_{60}^{3-} in benzonitrile solution. A novel

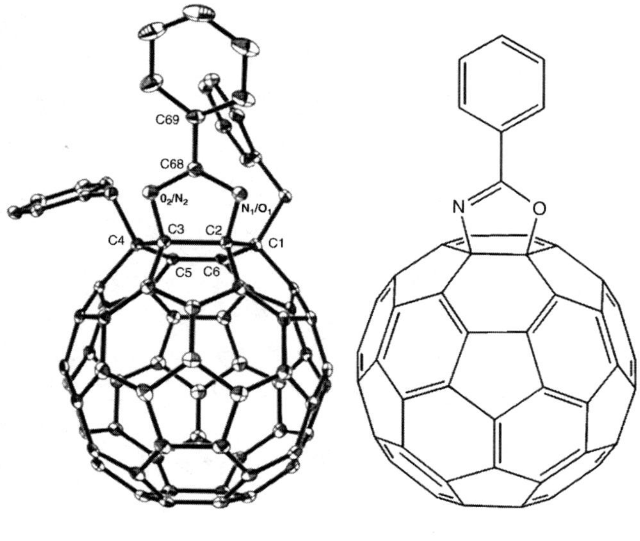

12 13

FIGURE 13.18. The structure of (**a**) 1,4-dibenzyl-2,3-cyclic phenylimidate C_{60} (**12**), and (**b**) [6,6] cyclic phenylimidate C_{60} (**13**). Reprinted from reference 135 with permission. Copyright © 2008, American Chemical Society.

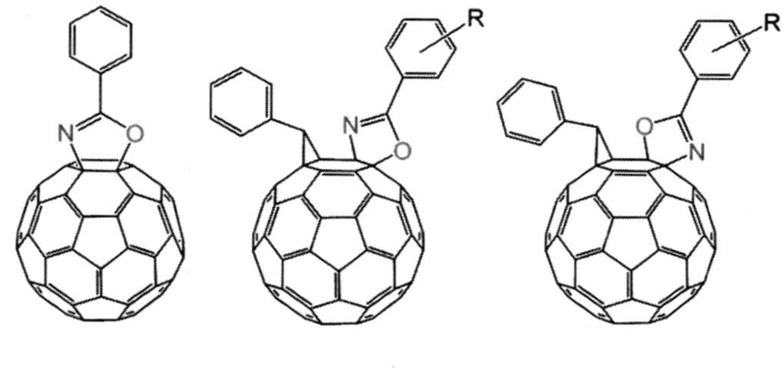

13 14 15

a: R = H a: R = H

b: R = *m*-OCH_3 b:R = *m*-OCH_3

FIGURE 13.19. Possible regioisomers from the electrosysnthetic reaction of $[C_{61}HPh]^{3-}$ in various nitrile solvents. Reprinted from reference 136 with permission. Copyright © 2009, American Chemical Society.

heterocyclic derivative of C_{60} was prepared by electrosynthesis. The procedure for generating $C_{60}{}^{3-}$ by electrolysis was similar to that for producing $C_{60}{}^{2-}$, except that the applied potential was not switched off after the theoretical number of coulombs required for a complete conversion of C_{60} to $C_{60}{}^{2-}$ had been reached. As the reduction potential continued to be applied, an extra irreversible anodic wave at −0.50 V versus SCE appeared. This indicated that a reaction involving anionic C_{60} in benzonitrile had occurred. The extra irreversible anodic wave became stronger the longer the reduction potential was applied. The potentiostat was switched off when the irreversible anodic wave at −0.50 V versus SCE became prominent while still retaining the features of the C_{60} moiety. After addition of excess $PhCH_2Br$ to the solution, the reaction was allowed to proceed for 40 min with stirring. In contrast to previously reported reactions between $C_{60}{}^{2-}$ and benzyl bromide [124], where the only major product is 1,4-$(PhCH_2)_2C_{60}$ (**3**), an unexpected *cis*-1 adduct of C_{60}, 1,4-dibenzyl-2,3-cyclic phenylimidate C_{60} (**12**) (Figure 13.18), was regioselectively formed. The formation of 1,4-$(PhCH_2)_2C_{60}$ (**3**), trace amounts of 1,2-$(PhCH_2)_2C_{60}$ (**10**), 1,2-$HPhCH_2C_{60}$ (**9**), [6,6]methanofullerene $C_{61}HPh$ (**8**), and unreacted C_{60} was also observed. Controlled experiments were performed and $C_{60}{}^{3-}$ was shown to be the reactive species. The different products generated are ascribed to different reactivities between $C_{60}{}^{2-}$ and $C_{60}{}^{3-}$ in benzonitrile. The source of the oxygen atom in the structure is still not clear and remains to be investigated. A related compound bearing the same heterocycle, [6,6] cyclic phenylimidate C_{60} (**13**), was obtained from the same anionic C_{60} benzonitrile solution, when it was reoxidized at 0 V versus SCE without adding benzyl bromide.

In the latter reaction, due to the high symmetry of the C_{60} molecule, there is no regioselectivity for the addition of oxygen and nitrogen atoms on the C_{60} sphere. Thus the reaction leads to the formation of only one regioisomer. In order to investigate the regioselectivity during addition of heteroatoms, a novel reaction was also conducted by Gao and co-workers [136] starting with $C_{61}HPh$ instead of pristine C_{60}. Various solvents were used, including benzonitrile (PhCN), *m*-methoxybenzonitrile (*m*-OCH_3PhCN), *m*-tolunitrile (*m*-CH_3PhCN), and *o*-tolunitrile (*o*-CH_3PhCN). Theoretically, the reaction was expected to give rise to two types of regioisomers **14** and **15** in Figure 13.19 due to the decrease in molecular symmetry [136]. However, only regioisomers **14a–d** were obtained, although the alternative isomers **15a–d** are slightly energetically favorable as predicted by DFT calculations. Such regioselectivity for the formation of fullerooxazoles can be well rationalized by the charge distributions on $C_{61}HPh^{3-}$, where the site for the oxygen addition has the largest negative charge. The reactive species for this reaction is $C_{60}{}^{3-}$, since only trianionic C_{60} is capable of initiating the reaction. Therefore $[C_{61}HPh]^{3-}$ was formed directly by CPE at the appropriate applied potential in each solvent, followed by bulk electrooxidation at 0 V versus SCE. Four fullerooxazole products **14a–d** were obtained regioselectively and fully characterized. A single-crystal structure for one of the products **14a** obtained in benzonitrile was resolved (Figure 13.20). The crystal structure combined with NMR data demonstrated the regioselectivity of the hetroatoms.

The electrochemical properties of 1,4-dibenzyl-2,3-cyclic phenylimidate C_{60} (**12**), [6,6] cyclic phenylimidate C_{60} (**13**), and 1,2-benzal-3-*N*-4-*O*-cyclic phenylimidate C_{60} (**14a**)

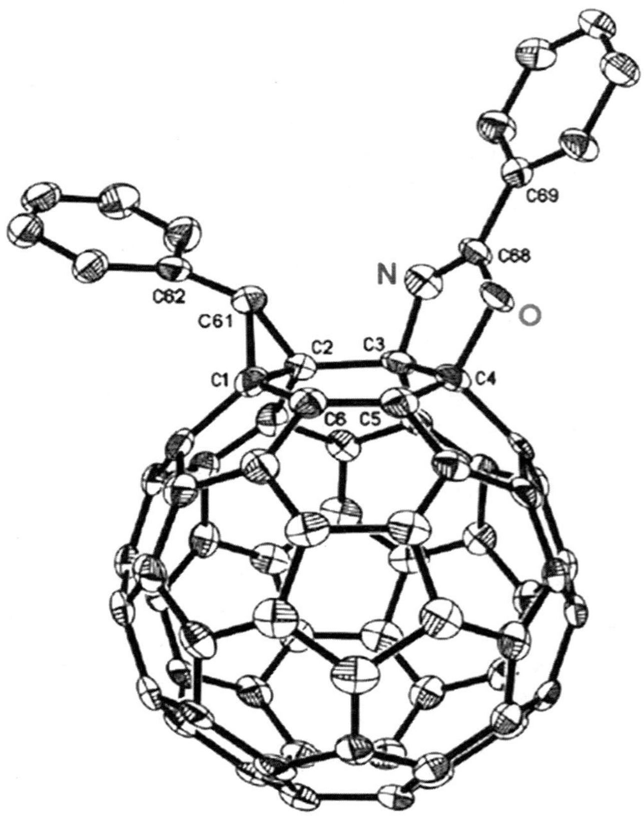

14a

FIGURE 13.20. Crystal structure of fullerooxazole **14a**. Hydrogen atoms and solvent molecules are omitted for clarity. Reprinted from reference 136 with permission. Copyright © 2009, American Chemical Society.

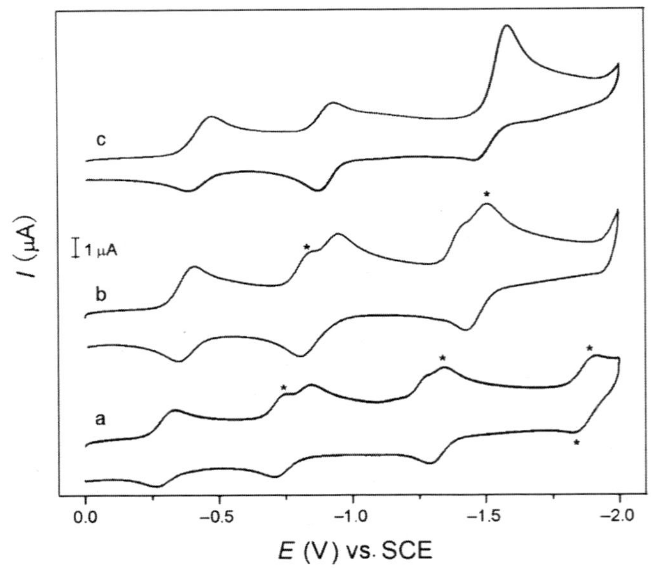

FIGURE 13.21. Cyclic voltammograms of compounds (**a**) **13**, (**b**) **14a**, and (**c**) **12** in DMF containing 0.1 M TBAP at a scan rate of 100 mV/s. Only the first scan for each compound is shown. Peaks marked with asterisks show the redox process of C_{60} in (**a**) and $C_{61}HPh$ in (**b**). Reprinted from reference 137 with permission. Copyright © 2009, American Chemical Society.

were also studied by cyclic voltammetry [137]. As shown in Figure 13.21, all three compounds undergo a reversible one-electron first reduction process, indicating that the monoanions are stable on the cyclic voltammetric time scale. The observed $E_{1/2}$ values are all cathodically shifted relative to the value of C_{60} (−0.26 V versus SCE) [138]. The values for **12** and **14a** are negatively shifted more than that for **13**, consistent with the fact that as the number of addends on the sphere of C_{60} increases the reduction potential is shifted more cathodically [7]. Interestingly, unlike monoanions, the dianions and trianions of these fullerooxazoles are unstable on the cyclic voltammetric timescale, as evidenced by the appearance of extra reduction waves during the first scan. These waves correspond to the reduction of C_{60} and $C_{61}HPh$ respectively [137]. These results clearly demonstrate that compound **12** is more stable electrochemically than **13** and **14a** under reductive conditions, as **13** and **14a** undergo retro-cycloaddition of the cyclic phenylimidate upon formation of the dianions. The stability of the mono- and dianionic fullerooxazoles **12, 13**, and **14a** was further evaluated by CPE, which was performed at potentials ~100–200 mV more cathodic than those of the respective E_{pc} values. The anionic solution was then electrochemically oxidized at 0 V versus SCE, and the recovered products were analyzed by HPLC. The results confirmed that compounds **13** and **14a** undergo retro-cycloaddition reactions that lead to the formation of C_{60} and $C_{61}HPh$, respectively, upon transfer of two electrons. Results also confirmed that compound **12** is much more electrochemically stable because no retro-cycloaddition reaction occurs under similar conditions. Examination of the crystal structure of compound **12** revealed that intramolecular hydrogen bonding is responsible for the electrochemical stability of **12** (Figure 13.22).

There are only two types of bonds in C_{60}, whereas there are eight types of bonds in C_{70} (see Figure 13.23) [139]. This makes the regiochemistry of C_{70} organofullerene derivatives much more complex than that of the C_{60}. The reactivity of C_{70} anions has not been as extensively studied as those of C_{60} anions. Nevertheless, C_{70} has been shown to have electrochemical properties and reactivity similar to C_{60}, and not surprisingly, electrosynthesis is also applicable for the preparation of C_{70} organofullerene derivatives. Applying the same procedure used to prepare derivatives of C_{60}, Kadish and Fukuzumi synthesized $(PhCH_2)_2C_{70}$ (**16**) from the dianion of C_{70} in 2000 [139]. The compound was obtained as a mixture of isomers, but no single isomer was isolated. The reaction

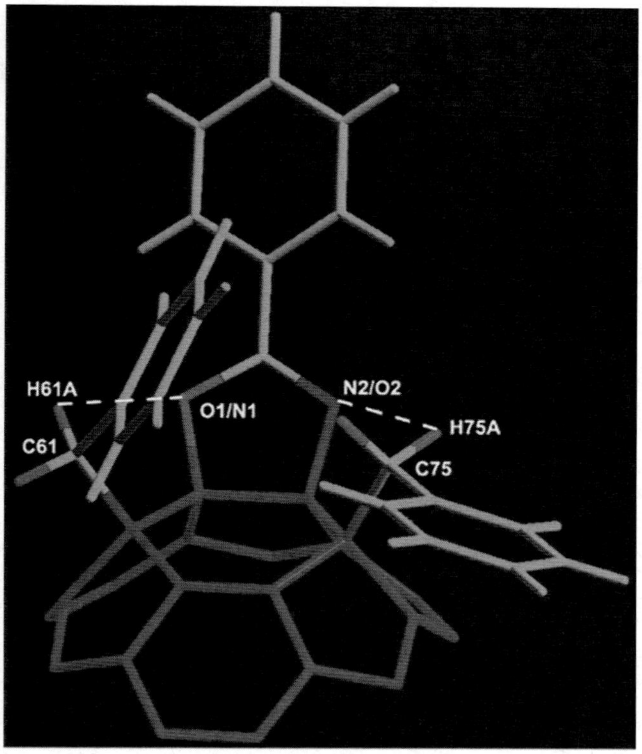

FIGURE 13.22. Expanded view of the crystal structure of **12**, with H61A· · · O1/N1 and H75A· · · N2/O2 distances shown as dashed lines. Reprinted from reference 137 with permission. Copyright © 2009, American Chemical Society.

mechanism for the electrogeneration of $(PhCH_2)_2C_{70}$ was consistent with that of the C_{60} analogue. The electrochemical properties of this mixture were explored by cyclic voltammetry. Two sets of electrochemical processes were found. The major electroactive isomer undergoes four reversible

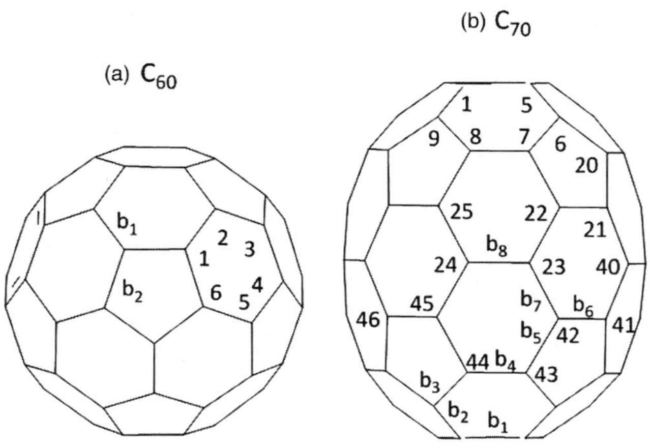

FIGURE 13.23. Schematic representation including the labeling of different types of bonds and the numbering of some carbon atoms of (**a**) C_{60} and (**b**) C_{70}.

one-electron reductions at $E_{1/2} = 0.37, -0.84, -1.37$, and -1.84 V. Surprisingly, the first reduction is anodically shifted 50 mV with respect to the corresponding reduction of pristine C_{70}, in contrast with previous results for C_{60}, where funtionalization leads to cathodically shifted reduction waves with respect to the parent fullerene.

13.4.1 Regioselective Electrosynthesis of C_{60} Derivatives

Multiple studies have focused on the synthesis of C_{60} derivatives with two or more different functional groups at specific positions on the sphere in order to fine-tune the properties of the resulting material for specific nanodevice and molecular applications. Due to the high symmetry of C_{60}, many isomers are formed during multiadditions [140]. Several cycloaddition reactions show selective reactivity for the 30 double bonds on the sphere (denominated 6,6-bonds) over the remaining 60 single bonds (denominated 5,6-bonds). Using strategic protection–deprotection methodologies involving thermally labile and redox-active addends, regioselective multiadduct products have been isolated and characterized [141–146].

In 2000, Kessinger et al. [147] reported for the first time and in reasonably high yields (60% or higher) the selective removal of bis(ethoxycarbonyl)methano addend (Bingel addend) in the presence of other cycloaddition groups by controlled potential electrolysis (CPE). In this electrochemically controlled reaction, known as the retro-Bingel reaction [148], two e^- per Bingel addends were transferred per molecule during electrolysis. Electrolysis was conducted at the potential for the second fullerene-centered reduction step, and clearly the non-Bingel addend was stable under reduction conditions while the Bingel addend was not [147]. Based on these results, the Bingel addend was proposed as a suitable, easily removable protecting/directing group for the synthesis of other C_{60} regiospecific derivatives, which are not favorably produced by other means.

In 2006, Lukoyanova et al. [149] reported the selective removal of the two most common addends, pyrrolidino and Bingel, in the presence of each other by electrochemical retro-cycloaddition or retro-cyclopropanation, respectively, by CPE (Figure 13.24a). The pyrrolidino addend was shown to be unstable under oxidative conditions and therefore could be removed in the presence of the Bingel addend to produce pure bis(ethoxycarbonyl) methanofullerene in high yields (60%) (Figure 13.24b). Likewise, the Bingel addend was removed by retrocyclopropanation in the presence of the pyrrolidino addend, producing N-methyl pyrrolidinofullerene as a major product (89%). These results provide an excellent electrochemical strategy for the selective elimination of 1,3-dipolar addends or Bingel addends from mixed derivatives of C_{60}.

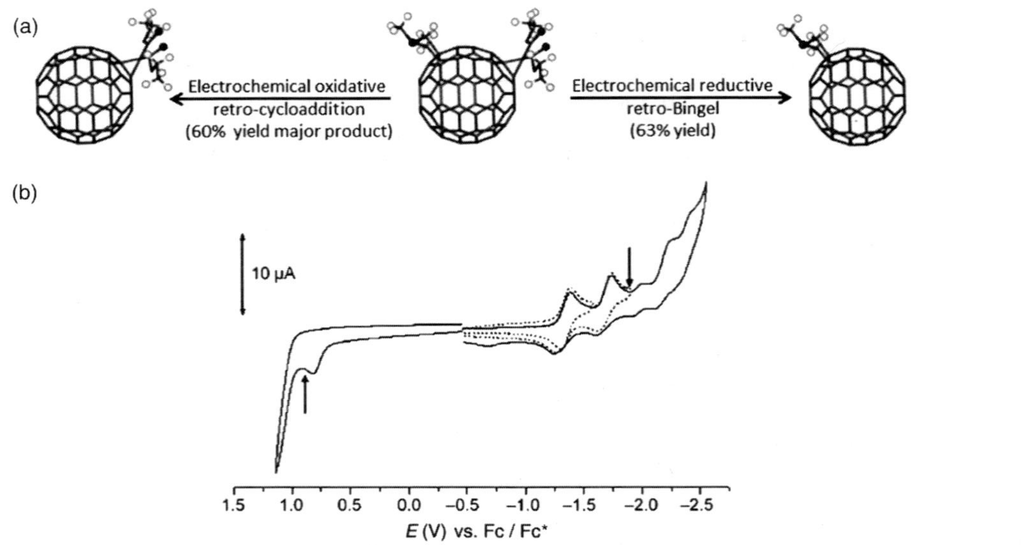

FIGURE 13.24. (**a**) Selective electrochemical retro-cycloaddition under oxidative CPE and retro-Bingel under reductive CPE of a hybrid bis-adduct of C_{60}. (**b**) Cyclic voltammetry of the parent hybrid bis-adduct in 0.1 M $TBAPF_6$ in dichloromethane (arrows indicate selected potential at which reductive or oxidative electrolysis was performed). Reprinted with permission from reference 149. Copyright © 2006, Wiley–VCH Verlag GmbH & Co. KGaA, Weinheim.

Implementing the previous methodology, Zhang et al. [150] reported in 2006 the synthesis of two *trans*-1 bis-adducts (compounds **18** and **20**, Figure 13.25) by using a protection-deprotection strategy that combined chemical and electrochemical addition/elimination steps. The last step of the synthesis of **18** and **20** involved the selective removal of Bingel addends from compounds **17** and **19** using the retro-Bingel process (Figure 13.25). The cyclic voltammetry of compounds **17** and **19** shows the irreversible reduction processes expected for the Bingel addends present in the materials (Figure 13.26a). After approximately $8e^-$ per molecule were transferred during the electrolysis of **17** ($2e^-$ per Bingel addend), the cyclic voltammogram showed two reversible reduction steps, suggesting that the Bingel addends were fully removed from the substrate and the new compound, **18,** was formed (Figure 13.26b). Similar results were obtained for compound **19**.

In 2011, Ortiz and Echegoyen [151] reported the reductive controlled potential electrolysis of a D_{2h}-*hexakis*-C_{60} derivative (Compound **21** in Figure 13.27a). The CPE was performed until $6e^-$ per molecule were discharged, after which two reversible reductive processes were observed by cyclic voltammetry (Figure 13.27b). However, after purification, compounds **22** to **24**, which correspond to the removal of

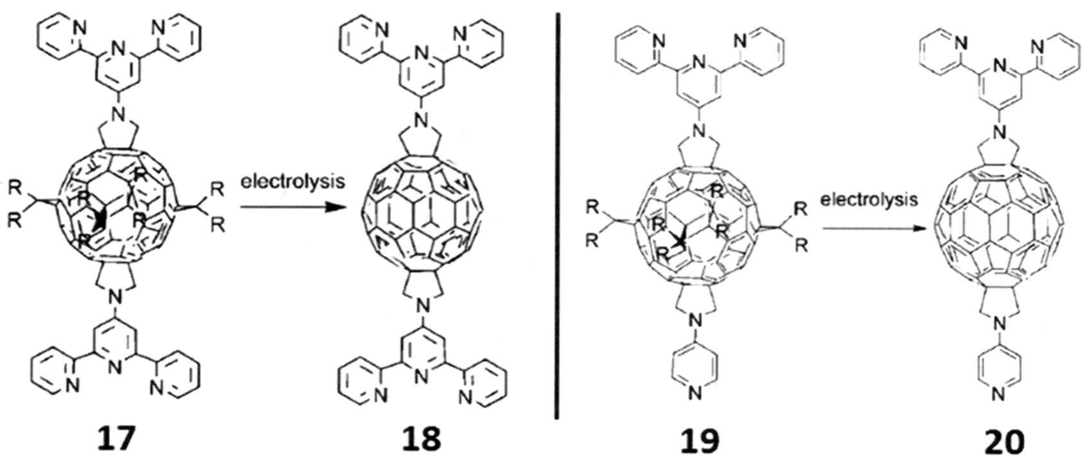

FIGURE 13.25. Structure of compounds **17**–**20**. Reprinted from reference 150 with permission. Copyright ©2006, Wiley–VCH Verlag GmbH & Co. KGaA, Weinheim.

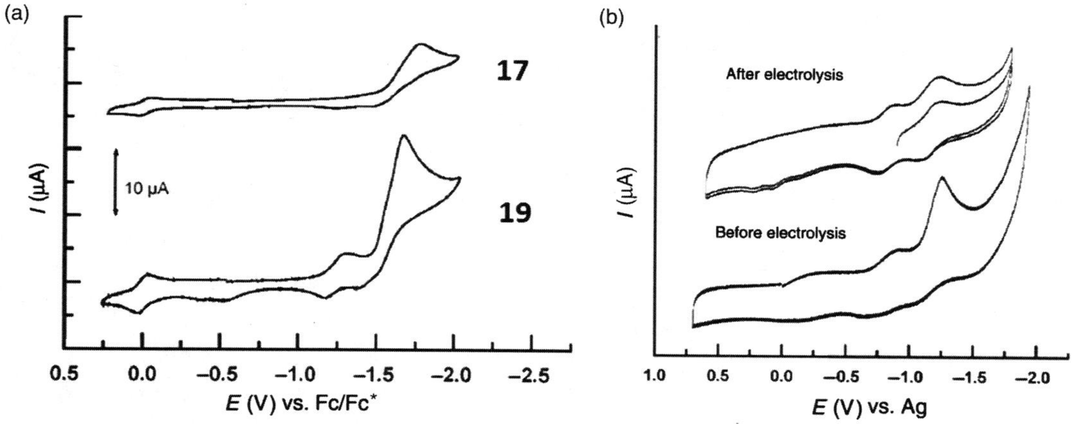

FIGURE 13.26. (**a**) Cyclic voltammograms of **17** and **19** in 0.1 M $TBAPF_6$ in dichloromethane. (**b**) Cyclic voltammograms of **17** before and after reductive CPE. Reprinted from reference 150 with permission. Copyright ©2006, Wiley–VCH Verlag GmbH & Co. KGaA, Weinheim.

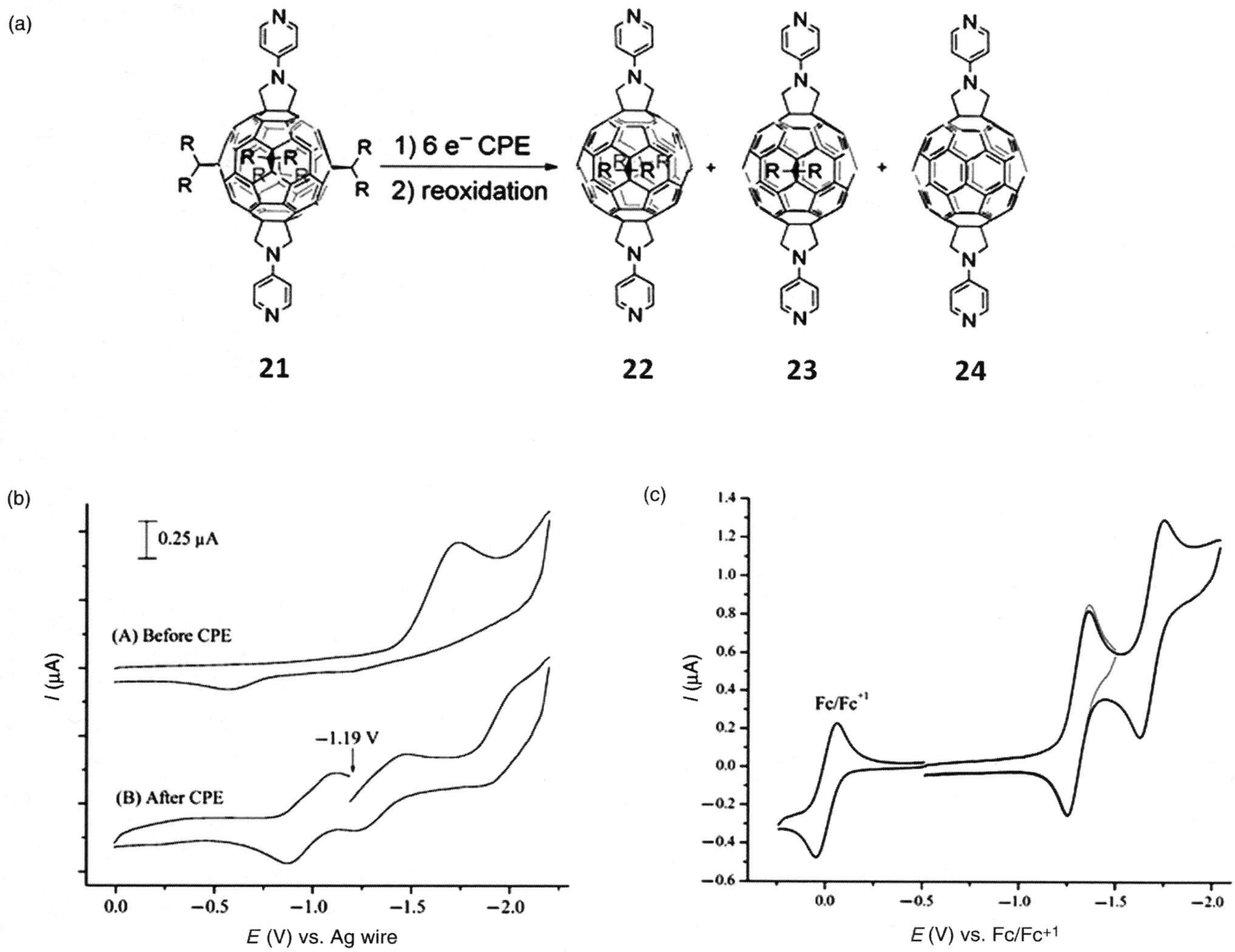

FIGURE 13.27. (**a**) Electrochemical retro-Bingel of **21** by CPE. (**b**) Cyclic voltammogram of **21** before and after reductive CPE. (**c**) Cyclic voltammogram of **22**. Reprinted from reference 151 with permission. Copyright © 2010, Royal Society of Chemistry.

two, three, and four Bingel addends of the four present in the starting material, were identified by MALDI-TOF MS. Compound **22** was identified as a single isomer with a highly symmetrical geometry in which all four addends are positioned on the equatorial belt of the C_{60} sphere. Two unexpected electrochemical properties were reported for compound **22**. First, it displayed two one-electron reversible reduction processes by cyclic voltammetry (Figure 13.27c), which is quite unexpected for a Bingel-addend fullerene derivative. Second, upon electrolysis at the first reduction potential, the usual rearrangement or "walk-on-the-sphere" expected for Bingel bis-adducts was not observed [152], indicating a high stability and isomeric preference for this intriguing compound. This material was proposed to be a potentially useful candidate for molecular electronic applications due to its chemical and electrochemical reversible behavior.

13.4.2 Electrosynthesis of Endohedral Fullerene Derivatives

Exohedral chemical functionalization of TNT–EMFs is crucial for the development of novel organometallofullerene materials for a variety of potential applications. Reactions that have been successful with empty-cage fullerenes have also been applied to various TNT–EMFs, such as Diels–Alder [89, 90], 1,3-dipolar [91–99], cyclopropanation [100–103], [2 + 2]-addition [104], and free radical additions reactions [100, 105, 108–110]. The reactivity of endohedral metallofullerene anions remained unexplored until 2010 when Echegoyen and co-workers [153] prepared the first derivative of $Lu_3N@I_h–C_{80}$ from the nucleophilic reaction of benzyl bromide with $[Lu_3N@I_h–C_{80}]^{2-}$ generated by CPE. The product was characterized experimentally and theoretically as the [6,6]-open [$Lu_3N@I_h–C_{80}$-CHPh] regioisomer **(25)** shown in Figure 13.28.

On the basis of many reports, the electrochemical behavior of TNT–EMF monoadducts depends mainly on the location of the addends [5]. [5, 6]-TNT–EMF adducts exhibit reversible cathodic electrochemical behavior [94], while, in general, [6,6]-adducts exhibit irreversible behavior [94, 101, 102]. $Sc_3N@I_h–C_{80}$ [6,6]-adducts are an exception. They display reversible reductive behavior [104, 154]. As a consequence of these observations, electrochemistry can serve as a complementary diagnostic tool to determine the position of functional groups on the cage surface. The electrochemical behavior of **25** was recorded by cyclic voltammetry as shown in Figure 13.29. Irreversible cathodic electrochemical behavior, with E_{pc} values of −1.49, −1.95, and −2.32 V versus Fc/Fc^+, was observed. The reduction potentials are cathodically shifted by 70, 150, and 60 mV, respectively, with respect to those observed for the pristine $Lu_3N@I_h–C_{80}$. Quasi-reversible anodic electrochemical behavior with E_{pa} values of +1.10 and +0.59 V versus Fc/Fc^+ was also

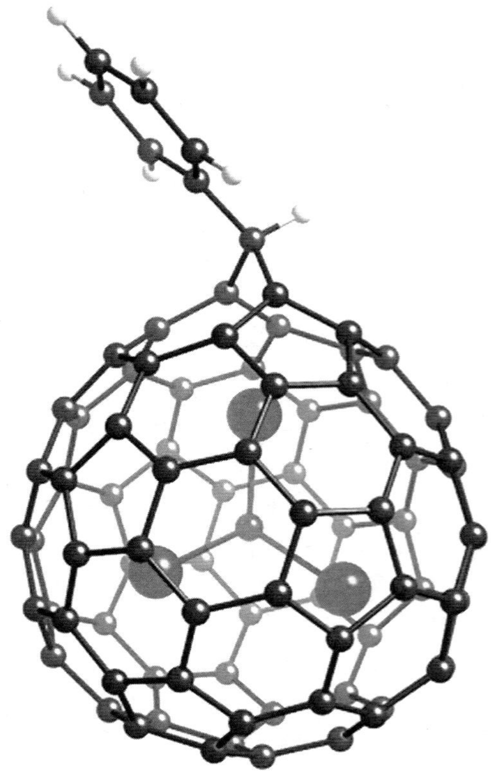

25

FIGURE 13.28. Representation of the structure of the [6,6]-[$Lu_3N@I_h–C_{80}$–CHPh] regioisomer (**25**). Reprinted from reference 153 with permission. Copyright © 2011, American Chemical Society.

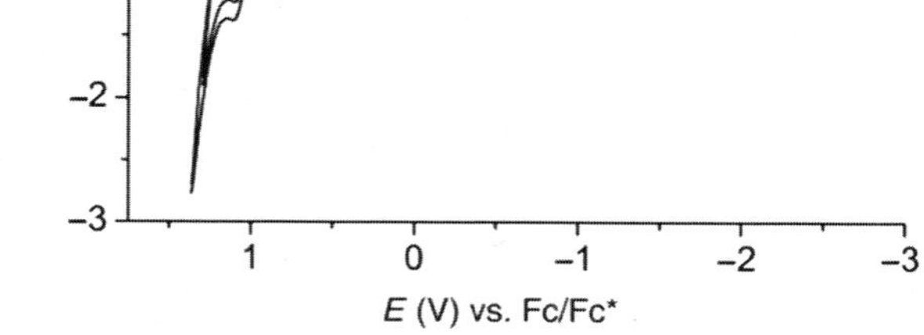

FIGURE 13.29. Cyclic voltammogram of [6,6]-[$Lu_3N@I_h–C_{80}$–CHPh] (**25**). The voltammogram was recorded in a 0.05 M solution of n-Bu_4NPF_6 in o-DCB at a scan rate of 100 mV/s. Reprinted from reference 153 with permission. Copyright © 2011, American Chemical Society.

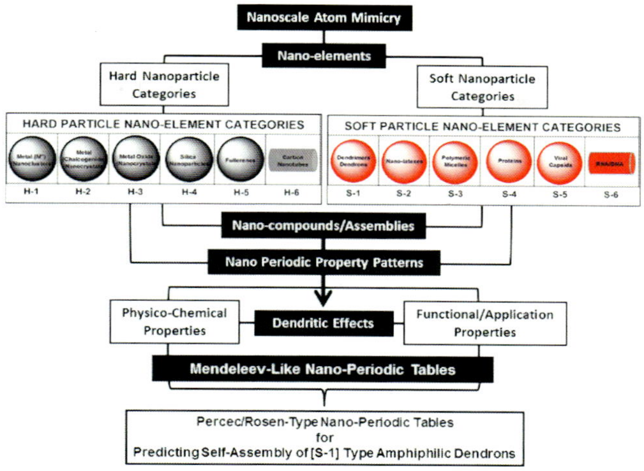

FIGURE 1.16. The first examples of Mendeleev-like nano-periodic tables have recently fulfilled these expected nano property pattern/trend predictions [2, 31]. Percec/Rosen [46] have reported the first three nano-periodic tables for predicting the self-assembly patterns for [S-1] type amphiphilic dendrons with predictive accuracies of 85% to >90% based on knowledge of the primary dendron CNDPs: namely, (a) size, (b) shape, (c) surface/apex chemistry, and (d) flexibility/rigidity [45].

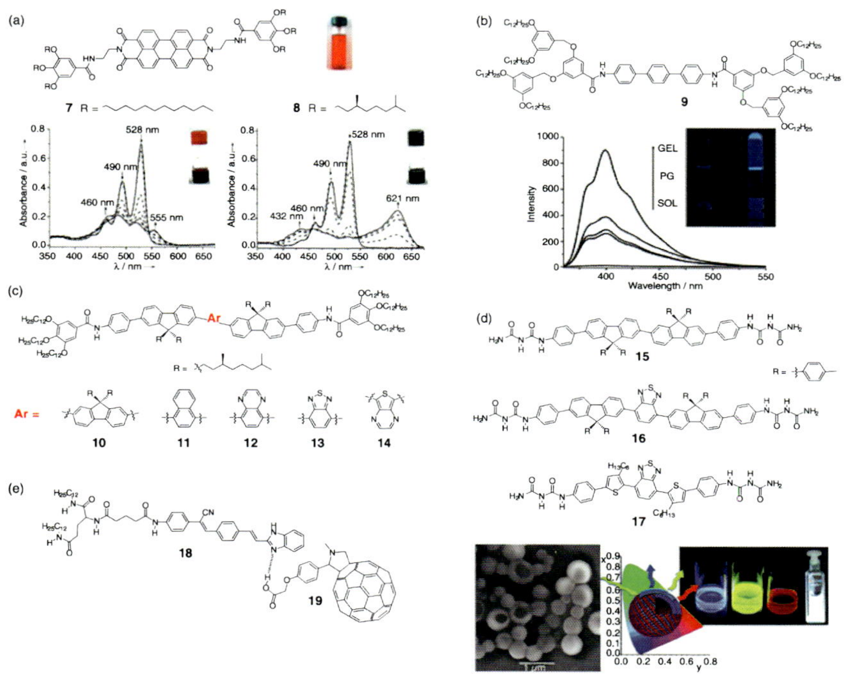

FIGURE 2.3. (**a**) Structure of PBIs **7** and **8**. Solvent-dependent UV–vis absorption spectra of PBIs **7** and **8** at a concentration of 10^{-5} M at 25°C. Arrows indicate the spectral changes upon increasing the methylcyclohexane/$CHCl_3$ ratio. The inserted pictures show the solution and gel colors in each case. (**b**) Structure of terphenylene **9**. Fluorescence spectra of **9** in xylene solution (SOL), partial gels (PG), and gel state at the same concentration (4.2 mM) excited at 350 nm. The inset shows the fluorescence images of the solution (*left*) and gel (*right*) in xylene taken under illumination with 365 nm UV light. (**c**) Structure of fluorene oligomers **10**–**14**. (**d**) Structure of oligomers **15**–**17**. SEM images of the spherical aggregates formed by compound **15** upon dropcasting from dilute anhydrous THF solution ([**15**] = 0.1 mM) onto a SiO_2 substrate. The presence of holes in some of the aggregates is indicative of their hollow nature. (**e**) Structure of the H-bonded complex **18**·**19**. Adapted from: (a) reference 39 with the permission of Wiley-VCH; (b) reference 16 with permission from the American Chemical Society; (c) reference 23; (d) reference 25 with the permission of Wiley-VCH; (e) reference 84.

Organic Nanomaterials: Synthesis, Characterization, and Device Applications, First Edition. Edited by Tomás Torres and Giovanni Bottari.

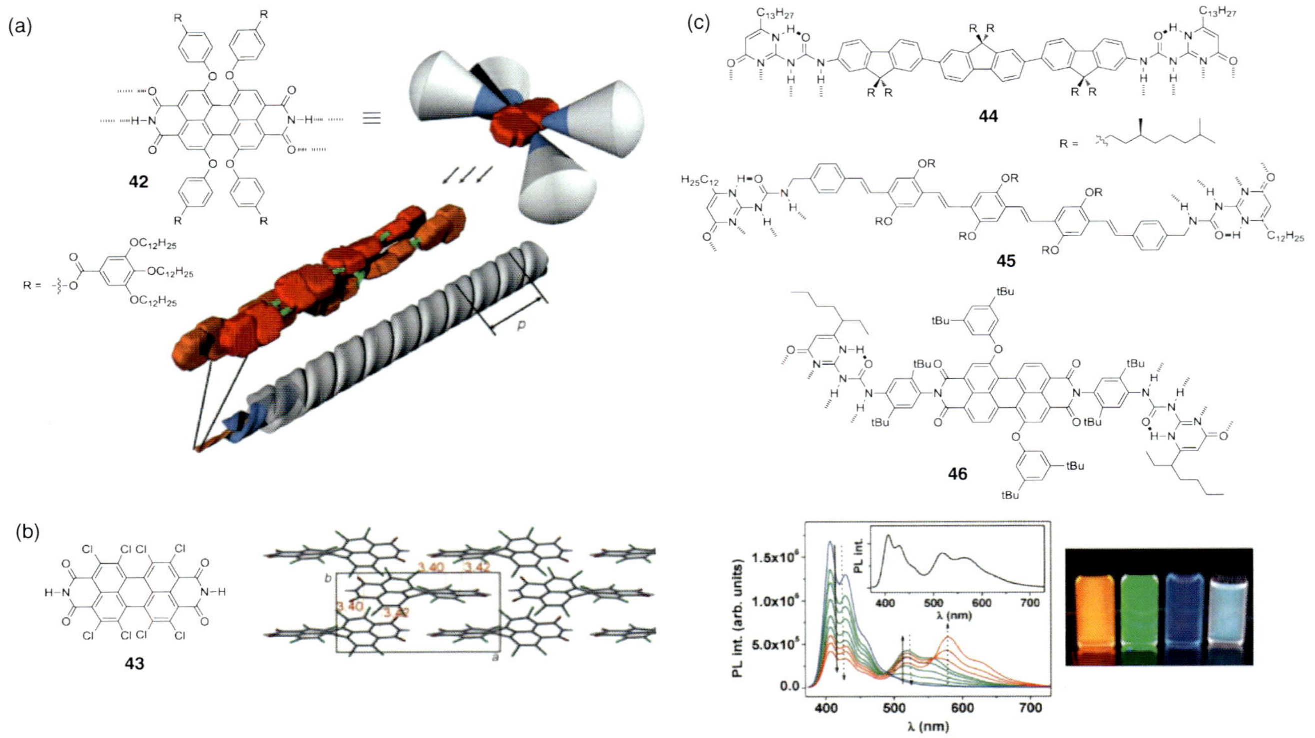

FIGURE 2.10. (**a**) Structure of PBI **43** and schematic representation of its self-assembly process into J-type helical aggregates by H-bonding between the imide groups and π–π stacking between the twisted perylene cores. The gray cones with an apex represent the bay substituents. (**b**) Structure of PBI **44** and packing arrangement in the crystalline state. (**c**) Structure of OF **45**, OPV **46**, and PBI **47** functionalized at the terminal positions by quadruple H-bonding Upy groups. Below on the left, a titration experiment is shown (pure **45**, successive addition of **46**, further addition of **47**). The solid arrows indicate spectral changes upon addition of **46** to **45**, and the dotted arrow represents the change upon addition of **47** to a mixture of **46** and **45**. The inset shows the PL spectrum corresponding to a solution containing **45**:**46**:**47** in a ratio of 59:33:8. On the right, the solutions of pure di-UPy chromophores **45**, **46**, **47** and a white emitting mixture in chloroform under UV irradiation are displayed, in that order. Adapted from: (a) reference 140; (b) reference 144 with permission from Wiley–VCH; (c) reference 150 with permission from the American Chemical Society.

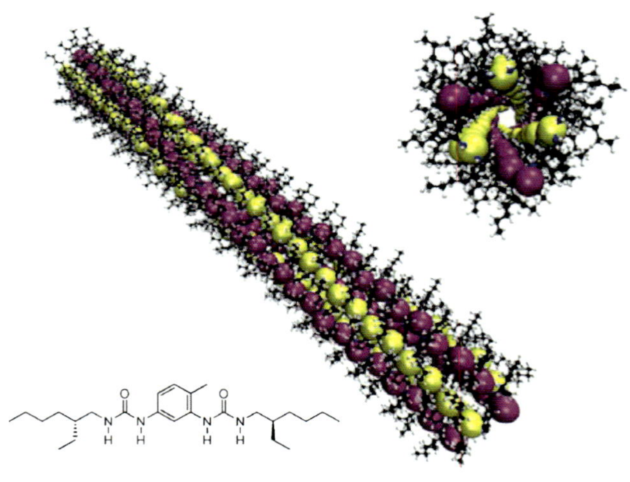

FIGURE 3.11. A chiral bis-urea which self-assembles to give chiral supramolecular nanotubes, whose molecular model is shown with different hydrogen-bonded chains in different colors.

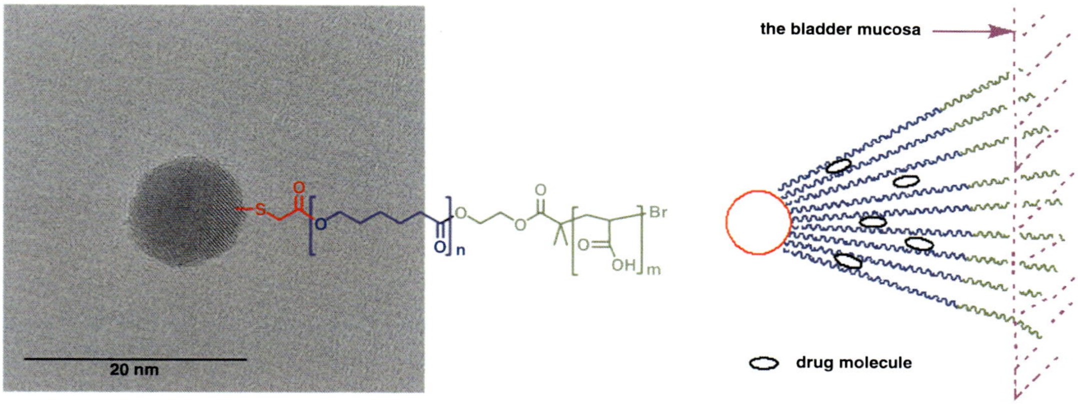

FIGURE 4.1. (**A**) TEM image of one GNP decorated with one amphiphilic PCL-*b*-PAA chain. (**B**) A small fraction of the GNP corona (loaded with drug molecules) adhering to bladder mucosa.

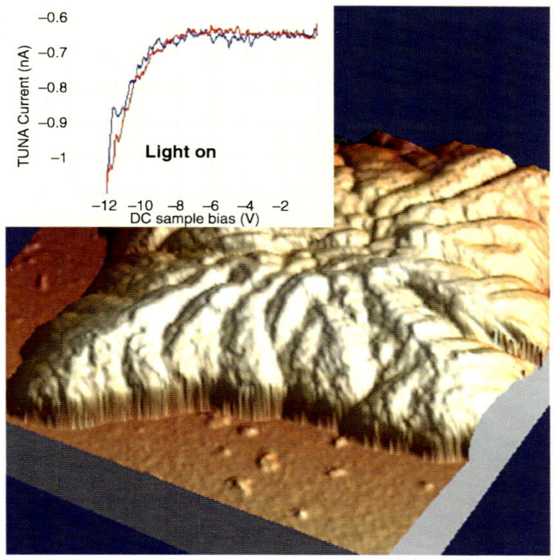

FIGURE 5.13. (**a**) AFM studies of a ZnTPPS/SnT(N-EtOH-4-Py) clover. (**b, c**) TUNA current–voltage curve for conduction through the clover from the p-doped conductive Si substrate without (**b**) and with (**c**) illumination with cool white light. Adapted from reference 20.

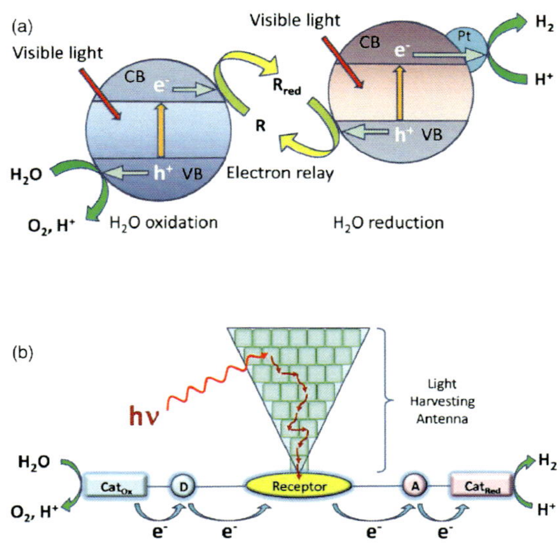

FIGURE 5.18. A two-semiconductor artificial photosynthesis system for light-driven water splitting (**a**) requires the nanotube to transport electrons and excitons. The biomimetic approach uses the porphyrin nanostructure as a light-harvesting component (**b**) and requires additional components (e.g., energy receptor and relay molecules, A = electron acceptor, D = electron donor) to split water. Adapted from reference 19.

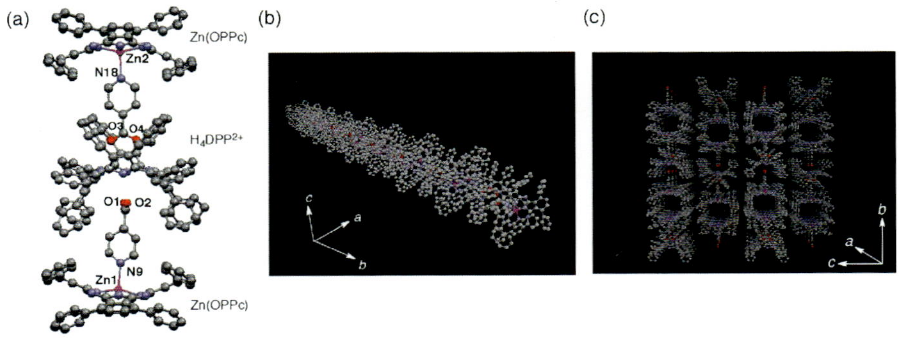

FIGURE 6.16. (**a**) Crystal structure of (H_4DPP){Zn(OPPc)(4-PyCOO)}$_2$. Hydrogen atoms are omitted for clarity. (**b**) One-dimensional nanowire structure of (H_4DPP){Zn(OPPc)(4-PyCOO)}$_2$ in the crystal directed to the crystallographic *b* axis. (**c**) Crystal packing of (H_4DPP){Zn(OPPc)(4-PyCOO)}$_2$. Solvent molecules of crystallization are omitted for clarity.

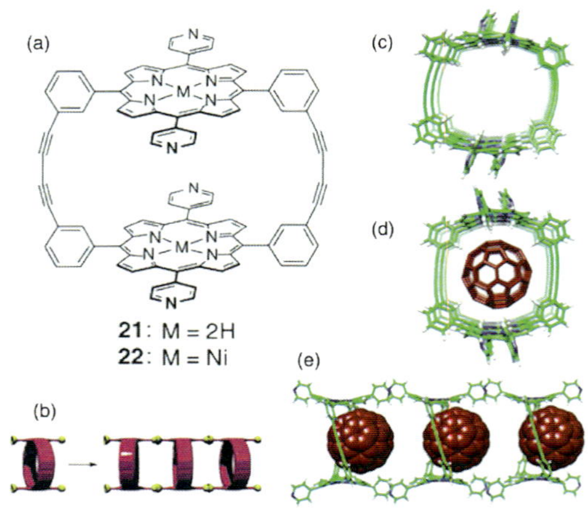

FIGURE 7.8. (**a**) Compounds **21** and **22**, (**b**) their self-assembly, (**c**) x-ray crystal structures of **22**, and (**d**, **e**) C_{60}·**22**. Copyright Wiley–VCH Verlag GmbH & Co. KGaA. Reproduced with permission from reference 8a.

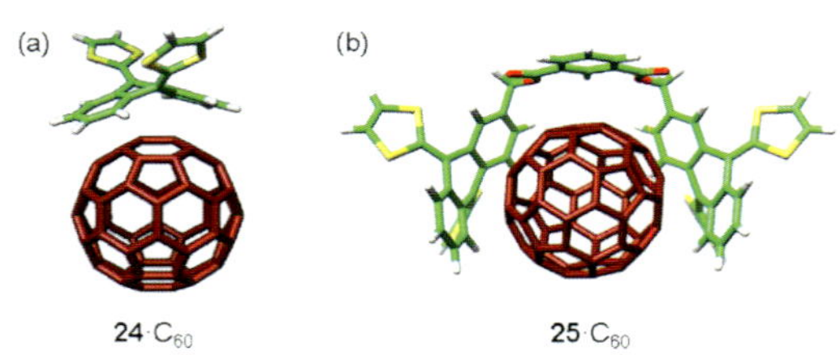

FIGURE 7.10. (**a**) Molecular model showing the shape complementarity between **24** and C_{60}. (**b**) Energy-minimized structure of the **25**·C_{60} complex, as predicted by calculations at the BH&H/6-31G** level.

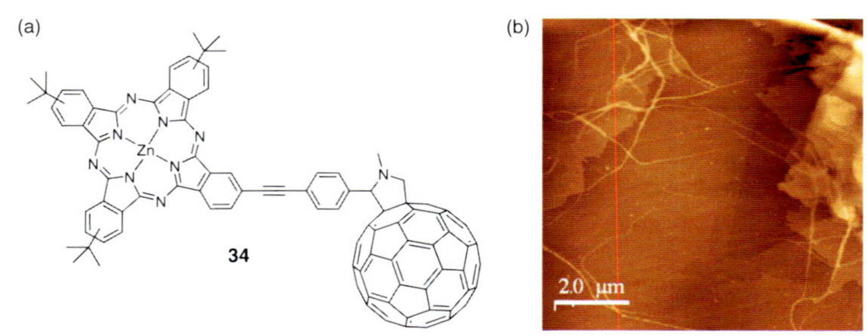

FIGURE 8.14. (**a**) Molecular structure of Pc–C_{60} conjugate **34**. (**b**) AFM topographic image of dyad **34** drop casted on HOPG. The image in (**b**) is reprinted with permission from reference 58. Copyright 2008, Wiley-VCH.

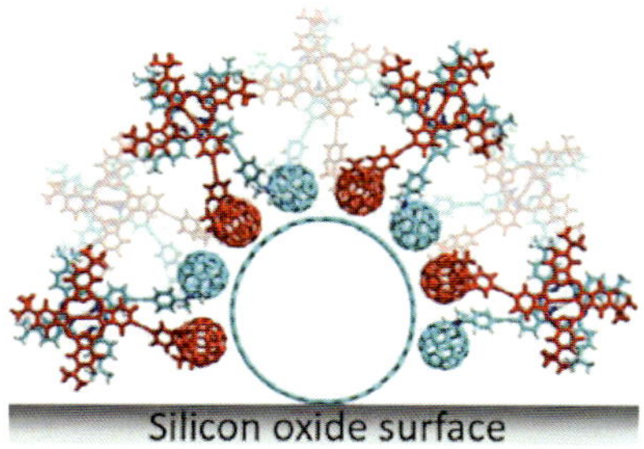

FIGURE 8.15. Frontal view of the proposed supramolecular organization of Pc–C_{60} dyad **34** on a 2.5-nm SWCNT on a silicon oxide surface. The image is reprinted with permission from reference 59. Copyright 2010, Royal Society of Chemistry.

FIGURE 10.1. Carbon nanostructures.

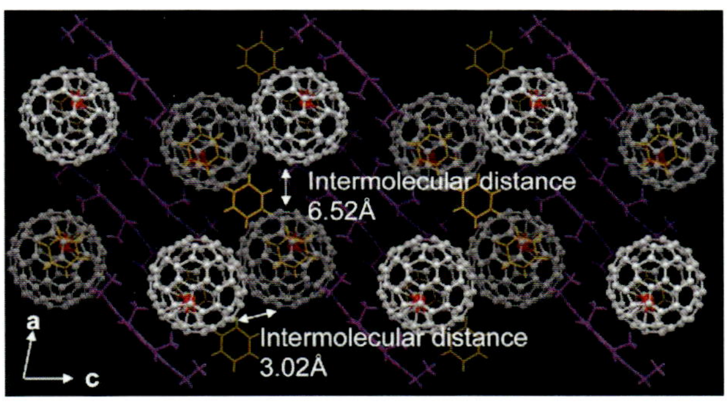

FIGURE 12.9. Crystal packing of La@C_{2v}-C_{82}·Ni^{II}(OEP)·1.5-benzene. The purple and orange colors denote porphyrin and benzene molecules, respectively.

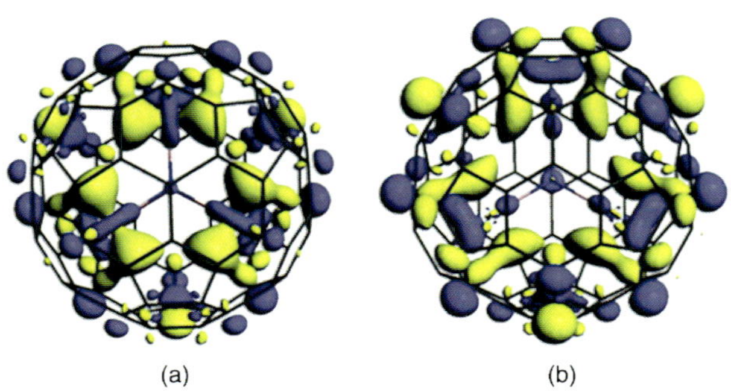

FIGURE 13.31. Representation of the HOMOs for (**a**) $[Sc_3N@I_h–C_{80}]^{2-}$ and (**b**) $[Lu_3N@I_h–C_{80}]^{2-}$. Reprinted from reference 153 with permission. Copyright © 2011, American Chemical Society.

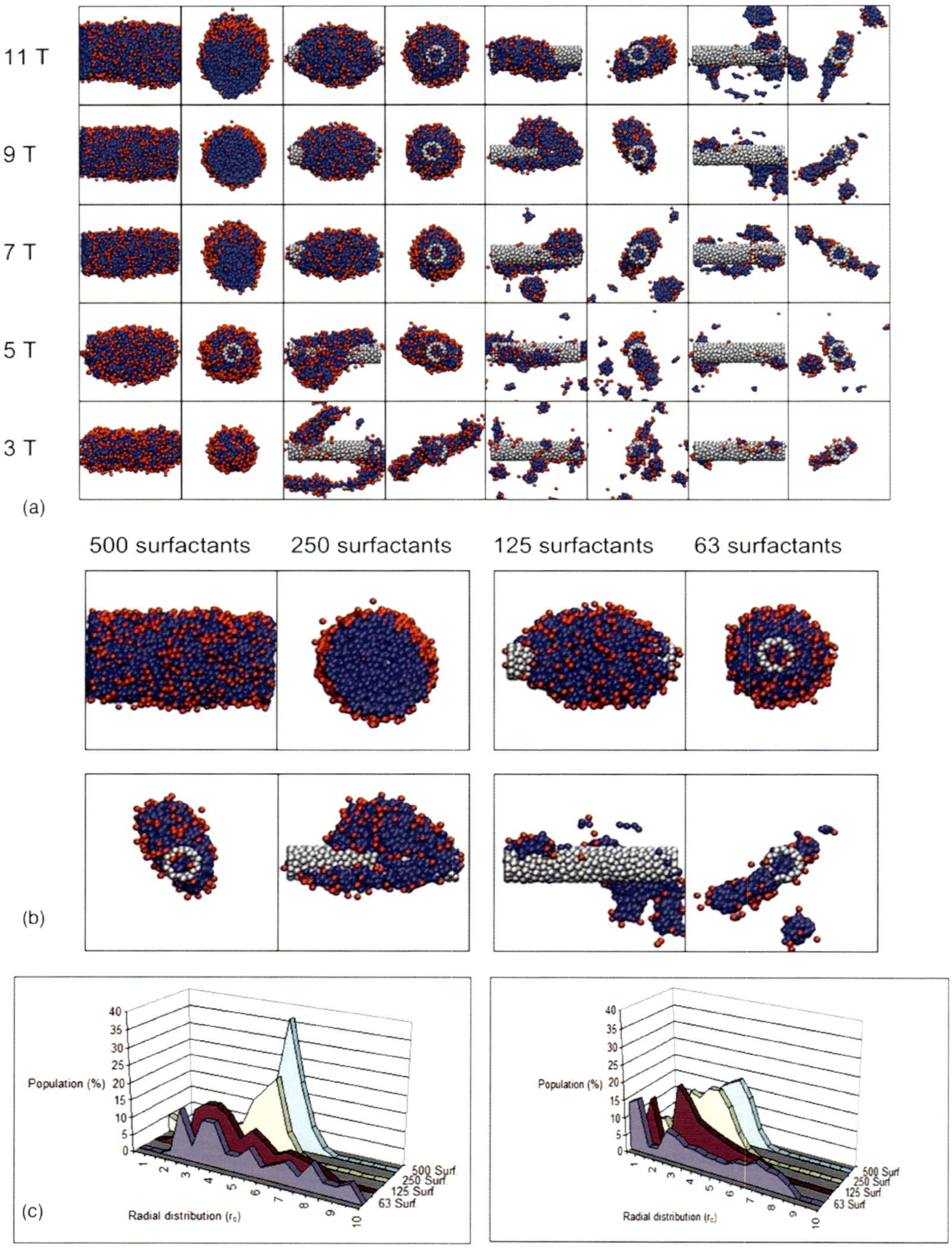

FIGURE 14.14. (**a**) Snapshots of equilibrium morphologies of surfactant assemblies around the CNT. *Top to bottom*: Surfactant chain length of 11, 9, 7, 5, and 3 beads. *Left to right*: 500, 250, 125, and 63 surfactants in the simulation box. Water molecules are removed for clarity. (**b**) Typical morphologies from two points of view obtained for the adsorption of surfactants on the CNT. *Top left*: Cylindrical micelle. *Top right*: Hemimicelle. *Bottom left*: Random adsorption. *Bottom right*: Adsorption of micelles. (**c**) The radial distribution of the heads (*left*) and of the tails (*right*) of the surfactants around the axis of the CNT. Reprinted from reference 68, with permission. Copyright 2009, Wiley–VCH.

FIGURE 15.11. Images showing the changing temperature exhibited by compound **4b**.

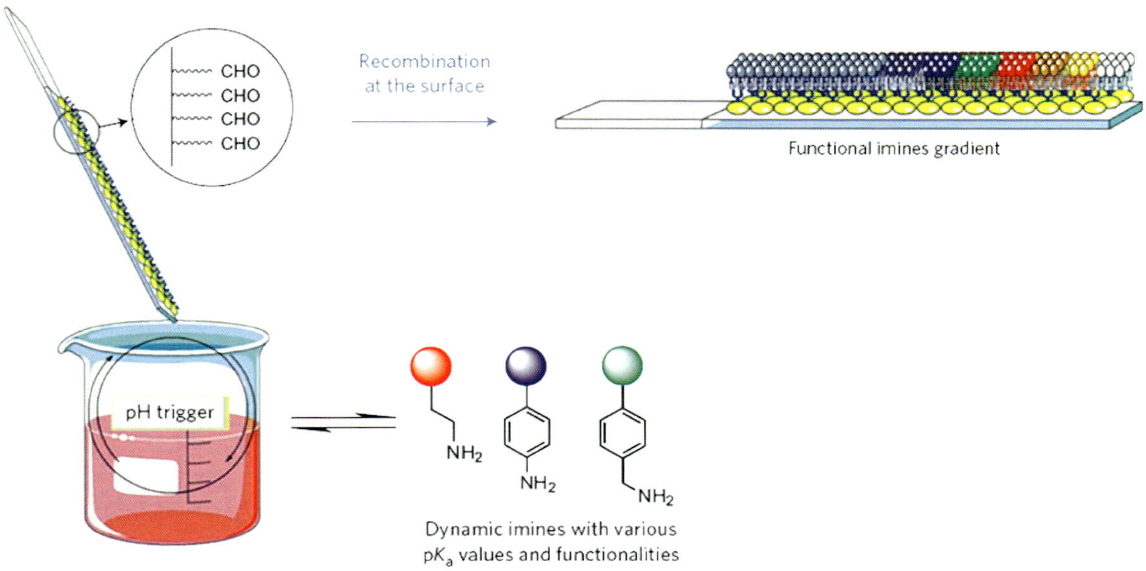

FIGURE 16.6. An aldehyde-terminated surface is progressively removed from a solution of various functional amines as the pH is varied. Each amine combines a single functionality associated with a unique physical property for a given pK_a, which thus leads to a gradient of functional imines. Reprinted from reference 123b, with permission.

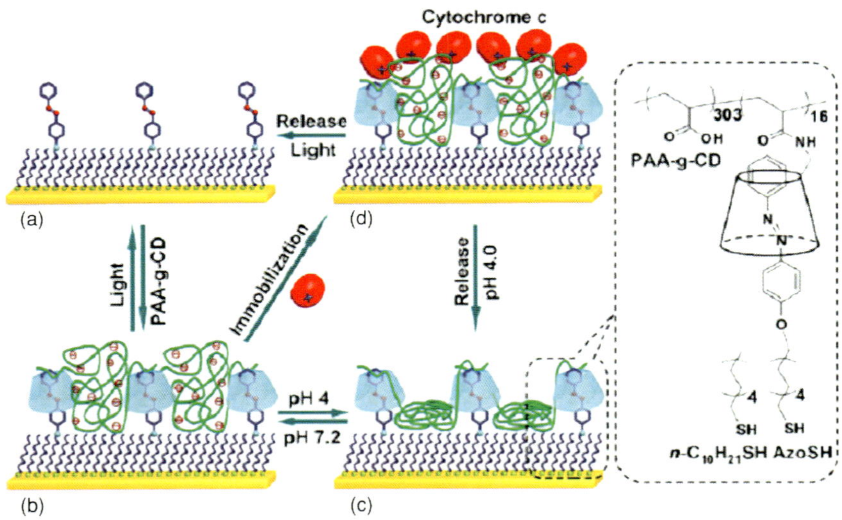

FIGURE 16.12. The light and pH-sensitive monolayer used for reversible immobilization of cytochrome c. Reproduced from reference 175, with permission.

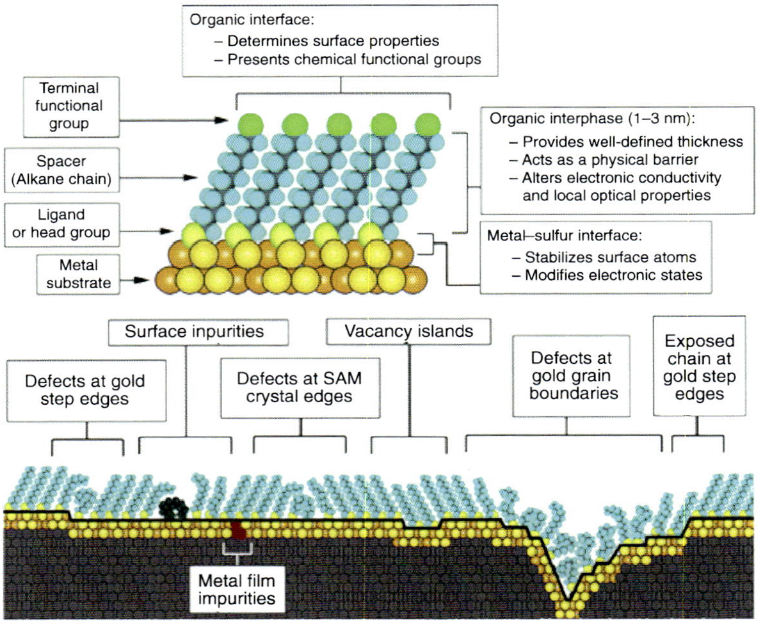

FIGURE 17.5. Detailed structure of a self-assembled monolayer (SAM) [57]. Reproduced with permission.

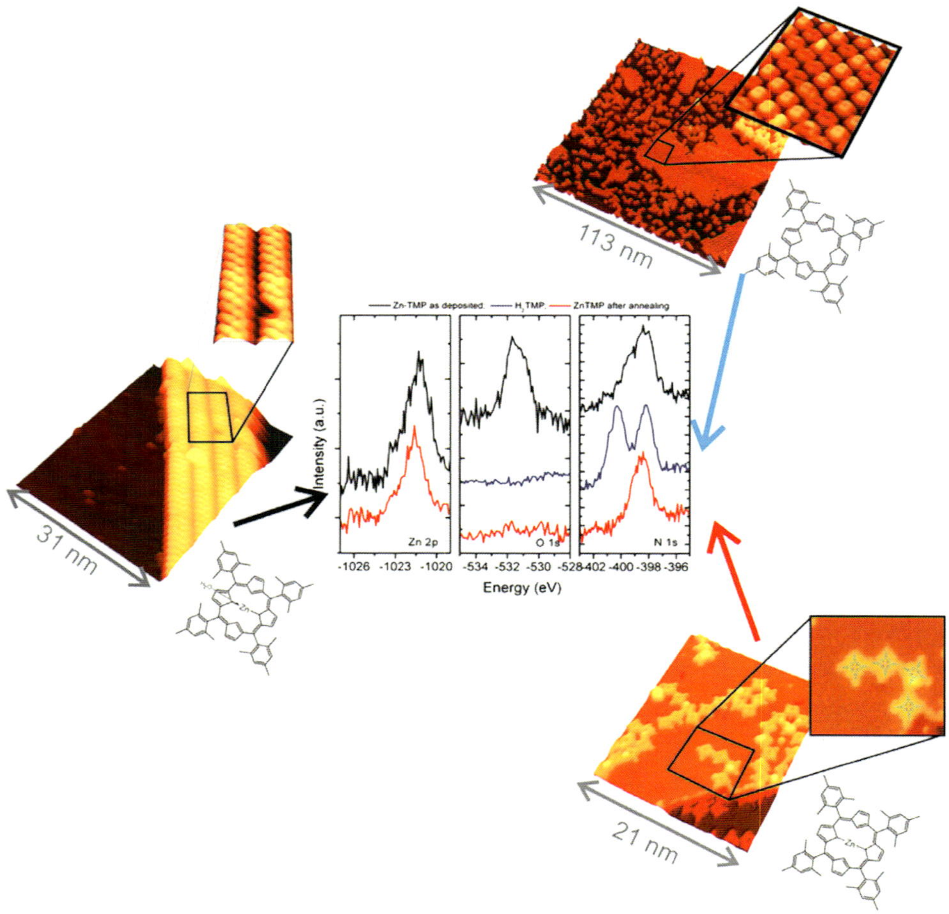

FIGURE 19.8. Correlation between XPS spectra on self-assembled tetramesitylporphyrin (TMP) derivatives on Cu(100) and characteristic STM images. Upon deposition of Zn-TMP, a clear oxygen peak can be found in the XPS spectra, a situation in which STM images reveal nanorod formation (*left inset*). Deposition of base-free H_2-TMP, however, does not show either oxygen in the XPS spectra or nanorod formation (*top inset*). Similarly, water detachment achieved by annealing to 525 K also leads to nanorod dissociation.

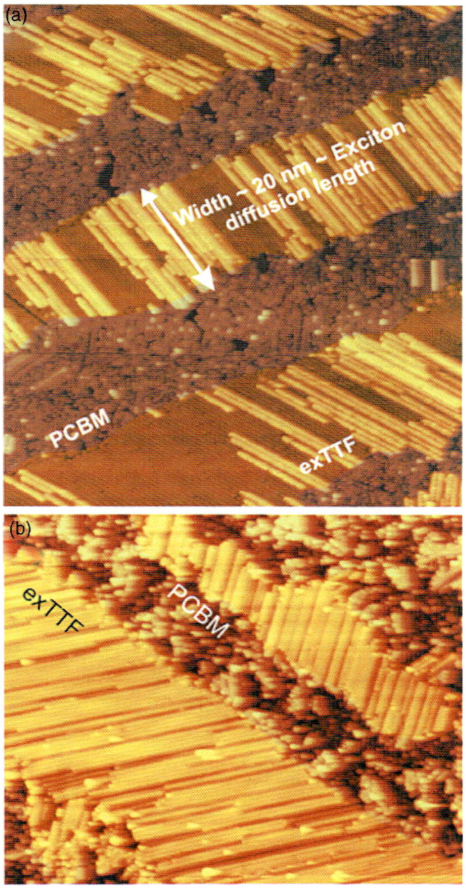

FIGURE 19.19. Lateral nanoscale organic donor/acceptor superlattice. (**a**) Nanoscale segregation of electron-donor/electron-acceptor molecules on Au(111) (118 nm × 132 nm). Since both PCBM and exTTF molecules are imaged with similar heights, a semitransparent color has been superimposed on the disordered PCBM areas to enhance visibility. The width of the exTTF stripes is about 20 nm, of the same order as typical exciton diffusion lengths, and there exists a large acceptor/donor interface, as required for optimum solar-cell performance. (**b**) (118 nm × 115 nm) The selective adsorption of excess exTTF molecules on areas previously covered by exTTF molecules implies that this morphology can be extended beyond the first monolayer.

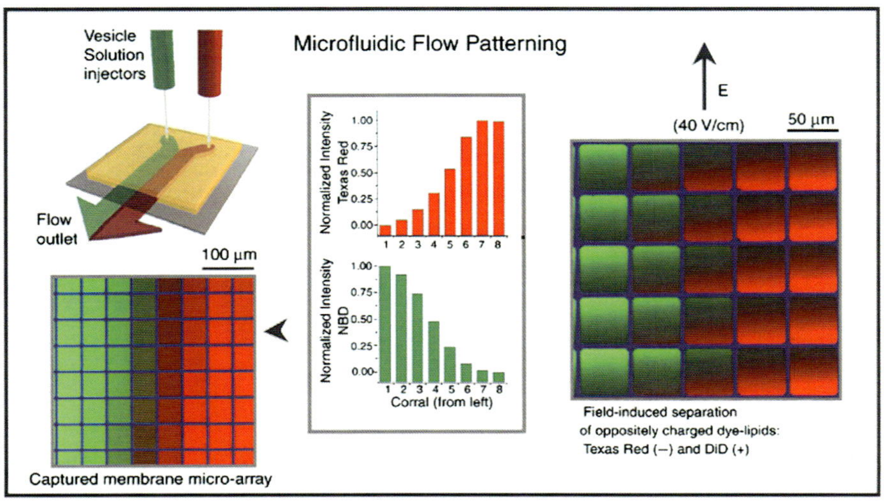

FIGURE 20.14. Addressing by laminar flow in a microfluidic channel. Diffusive mixing in a microchannel under laminar flow conditions provides a concentration gradient of different dye-labeled vesicles. The concentration of vesicles in the gradient is reflected in the surface concentration of each membrane in the resultant array. The array shown is a mixture of Texas Red-labeled lipids and DiD-labeled lipids. Since the dyes have opposite charge, they can be separated in an electric field [7].

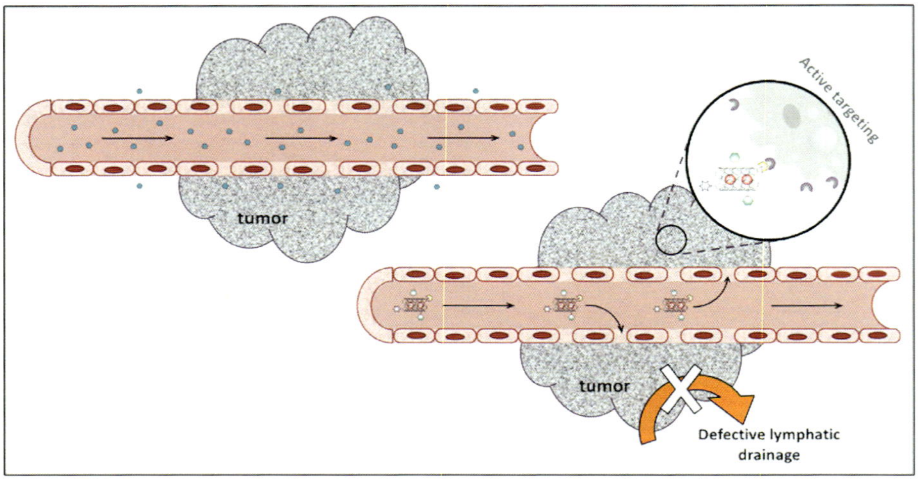

FIGURE 21.5. Enhanced permeability and retention effect for a model CNT-based drug delivery system.

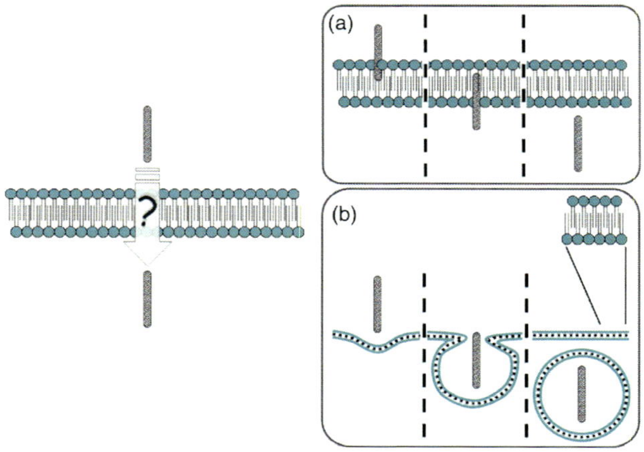

FIGURE 21.6. Nano-needle (**a**) and endocytosis (**b**) penetration mechanism.

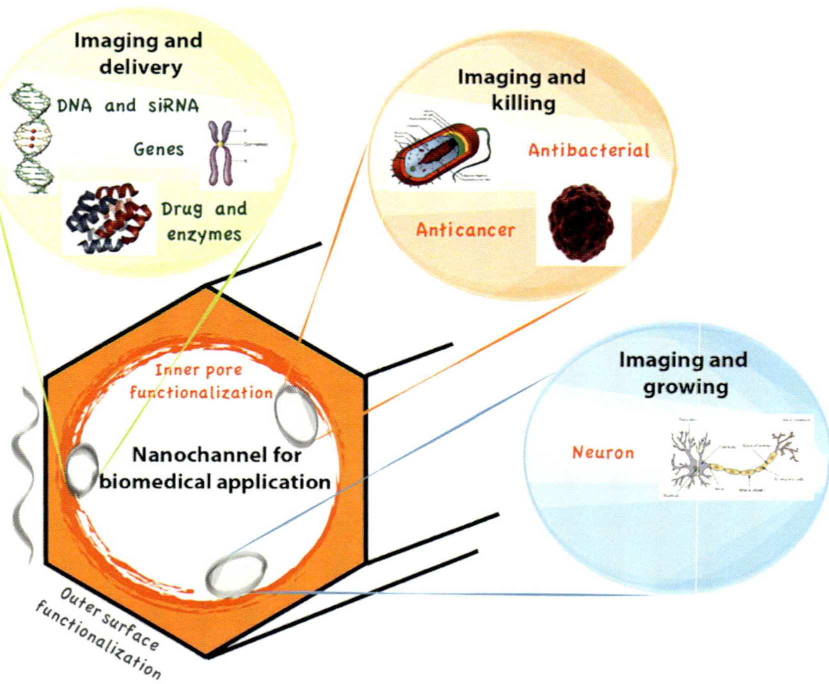

FIGURE 22.2. Schematic illustration of multifunctional nanocontainers and their potential bio applications.

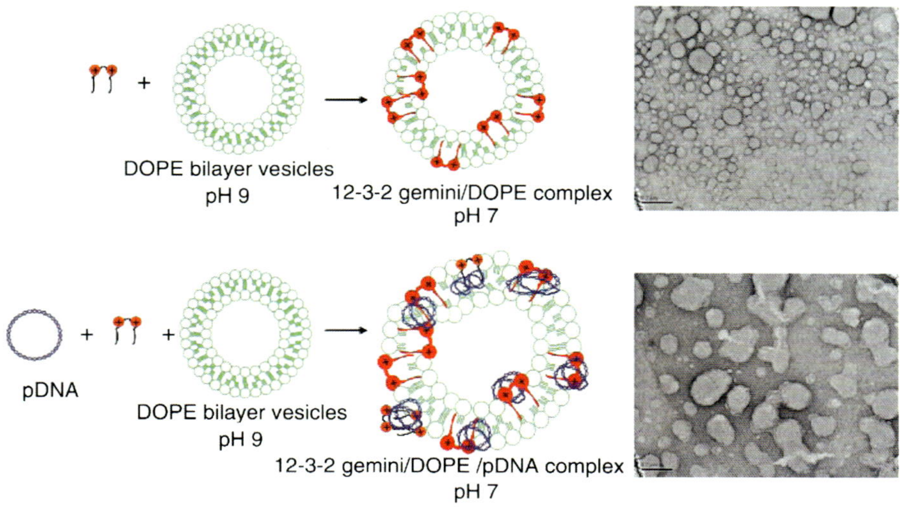

FIGURE 23.5. Schematic illustration of gemini nanoparticle formation and comparative electron micrographs of the structural morphology of gemini nanoparticles and plasmid–gemini nanoparticles prepared from 12–3–12/DOPE (**A**) and 12–3–12/DOPE/plasmid (**B**) by negative staining. Complexation of the plasmid DNA in the gemini/DOPE nanoparticles results in slightly larger particle size distribution and "fuzzy" particles, which is indicative of the presence of an additional polymorphic phase [92] and confirms our previous results on the structural features of these nanoparticles by small-angle x-ray scattering [93].

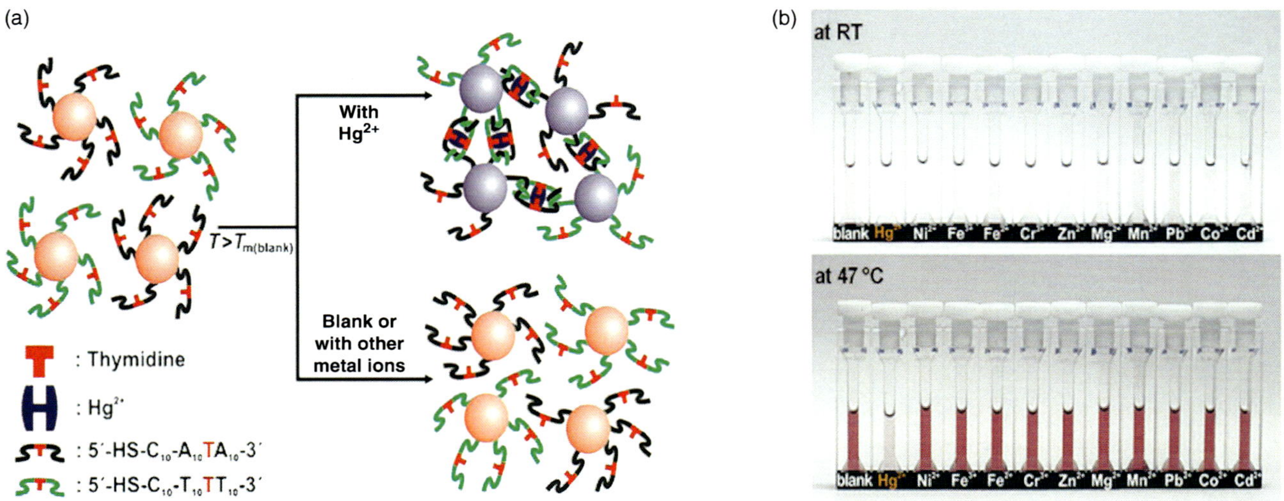

FIGURE 24.2. (**a**) Scheme of the colorimetric detection of Hg(II) ions using hybrid DNA–Au NPs. (**b**) Color change of the aggregates in the presence of various representative metal ions upon heating from room temperature to 47°C. Reproduced from reference 37 with permission from the Wiley-VCH.

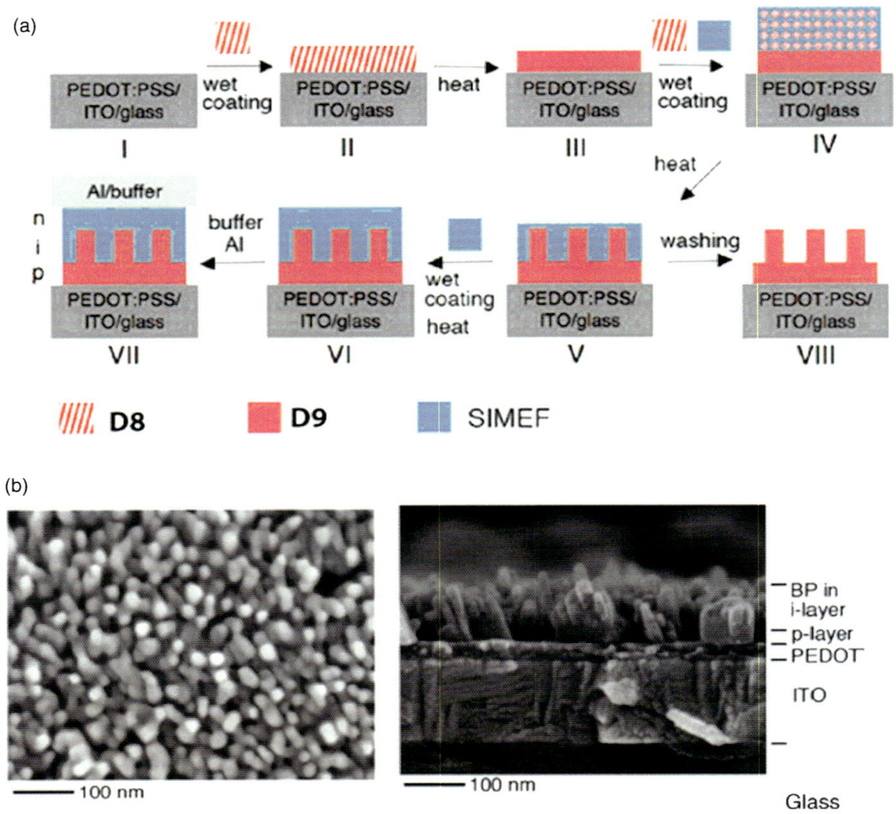

FIGURE 25.26. (**a**) Technology for the solar cell construction based on SIMEF and **D9/D10**. (**b**) Top view (*left*) and side view (*right*) SEM images of the exposed porphynin **D10** network after removing SIMEF material. Reproduced from reference 49, with permission of the American Chemical Society.

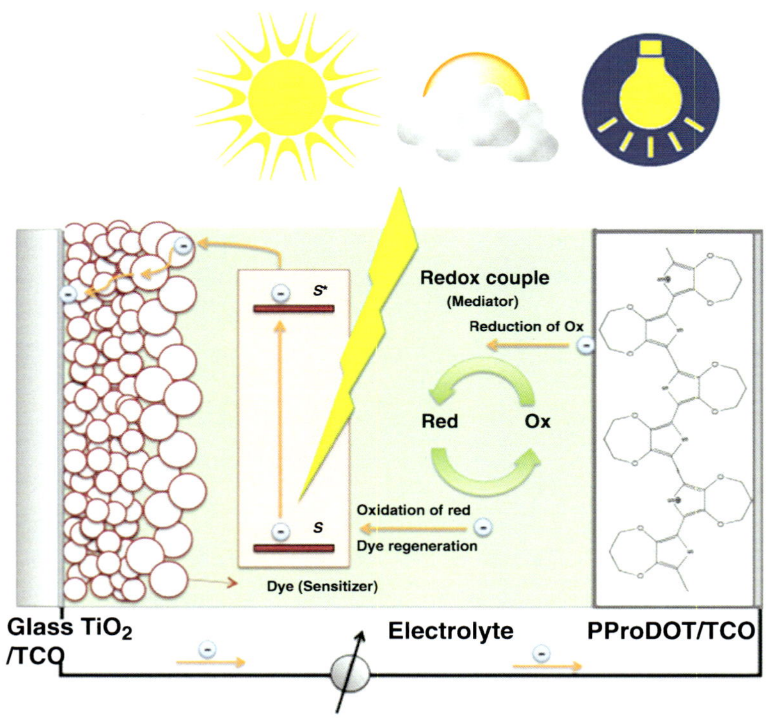

FIGURE 26.17. Schematic diagram of DSSCs illustrating PProDOT as a counterelectrode. The diagram is for illustration and is not to scale.

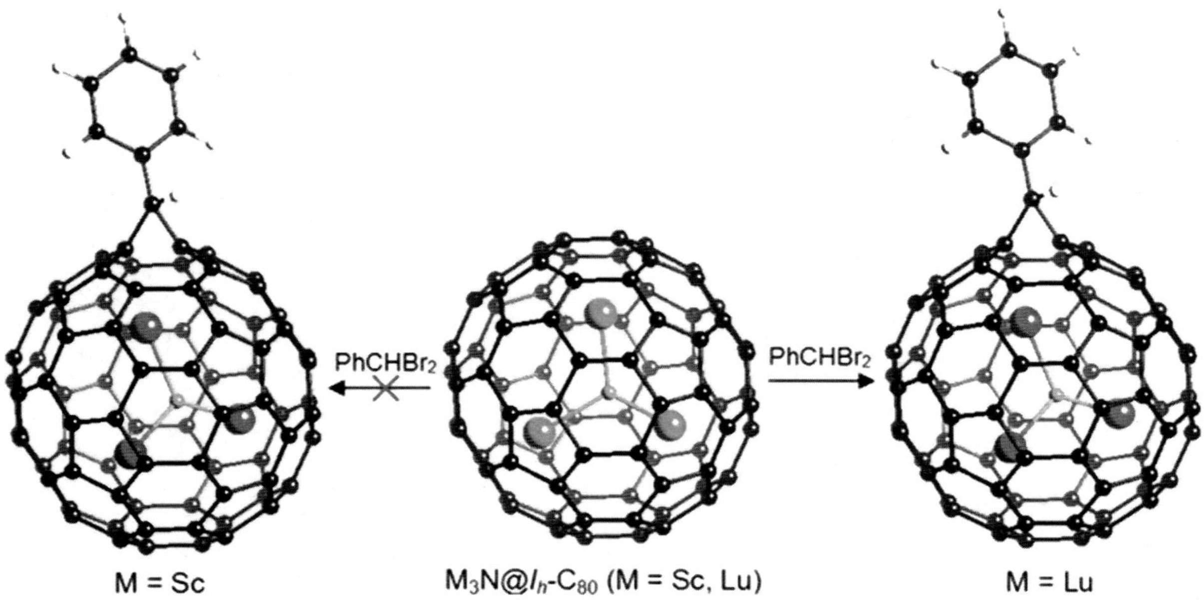

FIGURE 13.30. Reactions of $Sc_3N@I_h–C_{80}$ and $Lu_3N@I_h–C_{80}$ with benzyl bromide under CPE conditions. Reprinted from reference 153 with permission. Copyright © 2011, American Chemical Society.

observed. Notably, the first oxidation and reduction potentials are comparable to those observed for $Lu_3N@I_h–C_{80}$–PCBM and $Lu_3N@I_h–C_{80}$–PCBH, which were shown to be excellent acceptor materials in organic photovoltaic devices [155, 156]. As reported, the increase in the reduction potential is expected to lead to high open-circuit voltages (Voc) and consequently to relatively high efficiencies in OPV devices [155–159].

$[Sc_3N@I_h–C_{80}]^{2-}$ was also produced by CPE under the same conditions. Surprisingly, $[Sc_3N@I_h–C_{80}]^{2-}$ did not react with benzyl bromide (Figure 13.30). The $Sc_3N@I_h–C_{80}$ dianion is clearly much less reactive toward electrophiles than the $Lu_3N@I_h–C_{80}$ dianion. DFT calculations of both dianionic metallofullerenes showed that their electronic structures are very different. The HOMO is mainly localized on the fullerene cage for $[Lu_3N@I_h–C_{80}]^{2-}$, whereas for $[Sc_3N@I_h–C_{80}]^{2-}$, it is mostly localized on the metal cluster (Figure 13.31). Therefore, $[Lu_3N@I_h–C_{80}]^{2-}$ is more electron-rich consequently more reactive than $[Sc_3N@I_h–C_{80}]^{2-}$. The computational studies provided a clear explanation for the different reactivities and the strong influence of the endohedral cluster on the exohedral chemistry.

13.5 CARBON NANO-ONIONS (CNOs)

Multishell fullerenes are commonly known as onion-like structures or simply carbon nano-onions. They consist of a hollow spherical fullerene core surrounded by larger concentric fullerene layers forming quasi-spherical multilayers [160, 161]. Although they were discovered soon after carbon nanotubes, it has not been until recently that synthetic methodologies have been developed to produce them in sufficient amounts [162, 163]. Hence, research on these promising materials has been somewhat limited [37, 160–169]. Annealing of nanodiamond powder seems to be the most effective way to prepare large quantities of CNOs [167]. This method yields a distribution of sizes (5–6 nm) and shell numbers (6–8) [37, 61]. The nomenclature of CNOs is analogous to that of endohedral fullerenes, where the number of outer-cage carbons follow the number of inner-shell carbons using the

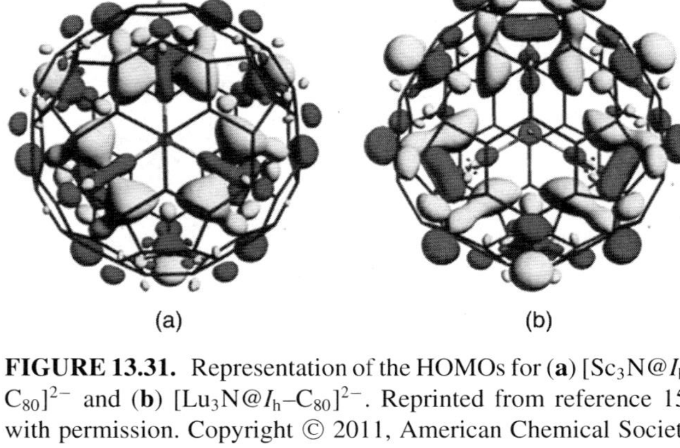

FIGURE 13.31. Representation of the HOMOs for (**a**) $[Sc_3N@I_h–C_{80}]^{2-}$ and (**b**) $[Lu_3N@I_h–C_{80}]^{2-}$. Reprinted from reference 153 with permission. Copyright © 2011, American Chemical Society. See color insert.

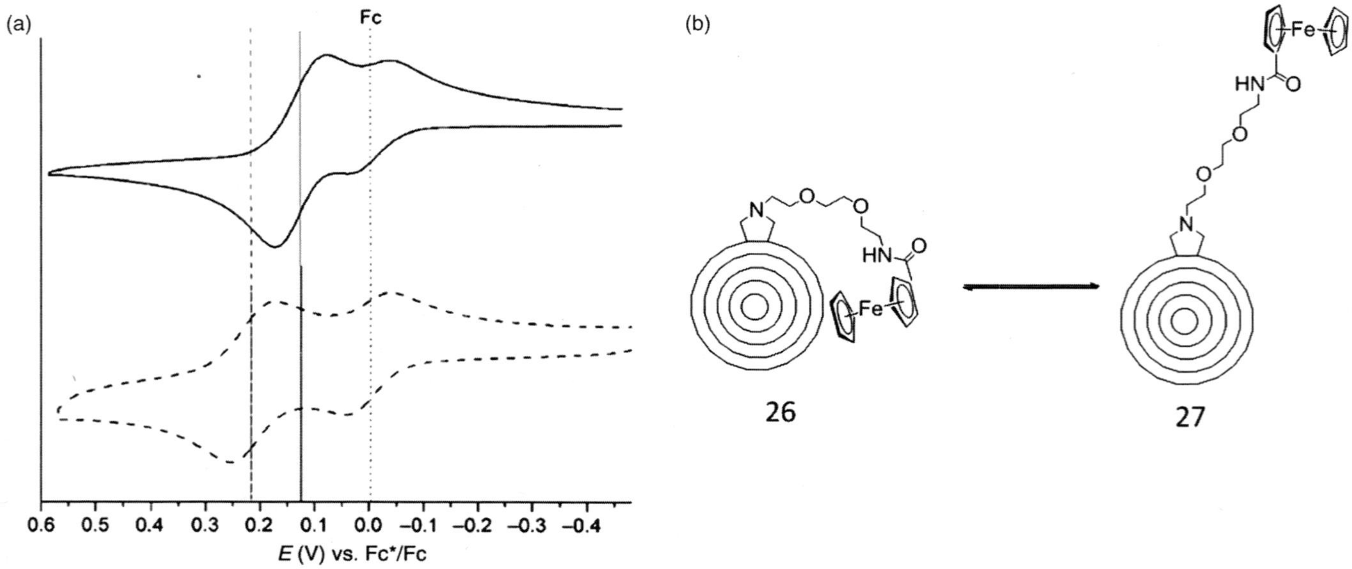

FIGURE 13.32. (**a**) Cyclic voltammograms of ferrocene functionalized CNO (*bottom*) and reference ferrocene moeity (*top*) in THF (0.1M NBu_4PF_6) at a scan rate of 0.1 Vs^{-1}. (**b**) Schematic representation of "face-to-face" conformation (*left*) and "loose" conformation (*right*) of ferrocene functionalized CNO. Reprinted from reference 166 with permission. Copyright ©2009, Wiley–VCH Verlag GmbH & Co. KGaA, Weinheim.

@ symbol; for example C_{60}@C_{240} denotes a C_{60} cage inside a C_{240} cage to make a two-shell CNO. To date, there are no reported methodologies to prepare and isolate *small*-CNOs (e.g., 2–3 shells) or size-homogeneous samples of CNOs. Nonetheless, research on these multilayered fullerenes is gaining strength thanks to their potential use as capacitor active materials [168] and as scaffold nanoparticles.

Since pristine CNOs are insoluble in protic and aprotic solvents, their electrochemical studies have only been possible after surface functionalization [164]. These measurements therefore reflect the electronic properties of both the CNOs and the functional groups attached to them [166]. It has been found that after oxidation with diluted HNO_3, CNOs are electrochemically active at negative potentials due to the presence of a variety of different oxygen-containing functional groups (e.g., carbonyl and carboxylate) [164, 166]. Cioffi et al. also reported a ferrocene functionalized CNO derivative that showed an anodic shift (about 60 mV) with respect to the oxidation potential of the unattached ferrocene moiety (Figure 13.32a). This shift was attributed to a possible through-space interaction between the attached ferrocene moiety and the CNO core (Figure 13.32b) [166].

Other electrochemical studies have been focused on the capacitive properties of pristine or composite-based (i.e., polymer/CNO mixtures) films of CNOs deposited on the surface of a working electrode (e.g., glassy carbon) [165, 169] or conductive substrate [168]. The composites create a nanostructured polymeric matrix that couples, and consequently modifies the electronic properties of both components, as previously studied with carbon nanotube-based materials [164, 170–174]. The cyclic voltammograms of such films show pseudo-rectangular cathodic and anodic profiles (Figure 13.33), expected for double-layer capacitive materials [164, 165]. These extraordinary results led to further studies to determine the potential application of CNOs as supercapacitors [175].

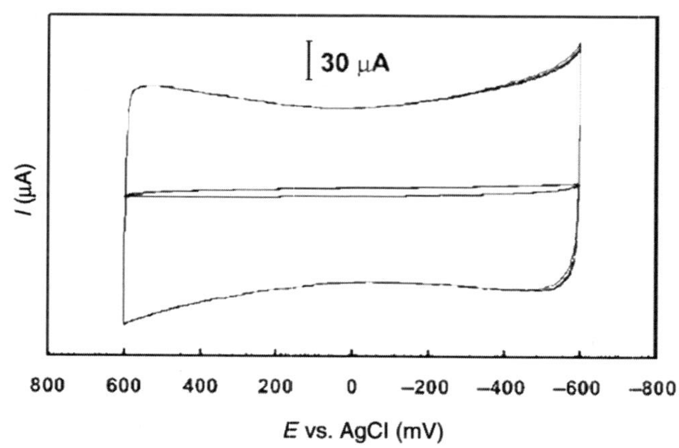

FIGURE 13.33. Cyclic voltammogram of CNOs/Chit (1:1.5) composite deposited on glassy carbon disc electrode in 0.1M H_2SO_4 in water. Sweep rate of 100 mV s^{-1}. Reprinted from reference 164 with permission. Copyright © 2010, Royal Society of Chemistry.

13.6 FULLERENE-BASED COMPOUNDS FOR POTENTIAL PHOTOVOLTAIC APPLICATIONS

13.6.1 C_{60}-Based Dyads and Triads

The remarkable electron acceptor properties of ground-state C_{60}, combined with its reasonably high absorption throughout the visible spectral region and its ability for rapid photoinduced charge separation, have made it a most desirable acceptor moiety for the construction of covalent donor–acceptor dyads. C_{60} and its derivatives are among the most intensely studied organic materials for photovoltaics applications [176, 177].

The number of potential donor moieties that can be attached to fullerenes for the preparation of these dyads is very large. In this chapter, we focus on the most frequently used donor moieties as shown in Figure 13.34.

13.6.1.1 Ferrocene-Based Donors. Ferrocene (Fc) is commonly used as an electron donor because the ferrocenyl centers can be easily and reversibly oxidized [178]. A series of covalently linked fullerene/Fc-based donor–bridge–acceptor dyads were reported by Guldi et al. [179] in 1997 (**29–33** in Figure 13.35). The fluorescence of dyads **29–33** in methylcyclohexane at 77 K was substantially quenched, relative to the reference compound, *N*-methylfulleropyrrolidine **28**, indicating intramolecular quenching of the fullerene excited singlet state. Picosecond-resolved photolysis of dyads **29–33** in toluene showed light-induced formation of the excited singlet state, which in turn underwent rapid intramolecular quenching. These results indicate that the electron transfer from the ferrocenyl moiety causes bleaching of the fullerene singlet excited state. The different spacers between C_{60} and Fc may be responsible for two different quenching mechanisms: (a) through-bond electron transfer for dyads **29**, **32**, and **33** (b) and formation of a transient intramolecular exciplex for dyads **30** and **31**. No electrochemical properties for these fullerene/ferrocene-based dyads were discussed in this study.

In 2000, fullerene/Fc-based triads were prepared by Herranz et al. [180] (see Figure 13.36). These C_{60}-based triads are constituted by a fulleropyrrolidine moiety and two different electroactive units: two donors consisting of a Fc and a tetrathiafulvalene (TTF) as in **34**, or two donors consisting of a Fc and a π-extended TTF (ext-TTF) as in **35a** and **35b**. In the case of **36** and **37**, the fulleropyrrolidine is connected to one donor (Fc) and one other acceptor: anthraquinone (AQ) in **36** and tetracyanoanthraquinodimethane (TCAQ) in **37** (see Figure 13.40). Photophysical studies showed intramolecular electron transfer (ET) processes from the stronger electron donor (i.e., TTF or extended TTF) to the fullerene singlet excited state, rather than from the ferrocene donor in **34** and **35a,b**. No evidence for a subsequent ET from Fc to $TTF^{\bullet+}$ or

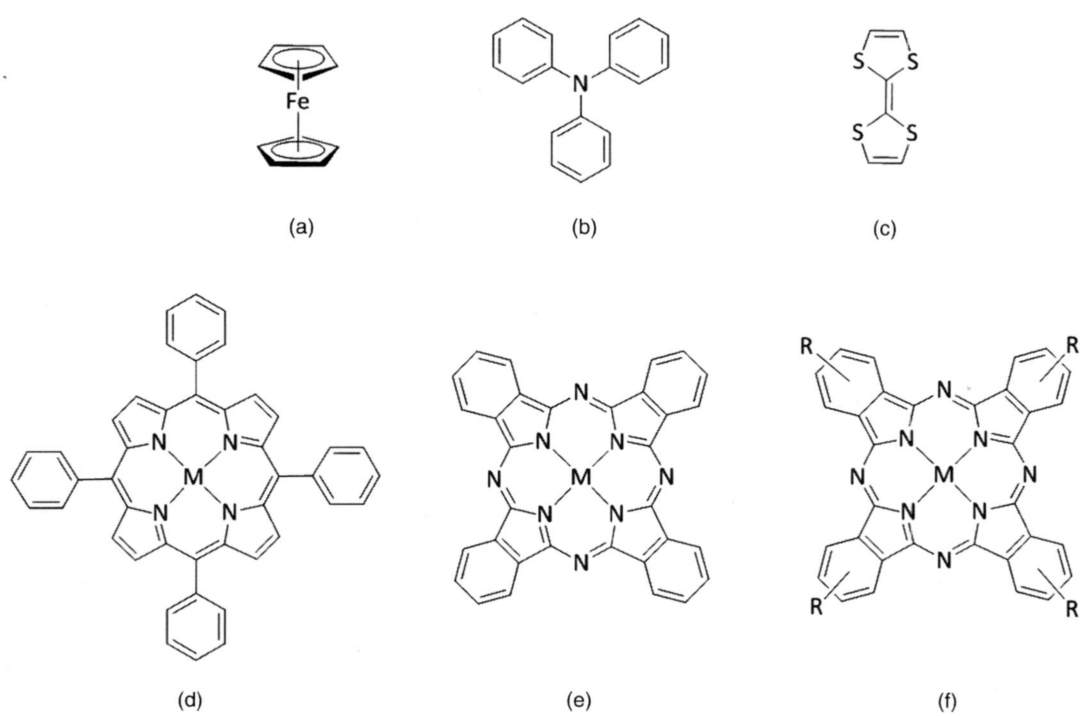

FIGURE 13.34. Commonly used donor moieties for the construction of donor–(fullerene) acceptor systems. (**a**) Ferrocene, (**b**) triphenylamine (TPA), (**c**) tetrathiafulvalene (TTF), (**d**) porphyrin, (**e**) phthalocyanine, and (**f**) subphthalocyanine.

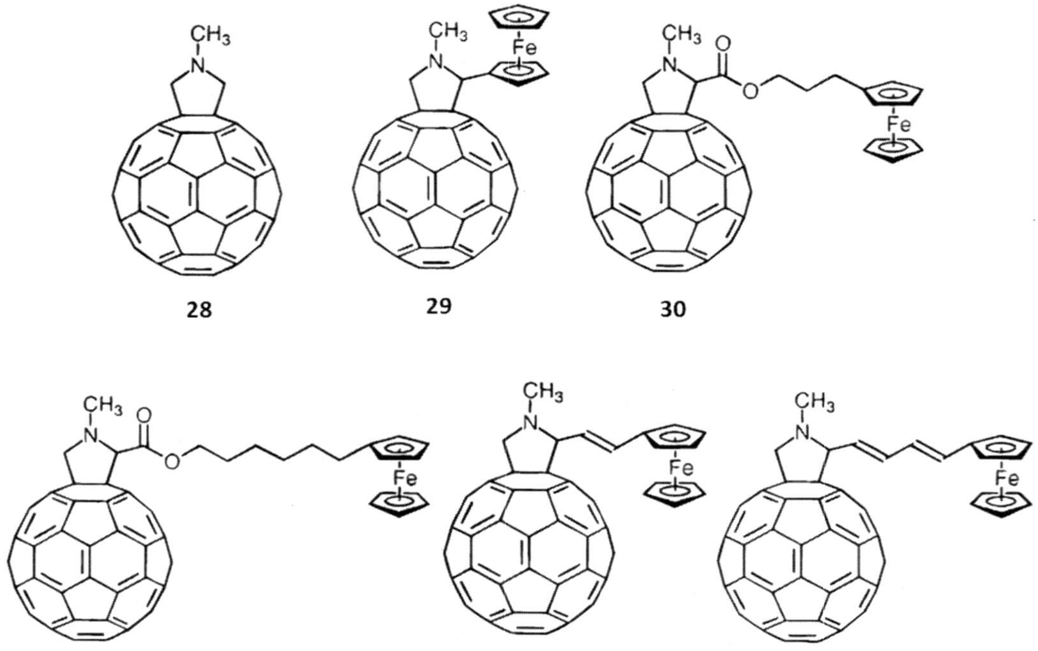

FIGURE 13.35. Molecular structures of fullerene/ferrocene-based donor–bridge–acceptor dyads. Reprinted from reference 179 with permission. Copyright © 1997, American Chemical Society.

π-extended $TTF^{\bullet+}$ was observed. In these triads, intramolecular ET from the TTF and π-extended TTF moieties to the fullerene singlet excited state yielded the respective charge separated radical pairs. An alternative ET from the Fc donor was not observed, due to the unfavorable free energy changes associated with the ET process. In contrast, the singlet lifetimes in triads **36** and **37**, bearing C_{60} and one other acceptor, reveal that only a reaction between Fc as donor and C_{60} as acceptor occurs.

The electrochemical behavior of these triads was studied by cyclic voltammetry. Redox potentials are listed in Table 13.4. Triads **34** and **35a** display four reduction waves corresponding to the first four reduction steps of the fullerene moiety. As expected, all of them are cathodically shifted compared to those of pristine C_{60} [178]. Two oxidation steps were also observed for triad **34**. The first one, at 0.49 V versus SCE, is quasi-reversible and corresponds to the formation of the radical cation of the TTF unit. The second one at 0.73 V versus SCE is very broad and was ascribed to the superimposed features of the second oxidation step of the TTF moiety and the first oxidation step of the Fc unit (see Table 13.4). While triad **35a** shows two oxidation waves at 0.55 V (two electrons) and 0.75 V (one electron), triad **35b** undergoes three reduction steps, all of which are cathodically shifted compared to those of the parent C_{60}. It displays one broad oxidation wave at 0.70 V, which corresponds to

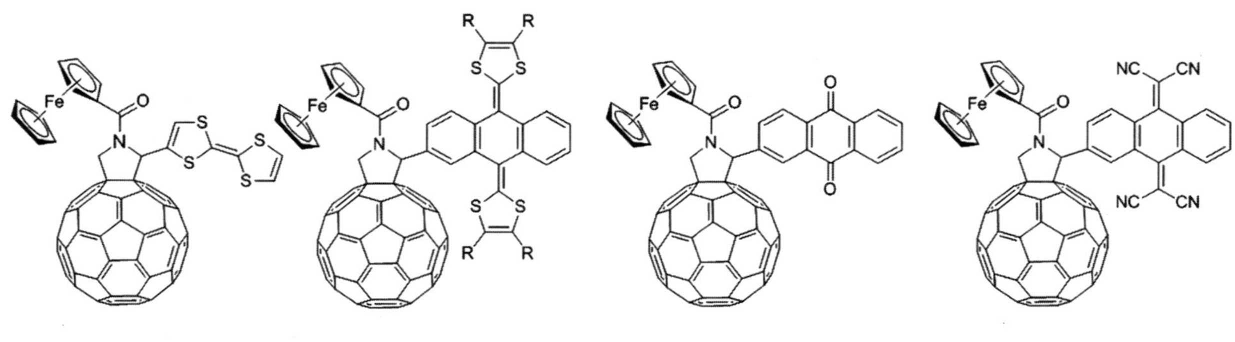

FIGURE 13.36. Fullerene/ferrocene–based triads.

TABLE 13.4. Redox Potentials of the C_{60} Triads 34, 35a,b, 36, 37, and C_{60} as Reference Compounds

Compound	E^1_{ox}	E^2_{ox}	E^1_{red}	E^2_{red}	E^3_{red}	E^4_{red}
34[a]	0.49	0.73 (broad)	−0.62	−1.25	−1.65	−1.96
35a[a]	0.55 (2e⁻)	0.75	−0.63	−1.07	−1.64	−2.14
35b[a]	0.70 (3e⁻) (broad)		−0.66	−1.06	−1.64	
36[b]	0.62		−0.59	−0.95	−1.65	
37[b]	0.61		−0.60	−1.00	−1.61	
C_{60}[a]			−0.60	−1.00	−1.52	−1.93

[a]Potentials in V versus SCE; Tol/MeCN (4:1) as solvent.
[b]Potentials in V versus SCE; Tol/MeCN (5:1) as solvent; scan rate 200 mV/s; 0.1 mol/L $Bu_4N^+ClO_4^-$ as supporting electrolyte; GCE as working electrode.

the simultaneous transfer of three electrons. The anodic shift of the oxidation potential is indicative of the substitution of the 1,3-dithiole rings. Introduction of SMe groups on the 1,3-dithiole ring results in a poorer electron donor system than the parent unsubstituted TTF (see Table 13.4). For triads **36** and **37**, three one-electron reduction steps that correspond to the [60] fullerene moiety are observed. The reduction potential values are quite close to those of the parent C_{60} due to the presence of the electron-acceptor unit bound to the pyrrolidine ring. This study revealed that although the electroactive units in the triads preserve their identity, a noticeable electronic interaction between these units is present.

In 2007, two new triads based on *N*-methylfulleropyrolidine, oligothienylenevinylenes (*n*TV), and ferrocene (Fc) were synthesized, namely C_{60}–*n*TV–Fc ($n = 2, 4$). The structures are displayed in Figure 13.37 [181]. In both polar and nonpolar solvents, photoinduced charge-separation (CS) processes take place from the singlet excited states of C_{60} and *n*TV. The charge-separation process for both triads and the reference compound C_{60}–2TV were confirmed by fluorescence quenching. However, the lifetimes of the CS states of C_{60}–2TV–Fc were shorter than those of C_{60}–4TV–Fc. These results suggest that the introduction of the Fc donor moiety connected to C_{60} via the longer chain of TV effectively increases the ability and efficiency of the charge-separation processes.

The electrochemical characteristics of C_{60}–*n*TV–Fc triads were examined by using cyclic voltammetry (CV) and Osteryoung square-wave voltammetry (OSWV). As an example, the cyclic voltammogram of **38** is shown in Figure 13.38. On the cathodic side, both **38** and **39** show three reversible reduction waves. The first two waves can be assigned to the fullerene moiety, while the third one is attributed to the simultaneous third reduction step of the fullerene cage and the reduction of the *n*TV–Fc moiety. On the anodic side, one reversible oxidation wave appears at 0.04 V for **38** and at 0.09 V for **39**. Therefore, considering the similar redox potential values for ferrocene and *n*TV–Fc, these low oxidation potential waves can be assigned to the ferrocene [181]. The experimentally measured HOMO–LUMO gaps are low: 1.11 eV for **38** and 1.09 eV for **39**, making these systems good candidates to act as molecular rectifiers [182].

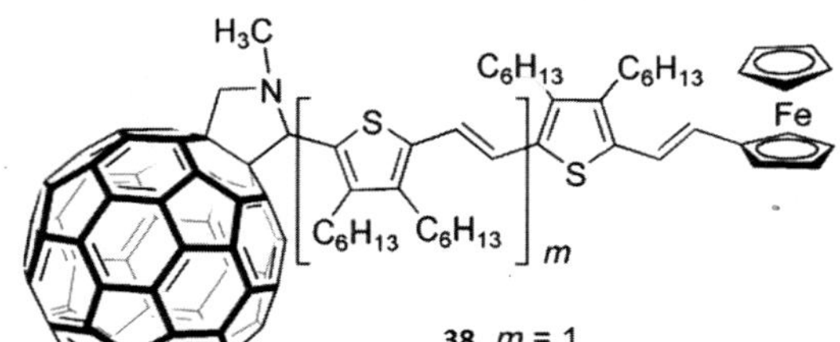

38 *m* = 1

39 *m* = 3

FIGURE 13.37. Structures of C_{60}-2TV-Fc (**38**) and C_{60}-4TV-Fc (**39**). Reprinted from reference 181 with permission. Copyright © 2007, Wiley–VCH Verlag GmbH & Co. KGaA, Weinheim.

13.6.1.2 Triphenylamine-Based Donors.

Triphenylamine (TPA) and its derivatives have been successfully employed as donor systems in the construction of small molecular donor-solution-processable organic solar cells [183–186], TiO_2 dye-sensitized solar cells [187, 188], and fullerene donor–acceptor conjugates [189–191]. Although TPA derivatives are not good absorbers in the visible range,

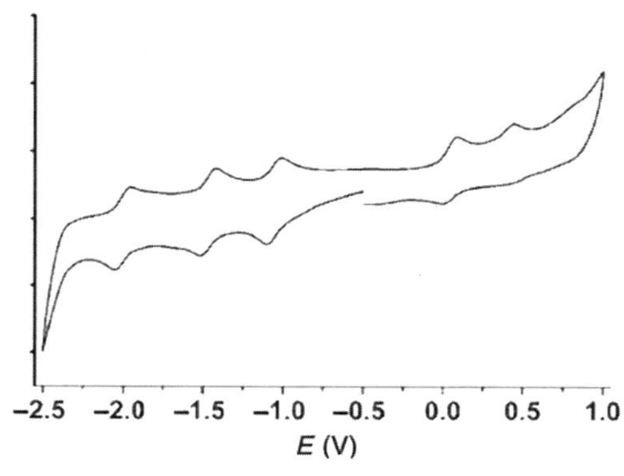

FIGURE 13.38. Cyclic voltammogram of C_{60}-2TV-Fc (**38**) in *o*-DCB/acetonitrile 4:1 containing 0.1 m (*n*-Bu)$_4$NClO$_4$. Scan rate 100 mV/s. Reprinted from reference 181 with permission. Copyright © 2007, Wiley–VCH Verlag GmbH & Co. KGaA, Weinheim.

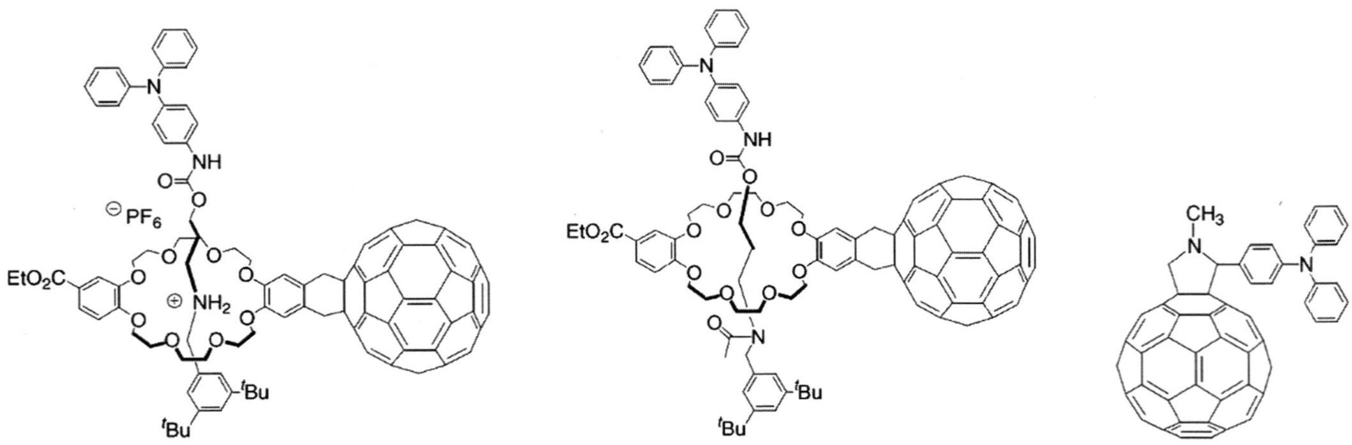

40 **41** **42**

FIGURE 13.39. Structures of fullerene–triphenylamine dyads. Reprinted from reference 194 with permission. Copyright © 2004, American Chemical Society.

they are good donors and have excellent hole transporting properties [192]. They have a propeller structure that is advantageous for solution processability because after evaporation of the solvent they produce highly homogeneous thin films [184]. Finally, they have the ability to form and stabilize cation radicals [193]. Based on all these properties, fullerene–TPA dyads, triads, and tetrads have been studied by several groups [189–191, 194–197].

Photoinduced electron-transfer processes for systems involving [C_{60}]fullerene and TPA moieties were explored in 2004 by Sandanayaka et al. [194]. In one system the donor and acceptor moieties were tethered by rotaxane structures, and in the other they were covalently connected (see Figure 13.39). Results indicate that in the rotaxane-tethered systems (**40**, **41**), an efficient, photoinduced CS process takes place via $^3C_{60}{}^*$ and produces long-lived CS states lasting 360 and 290 ns, respectively, in benzonitrile. However, in the covalently bonded system (**42**), fast charge separation occurs in the excited singlet state (via $^1C_{60}{}^*$), followed by fast charge recombination.

In connection with the covalently bonded C_{60}-TPA dyad (**42**), Pinzón et al. [197] demonstrated the effect of the TPA donor system linkage position (N-substituted versus 2-substituted pyrrolidine) in the formation of the photoinduced CS states. It was found that when the TPA donor is connected to the pyrrolidine nitrogen atom as in **43** (Figure 13.40), the resulting dyad produces a significantly longer-lived radical pair (k_{CR}/PhCN $= 1.9 \pm 0.5 \times 10^9$ s^{-1}) than the corresponding 2-substituted isomer (**42**) (k_{CR}/PhCN $= 6.5 \pm 0.5 \times 10^9$ s^{-1}) because by linking TPA to the nitrogen of the pyrrolidine ring, the donor–acceptor separation increases, which exerts a significant impact on the charge separation dynamics, namely a notable slow down.

Studies of C_{60}-TPA triads **44a,b** have also been reported. These triads are composed of three triphenylamine units covalently linked at meso positions of the porphyrin ring and one fulleropyrrolidine connected at the fourth meso position (see Figure 13.41). The TPA units were included to act as energy-transferring antenna units and to enhance the electron-donating ability of both the free-base and zinc(II) porphyrin derivatives [190]. Spectral and computational studies revealed appreciable electronic interactions between the

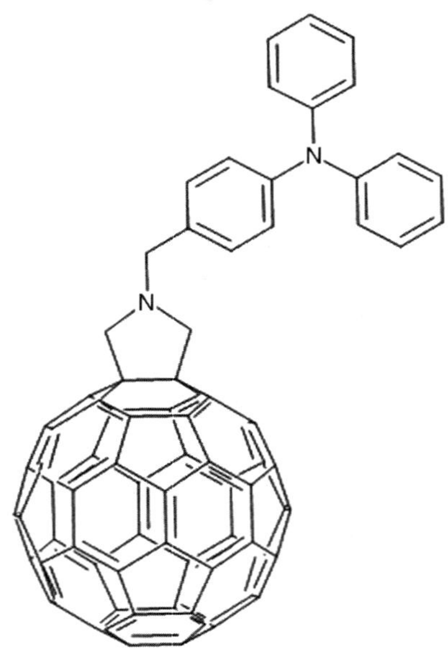

43

FIGURE 13.40. Structure of TPA donor connected to C_{60} via the pyrrolidine nitrogen atom.

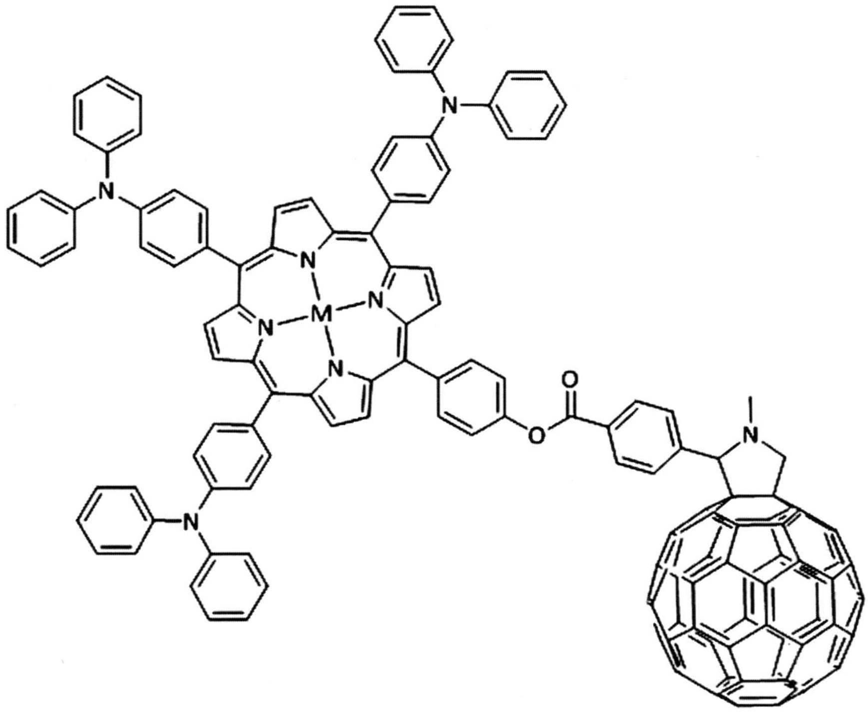

44a M = 2H
44b M = Zn

FIGURE 13.41. Molecular structure of C_{60}–TPA–porphyrin triads.

porphyrin π-system and the meso-substituted triphenylamine entities. These moieties acted together as an electron donor while the fullerene moiety acted as electron acceptor in the conjugates. Remarkably, charge-separated states with lifetimes on the order of a few microseconds were observed as a result of the delocalization of the porphyrin π-cation radical to the triphenylamine entities. This study presented a new route to generate long-lived charge-separated states.

Cyclic voltammetric studies were conducted to evaluate the redox potentials of the conjugates and the energetics of the electron-transfer processes. The first oxidation and the first reduction were found to be reversible for both triads. Additional peaks appeared when scanning the potential further into the anodic direction, indicating that films formed on the electrode surface. Figure 13.42 shows the cyclic voltammogram of conjugate **44a**. The HOMO–LUMO gaps for **44a**, calculated from the first oxidation to the first reduction of the fullerene-centered electrochemical steps, are 1.53 and 1.40 V in *o*-DCB and PhCN, respectively. These values are smaller by about 100 mV than those reported earlier for tetraphenylporphyrin–fullerene derivatives [198, 199]. The smaller values of the HOMO–LUMO gaps for **44a**,**b** are attributed to the presence of the triphenylamine entities.

C_{60}-tetrads containing a triphenylamine central building block were investigated recently by Seok et al. [196], who synthesized bisferrocene- and bisfullerene-substituted compounds. One tetrad, C_{60}–TPA–$(Fc)_2$ (**45**), has a TPA linked to one C_{60} moiety and two Fc moieties. Another tetrad, $(C_{60})_2$–TPA–Fc (**46**), has a TPA linked with two C_{60} moieties and one ferrocene unit, as shown in Figure 13.43. The electrochemical properties of these two compounds were

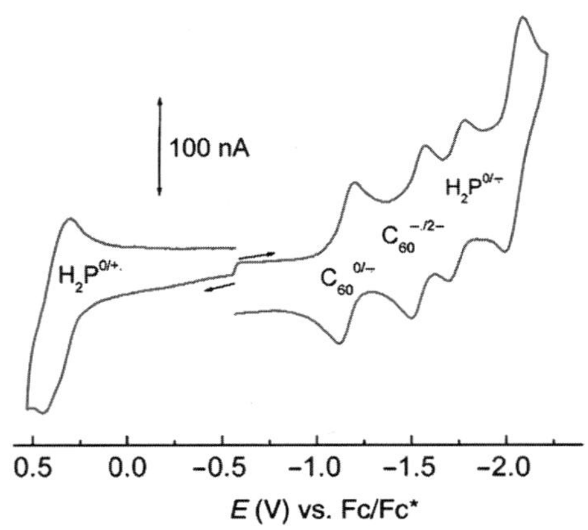

FIGURE 13.42. Cyclic voltammogram of **44a** in o-DCB, 0.1 M $(n\text{-}Bu_4N)ClO_4$. Scan rate 100 mV/s. Reprinted from reference 190 with permission. Copyright © 2007, American Chemical Society.

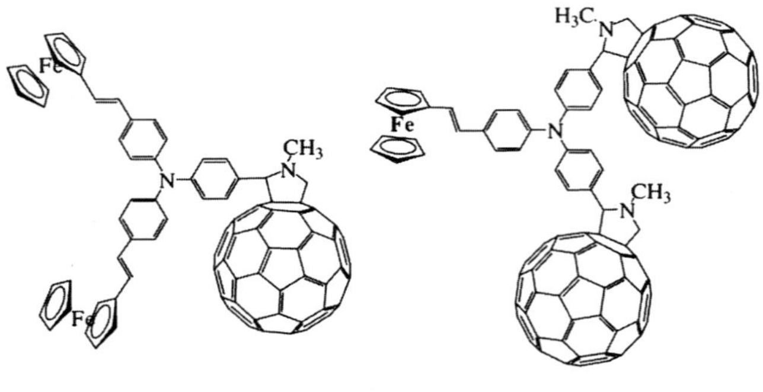

C_{60}-TPA-(Fc)2 (C_{60})2-TPA-Fc

45 **46**

FIGURE 13.43. Molecular structures of compounds **45** and **46**. Reprinted from reference 196 with permission. Copyright © 2009, Elsevier.

probed by cyclic voltammetry. As a general feature, both compounds exhibit three reversible one-electron reduction waves, which are attributed to three reduction potentials of the C_{60} cage. In the anodic region, a reversible oxidation observed at +0.07 V versus. Fc/Fc^+ corresponds to Fc ($E_{ox}{}^{Fc}$), while the other reversible oxidation at +0.80 V is assigned to TPA ($E_{ox}{}^{TPA}$). Photoinduced electron-transfer studies by time-resolved spectroscopic techniques showed that the ratio of Fc-donor to C_{60}-acceptor affects the charge separation efficiency via the excited singlet state of C_{60}. The CS process in C_{60}–TPA–$(Fc)_2$ turned out to be more efficient than that in $(C_{60})_2$–TPA–Fc.

13.6.1.3 TTF and exTTF-Based Donors. Upon oxidation at very low redox potentials, TTF forms stable radical cation and dication, which result in the formation of dithiolium aromatic rings (see Figure 13.44). This gain in aromatization energy is the reason for TTF's excellent donor properties [226, 227]. Functional TTF derivatives (π-extended TTFs, i.e., *ex*TTF) based on the electron-donor and charge-transfer abilities of TTF, have been synthesized as materials for different applications, which include electrochemical switches, sensors, surface modification agents, and so on [200–202].

Contrary to other donor systems, *ex*TTF is a ground-state donor system. The driving force for this process is the aromatization that *ex*TTF undergoes after oxidation at relatively

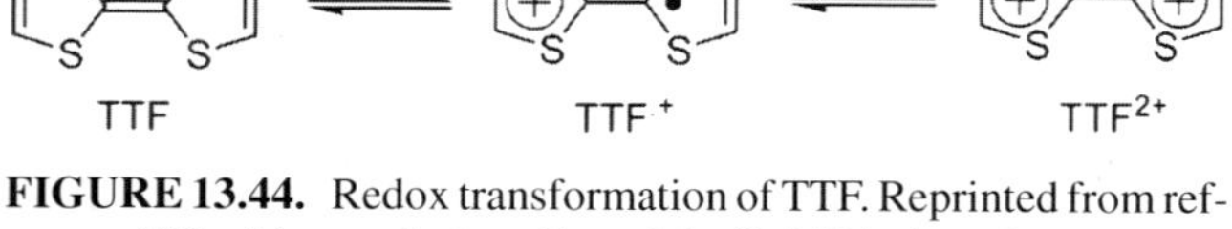

FIGURE 13.44. Redox transformation of TTF. Reprinted from reference 203 with permission. Copyright © 2004, American Chemical Society.

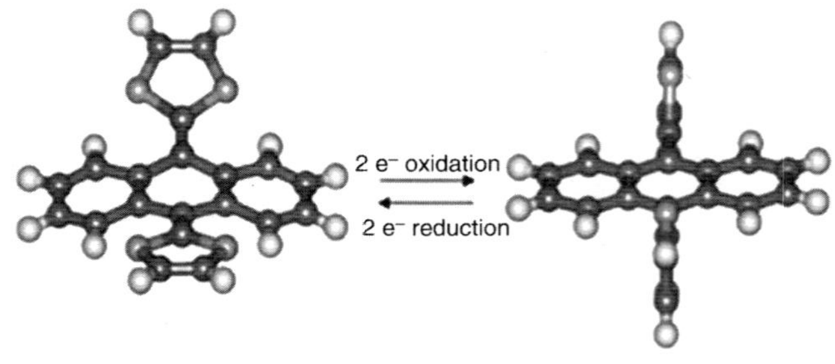

47 **48**

FIGURE 13.45. Geometry changes of *ex*TTF after redox reactions. **47**, ground state, **48**, oxidized state.

low potentials, which yields thermodynamically stable dicationic species having the dithiolium cation rings orthogonal to the anthracene core [202] (see Figure 13.45). These conformational changes may impact the reorganization energies and hence the lifetime of the charge separated states of the donor–acceptor systems built with *ex*TTF. However, the main reason for using *ex*TTF or generally TTF systems is because of the enhanced conductance properties.

Donor–acceptor dyads with fullerene as an acceptor and TTF or *ex*TTF as a donor have been extensively investigated in the past two decades [203]. The first TTF–fullerene dyads were prepared almost simultaneously by Prato et al. (**49a**) [204] and Martín et al. (**49a–d**) [205] (see Figure 13.46).

Since then, many fullerene–TTF conjugates—dyads, triads, tetrads, and pentads—have been synthesized and their

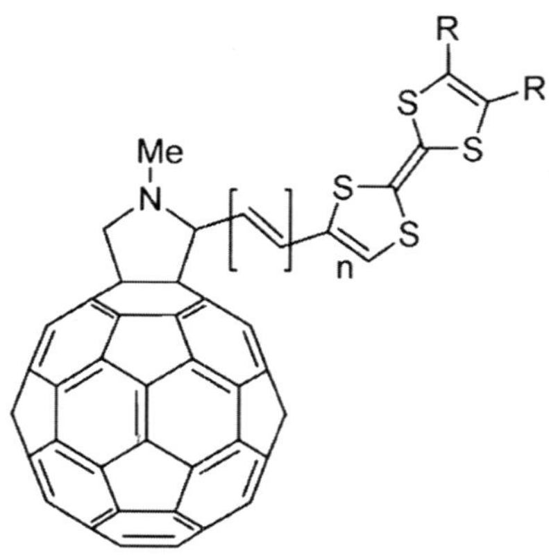

49a $n = 0$, R= H
49b $n = 0$, R-R = SCH_2CH_2S
49c $n = 0$, R= H
49d $n = 0$, R-R = SCH_2CH_2S

FIGURE 13.46. Fullerene–TTF dyads. Reprinted from reference 203 with permission. Copyright © 2004, American Chemical Society.

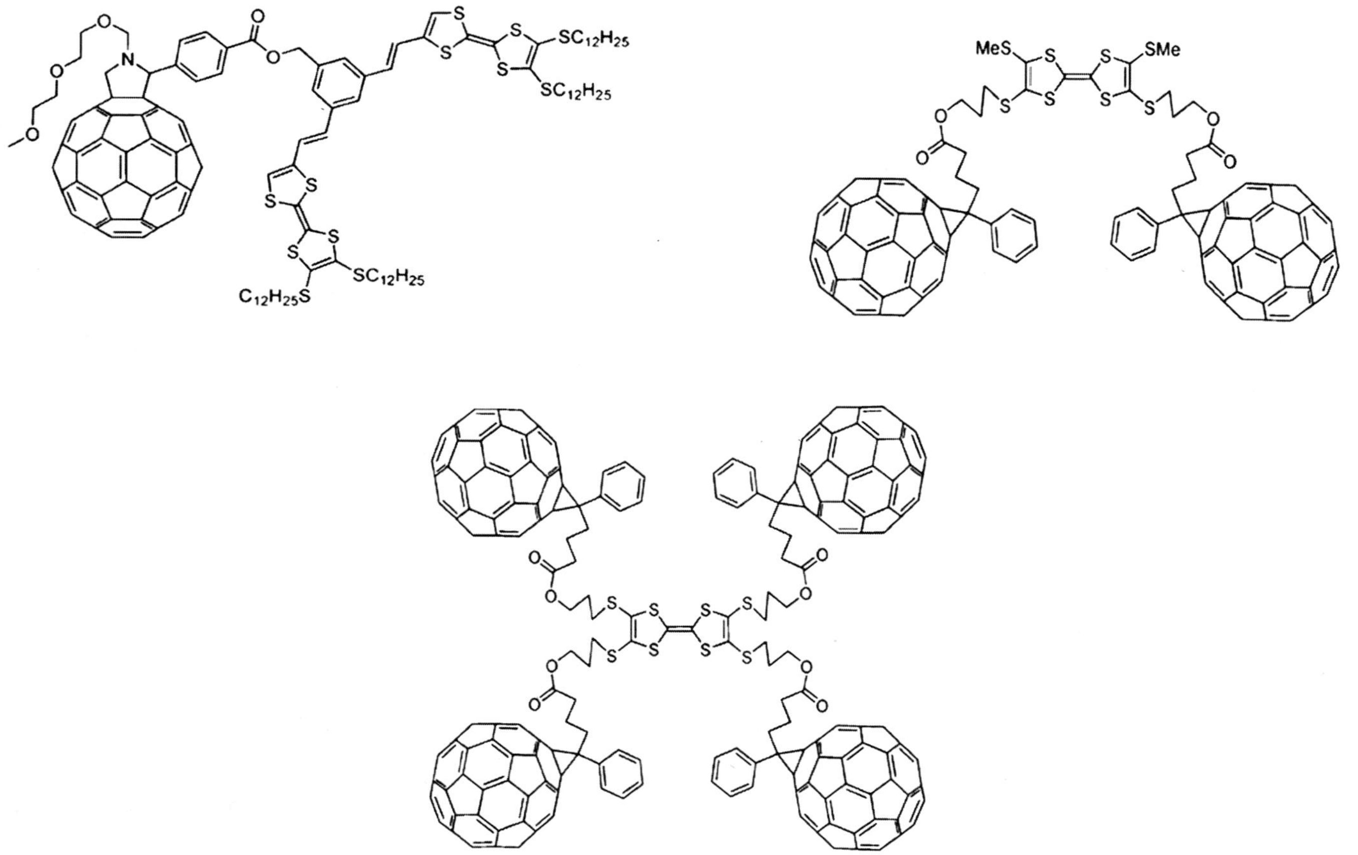

FIGURE 13.47. Examples of structures of fullerene–TTF triad and multiads. Reprinted from reference 203 with permission. Copyright © 2004, American Chemical Society.

electronic properties probed. The cyclic voltammetry of dyads generally shows two TTF-centered reversible oxidation waves and up to five reversible (or quasi-reversible) fullerene-centered reduction steps, giving rise to eight possible redox states. Electrochemical HOMO–LUMO gaps in the range of ~1–1.2 eV have been measured for these compounds. An excellent review has been published, which summarizes the electrochemical data for these dyads, triads, tetrads, and pentad [203].

Several examples of structures of fullerene–TTF triads and multiads are displayed in Figure 13.47. Comparing the electrochemical data for triads, tetrads, and pentads to those of the corresponding dyads, led to the conclusion that multiple substitution of the fullerene core decreases its reduction potential and increases the HOMO–LUMO gap.

The two-electron donor 9,10-bis(1,3-dithiol-2-ylidene)-9,10-dihidroanthracene (TTFAQ) (see Figure 13.48) has also been incorporated into fullerene dyads and triads, mostly by Martín and co-workers [206–212]. Some examples of structures for fullerene–TTFAQ conjugates are displayed in Figure 13.49.

Triads of C_{60}–TTFAQ-TTFAQ (**50a,b**) [208], C_{60}–porphyrin–TTF (**51a,b**) [213, 214], and C_{60}–porphyrin–*ex*TTF (**52a,b**) [213] were prepared in an attempt to control short- and long-range electron transfer processes (see Figure 13.50). Results demonstrated that donor–acceptor distances between 9.5 and 15.3 Å prevent substantial charge transfer in the ground state. Only upon photoexcitation of the fullerene chromophore, rapid and efficient charge transfer was observed for **50a,b**, which produced $C_{60}^{\bullet-}$–(exTTF)$^{\bullet+}$–(exTTF) (**50a,b**). Electrochemistry measurements show four quasi-reversible reduction waves for **50a,b**, whose potentials, as expected, are shifted to more negative values compared to C_{60}. On the oxidation side, they exhibit a broad quasi-reversible oxidation wave to form the tetracation

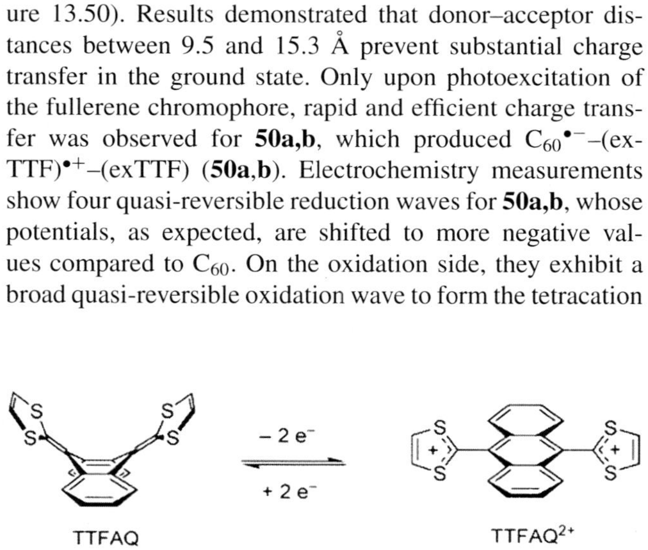

FIGURE 13.48. Two-electron oxidation of TTFAQ. Reprinted from reference 203 with permission. Copyright © 2004, American Chemical Society.

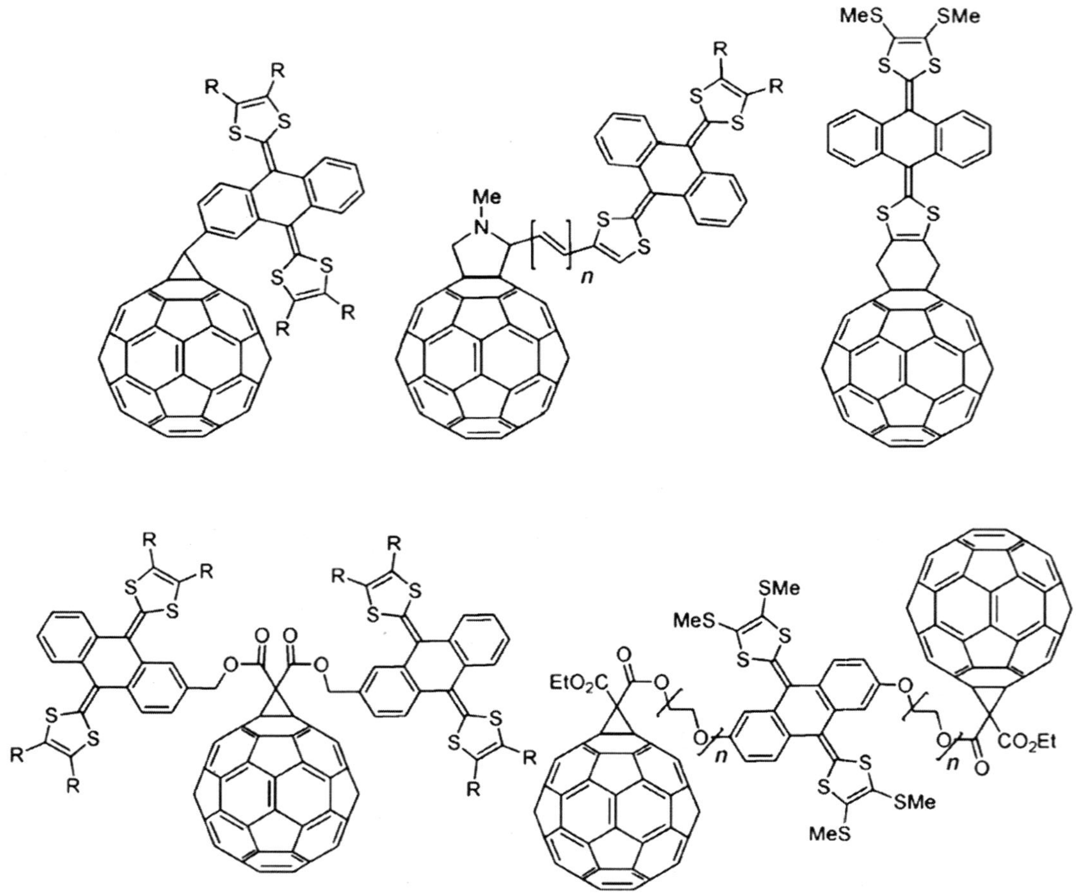

FIGURE 13.49. Structures of Fullerene–TTFAQ compounds. Reprinted from reference 203 with permission. Copyright © 2004, American Chemical Society.

species. Triad **50a**, bearing two identical π-extended TTFs (R = H), shows the broad oxidation peak centered at 0.41 V versus SCE. Similarly, triad **50b**, which bears two different π-extended TTFs (R = SMe), shows a broad oxidation peak centered at 0.43 V involving the oxidation of both *ex*TTF units.

Photoinduced electron-transfer studies on C_{60}–porphyrin–TTF (**51a,b**) and C_{60}–porphyrin–*ex*TTF (**52a,b**) indicated that the π-extended system (**52a,b**) exhibits many desirable characteristics as an electron donor in supramolecular systems [213, 214]. In addition, the π-extended TTF system has nearly identical first and second oxidation potentials, which make it a suitable two-electron donor in sequential multiphoton photoinduced electron-transfer experiments.

Electrochemical measurements of **51a,b** and **52a,b** were carried out to estimate the energies of the charge-separated states. Their potentials turned out to be very similar to those measured for model porphyrins and fullerenes, indicating that linking the moieties in the triad does little to perturb the individual redox centers [215].

In an effort to tune the electron-transfer process through molecular wires, a novel series of C_{60}–wire–*ex*TTF triads involving oligo-PPVs (polyphenylene vinylenes) as bridges were investigated by Martín and co-workers and Guldi and co-workers (see Figure 13.51) [216–218]. The conjugation length of the bridge oligomers was increased systematically. The point of this design was to optimize the coupling between the *ex*TTF moiety and the oligomeric bridge via full conjugation. The effect of the bridge over distances of 40 Å and beyond was probed. The results showed that the energies of the C_{60} HOMOs match especially those of the long oligo-PPVs. This facilitates electron/hole injection into the wire. The paraconjugation of the oligo-PPVs to the *ex*TTF electron donor leads to donor–acceptor coupling constants (V) of ~5.5 cm^{-1} and assists charge-transfer reactions that reveal a rather weak distance dependence.

The electrochemical measurements for **53–56** show the first four reduction steps of the fullerene core at −0.70, −1.10, −1.65, −2.10 V versus SCE [7] and the reduction of the oligo-PPV moiety at −1.9 V [219]. On the oxidation side, a two electron quasi-reversible oxidation wave at ~0.45 V to

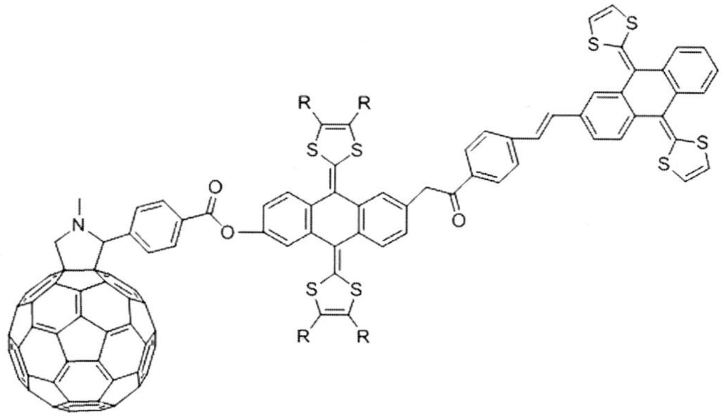

50a (R = H); **50b** (R = Sme)

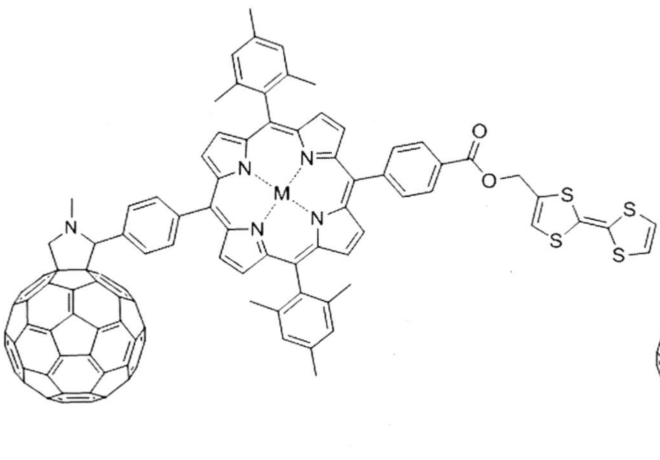

51a (M = 2H); **51b** (M = Zn)

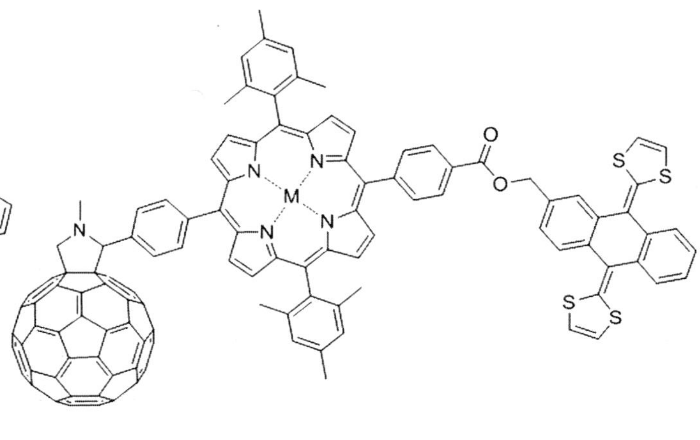

52a (M = 2H); **52b** (M = Zn)

FIGURE 13.50. Molecular structures of C_{60}–TTFAQ–TTFAQ (**50**), C_{60}–porphyrin–TTF (**51**), and C_{60}–porphyrin–*ex*TTF (**52**).

form the dication of the *ex*TTF moiety is observed [206]. A second oxidation wave is also observed in the range of 0.80–1.72 V versus SCE, which involves the oligo-PPV fragment. As the conjugation length of the oligomer increases, this second oxidation wave is cathodically shifted.

Another series of C_{60}–wire–*ex*TTF system **57a,b,c** incorporating *p*-phenyleneethynylene were prepared by Martín and co-workers (see Figure 13.52) [217]. In contrast with previous observations involving *p*-phenylenevinylene systems, the results revealed that a simple exchange of C–C double bonds (oligo-*p*-phenylenevinylenes, *o*PPV) for C–C triplet bonds (oligo-*p*-phenyleneethynylenes, *o*PPE) leads to a significantly altered long-range electron transfer (i.e., charge separation and charge recombination) in electron donor–acceptor conjugates due to the significant different electronic structure between the *o*PPE and *o*PPV systems. For the C_{60}–*o*PPV–*ex*TTF system, the HOMO partly localized on the *o*PPV bridge, whereas the HOMO in the C_{60}–*o*PPE–*ex*TTF system is completely localized on the exTTF moiety. Therefore, electron transfer is facilitated through better orbital overlap between the exTTF and the *o*PPV in comparison to the *o*PPE system.

Fluorene-based oligomers (see Figure 13.53) have emerged as a most unusual class of π-conjugated systems showing wire-like behavior; their energy levels change only slightly when the length of the bridge is increased [220]. Fluorene-based oligomers are able to act as efficient cables in C_{60}–*ex*TTF donor–acceptor systems, showing a remarkably low attenuation factor with β value of 0.09 $Å^{-1}$ [218].

Electrochemical measurements of **58a,b** indicate the absence of significant electronic interactions since the redox potentials determined by cyclic voltammetry for compounds **58a** (E^1_{ox} = 215 mV; E^1_{red} = −874 mV) and **58b** (E^1_{ox} = 281 mV; E^1_{red} = −889 mV) are comparable to those measured for precursor I-*ex*TTF (E^1_{ox} = 232 mV) and *N*-methylpyrrolidinofullerene (E^1_{red} = −863 mV).

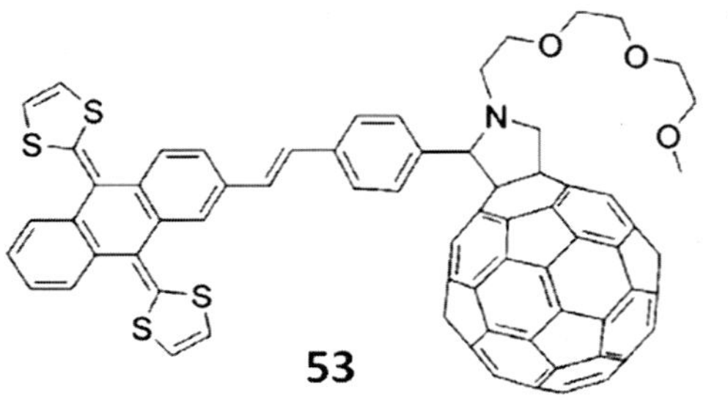

53

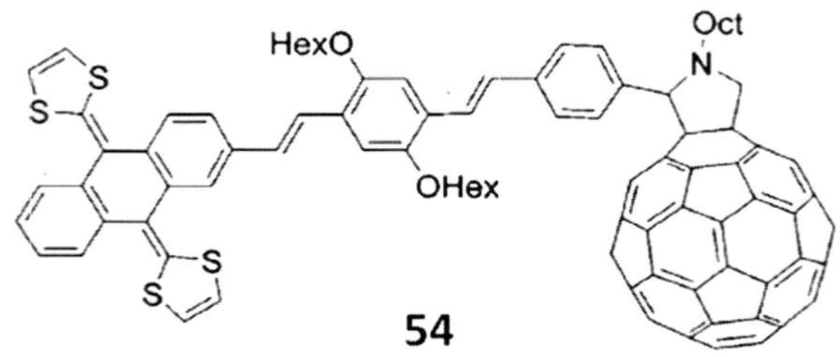

54

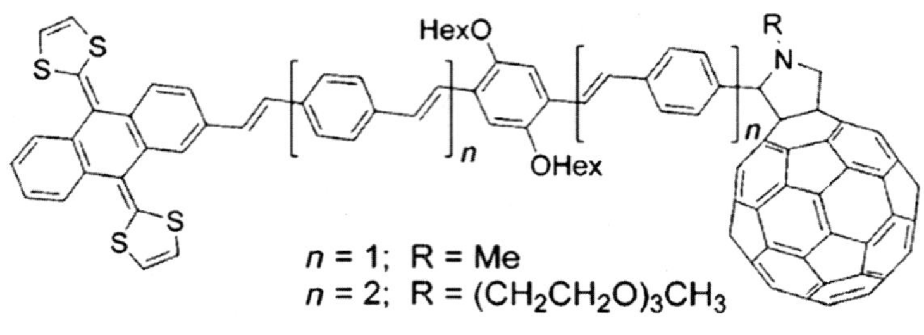

55

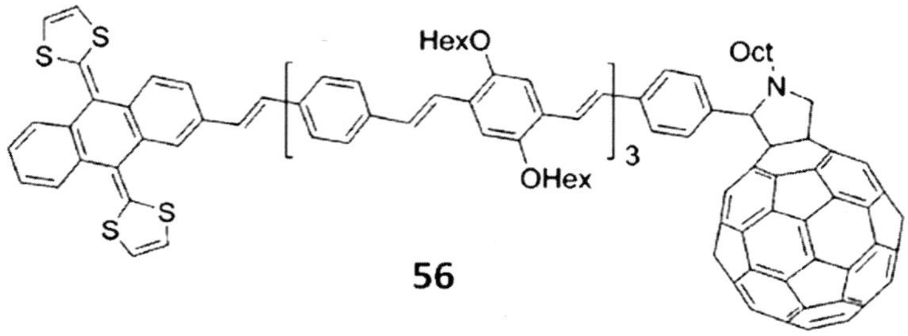

56

FIGURE 13.51. C_{60}–wire–*ex*TTF triads **53–56**. Reprinted from reference 216 with permission. Copyright © 2004, American Chemical Society.

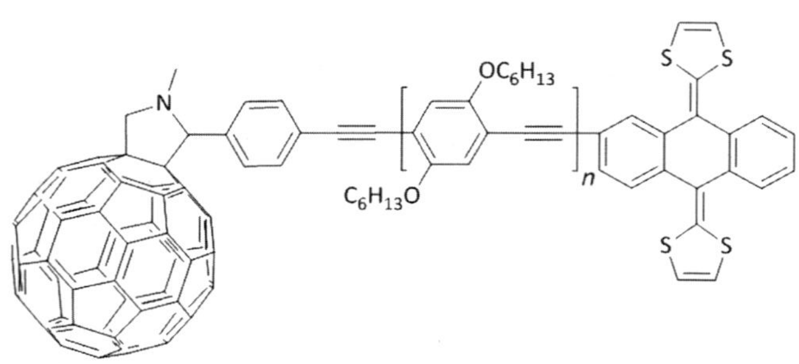

n=0
n=1
n=2
57a,b,c

FIGURE 13.52. C_{60}–wire–*ex*TTF triads involving *p*-phenyleneethynylene bridge.

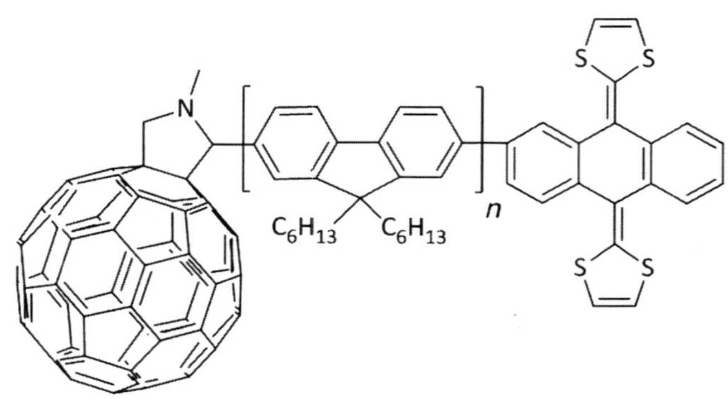

58a *n* = 1, **58b** *n* = 2

FIGURE 13.53. C_{60}–wire–*ex*TTF triads with fluorene-based bridges.

59

60

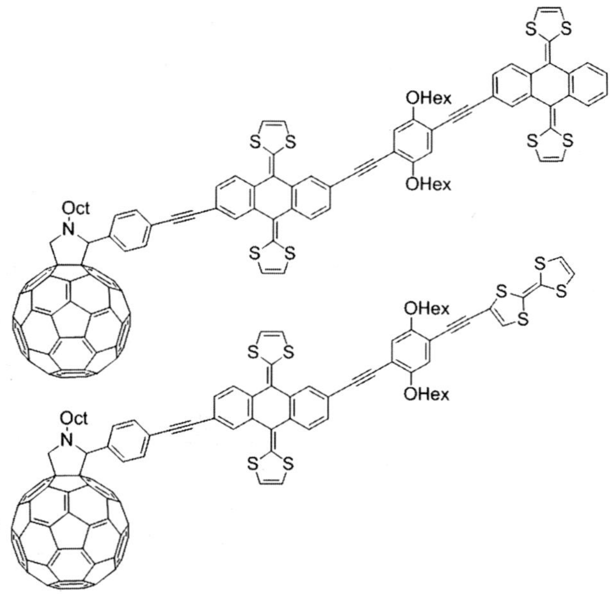

FIGURE 13.54. C_{60}–*ex*TTF–*ex*TTF (**59**) and C_{60}–*ex*TTF–TTF (**60**).

Very recently, rigid and soluble electron donor–acceptor conjugates combining *ex*TTF and/or TTF as donors and C_{60} as acceptor (**59** and **60** shown in Figure 13.54) have been synthesized by Martín and co-workers [221]. Their electrochemical behavior, along with their corresponding aldehydes, was investigated. Both **59** and **60** behave similarly to their corresponding aldehydes in the oxidation range. **59** exhibits an electrochemically irreversible oxidation wave associated with a four-electron transfer, which gives rise to the formation of the tetracationic species. Three oxidation steps are observed for **60**. Under reductive conditions, three one-electron quasi-reversible reduction waves are assigned to C_{60} (**59**: −0.84, −1.41, −2.05 V; **60**: −0.86, −1.42, −2.13 V versus $Ag/AgNO_3$). As expected, these potential values are cathodically shifted relative to pristine C_{60} (−0.72, −1.12, −1.60 V versus $Ag/AgNO_3$).

13.6.1.4 Porphyrin-Based Donors. Porphyrin is one of the building blocks frequently employed as an electron donor in artificial photosynthetic models. In this respect, photoinduced electron transfer systems comprising such macrocycles and C_{60} have been shown to be excellent combinations for revealing basic photophysical properties of donor-linked C_{60} systems. Gust and co-workers [222] described the first preparation and photophysical properties of C_{60}-linked porphyrins. Early studies about these C_{60}-porphyrin dyads and triads have been collected in reviews by Imahori and Sakata [223, 224] and by by Martín et al. [225]. In the past decade, photoinduced electron transfer in porphyrin–fullerene dyads, triads, and tetrads and their electrochemical properties have been studied by several groups [199, 226–249]. Guldi et al. [228, 235] investigated the effect of orientation of the donor and acceptor in electron transfer in these dyads, specifically parallel (face to face, Figure 13.55a) versus perpendicular (edge to face, Figure 13.55b) alignment. The different orientations have a strong effect on the back-electron-transfer (BET) dynamics. In THF, the lifetime of the $C_{60}{}^{\bullet -}$-$ZnTPP^{\bullet +}$

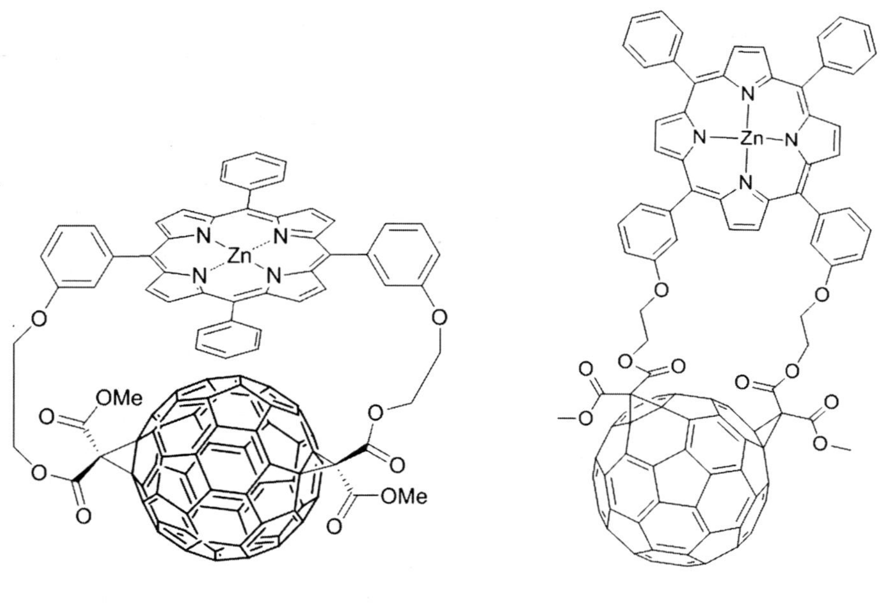

61a **61b**

FIGURE 13.55. Molecular structures of parallel (face to face, **61a**) versus perpendicular (edge to face, **61b**) alignment of porphyrin and fullerene.

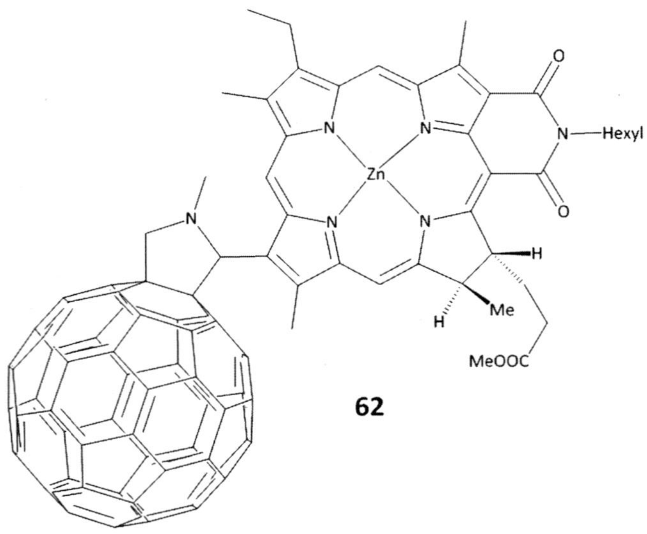

62

FIGURE 13.56. Molecular structure of zinc chlorin–fullerene dyad.

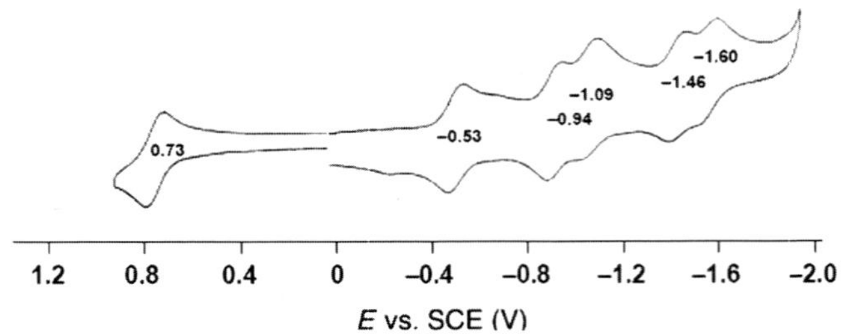

FIGURE 13.57. Cyclic voltammogram of ZnCh–C_{60} in the presence of 0.10 M n-Bu_4NClO_4 in deaerated PhCN at 298 K. Reprinted from reference 227 with permission. Copyright © 2004, Wiley–VCH Verlag GmbH & Co. KGaA, Weinheim.

Recent investigations focusing on ZnP–C_{60} electron donor–acceptor systems were summarized in a comprehensive review by Guldi et al. in 2009 [251], to which the reader is referred. Different types of C_{60}–wire–donor systems, such as C_{60}–wire–ZnP, C_{60}–wire–*ex*TTF, and C_{60}–wire–Fc, are discussed.

radical pair in the *trans*-2 dyad (**61a**) is on the order of a few hundred picoseconds with the back electron transfer rate constant of $K_{BET} = 2.6 \times 10^9$, while **61b** gives a lifetime on the order of microseconds with $K_{BET} = 3.8 \times 10^5$. Modification of the relative orientation of donor (ZnTPP) and acceptor (C_{60}), along with its corresponding alteration of the π-π interactions, changes the lifetime of the charge-separated state by four orders of magnitude. Electroanalytical investigations of **61a** using cyclic voltammetry and differential pulse voltammetry reveal two oxidative and six reductive electron transfer processes which are also present in either the parent porphyrin or the *trans*-2-bis(diethylmalonate) C_{60}. No electrochemical studies of **61b** were reported.

The lifetime of the CS state has recently been reported to be extended by decreasing the distance between the donor and acceptor [231, 250]. Ohkubo et al. [227] recently reported a zinc chlorin–fullerene dyad, **62,** which has an extremely short donor–acceptor distance (ZnCh–C_{60}, Figure 13.56). The lifetime of the CS state at $-150°$C is as long as 120 s, which is the longest CS lifetime reported for linked donor–acceptor systems so far.

The cyclic voltammogram of ZnCh–C_{60} exhibits three C_{60}-centered, one-electron reductions (−0.53, −1.09, and −1.60 V versus SCE), two ZnCh-centered, one-electron reductions (−0.94 and −1.46 V versus SCE), and a single ZnCh-centered, one-electron oxidation (0.73 V versus SCE) (Figure 13.57). These findings support prior electrochemical measurements of porphyrin–fullerene donor–acceptor systems, which generally display three or four reductions corresponding to the fullerene moiety, two reductions ascribed to the porphyrin, and one or two oxidations centered on the porphyrin [199, 215, 242].

13.6.1.5 Phthalocyanine-Based Donors. Among the electron donors phthalocyanines (Pcs) and sub-Pcs, aromatic macrocycles analogous to porphyrins (Figure 13.34e,f)) have a prominent position because they possess unique physicochemical properties, such as excellent semiconductivity, photoconductivity, chemical stability, and optical absorption in the UV–vis region with high extinction coefficients in the red–near-infrared spectral region. These properties make them valuable building blocks in donor–acceptor systems for their incorporation in photovoltaic and artificial photosynthetic devices [252–254]. A wide range of covalently linked D–A sub-Pc and Pc-based conjugates, incorporating fullerenes as the electron acceptor units, have been synthesized and their photophysical properties studied. The first molecular system containing both a Pc and a fullerene covalently linked dyad (**64**, see Figure 13.58) was prepared through a Diels–Alder reaction reported by Hanack, Hirsch, and co-workers [255]. Electrochemical measurements of **64** showed five reversible reduction peaks, which were assigned to be either fullerene- or Pc-based. However, these measurements showed that the reduction potentials centered on the Pc or the fullerene moieties do not change significantly compared to those of the respective parent compounds (**63** and the C_{60} fullerene monoadduct without the Pc moiety).

A Pc molecule extensively used to prepare Pc–C_{60} systems is tri-*tert*-butyliodo-Pc (**65**), either metalated or in its free base form (see Figure 13.59a), which has established itself as an interesting building block for the preparation of such D–A systems. The *tert*-butyl groups on the periphery of the macrocycle provide this molecule with high solubility and help reduce the strong self-aggregation tendency typical of Pcs. Another very important feature of **65** is the presence

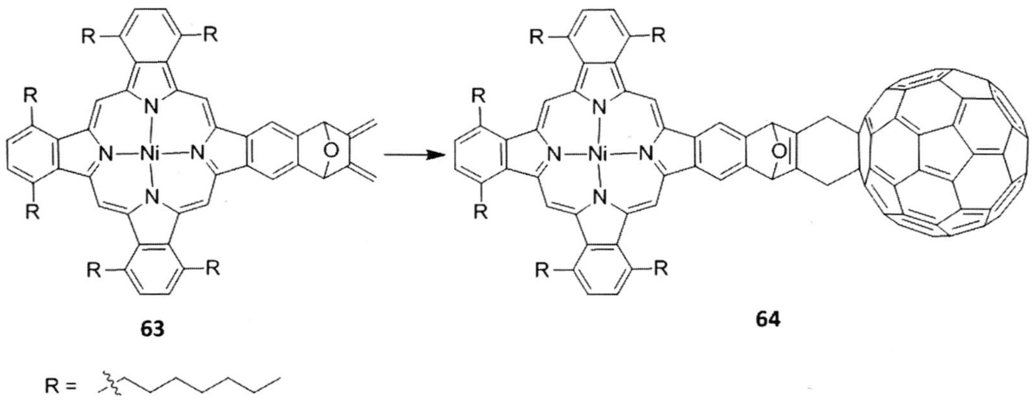

FIGURE 13.58. Synthesis of Pc–C_{60} fullerene dyad **64**.

of an iodine atom directly connected to the Pc framework, which allows for easy functionalization of this macrocycle with any functional group. Pc–C_{60} dyads incorporating **65** as the electron donor (see Figure 13.59b) were synthesized and their electrochemical properties explored by Echegoyen, Torres, and co-workers [256]. A thorough investigation of their physicochemical properties was carried out by Guldi, Torres, and co-workers [257]. Electrochemical investigation showed that the Pc-based oxidation and reduction potentials were positively shifted in **66a,b** with respect to reference compounds (Pcs and a fullerene derivative without the Pc moiety). Conversely, the C_{60}-based reduction potentials for **66a,b** were negatively shifted with respect to those of a fullerene derivative reference compound, thus suggesting some degree of ground-state intra- and/or intermolecular interaction between the electron-donating Pc to the electron-accepting fullerene moiety.

In 2009, Pc–C_{60} dyads in which Pcs were connected to the same 1-(3-carboxypropyl)-1-phenyl-[6,6]–C_{61} (PCBM) through different linkers were also reported (Figure 13.60) [258]. Cyclic voltammetry measurements of **67a** and **67d** in CH_2Cl_2 demonstrated essentially identical results. The wave potentials do not change upon elongation of the spacer between the donor and the acceptor units, suggesting the absence of intramolecular charge transfer in the ground state for these dyads.

Other phthalocyanine–fullerene dyads with double-bridge synthesized through the Bingel–Hirsch synthetic strategy were constructed [259]. Multiple Pc donors and/or C_{60} fullerene acceptor units have been assembled to form triads, tetrads, and multiads [260–268]. Very recently, an excellent review of phthalocyanie–fullerene donor–acceptor systems [269] has appeared to which the readers are referred. C_{60} fullerene-based donor–acceptor systems have been explored

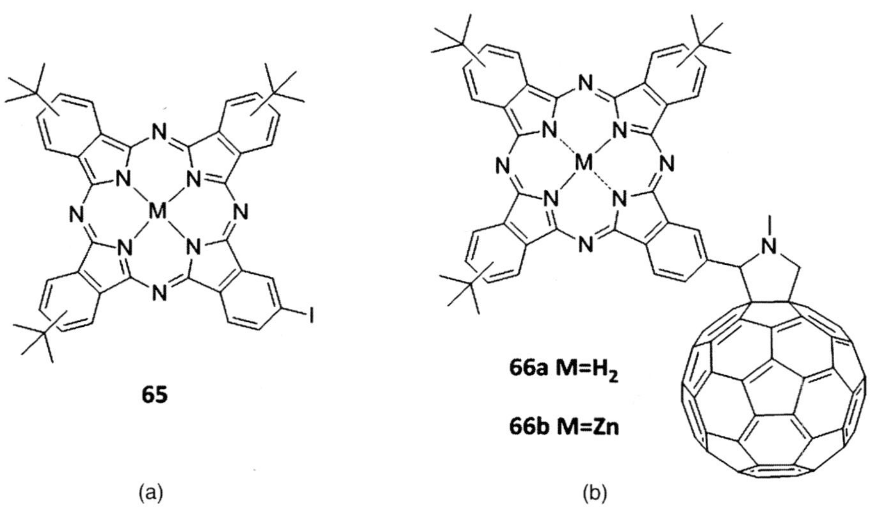

FIGURE 13.59. Molecular structure of (**a**) tri-*tert*-butyliodo-Pc (**65**) and (**b**) Pc–C_{60} dyads (**66**). M = H_2 or metal.

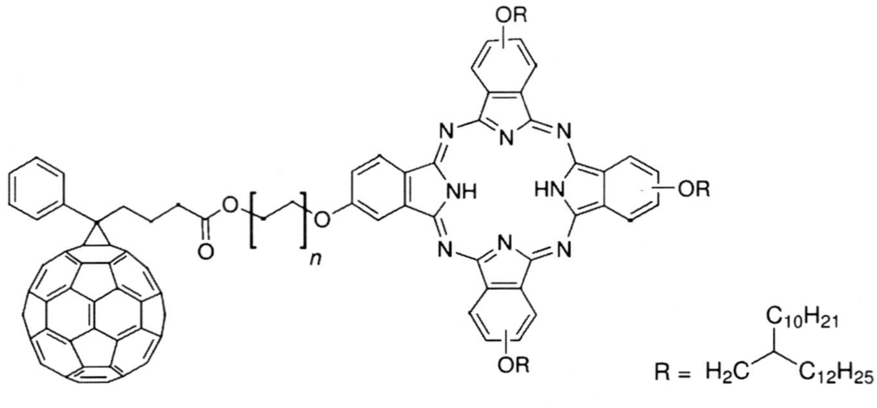

67, *n* = 3 (a), 4 (b), 5 (c), 6 (d)

FIGURE 13.60. Phthalocyanine–C_{60} dyads **67**.

extensively. The discovery of endohedral metallofullerenes (EMFs) has promoted their use in dyads.

13.6.2 Endohedral Metallofullerene Based Dyads

Until recently the construction of donor–acceptor systems using fullerenes as the acceptor moiety has been dominated by the use of C_{60}. However, the interesting properties displayed by endohedral metallofullerenes make them attractive for these applications. Among the endohedral fullerenes, carbon cages that encapsulate the trimetallic nitride cluster known as trimetallic nitride-templated (TNT) endohedral metallofullerenes (EMF) have been the focus of considerable effort since their discovery in 1999 by the process developed by Dorn and co-workers [270]. This method has allowed the preparation of TNT EMFs in high yield; and their structural, physical, and chemical properties have been explored in great detail [6]. $Sc_3N@C_{80}$ has drawn special attention because it can be isolated in large quantities. A fundamental aspect of $Sc_3N@C_{80}$ chemistry that remains unexplored is its applicability in electron donor–acceptor systems [198, 271, 272]. C_{60} and its derivatives exhibit uniquely low reorganization energies in charge-transfer reactions and show remarkable electron mobility features, acting as outstanding electron acceptors in different nanoconjugates [225, 273–276]. None of the trimetallic nitride endohedrals has ever been explored in this context despite the fact that early studies of their electrochemical studies revealed that, while preserving a remarkable electron-accepting ability similar to that of C_{60}, they possess much larger absorption coefficients than C_{60} in the visible region of the electromagnetic spectrum and low HOMO–LUMO energy gaps [6].

In the case of $Sc_3N@I_h–C_{80}$, theoretical calculations have established that the LUMO orbital has a strong contribution from the metal cluster [277], suggesting that the encapsulated cluster has a strong influence on the charge separation process. The difference in electrochemical behavior observed depending upon the functionalization position on the surface of the cage [94] further opens the possibility to either control or modulate the behavior of the resulting dyads. The preparation of covalently linked donor–acceptor systems using some common electron donors and TNT-EMFs as the electron acceptors has recently been accomplished using ferrocene [278].

The first ferrocene-based dyads donor–acceptor systems with $M_3N@I_h–C_{80}$ (M = Sc, Y) using the 1,3-dipolar cycloaddition reaction (Prato reaction) were prepared by Echegoyen and co-workers [278]. The differential reactivity between $Sc_3N@I_h–C_{80}$ and $Y_3N@I_h–C_{80}$ [93, 94, 96] made it possible for them to obtain two different fulleropyrrolidine dyads: a [5,6] in the case $Sc_3N@I_h–C_{80}$ and a [6,6] in the case of $Y_3N@I_h–C_{80}$. The structures are shown in Figure 13.61. Both compounds (**68** and **69**) were characterized

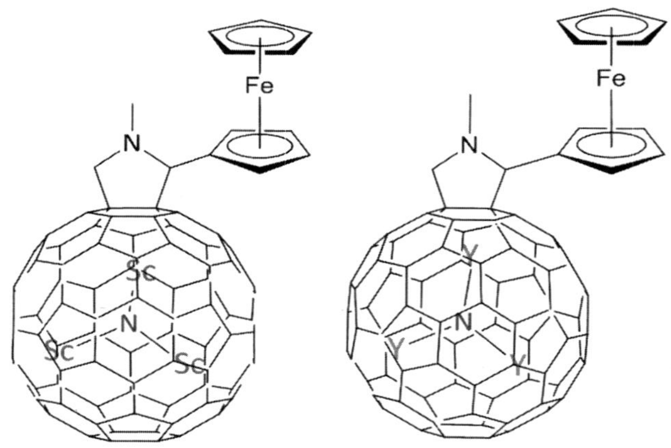

68 **69**

FIGURE 13.61. Structures of *N*-methyl-2-ferrocenyl-[5, 6]-$Sc_3N@I_h–C_{80}$–fulleropyrrolidine (**68**) and *N*-methyl-2-ferrocenyl-[6,6]-$Y_3N@I_h–C_{80}$–fulleropyrrolidine (**69**).

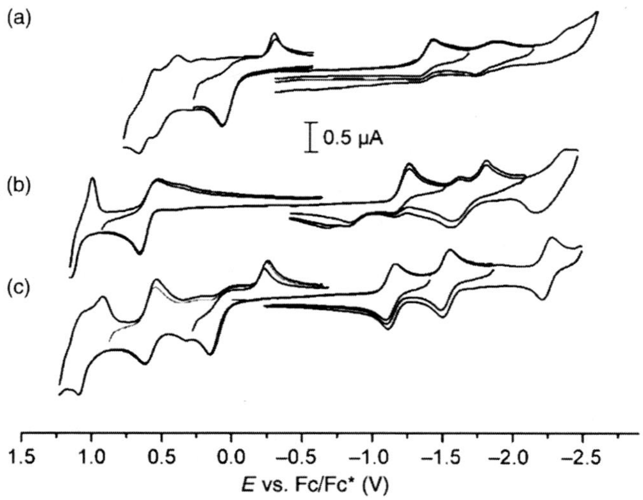

FIGURE 13.62. Cyclic voltammograms of (**a**) **69**, (**b**) $Sc_3N@I_h$–C_{80}, and (**c**) **68**. Recorded in *o*-DCB containing 0.05M *n*-Bu_4NPF_6 as supporting electrolyte, at a scan rate of 100 mV s^{-1}. Reprinted from reference 279 with permission. Copyright © 2009, Wiley–VCH Verlag GmbH & Co. KGaA, Weinheim.

by MALDI-TOF mass spectra and a variety of NMR techniques.

The solution electrochemistry was investigated for both compounds, and the resulting cyclic voltammograms of *N*-methyl-2-ferrocenyl-[5,6]-$Sc_3N@I_h$–C_{80}–fulleropyrrolidine (**68**) along with $Sc_3N@I_h$–C_{80} are shown in Figures 13.62c and 13.62b; respectively. For **68**, three one-electron reversible reductions at −1.14, −1.53, and −2.25 V versus. Fc/Fc^+ and three oxidations at +0.15, +0.61, and +1.09 versus. Fc/Fc^+ were observed. The processes in the negative potential range are assigned the fullerene cage. The first oxidation process at +0.15 V versus Fc/Fc^+ corresponds to the oxidation of the ferrocenyl moiety and the processes at +0.61 and +1.09 versus Fc/Fc^+ to oxidations of the fullerene cage [278].

The electrochemical properties of compound **69** were also investigated by cyclic voltammetry in *o*-DCB solution. The reduction waves of the pyrrolidinofullerene ferrocenyl monoadduct of $Y_3N@C_{80}$ show different reversibility from that observed in *N*-methyl-2-ferrocenyl-[5, 6]-$Sc_3N@I_h$–C_{80}–fulleropyrrolidine as shown in Figure 13.62. This indicates the formation of different regioisomer of [6,6]fulleropyrrolidine of $Y_3N@C_{80}$. The oxidation at +0.07 V corresponds to the oxidation of the ferrocenyl addend. Two pairs of oxidations were observed at +0.57 and +0.66 V; one of them can be attributed to the cage-based oxidation, while the other is possibly a result of electrochemical retrocycloaddtion reaction [279].

To determine if this $Sc_3N@C_{80}$–ferrocene electron donor–acceptor conjugate is indeed a promising material for photovoltaic applications, its electron-transfer interactions in the photoexcited state was tested. Not only did the results confirm that a photoinduced electron transfer occurred within this novel system, but, more significantly, the radical ion pair state formed turned out to be more stable than the corresponding C_{60}–ferrocene conjugate. Metal nitride cluster fullerene-based dyads are now envisioned as new promising materials for solar-energy conversion applications [278].

The *ex*TTF-based dyads with TNT–EMFs were synthesized using the Bingel reaction by Echegoyen's group and his collaborators [279]. The reaction is displayed in Figure 13.63.

Compound **70** was unstable and readily underwent conversion to **72** (see Figure 13.64) within a few hours. The chemical instability of **70** in solution contrasts with that of the analogue C_{60} derivative, which only decomposes after a few weeks. Compound **72**, containing the anthraquinone moiety, is not interesting for donor–acceptor systems because the anthraquinone is also a good electron acceptor.

The electrochemical behavior of both **70** and **72** was investigated by cyclic voltammetry in *o*-DCB solution.

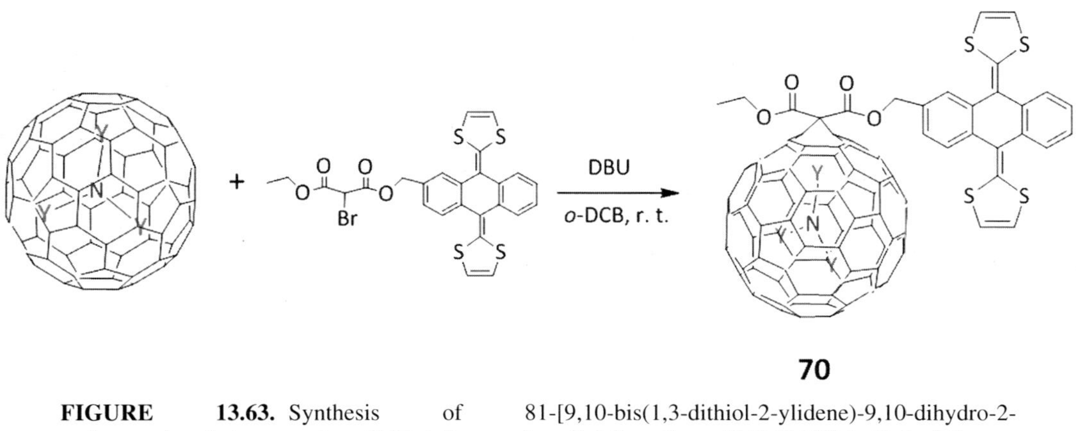

70

FIGURE 13.63. Synthesis of 81-[9,10-bis(1,3-dithiol-2-ylidene)-9,10-dihydro-2-anthracenylmethyl oxycarbonyl]-81-(ethoxycarbonyl)-1,2-methano-[6,6]-$Y_3N@I_h$–C_{80}] fullerene (**70**).

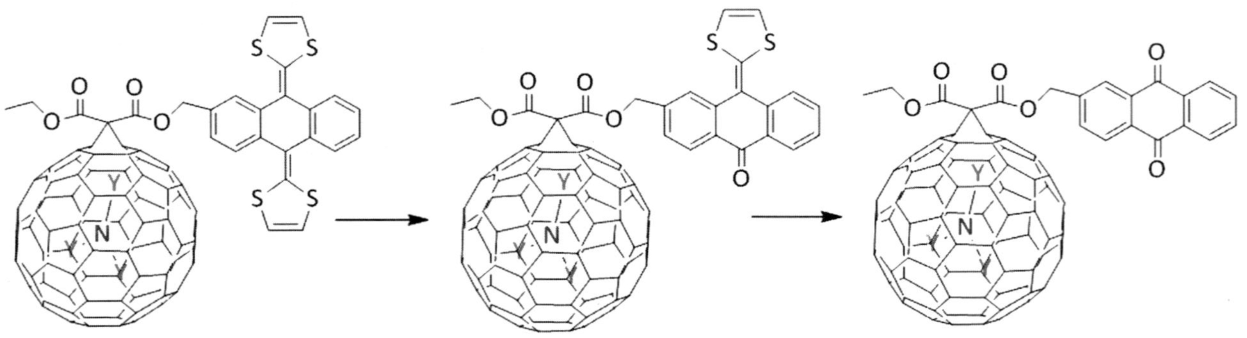

70 71 72

FIGURE 13.64. Conversion of compound **70** into compound **72**.

Figures 13.65a and 13.65b show representative CVs for the exTTF-based dyad **70** and the anthraquinone-$Y_3N@C_{80}$ Bingel derivative **72**. Dyad **70** exhibits an irreversible reductive behavior, which is the typical electrochemical behavior of [6,6]-methanofullerene derivatives of $Y_3N@C_{80}$ [94]. On the oxidation side, quasi-reversible processes of the respective parent metallofullerene and exTTF are observed at +0.90 and +0.23 V. The cyclic voltammogram of anthraquinone-$Y_3N@C_{80}$ Bingel adduct (**72**) shows two reduction and one oxidation steps, corresponding to the reductions of the anthraquinone moiety and the oxidation of the $Y_3N@C_{80}$ cage, respectively. The first reduction process at −1.34 V appears to be electrochemically reversible. The second reduction step at −1.72 V and the oxidation at +0.67 are electrochemically quasi-reversible. The shoulders observed at about −1.23 and −1.98 V probably correspond to the reduction of the C_{80} cage, since the reductive processes of both fragments occur almost at the same potential. The absence of the oxidation process corresponding to the exTTF addend, along with the presence of the two reduction waves of the anthraquinone addend, clearly demonstrated the conversion of exTTF into the anthraquinone adduct of $Y_3N@C_{80}$. The presence of the two reduction steps corresponding to anthraquinone addend and the absence of the oxidation process of the exTTF addend indicate the conversion of exTTF into the anthraquinone adduct.

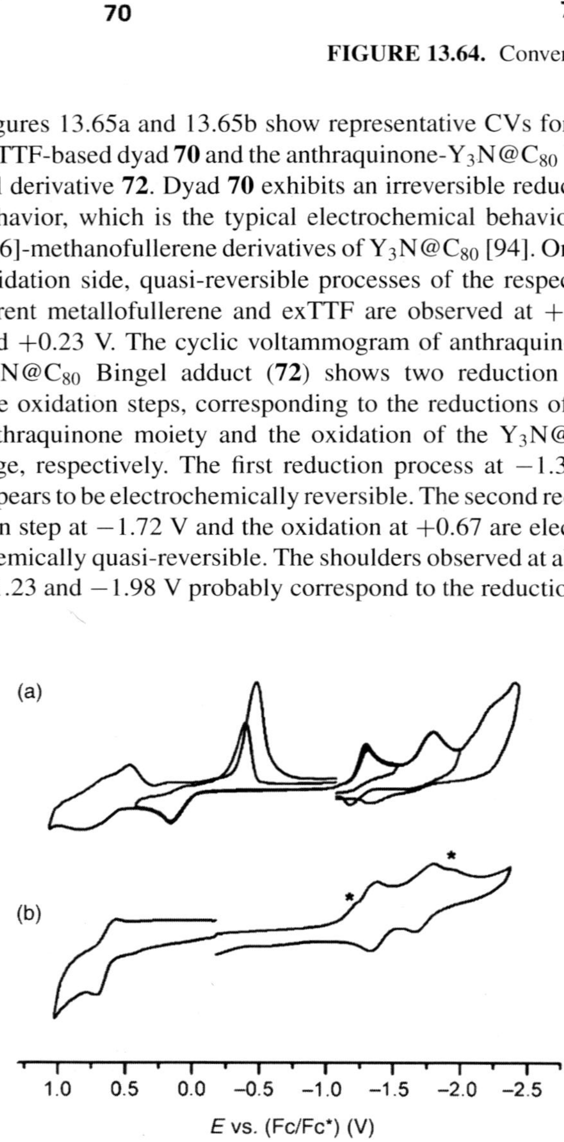

FIGURE 13.65. Cyclic voltammograms of (**a**) **70** and (**b**) **72**. Recorded in *o*-DCB containing 0.05M *n*-Bu_4NPF_6 as supporting electrolyte, at a scan rate of 100 mV s^{-1}. * indicates fullerene-based reduction processes. Reprinted from reference 279 with permission. Copyright © 2009, Wiley–VCH Verlag GmbH & Co. KGaA, Weinheim.

As indicated in Section 13.6.1.5, phthalocyanines (Pc), have been successfully used in the construction of donor–acceptor systems with fullerenes [280–284] and as TiO_2 photosensitizers for the construction of dye sensitized solar cells [285–288].

An attempt to prepare a Pc-substituted pyrrolidine-containing $Sc_3N@C_{80}$ dyad was carried out by Echegoyen and co-workers [279] and resulted in only a very limited amount of the Pc dyad. However, the Y analogue of this dyad was isolated (**73** in Figure 13.66). Additionally,

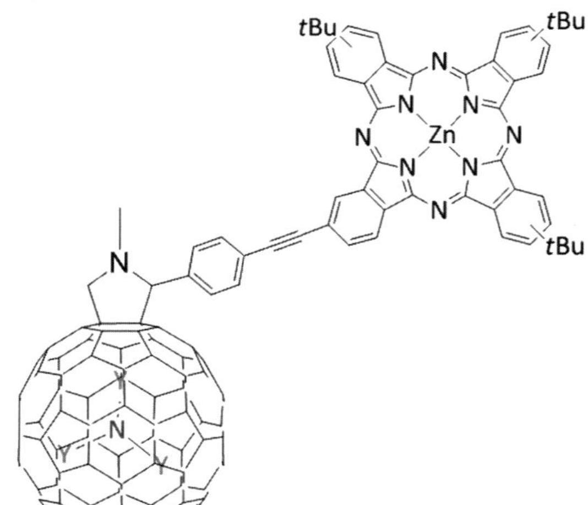

73

FIGURE 13.66. Synthesis of pyrrolidine-based TNT–EMFs substituted with phthalocyanine **73**.

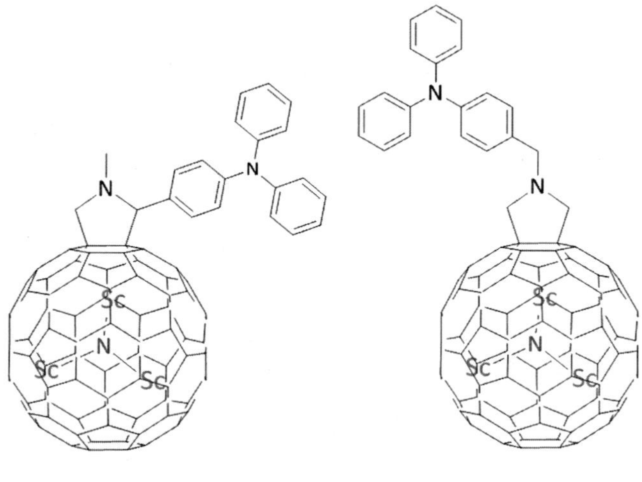

74 **75**

FIGURE 13.67. Molecular structures of triphenylaminofulleropyrrolidine electron donor–acceptor conjugates.

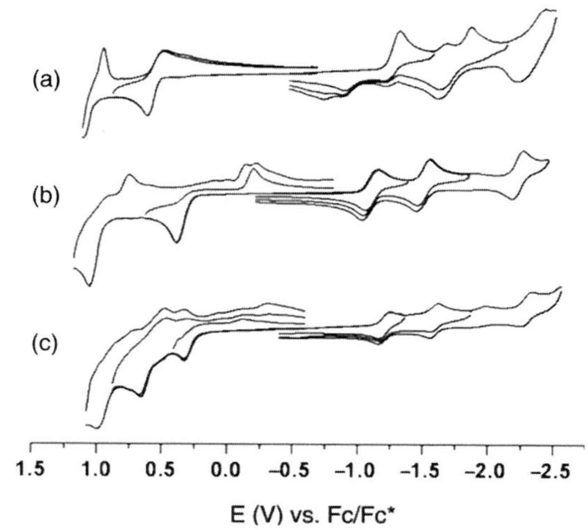

FIGURE 13.68. Cyclic voltammograms recorded on a GC electrode (1 mm) in a 0.05 M solution of tetra(*n*-butyl)ammoniumhexafluorophosphate in *o*-DCB as supporting electrolyte and a scan rate of 100 mV s^{-1}. (**a**) Pristine I_h-$Sc_3N@C_{80}$, (**b**) compound **74**, and (**c**) compound **75**. Reprinted from reference 197 with permission. Copyright © 2009, American Chemical Society.

the $Y_3N@C_{80}$ malonate fulleroid dyad (**73**) was prepared by following the Bingel–Hirsch protocol. Unfortunately, **73** decomposes under ambient conditions, giving rise to $Y_3N@C_{80}$. Consequently, it was not possible to study the electrochemical properties of **73**.

TPA-based dyads with endohedral metallofullerenes were synthesized and studied by Echegoyen and co-workers [197] in order to determine the effect of the linkage position of the donor system—that is, 2-substituted (**74**) versus N-substituted (**75**) pyrrolidine (Figure 13.67)—on the formation of photoinduced charge separated states and the thermal stability toward the retro-cycloaddition reaction.

The electrochemical properties of **74** and **75** were studied by cyclic voltammetry (CV) [197]. Their redox potentials were compared to those of $Sc_3N@C_{80}$. As observed in Figure 13.68, traces b and c, both **74** and **75** exhibit the typical three reversible reductions of [5,6]-$Sc_3N@I_h$–C_{80} fulleropyrrolidines [94]. In the anodic scan, **74** displays two irreversible oxidation processes; the first process is probably related to the oxidation of the triphenylamine [192], but it may also be the result of a pyrrolidine-based oxidation [149]. The second oxidation process is centered on the fullerene cage. In the case of **75**, three irreversible processes are observed. The first oxidation process at +0.32 V is related to either the oxidation of the triphenylamine group [192] or the pyrrolidine [94], and the other processes at +0.63 V and +0.99 V correspond to the oxidation of the fullerene cage.

It was found that when the donor is connected to the pyrrolidine nitrogen atom, the resulting dyad produces a significantly longer lived radical pair (rate constant of charge recombination is $<3.3 \pm 0.5 \times 10^8$) than the corresponding 2-substituted isomer (rate constant of charge recombination is $4.5 \pm 0.5 \times 10^8$) for I_h–$Sc_3N@C_{80}$ dyads. Even more remarkable is the fact that both **74** and **75** are longer-lived than the non-endohedral C_{60} analogues. Notably, the N-substituted TPA–I_h–$Sc_3N@C_{80}$ dyad, **75**, has much better thermal stability than the 2-subtituted one.

Synthetic meso-substituted porphyrins have emerged as important building blocks for the construction of fullerene-based donor–acceptor systems [289–292], and they are relatively easy to prepare. They have excellent photophysical properties such as high molar extinction coefficients in the visible range and fast energy transfer from their excited states toward neighbor acceptor molecules [293]. On the other hand, the highly conjugated *p*-phenyleneethynylene (PPE) is known for its behavior as a molecular wire that facilitates the charge transfer between donor and fullerene acceptor systems. The energies of the C_{60} HOMOs match those of PPE, which facilitates electron/hole injection into the wire [230].

Recently, Martin, Echegoyen, and Guldi's research groups reported an excellent dyad system in which $Sc_3N@C_{80}$ was used as electron acceptor while a metalloporphyrin was used as an electron donor. The acceptor $Sc_3N@C_{80}$ and donor metalloporphyrins were connected by a long spacer with center-to-center distances of up to 45 Å (Figure 13.69) [294]. For the first time, the long-range charge-transfer processes of electron-acceptor conjugates possessing zinc porphyrins (ZnP) as chromophore/electron donor and $Sc_3N@C_{80}$ as electron acceptor were studied. Radical ion pair states with lifetimes of 2.6 μs were achieved.

The cyclic voltammograms of **76** and **77** are shown in Figure 13.70. These dyads display not only the three typical reduction processes observed for [5,6]-$Sc_3N@C_{80}$

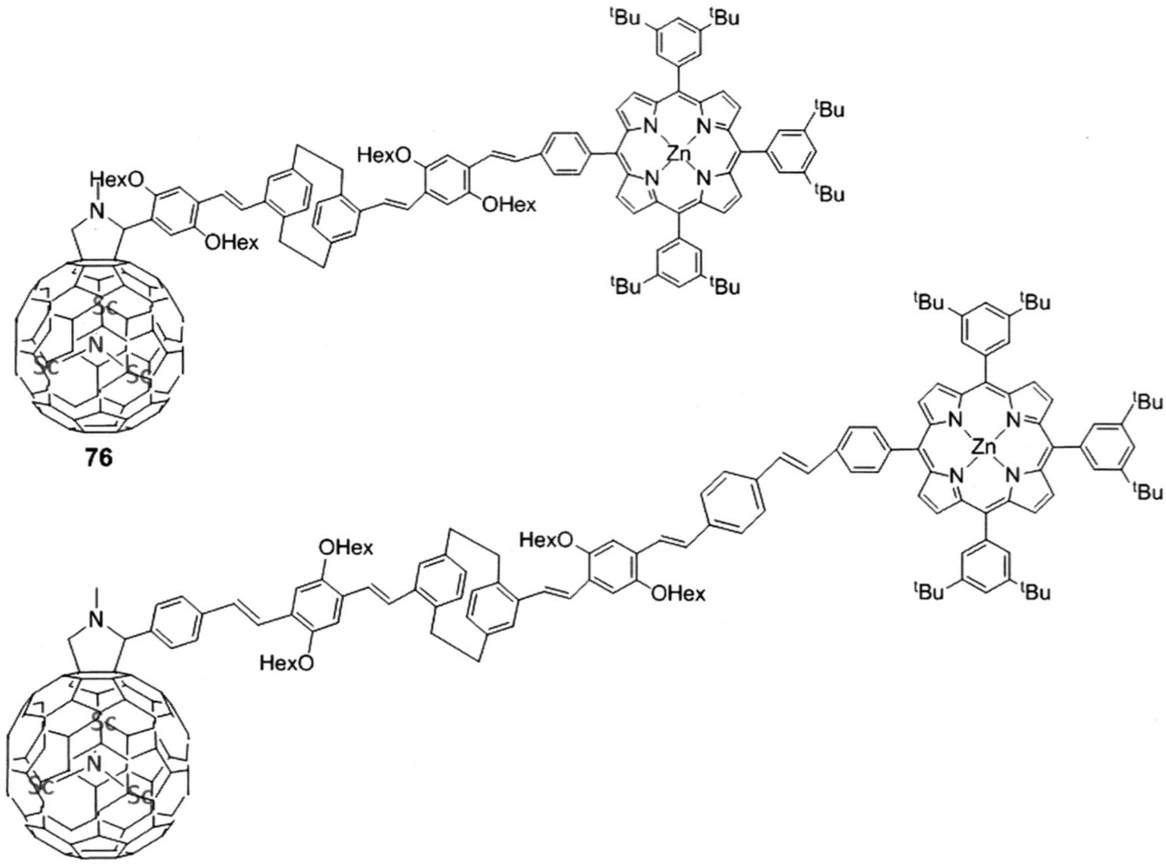

FIGURE 13.69. Molecular structures of **76** and **77**.

fulleropyrrolidines, but also one additional reduction process that is porphyrin-based [295]. The reversible reduction processes at −1.22/−1.18, −1.62/−1.56, and −2.32/−2.27 V are based on $Sc_3N@C_{80}$, while the reduction of ZnP occurs at −2.00/−1.98 V. The anodic scan shows three one-electron oxidation processes at +0.25/+0.28, +0.44/+0.40, and +0.57/+0.55 V versus Fc/Fc^+ that correlate with the first oxidation of ZnP, the first oxidation of $Sc_3N@C_{80}$, and the second oxidation of ZnP, respectively.

13.7 SUMMARY AND OUTLOOK

In this chapter, we have provided an update on the electrochemical characterization, electrosynthesis, and electronic properties of various carbon materials, some of which hold promise for optoelectronic applications. Emphasis was placed on fullerenes of various sizes and shapes, their endohedral counterparts and their functionalized derivatives. Functionalization of fullerenes is critical to the development of new materials with sufficient solubilities and fine-tuned electronic properties; and, to this end, many common synthetic strategies have been successfully applied, such as Diels–Alder, 1,3-dipolar cycloaddition, addition–elimination, and free radical addition reactions. Clever alternative approaches using electrosynthesis have also been employed to prepare derivatives that are otherwise not accessible. Many of the same synthetic and electrosynthetic reactions have been effectively used to functionalize some endohedral metallofullerenes.

Fullerenes and their endohedral counterparts constitute remarkably robust, three-dimensional building blocks to construct a variety of donor–acceptor systems for potential applications in optoelectronic devices. Molecular heterojunction dyads, triads, and multiads containing fullerenes as the electron acceptor have been shown to be excellent candidates for efficient photovoltaic cells. However, low synthetic yields and isolation difficulties, especially for the very promising endohedral metallofullerenes, have slowed down the progress. To advance the development of fullerene science and its applications, more research is necessary to increase the yields, find effective ways to isolate, purify, and functionalize the most promising compounds, and to improve the design and efficiency of electronic devices based on

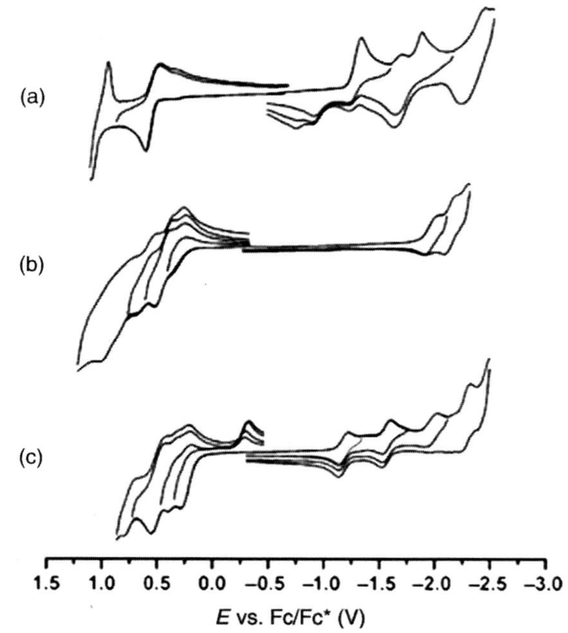

FIGURE 13.70. Cyclic voltammograms of (**a**) $Sc_3N@I_h$–C_{80}, (**b**) **76**, and (**c**) **77**. Recorded in tetra-*n*-butylammoniumhexafluorophosphate (0.05 M) containing *o*-DCB. Reprinted from reference 294 with permission. Copyright © 2010, Royal Society of Chemistry.

predictable and tunable 3D molecular architectures with the appropriate electronic properties.

Unquestionably, fullerenes, along with the many other forms of nanocarbon, hold tremendous potential for the development of light, flexible, and efficient optoelectronic devices, particularly those that involve solar energy conversion.

REFERENCES

1. Kroto, H. W., Heath, J. R., O'Brien, S. C., Curl, R. F., Smalley, R. E. (1985). C_{60}: Buckminsterfullerene. *Nature (London)* **318**, 162–163.
2. Wiles, P. G., Abrahamson, J. (1978). Carbon-fiber layers on arc electrodes. 1. Their properties and cool-down behavior. *Carbon* **16**, 341–349.
3. Iijima, S. (1991). Helical microtubules of graphitic carbon. *Nature (London)*, **354**, 56–58.
4. Novoselov, K. S., Geim, A. K., Morozov, S. V., Jiang, D., Zhang, Y., Dubonos, S. V., Grigorieva, I. V., Firsov, A. A. (2004). Electric field effect in atomically thin carbon films. *Science*, **306**, 666–669.
5. Chaur, M. N., Melin, F., Ortiz, A. L., Echegoyen, L. (2009). Chemical, electrochemical, and structural properties of endohedral metallofullerenes. *Angewandte Chemie International Edition*, **48**, 7514–7538.
6. Dunsch, L., Yang, S. (2007). Metal nitride cluster fullerenes: Their current state and future prospects. *Small*, **3**, 1298–1320.
7. Echegoyen, L., Echegoyen, L. E. (1998). Electrochemistry of fullerenes and their derivatives. *Accounts of Chemical Research*, **31**, 593–601.
8. Echegoyen, L. E., Herranz, M. A., Echegoyen, L. (2006). *Carbon. Fullerenes*. Wiley–VCH Verlag GmbH & Co. KGaA, Weinheim, pp. 145–201.
9. Kadish, K. M., Ruoff, R. S., editors. (2000). *Fullerenes: Chemistry, Physics, and Technology*. John Wiley & Sons, Hoboken, NJ.
10. Kavan, L., Dunsch, L. (2008). Electrochemistry of carbon nanotubes. *Topics in Appllied Physics*, **111**, 567–603.
11. Dumitrescu, I., Unwin, P. R., MacPherson, J. V. (2009). Electrochemistry at carbon nanotubes: Perspective and issues. *Chemical—Communications (Cambridge, U.K.)*, 6886–6901.
12. Pumera, M. (2009). The electrochemistry of carbon nanotubes: Fundamentals and applications. *Chemistry—European Journal*, **15**, 4970–4978.
13. Dillon, A. C. (2010). Carbon nanotubes for photoconversion and electrical energy storage. *Chemical Reviews*, **110**, 6856–6872.
14. Eder, D. (2010). Carbon nanotube—Inorganic hybrids. *Chemical Reviews*, **110**, 1348–1385.
15. Hu, L., Hecht, D. S., Grüner, G. (2010). Carbon nanotube thin films: Fabrication, properties, and applications. *Chemical Reviews*, **110**, 5790–5844.
16. Allen, M. J., Tung, V. C., Kaner, R. B. (2009). Honeycomb carbon: A review of graphene. *Chemical Reviews*, **110**, 132–145.
17. Howard, J. B., McKinnon, J. T., Makarovsky, Y., Lafleur, A. L., Johnson, M. E. (1991). Fullerenes C_{60} and C_{70} in flames. *Nature (London)*, **352**, 139–141.
18. Goel, A., Hebgen, P., Vander, S. J. B., Howard, J. B. (2002). Combustion synthesis of fullerenes and fullerenic nanostructures. *Carbon*, **40**, 177–182.
19. McKinnon, J. T., Bell, W. L., Barkley, R. M. (1992). Combustion synthesis of fullerenes. *Combustion and Flame*, **88**, 102–112.
20. Takehara, H., Fujiwara, M., Arikawa, M., Diener, M. D., Alford, J. M. (2005). Experimental study of industrial scale fullerene production by combustion synthesis. *Carbon*, **43**, 311–319.
21. Song, X., Liu, Y., Zhu, J. (2006). The effect of furnace temperature on fullerene yield by a temperature controlled arc discharge. *Carbon*, **44**, 1584–1586.
22. Ahmad, B., Riaz, M., Ahmad, M., Nawaz, S., Ahmad, S. (2008). Synthesis and characterization of higher fullerene (C_{84}) in DC arc discharge using Cu as a catalyst. *Material Letters*, **62**, 3367–3369.
23. Gonzalez-Aguilar, J., Moreno, M., Fulcheri, L. (2007). Carbon nanostructures production by gas-phase plasma processes at atmospheric pressure. *Journal of Physics D: Applied Physics*, **40**, 2361–2374.
24. Kikuchi, K., Nakahara, N., Wakabayashi, T., Honda, M., Matsumiya, H., Moriwaki, T., Suzuki, S., Shiromaru, H., Saito, K.,

et al. (1992). Isolation and identification of fullerene family: The carbon seventy-six-, seventy-eight, eighty-two-, eighty-four, ninety-, and ninety-six-atom molecules (C_{76}, C_{78}, C_{82}, C_{84}, C_{90}, and C_{96}). *Chemical Physics Letters*, **188**, 177–180.

25. Kroto, H. W., Walton, D. R. M. (1993). *The Fullerenes: New Horizons for the Chemistry, Physics and Astrophysics of Carbon. Cambridge University Press*, New York.

26. Rohlfing, E. A., Cox, D. M., Kaldor, A. (1984). Production and characterization of supersonic carbon cluster beams. *Journal Chemical Physics*, **81**, 3322–3330.

27. Kerner, R., Penson, K. A., Bennemann, K. H. (1992). Model for the growth of fullerenes (C_{60}, C_{70}) from carbon vapor. *Europhysics Letters*, **19**, 363–368.

28. Brothers, E. N., Scuseria, G. E. (2008). Enhanced enthalpies of formation from density functional theory through molecular reference states. *Journal of Physical Chemistry, A*, **112**, 13706–13711.

29. Fowler, P. W., Manolopoulos, D. E. (1995). An atlas of fullerenes, Oxford University Press, Oxford.

30. Fowler, P. W., Myrvold, W. (2010). Non-ipr fullerenes with properly closed shells. *Physical Chemistry Chemical Physics*, **12**, 14822–14826.

31. Zhao, X., Goto, H., Slanina, Z. (2004). C_{100} IPR fullerenes: Temperature-dependent relative stabilities based on the gibbs function. *Chemical Physics*, **306**, 93–104.

32. Zhao, X., Slanina, Z., Goto, H. (2004). Theoretical studies on the relative stabilities of C_{96} IPR fullerenes. *Journal of Physical Chemistry A*, **108**, 4479–4484.

33. Thilgen, C., Herrmann, A., Diederich, F. (1997). The covalent chemistry of higher fullerenes: C-70 and beyond. *Angewandte Chemie International Edition*, **36**, 2269–2280.

34. Yang, H., Mercado, B. Q., Jin, H., Wang, Z., Jiang, A., Liu, Z., Beavers, C. M., Olmstead, M. M., Balch, A. L. (2011). Fullerenes without symmetry: Crystallographic characterization of C_1(30)-C_{90} and C_1(32)-C_{90}. *Chemical Communications (Cambridge, U.K.)*, **47**, 2068–2070.

35. Yang, H., Beavers, C. M., Wang, Z. M., Jiang, A., Liu, Z. Y., Jin, H. X., Mercado, B. Q., Olmstead, M. M., Balch, A. L. (2010). Isolation of a small carbon nanotube: The surprising appearance of D_{5h}(1)-C_{90}. *Angewandte Chemie International Editions*, **49**, 886–890.

36. Beavers, C. M., Chaur, M. N., Olmstead, M. M., Echegoyen, L., Balch, A. L. (2009). Large metal ions in a relatively small fullerene cage: The structure of $Gd_3N@C_2(22010)$-C_{78} departs from the isolated pentagon rule. *Journal of the American Chemical Society*, **131**, 11519–11524.

37. Palkar, A., Melin, F., Cardona, C. M., Elliott, B., Naskar, A. K., Edie, D. D., Kumbhar, A., Echegoyen, L. (2007). Reactivity differences between carbon nano onions (CNOs) prepared by different methods. *Chemistry—An Asian Journal*, **2**, 625–633.

38. Haufler, R. E., Conceicao, J., Chibante, L. P. F., Chai, Y., Byrne, N. E., Flanagan, S., Haley, M. M., O'Brien, S. C., Pan, C., et al. (1990). Efficient production of C_{60} (buckminsterfullerene), $C_{60}H_{36}$, and the solvated buckide ion. *Journal of Physical Chemistry*, **94**, 8634–8636.

39. Allemand, P. M., Koch, A., Wudl, F., Rubin, Y., Diederich, F., Alvarez, M. M., Anz, S. J., Whetten, R. L. (1991). Two different fullerenes have the same cyclic voltammetry. *Journal of the American Chemical Society*, **113**, 1050–1051.

40. Dubois, D., Kadish, K. M., Flanagan, S., Haufler, R. E., Chibante, L. P. F., Wilson, L. J. (1991). Spectroelectrochemical study of the C_{60} and C_{70} fullerenes and their mono-, di-, tri- and tetraanions. *Journal of American Chemistry Society*, **113**, 4364–4366.

41. Xie, Q., Perez-Cordero, E., Echegoyen, L. (1992). Electrochemical detection of $C_{60}{}^{6-}$ and $C_{70}{}^{6-}$: Enhanced stability of fullerides in solution. *Journal of the American Chemistry Society*, **114**, 3978–3980.

42. Konarev, D. V., Kuzmin, A. V., Simonov, S. V., Khasanov, S. S., Yudanova, E. I., Lyubovskaya, R. N. (2011). Structure and properties of ionic fullerene complex $Co^+(dppe)_2\cdot(C_{60}{}^{\bullet-})\cdot(C_6H_4Cl_2)_2$: Distortion of the ordered fullerene cage of $C_{60}{}^{\bullet-}$ radical anions. *Dalton Transactions*, **40**, 4453–4458.

43. Buergi, H. B., Blanc, E., Schwarzenbach, D., Liu, S., Lu, Y. J., Kappes, M. M., Ibers, J. A. (1992). The structure of fullerene C_{60}: Orientation defect in the low-temperature modification of C_{60}. *Angewandte Chemie, International Edition in English*, **31**, 640–643.

44. Paul, P., Xie, Z., Bau, R., Boyd, P. D. W., Reed, C. A. (1994). Ordered structure of a distorted $C_{60}{}^{2-}$ fulleride ion. *Journal of the American Chemical Society*, **116**, 4145–4146.

45. Himmel, K., Jansen, M. (1998). Synthesis and single-crystal structure analysis of $[Ba(NH_3)_7]C_{60}\cdot NH_3$. *Inorganic Chemistry*, **37**, 3437–3439.

46. Sariciftci, N. S., Smilowitz, L., Heeger, A. J., Wudl, F. (1992). Photoinduced electron-transfer from a conducting polymer to buckminsterfullerene. *Science*, **258**, 1474–1476.

47. Haddon, R. C., Brus, L. E., Raghavachari, K. (1986). Electronic structure and bonding in icosahedral carbon cluster (C_{60}). *Chemical Physics Letters*, **125**, 459–464.

48. Wang, L. S., Conceicao, J., Jin, C., Smalley, R. E. (1991). Threshold photodetachment of cold carbon sixty-atom mononegative ion. *Chemical Physics Letters*, **182**, 5–11.

49. Xie, Q., Arias, F., Echegoyen, L. (1993). Electrochemically-reversible, single-electron oxidation of C_{60} and C_{70}. *Journal of the American Chemical Society*, **115**, 9818–9819.

50. Saito, S., Oshiyama, A. (1991). Electronic and geometric structures of fullerene (C_{70}). *Physics Reviews B: Condensed Matter*, **44**, 11532–11535.

51. Cox, D. M., Behal, S., Disko, M., Gorun, S. M., Greaney, M., Hsu, C. S., Kollin, E. B., Millar, J., Robbins, J., et al. (1991). Characterization of C_{60} and C_{70} clusters. *Journal of the American Chemical Society*, **113**, 2940–2944.

52. Azamar-Barrios, J. A., Dennis, T. J. S., Sadhukan, S., Shinohara, H., Scuseria, G. E., Penicaud, A. (2001). Characterization of six isomers of [84]fullerene C_{84} by electrochemistry, electron spin resonance spectroscopy, and molecular energy levels calculations. *Journal of Physical Chemistry, A*, **105**, 4627–4632.

53. Anderson, M. R., Dorn, H. C., Stevenson, S. A., Dana, S. M. (1998). The voltammetry of C_{84} isomers. *Journal of Electroanalytical Chemistry*, **444**, 151–154.

54. Zalibera, M., Rapta, P., Popov, A. A., Dunsch, L. (2009). Charged states of four isomers of C_{84} fullerene: In situ esr and vis–nir spectroelectrochemistry and dft calculations. *Journal of Physical Chemistry C*, **113**, 5141–5149.

55. Kavan, L., Dunsch, L. (2007). Spectroelectrochemistry of carbon nanostructures. *ChemPhysChem*, **8**, 974–998.

56. Kavan, L., Dunsch, L. (2011). Spectroelectrochemistry of carbon nanotubes. *ChemPhysChem*, **12**, 47–55.

57. Rapta, P., Bartl, A., Gromov, A., Stasko, A., Dunsch, L. (2002). In situ ESR/Vis/NIR spectroelectrochemistry of [60]fullerene: The origin of esr "spikes" and the reactivity of pristine fullerene anions. *ChemPhysChem*, **3**, 351–356.

58. Zalibera, M., Rapta, P., Dunsch, L. (2007). In situ ESR/UV/Vis/NIR spectroelectrochemistry of an empty fullerene anion and cation: The c82:3 isomer. *Electrochemical Communications*, **9**, 2843–2847.

59. Zalibera, M., Popov, A. A., Kalbac, M., Rapta, P., Dunsch, L. (2008). The extended view on the empty $C_2(3)$-C_{82} fullerene: Isolation, spectroscopic, electrochemical, and spectroelectrochemical characterization and dft calculations. *Chemistry—A European Journal*, **14**, 9960–9967.

60. Zalibera, M., Rapta, P., Dunsch, L. (2008). The power of in situ esr spectroelectrochemistry in the analysis of a C_{84} fullerene isomer. *Electrochemical Communications*, **10**, 943–946.

61. Akasaka, T., Wudi, F., Nagase, S. (2010). Chemistry of nanocarbons. Wiley, West Sussex, United Kingdom.

62. Chaur, M. N., Aparicio-Angles, X., Mercado, B. Q., Elliott, B., Rodriguez-Fortea, A., Clotet, A., Olmstead, M. M., Balch, A. L., Poblet, J. M., Echegoyen, L. (2010). Structural and electrochemical property correlations of metallic nitride endohedral metallofullerenes. *Journal of Physical Chemistry C*, **114**, 13003–13009.

63. Chen, N., Chaur, M. N., Moore, C., Pinzon, J. R., Valencia, R., Rodriguez-Fortea, A., Poblet, J. M., Echegoyen, L. (2010). Synthesis of a new endohedral fullerene family, $Sc_2S@C_{2n}$ ($n = 40$–50) by the introduction of SO_2. *Chemical Communications (Cambridge, U.K.)*, **46**, 4818–4820.

64. Dunsch, L., Yang, S., Zhang, L., Svitova, A., Oswald, S., Popov, A. A. (2010). Metal sulfide in a C_{82} fullerene cage: A new form of endohedral clusterfullerenes. *Journal of American Chemical Society*, **132**, 5413–5421.

65. Mercado, B. Q., Chen, N., Rodriguez-Fortea, A., Mackey, M. A., Stevenson, S., Echegoyen, L., Poblet, J. M., Olmstead, M. M., Balch, A. L. (2011). The shape of the $Sc_2(\mu_2\text{-}S)$ unit trapped in C_{82}: Crystallographic, computational, and electrochemical studies of the isomers, $Sc_2(\mu_2\text{-}S)@C_s(6)\text{-}C_{82}$ and $Sc_2(\mu_2\text{-}S)@C_{3v}(8)\text{-}C_{82}$. *Journal of the American Chemical Society*, **133**, 6752–6760.

66. Mercado, B. Q., Olmstead, M. M., Beavers, C. M., Easterling, M. L., Stevenson, S., Mackey, M. A., Coumbe, C. E., Phillips, J. D., Phillips, J. P., Poblet, J. M., Balch, A. L. (2010). A seven atom cluster in a carbon cage, the crystallographically determined structure of $Sc_4(\mu_3\text{-}O)_3@I_h\text{-}C_{80}$. *Chemical Communications (Cambridge, U.K.)*, **46**, 279–281.

67. Mercado, B. Q., Stuart, M. A., Mackey, M. A., Pickens, J. E., Confait, B. S., Stevenson, S., Easterling, M. L., Valencia, R., Rodriguez-Fortea, A., Poblet, J. M., Olmstead, M. M., Balch, A. L. (2010). $Sc_2(\mu_2\text{-}O)$ trapped in a fullerene cage: The isolation and structural characterization of $Sc_2(\mu_2\text{-}O)@C_s(6)\text{-}C_{82}$ and the relevance of the thermal and entropic effects in fullerene isomer selection. *Journal of the American Chemistry Society*, **132**, 12098–12105.

68. Valencia, R., Rodriguez-Fortea, A., Stevenson, S., Balch, A. L., Poblet, J. M. (2009). Electronic structures of scandium oxide endohedral metallofullerenes, $Sc_4(\mu_3\text{-}O)_n@I_h\text{-}C_{80}$ ($n = 2, 3$). *Inorganic Chemistry (Washington, DC, U.S.)*, **48**, 5957–5961.

69. Murata, M., Murata, Y., Komatsu, K. (2006). Synthesis and properties of endohedral C_{60} encapsulating molecular hydrogen. *Journal of the American Chemical Society*, **128**, 8024–8033.

70. Murata, M., Murata, Y., Komatsu, K. (2008). Surgery of fullerenes. *Chemical Communications (Cambridge, U.K.)*, 6083–6094.

71. Ramachandran, C. N., Roy, D., Sathyamurthy, N. (2008). Host–guest interaction in endohedral fullerenes. *Chemical Physics Letters*, **461**, 87–92.

72. Riahi, S., Pourhossein, P., Zolfaghari, A., Ganjali, M. R., Jooya, H. Z. (2009). Encapsulation of hydrogen molecule in fullerene (C_{60}). *Fullerenes, Nanotubes. Carbon Nanostructures*, **17**, 159–170.

73. Chaur, M. N., Melin, F., Ortiz, A. L., Echegoyen, L. (2009). Chemical, electrochemical, and structural properties of endohedral metallofullerenes. *Angewandte Chemie International Edition*, **48**, 7514–7538.

74. Chaur, M. N., Athans, A. J., Echegoyen, L. (2008). Metallic nitride endohedral fullerenes: Synthesis and electrochemical properties. *Tetrahedron*, **64**, 11387–11393.

75. Chaur, M. N., Valencia, R., Rodriguez-Fortea, A., Poblet, J. M., Echegoyen, L. (2009). Trimetallic nitride endohedral fullerenes: Experimental and theoretical evidence for the $M_3N^{6+}@C_{2n}{}^{6-}$ model. *Angewandte Chemie International Edition*, **48**, 1425–1428.

76. Boltalina, O. V., Ponomarev, D. B., Sidorov, L. N. (1998). Thermochemistry of fullerene anions in the gas phase. *Mass Spectrometry Reviews*, **16**, 333–351.

77. Ettl, R., Chao, I., Diederich, F., Whetten, R. L. (1991). Isolation of C_{76}, a chiral (D_2) allotrope of carbon. *Nature (London)*, **353**, 149–153.

78. Kobayashi, K., Nagase, S., Yoshida, M., Osawa, E. (1997). Endohedral metallofullerenes. Are the isolated pentagon rule and fullerene structures always satisfied? *Journal of the American Chemical Society*, **119**, 12693–12694.

79. Yang, S., Rapta, P., Dunsch, L. (2007). The spin state of a charged non-IPR fullerene: The stable radical cation of $Sc_3N@C_{68}$. *Chemcal Communications (Cambridge, U.K.)*, 189–191.

80. Rapta, P., Popov, A. A., Yang, S., Dunsch, L. (2008). Charged states of $Sc_3N@C_{68}$: An in situ spectroelectrochemical study of the radical cation and radical anion of a non-ipr fullerene. *Journal of Physical Chemistry, A*, **112**, 5858–5865.
81. Popov, A. A., Dunsch, L. (2008). Hindered cluster rotation and Sc-45 hyperfine splitting constant in distonoid anion radical $Sc_3N@C_{80}^{-}$, and spatial spin-charge separation as a general principle for anions of endohedral fullerenes with metal-localized lowest unoccupied molecular orbitals. *Journal American Chemical Society*, **130**, 17726–17742.
82. Vargova, A., Popov, A., Rapta, P., Sun, B., Zhang, L., Dunsch, L. (2010). Electrochemical tuning of spin states of the endohedral metallofullerene $Y@C_{82}$ as probed by esr spectroelectrochemistry. *ChemPhysChem*, **11**, 1650–1653.
83. Roncali, J. (2005). Linear π-conjugated systems derivatized with C_{60}-fullerene as molecular heterojunctions for organic photovoltaics. *Chemical Society Reviews*, **34**, 483–495.
84. Accorsi, G., Armaroli, N. (2010). Taking advantage of the electronic excited states of [60]-fullerenes. *Journal of Physical Chemistry, C*, **114**, 1385–1403.
85. Sanchez, L., Herranz, M. A., Martin, N. (2005). C_{60}-based dumbbells: Connecting C_{60} cages through electroactive bridges. *Journal of Material Chemistry*, **15**, 1409–1421.
86. Segura, J. L., Giacalone, F., Gomez, R., Martin, N., Guldi, D. M., Luo, C., Swartz, A., Riedel, I., Chirvase, D., Parisi, J., Dyakonov, V., Serdar, S. N., Padinger, F. (2005). Design, synthesis and photovoltaic properties of [60]fullerene based molecular materials. *Materials Science and Engineering, C*, ***C*25**, 835–842.
87. Bendikov, M., Wudl, F., Perepichka, D. F. (2004). Tetrathiafulvalenes, oligoacenenes, and their buckminsterfullerene derivatives: The brick and mortar of organic electronics. *Chemical Reviews (Washington, DC, U.S.)*, **104**, 4891–4945.
88. Sastre-Santos, A., Parejo, C., Martin-Gomis, L., Ohkubo, K., Fernandez-Lazaro, F., Fukuzumi, S. (2011). C_{60} dimers connected through pleiadene bridges: Fullerenes talking to each other. *Journal of Materials Chemistry*, **21**, 1509–1515.
89. Iezzi, E. B., Duchamp, J. C., Harich, K., Glass, T. E., Lee, H. M., Olmstead, M. M., Balch, A. L., Dorn, H. C. (2002). A symmetric derivative of the trimetallic nitride endohedral metallofullerene, $Sc_3N@C_{80}$. *Journal of the American Chemical Society*, **124**, 524–525.
90. Lee, H. M., Olmstead, M. M., Iezzi, E., Duchamp, J. C., Dorn, H. C., Balch, A. L. (2002). Crystallographic characterization and structural analysis of the first organic functionalization product of the endohedral fullerene $Sc_3N@C_{80}$. *Journal of the American Chemical Society*, **124**, 3494–3495.
91. Cardona, C. M., Kitaygorodskiy, A., Ortiz, A., Herranz, M. A., Echegoyen, L. (2005). The first fulleropyrrolidine derivative of $Sc_3N@C_{80}$: Pronounced chemical shift differences of the geminal protons on the pyrrolidine ring. *Journal of Organic Chemistry*, **70**, 5092–5097.
92. Cai, T., Ge, Z. X., Iezzi, E. B., Glass, T. E., Harich, K., Gibson, H. W., Dorn, H. C. (2005). Synthesis and characterization of the first trimetallic nitride templated pyrrolidino endohedral metallofullerenes. *Chemical Communications*, 3594–3596.
93. Cardona, C. M., Kitaygorodskiy, A., Echegoyen, L. (2005). Trimetallic nitride endohedral metallofullerenes: Reactivity dictated by the encapsulated metal cluster. *Journal of the American Chemical Society*, **127**, 10448–10453.
94. Cardona, C. M., Elliott, B., Echegoyen, L. (2006). Unexpected chemical and electrochemical properties of $M_3N@C_{80}$ (m = sc, y, er). *Journal of the American Chemical Society*, **128**, 6480–6485.
95. Campanera, J. M., Bo, C., Poblet, J. M. (2006). Exohedral reactivity of trimetallic nitride template (TNT) endohedral metallofullerenes. *Journal of Organic Chemistry*, **71**, 46–54.
96. Chen, N., Fan, L. Z., Tan, K., Wu, Y. Q., Shu, C. Y., Lu, X., Wang, C. R. (2007). Comparative spectroscopic and reactivity studies of $Sc_{3-x}Y_xN@C_{80}$ ($x = 0$–3). *Journal of Physical Chemistry, C*, **111**, 11823–11828.
97. Chen, N., Zhang, E. Y., Tan, K., Wang, C. R., Lu, X. (2007). Size effect of encaged clusters on the exohedral chemistry of endohedral fullerenes: A case study on the pyrrolidino reaction of $Sc_xGd_{3-x}N@C_{80}$ ($x = 0$–3). *Organic Letters*, **9**, 2011–2013.
98. Cai, T., Slebodnick, C., Xu, L., Harich, K., Glass, T. E., Chancellor, C., Fettinger, J. C., Olmstead, M. M., Balch, A. L., Gibson, H. W., Dorn, H. C. (2006). A pirouette on a metallofullerene sphere: Interconversion of isomers of *n*-tritylpyrrolidino i-h $Sc_3N@C_{80}$. *Journal of the American Chemical Society*, **128**, 6486–6492.
99. Cai, T., Xu, L. S., Anderson, M. R., Ge, Z. X., Zuo, T. M., Wang, X. L., Olmstead, M. M., Balch, A. L., Gibson, H. W., Dorn, H. C. (2006). Structure and enhanced reactivity rates of the D_{5h} $Sc_3N@C_{80}$ and $Lu_3N@C_{80}$ metallofullerene isomers: The importance of the pyracylene motif. *Journal of the American Chemical Society*, **128**, 8581–8589.
100. Shu, C. Y., Cai, T., Xu, L. S., Zuo, T. M., Reid, J., Harich, K., Dorn, H. C., Gibson, H. W. (2007). Manganese(iii)-catalyzed free radical reactions on trimetallic nitride endohedral metallofullerenes. *Journal of the American Chemical Society*, **129**, 15710–15717.
101. Pinzon, J. R., Zuo, T. M., Echegoyen, L. (2010). Synthesis and electrochemical studies of bingel-hirsch derivatives of $M_3N@I_h$-C_{80} (m = Sc, Lu). *Chemistry—A European Journal*, **16**, 4864–4869.
102. Lukoyanova, O., Cardona, C. M., Rivera, J., Lugo-Morales, L. Z., Chancellor, C. J., Olmstead, M. M., Rodriguez-Fortea, A., Poblet, J. M., Balch, A. L., Echegoyen, L. (2007). "Open rather than closed" malonate methano-fullerene derivatives. The formation of methanofulleroid adducts of $Y_3N@C_{80}$. *Journal of the American Chemical Society*, **129**, 10423–10430.
103. Chaur, M. N., Melin, F., Athans, A. J., Elliott, B., Walker, K., Holloway, B. C., Echegoyen, L. (2008). The influence of cage size on the reactivity of trimetallic nitride metallofullerenes: A mono- and bis-methanoadduct of $Gd_3N@C_{80}$ and a monoadduct of $Gd_3N@C_{84}$. *Chemical Communications*, 2665–2667.
104. Li, F. F., Pinzon, J. R., Mercado, B. Q., Olmstead, M. M., Balch, A. L., Echegoyen, L. (2011). [2 + 2] cycloaddition reaction to $Sc_3N@I_h$-C_{80}. The formation of very stable [5,6]-

and [6,6]-adducts. *Journal of the American Chemical Society*, **133**, 1563–1571.

105. Iiduka, Y., Ikenaga, O., Sakuraba, A., Wakahara, T., Tsuchiya, T., Maeda, Y., Nakahodo, T., Akasaka, T., Kako, M., Mizorogi, N., Nagase, S. (2005). Chemical reactivity of $Sc_3N@C_{80}$ and $La_2@C_{80}$. *Journal of the American Chemical Society*, **127**, 9956–9957.
106. Feng, L., Tsuchiya, T., Wakahara, T., Nakahodo, T., Piao, Q., Maeda, Y., Akasaka, T., Kato, T., Yoza, K., Horn, E., Mizorogi, N., Nagase, S. (2006). Synthesis and characterization of a bisadduct of $La@C_{82}$. *Journal of the American Chemical Society*, **128**, 5990–5991.
107. Feng, L., Wakahara, T., Nakahodo, T., Tsuchiya, T., Piao, Q., Maeda, Y., Lian, Y., Akasaka, T., Horn, E., Yoza, K., Kato, T., Mizorogi, N., Nagase, S. (2006). The bingel monoadducts of $La@C_{82}$: Synthesis, characterization, and electrochemistry. *Chemistry—A European Journal*, **12**, 5578–5586.
108. Wakahara, T., Iiduka, Y., Ikenaga, O., Nakahodo, T., Sakuraba, A., Tsuchiya, T., Maeda, Y., Kako, M., Akasaka, T., Yoza, K., Horn, E., Mizorogi, N., Nagase, S. (2006). Characterization of the bis-silylated endofullerene $Sc_3N@C_{80}$. *Journal of the American Chemical Society*, **128**, 9919–9925.
109. Shustova, N. B., Popov, A. A., Mackey, M. A., Coumbe, C. E., Phillips, J. P., Stevenson, S., Strauss, S. H., Boltalina, O. V. (2007). Radical trifluoromethylation of $Sc_3N@C_{80}$. *J Am Chemical Society*, **129**, 11676–11677.
110. Shu, C., Slebodnick, C., Xu, L., Champion, H., Fuhrer, T., Cai, T., Reid, J. E., Fu, W., Harich, K., Dorn, H. C., Gibson, H. W. (2008). Highly regioselective derivatization of trimetallic nitride templated endohedral metallofullerenes via a facile photochemical reaction. *Journal of the American Chemical Society*, **130**, 17755–17760.
111. Reed, C. A., Bolskar, R. D. (2000). Discrete fulleride anions and fullerenium cations. *Chemical Reviews*, **100**, 1075–1120.
112. Caron, C., Subramanian, R., D'Souza, F., Kim, J., Kutner, W., Jones, M. T., Kadish, K. M. (1993). Selective electrosynthesis of dimethylfullerene [$(CH_3)_2C_{60}$]: A novel method for the controlled functionalization of fullerenes. *Journal of the American Chemical Society*, **115**, 8505–8506.
113. Boulas, P. L., Zuo, Y., Echegoyen, L. (1996). A new route for the selective synthesis of [6,6]-methanofullerenes. Electrosynthesis of $C_{61}HCMe_3$ and $C_{61}HCN$. *Chemical Communications*, 1547–1548.
114. Subramanian, R., Kadish, K. M., Vijayashree, M. N., Gao, X., Jones, M. T., Miller, M. D., Krause, K. L., Suenobu, T., Fukuzumi, S. (1996). Chemical generation of $C_{60}{}^{2-}$ and electron transfer mechanism for the reactions with alkyl bromides. *The Journal of Physical Chemistry*, **100**, 16327–16335.
115. Fukuzumi, S., Suenobu, T., Hirasaka, T., Arakawa, R., Kadish, K. M. (1998). Formation of C_{60} adducts with two different alkyl groups via combination of electron transfer and SN_2 reactions. *Journal of the American Chemical Society*, **120**, 9220–9227.
116. Kadish, K. M., Gao, X., Caemelbecke, E. V., Hirasaka, T., Suenobu, T., Fukuzumi, S. (1998). Synthesis and spectroscopic and electrochemical characterization of di- and tetrasubstituted C_{60} derivatives. *The Journal of Physical Chemistry A*, **102**, 3898–3906.
117. Zhu, Y.-H., Song, L.-C., Hu, Q.-M., Li, C.-M. (1999). Synthesis and isolation of σ-bonded fullerene metal derivatives from reactions of fullerene dianion $C_{60}{}^{2-}$ with organometal halides. *Organic Letters*, **1**, 1693–1695.
118. Kadish, K. M., Gao, X., Van Caemelbecke, E., Suenobu, T., Fukuzumi, S. (2000). Electrosynthesis and structural characterization of two $(C_6H_5CH_2)_4C_{60}$ isomers. *Journal of the American Chemical Society*, **122**, 563–570.
119. Allard, E., Delaunay, J., Cheng, F., Cousseau, J., Ordúna, J., Garín, J. (2001). Novel C_{60}-based building blocks derived from $C_{60}{}^{2-}$ anion. *Organic Letters*, **3**, 3503–3506.
120. Cheng, F., Murata, Y., Komatsu, K. (2002). Synthesis, x-ray structure, and properties of the singly bonded C_{60} dimer having diethoxyphosphorylmethyl groups utilizing the chemistry of $C_{60}{}^{2-}$. *Organic Letters*, **4**, 2541–2544.
121. Meier, M. S., Bergosh, R. G., Gallagher, M. E., Spielmann, H. P., Wang, Z. (2002). Alkylation of dihydrofullerenes. *The Journal of Organic Chemistry*, **67**, 5946–5952.
122. Wang, Z., Meier, M. S. (2003). Monoalkylation of C_{60} and C_{70} with Zn and active alkyl bromides. *The Journal of Organic Chemistry*, **68**, 3043–3048.
123. Allard, E., Delaunay, J., Cousseau, J. (2003). $C_{60}{}^{2-}$ chemistry: C_{60} adducts bearing two ester, carbonyl, or alcohol groups. *Organic Letters*, **5**, 2239–2242.
124. Zheng, M., Li, F., Shi, Z., Gao, X., Kadish, K. M. (2007). Electrosynthesis and characterization of 1,2-dibenzyl C_{60}: A revisit. *The Journal of Organic Chemistry*, **72**, 2538–2542.
125. Chen, J., Cai, R.-F., Huang, Z.-E., Wu, H.-M., Jiang, S.-K., Shao, Q.-F. (1995). Isomerically pure organo[60]fullerenes from $C_{60}{}^{2-}$ salt: Synthesis and characterization of 1-benzyl-2-hydro[60]fullerene. *Journal of the Chemical Society, Chemical Communications*, 1553–1554.
126. Mangold, K. M., Kutner, W., Dunsch, L., Fröhner, J. (1996). Derivatization of fullerenes by electrosynthesis. *Synthetic Metals*, **77**, 73–76.
127. Kadish, K. M., Gao, X., Caemelbecke, E. V., Suenobu, T., Fukuzumi, S. (2000). Effect of addition pattern on the electrochemical and spectroscopic properties of neutral and reduced 1,2- and 1,4-$(C_6H_5CH_2)_2C_{60}$ isomers. *The Journal of Physical Chemistry A*, **104**, 3878–3883.
128. Cardullo, F., Seiler, P., Isaacs, L., Nierengarten, J.-F., Haldimann, R. F., Diederich, F., Mordasini-Denti, T., Thiel, W., Boudon, C., Gisselhrecht, J.-P., Gross, M. (1997). Bis-through tetrakis-adducts of C_{60} by reversible tether-directed remote functionalization and systematic investigation of the changes in fullerene properties as a function of degree, pattern, and nature of functionalization. *Helvetica Chimica Acta*, **80**, 343–371.
129. Yang, W.-W., Li, Z.-J., Gao, X. (2010). Reaction of $C_{60}{}^{2-}$ with organic halides revisited in dmf: Proton transfer from water to RC_{60^-} and unexpected formation of 1,2-dihydro[60]fullerenes. *The Journal of Organic Chemistry*, **75**, 4086–4094.

130. Cosa, G., Lukeman, M., Scaiano, J. C. (2009). How drug photodegradation studies led to the promise of new therapies and some fundamental carbanion reaction dynamics along the way. *Accounts Chem Res*, **42**, 599–607.

131. Yang, W.-W., Li, Z.-J., Gao, X. (2011). Formation of singly bonded $PhCH_2C_{60}$-$C_{60}CH_2Ph$ dimers from 1,2-$(PhCH_2)HC_{60}$ via electroreductive C_{60}-H activation. *The Journal of Organic Chemistry*, **76**, 6067–6074.

132. Guarr, T. F., Meier, M. S., Vance, V. K., Clayton, M. (1993). Electrochemistry of the $C_{60}H_2$ fullerene. *Journal of the American Chemical Society*, **115**, 9862–9863.

133. Khaled, M. M., Carlin, R. T., Trulove, P. C., Eaton, G. R., Eaton, S. S. (1994). Electrochemical generation and electron paramagnetic resonance studies of $C_{60}-$, $C_{60}{}^{2-}$, and $C_{60}{}^{3-}$. *Journal of the American Chemical Society*, **116**, 3465–3474.

134. Beulen, M. W. J., Echegoyen, L. (2000). Instability of [60]fullerene anions in dichloromethane: A synthetic avenue to C > (CH) methanofullerenes. *Chemical Communications*, 1065–1066.

135. Zheng, M., Li, F.-f., Ni, L., Yang, W.-w., Gao, X. (2008). Synthesis and identification of heterocyclic derivatives of fullerene C_{60}: Unexpected reaction of anionic C_{60} with benzonitrile. *The Journal of Organic Chemistry*, **73**, 3159–3168.

136. Li, F.-F., Yang, W.-W., He, G.-B., Gao, X. (2009). Formation of fullerooxazoles from $C_{61}HPh^{3-}$: The regioselectivity of heteroatom additions. *The Journal of Organic Chemistry*, **74**, 8071–8077.

137. Li, F.-F., Gao, X., Zheng, M. (2009). Why [6,6]- and 1,2-benzal-3-N-4-O-cyclic phenylimidate C_{60} undergo electrochemically induced retro-addition reactions while 1,4-dibenzyl-2,3-cyclic phenylimidate C_{60} does not? C−H···x (x = N, O) intramolecular interactions in organofullerenes. *The Journal of Organic Chemistry*, **74**, 82–87.

138. Dubois, D., Moninot, G., Kutner, W., Jones, M. T., Kadish, K. M. (1992). *J. Physical Chemistry*, **96**, 7173.

139. Kadish, K. M., Gao, X., Gorelik, O., Van Caemelbecke, E., Suenobu, T., Fukuzumi, S. (2000). Electrogeneration and characterization of $(C_6H_5CH_2)_2C_{70}$. *The Journal of Physical Chemistry A*, **104**, 2902–2907.

140. Hirsch, A. (1994). *The Chemistry of the Fullerenes*, G. Thieme Verlag, Stuttgart.

141. Diederich, F., Kessinger, R. (1999). Templated regioselective and stereoselective synthesis in fullerene chemistry. *Accounts of Chemical Research*, **32**, 537–545.

142. Chronakis, N., Hirsch, A. (2006). Macrocyclic malonates. A new family of tethers for the regio- and diastereoselective functionalization of [60]fullerene. *Comptes Rendu Chimie*, **9**, 862–867.

143. Thilgen, C., Diederich, F. (2006). Tether-directed remote functionalization of fullerenes C_{60} and C_{70}. *Comptes Rendu Chimie*, **9**, 868–880.

144. Isaacs, L., Haldimann, R. F., Diederich, F. (1994). Spacer-controlled long-range functionalization of buckminsterfullerenes regiospecific formation of a hexaadduct. *Angewandte Chemie*, **106**, 2434–2437 (see also *Angewandte Chemie International Edition English*, 1994, 2433(2422), 2339–2442).

145. Schwenninger, R., Mueller, T., Kraeutler, B. (1997). Concise route to symmetrical multiadducts of [60]fullerene: Preparation of an equatorial tetraadduct by orthogonal transposition. *Journal of the American Chemical Society*, **119**, 9317–9318.

146. Herranz, M. A., Diederich, F., Echegoyen, L. (2004). Electrochemically induced retro-cyclopropanation reactions. *European Journal of Organic Chemistry*, 2299–2316.

147. Kessinger, R., Fender, N. S., Echegoyen, L. E., Thilgen, C., Echegoyen, L., Diedrich, F. (2000). Selective electrolytic removal of bis(alkoxycarbonyl)methano addends from C_{60} bis-adducts and electrochemical stability of C_{70} derivatives. *Chemistry—A European Journal*, **6**, 2184–2192.

148. Kessinger, R., Crassous, J., Herrmann, A., Ruttimann, M., Echegoyen, L., Diederich, F. (1998). Preparation of enantiomerically pure C_{76} with a general electrochemical method for the removal of di(alkoxycarbonyl)methano bridges from methanofullerenes: The retro-Bingel reaction. *Angewandte Chemie International Edition*, **37**, 1919–1922.

149. Lukoyanova, O., Cardona, C. M., Altable, M., Filippone, S., Martin, D. A., Martin, N., Echegoyen, L. (2006). Selective electrochemical retro-cycloaddition reaction of pyrrolidinofullerenes. *Angewandte Chemie International Edition English*, **45**, 7430–7433.

150. Zhang, S., Lukoyanova, O., Echegoyen, L. (2006). Synthesis of fullerene adducts with terpyridyl- or pyridylpyrrolidine groups in trans-1 positions. *Chemistry—A European Journal*, **12**, 2846–2853.

151. Ortiz, A. L., Echegoyen, L. (2011). Unexpected and selective formation of an *(e,e,e,e)*-tetrakis-[60]fullerene derivative via electrolytic retro-cyclopropanation of a D_{2h}-hexakis-[60]fullerene adduct. *Journal of Material Chemistry*, **21**, 1362–1364.

152. Kessinger, R., Gomez-Lopez, M., Boudon, C., Gisselbrecht, J. P., Gross, M., Echegoyen, L., Diederich, F. (1998). Walk on the sphere: Electrochemically induced isomerization of C_{60} bis-adducts by migration of di(alkoxycarbonyl)methano bridges. *Journal of the American Chemical Society*, **120**, 8545–8546.

153. Li, F. F., Rodriguez-Fortea, A., Poblet, J. M., Echegoyen, L. (2011). Reactivity of metallic nitride endohedral metallofullerene anions: Electrochemical synthesis of a $Lu_3N@I_h$-C_{80} derivative. *Journal of the American Chemical Society*, **133**, 2760–2765.

154. Chen, N., Pinzón, J. R., Echegoyen, L. (2011). Influence of the encapsulated clusters on the electrochemical behaviour of endohedral fullerene derivatives: Comparative study of n-tritylpyrrolidino derivatives of $Sc_3N@I_h$-C_{80} and $Lu_3N@I_h$-C_{80}. *ChemPhysChem*, **12**, 1422–1425.

155. Ross, R. B., Cardona, C. M., Guldi, D. M., Sankaranarayanan, S. G., Reese, M. O., Kopidakis, N., Peet, J., Walker, B., Bazan, G. C., Van Keuren, E., Holloway, B. C., Drees, M. (2009). Endohedral fullerenes for organic photovoltaic devices. *Nature Materials*, **8**, 208–212.

156. Shu, C., Xu, W., Slebodnick, C., Champion, H., Fu, W., Reid, J. E., Azurmendi, H., Wang, C., Harich, K., Dorn, H. C., Gibson, H. W. (2009). Syntheses and structures of phenyl-c81-butyric acid methyl esters (PCBMs) from $M_3N@C_{80}$. *Organic Letters*, **11**, 1753–1756.

157. Ross, R. B., Cardona, C. M., Swain, F. B., Guldi, D. M., Sankaranarayanan, S. G., Van Keuren, E., Holloway, B. C., Drees, M. (2009). Tuning conversion efficiency in metallo endohedral fullerene-based organic photovoltaic devices. *Advanced Functional Materials*, **19**, 2332–2337.

158. Kooistra, F. B., Knol, J., Kastenberg, F., Popescu, L. M., Verhees, W. J. H., Kroon, J. M., Hummelen, J. C. (2007). Increasing the open circuit voltage of bulk-heterojunction solar cells by raising the lumo level of the acceptor. *Organic Letters*, **9**, 551–554.

159. Lenes, M., Wetzelaer, G.-J. A. H., Kooistra, F. B., Veenstra, S. C., Hummelen, J. C., Blom, P. W. M. (2008). Fullerene bisadducts for enhanced open-circuit voltages and efficiencies in polymer solar cells. *Advanced Materials*, **20**, 2116–2119.

160. Iijima, S. (1980). Direct observation of the tetrahedral bonding in graphitized carbon black by high resolution electron microscopy. *Journal of Crystal Growth*, **50**, 675–683.

161. Iijima, S. (1987). The 60-carbon cluster has been revealed. *Journal of Physical Chemistry*, **91**, 3466–3467.

162. Kuznetsov, V. L., Chuvilin, A. L., Butenko, Y. V., Mal'kov, I. Y., Titov, V. M. (1994). Onion-like carbon from ultra-disperse diamond. *Chemical Physics Letters*, **222**, 343–348.

163. Sano, N., Wang, H., Chhowalla, M., Alexandrou, I., Amaratunga, G. A. (2001). Synthesis of carbon 'onions' in water. *Nature*, **414**, 506–507.

164. Breczko, J., Winkler, K., Plonska-Brzezinska, M. E., Villalta-Cerdas, A., Echegoyen, L. (2010). Electrochemical properties of composites containing small carbon nano-onions and solid polyelectrolytes. *Journal of Material Chemistry*, **20**, 7761–7768.

165. Bushueva, E. G., Galkin, P. S., Okotrub, A. V., Bulusheva, L. G., Gavrilov, N. N., Kuznetsov, V. L., Moiseekov, S. I. (2008). Double layer supercapacitor properties of onion-like carbon materials. *Physica Status Solidi B*, **245**, 2296–2299.

166. Cioffi, C. T., Palkar, A., Melin, F., Kumbhar, A., Echegoyen, L., Melle-Franco, M., Zerbetto, F., Rahman, G. M. Á., Ehli, C., Sgobba, V., Guldi, D. M., Prato, M. (2009). A carbon nano-onion-ferrocene donor–acceptor system: Synthesis, characterization and properties. *Chemistry—A European Journal*, **15**, 4419–4427.

167. Kuznetsov, V. L., Chuvilin, A. L., Moroz, E. M., Kolomiichuk, V. N., Shaikhutdinov, S. K., Butenko, Y. V. (1994). Effect of explosion conditions on the structure of detonation soots: Ultradisperse diamond and onion carbon. *Carbon*, **32**, 873–882.

168. Pech, D., Brunet, M., Durou, H., Huang, P., Mochalin, V., Gogotsi, Y., Taberna, P.-L., Simon, P. (2010). Ultrahigh-power micrometre-sized supercapacitors based on onion-like carbon. *Nature Nanotechnology*, **5**, 651–654.

169. Plonska-Brzezinska, M. E., Palkar, A., Winkler, K., Echegoyen, L. (2010). Electrochemical properties of small carbon nano-onion films. *Electrochemical and Solid-State Letters*, **13**, K35–K38.

170. Xiao, Q., Zhou, X. (2003). The study of multiwalled carbon nanotube deposited with conducting polymer for supercapacitor. *Electrochimica Acta*, **48**, 575–580.

171. Hughes, M., Chen, G. Z., Shaffer, M. S. P., Fray, D. J., Windle, A. H. (2002). Electrochemical capacitance of a nanoporous composite of carbon nanotubes and polypyrrole. *Chemcal Materials*, **14**, 1610–1613.

172. Hassanien, A., Gao, M., Tokumoto, M., Dai, L. (2001). Scanning tunneling microscopy of aligned coaxial nanowires of polyaniline passivated carbon nanotubes. *Chemical Physics Letters*, **342**, 479–484.

173. Pieta, P., Grodzka, E., Winkler, K., Warczak, M., Sadkowski, A., Zukowska, G. Z., Venukadasula, G. M., D'Souza, F., Kutner, W. (2009). Conductive, capacitive, and viscoelastic properties of a new composite of the C_{60}-Pd conducting polymer and single-wall carbon nanotubes. *Journal of Physical Chemistry, B*, **113**, 6682–6691.

174. Grodzka, E., Pieta, P., Dluzewski, P., Kutner, W., Winkler, K. (2009). Formation and electrochemical properties of composites of the C_{60}-Pd polymer and multi-wall carbon nanotubes. *Electrochimica Acta*, **54**, 5621–5628.

175. Pech, D., Brunet, M., Durou, H., Huang, P., Mochalin, V., Gogotsi, Y., Taberna, P.-L., Simon, P. (2010). Ultrahigh-power micrometre-sized supercapacitors based on onion-like carbon. *Nature Nanotechnology*, **5**, 651–654.

176. Brabec, C. J., Sariciftci, N. S., Hummelen, J. C. (2001). Plastic solar cells. *Advanced Functional Materials*, **11**, 15–26.

177. Wudl, F. (2002). Fullerene materials. *Journal of Materials Chemistry*, **12**, 1959–1963.

178. Maggini, M., Karlsson, A., Scorrano, G., Sandona, G., Farnia, G., Prato, M. (1994). Ferrocenyl fulleropyrrolidines: A cyclic voltammetry study. *Journal of the Chemical Society, Chemical Communications*, 589–590.

179. Guldi, D. M., Maggini, M., Scorrano, G., Prato, M. (1997). Intramolecular electron transfer in fullerene/ferrocene based donor–bridge–acceptor dyads. *Journal of the American Chemical Society*, **119**, 974–980.

180. Herranz, M. Á., Illescas, B., Martín, N., Luo, C., Guldi, D. M. (2000). Donor/acceptor fulleropyrrolidine triads. *The Journal of Organic Chemistry*, **65**, 5728–5738.

181. Oswald, F., Islam, D. M. S., Araki, Y., Troiani, V., de la Cruz, P., Moreno, A., Ito, O., Langa, F. (2007). Synthesis and photoinduced intramolecular processes of fulleropyrrolidine–oligothienylenevinylene–ferrocene triads. *Chemistry—A European Journal* **13**, 3924–3933.

182. Matt, G. J., Sariciftci, N. S., Fromherz, T. (2004). Anomalous charge transport behavior of fullerene based diodes. *Applied Physics Letters*, **84**, 1570.

183. Alévêque, O., Leriche, P., Cocherel, N., Frère, P., Cravino, A., Roncali, J. (2008). Star-shaped conjugated systems derived from dithiafulvenyl-derivatized triphenylamines as active materials for organic solar cells. *Solar Energy Materials and Solar Cells*, **92**, 1170–1174.

184. Wu, G., Zhao, G., He, C., Zhang, J., He, Q., Chen, X., Li, Y. (2009). Synthesis and photovoltaic properties of a star-shaped molecule with triphenylamine as core and benzo[1,2,5]thiadiazol vinylene as arms. *Solar Energy Materials and Solar Cells*, **93**, 108–113.

185. He, C., He, Q., Yang, X., Wu, G., Yang, C., Bai, F., Shuai, Z., Wang, L., Li, Y. (2007). Synthesis and photovoltaic properties of a solution-processable organic molecule containing triphenylamine and dcm moieties. *The Journal of Physical Chemistry C*, **111**, 8661–8666.

186. Roquet, S., Cravino, A., Leriche, P., Alévêque, O., Frère, P., Roncali, J. (2006). Triphenylamine–thienylenevinylene hybrid systems with internal charge transfer as donor materials for heterojunction solar cells. *Journal of the American Chemical Society*, **128**, 3459–3466.

187. Shi, J., Wang, L., Liang, Y., Peng, S., Cheng, F., Chen, J. (2010). All-solid-state dye-sensitized solar cells with alkyloxy-imidazolium iodide ionic polymer/SiO_2 nanocomposite electrolyte and triphenylamine-based organic dyes. *The Journal of Physical Chemistry C*, **114**, 6814–6821.

188. Marinado, T., Nonomura, K., Nissfolk, J., Karlsson, M. K., Hagberg, D. P., Sun, L., Mori, S., Hagfeldt, A. (2009). How the nature of triphenylamine-polyene dyes in dye-sensitized solar cells affects the open-circuit voltage and electron lifetimes. *Langmuir*, **26**, 2592–2598.

189. El-Khouly, M. E., Shim, S. H., Araki, Y., Ito, O., Kay, K.-Y. (2008). Effect of dual fullerenes on lifetimes of charge-separated states of subphthalocyanine–triphenylamine–fullerene molecular systems. *The Journal of Physical Chemistry B*, **112**, 3910–3917.

190. D'Souza, F., Gadde, S., Islam, D. M. S., Wijesinghe, C. A., Schumacher, A. L., Zandler, M. E., Araki, Y., Ito, O. (2007). Multi-triphenylamine-substituted porphyrin–fullerene conjugates as charge stabilizing "antenna–reaction center" mimics. *The Journal of Physical Chemistry A*, **111**, 8552–8560.

191. Chen, Y., El-Khouly, M. E., Zhuang, X.-D., He, N., Araki, Y., Lin, Y., Ito, O. (2007). Synthesis and photoinduced electron-transfer process of a novel triphenylamine-substituted polyfluorene–C_{60} triad. *Chemistry—A European Journal*, **13**, 1709–1714.

192. Yeh, S. J., Tsai, C. Y., Huang, C.-Y., Liou, G.-S., Cheng, S.-H. (2003). Electrochemical characterization of small organic hole-transport molecules based on the triphenylamine unit. *Electrochemistry Communications*, **5**, 373–377.

193. Saeki, A., Seki, S., Satoh, N., Yamamoto, K., Tagawa, S. (2008). Long-lived hole stabilized at a triphenylamine core and shielded by rigid phenylazomethine dendrons: A pulse radiolysis study. *The Journal of Physical Chemistry B*, **112**, 15540–15545.

194. Sandanayaka, A. S. D., Sasabe, H., Araki, Y., Furusho, Y., Ito, O., Takata, T. (2004). Photoinduced electron-transfer processes between [C_{60}]fullerene and triphenylamine moieties tethered by rotaxane structures. Through-space electron transfer via excited triplet states of [60]fullerene. *The Journal of Physical Chemistry A*, **108**, 5145–5155.

195. Araki, Y., Ito, O. (2008). Factors controlling lifetimes of photoinduced charge-separated states of fullerene-donor molecular systems. *Journal of Photochemistry and Photobiology C: Photochemistry Reviews*, **9**, 93–110.

196. Seok, J. H., Park, S. H., El-Khouly, M. E., Araki, Y., Ito, O., Kay, K.-Y. (2009). Photoinduced processes of newly synthesized bisferrocene- and bisfullerene-substituted tetrads with a triphenylamine central block. *Journal of Organometallic Chemistry*, **694**, 1818–1825.

197. Pinzón, J. R., Gasca, D. C., Sankaranarayanan, S. G., Bottari, G., Torres, T. s., Guldi, D. M., Echegoyen, L. (2009). Photoinduced charge transfer and electrochemical properties of triphenylamine I_h-$Sc_3N@C_{80}$ donor–acceptor conjugates. *Journal of the American Chemical Society*, **131**, 7727–7734.

198. Gust, D., Moore, T. A., Moore, A. L. (2000). Mimicking photosynthetic solar energy transduction. *Accounts of Chemical Research*, **34**, 40–48.

199. D'Souza, F., Deviprasad, G. R., El-Khouly, M. E., Fujitsuka, M., Ito, O. (2001). Probing the donor–acceptor proximity on the physicochemical properties of porphyrin–fullerene dyads: "Tail-on" and "tail-off" binding approach. *Journal of the American Chemical Society*, **123**, 5277–5284.

200. Bryce, M. R. (1999). Tetrathiafulvalenes as π-electron donors for intramolecular charge-transfer materials. *Advanced Materials*, **11**, 11–23.

201. Nielsen, M. B., Lomholt, C., Becher, J. (2000). Tetrathiafulvalenes as building blocks in supramolecular chemistry ii. *Chemical Society Reviews*, **29**, 153–164.

202. Segura, J. L., Martín, N. (2001). New concepts in tetrathiafulvalene chemistry. *Angewandte Chemie International Edition*, **40**, 1372–1409.

203. Bendikov, M., Wudl, F., Perepichka, D. F. (2004). Tetrathiafulvalenes, oligoacenenes, and their buckminsterfullerene derivatives: The brick and mortar of organic electronics. *Chemical Reviews*, **104**, 4891–4946.

204. Prato, M., Maggini, M., Giacometti, C., Scorrano, G., Sandona, G., Farnia, G. (1996). Synthesis and electrochemical properties of substituted fulleropyrrolidines. *Tetrahedron*, **52**, 5221–5234.

205. Martín, N., Sánchez, L., Seoane, C., Andreu, R., Garín, J., Orduna, J. (1996). Semiconducting charge transfer complexes from [60]fullerene–tetrathiafulvalene (C_{60}-TTF) systems. *Tetrahedron Letters*, **37**, 5979–5982.

206. Martín, N., Pérez, I., Sánchez, L., Seoane, C. (1997). Synthesis and properties of the first highly conjugated tetrathiafulvalene analogues covalently attached to [60]fullerene. *The Journal of Organic Chemistry*, **62**, 5690–5695.

207. Herranz, M. A., Martín, N., Sánchez, L., Seoane, C., Guldi, D. M. (2000). New [pi]-extended tetrathiafulvalene-containing fulleropyrrolidine dyads endowed with vinyl spacers. *Journal of Organometallic Chemistry*, **599**, 2–7.

208. Sánchez, L., Pérez, I., Martín, N., Guldi, D. M. (2003). Controlling short- and long-range electron transfer processes in molecular dyads and triads. *Chemistry—A European Journal*, **9**, 2457–2468.

209. Herranz, M. A., Martín, N. (1999). A new building block for Diels–Alder reactions in π-extended tetrathiafulvalenes:

Synthesis of a novel electroactive C_{60}-based dyad. *Organic Letters*, **1**, 2005–2007.

210. Martin, N., Sanchez, L., Guldi, D. M. (2000). Stabilisation of charge-separated states via gain of aromaticity and planarity of the donor moiety in C_{60}-based dyads. *Chemical Communications*, 113–114.

211. Herranz, M. A., Martin, N., Ramey, J., Guldi, D. M. (2002). Thermally reversible C_{60}-based donor–acceptor ensembles. *Chemical Communications*, 2968–2969.

212. González, S., Martín, N., Guldi, D. M. (2002). Synthesis and properties of Bingel-type methanofullerene—π-extended-ttf diads and triads. *The Journal of Organic Chemistry*, **68**, 779–791.

213. Kodis, G., Liddell, P. A., de la Garza, L., Moore, A. L., Moore, T. A., Gust, D. (2002). Photoinduced electron transfer in [pi]-extended tetrathiafulvalene–porphyrin–fullerene triad molecules. *Journal of Materials Chemistry*, **12**, 2100–2108.

214. Liddell, P. A., Kodis, G., de la Garza, L., Bahr, J. L., Moore, A. L., Moore, T. A., Gust, D. (2001). Photoinduced electron transfer in tetrathiafulvalene–porphyrin–fullerene molecular triads. *Helvetica Chimica Acta*, **84**, 2765–2783.

215. Kuciauskas, D., Lin, S., Seely, G. R., Moore, A. L., Moore, T. A., Gust, D., Drovetskaya, T., Reed, C. A., Boyd, P. D. W. (1996). Energy and photoinduced electron transfer in porphyrin–fullerene dyads. *The Journal of Physical Chemistry*, **100**, 15926–15932.

216. Giacalone, F., Segura, J. L., Martín, N., Guldi, D. M. (2004). Exceptionally small attenuation factors in molecular wires. *Journal of the American Chemical Society*, **126**, 5340–5341.

217. Atienza, C., Martin, N., Wielopolski, M., Haworth, N., Clark, T., Guldi, D. M. (2006). Tuning electron transfer through *p*-phenyleneethynylene molecular wires. *Chemical Communications*, 3202–3204.

218. Atienza-Castellanos, C., Wielopolski, M., Guldi, D. M., van der Pol, C., Bryce, M. R., Filippone, S., Martin, N. (2007). Determination of the attenuation factor in fluorene-based molecular wires. *Chemical Communications*, 5164–5166.

219. Schenk, R., Gregorius, H., Meerholz, K., Heinze, J., Muellen, K. (1991). Novel oligo(phenylenevinylenes): Models for the charging of extended π chains. *Journal of the American Chemical Society*, **113**, 2634–2647.

220. Goldsmith, R. H., Sinks, L. E., Kelley, R. F., Betzen, L. J., Liu, W., Weiss, E. A., Ratner, M. A., Wasielewski, M. R. (2005). Wire-like charge transport at near constant bridge energy through fluorene oligomers. *Proceedings of the National Academy of Sciences of the United States of America*, **102**, 3540–3545.

221. Illescas, B. M., Santos, J., Wielopolski, M., Atienza, C. M., Martin, N., Guldi, D. M. (2009). Electron transfer through exttf bridges in electron donor–acceptor conjugates. *Chemical Communications*, 5374–5376.

222. Liddell, P. A., Sumida, J. P., Macpherson, A. N., Noss, L., Seely, G. R., Clark, K. N., Moore, A. L., Moore, T. A., Gust, D. (1994). Preparation and photophysical studies of porphyrin-C_{60} dyads. *Photochemistry and Photobiology*, **60**, 537–541.

223. Imahori, H., Sakata, Y. (1997). Donor-linked fullerenes: Photoinduced electron transfer and its potential application. *Advanced Materials*, **9**, 537–546.

224. Imahori, H., Sakata, Y. (1999). Fullerenes as novel acceptors in photosynthetic electron transfer. *European Journal of Organic Chemistry*, **1999**, 2445–2457.

225. Martín, N., Sánchez, L., Illescas, B., Pérez, I. (1998). C_{60} based electroactive organofullerenes. *Chemical Reviews*, **98**, 2527–2548.

226. Imahori, H., Yamada, H., Guldi, D. M., Endo, Y., Shimomura, A., Kundu, S., Yamada, K., Okada, T., Sakata, Y., Fukuzumi, S. (2002). Comparison of reorganization energies for intra- and intermolecular electron transfer. *Angewandte Chemie International Edition*, **41**, 2344–2347.

227. Ohkubo, K., Kotani, H., Shao, J., Ou, Z., Kadish, K. M., Li, G., Pandey, R. K., Fujitsuka, M., Ito, O., Imahori, H., Fukuzumi, S. (2004). Production of an ultra-long-lived charge-separated state in a zinc chlorin–C_{60} dyad by one-step photoinduced electron transfer. *Angewandte Chemie International Edition*, **43**, 853–856.

228. Guldi, D. M., Luo, C., Prato, M., Dietel, E., Hirsch, A. (2000). Charge-transfer in a [pi]-stacked fullerene porphyrin dyad: Evidence for back electron transfer in the 'marcus-inverted' region. *Chemical Communications*, 373–374.

229. Dietel, E., Hirsch, A., Eichhorn, E., Rieker, A., Hackbarth, S., Roder, B. (1998). A macrocyclic [60]fullerene-porphyrin dyad involving π–π stacking interactions. *Chemical Communications*, 1981–1982.

230. de la Torre, G., Giacalone, F., Segura, J. L., Martín, N., Guldi, D. M. (2005). Electronic communication through π-conjugated wires in covalently linked porphyrin/C_{60} ensembles. *Chemistry—A European Journal*, **11**, 1267–1280.

231. Fukuzumi, S., Ohkubo, K., Imahori, H., Shao, J., Ou, Z., Zheng, G., Chen, Y., Pandey, R. K., Fujitsuka, M., Ito, O., Kadish, K. M. (2001). Photochemical and electrochemical properties of zinc chlorin–C_{60} dyad as compared to corresponding free-base chlorin–C_{60}, free-base porphyrin–C_{60}, and zinc porphyrin–C_{60} dyads. *Journal of the American Chemical Society*, **123**, 10676–10683.

232. Imahori, H., Guldi, D. M., Tamaki, K., Yoshida, Y., Luo, C., Sakata, Y., Fukuzumi, S. (2001). Charge separation in a novel artificial photosynthetic reaction center lives 380 ms. *Journal of the American Chemical Society*, **123**, 6617–6628.

233. Schuster, D. I., Cheng, P., Jarowski, P. D., Guldi, D. M., Luo, C., Echegoyen, L., Pyo, S., Holzwarth, A. R., Braslavsky, S. E., Williams, R. M., Klihm, G. (2004). Design, synthesis, and photophysical studies of a porphyrin–fullerene dyad with parachute topology; charge recombination in the Marcus inverted region. *Journal of the American Chemical Society*, **126**, 7257–7270.

234. Kodis, G., Terazono, Y., Liddell, P. A., Andréasson, J., Garg, V., Hambourger, M., Moore, T. A., Moore, A. L., Gust, D. (2006). Energy and photoinduced electron transfer in a wheel-shaped artificial photosynthetic antenna-reaction center complex. *Journal of the American Chemical Society*, **128**, 1818–1827.

235. Guldi, D. M., Luo, C., Prato, M., Troisi, A., Zerbetto, F., Scheloske, M., Dietel, E., Bauer, W., Hirsch, A. (2001). Parallel (face-to-face) versus perpendicular (edge-to-face) alignment of electron donors and acceptors in fullerene porphyrin dyads: The importance of orientation in electron transfer. *Journal of the American Chemical Society*, **123**, 9166–9167.

236. Imahori, H. (2004). Porphyrin–fullerene linked systems as artificial photosynthetic mimics. *Organic & Biomolecular Chemistry*, **2**, 1425–1433.

237. Harriman, A. (2004). Unusually slow charge recombination in molecular dyads. *Angewandte Chemie International Edition*, **43**, 4985–4987.

238. Imahori, H., Yamada, K., Hasegawa, M., Taniguchi, S., Okada, T., Sakata, Y. (1997). A sequential photoinduced electron relay accelerated by fullerene in a porphyrin–pyromellitimide–C_{60} triad. *Angewandte Chemie International Edition in English*, **36**, 2626–2629.

239. Schumacher, A. L., Sandanayaka, A. S. D., Hill, J. P., Ariga, K., Karr, P. A., Araki, Y., Ito, O., D'Souza, F. (2007). Supramolecular triad and pentad composed of zinc–porphyrin(s), oxoporphyrinogen, and fullerene(s): Design and electron-transfer studies. *Chemistry—A European Journal*, **13**, 4628–4635.

240. Imahori, H., Tamaki, K., Guldi, D. M., Luo, C., Fujitsuka, M., Ito, O., Sakata, Y., Fukuzumi, S. (2001). Modulating charge separation and charge recombination dynamics in porphyrin–fullerene linked dyads and triads: Marcus-normal versus inverted region. *Journal of the American Chemical Society*, **123**, 2607–2617.

241. Liddell, P. A., Kodis, G., Andréasson, J., de la Garza, L., Bandyopadhyay, S., Mitchell, R. H., Moore, T. A., Moore, A. L., Gust, D. (2004). Photonic switching of photoinduced electron transfer in a dihydropyrene–porphyrin–fullerene molecular triad. *Journal of the American Chemical Society*, **126**, 4803–4811.

242. Schuster, D. I., Li, K., Guldi, D. M., Palkar, A., Echegoyen, L., Stanisky, C., Cross, R. J., Niemi, M., Tkachenko, N. V., Lemmetyinen, H. (2007). Azobenzene-linked porphyrin–fullerene dyads. *Journal of the American Chemical Society*, **129**, 15973–15982.

243. Poddutoori, P. K., Sandanayaka, A. S. D., Hasobe, T., Ito, O., van der Est, A. (2010). Photoinduced charge separation in a ferrocene–aluminum(III) porphyrin–fullerene supramolecular triad. *The Journal of Physical Chemistry B*, **114**, 14348–14357.

244. Guldi, D. M., Imahori, H., Tamaki, K., Kashiwagi, Y., Yamada, H., Sakata, Y., Fukuzumi, S. (2004). A molecular tetrad allowing efficient energy storage for 1.6 s at 163 k. *The Journal of Physical Chemistry A*, **108**, 541–548.

245. Vail, S. A., Krawczuk, P. J., Guldi, D. M., Palkar, A., Echegoyen, L., Tomé, J. P. C., Fazio, M. A., Schuster, D. I. (2005). Energy and electron transfer in polyacetylene-linked zinc–porphyrin–[60]fullerene molecular wires. *Chemistry—A European Journal*, **11**, 3375–3388.

246. Tashiro, K., Sato, A., Yuzawa, T., Aida, T., Yamanaka, K.-i., Fujitsuka, M., Ito, O. (2006). Long-range photoinduced electron transfer mediated by oligo-*p*-phenylenebutadiynylene conjugated bridges. *Chemistry Letters*, **35**, 518–519.

247. Ikemoto, J., Takimiya, K., Aso, Y., Otsubo, T., Fujitsuka, M., Ito, O. (2002). Porphyrin–oligothiophene–fullerene triads as an efficient intramolecular electron-transfer system. *Organic Letters*, **4**, 309–311.

248. Nakamura, T., Fujitsuka, M., Araki, Y., Ito, O., Ikemoto, J., Takimiya, K., Aso, Y., Otsubo, T. (2004). Photoinduced electron transfer in porphyrin–oligothiophene–fullerene linked triads by excitation of a porphyrin moiety. *The Journal of Physical Chemistry B*, **108**, 10700–10710.

249. Oswald, F., Islam, D. M. S., Araki, Y., Troiani, V., Caballero, R., Cruz, P. d. l., Ito, O., Langa, F. (2007). High effectiveness of oligothienylenevinylene as molecular wires in zn-porphyrin and C_{60} connected systems. *Chemical Communications*, 4498–4500.

250. Kashiwagi, Y., Ohkubo, K., McDonald, J. A., Blake, I. M., Crossley, M. J., Araki, Y., Ito, O., Imahori, H., Fukuzumi, S. (2003). Long-lived charge-separated state produced by photoinduced electron transfer in a zinc imidazoporphyrin-C_{60} dyad. *Organic Letters*, **5**, 2719–2721.

251. Guldi, D. M., Illescas, B. M., Atienza, C. M., Wielopolski, M., Martin, N. (2009). Fullerene for organic electronics. *Chemical Society Reviews*, **38**, 1587–1597.

252. El-Khouly, M. E., Ito, O., Smith, P. M., D'Souza, F. (2004). Intermolecular and supramolecular photoinduced electron transfer processes of fullerene–porphyrin/phthalocyanine systems. *Journal of Photochemistry and Photobiology C: Photochemistry Reviews*, **5**, 79–104.

253. D'Souza, F., Ito, O. (2005). Photoinduced electron transfer in supramolecular systems of fullerenes functionalized with ligands capable of binding to zinc porphyrins and zinc phthalocyanines. *Coordination Chemistry Reviews*, **249**, 1410–1422.

254. D'Souza, F., Ito, O. (2009). Supramolecular donor-acceptor hybrids of porphyrins/phthalocyanines with fullerenes/carbon nanotubes: Electron transfer, sensing, switching, and catalytic applications. *Chemical Communications*, 4913–4928.

255. Linssen, T. G., Durr, K., Hanack, M., Hirsch, A. (1995). A green fullerene: Synthesis and electrochemistry of a diels-alder adduct of [60]fullerene with a phthalocyanine. *Journal of the Chemical Society, Chemical Communications*, 103–104.

256. Gouloumis, A., Liu, S.-G., Sastre, Á., Vázquez, P., Echegoyen, L., Torres, T. (2000). Synthesis and electrochemical properties of phthalocyanine–fullerene hybrids. *Chemistry—A European Journal*, **6**, 3600–3607.

257. Guldi, D. M., Zilbermann, I., Gouloumis, A., Vázquez, P., Torres, T. (2004). Metallophthalocyanines: Versatile electron-donating building blocks for fullerene dyads. *The Journal of Physical Chemistry B*, **108**, 18485–18494.

258. Geerts, Y. H., Debever, O., Amato, C., Sergeyev, S. (2009). Synthesis of mesogenic phthalocyanine–C_{60} donor–acceptor dyads designed for molecular heterojunction photovoltaic devices. *Beilstein Journal of Organic Chemistry*, **5**.

259. Isosomppi, M., Tkachenko, N. V., Efimov, A., Vahasalo, H., Jukola, J., Vainiotalo, P., Lemmetyinen, H. (2006).

Photoinduced electron transfer of double-bridged phthalocyanine–fullerene dyads. *Chemical Physics Letters*, **430**, 36–40.

260. Gouloumis, A., de la Escosura, A., Vázquez, P., Torres, T., Kahnt, A., Guldi, D. M., Neugebauer, H., Winder, C., Drees, M., Sariciftci, N. S. (2006). Photoinduced electron transfer in a new bis(C_{60})–phthalocyanine triad. *Organic Letters*, **8**, 5187–5190.
261. Quintiliani, M., Kahnt, A., Vazquez, P., Guldi, D. M., Torres, T. (2008). Implementing a tripodal relay station in a phthalocyanine–[60]fullerene conjugate. *Journal of Materials Chemistry*, **18**, 1542–1546.
262. Kahnt, A., Quintiliani, M., Vázquez, P., Guldi, D. M., Torres, T. (2008). A bis(C_{60})–bis(phthalocyanine) nanoconjugate: Synthesis and photoinduced charge transfer. *ChemSusChem*, **1**, 97–102.
263. Quintiliani, M., García-Frutos, E. M., Vázquez, P., Torres, T. (2008). Synthesis of diiodo butadiynyl-bridged bisphthalocyaninatozinc(ii) complexes. *Journal of Inorganic Biochemistry*, **102**, 388–394.
264. Kim, K. N., Choi, C. S., Kay, K.-Y. (2005). A novel phthalocyanine with two axial fullerene substituents. *Tetrahedron Letters*, **46**, 6791–6795.
265. El-Khouly, M. E., Kang, E. S., Kay, K.-Y., Choi, C. S., Aaraki, Y., Ito, O. (2007). Silicon-phthalocyanine-cored fullerene dendrimers: Synthesis and prolonged charge-separated states with dendrimer generations. *Chemistry—A European Journal*, **13**, 2854–2863.
266. Martín-Gomis, L., Ohkubo, K., Fernández-Lázaro, F., Fukuzumi, S., Sastre-Santos, Á. (2007). Synthesis and photophysical studies of a new nonaggregated C_{60}–silicon phthalocyanine–C_{60} triad. *Organic Letters*, **9**, 3441–3444.
267. Martín-Gomis, L., Ohkubo, K., Fernández-Lá;zaro, F., Fukuzumi, S., Sastre-Santos, A. N. (2008). Adiabatic photoinduced electron transfer and back electron transfer in a series of axially substituted silicon phthalocyanine triads. *The Journal of Physical Chemistry C*, **112**, 17694–17701.
268. El-Khouly, M. E., Kim, J. H., Kay, K.-Y., Choi, C. S., Ito, O., Fukuzumi, S. (2009). Synthesis and photoinduced intramolecular processes of light-harvesting silicon phthalocyanine–naphthalenediimide–fullerene connected systems. *Chemistry—A European Journal*, **15**, 5301–5310.
269. Bottari, G., de la Torre, G., Guldi, D. M., Torres, T. S. (2010). Covalent and noncovalent phthalocyanine–carbon nanostructure systems: Synthesis, photoinduced electron transfer, and application to molecular photovoltaics. *Chemical Reviews*, **110**, 6768–6816.
270. Stevenson, S., Rice, G., Glass, T., Harich, K., Cromer, F., Jordan, M. R., Craft, J., Hadju, E., Bible, R., Olmstead, M. M., Maitra, K., Fisher, A. J., Balch, A. L., Dorn, H. C. (1999). Small-bandgap endohedral metallofullerenes in high yield and purity. *Nature*, **401**, 55–57.
271. Wasielewski, M. R. (2006). Energy, charge, and spin transport in molecules and self-assembled nanostructures inspired by photosynthesis. *The Journal of Organic Chemistry*, **71**, 5051–5066.
272. Guldi, D. M. (2007). Nanometer scale carbon structures for charge-transfer systems and photovoltaic applications. *Physical Chemistry and Chemical Physics*, **9**, 1400–1420.
273. Guldi, D. M. (2000). Fullerenes: Three dimensional electron acceptor materials. *Chemical Communications*, 321–327.
274. Imahori, H. (2004). Giant multiporphyrin arrays as artificial light-harvesting antennas. *The Journal of Physical Chemistry B*, **108**, 6130–6143.
275. Segura, J. L., Martin, N., Guldi, D. M. (2005). Materials for organic solar cells: The C_{60}/π-conjugated oligomer approach. *Chemical Society Reviews*, **34**, 31–47.
276. Martín, N., Sánchez, L., Herranz, M. a. Á., Illescas, B., Guldi, D. M. (2007). Electronic communication in tetrathiafulvalene (TTF)/C_{60} systems: Toward molecular solar energy conversion materials? *Accounts of Chemical Research*, **40**, 1015–1024.
277. Campanera, J. M., Bo, C., Poblet, J. M. (2005). General rule for the stabilization of fullerene cages encapsulating trimetallic nitride templates. *Angewandte Chemie International Edition*, **44**, 7230–7233.
278. Pinzón, J. R., Plonska-Brzezinska, M. E., Cardona, C. M., Athans, A. J., Gayathri, S. S., Guldi, D. M., Herranz, M. Á., Martín, N., Torres, T., Echegoyen, L. (2008). $Sc_3N@C_{80}$-ferrocene electron-donor/acceptor conjugates as promising materials for photovoltaic applications. *Angewandte Chemie International Edition*, **47**, 4173–4176.
279. Pinzón, J. R., Cardona, C. M., Herranz, M. Á., Plonska-Brzezinska, M. E., Palkar, A., Athans, A. J., Martín, N., Rodríguez-Fortea, A., Poblet, J. M., Bottari, G., Torres, T., Gayathri, S. S., Guldi, D. M., Echegoyen, L. (2009). Metal nitride cluster fullerene $M_3N@C_{80}$ (M = Y, Sc) based dyads: Synthesis, and electrochemical, theoretical and photophysical studies. *Chemistry—A European Journal*, **15**, 864–877.
280. Guldi, D. M., Gouloumis, A., Vázquez, P., Torres, T., Georgakilas, V., Prato, M. (2005). Nanoscale organization of a phthalocyanine–fullerene system: Remarkable stabilization of charges in photoactive 1-D nanotubules. *Journal of the American Chemical Society*, **127**, 5811–5813.
281. de la Escosura, A., Martínez-Díaz, M. V., Guldi, D. M., Torres, T. (2006). Stabilization of charge-separated states in phthalocyanine–fullerene ensembles through supramolecular donor–acceptor interactions. *Journal of the American Chemical Society*, **128**, 4112–4118.
282. Bottari, G., Olea, D., Gómez-Navarro, C., Zamora, F., Gómez-Herrero, J., Torres, T. (2008). Highly conductive supramolecular nanostructures of a covalently linked phthalocyanine–C_{60} fullerene conjugate. *Angewandte Chemie International Edition*, **47**, 2026–2031.
283. de la Torre, G., Claessens, C. G., Torres, T. (2007). Phthalocyanines: Old dyes, new materials. Putting color in nanotechnology. *Chemical Communications*, 2000–2015.
284. de la Torre, G., Vázquez, P., Agulló-López, F., Torres, T. (2004). Role of structural factors in the nonlinear optical properties of phthalocyanines and related compounds. *Chemical Reviews*, **104**, 3723–3750.

285. Morandeira, A., López-Duarte, I., Martínez-Díaz, M. V., O'Regan, B., Shuttle, C., Haji-Zainulabidin, N. A., Torres, T., Palomares, E., Durrant, J. R. (2007). Slow electron injection on ru–phthalocyanine sensitized TiO_2. *Journal of the American Chemical Society*, **129**, 9250–9251.

286. O'Regan, B. C., López-Duarte, I., Martínez-Díaz, M. V., Forneli, A., Albero, J., Morandeira, A., Palomares, E., Torres, T., Durrant, J. R. (2008). Catalysis of recombination and its limitation on open circuit voltage for dye sensitized photovoltaic cells using phthalocyanine dyes. *Journal of the American Chemical Society*, **130**, 2906–2907.

287. Reddy, P. Y., Giribabu, L., Lyness, C., Snaith, H. J., Vijaykumar, C., Chandrasekharam, M., Lakshmikantam, M., Yum, J.-H., Kalyanasundaram, K., Grätzel, M., Nazeeruddin, M. K. (2007). Efficient sensitization of nanocrystalline TiO_2 films by a near-ir-absorbing unsymmetrical zinc phthalocyanine. *Angewandte Chemie International Edition*, **46**, 373–376.

288. Cid, J.-J., Yum, J.-H., Jang, S.-R., Nazeeruddin, M. K., Martínez-Ferrero, E., Palomares, E., Ko, J., Grätzel, M., Torres, T. (2007). Molecular cosensitization for efficient panchromatic dye-sensitized solar cells. *Angewandte Chemie International Edition*, **46**, 8358–8362.

289. Kira, A., Umeyama, T., Matano, Y., Yoshida, K., Isoda, S., Park, J. K., Kim, D., Imahori, H. (2009). Supramolecular donor–acceptor heterojunctions by vectorial stepwise assembly of porphyrins and coordination-bonded fullerene arrays for photocurrent generation. *Journal of the American Chemical Society*, **131**, 3198–3200.

290. Megiatto, J. D., Schuster, D. I., Abwandner, S., de Miguel, G., Guldi, D. M. (2010). [2]catenanes decorated with porphyrin and [60]fullerene groups: Design, convergent synthesis, and photoinduced processes. *Journal of the American Chemical Society*, **132**, 3847–3861.

291. Kuramochi, Y., Satake, A., Itou, M., Ogawa, K., Araki, Y., Ito, O., Kobuke, Y. (2008). Light-harvesting supramolecular porphyrin macrocycle accommodating a fullerene–tripodal ligand. *Chemistry—A European Journal*, **14**, 2827–2841.

292. Ozawa, H., Kawao, M., Uno, S., Nakazato, K., Tanaka, H., Ogawa, T. (2009). A photo-responsive molecular wire composed of a porphyrin polymer and a fullerene derivative. *Journal of Materials Chemistry*, **19**, 8307–8313.

293. Fukuzumi, S., Kojima, T. (2008). Photofunctional nanomaterials composed of multiporphyrins and carbon-based π-electron acceptors. *Journal of Materials Chemistry*, **18**, 1427–1439.

294. Wolfrum, S., Pinzon, J. R., Molina-Ontoria, A., Gouloumis, A., Martin, N., Echegoyen, L., Guldi, D. M. (2011). Utilization of $Sc_3N@C_{80}$ in long-range charge transfer reactions. *Chemical Communications*, **47**, 2270–2272.

295. Li, K., Schuster, D. I., Guldi, D. M., Herranz, M. Á., Echegoyen, L. (2004). Convergent synthesis and photophysics of [60]fullerene/porphyrin-based rotaxanes. *Journal of the American Chemical Society*, **126**, 3388–3389.

14

SOLVATING INSOLUBLE CARBON NANOSTRUCTURES BY MOLECULAR DYNAMICS

MATTEO CALVARESI AND FRANCESCO ZERBETTO

Due to their unique physical properties [1], carbon nanotubes (CNTs) have been proposed for a wide range of potential applications [2–4]. However, their insolubility in both water and organic solvents hinders the path toward practical applications of this unique class of materials. Pristine nanotubes tend to assemble in bundles and ropes (Figure 14.1) that contain hundreds of close-packed CNTs, tightly interlaced and characterized by van der Waals attraction energies of 500 eV μm^{-1} [5].

The high interaction energy renders the dispersion of CNTs a challenge. Currently, the main approaches used to disperse nanotubes are either mechanical or chemical. The mechanical approach consists of ultrasonication and high-shear mixing. These processes are time-consuming and have low efficiency. They can result in the fragmentation of CNTs and subsequent decrease of their aspect ratio and can lead to poor stability of the dispersion [6]. The chemical approach [7–9] entails both covalent and noncovalent methods and is aimed at improving the chemical compatibility of CNTs with the target medium and at reducing their tendency to agglomerate. Covalent methods involve chemical modification (or chemical functionalization) of the CNT walls with various chemical moieties to improve solubility in solvents [10–12]. This aggressive approach may introduce defects in the CNT walls and modify the π-electron conjugation, which may result in the deterioration of their peculiar properties. Noncovalent functionalization, instead, is based on noncovalent interactions such as the physical adsorption of molecules on the CNT surface [13]. It is attractive because the π-electron cloud of the graphene sheet of the CNT is not disturbed and the characteristic properties of the CNT are preserved. To exfoliate the CNTs bundles, the tube surface can be modified, via van der Waals forces and π–π interactions, by adsorption or wrapping of polynuclear aromatic compounds, surfactants, polymers, or biomolecules.

A full understanding of the mechanisms that lead to the dispersion of CNTs is still lacking. Several models attempt to describe the phenomenon, but detailed interpretation of the experimental results is often difficult since direct evidence of the structures is not trivial to obtain and different experiments have led to contradictory conclusions.

Computer simulations are a convenient tool to study the solvation of CNTs because they can give a microscopic picture of the process. They avoid some of the experimental difficulties associated with the observation of the structures and ultimately afford a theoretical understanding of the effects that play a role in the dispersion of CNTs in solution. Molecular dynamics simulation can describe in detail the interface and morphology of molecules adsorbed on carbon surfaces and, more importantly, the interaction mechanisms that take place in different supramolecular aggregates [14]. Understanding the self-assembly mechanisms will guide further improvement of carbon nanostructure dispersion, sorting, and separation and facilitate the design of novel carbon-based nanomaterials.

14.1 CNT IN LIQUIDS

Owing to their high molecular weight, CNTs were initially thought insoluble. This disadvantage was circumvented by coating the nanotubes with a dispersant phase. A desirable

Organic Nanomaterials: Synthesis, Characterization, and Device Applications, First Edition. Edited by Tomás Torres and Giovanni Bottari.

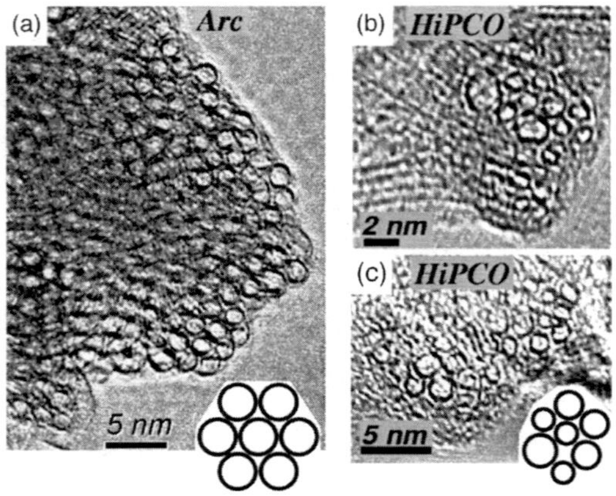

FIGURE 14.1. HRTEM image of as-grown SWNT bundle cross section of (**a**) ARC and (**b, c**) HiPCO materials. Reprinted with permission from *J. Phys. Chem. B*, 2005, **109**, 23358–23365. Copyright 2005, American Chemical Society.

scenario involved the discovery of a solvent where nanotubes would be thermodynamically soluble (i.e., where the free energy of mixing is negative) without the aid of a third, dispersant phase. Alternatively, it was thought to be advantageous to identify a solvent where nanotubes could be dispersed either individually or in small bundles that remained kinetically stable for reasonably long periods of time.

14.1.1 CNT in Water

To understand the dispersion process of CNTs in water, it is fundamental to know the structural properties of water surrounding the tube. MD simulations provided an atomistic picture of the structure of the CNT–water interface (Figure 14.2) [15]. The water at the carbon–water interface has a HOH plane nearly tangential to the interface.

The radial density profiles of the hydrogen and oxygen atoms are shown in Figure 14.2b. The maxima in the oxygen and hydrogen profiles at ~3.20 Å (corresponding to the value of σ_{CO}) nearly coincide, which indicates that the plane of the water molecules is approximately tangential to the cylindrical CNT–water interface. The orientation of the water molecules at the CNT–water interface can be inferred also from the orientation of the water dipole moment (Figure 14.2c). The water molecules in closest proximity to the CNT (at $r = 2.85$–3.37 Å) display a preference for angles of 94°–95°, indicating that the dipole moment is nearly tangential to the plane of the CNT. At a radial distance of 4.90 Å, which corresponds to a local minimum of density, the dipole moment has turned to an angle of 75° and therefore points in the direction of the bulk. The bulk properties are finally reached at $r = 9.00$ Å.

Analysis of the hydrogen bond population, expressed as the average number of hydrogen bonds per water molecule, showed that the average number of hydrogen bonds decreases from a value of 3.73 in the bulk phase to a value of 2.89 at the carbon–water interface [15].

The results of the CNT–water MD simulation can be used to analyze the energetic trends associated with the process of introducing a CNT from the gas phase into the aqueous solution. The total energy change in this process, denoted $\Delta E_{g\rightarrow w}$, has three main contributions:

$$\Delta E_{g\rightarrow w} = \Delta E_{CC} + \Delta E_{CW} + \Delta E_{WW} \qquad (14.1)$$

where ΔE_{CC} represents the change in energy due to the interactions between carbon atoms. This quantity, expected to be small, is primarily due to the presence of water molecules around the CNT that slightly alters the C–C interactions. ΔE_{CW} is the energy arising from the interactions between the carbon atoms of the nanotube and the surrounding water molecules. ΔE_{WW} represents the energy required to create a cavity in water to accommodate the CNT. This contribution is due to interactions between water molecules and is related to the surface energy.

The energy changes calculated by MD (in kJ mol^{-1} per length of the CNT in Å) are $\Delta E_{CC} = 0.59$, $\Delta E_{CW} = -17.33$, and $\Delta E_{WW} = 45.40$, which yields $\Delta E_{g\rightarrow w} = 28.66$ kJ mol^{-1} [15]. These results show that the energy contribution coming from C–C interactions is positive and quite small with respect to other contributions, as expected. The interactions between carbon atoms and the surrounding water molecules are attractive and fairly large. However, ΔE_{WW} has a positive value and is the largest contribution. The net result indicates that the process of introducing a CNT in an aqueous solution is energetically unfavorable. Changing the values of the three terms is chemically possible and leads to the solvation of CNTs. When a pair of CNTs in water are considered [16], it is found that the system displays drying, as evidenced by the expulsion of the interstitial water, when the two CNTs are separated by less than 9–10 Å.

Constrained molecular dynamics simulations for two carbon nanotubes in water allowed the extraction of the potential of mean force (PMF) that governs the drying transition. At equilibrium, the constraining force (F) balances the hydration force ($F = -F_{hyd}$). The first equilibrium configuration is found at the van der Waals contact point at a separation of approximately 3.2 Å. At shorter separation distances, the constraining force is negative due to van der Waals repulsion between the nanotubes. The attractive part of the force reaches its maximum at a tube spacing of 5.0 Å (Figure 14.3Ib and Figure 14.3IIa) which is approximately 1 Å shorter than the spacing required to host one layer of water. After the maximum is reached, the force decreases rapidly at larger distances and reaches a low but positive value for a tube separation of 6–7 Å, which allows accommodation of one unstable layer of water. Finally, a weak depression in the force is

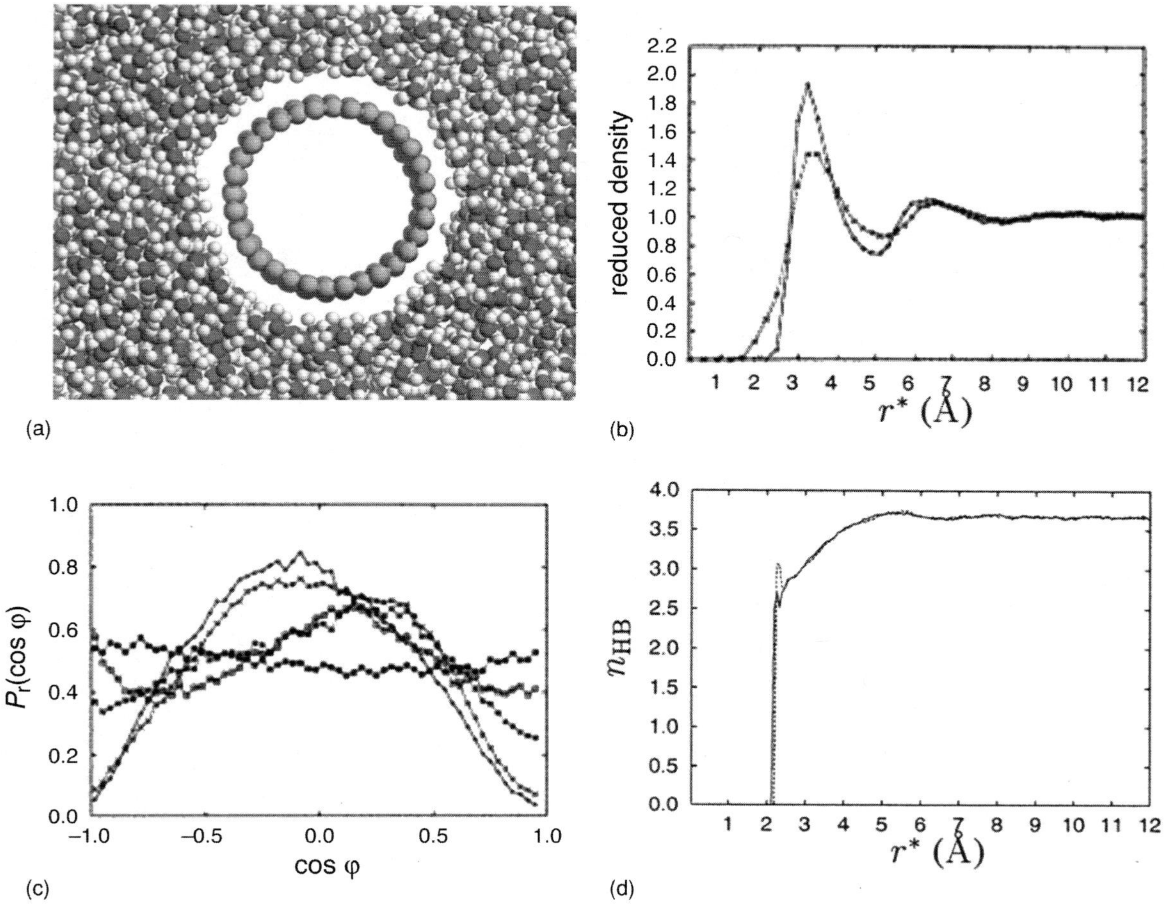

FIGURE 14.2. (**a**) Snapshot of the atoms for the simulation of a carbon nanotube in water. (**b**) Water radial density profile. Oxygen density profile ($\rho_O/\rho_O{}^0$) (– + – line) and hydrogen density profile ($\rho_H/\rho_H{}^0$) (– × –) line, where $\rho_O{}^0$ and $\rho_H{}^0$ are the bulk oxygen and hydrogen densities. (**c**) Orientation of the water dipole moment at different radial distances from the carbon nanotube wall. 2.85 Å (– + –); 3.37 Å (first peak) (– × –); 4.34 Å (— ∗ —); 4.79 Å (first minimum)(– ○ –); 8.89 Å (bulk phase) (– ■ –). (**d**) Radial profile of the hydrogen bond population, expressed as the average number of hydrogen bonds per water molecule (n_{HB}). Reprinted from reference 15, with permission. Copyright 2001, American Chemical Society.

observed at 9–10 Å, corresponding to a metastable interstitial region with two layers of water. For the 12.5-Å-diameter carbon nanotubes that are considered in the study, the magnitude of the PMF per unit length is 17 kJ mol^{-1} $Å^{-1}$. The main contribution to the PMF is the van der Waals attraction between the carbon nanotubes, as is demonstrated by measuring the force between two rigid carbon nanotubes in vacuum (Figure 14.3IIa) [16]. However, the deformation of the carbon nanotubes in water reduces the maximum attraction between the tubes, and the point of the maximum attraction occurs at a larger separation. The snapshots of the simulations shown in Figure 14.3Ic–e reveal an interstitial vapor phase indicative of cavitation for the metastable pre-drying state and an apparently stable water layer at larger tube spacing (Figure 14.3If) [16].

14.1.2 CNT in Organic Solvents

Stable suspensions of single-walled nanotubes (CNTs) have been prepared in a variety of common solvents [17–21]. It has now become increasingly clear that nanotubes can be suspended and even exfoliated in concentrations as high as 3.5 mg/ml [20].

In a solvent where nanotubes are thermodynamically soluble, the free energy of mixing, ΔG_{Mix}, is negative:

$$\Delta G_{\mathrm{Mix}} = \Delta H_{\mathrm{Mix}} - T\Delta S_{\mathrm{Mix}} \qquad (14.2)$$

where ΔH_{Mix} and ΔS_{Mix} are the enthalpy and entropy of mixing [22]. For most molecular combinations, ΔH_{Mix} is small and positive and solvation is usually driven by ΔS_{Mix}.

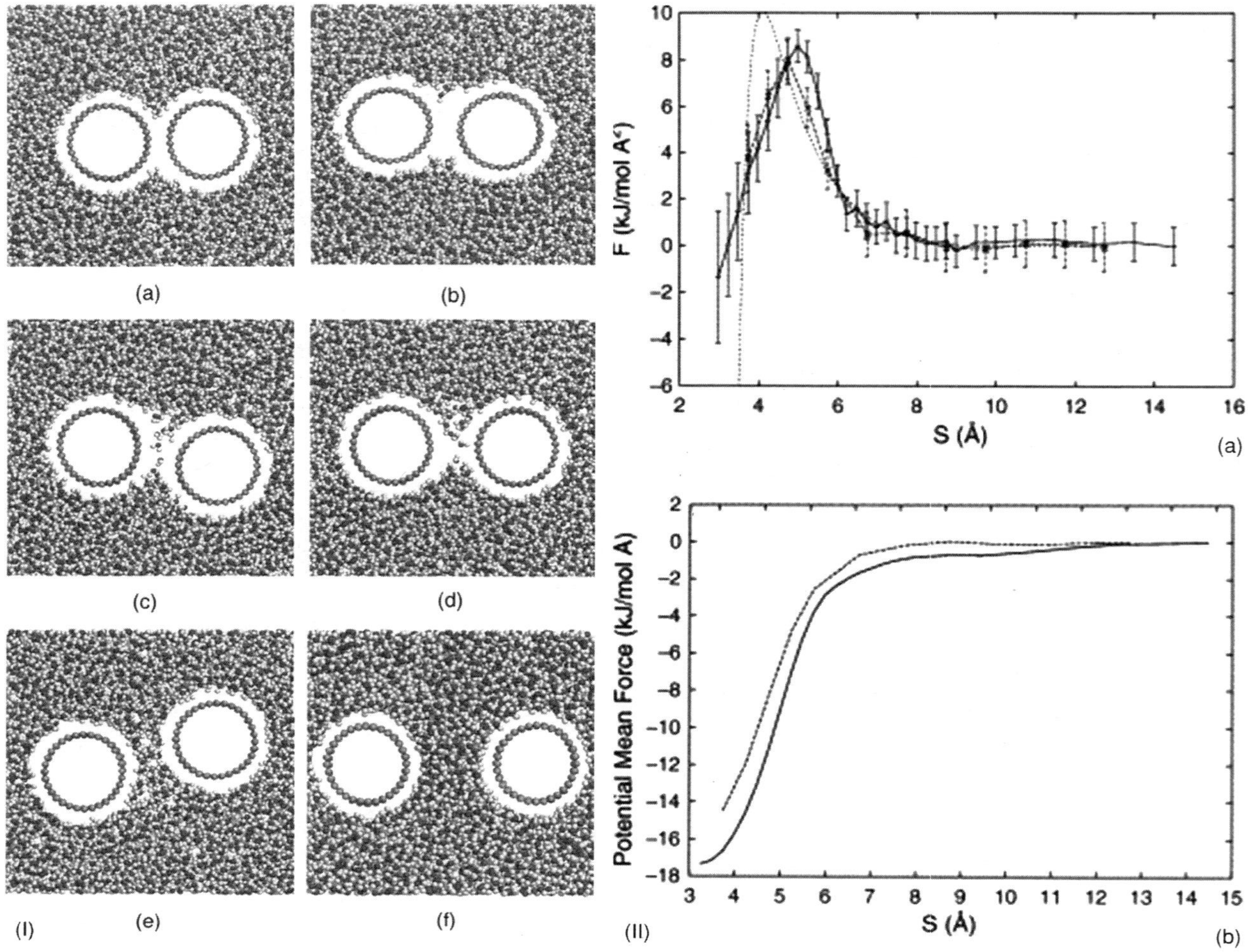

FIGURE 14.3. (**I**) Snapshot of the simulation of two carbon nanotubes in water. The center-of-mass distance and the minimum carbon–carbon spacing in brackets is (**a**) 3.48 Å, (**b**) 5.48 Å, (**c**) 6.98 Å, (**d**) 7.98 Å, (**e**) 9.48 Å, and (**f**) 14.48 Å. (**II**) (**a**) Constraining force per unit length of the carbon nanotube for the periodic system (——), for a slab (– –), and for rigid carbon nanotubes in vacuum (· · ·). The error bars in (**a**) indicate the standard deviation of the constraining force. (**b**) The Corresponding potential of mean force per unit length of the carbon nanotube. The reaction coordinate is the tube spacing (S). The potential is set to zero at the maximum tube spacing. Reprinted from reference 16, with permission. Copyright 2004, Elsevier.

However, the large molecular weight and high rigidity of nanotubes lead to an extremely small entropy of mixing. Due to their large mutual attraction, ΔH_{Mix} is generally expected to be positive for all conceivable solvent–nanotube mixtures, resulting in a positive ΔG_{Mix}, therefore prohibiting nanotube solvation [20].

Recently, Coleman and co-workers [20–22] proposed that CNT–*N*-methyl-pyrrolidone (NMP) interaction leads to an enthalpy of mixing that is approximately zero and hence gives a negative free energy of mixing. They showed that nanotubes spontaneously exfoliate when CNT–NMP dispersions are diluted. A dynamic equilibrium exists, which is characterized by significant populations of pristine individual nanotubes and small bundles. A detailed thermodynamic analysis of the CNT–NMP system led to the consideration that ΔH_{Mix} is minimized when the nanotube surface energy matches that of the solvent:

$$\frac{\Delta H_{\mathrm{mix}}}{V_{\mathrm{mix}}} \approx \frac{2}{R_{\mathrm{Bundle}}} (\delta_{\mathrm{NT}} - \delta_{\mathrm{sol}})^2 \phi \qquad (14.3)$$

where R_{Bundle} is the bundle radius, $\delta_i = \sqrt{E^i_{\mathrm{Sur}}}$, and ϕ is the nanotube volume fraction. The enthalpy of mixing depends on the balance of nanotube and solvent surface energies. Successful solvents have a surface energy close to the values of nanotube/graphite surface energy, which is estimated to be $\sim$70 mJm^{-2} [20]. A measurement of the Flory–Huggins

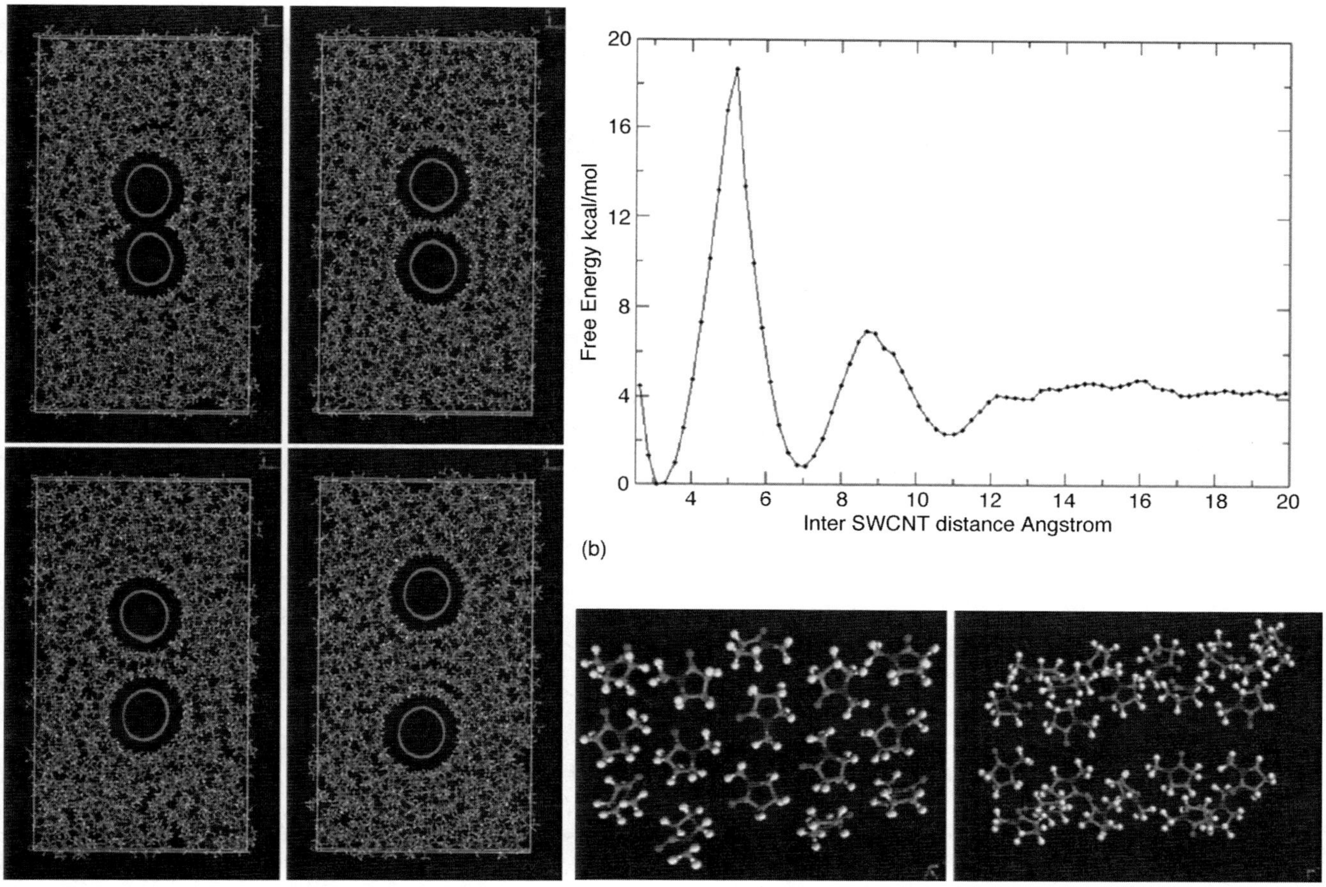

(a) (c) (d)

FIGURE 14.4. (**a**) Snapshot of the CNT–NMP systems at inter-CNT separations of 4, 7, 9, and 17 Å. (**b**) Landau free energy versus inter-CNT separation. Snapshot of the solvent molecules between two nearby CNTs: (**c**) inter-CNT distance of 7 Å; (**d**) inter-CNT distance of 5 Å that is just before the rupture of the inter-CNT film, where a cavity/hole starts appearing. Reprinted from reference 27, with permission. Copyright 2010, Scientific Research Publishing.

parameter showed that the enthalpy of mixing is negative and confirmed true solubility of CNTs in NMP [20]. However, some bundles are always observed in solvent dispersion. Their presence should not occur for a true solution. The presence of bundles implies that the exact mechanism of nanotube dispersion and exfoliation remains partly unclear.

An alternative kinetic explanation for the solvation of CNT in NMP was based on the understanding of the free energy landscape associated with transitions from isolated CNTs to bundles in solution [23]. Extensive molecular dynamics simulations indicated that bundled pairs of CNTs are more stable than pairs of isolated tubes [23]. They also showed the existence of a free energy barrier that explains (i) the long-lived transient aggregates of CNTs in NMP observed in experiment and (ii) why no spontaneous CNTs dispersion has been reported simply by immersion into NMP, or similar solvents, without intense sonication [23].

The PMF calculated for increasing intertube distances in NMP (Figures 14.4a and 14.4b) showed that the bundle is thermodynamically the more stable situation. However, there is a substantial free energy barrier of ~18 kcal/mol between 5 Å and 7 Å. From the comparison of the profile of PMF with the previous one calculated in water (Figure 14.2) emerges the ability of NMP to solvate CNT. The area of the molecular layer of solvent separating the two CNTs at a distance of 7 Å is approximately 12 × 30 Å^2. Using the tabulated surface tension of NMP (40 dynes/cm), the energy required to create such a surface is 20.6 kcal/mol. This means that the free energy barrier between two bundled and two isolated tubes is proportional to the surface tension of NMP and is likely to be responsible for the long-lived transients reported in the experiments. A pair of aggregated CNTs in NMP can only be separated by the injection of energy (e.g., intense ultrasound), which is consistent with experiments and leads to a possibly metastable state [24,25]. It is interesting to look at the nature of the single layer of solvent molecules at the first maximum and at the second minimum of the free energy in Figures 14.4c and 14.4d. When the tubes are 7 Å apart (Figure 14.4a), the solvent molecules are highly ordered, with the oxygen atoms, each bearing a negative partial charge, avoiding each other, and in contact with the hydrogen atoms. In contrast, when the tubes are separated by 5 Å (Figure 14.4d, corresponding to the maximum of free energy), the appearance of a hole or cavity is evident. At distances slightly less than 5 Å the molecular layer of solvent is no longer stable. These findings underline the role of solvent molecules in the solvation of CNT. Only MD can provide this level of atomistic description.

Understanding solvation processes of CNT cannot be limited to the understanding of how certain molecules such as NMP solvate CNTs, but requires also understand why some solvents do not solvate CNTs [17]. A relevant question is why hydrophobic nanotubes are not soluble in nonpolar solvents such toluene [17]. The results obtained by MD simulations show the role of the configurational entropy of solvent in the solvation of CNTs [26]. The total energy change accompanying the solvation of CNT in toluene has three contributions:

$$\Delta E_{T+\text{CNT}} = \Delta E_T + \Delta E_{\text{CNT}} + \Delta E^{\text{Int}}_{T+\text{CNT}} \qquad (14.4)$$

ΔE_T is the change in the solvent energy for the creation of a cylindrical cavity able to accommodate the CNT. ΔE_T has two contributions itself, one associated with a change in the bulk structure of the solvent caused by the presence of the solute (CNT) and the other associated with the surface energy of the cylindrical cavity. ΔE_{CNT} is the change in the CNT energy due to solvent–solute interactions. $\Delta E^{\text{Int}}_{T+\text{CNT}}$ is the energy of the interaction between the atoms/molecules of the solvent and the solute. Differently from solvation in water, the total enthalpic energy change associated with the introduction of a CNT into particular organic solvents—for example, toluene—is small but negative. This would imply that CNTs should be soluble in toluene. The largest contribution to solvation energy arises from the CNT–toluene interactions that overcome the energy cost accompanying the creation of a cylindrical cavity in the toluene solvent. However, to obtain an accurate description of the solvation process, one must also consider the entropy change.

Aromatic solvent molecules surrounding the SWCNT form a well-defined first solvation shell where the molecules tend to align in such a way that the aromatic plane is parallel to the CNT axis. The interactions between CNT and aromatic molecules result in major reorganization of the solvent molecules, and the associated conformation gives rise to a substantial decrease in the configurational entropy of solvent, larger than the negative enthalpic value, which, in turn, leads to a small but positive solvation Gibbs free energy. The reduced configurational entropy associated with π-stacking of solvent molecules at the CNT surface can prohibit nanotubes dispersion in otherwise "promising" solvents.

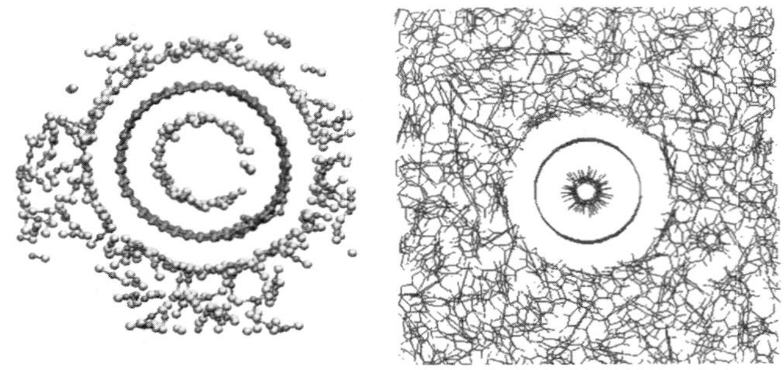

(a) (b)

FIGURE 14.5. Snapshots of (**a**) CS_2 and (**b**) benzene configuration in the presence of a CNT. (**a**) Reprinted from reference 27, with permission. Copyright 2010, Scientific Research Publishing. (**b**) Reprinted from reference 28, with permission. Copyright 2011, The Royal Society of Chemistry.

All simulations of CNT–solvent systems showed the presence of the same layered solvent molecules around the carbon nanotube. In CS_2, the solvent atoms become configured similarly to the cylindrical shape of CNT, especially in the first solvation shell (Figure 14.5) [27]. Benzene molecules, irrespective of the nanotube diameter, form cylindrical solvation shells outside the nanotubes [28]. The radial distribution exhibited distinctive cylindrical shell-like distributions, of which the first minimum is located at 0.6 nm from the exterior wall. In the first external solvation shell, the benzene molecular planes near the CNTs are oriented parallel to the nanotube surface, forming a π-stacked structure between the two. As expected, the benzene orientation becomes isotropic when sufficiently far from the surface of the nanotube.

14.1.3 CNT in Ionic Liquid

Room-temperature ionic liquids (RTILs) based on bulky and asymmetric organic cations have received intensive scrutiny recently [29]. RTILs, usually liquid at or near room temperature, are nonvolatile, nonflammable, and thermally stable. As such, they provide an environmentally benign "green" alternative to organic solvents for chemical synthesis, extractions, and biocatalysis [29]. Aida and co-workers [30, 31] found that RTILs are capable of dispersing CNTs. Specifically, when a powder of ground CNTs is mixed with an excess amount of imidazolium-based RTILs, gelatinous materials ("bucky gels") are formed. They can be processed into different shapes without disrupting nanotube structures. Despite rapidly growing interest in composite systems of carbon nanotubes and RTILs, they are not well understood at the molecular level. An initial attempt to gain theoretical understanding of these composite systems, studying solvation of small

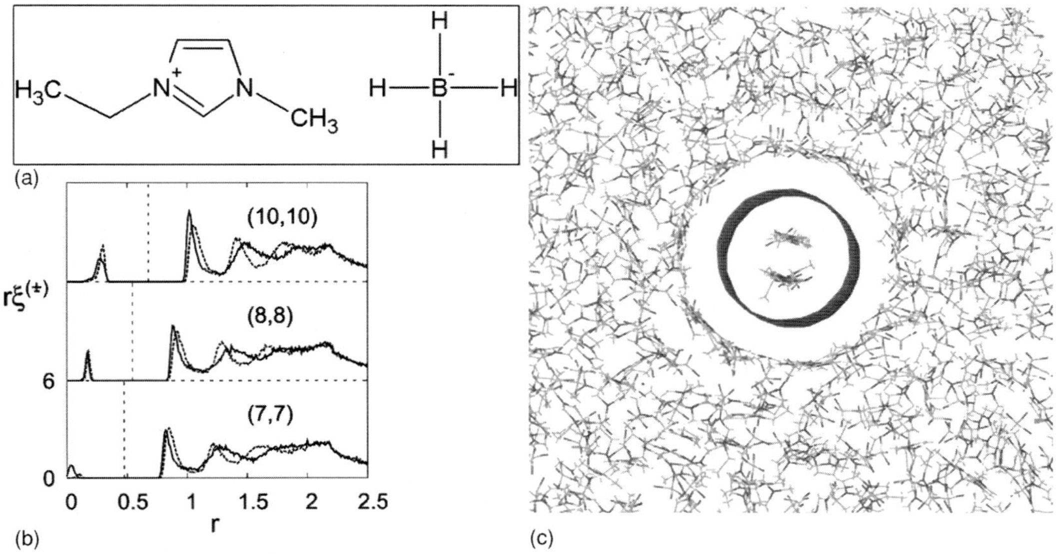

FIGURE 14.6. (**a**) 1-Ethyl-3-methylimidazolium tetrafluoroborate ($EMI^+BF_4^-$) structure. (**b**) Radial distributions of EMI^+ (——) and BF_4^- (···) around a CNT; *r* is the distance (nm) from the nanotube axis, the dashed vertical lines denote the positions of nanotube walls. (**c**) Snapshot of $EMI^+BF_4^-$ in the presence of a CNT. Reprinted from reference 32, with permission. Copyright 2009, American Chemical Society.

carbon nanotubes in 1-ethyl-3-methylimidazolium tetrafluoroborate ($EMI^+BF_4^-$), was done with MD simulations [32] (Figure 14.6).

Solvation structures show similar characteristics regardless of the nanotube diameter. Cation and anion distributions form smeared-out cylindrical shell-like structures (see Figure 14.6) around the nanotubes, with a primary and a secondary density peaks located ~0.35 and ~0.8 nm from the exterior nanotube surface [32]. The position of the first peak of EMI^+ is slightly closer to the nanotube wall than that of BF_4^-. The imidazole ring of EMI^+ in the first external solvation shell is mainly parallel to the CNT surface. This stacking orientation allows bulky cations to approach the nanotube surface more closely than the corresponding anions. Cation orientation in the second external solvation shell is essentially isotropic with respect to the radial direction [32]. The transition from π-stacking to isotropic configurations occurs over a distance of a ~0.4 nm. Rapid variations of ring orientations similar to these were also found for dimethylimidazolium cations confined inside two parallel plates [33]. This flexibility in ring orientations and the resulting easy adoption of π-stacking can play a significant role in enabling imidazolium-based RTILs to disperse carbon nanotubes.

14.2 NONCOVALENT FUNCTIONALIZATION OF CNTs

The most common method to separate nanotubes from their bundles is liquid phase exfoliation and stabilization of nanotubes. Such stabilization cannot occur unless the attractive inter-nanotube potential is balanced by a repulsive potential [21]. Various methods to provide this repulsive potential have been explored. Coating the nanotubes with a dispersant phase—usually amphiphilic molecules, surfactants, biomolecules, or polymers—can result in weak intertube repulsion and a metastable suspension of the coated tubes in the solvent of choice. Early on, nanotubes were treated using a variety of stabilization and processing techniques typical of colloid science. The two main techniques were:

(i) Electrostatic Stabilization. Stabilization of colloids often relies on the presence of a surface charge, which then attracts a layer of counterions from the liquid to form an electric double layer. Because of the diffusive nature (due to Brownian motion) of the counterions, the overall result is an effective surface charge, resulting in Coulomb repulsion between nearby charged colloids [21, 34]. It is possible to apply this process to carbon nanotubes, introducing a temporary and removable surface charge, by allowing molecules to adsorb onto nanotubes via their hydrophobic part. Usually an ion becomes dissociated from the hydrophilic head groups and acts as the counterion. The adsorbed molecular ions then interact with the solvent, usually water. The important point is that the diffuse cloud of counterions is spatially separated from the tail-group molecular ions. The presence of the double layer implies that the attractive van der Waals potential due to the nanotubes is balanced by the repulsive potential associated with the double layer.

(ii) Steric Stabilization. Colloids can also be stabilized by attachment of polymer chains or other linear molecules. The stabilization is entropic; when the colloids approach, the chains attached to the tubes interact. This interaction results in the reduction of the number of conformations available to the chains, thus lowering their entropy and increasing the free energy of the system. This process acts as a repulsive force and is known as steric stabilization [21, 35]. This mechanism can be realized by attaching polymers or molecules to nanotubes either covalently or by physical adsorption.

14.2.1 Molecules

14.2.1.1 Polynuclear Aromatic Molecule. Condensed aromatic derivatives carrying a hydrophilic or a hydrophobic moiety can dissolve CNTs in aqueous or in organic media. Nakashima and co-workers [36] studied the solvation of CNTs in aqueous environment with phenyl, naphthalene, phenanthrene, and pyrene–ammonium amphiphiles. Phenyl- and naphthalene-based amphiphiles were not able to disperse CNTs in water. On the contrary, the pyrene derivative was a more efficient solvating agent when compared to the phenanthrene one. Polynuclear aromatic molecules, such as pyrene, perylene, porphyrin, and their derivatives, can adsorb to the surface of CNTs through strong π–π stacking interactions utilizing their aromatic part as an anchor, opening the way for the solvation of CNTs [14, 37–39] (Figure 14.7).

Pyrene is probably the most studied molecule and will be used as the paradigm of this class of compounds. MD simulations showed that the pyrene–CNT complex has lower potential energy than the pyrene dimer. Therefore, pyrene molecules in solution interact with nanotubes and form complexes rather than assemble by themselves [40]. Experimentally to solvate CNT with pyrene, a 1.3:1 mole ratio of CNT unit to pyrene molecules is used, where a CNT unit consists of four hexagonal rings (as does the pyrene structure). This means that the pyrene molecules fully cover the external surface of the nanotubes. Adsorption models showed that five-pyrene molecules adsorb on the (5, 5) nanotube sidewall surface and form a pentagonal structure. Since the diameter of a (10, 0) nanotube is larger than that of a (5, 5) nanotube, six-pyrene molecules are required to form a regular hexagonal arrangement about a (10, 0) nanotube. In both (5, 5) and (10, 0) nanotube systems, the optimized geometries are not eclipsed conformations, and there is a shift of the pyrene molecules along the nanotube axis (Figure 14.8). The shift extent is approximately 1/3 to 1/2 of the molecular size [40]. MD simulation showed also that polynuclear aromatic molecules adsorbed on a tube are mobile at room temperature [41]. A decrease in temperature leads to localization of the adsorbed molecules. Adsorbed polynuclear aromatic molecules prefer lying flat on the tube surface and this trend is more pronounced at lower temperatures [41]. One of the major difficulties in properly evaluating the CNT–pyrene interaction is the high mobility that pyrene shows on the surface of CNT. Upon adsorption, small horizontal movements of the organic moiety readily break the π–π stacking interactions, which, however, are immediately re-formed a fraction of an angstrom away. Determination of the size of the interactions between CNT and pyrene is, however, fundamental to understand the dispersion of CNTs. For example, as the radius of CNTs decreases, one might qualitatively expect weaker interactions with the planar and rigid structure of pyrene. To address this issue, free energies were calculated by integrating the potential mean force (PMF) curves, which were obtained by molecular dynamics simulations at 298 K [38].

The optimized energy, the enthalpy, and the free energy at 298 K show similar trends: From the narrow (5,5) to the wider (17,17) CNTs, the radius varies by more than a factor of 3, while the various types of energy of interaction change by 1.5 times [38]. The several kcal mol^{-1} of difference suggests the tendency of pyrene molecule to interact with larger nanotubes than with those of narrower diameters. This tendency can explain the preferential solvation of larger diameter nanotubes by water-soluble pyrene derivatives [42].

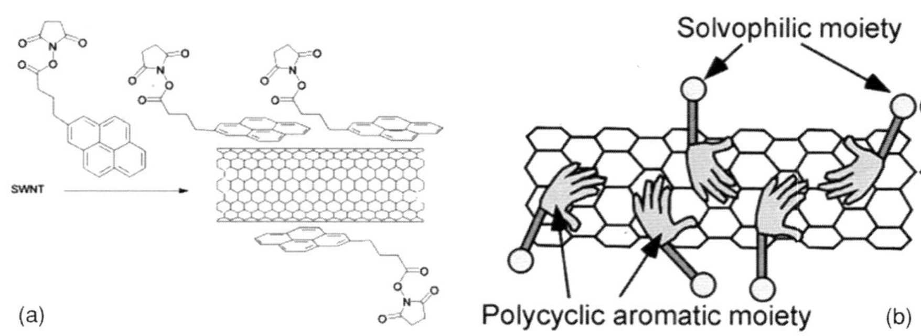

FIGURE 14.7. (**a**) Noncovalent anchoring of aromatic molecules to the sidewalls of CNTs. (**b**) Strategy to solvate CNTs by physical adsorption of polycyclic aromatic molecules carrying a solvating moiety onto the surfaces of CNTs. (**a**) Reprinted from reference 8, with permission. Copyright 2003, Wiley–VCH. (**b**) Reprinted from reference 36, with permission. Copyright 2006, Wiley–VCH.

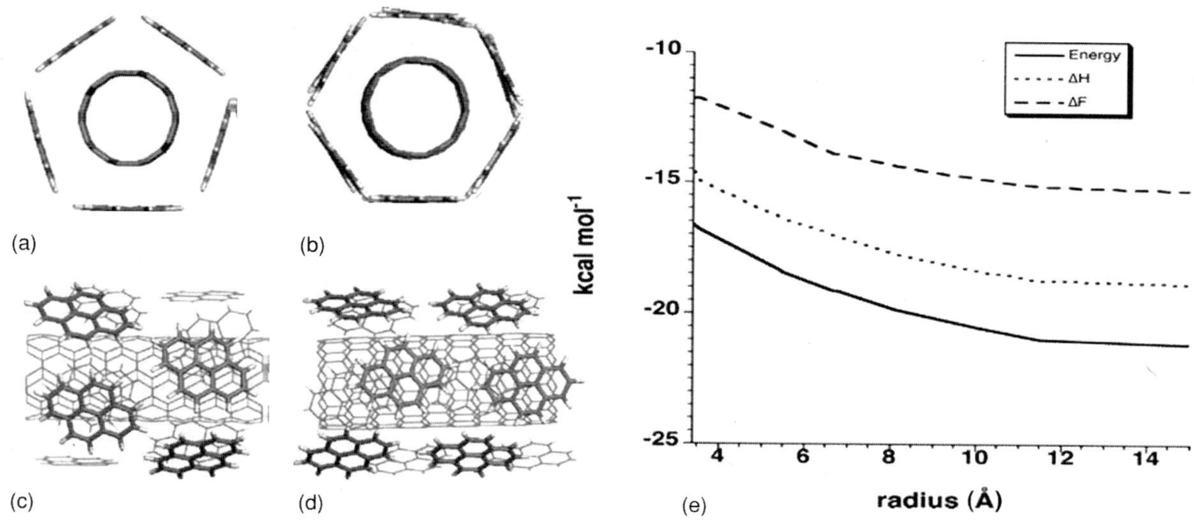

FIGURE 14.8. Optimized geometries of pyrene molecules interacting with (**a**) and (**c**) a (5, 5) nanotube, and with (**b**) and (**d**) a (10,0) nanotube. a, and b) Top view. (**c**) and (**d**) Side view. (**e**) Calculated potential energy, enthalpy, and free energy at room temperature versus CNT radius. (a,b,c,d) Reprinted from reference 40, with permission. Copyright 2007, American Scientific Publishers. (**e**) Reprinted from reference with permission. Copyright 2006, American Chemical Society.

Polynuclear aromatic molecules are also the fundamental components of a novel strategy for diameter-selective separation of CNTs. Komatsu and co-workers [43] synthesized pyrene nanotweezers that consist of two 1- or 2-pyrenes and 3,6-carbazolylenes with various N-substituents (Figure 14.9).

1-Pyrene nanotweezers selectively solvate and extract CNTs with diameters ranging from 0.84 nm to 0.97 nm, while 2-pyrene nanotweezers are not able to extract CNTs at all. Calculations supported the hypothesis that the minimal deformation of the 1-pyrene nanotweezers upon complexation allows selective complexation with specific nanotubes. When the deformation energy of the tweezer is too large, it destabilizes the complex and no CNTs are extracted.

14.2.2 Amphipilic Molecules

A simple and non-destructive method for nanotube solvation is based on noncovalent interactions of amphiphilic molecules with nanotube surfaces: The hydrophilic part of

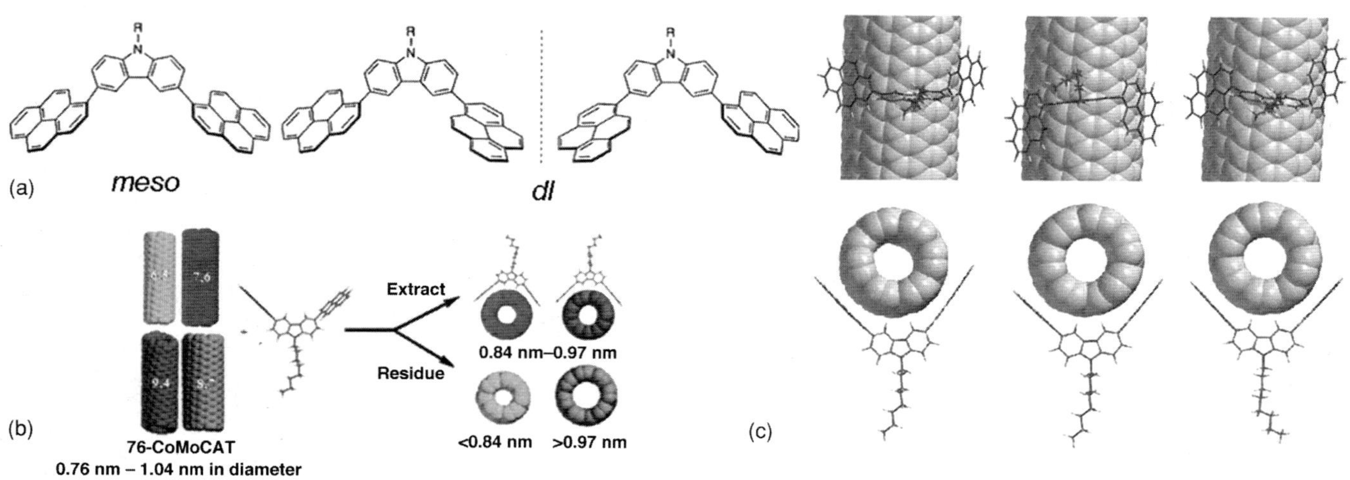

FIGURE 14.9. (**a**) Conformational stereoisomers of 1-pyrene nanotweezers, *meso* and *dl*. (**b**) Selective extraction of CNTs by 1-pyrene nanotweezers. (**c**) Computer-generated complex structures of *meso*-3, and *dl*-3 with (7,6) CNTs. Reprinted from reference 43, with permission. Copyright 2011, The Royal Society of Chemistry.

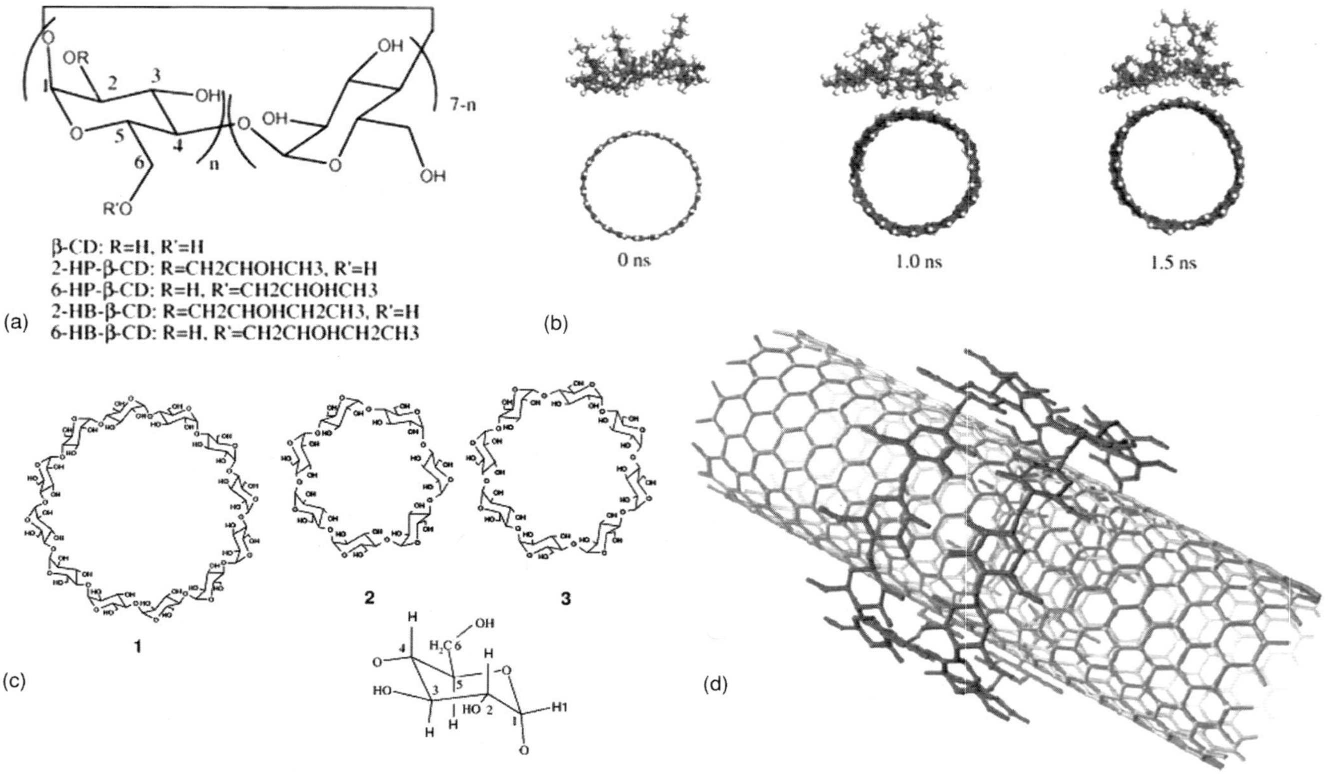

FIGURE 14.10. (**a**) Structures of β-CD, HB-β-CD, and HP-β-CD ($n = 3$). (**b**) Snapshots of the MD simulations of the interaction of 6-HB-β-CD and CNT. (**c**) η- **1**, β- **2**, and γ-**3** cyclodextrins and the atom numbering in the glucopyranosic unit. (**d**) Two η-CyDs **1** in the head-to-head arrangement threaded on CNT as modeled by MD. (a, b) Reprinted from reference 52, with permission. Copyright 2010, Elsevier. (c, d) Reprinted from reference 53, with permission. Copyright 2003, The Royal Society of Chemistry.

the molecules interact with the solvent and hydrophobic parts are adsorbed onto the nanotube surface, thus solvating CNTs and preventing them from aggregation into bundles and ropes.

Hu et al. [44] reported a noncovalent approach for the exfoliation and dissolution of CNTs in water by a rigid, planar, and conjugated diazo dye, namely Congo red (CR). Bromocresol green (BCG), bromophenol blue (BPB) in acidic and basic forms [45], and Disperse Orange 3 (DO3) [46] were also used efficiently as dispersing agents.

Supramolecular interaction and solvation of CNTs in aqueous media was observed with macrocyclic host molecules, such as calixarenes [47], cucurbiturils [48], and cyclodextrins [49, 50]. The interaction between cyclodextrins and CNT has been subject to a computational analysis [51, 52]. Cyclodextrins (CDs) are macrocyclic oligosugars commonly composed of 6, 7, or 8 glucosidic units and named α-, β-, and γ-cyclodextrin. They have a hydrophobic inner cavity and a hydrophilic external surface. Combining experimental results with calculations [51], a formation mechanism was presented where the driving forces for the formation of CNTs/CDs complexes originates from (i) van der Waals forces between CNTs and CDs adsorbed on the surface of CNTs and (ii) hydrogen-bonding interaction between adjacent CD molecules. Both interactions drive CD molecules to arrange rigidly onto the surface of CNTs. CD molecules have two different hydroxyl groups at the two ends of its cavity. Both the primary tail of 0.78-nm diameter and the secondary head of 1.53-nm diameter can interact with CNTs. Calculations demonstrated the attachment of the head (-40.4 kcal mol^{-1}) and not of the tail (-29.6 kcal mol^{-1}) of the CD molecules to the surface of CNTs [51]. Breakdown of the energy showed that the binding energy is mainly from the contribution of the nonbonding energy of the van der Waals forces. Van der Waals contributions between CNTs and CDs are larger than intertube interactions that are due to the bundling of CNTs. High-resolution transmission electron microscopy (HRTEM) showed CD cavities perpendicular rather than parallel to the CNT surface [51], in disagreement with the calculations. Subsequent molecular dynamics simulations of CNT interacting with β-cyclodextrin and four cyclodextrin derivatives revealed the influence of the substituent groups, as well as their positions in the CNT/CDs adducts in the interaction process [52] (Figure 14.10).

Functionalized CDs can more easily and completely wrap around the surface of CNTs. CDs with longer C-2 side chains have the strongest interactions with CNTs. The attractive interactions between CNTs and CDs monotonically increase with CNT radius. Calculations predict that with appropriately functionalized CDs it is possible to enhance the formation of CD–CNT complexes. Even more interesting is the complexation of CNT with 12-membered cyclodextrins, η-cyclodextrins, that enables not only their solvation in water but also their partial separation with respect to the diameter [53]. Both the NMR spectra as well as MM and MD simulations described the complexes as a poly-(pseudo-rotaxane) structures, excluding the possibilities that η-CDs stick to the external CNTs sides [53]. With its inner diameter of ~1.8 nm, η-cyclodextrin can host CNTs with outer diameters of ~1.2 nm. MM simulations of a system consisting of η-CD placed coaxially at one end CNT have shown that during the energy minimization the nanotube threads the η-CD. MD simulations of the CNT with two η-CD molecules threaded around it in the head-to-head arrangement exhibited concerted motions of the CDs, most probably due to the presence of a large system of 12 hydrogen bonds connecting the macrocycles [53].

Another very important class of amphiphilic molecules used to solvate CNTs are biological detergents such as steroid [54] and sugar biosurfactants [54, 55]. Bile salt biosurfactants have been utilized extensively to disperse individual CNTs in aqueous solution, and their role has been investigated computationally. The most common bile salt is sodium cholate (SC). Bile salts, unlike conventional linear surfactants, are rigid faced amphiphiles, referred to as "two-faced detergents." These surfactants possess a quasi-planar, slightly warped but rigid steroid ring with a hydrophilic face formed by the hydroxyl groups and the charged carboxylate group, as well as a hydrophobic face formed by the methyl groups and the tetracyclic carbon backbone (see Figure 14.11I). As a result of their chemical structure, bile salts act as very effective dispersants of biological molecules in living cells, including fat-soluble vitamins, bilirubin, and cholesterol. Due to the slightly bent but rigid steroid ring found in bile salts, these surfactants can very effectively accommodate the curvature of the CNT surface and enhance the stability of a dispersion of CNTs in aqueous solutions.

MD simulations described the variation of the density of SC-CNT assemblies when cholate ions where adsorbed onto CNTs characterized by different diameters [56, 57] and investigated the surface morphology of adsorbed SC as a function of surface coverage [58]. Both at low (75 mM) and at high (125 mM) SC concentration, the cholate ions wrap around the CNT to form a ring with the hydrophobic faces inward and the hydrophilic faces outward. The organization of the cholate ions on the CNT surface appears when comparing the radial distribution functions (RDF) of the cholate ions and of the charged carboxylate groups (Figure 14.11III). The distribution of the angles between the principal axis of the cholate ions and the cylindrical axis of the CNT (Figure 14.11IV) showed that for both SC surface coverages, the cholate ions prefer to orient almost parallel to the cylindrical axis of the CNT. The cholate ions also have a small tendency to orient perpendicular to the cylindrical axis of the CNT (the angle distribution profile also exhibits a smaller peak at ~90°). The interactions between two parallel CNTs with cholate ions adsorbed at approximately the saturated SC surface coverage were quantified by calculating the potential of mean force (PMF) per unit length of nanotube, as a function of the intertube separation, d (Figure 14.11Va). The PMF profile exhibits a primary, long-range repulsive potential energy barrier possessing a maximum of ~28 kJ mol^{-1} nm^{-1}.

Comparing the PMF results for SC with those of linear surfactants such as SDS [59], it appears than that SC is more effective than SDS at stabilizing aqueous dispersions of individual CNTs. Specifically, (i) the simulated potential energy barrier induced by SC at 2.0 nm is 40% higher than that induced by SDS at 2.4 nm [59], and (ii) the simulated attractive energy well induced by SC at 1.7 nm is 50% smaller than that induced by SDS at the same intertube separation [59]. A shallower attractive energy well still enhances the dispersion stability based on the theory discussed in reference 60, where the integration of the exponential of the PMF between two SC-coated CNTs was found inversely proportional to the coagulation rate of these two colloidal particles.

Another interesting molecule used to solvate CNTs is the flavin mononucleotide (FMN), a common redox cofactor (Figure 14.12). This molecule can perform chirality selection, wrapping around CNT in a helical pattern [61, 62]. MM calculations showed that the cooperative hydrogen bonding between adjacent flavin moieties results in the formation of a helical ribbon, which organizes around CNTs through concentric π–π interactions between the flavin mononucleotide and the graphene wall [61, 62]. The strength of the helical flavin mononucleotide assembly strongly depends on nanotube chirality. MD simulations [62] confirmed that FMN molecules adsorb with the isoalloxazine group on the CNT surface, yielding helical structures that resemble those proposed on the basis of energy minimization techniques. However, the structures observed in MD simulations were less compact. Change in the CNT diameter affects the orientation of FMN molecules and hence may enhance the separation of CNTs of different diameters. The calculation of the effective pair potential of mean force (PMF) between aqueous (6, 6) CNTs in the presence of FMN at two surface densities showed that increasing the surface density of FMN molecules increases the repulsive barrier between CNTs and confirmed the experimental observations according to which FMN surfactants are superior to SDS in stabilizing aqueous dispersions of CNTs, for reasons similar to those just discussed for the cholate ions.

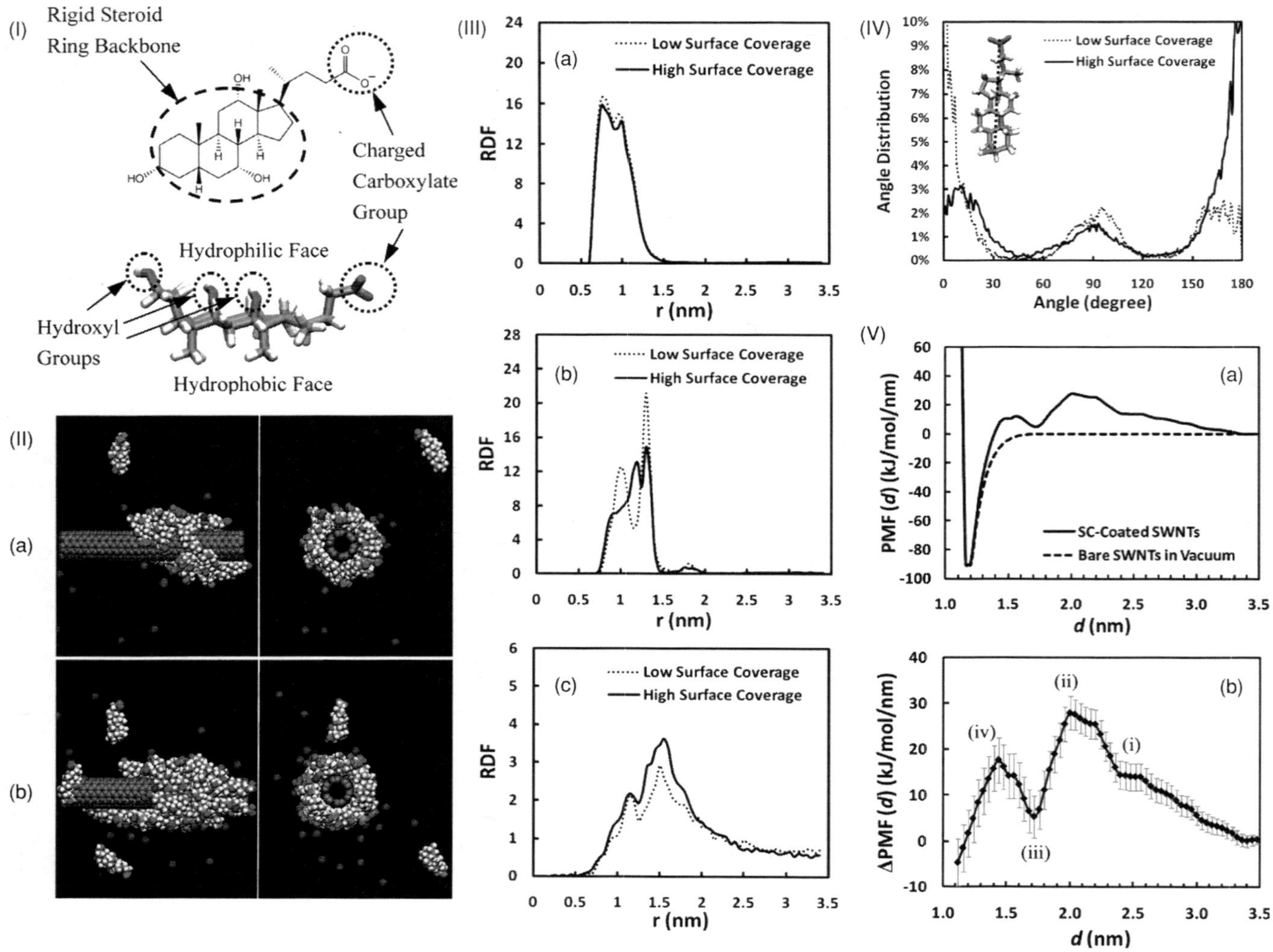

FIGURE 14.11. (**I**) Schematic (*top*) and spatial (*bottom*) chemical structures of sodium cholate, the hydroxyl groups (OH), and the charged carboxylate group (COO-). Color code: red, oxygen; light green, carbon; white, hydrogen. (**II**) Snapshots of a (6, 6) CNT in aqueous SC solutions at two different SC concentrations: (**a**) 75 mM and (**b**) 125 mM. Water molecules are not shown for clarity. (**III**) Radial distribution functions (RDFs) relative to the cylindrical axis of a CNT ("*r*" is measured radially from the CNT axis): (**a**) RDF of the cholate ions, (**b**) RDF of the carbon atoms of the charged carboxylate groups, and (**c**) RDF of the sodium counterions. The RDFs are plotted for low (corresponding to a total SC concentration of 75 mM) and high (corresponding to a total SC concentration of 125 mM) SC surface coverages. (**IV**) Simulated distribution profiles of the angle, θ, between the principal axis of the cholate ions and the cylindrical axis of the CNT. The bias resulting from variations in solid angle has been removed by a weighting factor of $(1/\sin\theta)$. In the SC molecular structure shown, the dotted line connecting the carbon atom in the carboxylate group with the carbon atom at the end of the steroid ring defines the principal axis of the cholate ion (refer also to part I). (**V**) (**a**) Simulated potential of mean force (PMF) corresponding to (i) two parallel CNTs coated with cholate ions (solid line) and (ii) two parallel bare CNTs in vacuum (dashed line), as a function of the intertube separation, d. Note that the two PMF profiles overlap for $d = 1.2$ nm, (**b**) Net contribution of SC to the PMF profile, ΔPMF, corresponding to two parallel CNTs coated with cholate ions. Reprinted from reference 58, with permission. Copyright 2010, American Chemical Society.

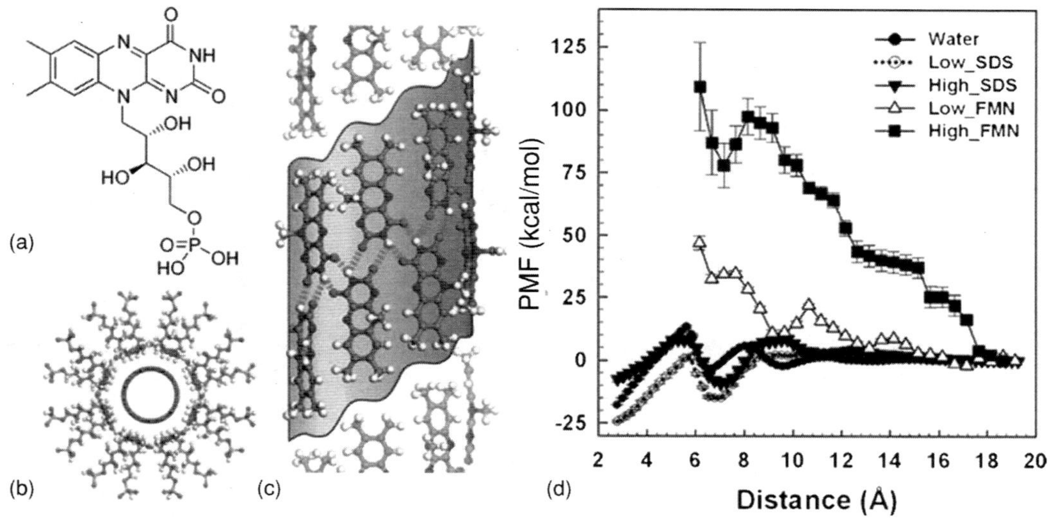

FIGURE 14.12. (**a**) Chemical structures of flavin mononucleotide (FMN). (**b**) The long *d*-ribityl phosphate side groups of FMN provide aqueous solubilization; (**c**) Top view of isoalloxazine moieties wrapped in an 8_1 helical pattern. The helical ribbon (shaded structure) is stabilized by (i) four H bonds (red dashed lines) between adjacent isoalloxazine moieties and (ii) charge-transfer interactions with the underlying graphene side walls. (**d**) Potential of mean force between two rigid (6, 6) CNTs in water, in aqueous SDS, and in aqueous FMN. For both surfactants, results are reported at two surface coverages [low (0.98 nm^2/SDS and 1.24 nm^2/FMN) and high (0.44 nm^2/SDS and 0.94 nm^2/FMN)]. (b,c) Reprinted from reference 61, with permission. Copyright 2008, Nature Publishing Group. (d) Reprinted from reference 62, with permission. Copyright 2010, American Chemical Society.

14.2.3 Surfactants

Surfactants are amphiphilic molecules with a hydrophilic head and a hydrophobic tail. Surfactants have played a very important role in the procedures used to prepare aqueous dispersions of CNTs [63–66]. The surfactant-aided dispersion of CNTs involves the application of an external energy input (ultrasound) to separate bundled nanotubes [6]. Subsequently, the separated CNTs provide new adsorption sites for the surfactant molecules. These amphiphilic molecules orient themselves to adsorb on the CNT surface with their hydrophobic tail, while the hydrophilic head is oriented toward the solution. The repulsive potential energy resulting from the adsorption of surfactant molecules (electrostatic for ionic surfactants and steric for nonionic surfactants) further enhances the separation process. There are several models that describe the adsorption, as depicted in Figure 14.13: hemimicelle adsorption of surfactants onto CNTs [64, 65], encapsulation of a CNT inside a cylindrical micelle of surfactants [63], and random adsorption of surfactants onto the tube surface [67]. MD simulations provided a picture of the adsorption process of surfactants on the CNTs that described the self-assembly process [68–72] and gave an atomistic model of the interaction mechanism between CNTs and these amphiphilic molecules [59, 63, 73–77].

Sodium dodecyl sulfate (SDS) is generally considered as the ideal model of surfactant. In the following, the analysis will be carried out for this surfactant. DPD [68] and coarse-grained simulations [69–72] showed that all the manners of surfactant–CNT interactions suggested in the literature (cylindrical micelle, hemicelles, random adsorption) are possible and smoothly change one into the other as the concentration or the nature of the surfactant changes (Figure 14.14a).

At low concentration of surfactants, the nanotube is only occasionally decorated by micelles, in agreement with the images reported by Smalley and co-workers [66]. Increasing the concentration, the appearance of the different behaviors depends on the length of the surfactant. For short surfactants, the nanotube is decorated by micelles; for longer surfactants, there is random adsorption on the surface, as suggested by Krishnamoorti and co-workers [67]. For further increases of the concentration, the system develops a rich variety of structures depending on the length of the surfactant, including the appearance of cylindrical micelles adsorbed onto the wall of the CN, or of adsorbed hemimicelles, as observed Mioskowski et al. [65]. At even higher concentration the coverage of the CNT is complete. For the shorter surfactants a monolayer covering the tube appears, while longer surfactants encapsulate the CNT inside a cylindrical micelle. The possibility for surfactants to self-assemble in a cylindrical micelle requires the fulfillment of strict geometrical parameters, as demonstrated by Matarredona et al. [78], that depend on the length of the chain.

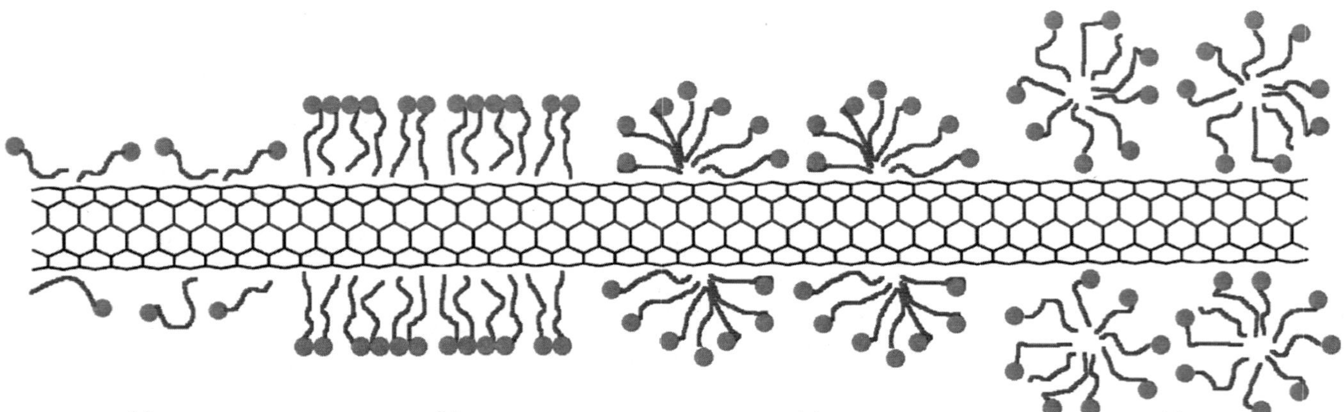

FIGURE 14.13. Schematic representations of surfactant interactions with a CNT. For clarity, only a few surfactants are shown and the CNT is in full view. (**a**) Random adsorption of surfactant molecules. (**b**) Encapsulation in a cylindrical surfactant micelle. (**c**) Hemimicellar adsorption. (**d**) Adsorption of micelles. Reprinted from reference 68, with permission. Copyright 2009, Wiley–VCH.

The DPD simulations [68] displayed also the dynamics of surfactant self-assembly around nanotube. For all surfactant concentrations, the coverage of the nanotube arises from collisions between the preformed micelles and the CNT. At low surfactant numbers, the micelles interact dynamically and weakly with the CNT. At higher surfactants numbers, the micelles adsorb on the CNT and dynamically cover it. The micelles spread randomly on the surface and remain in a dynamical equilibrium on it. At the highest surfactants numbers, they start to self-assemble in stable superstructures. The hemimicelles present in the case of medium–high densities of surfactants are replaced by cylindrical micelles at the higher densities. This is in agreement with the concentration-dependent study of Matarredona et al. [78], who observed two stable plateaus that were assigned to random face-on adsorption of the surfactants and to their ordered edge-on adsorption.

There are differences in the surfactant adsorption on carbon nanotubes and their bundles [71] because of the heterogeneity of the bundle surface and the difference in diameter of bundles compared to that of individual tubes. Whereas aggregation dominates adsorption on individual tubes, on bundles it is largely a Langmuir-type process. High adsorption energy sites on the outer surface of bundles, where surfactant molecules can interact with two tubes simultaneously (grooves), dominate at low coverage. They also cause adsorption on bundles to become significant before adsorption on individual tubes starts. The difference in the adsorption mechanisms leads to a crossover point at higher concentrations, when the adsorbed amount on individual tubes becomes larger than that for the bundles [72]. Geometric factors are of primary importance, both for surfactants and for nanotubes. Striolo and co-workers [74] demonstrated with MD calculations that the nanotube diameter is the primary factor that determines the morphology of the aggregates between CNT and surfactants. There is a competition between (a) the energy gained by the surfactants when they wrap around a nanotube and (b) the enthalpic penalty due to bending of the surfactant molecule. This aspect is important for improving separation techniques [79] and for understanding the driving forces responsible for determining the morphology of the aggregates.

MD simulations for SDS surfactants adsorbed on (6, 6), (12, 12), and (20, 20) CNTs suggested that the morphology of adsorbed aggregates depends on the surface coverage, but also, and more significantly, on the CNTs diameter (Figure 14.15).

At low surface coverage (Figure 14.15a) SDS surfactants on (6, 6) CNTs form "rings" where the surfactants lie parallel to the nanotube axis. As the CNTs diameter increases the SDS surfactants still lie predominantly flat on the nanotube surface, but the surface coverage appears becomes more uniform. The orientation of the adsorbed surfactants also changes as the nanotube diameter increases. An important factor is the rigidity of the SDS molecule. The SDS surfactants can lie along the CNT axis or they can wrap the nanotubes. Entropically, both possibilities can occur. However, when a SDS molecule wraps around a narrow tube, it has to bend and overcome an energy barrier. For (6, 6) CNT, the simulations show that the advantage of wrapping the nanotubes is not sufficient to balance the energetic penalty encountered to bend the SDS molecule. As the nanotube diameter increases, it becomes easier to wrap SDS around the CNTs not only because a smaller deformation of the SDS molecule is necessary, but also because the number of surfactant tail/carbon atoms contacts increases. It is worth pointing out that some small regions of the (6, 6) CNT surface remain exposed to water even at large surfactant surface density.

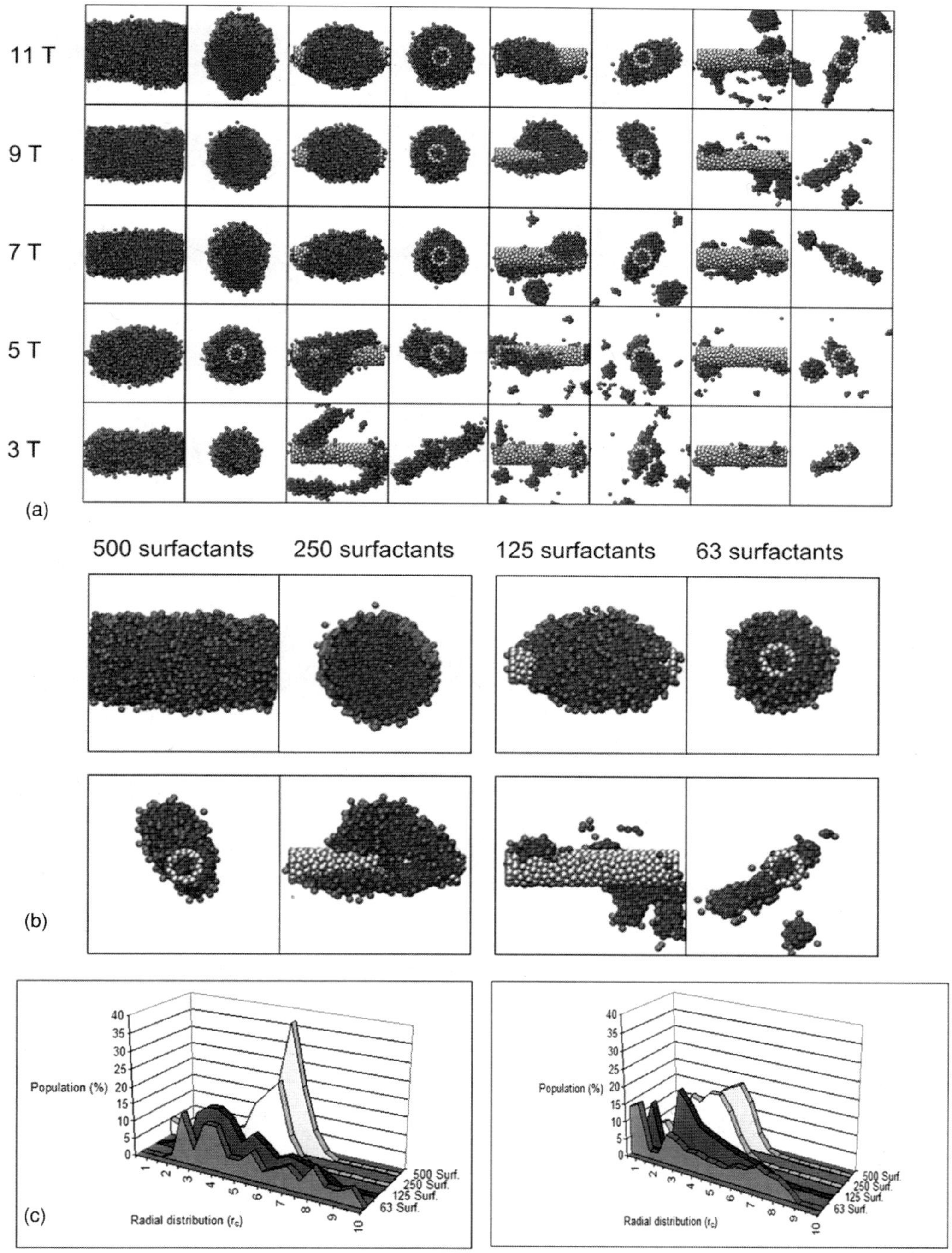

FIGURE 14.14. (**a**) Snapshots of equilibrium morphologies of surfactant assemblies around the CNT. *Top to bottom*: Surfactant chain length of 11, 9, 7, 5, and 3 beads. *Left to right*: 500, 250, 125, and 63 surfactants in the simulation box. Water molecules are removed for clarity. (**b**) Typical morphologies from two points of view obtained for the adsorption of surfactants on the CNT. *Top left*: Cylindrical micelle. *Top right*: Hemimicelle. *Bottom left*: Random adsorption. *Bottom right*: Adsorption of micelles. (**c**) The radial distribution of the heads (*left*) and of the tails (*right*) of the surfactants around the axis of the CNT. Reprinted from reference 68, with permission. Copyright 2009, Wiley–VCH. See color insert.

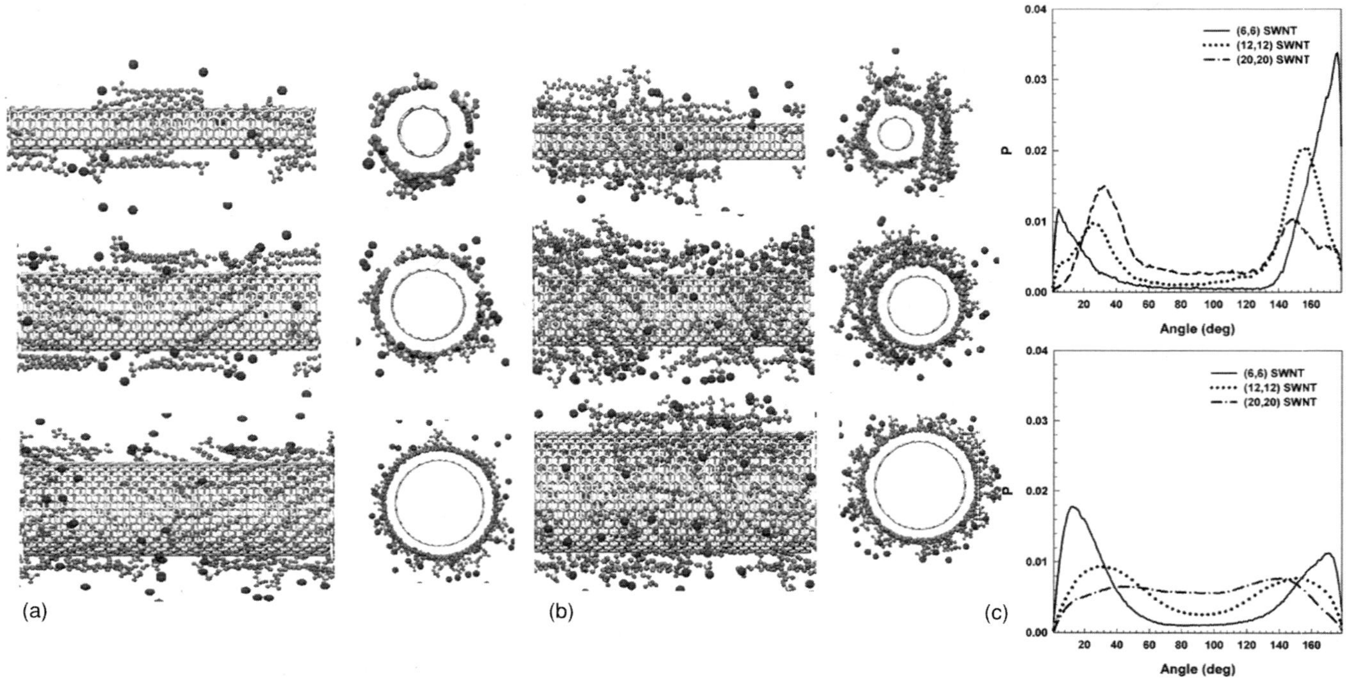

FIGURE 14.15. (**a**) Snapshots for (6, 6) (*top*), (12, 12) (*center*), and (20, 20) CNTs (*bottom*) covered by SDS surfactants at a surface density of 0.98 nm^2 per headgroup. Blue spheres are Na^+ ions. Cyan spheres are either CH_2 or CH_3 groups. Water molecules are not shown for clarity. (**b**) Same as (**a**), but for SDS surfactants at a surface density of 0.44, 0.49, and 0.81 nm^2 per headgroup on (6, 6), (12, 12), and (20, 20) CNTs. (**c**) Probability density for SDS surfactants orientation with respect to the CNTs axis (*top panel*) and for SDS–SDS relative orientation (*bottom panel*). Results are obtained for the systems shown in (**a**), in which the surfactant surface density is 0.98 nm^2 per headgroup. Reprinted from reference 74, with permission. Copyright 2009, American Chemical Society.

The exposure probably occurs because the adsorbed SDS surfactants find it more favorable to maximize the SDS/SDS interactions than spread evenly on the CNTs surface.

On (12, 12) CNTs, the SDS surfactants form a continuous monolayer of adsorbed surfactants in contact with the nanotube surface. The excess SDS molecules agglomerate to form a multilayered structure. On the (20,20) CNTs the surface coverage was not complete. The surfactants would in some cases prefer to agglomerate together rather than spread over the entire available surface.

PMF between two SDS-coated CNTs in the aqueous environment were calculated by MD simulations [59], in which the (6, 6) CNTs are fully covered with SDS molecules (2.8 molecules/nm^2) [80]. Two kinds of PMFs based on different interacting objects were simulated: (1) the bare nanotubes, considering the adsorbed surfactant as a part of solution medium (denoted as "bare PMF"); and (2) the united nanotubes, allowing for the whole entity composed of the bare nanotube itself and the covered surfactants (defined as "united PMF").

The bare PMF (Figure 14.16Ia) features a long-range repulsive free energy barrier with a maximum of 20 kJ mol^{-1} nm^{-1}, centered between 16 and 28 Å of intertube separation, and a steep attraction minimum of -34 kJ mol^{-1} nm^{-1} at around $d = 8$ Å. The free energy barrier hinders CNT from approaching the highly attractive region and prevents aggregation. Various contributions to the PMF are displayed in Figure 14.16Ib (bottom).

Figure 14.16II shows the evolution of the system at decreasing intertube distances until the two independent surfactant/nanotube aggregations assemble into a single supramolecular micelle encapsulating the two CNTs with only a single surfactant tail layer between. When the intertube spacing is reduced below 8 Å, the intertube surfactants are almost completely squeezed out of the intertube region, as shown in the Figure 14.16IId.

MD simulations also investigated the role of surface coverage in the intertube interactions. The PMF between two SDS/CNT aggregates was calculated at the low surfactant coverage (1.0 molecules/nm^2). As shown in Figure 14.16IIIa, the PMF for the low SDS packing density displays a free-energy barrier at 16.0 Å and a minimum at 8.0 Å. The free-energy barrier of the PMF for the low SDS coverage decreases significantly, while the minimum is much steeper, -85 kJ mol^{-1} nm^{-1}. Surfactant adsorption at low packing density may not prevent nanotubes from re-aggregation and is not sufficient for an efficient dispersion of CNT, as was observed experimentally CNT [64, 80]. The decomposition

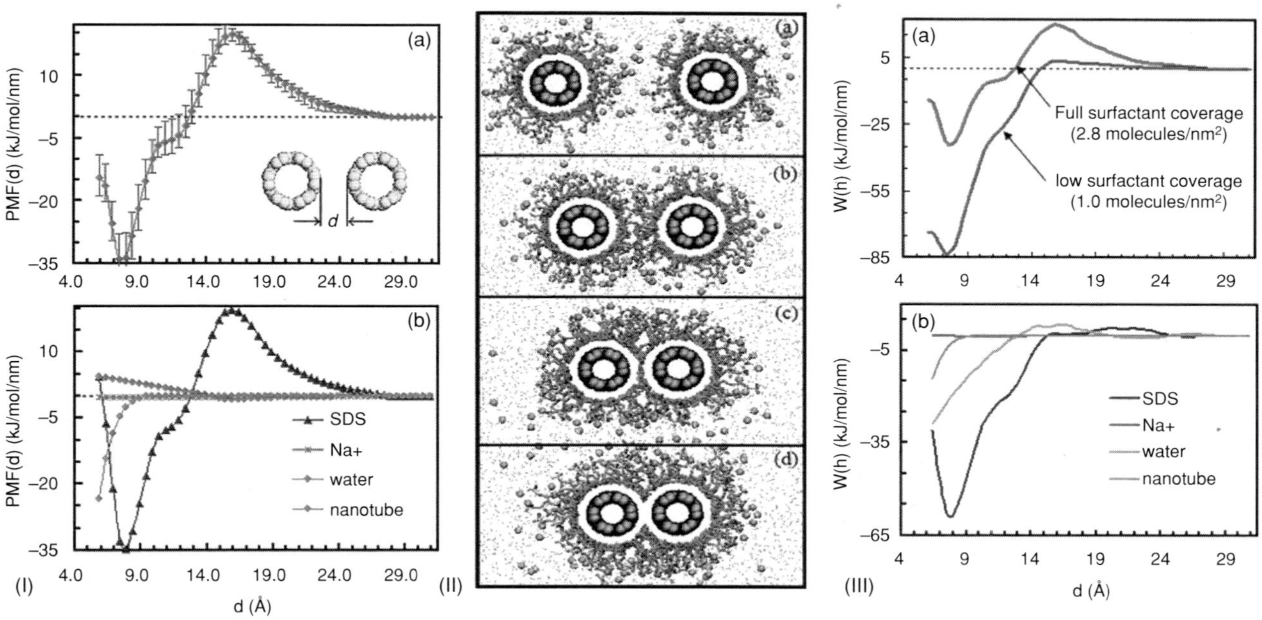

FIGURE 14.16. (**I**) (**a**) Potential of mean force, PMF(d), for the two bare carbon nanotubes at surfactant coverage of 2.8 molecules/nm^2 as a function of the intertube separation, d, obtained from MD simulations. (**b**) Water-induced, Na^+-induced, adsorbed SDS-induced, and tube–tube interaction potential contributions to the PMF. (**II**) Front view of snapshots obtained at the intertube separations of (**a**) 26.0 Å, (**b**) 16.0 Å, (**c**) 8.0 Å, and (**d**) 6.0 Å. (**III**) (**a**) Potential of mean force, PMF, for the two bare carbon nanotubes at the low surfactant coverage of 1.0 molecules/nm^2 as function of the intertube separation. (**b**) Water-induced, Na^+-induced, adsorbed SDS-induced, and tube–tube interaction potential contributions to the PMF. Reprinted from reference 59, with permission. Copyright 2010, American Chemical Society.

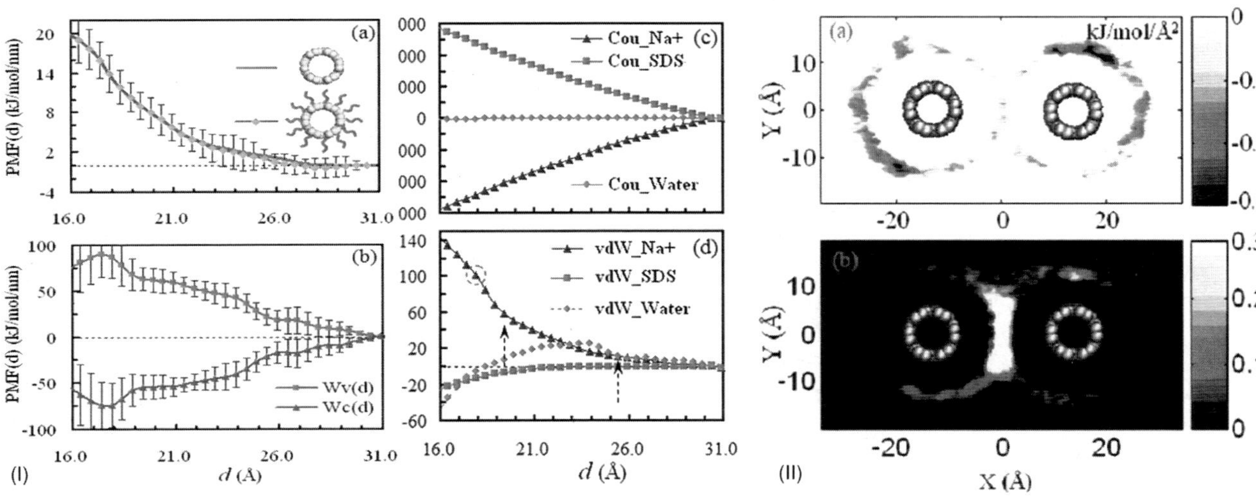

FIGURE 14.17. (**I**) (**a**) The united potential of mean force, PMF. The bare PMF (red line) is also shown for comparison. (**b**) The contributions from the Coulombic (Wc(d)) and vdW interactions (Wv(d)) to the united PMF (PMF(d)) Wc(d) + Wv(d)). (**c**) The contributions from water, Na^+, and the adsorbed surfactants to the Coulombic portion of the united PMF. (**d**) The induced contribution from water, Na^+, and the adsorbed surfactants to the vdW portion of the united PMF. (**II**) The two-dimensional strength maps of the surfactant–Na^+ vdW forces in the case of the intertube separation of 18 Å. (**a**) Na^+-induced vdW attractive interaction ($FC(d) < 0$) to the PMF, mainly arising from the outer surrounding sodium ions. (**b**) Na^+-induced repulsive vdW interaction ($FC(d) > 0$) to the PMF, arising from the intertube ion contribution. The bar shows the force–strength scale. Reprinted from reference 59, with permission. Copyright 2010, American Chemical Society.

of the PMF, with the various contributions to the PMF for the lower coverage shows that unlike the situation that occurs for full coverage, the water-induced contributions are significant.

To reveal the origin of the repulsive PMF generated by the adsorbed surfactants in the intertube separation range of 16–28 Å, the free energy between the two united entities, each composed of the nanotube and the covering surfactants, is presented in Figure 14.17. The bare PMF and united PMF, which are based on two different interacting objects, have almost identical repulsive features. The repulsive feature in the PMF profile is not due to the Coulombic interaction but to the vdW interaction. The sodium ions reside in the negatively charged pockets formed by the surfactant head groups. The sodium ion layer has a very strong electrostatic attractive interaction with the head segments of surfactants. A similar counterion-induced attraction has been experimentally reported in balancing the electrostatic repulsion between negatively charged DNAs [81]. Experimental investigations [82, 83] have shown that salt addition to the aqueous dispersion of nanotubes in the presence of SDS allows manipulation of the repulsive forces between the nanotubes that can lead to the selective aggregation of nanotubes.

14.3 CONCLUSION

MD simulations provide guidelines to exfoliate, solvate, and stabilize carbon nanotubes in solution. They suggest that it is necessary to employ dispersing agents that (1) strongly adsorb on the nanotube surface, (2) present hydrophilic groups, which function better if they are rigid, and (3) are not very mobile on the nanotube surface. Nanotube diameter and chirality sorting can be obtained if the dispersing agents show aggregates with a structure that depends on the nanotube geometry.

REFERENCES

1. Saito, R., Dresselhaus, G., Dresselhaus, M. S. (1998). *Physical Properties of Carbon Nanotubes*. Imperial College Press, London.
2. Dai, H. J., Hafner, J. H., Rinzler, A. G., Colbert, D. T., Smalley, R. E. (1996). Nanotubes as nanoprobes in scanning probe microscopy. *Nature*, **384**, 147–150.
3. Liu, C., Fan, Y. Y., Liu, M., Cong, H. T., Cheng, H. M., Dresselhaus, M. S. (1999). Hydrogen storage in single-walled carbon nanotubes at room temperature. *Science*, **286**, 1127–1129.
4. Baughman, R. H., Zakhidov, A. A., de Heer, W. A. (2002). Carbon nanotubes—The route toward applications. *Science*, **297**, 787–792.
5. Girifalco, L. A., Hodak, M., Lee, R. S. (2000). Carbon nanotubes, buckyballs, ropes, and a universal graphitic potential. *Physical Review B*, **62**, 13104–13110.
6. Lu, K. L., Lago, R. M., Chen, Y. K., Green, M. L. H., Harris, P. J. F., Tsang, S. C. (1996). Mechanical damage of carbon nanotubes by ultrasound. *Carbon*, **34**, 814–816.
7. Tasis, D., Tagmatarchis, N., Georgakilas, V., Prato, M. (2003). Soluble carbon nanotubes. *Chemistry European Journal*, **9**, 4000–4008.
8. Tasis, D., Tagmatarchis, N., Bianco, A., Prato, M. (2006). Chemistry of carbon nanotubes. *Chemistry Reviews*, **106**, 1105–1136.
9. Karousis, N., Tagmatarchis, N. (2010). Current progress on the chemical modification of carbon nanotubes. *Chemistry Reviews*, **110**, 5366–5397.
10. Hirsch, A. (2002). Functionalization of single-walled carbon nanotubes. *Angewandte Chemie International Edition*, **41**, 1853–1859.
11. Peng, X., Wong, S. S. (2009). Functional covalent chemistry of carbon nanotube surfaces. *Advanced Materials*, **21**, 625–642.
12. Singh, P., Campidelli, S., Giordani, S., Bonifazi, D., Bianco, A., Prato, M. (2009). Organic functionalisation and characterisation of single-walled carbon nanotubes. *Chemical Society Reviews*, **38**, 2214–2230.
13. Zhao, Y.-L., Stoddart, J. F. (2009). Noncovalent functionalization of single-walled carbon nanotubes. *Accounts on Chemical Research*, **42**, 1161–1171.
14. Calvaresi, M., Hoefinger, S., Zerbetto, F. (2012). Probing the structure of lysozyme–carbon-nanotube hybrids with molecular dynamics. *Chemistry—A European Journal*, **18**, 4308–4313.
15. Walther, Journal of H., Jaffe, R., Halicioglu, T., Koumoutsakos, P. (2001). Carbon nanotubes in water: Structural characteristics and energetics. *Journal of Physical Chemistry B*, **105**, 9980–9987.
16. Walther, J. H., Jaffe, R. L., Kotsalis, E. M., Werder, T., Halicioglu, T., Koumoutsakos, P. (2004). Hydrophobic hydration of C_{60} and carbon nanotubes in water. *Carbon*, **42**, 1185–1194.
17. Bergin, S. D., Sun, Z., D.Rickard, P. V.Streich, J. P.Hamilton, J. N.Coleman. (2009). Multicomponent solubility parameters for single-walled carbon nanotube–solvent mixtures. *ACS Nano*, **3**, 2340–2350.
18. Bahr, J. L., Mickelson, E. T., Bronikowski, M. J., Smalley, R. E., Tour, J. M. (2001). Dissolution of small diameter single-wall carbon nanotubes in organic solvents? *Chemical Communications*, **2**, 193–194.
19. Ausman, K. D., Piner, R., Lourie, O., Ruoff, R. S., Korobov, M. (2000). Organic solvent dispersions of single-walled carbon nanotubes: Toward solutions of pristine nanotubes. *Journal of Physical Chemistry B*, **104**, 8911–8915.
20. Bergin, S. D., Nicolosi, V., Streich, P. V., Giordani, S., Sun, Z., Windle, A. H., Ryan, P., Peter, N., Niraj, P., Wang, Z.-T. T., Carpenter, L., Blau, W. J., Boland, J. J., Hamilton, J. P., Coleman, J. N. (2008). Towards solutions of single-walled carbon nanotubes in common solvents. *Advanced Materials*, **20**, 1876–1881.
21. Coleman, J. N. (2009). Liquid-phase exfoliation of nanotubes and graphene. *Advanced Functional Materials*, **19**, 3680–3695.
22. Dill, K. A., Bromberg, S. (2003). *Molecular Driving Forces: Statistical Thermodynamics in Chemistry & Biology*. Garland Science, New York.

23. Mac Kernan, D., Blau, W. J. (2008). Exploring the mechanisms of carbon-nanotube dispersion aggregation in a highly polar solvent. *EPL*, **83**, 66009.
24. Furtado, C. A., Kim, U. J., Gutierrez, H. R., Pan, L., Dickey, E. C., Eklund, P. C. (2004). Debundling and dissolution of single-walled carbon nanotubes in amide solvents. *Journal of American Chemical Society*, **126**, 6095–6105.
25. Giordani, S., Bergin, S. D., Nicolosi, V., Lebedkin, S., Kappes, M. M., Blau, W. J., Coleman, J. N. (2006). Debundling of single-walled nanotubes by dilution: Observation of large populations of individual nanotubes in amide solvent dispersions. *Journal of Physical Chemistry B*, **110**, 15708–15718.
26. Grujicic, M., Cao, G., Roy, W. N. (2004). Atomistic simulations of the solubilization of single-walled carbon nanotubes in toluene. *Journal of Materials Science*, **39**, 2315–2325.
27. Kholmurodov, K., Aru, G., Yasuoka, K. (2010). Molecular dynamics simulations of the interaction of carbon nanotube and a carbon disulfide solvent. *Natural Science*, **2**, 902–910.
28. Shim, Y., Jung, Y. J., Kim, H. J. (2011). Carbon nanotubes in benzene: Internal and external solvation. *Physical Chemistry Chemical Physical*, **13**, 3969–3978.
29. Welton, T. (1999). Room-temperature ionic liquids. Solvents for synthesis and catalysis. *Chemical Reviews*, **99**, 2071–2084.
30. Fukushima, T., Kosaka, A., Ishimura, Y., Yamamoto, T., Takigawa, T., Ishii, N., Aida, T. (2003). Molecular ordering of organic molten salts triggered by single-walled carbon nanotubes. *Science*, **300**, 2072–2074.
31. Fukushima, T., Aida, T. (2007). Ionic liquids for soft functional materials with carbon nanotubes. *Chemistry—A European Journal*, **13**, 5048–5058.
32. Shim, Y., Kim, H. J. (2009). Solvation of carbon nanotubes in a room-temperature ionic liquid. *ACS Nano*, **3**, 1693–1702.
33. Pinilla, C., Del Popolo, M. G., Lynden-Bell, R. M., Kohanoff, J. (2005). Structure and dynamics of a confined ionic liquid. Topics of relevance to dye-sensitized solar cells. *Journal of Physical Chemistry B*, **109**, 17922–17927.
34. Hunter, R. J. (1994). *Introduction to Modern Colloid Science*, Oxford Science Publications, Oxford.
35. Israelachvili, J. (1991). *Intermolecular and Surface Forces*, Academic Press, London.
36. Tomonari, Y., Murakami, H., Nakashima, N. (2006). Solubilization of single-walled carbon nanotubes by using polycyclic aromatic ammonium amphiphiles in water—Strategy for the design of high-performance solubilizers. *Chemistry—A European Journal*, **12**, 4027–4034.
37. Chen, R. J., Zhang, Y., Wang, D., Dai, H. (2001). Single-walled carbon nanotubes for protein immobilization. *Journal of the American Chemical Society*, **123**, 3838–3839.
38. Ehli, C., Rahman, G. M. A., Jux, N., Balbinot, D., Guldi, D. M., Paolucci, F., Marcaccio, M., Paolucci, D., Melle-Franco, M., Zerbetto, F., Campidelli, S., Prato, M. (2006). Interactions in single wall carbon nanotubes/pyrene/porphyrin nanohybrids. *Journal of the American Chemical Society*, **128**, 11222–11231.
39. Backes, C., Schmidt, C. D., Rosenlehner, K., Hauke, F., Coleman, J. N., Hirsch, A. (2010). Nanotube surfactant design: The versatility of water-soluble perylene bisimides. *Advanced Materials*, **22**, 788–802.
40. Zhang, Y., Yuan, S., Zhou, W., Xu, J., Li, Y. (2007). Spectroscopic evidence and molecular simulation investigation of the interaction between pyrene molecules and carbon nanotubes. *Journal of Nanoscience and Nanotechnology*, **7**, 2366–2375.
41. Wesołowski, R. P., Furmaniak, S., Terzyk, A. P., Gauden, P. A. (2011). Simulating the effect of carbon nanotube curvature on adsorption of polycyclic aromatic hydrocarbons. *Adsorption*, **17**, 1–4.
42. Backes, C., Mundloch, U., Ebel, A., Hauke, F., Hirsch, A. (2010). Dispersion of HiPco® and CoMoCAT® single-walled nanotubes (CNTs) by water soluble pyrene derivatives—Depletion of small diameter CNTs. *Chemistry—A European Jouranl*, **16**, 3314–3317.
43. Mustafizur Rahman, A. F. M., Wang, F., Matsuda, K., Kimura, T., Komatsu, N. (2011). Diameter-based separation of single-walled carbon nanotubes through selective extraction with dipyrene nanotweezers. *Chemical Science*, **2**, 862–867.
44. Hu, C., Chen, Z., Shen, A., Shen, X., Li, J., Hu, S. (2006). Water-soluble single-walled carbon nanotubes via noncovalent functionalization by a rigid, planar and conjugated diazo dye. *Carbon*, **44**, 428–434.
45. Zhang, W., Silva, S. R. P. (2010). The effects of phenolic hydrogens and methyl substitute groups in organic dyes on their dispersion of multiple-walled carbon nanotubes. *Carbon*, **48**, 2063–2071.
46. Costanzo, G. D., Ledesma, S., Mondragon, I., Goyanes, S. (2010). Stable solutions of multiwalled carbon nanotubes using an azobenzene dye. *Journal of Physical Chemistry C*, **114**, 14347–14352.
47. Ogoshi, T., Yamagishi, T., Nakamoto, Y. (2007). Supramolecular single-walled carbon nanotubes (SWCNTs) network polymer made by hybrids of SWCNTs and water-soluble calix[8]arenes. *Chemical Communications*, **45**, 4776–4778.
48. Ogoshi, T., Inagaki, A., Yamagishi, T., Nakamoto, Y. (2008). Defection-selective solubilization and chemically-responsive solubility switching of single-walled carbon nanotubes with cucurbit[7]uril. *Chemical Communication*, **19**, 2245–2247.
49. Chen, J., Dyer, M. J., Yu, M. F. (2001). Cyclodextrin-mediated soft cutting of single-walled carbon nanotubes. *Journal of the American Chemical Society*, **123**, 6201–6202.
50. Chambers, G., Carroll, C., Farrell, G. F., Dalton, A. B., McNamara, M., Panhuis, M. I. H., Byrne, H. J. (2003). Characterization of the interaction of gamma cyclodextrin with single-walled carbon nanotubes. *Nano Letters*, **3**, 843–846.
51. Liu, K., Fu, H., Xie, Y., Zhang, L., Pan, K., Zhou, W. (2008). Assembly of â-cyclodextrins acting as molecular bricks onto multiwall carbon nanotubes. *Journal of Physical Chemistry C*, **112**, 951–957.
52. Pang, J., Xu, G., Bai, Y., Yuan, S., He, F., Wang, Y., Sun, H., Hao, A. (2010). Molecular dynamics simulations of the interactions between β-cyclodextrin derivatives and single-walled carbon nanotubes. *Computational Materials Science*, **50**, 283–290.
53. Dodziuk, H., Ejchart, A., Anczewski, W., Ueda, H., Krinichnaya, E., Dolgonos, G., Kutner, W. (2003). Water solubilization, determination of the number of different types of single-wall carbon nanotubes and their partial separation with respect

to diameters by complexation with η-cyclodextrin. *Chemical Communications*, **8**, 986–987.

54. Ishibashi, A., Nakashima, N. (2006). Individual dissolution of single-walled carbon nanotubes in aqueous solutions of steroid or sugar compounds and their Raman and near-IR spectral properties. *Chemistry—A European Journal*, **12**, 7595–7602.
55. Gorityala, B., K., Ma, J., Wang, X., Chenb, P., Liu, X.-W. (2010). Carbohydrate functionalized carbon nanotubes and their applications. *Chemical Society Reviews*, **39**, 2925–2934.
56. Carvalho, E. J. F., dos Santos, M. C. (2010). Role of surfactants in carbon nanotubes density gradient separation. *ACS Nano*, **4**, 765–770.
57. Quintilla, A., Hennrich, F., Lebedkin, S., Kappes, M. M., Wenzel, W. (2010). Influence of endohedral water on diameter sorting of single-walled carbon nanotubes by density gradient centrifugation. *Physical Chemistry Chemical Physics*, **12**, 902–908.
58. Lin, S., Blankschtein, D. (2010). Role of the bile salt surfactant sodium cholate in enhancing the aqueous dispersion stability of single-walled carbon nanotubes: A molecular dynamics simulation study. *Journal of Physical Chemistry B*, **114**, 15616–15625.
59. Xu, Z. J., Yang, X. N., Yang, Z. (2010). A molecular simulation probing of structure and interaction for supramolecular sodium dodecyl sulfate/single-wall carbon nanotube assemblies. *Nano Letters*, **10**, 985–991.
60. Russel, W. B., Saville, D. A., Schowalter, W. R. (1989). *Colloidal Dispersions*, Cambridge University Press, Cambridge.
61. Ju, S.-Y., Doll, J., Sharma, I., Papadimitrakopoulos, F. (2008). Selection of carbon nanotubes with specific chiralities using helical assemblies of flavin mononucleotide. *Nature Nanotechnology*, **3**, 356–362.
62. Tummala, N. R., Morrow, B. H., Resasco, D. E., Striolo, A. (2010). Stabilization of aqueous carbon nanotube dispersions using surfactants: Insights from molecular dynamics simulations. *ACS Nano*, **4**, 7193–7204.
63. O'Connell, M. J., Bachilo, S. M., Huffman, C. B., Moore, V. C., Strano, M. S., Haroz, E. H., Rialon, K. L., Boul, P. J., Noon, W. H., Kittrell, C., Ma, J. P., Hauge, R. H., Weisman, R. B., Smalley, R. E. (2002). Band gap fluorescence from individual single-walled carbon nanotubes. *Science*, **297**, 593–596.
64. Islam, M. F., Rojas, E., Bergey, D. M., Johnson, A. T., Yodh, A. G. (2003). High weight fraction surfactant solubilization of single-wall carbon nanotubes in water. *Nano Letters*, **3**, 269–273.
65. Richard, C., Balavoine, F., Schultz, P., Ebbesen, T. W., Mioskowski, C. (2003). Supramolecular self-assembly of lipid derivatives on carbon nanotubes. *Science*, **300**, 775–778.
66. Moore, V. C., Strano, M. S., Haroz, E. H., Hauge, R. H., Smalley, R. E., Schmidt, J., Talmon, Y. (2003). Individually suspended single-walled carbon nanotubes in various surfactants. *Nano Letters*, **3**, 1379–1382.
67. Yurekli, K., Mitchell, A., Krishnamoorti, R. (2004). Small-angle neutron scattering from surfactant-assisted aqueous dispersions of carbon nanotubes. *Journal of the American Chemical Society*, **126**, 9902–9903.
68. Calvaresi, M., Dallavalle, M., Zerbetto, F. (2009). Wrapping nanotubes with micelles, hemimicelles, and cylindrical micelles. *Small*, **5**, 2191–2198.
69. Wallace, E. J., Sansom, M. S. P. (2007). Carbon nanotube/detergent interactions via coarse-grained molecular dynamics. *Nano Letters*, **7**, 1923–1928.
70. Wallace E. J., Sansom, M. S. P. (2009). Carbon nanotube self-assembly with lipids and detergent: A molecular dynamics study. *Nanotechnology*, **20**, 045101.
71. Angelikopoulos, P., Gromov, A., Leen, A., Nerushev, O., Bock, H., Campbell, E. E. B. (2010). Dispersing individual single-wall carbon nanotubes in aqueous surfactant solutions below the cmc. *Journal of Physical Chemistry C*, **114**, 2–9.
72. Angelikopoulos, P., Bock, H. (2009). The differences in surfactant adsorption on carbon nanotubes and their bundles. *Langmuir*, **26**, 899–907.
73. Qiao, R., Ke, P. C. (2006). Lipid–carbon nanotube self-assembly in aqueous solution. *Journal of the American Chemical Society*, **128**, 13656–13657.
74. Tummala, N. R., Striolo, A. (2009). SDS surfactants on carbon nanotubes: Aggregate morphology. *ACS Nano*, **3**, 595–602.
75. Wang, H., Michielssens, S., Moors, S. L. C., Ceulemans, A. (2009). Molecular dynamics study of dipalmitoylphosphatidylcholine lipid layer self-assembly onto a single-walled carbon nanotube. *Nano Research*, **2**, 945–954.
76. Ke, P. C. (2007). Fiddling the string of carbon nanotubes with amphiphiles. *Physical Chemistry Chemical Physics*, **9**, 439–447.
77. Uddin, N. M., Capaldi, F., Farouk, B. (2010). *Journal of Engineering Materials Technology*, **132**, 021012.
78. Matarredona, O., Rhoads, H., Li, Z., Harwell, J. H., Balzano, L., Resasco, D. E. (2003). Dispersion of single-walled carbon nanotubes in aqueous solutions of the anionic surfactant NaDDB. *Journal of Physical Chemistry B*, **107**, 13357–13367.
79. Arnold, M. S., Green, A. A., Hulvat, J. F., Stupp, S. I., Hersam, M. C. (2006). Sorting carbon nanotubes by electronic structure using density differentiation. *Nature Nanotechnology*, **1**, 60–65.
80. Grossiord, N., Schoot, P. v. d., Meuldijk, J., Koning, C. E. (2007). Determination of the surface coverage of exfoliated carbon nanotubes by surfactant molecules in aqueous solution. *Langmuir*, **23**, 3646–3653.
81. Besterman, K.Eijk, K., Lemay, S. G. (2007). Charge inversion accompanies DNA condensation by multivalent ions. *Nature Physics*, **3**, 641–644.
82. Niyogi, S., Boukhalfa, S., Chikkannanavar, S. B., McDonald, T. J., Heben, M. J., Doorn, S. K. (2007). Selective aggregation of single-walled carbon nanotubes via salt addition. *Journal of the American Chemical Society*, **129**, 1898–1899.
83. Niyogi, S., Densmore, C. G., Doorn, S. K. (2009). Electrolyte tuning of surfactant interfacial behavior for enhanced density-based separations of single-walled carbon nanotubes. *Journal of the American Chemical Society*, **131**, 1144–1153.

15

INORGANIC CAPSULES: REDOX-ACTIVE GUESTS IN METAL CAGES

ANDREW MACDONELL AND LEROY CRONIN

15.1 INTRODUCTION

Inorganic capsules can be considered as both an offshoot of and an alternative to their organic counterparts. Whether in pure inorganic assemblies or combined with organic ligands, the versatile coordinations of the transition metals allow for a satisfying variety of architectures and the phenomenon of self-assembly allows complex structures to form in one-pot reactions with near-quantitative yields. Starting with a review of transition metals as components in capsule structures and looking at the development of this field, the focus of this chapter will then shift to polyoxometalates, inorganic molecules composed primarily of transition metals held together by oxygen bridges, giving an introduction to this unique chemical family followed by a discussion regarding how self-assembled architectures can provide an alternative form of encapsulation.

15.1.1 Transition Metals in Capsule Formation

A fundamental problem in the synthesis of organic capsules is the complexity of the target molecules. Due to the linear nature of hydrogen bonds, one of the principal intramolecular forces utilized in the design of organic capsules, the curvature of the capsule, has to be introduced elsewhere, which can demand lengthy and complex syntheses [1]. One means of achieving complex architectures without intensive synthetic effort is to make use of the varying coordination geometries of transition metals. By employing rigid polydentate ligands with more than one coordination site, it is possible to direct the self-assembly of complex three-dimensional polyhedral structures, in which the metals act as vertices and the linkers act as edges (with two coordination sites) or faces (with more than two) [2]. Some example structures are shown in Figure 15.1.

The structures produced by this method can be synthesized in one-pot reactions and at near-quantitative yield, although some synthetic effort may be involved in the synthesis of the organic ligands.

A comprehensive overview of transition metal capsules would require more space than this chapter is able to give. However, a general impression of the achievements of the field will be attempted. Two interesting results will be looked at: The first is a capsule functioning as an enzyme mimic which ultimately achieves enzyme-like rate acceleration, and the second is an investigation into the use of crystalline capsules as hosts for reactions.

15.1.2 Capsules for Catalysis

The work performed by the Raymond group (supramolecular chemistry) in conjunction with the Bergman group (catalysis) has provided a thorough exploration of a specific capsule system used to catalyze a number of simple organic reactions [3]. The capsule is shown on the left in Figure 15.1 with the formula $[Ga_4L_6]^{12-}$ (L = *N*,*N*′-bis(2,3-hydroxybenzoyl)-1,5-diaminonaphthalene). The naphthalene-based ligands form the edges of the tetrahedral structure, meeting at the four gallium-atom vertices, where three bidentate ligands occupy the octahedral coordination sphere of the gallium. These capsules form without a templating guest, are water-soluble and stable to guest exchange, and provide a hydrophobic cavity of up to 450 Å [3].

Organic Nanomaterials: Synthesis, Characterization, and Device Applications, First Edition. Edited by Tomás Torres and Giovanni Bottari.

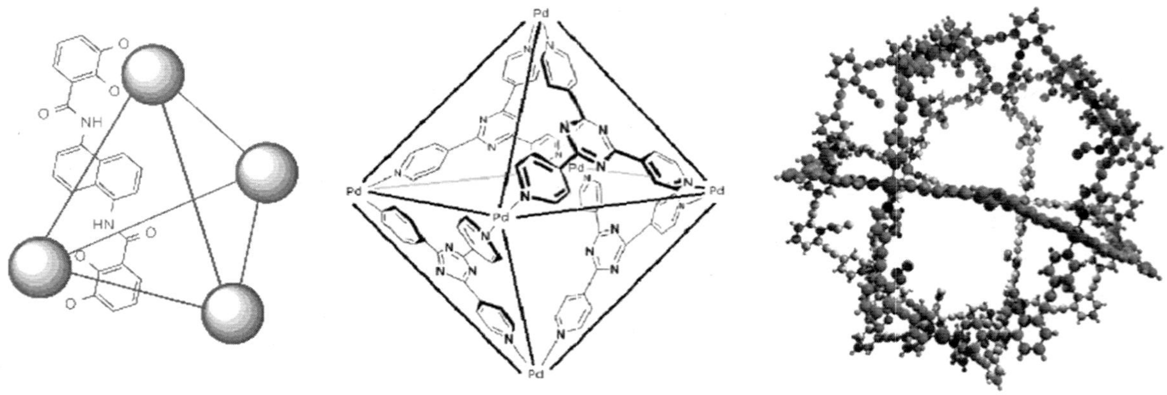

FIGURE 15.1. Three transition metal capsules. (**Left**) Gadolinium metal centers with a naphthalene-based linear ligand (Raymond). (**Center**) Palladium metal centers with triazole-based trigonal planar ligands (Fujita). (**Right**) Palladium centers with curved ligands bearing pendant glucose groups (Fujita).

The observation that water-reactive species could be stabilized within the hydrophobic cavity [4] prompted further research into encapsulating protonated guests. This revealed that protonated amines could be preserved inside the capsules even above pH values that would deprotonate a free amine, suggesting a strong stabilization [5]. It was reasoned that these capsules, following from the theory of transition state stabilization in enzyme catalysis [6], could catalyze reactions with protonated transition states in their rate-determining step.

Orthoformate and acetal hydrolysis, the mechanisms of which are shown in Figure 15.2 (both in solution and encapsulated within Ga_4L_6), were both investigated as model systems for this catalytic effect.

Both are acid-catalyzed systems and both result in a product with significantly different binding properties to the substrate, which decreases the likelihood of product inhibition and improves turnover. One of the more intuitive measures of catalytic activity is k_{cat}/k_{uncat}, which shows how many times faster the catalyzed reaction is compared to the uncatalyzed reaction under the same conditions. For the orthoformate reactions, this increase was between 150 and 3900 times, depending on the shape, size, and hydrophobicity of the compounds [7]. For the two acetal hydrolysis reactions, the increases were 190 and 980 [8]. While these results are a confirmation of the initial theory, the rate accelerations obtained are small compared to enzymatic catalysis [9]. However, using the same system to catalyze a different reaction, the Nazarov cyclization of pentadienols (shown in Figure 15.3 with the additional Diels–Alder step required to overcome the problem of product inhibition), two of the substrates studied showed rate accelerations on the order of 10^6 (2.1 and 1.7×10^6) [10]. This led the group to suggest that this combination of constrictive binding and functional group activation could be a general strategy for achieving enhanced reactivity.

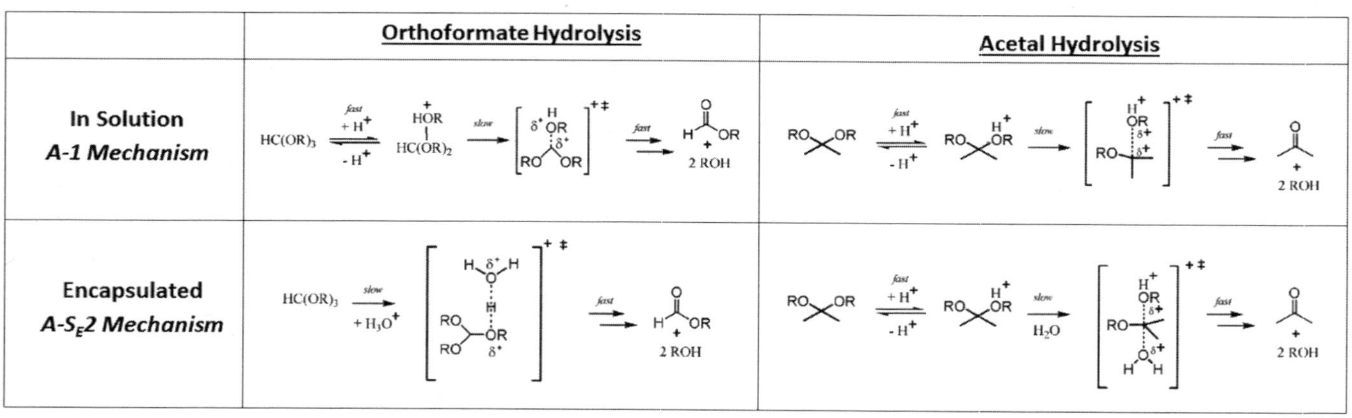

FIGURE 15.2. Scheme showing the different mechanisms of orthoformate and acetal hydrolysis both in solution without encapsulation and within the $[Ga_4L_6]^{12-}$ capsule.

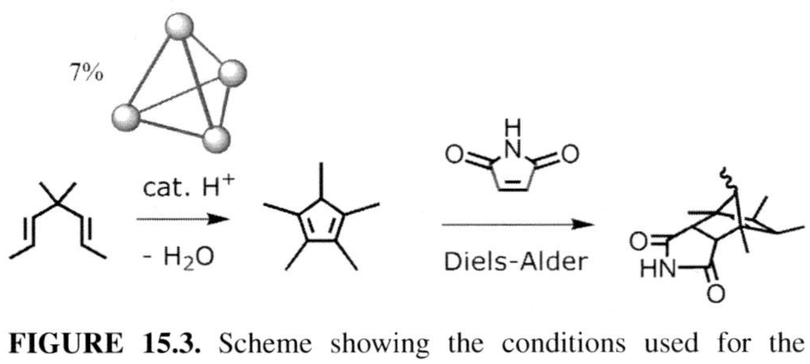

FIGURE 15.3. Scheme showing the conditions used for the Nazarov cyclization of Pentadienols, including the Diels–Alder reaction step required to avoid product inhibition.

15.1.3 Crystalline Capsules

A novel application of transition metal capsules has been pioneered by the Fujita group, in which they use crystallized molecular capsules to contain reactions, allowing the encapsulated guests to be observed via crystallography. If anything more than the slightest of chemical transformations are attempted on the component molecules of crystalline structures, the associated reorganization of the molecule will disrupt the regular repeating structure of the crystal, leading to an amorphous product [11]. However, if a molecular capsule forms the crystal, reactions can occur within their cavities without influencing the external crystal structure. Fujita's group, as a proof of concept, used an octahedral capsule with alternate closed and open faces formed from square-planar palladium (the vertices) and tris(4-pyridyl)triazine ligands (the faces) [12] as shown in the center of Figure 15.1. This was self-assembled and mixed in solution with acenaphthylene, forming a capsule–(acenapththylene)$_2$ complex from which crystals were formed by slow evaporation of the solvent [13]. While x-ray diffraction showed the acenaphthylene disordered over three positions within the cage, ultraviolet irradiation led to the [2 + 2] photodimerization, with the syn-product observed via x-ray diffraction [14]. A similar system, with encapsulated Cp'Mn(CO)$_3$ undergoing photodissociation of carbonyl ligands, was also investigated, showing that labile and otherwise non-isolable compounds could be clearly observed [15].

15.2 POLYOXOMETALATES

In addition to the numerous examples of organic–inorganic capsules available, there is also a family of entirely inorganic molecules which are capable of self-assembling into an astonishingly wide variety of clusters. This family, known as the polyoxometalates, contains a number of subsets capable of functioning as entirely inorganic capsule structures. Polyoxometalates, or POMs as they are commonly known, are discrete inorganic metal oxide molecules. Though their architectures and composition vary greatly, they share some distinguishing common features; they are all composed of high-oxidation-state, early-row transition metals (normally, though not exclusively, Mo and W) linked together by oxygen bridges to form polyhedral units in which the metal defines the center and the oxygen atoms define the vertices. These polyhedral units then link together, by sharing either one, two, or three oxygen vertices, referred to as corner-, edge- and face-sharing, respectively [16]. Depending on the nature of the metals and the reaction conditions, this linking of polyhedral units will lead to any one of a number of complex POM architectures based upon these basic building blocks. The resulting structures may have anywhere between 2 and 368 metal centers, with structures as diverse as wheels, crowns, spheres, and stars (see Figure 15.4) [17].

15.2.1 Synthesis and Assembly

POMs are normally assembled by acidifying a solution of the transition metal oxyanion, leading to the self-assembly of a variety of structures depending on specific pH, temperature, and the presence of other reagents. Somewhat surprisingly, despite the ubiquity of POMs in recent publications and the number of research groups regularly working with them, the actual formation mechanism has not been extensively researched. Initial theoretical speculation looking at tungsten POMs suggested that protonated WO_4 units would come together directly, followed by dehydration reactions to produce larger assemblies. This process was then elaborated to lead to the formation of larger POMs, despite limited

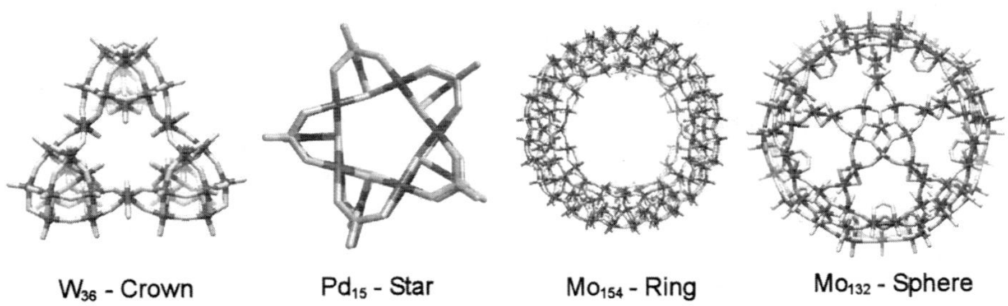

FIGURE 15.4. Crystal structures showing the variety of POM architectures, showing the metal, oxygen, and heteroatom components.

experimental evidence [18]. However, recent work employing DFT calculations and ESIMS experiments suggests that this theory does not fit as well as an alternative method, proposing a more complex mechanism of protonation and dimerisation, supported by MS identification of intermediate peaks [19].

In aqueous solution, POMs form at low pH and they require these conditions to remain stable; raising the pH will often cause the POM to degrade. However, raising the pH under strict conditions can cause the selective removal of metal centers, leading to so-called *lacunary* POMs. This *lacuna* can then be filled by a different transition metal, giving a mixed-addenda polyoxoanion, allowing new properties to be imbued and tuned. Lacunary structures can also serve as building blocks for larger POM structures.

15.2.2 Isopolyanions and Heteropolyanions

The many structures of polyoxometalates can be divided easily into two principal subgroups: isopolyanions, which are composed purely of metal and oxygen atoms, and heteropolyanions, which are composed of metal and oxygen atoms plus at least one heteroatom. Their formulas may be expressed as follows:

Isopolyanions (**IPA**s): $[M_nO_y]^{p-}$
Heteropolyanions (**HPA**s): $[X_aM_nO_y]^{q-}$, where $a \leq n$

The metal atoms, M, are considered to be peripheral to the heteroatoms and are referred to as *addenda* using Pope's terminology [16]. These addenda atoms are restricted to metals which can form strong d–π p–π bonds with oxygen, but the heteroatoms have no such restriction, with at least 65 elements represented. In the context of polyoxometalates, isopolyoxometalates (*iso*-POMs) and heteropolyoxometalates (*hetero*-POMs) are often used interchangeably with IPAs and HPAs.

The heteroatoms can be further divided into primary ("central") and secondary ("peripheral") heteroatoms. The primary heteroatoms are those which are crucial to the structure, normally (though not always) positioned in the center of the cluster, while the secondary heteroatoms may be removed from the structure to leave an independently stable HPA, an example of which would be the Cr(III) in $[SiW_{11}Cr(H_2O)O_{39}]^{5-}$ which can be removed to give $[Cr(H_2O)]^{3+}$ and $[SiW_{11}O_{39}]^{8-}$. The primary heteroatoms can direct both the chemistry and architecture of the clusters they form part of, with a general formula of $[XO_y]^{n-}$ but normally forming tetrahedral $[MO_4]^{n-}$ structures with M most commonly being Si, Ge, P, As or S. Metallic heteroatoms also occur, such as tetrahedral Co(III) or the $[XO_6]$ found at the center of the Anderson structure $[XM_6O_{24}]^{n-}$, where X = Mn(III), Te(VI), or Ni(II). The most common HPAs, however, are the Keggin and Wells–Dawson structures, which are of particular interest because the primary heteroatoms are entirely encapsulated within the metal cage.

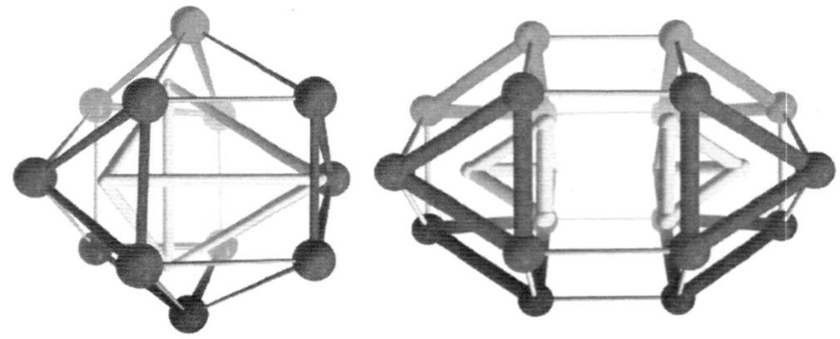

FIGURE 15.5. Schematic representation of the structures of the Keggin (**Left**) and Dawson (**right**) clusters. The triangles represent $[M_3O_9]$ units with the gray bonds representing their shared bridging oxygens. The tetrahedrons represent the heteroanions.

15.2.3 Keggin and Wells–Dawson Structures

The Keggin ($[XM_{12}O_{40}]^{n-}$ is one of the simplest heteropolyanionic structures, with a central heteroatom surrounded by 12 metal centers and 40 oxygen atoms (4 oxygens linking to the heteroatom, 24 bridging between metal centers, and 12 terminal oxo ligands on the metal centers). A simple means of understanding the structure is to consider the 12 metal centers making 4 triangles made of edge-sharing $[M_3O_9]$ units. The heteroatom and its four oxygens form a tetrahedron with a corner in the center of each of the four triangles. These triangles are then linked together at the corners to form a cage, as is shown on the left of Figure 15.5. The Keggin formula can be altered to better represent this structure as $[(XO_4)(M_3O_9)_4]^{n-}$ (all future Keggin–Dawson-based structures will provide this alternative structural formula). The five isomers of this structure are achieved by rotating each of these triangles by 45°. This increases edge-sharing in preference to corner sharing, bringing the metal centers closer together and increasing repulsion, making the isomers progressively less energetically favorable.

Slightly altering the reaction conditions for the Keggin anion can lead to the formation of the Wells–Dawson heteropolyanion ($[X_2M_{18}O_{62}]^{n-}$). Its 62 oxygen atoms can be divided into 8 bonding to the heteroatoms, 36 bridging between metal centers, and 18 terminal oxo ligands. Its structure is equivalent to two Keggin clusters, each with one $[M_3O_9]$ triangle unit removed, bound together by six equatorial oxo ligands at the position of the missing triangular units (see Figure 15.5). Similarly to the Keggin structure, its formula can be altered to $[\{(XO_4)(M_3O_9)_3\}_2]^{n-}$. It has six isomers and, for both the Keggin and Dawson structures, the combination of isomerization and lacunary structures can result in exposed oxo ligands attached to the heteroatom, able to coordinate directly to transition metals, which is of importance in catalysis. Both the Keggin and Dawson structures will be revisited in more detail.

15.2.4 Redox-Active Guests

For heteropolyanions possessing exposed heteroatoms, such as in the Anderson structure, these may play a direct role in the molecule's reactivity. However, for the Keggin–Dawson structure the heteroatoms are entirely surrounded by the metal atoms they bond to and they are often referred to as being "encapsulated" within the "cage" of the polyoxometalate framework. Though this prevents them from reacting directly with their chemical environment, there are other means by which the heteroatoms can be active in the chemical properties of the molecule.

One of the distinguishing characteristics of POMs is their ability to accept and donate electrons without significant change to their overall structure. Although POMs usually form with the addenda metal atoms in their highest oxidation state (d^1 or d^0), they are easily reduced, forming mixed-valence addenda species. Considering that these clusters are made up of many metal centers and that each metal center can potentially accept or donate multiple electrons, it is easy to imagine why POMs have been referred to as "electron reservoirs." If it were proven possible to link the redox potential of the cage with an encapsulated heteroatom functioning as a dopant, it could lead to molecules exhibiting properties not observed in their bulk analogues, which could form part of single-molecule electronic devices. This idea shall be returned to in the discussion of Wells–Dawson clusters.

15.2.5 Cation Exchange and Cation-Directed Synthesis

Since POMs are anionic, this charge must be balanced with cations. Normal synthetic conditions typically lead to small alkali metal cations or protons, but these can often be exchanged for bulkier organic equivalents. Cation exchange allows several properties of the POM to be altered or tuned, such as solubility or rate of crystallisation. Recently, however, it has been explored as a means of constructing new POM structures.

The popular "building-block" concept advocates the use of small POM units (building blocks), carefully chosen for their shape and charge, as a means of constructing large, complex clusters through controlled aggregation, as was demonstrated by Müller, who showed that, by varying reaction conditions and using building blocks such as the pentagonal $\{Mo(Mo)_5\}$ unit with suitable linkers, it was possible to construct huge mixed-valence clusters with varied structures, such as the spherical icosohedral $\{Mo_{132}\}$, big wheel $\{Mo_{154}/Mo_{176}\}$, capped cyclic $\{Mo_{248}\}$, and basket-shaped $\{Mo_{116}\}$ architectures [20]. In order to extend this concept,

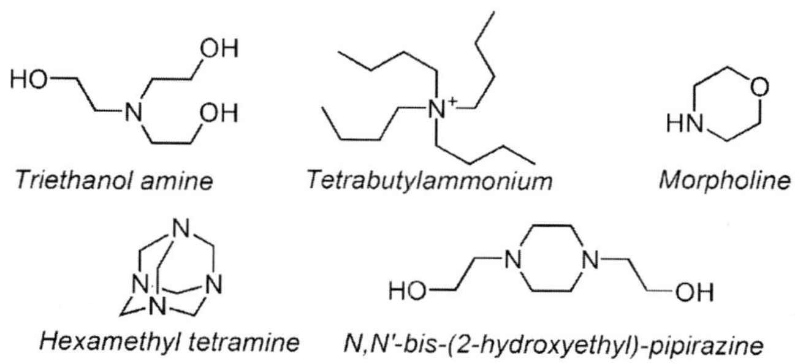

FIGURE 15.6. Some of the bulky organic cations used in the "shrink-wrapping" strategy.

a range of suitable low-symmetry clusters (i.e., with high negative charge and high nucleophilicity), such as Müller's pentagonal $\{Mo(Mo)_5\}$ unit, must be made available. However, the basic POM structures tend to be spherical or near-spherical (with high symmetry) in form due to the tendency of these highly charged, potent nucleophiles to aggregate into uniform, stable structures.

The Cronin group at Glasgow University, UK, has investigated a means of preventing this aggregation by a form of "reverse-templating" with bulky organic counterions [21]. It was found that bulky organic amine cations used in synthesis helped to isolate POM species in one-pot reactions, trapping and stabilizing them, thereby preventing their aggregation into more stable clusters and leading not only to novel compounds, but to entirely novel structures. In addition, they can also serve to direct the self-assembly of these building blocks into extended structures [22]. The amines used, shown in Figure 15.6, were primarily hexamethylene tetramine ($HMTA/HMTAH^+$), triethanol amine ($TEA/TEAH^+$), *N,N*-bis-(2-hydroxyethyl)-pipirazine ($BHEP/BHEPH^+$), and morpholine, capable of acting as encapsulating cations in the synthesis, but also functioning as buffers and even redox reagents in some cases. The technique is frequently referred to as "shrink-wrapping," due to the remarkably small distances between the POM surface and the cations, or "inverse templating," comparing it to the role of classical central templates, for example, in the formation of aluminosilicates or aluminophosphates [23].

This technique has led to a wealth of new discoveries. The first cluster to be isolated in this way using ($HMTAH^+$) was an isopolyoxomolybdate, with the formula $(C_6H_{14}N_4)_{10}[H_2Mo_{16}O_{52}]\cdot 34H_2O$. In contrast to other polyoxomolybdates of similar nuclearity, this structure is flat; four of the 12 Mo centers are one-electron reduced and the overall structure of the molecule has a central Mo_{12} cluster, with two Mo_2 "wings" extending on either side, as shown in Figure 15.7. The role of "shrink-wrapping" can be seen in the very close interactions between the cluster and the cations: The presence of 18 short hydrogen-bonded, cluster-surface, oxygen-to-cation interactions in the range of 2.581(7)

FIGURE 15.7. Crystal structures for each of the first three structures to come from the "shrink-wrap" technique. (**Left**) The $[H_2Mo_{16}O_{52}]^{10-}$ "bat-wing" cluster. (**Center**) The $[H_{12}W_{36}O_{120}]^{12-}$ "Celtic ring" cluster. (**Right**) The $[H_4W_{19}O_{62}]^{6+}$ "tungsten-centered Dawson" cluster.

$\leq d[E(\text{-H})\ldots O(MO)] \leq 3.140(5)$ Å shows that the cluster and cations are extensively linked via hydrogen bonds [24].

Applying this technique to tungsten systems using $(TEAH^+)$ led to the discovery of a new isopolyoxotungstate $[H_{12}W_{36}O_{120}]^{12-}$, $\{W_{36}\}$, composed of three $[H_4W_{11}O_{38}]^{6-}$ units linked together by three $[WO_6]$ bridges to form a ring structure which maps well onto the structure of an 18-crown-6 crown ether, as shown in Figure 15.7. Based on this, the possibility of the central cavity taking part in host–guest chemistry was investigated, resulting in the successful synthesis of several $\{W_{36}\text{-M}\}$ ($M = K^+, Rb^+, Cs^+, NH_4{}^+, Sr_2{}^+$ and Ba^{2+}) complexes. Further extension to a range of protonated aliphatic and aromatic guests showed not only that they were successfully bound in the cavity, but that the protruding components of these structures, the organic "tail", could direct the formation of the crystalline structures [25].

The next structure to be isolated is more relevant to the topic of inorganic capsules. While still being an isopolyoxometalate, the $(TEAH^+)_6[H_4W_{19}O_{62}]$, $\{W_{19}\}$ cluster has the external structure of a Dawson polyanion with a triangular-prismatic or octahedral (depending on the isomer) $[WO_6]^{6-}$ anion acting as the central template, replacing the normal two tetrahedral heteroanions (see Figure 15.7). The $[WO_6]^{6-}$ anion is positioned in the middle of the cluster, between two μ_3-oxo ligands; and between these oxo ligands and the central anion are two tetrahedral "voids" in the positions usually occupied by the two heteroanions [26].

15.3 THE WELLS–DAWSON CLUSTER $[X_2M_{18}O_{62}]^{n-}$

The general structure of the Wells–Dawson cluster has already been mentioned. However, there are several significant variations upon this structure, stemming principally from the encapsulated heteroatoms. In order to fully understand the novelty of these systems, the structure of the classic Dawson must first be revisited in more depth.

As has been mentioned, the Wells–Dawson cluster has the formula $[X_2M_{18}O_{62}]^{n-}$ or $[\{(XO_4)(M_3O_9)_3\}_2]^{n-}$, equivalent to two trilacunary Keggin structures joined symmetrically. The cage formed is generally oval in shape with terminal oxo ligands on the exterior of the cluster extending out from each of the metal centers. The internal structure relies upon the (XO_4) heteroanion template; one μ_4-oxo ligand coordinates to the three metal centers in the "cap" at the narrow end of the oval, while the remaining three μ_3-oxo ligands coordinate to two metal centers in the "belt" portion, at the middle of the oval. There are six possible isomers for the Dawson cage—the first three represent different rotations of the two caps:

α	Neither cap rotated	D_{3h} symmetry
β	One cap rotated	C_{3v} symmetry
γ	Both caps rotated	D_{3h} symmetry

For each of these, it is then possible to rotate the molecule around its equator, leading to:

α*	Neither cap rotated	D_{3h} symmetry
β*	One cap rotated	C_{3v} symmetry
γ*	Both caps rotated	D_{3h} symmetry

The α-isomer is the most common structure followed by the γ*, while β and γ have not yet been observed crystallographically (although their presence has been inferred using other spectroscopic methods [27]) and the α*-isomer has not yet been experimentally observed. In structures with missing or shifted heteroanions (which will be encountered in the following sections), additional oxygen atoms are found on the (W_3O_9) units, forming (W_3O_{10}) triangles, where the extra oxygen is located above the center of the triangle and coordinates to the three metal centers, replacing those usually provided by the heteroanion. These extra oxygens are often protonated.

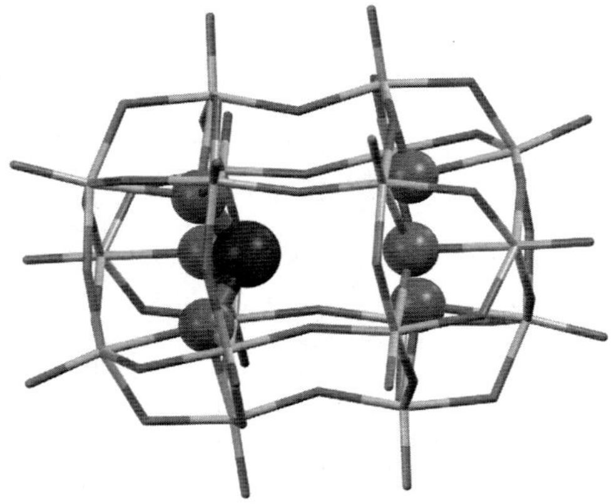

FIGURE 15.8. Crystal structure of the $[H_2AsW_{18}O_{60}]^{7-}$ cluster, **1**.

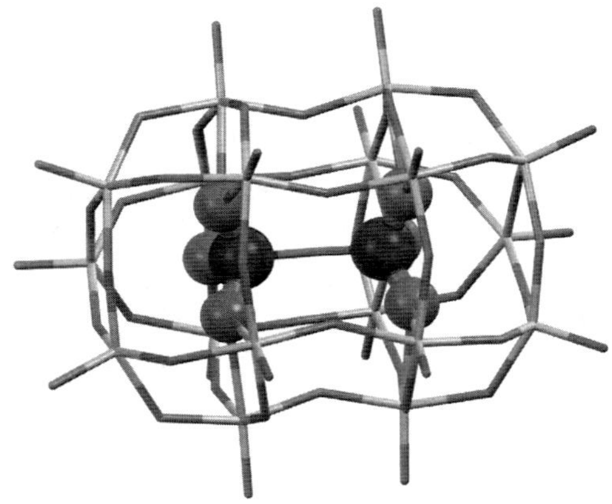

FIGURE 15.9. Crystal structure of the $[P_2Mo_{18}O_{61}]^{4-}$ cluster, **2**.

15.3.1 Single-Pyramidal Dawson $[H_x(XO_3)M_{18}O_{56}]^{n-}$

A significant difference from the oval-cage and tetrahedral-heteroanion form of the classic Dawson is found for structures incorporating pyramidal heteroanions. These include $[H_2AsW_{18}O_{60}]^{7-}$ (reformulated as $[\{(AsO_3)(W_3O_9)_3\}\{(H_2)(W_3O_{10})_3\}]^{7-}$) [28] (**1**) and $[H_3BiW_{18}O_{60}]^{7-}$ (reformulated as $[\{(BiO_3)(W_3O_9)_3\}\{(H_3)(W_3O_{10})_3\}]^{6-}$) [29] (**2**), both of which contain only one heteroanion, leaving the remaining void to be stabilized by protonated oxygens, in a fashion similar to that of the the Keggin-based metatungstate structure $[H_2W_{12}O_{40}]^{6-}$ ($[(H_2)(W_3O_{10})_4]^{6-}$). The As structure, which is representative of the monopyramidal structures, is depicted in Figure 15.8. Since the lone pairs of the pyramidal heteroatoms point into the center of the cluster, it has been suggested that the electronic repulsion this would generate, given the ionic radii of the heteroatoms, would make it impossible to fit two AsO_3 or BiO_3 units into the same cluster, providing a reason why they form these single-heteroatom species.

15.3.2 Pyrophosphate Dawson $[(P_2O_7)Mo_{18}O_{54}]^{4-}$

The next elaboration of the Dawson structure was synthesized in 1990 by Himeno *et al.* [30] and had its x-ray structure derived in 1994 by Kortz and Pope [31]; the $[P_2Mo_{18}O_{61}]^{4-}$ (or reformulated as $[(P_2O_7)\{(Mo_3O_9)_3\}_2]^{4-}$) (**3**) structure represented a new kind of Dawson structure, in which the two heteroanion positions are filled by a linked pyrophosphate group, as shown in Figure 15.9.

This was significant for a number of reasons: The internal structure of the Dawson is different, with each μ_4-oxo ligand of the pyrophosphate bridging one metal center from the cap and two from the belt; the bond lengths between the Mo atoms of the "caps" and the oxygens of the pyrophosphate groups are unusually long (2.64–2.72 Å), although this is also true (to a lesser extent) of the Keggin and Dawson structures with phosphate heteroanions (~2.4 Å), lending these compounds a more clathrate character; the corner-sharing in preference to edge-sharing which occurs in the "caps" of the pyrophosphate Dawson increases the distance between the Mo centers, which, given that the belt region has a similar size to the regular Dawson structure, gives the overall compound an hourglass or peanut-shaped structure; the pyrophosphate contained within the capsule was the first to be unambiguously assigned as having a linear, eclipsed P–O–P structure, all previous P–O–P structures being either nonlinear or having data which could be otherwise interpreted. Two reduced forms of the cluster were also investigated: the dark green one-electron-reduced compound and the dark blue two-electron-reduced compound, which showed mixed-valence electron mobility (a measure of the temperature at which an electron in a mixed-valent compound becomes "trapped" at a metal center) somewhere between the equivalently reduced phosphate Dawson cluster and phosphate Keggin cluster.

15.3.3 Double-Pyramidal Dawson $[(XO_3)_2M_{18}O_{54}]^{6-}$

A further alteration of the classic Dawson was achieved by the Cronin group in 2004, when, using the aforementioned "shrink-wrapping" technique, the two-electron-reduced $[(SO_3)_2Mo_{18}O_{54}]^{6-}$ (reformulated as $[\{(SO_3)(Mo_3O_9)_3\}_2]^{6-}$) (**4a**) was isolated using ($TEAH^+$) as the counterion [32]. This was the first example of the pyramidal SO_3 anion acting as a heteroanion in a Dawson structure and, while pyramidal structures have been incorporated into Dawson clusters before, this was also the first example of two nonlinked pyramidal heteroanions contained within a Dawson cluster, as can be seen in Figure 15.10.

Both the internal and external structure of the cage is similar to the pyrophosphate Dawson, although the lack of an oxygen bridge between the heteroatoms leads to a slightly tighter O_6 "belt" region. Two fully oxidized equivalents

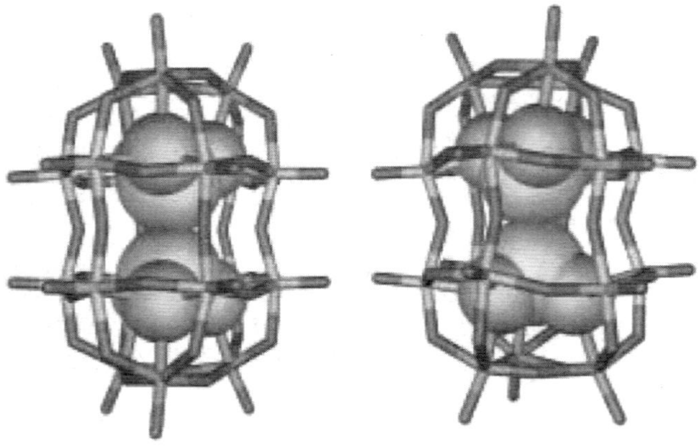

FIGURE 15.10. Crystal structure of the sulfite Dawson with the reduced $[\alpha\text{-}(SO_3)_2Mo_{18}O_{54}]^{6-}$ cluster **4a** (**left**) and the fully oxidized $[\beta\text{-}(SO_3)_2Mo_{18}O_{54}]^{4-}$ cluster **4c** (**right**).

were also synthesized, $[(SO_3)_2Mo_{18}O_{54}][Mo_6O_{19}](Bu_4N)_6$ (**4b**) and $[(SO_3)_2Mo_{18}O_{54}](C_2H_3N)(Bu_4N)_4$ (**4c**), in order to better understand the consequences of sulfite encapsulation. Compound **4b** crystallized with the $[Mo_6O_{19}]^{2-}$ Lindqvist anion, while compound **4c** was the β-isomer with a staggered arrangement of the two SO_3 groups (both compounds **4a** and **4b** were α-isomers, with an eclipsed arrangement). Both of the fully oxidized clusters were found to exhibit thermochromic behavior unprecedented for discrete POM clusters, changing from pale yellow at 77 K to deep red at 500 K in a fully reversible process (Figure 15.11). This observation, the ~8-nm shift in the HOMO–LUMO centered absorption band, was explained as a consequence of two factors: (a) a slight expansion of the POM framework with increasing temperature, which decreases the LUMO energy, and (b) a broadening of the HOMO energies due to the varying S . . . S distances at higher temperatures [33].

One of the reasons why the fully oxidized disulfite Dawson cluster was thought to be an interesting synthetic target was that the two sulfite clusters could potentially form a bond, leading to the formation of the dithionate $S_2O_6^{2-}$ anion, giving a structure similar to the pyrophosphate Dawson, but without the bridging oxygen atom between the heteroatoms (sulfur is the only main group element which can form an $X_2O_6^{n-}$ ion with an X–X single bond). The formation of the $S_2O_6^{2-}$ is an oxidative process, and it was hoped that stimulating this process could lead to the reduction of the surrounding Dawson cage (a system that was discussed in Section 15.2.4), giving a mixed-valent cage as is found in **4a**. Despite the S . . . S distances in **4a** (3.301(2) Å), **4b** (3.229(2) Å), and **4c** (3.271(5) Å) all being shorter than the sum of the van der Walls radii for two sulfur atoms (~3.6 Å), it appeared that the structural changes required to achieve the ~2.15-Å S–S single bond were too great, and this transformation was not observed in either solid-state or solution phase.

However, when molecules of **4c**, the staggered β-isomer, were isolated on a gold surface, the formation of the dithionate *was* observed [34]. By radically changing its environment, it became possible to convert the molecule between two distinct electronic states by raising and lowering the temperature. In order to prove the theory that the electrons reducing the metal cage were emanating from the sulfite/dithionate conversion (as opposed to some external source, such as the metal surface), a control cluster $[S_2Mo_{18}O_{62}]^{n-}$ (or $[\{(SO_4)(M_3O_9)_3\}_2]^{n-}$) (**5**) was used. This sulfate Dawson has the same charge as **4c**, its sulfite equivalent, and is actually easier to electrochemically reduce, but the sulfate groups are chemically inert and so cannot provide electrons to the metal cage. Hence, any reduction seen in the sulfite group which is not mirrored in the sulfate group would imply that the reducing electrons must come from the interior of the cluster. This was exactly what was observed when both clusters on a gold surface were analyzed with valence-level photoemission spectroscopy (for the valence-level electronic structure of the cluster layer) and core-level x-ray photoelectron spectroscopy (for the oxidation state of the Mo centers) at both 77 K and 298 K. Compound **5** did not show significant changes with the change in temperature, but the sulfite cluster **4c** showed changes consistent with the proposed reversible intramolecular redox process. The core-level spectrum for **4c** is shown in Figure 15.12, both at 77 K and 298 K, showing the evolution of a shoulder peak at 298 K indicating the presence of Mo^{V}. This shoulder was not observed for compound **5**.

In addition, Figure 15.13 shows the valence-level spectrum showing a band that can be attributed to the Mo $4d^1$ electron, consistent with the core-level spectrum, as well as bands associated with tetrahedral sulfur compounds which appear in the control at both temperatures, but only in the 298 K spectrum for compound **4c**, supporting the hypothesis that the formation of a tetrahedral dithionate-like anion causes the observed reduction of the cage.

In order to understand the effect that the gold surface plays in activating this reaction, density function theory (DFT)

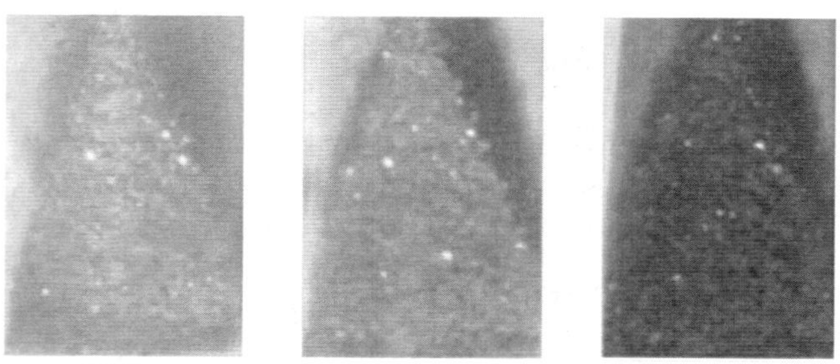

FIGURE 15.11. Images showing the changing temperature exhibited by compound **4b**. See color insert.

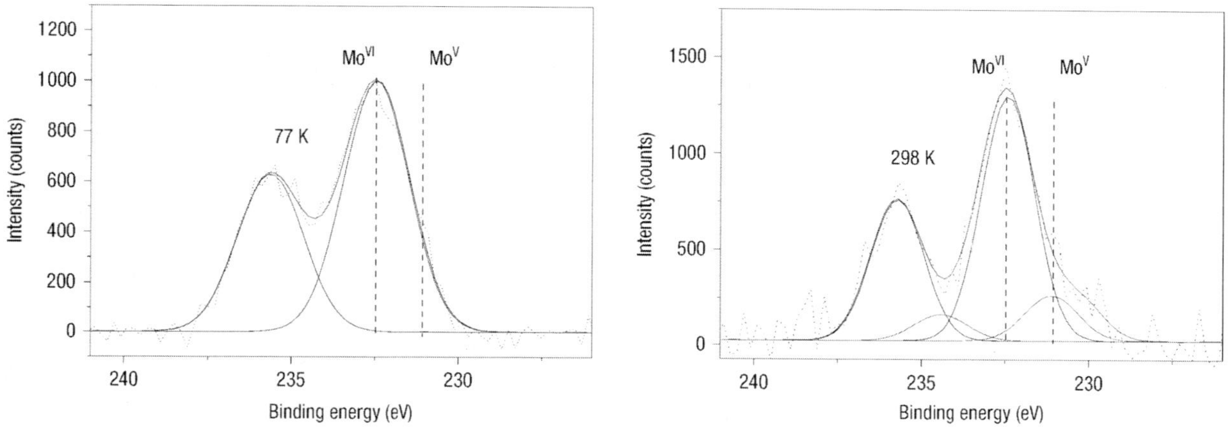

FIGURE 15.12. The Mo 3*d* Core-Level Spectra for sulfite Dawson compound **4c** at 77 K and 298 K. At 77 K the observed curve corresponds to the predicted peaks for Mo^{VI}, but at 298 K the presence of a shoulder peak at ~230 eV indicates the formation of Mo^{V}. The equivalent spectra for the sulfate Dawson do not show the development of this shoulder peak.

calculations were carried out to model the potential electronic effects underpinning the observed electronic changes. Adding a single reaction coordinate by shifting the two S centers of the sulfite Dawson by a displacement Δx (to simulate thermal activation) while shifting their oxo ligands by $\Delta x/2$ toward each other (along the main symmetry axis of the cluster), it was found that, in the gas phase, no intramolecular electron transfer nor formation of the S–S bond occurred. However, when the fluctuations in charge density (the *image charge*) induced in the polarizable Au surface by the cluster itself were modeled by four single positive charges forming a rectangle on one arbitrary side of the cluster, the same decrease in S–S distance *did* lead to the formation of a bond and the transfer of electrons to the cage.

Based on this reasoning, the same set of experiments, using both compound **5** and **4c**, were attempted using alternative surfaces: (a) an Au surface coated with a monolayer of

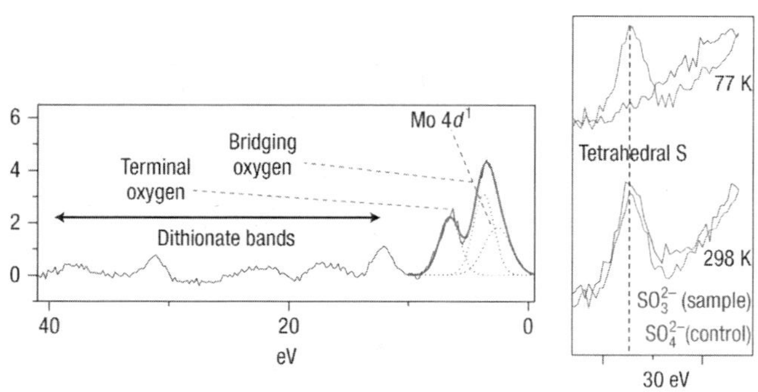

FIGURE 15.13. The valence level spectrum for compound **4c**, showing the presence of the Mo 4d^1 band (**left**) and the change from pyramidal sulfur to tetrahedral sulfur observed for the SO_3^{2-} sample at 298 K (**right**).

cystein and (b) a surface of highly oriented pyrolytic graphite (HOPG). The experiments showed no reaction for either molecule on the cystein-coated Au surface and no reaction for the control on the HOPG. However, compound **4c** on HOPG was found to be in its activated state even at room temperature, and the redox process was irreversible. These results show unambiguously that the surface effect is real: For the untreated gold, the image charges are located within the metal mirror images of their corresponding cluster charges (where the gold surface is the mirror plane), leading to an intermediate stabilization resulting in a reversible formation of the S–S bond with changing temperature; for the cystein-gold surface the image charges are weaker and further away due to the obstruction caused by the cystein molecules, leading to a surface-stabilization too weak to allow S–S bond formation; in contrast to the gold surfaces, however, HOPG localizes the image charges on the surface graphene sheet, bringing them closer to the cluster and increasing the local field felt by the internal S atoms. The effect is sufficiently strong that both Mo^{V} and metallic Mo^{0} species were observed by XPS studies at room temperature, a similar effect having been noted for the untreated gold surface when heated to decomposition at ~500 K. The control molecule did not experience any redox activity for any of the surface experiment, which leads to the remarkable conclusion that the HOPG surface was capable of inducing an internal rearrangement in a POM cluster without directly transferring charge to it (since, if this were the case, the more redox-active sulfate Dawson would also have been reduced).

After the synthesis of the molybdenum sulfite Dawson, attempts were made to create a tungsten analogue, leading to the formation of two cluster structures [35]: the isostructural equivalent to the molybdenum sulfite cluster, $[W^{VI}{}_{18}O_{54}(SO_3)_2]^{4-}$ (or $[\{(SO_3)(W_3O_9)_3\}_2]^{4-}$) (**6**), and a new

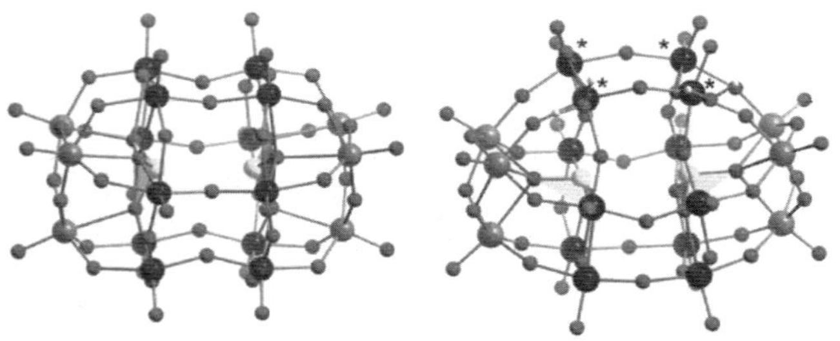

FIGURE 15.14. Crystal structures of the W sulfate clusters, with the $[W^{VI}{}_{18}O_{54}(SO_3)_2]^{4-}$ cluster **6** (**left**) and the $[W^{VI}{}_{18}O_{56}(SO_3)_2(H_2O)_2]^{8-}$ cluster **7** (**right**), with the metal centers not bound to the heteroanion indicated by a *.

structure, $[W^{VI}{}_{18}O_{56}(SO_3)_2(H_2O)_2]^{8-}$ (or $[\{(SO_3)(W_3O_9)_2(W_3O_{10}(H_2O))\}_2]^{8-}$) (**7**), shown in Figure 15.14.

Structure **7**, in addition to being the first closed *hetero*-POM found to include water ligands, showed some unique electronic properties. In a process similar to the Mo sulfite Dawsons on surfaces, the cluster undergoes an intramolecular redox reaction with the sulfite centers becoming oxidized and the surrounding cluster becoming reduced when the compound is heated to 400°C. However, in contrast to the dithionate formation experienced by the Mo sulfite Dawsons, this reaction involves the oxidation of the sulfite centers to sulfates, involving a rearrangement of the internal oxygens in the POM cage and the loss of the two water ligands, forming a $[W^{VI}{}_{14}W^{V}{}_4O_{54}(SO_4)_2]^{8-}$ (**8**) mixed-valence cluster. Although electron-transfer reactions and structural rearrangements are well known for *hetero*-POMs, this was the first example of a reaction combining the two and is also the first example of a fully characterized unimolecular reaction involving a *hetero*-POM.

Structure **7** is capable of this rearrangement due to its unique internal arrangement. Compared to structure **6**, isostructural to the Mo sulfite Dawson, in which the SO_3 groups are centrally aligned and each oxygen binds to three of the nine metal centers, in structure **7** the SO_3 groups are tilted so that, while one oxygen binds to three metal centers, the remaining two only bind two metal centers. This means that two of the nine metal centers in each half are not bound to the heteroanion, giving a total of four metal centers (4W*) not bound to heteroanions in the cluster which neighbor each other in an equatorial position. These four W* centers, combined with the tilted sulfite groups, reduce the overall symmetry of the cluster from D_{3h} (as is the case for structure **6**) to C_{2v}. The different orientation of the sulfite groups is also represented in the S... S distance of 3.61(2) Å, which is significantly longer than the 3.19(1) Å distance in structure **6**. The four W* metal centers in structure **7** each possess an additional terminal ligand (all metal centers having one terminal oxo ligand to begin with); two of these ligands are terminal oxo ligands (W* = O) while the other two are water ligands. This arrangement results in the bridging oxo ligands between the W* atoms of each W_9 half pointing inwards toward the vacant site of the sulfite groups.

All these factors result in the structure of **7** being essentially prearranged for internal reorganization with a concurrent redox reaction. When heated to 400°C the sulfite groups are oxidized to sulfate, acting as embedded reducing reagents and contributing up to four electrons to the W_{18} cage, leading to the formation of the intense blue mixed-valent cluster **8** with the loss of the two coordinated water ligands from the W* centers. Since the reduction of the shell is balanced by the oxidation of the heteroanions, the overall charge of the cluster remains constant at 8-.

15.3.4 Octahedral Dawson $[(XO_6)M_{18}O_{54}]^{6-}$

15.3.4.1 Octahedral Heteroatoms ($M_{18}O_{54}(IO_6)$, $M_{18}O_{54}(TeO_6)$). The discovery of the $\{W_{19}\}$ *iso*-POM (the Dawson-like structure with an octahedral or trigonal prismatic WO_6 template discussed in Section 15.2.5) suggested that it may be possible to form such a structure with an octahedral or trigonal prismatic heteroanion. This led to the synthesis of the $[H_3(IO_6)W_{18}O_{56}]^{6-}$ (or $[(IO_6)(W_3O_9)_4(W_3O_{10})_2(H_3)]^{6-}$) (**9**) cluster, shown in Figure 15.15, as a potassium (**9a**), tetrapropylammonium (TPA) (**9b**), and tetrabutylammonium (TBA) (**9c**) salt. This represents the first example of a Dawson-type cluster templated by an XO_6 heteroanion and the first crystallographically characterized tunstatoperiodate [36].

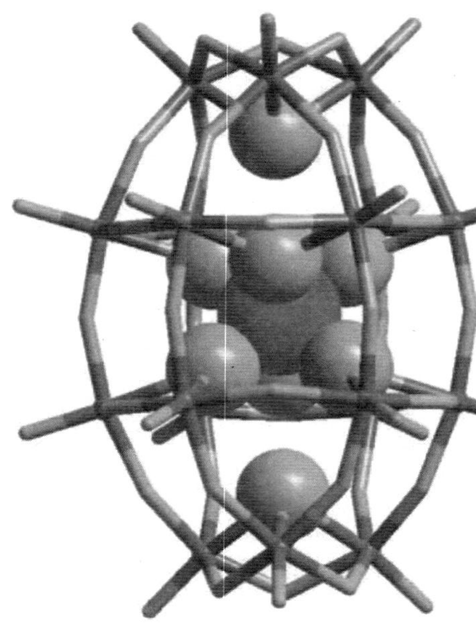

FIGURE 15.15. Crystal structure of the $[H_3(IO_6)W_{18}O_{56}]^{6-}$ cluster **9**.

Similar in structure to the $\{W_{19}\}$ cluster, the periodate cluster has two vacant positions where the two conventional heteroanions are usually found and the cap units, normally $[W_3O_9]$, both possess additional oxygen atoms usually provided by the heteroanion, giving $[W_3O_{10}]$. These extra oxygens are either singly or doubly protonated. The main difference between the $\{W_{19}\}$ structure and the periodate is the relative orientation of the cage and the heteroanion: The D_{3d} symmetry of the IO_6 anion does not match the D_{3h} symmetry of the cage, resulting in the overall cluster having a C_{3v} symmetry, representing the β^* isomer (the first β^* Dawson isomer to be identified crystallographically).

Since the only difference between the two clusters is the differing symmetry and the identity of the central heteroatom, it would be exceptionally difficult to confirm by crystallography alone that the supposed periodate cluster was not simply the $\{W_{19}\}$ cluster. In order to confirm the identity of the periodate cluster, high-resolution electrospray and cryospray mass spectrometry (MS) was performed on the **9b** salt and the **9a** salt after cation exchange to give the TPA/K salt, both in acetone, giving a series of peaks that could be attributed to the periodate cluster. This, combined with elemental analysis, confirms the presence of pure cluster **9** without $\{W_{19}\}$ impurities.

Syntheses were attempted for the tellurium analogue of the $[H_3(IO_6)W_{18}O_{56}]^{6-}$ cluster, but conventional synthetic methods did not yield the desired product. Based on the use of mass spectrometry to identify the cluster in the periodate synthesis, a similar procedure was performed on a variety of reaction mixtures, utilizing different cation systems, for the tellurium cluster, revealing peaks corresponding to a tellurium analogue of the periodate Dawson in the dimethylammonium ($DMAH^+$) and tetrabutylammonium (TBA) systems. Crystals were eventually obtained from both these reaction mixtures giving the $[H_3(Te^{VI}O_6)W_{18}O_{56}]^{7-}$ (**10**) cluster as its $DMAH^+$ (**10a**) and TBA (**10b**) salts. This was the first example of tellurium being incorporated into a Dawson-type structure [37].

The structure of cluster **10** is similar to the periodate cluster **9**, except for the symmetry of the two molecules. Cluster **10** adopts the more common γ^* conformation in contrast to cluster **9**'s far rarer β^* conformation. However, cluster **10** possesses the same voids found in cluster **9** at the sites where the two heteroanions are normally found in the classic Wells–Dawson structure and also shares the additional oxygens found on the caps. Cluster **10** shows some very unusual electronic activity. On exposure in acidified solution to the reducing agent $Na_2S_2O_4$, the usual intense blue colour (stemming from the reduction of W atoms in the cluster shell) was quickly replaced by a pale yellow one. Crystals obtained from this solution showed that a complete internal arrangement had occurred, with a marked shift in the Te position (1.10 Å toward the cap) and the conversion of the octahedral $Te^{VI}O_6$ into the pyramidal $Te^{IV}O_3$ heteroanion, accompanied

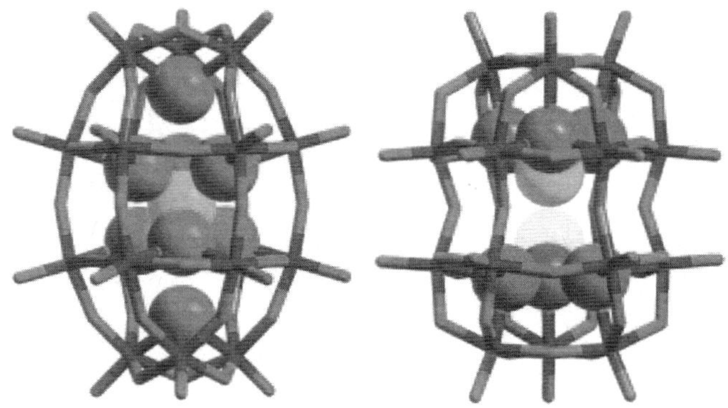

FIGURE 15.16. Crystal structures for the $[H_3(Te^{VI}O_6)W_{18}O_{56}]^{7-}$ cluster **10** (**left**) and the product of its reaction with $Na_2S_2O_4$, the $[H_3(Te^{IV}O_3)W_{18}O_{57}]^{5-}$ cluster **11** (**right**).

by the loss of both additional oxygens on the capping triads. This resulted in a final structure of $[H_3(Te^{IV}O_3)W_{18}O_{57}]^{5-}$ (**11**), with a fully oxidized cage, very similar in structure to its Sb analogue $[H_2(Sb^{III}O_3)W_{18}O_{57}]^{7-}$. Both structures can be seen in Figure 15.16.

Further investigations into this rearrangement/redox process showed that by adjusting the pH in solution of the compound to pH 2, nanosized crown-like tetrameric clusters could be formed, consisting of two **11** clusters linked by two $\{W_{11}\}$ units via adjacent distorted W centers on the **11** clusters. This shows that the presence of the pyramidal $Te^{IV}O_3$ heteroanion activates the cage surface to a sufficient extent that it forms nanoscale structures without the addition of other transition metal electrophiles.

15.4 THE KEGGIN CLUSTER

While the majority of this chapter is devoted to the Wells–Dawson cluster, since the presence of two internal heteroatoms presents more opportunities for engineering intramolecular reactions, the Keggin clusters have also been incorporated into systems in which its single encapsulated heteroatom plays an important part in the chemical properties of the materials.

15.4.1 The Keggin-Net

A great deal of interest has recently been generated in extended modular frameworks incorporating inorganic building blocks [38], such as metal–organic frameworks (MOFs) and coordination polymers [39], which allow for carefully directed assembly of porous structures, and zeolites [40], which offer robust structures with chemical functionality. An ideal system, which combined the controlled assembly of MOFs with the stability and functionality of zeolites and similar systems, would allow for the inclusion of specific

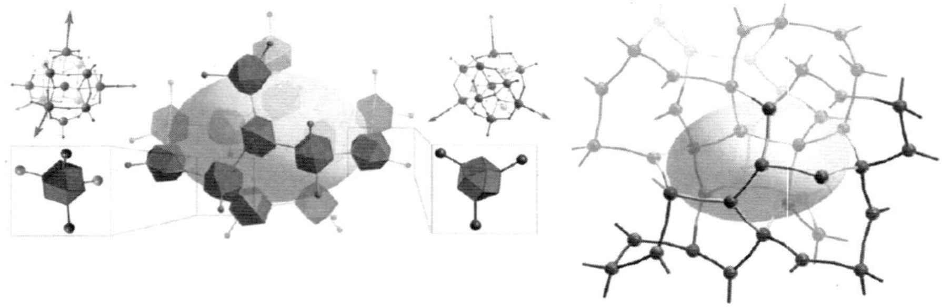

FIGURE 15.17. Two representations of the cavities found within the Keggin-net structure. The image on the left emphasizes how the alternating nodes connect around the cavity, while the image on the right emphasizes rings that make up the 3D structure.

guests at "active sites," thereby triggering chemical reactions (e.g., redox reactions) that could switch the nature of the framework. We have already seen that POM structures are highly redox active and that the incorporated heteroatoms in the Keggin and Dawson structures can contribute to this redox activity, so, if a redox active extended modular framework is the goal, one using *hetero*-POM clusters as building blocks would seem like a good synthetic target.

The Keggin-based framework $[(C_4H_{10}NO)_{40}(W_{72}Mn^{III}_{12}O_{268}Si_7)_n]$ (**12$_{ox}$**), or Keggin-Net, formed from lacunary Keggin clusters [γ-$SiW_{10}O_{36}$] and Mn^{II} in the presence of morpholinium cations and potassium permanganate, is an example of such a POM-based framework [41]. It is a material that can undergo reversible redox processes; the inclusion of a redox reagent couples with a controlled redox reaction in the framework itself, giving the reduced cluster **12$_{red}$**. This reduction occurs at the Mn^{III} heteroatoms, switching these to Mn^{II} and thereby retaining long-range order by cooperative structural changes within the W–O–Mn linkages that connect the Keggin units. Both **12$_{ox}$** and **12$_{red}$** can be dissolved in water and recrystallized to perfectly reform **12$_{ox}$** (with **12$_{red}$** being oxidized upon dissolution). The structure was the first to be formed purely of POM units, without the need for external linkers such as transition metal electrophiles. The lacunary Keggin clusters are linked directly by oxygen bridges between W and Mn atoms (W–O–Mn), and the overall structure is made up of alternating units of trigonal-planar three-linked (connected to three other Keggin units) and tetrahedral four-linked (connected to four other Keggin units) clusters cross-linked into an infinite 3D framework. These can be thought of as an equal distribution of trivacant $\{SiW_9O_{37}\}$ units and tetravacant (SiW_8O_{36}) units linked together by W–O–Mn bridges, which fix these units together.

The topology of the framework is identical to that of cubic germanium nitride, Ge_3N_4, in which the germanium is equivalent to the tetrahedral nodes and the nitrogen is equivalent to the trigonal nodes [42]. Both structures have puckered eight-membered ellipsoidal rings that are composed of four trigonal and four tetrahedral nodes. For Ge_3N_4 the dimensions of the ring are 2.97 Å × 4.61 Å; for the **12$_{ox}$** structure, the equivalent rings have dimensions of 9.45 Å × 12.93 Å, showing the effect of replacing atomic nodes in classic materials with nanosized clusters. Extrapolating the **12$_{ox}$** structure, it may be seen that four of these puckered rings form a larger 3D cavity, as can be seen in Figure 15.17. These cavities have nanoscale dimensions of 2.7 × 2.4 × 1.3 nm and accommodate both the charge balancing cations and a large number of solvent molecules.

Based on the assumption that it should be possible to alter the heteroatom of the [γ-$SiW_{10}O_{36}$] Keggin unit, a germanium-centered equivalent to the silicon-centerd Keggin-net was produced $[(C_4H_{10}NO)_{40}(W_{72}Mn^{III}_{12}O_{268}Ge_7)_n]$ (**13$_{ox}$**) [43]. This compound had reactivity very similar to that of compound **12**, with a slightly faster reduction time to **13$_{red}$** and a slower reoxidation time back to **13$_{ox}$**. The structural features are also almost identical, with the heteroatom-oxygen distance (1.6 Å for Si–O and 1.75 Å for Ge–O) being the only real distinction, leading to the difference in reactivity of the two frameworks. Adding another level of variability, further work resulted in the synthesis of Keggin-net structures using both Si and Ge heteroatoms, but with cobalt in place of manganese linking the lacunary units together, leading to $[W_{72}Co^{II}_{12}O_{268}Si_7]_n$ (**14$_{red}$**) and $[W_{72}Co^{II}_{12}O_{268}Ge_7]_n$ (**15$_{red}$**). In contrast to the Mn structures that crystallized in their 3+ oxidation state and could be reduced to 2+, the Co structures crystallize as 2+ and can be oxidized to 3+. Unlike their Mn counterparts, compounds **14** and **15** are not stable in their modified state and will, if left in solution, reduce back to their native 2+ state.

The availability of these four native structures and their four reduced/oxidized states led somewhat naturally to the idea of forming a framework alloy [44]: a structure based on two different frameworks, A and B, which share the same structure and could be combined to form a framework of AB units, also sharing the same structure. Since the difference in redox activity between the different heteroatoms is slight, the heterometals, Mn and Co, were selected for use, since their redox behavior is completely opposite, one undergoing oxidation while the other undergoes reduction. This was attempted by preparing reaction mixtures of

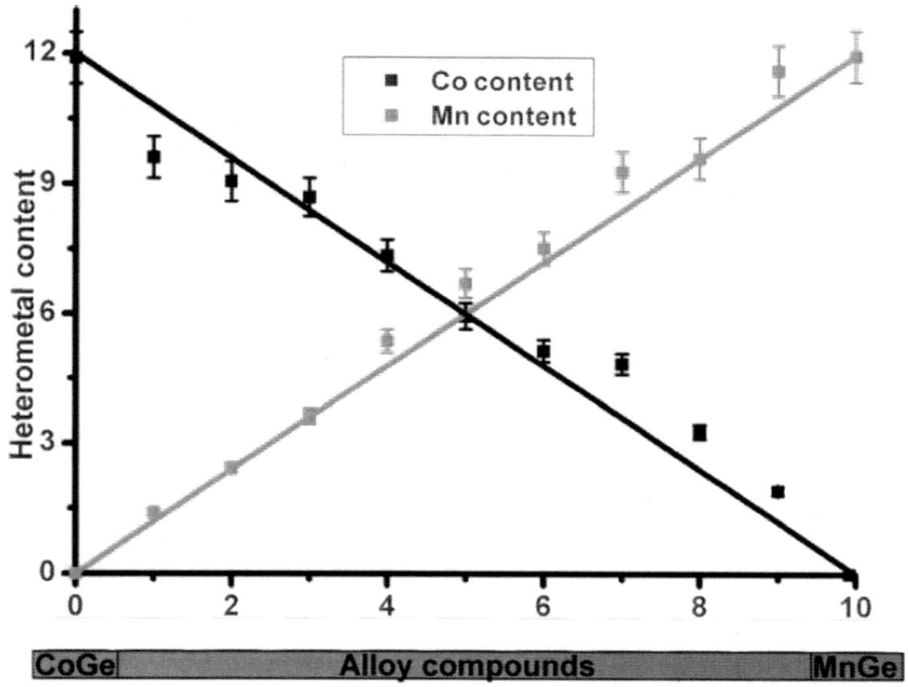

FIGURE 15.18. Graph showing the FAAS results for the Co/Mn Keggin-net alloy. It shows that the metal content of the molecular alloys varies linearly with the ratio of the reaction mixtures of **13** and **15**.

$[W_{72}Mn^{III}{}_{12}O_{268}Ge_7]_n$, **13**, and $[W_{72}Co^{II}{}_{12}O_{268}Ge_7]_n$ **14**, using Ge since it gave a far greater yield, and then combining these in varying ratios (9:1, 8:2, etc), resulting in a total of nine different framework alloy compounds. These could be visually identified since they showed a color variant going from the pure Mn^{III} cluster (purple) to the pure Co^{II} cluster (brown). A full spectrum of analysis was performed in order to confirm that the compounds were true alloys and not simply discrete mixtures of single crystals of pure compounds; in particular, flame atomic absorption spectroscopy (FAAS) was used, which confirmed the heterometal content to vary linearly for the 10% Mn/Co mixing steps, as shown in Figure 15.18. Investigations of the 5:5 alloy via cyclic voltammetry showed that the compound possessed some unexpected properties; while the second reduction wave lies at −0.97 V, the intermediate value of the two pure compounds, the first is at −0.893 V, far removed from the intermediate value of −0.76 V.

15.5 CONCLUSION

For readers who are unfamiliar with inorganic synthesis but who work extensively with organic capsules, hopefully this will have provided an insight into the possibilities presented by transition metals, both as structural components and as redox active materials. Much current work in POM synthesis focuses on the interface between polyoxometalates and organic structures, developing methods of grafting the two together to form hybrid organic-inorganic structures. The potential that we have seen of these electronically fascinating clusters can be both explored and channeled by extending them with the subtlety and finesse of organic synthesis.

REFERENCES

1. Vriezema, D. M., Comellas Aragonès, M., Elemans, J. A. A. W., Cornelissen, J. J. L. M., Rowan, A. E., Nolte, R. J. M. (2005). Self-assembled nanoreactors. *Chemical Reviews*, **105**(4), 1445–1490.
2. Leininger, S., Olenyuk, B., Stang, P. J. (2000). Self-assembly of discrete cyclic nanostructures mediated by transition metals. *Chemical Reviews*, **100**(3), 853–908.
3. (a) Caulder, D. L., Powers, R. E., Parac, T. N., Raymond, K. N. (1998). The self-assembly of a predesigned tetrahedral M_4L_6 supramolecular cluster. *Angewandte Chemie International Edition*, **37**(13–14), 1840–1843. (b) Davis, A. V., Fiedler, D., Ziegler, M., Terpin, A., Raymond, K. N. (2007). Resolution of chiral, tetrahedral M_4L_6 metal ligand hosts. *Journal of the American Chemical Society*, **129**(49), 15354–15363. (c) Davis, A. V., Raymond, K. N. (2005). The big squeeze: guest exchange in an M_4L_6 supramolecular host. *Journal of the American Chemical Society*, **127**(21), 7912–7919. (d) Pluth, M. D., Johnson, D. W., Szigethy, G. Z., Davis, A. V., Teat, S. J., Oliver, A. G., Bergman, R. G., Raymond, K. N. (2008). Structural consequences of anionic host–cationic guest interactions in a supramolecular assembly. *Inorganic Chemistry*, **48**(1), 111–120.
4. (a) Brumaghim, J. L., Michels, M., Pagliero, D., Raymond, Kenneth N. (2004). Encapsulation and stabilization of

reactive aromatic diazonium ions and the tropylium ion within a supramolecular host. *European Journal of Organic Chemistry*, **2004**(24), 5115–5118. (b) Brumaghim, J. L., Michels, M., Raymond, Kenneth N. (2004). Hydrophobic chemistry in aqueous solution: Stabilization and stereoselective encapsulation of phosphonium guests in a supramolecular host. *European Journal of Organic Chemistry*, **2004**(22), 4552–4559. (c) Dong, V. M., Fiedler, D., Carl, B., Bergman, R. G., Raymond, K. N. (2006). Molecular recognition and stabilization of iminium ions in water. *Journal of the American Chemical Society*, **128**(45), 14464–14465.

5. Pluth, M. D., Bergman, R. G., Raymond, K. N. (2007). Making amines strong bases: Thermodynamic stabilization of protonated guests in a highly-charged supramolecular host. *Journal of the American Chemical Society*, **129**(37), 11459–11467.
6. Kirby, A. J. (1994). Enzyme mimics. *Angewandte Chemie International Edition*, **33**(5), 551–553.
7. (a) Pluth, M. D., Bergman, R. G., Raymond, K. N. (2007). Acid catalysis in basic solution: A supramolecular host promotes orthoformate hydrolysis. *Science*, **316**(5821), 85–88. (b) Pluth, M. D., Bergman, R. G., Raymond, K. N. (2008). Supramolecular catalysis of orthoformate hydrolysis in basic solution. *Journal of the American Chemical Society*, **130**(34), 11423–11429.
8. (a) Pluth, M. D., Bergman, R. G., Raymond, K. N. (2007). Catalytic deprotection of acetals in basic solution with a self-assembled supramolecular "nanozyme". *Angewandte Chemie International Edition*, **46**(45), 8587–8589. (b) Pluth, M. D., Bergman, R. G., Raymond, K. N. (2008). The acid hydrolysis mechanism of acetals catalysed by a supramolecular assembly in basic solution. *The Journal of Organic Chemistry*, **74**(1), 58–63.
9. Radzicka, A., Wolfenden, R. (1995). A proficient enzyme. *Science*, **267**(5194), 90–93.
10. Hastings, C. J., Pluth, M. D., Bergman, R. G., Raymond, K. N. (2010). Enzyme-like catalysis of the nazarov cyclization by supramolecular encapsulation. *Journal of the American Chemical Society*, **132**(20), 6938–6940.
11. Kohlschütter, V., Tüscher, J. L. (1920). Zur Kenntnis topochemischer Reaktionen. Uber Bildung und Verhalten von Kupferhydroxyd. *Zeitschrift für anorganische und allgemeine Chemie*, **111**(1), 193–236.
12. Fujita, M., Oguro, D., Miyazawa, M., Oka, H., Yamaguchi, K., Ogura, K. (1995). Self-assembly of ten molecules into nanometer-sized organic host frameworks. *Nature*, **378**(6556), 469–471.
13. Yoshizawa, M., Takeyama, Y., Kusukawa, T., Fujita, M. (2002). Cavity-directed, highly stereoselective [2 + 2] photodimerization of olefins within self-assembled coordination cages. *Angewandte Chemie International Edition*, **41**(8), 1347–1349.
14. Takaoka, K., Kawano, M., Ozeki, T., Fujita, M. (2006). Crystallographic observation of an olefin photodimerization reaction that takes place via thermal tumbling within a self-assembled host. *Chemical Communications*, (15), 1625–1627.
15. Kawano, M., Kobayashi, Y., Ozeki, T., Fujita, M. (2006). Direct crystallographic observation of a coordinatively unsaturated transition-metal complex in situ generated within a self-assembled cage. *Journal of the American Chemical Society*, **128**(20), 6558–6559.
16. Pope, M. T. (1983). *Heteropoly and Isopoly Oxometalates*, Vol. 8. Springer-Verlag, Berlin.
17. (a) Miras, H. N., Cooper, G. J. T., Long, D.-L., Bogge, H., Müller, A., Streb, C., Cronin, L. (2010). Unveiling the transient template in the self-assembly of a molecular oxide nanowheel. *Science*, **327**(5961), 72–74. (b) Long, D.-L., Abbas, H., Kögerler, P., Cronin, L. (2004). A high-nuclearity "Celtic-ring" isopolyoxotungstate, $[H_{12}W_{36}O_{120}]^{12-}$, that captures trace potassium ions. *Journal of the American Chemical Society*, **126**(43), 13880–13881. (c) Müller, A., Kögerler, P., Dress, A. W. M. (2001). Giant metal-oxide-based spheres and their topology: From pentagonal building blocks to keplerates and unusual spin systems. *Coordination Chemical Reviews*, **222**, 193–218. (d) Xu, F., Scullion, R. A., Yan, J., Miras, H. N., Busche, C., Scandurra, A., Pignataro, B., Long, D.-L., Cronin, L. (2011). A supramolecular heteropolyoxopalladate $\{Pd_{15}\}$ cluster host encapsulating a $\{Pd_2\}$ dinuclear guest: $[Pd^{II}{}_2\{H_7Pd^{II}{}_{15}O_{10}(PO_4)_{10}\}]^{9-}$. *Journal of the American Chemical Society*, **133**(13), 4684–4686.
18. (a) Kepert, D. L. (2007). *Isopolytungstates in Progress in Inorganic Chemistry*. John Wiley & Sons, Hoboken, NJ, pp. 199–274. (b) Tytko, K.-H., Glemser, O. (1976) Isopolymolybdates and isopolytungstates. In: Emeléus, H. J., Sharpe, A. G., eds. *Advances in Inorganic Chemistry*, Academic Press, Vol. 19, pp. 239–315.
19. Vilà-Nadal, L., Rodriguez-Fortea, A., Yan, L. K., Wilson, E. F., Cronin, L., Poblet, J. M. (2009). Nucleation mechanisms of molecular oxides: A study of the assembly-dissassembly of $[W_6O_{19}]^{2-}$ by theory and mass spectrometry. *Angewandte Chemie International Edition*, **48**(30), 5452–5456.
20. (a) Müller, A., Kögerler, P., Kuhlmann, C. (1999). A variety of combinatorially linkable units as disposition: From a giant icosahedral keplerate to multi-oxide based network structures. *Chemical Communications*, (15), 1347–1358. (b) Cronin, L., Kögerler, P., Müller, A. (2000). Controlling growth of novel solid-state materials via discreet molybdenum-oxide-based building blocks as synthons. *Journal of Solid State Chemistry*, **152**(1), 57–67. (c) Cronin, L., Beugholt, C., Krickemeyer, E., Schmidtmann, M., Bögge, H., Kögerler, P., Luong, T. K. K., Müller, A. (2002). "Molecular symmetry breakers" generating metal-oxide-based nano-object fragments as synthons for complex structures: $[\{Mo_{128}Eu_4O_{388}H_{10}(H_2O)_{81}\}_2]^{20-}$], a giant-cluster dimer. *Angewandte Chemie International Edition*, **41**(15), 2805–2808.
21. Song, Y.-F., Long, D.-L., Ritchie, C., Cronin, L. (2011). Nanoscale polyoxometalate-based inorganic/organic hybrids. *The Chemical Record*, **11**(3), 158–171.
22. Abbas, H., Pickering, A. L., Long, D.-L., Kögerler, P., Cronin, L. (2005). Controllable growth of chains and grids from polyoxomolybdate building blocks linked by silver(I) dimers. *Chemistry—A European Journal*, **11**(4), 1071–1078.
23. Lewis, D. W., Willock, D. J., Catlow, C. R. A., Thomas, J. M., Hutchings, G. J. (1996). De novo design of

structure-directing agents for the synthesis of microporous solids. *Nature*, **382**(6592), 604–606.

24. Ritchie, C., Burkholder, E. M., Long, D.-L., Adam, D., Kögerler, P., Cronin, L. (2007). Exploiting the multifunctionality of organocations in the assembly of hybrid polyoxometalate clusters and networks. *Chemical Communications*, (5), 468–470.
25. (a) Long, D.-L., Abbas, H., Kögerler, P., Cronin, L. (2004). A high-nuclearity "Celtic-ring" isopolyoxotungstate, $[H_{12}W_{36}O_{120}]^{12-}$, that captures trace potassium ions. *Journal of the American Chemical Society*, **126**(43), 13880–13881. (b) Long, D.-L., Brücher, O., Streb, C., Cronin, L. (2006). Inorganic crown: The host–guest chemistry of a high nuclearity "celtic-ring" isopolyoxotungstate $[H_{12}W_{36}O_{120}]^{12-}$. *Dalton Transactions*, (23), 2852–2860. (c) Streb, C., McGlone, T., Brücher, O., Long, D.-L., Cronin, L. (2008). Hybrid host–guest complexes: Directing the supramolecular structure through secondary host-guest interactions. *Chemistry—A European Journal*, **14**(29), 8861–8868. (d) McGlone, T., Streb, C., Long, D.-L., Cronin, L. (2009). Guest-directed supramolecular architectures of $\{W_{36}\}$ polyoxometalate crowns. *Chemistry—An Asian Journal*, **4**(10), 1612–1618.
26. Long, D.-L., Kögerler, P., Parenty, A. D. C., Fielden, J., Cronin, L. (2006). Discovery of a family of isopolyoxotungstates $[H_4W_{19}O_{62}]^{6-}$: Encapsulating a $\{WO_6\}$ moiety within a $\{W_{18}\}$ Dawson-like cluster cage. *Angewandte Chemie International Edition*, **45**(29), 4798–4803.
27. Contant, R., Thouvenot, R. (1993). A reinvestigation of isomerism in the dawson structure: Syntheses and ^{183}W NMR structural characterisation of three new polyoxotungstates $[X_2W_{18}O_{62}]^{6-}$ (X = PV, AsV). *Inorganica Chimica Acta*, **212**(1–2), 41–50.
28. Jeannin, Y., Martin-Frere, J. (1979). X-ray study of $(NH_4)_7[H_2AsW_{18}O_{60}]\cdot 16H_2O$: First example of a heteropolyanion containing protons and arsenic(III). *Inorganic Chemistry*, **18**(11), 3010–3014.
29. Ozawa, Y., Sasaki, Y. (1987). Synthesis and crystal structure of $[(CH_3)_4N]_6[H_3BiW_{18}O_{60}]$. *Chemical Letters*, 923–926.
30. Himeno, S., Saito, A., Hori, T. (1990). Synthesis and characterization of 18-molybdopyrophosphate complex. *Bulletin of the Chemical Society of Japan*, **63**, 1602.
31. Kortz, U., Pope, M. T. (1994). Polyoxometalate–diphosphate complexes. 2. Structure of 18-molybdopyrophosphate, $[(P_2O_7)Mo_{18}O_{54}]^{4-}$, which encloses a linear, eclipsed conformation of the pyrophosphate anions, and preliminary characterization of Its one- and two-electron heteropoly blues. *Inorganic Chemistry*, **33**(25), 5643–5646.
32. Long, D.-L., Kögerler, P., Cronin, L. (2004). Old clusters with new tricks: Engineering S–S interactions and novel physical properties in sulfite-based dawson clusters. *Angewandte Chemie International Edition*, **43**(14), 1817–1820.
33. Tsunashima, R., Long, D.-L., Endo, T., Noro, S.-I., Akutagawa, T., Nakamura, T., Cabrera, R. Q., McMillan, P. F., Kögerler, P., Cronin, L. (2011). Exploring the termochromism of sulfite-embedded polyoxometalate capsules. *Physical Chemistry Chemical Physics*, **13**(16), 7295–7297.
34. Fleming, C., Long, D. L., Mcmillan, N., Johnston, J., Bovet, N., Dhanak, V., Gadegaard, N., Kögerler, P., Cronin, L., Kadodwala, M. (2008). Reversible electron-transfer reactions within a nanoscale metal-oxide cage mediated by metallic substrates. *Nature Nanotechnology*, **3**(4), 229–233.
35. (a) Long, D.-L., Abbas, H., Kögerler, P., Cronin, L. (2005). Confined electron-transfer reactions within a molecular metal-oxide "trojan horse". *Angewandte Chemie International Edition*, **44**(22), 3415–3419. (b) Cabrera, R. Q., Long, D.-L., Cronin, L., McMillan, P. F. (2010). In situ investigations of the polyoxometalate trojan horse compound $K_7Na[(W^{VI}{}_{18}O_{56}(SO_3)_2(H_2O)_2]\cdot 20H_2O$ under high temperature and high pressure conditions. *CrystEngComm*, **12**(9), 2568–2572.
36. Long, D.-L., Song, Y. F., Wilson, E. F., Kögerler, P., Guo, S. X., Bond, A. M., Hargreaves, J. S. J., Cronin, L. (2008). Capture of periodate in a $\{W_{18}O_{54}\}$ cluster cage yielding a catalytically active polyoxometalate $[H_3W_{18}O_{56}(IO_6)]^{(6-)}$ embedded with high-valent iodine. *Angewandte Chemie International Edition*, **47**(23), 4384–4387.
37. Yan, J., Long, D.-L., Wilson, E. F., Cronin, L. (2009). Discovery of heteroatom-"embedded" Te subset of $\{W_{18}O_{54}\}$ nanofunctional polyoxometalates by use of cryospray mass spectrometry. *Angewandte Chemie International Edition*, **48**(24), 4376–4380.
38. Chae, H. K., Siberio-Pérez, D. Y., Kim, J., Go, Y., Eddaoudi, M., Matzger, A. J., O'Keeffe, M., Yaghi, O. M. (2004). A route to high surface area, porostiy and inclusion of large molecules in crystals. *Nature*, **427**(6974), 523–527.
39. (a) Serre, C., Mellot-Draznieks, C., Surblé, S., Audebrand, N., Filinchuk, Y., Férey, G. (2007). Role of solvent–host interactions that lead to very large swelling of hybrid frameworks. *Science*, **315**(5820), 1828–1831. (b) Hasegawa, S., Horike, S., Matsuda, R., Furukawa, S., Mochizuki, K., Kinoshita, Y., Kitagawa, S. (2007). Three-dimensional porous coordination polymer functionalized with amide groups based on tridentate ligand: Selective sorption and catalysis. *Journal of the American Chemical Society*, **129**(9), 2607–2614.
40. Yuan, H.-M., Chen, J.-S., Zhu, G.-S., Li, J.-Y., Yu, J.-H., Yang, G.-D., Xu, R.-R. (2000). The first organo-templated cobalt phosphate with a zeolite topology. *Inorganic Chemistry*, **39**(7), 1476–1479.
41. Ritchie, C. I., Streb, C., Thiel, J., Mitchell, S. G., Miras, H. N., Long, D.-L., Boyd, T., Peacock, R. D., McGlone, T., Cronin, L. (2008). Reversible redox reactions in an extended polyoxometalate framework solid. *Angewandte Chemie International Edition*, **47**(36), 6881–6884.
42. Kroll, P. (2003) *J. SolidState Chem.*, **176**, 530–537.
43. Thiel, J., Ritchie, C., Streb, C., Long, D.-L., Cronin, L. (2009). Heteroatom-controlled kinetics of switchable polyoxometalate frameworks. *Journal of the American Chemical Society*, **131**(12), 4180–4181.
44. Thiel, J., Ritchie, C., Miras, H. N., Streb, C., Mitchell, S. G., Boyd, T., Ochoa, M. N. C., Rosnes, M. H., McIvet, J., Long, D.-L., Cronin, L. (2010). *Angewandte Chemie International Edition*, **49**(39), 6984–6988.

16

STIMULI-RESPONSIVE MONOLAYERS

Francesca A. Scaramuzzo, Mario Barteri, Pascal Jonkheijm, and Jurriaan Huskens

16.1 INTRODUCTION

The use of self-assembled monolayers (SAMs) for the functionalization of surfaces is extremely widespread among the scientific community, because they are easy to prepare and thermodynamically stable [1].

In general the concept of self-assembly concerns the spontaneous formation of complex hierarchical structures from defined starting building blocks. Such a phenomenon is quite common in nature, as seen, for example, in the formation of Fe–S clusters, membranes from lipid molecules or phages [2]. Even though different kinds of organic films have been observed and described for more than 200 years [3–6], a pioneering paper on the preparation of mono-molecular surfactant films on metal surfaces directly from solution was published in 1946 by Zisman and co-workers [1]. Actually, SAMs are defined as ordered molecular assemblies formed by the adsorption of a one-molecule-thick layer on a surface. In the last 30 years, various types of monolayers such as fatty acids on metal oxides [7], silanes on oxides [8–10], and sulfur-containing molecules (alkanethiols, dialkyl disulfides and dialkyl sulfides) on gold [11–14] have been described. In particular, the last two mentioned are the two families of SAMs which have been studied most extensively, although only monolayers on gold can be strictly considered to be self-assembled structures. As far as the formation of a stable monolayer on Au is concerned, it has been shown that the adsorption process follows a Langmuir adsorption isotherm. For the enthalpy of adsorption, it is necessary to take into account the binding force between the anchoring sulfur group and gold (the binding energy of a S–Au bond is about 45 kcal/mol) [15], as well as the interchain interactions (van der Waals interactions are on the order of a few kcal/mol per CH_2) [16]. The entropy plays an important role as well: For adsorption of 1-octadecanethiol on gold at 293 K, for example, it has been demonstrated that $\Delta S = -48 \pm 1$ cal $mol^{-1}K^{-1}$. This means that the entropy of adsorption is about four times larger than for a typical liquid-to-solid phase transition [17]. Electrochemical studies have demonstrated that alkanethiols adsorb as alkanethiolates [16], and recently it has been shown that, on Au nanoparticles, the process leads to the formation of H_2 [18]. Moreover, dialkyl disulfides form films that reveal the same structural characteristics as the corresponding alkanethiol monolayers [19]. For this reason, despite the fact that the literature is not totally coherent, it was concluded that during SAM formation the S–S bond of dialkyl disulfides breaks [20–22], leading to thiolate monolayers similarly to the adsorption of thiols. In comparison to alkanethiols and dialkyl disulfides, thioethers have been less studied [23]. However, both infrared [24] and electrochemical [25] data suggested a relatively less ordered packing compared to alkanethiols. Moreover, since their interaction with gold is weaker, they lead to films of inferior stability [26]. Inferior binding uniformity, packing density, and molecular alignment have recently been observed also for SAMs made of thioether-based tripodal ligands with respect to the corresponding SAMs made of thiol-based tripodal ligands [27].

By changing the chemical structure of the adsorbate, any functional group can be introduced as a tail group, thus yielding SAMs with a huge variety of molecular properties (e.g., reactivity, wettability, optical absorption) [28]. This is undoubtedly interesting both for nanotechnology applications and for fundamental studies of interfacial reactions. Moreover, in recent years, there is a growing interest in so-called smart SAMs, which give rise to responsive

Organic Nanomaterials: Synthesis, Characterization, and Device Applications, First Edition. Edited by Tomás Torres and Giovanni Bottari.

surfaces or, in other words, surfaces able to dynamically change their properties in response to external stimuli [29], which can be energy based (light, temperature, ultrasonic radiation, electric and magnetic fields) or chemically based (pH, ionic strength, solvents, metal ions, chemical and biochemical analytes, biomolecules) [30]. Nowadays there are many available techniques to characterize SAMs, which are based on different physical principles and stimulation; however, it is important to have a clear readout using different methods, since this allows us to have more information, often complementary to each other. In subsequent steps, having a SAM that is sensitive to external stimuli would be highly desirable for the interaction with other molecules (e.g., reaction monitoring), in order to reach high levels of control over adsorption processes, interaction, and reactivity on surfaces.

Ideally, a smart SAM would (i) give a clear readout dependent on surface coverage and molecular conformation on the surface, (ii) allow to monitor (post)immobilization of (bio)molecules, and (iii) allow to reversibly switch the assembly/disassembly process of the molecules on the surface. This chapter presents an overview of the literature on smart monolayers which respond to different stimuli both for their characterization and for their interaction with other molecules. We would like to focus on both interactions with different organic molecules and biomolecules. Particular emphasis is given to applications regarding cell-surface interactions, because they are most interesting for possible use in bionanotechnology.

The majority of responsive layers described in literature are sensitive to one specific stimulus. Upon applying the stimulus, a change generally detectable with a combination of conventional techniques (contact angle, UV–vis spectrometry, electrochemistry) occurs in the surface properties. In the coming sections, these monolayers are grouped and described on the basis of the stimulus they are responsive to. Thereafter, a few cases with systems that respond to multiple stimuli are covered.

16.2 LIGHT-RESPONSIVE MONOLAYERS

Light-responsive layers can be profitably used to design optical memories and switching devices. Initially they were fabricated and investigated at the air–water interface [31,32], but there are also some examples of (a) electrochemical transduction of optical signals rising from the photostimulation of SAMs on gold [33,34] and (b) correlation of electrochemical signals to the optical properties of the monolayers [35]. In the last decade, the interest for photoswitchable SAMs has grown and have been deeply investigated for applications of potential interest in nanotechnologies. The influence of light on a monolayer can be exploited for (i) photosensing, (ii) photoswitching, or (iii) photoreactions.

16.2.1 Photosensing

Fluorescence is considered quite a desirable property in a monolayer, because it allows a rapid and sensitive characterization, with minimal impact on the sample [36].

In biosensing, the use of fluorescent layers to monitor enzyme activity is not new: in the early 1990s there were examples of membranes decorated with fluorescent molecules and associated to enzymes, whose activity determined the formation of quenchers of the membrane fluorophore [37]. Focusing more strictly on SAMs, mixed monolayers were fabricated using biotin-containing and fluorescent BODIPY-containing molecules [38], which were anchored on amino-terminated alkyl chains on quartz. When the biotin on the surface was put in contact with streptavidin and an anti-biotin protein, the interaction with these macromolecules quenched the fluorescence of BODIPY nearby.

More recently, dansyl chromophores have been immobilized on an epoxy-terminated self-assembled monolayer on glass slides. The surface thus obtained was suitable for the detection of nitroaromatics in aqueous solution, because of the formation of a charge-transfer complex between electron-poor nitroderivatives and electron-rich dansyl resulting in fluorescence quenching. Such a sensor proved to be highly sensitive for nitrobenzene more than for other compounds of the same class [39].

Even though in most cases fluorescence is exploited for (bio)sensing, it can also be used to detect the presence and the reactivity of a monolayer, whose emissive properties usually change after interaction with different analytes. A fluorophore suitable for both purposes is coumarin. On surfaces, for example, a coumarin coupled to a peptide has been used to design a light-sensing microarray for the determination of protease activity. In short, Salisbury et al. [40] linked nonfluorescent peptidyl coumarins to a glass microarray via a chemoselective oxime forming reaction. The arrays thus fabricated were subjected to proteolysis by a variety of serine proteases: as a consequence, the peptides were released in the bulk, while the surface became fluorescent upon the exposure of a 7-amino coumarin. Differences in the extent of cleavage by the tested proteases were compared to the fluorescence emission of the corresponding substrates, which led to the quantitative evaluation of the substrate specificity of the proteases [40]. Similarly, Zhu et al. [41] described the detection and evaluation of the catalytic activity to different classes of hydrolytic enzymes using this coumarin fluorescence response. Furthermore, upon coupling of a PNA-linked azidocoumarin to acetylene-functionalized glass slides via a 1,3-Huisgen dipolar cycloaddition, the emission of triazolocoumarin was monitored. The fluorescence emission of the surface after the reaction is a clear and direct indication of the successful immobilization process (Figure 16.1) [42].

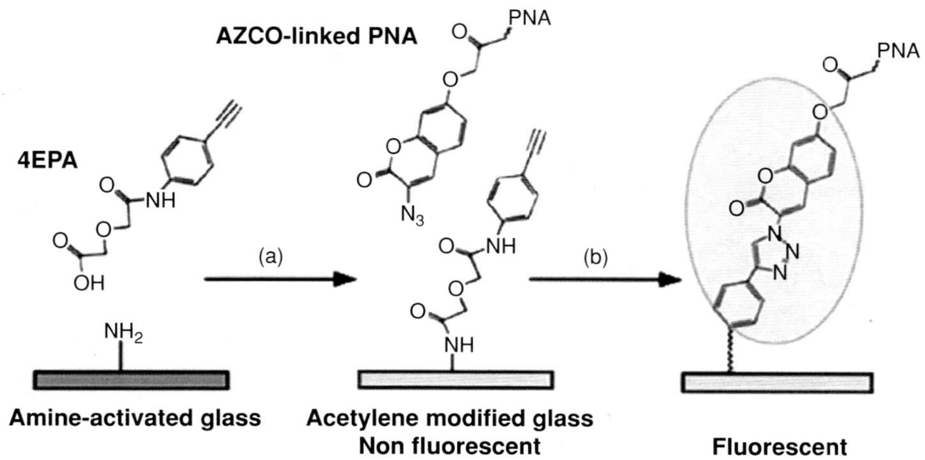

FIGURE 16.1. Fluorogenic immobilization strategy of PNA probes via 1,3-Huisgen dipolar cycloaddition. Reprinted from reference 42, with permission.

16.2.2 Photoswitching

The light irradiation-induced reversible transformation of a chemical species between two forms with different absorption spectra is named photochromism [43]. Light-driven switches, characterized by a change in color, are usually coupled with changes in structure, oxidation/reduction potential, refractive index, and dielectric constant [44]. If the photochemical reaction induces the formation of a thermal unstable isomer, the chromophore is of the so-called T-type (thermally reversible type), such as azobenzene [45] and spiropyran [46] derivatives, whereas if the formed isomer can be converted to the original state photochemically but not thermally, the chromophore is of the so-called P-type (photochemically reversible type), such as diarylethenes [47].

One of the most used chromophores in light-responsive layers is azobenzene, which undergoes a *cis/trans* isomerization under UV–vis excitation: Irradiation at 320–380 nm induces the formation of an excess of *cis* isomer, while under visible light the molecule switches to the *trans* isomer. One of the first azobenzene-based SAMs was described in 1999, when it was demonstrated that an *O*-carboxymethylated calix-[4]resorcinarene monolayer on aminosilylated silica can undergo a photoinduced reversible *cis-trans* isomerization. Upon irradiation with UV (365 nm) light, the formation of about 90% of *cis* isomer was observed. Subsequent irradiation of the *cis*-rich monolayer with blue light (436 nm) determined a conformational change of the molecules and the formation of the *trans*-rich monolayer [48]. One year later, the same group of scientists showed that an asymmetrical photo-irradiation of the previously described monolayer caused a gradient in surface free energy, exploitable for the motion of droplets on the surface [49].

An aminoazobenzene derivative, easily obtained from methyl red by a one-step synthesis, was used to fabricate monolayers on glass that showed high efficiency in the photoswitching of nematic liquid crystal cells [50]. In addition to this, a different kind of azobenzene derivative was assembled as a monolayer on a Nb thin metal film and was used to reversibly phototune the superconductive surface properties. It was found that UV–vis light caused photoisomerization of the monolayer and, as a consequence, changes in the electronic density of the surface [51].

The *cis–trans* isomerism of azo- and aminoazobenzene has been widely used also to tune the interaction between surfaces and (bio)macromolecules. For example, a monolayer made of photoresponsive peptide constituted of basic amino acids and an azobenzene chromophore on gold has been used to bind an RNA aptamer. The light-dependent *cis/trans* isomerization of the azobenzene group caused a change in the mutual orientation of two guanidinium groups of the peptide. In the absence of irradiation, the monolayer was able to bind with good affinity a certain RNA sequence. Irradiation of the monolayer at 360 nm for 5 min determined the formation of a high excess of *cis* isomer and a decrease of more than 90% of the binding within minutes. However, the process is reversible, as photo-irradiation with 430-nm light determined again the binding of the RNA to the surface [52]. Moreover, azobenzene has been used to photoregulate the binding to a gold surface of the protein α-chymotripsin [53]. As shown in Figure 16.2, in this case the surface was functionalized using a phenylalanine-based trifluoromethylketone inhibitor containing an azobenzene core, an oligoethylene glycol tether, and a terminal alkyne, coupled through a "click" Huisgen 1,3-dipolar cycloaddition to a dextran polymer matrix. The reversible photoswitched binding of α-chymotripsin to the layer thus formed was monitored via SPR. The *cis*-azobenzene isomer, obtained after irradiating the system at 320–380 nm, showed an improved binding of the protein to the SAM. On the other hand, irradiating with visible light isomerized the azobenzene back to the *trans*

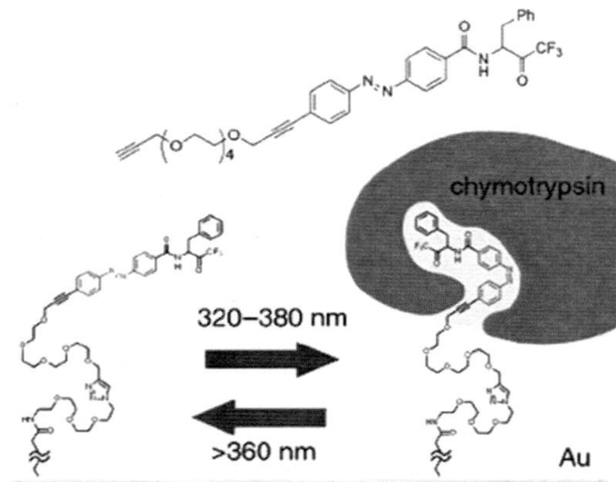

FIGURE 16.2. Light-induced azobenzene isomerization and interaction with α-chymotripsin. Reprinted from reference 53, with permission.

isomer, yielding a SAM less prone to protein binding. As the approach used is efficient and relatively simple, and considering the fact that the inhibitor design is modular, this method can be considered of general use for the whole protease family. Azobenzene-based monolayers have also been used to achieve a reversible photo-control of the cell–surface interaction. An RGD peptide, well known for its property of cell binding, was included in the monolayer containing the azobenzene group and anchored on Au. The *cis*-azobenzene isomer masked the peptide, while the *trans* isomer allowed it to be exposed toward the lattice and thus available for cell adhesion [54].

The reversible isomerization of spiropyran to merocyanine was exploited in 1997 by Willner's group who, besides describing the photoisomerization of the molecule on the surface [55], studied the tuning of the activity of an immobilized enzyme by using a photoisomerizable semi-synthetic factor [56].

The use of photosensitive biomaterials is considered highly attractive in nanotechnology, since it allows designing and fabricating optobioelectronic devices [57]. In an example, a nitrospiropyran flavin adenine dinucleotide (FAD) was assembled to a glucose oxidase, which was then covalently bound to an organic layer on gold, thus forming a photoisomerizable glucose oxidase SAM on Au, potentially usable as a biocatalyst. FAD is a redox cofactor involved in biochemical processes with one or two-electrons transfer, such as occurring in the Krebs cycle. The nitrospyran part of the semisynthetic cofactor exhibited reversible photoisomerization: Upon irradiation between 360 and 380 nm, a nitromerocyanine was formed, while irradiation at 475 nm restored the nitrospyran. The nitrospyran-FAD form of the monolayer was inactive, while the isomerized form obtained upon UV irradiation could catalyze the oxidation of glucose.

As already shown for spiropyran, diarylethenes also undergo a ring open-closure process upon UV–vis irradiation. The pioneering work of Irie [47] on photosensitive diarylethene-based compounds led to the fabrication of photosensitive monolayers both on nanoparticles and on surface [58–61]. One of the most recent examples in this sense is represented by a monolayer containing thiophene-substituted diarylethene and a viologen moiety on Si(111). The photoswitching of the diarylethene unit caused a change in the electron transfer between the viologen and the surface [62], explained by the different conductivities of the diarylethene in the open and in the closed forms [63,64]. The ratio between the open and closed forms of a dithienylethene molecule on gold was measured via a break-junction technique and was in good agreement with *ab initio* calculations [65].

Recently, also two monolayers of stilbene-derivative chromophores (i.e., a 1-cyano-1-phenyl-2-[4′-(10-undecenyloxy)phenyl]-ethylene and a 1-cyano-1-(4-Cl-phenyl)-2-[4′-(10-undecenyloxy)phenyl]-ethylene monolayers) on Si(100) were shown to give a reversible *cis–trans* photoswitching under UV irradiation, without suffering from steric hindrance that could, in principle, lower the yield of the process on the surface [66]. Examples of photoswitches have also been obtained in light-driven molecular machines on surfaces. In recent years, research was focused on immobilization of these systems on nanoparticles; nevertheless, there are also interesting examples on planar surfaces [67]. Feringa et al obtained a photoinduced controlled rotation with a second-generation motor attached to a quartz surface [68]. Light-excitation of a two-station rotaxane physisorbed on a modified gold surface caused a shuttle movement able to induce changes in surface tension [69]. The photoactive components in this case were a fumaramide and a tetrafluorosuccinimide. A scheme of the system is represented in Figure 16.3.

Interesting photoswitchable molecular machines have been fabricated using β-cyclodextrins. For example, the host–guest interaction between an immobilized heptathioether-functionalized β-cyclodextrin on a gold electrode and a bipyridinium azobenzene could be tuned with light and transduced to the conductive surface [70]. Moreover, a rotaxane having as axle an azobenzene was immobilized on Au, and as a rotor a ferrocene-modified β-cyclodextrin, which is able to transduce an optical signal into an electronic signal by means of photocontrolled ring shuttling due to azobenzene isomerization [71].

16.2.3 Photoinduced Reactivity

Light can also causes changes in the monolayer structure, inducing bond-forming or bond-breaking reactions between the adsorbate molecules.

For example, it was shown that upon irradiation at 350 nm a coumarin-derivatized SAM on Au dimerized, forming the *syn* H–H dimer [72]. The reaction is reversible, because a subsequent irradiation at 254 nm causes the photocleavage of

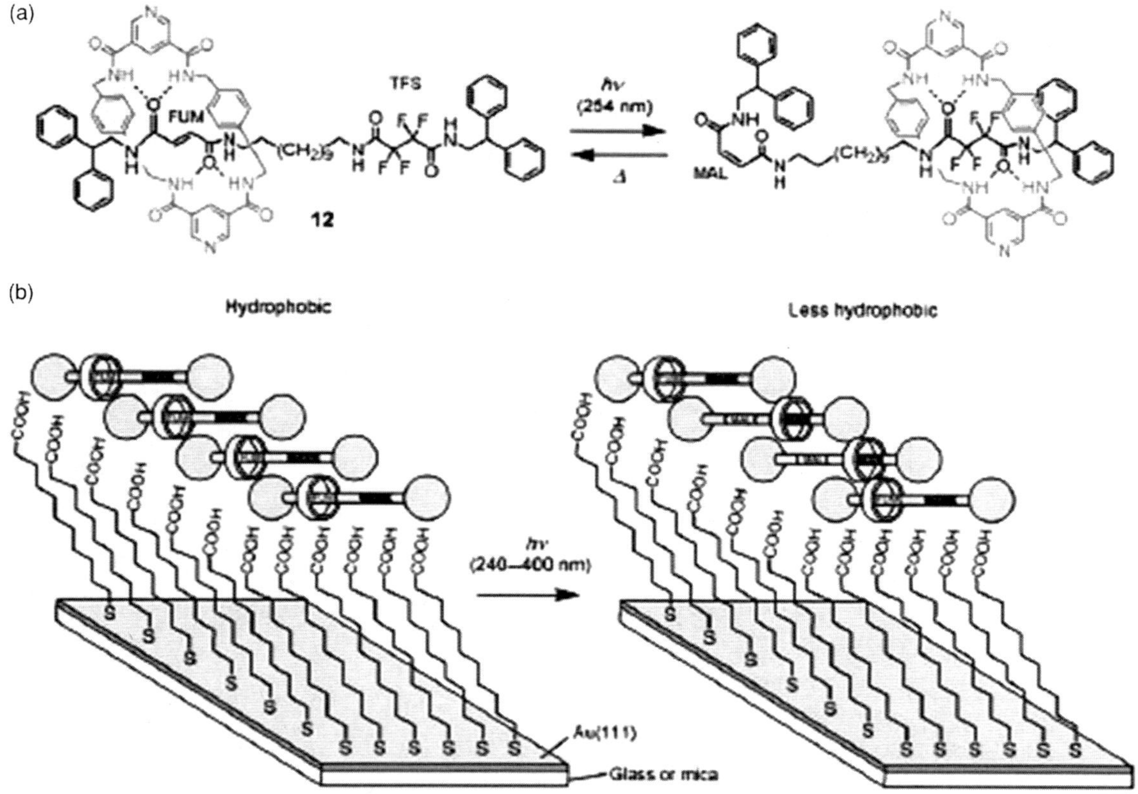

FIGURE 16.3. Scheme of the photoswitchable fumaramide-based rotaxane used to tune the surface wettability properties. Reprinted from reference 69, with permission.

the dimer. At the same irradiation wavelength, the reversible dimerization of the anthryl groups of a 10-thiodecyl-2-anthryl ether monolayer was observed [73]. Similarly, a SAM of pyrimidine on Au was shown to give dimerization and the reverse reaction, along with concomitant surface wettability changes, after irradiation at 280 and 240 nm, respectively [74].

An interesting light-responsive monolayer on silica is constituted of phosphomolybdic acid molecules. The monolayer, which usually shows the typical phosphomolybdate anion absorbance peaks at 196 and 310 nm, after 5 min of UV-irradiation changed color, and the process was monitored with a microscope equipped with a color CCD camera. Moreover, in the absorption spectrum a new peak at 780 nm appeared, indicating the presence of photoreduced Mo^{5+} species in the monolayer after irradiation [75]. It has been reported that a SAM of a flutamide derivative on Au was able to release NO quantitatively only as an effect of irradiation at 380–480 nm [76]. Such an application is interesting, since NO plays an important role in many different physiological processes and metabolic pathways [77, 78].

The possibility to induce photoreactions in the monolayer which can facilitate cell–surface interactions has also been exploited. Because it is well known that the 2-nitrobenzyl group is photocleavable, layers containing this group have been used for dynamic photocontrol of cell adhesion [79]. Nakanishi et al. [80] have fabricated a 1-(2-nitrophenyl)ethyl-5-trichlorosilylpentanoate monolayer on glass coverslips, which they covered with bovine serum albumin (BSA), thus preventing cell adhesion. When this layer was irradiated with 365-nm UV light, the 2-nitrobenzyl group was photocleaved and BSA diffused into the bulk. As a consequence of the photocleavage, the hydrophilicity of the substrate increased, since polar carboxylic groups are exposed to the surface, now suitable for fibronectin and cell immobilization. The same system has also been used in a microarray format to induce cell migration and to quantitatively evaluate the extension rate of cell proliferation [81]. The system and its application in cell migration studies are represented in Figure 16.4. The introduction of an RGD peptide in the 2-nitrobenzyl containing monolayer facilitated the cell adhesion through interaction with the integrins of the extracellular matrix [82].

Other well-studied photosensitive molecules at the surface are porphyrins, used for the development of donor–acceptor systems on different substrates. It has been reported that the successive assembly of Zn porphyrins and porphyrin–fullerene dyads on ITO electrodes led to the formation of well-ordered, photoactive multilayer structures that are very promising for applications in photovoltaic devices [83]. High

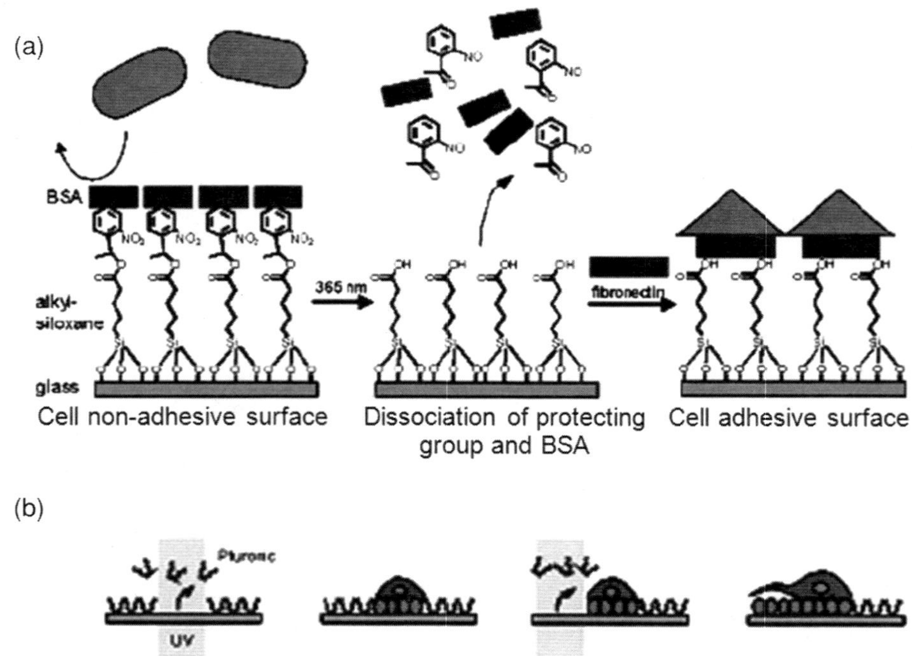

FIGURE 16.4. (**a**) Schematic representation of cell adhesion with a 2-nitrobenzyl-based photosensitive monolayer and (**b**) illustration of cell migration induction. *First picture*: UV illumination and pluronic desorption. *Second picture*: Fibronectin adsorption and cell adhesion. *Third picture*: UV illumination and pluronic desorption. *Fourth picture*: Fibronectin adsorption and cell migration. Reprinted from reference 81, with permission.

photocurrent generation efficiency has been achieved also using a fullerene metal cluster–porphyrin dyads in the presence of diazabicyclooctane [84]. Porphyrin-modifed Au nanoclusters have been used for the photocatalytic reduction of hexyl viologen [85], and very recently Akiyama et al. [86] showed that the excitation in the near-infrared region of a SAM of a porphyrin derivative on a Au nanostructure obtained by electrodeposition on ITO gave a remarkable enhancement of photocurrent.

16.3 TEMPERATURE-RESPONSIVE LAYERS

Temperature-sensitive hairpin-forming DNAs have been used to form SAMs on gold electrodes. When these DNA-functionalized surfaces contain an electroactive molecule (i.e., thionine coupled via a 1,4-phenylenediisothiocyanate to the DNA strand), the SAMs thus obtained can be used as working electrodes that give a temperature-dependent response in redox current [87]. A scheme of a typically modified electrode typically used is shown in Figure 16.5a. However, the majority of the examples on thermally switchable surfaces employ polymers that are able to modify their wettability properties in response to temperature changes.

Thermoresponsive polymers have a characteristic low critical solution temperature (LCST). At temperatures below the LCST these polymers are soluble in water, whereas at temperatures above the LCST they undergo a phase transition and aggregation [88]. Typical examples of temperature responsive polymers are poly(*N*-isopropylacrylamide) (PNIPAM) and poly(acrylic acid) (PAA). The changes in the PNIPAM layer structure and concomitant surface wettability as a temperature function are shown in Figure 16.5b.

The use of this kind of layer has been shown to be helpful for biomolecular separation, cell culturing, and tissue engineering applications [89, 90]. Controlled cell desorption due to temperature change is an interesting topic, since it offers an alternative to the proteolitic enzymatic treatments with trypsin [91] or dispase [92], which damage the cells by hydrolyzing the adhesion proteins on the external part of their membrane. On the other hand, temperature-dependent cell lift-off is mild and does not destroy the extracellular matrix [93]. The main physical rationale behind the use of some polymers as temperature-responsive layers for cell detachment is that their LCST is just under the typical culture temperature (37°): LCST of PNIPAM, for example, is 32°C in water. At higher temperature the layer is slightly hydrophobic, which facilitates cell adhesion. However, below 32°C it undergoes a rapid hydration phase transition, leading to a spontaneous cell release. Above the LCST, the PNIPAM chains are only partially dehydrated, and thus other variables such as swelling ratio, molecular mobility, chain density, and

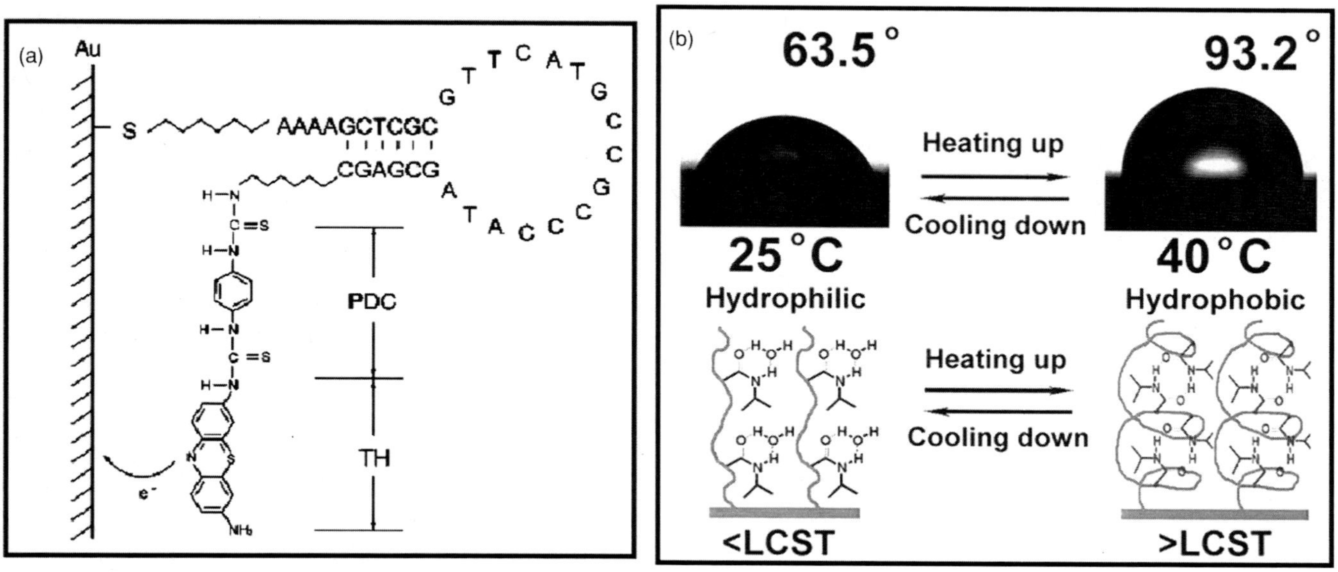

FIGURE 16.5. (**a**) The structure of hairpin oligonucleotide temperature-responsive modified electrode and (**b**) the structural change in a PNIPAM layer upon change of temperature and the subsequent change in surface wettability detected through contact angle measurements. Reprinted from references 87 and 88, with permission.

concentration of the hydrophobic groups must be assessed as well for cell adhesion and proliferation [94]. It has been demonstrated that PNIPAM on polystyrene culture dishes interacts with the fibronectin matrix allowing for attachment and proliferation of endothelial cells at 37°C, while at 20°C it causes the release of a monolayer sheet of the same cell culture [95]. Moreover, poly(*N*-isopropylacrylamide-co-2-carboxyisopropylacrylamide) copolymer on polystyrene dishes permits the immobilization of the Arg-Gly-Asp-Ser (RGDS) peptide, a well-known ligand for cell integrins. By tuning the temperature of this system, it is possible to achieve an "on–off" affinity control between the integrins and the ligand [96]. However, the most relevant applications in this field concern biomedical applications such as transplantations and tissue regeneration. As an example, PNIPAM-grafted tissue culture dishes have been used for cultivation of lung cells, which kept their integrity and functionality after temperature-based harvesting. Being possible to maintain these cells for up to 70 days in culture, the system represents a good starting point for the development of a hybrid artificial lung [97]. Another study was performed using human corneal endothelial cells. These cells grown on PNIPAM layers were able to reach a density of at least 3000 cells/mm^2, sufficient for routine in vitro and in vivo applications. The temperature-response-based-detachment from the dish maintains the cells in their sheets-like assembly, free of the contaminants characteristic of enzymatic detachment methods [98]. Kitano et al. [99] recently fabricated a polymer brush made of poly(2-(2-methoxyethoxy)ethylmethacrylate) with ω-methoxy-di(ethylene glycol) side chains (PMDM) on gold. Heating caused a coil-globule transition (i.e., a shrinkage) of the PMDM chains, which caused a variation in refractive index and the increase of absorbance at 550 nm. Due to its thermal responsiveness, a SH-terminated PNIPAM monolayer on gold is able to influence the absorption and desorption of bovine serum albumin (BSA) immobilized on an AFM tip as a function of the temperature, as proven by quartz crystal balance measurements [100]. Densely grafted PNIPAM brushes of different lengths have been prepared both on glass coverslips or fused silica capillaries [101] and on silica bead surfaces [102] using the surface-initiated atom transfer radical polymerization technique. The surfaces thus obtained show temperature-tunable hydrophobic properties, and for this reason these have been used for chromatographic separation of steroids and peptides. In this kind of application the chain length plays a key role: If it is too short, the temperature increase does not cause sufficient dehydration and aggregation, so that the steroid separation is not efficient. This method provides an advantage in purification methods, since it allows an effective separation without changing the pH of the mobile phase. Huber et al. [103] have described a microfluidic device, the channel of which had been functionalized with a SAM of PNIPAM, able to induce reversible protein adsorption. Moreover, it has been demonstrated that temperature-sensitive SAMs of PNIPAM and PEG copolymers allow a fine control of the adhesion of different cell types both in microfluidic channels [104] and on gold surfaces [105]. Temperature increase can also be the driving force of covalent assembly of SAM-functionalized particles: By alternatingly heating and cooling a system made of furan-modified Au nanoparticles and maleimide bearing oligo(*p*-phenylenevinylene), a thermally reversible Diels–Alder reaction was induced, leading to ordered monolayer protected gold cluster self-assembly [106].

More recently, a mixed monolayer made of a carboxyl group-terminated PNIPAM and different aza-amino compounds bound to aminated polymethacrylate beads was described. Actually, in this case PNIPAM is the only responsive element, while the second component is exploited for its ability to bind Cibacron Blue F3G-A, a well-known ligand for purification of albumin. By changing the polymer state as a function of temperature, using such a modified surface allows a temperature-controlled molecular recognition. The system, based on masking and forced-releasing effects, could be used in chromatographic matrices for protein purification [107]. An analogous surface has been obtained using a modified oligo(ethylene glycol) and a biotinylated disulfide [108]. Oligo(ethylene glycol) is a thermosensitive element, because its hydrodynamic diameter shrinks when the temperature increases up to 45°C. Its property is exploited to control the immobilization of streptavidin on the second monolayer component—that is, the biotinylated chain. The choice of an oligo(ethylene glycol) derivative is indeed useful to achieve selective protein immobilization, because it is well known to prevent nonspecific protein adsorption [109].

16.4 pH-RESPONSIVE MONOLAYERS

The use of pH change as external stimulus is considered highly important, considering the huge number of pH gradients that exist both in vitro and in vivo.

In pH-responsive monolayers, pH changes induce changes in structure, barrier properties, and surface properties, such as wettability [110]. Interesting examples of pH-sensitive monolayers have been obtained by fabricating mixed monolayers consisting of two kinds of molecules, one pH-responsive (the sensing one) and another electrochemically active. In this case the main advantage is that it is possible to use the redox signal as an internal readout, so that the presence of an electrochemically inactive compound can be electrochemically detected [111]. It has been observed for a long time that lowering the pH of an electrolytic aqueous $Fe(CN)_6^{3-}$ solution from 5.3 to 2.9 improves the electron transfer mediated by a self-assembled glutathione monolayer modified gold electrode [112]. When a gold electrode is modified with a mixed ferrocene–thiol/carboxylic acid–thiol monolayer or, as an alternative, with a ferrocene–acid sulfide monolayer, a shift of the cathodic signal using cyclic voltammetry (CV) measurements was shown upon changing the pH [113].

Also, pH-switchable layers were fabricated using bipolar α,ω-bis(4-aminophenoxy)alkanes and ω-mercapto-carboxylic acid monolayers on gold. Amidines form cyclic, hydrogen-bonded ion pairs with oxoacids, but in acidic conditions the interaction is lost. As a consequence, at low pH no self-assembly is observed. The layer stability is strongly pH-dependent, being influenced by the degree of ionization of the surface acidic group. Disassembly and reassembly of the layer could be reversibly observed when cycling the pH between 3 and 9. Furthermore, the positively charged amidinium surface thus obtained was used for the selective adsorption of phosphate biomolecules such as ATP and oligonucleotides. Both the amidine assembly/disassembly cycles and the biomolecule adsorption caused significant changes in the layer thickness, as witnessed by ellipsometry [114]. The same system has been used for reversible protein immobilization; at high pH it was possible to deposit a second amidine layer, which was coated with fibrinogen and lysozyme. Decreasing the pH caused the detachment of the complex amidine–protein, leaving only one anchored amidine layer. Ellipsometry and neutron reflectivity gave information about the layer thickness, and infrared reflection–absorption spectroscopy (IRRAS) indicated the presence or absence of typical amidine peaks [115]. Another example of a pH-responsive SAM was described by Jiang et al. [116] who immobilized 2-(11-mercaptoundecanamido)benzoic acid on gold: in this case, hydrogen bonding can occur between the amide groups of adjacent chains. The presence of this kind of interaction gave rise to different stretching and bending motions of the molecular functional groups: these changes were monitored as a function of pH using IRRAS. In addition to this, the pH influenced the surface wettability.

The wettability change of a functionalized surface in case of pH sensitivity can be substantial: Yu et al. [117] showed that mixed monolayers-containing both alkyl and carboxylic groups on gold can undergo a wide contact angle variation (from 154° to less than 30° in 1 s) by applying respectively an acid or a base water droplet on the surface. To achieve such a remarkable result, both the monolayer composition and the surface roughness had to be suitably tuned.

It has been reported that pH-sensitive monolayer-modified Au electrodes can affect the electrochemical response of a substrate. The negatively charged pyrroloquinoline quinone (PQQ), for example, undergoes a quasi-reversible redox process at a protonated monolayer-modified electrode, which shifts to an irreversible electrochemical behavior in the presence of a negatively charged monolayer. Such a pH-switched electroactivity, proven with cystamine, 4-aminothiophenol and 4-pyridyldisulfide-modified electrodes, has been attributed to the electrostatic attraction of PQQ to the positively charged monolayer interface which facilitates the interfacial electron transfer [118]. This is in perfect agreement with the general consideration that electrochemical processes can be controlled by the electrical charge associated with the electrode surface [119]. The pH value at which the transition is observed is related to the monolayer composition (i.e., to its pK_a).

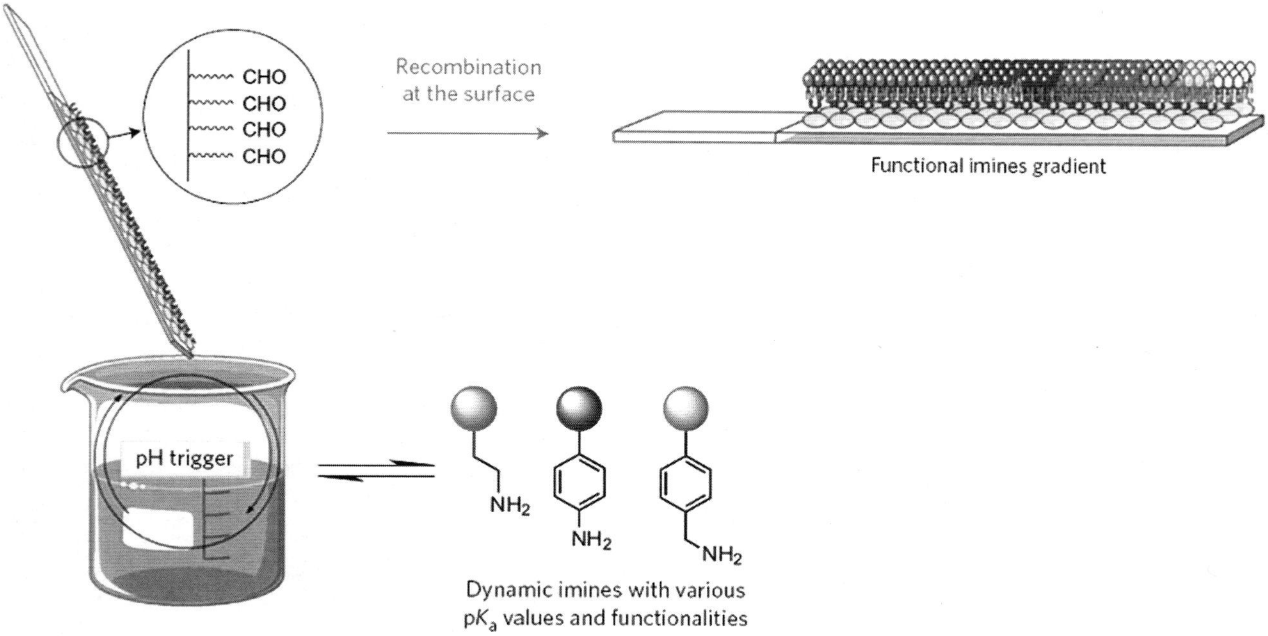

FIGURE 16.6. An aldehyde-terminated surface is progressively removed from a solution of various functional amines as the pH is varied. Each amine combines a single functionality associated with a unique physical property for a given pK_a, which thus leads to a gradient of functional imines. Reprinted from reference 123b, with permission. See color insert.

Carboxyl-terminated thiol monolayers have been used to prepare pH-responsive membranes, since they are able to influence the flux of benzenesulfonate anions through a polycarbonate track-etched membrane [120]. In addition to this, it has been demonstrated that the rate constants for electron tunneling through wild-type and modified *Pseudomonas aeruginosa* azurins adsorbed on mixed monolayers of alkanethiols and hydroxyalkanethiols of different lengths on gold are higher at pH 11 than at pH 4.6. This has been explained by enhanced coupling of a negatively charged SAM to the Asn47 residue, due to the formation of hydrogen bonds [121].

The choice of a specific molecule for the fabrication of a pH responsive monolayer can also be based on the results of molecular dynamics experiments. For example, the conformational properties of a synthetic peptide were investigated with simulations at different pH values. After theoretical demonstration that the peptide undergoes a conformational change going from α-helix at pH 2 to random coil at pH 7, this was established as an interesting molecule to obtain a pH-sensitive SAM on gold. This hypothesis was then confirmed by electrochemical measurements performed with a SAM of the peptide-modified electrode [122].

Parallel to the use of conventional techniques for pH-driven changes of monolayer features, there is a growing interest in new pH response detection ways. An interesting example in this field concerns the formation of organosilane monolayers on silicon oxide surfaces. A multichannel electrode with active and passive sites for pH responses was designed by patterning monolayers of pH-sensitive and pH-nonsensitive silanes on the same substrate. This kind of array can be used in nanoelectronic applications: the active part is suitable to act as an electrode for an ion-sensitive field effect transistor (FET), and the inactive one is usable as a reference electrode [123a].

Giuseppone and co-workers [123b] recently fabricated an aldehyde-SAM on quartz, which was then immersed in a solution of different amines of various pK_a, thus forming the corresponding imines (Figure 16.6). Layer dynamics was demonstrated by performing a transimination reaction at the surface; this showed that by varying the pH, the surface reactivity was tuned toward different amines. In this way a gradient of immobilized functional imines was installed [123b]. The method has been successfully extended to proteins. Such work provides a general and predictable method to control the reactivity at the surface and opens new perspectives in terms of design of responsive tools.

16.5 ELECTROCHEMICALLY RESPONSIVE MONOLAYERS

Application of a negative potential (< -1.2 V) to gold substrates results in partial or complete desorption of the SAM, which was used by Whitesides to detach patterned cells [124]. The same strategy has also been used to release nanoparticles, proteins, and nucleic acids from the surface [125, 126].

For example, Huang et al. [126] fabricated dynamic nanoarrays of thiolated DNA sequences. After hybridization with complementary DNA-biotin strands to the HS-DNA at the electrode surface and subsequent incubation with streptavidin, these assemblies were released by employing −1.6 V for 2 s. When focusing on nondestructive electrochemically responsive SAMs, such SAMs should (i) be used to detect or induce a change in the conformation of the molecular SAM structure under the application of a potential or (ii) contain an electroactive moiety, which is used for chemical modifications.

The possibility to anchor electroactive units on a surface is highly convenient, in the perspective of designing and fabricating nanostructured materials with tailored functions and properties [127]. Usually, the confinement of a redox center on a surface can be proven by cyclic voltammetry measurements: The peak current varies linearly with the scan rate, and the slope of the plot is related to the number of electrons involved, the electrode area, and the surface coverage. As a consequence, cyclic voltammetry allows quantifying the amount of the electroactive species adsorbed on the surface [128]. The most common moieties used to design electrochemically responsive SAMs are ferrocene, quinone, azobenzene, and viologen. However, there are also less common examples, such as salophenes [129], fullerenes [130], and others. Since it is difficult to make densely packed SAMs of these bulky species, the general approach is to use mixed monolayers of chains bearing the group of interest and alkanethiols. However, the length of the chain is crucial: As has been observed for mixed monolayers containing hydroquinone, the diluting chain must not exceed the length of the hydroquinone chain, since this causes a decrease of the redox activity [131a]. An alternative approach is the functionalization with long-chain thioethers, as recently demonstrated for adamantane-based tripodal ligands bearing a redox-active ferrocene [131b]. As far as the immobilization of redox-active proteins is concerned, a well-known example is given by cytochrome c. Monolayers of cytochrome c give electron transfer to a gold electrode [132] at E° between 0 and −60 mV versus SCE. Recently, the influence of the metal substrate on the electrochemical response of this protein has been studied, showing that, using the same alkanethiol as a linker, the electron transfer on gold is about two times faster than on silver [133]. Other interesting examples in this field concern copper proteins, such as azurines and plastocyanines. Once they are immobilized on gold (directly or at modified surfaces), they show a robust electrochemical response that allows to perform molecular-level tunneling and topographic analyses [134].

The remainder of this section is divided into two parts: The first one describes electrochemically detected/induced conformational changes of different kinds of SAMs, whereas the second one focuses on examples of electroactive SAMs used for chemical modifications.

16.5.1 Electroactive Sensing/Switching

Treatment with an oxidizing or reducing potential may permit to switch the molecular conformation of ordered monolayers, thus changing the surface properties. Such a concept was introduced in 2003 by Langer and co-workers [135], who demonstrated that, when applying an electric potential, a low-density SAM of 16-mercaptohexadecanoic acid on Au undergoes a stretching/bending process due to attraction or repulsion toward the electrode. The process, which is also associated to a conformational change, was followed using sum-frequency generation spectroscopy and contact angle measurements.

Exploiting the same property, a potential-controlled "molecular arm" was fabricated immobilizing a bipyridinium derivative on Au [136]. In this case, when the applied potential was 0.3 V versus SCE, the electrostatic repulsion between the positively charged electrode and head groups of the monolayer caused a stretching of the immobilized molecules. Switching the potential to −0.2 V did not yield electron transfer (which happens only at −0.7 V) but, instead, resulted in an electrostatic attraction between the negatively charged electrode and the bipyridinium units, which bend toward the surface. This motion causes a change in the wettability of the surface: At positive potentials the charged pyridinium ring faces the exterior of the monolayer, making the surface hydrophilic, whereas at negative potentials the alkyl chain is more exposed and the surface shows an increased hydrophobicity.

The potential-controlled chain-bending has been demonstrated to play a key role in obtaining a selective protein adsorption on this kind of functionalized surface [137], which, for this reason, has also been used in microfluidic devices in order to obtain protein separation under electrical control [138].

The electrochemical response of monolayers of supramolecular structures has been intensively studied. Electrochemical-responsive supramolecular systems have been realized using host–guest complexes based on calixarenes filled with cobaltocenium or ferrocenium [139]. Tetra-tosylurea calix[4]arene functionalized with sulfide groups for the binding on gold formed heterodimers with tetra-urea calix[4]arenes. The inclusion of ferrocenium as a guest made it possible to monitor the behavior of the electroactive encapsulated species, which shows good reversibility [140].

Our group demonstrated that multivalent ferrocene dendrimers can be reversibly adsorbed and removed from β-cyclodextrin SAMs by electrochemical oxidization of the ferrocene end groups [141,142]. More recently, the reversible attachment of β-cyclodextrin-functionalized nanoparticles on ferrocenyl-functionalized poly(propylene imine) dendrimers at a β-cyclodextrin SAM was achieved in a similar way in our group [143], as schematically shown in

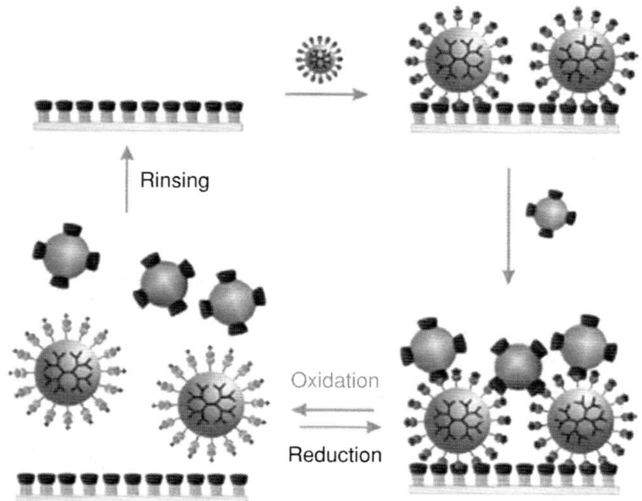

FIGURE 16.7. Reversible adsorption of β-cyclodextrin-functionalized nanoparticles on ferrocenyl-functionalized poly (propylene imine) dendrimers at a β-cyclodextrin SAM. Permission from [143].

Figure 16.7. In this case, the ferrocenyl-functionalized dendrimers act as reversible molecular glue between the surface and the nanoparticles.

A large body of work on electrochemically responsive supramolecular layers has been carried out using rotaxanes and catenane systems. Willner and co-workers [144, 145] have fabricated a rotaxane monolayer on gold consisting of a cyclophane cyclobis(paraquat-*p*-phenylene) (CBPQT) and an axle including a σ-donor diiminobenzene unit and an adamantyl unit as stopper. Normally, the cyclophane forms a σ-donor–acceptor complex with the diiminobenzene unit of the axle, but if it is electrochemically reduced, it acts as a molecular shuttle, being driven toward the electrode. An evolution of this system has been realized replacing the adamantyl stopper by glucose oxidase, giving the supramolecular structure schematically reported in Figure 16.8. This system proved to be effective in the bioelectrocatalytic oxidation of glucose, and it is the first example of an interlocked rotaxane for electrical contacting of redox active enzymes [146]. Stoddart and co-workers [147] immobilized on gold a disulfide-tethered rotaxane, the axle of which was characterized by a tetrathiafulvalene (TTF) unit and a (dioxynaphthalene) DNP unit. Also in this case the interchained CBPQT acted as a molecular shuttle along the axle when a potential between −100 and +900 mV versus Ag^+/AgCl was applied. Applying an oxidizing potential (+800 mV), CBPQT was attracted toward the DNP as a consequence of the TTF oxidation. Upon application of a reducing potential, the TTF became neutral again, so the chain was supposed to move back. However, starting from the second oxidation cycle a metastable state was observed, where the chain remained on the DNP, the population of which could be tuned by the scan rate [147].

Conformations of DNAs have been monitored electrochemically as well. For example, Plaxco and co-workers [148] have described an aptamer-based biosensor for the detection of the enzyme thrombin. The DNA used was modified with methylene blue, for the electron transfer with the surface. While in the unbound state the aptamer is in the unfolded state, in case of interaction with thrombin it is in a G-quartet conformation and assumes a rigidity that forces the electrochemically active unit far from the electrode, thus

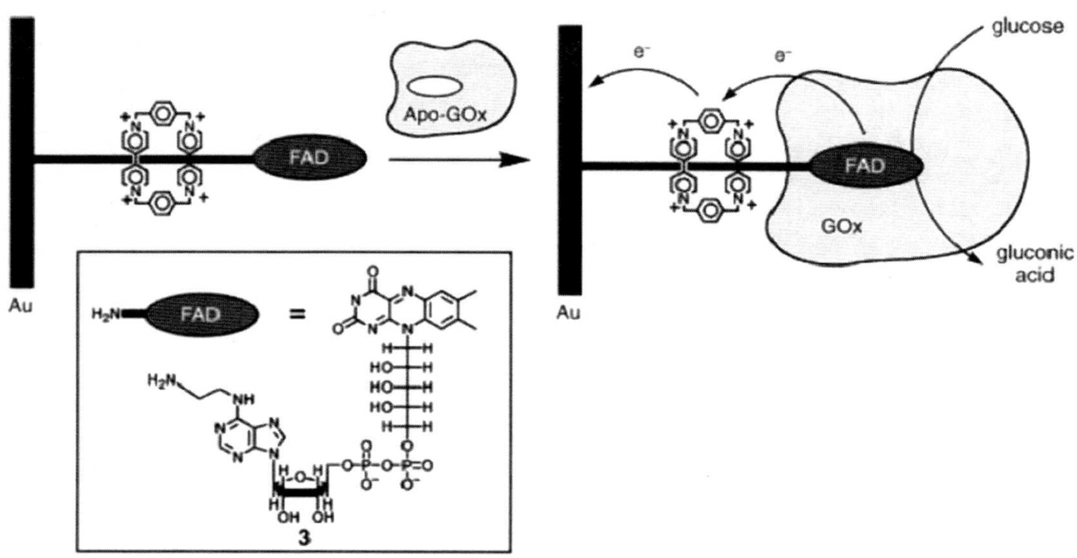

FIGURE 16.8. Scheme of an interlocked rotaxane with a gluocose oxidase stopper. Reprinted from reference 146, with permission.

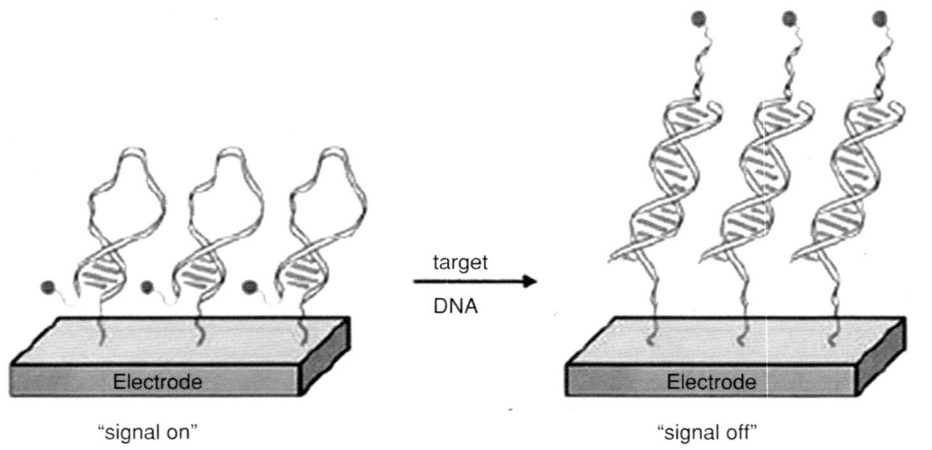

FIGURE 16.9. Schematic representation of a labeled hairpin for electrochemical DNA hybridization detection. Reprinted from reference 153, with permission.

inhibiting the electron transfer [149]. Plaxco's group also developed an aptamer-based sensor for cocaine detection in adulterated samples and biological fluids. This sensor can be easily regenerated by rinsing at room temperature and is unaffected by nonspecific contaminants [150].

A ferrocene-tagged DNA stem–loop structure SAM has been proven to be highly sensitive for the detection of DNA hybridization. Hybridization induces a conformational change in the immobilized molecules [151], with concomitant alteration of the electron-transfer tunneling distance between them and the electrode [152]. In the same way, an electrochemically active anthraquinone-capped hairpin DNA monolayer on Au has been used for the investigation of electron transfer through double-helical DNA on a solid surface (Figure 16.9) [153].

In the opposite way, a ferrocene-modified aptamer has been described which, in the initial state, forms a duplex and cannot do electron transfer. Such an aptamer is able to react with ATP, changing its tertiary structure. In this case the nucleic acid molecule is in a floppy state and the ferrocene is closer to the surface, so that the electron transfer becomes possible and the sensor switches from an OFF to an ON state [154]. Hybridization of DNA can also occur between a single-strand DNA and a target ss-DNA immobilized on gold nanoparticles and labeled with a ruthenium bis(2,2′-bipyridine)(2,2′-bipyridine-4,4-dicarboxylic acid)-*N*-hydroxysuccinimide ester. When the two filaments interact, forming a double helix, this head group causes the electrogeneration of a chemiluminescence, thus allowing an easier detection of the binding [155]. It has been demonstrated that, while the probe density does not affect the hybridization specificity, higher probe density sensors give optimal signal suppression [156]. Moreover, the DNA does not uniformly cover the surface: if the molecule is also labeled with a fluorescent tag, such surface heterogeneity can be detected using electrofluorescence microscopy [157].

16.5.2 Electroactive SAMs for Chemical Modifications

The presence of a redox center in a monolayer can be exploited to further modify the surface after an electrochemical stimulus. The reduction of SAMs of nitro derivatives, for example, has been used to obtain surface amino groups, which can then bind other functionalities and thus modify the surface. 4-Nitrobenzenediazonium tetrafluoroborate was used to functionalize carbon nanotubes [158] and conductive nanocrystalline diamond thin films [159]. The nitro groups were reduced to amines by applying a voltage of -1.4 V versus Ag/Ag^+. Since the reaction is almost irreversible, it can be highly useful to obtain surface amino groups that can be chemically modified in a further step. Another application has recently been described by Stoddart and co-workers [160]: When a monolayer of 4-nitrophenol on gold electrode surfaces was selectively and electrochemically reduced, it was possible to obtain exposed amino groups suitable for protein immobilization. In this way it is possible to obtain the desired geometry to design complex protein structures.

A 4-(1,4.dihydroxybenzene)butyl phosphonic acid has been used to selective functionalize In_2O_3 nanowires. After the oxidation, which led to the formation of a quinone, the monolayer reacted with a thiol-terminated single-strand DNA. The system thus obtained is potentially useful as a biosensor for oligonucleotides [161a]. Recently, a 4-thiouracil SAM has been used to obtain an electrochemically triggered Michael addition with various hydroquinoid compounds. The redox-active monolayer obtained in this way can differently interact with Ni^{2+} and Cu^{2+} ions, thus opening new perspectives in sensoring and analysis of metal ions [161b]. An electrochemically oxidized quinone anchored on a surface has been made to react chemoselectively also with an aminooxy acetic acid to give the corresponding oxime [162]. The oxime is also electrochemically active, but at a different potential, so the obtained surface was still available

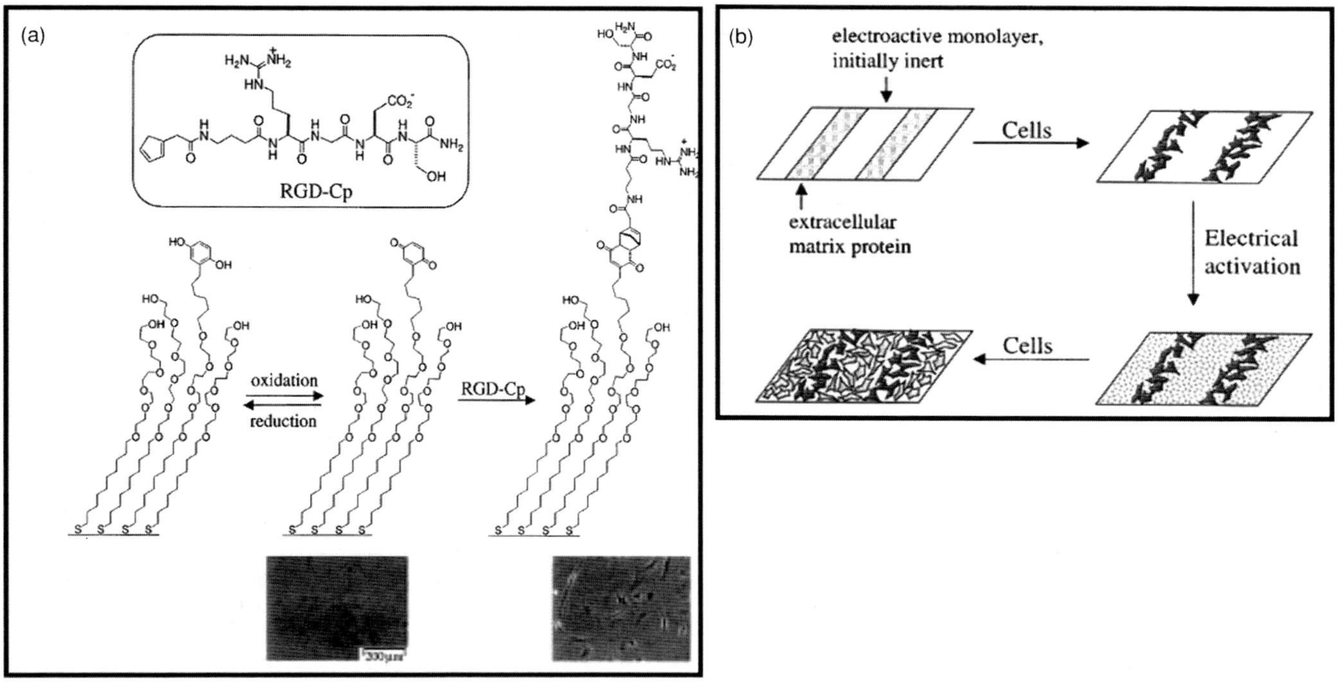

FIGURE 16.10. (**a**) Quinone-based electroactive monolayer for controlled cell adhesion, release, and migration. (**b**) An effective strategy for patterning two different cell types on a substrate using a quinone-based electroactive monolayer. Reprinted from references 166 and 164, with permission.

for further redox reactions. The authors used this system to bind an aminooxy-terminated peptides; using the RGD-oxyamine peptide, it was even possible to anchor cells.

Mrksich and co-workers [163] were able to electrochemically switch cell–surface interactions on SAMs on gold. As a first step, on a SAM background of penta(ethylene glycol) groups the hydroquinone head groups were oxidized to benzoquinone, which can undergo a Diels–Alder reaction with cyclopentadiene (Cp) [164]. Tethering in this way a Cp-peptide containing the RGD sequence (RGD-Cp) at those areas of the SAM where hydroquinone was patterned by micro-contact printing, 3T3 fibroblasts were attached and spread on the RGD-functionalized patterns (Figure 16.10a). In addition to electrochemical activation of cell adhesion, Mrksich and co-workers [165] also developed a SAM presenting RGD peptides anchored via a quinone ester. Upon electrochemical reduction of this quinone to the corresponding hydroquinone, cyclization of the lactone occurred, leading to the release of the RGD peptide and of the cells. The authors showed that it is possible to pattern a surface with this kind of electroactive layer either to induce migration of cells [164] or to obtain controlled spatial functionalization of surfaces with multiple cell types [166], schematically shown in Figure 16.10b.

The incorporation in the monolayer of an *O*-silyl hydroquinone moiety, combining two dynamic properties on the same substrate, gives additional flexibility to the whole system [167]. The *O*-silyl hydroquinone allows the selective release of the RGD peptide from the substrate upon oxidation at 650 mV, yielding the corresponding benzoquinone. Patterning the substrate with distinct regions that consist of two different redox-active molecules (the quinone ester and the *O*-silyl hydroquinone) which respond to different potentials (−650 and + 650 mV respectively), the RGD-ligand release can be selectively induced on different areas (Figure 16.11a). In this way, control over cell adhesion can be achieved [168]. Yeo and Mrksich [169] described the use of an electroactive protecting group that is removed by applying an oxidative potential (900 mV) yielding aldehyde moieties on the SAM and the release of the protecting group itself into the solution. The authors showed the coupling of aminobiotin to this aldehyde SAM, which was then able to capture streptavidin, as identified by mass spectrometric analysis. The general strategy followed is shown in Figure 16.11b.

A chemical and electrochemical control of the surface properties was achieved using a quinone functionalized monolayer on Au [170]. The conversion of 1,4-benzoquinone to 1,4-hydroquinone monolayer, obtained by applying a potential and simultaneously immersing the SAM in a solution of ascorbic acid, led to an increase of the surface hydrophilicity (i.e., a decreasing of the contact angle value). This phenomenon was enhanced in the presence of methyl viologen, which forms a donor–acceptor complex with the reduced quinine.

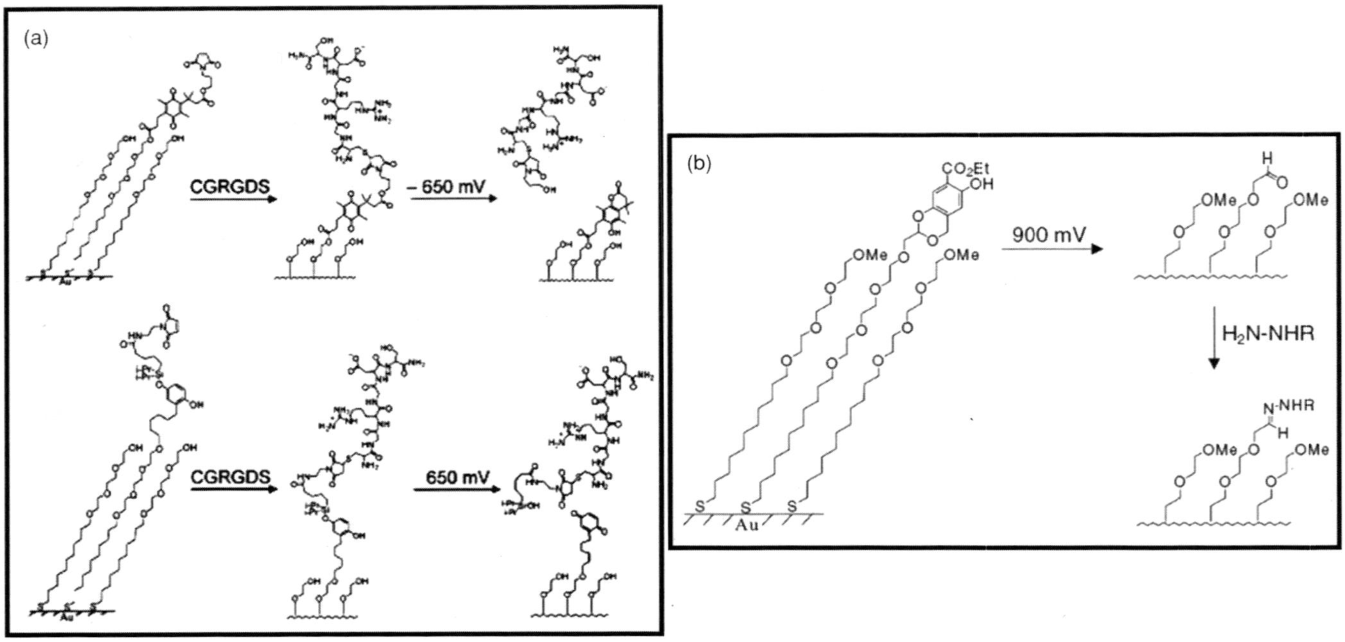

FIGURE 16.11. (**a**) Molecular strategies used to prepare dynamic substrates that can release the RGD-containing ligand in response to opposite applied potentials. (**b**) Scheme of the electrodynamical oxidation of a 4-*H*-benzo[*d*][1, 3]dioxinol functionalized SAM to yield an aldehyde group, which can further react with amines. Reprinted from references 167 and 169, with permission.

4-Hydroxy-(3-mercaptopropyl)phenyl valerate coupled to a maleimide-presenting monolayer on gold was subsequently treated with the enzyme cutinase and an oxidizing potential, upon which it switched from a redox inactive to a redox active surface. When the enzyme is artificially introduced on a cell membrane, and the engineered cells were immobilized on the monolayer, it was possible to transduce the cellular activity into an electrical output [171].

16.6 MULTI-RESPONSIVE MONOLAYERS

Recently there has been a growing interest in monolayers that respond to more than one stimulus. Different types of stimulati can cause changes in the same or in different properties. The ideal mode to obtain a multi-responsive layer is to use a molecule that can change one or more specific property while stimulating it in different ways. As an alternative, it is also possible to design molecules constituted by different building blocks, each one responsive to the change of one lattice property. In this way, for example, it has been possible to fabricate multi-responsive block copolymers such as *N*-isopropylacrylamide P(NIPAM-co-AAc), thermoresponsive like PNIPAM, and pH-responsive ones such as PAA [172].

In 1996, Willner and co-workers [173] described how the electrochemical response of a phenoxynaphthacenequinone is influenced by light irradiation and pH conditions. The *trans*-quinone, obtainable under irradiation at 430 nm, in a mixed monolayer with 1-tetradecanethiol on gold, exhibits electrochemical activity, undergoing a double-electron redox process. The photoisomerization, obtainable under irradiation between 305 and 320 nm, leads to the formation of an electrochemically inactive form. As the process is reversible, it is possible to switch on and off the electroactivity of the monolayer. Moreover, the electrochemical activity of the *trans*-isomer monolayer is pH-dependent, as E° decreases linearly upon increasing the pH value.

Both UV- and pH-responsive as well is the malachite green derivative-based monolayer on Au, described by Jiang et al. [174]. As already mentioned, light irradiation can also cause changes in properties different from the UV–vis absorption or the isomeric form, such as the charge state and wettability, typically pH-tunable features. The electrically neutral malachite green group can be ionized into its corresponding delocalized triphenylmethyl cation and a cyanide anion: the latter is released in solution, while the former acts as a reversible charge generator. As a consequence, the surface shows superhydrophilic properties at pH 1 and superhydrophobic properties at pH 13, with subsequent change in the contact angle value. Very similar results were obtained by stimulating the malachite-green monolayer with UV irradiation.

Another example of light or pH dual-responsive surface has been reported for reversible immobilization of cytochrome c on Au [175]. In this case, the authors used the photo-controlled reversible host–guest interaction between

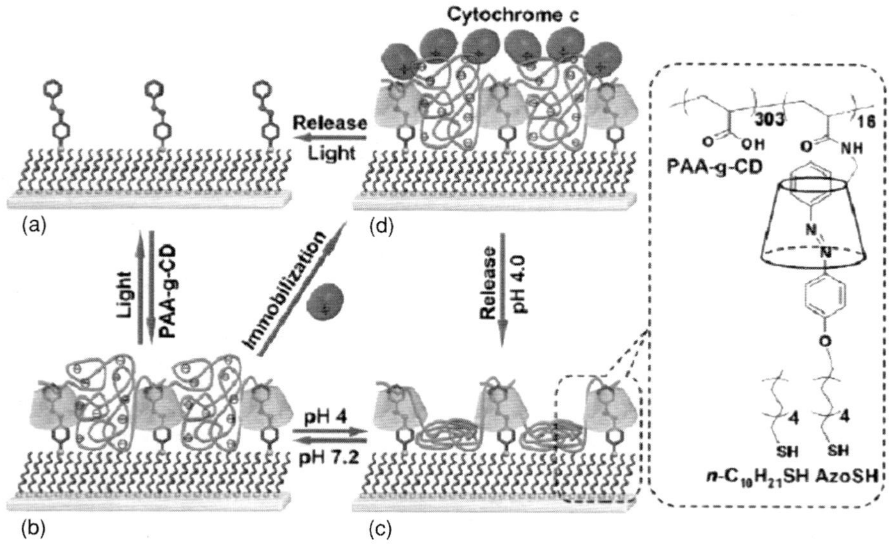

FIGURE 16.12. The light and pH-sensitive monolayer used for reversible immobilization of cytochrome c. Reproduced from reference 175, with permission. See color insert.

an azobenzene containing SAM and the pH responsive poly(acrylic acid) polymer grafted with β-cyclodextrin moieties (PAA-g-CD). This biocompatible polymer was chosen for its ability to adsorb large amounts cytochrome c at physiological pH (7.2) and to release it at pH 4. Such a phenomenon is possible because under pH 10, cytochrome is positively charged; on the other hand, PAA is negatively charged at pH 7.2 and neutral at pH 4. A scheme of the system is reported in Figure 16.12.

The light-driven reversible attachment of PAA-g-CD on an azobenzene-containing SAM can be detected through cyclic voltammetry measurements, since the negative charges of PAA-g-CD polymer on the azo-SAM influences the passage of charges on a modified gold electrode. Under acidic pH conditions, the carboxylated groups of PAA are hardly ionized; as a consequence the polymer, which is in the electroneutral and relaxed state, releases the cytochrome c. The negative charged state of the polymer can be easily restored by switching the pH back to 7.2.

16.7 CONCLUSIONS AND FUTURE PERSPECTIVES

The examples described here show the high importance of smart SAMs for surface characterization, post-immobilization monitoring, and switching processes. Even though a huge number of results are already available for each type of stimulus, it appears obvious that it would be highly desirable to have a SAM responsive to two or more stimuli. This type of application would permit to improve the systems versatility and the clearness of readouts.

REFERENCES

1. Bigelow, W. C., Pickett, D. L., Zisman, W. A. (1946). Oleophobic monolayers. I. Films adsorbed from solution in nonpolar liquids. *Journal of Colloid and Interface Science*, **1**, 513–518.
2. Schreiber, F. (2000). Structure and growth of self-assembling monolayers. *Progress Surface Science*, **65**, 151–257.
3. Franklin, B. (1774). On the stilling of waves by means of oil. *Philosophical Transactions of the Royal Society London*, **64**, 445–460.
4. Pockels, A. (1891). Prof. Van der Waals on the continuity of the liquid and gaseous states. *Nature*, **43**, 437–439.
5. Langmuir, I. (1917). The constitution and fundamental properties of solids and liquids. II. Liquids. *Journal of the American Chemical Society*, **39**, 1848–1906.
6. Blodgett, K. (1935). Films built by depositing successive monomolecular layers on a solid surface. *Journal of the American Chemical Society*, **57**, 1007–1022.
7. Allara, D. L., Nuzzo, R. G. (1985). Spontaneously organized molecular assemblies. 1. Formation, dynamics, and physical properties of *n*-alkanoic acids adsorbed from solution on an oxidized aluminum surface. *Langmuir*, **1**, 45–52.
8. Sagiv, J. (1980). Organized monolayers by adsorption. 1. Formation and structure of oleophobic mixed monolayers on solid surfaces. *Journal of the American Chemical Society*, **102**, 92–98.
9. Finklea, H. O., Robinson, L. R., Blackburn, A., Ritcher, B., Allara, D. L., Bright, T. (1986). Formation of an organized monolayer by solution adsorption of octadecyltrichlorosilane on gold: Electrochemical properties and structural characterization. *Langmuir*, **2**, 239–244.
10. Sabatani, E., Rubinstein, I., Maoz, R., Sagiv, J. (1987). Organized self-assembling monolayers on electrodes: Part I.

Octadecyl derivatives on gold. *Journal of Electroanalytical Chemistry*, **219**, 365–371.

11. Nuzzo, R. G., Allara, D. L. (1983). Adsorption of bifunctional organic disulfides on gold surfaces. *Journal of the American Chemical Society*, **105**, 4481–4483.
12. Bain, C. D., Troughton, E. B., Tao, Y. T., Evall, J., Whitesides, G. M., Nuzzo, R. G. (1989). Formation of monolayer films by the spontaneous assembly of organic thiols from solution onto gold. *Journal of the American Chemical Society*, **111**, 321–335.
13. Huisman, B. H., Rudkevich, D. M., van Veggel, F. C. J. M., Reinhoudt, D. N. (1996). Self-assembled monolayers of carceplexes on gold. *Journal of the American Chemical Society*, **118**, 3523–3524.
14. van der Vegte, E. W., Subbotin, A., Hadziioannou, G., Ashton, P. R., Preece, J. A. (2000). Nanotribological properties of unsymmetrical *n*-dialkyl sulfide monolayers on gold: Effect of chain length on adhesion, friction, and imaging. *Langmuir*, **16**, 3249–3256.
15. Bishop, A. R., Nuzzo, R. G. (1996). Self-assembled monolayers: Recent developments and applications. *Current Opinion on Colloid Interface Science*, **1**, 127–136.
16. Nuzzo, R. G., Zegarski, B. R., Dubois, L. H. (1987). Fundamental studies of the chemisorption of organosulfur compounds on gold(111). Implications for molecular self-assembly on gold surfaces. *Journal of the American Chemical Society*, **109**, 733–740.
17. Schessler, H. M., Karpovich, D. S., Blanchard, G. J. (1996). Quantitating the balance between enthalpic and entropic forces in alkanethiol/gold monolayer self assembly. *Journal of the American Chemical Society*, **118**, 9645–9651.
18. Petroski, J., Chou, M., Creutz, C. (2009). The coordination chemistry of gold surfaces: Formation and far-infrared spectra of alkanethiolate-capped gold nanoparticles. *Journal of Organometallic Chemistry*, **694**, 1138–1143.
19. Biebuyck, H. A., Bain, C. D., Whitesides, G. M. (1994). Comparison of organic monolayers on polycrystalline gold spontaneously assembled from solutions containing dialkyl disulfides or alkanethiols. *Langmuir*, **10**, 1825–1831.
20. Biebuyck, H. A., Whitesides, G. M. (1993). Interchange between monolayers on gold formed from unsymmetrical disulfides and solutions of thiols: Evidence for sulfur–sulfur bond cleavage by gold metal. *Langmuir*, **9**, 1766–1770.
21. Hagenhoff, B., Benninghoven, A., Spinke, J., Liley, M., Knoll, W. (1993). Time-of-flight secondary ion mass spectrometry investigations of self-assembled monolayers of organic thiols, sulfides, and disulfides on gold surfaces. *Langmuir*, **9**, 1622–1624.
22. Badia, A., Demers, L., Dickinson, L., Morin, F.G, Lennox, R. B., Reven, L. (1997). Gold–sulfur interactions in alkylthiol self-assembled monolayers formed on gold nanoparticles studied by solid-state NMR. *Journal of the American Chemical Society*, **119**, 11104–11105.
23. Troughton, E. B., Bain, C. D., Whitesides, G. M., Nuzzo, R. G., Allara, D. L., Porter, M. D. (1988). Monolayer films prepared by the spontaneous self-assembly of symmetrical and unsymmetrical dialkyl sulfides from solution onto gold substrates: Structure, properties, and reactivity of constituent functional groups. *Langmuir*, **4**, 365–385.
24. Strong, L., Whitesides, G. M. (1988). Structures of self-assembled monolayer films of organosulfur compounds adsorbed on gold single crystals: Electron diffraction studies. *Langmuir*, **4**, 546–558.
25. Zhang, M., Anderson, M. R. (1994). Investigation of the charge transfer properties of electrodes modified by the spontaneous adsorption of unsymmetrical dialkyl sulfides. *Langmuir*, **10**, 2807–2813.
26. Lavrich, D. J., Wetterer, S. M., Bernasek, S. L., Scoles, G. (1998). Physisorption and chemisorption of alkanethiols and alkyl sulfides on Au(111). *Journal of Physical Chemistry B*, **102**, 3456–3465.
27. Weidner, T., Ballav, N., Siemeling, U., Troegel, D., Walter, T., Tacke, R., Castner, D. G., Zharinikov, M. (2009). Tripodal binding units of self-assembled monolayers on gold: A comparison of thiol and thioether headgroups. *Journal of Physical Chemistry C*, **113**, 19609–19617.
28. Love, J. C., Estroff, L. A., Kriebel, J. K., Nuzzo, R. G., Whitesides, G. M. (2005). Self-assembled monolayers of thiolates on metals as a form of nanotechnology. *Chemical Reviews*, **105**, 1103–1170.
29. Sun, A., Lahann, J. (2009). Dynamically switchable biointerfaces. *Soft Matter*, **5**, 1555–1561.
30. Cole, M. A., Voelcher, N. H., Tissen, H., Griesser, H. J. (2009). Stimuli-responsive interfaces and systems for the control of protein–surface and cell–surface interactions. *Biomaterials*, **30**, 1827–1850.
31. Higuchi, M., Minoura, N., Kinoshita, T. (1995). Photoresponsive behavior of a monolayer composed of an azobenzene containing polypeptide in the main chain. *Colloid and Polymer Science*, **273**, 1022–1027.
32. Seki, T., Sekizawa, H., Ichimura, K. (1999). Non-linear characteristics of the photoinduced deformation in azobenzene-containing monolayer at the air–water interface. *Polymer Journal*, **31**, 1079–1082.
33. Katz, E., Lion-Dagan, M., Willner, I. (1995). Control of electrochemical processes by photoisomerizable spiropyran monolayers immobilized onto Au electrodes: Amperometric transduction of optical signals. *Journal of Electroanalytical Chemistry*, **382**, 25–31.
34. Katz, E., Willner, B., Willner, I. (1997). Light-controlled electron transfer reactions at photoisomerizable monolayer electrodes by means of electrostatic interaction: Active interfaces for the amperometric transduction of recorded optical signals. *Biosensors and Bioelectronics*, **12**, 703–719.
35. Koide, Y., Terasaki, N., Akiyama, T., Yamada, S. (1999). Effects of spacer-chain length on the photoelectrochemical responses of monolayer assemblies with ruthenium tris(2,2′-bipyridine)–viologen linked disulfides. *Thin Solid Films*, **350**, 223–227.
36. Zimmerman, R., Basabe-Desmonts, L., van der Baan, F., Reinhoudt, D. N., Crego-Calama, M. (2005). A combinatorial

approach to surface-confined cation sensors in water. *Journal of Materials Chemistry*, **15**, 2772–2777.

37. Brennan, J. D., Brown, S., Forster, D., Kallury, R. K., Krull, U. J. (1991). Fluorescence transduction of an enzyme–substrate reaction by covalently immobilized monolayers of amphiphiles. *Analytica Chimica Acta*, **255**, 73–82.
38. Sekar, M. M. A., Hampton, P. D., Buranda, T., Lopez, G. P. (1999). Multifunctional monolayer assemblies for reversible direct fluorescence transduction of protein-ligand interactions at surfaces. *Journal of the American Chemical Society*, **121**, 5135–5141.
39. Li, H., Kang, J., Ding, L., Lü, F., Fang, Y. (2008). A dansyl-based fluorescent film: Preparation and sensitive detection of nitroaromatics in aqueous phase. *Journal of Photochemistry and Photobiology A*, **197**, 226–231.
40. Salisbury, C. M., Maly, D. J., Ellman, J. A. (2002). Peptide microarrays for the determination of protease substrate specificity. *Journal of the American Chemical Society*, **124**, 14868–14870.
41. Zhu, Q., Uttamchandani, M., Li, D., Lesaicherre, M. L., Yao, S. Q. (2003). Enzymatic profiling system in a small-molecule microarray. *Organic Letters*, **5**, 1257–1260.
42. Lim, S. Y., Chung, W. Y., Lee, H. K., Park, M. S., Park, H. G. (2008). Direct and nondestructive verification of PNA immobilization using click chemistry. *Biochemical and Biophysical Research Communications*, **376**, 633–636.
43. Feringa, B. L., Jager, W. F., de Lange, B. (1993). Organic materials for reversible optical data storage. *Tetrahedron*, **49**, 8267–8310.
44. Dürr, H., Bouas-Laurent, H. (2003). *Photochromism: Molecules and Systems*. Elsevier, Amsterdam.
45. Zhao, Y., Ikeda, T. (2009). *Smart Light-Responsive Materials: Azobenzene-Containing Polymers and Liquid Crystals*. Wiley-Interscience, Hoboken, NJ.
46. Sahyun, M. R. V. (1972). Characterization of the photochemical processes involved in the photochromism of spiropyrans. *Photograhphic Science and Engineering*, **16**, 63–68.
47. Irie, M. (2000). Diarylethenes for memories and switches. *Chemical Reviews*, **100**, 1685–1716.
48. Oh, S. K., Nakagawa, M., Ichimura, K. (1999). Self-assembled monolayers derived from calix[4]resorcinarenes exhibiting excellent desorption-resistance and their applicability to surface energy photocontrol. *Chemistry Letters*, **4**, 349–350.
49. Ichimura, K., Oh, M. Nakagawa, S. K. (2000). Light-driven motion of liquids on a photoresponsive surface. *Science*, **288**, 1624–1626.
50. Yi, Y., Farrow, M. J., Korblova, E., Walba, D. M., Furtak, T. E. (2009). High-sensitivity aminoazobenzene chemisorbed monolayers for photoalignment of liquid crystals. *Langmuir*, **25**, 997–1003.
51. Ikegami, A., Suda, M., Watanabe, T., Einaga, Y. (2010). Reversible optical manipulation of superconductivity. *Angewandte Chemie International Edition*, **49**, 372–374.
52. Hayashi, G., Hagihara, M., Dohno, C., Nakatani, K. (2007). Photoregulation of a peptide-RNA interaction on a gold surface. *Journal of the American Chemical Society*, **129**, 8678–8679.
53. Pearson, D., Downard, A. J., Muscroft-Taylor, A., Abell, A. D. (2007). Reversible photoregulation of binding of α-chymotrypsin to a gold surface. *Journal of the American Chemical Society*, **129**, 14862–14863.
54. Liu, D., Xie, Y., Shao, H., Jiang, X. (2009). Using azobenzene-embedded self-assembled monolayers to photochemically control cell adhesion reversibly. *Angewandte Chemie International Edition*, **48**, 4406–4408.
55. Donor, A., Katz, E., Tao, G., Willner, I. (1997). Photochemically-, chemically-, and pH-controlled electrochemistry at functionalized spiropyran monolayer electrodes. *Langmuir*, **13**, 1783–1790.
56. Blonder, R., Katz, E., Willner, I., Wray, V., Bückmann, A. F. (1997). Application of a nitrospiropyran-FAD-reconstituted glucose oxidase and charged electron mediators as optobioelectronic assemblies for the amperometric transduction of recorded optical signals: Control of the "on"–"off" direction of the photoswitch. *Journal of the American Chemical Society*, **119**, 11747–11757.
57. Willner, I., Rubin, S. (1996). Control of the structure and functions of biomaterials by light. *Angewandte Chemie International Edition*, **35**, 367–385.
58. Matsuda, K., Ikeda, M., Irie, M. (2004). Photochromism of diarylethene-capped gold nanoparticles. *Chemistry Letters*, **33**, 456–457.
59. Ikeda, M., Tanifuji, N., Yamaguchi, H., Irie, M., Matsuda, K. (2007). Photoswitching of conductance of diarylethene-Au nanoparticle network. *Chemical Communications*, 1355–1357.
60. Sakano, T., Yamaguchi, H., Tanifuji, N., Irie, M., Matsuda, K. (2008). Percolation-type photoswitching behavior in conductance of diarylethene–silver nanoparticle networks. *Chemistry Letters*, **37**, 634–635.
61. Matsuda, K., Yamaguchi, H., Sakano, T., Ikeda, M., Tanifuji, N., Irie, M. (2008). Conductance photoswitching of diarylethene-gold nanoparticle network induced by photochromic reaction. *Journal of Physical Chemistry C*, **112**, 17005–17010.
62. Masuda, T., Irie, M., Uosaki, K. (2009). Photoswitching of electron transfer property of diarylethene–viologen linked molecular layer constructed on a hydrogen-terminated Si(111) surface. *Thin Solid Films*, **518**, 591–595.
63. Dulic, D., van der Molen, S. J., Kudernac, T., Jonkman, H. T., de Jong, J. J. D., Bowden, T. N., van Esch, J., Feringa, B. L., van Wees, B. J. (2003). One-way optoelectronic switching of photochromic molecules on gold. *Physical Review Letters*, **91**, 207402-1–207402-4.
64. van der Molen, S. J., Van der Vegte, H., Kudernac, T., Amin, I., Feringa, B. L., van Wees, B. J. (2006). Stochastic and photochromic switching of diarylethenes studied by scanning tunnelling microscopy. *Nanotechnology*, **17**, 310–314.
65. He, J., Chen, F., Liddell, P. A., Andréasson, J., Straigh, S. D. T, Gust, D., Moore, T. A., Moore, A. L., Li, J., Sankey, O. F., Lindsay, S. M. (2005). Switching of a photochromic

molecule on gold electrodes: Single-molecule measurements. *Nanotechnology*, **16**, 695–702.

66. Gulino, A., Lupo, F., Condorelli, G. G., Fragalà, M. E., Amato, M. E., Scarlata, G. (2008). Reversible photoswitching of stimuli-responsive Si(100) surfaces engineered with an assembled 1-cyano-1-phenyl-2-[4′-(10-undecenoyloxy)phenyl]-ethylene monolayer. *Journal of Materials Chemistry*, **18**, 5011–5018.
67. Balzani, V., Credi, A., Venturi, M. (2008). Molecular machines working on surfaces. *ChemPhysChem*, **9**, 202–220.
68. Pollard, M. M., Lubonska, M., Rudolf, P., Feringa, B. L. (2007). Controlled rotary motion in a monolayer of molecular motors. *Angewandte Chemie International Edition*, **46**, 1278–1280.
69. Berna, J., Leigh, D. A., Lubomska, M., Mendoza, S. M., Pérez, E. M., Rudolf, P., Teobaldi, G., Zerbetto, F. (2005). Macroscopic transport by synthetic molecular machines. *Nature Materials*, **4**, 704–710.
70. Popovic, Z., Otter, M., Calzaferri, G., De Cola, L. (2007). Self-assembling living systems with functional nanomaterials. *Angewandte Chemie International Edition*, **46**, 6188–6191.
71. Willner, I., Pardo-Yssar, V., Katz, E., Ranjit, K. T. (2001). A photoactivated 'molecular train' for optoelectronic applications: Light-stimulated translocation of a β-cyclodextrin receptor within a stoppered azobenzene-alkyl chain supramolecular monolayer assembly on a Au-electrode. *Journal of Electroanalytical Chemistry*, **497**, 172–177.
72. Li, W., Lynch, V., Thompson, H., Fox, M. A. (1997). Self-assembled monolayers of 7-(10-thiodecoxy)coumarin on gold: Synthesis, characterization, and photodimerization. *Journal of the American Chemical Society*, **119**, 7211–7217.
73. Fox, M. A., Wooten, M. D. (1997). Characterization, adsorption and photochemistry of self-assembled monolayers of 10-thiodecyl-2-anthryl ether on gold. *Langmuir*, **13**, 7099–7105.
74. Abbott, S., Ralston, J., Reynolds, G., Hayes, R. (1999). Reversible wettability of photoresponsive pyrimidine-coated surfaces. *Langmuir*, **15**, 8923–8928.
75. Zhang, G., He, T., Ma, Y., Chen, Z., Yang, W., Yao, J. (2003). Fabrication of photochromic phosphomolybdic acid monolayer film. *Physics Chemitry and Chemical Physical*, **5**, 2751–2753.
76. (a) Sortino, S., Petralia, S., Compagnini, G., Conoci, S., Condorelli, G. (2002). Light-controlled nitric oxide generation from a novel self-assembled monolayer on a gold surface. *Angewandte Chemie International Edition*, **41**, 1914–1916. (b) Sortino S. (2010). Light-controlled nitric oxide delivering molecular assemblies. *Chemical Society Reviews*, **39**, 2903–2913.
77. Culotta, E., Koshland, D. E. (1992). NO news is good news. *Science*, **258**, 1862–1865.
78. Maragos, C. M., Wang, J. M., Hrabie, J. A., Oppenheim, J. J., Keefer, L. K. (1993). Nitric oxide/nucleophile complexes inhibit the in vitro proliferation of A375 melanoma cells via nitric oxide release. *Cancer Research*, **53**, 564–568.
79. Ohmuro-Matsuyama, Y., Tatsu, Y. (2008). Photocontrolled cell adhesion on a surface functionalized with a caged arginine-glycine-aspartate peptide. *Angewandte Chemie International Edition*, **47**, 7527–7529.
80. Nakanishi, J., Kikuchi, Y., Takarada, T., Nakayama, H., Yamaguchi, K., Maeda, M. (2004). Photoactivation of a substrate for cell adhesion under standard fluorescence microscopes. *Journal of the American Chemical Society*, **126**, 16314–16315.
81. Nakanishi, J., Kikuchi, Y., Inoue, S., Yamaguchi, K., Takarada, T., Maeda, M. (2007). Spatiotemporal control of migration of single cells on a photoactivatable cell microarray. *Journal of the American Chemical Society*, 6694–6695.
82. Petersen, S., Alonso, J. M., Specht, A., Duodu, P., Goeldner, M., del Campo, A. (2008). Phototriggering of cell adhesion by caged cyclic RGD peptides. *Angewandte Chemie International Edition*, **47**, 3192–3195.
83. Isosomppi, M., Tkachenko, N. V., Efimov, A., Kaunisto, K., Hosomizu, K., Imahori, H., Lemmetyinen, H. (2005). Photoinduced electron transfer in multilayer self-assembled structures of porphyrins and porphyrin-fullerene dyads on ITO. *Journal of Materials Chemistry*, **15**, 4546–4554.
84. Cho, Y. J., Ahn, T. K., Song, H., Kim, K. S., Lee, C. Y., Seo, W. S., Lee, K., Kim, S. K., Kim, D., Park, J. T. (2005). Unusually high performance photovoltaic cell based on a [60]fullerene metal cluster-porphyrin dyad SAM on an ITO electrode. *Journal of the American Chemical Society*, **127**, 2380–2381.
85. Fukuzumi, S., Endo, Y., Kashiwagi, Y., Araki, Y., Ito, O., Imahori, H. (2003). Novel photocatalytic function of porphyrin-modified gold nanoclusters in comparison with the reference porphyrin compound. *Journal of Physical Chemistry B*, **107**, 11979–11986.
86. Akiyama, T., Aiba, K., Hoashi, K., Wang, M., Sugawa, K., Yamada, S. (2010). Enormous enhancement in photocurrent generation using electrochemically fabricated gold nanostructures. *Chemical Communications*, **46**, 306–308.
87. Mao, Y., Luo, C., Ouyang, Q. (2003). Studies of temperature-dependent electronic transduction on DNA hairpin loop sensor. *Nucleic Acids Research*, **31**, e108–e113.
88. Heskins, M., Guillet, J. E., James, E. (1968). Solution properties of poly(*N*-isopropylacrylamide). *Journal of Macromolecular Science Chemistry*, **A2**, 1441–1455.
89. Yamada, N., Okano, T., Sakai, H., Karikusa, F., Sawasaki, Y., Sakurai, Y. (1990). Thermo-responsive polymeric surfaces; control of attachment and detachment of cultured cells. *Die Makromolekulare Chemie, Rapid Communications*, **11**, 571–576.
90. Nagase, K., Kobayashi, J., Okano, T. (2009). Temperature-responsive intelligent interfaces for biomolecular separation and cell sheet engineering. *Journal of Royal Society Interface*, **6**, S293–S309.
91. Scherer, W. F., Syverton, J. T., Gey, G. O. (1953). Studies on the propagation in vitro of poliomyelitis viruses IV. Viral multiplication in a stable strain of human malignant epithelial

cells (strain hela) derived from an epidermoid carcinoma of the cervix. *Journal of Experimental Medicine*, **97**, 695–710.

92. Frangakis, M. V., Kimelberg, H. K. (1984). Dissociation of neonatal rat brain by dispase for preparation of primary astrocyte cultures. *Neurochemistry Research*, **9**, 1689–1698.

93. Canavan, H. E., Cheng, X., Graham, D. J., Ratner, B. D. (2005). Cell sheet detachment affects the extracellular matrix: A surface science study comparing thermal liftoff, enzymatic, and mechanical methods. *Journal of Biomedical Materials Research*, **75A**, 1–13.

94. da Silva, R. M. P., Mano, J. F., Reis, R. L. (2007). Smart thermoresponsive coatings and surfaces for tissue engineering: Switching cell-material boundaries. *Trends in Biotechnology*, **25**, 577–583.

95. Kushida, A., Yamato, M., Konno, C., Kikuchi, A., Sakurai, Y., Okano, T. (1999). Decrease in culture temperature releases monolayer endothelial cell sheets together with deposited fibronectin matrix from temperature-responsive culture surfaces. *Journal Biomedical Materials Research*, **45**, 355–362.

96. Ebara, M., Yamato, M., Aoyagi, T., Kikuchi, A., Sakai, K., Okano, T. (2004). Temperature-responsive cellculture surfaces enable "on–off" affinity control between cell integrins and RGDS ligands. *Biomacromolecules*, **5**, 505–510.

97. Nandkumar, M. A., Yamato, M., Kushida, A., Konno, C., Kikuchi, A., Sakurai, Y., Okano, T. (2002). Two-dimensional cell sheet manipulation of heterotypically co-cultured lung cells utilizing temperature-responsive culture dishes results in long-term maintenance of differentiated epithelial cell functions. *Biomaterials*, **23**, 1121–1130.

98. Sumide, T., Nishida, K., Yamato, M., Ide, T., Hayashida, Y., Watanabe, K., Yang, J., Konno, C., Kikuchi, A., Maeda, N., Watanabe, H., Okano, T., Tano, Y. (2006). Functional human corneal endothelial cell sheets harvested from temperature-responsive culture surfaces. *FASEB*, **20**, 392–414.

99. Kitano, H., Kago, H., Matsuura, K. (2009). Temperature-responsive polymer brush constructed on a colloidal gold monolayer. *Journal of Colloid Interface Science*, **331**, 343–350.

100. Cho, E. C., Kim, Y. D., Cho, K. (2004). Thermally responsive poly(*N*-isopropylacrylamide) monolayer on gold: Synthesis, surface characterization, and protein interaction/adsorption studies. *Polymer*, **45**, 3195–3204.

101. Idota, N., Kikuchi, A., Kobayashi, J., Akiyama, Y., Sakai, K., Okano, T. (2006). Thermal modulated interaction of aqueous steroids using polymer-grafted capillaries. *Langmuir*, **22**, 425–430.

102. Nagase, K., Kobayashi, J., Kikuchi, A., Akiyama, Y., Kanazawa, H., Okano, T. (2007). Interfacial property modulation of thermoresponsive polymer brush surfaces and their interaction with biomolecules. *Langmuir*, **23**, 9409–9415.

103. Huber, D. L., Manginell, R. P., Samara, M. A., Kim, B. I., Bunker, B. C. (2003). Programmed adsorption and release of proteins in a microfluidic device. *Science*, **301**, 352–354.

104. Ernst, O., Lieske, A., Jager, M., Lankenau, A., Duschl, C. (2007). Control of cell detachment in a microfluidic device using a thermo-responsive copolymer on a gold surface. *Lab Chip*, **7**, 1322–1329.

105. Ernst, O., Lieske, A., Holländer, A., Lankenau, A., Duschl, C. (2008). Tuning of thermo-responsive self-assembly monolayers on gold for cell-type-specific control of adhesion. *Langmuir*, **24**, 10259–10264.

106. Liu, X., Liu, H., Zhou, W., Zheng, H., Yin, X., Li, Y., Guo, Y., Zhu, M., Ouyang, C., Zhu, D., Xia, A. (2010). Thermoreversible covalent self-assembly of oligo(*p*-phenylenevinylene) bridged gold nanoparticles. *Langmuir*, **26**, 3179–3185.

107. Yoshizako, K., Akiyama, Y. (2002). Regulation of protein binding toward a ligand on chromatographic matrixes by masking and forced-releasing effects using thermoresponsive polymer. *Analytical Chemistry*, **74**, 4160–4166.

108. Zareie, H. M., Boyer, C., Bulmus, V., Nateghi, E., Davis, T. P. (2008). Temperature-responsive self-assembled monolayers of oligo(ethylene glycol): Control of biomolecular recognition. *ACS Nano*, **2**, 757–765.

109. Prime, K. L., Whitesides, G. M. (1991). Self-assembled organic monolayers: Model systems for studying adsorption of proteins at surfaces. *Science*, **252**, 1164–1167.

110. Bain, C. D., Whitesides, G. M. (1989). A study by contact angle of the acid–base behavior of monolayers containing α-mercaptocarboxylic acids adsorbed on gold: An example of reactive spreading. *Langmuir*, **5**, 1370–1378.

111. Ullman, A. (1991). *An Introduction to Ultrathin Organic Films. From Langmuir–Blodgett to Self-Assembly*. Academic Press, New York.

112. Takehara, K., Ide, Y., Aihara, M., Obuchi, E. (1992). An ion-gate response of the glutathione monolayer assembly formed on a gold electrode. Part 1. The effect of pH, K^+ and Ca^{2+}. *Bioelectrochemistry and Bioenergetics*, **29**, 103–111.

113. Beulen, M. W. J., van Veggel, F. C. J. M., Reinhoudt, D. N. (1999). Coupling of acid–base and redox functions in mixed sulphide monolayer on gold. *Chemical Communications*, 503–504.

114. Sellergren, B., Swietlow, A., Arnebrant, T., Unger, K. (1996). Consecutive selective adsorption of pentamidine and phosphate biomolecules on a self assembled layer: Reversible formation of a chemically selective coating. *Analytical Chemistry*, **68**, 402–407.

115. Auer, F., Schubert, D. W., Stamm, M., Arnebrant, T., Swietlow, A., Zizlsperger, M., Sellergren, B. (1999). Switchable assembly of stable, ordered molecular layers. *Chemistry—A European Journal*, **5**, 1150–1159.

116. Jiang, Y., Wang, Z., Xu, H., Chen, H., Zhang, X., Smet, M., Dehaen, W., Hirano, Y., Ozaki, Y. (2006). Investigation into pH-responsive self-assembled monolayers of acylated anthranilate-terminated alkanethiol on a gold surface. *Langmuir*, **22**, 3715–3720.

117. Yu, X., Wang, Z., Jiang, Y., Shi, F., Zhang, X. (2005). Reversible pH-responsive surface: From superhydrophobicity to superhydrophilicity. *Advanced Materials*, **17**, 1289–1293.

118. Katz, E., Lion-Dagan, M., Willner, I. (1996). pH-switched electrochemistry of pyrroloquinoline quinone at Au electrodes

modified by functionalized monolayers. *Journal of Electroanalytical Chemistry*, **408**, 107–112.

119. Delahay, P. (1965). *Double Layer and Electrode Kinetics*. Interscience, New York.

120. Hou, Z., Abbott, N. L., Stroeve, P. (2000). Self-assembled monolayers on electroless gold impart pH-responsive transport of ions in porous membranes. *Langmuir*, **16**, 2401–2404.

121. Yokoyama, K., Leigh, B. S., Sheng, Y., Niki, K., Nakamura, N., Ohno, H., Winkler, J. R., Gray, H. B., Richards, J. H. (2008). Electron tunnelling through Pseudomonas aeruginosa azurins on SAM gold electrodes. *Inorganica Chimica Acta*, **361**,1095–1099.

122. Doneux, T., Bouffier, L., Mello, L. V., Rigden, D. J., Kejnovská, I., Fernig, D. G., Higgins, S. J., Nichols, R. J. (2009). Molecular dynamics and electrochemical investigations of a pH-responsive peptide monolayer. *Journal of Physical Chemistry C*, **113**, 6792–6799.

123. (a) Niwa, D., Yamada, Y., Homma, T., Osaka, T. (2004). Formation of molecular templates for fabricating on-chip biosensing devices. *Journal of Physical Chemistry B*, **108**, 3240–3245. (b) Tauk, L., Schröder, A. P., Decher, G., Giuseppone, N. (2009). Hierarchical functional gradients of pH-responsive self-assembled monolayers using dynamic covalent chemistry on surfaces. *Nature Chemistry*, **1**, 649–656.

124. Jiang, X., Ferrigno, R., Mrksich, M., Whitesides, G. M. (2003). Electrochemical desorption of self-assembled monolayers noninvasively releases patterned cells from geometrical confinements. *Journal of the American Chemical Society*, **125**, 2366–2367.

125. Mali, P., Bhattacharjee, N., Searson, P. C. (2006). Electrochemically programmed release of biomolecules and nanoparticles. *Nano Letters*, **6**, 1250–1253.

126. Huang, S., Schopf, E., Chen, Y. (2007). Dynamically configurable biomolecular nanoarrays. *Nano Letters*, **7**, 3116–3121.

127. Ceroni, P., Credi, A., Venturi, M. (2010). *Electrochemistry of Functional Supramolecular Systems*. John Wiley & Sons, New York.

128. Bard, A. J., Faulkner, L. R. (2001). *Electrochemical Methods: Fundamentals and Applications*. John Wiley & Sons, New York.

129. Beulen, M. W. J., van Veggel, F. C. J. M., Reinhoudt, D. N. (2000). Self-assembled monolayers of metallosalophenes on gold. *Israel Journal of Chemistry*, **40**, 73–80.

130. Cho, Y. J., Song, H., Lee, K., Kim, K., Kwak, J., Kim, S., Park, J. T. (2002). The first observation of four-electron reduction in [60]fullerene-metal cluster self-assembled monolayers (SAMs). *Chemical Communications*, 2966–2967.

131. (a) Larsen, A. G., Gothelf, K. V. (2005). Electrochemical properties of mixed self-assembled monolayers on gold electrodes containing mercaptooctylhydroquinone and alkylthiols. *Langmuir*, **21**, 1015–1021. (b) Weidner, T., Zharnikov, M., Hoβbach, J., Castner, D. G., Siemeling, U. (2010). Adamantane-based tripodal thioether ligands functionalized with a redox-active ferrocenyl moiety for self-assembled monolayers. *Journal of Physical Chemistry C*, **114**, 14975–14982.

132. Niwa, M., Fukui, H., Higashi, N. (1993). Irreversible adsorption and electron transfer reaction of cytochrome c on poly(methacrylic acid) segment-carrying triblock-polyion-complex monolayer electrodes. *Macromolecules*, **26**, 5816–5818.

133. Millo, D., Ranieri, A., Gross, P., Ly, H. K., Borsari, M., Hildebrandt, P., Wuite, G. J. L., Gooijer, C., ven der Zwan, G. (2009). Electrochemical response of cytochrome c immobilized on smooth and roughened silver and gold surfaces chemically modified with 11-mercaptounodecanoic acid. *Journal of Physical Chemistry C*, **113**, 2861–2866.

134. Andolfi, L., Bruce, D., Cannistraro, S., Canters, G. W., Davis, J. J., Hill, H. A. O., Crozier, J., Verbeet, M. P.., Wrathmell, C. L., Astier, Y. (2004). The electrochemical characteristics of blue copper protein monolayers on gold. *Journal of Electroanalytical Chemistry*, **565**, 21–28.

135. Lahann, J., Mitragotri, S., Tran, T. N., Kaido, H., Sundaram, J., Choi, I. S., Hoffer, S., Somorjai, G. A., Langer, R. (2003). A reversibly switching surface. *Science*, **299**, 371–374.

136. Wang, X., Kharitonov, A. B., Katz, E., Willner, I. (2003). Potential-controlled molecular machinery of bipyridinium monolayer-functionalized surfaces: An electrochemical and contact angle analysis. *Chemical Communications*, 1542–1543.

137. Liu, Y., Mu, L., Liu, B., Zhang, S., Yang, P., Kong, J. (2004). Controlled protein assembly on a switchable surface. *Chemical Communications*, 1194–1195.

138. Mu, L., Liu, Y., Cai, S., Kong, J. (2007). A smart surface in a microfluidic chip for controlled protein separation. *Chemistry—A European Journal*, **13**, 5113–5120.

139. Xu, S., Podoprygorina, G., Böhmer, V., Ding, Z., Rooney, P., Rangan, C., Mittler, S. (2007). Tetraurea calix[4]arenes with sulfur functions: Synthesis, dimerization to capsules, and self-assembly on gold. *Organic and Biomolecular Chemistry*, **4**, 558–568.

140. Xu, S., Podoprygorina, G., Moon, C., Böhmer, V., Ding, Z., Mittler, S. (2008). Electrochemical response of ferrocenium filled calix[4]arene capsules in solution and immobilized on gold. *Electrochimica Acta*, **53**, 7981–7987.

141. Nijhuis, C. A., Huskens, J., Reinhoudt, D. N. (2004). Binding control and stoichiometry of ferrocenyl dendrimers at a molecular printboard. *Journal of the American Chemical Society*, **126**, 12266–12267.

142. Nijhuis, C. A., Yu, F., Knoll, W., Huskens, J., Reinhoudt, D. N. (2005). Multivalent dendrimers at molecular printboards: Influence of dendrimer structure on binding strength and stoichiometry and their electrochemically induced desorption. *Langmuir*, **21**, 7866–7876.

143. Ling, X. Y., Reinhoudt, D. N., Huskens, J. (2008). Reversible attachment of nanostructures at molecular printboards through supramolecular glue. *Chemical Materials*, **20**, 3574–3578.

144. Katz, E., Lioubashevsky, O., Willner, I. (2004). Electromechanics of a redox-active rotaxane in a monolayer assembly on an electrode. *Journal of the American Chemical Society*, **126**, 15520–15532.

145. Katz, E., Baron, R., Willner, I., Richke, N., Levine, R. D. (2005). Temperature-dependent and friction-controlled electrochemically induced shuttling along molecular strings associated with electrodes. *ChemPhysChem*, **6**, 2179–2189.

146. Katz, E., Sheeney-Haj-Ichia, L., Willner, I. (2004). Electrical contacting of glucose oxidase in a redox-active rotaxane configuration. *Angewandte Chemie International Edition*, **43**, 3292–3300.

147. Tseng, H. R., Wu, D., Fang, N. X., Zhang, X., Stoddart, J. F. (2004). Electrochemically controlled nanoscale machine on gold surfaces. *ChemPhysChem*, **5**, 111–116.

148. Xiao, Y., Lubin, A. A., Heeger, A. J., Plaxco, K. W. (2005). Label-free electronic detection of thrombin in blood serum by using an aptamer-based sensor. *Angewandte Chemie International Edition*, **44**, 5456–5459.

149. Radi, A. E., Sanchez, J. L. A., Baldrich, E., O'Sullivan, C. K. (2006). Reagentless, reusable, ultrasensitive electrochemical molecular beacon aptasensor. *Journal of the American Chemical Society*, **128**, 117–124.

150. Baker, B. R., Lai, R. Y., Wood, M. S., Doctor, E. H., Heeger, A. J., Plaxco, K. W. (2006). An electronic, aptamer-based small-molecule sensor for the rapid, label-free detection of cocaine in adulterated samples and biological fluids. *Journal of the American Chemical Society*, **128**, 3138–3139.

151. Immoos, C. E., Lee, S. J., Grinstaff, M. W. (2004). Conformationally gated electrochemical gene detection. *ChemBioChem*, **5**, 1100–1103.

152. Fan, C., Plaxco, K. W., Heeger, A. J. (2003). Electrochemical interrogation of conformational changes as a reagentless method for the sequence-specific detection of DNA. *Proceedings of the National Academy of Sciences USA*, **100**, 9134–9137.

153. Nakamura, M., Ueda, M., Watanabe, S., Kumamoto, S., Yamana, K. (2007). Syntheses of antraquinone capped hairpin DNAs and electrochemical redox responses from their self-assembled monolayers on gold electrode. *Tetrahedron Letters*, **48**, 6159–6162.

154. Zuo, X., Song, S., Zhang, J., Pan, D., Wang, L., Fan, C. (2007). A target-responsive electrochemical aptamer switch (TREAS) for reagentless detection of nanomolar ATP. *Journal of the American Chemical Society*, **129**, 1042–1043.

155. Wang, H., Zhang, C., Li, Y., Qi, H. (2006). Electrogenerated chemiluminescence detection for deoxyribonucleic acid hybridization based on gold nanoparticles carrying multiple probes. *Analytica Chimica Acta*, **575**, 205–211.

156. Ricci, F., Lai, R. Y., Heeger, A. J., Plaxco, K. W., Sumner, J. J. (2007). Effect of molecular crowding on the response of an electrochemical DNA sensor. *Langmuir*, **23**, 6827–6834.

157. Murphy, J. N., Cheng, A. K. H., Yu, H. Z., Bizzotto, D. (2009). On the nature of DNA self-assembled monolayers on Au: Measuring surface heterogeneity with electrochemical in situ fluorescence microscopy. *Journal of the American Chemical Society*, **131**, 4042–4050.

158. Lee, C. S., Baker, S. E., Marcus, M. S., Yang, W., Eriksson, M. A., Hamers, R. J. (2004). Electrically addressable bimolecular functionalization of carbon nanotube and carbon nanofiber electrodes. *Nano Letters*, **4**, 1713–1716.

159. Yang, W., Baker, S. E., Butler, J. E., Lee, C. S., Russell, J. N., Shang, L., Sun, B., Hamers, R. J. (2005). Electrically addressable biomolecular functionalization of conductive nanocrystalline diamond thin film. *Chemical Materials*, **17**, 938–940.

160. Mendes, P. M., Christman, K. L., Parthasarathy, P., Schopf, E., Ouyang, J., Yang, Y., Preece, J. A., Maynard, H. D., Chen, Y., Stoddart, J. F. (2007). Electrochemically controllabe conjugation of proteins on surfaces. *Bioconjugate Chemistry*, **18**, 1919–1923.

161. (a) Curreli, M., Li, C., Sun, Y., Lei, B., Gundersen, M. A., Thompson, M. E., Zhou, C. (2005). Selective functionalization of In_2O_3 nanowire mat devices for biosensing applications. *Journal of the American Chemical Society*, **127**, 6922–6923. (b) Tian, Y., Ye, S., Ran, Q., Xian, Y., Xu, J., Peng, R., Jin, L. (2010). Generation of surface-confined catechol terminated SAMs via electrochemically triggered Michael addition: Characterization, electrochemistry and complex with Ni(II) and Cu(II) cations. *Physical Chemistry Chemical Physics*, **12**, 13287–13295.

162. Chan, E. W. L., Yousaf, M. N. (2006). Immobilization of ligands with precise control of density to electroactive surfaces. *Journal of the American Chemical Society*, **128**, 15542–15546.

163. Yousaf, M. N., Houseman, B. T., Mrksich, M. (2001). Turning on cell migration with electroactive substrates. *Angewandte Chemie International Edition*, **40**, 1093–1096.

164. Yousaf, M. N., Mrksich, M. (1999). Diels–Alder reaction for the selective immobilization of protein to electroactive self-assembled monolayers. *Journal of the American Chemical Society*, **121**, 4286–4287.

165. Yeo, W. S., Hodneland, C. D., Mrksich, M. (2001). Electroactive monolayer substrates that selectively release adherent cells. *ChemBioChem*, **2**, 590–593.

166. Yousaf, M. N., Houseman, B. T., Mrksich, M. (2001). Using electroactive substrates to pattern the attachment of two different cell populations. *Proceedings of the National Academy of Sciences USA*, **98**, 5992–5996.

167. Yeo, W. S., Yousaf, M. N., Mrksich, M. (2003). Dynamic interfaces between cells and surface: Electroactive substrates that sequentially release and attach cells. *Journal of the American Chemical Society*, **125**, 14994–14995.

168. Yeo, W. S., Mrksich, M. (2006). Electroactive self-assembled monolayers that permit orthogonal control over the adhesion of cells to patterned substrates. *Langmuir*, **22**, 10816–10820.

169. Yeo, W. S., Mrksich, M. (2004). Electroactive substrates that reveal aldehyde groups for bio-immobilization. *Advanced Materials*, **16**, 1352–1356.

170. Wieckowska, A., Braunschweig, A. B., Willner, I. (2007). Electrochemical control of surface properties using a quinone-functionalized monolayer: Effects of donor–acceptor complexes. *Chemical Communications*, 3918–3920.

171. Collier, J. H., Mrksich, M. (2006). Engineering a biospecific communication pathway between cells and electrodes.

Proceedings of the National Academy Sciences USA, **103**, 2021–2025.

172. Zhou, J., Wang, G., Hu, J., Lu, X., Li, J. (2006). Temperature, ionic strength and pH induced electrochemical switching of smart polymer interfaces. *Chemical Communications*, 4820–4822.

173. Doron, A., Portnoy, M., Lion-dagan, M., Katz, E., Willner, I. (1996). Amperometric transduction and amplification of optical signals recorded by a phenoxynaphthacenequinone monolayer electrode: Photochemical and pH-gated electron transfer. *Journal of the American Chemical Society*, **118**, 8937–8944.

174. Jiang, Y., Wan, P., Smet, M., Wang, Z., Zhang, X. (2008). Self-assembled monolayers of a malachite green derivative: Surface pH- and UV-responsive wetting properties. *Advanced Materials*, **20**, 1972–1977.

175. (a) Wan, P., Wang, Y., Jiang, Y., Xu, H., Zhang, X. (2009). Fabrication of reactivated biointerface for dual-controlled reversible immobilization of cytochrome c. *Advanced Materials*, **21**, 4362–4365. (b) Wan, P., Chen, Y., Xing, Y., Chi, L., Zhang, X. (2010). Combining host-guest systems with nonfouling material for the fabrication of a biosurface: Toward nearly complete and reversible resistance of cytochrome c. *Langmuir*, **26**, 12515–12517.

17

SELF-ASSEMBLED MONOLAYERS AS MODEL BIOSURFACES

Anna Laromaine and Charles R. Mace

17.1 INTRODUCTION

In this chapter, we discuss organic monolayer films that have been developed as a powerful and flexible approach with which to study the interactions between surfaces and adsorbates. The methods used to fabricate these films experimentally correlate with models that predict the packing of molecules at a surface, which result in control over the interactions that lead to a precise atomic structure (e.g., the orientation of functional groups that decorate a surface). At higher scales, these interactions facilitate the spontaneous production of desired surface properties from a bulk material.

Self-organization is ubiquitous in nature: Molecules self-organize, creating functional materials that orchestrate life. Multiple examples exist in the literature highlighting the power and the importance of understanding and controlling these organizations of molecules to build materials that are relevant to biology; these organizations include bones, collagen structures, protein complexes [1], and biofilms [2], but this classification can also include biochips [3].

We present here an overview of self-assembled monolayers (SAMs) that have been investigated as representative models of biological surfaces. We will explore the reasons why SAMs have become so important to the study of biological interactions and describe particular examples of SAMs that model biological interactions.

17.1.1 Scales of Molecular Interactions and Examples

Biology utilizes molecular systems composed of long- and short-ranged structures that occur between the scale of nanometers (e.g., intermolecular) and micrometers (e.g., intercellular), called the *mesoscale*. Examples of complexity arising from molecular organization at the mesoscale include protein folding, amphiphile aggregation into micelles, or the nucleation of hydroxyapatite to form bone nodules [4].

Figure 17.1 shows some examples of the organization of molecules at different scales, ranging from the nanoscale to the macroscale. In Figure 17.1A, an amphiphilic molecule (e.g., a phospholipid) composed of a hydrophobic chain and a hydrophilic head group can assemble into a number of geometries driven by hydrophobic and hydrophilic interactions; for example, two layers of phospholipids, interacting with intercalating proteins and carbohydrates, organize to form a lipid bilayer membrane. These types of membranes are present in living organisms and in viruses and are a basic component required for life. Unicellular organisms, such as *Escherichia coli* (Figure 17.1C), contain multiple additional layers of organic thin films within their outer cellular enclosures (e.g., organelles) that are themselves ordered and functional structures. Cells grow and interact to create tissues with extraordinary complexity. Our bones, skin, and digestive systems are composed of multiple layers of cells that organize to create functional tissues. For instance, the human intestine is composed of layers with different nanoscale and microscale features that aid digestive function. Microvilli are small protuberances that enlarge the surface area of our intestine, increasing the capacity of diffusion and absorption of nutrients along our digestive system. Bacteria are also present in our digestive tract, whereby they assist symbiotically in the digestion of food. Two examples where self-assembled monolayers have found use as a model system are (a) the design of a surface that allows the formation of a biofilm in

Organic Nanomaterials: Synthesis, Characterization, and Device Applications, First Edition. Edited by Tomás Torres and Giovanni Bottari.

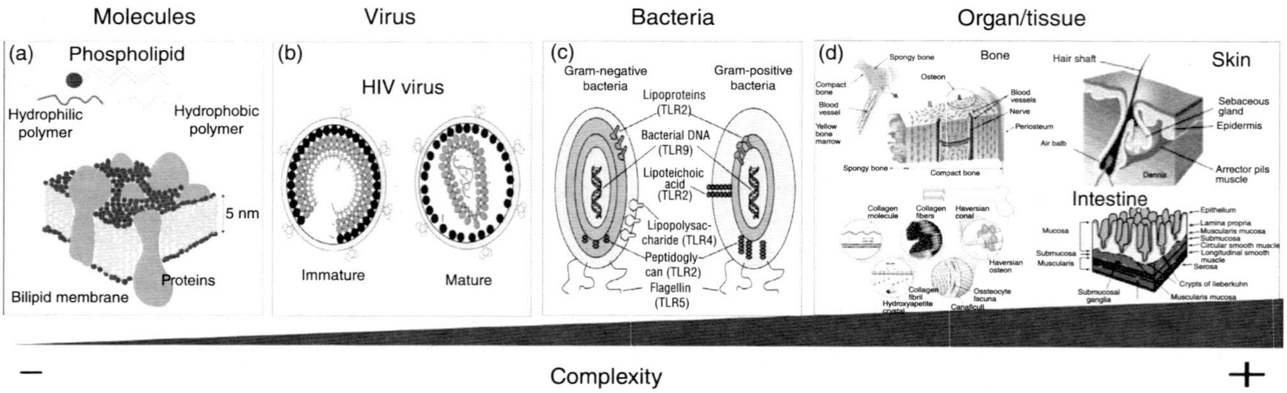

FIGURE 17.1. Examples of complexity arising from molecular organization on multiple size scales: (**A**) nanoscale (e.g., molecules), (**B**) mesoscale (e.g., viruses) [5], (**C**) microscale (e.g., bacteria and unicellular organisms) [6], and (**D**) macroscale (e.g., organs or other multicellular systems) [7–9]. Panels B–D are reproduced with permission.

the digestive tract and (b) the prevention of biofilm formation in bone implants.

17.2 ORGANIC MONOLAYER FILMS

Organic monolayer films have emerged as useful materials with which to study interfaces. Interfaces in biology present a complex environment—consisting of proteins, lipids, oligosaccharides, peptides, ions, and salts, among others—and are a rich area of experimentation.

Interfaces are the point, area, or surface along which two substances or materials interact. The composition of each surface dictates how materials interact and organize to produce complex and functional materials. Often, the molecular structure of the surface at the interface drives the interactions between materials, rather than the bulk properties of the materials themselves. Examples of such interfaces exist in all areas of our lives, either in common devices (e.g., glucose sensors) or in our own biological structure. The control over these interfacial interactions is a principal challenge in the development of materials, and their importance demands reproducible and reliable methods of fabrication to allow rigorous investigations of molecular interactions.

The experimental fabrication and study of these films, further supported by theoretical models, allow superior control over the molecular interactions that lead to a precise atomic structure of the surface of a material. At higher scales, this precision permits the reproduction of a desired surface functionality from a bulk material. Single monolayers of organic molecules can be produced using a number of methods, including (i) atomic layer deposition (ALD), (ii) the Langmuir–Blodgett techniques, (iii) layer-by-layer deposition (LbL), and (iv) self-assembly. We will briefly discuss each method. In this chapter, however, we focus our discussion on self-assembled monolayers because they offer easy methods of fabrication, flexibility of design, and the possibility to combine biological and organic molecules.

17.2.1 Atomic Layer Deposition

Atomic layer deposition (ALD) is a modification of bulk chemical vapor deposition (CVD) techniques [10–12]. Typically, ALD deposits layers of inorganic molecules on a surface using cycles of a process that rely on chemical reactions. In brief, a volatile precursor (a) is pulsed at high pressure into a reactor where it chemisorbs on a surface or a structure until saturation is achieved; the precursor (a) is then purged and a second precursor (b), usually water, is introduced. The second precursor (b), too, chemisorbs to the surface until saturation is achieved, whereby it is purged from the system (Figure 17.2). This process is repeated cyclically (each cycle usually adds a sub-monolayer quantity of the material), and it is repeated until the desired layer thickness is acquired. The performance of ALD varies with the temperature (around 200°C, typically), the reactor chamber pressure, and the flow rate of the gas precursors [13]. Layer formation by ALD is most often performed using solid inorganic surfaces, although substrates such as wires, tubes, laminates, and ionic liquids are emerging as candidate surfaces [14].

ALD is a suitable method to fabricate inorganic thin films used in electronic and optical coatings. The pressures and temperatures used for the ALD process, however, may alter the physical and chemical properties of the resulting layers, particularly for monolayers grown on soft or porous materials. This inconvenience makes the ALD technique not suitable to obtain films of biomaterials or biological species [15].

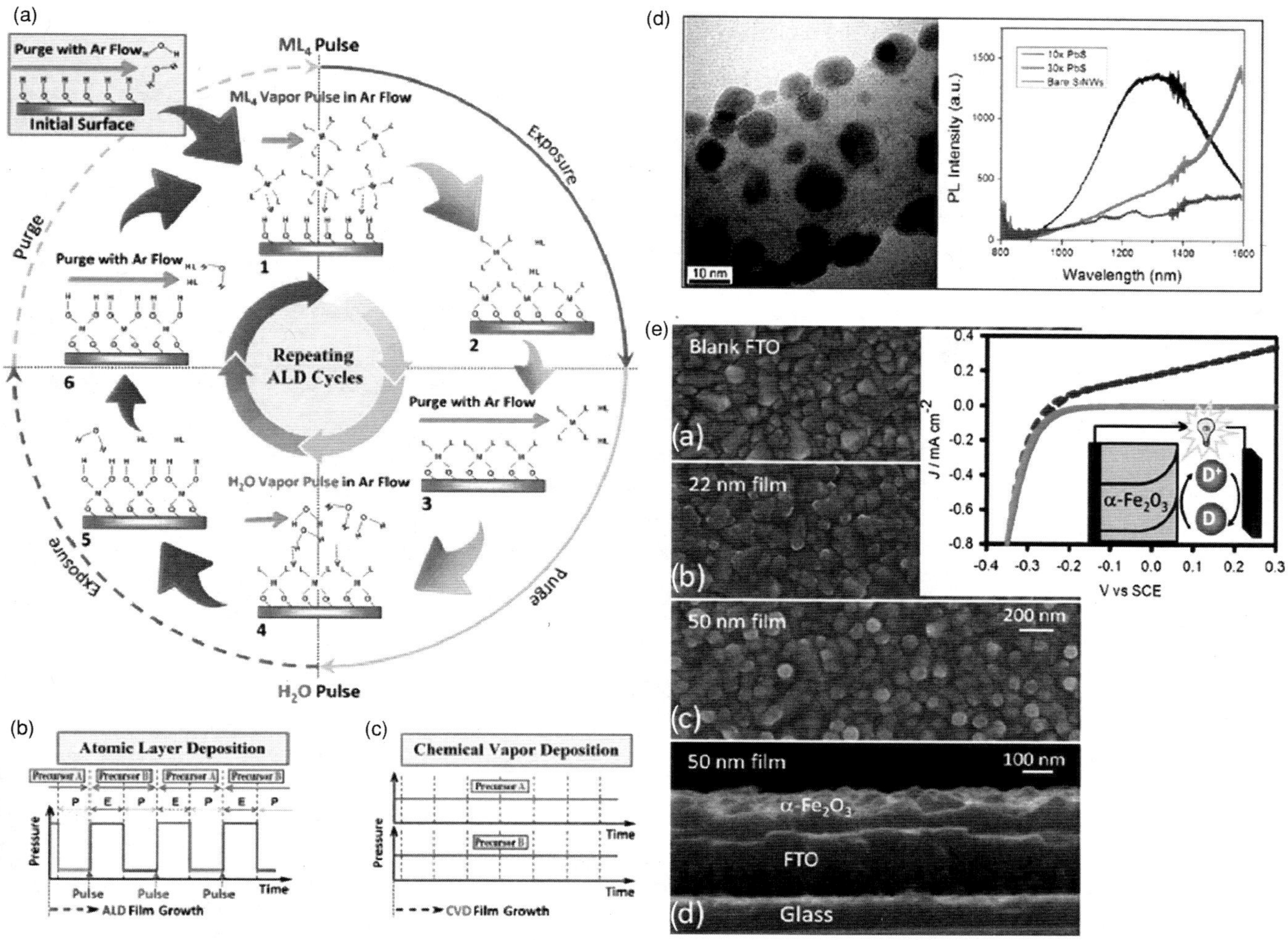

FIGURE 17.2. (**A–C**) Comparison of the atomic layer deposition (ALD) cyclic process and chemical vapor deposition (CVD) technique [15]. In panels D and E, we present some examples of surfaces fabricated by ALD published in 2011. (**D**) Atomic layer deposition of lead sulfide quantum dots on nanowire surfaces [16]. (**E**) Photoelectrochemical investigation of ultrathin film iron oxide solar cells prepared by atomic layer deposition [17]. Reproduced with permission.

17.2.2 Langmuir–Blodgett Technique

The Langmuir–Blodgett technique (LB) relies on the formation of monolayers of amphiphilic molecules on the surface of a liquid. In brief, molecules spread at the air–water interface, and then the placement of a barrier at the interface compresses them mechanically until they spontaneously pack and adopt a common orientation. The hydrophobic groups of the amphiphilic molecules position away from the aqueous phase, while the hydrophilic groups orient toward the aqueous phase (Figure 17.3a,b). After assembly, monolayers are transferred to a solid substrate to provide mechanical stability to the films in order to facilitate their use and their study. The transfer of the monolayer to a surface can be performed vertically (Langmuir–Blodgett, Figure 17.3d) or horizontally (Langmuir–Schaefer, Figure 17.3c) [18,19], although the former is the most commonly used approach.

Despite the challenge of obtaining intact films, LB has found exceptional utility in the formation of bilayer lipid membranes [20]. Bilayer lipid membranes prepared by LB have been used to study the properties of membrane proteins [20–22], enzymes [23–25], and transmembrane ion currents [26].

17.2.3 Layer-by Layer Technique

The layer-by-layer technique (LbL) assembles thin films by introducing alternating layers of oppositely charged materials (e.g., polycations (P) and polyanions (N)) on a surface. In brief, the immersion of a substrate into an aqueous solution of a P, followed by a wash step and a second immersion in an aqueous solution of an N, deposits a layer of PN on the surface of the substrate; a single PN layer can have a thickness

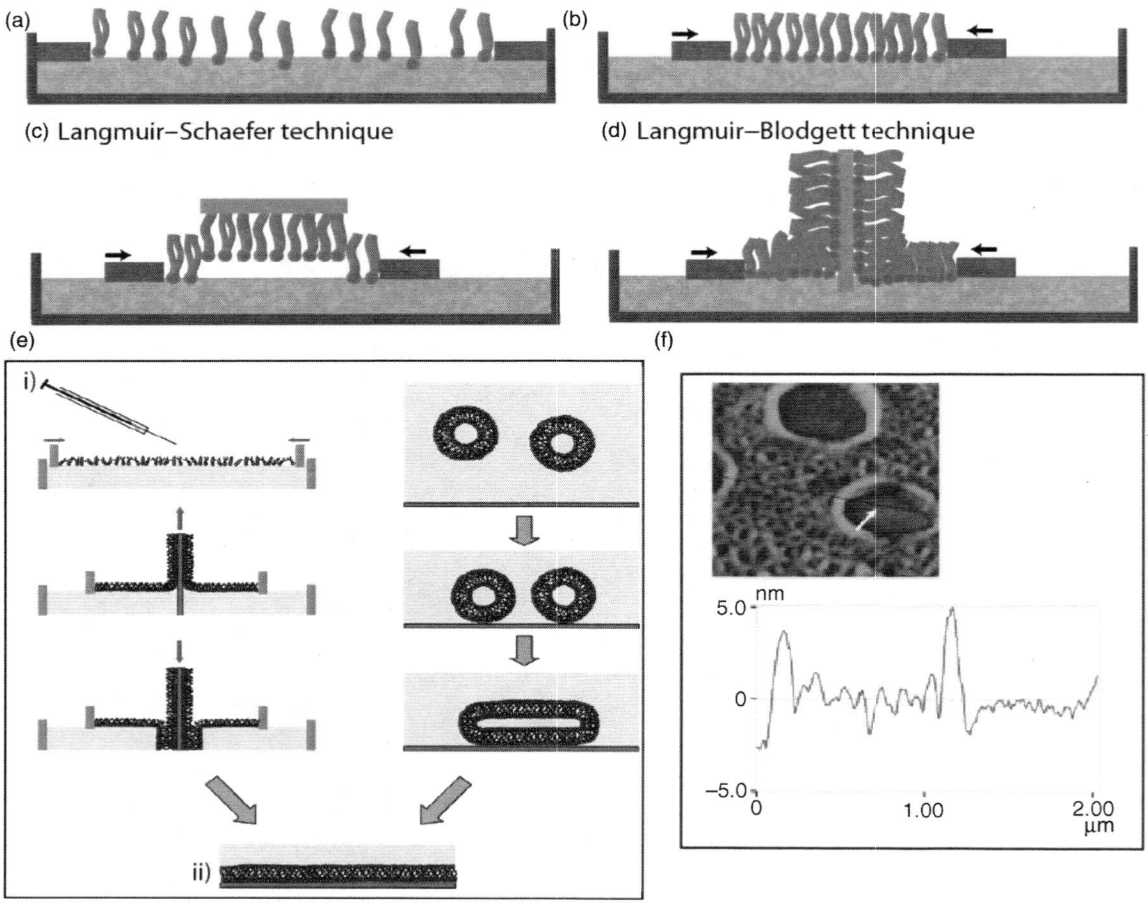

FIGURE 17.3. Schematic representation of the formation of films (**a**, **b**) and then two methods of obtaining a film: horizontally, called the Langmuir–Schaefer technique (**c**); or vertically, called the Langmuir–Blodgett technique (**d**). (**e**) Preparation of bilayers for AFM studies. Two methods commonly used for preparing supported lipid bilayers: (i) Langmuir–Blodgett technique and (ii) fusion of lipid vesicles [27]. (**f**) This image show the network structure of collagen layers as LB films on bare mica supports [28]. Panel e and f are reproduced with permission.

of angstroms to nanometers, and hundreds of layers can be deposited on a substrate.

Repetition of this simple process deposits a multilayer film on the surface, giving rise to complex and multifunctional thin films. The charged layers of an LbL can be used as an anchor to interact, via electrostatic interactions or hydrogen-bonding, with additional building blocks to broaden the ultimate chemical and physical properties of the LbL film. Figure 17.4 illustrates schematically how a multilayer is formed using the LbL technique and some examples of LbL films formed using different molecules. Since the introduction of LbL by Decher in 1997, thin films composed of DNA, proteins, metals, polymers, or inorganic nanoparticles have been produced [29–32].

Dip-coating, spin-coating, spray-coating, and flow-based methods are experimentally simple techniques that can be used to deposit layers on substrates in the production of LbL films; however, the reproducibility of the manufacture of the LbL films depends upon a number of conditions including immersion time and the solution temperature, concentration, and pH [29, 33, 34]. Automated instrumentation has been developed to control many of these conditions. Characterization of LbL films are typically performed by optical techniques such as dual polarization interferometry and ellipsometry, or by quartz-crystal microbalance.

17.2.4 Self-Assembled Monolayers (SAMs)

In comparison to LB, the formation of self-assembled monolayers is a more effective and robust method to modify the properties of a surface. SAMs can be prepared from organic, inorganic, and biological molecules, or mixtures of molecular species with broad chemical properties. The high degree of control of the properties of SAMs, including the molecular

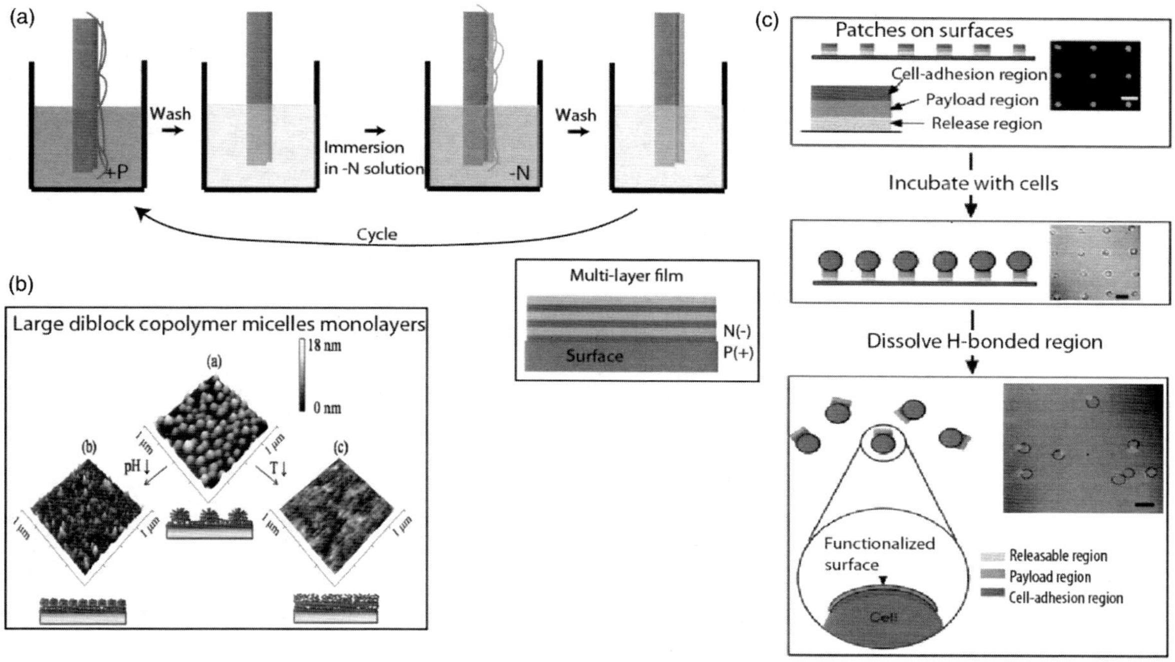

FIGURE 17.4. (**a**) Schematic representation of the cyclic steps applied toward the formation of surface layers using layer-by-layer (LbL) technique. (**b**) Monolayers of large diblock copolymer micelles deposited on a surface and their resulting RMS roughness compared using AFM [35]. (**c**) Process of incubation of an array of surface-bound patches with B-cells spaced 50 μm apart (green fluorescence corresponds to the fluorescein in the payload region). After decreasing the temperature, from 37°C to 4°C for 30 min, the patches released from the surface while remaining attached to the cell membrane [36]. Reproduced with permission.

packing density, affords the application of SAMs toward a wide range of problems, which include molecular electronics [37–44], catalysis [45–47], and biomedicine [48–52].

Since SAMs form spontaneously, their fabrication, in concept, is a simple process and can be performed either in the liquid phase (e.g., through a dip-coating process), in supercritical solvents (e.g., ethanol–supercritical carbon dioxide mixtures) [53], or in the gas phase (e.g., through evaporative methods). The type of substrates, the interaction time, the concentration and purity of the adsorbates, and the cleanliness of the surface are variables that must be considered to achieve a functional SAM.

The strong and specific chemical interactions between the substrate and the adsorbate (e.g., noble metals and thiols) and between adsorbed species (e.g., alkyl chains) influence the spontaneous formation of well-ordered, close-packed monolayers. The thickness of a SAM is on the order of angstroms to nanometers, but the choice of the adsorbate can be used to manipulate the properties of the SAM at the atomic level. This tunability affords remarkable control over the chemical and physical properties of the interface that results between the SAM and the external environment—for example, hydrophobicity [54, 55] or electron tunneling barrier [41, 56].

This chapter is dedicated to review, with examples, how the structures of self-assembled monolayers have been varied in a controlled manner to obtain information on interactions present between biomolecules, cells, and organisms (Figure 17.5). We include references to the literature that detail the generation of self-assembled monolayers and the variables that control their formation [57–59].

17.3 SELF-ASSEMBLED MONOLAYERS

The reactivity between the substrate and the adsorbate and the properties of the adsorbates define the final properties of the SAMs; therefore, their careful design allows the study of different interfaces and mesoscale systems.

17.3.1 Substrates

Substrates refer to the surface onto which SAMs are formed or those surfaces that support SAMs. Substrates can range from planar or rough surfaces to highly curved surfaces [60–62]. The nature of the substrate is typically a thin film of a noble metal (e.g., Au or Pt), oxides of transition metals (e.g., TiO_2), post-transition metals (e.g., Al_2O_3), metalloids (e.g.,

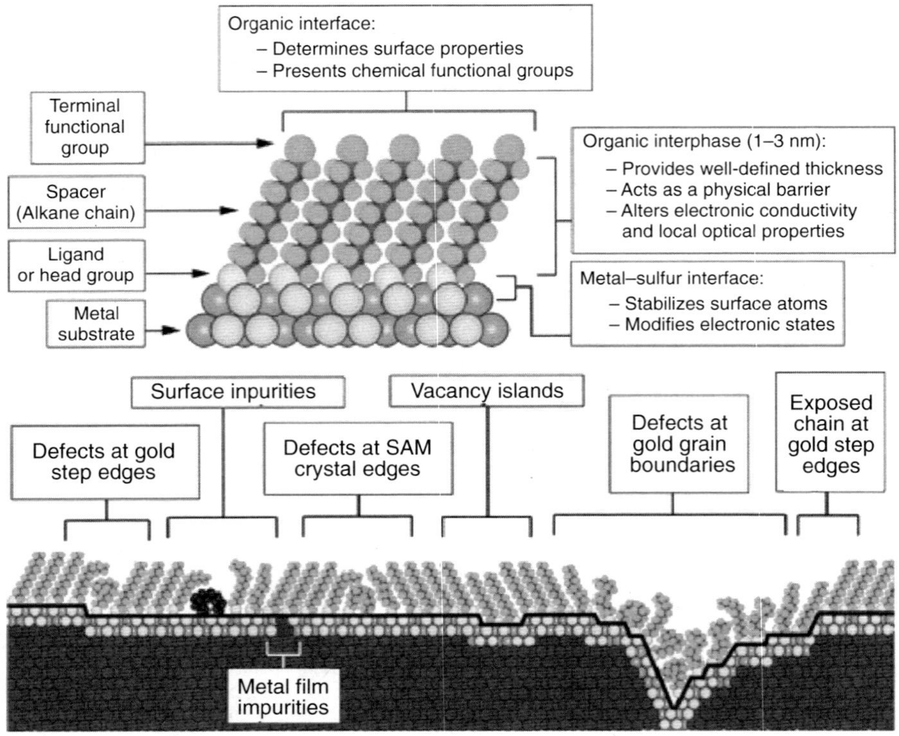

FIGURE 17.5. Detailed structure of a self-assembled monolayer (SAM) [57]. Reproduced with permission. See color insert.

SiO_2), or alloys (e.g., nitinol). The choice of the substrate will also determine the chemistry used to assemble the applied adsorbate and the final application of the SAM.

The most common substrates for SAMs are films of Au and SiO_2, although others are known. Their functionalization chemistries, based on thiols and silanes, respectively, are well-defined, thoroughly characterized, and highly reproducible [63–65].

An important feature that defines the quality of the SAMs is the smoothness of the starting substrate. Therefore, the support on which the substrates are deposited strongly affects the packing density, number and types of defects formed, and the functional properties of the resulting SAM. Glass surfaces are used frequently because they are commercially available, inexpensive, and integrated readily into common metal deposition processes (e.g., by evaporation). The use of many glasses to prepare uniform SAMs is limited by surface roughness. As a result, ultraflat silicon wafers, thermally grown oxides, and mica supports are the preferred supports to improve the smoothness of the thin SAM film.

To further improve the smoothness of the final thin films, strategies that strip off a sacrificial layer have been reported recently. These template stripping techniques produce ultrasmooth metals on a variety of supports: A metal is e-beam evaporated on a silicon wafer and then, using adhesive or solder, a glass support or flexible polymer is attached to the evaporated metal [66]. Using mechanical methods, the sandwich (metal and glass support) is removed from the silicon wafer. This method exploits the original ultraflat surfaces to produce higher-quality films than can be prepared by traditional approaches to deposition: In gold, an RMS of 0.6 nm can be produced after template stripping, compared to an RMS of 4.5 nm using deposition methods.

Techniques to improve the smoothness of the substrate are highly desirable, as the substrate quality is a critical component for the quality of a resultant SAM. Template stripping is advantageous over other methods because it is a fast, inexpensive, and reliable method to produce reproducible substrates.

17.3.2 Adsorbates

Adsorbates are classified broadly as the molecules that assemble on the surface of substrates. Adsorbate molecules are composed of three distinctive parts: (i) head group, (ii) spacer, and (iii) terminal group. The head group defines the specificity of the interaction between the adsorbate and the substrate and confers stability to the SAM, while the spacer and terminal groups can have common chemical properties that are independent of unique head groups.

(i) As examples of head groups, we describe thiols and silanes and their interactions with appropriate surfaces. For example, the strong interactions of thiols/thiolates on surfaces of gold or other noble metals have been explored widely as chemical interactions that form SAMs. The specificity of the thiol–gold interaction and the relative ease by which thiolated molecules can be synthesized have both contributed to their use toward the production and study of self-assembled monolayers.

The final composition and structural nature of the bonds of the thiol–gold bond on the surface of a gold film is complex and a continuing topic of study. We refer the reader to a number of references exploring the nature of these chemical bonds [58, 67, 68].

Silanes are a class of chemicals that form self-assembled monolayers on glass or metal oxide surfaces [69]. Their use has become commonplace, and they follow one of two routes of attachment: displacement of a silyl chloride or an alkoxy group. General methods exist for the evaporative deposition of silanes or through the use of anhydrous solvent. These approaches ensure that the reactive halide or alkoxy groups have not hydrolyzed due to the presence of water. Hydrolized silanes, degraded to silanols, are able to coordinate to silicon dioxide surfaces through a network of hydrogen bonds, which can later be condensed to covalent bonds by heat. However, there is a possibility of polymerization and the formation of multilayers of silanes under these conditions.

(ii) The spacer group of an adsorbate is typically an alkane chain. The composition of the spacer group defines the thickness of the SAM, the packing of the SAM, and the steric effects within the layer. Linear, branched, or cyclic alkanes may confer different structural effects to the SAM that is reflected by the final properties observed at the interface; for example, terminal group orientation is dominated by odd/even effects of an alkyl spacer group [38, 70–72]. Additional groups within the alkane chain may also be used to modify the properties of the SAM. For example, the use of ferrocenes [73–75] or hydroquinone confer additional electronic and optical properties to the SAM that can be used to produce electrochemical switches capable of interacting with adhered cells [76, 77]. The ability to include multiple chemical groups within the SAM through the formation of mixed monolayers also increases the structural and chemical diversity available to SAMs.

(iii) The terminal functional group of the adsorbate influences the final properties of the monolayers. For example, aromatic or charged moieties employed as terminal groups would bestow hydrophobicity or hydrophilicity to the surface of the SAM, respectively. Terminal functional groups allow further modifications to the SAM surface through the use of chemical reactions: a prominent and useful example is the reaction of carboxy-terminated SAMs with the amine groups of antibodies to produce biofunctional SAMs via amide formation [78]. A number of other useful reactive chemical groups that have been incorporated into SAMs that result in active or functional surfaces include amines, aldehydes, *N*-hydroxysuccinimide, maleimide, aminophenyl groups [79], hyaluronic acid [80] or other carbohydrates, polymers, and dialkyldithiocarbamate salts [81].

Particles have also been incorporated successfully into SAMs as a terminal group. Surfaces modified with particles provide opportunities for novel applications (e.g., catalysis). Gold nanoparticles, if assembled onto gold substrates, require a bifunctional tether to conjugate them to the surface.

For tables with combinations of adsorbates and substrates, we refer the reader to the following articles and reviews [57, 84, 85]:

17.3.3 Types of SAMs

SAMs can be produced with varying degrees of surface complexity: The condition whereby a single type of adsorbate is present in the monolayer is referred to as a uniform SAM (Figure 17.4A), whereas a monolayer comprising a mixture of adsorbates is defined as a mixed SAM (Figure 17.4B). Self-assembly is controlled by the thermodynamic equilibrium between adsorbates on the substrate surface and the precursors that are in the solution-phase. Therefore, the concentration and purity of the reagents affect the formation of the SAMs, as well as the structural dynamics within the SAM. The cleanliness, topography, and crystallinity of the substrate also influence equilibrium by determining the surface area available for SAM formation. For information about the current understanding of the structure of SAM formation, we refer the reader to a series of reviews present in the literature [45, 63, 67, 86–90].

Uniform SAMs allow the researcher to decouple the effects of multiadsorbate interactions, study the interface between well-defined surfaces, and assess explicitly the factors that influence the adsorbed molecule. SAMs allow us to integrate a variety of chemical compositions of the monolayers and also different topographies of the surface.

Mixed SAMs offer a sophisticated surface by increasing the chemical and physical complexity of the environment, which is a more accurate representation of interfaces found commonly in nature. The technique most often employed to generate a mixed SAM is immersion of a substrate into a

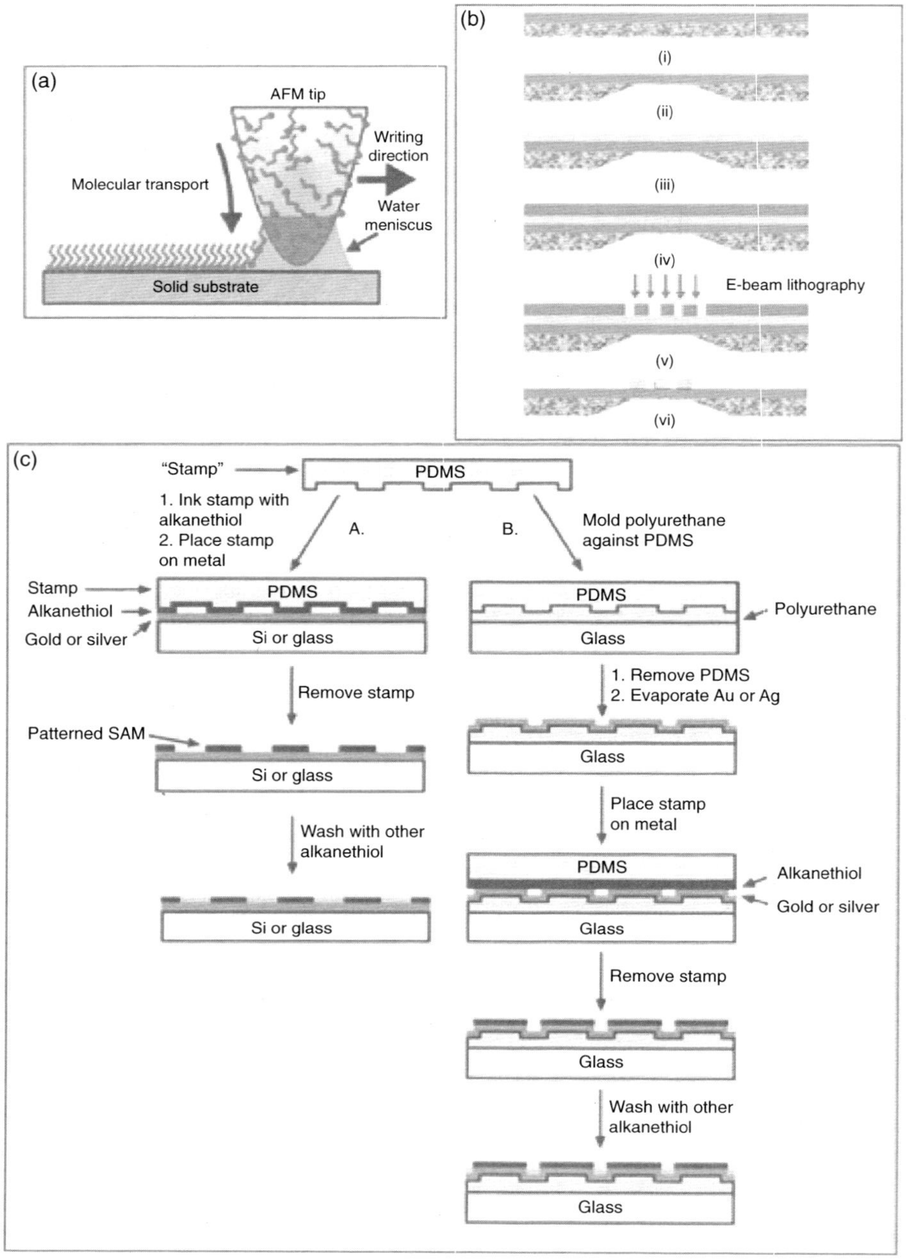

FIGURE 17.6. Brief schematics of techniques that can be used to create mixed SAMs: (**A**) Dip-pen lithography [121], (**B**) e-beam lithography [122], and (**C**) Microcontact printing or molding [93]. Reproduced with permission.

solution containing multiples adsorbates. This process, however, provides poor control over the position of the adsorbates on the substrate surface, leading to islands rather than a homogeneously distributed mixture. To increase the spatial control over the formation of the mixed SAM, different patterning techniques have been explored. Some examples are (i) microcontact printing [52,91–98], (ii) patterning by topography [99,100], (iii) electrochemical deposition [101–105], (iv) nanoimprinting [106–109], (v) electrospinnning [110, 111], (vi) dip-pen lithography [112–117], and (vii) e-beam lithography [118–120] (Figure 17.6).

The facile generation of SAMs, the flexibility of designs via patterning at the micro- and nanoscale, and their broad utility have contributed to the expansion of this methodology. It follows, then, that the ability to characterize the structure and composition of SAMs, both at the atomic level and as

TABLE 17.1. Some Techniques Used to Characterize SAMs

Atomic force microscopy (AFM): Imaging technique with subnanometer resolution, widely used in surface science; it also allows the manipulation of matter at the nanoscale.

Auger electrons spectroscopy (AES): Quantifies the energies of electrons emitted after excitation of a surface and, therefore, the chemical composition of a surface.

Contact angle goniometry: Measures the contact angle between a drop of liquid and the surface; provides hydrophobicity/hydrophilicity information.

Electrochemistry: Studies the chemical reactions that take place at the interface of a metal or semiconductor with other molecules involved in electron transfer mechanisms.

Ellipsometry: Optical technique that investigates the dielectric properties of films, commonly used to assess film thickness changed from a few angstroms to micrometers.

Grazing incidence x-ray diffraction (GID): Studies surfaces using x-ray illumination at small incident angles.

Helium atom scattering (HAS): Provides information of the surface structure and lattice dynamics of the material. Interesting because the information from scattered helium is only from the very surface, in contrast to electron scattering.

High-resolution electron energy loss spectroscopy (HREELS): Gives information about the small energy losses in the electronic and vibrational modes of surfaces or molecules absorbed to a surface.

Nuclear magnetic resonance spectroscopy (NMR): Provides information of the quantum and mechanical magnetic properties of the atoms on the surface of a material. *Magnetic resonance imaging* (*MRI*) is commonly used for medical applications.

Quartz crystal microbalance (QCM): Measures the mass change on a surface per unit area as a function of the frequency change of a quartz-crystal resonator.

Raman spectroscopy: Evaluates low-frequency modes in a system. Gives complementary information to *infrared spectroscopy* (IR).

Reflectance adsorption infrared spectroscopy (RAIRS): Vibrational spectroscopic technique used to identify species generated upon molecular absorption.

Scanning electron microscopy (SEM): Obtains information on the topography, composition, and electrical conductivity of surfaces.

Scanning tunneling microscopy (STM): Technique used to image surfaces with subnanometer resolution (usually 0.1 nm lateral and 0.01 nm in depth).

Small angle x-ray scattering (SAXS): Allows the acquisition of information of size, shape of molecules, and partially ordered materials at very low angles of elastic scattering of x rays.

Surface enhanced-Raman spectroscopy (SERS): A variant of Raman spectroscopy that can detect single molecules on a surface.

Surface plasmon resonance spectroscopy (SPR): Measures the absorbance of light due to the oscillation of electrons on a surface, important technique in color-based biosensors and lab-on-a-chip sensors.

Transmission electron microscopy (TEM): Used to measure the thickness and composition of a material due to the absorption or scatter of electrons. Modulations in chemical identity, crystal orientation, electronic structure and electron phase can be observed.

X-ray absorption near edge spectroscopy (XANES): Used to study the orientation of molecules with σ and π-bonds on a surface.

a bulk *mesoscale* film, is paramount to ultimate utility and adoption of these materials in a large number of applications and fields of use.

17.3.4 Characterization of SAMs

There are multiple techniques that have been applied to the characterization of self-assembled monolayers. Techniques to characterize SAMs provide information about the atomic structure, purity, and uniformity of the SAM in isolated segments and as intact films. These techniques can be applied to SAMs formed on planar surfaces or curved substrates.

In Table 17.1, we provide a list of some techniques used to characterize SAMs, although readers should bear in mind that it is not a complete list and that these and other techniques are in continuous development. We refer the readers to excellent reviews on characterization of SAMs [58, 123, 124].

17.4 BIOLOGICAL SURFACES

In the next sections, we will focus our discussion on the use of self-assembled monolayers, and their advantages and disadvantages, in biological applications and how a chemical understanding of SAMs assisted the development of new materials used to describe surface interactions and phenomena in biology.

SAMs Allow Reproducibility. Biological experimentation is highly sensitive to variations in experimental design. Variables such as temperature, pH, humidity, air quality, and the cleanliness of the environment, to name a few, are difficult parameters to control even though they are known to affect strongly the outcome of the biological experiments.

Experimental reproducibility in any discipline is important, but it is vital for biological applications because the observed effects are often small. To account for variations within an experiment, we measure a number (n) of independent samples and conduct replicate experiments or observations. Statistical analyses of sample populations composed of large values of n impart confidence to the experimental results and to the effects observed, if any [125]. The distribution of data for a population is represented by the mean, M, and the standard deviation from the mean. The statistical reliability that M is an accurate value for the true mean of the population increases as n increases.

Robust procedures and techniques are required to produce a number of independent experiments that replicate reliably experimental conditions. SAMs are considered to be one such methodology to produce monolayers on surfaces with high reproducibility.

Biological experimentation is usually reported as comparisons of experiments and controls with respect to a set of varied parameters. A large experimental population strengthens the tested hypothesis and the fidelity of the results that are obtained as a result of experimentation. Parallelization is an ideal strategy for biological experimentation, and SAMs offer the possibility to achieve multiple, varied surfaces easily and reproducibly. Several patterning techniques described previously allow the generation of multiple conditions on a single substrate, either replicating instances of a unique SAM or replicating a number of SAMs with varied characteristics within the same surface. The ability to perform multiple experiments on the same substrate minimizes the effects of environmental factors and inter-assay variations.

17.4.1 Inert Surfaces

The ability of a surface to resist the adsorption of molecules and proteins is usually defined as *inertness*. Many fields of research—including bioanalysis, tissue engineering, proteomics, protein chromatography, clinical diagnostics, biomedical materials, cellular adhesion, and drug delivery—demand a clear understanding of adsorption of proteins or cellular material. Many systems are designed to perform a desired action; therefore, any unwanted adsorption can cause a failure and a modification of the intended function.

The use of SAMs is ideal to study surfaces that resist the nonspecific adsorption of proteins, provide resistance to the adhesion of bacteria, or, on the other hand, promote adhesion. SAMs terminated with hexa(ethylene glycol) [126, 127], mannitose [128], and PNIPAAm (a complex polymer with variable hydrophilic and hydrophobic properties) [129, 130] are some examples of terminal surface layers that are described to effectively resist protein adsorption. To different degrees, the adsorption of proteins on SAMs can be tailored from the mixture of hydrophobic and hydrophilic SAMs.

Prime and Whitesides [131, 132] described how SAMs terminated with ethylene glycol (EG), and their arrangement as crystalline helical and amorphous forms, act as protein-resistant surfaces [133]. SAMs of oligomers of EG, produced using thiolated derivatives (i.e., HS-$(CH_2)_n$-EG_3OH), were established as protein resistance surfaces, and the mechanism by which these surfaces resisted adhesion was described by De Gennes and co-workers [134]. SAMs of ethylene oxide (EO; HS-$(CH_3)_{11}(OCH_2\ CH_2)_nOH$) also demonstrated protein resistance. The experiments that compared EG to EO surfaces were used further to establish rules for the synthesis and design of SAMs, as EO monolayers demonstrated that SAMs could be prepared without the incorporation of long alkane chains into the adsorbing molecules. Those rules were then applied to develop novel bacterial- and protein-resistant SAMs based on the polyethylene glycol (PEG) molecule and its many derivatives [2, 135–137]. Unfortunately the protein-resistant properties of PEG do not persist for significant periods of time because the oxidation of PEG in the presence of oxygen forms aldehydes, which can react chemically with undesired species [138]. As a consequence, the investigation of new SAMs to create protein-resistant surfaces was required. Based on Whitesides and co-workers' rules [133, 136], a number of SAMs with resistant properties were found, including those produced from peptides and peptoids [139, 140], carbohydrates, and glycerol derivatives [141, 142].

Kosmotropic molecules stabilize and structure the interactions between water molecules, favoring intermolecular interactions, and have therefore been explored as protein-resistant surfaces. SAMs of kosmotropes such as taurine, betaine, and trimethylamine *N*-oxide (TMAO) have been shown to prevent protein adsorption. The mechanism by which these molecules exclude proteins from a surface is not yet understood completely; some explanations allude to gradients in the concentration of ordered water originating at the surface [143], which then exclude the binding of proteins or other species [144] (Figure 17.7).

The connection between protein resistance, ordered surface waters, osmolality, and kosmotropicity must be considered in order to achieve protein-resistant surfaces. We suggest the reader to consider the following review on surface coatings that resist protein adsorption [138].

17.4.2 Proteins

Physical adsorption of proteins on a surface is a direct and simple immobilization method that has been explored as a nonchemical means to create functional surfaces for bioanalytical applications. Adsorption has disadvantages, however, as the mechanism for adsorption can (i) denature and inactivate proteins, (ii) produce heterogeneous surface compositions and irregular protein densities, and (iii) lead to irreproducible results because the orientation of adsorbed proteins is not controlled. Additionally, if surfaces coated with proteins

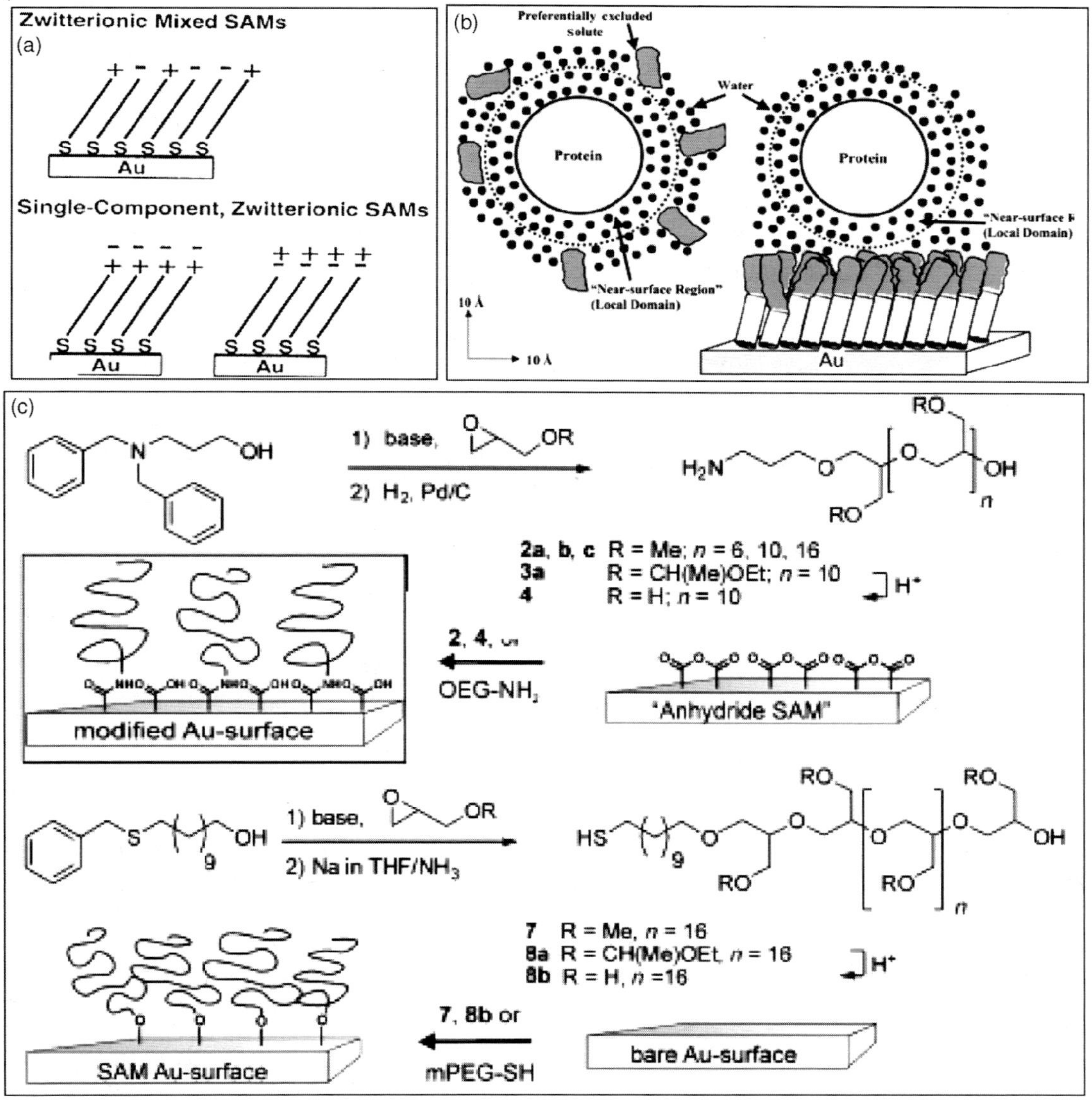

FIGURE 17.7. Examples of surfaces used to prevent the nonspecific adsorption of proteins. (**A**) Zwitterionic SAMs [145]. (**B**) Surface containing a kosmotropic molecule and the purported mechanism by which the SAM excludes proteins [143]. (**C**) Example of dendritic polyglycols [146]. Schematics are reproduced with permission.

interact with cells, bacteria, or other organisms, the interacting species often remodels the initial surface, further complicating the control over the composition of the surface [147]. Self-assembled monolayers offer a more robust method than physical adsorption to functionalize surfaces with proteins: Terminal groups presented by SAMs offer the means to covalently attach proteins to a surface in a structurally defined manner and in nondenaturing conditions. SAMs control the density of protein linkages and the environment of the proteins immobilized on the surface, provide uniform and high coverage structures, and reduce the random orientations of immobilized proteins, all of which contribute to the reproducibility and stability of the protein monolayers.

The attachment of a protein to a surface can be approached by modifying chemically either the protein or the substrate to facilitate specific interactions. The preferred method will be dictated by the choice of the substrate and adsorbate. For example, surface–protein interactions can be mediated by thiol–gold bonds by exploiting surface cysteines or the incorporation of a terminal cysteine through molecular biological manipulations. Most common reactions to incorporate the thiol group in proteins include reactions with the amines of proteins with *N*-succinimidyl-3,2-pyridyldithio propionate (SPDP), 1,4-dimercapto-2,3-butanediol, or 2-iminothiolane. An alternative route is to produce a SAM with terminal carboxylic acid groups to facilitate any number of esterification reactions with the amines of proteins, including *N*-hydroxysuccinimide-activated esters.

17.4.3 Bacterial Cells

Prokaryote microorganisms, such as bacteria, are decisive in many aspects of our lives. They are responsible for fixation of nitrogen from the atmosphere, production of antibiotics, and many symbiotic interactions with host organisms (e.g., human intestinal flora). There are, of course, a number of bacterial species that are pathogenic to humans or other forms of life.

Bacterial adhesion on surfaces is a complex process that involves multiple steps, including an initial weak association of bacteria with the surface, fostered by multivalent nonspecific interactions, followed by robust adhesion. Each of these steps, too, are composed of multiple stages, and their continued study is an active area of research [148]. To learn more about mechanistic biofilm formation, we refer the readers to comprehensive reviews [149–153].

Multiple factors control the adhesion of bacteria on surfaces [154]. These include: (i) environmental factors, namely, the temperature and pH of the medium, the time of exposure, the concentration of bacteria, and the presence of biocides or antibiotics in the medium [155]; (ii) the physicochemical characteristics of the bacteria, namely, the deformability of the bacterium and the types of molecules present on the surface of a bacterium (e.g., glycosylated or charged proteins); and (iii) the properties of the material surface, namely, the chemical composition, roughness, and wettability. The secretion of biological molecules by bacteria or the extracellular polymeric substance (EPS) matrix can also contribute to the adhesion of bacteria on a surface. EPS are glycosylated polymers present on the surface of the organism, and they have been shown to adhere to a wide variety of particulate materials such as clay, dead cells, and precipitated materials. The EPS, therefore, is used to promote the attachment or growth of organisms on a number of surfaces [156]. Approaches to prevent the adhesion of EPS are similar to those investigations of the prevention of nonspecific adsorption and can be performed using SAMs [157].

Although the resistance of a surface to the adhesion of bacteria or cells was thought to correlate to the ability of a surface to be coated by proteins, the use of SAMs to study adhesion demonstrated the unique function of some molecules to prevent bacterial adhesion by isolating individual factors that affect adhesion. Self-assembled monolayers have been instrumental to show that the type of microbe [129, 158], surface roughness [159–162]), hydrophobicity [160–165], electrostatic interactions [166–171], hydrogen bonding [172, 173], and Lewis acid–base [150, 173], can strongly affect attachment.

We use the remainder of this section to discuss how the hydrophobicity of surfaces affects the attachment of different bacterial strains and how the use of novel molecular coatings on surfaces prevents biofouling. In both cases, these studies were made possible by SAMs. To further describe information regarding other parameters, we would refer the readers to excellent reviews [138, 148, 174].

SAMs prepared from polyethylene glycol (PEG) and its derivatives are among the most explored chemicals used to fabricate surfaces that resist bacterial contamination because of their superior resistance to proteins [133, 175–178].

SAMs of other polymers, such as poly(*N*-isopropylacrylamide) (PNIPAAm) on gold and silica substrates, have been assessed for their ability to promote resistance to adhesion with different strains of bacteria (b) and yeast (y). PNIPAAm has been polymerized in situ on a gold substrate functionalized with a carboxy-terminated alkanethiol and the free-radical initiator 2,2-azobis(2-amidopropane) hydrochloride (ABAH). Other polymers, such as those comprising subunits of ethylene glycol, can also be grown in situ from surface-bound precursors [175, 179].

Biological microorganisms, due to their unique molecular markers, are characterized by a wide range of surface energies and, therefore, wettabilities: *Pseudomonas aeruginosa* (b) and *Candida tropicalis* (y) are relatively hydrophobic, whereas *Staphylococcus epidermidis* (b) and *Candida albicans* (y) are hydrophilic. As an example, SAMs of PEO with low surface energies (i.e., a hydrophobic surface) support interactions with hydrophobic microbes rather than those microbes whose surfaces are hydrophilic [175]. Likewise, studies using *Staphylococcus epidermis* and SAMs of PNIPAAm confirmed the preferential interaction of materials and microbes with hydrophilic surfaces [179].

SAMs that provide microbiocidal properties to surfaces include mannitol-terminated SAMs [180], surfactants attached to zwitterionic SAMs [145, 181, 182], glycerol saccharides [142], antimicrobial peptides [183], and silver coatings on titanium substrates for implants [184] (Figure 17.8).

17.4.4 Mammalian Cells

Cells function best in environments replete with the conditions required for survival. Some cells grow and propagate preferentially when adhered to surfaces. Understanding the process by which cells adhere to the surface of a substrate is of great importance in the study of cell growth, proliferation, migration, and the formation of tissues.

SAMs have been used to interrogate the time-dependent effects of cell adhesion on surfaces, both to changes in the surface and changes to the morphology or biochemistry of the cell. The study of these effects caused by the interactions between cells and surfaces provide invaluable information toward the efforts to develop smart biomaterials or sensing technologies. SAMs patterned onto surfaces allow multiple experiments to be performed in parallel and under controlled

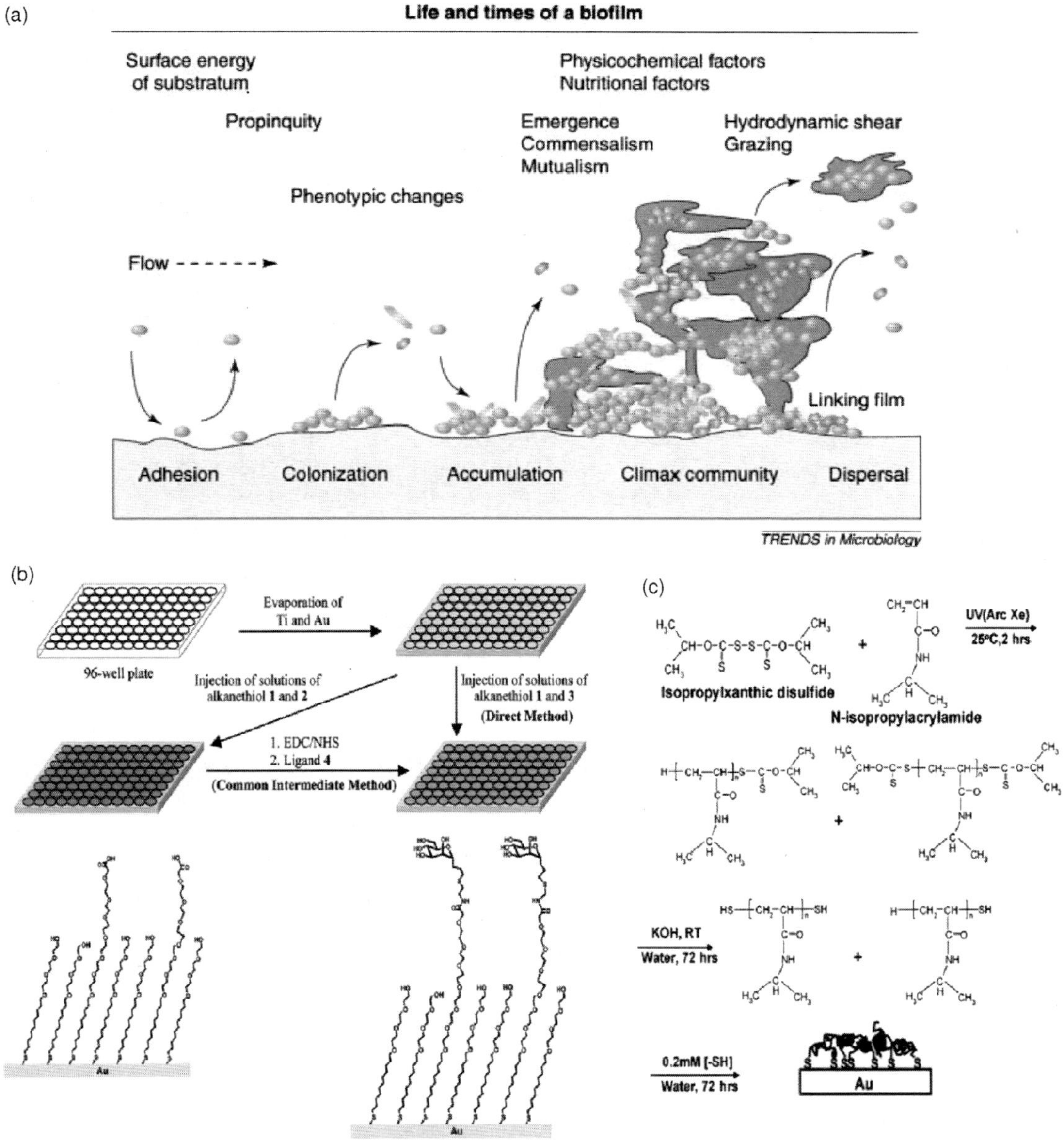

FIGURE 17.8. (**A**) Diagram showing the initiation and development of a biofilm, depicted as a developmental cycle, the various stages of which are determined by a range of physical, biological, and environmental factors [185]. Different approaches to the functionalization of surfaces to prevent bacterial adhesion, using PEG (**B**) [186] and PNIPAAm (**C**) as examples [187]. Schematics reproduced with permission.

conditions, which are absolute requirements for biological experimentation. In this section we focus on SAMs that promote cell adhesion, and we discuss how specific experiments have helped elucidate some of the mechanisms of cell adhesion.

The response of different cell lines to functional groups displayed on a substrate has been surveyed, and the correlation between cell spreading and attachment due to surface wettability were observed. Table 17.2 summarizes the responses of different cell lines exposed to various functional groups that were grafted on a surface using SAMs [188]. This table is far from comprehensive, but we use it to indicate the wide variety of applications and experimental results that can be found in the literature.[1] In Table 17.2, we include

[1]Search at ISI Web of Knowledge (14th November 2011) with keywords *surface chemistry* and *cells*: in 2009 (390 articles), 2010 (455), and 2011 (450).

TABLE 17.2. Some Functional Groups Incorporated Onto a Variety of Surface as SAMs, the Cell Lines Cultured on the SAM, and the Resulting Response of the Cell Line

Terminus	Substrate	Cells and Response Studied
CH_3	Au	Human fibroblast adhesion [190]
		Human neuroblastoma adhesion [191]
		Mammalian endothelial adhesion [192]
		Neutrophile adhesion [193]
		Myoblast proliferation differentiation [194]
	Si	Human fibroblast adhesion [195]
		Mesenchymal stem cells—adhesion [196]
		Endothelial cells—adhesion [197]
		Osteoblasts adhesion, mineralization [198]
	Ag	Mammalian endothelial cells adhesion [192]
OH	Au	Human fibroblast cells adhesion [190]
		Mammalian endothelial cells adhesion [192]
		Myoblasts cells proliferation and differentiation [194]
	Si	Endothelial cells adhesion [197]
		Osteoblast adhesion mineralization [198]
CF_3	Si	Canine endothelial cells adhesion [197]
NH_2	Au	Myoblasts proliferation and differentiation [194]
		Osteoblasts adhesion mineralization [199]
	Si	Human fibroblast cells adhesion [190]
		Erythroleukemia cells attachment [200]
		Mesenchymal stem cells adhesion, proliferation, mRNA expression [201]
		Myoblasts proliferation and differentiation [194]
COOH	Au	Neuronal growth adhesion [202]
	Si	Human fibroblast adhesion [195]
	Ti	Neuroblastoma and fibroblasts adhesion [203]
EDA	Si	Endothelial cells—differentiation and adhesion [204]
PC	Au	Neutrophiles [193]
RGDS,	Au	Human neuroblastoma [191]
RGDSP,		Neuronal cells [205]
TYRRKY		Endothelial adhesion [147]
	Si	Osteoblasts adhesion [206]
	Ti	MC3T3-E1 pre-osteoblasts adhesion and mineralization [207, 208]
PEG	Ti-S	Osteoblast adhesion and osseointegration [207]
FN/BMP-2		Antifouling properties [209, 210]
SBP	Au-S	Osteoblast adhesion [82]
FN	Au	Keratocytes adhesion [211]
		Osteoblasts adhesion [212]

Abbreviations: PEG, poly(ethylene)glycol; poly(OEGMA), poly(oligo(ethylene glycol) methacrylate); FN, fibronectin; BMP-2, bone morphogenetic protein-2; SBP, soy-bean peroxidase; EDA, *N*-(2-aminoethyl)-3-aminopropyl trimethoxysilane; PC, phosphorylcholine.

the following experimental details: the substrate employed (mainly Au or Si) [189], the terminal group exposed to the surface of the SAM (e.g., small molecules and proteins, via additional chemical modifications), the cell line under investigation, and the response of the cell to the surface.

The extracellular matrix (ECM) is a material that mediates cell adhesion in tissues; the ECM is composed of a number of protein complexes that present biochemical and mechanical cues to cells. The study of cell adhesion to the ECM and, specifically, the determination of peptide motifs of proteins that promote cellular adhesion have been guided through the use of SAMs.

Substrates grafted with complex SAMs of proteins such as fibronectin, integrins, nephronectin, laminin A, and heparin sulfate have been cultured with cells to elucidate peptide domains that promote cell adhesion. Peptide domains found in these studies include Arg-Gly-Asp (RGD) [191, 204, 213], Arg-Gly-Asp-Ser (RGDS) [191], Pro-His-Ser-Arg-Asp (PHSRN) [214, 215], Ile-Lys-Val-Ala-Val (IKVAV) [216], and Phe-Glu-Ile (FEI) [147]. Each peptide sequence is currently applied in studies toward the understanding of protein binding and cellular adhesion, and they have been used in a number of applications in cell biology, the development of cell sensors, and cell-based assays.

SAMs have additionally found use as a tool for researchers studying the propagation and differentiation of primary cells and stem cells. Surfaces grafted with proteins or peptides and surfaces that vary in topography and stiffness have been found to direct the development of stem cells and primary cells [217–222]. SAMs have also been instrumental in the elucidation of the effects of external geometric factors and mechanical tensile forces on the proliferation, differentiation, and movement of cells. The use of SAMs as model surfaces allowed Ingber [223] to formulate the tensegrity theory for adherent cells.

The adhesion and growth properties of endothelial cells, fibroblasts, or muscle cells were assessed on islands of SAMs that were 5–50 μm in size, circular or polygonal in shape, and comprised of molecules to promote or resist adhesion. Islands of hexadecanethiol ($HS(CH_2)_{15}CH_3$; the adhesive component) and tri(ethylene glycol)-terminated alkanethiol ($HS(CH_2)_{11}(OCH_2CH_2)_3OH$; the resistive component) were grafted onto gold substrates using microprinting techniques [92]. Cells cultured on these uniform SAMs adopted the precise size and shape of the island, cells cultured on polygonal islands demonstrated motile processes preferentially from their corners, and cells cultured on circular islands exhibited no bias. Islands of 20 μm activated an apoptotic mechanism in cells, whereas the cells placed in 50-μm islands grew with no adverse effects (Figure 17.9) [224, 225].

Following these initial experiments, Ingber et al. defined the tensegrity theory, which describes how the tensegrity structure of cells—microtubules, microfilaments, and intermediate filaments—respond to mechanical forces and influence cellular processes and the morphogenesis and vascularization of tissues. The prospective applications of these findings in clinical and medical therapies expanded the need for the investigations of surfaces with a variety of topographies [226] and mechanical properties [227, 228].

17.4.5 Implants

In our aging society, tissue regeneration strategies and the development of materials to survive extended implantation are a necessity. Hip, knee, and dental implants are the most used implants in the world to replace and restore function to damaged tissue. The optimization and development of novel implants to improve mobility and regeneration are active fields of research.

The successful integration of an implant with the surrounding tissue involves multiple factors from the characteristics of the patient, the surgical procedure utilized, the recovery process, the duration of the implantation, and the materials used to fabricate the implanted material. Since some circumstances are out of the control of researchers, the focus on the development of novel devices with optimal integration capabilities is placed largely at the tissue-implant interface.

The tissue–implant interface ideally should integrate the foreign implant within the region, inducing controlled, guided, and rapid healing of the area and minimizing any mismatch between the insert and the native tissue. Considerations of surface chemistry contribute to the design and optimization of the interfacial area. Due to the great societal impact of proper implantation procedure and materials, there are many researchers leading research programs on this topic.[2]

The surface of the implanted material controls the interaction between the native tissue and the foreign implant. The interface can facilitate the adhesion and spreading of cells and, therefore, the successful integration of the implant. The composition, surface chemistry, surface energy, charge transfer capabilities, stiffness, and topography of the surface of the implant material all perform important roles in integration [231–233].

Those, as well as other, studies highly impacted the development of commercial implants and the tissue regeneration field itself, resulting in a new direction to produce implants that contain a coating (e.g., chemical functionalization) or exhibit features (e.g., modification of topography) to improve their integration into the damaged area [234]. Commercial implants, such as those from Merete Co., contain bioactive calcium to promote osteointegration. Other implantable devices contain vancomycin or gentamycin as antiseptics to resist biofouling [235]. Bioretec also increases the integration of their materials by incorporating grooves into the bioresorbable implantable pin made of poly(lactic-co-glycolic acid) [236].

In this section we describe investigations that used SAMs to develop successful or promising metallic implants for bones, focusing on the elucidation of chemical groups, material properties, and topography that promoted integration. SAMs influence osteointegration, but are also explored as vehicles for drug delivery or their antifouling or antiseptic properties when incorporated in implants.

The study of SAMs for implants expanded the range of metallic surfaces onto which SAMs are grafted. Since metallic surfaces are needed to achieve successful tissue supports, they must also approximate the properties of native tissue. The human osseous tissue withstands pushing forces, but pulling or torsional forces damages this tissue easily, since it has relatively high compressive and tensile strengths and low shear stress strengths [237]. Any implant for bone regeneration needs to duplicate osseous tissues characteristics, and

[2]Search at ISI Web of Knowledge (14th November 2011) with keywords *implants* and *self-assembled monolayers:* in 2011 (33 articles), 2010 (29), 2009 (24), 2008 (29).

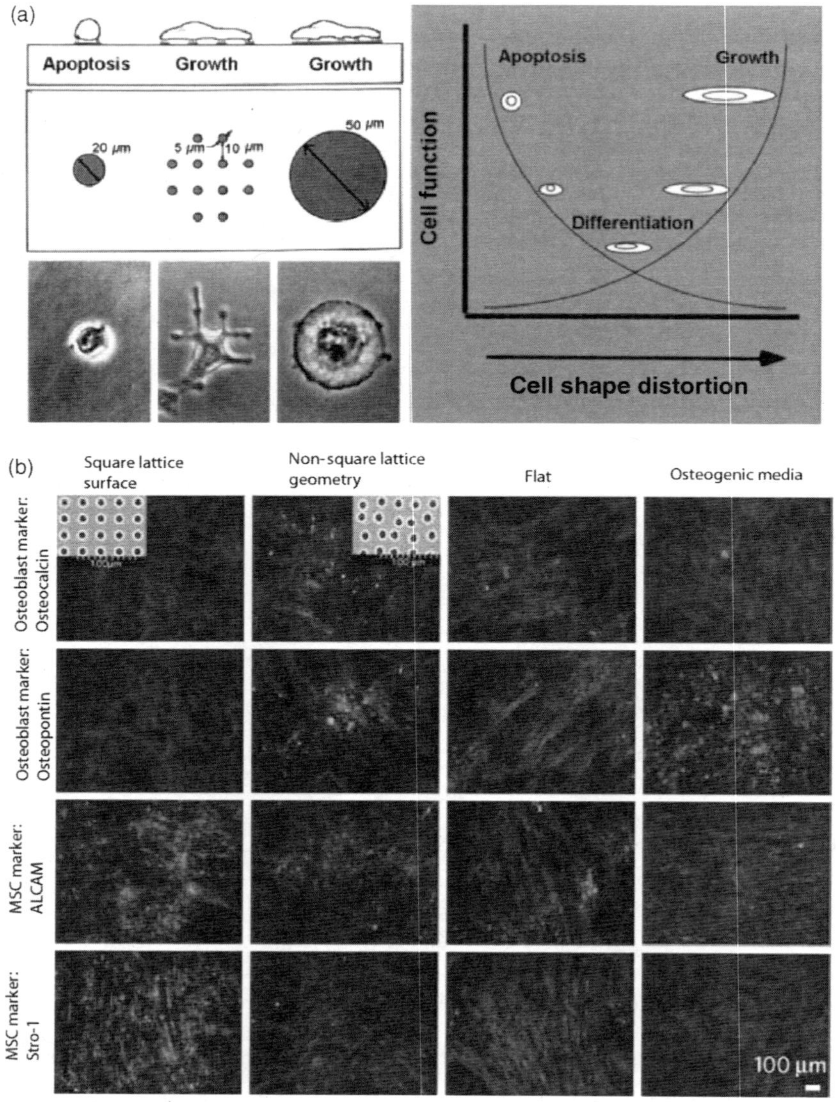

FIGURE 17.9. (**A**) Schematic describing the experiments performed by Ingber and co-workers [229] that led to the postulation of the tensegrity theory, whereby the size, shape, and distribution of islands of SAMs control cell growth. (**B**) Recent findings demonstrating how mesenchymal stem (MS) cells can be directed and aligned due to features patterned on the surface of a material [230]. Images reproduced with permission.

titanium has emerged as a good candidate material for bone implants.

According to at the American Food Drug Administration website (FDA, www.fda.com), a number of metal substrates have been approved for implantation in humans (e.g., dental prosthetics and orthopedic). These metals include titanium, titanium alloys, stainless steel, nickel, chromium, cobalt alloys, (Co–Cr–Mo) alloy, and tantalum. Metal oxides are also of interest as implant candidates because metals oxidize in vivo to form thin layers of metal oxides. Materials coating the metal oxides have been shown to promote osteointegration—for example, SAMs of polymethylmethacrylate (PMMA) or hydroxyapatite (HA) [234].

Metallic implants constitute only a segment of the implants used in orthopedics. Other materials—ceramics (typically made by pressing and heating aluminum oxide and zirconium oxide), composite materials, Trabecular Metal™ Material (tantalum over carbon), bioadsorbable plastics, polytetrafluoroethylene (PTFE), and vitreous carbon and silicone—are also incorporated in commercial prostheses (Table 17.3).

Combining chemical modifications or alterations to the topography of an implanted material have also been shown

TABLE 17.3. Brief List of Metallic Surfaces Studied as Medical Implants, the Molecules Grafted as SAMs Onto the Metallic Surfaces to Improve Tissue Integration, and the Types of Applications of the Implant

Metallic Surface	SAM	Suggested Application of the Implants
Ti or TiO_2 (as native oxide)	Phosphates [238, 239] RGD-phosphonates [240] OETS [241] OH [203] COOH [203, 242] RGD [243] BMP-2 [209] Hydroxyapatite [244]	Mainly for bone implants
Stainless steel	OH/ Ibuprofen Carbohydrates [246] Hyaluran [247] Phosphonates [248]	Coronary artery stent [245] NS
NiTi alloy (nitinol), a shape memory alloy	Phosphonate Amphiphilic peptides	Surgical implants [249] Bone plates and stents [250]
Co–Cr–W–Ni alloy	Silanes	NS [251]
Sapphire	Heparin	Neuroprosthetic [252]
Au	Phosphonates RGD	NS [243]
Zr	phosphonates [253]	NS

NS, not specified; OETS, 7-oct-1-enyltrichlorosilane.

to impact adoption. Features such as grooves, islands, lines, and meshes (both isotropic or anisotropic) can be patterned onto materials to achieve enhanced bone–implant contact. Excellent reviews about this topic are available [218, 254–258].

Assessing how the topography of an implant affects integration with the surrounding tissue is of great importance to the adhesion of materials with osteoblasts and osteoclasts, the cell lines studied for osteo-integration (Figure 17.10).

Integration of implants involves an initial stage where proteins first adhere to the surface prior to the docking of cells. Therefore, integration is a complex process that combines protein adhesion, topography, and cell adhesion. Although it is difficult to separate the effects of protein–cell interactions and topographical features on the efficacy of implant adoption, experiments have demonstrated that osteoblasts adhere preferentially to rough surfaces rather than smooth surfaces even if the same protein coating is performed. It is important to note that roughness does not promote adhesion to *all* cell lines; for example, smooth muscle cells prefer smooth surfaces as attachment sites [259]. The mechanisms used by cells to recognize surface features for adhesion are not completely understood, although the filapodia of cells likely play a role [254, 260].

A mathematical model described by Hansoon and Norton described the relationship between surface roughness and the interaction between bone and an implant. The result of the model identified that pits which are 1.5 μm in depth and 3–5 μm in diameter would positively influence the integration [261]. Tables summarizing the responses of osteoblasts on surfaces with different topographical features in vitro and in vivo are useful for the design of optimal implants and predicted their expected behavior [257]. Software programs to evaluate the integration of the implants and SAMs have also been developed to assist the realization of functional devices [262].

Multiple methods have been used to achieve these topographies experimentally, including etching by hydrofluoric acid, microabrasion, metal deposition, and the use of patterned self-assembled monolayers [232, 263–265]. Table 17.3 summarizes some examples of SAMs on a variety of metallic surfaces; although this list is far from complete, it highlights a number of salient cases.

Chemical patterning of the implant results in the display of different molecules on the interfacial region and also imparts a molecular topography to the surface. SAMs have been explored in combination with other topographical methods described here as a means to induce additional features on the material surface. Titanium and titanium oxide implants, or those coated with a layer of gold, are common materials for these purposes, since the chemistries used to functionalize them are well established [266]. After functionalization, surfaces are often appended with adhesion proteins, such as the RGD peptide sequence of extracellular matrices, oligosaccharides, integrins, or collagen, among others. Patterned SAMs have also emerged as a suitable tool to confine cells, whereby micro-patterned surfaces promote the differentiation of cells or the isotropic behavior of cells—for

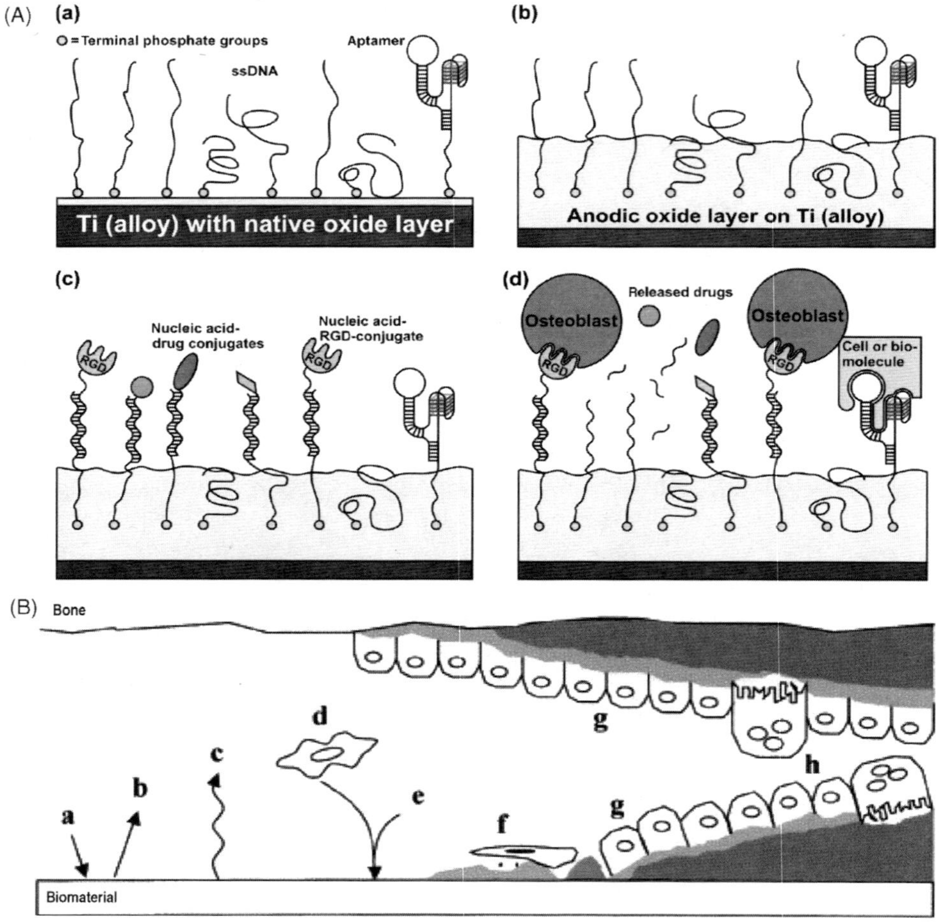

FIGURE 17.10. Integration of implants with native bone. Panel **A** shows a number of different approaches, modeled as a mixed layer for clarity, to the adhesion of osteoblasts onto a Ti surface: (**a**) Single-stranded anchors or functional nucleic acids (e.g., aptamers) are adsorbed onto the native oxide layer of titanium materials. (**b**) Anodic polarization leads to a partial entrapment of adsorbed nucleic acids in the thickened oxide layer. (**c**) Nucleic acid conjugates of bioactive molecules are hybridized to the single-stranded "anchor strands" (AS) fixed on the surface. (**d**) In a mammalian organism, osteoblasts can bind to RGD peptides on the surface. Aptamers can also be generated to bind specific cell types or molecules. These approaches lead to faster integration into the surrounding bone. Released drugs from the surface-bound conjugates may include antibiotics, antiphlogistics, or growth factors [268]. Panel **B** is drawing representing the events leading to healing at the bone-implant interface: (**a**) Protein adsorption from blood and tissue fluids, (**b**) protein desorption, (**c**) surface changes and material release, (**d**) inflammatory and connective tissue cells approach the implant, (**e**) possible targeted release of matrix proteins and selected adsorption of proteins such as BSP and OPN, (**f**) formation of *lamina limitans* and adhesion of osteogenic cells, (**g**) bone deposition on both the exposed bone and implant surfaces, (**h**) remodeling of newly formed bone [255]. Reproduced with permission.

example, observing the expression of osteogenic markers in stem cells or the preferential deposition of mineral [201,267].

17.5 CONCLUSIONS

This chapter described self-assembled monolayers (SAMs) and their applications in the study of interactions between synthetic surfaces and biomolecules, cells, and organisms. The use of SAMs to modify surfaces is an ideal approach for the study of surfaces, interfaces, and the interactions between biological species and materials. SAMs can be patterned onto a variety of materials—including noble metals, metal oxides, and plastics—and materials-specific chemical reactions offer an expansive toolbox by which to modify surfaces such that functional groups are presented to the interacting environment. SAMs can promote adhesion, resist nonspecific adsorption, and sustain the in vivo implantation of medical devices, and they are a robust and broadly useful research tool.

REFERENCES

1. Fratzl, P., Weinkamer, R. (2007). Nature's hierarchical materials. *Progress in Materials Science*, **52**(8), 1263–1334.
2. Ostuni, E., et al. (2001). Self-assembled monolayers that resist the adsorption of proteins and the adhesion of bacterial and mammalian cells. *Langmuir*, **17**(20), 6336–6343.
3. Jonkheijm, P., et al. (2008). Chemical strategies for generating protein biochips. *Angewandte Chemie International Edition*, **47**(50), 9618–9647.
4. Kumar, A., et al. (1995). Patterned self-assembled monolayers and mesoscale phenomena. *Accounts of Chemical Research*, **28**(5), 219–226.
5. Cortines, J. R., et al. (2011). A retroviral chimeric capsid protein reveals the role of the N-terminal β-hairpin in mature core assembly. *Journal of Molecular Biology*, **410**(4), 641–652.
6. Nduka, O. O., Parrillo, J. E. (2011). The pathophysiology of septic shock. *Critical Care Nursing Clinics of North America*, **23**(1), 41–66.
7. Khavkin, J., Ellis, D. A. F. (2011). Aging skin: histology, physiology, and pathology. *Facial Plastic Surgery Clinics of North America*, **19**(2), 229–234.
8. Nalla, R. K., et al. (2006). Role of microstructure in the aging-related deterioration of the toughness of human cortical bone. *Materials Science and Engineering: C*, **26**(8), 1251–1260.
9. Wallace, K. N., et al. (2005). Intestinal growth and differentiation in zebrafish. *Mechanisms of Development*, **122**(2), 157–173.
10. Kim, K. S., et al. (2010). Selective adhesion of intestinal epithelial cells on patterned films with amine functionalities formed by plasma enhanced chemical vapor deposition. *Applied Surface Science*, **257**(2), 398–403.
11. Spencer, N. D. (1998). Advanced surfaces: Their tailoring and analysis. *Chimia*, **52**(10), 598–606.
12. Kong, Z., et al. (2010). Study on chemical vapor deposited copper films on cyano and carboxylic self-assembled monolayer diffusion barriers. *Thin Solid Films*, **518**(17), 4852–4859.
13. Lee, S. M., Pippel, E., Knez, M. (2011). Metal infiltration into biomaterials by ALD and CVD: A comparative study. *Chemphyschem*, **12**(4), 791–798.
14. Knez, M., Niesch, K., Niinisto, L. (2007). Synthesis and surface engineering of complex nanostructures by atomic layer deposition. *Advanced Materials*, **19**(21), 3425–3438.
15. Lee, S.-M., Pippel, E., Knez, M. (2011). Metal infiltration into biomaterials by ALD and CVD: A comparative study. *Chemphyschem*, **12**(4), 791–798.
16. Dasgupta, N. P., et al. (2011). Atomic layer deposition of lead sulfide quantum dots on nanowire surfaces. *Nano Letters*, **11**(3), 934–940.
17. Klahr, B. M., Martinson, A. B. F., Hamann, T. W. (2011). Photoelectrochemical investigation of ultrathin film iron oxide solar cells prepared by atomic layer deposition. *Langmuir*, **27**(1), 461–468.
18. Myagkov, I. V., Novak, V. R. (1987). Polar langmuir-schaefer multilayers from azo dyes with an aliphatic radical. *Colloid Journal of the USSR*, **49**(5), 892–894.
19. Estillore, N. C., Park, J. Y., Advincula, R. C. (2010). Langmuir–schaefer (LS) macroinitiator film control on the grafting of a thermosensitive polymer brush via surface initiated-ATRP. *Macromolecules*, **43**(16), 6588–6598.
20. Girard-Egrot, A. P., Marquette, C. A., Blum, L. J. (2010). Biomimetic membranes and biomolecule immobillisation strategies for nanobiotechnology applications. *International Journal of Nanotechnology*, **7**(4–8), 753–780.
21. Yokoi, H., Hayashi, S., Kinoshita, T. (2003). Polypeptide membranes at an interface. *Progress in Polymer Science*, **28**(3), 341–357.
22. Belegrinou, S., et al. (2009). Biomimetic supported membranes from amphiphilic block copolymers. *Soft Matter*, **6**(1), 179–186.
23. Ariga, K., et al. (2008). Supramolecular chemistry as a versatile tool for advanced sciences in nanospace. *Advanced Science Letters*, **1**(1), 28–58.
24. Badia, A., et al. (2008). Nanostructured phospholipid membranes. *International Journal of Nanotechnology*, **5**(9–12), 1371–1395.
25. Girard-Egrot, A. P., Godoy, S., Blum, L. J. (2005). Enzyme association with lipidic langmuir-blodgett films: Interests and applications in nanobioscience. *Advances in Colloid and Interface Science*, **116**(1–3), 205–225.
26. Le Pioufle, B., et al. (2008). Lipid bilayer microarray for parallel recording of transmembrane ion currents. *Analytical Chemistry*, **80**(1), 328–332.
27. El Kirat, K., Morandat, S., Dufrêne, Y. F. (2010). Nanoscale analysis of supported lipid bilayers using atomic force microscopy. *Biochimica et Biophysica Acta (BBA)—Biomembranes*, **1798**(4), 750–765.
28. Chen, Q., et al. (2007). Network structure of collagen layers absorbed on LB film. *Journal of Colloid and Interface Science*, **316**(1), 1–9.
29. Evtugyn, G. A., Hianik, T. (2011). Layer-by-layer polyelectrolyte assembles involving DNA as a platform for DNA sensors. *Current Analytical Chemistry*, **7**(1), 8–34.
30. Machado, G., et al. (2011). Structural control of gold nanoparticles self-assemblies by layer-by-layer process. *Nanoscale*, **3**(4), 1717–1723.
31. Such, G. K., Johnston, A. P. R., Caruso, F. (2011). Engineered hydrogen-bonded polymer multilayers: From assembly to biomedical applications. *Chemical Society Reviews*, **40**(1), 19–29.

32. Tang, Z., et al. (2007). Biomedical applications of layer-by-layer assembly: From biomimetics to tissue engineering. *Advanced Materials*, **19**(7), 906–906.

33. Ariga, K., et al. (2011). Nanoarchitectonics: A conceptual paradigm for design and synthesis of dimension-controlled functional nanomaterials. *Journal of Nanoscience and Nanotechnology*, **11**(1), 1–13.

34. Fendler, J. H. (1996). Self-assembled nanostructured materials. *Chemistry of Materials*, **8**(8), 1616–1624.

35. Xu, L., Zhu, Z., Sukhishvili, S. A. (2011). Polyelectrolyte multilayers of diblock copolymer micelles with temperature-responsive cores. *Langmuir*, **27**(1), 409–415.

36. Swiston, A. J., et al. (2008). Surface functionalization of living cells with multilayer patches. *Nano Letters*, **8**(12), 4446–4453.

37. Arya, S. K., et al. (2009). Recent advances in self-assembled monolayers based biomolecular electronic devices. *Biosensors & Bioelectronics*, **24**(9), 2810–2817.

38. Heimel, G., et al. (2008). Odd–even effects in self-assembled monolayers of omega-(biphenyl-4-yl)alkanethiols: A first-principles study. *Langmuir*, **24**(2), 474–482.

39. Bhagirath, N., Kanhere, E., Bhattacharya, E. (2010). MOS capacitor for simple thiol based biosensor applications. *Thin Solid Films*, **519**(3), 982–986.

40. Walther, M. E., et al. (2010). Photoinduced processes in fluorene-bridged rhenium-phenothiazine dyads—comparison of electron transfer across fluorene, phenylene, and xylene bridges. *European Journal of Inorganic Chemistry*, **30**, 4843–4850.

41. Seo, K., Lee, H. (2009). Molecular electron transport changes upon structural phase transitions in alkanethiol molecular junctions. *Acs Nano*, **3**(9), 2469–2476.

42. Shaw, J., et al. (2011). Integration of self-assembled redox molecules in flash memory devices. *Ieee Transactions on Electron Devices*, **58**(3), 826–834.

43. DiBenedetto, S. A., et al. (2009). Charge conduction and breakdown mechanisms in self-assembled nanodielectrics. *Journal of the American Chemical Society*, **131**(20), 7158–7168.

44. Liu, H. M., et al. (2009). Asymmetric electron transport realized by decoupling between molecule and electrode. *Physical Chemistry Chemical Physics*, **11**(44), 10323–10330.

45. Ma, Z., Zaera, F. (2006). Organic chemistry on solid surfaces. *Surface Science Reports*, **61**(5), 229–281.

46. Tam-Chang, S. W., Iverson, I. (1999). Applications of self-assembled monolayers (SAMs) of alkanethiolates on gold. In: Debrowski, A., ed. *Adsorption and Its Applications in Industry and Environmental Protection*, Vol. I: *Applications in Industry*, pp. 917–950, Elsevier Science.

47. Zaupa, G., et al. (2011). Catalytic self-assembled monolayers on Au nanoparticles: The source of catalysis of a transphosphorylation reaction. *Chemistry—A European Journal*, **17**(17), 4879–4889.

48. Fischer, M., Sperling, C., Werner, C. (2010). Synergistic effect of hydrophobic and anionic surface groups triggers blood coagulation in vitro. *Journal of Materials Science—Materials in Medicine*, **21**(3), 931–937.

49. Castner, D. G., Ratner, B. D. (2002). Biomedical surface science: Foundations to frontiers. *Surface Science*, **500**(1–3), 28–60.

50. Cheng, M. M. C., et al. (2006). Nanotechnologies for biomolecular detection and medical diagnostics. *Current Opinion in Chemical Biology*, **10**(1), 11–19.

51. Patel, D. N., Bailey, S. R. (2007). Nanotechnology in cardiovascular medicine. *Catheterization and Cardiovascular Interventions*, **69**(5), 643–654.

52. Mrksich, M., Whitesides, G. M. (1996). Using self-assembled monolayers to understand the interactions of man-made surfaces with proteins and cells. *Annual Review of Biophysics and Biomolecular Structure*, **25**, 55–78.

53. Yan, D., Jennings, G. K., Weinstein, R. D. (2002). Controlling the properties of *n*-alkanethiolate self-assembled monolayers on gold by using supercritical carbon dioxide–ethanol mixtures as solvents. *Industrial & Engineering Chemistry Research*, **41**(18), 4528–4533.

54. Wang, J., Wu, H., Angnes, L. (1993). Online monitoring of hydrophobic compounds at self-assembled monolayer modified amperometric flow detectors. *Analytical Chemistry*, **65**(14), 1893–1896.

55. Acharya, H., et al. (2010). Mapping hydrophobicity at the nanoscale: Applications to heterogeneous surfaces and proteins. *Faraday Discussions*, **146**, 353–365.

56. Pakoulev, A. V., Burtman, V. (2009). Temperature dependent barrier crossover regime in tunneling single molecular devices based on the matrix of isolated molecules. *Journal of Physical Chemistry C*, **113**(51), 21413–21421.

57. Love, J. C., et al. (2005). Self-assembled monolayers of thiolates on metals as a form of nanotechnology. *Chemical Reviews*, **105**(4), 1103–1169.

58. Schreiber, F. (2000). Structure and growth of self-assembling monolayers. *Progress in Surface Science*, **65**(5–8), 151–256.

59. Ulman, A. (1996). Formation and structure of self-assembled monolayers. *Chemical Reviews*, **96**(4), 1533–1554.

60. Osman, M. A., et al. (2002). Structure and molecular dynamics of alkane monolayers self-assembled on mica platelets. *Journal of Physical Chemistry B*, **106**(3), 653–662.

61. Jackman, R. J., Wilbur, J. L., Whitesides, G. M. (1995). Fabrication of submicrometer features on curved substrates by microcontact printing. *Science*, **269**(5224), 664–666.

62. Browne, K. P., Grzybowski, B. A. (2011). Controlling the properties of self-assembled monolayers by substrate curvature. *Langmuir*, **27**(4), 1246–1250.

63. Delamarche, E., et al. (1996). Golden interfaces: The surface of self-assembled monolayers. *Advanced Materials*, **8**(9), 719–&.

64. Aswal, D. K., et al. (2006). Self assembled monolayers on silicon for molecular electronics. *Analytica Chimica Acta*, **568**(1–2), 84–108.

65. Yan, L., Huck, W. T. S., Whitesides, G. M. (2004). Self-assembled monolayers (SAMS) and synthesis of

planar micro- and nanostructures. *Journal of Macromolecular Science-Polymer Reviews*, **C44**(2), 175–206.

66. Weiss, E. A., et al. (2007). Si/SiO_2-templated formation of ultraflat metal surfaces on glass, polymer, and solder supports: Their use as substrates for self-assembled monolayers. *Langmuir*, **23**(19), 9686–9694.
67. Kluth, G. J., Carraro, C., Maboudian, R. (1999). Direct observation of sulfur dimers in alkanethiol self-assembled monolayers on Au(111). *Physical Review B*, **59**(16), 10449–10452.
68. Duwez, A. S. (2004). Exploiting electron spectroscopies to probe the structure and organization of self-assembled monolayers: A review. *Journal of Electron Spectroscopy and Related Phenomena*, **134**(2–3), 97–138.
69. Ciampi, S., Harper, J. B., Gooding, J. J. (2010). Wet chemical routes to the assembly of organic monolayers on silicon surfaces via the formation of Si–C bonds: Surface preparation, passivation and functionalization. *Chemical Society Reviews*, **39**(6), 2158–2183.
70. Thuo, M. M., et al. (2011). Odd-even effects in charge transport across self-assembled monolayers. *Journal of the American Chemical Society*, **133**(9), 2962–2975.
71. Tao, Y.-T., et al. (2011). Odd–even modulation of electrode work function with self-assembled layer: Interplay of energy barrier and tunneling distance on charge injection in organic light-emitting diodes. *Organic Electronics*, **12**(4), 602–608.
72. Vervaecke, F., et al. (2011). Odd–even effects in ion-beam-induced desorption of biphenyl-substituted alkanethiol self-assembled monolayers. *Chemphyschem*, **12**(1), 140–144.
73. Imahori, H., et al. (2004). Vectorial electron relay at ITO electrodes modified with self-assembled monolayers of ferrocene–porphyrin–fullerene triads and porphyrin–fullerene dyads for molecular photovoltaic devices. *Chemistry—A European Journal*, **10**(20), 5111–5122.
74. Doering, K., et al. (2010). Synthesis, electrochemical behavior, and self-assembly of metallocene-functionalized thiollu-orenes. *European Journal of Inorganic Chemistry*, **25**, 3952–3960.
75. Dubacheva, G. V., et al. (2011). Redox strategy for reversible attachment of biomolecules using bifunctional linkers. *Chemical Communications*, **47**(12), 3565–3567.
76. Dillmore, W. S., Yousaf, M. N., Mrksich, M. (2004). A photochemical method for patterning the immobilization of ligands and cells to self-assembled monolayers. *Langmuir*, **20**(17), 7223–7231.
77. Petrangolini, P., et al. (2010). An electrochemical scanning tunneling microscopy study of 2-(6-Mercaptoalkyl)hydroquinone molecules on Au(111). *Journal of the American Chemical Society*, **132**(21), 7445–7453.
78. Baccar, H., et al. (2010). Surface plasmon resonance immunosensor for bacteria detection. *Talanta*, **82**(2), 810–814.
79. Zhang, X., et al. (2010). A new strategy for the preparation of maleimide-functionalised gold surfaces. *Electrochemistry Communications*, **12**(10), 1403–1406.
80. Tanaka, N., et al. (2010). Self-assembly immobilization of hyaluronan thiosemicarbazone on a gold surface for cell culture applications. *Carbohydrate Polymers*, **82**(1), 100–105.
81. Weinstein, R. D., et al. (2007). Characterization of self-assembled monolayers from lithium dialkyldithiocarbamate salts. *Langmuir*, **23**(5), 2887–2891.
82. Aina, V., et al. (2011). Novel bio-conjugate materials: Soybean peroxidase immobilized on bioactive glasses containing Au nanoparticles. *Journal of Materials Chemistry*, **21**(29), 10970–10981.
83. Li, X. M., Huskens, J., Reinhoudt, D. N. (2004). Reactive self-assembled monolayers on flat and nanoparticle surfaces, and their application in soft and scanning probe lithographic nanofabrication technologies. *Journal of Materials Chemistry*, **14**(20), 2954–2971.
84. Li, J., Thiara, P. S., Mrksich, M. (2007). Rapid evaluation and screening of interfacial reactions on self-assembled monolayers. *Langmuir*, **23**(23), 11826–11835.
85. Prats-Alfonso, E., Albericio, F. (2011). Functionalization of gold surfaces: Recent developments and applications. *Journal of Materials Science*, **46**(24), 7643–7648.
86. Sandig, N., Zerbetto, F. (1999). Molecules on gold. *Chemical Communications*, **46**(5), 667–676.
87. Jung, D. R., Czanderna, A. W. (1994). Chemical and physical interactions at metal self-assembled organic monolayer interfaces. *Critical Reviews in Solid State and Materials Sciences*, **19**(1), 1–54.
88. Nishida, N., et al. (1996). Thermal desorption spectroscopy of alkanethiol self-assembled monolayer on Au(111). *Japanese Journal of Applied Physics Part 1—Regular Papers Short Notes & Review Papers*, **35**(11), 5866–5872.
89. Van Vaeck, L., Adriaens, A., Gijbels, R. (1999). Static secondary ion mass spectrometry: (S-SIMS) Part 1. Methodology and structural interpretation. *Mass Spectrometry Reviews*, **18**(1), 1–47.
90. Heimel, G., et al. (2008). The interface energetics of self-assembled monolayers on metals. *Accounts of Chemical Research*, **41**(6), 721–729.
91. Weibel, D. B., DiLuzio, W. R., Whitesides, G. M. (2007). Microfabrication meets microbiology. *Nature Reviews Microbiology*, **5**(3), 209–218.
92. Chen, C. S., et al. (1998). Micropatterned surfaces for control of cell shape, position, and function. *Biotechnology Progress*, **14**(3), 356–363.
93. Kane, R. S., et al. (1999). Patterning proteins and cells using soft lithography. *Biomaterials*, **20**(23–24), 2363–2376.
94. Xia, Y. N., Whitesides, G. M. (1998). Soft lithography. *Annual Review of Materials Science*, **28**, 153–184.
95. Park, T. H., Shuler, M. L. (2003). Integration of cell culture and microfabrication technology. *Biotechnology Progress*, **19**(2), 243–253.
96. Mrksich, M., Whitesides, G. M. (1995). Patterning self-assembled monolayers using microcontact printing—A new technology for biosensors. *Trends in Biotechnology*, **13**(6), 228–235.
97. Quist, A. P., Pavlovic, E., Oscarsson, S. (2005). Recent advances in microcontact printing. *Analytical and Bioanalytical Chemistry*, **381**(3), 591–600.

98. Castellana, E. T., Cremer, P. S. (2006). Solid supported lipid bilayers: From biophysical studies to sensor design. *Surface Science Reports*, **61**(10), 429–444.

99. Takayama, S., et al. (2000). Patterning the topographical environment for mammalian cell culture using laminar flows in capillaries. In: Dittmar, A. B. D., eds. *1st Annual International Ieee-Embs Special Topic Conference on Microtechnologies in Medicine & Biology, Proceedings*, pp. 322–325.

100. Gooding, J. J., et al. (2003). Self-assembled monolayers into the 21(st) century: Recent advances and applications. *Electroanalysis*, **15**(2), 81–96.

101. Panasyuk, T. L., et al. (1999). Electropolymerized molecularly imprinted polymers as receptor layers in a capacitive chemical sensors. *Analytical Chemistry*, **71**(20), 4609–4613.

102. Kelley, S. O., et al. (1998). Orienting DNA helices on gold using applied electric fields. *Langmuir*, **14**(24), 6781–6784.

103. Holdcroft, S. (2001). Patterning pi-conjugated polymers. *Advanced Materials*, **13**(23), 1753–1765.

104. Schreiber, F. (2004). Self-assembled monolayers: From "simple" model systems to biofunctionalized interfaces. *Journal of Physics-Condensed Matter*, **16**(28), R881–R900.

105. Zhang, X., et al. (2008). Superhydrophobic surfaces: From structural control to functional application. *Journal of Materials Chemistry*, **18**(6), 621–633.

106. Menard, E., et al. (2007). Micro- and nanopatterning techniques for organic electronic and optoelectronic systems. *Chemical Reviews*, **107**(4), 1117–1160.

107. Michel, B., et al. (2001). Printing meets lithography: Soft approaches to high-resolution printing. *Ibm Journal of Research and Development*, **45**(5), 697–719.

108. Stewart, M. E., et al. (2006). Quantitative multispectral biosensing and 1D imaging using quasi-3D plasmonic crystals. *Proceedings of the National Academy of Sciences of the United States of America*, **103**(46), 17143–17148.

109. Jung, G. Y., et al. (2005). Vapor-phase self-assembled monolayer for improved mold release in nanoimprint lithography. *Langmuir*, **21**(4), 1158–1161.

110. Iskandar, F. (2009). Nanoparticle processing for optical applications—A review. *Advanced Powder Technology*, **20**(4), 283–292.

111. Kim, G. M., et al. (2005). One-dimensional arrangement of gold nanoparticles by electrospinning. *Chemistry of Materials*, **17**(20), 4949–4957.

112. Salaita, K., Wang, Y. H., Mirkin, C. A. (2007). Applications of dip-pen nanolithography. *Nature Nanotechnology*, **2**(3), 145–155.

113. Salaita, K., Wang, Y., Mirkin, C. A. (2007). Applications of dip-pen nanolithography. *Nature Nanotechnology*, **2**(3), 145–155.

114. Piner, R. D., et al. (1999). "Dip-pen" nanolithography. *Science*, **283**(5402), 661–663.

115. Ginger, D. S., Zhang, H., Mirkin, C. A. (2004). The evolution of dip-pen nanolithography. *Angewandte Chemie-International Edition*, **43**(1), 30–45.

116. Geissler, M., Xia, Y. N. (2004). Patterning: Principles and some new developments. *Advanced Materials*, **16**(15), 1249–1269.

117. Lee, K. B., et al. (2002). Protein nanoarrays generated by dip-pen nanolithography. *Science*, **295**(5560), 1702–1705.

118. Golzhauser, A., et al. (2001). Chemical nanolithography with electron beams. *Advanced Materials*, **13**(11), 803–806.

119. Golzhauser, A., et al. (2000). Nanoscale patterning of self-assembled monolayers with electrons. *Journal of Vacuum Science & Technology B*, **18**(6), 3414–3418.

120. Xia, Y. N., Zhao, X. M., Whitesides, G. M. (1996). Pattern transfer: Self-assembled monolayers as ultrathin resists. *Microelectronic Engineering*, **32**(1–4), 255–268.

121. Ginger, D. S., Zhang, H., Mirkin, C. A. (2004). The evolution of dip-pen nanolithography. *Angewandte Chemie International Edition*, **43**(1), 30–45.

122. Jiang, W., et al. (2011). Fabrication of submicron photon sieve using E-beam lithography and X-ray lithography. *Microelectronic Engineering*, **88**(10), 3178–3181.

123. Sandhyarani, N., Pradeep, T. (2003). Current understanding of the structure, phase transitions and dynamics of self-assembled monolayers on two- and three-dimensional surfaces. *International Reviews in Physical Chemistry*, **22**(2), 221–262.

124. Ferretti, S., et al. (2000). Self-assembled monolayers: A versatile tool for the formulation of bio-surfaces. *Trac—Trends in Analytical Chemistry*, **19**(9), 530–540.

125. Cumming, G., Fidler, F., Vaux, D. L. (2007). Error bars in experimental biology. *The Journal of Cell Biology*, **177**(1), 7–11.

126. Herrwerth, S., et al. (2003). Factors that determine the protein resistance of oligoether self-assembled monolayers—Internal hydrophilicity, terminal hydrophilicity, and lateral packing density. *Journal of the American Chemical Society*, **125**(31), 9359–9366.

127. Ostuni, E., Yan, L., Whitesides, G. M. (1999). The interaction of proteins and cells with self-assembled monolayers of alkanethiolates on gold and silver. *Colloids and Surfaces B-Biointerfaces*, **15**(1), 3–30.

128. Kitano, H., et al. (2009). Anti-biofouling properties of a telomer brush with pendent glucosylurea groups. *Langmuir*, **25**(16), 9361–9368.

129. Ista, L. K., et al. (2001). Synthesis of poly(*N*-isopropylacrylamide) on initiator-modified self-assembled monolayers. *Langmuir*, **17**(9), 2552–2555.

130. Ista, L. K., Perez-Luna, V. H., Lopez, G. P. (1999). Surface-grafted, environmentally sensitive polymers for biofilm release. *Applied and Environmental Microbiology*, **65**(4), 1603–1609.

131. Prime, K. L., Whitesides, G. M. (1991). Self-assembled organic monolayers—Model systems for studying adsorption of proteins at surfaces. *Science*, **252**(5009), 1164–1167.

132. Prime, K. L., Whitesides, G. M. (1993). Adsorption of proteins onto surfaces containing end-attached oligo(ethylene oxide): A model system using self-assembled monolayers. *Journal of the American Chemical Society*, **115**, 10714–10721.

133. Ostuni, E., et al. (2001). A survey of structure–property relationships of surfaces that resist the adsorption of protein. *Langmuir*, **17**(18), 5605–5620.

134. Jeon, S. I., et al. (1991). Protein surface interactions in the presence of polyethylene oxide 1. Simplified theory. *Journal of Colloid and Interface Science*, **142**(1), 149–158.

135. Hucknall, A., Rangarajan, S., Chilkoti, A. (2009). In pursuit of zero: Polymer brushes that resist the adsorption of proteins. *Advanced Materials*, **21**(23), 2441–2446.

136. Chapman, R. G., et al. (2000). Surveying for surfaces that resist the adsorption of proteins. *Journal of the American Chemical Society*, **122**(34), 8303–8304.

137. Morra, M. (2000). On the molecular basis of fouling resistance. *Journal of Biomaterials Science—Polymer Edition*, **11**(6), 547–569.

138. Banerjee, I., Pangule, R. C., Kane, R. S. (2011). Antifouling coatings: Recent developments in the design of surfaces that prevent fouling by proteins, bacteria, and marine organisms. *Advanced Materials*, **23**(6), 690–718.

139. Statz, A. R., et al. (2005). New peptidomimetic polymers for antifouling surfaces. *Journal of the American Chemical Society*, **127**(22), 7972–7973.

140. Chelmowski, R., et al. (2008). Peptide-based SAMs that resist the adsorption of proteins. *Journal of the American Chemical Society*, **130**(45), 14952–14953.

141. Siegers, C., Biesalski, M., Haag, R. (2004). Self-assembled monolayers of dendritic polyglycerol derivatives on gold that resist the adsorption of proteins. *Chemistry—A European Journal*, **10**(11), 2831–2838.

142. Calderon, M., et al. (2010). Dendritic polyglycerols for biomedical applications. *Advanced Materials*, **22**(2), 190–218.

143. Kane, R. S., Deschatelets, P., Whitesides, G. M. (2003). Kosmotropes form the basis of protein-resistant surfaces. *Langmuir*, **19**(6), 2388–2391.

144. Dilly, S. J., et al. (2006). Novel tertiary amine oxide surfaces that resist nonspecific protein adsorption. *Langmuir*, **22**(19), 8144–8150.

145. Weinhart, M., et al. (2010). Linear poly(methyl glycerol) and linear polyglycerol as potent protein and cell resistant alternatives to poly(ethylene glycol). *Chemistry—An Asian Journal*, **5**(9), 1992–2000.

146. Holmlin, R. E., et al. (2001). Zwitterionic SAMs that resist nonspecific adsorption of protein from aqueous buffer. *Langmuir*, **17**(9), 2841–2850.

147. Sanchez-Cortes, J., Mrksich, M. (2011). Using self-assembled monolayers to understand alpha 8 beta 1-mediated cell adhesion to RGD and FEI motifs in nephronectin. *Acs Chemical Biology*, **6**(10), 1078–1086.

148. Kokare, C. R., et al. (2009). Biofilm: Importance and applications. *Indian Journal of Biotechnology*, **8**(2), 159–168.

149. Simões, M., Simões, L. C., Vieira, M. J. (2010). A review of current and emergent biofilm control strategies. *LWT—Food Science and Technology*, **43**(4), 573–583.

150. Chmielewski, R. A. N., Frank, J. F. (2003). Biofilm formation and control in food processing facilities. *Comprehensive Reviews in Food Science and Food Safety*, **2**(1), 22–32.

151. Donlan, R. M., Costerton, J. W. (2002). Biofilms: Survival mechanisms of clinically relevant microorganisms. *Clinical Microbiology Reviews*, **15**(2), 167–193.

152. Hall-Stoodley, L., Stoodley, P. (2002). Developmental regulation of microbial biofilms. *Current Opinion in Biotechnology*, **13**(3), 228–233.

153. O'Toole, G., Kaplan, H. B., Kolter, R. (2000). Biofilm formation as microbial development. *Annual Review of Microbiology*, **54**, 49–79.

154. An, Y. H., Friedman, R. J. (1998). Concise review of mechanisms of bacterial adhesion to biomaterial surfaces. *Journal of Biomedical Materials Research*, **43**(3), 338–348.

155. Simoes, M., Simoes, L. C., Vieira, M. J. (2010). A review of current and emergent biofilm control strategies. *Lwt-Food Science and Technology*, **43**(4), 573–583.

156. Donlan, R. M. (2002). Biofilms: Microbial life on surfaces. *Emerging Infectious Diseases*, **8**(9), 881–890.

157. Omoike, A., Chorover, J. (2004). Spectroscopic study of extracellular polymeric substances from *Bacillus subtilis*: Aqueous chemistry and adsorption effects. *Biomacromolecules*, **5**(4), 1219–1230.

158. Callow, M. E., et al. (2000). Use of self-assembled monolayers of different wettabilities to study surface selection and primary adhesion processes of green algal (Enteromorpha) zoospores. *Applied and Environmental Microbiology*, **66**(8), 3249–3254.

159. Lichter, J. A., et al. (2008). Substrata mechanical stiffness can regulate adhesion of viable bacteria. *Biomacromolecules*, **9**(6), 1571–1578.

160. Ploux, L., et al. (2007). Quantitative and morphological analysis of biofilm formation on self-assembled monolayers. *Colloids and Surfaces B—Biointerfaces*, **57**(2), 174–181.

161. Emerson, R. J., et al. (2006). Microscale correlation between surface chemistry, texture, and the adhesive strength of *Staphylococcus epidermidis*. *Langmuir*, **22**(26), 11311–11321.

162. Ista, L. K., et al. (2009). Smart surfaces for the control of bacterial attachment and biofilm accumulation. *Smart Coatings Ii*, **1002**, 95–110.

163. Luo, X., et al. (2009). Immobilization of the hyperthermophilic hydrogenase from Aquifex aeolicus bacterium onto gold and carbon nanotube electrodes for efficient H(2) oxidation. *Journal of Biological Inorganic Chemistry*, **14**(8), 1275–1288.

164. Bayoudh, S., et al. (2008). Electrical detection and characterization of bacterial adhesion using electrochemical impedance spectroscopy-based flow chamber. *Colloids and Surfaces a—Physicochemical and Engineering Aspects*, **318**(1–3), 291–300.

165. Barberousse, H., et al. (2007). Adhesion of facade coating colonisers, as mediated by physico-chemical properties. *Biofouling*, **23**(1), 15–24.

166. Ubbink, J., Schar-Zammaretti, P. (2005). Probing bacterial interactions: Integrated approaches combining atomic force

microscopy, electron microscopy and biophysical techniques. *Micron*, **36**(4), 293–320.

167. Kalasin, S., Santore, M. M. (2009). Non-specific adhesion on biomaterial surfaces driven by small amounts of protein adsorption. *Colloids and Surfaces B-Biointerfaces*, **73**(2), 229–236.
168. Hu, Y. F., Zhang, J. D., Ulstrup, J. (2010). Interfacial electrochemical electron transfer processes in bacterial biofilm environments on Au(111). *Langmuir*, **26**(11), 9094–9103.
169. Eleta Lopez, A., et al. (2011). Influence of surface chemistry and protein concentration on the adsorption rate and S-layer crystal formation. *Physical Chemistry Chemical Physics*, **13**(25), 11905–11913.
170. Xi, F. N., et al. (2011). Discrimination and detection of bacteria with a label-free impedimetric biosensor based on self-assembled lectin monolayer. *Journal of Electroanalytical Chemistry*, **656**(1–2), 252–257.
171. Huda, S., et al. (2010). Antibacterial nanoparticle monolayers prepared on chemically inert surfaces by cooperative electrostatic adsorption (CELA). *ACS Applied Materials & Interfaces*, **2**(4), 1206–1210.
172. Jin, Y. G., et al. (2008). Monolayers of the lipid derivatives of isoniazid at the air/water interface and the formation of self-assembled nanostructures in water. *Colloids and Surfaces B-Biointerfaces*, **64**(2), 229–235.
173. Khan, M. M. T., et al. (2011). Experimental and theoretical examination of surface energy and adhesion of nitrifying and heterotrophic bacteria using self-assembled monolayers. *Environmental Science & Technology*, **45**(3), 1055–1060.
174. Hall-Stoodley, L., Costerton, J. W., Stoodley, P. (2004). Bacterial biofilms: From the Natural environment to infectious diseases. *Nature Reviews Microbiology*, **2**(2), 95–108.
175. Roosjen, A., et al. (2004). Microbial adhesion to poly(ethylene oxide) brushes: Influence of polymer chain length and temperature. *Langmuir*, **20**(25), 10949–10955.
176. Lee, H. J., et al. (2000). Platelet and bacterial repellence on sulfonated poly(ethylene glycol)-acrylate copolymer surfaces. *Colloids and Surfaces B-Biointerfaces*, **18**(3–4), 355–370.
177. Kenan, D. J., et al. (2006). Peptide-PEG amphiphiles as cytophobic coatings for mammalian and bacterial cells. *Chemistry & Biology*, **13**(7), 695–700.
178. Fernandez, I. C. S., et al. (2007). The inhibition of the adhesion of clinically isolated bacterial strains on multi-component cross-linked poly(ethylene glycol)-based polymer coatings. *Biomaterials*, **28**(28), 4105–4112.
179. Ista, L. K., Mendez, S., Lopez, G. P. (2010). Attachment and detachment of bacteria on surfaces with tunable and switchable wettability. *Biofouling*, **26**(1), 111–118.
180. Luk, Y. Y., Kato, M., Mrksich, M. (2000). Self-assembled monolayers of alkanethiolates presenting mannitol groups are inert to protein adsorption and cell attachment. *Langmuir*, **16**(24), 9604–9608.
181. Vaisocherova, H., et al. (2008). Ultralow fouling and functionalizable surface chemistry based on a zwitterionic polymer enabling sensitive and specific protein detection in undiluted blood plasma. *Analytical Chemistry*, **80**(20), 7894–7901.
182. Zhang, Z., et al. (2009). Polysulfobetaine-grafted surfaces as environmentally benign ultralow fouling marine coatings. *Langmuir*, **25**(23), 13516–13521.
183. Onaizi, S. A., Leong, S. S. J. (2011). Tethering antimicrobial peptides: Current status and potential challenges. *Biotechnology Advances*, **29**(1), 67–74.
184. Amalric, J., et al. (2009). Phosphonate monolayers functionalized by silver thiolate species as antibacterial nanocoatings on titanium and stainless steel. *Journal of Materials Chemistry*, **19**(1), 141–149.
185. Jenkinson, H. F., Lappin-Scott, H. M. (2001). Biofilms adhere to stay. *Trends in Microbiology*, **9**(1), 9–10.
186. Qian, X. P., et al. (2002). Arrays of self-assembled monolayers for studying inhibition of bacterial adhesion. *Analytical Chemistry*, **74**(8), 1805–1810.
187. Cho, E. C., Kim, Y. D., Cho, K. (2004). Thermally responsive poly(*N*-isopropylacrylamide) monolayer on gold: Synthesis, surface characterization, and protein interaction/adsorption studies. *Polymer*, **45**(10), 3195–3204.
188. Roach, P., et al. (2007). Modern biomaterials: A review-bulk properties and implications of surface modifications. *Journal of Materials Science—Materials in Medicine*, **18**(7), 1263–1277.
189. Schlecht, C. A., Maurer, J. A. (2011). Functionalization of glass substrates: Mechanistic insights into the surface reaction of trialkoxysilanes. *RSC Advances*, **1**, 1146–1448.
190. McClary, K. B., Ugarova, T., Grainger, D. W. (2000). Modulating fibroblast adhesion, spreading, and proliferation using self-assembled monolayer films of alkylthiolates on gold. *Journal of Biomedical Materials Research*, **50**(3), 428–439.
191. Lewandowska, K., et al. (1992). Cell- type-specific adhesion mechanisms mediated by fibronectin adsorbed to chemically derivatized substrata. *Journal of Biomedical Materials Research*, **26**(10), 1343–1363.
192. Mrksich, M., et al. (1997). Using microcontact printing to pattern the attachment of mammalian cells to self-assembled monolayers of alkanethiolates on transparent films of gold and silver. *Experimental Cell Research*, **235**(2), 305–313.
193. Tegoulia, V. A., et al. (2001). Surface properties, fibrinogen adsorption, and cellular interactions of a novel phosphorylcholine-containing self-assembled monolayer on gold. *Langmuir*, **17**(14), 4396–4404.
194. Lan, M. A., et al. (2005). Myoblast proliferation and differentiation on fibronectin-coated self assembled monolayers presenting different surface chemistries. *Biomaterials*, **26**(22), 4523–4531.
195. Faucheux, N., et al. (2004). Self-assembled monolayers with different terminating groups as model substrates for cell adhesion studies. *Biomaterials*, **25**(14), 2721–2730.
196. Curran, J. M., Chen, R., Hunt, J. A. (2005). Controlling the phenotype and function of mesenchymal stem cells in vitro by adhesion to silane-modified clean glass surfaces. *Biomaterials*, **26**(34), 7057–7067.

197. Margel, S., et al. (1993). Peptide, protein, and cellular interactions with self-assembled monolayer model surfaces. *Journal of Biomedical Materials Research*, **27**(12), 1463–1476.

198. Scotchford, C. A., et al. (2002). Protein adsorption and human osteoblast-like cell attachment and growth on alkylthiol on gold self-assembled monolayers. *Journal of Biomedical Materials Research*, **59**(1), 84–99.

199. Keselowsky, B. G., Collard, D. M., Garcia, A. J. (2005). Integrin binding specificity regulates biomaterial surface chemistry effects on cell differentiation. *Proceedings of the National Academy of Sciences*, **102**(17), 5953–5957.

200. Lee, M. H., et al. (2006). Effect of biomaterial surface properties on fibronectin-alpha(5)beta(1) integrin interaction and cellular attachment. *Biomaterials*, **27**(9), 1907–1916.

201. Curran, J. M., et al. (2011). The use of dynamic surface chemistries to control msc isolation and function. *Biomaterials*, **32**(21), 4753–4760.

202. Palyvoda, O., et al. (2008). Molecular organization in SAMs used for neuronal cell growth. *Langmuir*, **24**(8), 4097–4106.

203. Culp, L. A., Sukenik, C. N. (1998). Cell type-specific modulation of fibronectin adhesion functions on chemically-derivatized self-assembled monolayers. *Journal of Biomaterials Science—Polymer Edition*, **9**(11), 1161–1176.

204. Spargo, B. J., et al. (1994). Spatially controlled adhesion, spreading, and differentiation of endothelial-cells on self-assembled molecular monolayers. *Proceedings of the National Academy of Sciences of the United States of America*, **91**(23), 11070–11074.

205. Zhang, S. G., et al. (1999). Biological surface engineering: A simple system for cell pattern formation. *Biomaterials*, **20**, 1213.

206. Dee, K. C., Andersen, T. T., Bizios, R. (1999). Osteoblast population migration characteristics on substrates modified with immobilized adhesive peptides. *Biomaterials*, **20**(3), 221–227.

207. Park, J.-W., et al. (2011). Bone healing of commercial oral implants with RGD immobilization through electrodeposited poly(ethylene glycol) in rabbit cancellous bone. *Acta Biomaterialia*, **7**(8), 3222–3229.

208. Schwartz, J., et al. (2003). Cell attachment and spreading on metal implant materials. *Materials Science & Engineering C-Biomimetic and Supramolecular Systems*, **23**(3), 395–400.

209. Ren, X., et al. (2011). Fibronectin and bone morphogenetic protein-2-decorated poly(OEGMA-r-HEMA) brushes promote osseointegration of titanium surfaces. *Langmuir*, **27**(19), 12069–12073.

210. Lorenz, C., et al. (2011). Coating of titanium implant materials with thin polymeric films for binding the signaling protein BMP2. *Macromolecular Bioscience*, **11**(2), 234–244.

211. Bush, K. A., et al. (2009). Designing tailored biomaterial surfaces to direct keratinocyte morphology, attachment, and differentiation. *Journal of Biomedical Materials Research Part A*, **90A**(4), 999–1009.

212. Keselowsky, B. G., Collard, D. M., Garcia, A. J. (2004). Surface chemistry modulates focal adhesion composition and signaling through changes in integrin binding. *Biomaterials*, **25**(28), 5947–5954.

213. Houseman, B. T., Mrksich, M. (2001). The microenvironment of immobilized Arg-Gly-Asp peptides is an important determinant of cell adhesion. *Biomaterials*, **22**(9), 943–955.

214. Eisenberg, J. L., Piper, J. L., Mrksich, M. (2009). Using self-assembled monolayers to model cell adhesion to the 9th and 10th type III domains of fibronectin. *Langmuir*, **25**(24), 13942–13951.

215. Petrie, T. A., et al. (2006). Integrin specificity and enhanced cellular activities associated with surfaces presenting a recombinant fibronectin fragment compared to RGD supports. *Biomaterials*, **27**(31), 5459–5470.

216. Heller, D. A., et al. (2005). Patterned networks of mouse hippocampal neurons on peptide-coated gold surfaces. *Biomaterials*, **26**(8), 883–889.

217. Hudalla, G. A., Murphy, W. L. (2010). Immobilization of peptides with distinct biological activities onto stem cell culture substrates using orthogonal chemistries. *Langmuir*, **26**(9), 6449–6456.

218. Anselme, K., Ponche, A., Bigerelle, M. (2010). Relative influence of surface topography and surface chemistry on cell response to bone implant materials. Part 2: Biological aspects. *Proceedings of the Institution of Mechanical Engineers Part H—Journal of Engineering in Medicine*, **224**(H12), 1487–1507.

219. Kim, S. H., et al. (2010). Comparative investigation of aged human mesenchymal stem cells on defined self-assembled monolayer (SAM) surfaces. *Differentiation*, **80**, S22–S22.

220. Klim, J. R., et al. (2010). A defined glycosaminoglycan-binding substratum for human pluripotent stem cells. *Nature Methods*, **7**(12), 989–U72.

221. Votteler, M., et al. (2010). Stem Cell Microenvironments—Unveiling the secret of how stem cell fate is defined. *Macromolecular Bioscience*, **10**(11), 1302–1315.

222. Ayala, R., et al. (2011). Engineering the cell-material interface for controlling stem cell adhesion, migration, and differentiation. *Biomaterials*, **32**(15), 3700–3711.

223. Ingber, D. E. (2003). Tensegrity I. Cell structure and hierarchical systems biology.

224. Chen, C. S., et al. (1997). Geometric control of cell life and death. *Science*, **276**(5317), 1425–1428.

225. Ostuni, E., et al. (2009). Using self-assembled monolayers to pattern ECM proteins and cells on substrates. *Methods in Molecular Biology*, **522**, 183–194.

226. Raghavan, S., et al. (2010). Micropaterned dynamically adhesive substrates for cell migration. *Langmuir*, **26**(22), 17733–17738.

227. Wang, N., Ingber, D. E. (1994). Control of cytoskeletal mechanics by extracellular-matrix, cell-shape, and mechanical tension. *Biophysical Journal*, **66**(6), 2181–2189.

228. Guillame-Gentil, O., et al. (2010). Engineering the extracellular environment: Strategies for building 2D and 3D cellular structures. *Advanced Materials*, **22**(48), 5443–5462.

229. Chen, C. S., Brangwynne, C., Ingber, D. E. (1999). Squaring up to the cell-shape debate. *Trends in Cell Biology*, **9**(7), 283.

230. McMurray, R. J., et al. (2011). Nanoscale surfaces for the long-term maintenance of mesenchymal stem cell phenotype and multipotency. *Nature Materials*, **10**(8), 637–644.

231. Kasemo, B. (2002). Biological surface science. *Surface Science*, **500**(1–3), 656–677.

232. Schwartz, Z., et al. (2001). Osteoblast response to titanium surface roughness and 1α,25-(OH)2D3 is mediated through the mitogen-activated protein kinase (MAPK) pathway. *Journal of Biomedical Materials Research*, **56**(3), 417–426.

233. Shekaran, A., Garcia, A. J. (2011). Nanoscale engineering of extracellular matrix-mimetic bioadhesive surfaces and implants for tissue engineering. *Biochimica et Biophysica Acta—General Subjects*, **1810**(3), 350–360.

234. Gross, K. A., Saber-Samandari, S., Heemann, K. S. (2010). Evaluation of commercial implants with nanoindentation defines future development needs for hydroxyapatite coatings. *Journal of Biomedical Materials Research Part B: Applied Biomaterials*, **93B**(1), 1–8.

235. http://www.merete-medical.com/.

236. http://www.bioretec.com/.

237. Cowin, S. C. (2001). *Bone Mechanics Handbook*, 2nd ed. p. 980.

238. Gawalt, E. S., et al. (2002). Bonding organics to Ti alloys: Facilitating human osteoblast attachment and spreading on surgical implant materials. *Langmuir*, **19**(1), 200–204.

239. Tizazu, G., et al. (2011). Large area nanopatterning of alkylphosphonate self-assembled monolayers on titanium oxide surfaces by interferometric lithography. *Nanoscale*, **3**(6), 2511–2516.

240. Shannon, F. J., et al. (2008). A novel surface treatment for porous metallic implants that improves the rate of bony ongrowth. *Journal of Biomedical Materials Research Part A*, **86A**(4), 857–864.

241. Liu, Q., et al. (2002). The role of surface functional groups in calcium phosphate nucleation on titanium foil: A self-assembled monolayer technique. *Biomaterials*, **23**(15), 3103–3111.

242. Konovalova, T., Kispert, L., Konovalov, V. (1999). Surface modification of TiO_2 nanoparticles with carotenoids. EPR study. *Journal of Physical Chemistry B*, **103**(22), 4672–4677.

243. Heijink, A., et al. (2008). Self-assembled monolayer films of phosphonates for bonding RGD to titanium. *Clinical Orthopaedics and Related Research*, **466**(4), 977–984.

244. Sampathkumaran, U., De Guire, M. R., Wang, R. (2001). Hydroxyapatite coatings on titanium. *Advanced Engineering Materials*, **3**(6), 401–405.

245. Kaufmann, C. R., et al. (2010). Long-term stability of self-assembled monolayers on 316L stainless steel. *Biomedical Materials*, **5**(2), 025008.

246. Slaney, A. M., et al. (2011). Biocompatible carbohydrate-functionalized stainless steel surfaces: A new method for passivating biomedical implants. *ACS Applied Materials & Interfaces*, **3**(5), 1601–1612.

247. Pitt, W. G., et al. (2004). Attachment of hyaluronan to metallic surfaces. *Journal of Biomedical Materials Research Part A*, **68A**(1), 95–106.

248. Raman, A., et al. (2006). Formation of self-assembled monolayers of alkylphosphonic acid on the native oxide surface of SS316L. *Langmuir*, **22**(15), 6469–6472.

249. Petrovic, Z., et al. (2011). Modification of a nitinol surface by phosphonate self-assembled monolayers. *Journal of the Electrochemical Society*, **158**(10), F159–F165.

250. Sargeant, T. D., et al. (2008). Covalent functionalization of NiTi surfaces with bioactive peptide amphiphile nanofibers. *Biomaterials*, **29**(8), 1085–1098.

251. Mani, G., et al. (2009). Surface modification of cobalt–chromium–tungsten–nickel alloy using octadecyltrichlorosilanes. *Applied Surface Science*, **255**(11), 5961–5970.

252. Wang, A., et al. (2007). Short-and long-term neural biocompatibility of heparin coated sapphire implants. *Materials Science & Engineering C-Biomimetic and Supramolecular Systems*, **27**(2), 237–243.

253. Cinier, M., et al. (2009). Bisphosphonate adaptors for specific protein binding on zirconium phosphonate-based microarrays. *Bioconjugate Chemistry*, **20**(12), 2270–2277.

254. Lim, J. Y., Donahue, H. J. (2007). Cell sensing and response to micro- and nanostructured surfaces produced by chemical and topographic patterning. *Tissue Engineering*, **13**(8), 1879–1891.

255. Puleo, D. A., Nanci, A. (1999). Understanding and controlling the bone–implant interface. *Biomaterials*, **20**(23-24), 2311–2321.

256. Wilson, C. J., et al. (2005). Mediation of biomaterial–cell interactions by adsorbed proteins: A review. *Tissue Engineering*, **11**(1–2), 1–18.

257. Mendonça, G., et al. (2008). Advancing dental implant surface technology—From micron- to nanotopography. *Biomaterials*, **29**(28), 3822–3835.

258. Marquis, M.-E., et al. (2009). Bone cells—biomaterials interactions. *Frontiers in Bioscience*, **14**, 1023–1067.

259. Webster, T. J., et al. (2000). Specific proteins mediate enhanced osteoblast adhesion on nanophase ceramics. *Journal of Biomedical Materials Research*, **51**(3), 475–483.

260. Dalby, M. J., et al. (2004). Investigating filopodia sensing using arrays of defined nanopits down to 35 nm diameter in size. *International Journal of Biochemistry & Cell Biology*, **36**(10), 2005–2015.

261. Hansson, S., Norton, M. (1999). The relation between surface roughness and interfacial shear strength for bone-anchored implants. A mathematical model. *Journal of Biomechanics*, **32**(8), 829–836.

262. Agashe, M., et al. (2005). Molecular simulation to characterize the adsorption behavior of a fibrinogen gamma-chain fragment. *Langmuir*, **21**(3), 1103–1117.

263. Yoshinari, M., et al. (2002). Bone response to calcium phosphate-coated and bisphosphonate-immobilized titanium implants. *Biomaterials*, **23**(14), 2879–2885.

264. Sader, M. S., et al. (2005). Effect of three distinct treatments of titanium surface on osteoblast attachment, proliferation, and differentiation. *Clinical Oral Implants Research*, **16**(6), 667–675.

265. Boyan, B. D., et al. (1998). Titanium surface roughness alters responsiveness of MG63 osteoblast-like cells to 1 alpha,25-(OH)(2)D-3. *Journal of Biomedical Materials Research*, **39**(1), 77–85.

266. Beutner, R., et al. (2010). Biological nano-functionalization of titanium-based biomaterial surfaces: a flexible toolbox. *Journal of the Royal Society Interface*, **7**, S93–S105.

267. Matsuzaka, K., et al. (2004). Effects of multigrooved surfaces on osteoblast-like cells in vitro: Scanning electron microscopic observation and mRNA expression of osteopontin and osteocalcin. *Journal of Biomedical Materials Research Part A*, **68A**(2), 227–234.

268. Michael, J., et al. (2009). Oligonucleotide–RGD peptide conjugates for surface modification of titanium implants and improvement of osteoblast adhesion. *Bioconjugate Chemistry*, **20**(4), 710–718.

18

LOW-DIMENSIONALITY EFFECTS IN ORGANIC FIELD EFFECT TRANSISTORS

STEFANO CASALINI, TOBIAS CRAMER, FRANCESCA LEONARDI, MASSIMILIANO CAVALLINI, AND FABIO BISCARINI

18.1 INTRODUCTION

The organic field effect transistor (OFET) is a device where a thin film of an organic semiconductor (OS) bridges a channel between source and drain electrodes. The gate electrode, separated by a dielectric thin film from the organic semiconductor, controls the charge carrier density in the organic semiconductor by capacitive coupling. OFET responds with a current between source and drain to a voltage bias applied to the gate and drain electrodes, the source being grounded. Response of OFET is measured by a set of parameters that are extracted from the current voltage characteristics as a function of the gate voltage (transfer curves) or the drain voltage (output curves).

The OFET has been studied for more than two decades because of its potential applications in flexible circuits, RFID tags, wearable electronics, and back-panel active matrix displays. Although organic electronics is the main technology driver, OFET play a central role in the fundamental studies aimed to elucidate charge transport in organic semiconductors. OFET are widely used as an experimental gauge for probing charge mobility, carrier density, and doping levels in organic thin films and nanostructures.

Despite the apparent simplicity of the architecture, the device physics is more subtle and elusive. OFET response cannot be simply elicited from molecular design and crystal packing, as it is dominated by the interactions of the OS with the device interfaces. At the OS/metal electrodes, charge injection/extraction occurs; at the OS/gate dielectric interface, charge carriers are capacitively accumulated, depleted, trapped, and transported; charge carriers cross organic semiconductor domain boundaries and are scattered/trapped by morphological/structural defects; the outer OS surface is exposed to the environment. The OFET response is extremely sensitive to any change occurring at these interfaces, happening either spontaneously (like in the case of charge trapping and bias stress), accidentally (a parasitic dopant), or by design (specific interaction of an analyte with species adsorbed or grafted at the interfaces). The effect of these interfaces is intertwined in the OFET response, and it is difficult to experimentally disentangle it.

On one hand, the OFET inherent instability is detrimental to electronics applications; on the other hand, it makes the device interesting for exploring new paradigms of sensing and transduction. In the past five years, an increasing trend of publications with OFETs used as (bio-)sensors is observed [1]. Advantages of OFETs with respect to more robust and established devices as CMOS are: the ease of interface tailoring toward analytes, living cells, and tissues; the use of low-cost scalable fabrication (important for single shot sensing); technology transferrable on flexible substrates with tunable mechanical compliance; a library of biocompatible and biodegradable materials, the latter yet *in nuce* (both crucial features for implantable devices); fabrication of devices with minimal amounts of materials (semiconductors, conductors, dielectrics, recognition groups); upscaling and integration are simpler than for resistive or amperometric sensors.

This chapter looks into the OFET as a low-dimensional device and how this distinctive feature can be exploited for a rational design of sensing devices. Owing to the vastity

Organic Nanomaterials: Synthesis, Characterization, and Device Applications, First Edition. Edited by Tomás Torres and Giovanni Bottari.

of the OFET research, and to the fact that the rationale is more effectively inferred from devices where molecules and oligomers constitute the active layers, we decided to focus this chapter on OFET operating with molecular semiconductors, in the form of ultra-thin films, self-assembled monolayers, and molecular nanostripes. We intentionally left polymeric devices a bit on the side, despite their technological relevance.

The chapter is organized as follows: In Section 18.2, a brief description of OFET working principles is given; in Section 18.3, different approaches to unconventional fabrication and growth of OFETs are reviewed; in Section 18.4, the focus is on the charge injection response, when the OFET acts as a charge tunneling device; in Section 18.5, low-dimensional charge transport in OFETs is discussed; in Section 18.6, sensing principles and applications are reviewed. Conclusions are given in Section 18.7.

18.2 PHENOMENOLOGICAL DESCRIPTION OF OFETs

18.2.1 Basic Operation Principles

The active material in an OFET is an OS thin film consisting of π-conjugated molecules, deposited by either high or ultra-high vacuum sublimation [2] or by spin-coating, drop-casting, or printing techniques if soluble [3]. An OS acts as a *p*-channel (*n*-channel) material when holes (electrons) are the majority charge carriers that are induced upon application of a suitable voltage to a capacitively coupled gate electrode. The OS, *per se*, is not intrinsically *p*-type or *n*-type, although the majority of semiconductors exhibit larger hole than electron mobility. The devices specific response depends on the materials, device architecture, and measurements conditions [4,5]. Doping is often accidental or it arises from the chemistry of interfaces, whereas intentional doping is rarely adopted.

The OFET is a three-terminal device (Figure 18.1, top). Source and drain electrodes are directly connected to the semiconductor, and the gate electrode is separated from the organic semiconductor thin film by a dielectric layer. The channel resembles a metal–insulator–semiconductor (MIS) structure. The source electrode is grounded. When no voltage is applied to the gate ($V_{GS} = 0$ V), the device is in the "off" state and a small current flows upon drain bias V_{DS}, because of the low intrinsic charge carrier density in the organic semiconductor. When the gate electrode is biased (Figure 18.1, middle), a conductive channel between source and drain electrodes forms due to the accumulation of charge carriers in the semiconductor close to the dielectric–organic semiconductor interface. Charge transport occurs in the channel with source–drain voltage, and the transistor is in the "on" state (Figure 18.1, bottom).

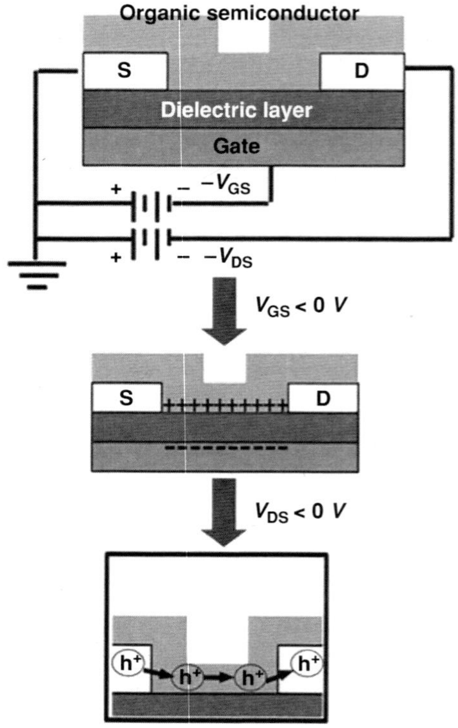

FIGURE 18.1. Schematic of *p*-channel thin-film transistor operation.

With no voltage applied to the drain ($V_{DS} = 0$), the density of charge carriers is uniform across the channel. For $V_{DS} \ll V_{GS} - V_{th}$ the uniform charge distribution is only perturbed and I_{DS} increases linearly with V_{DS} (Figure 18.2, top). When the drain potential is further increased up to $V_{DS} = V_{GS} - V_{th}$, a depletion region forms close to the drain electrode, the conductive channel is pinched off (Figure 18.2, middle), and the current saturates, thereby becoming V_{DS}-independent (Figure 18.2, bottom). The source–drain current can be described by Equation (18.1) and Equation (18.2) in the linear and the saturation regimes, respectively:

$$I_{DS} = \frac{W}{L} C_i \mu \left(V_{GS} - V_{th} - \frac{V_{DS}}{2} \right) V_{DS} \quad \text{for } V_{DS} < V_{GS} - V_{th} \tag{18.1}$$

$$I_{DS} = \frac{W}{2L} C_i \mu (V_{GS} - V_{th})^2 \quad \text{for } V_{DS} > V_{GS} - V_{th} \tag{18.2}$$

where W is the channel width, L is the channel length, C_i is the capacitance per unit area of the insulating layer, V_{th} is the threshold voltage, and μ is the field-effect mobility. Equations (18.1) and (18.2) hold upon the following assumptions: (i) the gradual channel approximation and (ii) the constant

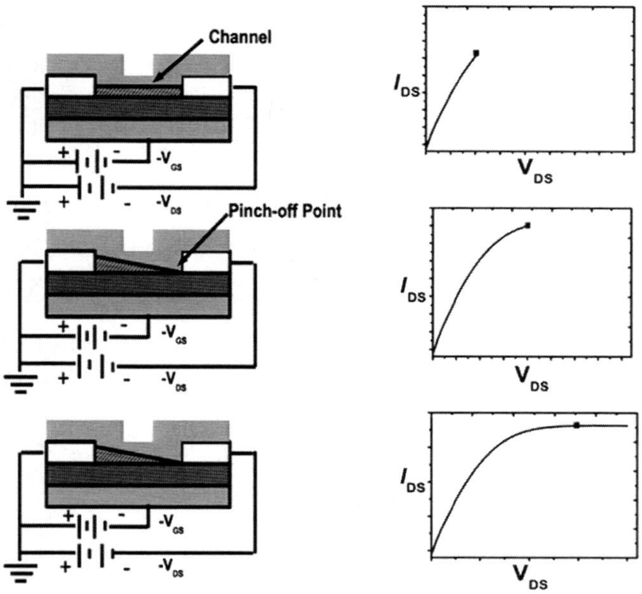

FIGURE 18.2. Illustration of operating regime and corresponding I–V characteristic: (**Top**) Linear regime. (**Middle**) Start of saturation regime at pinch off point. (**Bottom**) Saturation regime.

behavior of the mobility versus gate voltage. The former requires that the longitudinal electric field across the channel is lower than the transversal one; the latter considers the mobility as an intrinsic parameter of the organic semiconductor [6–8] and neglects the influence of trapping states. Figure 18.3a shows output characteristics I_{DS} versus V_{DS} of a p-channel transistor for different gate voltages. Figures 18.3b and 18.3c show the transfer characteristics I_{DS} versus V_{GS} for a fixed V_{DS}. Both linear and saturation regimes can be identified in the curves.

An OFET is a multiparameter device characterized by: charge mobility μ, threshold voltage V_{th}, on/off current ratio, subthreshold swing SS, and turn-on voltage V_{ON} [9]. These parameters are obtained from the characteristics as defined in Figure 18.3. The charge carrier mobility (in units of $cm^2/V{\cdot}s$) is the average velocity of the charge carriers along the channel normalized to the longitudinal field, and it can be extracted from the slope of the transfer curve in the linear regime (Figure 18.3b). In the saturation regime, μ is calculated from the curvature, or else from the slope of the linear trend in the plot of $\sqrt{I_{DS}}$ versus V_{GS} (Figure 18.3c)

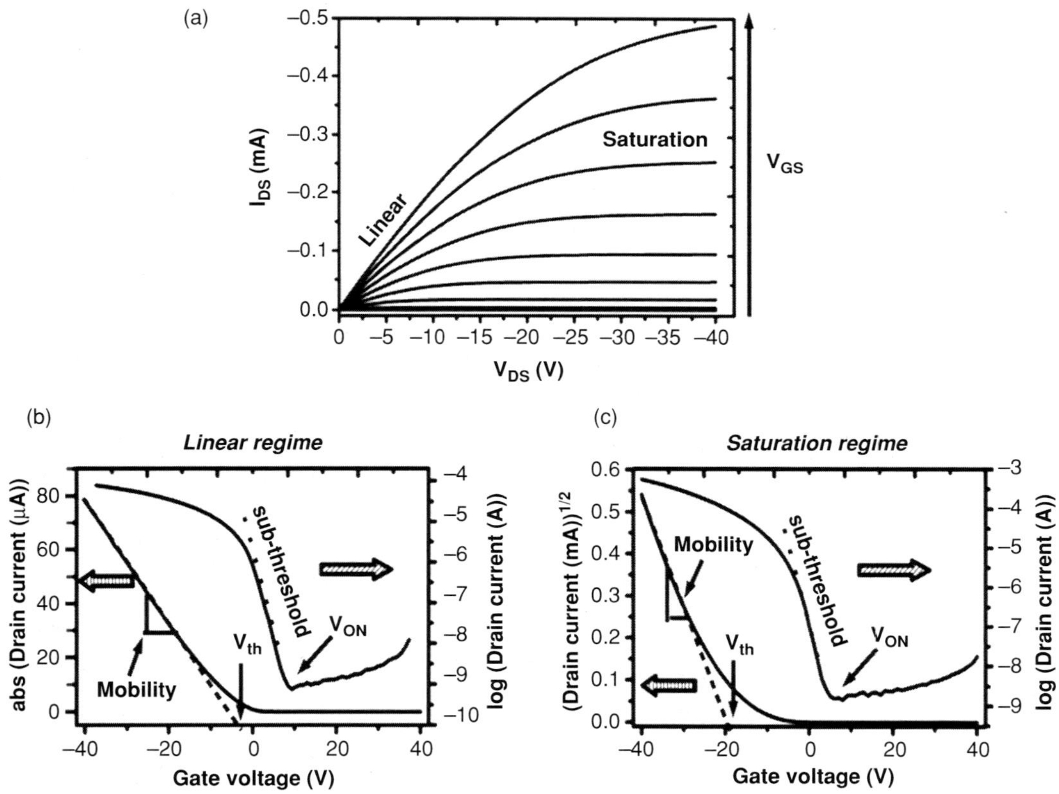

FIGURE 18.3. Current–voltage characteristic of a pentacene organic field effect transistor: (**a**) Output characteristic indicating the linear and saturation regimes. (**b**, **c**) Transfer characteristic in the linear regime and saturation regime, respectively; the linear graph indicates mobility (μ) and V_{th}, and the log-lin curves show the SS and V_{ON}.

[10, 11]. V_{th} is extracted as the intersection of the extrapolated linear part of the transfer characteristics and the V_{GS} axis (Figure 18.3c) [12]. The semilogarithmic plot (right axis of Figure 18.3b,c) is useful for extracting V_{ON}, SS, and the on/off ratio. V_{ON} is the voltage at which a I_{DS} increases above the noise level of the off-current. The subthreshold swing SS (expressed in mV/decade) is a measure of how rapidly the device switches from the off state to the on state and is extracted from the steep region of the I_{DS} trend [13]. The on/off ratio between the maximum and minimum value of the current I_{DS} is an estimate of the amplification. Both SS and on/off ratio are limited by the density of dopants and shallow traps.

Charge mobility in excess of 1 cm^2/(V·s) has been achieved with pentacene [14, 15] and rubrene [16] FETs. This value is comparable to the charge mobility of hydrogenated amorphous silicon (α-Si:H), which is currently used in back panel displays. Contrarily to silicon, OFETs suffer from bias stress which manifests itself as a continuous decrease of the current I_{DS} in time upon prolonged application of gate voltage. Bias stress causes hysteresis of the transfer characteristics upon a gate sweep cycle. It is ascribed to charge trapping that occurs at the interface between the OS film and the gate dielectric layer, and it still represents a problem in OS technology [17, 18].

OFETs architectures (Figure 18.4) are classified according to the position of the electrodes with respect to the organic semiconductor thin film. The specific architecture might have relevance on the values of the parameters extracted from the device characteristics. Test patterns on silicon oxide/Si wafer are usually fabricated into a BG/BC architecture. BG/TC devices exhibit the highest charge mobility and smallest contact resistance, as the result of better-defined interfaces [14].

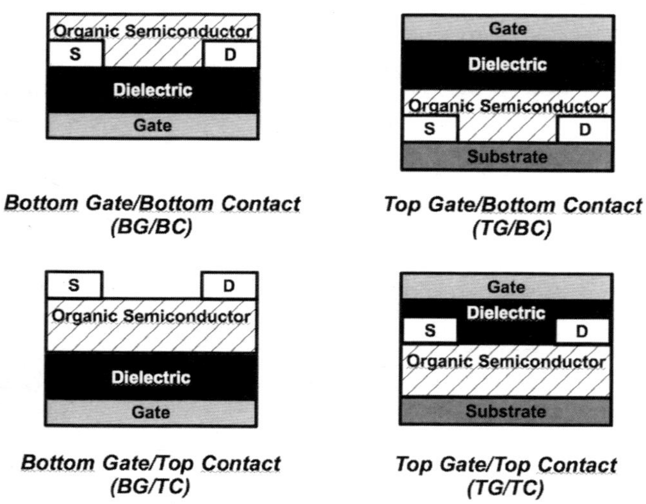

FIGURE 18.4. Basic OFET architectures.

18.2.2 Contact Resistance

In an ideal OFET device, the source–drain electrodes exhibit ohmic behavior [19]; and the contact resistance R_c, defined as the sum of the source and drain resistances, is negligible with respect to the resistance of the channel. This condition, necessary for measuring the intrinsic charge mobility, is often not fulfilled in OFETs. A heuristic approach to reconciliate the description by Eqs. (18.1) and (18.2) assigns the OS material close to the charge injection interfaces a different mobility with respect to the channel [20].

Here we introduce an explicit phenomenological description of R_c, and we discuss its microscopic origin in Section 18.4. The contact resistance R_c contributes to the effective resistance of the OFET, R_{eff}, which in the linear regime reads:

$$R_{eff} = \frac{V_{DS}}{I_{DS}} = 2R_c + R_{Ch} = \frac{2\rho_c \cdot \lambda}{W} + \frac{L}{W} \cdot \frac{1}{\mu \cdot C_i \cdot (V_{GS} - V_{th})} \tag{18.3}$$

The effective length scale λ arises from the interplay of energetic, structural, and morphological disorder at the metal/organic interface [21]. Here it is sufficient to treat it as a phenomenological parameter independent of the channel length. The resistivity of the contact ρ_c depends on the chemical nature and electronic structure of the junction. The other variables were introduced in Eqs. (18.1) and (18.2).

In Eq. (18.3) the channel resistance scales as L/W, whereas the contact resistance scales as $1/W$. It turns out that the effective resistance of the devices with smaller L will be affected more substantially by the contribution of R_c. This geometry dependence is a major barrier against downscaling and integration of OFETs. The contact resistance dominates for small L, so it is important to minimize ρ_c and λ when the charge mobility and the capacitance are both large. Equation (18.3) also shows that, when the channel contribution becomes negligible, the device resistance is governed by the charge injection interface. This regime, which is undesirable for circuits since marginal gain is obtained, allows one to explore the charge injection mechanism at the length scale λ, as described in Section 18.4. The resistivity ρ_c is associated with the chemical nature of the interface, because it is a rapidly varying property upon specific adsorption of chemical species. It can be exploited for sensing, as hinted in Section 18.6.

The contact resistance R_c is directly measured from (i) the transfer line method (TLM), (ii) gated four-probe measurements, (iii) local potentiometry using scanning Kelvin probe force microscopy (KPFM) [22, 23] or electrostatic force microscopy (EFM) [24].

TLM requires the measurement of the linear transfer characteristics from a set of devices with the same W and different L. Equation (18.3) shows that the intercept of R_{eff} versus L in the limit $L = 0$ yields twice R_c. In gated four-point probes, two additional electrodes measuring the potential are placed into the channel, with the transistor being operated in the linear regime, for guaranteeing uniform charge density in the conductive channel [25]. KPFM [22,23] and EFM [24] are based on a conductive tip, scanned across the operating device as in atomic force microscopy (AFM). Both techniques yield a map of the local electrostatic potential across the operating device. By measuring the voltage drop at the contacts on line scan profiles, the contact resistance is extracted and it is possible to estimate ρ_c. The contact resistances of the source and drain electrodes are usually not symmetrical (i.e., they deviate from the behavior of Schottky junctions [26]). The voltage drop is typically larger at the drain contact.

The contact resistance R_c depends also on the specific OFET architectures. The BG-BC configuration is affected by morphological disorder of the OS film close to the contact edges. On polycrystalline metal electrodes, with granular morphology, disorder arises from three-dimensional nucleation and growth of the OS film on the electrodes, as opposed to 2D nucleation and growth in the channel. The differences in growth on electrodes and channels yield discontinuity in the thin film across the interface. The 3D growth of OS nuclei on the electrodes can lead to recrystallization into large crystals protruding in the channel. This process depletes OS molecules from the interface lowering the effective W [27].

The topology of the contacts also influences R_c. The dominant contribution to charge injection in BG/BC is from the contact line between the source and drain contacts and the channel, with the electric field lines being denser there than on the rest of the electrode. This contact line is less accessible to the molecules during high vacuum sublimation, due to the shadowing of the metal contacts. In BG-TC the contact resistance R_c is lower than the BG-BC, as a consequence of the larger contact area arising from the in-depth diffusion of the metal atoms across the OS film [28] and the electric field between the contacts and the gate electrode which facilitates charge injection. However, since the conductive channel is far from the source and drain electrodes, the pathways for charge carriers to reach the charge injection interfaces depend on the OS film thickness and the coupling between stacked monolayers. This gives rise to the additional *access resistance* R_a. The TG-BC and TG-TC configurations require a pinhole-free robust dielectric/OS. They are mainly of interest in circuit design, and less relevant for sensing, so we will not deal with them further.

18.3 OFET FABRICATION

Most OFET studies were performed in BG/BC configuration using p^{++} Si wafer as common gate with SiO_2 dielectric thin film and metal electrodes fabricated by photolithography. These devices are reliable test beds for laboratory prototyping and materials assessment, although they are less relevant for technology. OFET manufacturing should be based on low-cost additive processing and flexibility of the substrate, and each OFET should be independently gated in circuit manufacturing. The need to match these requirements has stimulated the development of platforms for the fabrication of electrodes, dielectric, and OS on a variety of substrates.

18.3.1 Fabrication of Contacts

Electrodes can be fabricated by a variety of techniques: stencil printing of Au nanoparticles [29], inkjet printing [30–32], Ag electroless plating followed by microcontact patterning [33], lamination [34], microtransfer printing of Ag nanoparticles [35], metal transfer printing [36], and soft lithography [37,38].

Additive manufacturing using not only metal precursors, but also semiconductors and dielectrics, can be realized by inkjet printing and unconventional lithography. The feasibility of inkjet technology with subfemtoliter droplet volume and submicrometer resolution for electronic device applications was demonstrated [30]. Both p-channel and n-channel OFETs with source–drain contacts were prepared by subfemtoliter inkjet printing of nanoparticles, polymers, or metal precursors deposited directly on the substrate surface or onto the organic semiconductor layers, without the need for any photolithographic pre-patterning or surface pre-treatment.

Unconventional wet lithography, including micromolding in capillaries (MIMICs) [39] and lithographically controlled wetting (LCW) [40,41], exploits capillary forces in solutions confined between the substrate and a soft stamp. Both are additive methods for patterning soluble multifunctional materials [42]. They were used for fabricating OFET electrodes, as well as in combination OS nanostripes across the channel as shown in Figure 18.5 [43].

18.3.2 Fabrication of the Active Layer

OFETs applications depend on the precise control of the interfaces, the morphology of the film, and the structure of the active layers [44]. To date, the fabrication of the active layer in OFETs relies on thin film growth in high vacuum or deposition techniques. Thin-film technology, by either spin coating or physical deposition techniques, is widely used for producing multilayer architectures [42]. The control of thickness is limited to tens of nanometers, whereas the control of the lateral size and the position of the OS domains is poor. When more than one device is present on a test pattern, continuous films yield cross-talking. Another negative aspect is the loss of valuable OS material (>90%) during thin-film deposition.

As discussed in Section 18.5, the thickness of the OS films influences the local polarization around the charge

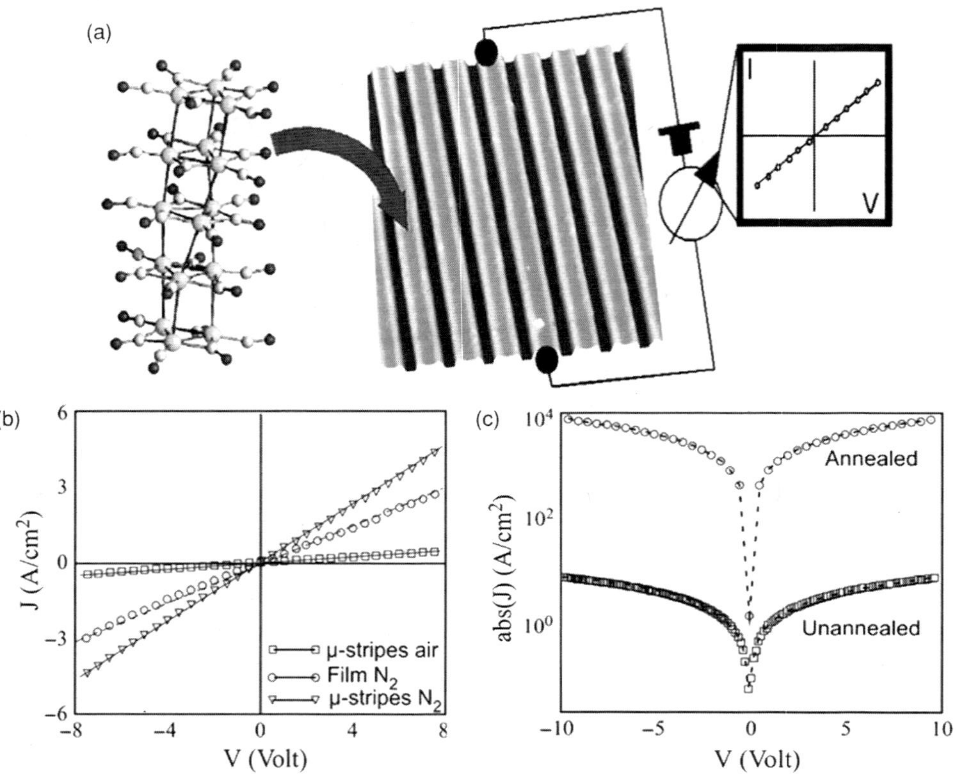

FIGURE 18.5. (**a**) AFM topography image of conductive wires of submicrometer width fabricated on glass by MIMIC using a platinum carbonic cluster $[NBu_4]_2[Pt_{15}(CO)_{30}]$ solution. (**b**) Current density–voltage characteristic of μ-stripes fabricated in air (□) and μ-stripes fabricated under nitrogen (Δ). (**c**) Comparison of current density–voltage characteristic of annealed and unannealed μ-stripes of $[NBu_4]_2[Pt_{15}(CO)_{30}]$ fabricated under nitrogen (current density is plotted in log scale). See reference 38.

carriers, and it can constitute an in-series capacitance coupled to the environment. High and ultra-high vacuum sublimation of the OS thin film allow a finer control of film thickness, down to sub-monolayer. The OFET performance depends on the continuity and the characteristic size of domains in each monolayer, the density of domain boundaries, and the stacking of monolayer terraces. Choosing an optimum thickness is crucial, because during the growth of the OS thin film the surface roughness increases, and competing phenomena such as secondary nucleation, growth transitions, dewetting, and recrystallization hamper the control of interfacial morphology. A quantitative understanding of morphological correlations with OFET parameters is still a research topic nowadays, and the control of the molecular organization at multiple length scales is a formidable barrier for the optimization of OFET response [45].

A strategy to overcome the limitations of thin-film technology is to adopt unconventional lithographic methods, where only the functional amount of a soluble material is deposited into spatially defined regions [46] (Figure 18.5). Characteristic timescales for self-organization depend on the volume and dimensionality. The capillary flow inside micrometric channels drives material distribution and crystal orientation [45, 47, 48]. Quasi-equilibrium conditions can be obtained in shorter timescales, compared to conventional thin-film wet methods, as the physical dimension of the system where the material self-organizes is reduced in size and dimensionality. This implies, among other effects, enhancement of surface nucleation with respect to bulk aggregation, or disappearance of diffusion limited processes. Furthermore, the presence of the stamp during the nanostructure formation can dramatically change the morphology and the structure of the thin deposit.

A representative example on how the nanostructuring of OFETs via wet lithography enhances charge transport properties has been reported using an alternate bis-fluorene-terthienyl block-oligomer [45]. This material normally does not form thin films suitable for OFET fabrication upon spin casting or vacuum sublimation. When deposited by lithographically controlled wetting [46], bis-fluorene-terthienyl block-oligomer forms oriented crystalline nanostripes exhibiting a charge mobility 500 times higher than

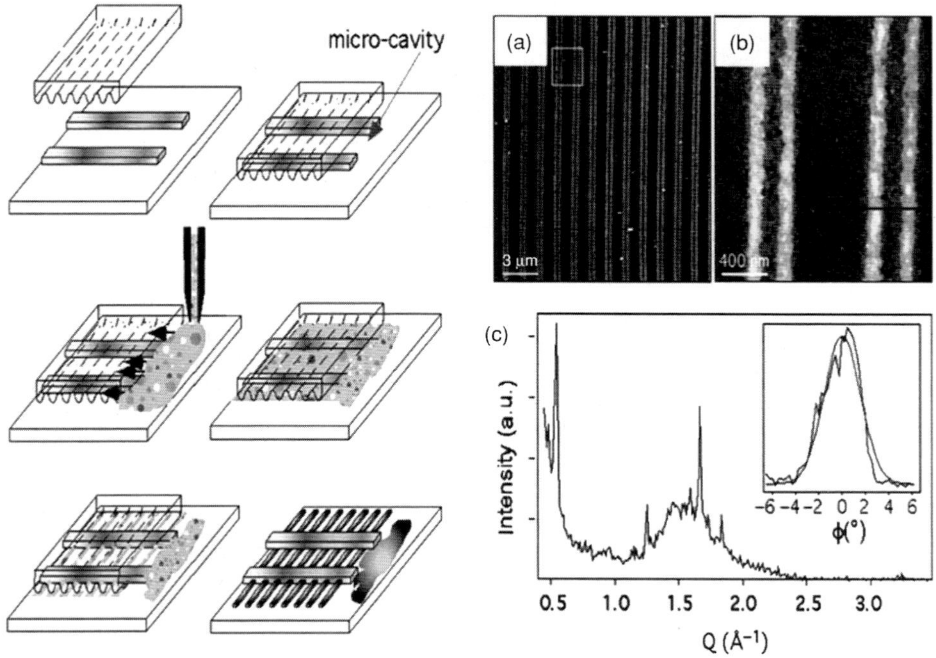

FIGURE 18.6. (**Left panel**) Schematic representation of the deposition process of F–T3–F molecules, showing the steps related to micromolding in capillaries (*top* and *middle*) and the steps related to lithographically controlled wetting (bottom). (**Right panel**) (**a**) Atomic force microscopy topography of printed nanostripes 20 × 20 μm^2 area of printed film; (**b**) Detail of (a). (**c**) grazing incidence x-ray diffraction of F-T3-F patterned by LCW. Q scan along the stripes direction; inset displays the in-plane rocking scan at $Q = 1.51$ Å^{-1}. $\varphi = 0$ corresponds to the direction of the stripes. See reference 45.

spin-coated thin film (Figure 18.6). A similar approach was used to fabricate ambipolar multi-stripe organic field effect transistors [49] and nanowires [50] in an additive dual-step procedure as shown in Figure 18.7.

18.3.3 Chemical Functionalization of the Interfaces

Common gate devices use dielectric films such as thermally grown SiO_2 and sputtered metal oxides [51, 52] on silicon wafers. Although the surface roughness of the dielectric thin films is <0.5 nm at saturation length scale, there are intrinsic surface defects related to the presence of hydroxyl groups. As a consequence, OS thin films give rise to dislocations within crystalline domains. An improvement of inorganic dielectric films for OFETs applications is their chemical functionalization by self-assembled monolayers (SAMs). This is an efficient way to tune the chemical properties of the interfaces, because SAMs form homogeneous surfaces at large scale and a variety of molecules can be designed in order to tune the surface energy and morphology which are two properties determining the diffusional barriers and the nucleation density of the growing OS thin film. The most common chemical functionalization for the channel is silanization, which was first introduced by A. Salleo et al. [53]. Precursors such as *n*-octadecyltrichlorosilane (OTS) condense in the presence of water traces with –OH groups of the SiO_2 surface, to form a SAM. Silanization or phosponization can be used with other metal oxides. The principal effects of SAM functionalization of the gate dielectric layer are: the changes in the molecular orientation of the OS molecules, the neutralization of polar defects and interfacial traps, and the modification of interface dipole and surface energy. The SAM functionalization does not change substantially the capacitance of the gate dielectric as it adds an in-series large capacitance. Its main benefits arise because the SAM decreases the energy disorder of localized states at the organic/dielectric interface both chemically, by passivating the hydroxyl groups at the channel surface, and morphologically, by smoothening the intrinsic roughness of the gate dielectric oxide that "imprints" structural defects in the organic semiconductor layer from the bottom [54]. SAMs also increase the desorption of water from the surface arising from moisture that is likely the source for charge trapping [55]. SAM can also help to decreases the nucleation density of the organic semiconductor thin film leading to larger domains, by enhancing the molecular diffusivity; it improves the inplane molecular packing, by promoting homeotropic orientation of rodlike

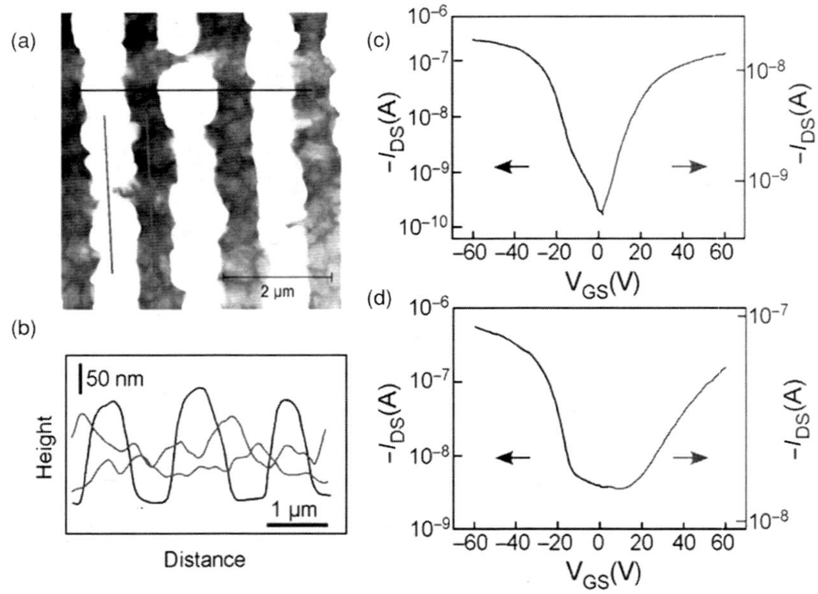

FIGURE 18.7. (**a**) Atomic force microscopy topography of pentacene thin film deposited on printed stripes of soluble perylene. In order to show the continuous film of pentacene in between the stripes, the palette of z range is limited from 0 to 25 nm. (**b**) Topographic line profiles of (a). (**c**) Transfer characteristics of ambipolar OFET based on multi-microwires measured in air. (**c**) Linear and (**d**) saturation transfer characteristics of ambipolar OFET. Black lines represent the p-current, red lines represent the n-current. See reference 49.

molecules with respect to the planar one. As a result of these effects in the devices, the charge carrier mobility increases, whereas V_{th} and SS decrease [56]. SAM functionalization of the channel can also dramatically change the OFETs characteristics, influencing the channel conductance and/or charge carrier density. As an example in 2005, Chua et al. [57] reported that the trapping of electrons at the semiconductor–dielectric interface by hydroxyl groups prevented most conjugated polymers from yielding n-channel FET conduction.

18.4 CHARGE INJECTION IN OFETs: THE ORGANIC–METAL INTERFACE

At the metal–organic semiconductor interface charges are injected from (to) the electrode to (from) the OS. This interface is associated with the contact resistance R_c introduced in Section 18.2.2. At the microscopic level, R_c arises from electronic and morphological effects: (1) the energy mismatch between the Fermi level of the metal and the HOMO/LUMO levels in the OS; (2) the structure and morphology of the OS at the contact interfaces [58]; (3) a reduced electronic coupling across the interface. In real OFET devices, these effects are largely caused by the OS–metal surface interactions.

Functionalization of metal electrodes with SAMs allows to improve electrical parameters, such as mobility, subthreshold slope, and on/off ratio [59–62]. SAMs not only act as a spacer between the metal and the OS first monolayer, but also tune the work function of the metal electrode [63–66], decrease the hole injection barrier by spacing the OS molecules apart from the metal, and enhance the morphological order of the OS film [59, 60, 62, 67]. These aspects can actively enhance the transmissivity of the interface. Phenomenologically, this implies that ρ_c and/or λ in Eq. (18.3) can be decreased by a suitable choice of SAM. Below, the electronic and morphological effects, as well as the implications in the choice of SAM, are discussed.

18.4.1 Alignment of Energy Levels at the Charge Injection Interface

The MO interfaces were described following the simple rule of the vacuum level alignment (termed Schottky–Mott limit), in which the hole/electron-injection barriers (ϕ_h/ϕ_e) are the differences between the ionization energy/electron affinity (IE/EA) of the OS and the metal work function ϕ_M, respectively. The breakdown of the vacuum level alignment occurs by the formation of an interfacial dipole moment Δ [58]. Upon physical adsorption of the organic molecules onto the metal electrode, it usually occurs a charge transfer due to the matching of the ϕ_M and the HOMO/LUMO and/or by the exchange repulsion (leading to the distortion of the "cushion" formed by the free metal electrons at the interface) [68]. Chemical adsorption leads to a covalent link of OS molecules onto the substrate [69] with the emergence of new hybrid resonances.

3,4,9,10-Perylenetetracarboxylic bisbenzimidazole (PTCBI) on Ag [70], hexadecafluoro copper phthalocyanine on Mg/Al, and perylene-3,4,9,10-tetracarboxylic-3,4,9,10-dianhydride (PTCDI) on Mg and In are representative for the charge transfer [71, 72]. Regarding the exchange repulsion, it was reported for pentacene, *p*-sexiphenyl, *N,N′*-diphenyl-*N*-*N′*-(3-methylphenyl)-1,1′-biphenyl-4,4′-diamine (TPD) on Au [68, 73–77]. The physical adsorption often induces a downward shift of the vacuum level yielding a counterintuitive increase of the hole injection barrier, except for such metals, like Sm, which have low Δ [78, 79].

Regarding the chemical reaction, many organic compounds are able to self-assemble onto the metal substrate because they have a head group that covalently binds the substrate, an aliphatic and/or π-conjugated backbone (inducing a variable extent of inter- and intramolecular interactions), and a tail group exposed to the outside of the surface. For instance, Au/Ag and thiol molecules are widely studied, and nowadays many self-assembly monolayers (SAMs) are available. The extent of the surface dipole depends on several factors such as (i) the electronegativity of the tail group [80], (ii) the overall orientation of the packed molecules, (iii) the nature of the bond between the head group and the metal [63], (iv) the length and polarizability of the backbone [63,64], and (v) the inter- and intramolecular interactions. If the surface dipole is oriented toward the outside (e.g., alkanethiol), the ϕ_M shows a downward shift, whereas for an opposite direction (e.g., perfluorinated thiols) an upward shift occurs. SAMs can be used for a fine modulation of the electrode work function, allowing a lowering/increase of the hole/electron energy barriers [66].

The direction of the dipole moment can be foreseen by simple physical–chemical arguments, whereas the quantification of its magnitude requires the use of ultraviolet photoemission spectroscopy (UPS) and inverse photoemission spectroscopy (IPES). Photoelectron spectroscopies probe the valence (HOMO) and conduction (LUMO) bands, and they measure directly their shift from the Fermi level of the bare metal [58, 81]. In ambient conditions the energy structure of the MO interfaces in the devices can be rather far from the one probed by ultra-high-vacuum UPS and IPES measurements. An adventitious contaminant layer [82], a specifically designed material [76], or a chemically defined layer like a SAM causes the weakening of the dipole moment at the metal surface, thus leading to a decrease of the charge-injection barrier.

18.4.2 Structural and Morphological Disorder at the Charge Injection Interface

As mentioned in Section 18.2.2, KPFM and EFM allow a direct extraction of the contact and channel resistances. By varying temperature, ionization potential, and electrode work function, the formation of "good" and "bad" contacts for BG-BC architecture of P3HT-based OTFTs has been demonstrated. In particular, the temperature dependence shows activation energies around 100 meV, which are too small to support the thermionic emission model [83] for charge injection. This evidence has been explained by assuming the presence of two depletion regions close to the source and drain electrodes, where the OS layer has a lower quality compared to the bulk [20].

The existence of an interfacial region of morphological disorder has been confirmed using pentacene FETs with electrodes functionalized with the whole homologue series of alkanethiols. The charge mobility first increases for short alkyl chains, reaches a maximum at eight methylene units, and then exponentially decreases at long chain lengths. The interpretation relies on the concept of the injection lengthscale, which arises from the interplay of the Schottky barrier modulation and the OS morphological disorder close to the electrodes. The latter is generated by planar adsorption and three-dimensional nucleation on the bare Au electrodes, as opposed to 2D nucleation and growth in the channel. This differential growth yields regions of discontinuity, as a result of the 3D growth of domains nucleated on the electrodes and protruding in the channel [27]. The minimum injection lengthscale at eight methylene units is the best compromise between (i) the formation of a compact SAM where homeotropic adsorption and 2D nucleation occur, and (ii) maximum transmissivity of charge carriers by tunneling, as discussed in the following.

18.4.3 OFET as Charge Tunneling Device

The entanglement between work-function modulation and morphological disorder at the injection interface results in the contact resistance. SAM functionalization intervenes on both phenomena. The dependence of the contact resistance as a function of the work-function modulation, has recently been demonstrated, whereas the mobility and threshold voltage do not show any clear behavior [67]. It has been also observed that a systematic increase of the SAM thickness (using the homolog series of alkanethiols) leads to the exponential decay of the charge mobility accompanied by odd–even fluctuations (Figure 18.8) [21]. This evidence indicates that with the SAM between Au and pentacene, charge carriers cross the SAM barrier by non-resonant through-bond tunneling. Odd–even effects demonstrate the sensitivity of OFET to the relative orientation of the last sigma bond in the alkyl chain relative to the pentacene molecules standing on the SAM.

The SAM-functionalized OFET approach has recently been extended toward the investigation of different aromatic SAMs showing $\beta \approx 0.2$ Å^{-1} [84]. This value, substantially smaller than the one measured for the alkanethiol homologue family, is consistent with the better matching of the

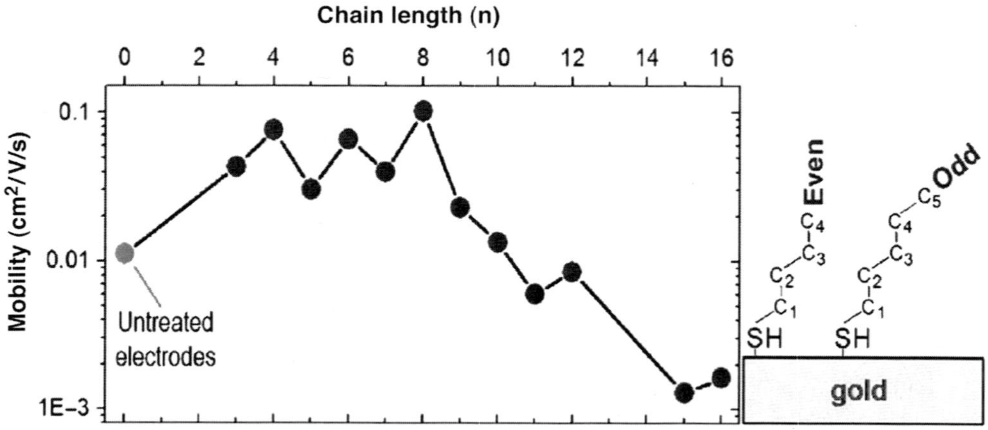

FIGURE 18.8. Charge carrier mobility μ behavior of pentacene OFET with alkanethiol monolayers of chain length n. The charge carrier mobility exhibits large fluctuations correlated with odd–even n. For $n < 8$, μ increases by one order of magnitude owing to the decrease of the hole injection barrier and the improved molecular order at the organic–metallic interface. For $n \geq 8$, decays exponentially with an inverse decay length $\beta = 0.6$ Å^{-1}. See reference 21.

resonances in the SAM with the Fermi level and HOMO level of pentacene.

These results point to the fact that SAM-functionalized OFETs behave as molecular charge tunneling devices, rather than transistors, when the SAM becomes resistive. Two important features of these devices make them attractive gauges for tunneling measurements through SAMs, as compared to large-area molecular junctions, nanopores, break junctions, and sessile Hg or eutectic drop electrodes in vertical junctions [85]: (i) the fabrication of OFET molecular junctions is easy and highly reproducible and (ii) the nature of majority carriers is controlled via the gate voltage.

18.5 LOW-DIMENSIONAL CHARGE TRANSPORT IN OFETs: DIELECTRIC/ORGANIC AND ORGANIC/ORGANIC INTERFACES

OFETs are low-dimensional devices, where charge accumulation and transport occurs only in the very first few molecular layers of the organic semiconductor in contact with the gate dielectric. The interplay of electronic and morphological effects in the charge transport layer can be used to rationalize the influence of adventitious dopants or analytes from the environment on the transistor response. The understanding of these basic transport phenomena should be used to design reproducible transducers and sensors based on ultra-thin film OFETs.

18.5.1 Low-Dimensional Charge Transport

The earliest claim of low dimensionality in OFET was by Dodabalapur et al. [86], who observed that charge mobility in T6 FETs with three different nominal thicknesses (5, 50, 100 nm) was invariant. First experimental demonstration of low-dimensional charge transport in organic semiconductors was by our group [87] on T6 OFETs with thicknesses ranging from 0.1 to 10 ML: charge mobility is contributed by the first two monolayers, and it is correlated to the onset of a layer-by-layer to island growth transition. This transition seems to be a universal feature of organic thin-film growth by sublimation in high vacuum; and it resembles a Stranski–Krastanov or, more precisely, a roughening transition [88]. Interestingly, the transition does not change the crystal structure of the OS monolayers [89], but instead corresponds to an acceleration of the nucleation and stacking of molecular terraces which is governed by the Schwoebel barrier [90].

The finite thickness of the charge transport layer was confirmed by other groups for pentacene [91, 92], T6 [93], and dihexyl-T4 [94, 95], with a variety of critical thicknesses ranging from 1 to 6 MLs. The dispersion of critical thickness for pentacene OFETs was explained by Shehu et al. [96] by means of an *in-situ* real-time experiment as the result of different deposition rates and, hence, different nucleation and growth modes (Figure 18.9). The rate dependence of the onset of current versus thickness is consistent with a percolative transport in 3D at low rate and 2D at higher rate. This indicates that the individual monolayers are better connected in-plane when growth occurs in a quasi layer-by-layer fashion. In the case of sexithienyl, only one or two active monolayers were reported [87, 91], whereas in the case of pentacene, up to three were reported [94].

Interestingly, the charge carrier distribution and the contributions of each monolayer to the OFET current are not monotonically decreasing with the thickness. This cannot be explained by electrostatic arguments solving the Poisson–Boltzmann equation on continuous semi-infinite OS films [97] or layered OS films [51]. Both models predict that only the very first monolayer would be substantially populated, with a Debye screening length of the OS close to 1 nm.

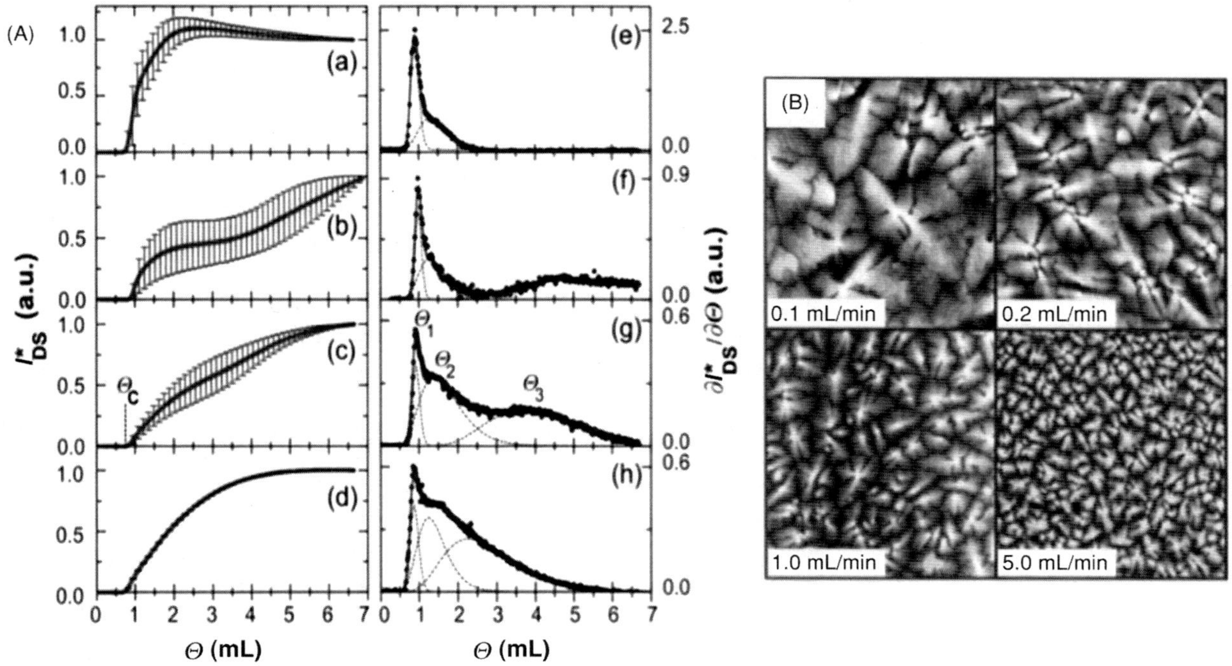

FIGURE 18.9. (**A**) (*Left*) Evolution of I_{DS} (normalized to $\Theta = 7$ ML) versus Θ measured in real time and in situ conditions for φ: (**a**) 0.1 ML/min, (**b**) 0.2 ML/min, (**c**) 1 ML/min, (**d**) 5 ML/min. The percolation threshold Θ_c is schematically indicated (dotted vertical segment on curve (c). Error bars are standard deviations (*right*). Derivative $\delta I_{DS}/\delta\Theta$ (**e**)–(**h**), circles); continuous line is best-fit by linear combination of three log-normal functions (dotted lines). (**B**) 5×5 μm^2 AFM images in the FET channel at $\Theta = 7$ ML [96].

Shehu et al. [96] demonstrated that the effective Debye length is substantially larger, and it scales with the correlation length of the OS thin film. The analysis in reference 96 also demonstrates that the first monolayer contributes less to the overall current with respect to the second and third ones, except for thin films grown at low rates.

To understand this, one should consider that the contribution from each monolayer is the product of the charge carrier density (which depends on the electrostatic screening in the material) and the charge mobility (which depends on the 2D percolation pathways across the monolayer). Both properties vary for each layer close to the gate dielectric and depend on the local thickness because the upper layers, although not connected, can polarize the charge carriers in the layers below. The monolayer that contributes most of the current is the one that maximizes this product. The outcome is critically determined by the growth conditions and the resulting morphology. The finding that monolayers more distant from the dielectric have the largest contribution to the current, can then be rationalized by an increased trap density in the first monolayer due to smaller domain size and more in-plane defects. Instead the second and higher monolayers offer more efficient charge transport paths which compensate the lower charge carrier density. In addition charge transfer between layers depends on stacking relations between adjacent layers.

Another important aspect of ultra-thin-film OFETs is that the threshold voltage, and hence the charge trap distribution, depends on the film thickness [98]. This might be understood to some extent in ambient operations, due to the capacitive coupling of carriers with water and oxygen molecules percolating through the active layer at a different rate depending on the thickness and the surface area. It is also possible that this thickness-dependent threshold voltage arises from the stacking of the OS monolayers to form uncomplete layers for charge transport. In this case the upper terraces, although scarcely populated or highly resistive, *de facto* play the role of an outer polarization of the domains where charge carriers are accumulated and transported. This also is consistent with the observation that the dielectric permittivity of the OS scales with film thickness from higher values for the first monolayers down to smaller values (2–3) at larger thicknesses. The experimental observation once again indicates that details of the film growth are important and determine the OFET parameters and their reproducibility and relevance.

The accessibility of molecules from the outer environment to the channel region of the OS ultra-thin film opens also the possibility to use such devices as transducers and sensors for molecules in the environment. The observation that there is a dominant monolayer determining the OFET response suggests that the diffusion of analytes must occur down to the dominant monolayer in order to yield the maximum response. As a consequence, ultra-thin-film OFETs are predicted to be more sensitive devices than thicker ones, and this should be used as design rule. Specific recognition upon functionalization of the OS/dielectric interface with designed SAMs was indeed suggested by Torsi [99] and demonstrated with recognition of biotin/streptavidin.

18.5.2 Polarization Affects the Charge Distribution and Mobility

The distribution and motion of charge carriers in the OS is determined by a superposition of electrostatic forces. On the one hand, external electric fields originating from the three electrodes control the charge carrier drift and accumulation. The gate field restricts the accumulation of charge carriers to the first few monolayers in contact with the gate dielectric. On the other hand, local electrostatic interactions of the excess charge with the underlying atomic lattice form the basis for nuclear polarization. Polarization leads to charge carrier localization (creation of polarons) which, in the proximity of interfaces, can give rise to various trapping effects. Especially in sensors, where the channel is in contact with a strongly polarizable medium like water, large effects on the current characteristics are expected. In the following we describe the basic concepts dealing with polarization in thin films for OFET-based sensors.

Organic semiconductors are low-permittivity materials, and the nuclear rearrangement upon introduction of an excess charge is often negligible in comparison to electronic polarization. In small-molecule semiconductors the major contribution to nuclear polarization comes from the rearrangement of molecular quadrupole in the neighborhood of the charge carrier [100]. Dipole polarization is typically not present due to the molecular inversion symmetry. The low polarization results in fast, coherent band-like transport in highly ordered crystalline phases at low temperatures [101].

Experimental and theoretical studies show that polarization effects gain importance in the vicinity of interfaces or grain boundaries in OFETs [102]. By systematically varying the gate dielectric and its permittivity, it has been demonstrated that polarization extends over the interface and deteriorates electrical characteristics of the semiconductor due to the formation of small interfacial polarons [100]. With increasing permittivity of the gate dielectric, the charge carrier is more strongly coupled to nuclear displacements in the dielectric. As a result, a decrease in mobility and a shift of its temperature dependence from a band-like regime to thermally activated transport were observed [103]. Molecular micro-electrostatic calculations elucidated the importance and complexity of polarization at pentacene grain boundaries [100]. The orientation of the semiconductor crystal phases and the presence of other polarizable phases or impurities in contrast to void determine whether trap states are created at grain boundaries, or whether charge carriers are repelled from them due to reduced polarization. In both cases, it can be anticipated that the mobility of the semiconductor film is reduced and transport has to occur in a thermally activated regime as trap states or barriers at grain boundaries have to be crossed.

An extreme case for polarization is represented by organic semiconductors operating under water. Rearrangement of the strong molecular dipoles at the liquid interface leads to stabilization of charge carriers in the semiconductor. Simulations for pentacene in direct contact with water predict a stabilization of 0.6 eV for positive charges in the interfacial polaronic trap state [102]. However, thermal fluctuations of the water molecules provoke two-dimensional diffusive hopping of the polaron parallel to the interface with mobilities of up to 0.6 $cm^2\ s^{-1}\ V^{-1}$. As a consequence, immersion of the OS ultra-thin film in water is expected to result in fast thermally activated hopping transport, although strong polarization effects are present. Polaron trapping is thus not expected to be a limiting factor for transistor-based sensors that operate under water.

Ultra-thin OFETs are largely affected by polarization that occurs at the different interfaces of the device. In the context of sensors, these effects might be exploited in multiparametric device analysis. Polarization broadens the energy landscape that a carrier has to cross in order to arrive at the drain electrode. As a consequence the transport is shifted into the thermally activated hopping regime; mobility, threshold voltage, and subthreshold slope are also influenced. The interaction is complex, and models have to include molecular micro-electrostatic details and dynamic aspects in order to predict changes in device characteristics. A simpler picture emerges for interfacial charging effects that appear when the sensor is coupled to ambient. They result mainly in a shift of threshold voltage as described in Section 18.6.

18.6 COUPLING THE CHANNEL TO AMBIENT: SENSING PRINCIPLES AND APPLICATIONS

The sensitivity of OFETs to the environment might be detrimental for stable operations in electronics, but is useful to achieve ultra-sensitive transduction of ionic and molecular analytes from the environment. Sensing is an emerging OFET application [104, 105].

18.6.1 Sensing Principles

The main use of transistors in electronics is the amplification of electronic signals acting on the gate electrode. Depending on the transistor specifications, small potential changes are already sufficient to switch between on and off state of the transistor and modulate strongly its current response. In an OFET sensor these potential changes have to arise from

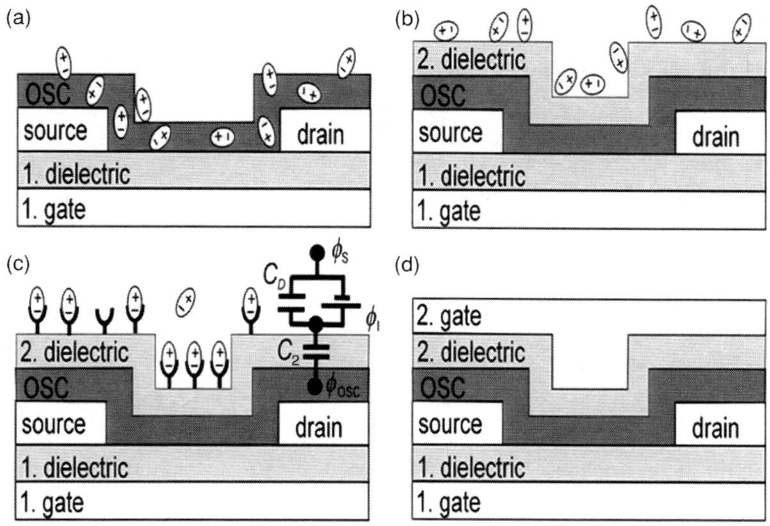

FIGURE 18.10. OTFT as sensor for dipolar molecules in solution. (**a**) Diffusion into the organic semiconductor (OSC). (**b**) Capping of the OSC with a second dielectric layer. (**c**) Specific binding of analyte molecules to the second dielectric and electric circuit model for the coupling between OSC and solvent. (**d**) Equivalent dual-gate transistor.

the specific recognition of an analyte at one of the transistors active interfaces. Therefore OFET sensors transduce a small chemical stimulus from the ambient into changes of the electronic response of the transistor.

Examples of sensing include the detection of molecules either from the gas or from the liquid phase. In both cases the analyte molecule has to diffuse into the semiconducting thin film (Figure 18.10a) or adsorb to one of its interfaces (Figure 18.10b,c) in order to interact with charge carriers. The final response of the device is then a consequence of a change in mobility or of accumulation/depletion of carriers, or both. A detailed physical understanding of the underlying sensing mechanism allows a rational design of sensors with high sensitivity and sensibility combined with reversibility for multiple uses. In the following we summarize the basic concepts. Details can be found in the literature [105–109].

Early it was recognized that ultra-thin transistors are sensitive to moisture and other reactive gas molecules. The high sensitivity was attributed to the close proximity of interfaces to the charge carrier channel. Due to the gate-induced field effect, the channel forms in the first few nanometers adjacent to the gate dielectric [96]. In ultra-thin films of semiconductors or FETs based on self-assembled monolayers, the interface to ambient is also only a few nanometers apart. In addition, small-molecule-based thin films have internal interfaces due to grain boundaries. Molecules that adsorb on these interfaces or start to penetrate into the film from these interfaces interact directly with charge carriers in the channel and change the device characteristics. The relevant interactions are complex and include polarization (see above) and other electrostatic interactions that induce changes in the density of states (DOS) of the semiconductor. A broadening of the DOS in combination with a localization of the transport states is expected [102]. In addition, depending on HOMO and LUMO levels of the analyte molecule, charge transfer to or from the semiconductor might appear. In this case the analyte acts as a strong dopant or trap state which induces a shift of the threshold voltage [110]. Often in experiments the combined effects on mobility as well as threshold voltage are observed (Figure 18.11).

OFET gas sensors have been explored in electronic noses, because of their scalable size and the variety of OS molecules that can be integrated in arrays for multiparametric fingerprinting approaches [111–113]. In the case of gas sensors the interface is not in contact with an electrical ground, and the situation corresponds to a floating second gate from a device point of view. Interfacial charges can build up due to charge transfer from the semiconductor. A well-characterized example is an OFET sensor for nitride oxide (NO) [114]. The selective adsorption of NO to the semiconductor surface was achieved by the deposition of a porphyrin complex on top of the device. The adsorbed NO molecule becomes a strong electron acceptor; and doping of the *p*-type semiconductor occurred already at ppb concentrations, leading to a positive shift in threshold voltage (Figure 18.12).

18.6.2 Dual Gate OFETs

Although the direct interaction between analyte and semiconductor offers large sensitivity, its drawback are the lack of selectivity and device stability, especially when operated in water as is necessary for biosensing applications. A second

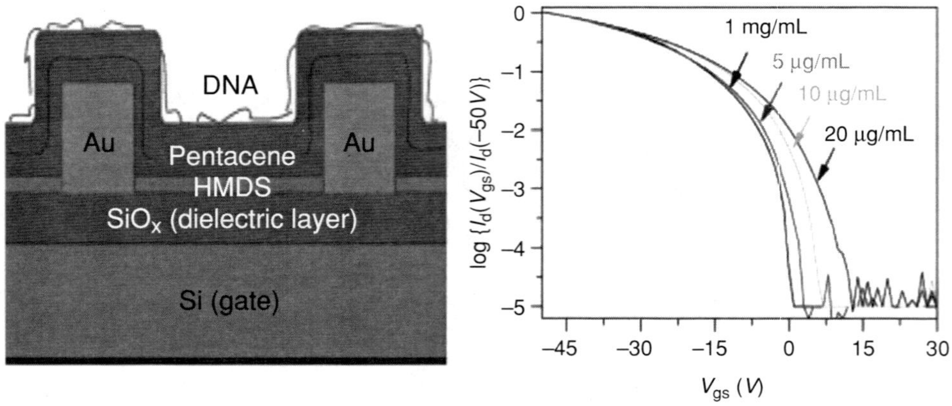

FIGURE 18.11. (**Left**) Schematic of the OFET used in reference 110. (**Right**) Transfer characteristics for the transistors at different DNA concentrations (increasing to the right) [110].

dielectric is introduced as a capping layer which prevents direct interactions between analyte, solvent, and semiconductor (Figure 18.10b) [98]. Selectivity can be obtained by adding receptor molecules to the capping layer that bind only specific analyte molecules to the interface (Figure 18.10c) [115]. As the analyte molecules are now separated from the conducting channel by a dielectric layer, only long-range electrostatic interactions and charge transfer may result in conductivity changes. Improved stability and selectivity come at the price of reduced sensitivity.

In general the sensing mechanism of devices with a second dielectric is equivalent to the operation of dual-gate transistors (Figure 18.10d) [116]. If a potential V_2 is applied to the additional second gate, changes in the charge carrier concentration occur: $\Delta\rho = -C_2V_2$, where C_2 is the capacitance of the second dielectric. As a consequence the threshold voltage V_{th}, as measured by performing a transfer scan with the conventional gate, shifts by $\Delta V_{th} = -(C_2V_2)/C_1$. The ratio of the capacitances of the two dielectric layers, termed capacitive coupling, is the critical parameter that determines the sensitivity to potential changes at the second gate. High sensitivity is possible by using ultra-thin layers with a high permittivity as second dielectric.

In sensors the interfacial potential ϕ_I corresponds to the potential V_2 in dual-gate transistors [117]. Depending on whether a sensor is operating in the gas or liquid phase, different origins of interfacial potentials have to be discussed. In liquids, ϕ_I builds up due to the specific adsorption of charged or polar molecules from the liquid to the surface [106]. The charge is counter-balanced either by intramolecular charges in the case of dipolar molecules (Figure 18.10c) or by counterions in the solution. In both cases a double-layer capacitance C_D forms which couples the interfacial charge to the potential of the grounded solution ϕ_S. An equivalent circuit is depicted in Figure 18.10c. The included potential source is driven by the chemical free energy, which is released upon adsorption of charged/polar molecules. Thus the dependence of ϕ_I on molecular concentrations follows in an ideal case the Nernst behavior: $\phi_I = \phi_{I,0} + RT/e \cdot \ln(c_{bulk}/c_{surface})$ [114]. The resulting logarithmic dependence of threshold voltage on concentration bulk was experimentally observed for a variety of analytes [107]. Especially for pH-sensing, where protons are adsorbed by the surface, the equation was proved

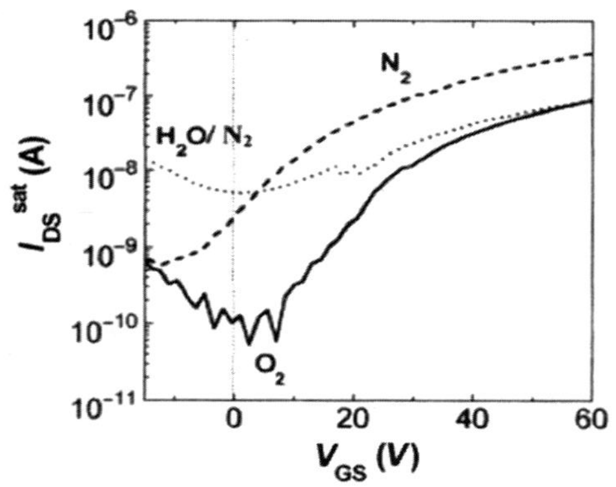

	N_2	O_2	H_2O/N_2
σ_{off} (s·cm^{-1})	3×10^{-8}	2×10^{-9}	1.5×10^{-7}
σ_{on} (s·cm^{-1})	2×10^{-4}	6.4×10^{-5}	$<3\times10^{-5}$
μ_{FET} (cm^2·V^{-1}·s)	1×10^{-4}	8×10^{-5}	4×10^{-5}
V_{th} (V)	-4	18	<-30
σ_{on}/σ_{off}	7×10^{3}	3×10^{4}	$<2\times10^{2}$

FIGURE 18.12. (**Top**) Transistor *I*–*V* transfer characteristics (i.e., I_{DS}^{sat} versus V_{GS}) measured in the saturated regime (i.e., at V_{DS} = 100 V) in an inert atmosphere of 100% N_2 (dashed line), in 100% O_2 (solid line) and wet nitrogen, that is, $H_2O = 2.1 \times 10^{-2}$ molar fraction in N_2 (dotted line) at room temperature. The gate bias ranges between −15 and 60 V. The vertical dotted line indicates the current values at zero gate bias. (**Bottom**) Values for the NTCDA bulk conductivity (σ_{off}) and for device parameters [111].

to be valid over orders of magnitude changes in concentration [117]. Thus, the dual-gate OFET can be assimilated to an ISFET (Figures 18.13 and 18.14) [117, 118].

The crucial role of the capacitance of the second dielectric, C_2, for sensitivity raises the question of how to maximize it without losing device stability. Recently, it has been shown that instabilities due to moisture are mainly a result of charge trapping at the first gate dielectric/OS interface [119]. The deterioration process can be circumvented when instead of the bottom gate the potential of the liquid itself is used to gate the semiconductor as is done in electrochemical transistors [120]. Redox reactions of charge carriers with water or oxygen can be excluded by the correct choice of the *p*-type semiconductor, and devices show stable operation even in the complete absence of a capping layer.

In an ultra-thin-film OFET exposed to an aqueous environment, C_2 equals the capacitance C_D of the diffuse double layer at the OS–water interface (just a few angstroms thick),

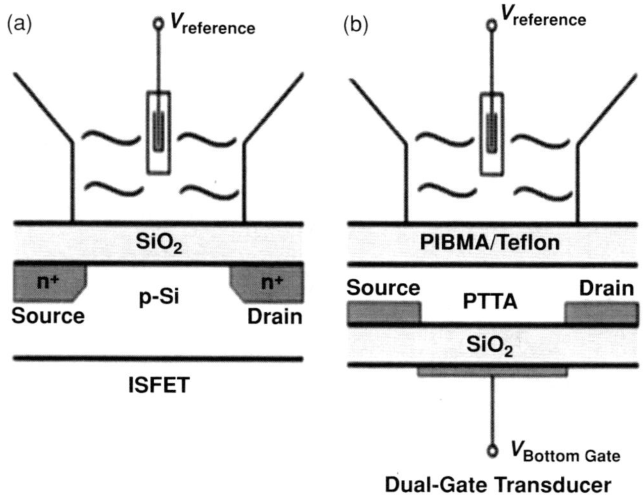

FIGURE 18.13. Schematic representation of an ISFET (**a**) and of a dual-gate transducer (**b**).

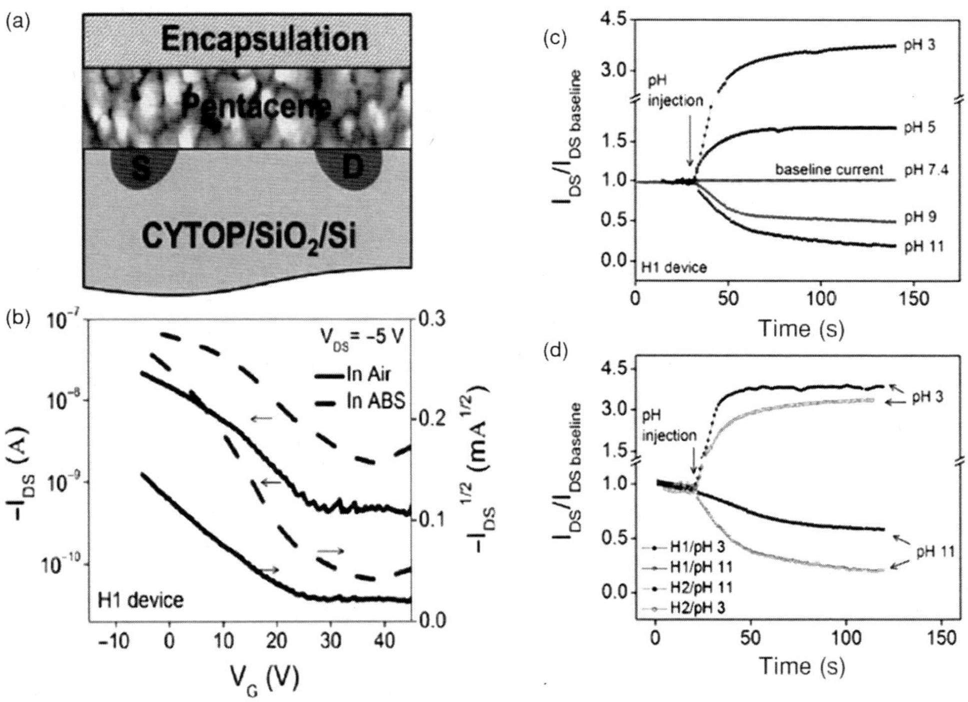

FIGURE 18.14. (**Left panel**) (**a**) Schematic representation of the OTFT as a biochemical sensor investigated in reference 115. Fabrication procedure for a bottom contact OTFT with 30 nm pentacene on 15 nm CYTOP/200 nm SiO_2 and source-drain electrodes with a channel W/L of 10. The entire device was passivated with PE-CVD-based ppPFDMCH layer and/or spin-coated CYTOP layer with variable thickness. (**b**) [H1-sensor] transfer characteristics (I_{DS} versus V_{GS}) at V_{DS} of -5 V in ambient (solid curves) and in PBS (dashed curves). **Right panel**: (**c**) pH detection in PBS of a representative OTFT with 30 nm pentacene on 15 nm CYTOP/200 nm SiO_2 and W/L of 10 measured at constant bias ($V_{GS} = -2$ V and $V_{DS} = -5$ V) with a flow rate of 300 μL/min. (**d**) [H1-device] current response ($I_{DS}/I_{\mathrm{DS\,baseline}}$) with time upon exposure to the solutions of four different pH (3, 5, 9, and 11) and (**b**) current response ($I_{DS}/I_{\mathrm{DS\,baseline}}$) with time upon exposure to solution of pH 3 and pH 11 to H1 (black curves) and H_2 devices [118].

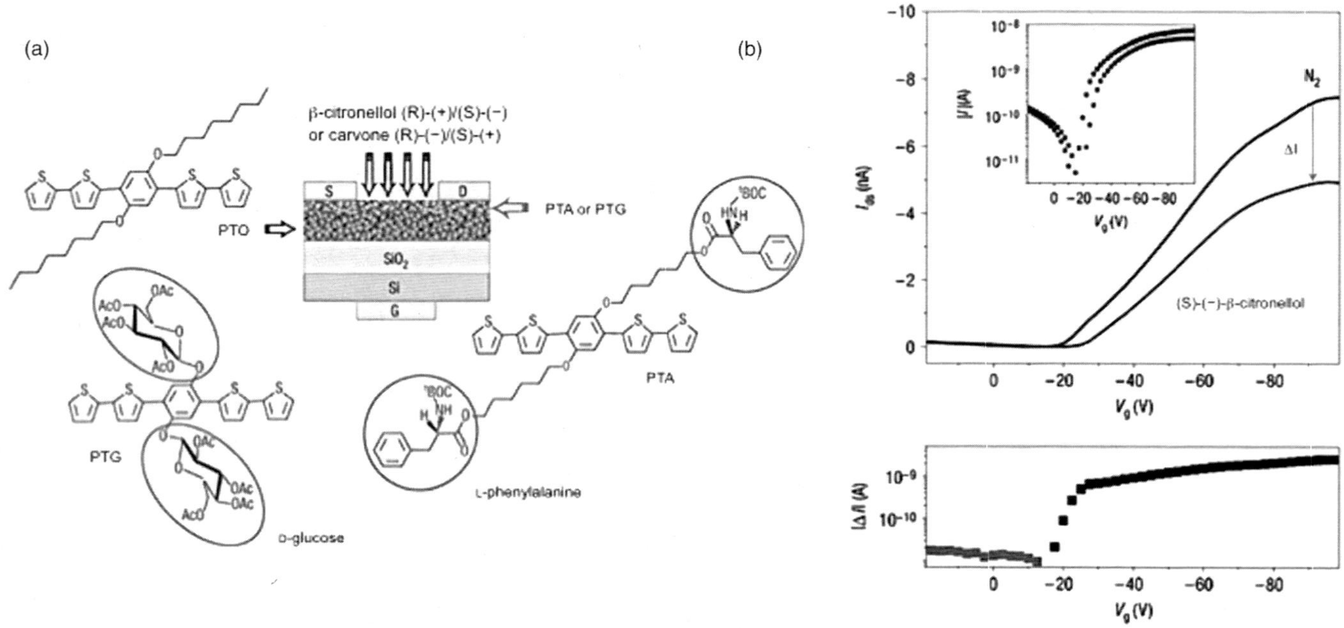

FIGURE 18.15. (**a**) Bilayer OTFT chiral sensor structure. (**b**) PTO–PTA-bilayer OTFT sensor response to (S)-(−)-β-citronellol vapours. The I_{DS}–V_{GS} transfer characteristics were measured by biasing the device in a common-source configuration by sweeping V_{GS} (from positive to negative potentials) at a fixed V_{DS}, namely −50 V. The current, I_{DS}, flowing in the channel region was measured in a nitrogen flux and in a flux of 30.6 ppm of (S)-(−)-β-citronellol in N_2. Each of the two steps lasted for 45 s, and in both cases the total flux was 200 mL min^{-1}. Before starting each run, the unbiased device was conditioned by exposure to the analyte atmosphere for 45 s [124].

which is constituted by mobile charge carriers in the semiconductor and solute counterions [120]. For water, this has been measured in OFETs [120, 121] to be up to 12 μF/cm^2 [122] compared to 10–20 nF/cm^2 for a technological dielectric such as thermal silicon oxide. The large capacitive coupling (500–1000 times) enhances the sensitivity of the OFET to external perturbations of the interfacial potential, and it opens interesting scenarios to the OFET as transducer of cellular signals and other bio-mediated phenomena occurring in the environment.

18.6.3 OFET Sensors and Transducers in Liquids

An open question is whether the electrostatic interactions described in Section 18.6.1 can give rise to a specific response for a given OS. The functionalization of the OS thin film has been envisioned as a possible scheme to enhance chemical–physical selectivity toward a specific analyte [123]. Torsi et al. [124] demonstrated chiral discrimination of natural enantiomers and racemic mixtures using OFETs where the OS is terminated with a chiral pending group. Importantly, they showed that the sensitivity can be enhanced by three orders of magnitude turning on the gate voltage (Figure 18.15).

It has been demonstrated that the interplay between the channel length and the morphological features of the thin layer affects the performance of OFET-based e-noses [113]. OFETs functionalized with the top gate dielectric have been used for immunoassays with antibodies [115], as an example of specific sensing with OFETs.

Another important class of organic device sensors, termed *organic electrochemical transistors* (OECT), was investigated for glucose sensing (by exploiting the catalytic glucose oxidation (Figure 18.16) [125], for controlling cell adhesion via generation of longitudinal electrochemical gradients [126], and, in a modified layout termed *electrochemical ion pumps*, to deliver ionic species between two solutions separated in space [127].

Another scheme for sensing, in view of the arguments discussed in Section 18.5, is to use SAMs or mixed SAMs where one component has a tail designed for the recognition of analytes. Transduction relies on the change of transmissivity due to subtle changes of thickness and/or electronic structure at the charge injection interface. Along this line, a few sensor configurations have been reported.

Interfacing living cells and tissues to electronics, so-called bioelectronics, has been pursued for almost three decades using CMOS technology. Aims are to monitor neural signals; stimulate cell recruitment, adhesion, and proliferation;

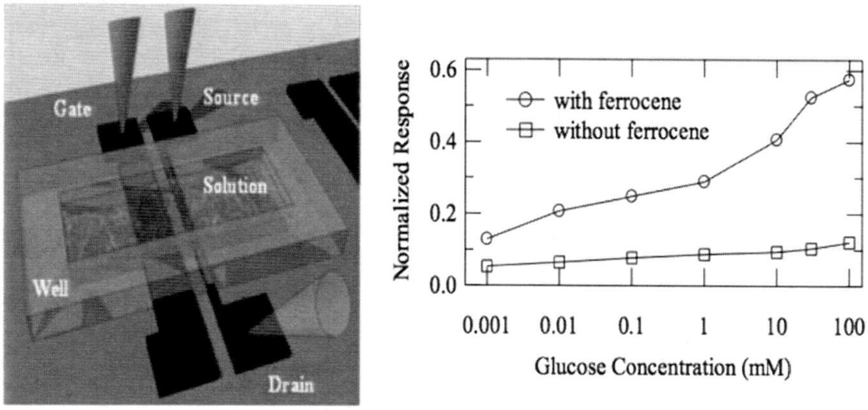

FIGURE 18.16. (**Left**) Layout of a finished device used in [125]. (**Right**) Normalized response to glucose concentration for OECTs preloaded with a mixture with (open circles) and without (open squares) ferrocene mediator. The lines are guides to the eye [125].

and drive microfluidic devices for drug delivery. Applications are envisioned in biodiagnostics, loco-regional therapy, and regenerative medicine. Organic electronics offers some advantages with respect to silicon technology: higher sensitivity, tailoring the interface with the biological environment, choice and design of biocompatible materials, and processes easily aligned with sterility requirements. A fundamental feature that makes OFET the cornerstone device of an interfaceable or implantable electronic platform is the mechanical compliance that devices in a living body must possess. The ability to adapt to the bodily motion and the mechanical deformation is a key element to minimize the effects of the immune response of the organism. Organic electronics can fill this gap more easily than silicon electronics. Moreover, it would be desirable to use biodegradable substrates and active and passive materials. This requires still a substantial effort in materials research, as only a few cases of organic semiconductors and dielectrics have been reported [128].

A wide palette of organic semiconductors can be interfaced to cells without exhibiting cytotoxicity in a few weeks' timescale. Different groups demonstrated that different neural cell lineages adhere and proliferate on OS thin films *in vitro* (Figure 18.17) [98, 126, 129] and can be cultured for several days without the need of a protein adhesion layer.

The mechanical forces exerted by the cells on the OS thin film were shown not to destroy the continuity of the OS thin film [130]. Both primary neurons and stem cells, the latter differentiated into neurons upon chemical stimulation, adhere to pentacene and form neural networks. The ability of the cells to adhere to the OS thin film is sensitive to the local

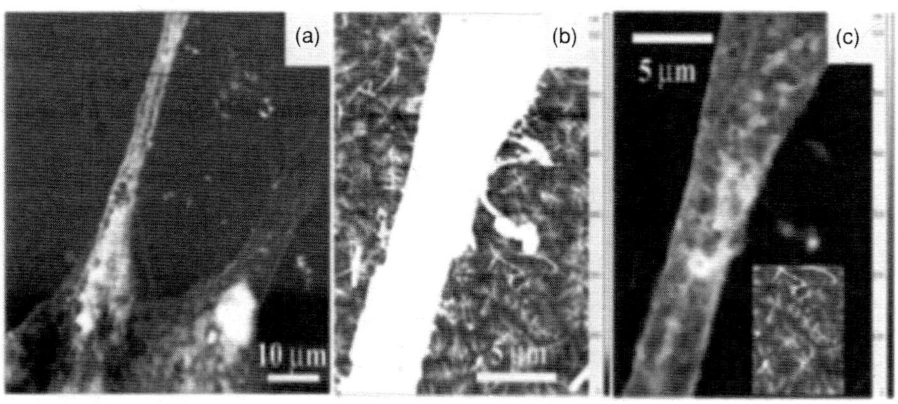

FIGURE 18.17. (**a**) AFM image of a differentiated NE-4C-derived neuron (*arrows*). (**b**) The magnified image shows that the morphology of the pentacene thin film is neither damaged nor modified by the interactions with the cell. (**c**) Same image as (b), with a different height scale to highlight the axon morphology. Inset shows the unaltered pentacene film from (b). The change in height scale highlights the axons [130].

curvature of the thin-film surface [131], which is a function of both roughness and correlation length [132]. Both these properties scale non-monotonically with the film thickness. It is observed that astrocytoma cell adhesion doubles when the film thickness is below 10 nm, and specifically is maximum for 4 ML thick films, corresponding to about 3 ML of pentacene. Once the cells have adhered, the proliferation rate remains the same independently of the film morphology. Thus, a small OS thin-film thickness represents the optimum compromise for charge transport, sensitivity, and cell adhesion.

It is also possible to define dielectric interfaces between the OS thin film and the cells in dual-gate configurations. These dielectric thin films can be artificial as PTrFE/PVDF thin films or bio-derived, as silk fibroin. Neuronal cell signals, triggered by chemical stimuli, are shown to be detectable in the spectral features of the electrical noise of the OFET. It is also shown that neuronal cells exhibit a standard electrophysiological response on silk fibroin films [133] and that this support can be used for interfacing conductive electrodes to the brain of an animal model [134].

18.7 CONCLUSION AND OUTLOOK

Because in the last decade the OFETs performance and stability have considerably improved, there is a growing interest in new OFETs applications: all-organic active matrix displays [32], logic elements [135], inverters [136], organic light emitting transistors (OLETs) [137], and sensors [138, 139]. These applications require more complex layouts and a finer control of the structure of the OS layers [44] but especially of their interfaces with a variety of functional materials. Examples include: the integration of metal nanoparticles [140] in the OS layer, which causes the device to mimic the response of a neural synapsis; the dual-gate dielectric OFETs, where a second gate electrode is deposited on top of the OS in a sandwich [141]; OLET [137], where a high emissive yield is obtained by a multi-layered system; split-gate field effect transistor [142] to enhance transport for specific carrier species and separate control of carrier polarity; and organic electrochemical transistor, which can be used for sensing [104] and drive cell adhesion by the control of the electrochemical potential across the channel [126, 143]. Sensing can be implemented with a variety of device layouts and response principles.

We have discussed how the response of OFETs is governed by interfaces. Different phenomena at nanometer length scales contribute to the physics of the device: charge transport, charge injection, interactions within the OS and with the outer environment. The technological control of such chemical-physical phenomena is of uttermost importance for the standardization of the devices. The low-dimensional character of the phenomena clearly indicate the need to move from a thin-film transistor technology to a technology based on ultra-thin films and patterned self-organized monolayers and nanostructures.

Charge transport is confined to the very first few monolayers in contact with the gate dielectrics. The growth conditions and the chemistry of the channel surface are extremely important for determining the morphology, but also the polarization of the OS active layers. Restricting the growth to a thickness corresponding to that of the charge transport layer, allows not only to avoid undesired phenomena competing with the desired layered growth, but also to optimize the polarization of the OS which contributes to the carrier density and charge mobility. In a sensing scheme, ultra-thin-film OFETs have to present the dominant monolayer as much as possible to the exposition to the outer environment, in order to gain sensitivity.

The modulation of the electronic and structural properties of the charge injection interface by surface chemistry and SAM-functionalization has been shown to induce a crossover in the OFET response from channel-limited to injection-limited transport. Controlling the charge injection interface is essential for making better electronics, by minimization of the contact resistance. Even more interestingly, the OFET can be used for studying charge transport phenomena at the contact/semiconductor interface down to molecular length scales, including a horizontal metal/SAM/semiconductor junction. The OFET then takes the role of a device capable of bridging organic electronics, governed by micro-mesoscopic charge transport, and molecular electronics, governed by nanoscale charge transport.

The OFET as a sensor and transducer is a consequence of the sensitivity to the environment of the different device interfaces. As a consequence of the extreme sensitivity of the charge injection interface to any modification of the chemistry of the contact surface, a new paradigm of label-free sensing can be envisioned with the charge injection interface as the sensing component of the device. Similarly, capacitive coupling of analytes to the OS might be achieved using thin dielectric layers, or even no layer. Problems of stability and repeatability have to be addressed for the future, as well as for standardization of the OFET response, but this appears as dominant in the near future, especially in view of implantable devices.

REFERENCES

1. Yaghmazadeh, O., Cicoira, F., Bernards, D., Yang, S. Y., Bonnassieux, Y., Malliaras, G. G. (2011). Optimization of organic electrochemical transistor for sensor applications. *Journal of Polymer Science Part B—Polymer Physics*, **49**, 34–39.
2. Tsumura, A., Koezuka, H., T.Ando (1986). Macromolecular electronic device: Field-effect transistor with a polythiophene thin film. *Applied Physics Letters*, **49**, 1210–1212.

3. Bao, Z., Rogers, J. A., Katz, H. E. (1999). Printable organic and polymeric semiconducting materials and devices. *Journal of Materials Chemistry*, **9**, 1895–1904.
4. Sirringhaus, H. (2005). Device physics of solution-processed organic field-effect transistors. *Advanced Materials*, **7**, 2411–2425.
5. Facchetti, A. (2007) Semiconductors for organic transistors. *Materials Today*, **10**, 28–37.
6. Newman, C. R., Frisbie, C. D., da Silva Filho, D. A., Brédas, J.-L., Ewbank, P. C., Mann, K. R. (2004). Introduction to organic thin film transistors and design of *n*-channel organic semiconductors. *Chemistry of Materials*, **16**, 4436–4451.
7. Di, C.-an, Yu, G., Liu, Y., Zhu, D., (2007). High-performance organic field-effect transistor: Molecular design, device fabrication, and physical properties. *Journal of Physical Chemistry B*, **111**, 14083–14096.
8. Brown, A. R., Jarret, C. P., de Leeuw, D. M., Matters, M. (1997). Field-effect transistor made from solution-processed organic semiconductors. *Synthetic Metals*, **88**, 37–55.
9. Zaumseil, J., Sirringhaus, H., (2007). Electron and ambipolar transport in organic field-effect transistors. *Chemical Reviews*, **107**, 1296–1323.
10. Dimitrakopoulos, C. D., Malenfant, P. R. L., (2002). Organic thin film transistors for large area electronics. *Advanced Materials*, **14**, 99–117.
11. Braga, D., Horowitz, G., (2009). High-performance organic field-effect transistors. *Advanced Materials*, **21**, 1473–1486.
12. Horowitz, G., Hajlaoui, R., Bouchriha, H., Bourguiga, R., Hajlaoui, M. (1998). The concept of "threshold voltage" in organic field-effect transistors. *Advanced Materials*, **10**, 923–927.
13. Colinge, J., Flandre, D., Vandewiele, F. (1994). Subthreshold slope of long-channel, accumulation-mode p-channel SOI MOSFETs. *Solid-State Electronics*, **37**, 289–294.
14. Lin, Y., Gundlach, D. J., Nelson, S. F., Jackson, T. N. (1997). Stacked pentacene layer organic thin-film transistors with improved characteristics. *IEEE Electron Device Letters*, **18**, 606–608.
15. Klauk, H., Halik, M., Zschieschang, U., Schmid, G., Radlik, W., Weber, W. (2002). High-mobility polymer gate dielectric pentacene thin film transistors. *Journal of Applied Physics*, **92**, 5259–5263.
16. Stingelin-Stutzmann, N., Smits, E., Wondergem, H., Tanase, C., Blom, P., Smith, P., de Leeuw, D. (2005). Organic thin-film electronics from vitreous solution-processed rubrene hypereutectics. *Nature materials*, **4**, 601–606.
17. Gomes, H. L., Stallinga, P., Dinelli, F., Murgia, M., Biscarini, F., de Leeuw, D., Muck, T., Geurts, J., Molenkamp, L. W. (2004). Bias-induced threshold voltages shifts in thin-film organic transistors. *Applied Physics Letters*, **84**, 3184–3186.
18. Ryu, K. K., Nausieda, I., He, D. D., Akinwande, A. I.Bulovic, V., Sodini, C. G., (2010). Bias-stress effect in pentacene organic thin-film transistors. *IEEE Transactions on Electron Devices*, **57**, 1003–1008.
19. Shen, Y., Hosseini, A. R., Wong, M. H., Malliaras, G. G., (2004). How to make ohmic contacts to organic semiconductors. *Chemphyschem: A European Journal of Chemical Physics and Physical Chemistry*, **5**, 16–25.
20. Li, T., Ruden, P. P., Campbell, I. H., Smith, D. L. (2003). Investigation of bottom-contact organic field effect transistors by two-dimensional device modeling. *Journal of Applied Physics*, **93**, 4017–4022.
21. Stoliar, P., Kshirsagar, R., Massi, M., Annibale, P., Albonetti, C., de Leeuw, D., Biscarini, F. (2007). Charge injection across self-assembly monolayers in organic field-effect transistors: Odd–even effects. *Journal of the American Chemical Society*, **129**, 6477–6484.
22. Bürgi, L., Sirringhaus, H., Friend, R. H. (2002). Noncontact potentiometry of polymer field-effect transistors. *Applied Physics Letters*, **80**, 2913–2915.
23. Liscio, A., Palermo, V., Gentilini, D., Nolde, F., Müllen, K., Samorì, P. (2006). Quantitative measurement of the local surface potential of π-conjugated nanostructures: A Kelvin probe force microscopy study. *Advanced Functional Materials*, **16**, 1407–1416.
24. Annibale, P., Albonetti, C., Stoliar, P., Biscarini, F. (2007). High-resolution mapping of the electrostatic potential in organic thin-film transistors by phase electrostatic force microscopy. *The Journal of Physical Chemistry A*, **111**, 12854–12858.
25. Zaumseil, J., Baldwin, K. W., Rogers, J. A. (2003). Contact resistance in organic transistors that use source and drain electrodes formed by soft contact lamination. *Journal of Applied Physics*, **93**, 6117–6124.
26. Puntambekar, K. P., Pesavento, P. V., Frisbie, C. D. (2003). Surface potential profiling and contact resistance measurements on operating pentacene thin-film transistors by Kelvin probe force microscopy. *Applied Physics Letters*, **83**, 5539–5541.
27. Asadi, K., Wu, Y., Gholamrezaie, F., Rudolf, P., Blom, P. W. M. (2009). Single-layer pentacene field-effect transistors using electrodes modified with self-assembled monolayers. *Advanced Materials*, **21**, 4109–4114.
28. Watkins, N. J., Yan, L., Gao, Y. (2002). Electronic structure symmetry of interfaces between pentacene and metals. *Applied Physics Letters*, **80**, 4384–4386.
29. Wu, Y., Li, Y., Ong, B. S., Liu, P., Gardner, S., Chiang, B. (2005). High-performance organic thin-film transistors with solution-printed gold contacts. *Advanced Materials*, **17**, 184–187.
30. Sirringhaus, H., Kawase, T., Friend, R. H., Shimoda, T., Inbasekaran, M., Wu, W., Woo, E. P. (2000). High-resolution inkjet printing of all-polymer transistor circuits. *Science*, **290**, 2123–2126.
31. Sekitani, T., Noguchi, Y., Zschieschang, U., Klauk, H., Someya, T. (2008). Organic transistors manufactured using inkjet technology with subfemtoliter accuracy. *Proceedings of the National Academy of Sciences of the United States of America*, **105**, 4976–4980.

32. Gelinck, G., Heremans, P., Nomoto, K., Anthopoulos, T. D. (2010). Organic transistors in optical displays and microelectronic applications. *Advanced Materials*, **22**, 3778–3798.
33. Tate, J., Rogers, J. A., Jones, C. D. W., Vyas, B., Murphy, D. W., Li, W., Bao, Z., Slusher, R. E., Dodabalapur, A., Katz, H. E. (2000). Anodization and microcontact printing on electroless silver: Solution-based fabrication procedures for low-voltage electronic systems with organic active components. *Langmuir*, **16**, 6054–6060.
34. Loo, Y.-L., Someya, T., Baldwin, K. W., Bao, Z., Ho, P., Dodabalapur, A., Katz, H. E., Rogers, J. A. (2002). Soft, conformable electrical contacts for organic semiconductors: High-resolution plastic circuits by lamination. *Proceedings of the National Academy of Sciences of the United States of America*, **99**, 10252–10256.
35. Blümel, A., Klug, A., Eder, S., Scherf, U., Moderegger, E., List, E. J. W. (2007). Micromolding in capillaries and microtransfer printing of silver nanoparticles as soft-lithographic approach for the fabrication of source/drain electrodes in organic field-effect transistors. *Organic Electronics*, **8**, 389–395.
36. Wang, Z., Yuan, J., Zhang, J., Xing, R., Yan, D., Han, Y. (2003). Metal transfer printing and its application in organic field-effect transistor fabrication. *Advanced Materials*, **15**, 1009–1012.
37. Hu, J., Beck, R. G., Deng, T., Westervelt, R. M., Maranowski, K. D., Gossard, A. C., Whitesides, G. M. (1997). Using soft lithography to fabricate GaAs/AlGaAs heterostructure field effect transistors. *Applied Physics Letters*, **71**, 2020–2022.
38. Greco, P., Cavallini, M., Stoliar, P., Quiroga, S. D., Dutta, S., Zacchini, S., Iapalucci, M. C., Morandi, V., Milita, S., Merli, P. G., Biscarini, F. (2008). Conductive sub-micrometric wires of platinum-carbonyl clusters fabricated by soft-lithography. *Journal of the American Chemical Society*, **130**, 1177–1182.
39. Kim, E., Xia, Y., Whitesides, G. M. (1995). Polymer microstructures formed by moulding in capillaries. *Nature*, **376**, 581–584.
40. Cavallini, M., Biscarini, F. (2003). Nanostructuring conjugated materials by lithographically controlled wetting. *Nano Letters*, **3**, 1269–1271.
41. Cavallini, M., Facchini, M., Massi, M., Biscarini, F. (2004). Bottom–up nanofabrication of materials for organic electronics. *Synthetic Metals*, **146**, 283–286.
42. Liu, S., Wang, W. M., Briseno, A. L., Mannsfeld, S. C. B., Bao, Z. (2009). Controlled deposition of crystalline organic semiconductors for field-effect-transistor applications. *Advanced Materials*, **21**, 1217–1232.
43. Serban, D. A., Greco, P., Melinte, S., Vlad, A., Dutu, C. A., Zacchini, S., Iapalucci, M. C., Biscarini, F., Cavallini, M. (2009). Towards all-organic field-effect transistors by additive soft lithography. *Small*, **5**, 1117–1122.
44. Hwang, D. K., Fuentes-Hernandez, C., Kim, J. B., Potscavage, W. J., Kippelen, B. (2011). Flexible and stable solution-processed organic field-effect transistors. *Organic Electronics*, **12**, 1108–1113.
45. Cavallini, M., Stoliar, P., Moulin, J. F., Surin, M., Leclere, P., Lazzaroni, R., Breiby, D. W., Andreasen, J. W., Nielsen, M. M., Sonar, P., Grimsdale, A. C., Mullen, K., Biscarini, F. (2005). Field-effect transistors based on self-organized molecular nanostripes. *Nano Letters*, **5**(12), 2422–245.
46. Cavallini, M., Albonetti, C., Biscarini, F. (2009). Nanopatterning soluble multifunctional materials by unconventional wet lithography. *Advanced Materials*, **21**, 1043–1053.
47. Cavallini, M., Bergenti, I., Milita, S., Ruani, G., Salitros, I., Qu, Z. R., Chandrasekar, R., Ruben, M. (2008). Micro- and nanopatterning of spin-transition compounds into logical structures. *Angewandte Chemie (International ed. in English)*, **47**, 8596–8600.
48. Cavallini, M. (2009). Inhomogeneous thin deposits: a strategy to exploit their functionality. *Journal of Materials Chemistry*, **19**, 6085–6092.
49. Cavallini, M., D'Angelo, P., Criado, V. V., Gentili, D., Shehu, A., Leonardi, F., Milita, S., Liscio, F., Biscarini, F. (2011). Ambipolar multi-stripe organic field-effect transistors. *Advanced Materials*, **23**, 5091–5097.
50. Zhang, F., Nyberg, T., Inganäs, O. (2002). Conducting polymer nanowires and nanodots made with soft lithography. *Nano Letters*, **2**, 1373–1377.
51. Kalb, W., Lang, P., Mottaghi, M., Aubin, H., Horowitz, G., Wuttig, M. (2004). Structure–performance relationship in pentacene/Al_2O_3 thin-film transistors. *Synthetic Metals*, **146**, 279–282.
52. Kang, S., Chung, K. B., Park, D. S., Kim, H. J., Choi, Y. K., Jang, M. H., Noh, M., Whang, C. N. (2004). Fabrication and characterization of the pentacene thin film transistor with a GdO gate insulator. *Synthetic Metals*, **146**, 351–354.
53. Salleo, A.Chabinyc, M. L., Yang, M. S., Street, R. A. (2002). Polymer thin-film transistors with chemically modified dielectric interfaces. *Applied Physics Letters*, **81**, 4383–4385.
54. Di, C.-an, Liu, Y., Yu, G. U. I.Zhu, D. (2009). Interface engineering: An effective approach transistors. *Accounts of Chemical Research*, **42**, 1573–1583.
55. Petrović, A., Pavlica, E., Bratina, G. (2009). Morphology of the metal–organic semiconductor contacts: The role of substrate surface treatment. *Interface Controlled Organic Thin Film*, **129**, 205–210.
56. Di Benedetto, S., Facchetti, A., Ratner, M., Marks, T. J. (2009). Molecular self-assembled monolayers and multilayers for organic and unconventional inorganic thin-film transistor applications. *Advanced Materials*, **21**, 1407–1433.
57. Chua, L. L., Zaumseil, J., Chang, J. C., Ou, E. C. W., Ho, P. K. H., Sirringhaus, H., Friend, R. H. (2005). General observation of n-type field-effect behaviour in organic semiconductors. *Nature*, **434**, 194–199.
58. Kahn, A., Koch, N., Gao, W. (2003). Electronic structure and electrical properties of interfaces between metals and conjugated molecular films. *Journal of Polymer Science Part B: Polymer Physics*, **41**, 2529–2548.
59. Gundlach, D. J., Jia, L., Jackson, T. N. (2001). Pentacene TFT with improved linear region characteristics using chemically modified source and drain electrodes. *IEEE Electron Device Letters*, **22**, 571–573.
60. Bock, C., Pham, D. V., Kunze, U., Käfer, D., Witte, G., Wöll, C. (2006). Improved morphology and charge carrier injection

in pentacene field-effect transistors with thiol-treated electrodes. *Journal of Applied Physics*, **100**, 1145171–1145177.

61. Mottaghi, M., Horowitz, G. (2006). Field-induced mobility degradation in pentacene thin-film transistors. *Organic Electronics*, **7**, 528–536.

62. Kymissis, I., Dimitrakopoulos, C. D., Purushothaman, S. (2001). High-performance bottom electrode organic thin-film transistors. *IEEE Transactions on Electron Devices*, **48**, 1060–1064.

63. Alloway, D. M., Graham, A. L., Yang, X., Mudalige, A., Colorado, R., Wysocki, V H., Pemberton, J. E., Lee, T. R., Wysocki, J., Armstrong, N. R. (2009). Tuning the effective work function of gold and silver using ω-functionalized alkanethiols: Varying surface composition through dilution and choice of terminal groups. *Journal of Physical Chemistry C*, **113**, 20328–20334.

64. Alloway, D. M., Hofmann, M., Smith, D. L., Gruhn, N. E., Graham, A. L., Colorado, R., Wysocki, V H., Lee, A. L., Lee, T. R., Armstrong, N. R. (2003). Interface dipoles arising from self-assembled monolayers on gold: UV–photoemission studies of alkanethiols and partially fluorinated alkanethiols. *The Journal of Physical Chemistry B*, **107**, 11690–11699.

65. Sushko, M. L., Shluger, A. L. (2009). Rough and fine tuning of metal work function via chemisorbed self-assembled monolayers. *Advanced Materials*, **21**, 1111–1114.

66. de Boer, B., Hadipour, A., Mandoc, M. M., van Woudenbergh, T., Blom, P. W. M. (2005). Tuning of Metal Work Functions with Self-Assembled Monolayers. *Advanced Materials*, **17**, 621–625.

67. Boudinet, D., Benwadih, M., Qi, Y., Altazin, S., Verilhac, J. M., Kroger, M., Serbutoviez, C., Gwoziecki, R., Coppard, R., Le Blevennec, G., Kahn, A., Horowitz, G. (2010). Modification of gold source and drain electrodes by self-assembled monolayer in staggered *n*- and *p*-channel organic thin film transistors. *Organic Electronics*, **11**, 227–237.

68. Witte, G., Lukas, S., Bagus, P. S., Wöll, C. (2005). Vacuum level alignment at organic/metal junctions: 'Cushion' effect and the interface dipole. *Applied Physics Letters*, **87**, 2635021–2635023.

69. Love, J. C., Estroff, L. A., Kriebel, J. K., Nuzzo, R. G., Whitesides, G. M. (2005). Self-assembled monolayers of thiolates on metals as a form of nanotechnology. *Chemical Reviews*, **105**, 1103–1169.

70. Hill, I. (2000). Organic semiconductor interfaces: electronic structure and transport properties. *Applied Surface Science*, **166**, 354–362.

71. Shen, C., Kahn, A., Schwartz, J. (2001). Role of metal–molecule chemistry and interdiffusion on the electrical properties of an organic interface: The Al–F16CuPc case. *Journal of Applied Physics*, **90**, 6236–6241.

72. Hirose, Y., Kahn, A., Aristov, V., Soukiassian, P., Bulovic, V., Forrest, S. (1996). Chemistry and electronic properties of metal–organic semiconductor interfaces: Al, Ti, In, Sn, Ag, and Au on PTCDA. *Physical Review B*, **54**, 13748–13758.

73. Ishii, H., Sugiyama, K., Ito, E., Seki, K. (1999). Energy level alignment and interfacial electronic structures at organic/metal and organic/organic interfaces. *Advanced Materials*, **11**, 605–625.

74. Baldo, M., Forrest, S. (2001). Interface-limited injection in amorphous organic semiconductors. *Physical Review B*, **64**, 1–17.

75. Crispin, X., Geskin, V., Crispin, A., Cornil, J., Lazzaroni, R., Salanek, W. R., Bredas, J. L. (2002). Characterization of the interface dipole at organic/metal interfaces. *Journal of the American Chemical Society*, **124**, 8131–8141.

76. Koch, N., Kahn, A., Ghijsen, J., Pireaux, J.-J., Schwartz, J., Johnson, R. L., Elschner, A., (2003). Conjugated organic molecules on metal versus polymer electrodes: Demonstration of a key energy level alignment mechanism. *Applied Physics Letters*, **82**, 70–72.

77. Vázquez, H., Dappe, Y. J., Ortega, J., Flores, F. (2007). Energy level alignment at metal/organic semiconductor interfaces: "Pillow" effect, induced density of interface states, and charge neutrality level. *The Journal of Chemical Physics*, **126**, 1447031–1447038.

78. Koch, N., Ghijsen, J., Johnson, R. L., Schwartz, J., Pireaux, J.-J. and Kahn, A., (2002). Physisorption-like interaction at the interfaces formed by pentacene and samarium. *The Journal of Physical Chemistry B*, **106**, 4192–4196.

79. Koch, N., Zojer, E., Rajagopal, A., Ghijsen, J., Johnson, R. L., Leising, G., Pireaux, J.-J. (2001). Electronic properties of the interfaces between the wide bandgap organic semiconductor para-sexiphenyl and samarium. *Advanced Functional Materials*, **11**, 51–58.

80. Campbell, I., Rubin, S., Zawodzinski, T. A., Kress, J. D., Martin, R. L., Smith, L., Barashkov, N. N., Ferraris, J. P. (1996). Controlling Schottky energy barriers in organic electronic devices using self-assembled monolayers. *Physical Review B*, **54**, 14321–14324.

81. Koch, N. (2008). Energy levels at interfaces between metals and conjugated organic molecules. *Journal of Physics: Condensed Matter*, **20**, 184008.

82. Wan, A, Hwang, J., Amy, F., Kahn, A. (2005). Impact of electrode contamination on the α-NPD/Au hole injection barrier. *Organic Electronics*, **6**, 47–54.

83. Bürgi, L., Richards, T. J., Friend, R. H., Sirringhaus, H. (2003). Close look at charge carrier injection in polymer field-effect transistors. *Journal of Applied Physics*, **94**, 6129–6137.

84. Casalini S., Shehu, A.Destri, S., Porzio, W., Pasini, M. C., Vignali, F., Borgatti, F., Leonardi, F., Albonetti, C., Biscarini, F. (*submitted* 2011). Molecular junctions with organic ultrathin film transistor. *Organic Electronics*.

85. Thuo, M. M., Reus, W., Nijhuis, C. A., Barber, J. R., Kim, C., Schulz, M. D., Whitesides, G. M. (2011). Odd-even effects in charge transport across self-assembled monolayers. *Journal of the American Chemical Society*, **133**, 2962–2975.

86. Dodabalapur, A., Torsi, L., Katz, H. E. (1995). Organic transistors: Two-dimensional transport and improved electrical characteristics. *Science*, **268**, 270–271.

87. Dinelli, F., Murgia, M., Levy, P., Cavallini, M., Biscarini, F., de Leeuw, D. M. (2004). Spatially correlated charge transport in organic thin film transistors. *Physical Review Letters*, **92**, 90–93.

88. Amar, J. G., Family, F., Lam, P.-M. (1994). Dynamic scaling of the island-size distribution and percolation in a model of submonolayer molecular-beam epitaxy. *Physical Review B*, **50**, 8781–8797.

89. Dinelli F., Moulin, J. F., Loi, M. A., Da Como, E., Massi, M., Murgia, M., Muccini, M., Biscarini, F., Wie, J., Kingshott, P. (2006). Effects of surface chemical composition on the early growth stages of α-sexithienyl films on silicon oxide substrates. *Journal of Physical Chemistry B*, **110**, 258–263.

90. Mayer, A., Ruiz, R., Zhou, H., Headrick, R., Kazimirov, A., Malliaras, G. (2006). Growth dynamics of pentacene thin films: Real-time synchrotron x-ray scattering study. *Physical Review B*, **73**, 1–5.

91. Ruiz, R., Papadimitratos, A., Mayer, A. C., Malliaras, G. G., (2005). Thickness dependence of mobility in pentacene thin-film transistors. *Advanced Materials*, **17**, 1795–1798.

92. Park, B.-N., Seo, S., Evans, P. G. (2007). Channel formation in single-monolayer pentacene thin film transistors. *Journal of Physics D: Applied Physics*, **40**, 3506–3511.

93. Kiguchi, M., Nakayama, M., Shimada, T., Saiki, K. (2005). Electric-field-induced charge injection or exhaustion in organic thin film transistor. *Physical Review B*, **71**, 1–5.

94. Muck, T., Wagner, V., Bass, U., Leufgen, M., Geurts, J., Molenkamp, L. (2004). In situ electrical characterization of DH4T field-effect transistors. *Synthetic Metals*, **146**, 317–320.

95. Muck, T., Fritz, J., Wagner, V. (2005). Better bottom contact properties in organic field-effect transistors with ultrathin layers. *Applied Physics Letters*, **86**, 2321011–2321013.

96. Shehu, A., Quiroga, S. D., D'Angelo, P., Albonetti, C., Borgatti, F., Murgia, M., Scorzoni, A., Stoliar, P., Biscarini, F. (2010). Layered distribution of charge carriers in organic thin film transistors. *Physical Review Letters*, **104**, 1–4.

97. Garnier, F., Hajlaoui, R., El Kassmi, A., Horowitz, G., Laigre, L., Porzio, W., Armanini, M., Provasoli, F. (1998). Dihexylquaterthiophene, A two-dimensional liquid crystal-like organic semiconductor with high transport properties. *Chemistry of Materials*, **10**, 3334–3339.

98. Göllner, M., Huth, M., Nickel, B. (2010). Pentacene thin-film transistors encapsulated by a thin alkane layer operated in an aqueous ionic environment. *Advanced materials*, **22**, 4350–4354.

99. Angione, M. D., Cotrone, S., Magliulo, M., Mallardi, A., Altamura, D., Giannini, C., Cioffi, N., Sabbatini, L., Fratini, E., Baglioni, P., Scamarcio, G., Palazzo, G., Torsi, L. (2012). Interfacial electronic effects in functional biolayers integrated into organic field effect transistors. Proceeding of the National Academy of Sciences of the United States of America, **109**, 6429–6434.

100. Verlaak, S., Heremans, P. (2007). Molecular microelectrostatic view on electronic states near pentacene grain boundaries. *Physical Review B*, **75**, 1–14.

101. de Boer, R. W. I., Jochemsen, M., Klapwijk, T. M., Morpurgo, A. F., Niemax, J., Tripathi, A. K., Pflaum, J. (2004). Space charge limited transport and time of flight measurements in tetracene single crystals: A comparative study. *Journal of Applied Physics*, **95**, 1196–1202.

102. Cramer, T., Steinbrecher, T., Koslowski, T., Case, D., Biscarini, F., Zerbetto, F. (2009). Water-induced polaron formation at the pentacene surface: Quantum mechanical molecular mechanics simulations. *Physical Review B*, **79**, 1–10.

103. Hulea, I. N., Fratini, S., Xie, H., Mulder, C. L., Iossad, N. N., Rastelli, G., Ciuchi, S., Morpurgo, A. F. (2006). Tunable Fröhlich polarons in organic single-crystal transistors. *Nature materials*, **5**(12), 982–986.

104. Yang, S. Y., et al. (2010). Electrochemical transistors with ionic liquids for enzymatic sensing. *Chemical Communications (Cambridge, England)*, **46**(42), 7972–7974.

105. Mabeck, J. T., Malliaras, G. G. (2006). Chemical and biological sensors based on organic thin-film transistors. *Analytical and Bioanalytical Chemistry*, **384**, 343–353.

106. Bergveld, P. (1996). The future of biosensors. *Sensors and Actuators A: Physical*, **56**, 65–73.

107. Lin, P., Yan, F. (2011). Organic thin-film transistors for chemical and biological sensing. *Advanced Materials*, 1–18.

108. Roberts, M. E., Mannsfeld, S. C. B., Queraltò, N., Reese, C., Locklin, J., Bao, Z. (2008). Water-stable organic transistors and their application in chemical and biological sensors. *Proceedings of the National Academy of Sciences of the United States of America*, **105**, 12134–12139.

109. Sokolov, A. N., Roberts, M. E., Bao, Z. (2009). Fabrication of low-cost electronic biosensors. *Materials Today*, **12**, 12–20.

110. Stoliar, P., Bystrenova, E., Quiroga, S. D., Annibale, P., Facchini, M., Spijkman, M., Setayesh, S., de Leeuw, D., Biscarini, F. (2009). DNA adsorption measured with ultra-thin film organic field effect transistors. *Biosensors & Bioelectronics*, **24**, 2935–2938.

111. Torsi, L., Dodabalapur, A., Sabbatini, L., Zambonin, P. G. (2000). Multi-parameter gas sensors based on organic thin-film-transistors. *Sensors and Actuators B: Chemical*, **67**, 312–316.

112. Zhu, Z.-T., Mason, J. T., Dieckmann, R., Malliaras, G. G. (2002). Humidity sensors based on pentacene thin-film transistors. *Applied Physics Letters*, **81**, 4643–4645.

113. Wedge, D. C., Das, A., Dost, R., Kettle, J., Madec, M. B., Morrison, J. J., Grell, M., Kell, D. B., Richardson, T. H., Yeates, S., Turner, M. L. (2009). Real-time vapour sensing using an OFET-based electronic nose and genetic programming. *Sensors and Actuators B: Chemical*, **143**, 365–372.

114. Andringa, A.-M., Spijkman, M., Smits, E. C. P., Mathijssen, S. G. J., van Hall, P. A., Setayesh, S., Willard, N. P., Borshchev, O. V., Ponomarenko, S. A., Blom, P. W. M., de Leeuw, D. (2010). Gas sensing with self-assembled monolayer field-effect transistors. *Organic Electronics*, **11**, 895–898.

115. Khan, H. U., Jang, J., Kim, J.-J., Knoll, W. (2011). In situ antibody detection and charge discrimination using aqueous stable pentacene transistor biosensors. *Journal of the American Chemical Society*, **133**, 2170–2176.

116. Spijkman, M.-J., Myny, K., Smits, E. C. P., Heremans, P., Blom, P. W. M., de Leeuw, D. M. (2011). Dual-gate thin-film transistors, integrated circuits and sensors. *Advanced Materials*, 3231–3242.

117. Spijkman, M.-J.Brondijk, J. J., Geuns, T. C. T., Smits, E. C. P., Cramer, T., Zerbetto, F., Stoliar, P., Biscarini,

F., Blom, P. W. M., de Leeuw, D. M. (2010). Dual-gate organic field-effect transistors as potentiometric sensors in aqueous solution. *Advanced Functional Materials*, **20**, 898–905.

118. Khan, H. U., Jang, J., Kim, J.-J., Knoll, W. (2011). Effect of passivation on the sensitivity and stability of pentacene transistor sensors in aqueous media. *Biosensors & Bioelectronics*, **26**, 4217–4221.

119. Sharma, A., Mathijssen, S. G. J., Kemerink, M., de Leeuw, D. M., Bobbert, P. A. (2009). Proton migration mechanism for the instability of organic field-effect transistors. *Applied Physics Letters*, **95**, 2533051–2533053.

120. Kergoat, L., Herlogsson, L., Braga, D., Piro, B., Pham, M.-C., Crispin, X., Berggren, M., Horowitz, G. (2010). A water-gate organic field-effect transistor. *Advanced Materials*, **22**, 2565–2569.

121. Buth, F., Kumar, D., Stutzmann, M., Garrido, J. A. (2011). Electrolyte-gated organic field-effect transistors for sensing applications. *Applied Physics Letters*, **98**, 1533021–1233023.

122. Cramer, T., Kyndiah, A., Murgia, M., Leonardi, F., Casalini, S., Biscarini, F. (2012). Double layer capacitance measured by organic field effect transistor operated in water. *Applied Physics Letters*, **100**, 143302.

123. Khan, H. U., Roberts, M. E., Knoll, W., Bao, Z. (2011). Pentacene based organic thin film transistors as the transducer for biochemical sensing in aqueous media. *Chemistry of Materials*, **23**, 1946–1953.

124. Torsi, L., Farinola, G. M., Marinelli, F., Tanese, M. C., Omar, H. O., Valli, L., Babudri, F., Palmisano, F., Zambonin, P. G., Naso, F. (2008). A sensitivity-enhanced field-effect chiral sensor. *Nature Materials*, **7**, 412–417.

125. Shim, N. Y., Bernards, D. A., Macaya, D. J., DeFranco, J. A., Nikolou, M., Owens, R. M., Malliaras, G. G. (2009). All-plastic electrochemical transistor for glucose sensing using a ferrocene mediator. *Sensors*, **9**, 9896–9902.

126. Bolin, M. H., Svennersten, K., Nilsson, D., Sawatdee, A., Jager, E. W. H., Richter-Dahlfors, A., Berggren, M. (2009). Active control of epithelial cell-density gradients grown along the channel of an organic electrochemical transistor. *Advanced Materials*, **21**, 4379–4382.

127. Berggren, M., Richter-Dahlfors, A. (2007). *Organic Bioelectronics. Advanced Materials*, **19**, 3201–3213.

128. Bettinger, C. J., Bao, Z. (2010). Organic thin-film transistors fabricated on resorbable biomaterial substrates. *Advanced Materials (Deerfield Beach, Fla.)*, **22**, 651–655.

129. Ghezzi, D., Antognazza, R., Lanzarini, E., Benfenati, F., Lanzani, G. (2011). A hybrid bioorganic interface for neuronal photoactivation. *Nature Communications*, **2**, 164–167.

130. Bystrenova, E., Jelitai, M., Tonazzini, I., Lazar, A. N., Huth, M., Stoliar, P., Dionigi, C., Cacace, M., Nickel, B., Madarasz, E., Biscarini, F. (2008). Neural networks grown on organic semiconductors. *Advanced Functionals Materials*, **18**, 1751–1756.

131. Tonazzini, I., Bystrenova, E., Chelli, B., Greco, P., Stoliar, P., Calò, A., Borgatii, F., D'Angelo, P., Martini, C., biscarini, F. (2010). Multiscale morphology of organic semiconductor thin films controls the adhesion and viability of human neural cells. *Biophysical Journal*, **98**, 2804–2812.

132. Biscarini, F., Samorí, P., Greco, O., Zamboni, R. (1997). Scaling behavior of anisotropic organic thin films grown in high vacuum. *Physical Review Letters*, **78**, 2389–2392.

133. Capelli, R., Amsden, J. J., Generali, G., Toffanin, S., Benfenati, V., Muccini, M., Kaplan, D. L., Omenetto, F. G., Zamboni, R. (2011). Integration of silk protein in organic and light-emitting transistors. *Organic Electronics*, **12**, 1146–1151.

134. Kim, D.-H., Lu, N., Ma, R., Kim, Y.-S., Kim, R.-H., Wang, S., Wu, J., Won, S. M., Tao, H., Islam, A., Yu, K. J., Kim, T.-i., Chowdhury, R., Ying, M., Xu, L., Li, M., Chung, H.-J., Keum, H., McCormick, M., Liu, P., Zhang, Y.-W., Omenetti, F. G., Huang, Y., Coleman, T., Rogers, J. A. (2011). Epidermal electronics. *Science*, **333**, 838–843.

135. Heremans, P., Gelinck, G. H., Müller, R., Baeg, K.-J., Kim, D.-Y., Noh, Y.-Y. (2011). Polymer and organic nonvolatile memory devices. *Chemistry of Materials*, **23**, 341–358.

136. Mukherjee, B., Shin, T. J., Sim, K., Mukherjee, M., Lee, J., Kim, S. H., Pyo, S. (2010). Periodic arrays of organic crystals on polymer gate dielectric for low-voltage field-effect transistors and complementary inverter. *Journal of Materials Chemistry*, **20**, 9047–9051.

137. Capelli, R., Toffanin, S., Generali, G., Usta, H., Facchetti, A., Muccini, M. (2010). Organic light-emitting transistors with an efficiency that outperforms the equivalent light-emitting diodes. *Nature materials*, **9**, 496–503.

138. Someya, T., Dodabalapur, A., Huang, J., See, K. C., Katz, H. E. (2010). Chemical and physical sensing by organic field-effect transistors and related devices. *Advanced Materials*, **22**, 3799–3811.

139. Wöbkenberg, P. H., Labram, J. G., Swiecicki, J.-M., Parkhomenko, K., Sredojevic, D., Gisselbrecht, J.-P., de Leeuw, D. M., Bradley, D. D. C., Djukic, J.-P.Anthopoulos, T. D. (2010). Ambipolar organic transistors and near-infrared phototransistors based on a solution-processable squarilium dye. *Journal of Materials Chemistry*, **20**, 3673–3680.

140. Alibart, F., Pleutin, S., Guerin, D., Novembre, C., Lenfant, S., Lmimouni, K., Gamrat, C., Vuillaume, D. (2010). An organic nanoparticle transistor behaving as a biological spiking synapse. *Advanced Functional Materials*, **20**, 330–337.

141. Maddalena, F., Spijkman, M.-J., Brondijk, J. J., Fonteijn, P., Brouwer, F., Hummelen, J. C., de Leeuw, D. M., Blom, P. W. M., de Boer, B. (2008). Device characteristics of polymer dual-gate field-effect transistors. *Organic Electronics*, **9**, 839–846.

142. Hsu, B. B. Y., Namdas, E. B., Yuen, J. D., Cho, S., Samuel, I. D. W., Heeger, A. J. (2010). Split-gate organic field effect transistors: Control over charge injection and transport. *Advanced Materials (Deerfield Beach, Fla.)*, **22**, 4649–4653.

143. Nilsson, D., Chen, M., Kugler, T., Remonen, T., Armgarth, M., Berggren, M. (2002). Bi-stable and dynamic current modulation in electrochemical organic transistors. *Advanced Materials*, **14**, 51–54.

19

THE GROWTH OF ORGANIC NANOMATERIALS BY MOLECULAR SELF-ASSEMBLY AT SOLID SURFACES

José M. Gallego, Roberto Otero, and Rodolfo Miranda

In recent years an enormous progress has been made in growing ultrathin organic films with a wide range of exciting electronic and optoelectronic properties, which has allowed their use as components in electronic devices, sensors, displays, and logic elements. Recent reports of functional systems incorporating molecular films include light-emitting diodes, nonlinear optical materials, field effect transistors, and photovoltaic devices. This growing interest in films based on molecular components, rather than the more traditional inorganic analogues, is based on the premise that optical and electronic properties can be systematically manipulated through molecular design. In many cases, however, many of these properties depend strongly upon film structure and orientation with respect to the substrate upon which they are deposited, which calls for a careful attention to the interface between the molecular overlayer and the substrate surface.

In this work we review several examples from our recent research demonstrating how the surface influence may affect the film morphology, or even give rise to new structures with potentially promising properties for the improvement or the development of new electronic and optolectronic devices.

19.1 INTRODUCTION: MOLECULAR SELF-ASSEMBLY

Self-assembly is a term used to describe processes in which a number of objects that are spatially arranged in a disordered pattern at some particular time will order themselves via local interactions [1]. When the constitutive components are molecules, the process is termed "molecular self-assembly." The distinction between "self-assembly" and "self-organization" is diffuse, although the first term is commonly used to refer to equilibrium situations, while self-organization only occurs far from equilibrium, in open systems, as it requires an external energy source.

Besides the fundamental interest in understanding and controlling the formation of ordered patterns out of disordered systems, the self-assembly of nanoscale objects, such as organic molecules, quantum dots, or nanotubes, is generally accepted to be the most promising method for large-scale fabrication of functional nanostructures—that is, nanoscale systems specifically designed to perform a given function. It is well known that, especially when we deal with nanoscale systems, structure determines functionality. For example, the efficiency of the proteins involved in the capture of light to trigger the photosynthetic reactions depends on a very particular spatial distribution of the pigments that enable the delocalization of the photoexcitation, thus lengthening their lifetimes and preventing fast exciton recombination [2]. In the case of molecular self-assembly, it is generally assumed that the design of the molecules can be done so as to show particular properties (electrical, optical, and catalytic) and, at the same time, a given chemical architecture that would enable them to assemble in specific ways to optimize the function of the self-assembled system.

Thus engineering functional nanoarchitectures requires a very precise control on the positioning of the molecular building blocks. Although several mechanisms to control such positioning down to individual nanoobjects exist in the literature (for example, by using the tip of a scanning tunneling microscope [3–5]), the serial character of these methods prevents the fabrication of the large number of systems that practical applications need. It is thus

Organic Nanomaterials: Synthesis, Characterization, and Device Applications, First Edition. Edited by Tomás Torres and Giovanni Bottari.

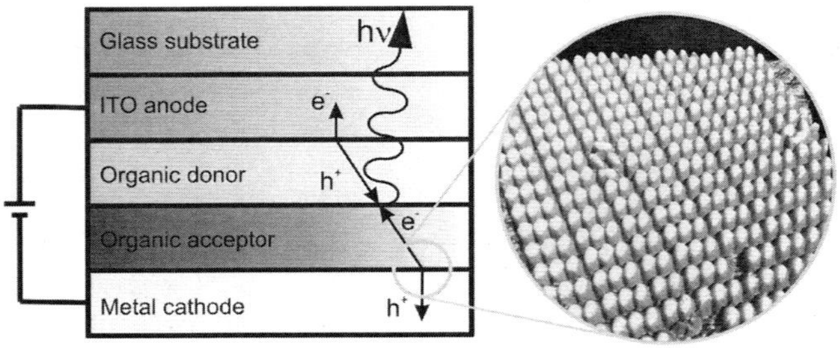

FIGURE 19.1. Organic/inorganic interfaces are very important for the functioning of many electronic and optoelectronic devices based on molecular materials. The figure shows a schematic representation of an organic light-emitting diode. Applying voltage to the active organic layer leads to hole injection at the donor interface, electron injection at the acceptor interface, and electron–hole recombination which leads to photon emission. Such injections processes are governed by the nature and strength of the molecule–metal interaction and the structure and packing density of the molecular layer. Copyright 2011, Wiley-VCH.

accepted nowadays that parallel fabrication approaches based on the self-ordering of the nanoobjects is the way to go in order to bring fundamental nanoscience into more applied nanotechnology.

In addition, for many applications it is often necessary that the self-assembled system is not freestanding, but deposited onto another solid, most commonly an inorganic solid, that would act as a support and which in many cases can also play an important functional role such as that of an electrode. It has been shown that the interface between the self-assembled organic layer and the electrode is very often of paramount importance for the functioning of the devices [6–8]. Details about molecular orientation and packing at the interface determine properties such as charge mobility and injection across the interface between the organic material and the electrode (see Figure 19.1) [4, 8, 9].

All the reasons mentioned above have motivated a very intense research effort to understand the adsorption and self-assembly of organic molecules on solid surfaces [10, 11]. Such research has experienced a tremendous boost by the use of a broad battery of powerful surface science techniques to grow and characterize the organic layers deposited on solid substrates. Growth techniques traditionally applied to fully inorganic systems have been extended to organic materials, in such a way that molecular films can be grown in ultra-high vacuum (10^{-10} torr) conditions by molecular beam epitaxy. The use of such ultra-clean environments is essential to quantify intermolecular interactions, which are much more difficult to understand in solution due to the presence of solvents [12].

But besides the importance of vacuum-based growth techniques in building well-defined self-assembled systems, the most important contribution of surface science to supramolecular chemistry has been the use of new characterization techniques that have provided a wealth of new data with unprecedented detail. For example, the extensive use of scanning tunneling microscopy to study the molecular arrangement of organic material on solid surfaces has offered real-space images of the self-assembled structures with molecular, and in some cases even submolecular, resolution [10, 11]. This has enabled the structural characterization of many systems, finding similarities and differences with the self-assembly of the same systems in solution phase. In some case the similarities are striking. For example, DNA base guanine was shown to self-assemble on relatively inert gold surfaces by forming quartets similar to those found in telomeric DNA, even in the absence of backbone, solvent, or counterions [13, 14]. The cleanliness of such systems together with the application of traditional vacuum-based surface science techniques, such as photoelectron spectroscopy, also allow for the characterization of self-assembled structures from the electronic and morphological points of view; thus, important contributions to the fundamental knowledge of the nonbonding interactions that hold supramolecular systems together are to be expected in the years to come.

In general, the 2D arrangement of organic molecules on solid surfaces is the result of a combination of intermolecular forces with molecule–substrate interactions [10, 11]. It is generally accepted that in the case of weak noncovalent intermolecular forces (such as van der Waals, or dispersive forces), molecule–substrate interactions determine adsorption geometry and conformation, while intermolecular interactions only affect the subsequent self-assembly of the adsorbates [11]. Only when the molecules have the possibility of forming strong directional bonds, like hydrogen or coordination bonds [13, 15, 16], and the corrugation of the potential energy for the adsorbed species is small compared to the energy gain from intermolecular interactions, supramolecular structures formed on surfaces are mainly determined by intermolecular forces (although even in this case the substrate influence may still be important).

In addition, since the role of the surface in determining the molecular arrangement cannot be generally neglected, investigations on the self-assembly of organic molecules adsorbed on metal surfaces offer new opportunities to grow supramolecular systems with unexpected molecular arrangements and thus functionalities. For example, the fact that adsorption sites and geometries are usually determined by molecule–substrate interactions can be exploited to steer the growth of molecular nanostructures with morphologies dictated by the substrate. When the surface is not homogeneous (because of the presence of regions with different chemical compositions [17, 18], different lattice parameters [19, 20], or different electronic structures), the effect known as site-selective adsorption might take place. If the special anchoring sites are distributed in an ordered fashion over the surface, the molecular nanostructures will thus mimic the substrate pattern, and the resulting morphology will not be dictated by

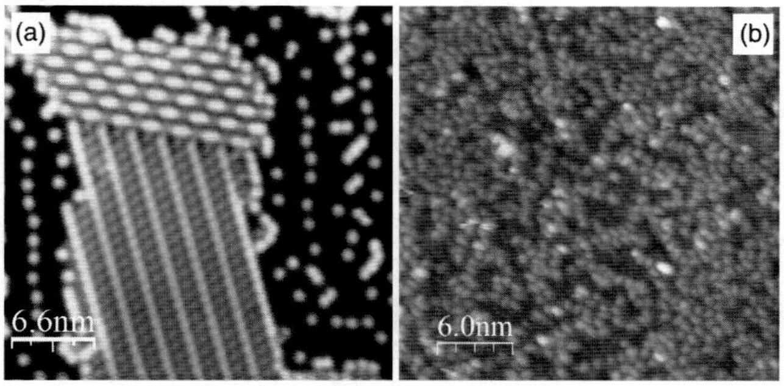

FIGURE 19.2. STM images taken after deposition (TTF + TCNQ) on (**a**) Au(111) and (**b**) Cu(100). Reprinted with permission from I. Fernández-Torrente, PhD Thesis.

FIGURE 19.3. (**a**) STM images taken after depositing TCNQ on Cu(100) at room temperature. (**b**) Electron density difference plot showing in dark gray (light gray) those molecular orbitals that have gained (lost) electrons upon deposition on the Cu(100) surface.

intermolecular interactions but rather by molecule–substrate interactions [21–23].

Care must be taken, however, since molecule–substrate interaction may be such that it deeply affects or even destroys the fundamental properties we are looking for in a particular organic molecule. For example, in recent years much effort has been invested in the co-assembly of electron donors and acceptors for optoelectronic or photovoltaic applications [24, 25], but reports on the successful formation of charge transfer complexes on metal surfaces has been rather scarce [26, 27]. As an example, Figure 19.2a shows an STM image taken after depositing submonolayer amounts of the electron donor tetrathiafulvalene (TTF) and the acceptor tetracyanoquinodimethane (TCNQ) on a Au(111) surface held at room temperature. The results show that the molecules self-assemble into mixed domains of alternating rows with a 1:1 stoichiometry [27]. Density functional theory (DFT) calculations and scanning tunneling spectroscopy (STS) measurements show that both molecules conserve their donor and acceptor characters on the Au(111) surface, with TCNQ taking ~1 electron from TTF. (The bulk charge-transfer complex TTF-TCNQ is an organic metal where the donor, TTF, transfers 0.6 e^- to the acceptor, TCNQ.)

However, attempts in our group to duplicate these results on a copper surface have been unsuccessful. Figure 19.2b shows an STM image taken after depositing TTF and TCNQ on Cu(100) at room temperature. For any surface temperature, post-annealing procedure, or order of deposition, we could never observe any indication of the formation of a compound with a well-defined stoichiometry, the reason being that, contrary to the Au(111) surface, TCNQ does not preserve its acceptor character when deposited on the Cu(100) surface [9]. Figure 19.3a shows a typical STM image taken after depositing a submonolayer amout of TCNQ on Cu(100) at room temperature. Under these conditions, TCNQ self-assembles in the form of two-dimensional compact islands, with a well-defined rectangular shape. However, DFT calculations in combination with photoemission spectroscopy and work function measurements have shown that upon adsorption there is a considerable amount of charge transfer (~1.9 e^-) from the surface to the molecule. As a consequence, the LUMO of the molecule is now almost completely occupied (see Figure 19.3b), which forbids charge transfer from further deposited TTF.

In the following, we will show several examples extracted from our recent research in which the surface influence affects or even dictates the morphology and structure of the growing organic film, giving rise to new structures with potentially new functionalities. The first two examples are related to photovoltaics: systems that could improve the performance of organic solar cells either by increasing the light harvesting capabilities (Section 19.2) or by optimizing the donor–acceptor phase segregation morphology in bulk heterojunction solar cells (Section 19.3). The third example (Section 19.4) is related to the use of solid surfaces to steer the growth of organic nanocrystals with a controlled size and shape. All these results demonstrate that atomistic studies on the growth of organic thin films under ultra-high vacuum (UHV) conditions can lead to the very accurate control over the film morphology that is needed in order to improve the efficiency of electronic and optoelectronic organic devices.

19.2 LIGHT-HARVESTING SYSTEMS: Zn-TMP/Cu(100)

Compared to nature's energy-conversion molecular devices, the efficiencies of man-made organic photovoltaic (PV) devices are still very poor. Photosynthetic organisms are provided with specific light-harvesting antenna complexes that allow them to convert efficiently solar energy into chemical energy even under diffuse or poor illumination conditions [28]. It has been suggested that all-organic or hybrid organic–inorganic solar cells functionalized with biological or bio-inspired antenna complexes would enhance their light-harvesting properties and thus their energy-conversion efficiencies [29].

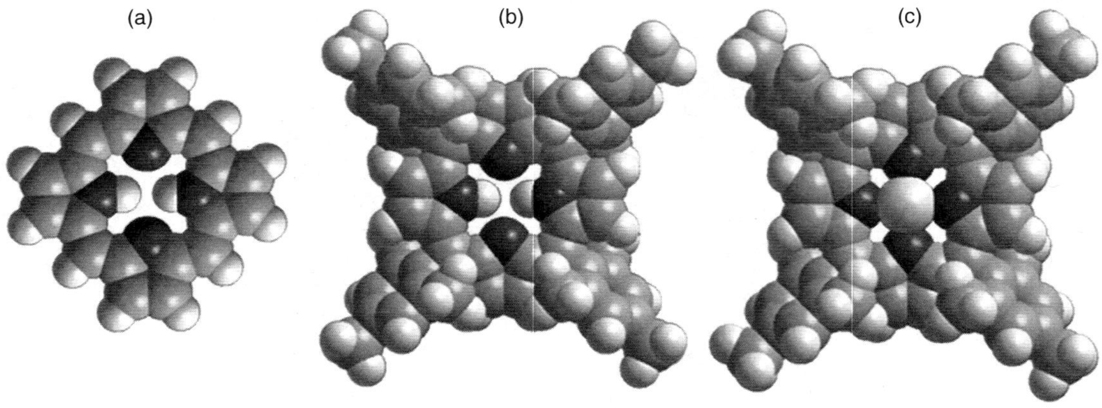

FIGURE 19.4. Space filling model of (**a**) porphyrin, (**b**) meso-tetramesitylporphyrin (H_2-TMP), and (**c**) zinc meso-tetramesitylporphyrin (Zn-TMP).

The most promising candidates of synthetic compounds for artificial photosynthetic components are porphyrins, which absorb visible light with high efficiency and are involved in a number of important biological electron-transfer systems. Porphyrins are heterocyclic macrocycles composed of four modified pyrrole subunits interconnected via methine (=CH-) bridges (Figure 19.4a). They have 18 π electrons in the inner 16-membered ring; that is, they obey Hückel's rule for aromaticity (aromatic systems contain $4n + 2\pi$ electrons). As a consequence, they usually have very intense absorption bands in the visible region and may be deeply colored. In addition, they can coordinate with metals in their central cavity, and it is also possible to add a great variety of functional groups in the meso positions. Taking the natural systems as a model, a large number of studies has been devoted to creating artificial, highly ordered arrays of porphyrins that can efficiently capture light.

Chlorosomes are the light-harvesting organelles of photosynthetic green bacteria [30], acting as sensitive antennas capable of gathering enough energy to support life even at distances of 50 m from the sea surface. They are usually placed close to the inner side of the cytoplasmic membrane, and they consist of a lipid monolayer enclosing the pigments that lead to light absorption, mostly bacteriochlorophyll *c* (BChl *c*). Unlike most other biological antenna systems, the nanoscale spatial arrangement of the BChl *c* molecules within the chlorosomes is not determined by interaction with protein scaffoldings [31]; instead, their arrangement (see Figure 19.5a) is the result of a self-assembly process due to (i) π–π interactions between the chlorin macrocycles, (ii) coordination of the central magnesium ion by the hydroxy group of a neighbor, and (iii) hydrogen bonding of the same hydroxy group to the carbonyl group of a third bacteriochlorophyll molecule [32–35].

The strategy previously followed in order to obtain self-assembled molecular nanostructures similar to the ones found in chlorosomes and having comparable optical properties usually implies the use of alternative synthetic dyes inspired in BChl *c* molecules [34, 36–38]. In particular, Zn-containing porphyrin derivatives with a hydroxyl group (as either a hydroxymethyl or a 1-hydroxyethyl) and acetyl substituents at the core's periphery can form extended one-dimensional chains in which Zn ions are linked to the oxygen atom of hydroxyethyl substituent while the carbonyl groups are displaced toward the pentacoordinated Zn of the subsequent metalloporphyrin approaching from the opposite face than the linked hydroxyl group (Figure 19.5b). For their use as sensitizing antennas in PV devices, these structures should form and be stable when deposited on a solid surface. However, the presence of reactive and unprotected hydroxyl and carbonyl substituents in the reported Zn-porphyrin analogues is likely to mediate strong molecule–surface interactions that could determine the geometry of the molecular adlayer. Previous works have also attempted to functionale the porphyrin core with polar groups for adsorption on ionic surfaces [39]. Thus the success of such an approach depends very strongly on the particular atomic arrangement at the solid surface and might not be directly exported to other technologically relevant surfaces.

We have used a different approach to fabricate long one-dimensional stacks of Zn-porphyrins on metal surfaces in which neighbouring subunits are linked via coordination bonds to the Zn atom at the porphyrin's core and to water molecules that act as bridging ligands. The porphyrin derivative used in our studies is the coordination compound zinc *meso*-tetramesitylporphyrin (Zn-TMP, Figure 19.4c). We show here variable-temperature scanning tunnelling microscopy (VT-STM) experiments demonstrating that Zn-TMP molecules self-assemble on Cu(100) and Au(111) by forming long nanorods that are only one molecule wide, in which the porphyrin core is not lying flat on the surface, but it is perpendicular to it [40].

Figure 19.6a and 19.6b show STM images recorded at 150 K of the molecular structures formed upon deposition

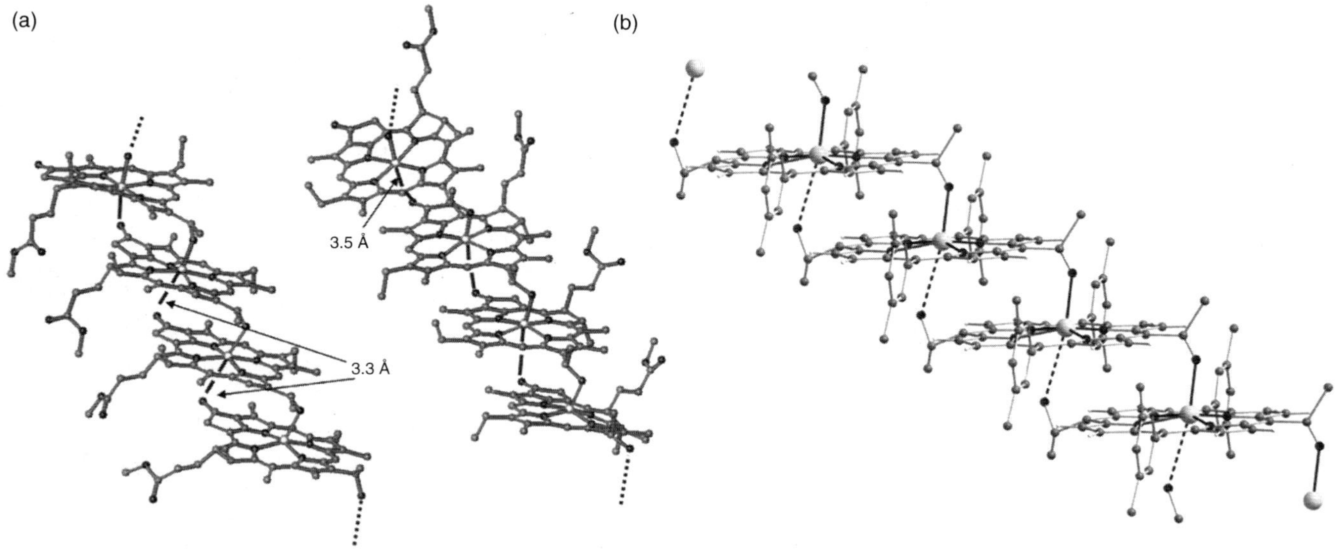

FIGURE 19.5. (**a**) BChl c modeled into stacks. The left stack is formed exclusively of α-ligated (3^1R)-BChl c, while the stack at left is β-ligated (3^1S)-BChl c. There are several possibilities for adjacent stacks to interdigitate with average Mg–Mg distances of ~2 nm with either α- or β-ligations and with both R and S stereochemistries of the 3^1 carbon atom. The weak >C=O···Mg ligation (3.3–3.5 Å long) is indicated by solid lines, while the stack extension is indicated by dotted lines [34]. Copyright 2005, American Chemical Society. (**b**) Crystal structure of the synthetic Zn-porphyrin derivative used to mimic the natural chlorosomal BChl c [36]. Copyright 2005, Wiley-VCH.

of Zn-TMP molecules on Au(111) and Cu(100), respectively. The formation of long (hundreds of nanometers) 1D rod-looking structures is clearly observed on both surfaces, independently of the substrate's chemical nature and structural symmetry, a clear indication that intermolecular interactions must be the main driving force for nanorod self-assembly. High-resolution images of the porphyrin nanorods (Figure 19.6c) reveal an ordered array of protrusions with a typical diameter of about 0.6 nm (i.e., significantly smaller that the size of a single molecule) and a typical height of

FIGURE 19.6. (**a**) STM image ($V_b = -2.1$ V, $I = 0.3$ nA) of the nanorods formed on the Au(111) surface upon Zn-TMP deposition. (**b**) STM image ($V_b = -2.1$ V, $I = 0.5$ nA) of the nanorods on Cu(100). (**c**) High-resolution STM images reveal a periodic corrugation along the tube direction which is identical for both metal surfaces, showing that the substrate's influence on nanorod formation is negligible.

about 0.9 nm. As discussed below, such protrusions can be identified with the topmost mesityl groups for standing up porphyrins. A planar adsorption geometry, in which the porphyrin ring is placed parallel to the surface, seems to be inconsistent with these observations.

On the basis of previous studies of ordered adlayers of porphyrin derivatives on a wide variety of substrates, a nonplanar adsorption geometry on metal surfaces is not expected at all [41–46]. Standing-up geometries have been reported only for the adsorption of porphyrins functionalized with polar groups on insulating surfaces, due to (a) the reduced dispersive interactions between surface and macrocycle and (b) the strong electrostatic interactions between the polar groups and ions at the surface [39]. This points directly toward strong intermolecular interactions as responsible for stabilizing the nonplanar adsorption geometry.

If the nanorods were held together by intermolecular interactions such as π-stacking, they should also be observed for free-base tetramesitylporphyrin (H_2-TMP or simply TMP). Since there are no previous STM studies on mesityl functionalized porphyrins, we have deposited, for the sake of comparison, TMP on the Cu(100) surface [47]. In this molecule the mesityl groups are rotated alternately an angle $\theta \sim \pm 61^\circ$ around the σ bonds, resulting in a rectangular symmetry. In this way, four methyl groups (one in each mesityl group) stay above the mean porphyrin plane, while another four remain below this plane (Figure 19.4b). In addition, the porphyrin core is nonplanar, but slightly saddle-shaped, with the pairs of C atoms forming the outer edge of each pyrrole ring alternately displaced by ~0.3 Å above and below the mean porphyrin plane.

Figure 19.7a shows a large-scale STM image of the Cu(100) surface after depositing ~0.2 ML of TMP. The surface steps appear completely covered with molecules, which indicates a high diffusion coefficient at room temperature. However, on the terraces the molecules remain mostly isolated which, taking into account the high molecular mobility, is a signal of weak intermolecular interactions. A closer look (Figure 19.7b) reveals that the molecules display a six-lobe shape, with rectangular symmetry and three lobes at each side of the main symmetry axis. The four external lobes have an apparent height of 2.1 Å, while the inner ones are 2.3 Å high. The STM images are then consistent with the average porphyrin plane being parallel to the copper surface, with the outer lobes corresponding to the four raised methyl groups and the central ones corresponding to the two pyrrole rings situated above the porphyrin core. Figure 19.7b also shows that the molecular main axis are parallel to the high symmetry directions of the Cu(100) surface. This indicates that the molecular orientation is indeed dictated by the molecule–substrate interaction. When increasing the coverage, large ordered islands with a height around 0.2–0.3 nm (i.e., about three times smaller than the height of the nanorods) grow in size until covering almost completely the copper surface (Figure 19.7c). Curiously, the STM images show the existence of a bimodal distribution of islands sizes [48].

Then, since free-base porphyrins do not form rods, but adsorb on Cu(100) with a planar geometry, it seems unlikely that dispersive interactions alone can account for the preference of the metallated units to bind to each other instead of binding to the substrate, which is to be expected since the electronic polarizability of a metal surface is necessarily larger than the polarizability of the π-clouds in the macrocycles. On the other hand, it is well known that columnar structures in metallic macrocycles are very often stabilized by coordination between neighboring metal centers through an organic ligand [49–55]. It seems thus natural to attribute this strong intermolecular interaction to axial coordination

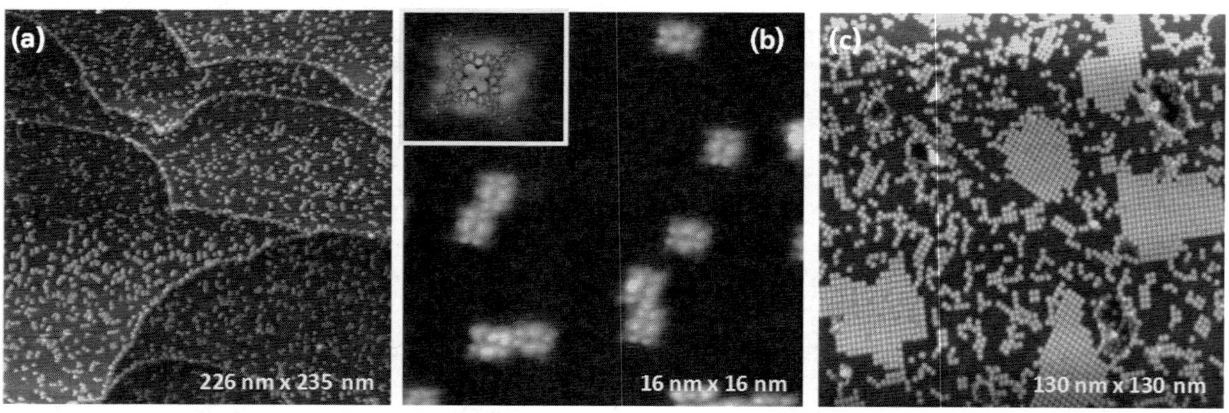

FIGURE 19.7. (**a, b**) STM images, taken at 150 K, of the Cu(100) surface after depositing ~0.2 ML of TMP with the substrate held at 300 K. The top inset in (**b**) shows an enlarged view of an isolated porphyrin, with an schematic drawing of the molecular structure superimposed. (**c**) STM image taken after depositing 0.45 ML of TMP on Cu(100). (**a**) $V_b = -3.5$ V; $I = 0.20$ nA; (**b**) $V_b = -2.1$ V; $I = 0.25$ nA; (**c**) $V_b = -3.0$ V; $I = 0.71$ nA. Reprinted with permission from reference [47]. Copyright 2008, American Chemical Society.

to yield shish-kebab polymers such as those described in references [50–52].

In order to check the presence of axial ligands, we have carried out x-ray photoemission spectroscopy (XPS) on the nanorods structures obtained for the metallated Zn-TMP species. Apart from the expected C 1s, N 1s, and Zn 2p core level peaks, at about 284.8 eV, 398.3 eV, and 1021.0 eV, binding energies respectively [56–58], we clearly detect an oxygen signal (O 1s centered at 531.4 eV binding energy) that is not present before deposition; that is, it cannot be attributed to sample contamination. Oxygen is not present in the chemical structure of Zn-TMP, but x-ray diffraction studies carried out with crystals of Zn-TMP did show a water molecule axially coordinated to the Zn atom [59]. On the contrary, in photoemission spectra taken on the free-base porphyrins, besides the C 1s peak and the two components expected in the N1s core level (corresponding to the two chemically different nitrogen species on the non-metallated porphyrin ring); no measurable traces of oxygen were found.

For the Zn-TMP molecules, ligand detachment from the adsorbed nanorods can be achieved by annealing the surface up to 575 K (higher than the Zn-TMP sublimation temperature), as confirmed by the absence of the oxygen peak in XPS spectra. At the same time the N 1s and Zn 2p core levels slightly shift to the reported values for other Zn porphyrins [57]. STM images reveal that, after this annealing treatment, the nanorods no longer exist on the surface. Instead, the molecules are imaged as rings with four little protrusions at 90° from each other. Superimposing a stick-and-ball model for the Zn-TMP molecules on the experimental STM images, we can ascribe the central ring to the porphyrin ring and the four protrusions to the peripheral mesityl groups. The STM images in the bottom inset of Figure 19.8 are very similar to the results obtained for other porphyrins on a surface after an annealing procedure [46, 60, 61]. After the annealing treatment, the molecules can be imaged by STM even at room temperature, but they cannot be pushed away with the STM tip from their adsorption positions. This enhanced stability could possibly be explained by the formation of a covalent bond by dehydrogenation of the methyl groups at the mesityl legs, as suggested in reference [46].

The possible effect of oxygen-containing axial ligands in bridging two neighboring Zn-TMP molecules has been also investigated by density functional theory calculations for water and other small molecules present in the atmosphere. A stable structure was found in which a water molecule is sandwiched between two Zn-TMP molecules rotated 45° around their C_4 axis with respect to each other, in such a way that the lone pairs in the oxygen atom of the water molecule can interact with both Zn centers (Figure 19.9a). The calculated binding energy is about 0.6 eV, indicating that the formation of the Zn-TMP–water–Zn-TMP complex is significantly exothermic.

Similar results were obtained for CO, but not for O_2, which is not able to bridge neighboring porphyrin molecules in a stable way, while CO_2 forces a nonlinear geometry that is not compatible with our observations. Finally, although CO shows a behavior similar to that shown by water, its scarcity in the atmosphere renders it an unlikely candidate to explain our observations.

Based upon the theoretical models described above, we propose the structure depicted in Figure 19.9b for the observed nanorods. Each porphyrin molecule is rotated by 45° around its C_4 axis with respect to its neighbors. The Zn-TMP molecules are linked to each other through coordination with bridging water molecules. Similar structures are known to occur in solution for a variety of metaloporphyrins and phthalocyanines with different ligands acting as linkers. An STM image of such a tube can be expected to show bright protrusions at the position of the topmost mesityl groups, along with lighter protrusions for the remaining mesityl groups pointing toward the top of the tube. This description fits well with the observed STM images (Figure 19.9b).

In summary, we have described the synthesis of shish-kebab-type coordination polymers of porphyrin derivatives on solid surfaces. The polymers have a straight conformation for over hundreds of nanometers. We suggest that the axial ligands that bridge the Zn atoms are water molecules that can bind to the upper and lower Zn atoms via the lone pairs at the oxygen atom, but other possibilities cannot be excluded. The described structure of the H_2O–Zn-TMP nanorods is reminiscent of the 1D antenna complexes of green bacteria. In both cases there is a 1D structure of stacked porphyrin-like macrocycles with overlapping π systems, linked by coordination between consecutive metal centers. There is ample evidence in the literature that this kind of aggregation leads to a red shift and a broadening in the light absorption spectra of the molecular dyes, both highly desirable properties for sunlight harvesting devices [34, 62]. Moreover, the nanorods have been synthesized on different metal surfaces, a clear indication that the driving force for self-assembly is not significantly affected by the surface and can thus be easily extended to a wider variety of substrates. We have thus demonstrated that, armed with state-of-the-art growth techniques and knowledge on surface assembly, building up solid-supported molecular nanostructures inspired in functional biological systems is indeed possible and is also a promising territory to explore in order to enhance device design and performances according to nature's wisdom.

19.3 OPTIMIZED GEOMETRIES FOR BULK HETEROJUNCTIONS SOLAR CELLS: PCBM–ExTTF/Au(111)

Another important aspect with respect to the efficiency of "plastic" solar cells concerns the morphology of the active

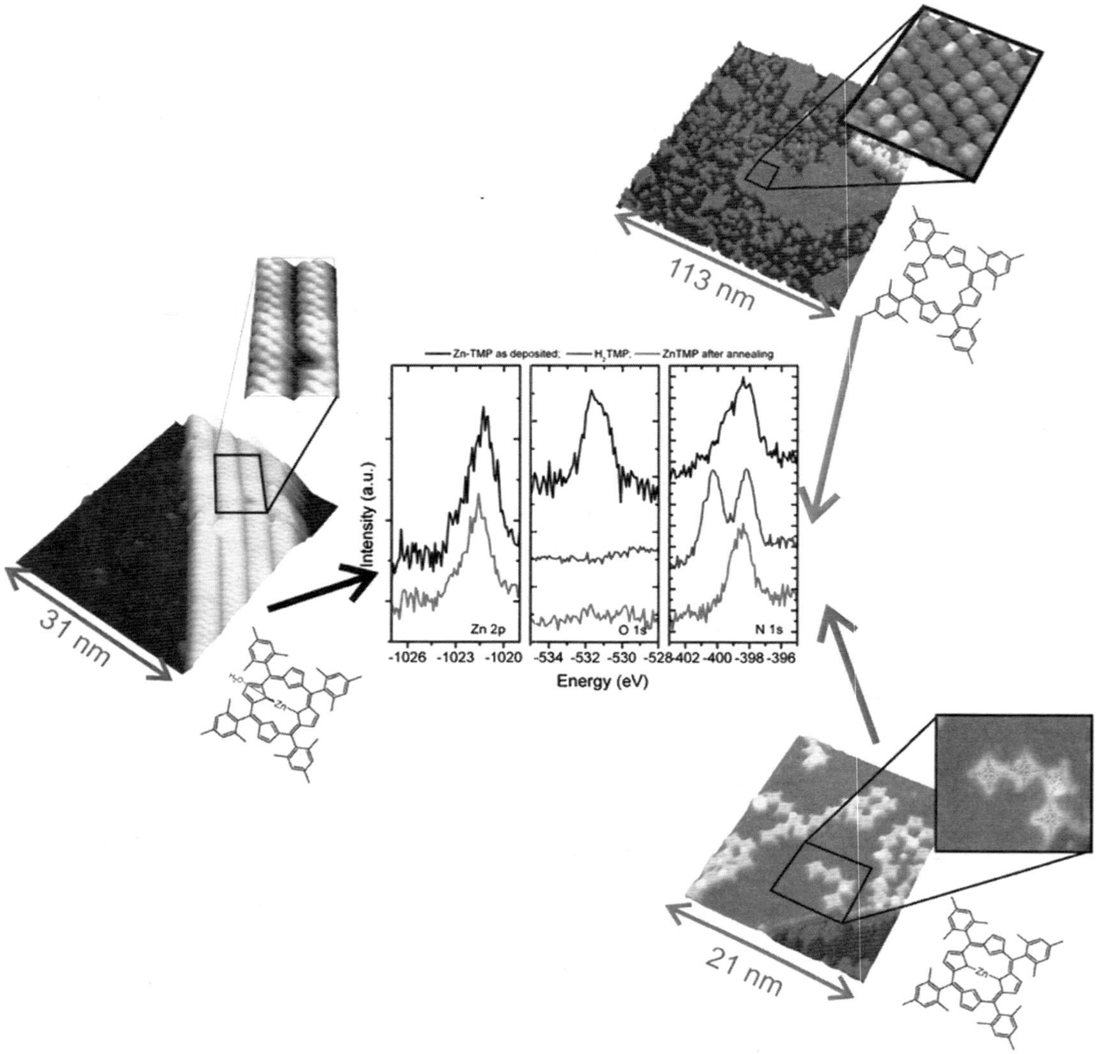

FIGURE 19.8. Correlation between XPS spectra on self-assembled tetramesitylporphyrin (TMP) derivatives on Cu(100) and characteristic STM images. Upon deposition of Zn-TMP, a clear oxygen peak can be found in the XPS spectra, a situation in which STM images reveal nanorod formation (*left inset*). Deposition of base-free H_2-TMP, however, does not show either oxygen in the XPS spectra or nanorod formation (*top inset*). Similarly, water detachment achieved by annealing to 525 K also leads to nanorod dissociation. (bottom inset). Reprinted with permission from reference [40]. Copyright 2011, Royal Society of Chemistry. See color insert.

layer. In particular, the so-called bulk-heterojunction solar cells (BHJ) are prepared by a blend mixture of electron-donor/electron-acceptor molecules [63–70]. The importance of blending donors and acceptors rests on the capability of their interfaces to dissociate the tightly bound excitons that are generated in organic materials upon photon excitation, the formation and recombination of which is one of the major factors limiting the efficiency of photovoltaic devices. In principle, a number of criteria must be satisfied by the morphology of the blend for optimum solar cell performance: first, electron-donor and acceptor domains must segregate into chemically homogeneous regions with typical sizes of the order of the exciton diffusion length (10–20 nm), in order to enhance the exciton dissociation probability and avoid wasteful radiative recombination events; and, second, donor (acceptor) domains must be continuously connected to the anode (cathode) to favor efficient charge transport [66, 68–70]. Figure 19.10 shows a schematic representation of a morphology that would satisfy these criteria [68].

Motivated by previous works, we have attempted to use the nanometer-scale pattern provided by the well-known $22 \times \sqrt{3}$ "herringbone" reconstruction of the Au(111) surface as a template to steer the growth of donor/acceptor species into 1D molecular nanostructures with sizes in good

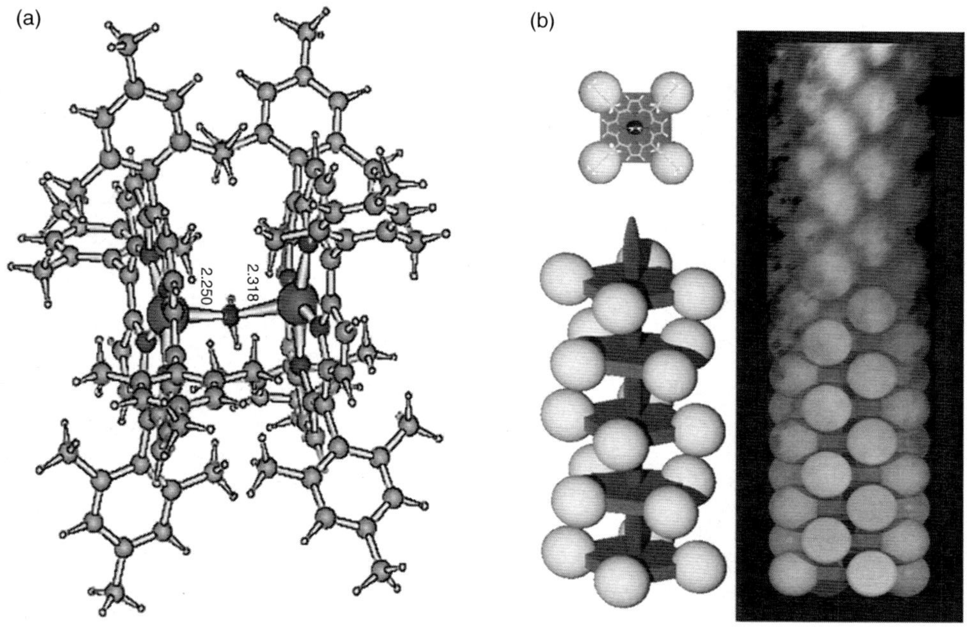

FIGURE 19.9. (**a**) Result of a DFT calculation showing the optimized geometry of a ZnTMP–H_2O–ZnTMP complex. (**b**) Schematic diagram showing the proposed structure for the nanorods. Each Zn-TMP molecule is rotated by 45° with respect to their common C_4 axis, in such a way that the mesityl groups fit between the neighboring mesityl groups of the adjacent Zn-TMP molecule. The Zn atoms are bound to each other through a bridging water molecule. Comparison with STM images reveal that the brightest central protrusions in the nanorods can be ascribed to the uppermost methyl groups whereas the lighter bumps correspond to the methyl groups pointing sideways. Reprinted with permission from reference [40]. Copyright 2011, Royal Society of Chemistry.

registry with exciton diffusion lengths. Our experiments show that, although this simple picture is far from reality and the substrate cannot be regarded as an static checkerboard upon adsorption of the molecular species, molecule–molecule and molecule–substrate interactions conspire to steer a lateral segregation of the donor/acceptor blends into a long-range ordered superlattice of the kind described in Figure 19.10 as a highly desirable morphology for improving solar cell efficiencies [71].

The molecular species considered here are the fullerene derivative PCBM (phenyl-C61-butyric acid methyl ester, Figure 19.10) as the electron acceptor [72] and exTTF (2-[9-(1,3-dithiol-2-ylidene)anthracen-10(9H)-ylidene]-1,3-dithiole, Figure 19.10) as the electron donor [73,74]. PCBM is currently the most common organic acceptor used in photovoltaic applications [63, 65, 70]. On the other hand, exTTF is one of the most important derivatives of the parent tetrathiafulvalene (TTF) [73, 75–78] and can be used in solar cells as the electron donor material [79]. It is a butterfly-shaped nonplanar electron donor [74] endowed with four sulfur atoms, located on the two dithiole rings, which are expected to bind strongly to the Au(111) surface. The planar conformation of the molecule is strongly hindered by the very short contacts between the sulfur atoms and the hydrogen atoms in peri positions. The most important interaction expected between PCBM and exTTF is the π–π interaction between the fullerene moiety and the benzene rings of exTTF, as is the case for exTTF-based tweezers, capable of binding efficiently to C_{60} molecules [80].

The surface of the chosen substrate, Au(111), reconstructs by increasing the atomic density with respect to a (111) plane in the bulk [19, 81]. The reconstruction has a primary structure by which the surface topmost layer of atoms is uniaxially compressed by 4.5% with respect to the underlying bulk Au lattice. The change in the lattice parameter at the surface makes it impossible for the surface atoms to preserve the fcc stacking characteristic of bulk Au, so that only a fraction of the atoms occupy fcc positions, another fraction occupies hcp sites, and fcc and hcp areas are separated by the so-called domain walls (Figure 19.11). In addition, an isotropic stress relief is obtained by the formation of stress domains in which the contraction alternates by 120° between two equivalent domains. These stress domains give rise to the periodic network of surface lattice dislocations that make up the herringbone pattern.

It has been previously shown that the "elbows" of the reconstruction (places where the dislocation lines change direction; see Figure 19.11) act as nucleation centers, and the secondary structure of the herringbone templates the growth of ordered arrays of inorganic and organic nanostructures

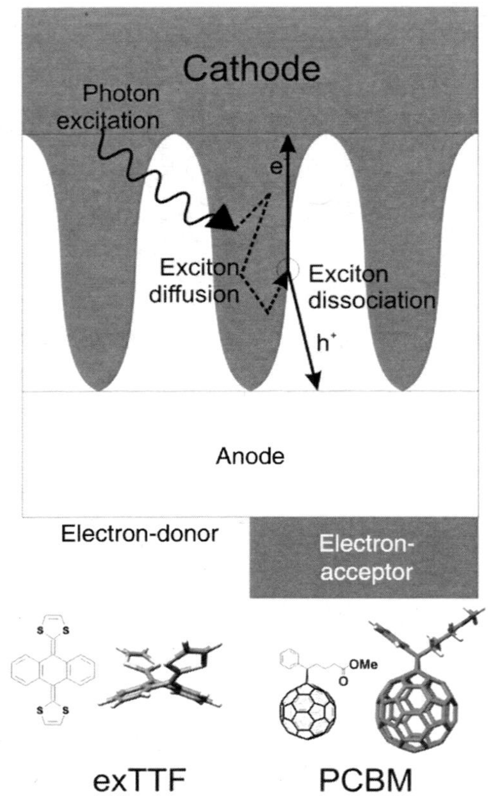

FIGURE 19.10. Schematic representation of the different steps in photoinduced charge separation in organic solar cells. Upon photon excitation, a tightly bound exciton is formed, which splits into free charge carriers after diffusing to a nearby donor/acceptor interface. Thus, maximizing the solar cell efficiency requires, us to maximize the area of the donor–acceptor interface, to decrease the size of pure donor and pure acceptor areas down to dimensions of the order of the exciton diffusion length to avoid radiative recombination, and to preserve electrical connection between the interfaces and the electrodes. The chemical structures of the electron-donor (exTTF) and electron-acceptor (PCBM) molecules used in this study are included. Reprinted with permission from reference [71]. Copyright 2007, American Chemical Society.

[81–89]. Thus, the inter-island distance reflects the inter-elbow distance, which is known to depend on preparation conditions and applied surface stress [90], but is generally in the range of 10–20 nm. If the herringbone secondary structure templated the growth of the donor and acceptor molecules on Au(111) in the same manner, it would result in a lateral superlattice with sizes of the order of the exciton diffusion length.

19.3.1 PCBM/Au(111)

Figure 19.12 shows STM images (taken at 170 K) of the Au(111) surface after depositing increasing amounts of PCBM molecules with the surface held at 300 K [91]. After the first molecules decorate the atomic steps of the surface (which implies a high room temperature diffusivity), the following molecules nucleate at the elbows of the herringbone reconstruction while simultaneously a number of finger-like zigzag structures appear on the surface (Figure 19.12a).

A closer look (see the inset in Figure 19.12c) reveals that these zigzag arrays appear exclusively on the fcc areas of the reconstruction and are actually double rows of PCBM molecules composed of shorter, straight line fragments, each one containing a small number of molecules separated around 10 Å. These short fragments do not run parallel to the fcc lines, but along a close-packed direction of the surface, thus forming an angle of 30° with the fcc lines. Further deposition (Figure 19.12b) causes the formation of parallel molecular chains hundreds of nanometers long and separated by only a few nanometers (~6 nm). These chains grow in length until they cover completely the fcc areas of the surface. Then, the growth proceeds along the lines joining the elbows of the reconstruction, giving rise to a highly organized 2D network of PCBM molecules resembling a nanosized "spider web" (Figure 19.12c).

For these coverages, only the fcc regions of the reconstruction are decorated with PCBM molecules. This behavior contrasts with that previously observed for pristine C_{60} on the same surface for room temperature deposition, which first nucleates at step edges and then produces compact hexagonal islands, disregarding the morphology of the Au(111) reconstruction [92]. The difference in site-sensitivity between C_{60} and PCBM implies an important role of the organic addend of the latter in the mechanism of site-selective adsorption and, thus, suggest an adsorption geometry in which this addend is in close contact to the surface.

Up to this stage the supramolecular ordering is the result of two combined effects: first, the particular interaction between the molecular tail and the surface reconstruction, leading to the impossibility for adsorbed PCBM molecules to sit on the dislocation lines, which results in an almost exclusive nucleation on the fcc areas of the surface; and second, the π–π interactions among the C_{60} cages, which causes the formation of double rows of molecules (further compact arrangement is impeded by the existence of the side tail and its attraction to the substrate surface). The reason for this preferential nucleation may be both steric—the fcc areas are wider than the hcp ones—and/or electronic—the charge density of the surface state electrons is different in the fcc regions, the hcp regions, and the dislocation lines of the reconstruction [93, 94]. Note that the corrugation of the Au(111) surface—that is, the height difference between the fcc areas and the dislocation lines—is only 0.2 Å, which discards a simple geometric effect.

When the density of deposited molecules exceeds that of the available fcc areas, the interactions between the organic addends of the fullerenes take over molecule–substrate interaction, removing the site-selectivity in the adsorption of

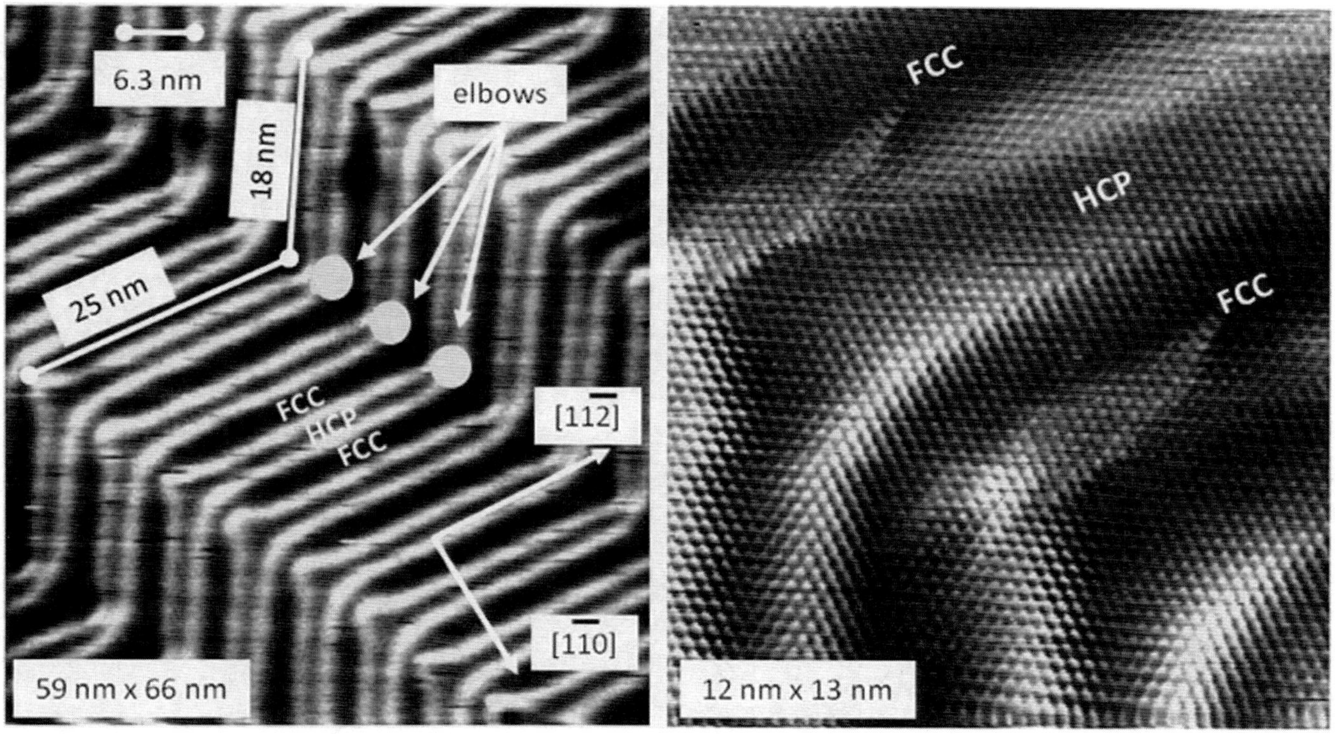

FIGURE 19.11. STM images of the Au(111) surface showing the structure of the herringbone reconstruction. Copyright 2011, Wiley-VCH.

PCBM and forcing the molecules to reorganize into a compact arrangement of double-row chains, equally spaced and running parallel to the close-packed directions of the Au(111) surface, with total disregard of the surface reconstruction (Figure 19.13a).

A high-resolution image of the compact double-row nanonetwork of PCBM is shown in Figure 19.13b. Again, the distance between molecules within a row is ~10 Å. In addition, the image reveals now some bright spots in the region separating adjacent double rows. In order to clarify the nature of the interaction between neighboring rows, we have performed DFT calculations of the monomer, several possible dimers, and different tetrameric structures for freestanding PCBM. Top and side views of the resulting symmetric, minimum-energy dimer structure compatible with a 2D geometry on a solid surface are shown in Figure 19.13c,d. Under this configuration, the bond between the two PCBM molecules is due mainly to the formation of two weak hydrogen bonds (C—H· · ·O) between the two tails, leading to an energy gain of 0.1 eV with respect to two isolated molecules. The separation between the centers of the C_{60} cages is 23.7 Å, to be compared with the experimental value of ~ 21 Å. The optimized conformation for a tetramer was also calculated. The final geometry (Figure 19.13e) is due to the formation of two additional hydrogen bonds between adjacent dimers. The comparison with the experimental data (see Figure 19.13b) is quite good. Hydrogen bonds such as those predicted by the calculations can only be formed if the hydrogen donor and acceptor groups at the PCBM tails face each other in the right geometry. This implies that, due to the large size of the C_{60} cages, the tails cannot be in contact with the surface any more (see Figure 19.13d). Note that this model naturally explains the extra features found in the STM images as arising from the organic tails that hold the rows together.

In summary, the vacuum deposition of the fullerene derivative PCBM on Au(111) leads to a coverage-dependent transition from substrate-controlled to (weak) hydrogen-bond-controlled self-assembly. At low coverages, due to the influence of the "side tail" of PCBM, the herringbone reconstruction of Au (111) acts as an efficient template that dictates the resulting structure: Starting with a preferential nucleation of PCBM at the elbows of the herringbone reconstruction the organization process continues with the formation of one-dimensional wires of PCBM molecules nucleated exclusively at the fcc areas of the reconstruction. Once the fcc areas are all occupied, new incoming molecules must sit on the energetically unfavorable dislocation lines. As the density of molecules increases, the molecular rows are now so close to each other that they must interact via the tails by hydrogen-bond formation between double rows, so that the PCBM tail does not touch the surface any longer, which modifies the adsorption geometry of the PCBM molecule, lifting the

FIGURE 19.12. STM images images of the Au(111) after depositing increasing amounts of PCBM. (**a**) 0.1 ML, (**b**) 0.3 ML, (**c**) 0.4 ML. The inset in (c) shows a close-up of one of the zigzag structures. Reprinted with permission from reference [91]. Copyright 2007, Wiley-VCH.

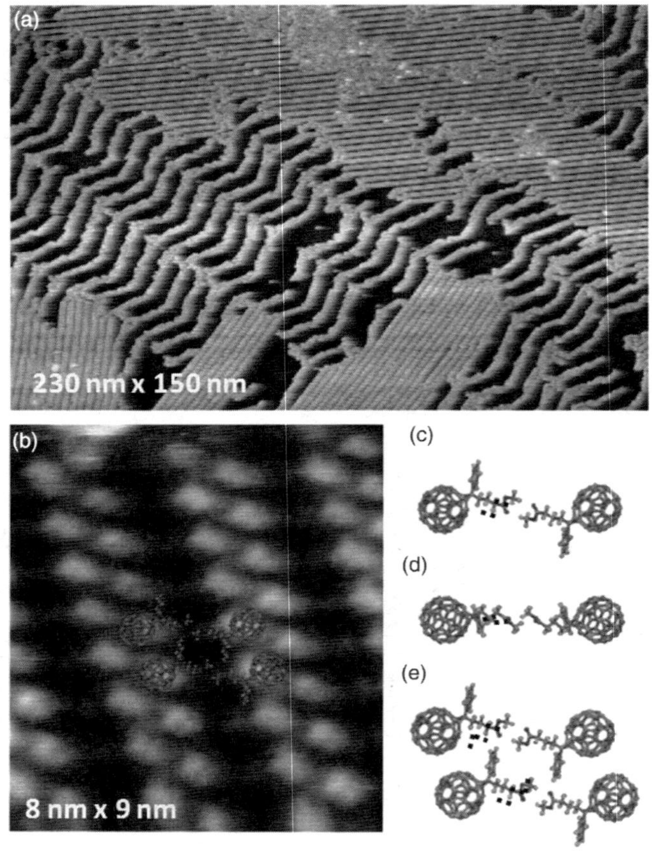

FIGURE 19.13. (**a**) Large-scale STM image of the Au(111) surface after depositing ~0.6 ML of PCBM, showing the coexistence of two different phases: the nanoscale spiderweb (created by the templating effect of the substrate surface) and the sets of parallel double rows connected by weak hydrogen. (**b**) A closer look of the parallel double rows phase. (**c**) Top view and (**d**) side views showing the optimized calculated structure for a PCBM dimer. (**e**) Optimized structure for a PCBM tetramer. The dotted lines mark the weak hydrogen bonds responsible for this conformation. Reprinted with permission from reference [91]. Copyright 2007, Wiley-VCH.

templating effect of the substrate reconstruction and giving rise to the formation of long, straight wires of double rows of PCBM.

19.3.2 Ex-TTF/Au(111)

The behavior of exTTF, when deposited on the clean Au(111) surface, is completely different. After depositing a small amount of exTTF at room temperature, no isolated molecule could be imaged with the STM, even at 140 K, possibly due to a high molecular diffusivity [95]. When increasing the coverage, however, stable islands start to form (Figure 19.14a). These islands are strongly elongated, with their long side parallel to the lines joining the elbows of the herringbone reconstruction. The minimum width of a stable island is around ~ 200 Å, and then increases with coverage up to ~ 700 Å, when the islands coalesce and the first monolayer is complete (Figure 19.14b).

Note that the gold reconstruction is still visible under the molecular layer, although the geometry is slightly different when compared to the clean gold surface. These first-layer islands appear composed of rows of molecules, making an angle of 10° (Figure 19.14e) with the reconstruction lines. A close-up look (Figure 19.14f) reveals the molecular structure of the islands. The unit cell of the exTTF monolayer (drawn in solid black) is rhombohedral, with sides 10.3 and 7.7 Å long, which form an angle of 65° between them.

The growth of exTTF can be continued beyond 1 ML. Second-layer islands grow in good registry with the first monolayer, and the islands are again elongated, aligned with the directions of the molecular rows of 1ML exTTF/Au(111) (Figure 19.14c). Molecules on the second and third monolayers also form well-ordered 1D-rows, but their lateral arrangement lacks the long-range order observed of the first monolayer (see Figure 19.14d). The ease with which second and even third layer exTTF molecules can be observed by STM implies a good electronic coupling and a strong bonding between consecutive layers.

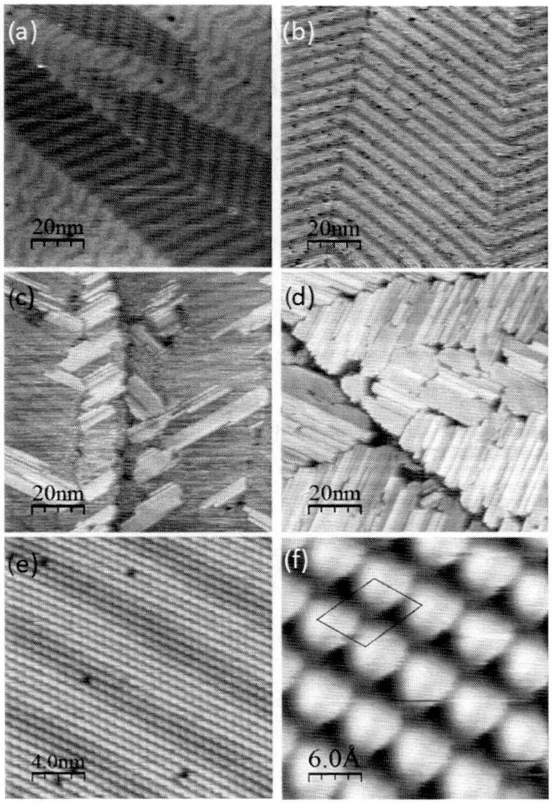

FIGURE 19.14. STM images of the Au surface after depositing (**a**) 0.5 ML (V_b = 0.38 V, I = 0.32 nA), (**b**) 1.0 ML (V_b = −1.24 V, I = 0.67 nA), (**c**) 1.2 ML (V_b = −0.88 V, I = 0.19 nA), (**d**) 2.1 ML (V_b = −1.47 V, I = 0.16 nA), of exTTF on Au(111) at room temperature. (**e, f**) Close-up STM images of the internal island structure (e) V_b = 1.24 V, I = 0.45 nA; (**f**) V_b = 1.24 V, I = 1.11 nA). The solid line in (f) marks the rhombohedral unit cell. Reprinted with permission from reference [95]. Copyright 2010, American Chemical Society.

To find out the adsorption conformation, DFT calculations were carried out. The results show that the molecule adopts a butterfly conformation, very similar to the one found in gas phase (Figure 19.15). The bond to the surface is dominated by the sulfur atoms; but due to geometry restrictions, the four sulfur atoms cannot be bonded to the surface simultaneously. Instead, only one of the two dithiole rings is close and almost parallel to the surface, while the other is almost perpendicular to it. The S atoms tend to be close to on-top positions, the average distance to the surface being around 3.0 Å. The calculated adsorption energy is 0.32 eV per exTTF molecule. The DFT calculations also predict a charge transfer of 0.3 e^- from the molecule to the gold surface, a result supported by the decrease in the work function while increasing the exTTF coverage. The predicted theoretical conformation is fully supported by photoemission (XPS) experiments [95].

As in similar systems [96, 97], the reconstruction of the Au(111) gold surface has a certain influence on the arrangement of the exTTF overlayer. On an unreconstructed hexagonal lattice, there are six symmetry-related possible orientations of a rhombohedral unit cell (Figure 19.16), which should give rise to six different molecular domains. All of them can be found in the exTTF layer in different regions of the sample surface, but in every case the orientation of the molecular domains is strongly determined by the direction of the underlying surface reconstruction. As can be seen in Figure 19.17, the overlayer unit cell changes its direction in accordance with the direction of the surface reconstruction, in such a way that the short side always makes an angle of ± 10° with the domain walls direction. Thus, the small structural differences between the different orientations produced by the anisotropic compression of the gold surface restricts the number of observed domains to only two (domains I and II in Figure 19.15b) for a given direction of the surface reconstruction.

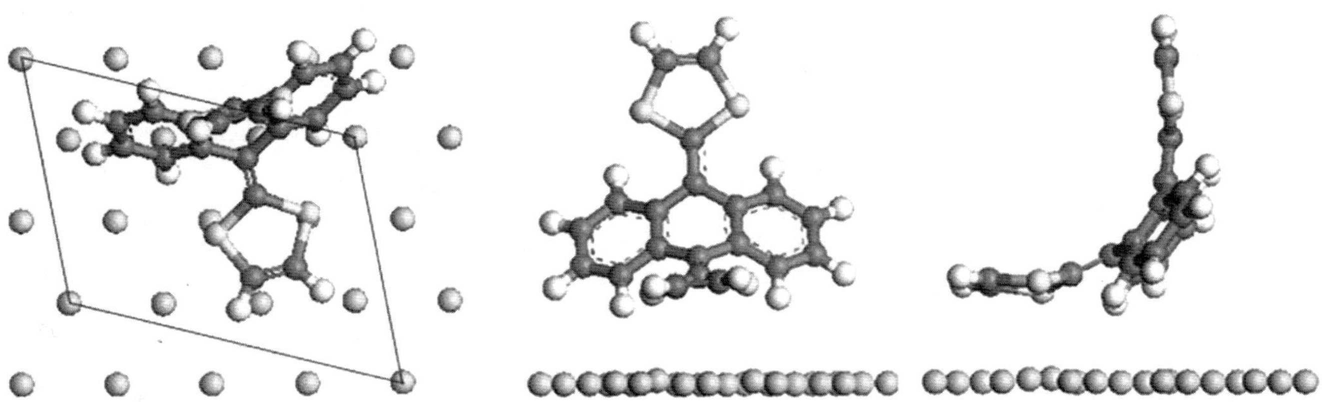

FIGURE 19.15. Top, front, and side views of the calculated minimum energy conformation for an exTTF molecule on the Au(111) unreconstructed surface. The solid line represents the unit cell of the complete monolayer. Reprinted with permission from reference [95]. Copyright 2010, American Chemical Society.

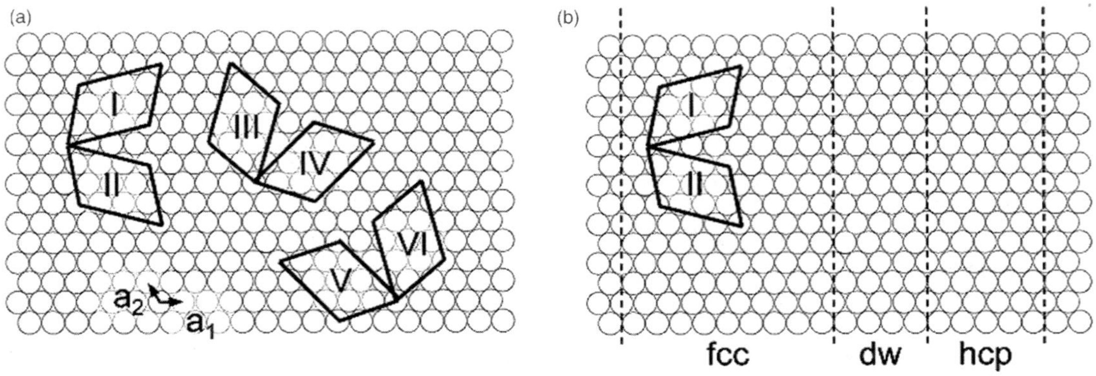

FIGURE 19.16. (**a**) The six possible orientations of the experimentally measured rhombohedral unit cell on an unreconstructed hexagonal surface. (**b**) Experimentally, however, for a given direction of the Au(111) surface reconstruction, only two domains are observed. If, for example, the domain walls run parallel to the dashed lines drawn in the figure, only domains I and II are present. Reprinted with permission from reference [95]. Copyright 2010, American Chemical Society.

Reciprocally, if the surface reconstruction has a clear effect on the arrangement of the exTTF molecules, the molecule–substrate interaction is large enough to cause alterations in the geometric structure of the herringbone reconstruction (Figure 19.18). Thus, the periodicity of the gold reconstruction below the exTTF overlayer is ~72.1 Å, instead of the 63.5 Å for the clean Au surface [98]. This means that now there is one extra atom in the surface layer every 25 atoms in the bulk, instead of every 22 atoms as in the clean surface. As a consequence, the hcp areas have vanished, there only being areas with fcc stacking and domain walls. The width of the domains has also increased, the reconstruction direction changing orientation every ~600 Å, instead of every ~ 250 Å as in the clean surface. This is thus an intermediate situation between those molecular systems that have little or no influence on the herringbone reconstruction and those where this is completely lifted.

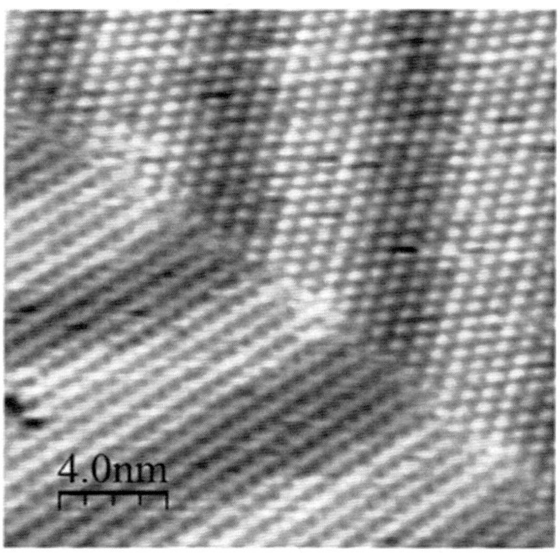

FIGURE 19.17. STM image of a domain wall between two molecular domains. Reprinted with permission from reference [95]. Copyright 2010, American Chemical Society.

In summary, the interaction of exTTF with the gold surface is dominated by the S–Au bonds. The molecule adopts a conformation very similar to the gas-phase conformation, with the two dithiole rings almost perpendicular to each other, and then only one in close contact with the gold surface. The molecule–substrate interaction is enough to cause distortions in the geometry of the gold herringbone reconstruction, probably due to the partial charge transfer (0.3 e^- per molecule) from exTTF to the substrate. Note that the modification of the secondary elbow structure of the herringbone reconstruction upon exTTF adsorption implies a breakdown of the surface templating assumption. Nevertheless, there is still a very important influence of the reconstruction on the exTTF stripe-island morphology: Instead of isotropic or hexagonal islands, as might be expected for a fcc(111) surface, it self-assembles into stripe islands in which the preferential growth direction is the one in which the elbows are aligned.

19.3.3 (PCBM + exTTF)/Au(111)

The influence of the herringbone reconstruction on the morphology of exTTF islands and the difference in bonding strength with Au(111) between PCBM and exTTF can be used to steer the nanoscale morphology of exTTF/PCBM blends into a lateral superlattice like the one described in Figure 19.10. The stronger interaction of exTTF with Au implies a dominating role for this molecule to form the striped structures, leaving only the space in between the exTTF stripes for the disordered PCBM close-packing. As shown in Figure 19.19, this expectation is fulfilled by the experiments. Figure 19.19a shows the result of adding, at room temperature, about 0.6 ML exTTF on top of the nanoscale spider-web PCBM structure depicted in Figure 19.12c

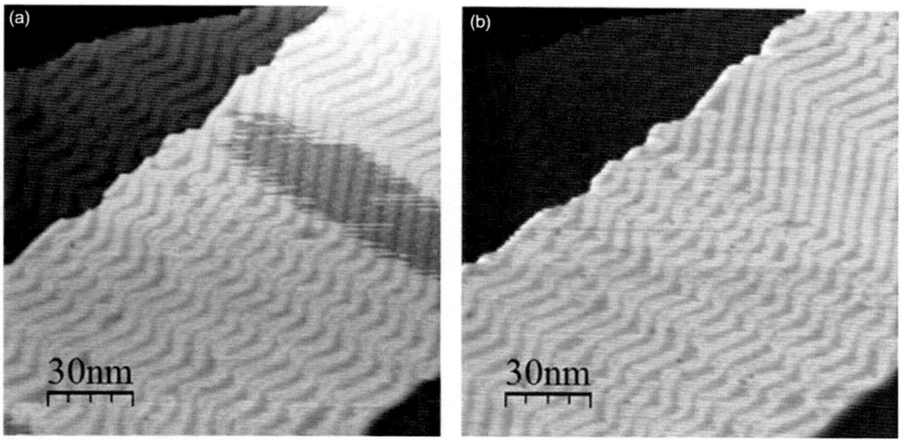

FIGURE 19.18. (**a, b**) Two different STM images of the same area of a partially covered gold surface. The exTTF island in panel (a) is not visible in the panel (b), but the changes in the surface reconstruction indicate the presence of a high concentration of exTTF molecules. (a) $V_b = -1.16$ V, $I = 0.49$ nA; b) $V_b = -0.82$ V, $I = 0.42$ nA. The image in (a) is reprinted with permission from reference [95]. Copyright 2010, American Chemical Society.

(coverage of about 0.5 ML). Two distinct areas can be distinguished in the large-scale STM images: (a) a well-ordered area with a molecular structure identical to that shown by exTTF and where the elbow-free herringbone reconstruction underneath can be recognized as a long-period corrugation of the molecular rows and (b) a disordered area reminiscent of the high-coverage PCBM phase observed in some areas in Figure 19.13a. The typical width of each area is about 20 nm. The ordered and disordered areas are assigned to (a) pure exTTF molecules with a local thickness in excess of 1 ML and (b) disordered PCBM molecules, respectively. Changing the order of the deposition—that is, depositing about 0.5 ML exTTF first and then a submonolayer amount of PCBM—does not substantially alter this result.

This peculiar nanoscale phase segregation can be understood as follows. First, exTTF and PCBM do not show a tendency to mix, in good agreement with previous results observed in solution [99]. Second, PCBM molecules are highly mobile at room temperature (RT), which is the temperature at which the PCBM/Au(111) substrate is held during deposition of exTTF. Since the stronger interaction of exTTF leads to self-assembled stripe islands that can be observed at RT, exTTF molecules start nucleating and forming the stripe islands, while PCBM molecules behave like a 2D gas on the Au(111) surface. As exTTF coverage increases, the space left for the PCBM 2D molecular gas decreases until the concentration of PCBM molecules is so large that it cannot behave as a gas any longer; that is, the density of PCBM molecules roughly equals the reciprocal of the molecular volume. The only area in which such a disordered PCBM 2D solid could stay is the area between adjacent exTTF stripe islands, which must then have a stripe morphology as well. Thus the joint effect of molecule–substrate interactions (leading to striped morphology of exTTF islands and to a high mobility of PCBM molecules at RT) and molecule–molecule interactions (leading to PCBM/exTTF segregation and to strongly bound exTTF islands that are stable at room temperature) leads to the formation of the lateral superlattice shown in Figure 19.19.

The presence of second-layer areas exclusively on the exTTF islands is a signal of the strong interaction among the exTTF molecules. As a consequence, the growth morphology and phase separation remain almost intact upon increasing the coverage. Figure 19.19b shows the result of depositing an additional 0.4 MLs of exTTF on the surface of Figure 19.19a. Although the width of the exTTF islands has increased slightly (which has caused some of the PCBM molecules to jump to the second level), the most notorious change is their growth in height, now being almost completely two layers high with a significant portion of third layer. Although with STM alone it is impossible to ascertain how far the template effect of the substrate surface will dominate the mixed film structure, these results indicate that phase separation is an intrinsic property of the blend morphology.

As discussed above, the characteristic width of the exTTF and PCBM stripes is about 20 nm, which compares well with typical exciton diffusion lengths. Both exTTF and PCBM stripes extend laterally over micron distances, and the morphology is statistically the same in all explored regions of the Au surface. These morphological features are ideal for the construction of highly efficient solar cells. The applicability of the method to other metallic substrates needs to be verified. It will rely on the possibility of employing nanostructured or reconstructed surfaces with similar ordered dislocation patterns, such Ag/Cu(111), Cu/Ru(0001), or Pt/Cu(111). We expect that similar and even more refined approaches will be exploited in the near future, further boosting the search for the optimum morphology of highly efficient organic solar cells.

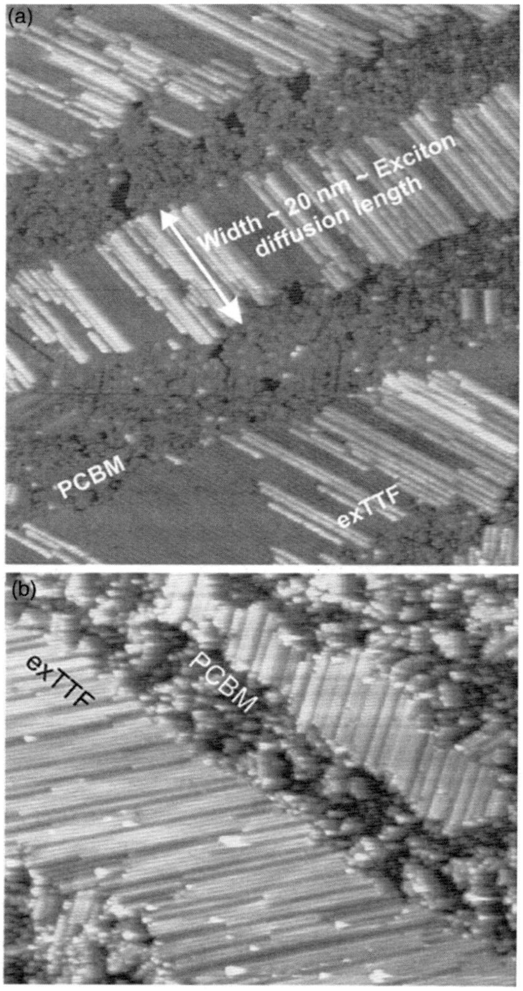

FIGURE 19.19. Lateral nanoscale organic donor/acceptor superlattice. (**a**) Nanoscale segregation of electron-donor/electron-acceptor molecules on Au(111) (118 nm × 132 nm). Since both PCBM and exTTF molecules are imaged with similar heights, a semitransparent filter has been superimposed on the disordered PCBM areas to enhance visibility. The width of the exTTF stripes is about 20 nm, of the same order as typical exciton diffusion lengths, and there exists a large acceptor/donor interface, as required for optimum solar-cell performance. (**b**) (118 nm × 115 nm) The selective adsorption of excess exTTF molecules on areas previously covered by exTTF molecules implies that this morphology can be extended beyond the first monolayer. Reprinted with permission from reference [71]. Copyright 2007, American Chemical Society. See color insert.

19.4 ORGANIC NANOCRYSTALS: SubPc/Cu(111)

Controlling the formation of nanosized organic particles of well-defined size and shape is one of the challenges facing modern chemistry [100–102]. As in the case of their inorganic counterparts [103–105], the optical properties of such organic nanostructures may be very different from the properties found for the same materials in bulk [100–102]. For example, organic nanoparticles show size-dependent absorption and fluorescence bands [106, 107] or single-photon emission [108]. The origin of size effects in organic nanoparticles, however, is very different from their inorganic counterparts: In metals, the size effect is associated with the collective behavior of the electrons within the nanobulk; in organic nanoparticles, electrons are confined in each molecule, and their mobility is limited and do not expand over large domains.

A substantial difference in the optical and electronic properties of organic nanomaterials from the properties of nanomaterials based on metals and inorganic substances is associated with the presence of weak intermolecular interactions such as the van der Waals forces and π–π conjugation in the former. This has delayed the development of studies of organic nanoparticles, compared to metal or semiconductor nanoparticles. Reports on the size effects in such organic nanoparticles are relatively recent, due to the low melting temperatures and lesser thermal stability of organic nanostructures as compared with those of inorganic compounds, which limits the methods for their synthesis and applications. In addition, a detailed understanding of the size effects is hindered by the difficulty in the synthesis of organic nanocrystals—that is, organic nanoparticles with an ordered molecular arrangement.

A possibility that remains mostly unexplored is the synthesis of such nanocrystals on solid surfaces. In the same way in which crystalline inorganic nanodots can be epitaxially grown on suitable substrates under conditions in which 3D Volmer–Weber growth takes place [109], an organic system could in principle be devised such that the growth of crystalline 3D islands sets in before the completion of the first monolayer. In practice, however, for organic adsorbates deposited on inorganic substrates, intermolecular interactions are much weaker than molecule–substrate interactions [10, 110], thus promoting a layer-by-layer growth mode, and preventing the fabrication of isolated 3D nanocrystals.

Among the large variety of metallo-macrocycles which can allow us to grow organic nanomaterials, phthalocyanines play a quite important role. Like porphyrins, phthalocyanines (Pcs) have attracted special attention because, due to their interesting optical and physicochemical properties, they have many applications in the areas of gas-sensing devices, photovoltaic applications, light-emitting diodes, organic field effect transistors, pigments and dyes, etc. [111]. Phthalocyanines are planar aromatic macrocycles composed of four isoindole units linked together through nitrogen atoms (see the inset in Figure 19.20). Their 42 π electrons are distributed over 32 carbon and 8 nitrogen atoms; but the electronic delocalization mainly takes place on the inner ring which is made of 16 atoms and 18 π electrons; the outer benzene rings maintain their electronic structure. Moreover, metal-phthalocyanines offer a huge variety of electronic properties by simple substitution of the central metal ion as well as the modification of its periphery by suitable

FIGURE 19.20. STM images of FePc on Au(111). Copyright 2011, Wiley-VCH.

functional/non-functional groups, leading to new complex and less-studied molecules.

On most metal surfaces, phthalocyanines' ultrathin layers are reported to adsorb in a planar configuration, forming large areas of nearly defect-free square crystalline structures. The weak adsorbate–substrate interaction allows a relatively high molecular diffusion, and the final square symmetry of the self-assembled structure is determined mostly by intermolecular interactions, consisting of weak van der Waals forces and steric repulsions, between cross-shaped molecules.

As an example, Figure 19.20 shows two STM images taken after depositing ~1 ML of FePc on Au(111). The molecules form an almost perfectly ordered layer with quasi-quadratic unit cell. The FePc molecules are recognized as a four-lobed cross structure with a protrusion at the center, which is consistent with its chemical structure and with enhanced tunneling through the half-filled $d_z{}^2$ orbital of Fe. The size and shape of the individual features indicates that the FePc molecules are adsorbed flat on the surface [112, 113].

The basic structure of the phthalocyanine macrocycle described before may be modified, giving rise to phthalocyanine derivatives such as subphthalocyanines (SubPc, from now on) [114–116], with only three isoindole units coordinated to a central boron atom (see Figure 19.21a). As a

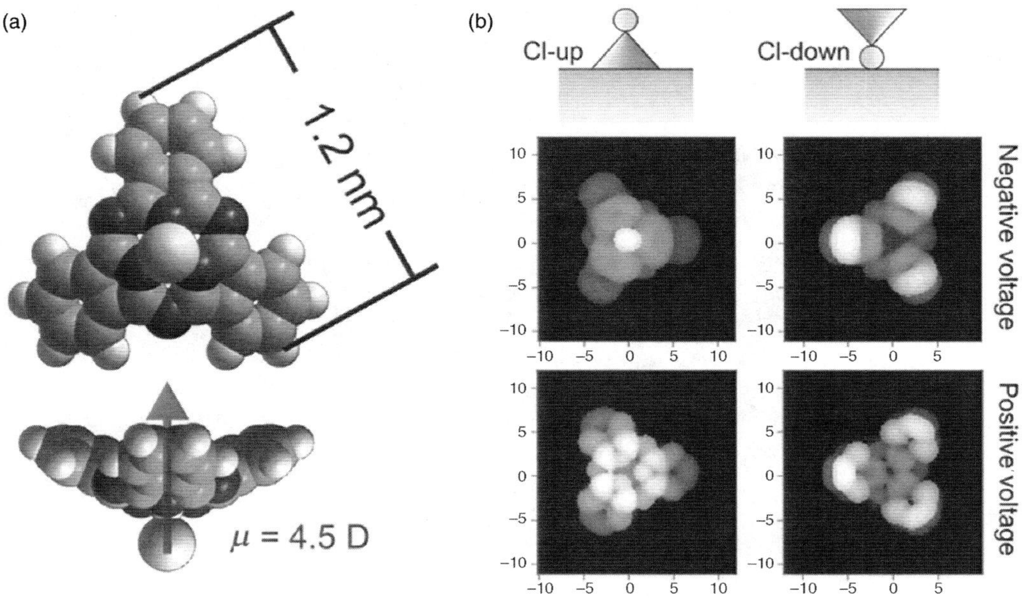

FIGURE 19.21. (**a**) CPK-model (top and side view) of a SubPc molecule, showing the nonplanar, conical shape. (**b**) Simulated STM images for Cl-up and Cl-down adsorption geometries for positive voltages and negative voltages [119]. Copyright 2011, Royal Society of Chemistry.

result of this modification, SubPc displays a cone-shaped structure, which does not prevent them from being aromatic molecules. As in the case of the phthalocyanines, the aromatic nature of the macrocycle is essentially located in the inner ring, where 14 π electrons are delocalized over 6 carbon and 6 nitrogen atoms, with the peripheral benzene rings mostly retaining their electronic structure. The axial ligand covalently attached to the boron atom (in this case a Cl atom) confers a strong dipolar character (μ ~4.5 D) to the overall molecule mostly oriented along the C_3 symmetry axis. Subphthalocyanines also show very attractive photophysical and electrochemical properties and have found applications in nonlinear optics, OLEDs, and other photovoltaic devices and multicomponent donor–acceptor systems [111, 117, 118].

Here we show that, upon deposition of cone-shaped, high-dipole-moment subphthalocyanines molecules on Cu(111), 3D growth is promoted versus layer-by-layer growth, and thus isolated triangular nanocrystallites up to 3 ML high appear on the surface before the completion of the first monolayer [119].

Figure 19.22 a,b shows representative STM images corresponding to a coverage of ~ 0.2 ML deposited on the copper surface at room temperature and recorded with the STM held at 150 K. Two different shapes can be associated with individual SubPc molecules: a bright circular spot and a darker trefoil shape. STM images of SubPc molecules on a variety of semiconducting [120] and metallic [121–123] surfaces have been previously reported, and the different shapes have been attributed to adsorption configurations in which the Cl atom is bonded to the surface (Cl-down) or the Cl atom is pointing in the opposite direction (Cl-up). By comparing with STM simulated theoretical images (Figure 19.21), we attribute the circular shape to the Cl-up conformation and the darker trefoil shape to the Cl-down geometry.

Cl-up molecules self-assemble by forming small patches with a honeycomb structure (Figure 19.22a,b), and the domains grow larger with postdeposition annealing treatments to 310 K (Figure 19.22c). This honeycomb network is aligned with the close-packed direction of the Cu(111) substrate, making a 9 × 9 unit cell superstructure. In this structure the closest SubPc-SubPc distance is about 1.3 nm.

In addition to the Cl-up and Cl-down structures described above, triangular islands with yet another kind of ordered structure can be found on the surface (Figure 19.23a). These islands are present on the surface at any coverage in excess of 0.2 ML, but they are not very common (about 1–2 islands in 100 nm^2). The islands are found with two different orientations with respect to the underlying close-packed directions of the Cu(111) surface: The angle between the island edges and the unit cell vectors of the substrate is ±15°, for each of these two orientational enantiomeric domains.

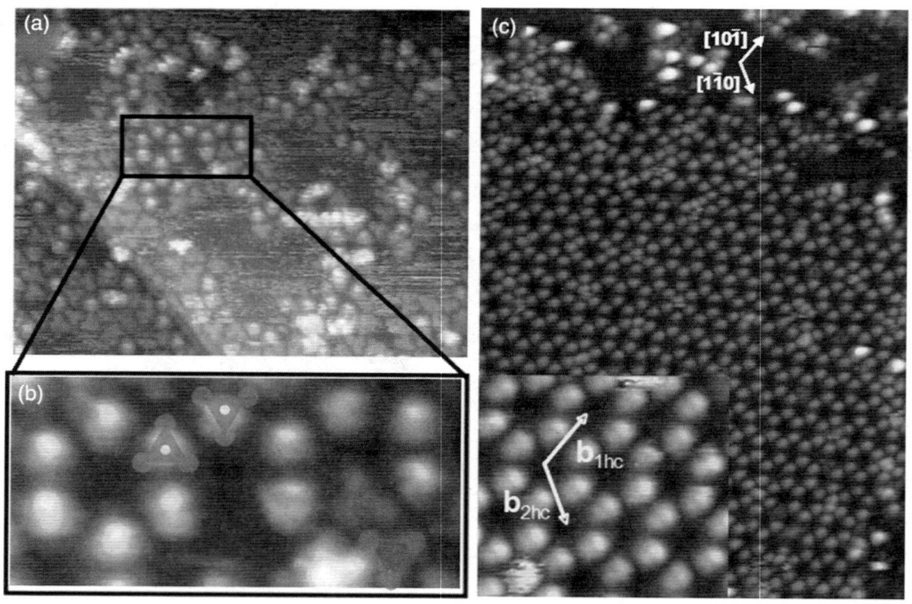

FIGURE 19.22. (**a**) STM image of 0.2 ML SubPc/Cu(111) (33.4 nm × 26.5 nm; V_b = −2,1V; I = −0,38 nA). (**b**) Zoom-in (10.8 nm × 5.5 nm) showing the two types of molecular-size features associated with the adsorption of SubPc: a bright protrusion and a trefoil shape. We interpret these features as corresponding to the coexistence of Cl-up and Cl-down adsorption geometries. (**c**) STM image (42.4 nm × 60 nm; V_b = −2.5 V; I = −0.43 nA) of a large honeycomb domain of Cl-up molecules obtained by heating the sample up to 310 K. The unit cell vectors (arrows) are parallel to the substrate's close-packed directions. The box encloses some Cl-down molecules. Reprinted with permission from reference [119]. Copyright 2011, Royal Society of Chemistry.

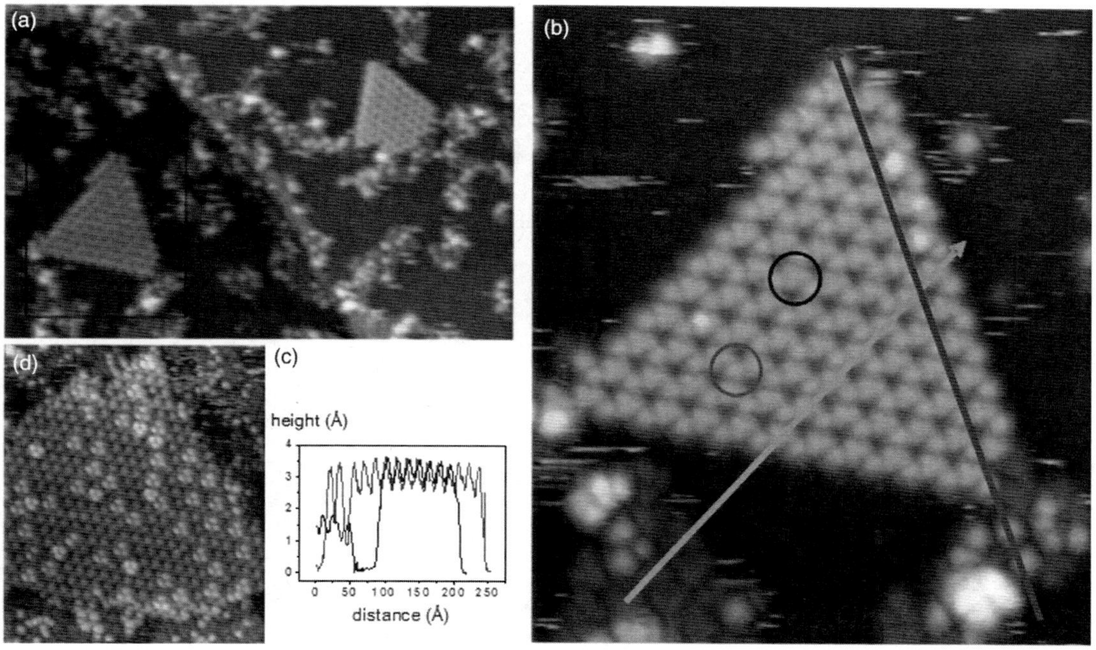

FIGURE 19.23. (**a**) STM image of 0.4 ML SubPc/Cu(111) (62.4 nm × 41.6 nm; $V_b = -2.9$ V; $I = -0.8$ nA). Two triangular islands can be found. (**b**) High-resolution STM images (22.3 nm × 26.1 nm) show two different molecular features, bright protrusions (light circle) and trefoil shapes (dark circle). The bright protrusions are identical in shape and size to the Cl-up molecules identified in Figure 19.22, but the trefoil features are 0.2 nm higher, as shown by the line profile in (**c**). (**d**) Even thicker islands can be found upon further deposition (33.4 nm × 39.2 nm). Reprinted with permission from reference [119]. Copyright 2011, Royal Society of Chemistry.

High-resolution STM images (see Figure 19.23b) of the triangular islands reveal that the molecular arrangement is made out of two different SubPc "species," which are imaged as bright protrusions (light circle) and trefoils (dark circle). The sublattice of bright protrusions and the sublattice of trefoil features are both hexagonal (and not honeycomb-like), aligned with the island edges and with a lattice parameter of 1.6 nm—that is, about 30% smaller than the unit cell of the honeycomb structure, but about 20% larger than the closest distance between molecules in the honeycomb. The island edges only contain one SubPc species: the one imaged as bright protrusions in the STM images (Figure 19.23b).

The height of the bright protrusions is the same as the height of the features previously identified as Cl-up molecules; in contrast, the height of the trefoil features in the triangular islands is about 0.2 nm higher than the Cl-down molecules in the first layer (see Figure 19.23c), even though they share a similar trefoil appearance in the STM images. This is consistent with (i) the bright protrusions corresponding to first-layer Cl-up molecules with a hexagonal arrangement and (ii) the trefoil features corresponding to second-layer Cl-down molecules sitting on the three-fold hollow sites of the first-layer Cl-up arrangement. As a consequence, the orientations of the dipole moments in the first and second layer are opposite to each other (as expected in order to maximize dipole–dipole interaction energy); that is, they are antiferroelectrically stacked. A similar antiferroelectric stacking has been found in 2-ML-thick films of the related ClAlPc and ClTiPc [124–126].

Further deposition of SubPc molecules on the sample containing nanocrystallites leads to a twofold effect. On the one hand, the disordered areas surrounding the triangular islands become more and more populated until there exist a 1-ML-thick connective tissue linking all the nanocrystallites. On the other hand, new and even brighter features can be observed on top of the bilayer islands (see Figure 19.23d). Such protrusions appear clustered in trimers; and they are sitting in the threefold hollow sites of the second layer, preferently those that do not have a first-layer Cl-up molecule underneath, in a fcc-like stacking; Such hcp stacking explains the preferential orientation of the trimers in the same direction as the island. Only a small fraction of trimers are oriented in the opposite direction, and in this case the molecules follow an hcp-like stacking, that is, the third-layer molecules are sitting on top of the threefold hollow sites with a first-layer Cl-up molecule underneath. The presence of another molecule directly underneath the third-layer molecule might explain the brighter appearance of inverted third-layer trimers versus normally oriented third-layer trimers.

We have thus identified a system for which the growth of 3D nanocrystallites sets in before completion of the first monolayer. Note, however, that the honeycomb structure of the 1-ML-thick Cl-up islands is different from the hexagonal arrangement of the Cl-up first layer in the nanocrystallites, indicating that the formation mechanism of the bilayer islands is not the adsorption of Cl-down SubPc molecules on previously formed 1-ML-thick Cl-up honeycomb structure. To seek alternative mechanisms for nanocrystallite formation, we need further information about their structure.

The orientation of the trefoil-shaped Cl-down molecules on the second layer around their C_3 axis with respect to the substrate lattice can be determined from our STM images. On the other hand, these images do not offer any clue as to the orientation of the underlying first-layer Cl-up molecules. In order to propose a structural model for the bilayer islands, we have performed theoretical calculations of a four-molecule cluster in the gas phase (molecular mechanics calculations with the MM+ force field from the HyperChem 7.0 package software). In this cluster we have placed three Cl-down molecules forming an equilateral triangle of 1.58 nm: Each of the Cl-down molecules is aligned with the direction of the supramolecular triangle, as observed for second-layer Cl-down molecules in our STM images (see Figure 19.24a).

A Cl-up molecule (representing one of the molecules in the first layer of the triangular islands) is placed along the axis defined by the triangle's baricenter and the dipole moment of the SubPc molecules. In the calculation the distance (z) between (a) the Cl atom of the Cl-up molecule and (b) the plane determined by the topmost hydrogen atoms of the second-layer Cl-down molecules is changed from 0 nm to 0.3 nm. We have calculated the bonding energy of the system as a function of the molecular orientation of the first layer with respect to the second (as determined by the angle θ defined in Figure 19.24a) for different interlayer distances z, and the resulting energy landscapes are plotted in Figure 19.24b.

Van der Waals (vdW) and π–π interactions as well as dipole–dipole interactions between both layers are the main bonding interactions expected to play a role in the stabilization of the nanocrystallites. The trend to higher bonding energies for decreasing interlayer distance in the range 0.3 nm > z > 0.05 nm can be ascribed to the strengthening of dipole–dipole interactions at short distances. Indeed, at a distance of 0.05 nm, the binding energy for the optimal angle configuration is about 16 kcal/mol (0.7 eV) larger than for a distance of 0.3 nm. A simple back-of-the-envelope calculation of the interaction energy between four dipole moments with the configuration of Figure 19.24a at the experimental distance yields a result of about 0.42 eV, even without taking into account polarization of the π-electron cloud, which can only enhance the dipole–dipole interaction. Such energy range is similar to that of multiple hydrogen-bonded networks on surfaces previously reported [13, 14, 16, 127, 128] that are stable at temperatures close to room temperature.

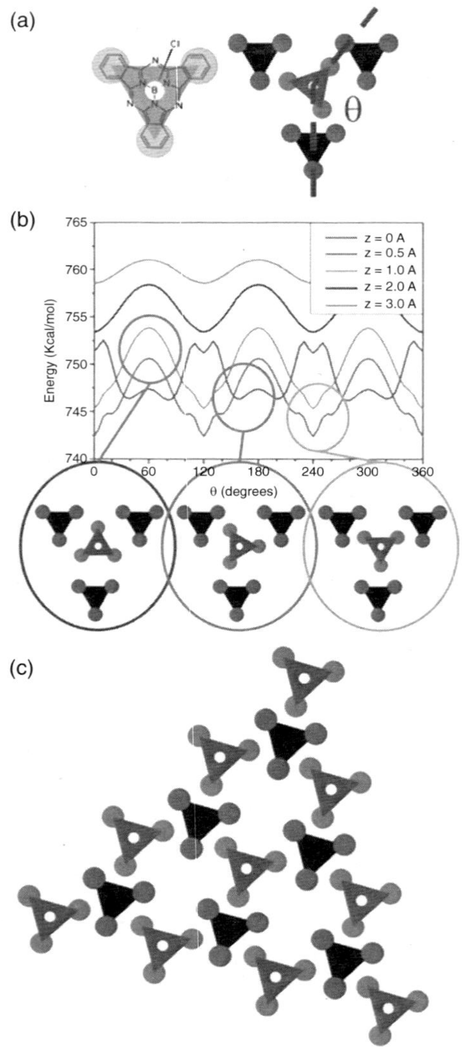

FIGURE 19.24. (**a**) Gas-phase model cluster for molecular mechanics calculations. In this schematic representation, light-gray triangles with a central white dot represents Cl-up molecules, whereas black triangles represent Cl-down molecules, with the circles representing the peripheral phenyl rings. The configuration of the second-layer Cl-down molecules is kept fixed to the experimentally observed configuration, whereas both the interlayer distance and the molecular orientation of the first-layer Cl-up molecule is changed systematically in order to obtain the energy landscape. (**b**) Angular dependence of the cluster's bonding energy for different interlayer distance. All the curves show the C_{3v} symmetry of the original cluster. The configurations of the maxima and minima in the curves are shown as insets. (**c**) Structural model for the triangular nanocrystallites based on experimental STM images and the cluster calculations described in (a) and (**b**). Reprinted with permission from reference [119]. Copyright 2011, Royal Society of Chemistry.

However, since the molecular dipoles are parallel to the B–Cl bond (i.e., out of plane), the dipole–dipole interaction is not expected to change significantly with rotations around the C_3 molecular axis. The molecular orientation in such a model cluster must rather be determined by the vdW and π–π interactions. This idea is confirmed by our calculations in Figure 19.24b: For every interlayer distance, except the smallest one, the minimum energy configuration is such that the first-layer Cl-up molecule shares the same orientation with the second-layer Cl-down molecules, thereby maximizing π–π interactions between the inner rings of the Cl-down and the outer rings of the three Cl-up molecules. Moreover, $\theta = 60^\circ$ is always a local energy maximum due to the lack of π–π interactions. Finally, for $z = 0$ nm, the inner rings of the Cl-down and the outer rings of the three Cl-up molecules get close enough for steric repulsion to take over. In this case the angle-averaged bonding energy increases, and the minimum energy configuration is characterized by a rotation to $\theta \approx 90^\circ$.

From our cluster calculations we propose the structural model depicted in Figure 19.24c for the triangular nanocrystallites. The second-layer (first-layer) molecules are depicted in darker (lighter) colors. The orientation of the second-layer molecules is the one obtained from our experimental STM images. This model explains the triangular shape of the nanocrystallites. Since both first-layer and second-layer molecules show hexagonal arrangements, the reason for the crystallites to be triangular instead of hexagonal is not immediately clear. However, close inspection of the model shows that the structure of a hypothetical step-edge at 60° with respect to the observed would be less stable than the observed one. In such a step-edge, first-layer molecules would be interacting with only one instead of two second-layer neighbors, thus rendering its overall configuration energetically unfavorable.

In summary, our results show that the strong dipolar and π–π interactions between adsorbed SubPc molecules, together with the coexistence of two adsorption geometries, Cl-up and Cl-down, on the Cu(111) surface direct the growth of nanocrystallites with well-defined shapes and sizes even before the first monolayer has been completed. On the basis of our results, we propose that the growth of nanocrystallites on solid surfaces versus single-height monolayers can be promoted by using nonplanar molecules provided with strong dipoles and whose adsorption geometry forces the dipoles to be directed out of the surface plane. This strategy is thus opposite to the one followed in most of the previous organic growth studies on solid surfaces, in which the molecular species rest flat on the surface, and their possible dipole moment is also parallel to the surface. As a result of the strong out-of-plane dipole–dipole interaction, such islands will tend to show an antiferroelectric stacking. We expect that a careful control of the 3D shape of the molecular adsorbates will lead to new ways to build 3D molecular nanostructures at solid surfaces, with tunable optical properties and, thus, a strong potential for optoelectronic device applications.

19.5 CONCLUSIONS AND OUTLOOK

In this review we have presented a number of case studies, selected among our recent work, on the self-assembly of organic molecules on solid surfaces to fabricate new materials with new structures at the nanoscale. The lesson to be learned from all these investigations is that, with very few exceptional situations, the surface is an active part of such arrangements, and its role must not be neglected in order to understand and design molecular nanoarchitectures on solid surfaces. Even when the molecules are weakly physisorbed on the surface, adsorption geometries, sites, and orientations are in most cases determined by the interaction between the molecular adsorbates and the solid substrate, thereby restricting considerably the freedom of organic molecules to self-assemble exclusively as they would if they were being arranged in solution. The role of the surface, of course, becomes more and more dominant as we move from weakly interacting adsorbates to strongly chemisorbed organic molecules, especially those containing heteroatoms such as oxygen, nitrogen, sulfur, and so on.

But besides the role of selecting adsorption geometries, solid surfaces might play an even more active role on the assembly of organic molecules. For example, inhomogeneous surfaces, such as surfaces with ordered arrays of dislocation networks, steps, or patches with different chemical composition, might act as patterns to guide the growth of molecular nanostructures whose morphology is not determined by intermolecular interactions, but by adsorption-site selectivity arising from the inhomogeneity of the molecule–substrate interaction.

Moreover, in those situations in which molecule–substrate interactions are very strong, even the substrate atoms may move out of the lattice positions in response to the new bonding configuration. Such adsorbate-induced surface reconstructions modify locally the substrate in the vicinity of the adsorbed molecules and are capable of mediating new interactions between the adsorbates leading to completely new self-assembled patterns that cannot be understood if the role of the substrate is not properly taken into account.

In summary, a change of point of view seems to be necessary: One should not think of the self-assembly of organic molecules *on* a solid surface, but rather one should think of the *co*assembly of organic molecules *and* a solid surface. This kind of Copernican revolution, according to which the surface is not a spectator but an actor in the assembly, seems in order if we are to satisfy the high scientific and technological expectations that are currently focused on the field of molecular nanoscience.

ACKNOWLEDGMENTS

Financial support from the Ministerio de Ciencia e Innovación (FIS2010-18847, MAT2009-13488 and Consolider-Ingenio en Nanociencia Molecular, ref. CSD2007–00010), Comunidad de Madrid (grant S2009/MAT-1726), and EU (SMALL PITN-GA-2009-23884) is gratefully acknowledged. R.O. thanks the Spanish Ministry for salary support through the "Ramón & Cajal" program.

REFERENCES

1. Lehn, J. M. (1995). *Supramolecular Chemistry: Concepts and Perspectives*. VCH–Wiley, Weinheim.
2. Cotterill, R. (2002). *Biophysics: An Introduction*. Chichester, West Sussex, England.
3. Eigler, D. M., Schweizer, E. K. (1990). Positioning single atoms with a scanning tunneling microscope. *Nature*, **344**, 524–526.
4. Otero, R., Rosei, F., Besenbacher, F. (2006). Scanning tunneling microscopy manipulation of complex organic molecules on solid surfaces. In: *Annual Review of Physical Chemistry*, pp. 497–525.
5. Bartels, L., Meyer, G., Rieder, K. H. (1997). Basic steps of lateral manipulation of single atoms and diatomic clusters with a scanning tunneling microscope tip. *Physical Review Letters*, **79**, 697–700.
6. Ishii, H., Sugiyama, K., Ito, E., Seki, K. (1999). Energy level alignment and interfacial electronic structures at organic metal and organic organic interfaces. *Advanced Materials*, **11**, 605–625.
7. Chua, L. L., Zaumseil, J., Chang, J. F., Ou, E. C. W., Ho, P. K. H., Sirringhaus, H., Friend, R. H. (2005). General observation of *n*-type field-effect behaviour in organic semiconductors. *Nature*, **434**, 194–199.
8. Dimitrakopoulos, C. D., Malenfant, P. R. L. (2002). Organic thin film transistors for large area electronics. *Advanced Materials*, **14**, 99.
9. Tseng, T. C., Urban, C., Wang, Y., Otero, R., Tait, S. L., Alcami, M., Ecija, D., Trelka, M., Gallego, J. M., Lin, N., Konuma, M., Starke, U., Nefedov, A., Langner, A., Woll, C., Herranz, M. A., Martin, F., Martin, N., Kern, K., Miranda, R. (2010). Charge-transfer-induced structural rearrangements at both sides of organic/metal interfaces. *Nature Chemistry*, **2**, 374–379.
10. Barth, J. V. (2007). Molecular architectonic on metal surfaces. *Annual Review of Physical Chemistry*, **58**, 375–407.
11. De Feyter, S., De Schryver, F. C. (2003). Two-dimensional supramolecular self-assembly probed by scanning tunneling microscopy. *Chemical Society Reviews*, **32**, 139–150.
12. De Feyter, S., Grim, P. C. M., Rucker, M., Vanoppen, P., Meiners, C., Sieffert, M., Valiyaveettil, S., Mullen, K., De Schryver, F. C. (1998). Expression of chirality by achiral coadsorbed molecules in chiral monolayers observed by STM. *Angewandte Chemie International Edition*, **37**, 1223–1226.
13. Otero, R., Schock, M., Molina, L. M., Laegsgaard, E., Stensgaard, I., Hammer, B., Besenbacher, F. (2005). Guanine quartet networks stabilized by cooperative hydrogen bonds. *Angewandte Chemie International Edition*, **44**, 2270–2275.
14. Otero, R., Lukas, M., Kelly, R. E. A., Xu, W., Laegsgaard, E., Stensgaard, I., Kantorovich, L. N., Besenbacher, F. (2008). Elementary structural motifs in a random network of cytosine adsorbed on a gold(111) surface. *Science*, **319**, 312–315.
15. Barth, J. V., Weckesser, J., Cai, C. Z., Gunter, P., Burgi, L., Jeandupeux, O., Kern, K. (2000). Building supramolecular nanostructures at surfaces by hydrogen bonding. *Angewandte Chemie International Edition*, **39**, 1230–1234.
16. Theobald, J. A., Oxtoby, N. S., Phillips, M. A., Champness, N. R., Beton, P. H. (2003). Controlling molecular deposition and layer structure with supramolecular surface assemblies. *Nature*, **424**, 1029–1031.
17. Kern, K., Niehus, H., Schatz, A., Zeppenfeld, P., George, J., Comsa, G. (1991). Long-range spatial self-organization in the adsorbate-induced restructuring of surfaces—Cu(110)–(2x1)O. *Physical Review Letters*, **67**, 855–858.
18. Leibsle, F. M., Flipse, C. F. J., Robinson, A. W. (1993). Structure of the Cu(100)-c(2x2)N surface—A scanning-tunneling-microscopy study. *Physical Review B*, **47**, 15865–15868.
19. Barth, J. V., Brune, H., Ertl, G., Behm, R. J. (1990). Scanning tunneling microscopy observations on the reconstructed Au(111) surface—Atomic-structure, long-range superstructure, rotational domains, and surface-defects. *Physical Review B*, **42**, 9307–9318.
20. Brune, H., Giovannini, M., Bromann, K., Kern, K. (1998). Self-organized growth of nanostructure arrays on strain-relief patterns. *Nature*, **394**, 451–453.
21. Otero, R., Naitoh, Y., Rosei, F., Jiang, P., Thostrup, P., Gourdon, A., Laegsgaard, E., Stensgaard, I., Joachim, C., Besenbacher, F. (2004). One-dimensional assembly and selective orientation of lander molecules on an O–Cu template. *Angewandte Chemie International Edition*, **43**, 2092–2095.
22. Ecija, D., Trelka, M., Urban, C., de Mendoza, P., Echavarren, A., Otero, R., Gallego, J. M., Miranda, R. (2008). Templated growth of an ordered array of organic bidimensional mesopores. *Applied Physics Letters*, **92**, 223117.
23. Sanchez, L., Otero, R., Gallego, J. M., Miranda, R., Martin, N. (2009). Ordering fullerenes at the nanometer scale on solid surfaces. *Chemical Reviews*, **109**, 2081–2091.
24. Peumans, P., Uchida, S., Forrest, S. R. (2003). Efficient bulk heterojunction photovoltaic cells using small-molecular-weight organic thin films. *Nature*, **425**, 158–162.
25. Delgado, J. L., Bouit, P. A., Filippone, S., Herranz, M. A., Martin, N. (2010). Organic photovoltaics: A chemical approach. *Chemical Communications*, **46**, 4853–4865.
26. Jackel, F., Perera, U. G. E., Iancu, V., Braun, K. F., Koch, N., Rabe, J. P., Hla, S. W. (2008). Investigating molecular charge transfer complexes with a low temperature scanning tunneling microscope. *Physical Review Letters*, **100**, 126102.
27. Gonzalez-Lakunza, N., Fernandez-Torrente, I., Franke, K. J., Lorente, N., Arnau, A., Pascual, J. I. (2008). Formation of dispersive hybrid bands at an organic–metal interface. *Physical Review Letters*, **100**, 156805.
28. Vangrondelle, R., Dekker, J. P., Gillbro, T., Sundstrom, V. (1994). Energy-transfer and trapping in photosynthesis. *Biochimica et Biophysica Acta—Bioenergetics*, **1187**, 1–65.
29. Das, R., Kiley, P. J., Segal, M., Norville, J., Yu, A. A., Wang, L. Y., Trammell, S. A., Reddick, L. E., Kumar, R., Stellacci,

F., Lebedev, N., Schnur, J., Bruce, B. D., Zhang, S. G., Baldo, M. (2004). Integration of photosynthetic protein molecular complexes in solid-state electronic devices. *Nano Letters*, **4**, 1079–1083.

30. Holzwarth, A. R., Griebenow, K., Schaffner, K. (1992). Chlorosomes, photosynthetic antennae with novel self-organized pigment structures. *Journal of Photochemistry and Photobiology a—Chemistry*, **65**, 61–71.
31. McDermott, G., Prince, S. M., Freer, A. A., Hawthornthwaitelawless, A. M., Papiz, M. Z., Cogdell, R. J., Isaacs, N. W. (1995). Crystal-structure of an integral membrane light-harvesting complex from photosynthetic bacteria. *Nature*, **374**, 517–521.
32. Balaban, T. S., Holzwarth, A. R., Schaffner, K., Boender, G. J., Degroot, H. J. M. (1995). CP-MAS C-13-NMR dipolar correlation spectroscopy of C-13-enriched chlorosomes and isolated bacteriochlorophyll-c aggregates of Chlorobium-tepidum—The self-organization of pigments is the main structural feature of chlorosomes. *Biochemistry*, **34**, 15259–15266.
33. van Rossum, B. J., Steensgaard, D. B., Mulder, F. M., Boender, G. J., Schaffner, K., Holzwarth, A. R., de Groot, H. J. M. (2001). A refined model of the chlorosomal antennae of the green bacterium *Chlorobium tepidum* from proton chemical shift constraints obtained with high-field 2-D and 3-D MAS NMR dipolar correlation spectroscopy. *Biochemistry*, **40**, 1587–1595.
34. Balaban, T. S. (2005). Tailoring porphyrins and chlorins for self-assembly in biomimetic artificial antenna systems. *Accounts of Chemical Research*, **38**, 612–623.
35. Elemans, J., Van Hameren, R., Nolte, R. J. M., Rowan, A. E. (2006). Molecular materials by self-assembly of porphyrins, phthalocyanines, and perylenes. *Advanced Materials*, **18**, 1251–1266.
36. Balaban, T. S., Linke-Schaetzel, M., Bhise, A. D., Vanthuyne, N., Roussel, C., Anson, C. E., Buth, G., Eichhofer, A., Foster, K., Garab, G., Gliemann, H., Goddard, R., Javorfi, T., Powell, A. K., Rosner, H., Schimmel, T. (2005). Structural characterization of artificial self-assembling porphyrins that mimic the natural chlorosomal bacteriochlorophylls c, d, and e. *Chemistry—A European Journal*, **11**, 2268–2275.
37. Tamiaki, H., Miyatake, T., Tanikaga, R., Holzwarth, A. R., Schaffner, K. (1996). Self-assembly of an artificial light-harvesting antenna: Energy transfer from a zinc chlorin to a bacteriochlorin in a supramolecular aggregate. *Angewandte Chemie International Edition in English*, **35**, 772–774.
38. Balaban, T. S., Linke-Schaetzel, M., Bhise, A. D., Vanthuyne, N., Roussel, C. (2004). Green self-assembling porphyrins and chlorins as mimics of the natural bacteriochlorophylls c, d, and e. *European Journal of Organic Chemistry*, **2004**, 3919–3930.
39. Maier, S., Fendt, L. A., Zimmerli, L., Glatzel, T., Pfeiffer, O. F., Diederich, F., Meyer, E. (2008). Nanoscale engineering of molecular porphyrin wires on insulating surfaces. *Small*, **4**, 1115–1118.
40. Trelka, M., Urban, C., Rogero, C., de Mendoza, P., Mateo-Marti, E., Wang, Y., Silanes, I., Ecija, D., Alcami, M., Yndurain, F., Arnau, A., Martin, F., Echavarren, A. M., Martin-Gago, J. A., Gallego, J. M., Otero, R., Miranda, R. (2011). Surface assembly of porphyrin nanorods with one-dimensional zinc-oxygen spinal cords. *CrystEngComm*, **13**, 5591–5595.
41. Jung, T. A., Schlittler, R. R., Gimzewski, J. K. (1997). Conformational identification of individual adsorbed molecules with the STM. *Nature*, **386**, 696–698.
42. Scudiero, L., Barlow, D. E., Hipps, K. W. (2000). Physical properties and metal ion specific scanning tunneling microscopy images of metal(II) tetraphenylporphyrins deposited from vapor onto gold (111). *Journal of Physical Chemistry B*, **104**, 11899–11905.
43. Qiu, X. H., Nazin, G. V., Ho, W. (2004). Mechanisms of reversible conformational transitions in a single molecule. *Physical Review Letters*, **93**, 196806.
44. Auwarter, W., Klappenberger, F., Weber-Bargioni, A., Schiffrin, A., Strunskus, T., Woll, C., Pennec, Y., Riemann, A., Barth, J. V. (2007). Conformational adaptation and selective adatom capturing of tetrapyridyl-porphyrin molecules on a copper (111) surface. *Journal of the American Chemical Society*, **129**, 11279–11285.
45. Grill, L., Dyer, M., Lafferentz, L., Persson, M., Peters, M. V., Hecht, S. (2007). Nano-architectures by covalent assembly of molecular building blocks. *Nature Nanotechnology*, **2**, 687–691.
46. Veld, M. I., Iavicoli, P., Haq, S., Amabilino, D. B., Raval, R. (2008). Unique intermolecular reaction of simple porphyrins at a metal surface gives covalent nanostructures. *Chemical Communications*, 1536–1538.
47. Ecija, D., Trelka, M., Urban, C., de Mendoza, P., Mateo-Marti, E., Rogero, C., Martin-Gago, J. A., Echavarren, A. M., Otero, R., Gallego, J. M., Mirandat, R. (2008). Molecular conformation, organizational chirality, and iron metalation of meso-tetramesitylporphyrins on copper(100). *Journal of Physical Chemistry C*, **112**, 8988–8994.
48. Miranda, R., Gallego, J. M. (2001). Bimodal island-size distributions in submonolayer growth. *Physical Review B*, **64**, 085426.
49. Marks, T. J. (1985). Electrically conductive metallomacrocyclic assemblies. *Science*, **227**, 881–889.
50. Martinsen, J., Stanton, J. L., Greene, R. L., Tanaka, J., Hoffman, B. M., Ibers, J. A. (1985). Metal-spine conductivity in a partially oxidized metallomacrocycle—(Phthalocyaninato)cobalt iodide. *Journal of the American Chemical Society*, **107**, 6915–6920.
51. Collman, J. P., McDevitt, J. T., Yee, G. T., Leidner, C. R., McCullough, L. G., Little, W. A., Torrance, J. B. (1986). Conductive polymers derived from iron, ruthenium, and osmium metalloporphyrins—The shish-kebab approach. *Proceedings of the National Academy of Sciences of the United States of America*, **83**, 4581–4585.
52. Hanack, M., Lang, M. (1994). Conducting stacked metallophthalocyanines and related-compounds. *Advanced Materials*, **6**, 819–833.
53. Ballester, P., Costa, A., Castilla, A. M., Deya, P. M., Frontera, A., Gomila, R. M., Hunter, C. A. (2005). DABCO-directed

self-assembly of bisporphyrins (DABCO = 1,4-diazabicyclo 2.2.2 octane). *Chemistry—A European Journal*, **11**, 2196–2206.

54. Kelley, R. F., Shin, W. S., Rybtchinski, B., Sinks, L. E., Wasielewski, M. R. (2007). Photoinitiated charge transport in supramolecular assemblies of a 1,7,*N*,*N*′-tetrakis(zinc porphyrin)-perylene-3,4:9,10-bis(dicarboximide). *Journal of the American Chemical Society*, **129**, 3173–3181.
55. Taylor, P. N., Anderson, H. L. (1999). Cooperative self-assembly of double-strand conjugated porphyrin ladders. *Journal of the American Chemical Society*, **121**, 11538–11545.
56. Kretschmann, A., Walz, M. M., Flechtner, K., Steinruck, H. P., Gottfried, J. M. (2007). Tetraphenylporphyrin picks up zinc atoms from a silver surface. *Chemical Communications*, 568–570.
57. Flechtner, K., Kretschmann, A., Bradshaw, L. R., Walz, M. M., Steinruck, H. P., Gottfried, J. M. (2007). Surface-confined two-step synthesis of the complex (ammine)(meso-tetraphenylporphyrinato)-zinc(II) on Ag(111). *Journal of Physical Chemistry C*, **111**, 5821–5824.
58. Gonzalez-Moreno, R., Sanchez-Sanchez, C., Trelka, M., Otero, R., Cossaro, A., Verdini, A., Floreano, L., Ruiz-Bermejo, M., Garcia-Lekue, A., Martin-Gago, J. A., Rogero, C. (2011). Following the metalation process of protoporphyrin IX with metal substrate atoms at room temperature. *Journal of Physical Chemistry C*, **115**, 6849–6854.
59. Song, H., Scheidt, W. R. (1990). Crystal-structure of the bis-pocket porphyrin aquo(tetramesitylporphinato)zinc(II). *Inorganica Chimica Acta*, **173**, 37–41.
60. Williams, F. J., Vaughan, O. P. H., Knox, K. J., Bampos, N., Lambert, R. M. (2004). First observation of capping/uncapping by a ligand of a Zn porphyrin adsorbed on Ag(100). *Chemical Communications*, 1688–1689.
61. Vaughan, O. P. H., Williams, F. J., Bampos, N., Lambert, R. M. (2006). A chemically switchable molecular pinwheel. *Angewandte Chemie International Edition*, **45**, 3779–3781.
62. Balaban, T. S., Bhise, A. D., Fischer, M., Linke-Schaetzel, M., Roussel, C., Vanthuyne, N. (2003). Controlling chirality and optical properties of artificial antenna systems with self-assembling porphyrins. *Angewandte Chemie International Edition*, **42**, 2140–2144.
63. Yu, G., Gao, J., Hummelen, J. C., Wudl, F., Heeger, A. J. (1995). Polymer photovoltaic cells—enhanced efficiencies via a network of internal donor–acceptor heterojunctions. *Science*, **270**, 1789–1791.
64. Halls, J. J. M., Walsh, C. A., Greenham, N. C., Marseglia, E. A., Friend, R. H., Moratti, S. C., Holmes, A. B. (1995). Efficient photodiodes from interpenetrating polymer networks. *Nature*, **376**, 498–500.
65. Brabec, C. J., Sariciftci, N. S., Hummelen, J. C. (2001). Plastic solar cells. *Advanced Functional Materials*, **11**, 15–26.
66. Nelson, J. (2002). Organic photovoltaic films. *Current Opinion in Solid State & Materials Science*, **6**, 87–95.
67. Peumans, P., Yakimov, A., Forrest, S. R. (2003). Small molecular weight organic thin-film photodetectors and solar cells. *Journal of Applied Physics*, **93**, 3693–3723.
68. Sylvester-Hvid, K. O., Rettrup, S., Ratner, M. A. (2004). Two-dimensional model for polymer-based photovoltaic cells: Numerical simulations of morphology effects. *Journal of Physical Chemistry B*, **108**, 4296–4307.
69. Yang, F., Shtein, M., Forrest, S. R. (2005). Controlled growth of a molecular bulk heterojunction photovoltaic cell. *Nature Materials*, **4**, 37–41.
70. Ma, W. L., Yang, C. Y., Gong, X., Lee, K., Heeger, A. J. (2005). Thermally stable, efficient polymer solar cells with nanoscale control of the interpenetrating network morphology. *Advanced Functional Materials*, **15**, 1617–1622.
71. Otero, R., Ecija, D., Fernandez, G., Gallego, J. M., Sanchez, L., Martin, N., Miranda, R. (2007). An organic donor/acceptor lateral superlattice at the nanoscale. *Nano Letters*, **7**, 2602–2607.
72. Hummelen, J. C., Knight, B. W., Lepeq, F., Wudl, F., Yao, J., Wilkins, C. L. (1995). Preparation and characterization of fulleroid and methanofullerene derivatives. *Journal of Organic Chemistry*, **60**, 532–538.
73. Yamashita, Y., Kobayashi, Y., Miyashi, T. (1989). *p*-Quinodimethane analogs of tetrathiafulvalene. *Angewandte Chemie International Edition in English*, **28**, 1052–1053.
74. Martin, N., Sanchez, L., Seoane, C., Orti, E., Viruela, P. M., Viruela, R. (1998). Synthesis, properties, and theoretical characterization of largely pi-extended tetrathiafulvalene derivatives with quinonoid structures. *Journal of Organic Chemistry*, **63**, 1268–1279.
75. Bryce, M. R. (1999). Tetrathiafulvalenes as pi-electron donors for intramolecular charge-transfer materials. *Advanced Materials*, **11**, 11–23.
76. Segura, J. L., Martin, N. (2001). New concepts in tetrathiafulvalene chemistry. *Angewandte Chemie International Edition*, **40**, 1372–1409.
77. Bendikov, M., Wudl, F., Perepichka, D. F. (2004). Tetrathiafulvalenes, oligoacenenes, and their buckminsterfullerene derivatives: The brick and mortar of organic electronics. *Chemical Reviews*, **104**, 4891–4945.
78. Canevet, D., Salle, M., Zhang, G. X., Zhang, D. Q., Zhu, D. B. (2009). Tetrathiafulvalene (TTF) derivatives: Key building-blocks for switchable processes. *Chemical Communications*, 2245–2269.
79. Martin, N., Sanchez, L., Herranz, M. A., Illescas, B., Guldi, D. M. (2007). Electronic communication in tetrathiafulvalene (TTF)/C-60 systems: Toward molecular solar energy conversion materials? *Accounts of Chemical Research*, **40**, 1015–1024.
80. Perez, E. M., Sanchez, L., Fernandez, G., Martin, N. (2006). exTTF as a building block for fullerene receptors. Unexpected solvent-dependent positive homotropic cooperativity. *Journal of the American Chemical Society*, **128**, 7172–7173.
81. Chambliss, D. D., Wilson, R. J., Chiang, S. (1991). Nucleation of ordered Ni island arrays on Au(111) by surface-lattice dislocations. *Physical Review Letters*, **66**, 1721–1724.

82. Yokoyama, T., Yokoyama, S., Kamikado, T., Mashiko, S. (2001). Nonplanar adsorption and orientational ordering of porphyrin molecules on Au(111). *Journal of Chemical Physics*, **115**, 3814–3818.
83. Voigtlander, B., Meyer, G., Amer, N. M. (1991). Epitaxial-growth of Fe on Au(111)—A scanning tunneling microscopy investigation. *Surface Science*, **255**, L529–L535.
84. Voigtlander, B., Meyer, G., Amer, N. M. (1991). Epitaxial-growth of thin magnetic cobalt films on Au(111) studied by scanning tunneling microscopy. *Physical Review B*, **44**, 10354–10357.
85. Cullen, W. G., First, P. N. (1999). Island shapes and intermixing for submonolayer nickel on Au(111). *Surface Science*, **420**, 53–64.
86. Fruchart, O., Klaua, M., Barthel, J., Kirschner, J. (1999). Self-organized growth of nanosized vertical magnetic Co pillars on Au(111). *Physical Review Letters*, **83**, 2769–2772.
87. Clair, S., Pons, S., Brune, H., Kern, K., Barth, J. V. (2005). Mesoscopic metallosupramolecular texturing by hierarchic assembly. *Angewandte Chemie International Edition*, **44**, 7294–7297.
88. Maksymovych, P., Sorescu, D. C., Dougherty, D., Yates, J. T. (2005). Surface bonding and dynamical behavior of the CH(3)SH molecule on Au(111). *Journal of Physical Chemistry B*, **109**, 22463–22468.
89. Canas-Ventura, M. E., Xiao, W., Wasserfallen, D., Mullen, K., Brune, H., Barth, J. V., Fasel, R. (2007). Self-assembly of periodic bicomponent wires and ribbons. *Angewandte Chemie International Edition*, **46**, 1814–1818.
90. Schaff, O., Schmid, A. K., Bartelt, N. C., de la Figuera, J., Hwang, R. Q. (2001). In-situ STM studies of strain-stabilized thin-film dislocation networks under applied stress. *Materials Science and Engineering a-Structural Materials Properties Microstructure and Processing*, **319**, 914–918.
91. Ecija, D., Otero, R., Sanchez, L., Gallego, J. M., Wang, Y., Alcami, M., Martin, F., Martin, N., Miranda, R. (2007). Crossover site-selectivity in the adsorption of the fullerene derivative PCBM on Au(111). *Angewandte Chemie International Edition*, **46**, 7874–7877.
92. Altman, E. I., Colton, R. J. (1992). Nucleation, growth, and structure of fullerene films on Au(111). *Surface Science*, **279**, 49–67.
93. Chen, W., Madhavan, V., Jamneala, T., Crommie, M. F. (1998). Scanning tunneling microscopy observation of an electronic superlattice at the surface of clean gold. *Physical Review Letters*, **80**, 1469–1472.
94. Burgi, L., Brune, H., Kern, K. (2002). Imaging of electron potential landscapes on Au(111). *Physical Review Letters*, **89**.
95. Urban, C., Ecija, D., Wang, Y., Trelka, M., Preda, I., Vollmer, A., Lorente, N., Arnau, A., Alcami, M., Soriano, L., Martin, N., Martin, F., Otero, R., Gallego, J. M., Miranda, R. (2010). Growth and structure of self-assembled monolayers of a TTF derivative on Au(111). *Journal of Physical Chemistry C*, **114**, 6503–6510.
96. Maksymovych, P., Dougherty, D. B. (2008). Molecular self-assembly guided by surface reconstruction: CH3SH monolayer on the Au(111) surface. *Surface Science*, **602**, 2017–2024.
97. Dretschkow, T., Lampner, D., Wandlowski, T. (1998). Structural transitions in 2,2'-bipyridine adlayers on Au(111)—An in-situ STM study. *Journal of Electroanalytical Chemistry*, **458**, 121–138.
98. Harten, U., Lahee, A. M., Toennies, J. P., Woll, C. (1985). Observation of a soliton reconstruction of Au(111) by high-resolution helium-atom diffraction. *Physical Review Letters*, **54**, 2619–2622.
99. Zhang, G. F., Maaloum, M., Muller, P., Benoit, N., Krafft, M. P. (2004). Surface micelles of semifluorinated alkanes in Langmuir–Blodgett monolayers. *Physical Chemistry Chemical Physics*, **6**, 1566–1569.
100. Asahi, T., Sugiyama, T., Masuhara, H. (2008). Laser fabrication and spectroscopy of organic nanoparticles. *Accounts of Chemical Research*, **41**, 1790–1798.
101. Zhao, Y. S., Fu, H. B., Peng, A. D., Ma, Y., Xiao, D. B., Yao, J. N. (2008). Low-dimensional nanomaterials based on small organic molecules: Preparation and optoelectronic properties. *Advanced Materials*, **20**, 2859–2876.
102. Utekhina, A. Y., Sergeev, G. B. (2011). Organic nanoparticles. *Russian Chemical Reviews*, **80**, 219–233.
103. Alivisatos, A. P. (1996). Semiconductor clusters, nanocrystals, and quantum dots. *Science*, **271**, 933–937.
104. Murray, C. B., Kagan, C. R., Bawendi, M. G. (2000). Synthesis and characterization of monodisperse nanocrystals and close-packed nanocrystal assemblies. *Annual Review of Materials Science*, **30**, 545–610.
105. Lounis, B., Orrit, M. (2005). Single-photon sources. *Reports on Progress in Physics*, **68**, 1129–1179.
106. Fu, H. B., Yao, J. N. (2001). Size effects on the optical properties of organic nanoparticles. *Journal of the American Chemical Society*, **123**, 1434–1439.
107. Xiao, D. B., Lu, X., Yang, W. S., Fu, H. B., Shuai, Z. G., Fang, Y., Yao, J. N. (2003). Size-tunable emission from 1,3-diphenyl-5-(2-anthryl)-2-pyrazoline nanoparticles. *Journal of the American Chemical Society*, **125**, 6740–6745.
108. Masuo, S., Masuhara, A., Akashi, T., Muranushi, M., Machida, S., Kasai, H., Nakanishi, H., Oikawa, H., Itaya, A. (2007). Photon antibunching in the emission from a single organic dye nanocrystal. *Japanese Journal of Applied Physics Part 2—Letters & Express Letters*. **46**, L268–L270.
109. Stangl, J., Holy, V., Bauer, G. (2004). Structural properties of self-organized semiconductor nanostructures. *Reviews of Modern Physics*, **76**, 725–783.
110. Kowarik, S., Gerlach, A., Schreiber, F. (2008). Organic molecular beam deposition: fundamentals, growth dynamics, and in situ studies. *Journal of Physics—Condensed Matter*, **20**, 184005.
111. de la Torre, G., Claessens, C. G., Torres, T. (2007). Phthalocyanines: Old dyes, new materials. Putting color in nanotechnology. *Chemical Communications*, 2000–2015.

112. Cheng, Z. H., Gao, L., Deng, Z. T., Jiang, N., Liu, Q., Shi, D. X., Du, S. X., Guo, H. M., Gao, H. J. (2007). Adsorption behavior of iron phthalocyanine on Au(111) surface at submonolayer coverage. *Journal of Physical Chemistry C*, **111**, 9240–9244.

113. Cheng, Z. H., Gao, L., Deng, Z. T., Liu, Q., Jiang, N., Lin, X., He, X. B., Du, S. X., Gao, H. J. (2007). Epitaxial growth of iron phthalocyanine at the initial stage on Au(111) surface. *Journal of Physical Chemistry C*, **111**, 2656–2660.

114. Claessens, C. G., Gonzalez-Rodriguez, D., Torres, T. (2002). Subphthalocyanines: Singular nonplanar aromatic compounds—Synthesis, reactivity, and physical properties. *Chemical Reviews*, **102**, 835–853.

115. Inokuma, Y., Osuka, A. (2008). Subporphyrins: emerging contracted porphyrins with aromatic 14 pi-electronic systems and bowl-shaped structures: Rational and unexpected synthetic routes. *Dalton Transactions*, 2517–2526.

116. Kobayashi, N. (2002). Phthalocyanines: Synthesis. In: Kadish, K., Guilard, R., Smith, K. M., eds. *The Porphyrin Handbook*. Academic Press, San Diego, CA, pp. 161–262.

117. Claessens, C. G., Gonzalez-Rodriguez, D., Torres, T., Martin, G., Agullo-Lopez, F., Ledoux, I., Zyss, J., Ferro, V. R., de la Vega, J. M. G. (2005). Structural modulation of the dipolar–octupolar contributions to the NLO response in subphthalocyanines. *Journal of Physical Chemistry B*, **109**, 3800–3806.

118. Diaz, D. D., Bolink, H. J., Cappelli, L., Claessens, C. G., Coronado, E., Torres, T. (2007). Subphthalocyanines as narrow band red-light emitting materials. *Tetrahedron Letters*, **48**, 4657–4660.

119. Trelka, M., Medina, A., Ecija, D., Urban, C., Groning, O., Fasel, R., Gallego, J. M., Claessens, C. G., Otero, R., Torres, T., Miranda, R. (2011). Subphthalocyanine-based nanocrystals. *Chemical Communications*, **47**, 9986–9988.

120. Yanagi, H., Schlettwein, D., Nakayama, H., Nishino, T. (2000). Site-specific physisorption and chemical reaction of subphthalocyanine molecules on silicon(111)-(7 × 7). *Physical Review B*, **61**, 1959–1964.

121. Yanagi, H., Ikuta, K., Mukai, H., Shibutani, T. (2002). STM-induced flip-flop switching of adsorbed subphthalocyanine molecular arrays. *Nano Letters*, **2**, 951–955.

122. Berner, S., de Wild, M., Ramoino, L., Ivan, S., Baratoff, A., Guntherodt, H. J., Suzuki, H., Schlettwein, D., Jung, T. A. (2003). Adsorption and two-dimensional phases of a large polar molecule: Sub-phthalocyanine on Ag(111). *Physical Review B*, **68**, art. no. 115410.

123. Yanagi, H., Ikuta, K. (2005). Stochastic switching of subphthalocyanine arrays triggered by scanning tunneling microscopy. *Surface Science*, **581**, 9–16.

124. Azuma, Y., Yokota, T., Kera, S., Aoki, M., Okudaira, K. K., Harada, Y., Ueno, N. (1998). Characterization of thin films of chloroaluminum phthalocyanine on MoS(2): HREELS, LEET and PIES study. *Thin Solid Films*, **327**, 303–307.

125. Kera, S., Yamane, H., Honda, H., Fukagawa, H., Okudaira, K. K., Ueno, N. (2004). Photoelectron fine structures of uppermost valence band for well-characterized ClAl-phthalocyanine ultrathin film: UPS and MAES study. *Surface Science*, **566**, 571–578.

126. Fukagawa, H., Yamane, H., Kera, S., Okudaira, K. K., Ueno, N. (2006). Experimental estimation of the electric dipole moment and polarizability of titanyl phthalocyanine using ultraviolet photoelectron spectroscopy. *Physical Review B*, **73**, 041302.

127. Weckesser, J., De Vita, A., Barth, J. V., Cai, C., Kern, K. (2001). Mesoscopic correlation of supramolecular chirality in one-dimensional hydrogen-bonded assemblies. *Physical Review Letters*, **87**, art. no. 096101.

128. Barth, J. V., Weckesser, J., Lin, N., Dmitriev, A., Kern, K. (2003). Supramolecular architectures and nanostructures at metal surfaces. *Applied Physics a—Materials Science & Processing*, **76**, 645–652.

20

BIOFUNCTIONALIZED SURFACES

MARISELA VÉLEZ

20.1 INTRODUCTION

Nature self-assembles functional organic materials at the nanoscale to sustain life. Living organism can be multicellular and large, but we know since the seventeenth century that living organisms, even the largest ones, are constituted by micrometer-sized structural units called cells. Primitive optical microscopes developed by Anton van Leeuwenhoek allowed their observation for the first time. He also identified that some living organism could be constituted by one single cell, indicating that those individual units had all the capabilities to sustain life. The more recent advent of the electron microscope, in the early twentieth century, provided deeper insight into the elements that constituted these unique structures. X-ray diffraction techniques developed at around the same time and applied a few decades later to the structural analysis of proteins have complemented our understanding of the complexity of the functional organic constituents of living cells. All this structural information, combined with biochemical and genetic analysis, confirmed that micrometer-sized cells are put together from myriads of smaller nanometer self-assembled parts.

All living organisms, from the largest animals or plants, such as whales or sequoias, to the smallest viruses or microorganisms, living at mild temperatures or at extreme conditions, are constituted by assemblies of nanometer units of basically four different chemical components: proteins (amino acid polymers), lipids (phospholipid assemblies), carbohydrates (sugar polymers), and nucleic acids (DNA and RNA, nucleotide polymers). These different materials self-assemble into "hybrid" structures of well-defined conformations that are able to develop highly specialized and complex functions. DNA–protein complexes, protein–sugar–lipid assemblies, and sugar–protein complexes are the elements that sustain the structure and function of all living processes.

So, we have much to learn from biology when it comes to self-assembling nanomaterials into hybrid complexes to develop highly specialized functions. We can simply try to understand how these specialized structures are assembled and use this knowledge to help us design new materials to achieve complex functions, different from the ones found in nature, or we can try to manipulate and combine the different materials found in nature to reproduce in a controlled manner specific functions. Whatever the aim we pursue, we first need to grasp and manipulate these small and delicate materials, keeping in mind that the degree of organization of the assemblies is essential for their function.

In biological systems, surfaces are very important. The fact that even large organisms are constituted by much smaller cells is related to the need to maintain a large surface-area-to-volume ratio. Only small volume units have enough surface area to sustain the living processes. This illustrates how important surfaces are for a living cell. Many essential processes take place there: cell recognition and communication, food exchange, and excretion, to name only a few. It is then easy to imagine how many specialized and interesting functions are governed by protein–lipid–sugar assemblies that are structured and oriented on surfaces.

In the fast growing field of nanotechnology, surfaces are also very important. Objects of reduced size also increase their surface-to-volume ratio; and it is the activity of this surface that, similarly to what happens with the cells, determines how different elements interact with each other. So, devices coupled to any biological process require strategies to interface biological molecules with inorganic surfaces. The interest is diverse: biosensor development, which takes

Organic Nanomaterials: Synthesis, Characterization, and Device Applications, First Edition. Edited by Tomás Torres and Giovanni Bottari.

advantage of the specificity of biomolecules to detect other biomolecules of interest; anchoring biological molecules in order to manipulate them with available single molecule tools; interrogating cell behavior by providing them with a tailored structure similar to the one found in their native surroundings; or modifying surfaces of different size and shape, from nanoparticles to scaffolds of diverse materials, for medical applications that go from facilitating tissue regeneration, improving implant materials to sophisticated therapeutic and diagnostic purposes.

It is therefore no surprise that the field of biofunctionalizing surfaces is broad and fast-growing, but still facing huge challenges. Biological organic materials are to be coupled to inorganic surfaces, in a mild way to preserve their functionality, ideally facilitating their assembly into more complex structures, all that to be achieved on diverse substrates: SiO_2, titanium, gold, plastic, or iron for magnetic nanoparticles, depending on the applications. This chapter will first provide a brief and general overview of some of the strategies available to modify different surfaces at the nanoscale with biological materials and then will briefy mention a few examples of applications in more specialized areas.

Experimental techniques used to characterize biofunctional surfaces are similar to the ones used to study monolayers of other organic materials: scanning probe microscopies (AFM, STM), electron microsopies, fluorescence, surface plasmon resonance (SPR), quartz crystal microbalance (QCM), or electrochemistry, to name some of them. Characterization techniques will not be addressed in this chapter, and the reader is referred to some recent reviews dedicated to this subject [1].

20.2 ASSEMBLING BIOLOGICAL MATERIALS ON INORGANIC SURFACES

Biologically relevant materials can be of different types: proteins, lipids, DNA, and carbohydrates. Each one of them is optimized to develop a different biological function, and therefore their structural and physical properties are very different. When looking for strategies to immobilize them, there are several issues to be considered depending on the material and the purpose of their immobilization. It might be of interest to orient the material on a solid support, to attach it covalently or not, or to deposit the material with defined spatial patterns. Furthermore, the type of substrate to be modified can be different: Glass, quartz, or transparent substrates are preferred if the aim is to attach cells that will be followed or analyzed using optical microscopy. Gold or other conductive materials are necessary if the biological material is to be coupled to an electrochemical or SPR biosensor. More specific materials such as iron oxide, silicon, carbon, or titanium could also be of interest if magnetic or semiconductor properties are required or if the interest is to develop medical applications. During recent years many different protocols have been developed [2] to attach the different biological materials to a large number of surfaces. I will summarize some of the most relevant strategies used for each of the different biological materials. The type of application pursued defines the requirements: Random noncovalent or covalent protein attachment might be enough for some biosensing applications, whereas in some cases a controlled orientation and spatial distribution are essential. Some applications might require spatial control to mimic and favor other interactions to allow for the formation of more complex molecular aggregates.

20.2.1 Random Noncovalent Attachments

Proteins can absorb on surfaces through ionic bonds and hydrophobic and polar interactions, producing random physisorption. The type of intermolecular force dominating the interaction will depend on the particular protein and surface involved. It is therefore relatively easy to produce a heterogeneous and randomly oriented protein layer adsorbed on a surface. Immobilization can also be achieved through electrostatic interactions. Surfaces that are modified to contain positively charged amine or negatively charged carboxy groups are most suitable for this approach. A clean gold surface can also be directly functionalized with thiol-containing molecules. Therefore, direct immobilization of proteins on gold surfaces can be achieved by simply exploiting the high affinity of cysteine residues in the protein for the gold surface, resulting in efficient chemisorptions [3,4]. An important limitation of this strategy is that proteins tend to lose activity when in direct contact with metal surfaces.

20.2.2 Functionalizing the Surface

Random or oriented covalent immobilization can be achieved after proper surface activation. The main method for functionalizing glass surfaces is using reactive silanol (Si-OH) (Figure 20.1). The silanol groups can be generated by pretreatment of the surface with, for example, piranha solution (H_2O_2/H_2SO_4) or oxygen plasma. Organofunctional silanes of the general structure $(RO)_3Si(CH_2)_nX$ or trichlorosilanes are then used to introduce a new functional group on the surface. A large variety of silane reagents are commercially available, bearing amine, thiol, carboxy, epoxide, and other functional groups for subsequent modification steps. Various protocols for silanization can be found in the literature: Employing deposition of silanes from organic solutions, using aqueous solutions, utilizing the gas phase, or using chemical vapor deposition [5–8].

Chemical modification of silicon surfaces is more complicated than modifying glass because silicon spontaneously oxidizes in air to produce and amorphous silica layer. Surface modification strategies for the formation of covalent

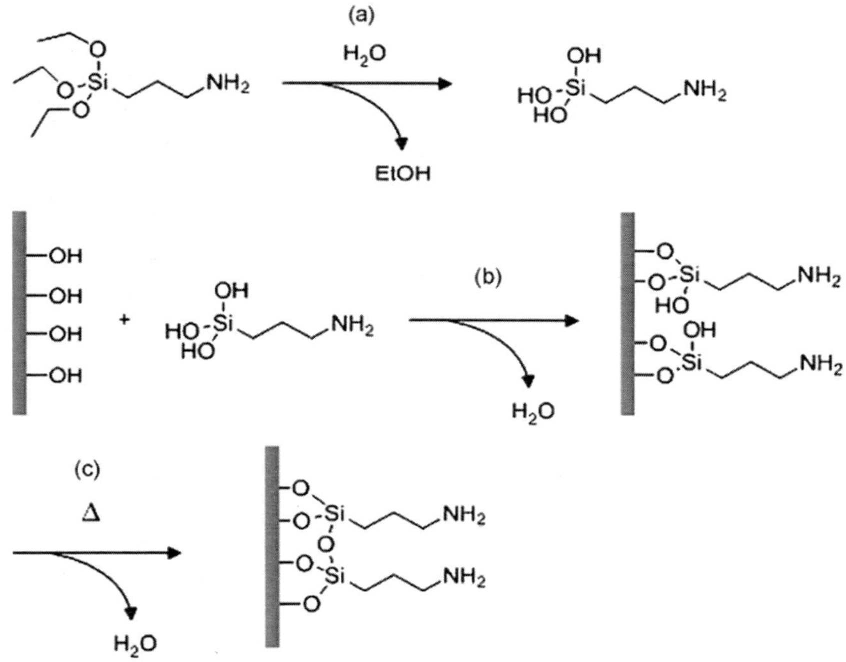

FIGURE 20.1. Proposed mechanism for the silanization with aminopropyltriethoxysilane (APTES): Hydrolysis of the reactive siloxanes (**a**), which can take place in solution or on the substrate surface, allows condensation with surface silanol groups (**b**). Thermal curing of the resulting film causes further cross-linking (**c**). Reprinted from reference [2], with permission.

silicon–carbon bonds require, first, a special pretreatment of the silicon surface to remove the oxide layer and, second, an activation of the silicon surface for subsequent reaction (Figure 20.2).

This activation can be achieved by treatment with HF, to generate a hydrogen-terminated Si (111) surface that can further react with unsaturated ω-functionalized alkenes upon ultraviolet irradiation or thermal activation [9], or by oxidation with plasma to further functionalize with organosilanes in analogy to glass slides. Various procedures have been developed to functionalize a range of alternative oxide surfaces that are of particular interest for specialized applications such as implants (titanium [10], tantalum, and niobium), electrical devices (indium tin oxide (ITO) and diamond [11]), and others, such as silicate minerals (mica [12]). Silane chemistry and electropolymerization [13] procedures have been applied in the case of ITO, while photoimmobilization has been used to activate diamond. To functionalize mica, (poly)electrolytes have been used in addition to silane chemistry while for titanium, tantalum, and niobium [14], self-assembled monolayer (SAM) formation using thiols and phosphonates has been reported [10, 15].

Covalent attachment of proteins to gold surfaces is also of great interest, particularly for biosensing applications. This metal substrate has the advantage of being easily functionalized with SAMs (self-assembled monolayers) of ω-functionalized thiols, disulfides, and sulfides [16]. The generation of SAMs on such surfaces strongly depends on the crystalline morphology of the underlying metal. Au(111) yields SAMs having the highest density and highest degree of regularity and is therefore most widely applied. The preparation of smooth gold surfaces can be carried out employing the so-called template stripping method of gold from, for example, mica [17]. The chemistry of the gold–thiol interface is well known and is much easier to control than organosilane chemistry. A typical thiol monolayer on gold is shown in Figure 20.3. Generation and applications of SAMs have recently been reviewed elsewhere [16].

The terminal groups of heterobifunctional thiol compounds are important for the potential interaction of the SAM with proteins, and thus a variety of functionalized thiols are commercially available and have been used in protein biochip applications [18]. The use of SAMs of mixed composition has been used to optimize the presentation of anchor molecules and to decrease steric hindrance of large binding proteins. The major drawbacks of using thiolates on surfaces are, however, their mobility on the solid surface, which limits the lifetime of the chips [16, 19] and their susceptibility to photooxidation [20].

More recently, mainly due to the increasing use of polymeric supports in standard microarray equipment and microfluidic chips, strategies for protein immobilization on polymeric surfaces has become of interest. Different chemical surface modification of polymeric materials such as

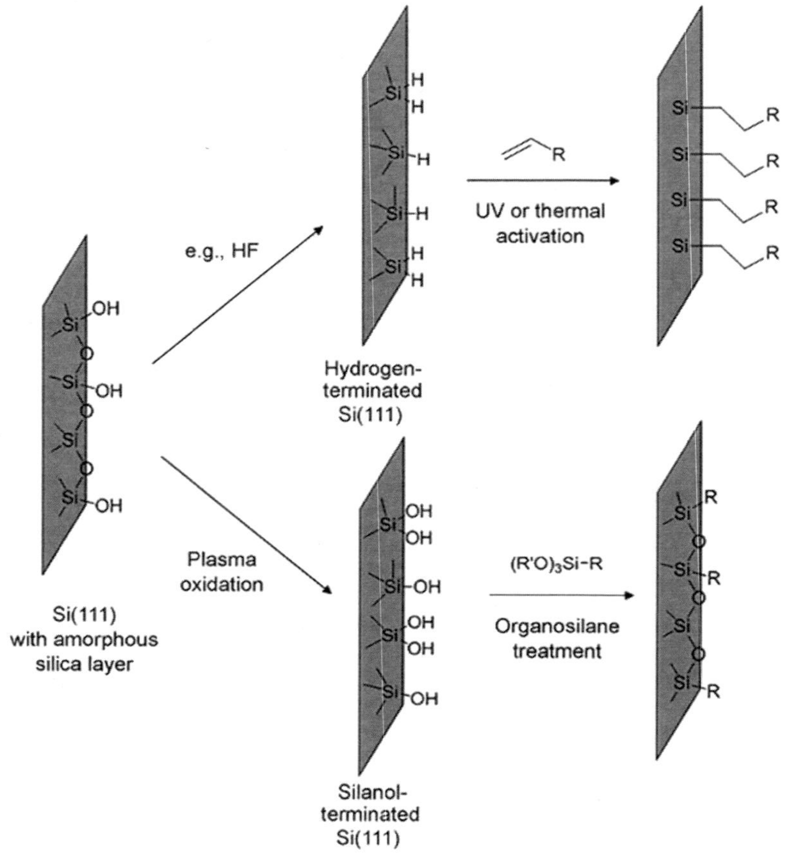

FIGURE 20.2. Examples of Si(111) functionalization: Si(111) spontaneously forms an amorphous silica layer in air. Treatment with HF, for example, produces a hydrogen-terminated silicon surface that can react further, for example with ω-functionalized alkenes. Treatment with oxygen plasma provides silanols at the Si(111) surface, which can then react with organosilanes [2].

poly(dimethylsiloxane) (PDMS), poly(methyl methacrylate) (PMMA), and polycarbonate (PC) allow protein immobilization [21–23]. All the above-described surface modifications techniques render the different surfaces active to bind proteins or the molecules of interest through covalent modifications. The reactivity of the surface is determined by the functional groups it displays.

*20.2.2.1 **Proteins.*** Proteins are constituted by a linear sequence of amino acids that fold in a three-dimensional

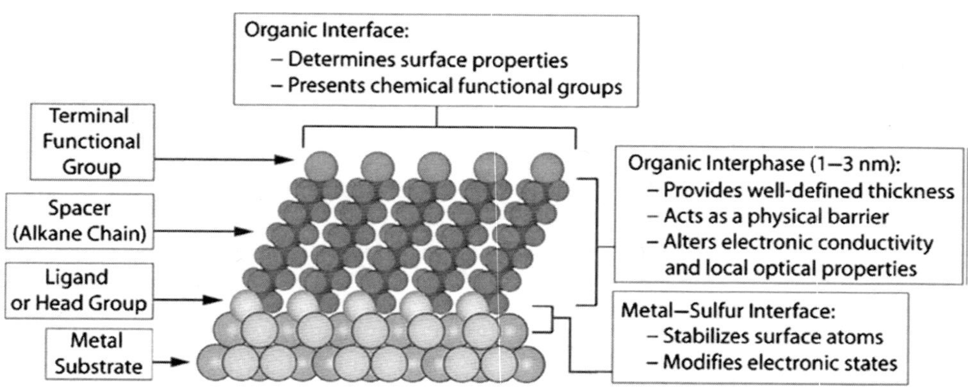

FIGURE 20.3. Ideal self-assembled monolayer (SAM) of terminally functionalized alkylthiolates bound to a Au(111) surface, showing the alkyl chains in the characteristically tilted orientation [16]. Reproduced with permission from ACS.

TABLE 20.1. Methods for Nonspecific Covalent Protein Immobilization

Surface Functional Groups		Protein Functional Groups	Product	
NHS ester [126–130]	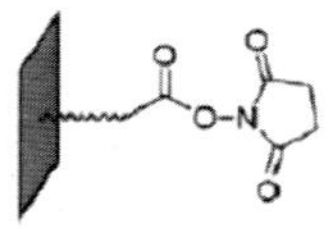	H_2NR	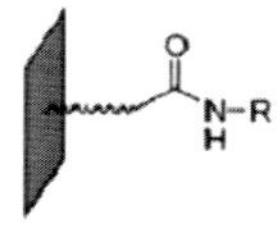	Amide
Aldehyde [7, 9, 131–138]	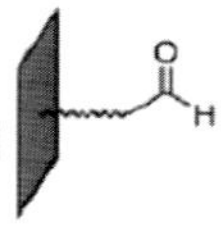	H_2NR	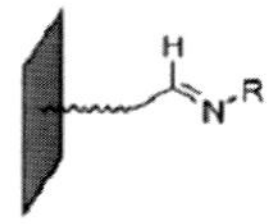	Imine
Isothiocyanate [99]	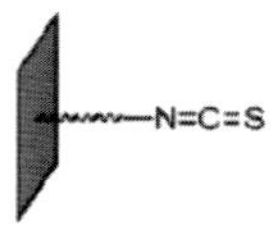	H_2NR	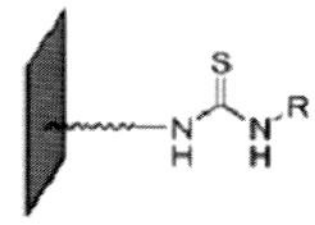	Thiourea
Epoxide [133, 139, 140]	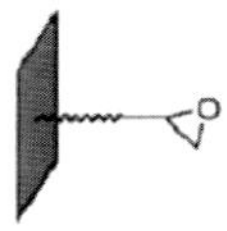	H_2NR	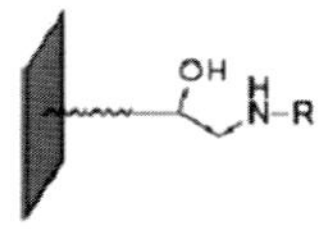	Aminoalcohol
Amine [121, 157] [a]	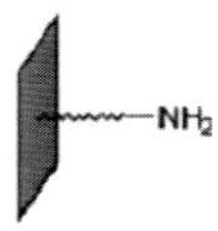	$HO(O)CCH_2R$	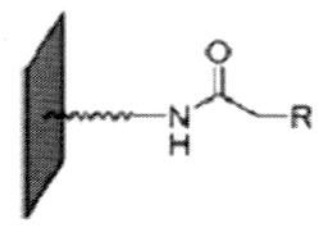	Amide

[a] With coupling reagent (e.g., CDI).
Source: Reproduced from reference [2] with permission.

structure maintained by weak interactions. Their specific function relies on the integrity of this spatial organization. The big challenge when attaching a protein to a solid surface is maintaining this native conformation to preserve its biological function. It is therefore important to bind the protein through functional groups or tags that do not compromise their function. In some cases, achieving just a random orientation of the protein on the surface is not enough, and controlling their orientation becomes crucial to retain the searched functionality [24, 25].

Proteins offer many functional groups, mainly in the amino acid side chains, that are suitable for immobilization purposes. Such functional groups can be used to covalently couple proteins to surfaces by a range of different reactions. Suitable complementary groups can be installed on the solid support.

Various chemically modified surfaces are commercially available for this purpose, and some of them are designed to suppress nonspecific adsorption. Table 20.1 shows some typical examples of compatible groups attached to surfaces and the functional groups they react with (NHS ester [26], aldehyde [27, 28], isothiocyanate [29], epoxide [30] and amine [31]).

Protein attachment can occur simultaneously through many residues, thereby restricting degrees of conformational freedom (and thus possibly activity) and also increasing heterogeneity in the population of immobilized proteins. So, to orient the proteins, by either noncovalent or covalent attachments, additional strategies have to be adopted. Orienting the site of interest of an immobilized protein away from the chip surface should facilitate interaction analysis, especially in the case of large interaction partners, such as other proteins. Noncovalent strategies adopted from established capture-reagent–fusion-protein pairs, which were originally developed for protein purification by column chromatography, have been adopted to immobilize fusion proteins onto surface-bound affinity tags to uniformly orient proteins on a chip surface. Many biologically active fusion proteins are available, including popular fusion tags such as glutathione *S*-transferase (GST)[32], maltose binding protein (MBP)[33], FLAG peptide [34], hexahistidine (His_6)[35], and dehalogenase [36]. The advantage over physisorption or covalent chemistry lies in the specificity and directionality of the supramolecular interaction and the tunability of the type and number of host–guest interactions. In addition to homogeneous and oriented attachment, the reversibility of immobilization can be very attractive from an economical point of view, because chip and sensor surfaces might be recyclable and suitable for repeated use.

The use of nickel nitrilotriacetic acid surfaces stands out as a common and versatile strategy (Figure 20.4). The extended use of engineered His_6 tag for purification makes these modified proteins readily available, and it is very convenient to use this same modification to immobilize the proteins on surfaces. Fluorescent proteins, antibodies, virus proteins, and growth factors can be immobilized on Ni-NTA chip surfaces

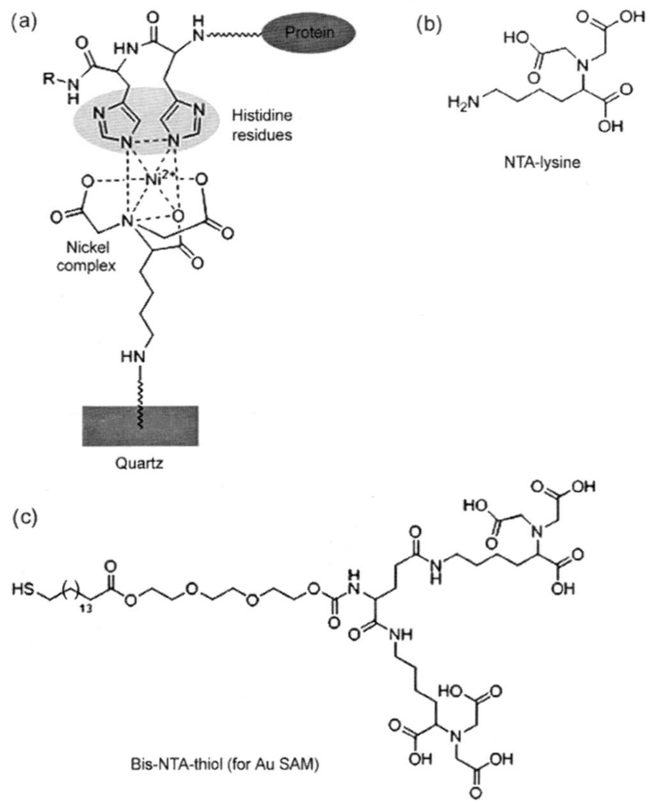

FIGURE 20.4. (**a**) Binding of a His6-tagged protein to a Ni-NTA-functionalized quartz surface. The protein binds through two of its histidine residues. (**b**) NTA-lysine, typically used for preparing Ni-NTA surfaces. (**c**) A multivalent, bis-NTA-thiol showing improved binding capabilities for His6-tagged proteins over single NTA groups [2].

[37–41]. Figure 20.4 illustrates how the tetradentate ligand NTA forms an hexagonal complex with different divalent metal ions (usually Ni^{2+}), leaving two binding positions available for binding to a His_6 sequence. It is possible to reverse the immobilization by the addition of ethylenediaminetetraacetate (EDTA) or imidazole, as is the case in conventional Ni-NTA affinity chromatography. NTA derivatives can be covalently bound through active esters (*N*′-(3-dimethylaminopropyl)-*N*-ethylcarbodiimide (EDC),NHS) or maleimide derivatives to dextran surfaces or glass slides, for example. NTA was installed on gold after reaction of maleimide NTA with *N*-succinimidyl-*S*-acetylthiopropionate or directly through thiolated NTA derivatives [37, 40, 42]. To overcome the relative low affinity of the His tag to the Ni-NTA complex ($K = 10^7\ M^{-1}$), potentially leading to unwanted dissociation of immobilized proteins, an increase in the surface density of NTA groups can increase the binding affinity of the His_6 tag to NTA receptors by several orders of magnitude due to the multivalency principle [43].

Another strategy adapted from affinity chromatography is the specific binding of biotin to the proteins avidin or streptavidin (SAv). SAv comprises four identical subunits, each of which binds one biotin molecule. Owing to the high binding affinity between streptavidin and biotin ($K = 10^{13}$–$10^{15}\ M^{-1}$), the formation of this complex can be regarded as nearly irreversible, on a scale nearly comparable to a covalent bond [44]. Biotin–SAv bond formation is very rapid and is not affected by pH value, temperature, organic solvents, enzymatic proteolysis, or other denaturing agents. This high-affinity biotin–SAv binding system has found many surface applications and has been reviewed recently [45]. One frequently used design principle is fabricating SAv monolayers in a stacked composition biotin–SAv–biotin. For instance, the biotin layer directs the order in the SAv layer, with two biotin-binding sites facing the surface-bound biotin layer and the other two sites facing outward for capturing biotinylated proteins. This order can be combined with good control over the surface density of biotin groups in the SAMs, for example using mixed SAM monolayers composed of two thiol species, one of which is biotinylated and the other one not [46] (Figure 20.5).

Biotinylation of proteins has been carried out historically by standard bioconjugation techniques using chemically activated biotin derivatives (Table 20.2). To avoid random biotinylation and subsequent inactivation of proteins, site-specific labeling of proteins using biotin ligase strategies [47] or tag-free intein-based methods have been developed [46, 48].

Another strategy to immobilize proteins makes use of nature's own protein capturing agents such as antibodies and antibody-binding proteins (e.g., protein A and protein G). Microarrays have been generated by printing an array of monoclonal or polyclonal antibodies, antibody fragments, or synthetic polypeptide ligands [49]. However, the lack of specificity due to their large area for interaction caused by their glycosilation can cause cross-reactivity between target proteins, and this lack of specificity can potentially lead to large numbers of false positives and negatives. Control over orientation can be achieved by using, for example, protein A, an available natural IgG binding protein. Detailed reviews on oriented immobilization of antibodies and the application of such chips for immunoassays have been published [49]. Current research is focused on further increasing control over the orientation of protein A attachment, which obviously influences the antibody orientation using several approaches [50–52].

More recently, a versatile approach has been taken to identify peptides with specific affinity to different inorganic materials [53, 54]. Biology is used as a guide to understand, engineer, and control peptide–material interactions and exploit them as a new design tool for novel materials and systems. Short peptides with specificity to a variety of practical materials can be selected adapting protocols of combinatorially designed peptide libraries. These genetically engineered peptides for inorganics (GEPI) can be used

TABLE 20.2. Typical Reagents Used for Protein Biotinylation

Name	Use	Structure
Biotin NHS	Amine-reactive	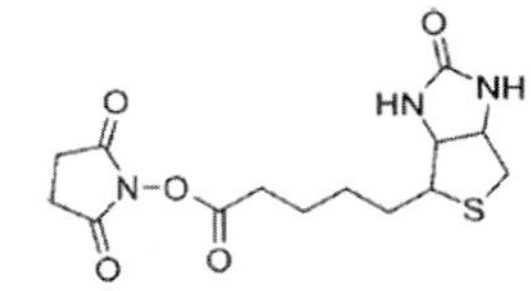
Biotin hydrazide	Carbohydrate-reactive	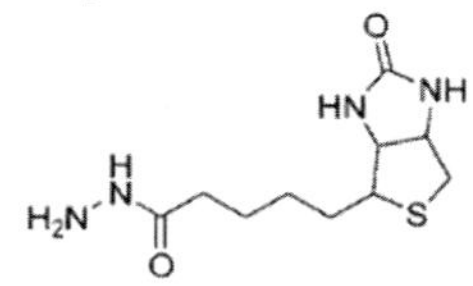
Pentylamine biotin	Carboxylic acid reactive, Pierce—used with coupling reagent EDC	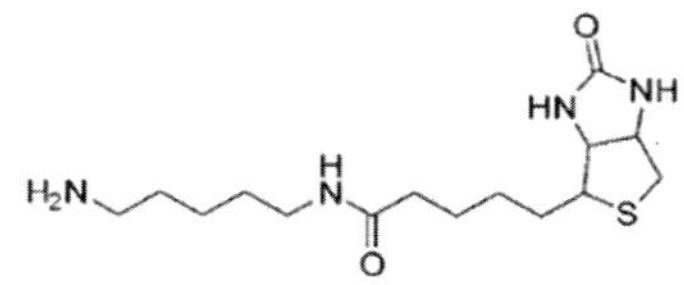
Biotin BMCC	Thiol-reactive, Pierce	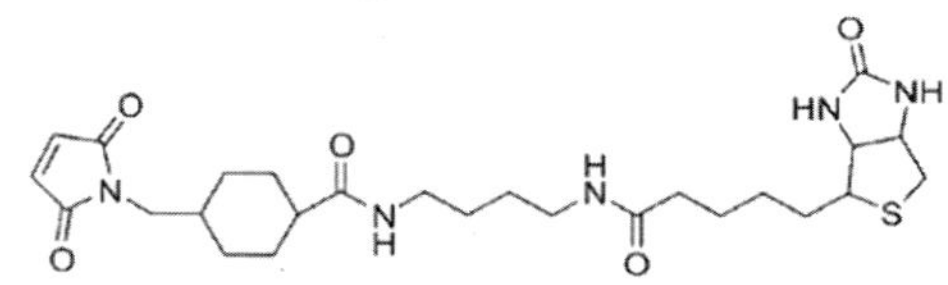

Source: Reproduced from reference [2] with permission.

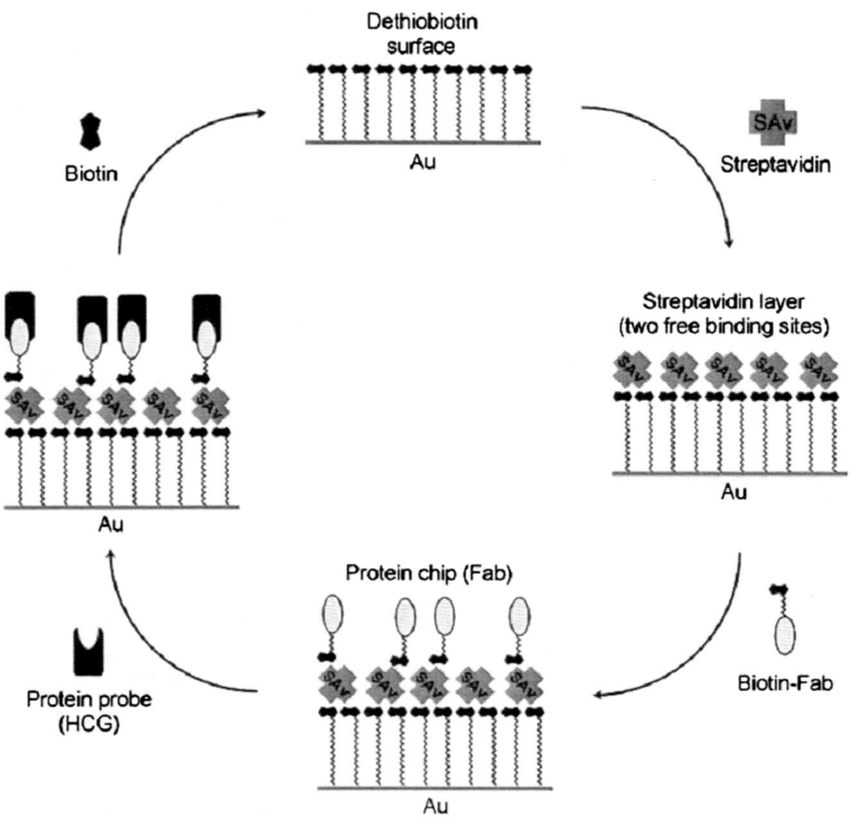

FIGURE 20.5. Creating protein chips using the biotin/SAv/biotin template approach: A SAM on Au displaying dethiobiotin is incubated with SAv, creating a SAv surface with two free biotin binding sites per SAv molecule. Incubation with a biotinylated target protein (here an antibody fragment, biotinylated anti-HCG Fab) results in the final protein chip, which was then used to detect a protein probe, human chorionic gonadotropin (HCG) [2].

to bind desired proteins to desired surfaces. This molecular biomimetic approach opens up new avenues for the design and utilization of multifunctional molecular systems in a wide range of applications from tissue engineering, disease diagnostics, and therapeutics to various areas of nanotechnology where integration is required among inorganic, organic, and biological materials.

20.2.2.2 DNA. Nucleic acids are formed by units called nucleotides, organic compounds made up of a nitrogenous base, a sugar, and a phosphate group. Some of the methods described above can also be applied to attach modified DNA molecules. For example, DNA can be biotinylated [55] and bound to a streptavidin surface.

The specificity of the base pairing of two complementary single strands can be used to direct the assembly of different molecules or complexes to a surface [55]. This oligonucleotide-directed immobilization provides exceptionally high stability and unique site selectivity and relies on well-established DNA chip production technology [56]. However, it is still a demanding task to incorporate oligonucleotides into large proteins. Syntheses that couple thiopyridyl- or maleimido-modified oligonucleotides to cysteine residues of proteins, succinimide-modified oligonucleotides to lysine residues of proteins, or aldehyde-modified oligonucleotides to hydrazine-modified antibodies have been developed to achieve this purpose [57–59].

DNA can also be used to create arbitrary two- and three-dimensional shapes at the nanoscale. The specificity of the interactions between complementary base pairs and a smart design of the sequence allow using DNA as a construction material to create nanoscale self-assembled structures on surfaces [60].

20.2.2.3 Lipids. Biological membranes are constituted mainly by different kinds of phospholipids, amphipathic molecules that self-assemble into a 5 to 7 nm-thick liquid-crystalline bilayer. This two-dimensional dynamic structure forms the cell envelope that defines and controls cellular organization in living systems. Although biological membranes are rich in proteins, it is essentially the lipid molecules that make them elastic two-dimensional fluids with diverse physical–chemical properties. Over the years, different experimental artificial platforms have been developed to perform biophysical studies of lipid bilayers and lipid–protein complexes in model systems; recently the field has expanded greatly, attracting the attention of different disciplines. There is, on the one hand, increasing interest in understanding fundamental aspects of how these chemically heterogeneous materials, displaying a very rich phase behavior and dynamics, produce such a large set of functions in biological membranes. On the other hand, it is well accepted that biomembranes are extremely interesting technological platforms for the development of biosensors. Several journals have published recently monographic issues dedicated to the field of supported lipid membranes and their applications [61,62]. A general overview of the different approaches currently used to deposit functional biomembranes on different surfaces will be presented here, and the readers are referred to more specialized review articles on the subject for further details on available characterization techniques or on more specialized applications [63–67].

Spreading a fluid lipid membrane on a solid support is a very attractive and convenient surface modification protocol useful for many applications that require interfacing biological materials with inorganic structures. Coupling these two types of materials is an essential step to read the activity of biological processes using a physical output, being it the quantification of a protein–protein interaction process or the opening or closing of an ion channel embedded in the membrane. Surface plasmons total internal reflection fluorescence, surface enhanced Raman [68, 69], electrochemical readings [70], or surface acoustic waves [71] all require arranging the biological materials in close proximity to an inorganic surface. A lipid bilayer immediately renders any underlying surface—gold, carbon, or any other metal oxide—friendlier to proteins. Additionally, the phospholipid composition of the membrane can be modified to include different binding motifs to provide specific anchors to the proteins. A large variety of phospholipid derivatives with some of the anchoring motives described earlier as covalent protein linkers are commercially available. Thus, one can incorporate phospholipids modified with biotin, NTA, cisteine, or maleimide groups into the membrane to serve as protein anchors.

Supporting lipid membranes on a solid surface makes them accessible to a wide variety of surface-specific analytical techniques, so this strategy has been used for many years to prepare model systems to study biophysical properties of lipid membranes. Figure 20.6 illustrates some of different methods that have been traditionally employed [69, 72].

Langmuir–Blodgett films (Figure 20.6b) were first used over 50 years ago. The technique consists on using a motorized stage to move the substrate between an aqueous and a gas phase. A lipid monolayer is held at a defined tension at the interface, which controls the packing density. Using a hydrophilic substrate and starting in solution, a lipid monolayer can be deposited by removal of the substrate [73]. Vertical reinsertion of the lipid monolayer formed through the interface deposits a (second) monolayer on top, which results in a lipid bilayer. If the substrate is instead taken through the interface horizontally the same result is achieved, but it is referred to as Langmuir–Schäfer deposition or sometimes as "tip-dip" when capillaries are used [74]. Although the contribution of Langmuir–Blodgett technology to the study of biomembranes has been important, the limited stability of the membranes, the difficulties encountered when incorporating

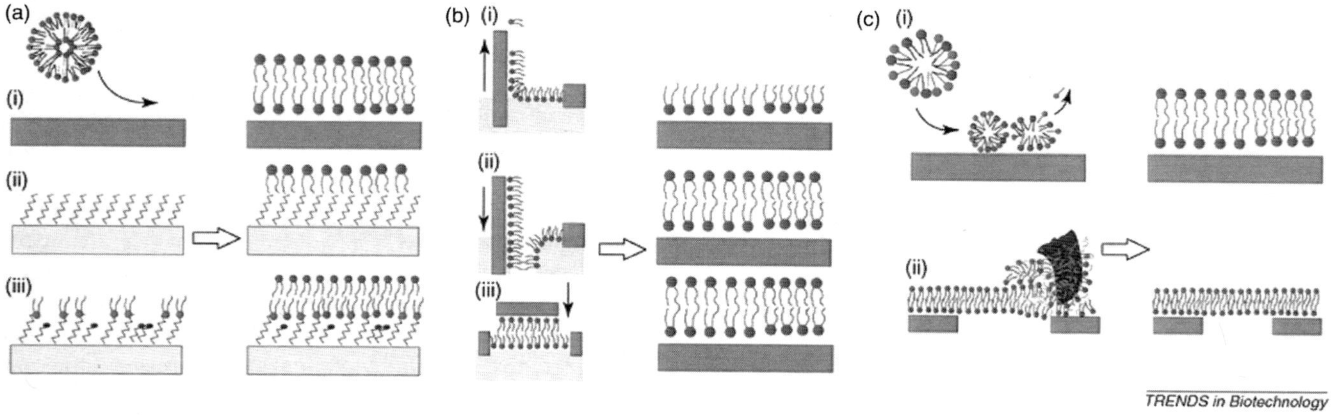

FIGURE 20.6. Methods used to assemble lipid membranes on sensor substrates (see text for a more detailed description). (**a**) Vesicle fusion: (i) on hydrophilic substrates and (ii) on hydrophobic surfaces. (iii) More complex tethers. (**b**) Langmuir–Blodgett techniques. (**c**) Detergent dialysis and painting. (i) Formation of micelles of lipids mixed with detergents (ii) Painting and solvent extraction: a bilayer at the substrate–aqueous solution interface [64].

proteins, and the demanding preparation protocols have limited their more extensive use, particularly in more applied fields as the development of biosensors.

In 1984 Brian and McConnell [75] first described a simpler procedure to form solid supported lipid membranes. They showed that lipid vesicles formed in solution spontaneously fused on hydrophilic surfaces to form lipid bilayers that, despite their stable attachment to solid substrates, retained the fluidity of biological membranes (Figure 20.6a). This simple method of preparing supported membranes has extended in recent years and has given rise to further optimizations, because techniques such as the quartz crystal microbalance have provided information about how the process takes place [76]. Vesicles fuse spontaneously to form bilayers on diferent hydrophilic substrates (e.g., silicon oxide and silicon nitride) [77–79], or monolayers when fused on hydrophobic surfaces (e.g., preformed thiol-alkyl monolayers) [80]. More complex tethers providing additional aqueous space under the self-assembled lipid layer, such as hydrophilic spacers with covalently bound lipids, can also be used to drive liposome fusion on the surface [81].

Another way of forming supported bilayers is shown in Figure 20.6c. Micelles of lipids mixed with detergents can be used to deposit the lipid material in aqueous solution at the solid interface. The detergent is continuously removed from the micelles by dialysis leading to decomposition of the micelles and the formation of a planar lipid bilayer [82]. Alternatively, if a drop of organic solvent containing dissolved lipids is added to a surface in an aqueous phase, the amphiphilic lipids will align at the solvent interface. When the solvent is extracted, the lipids at the interface fuse to form a bilayer at the substrate–aqueous solution interface [83].

Depending on the motivation for preparing the model systems or the analytical technique to be employed for its characterization, the solid substrate to deposit the lipids may vary. Gold or other conducting surfaces are required for electrochemical studies or for detection of interactions using surface plasmon resonance [84], whereas different oxides might be appropriate to couple the bilayers to other substrates used for acoustic sensing or other medical applications [38, 71, 85]. Glass or mica are, on the other hand, always convenient hydrophilic substrates whenever optical techniques needing transparent substrates are to be used.

The surface-supported bilayers can be also used as a substrate to study proteins. If the proteins are soluble, they can be oriented using lipids with specific binding motifs. But it is also of great interest to use these model systems to study membrane proteins. The selective transport of molecules and ions across the lipid bilayer to create and modify ion gradients is performed in the cells by integral membrane proteins. The fact that signal transduction events governed by these proteins guide many important processes makes them very interesting from a pharmacological point of view. Over 50% of all current drugs target membrane proteins [86]. It is thus essential to characterize membrane barrier properties and the changes that occur when these proteins are incorporated into a lipid membrane, activated or blocked. In these cases, the proximity of the solid surface to the membrane present in the preparations described above imposes important constrains.

Over the years, different approaches have been taken to overcome this limitation and provide membrane proteins with the appropriate environment to retain them fully functional. In the 1960s a method of preparing the latter called "Black lipid membranes" was described [72, 87] to study proteins that served as ionic channels. The name derives from their appearance by optical microscopy. When Mueller and Tien [87] observed the formation of the first black lipid membranes from extracted brain lipids, they noted

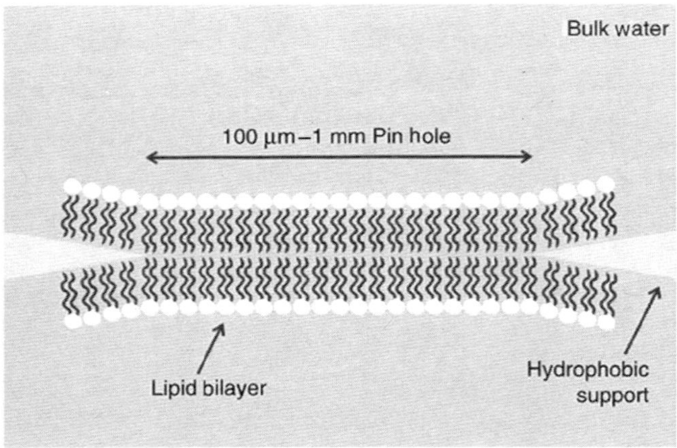

FIGURE 20.7. Illustration of a black lipid membrane. The phospholipid membrane spans a 100-μm to 1-mm pinhole in a hydrophobic support [123].

interference bands giving rise to color in the membrane. This interference effect disappeared during the thinning of the painted lipid mass and is thought to indicate the formation of a single bilayer membrane, as shown (Figure 20.7). An excellent resource on black lipid membranes is reference 69. All methods for producing black lipid membranes involve the formation of a membrane over a small aperture, usually less than 1 mm in diameter. The hole is formed in a hydrophobic material such as polyethylene or Teflon and is usually part of a wall separating two compartments that can be filled with aqueous solution, each containing a reference electrode. The result is a bilayer suspended over the aperture with an aqueous compartment on each side. These black lipid membranes have been used to investigate various biophysical processes. One of the most important ones is the formation of ion channels in phospholipid bilayers by peptides [88], proteins [89], antibiotics [90], and other pore-forming biomolecules. The membranes are suspended in solution, and there are no unwanted interferences of the membrane with an underlying support. The absence of such a support also means that transmembrane proteins suspended within the phospholipid bilayer remain fully mobile and active. However, this also limits the lifetime of the bilayer due to poor stability of the membrane. The methods of detection that can be employed with black lipid membranes are also typically limited to electrical conduction and simple light microscopy.

The strong motivation to create robust supported membrane systems compatible with the incorporation of functional membrane proteins and the detection of their activity has spurred the search for more robust alternatives to the black lipid bilayers. The idea behind is that in vitro membrane systems enabling direct incorporation and integrated electrochemical or voltage clamp measurements of membrane protein function (i.e. ion, charge, and liquid transport) could potentially revolutionize current technologies in drug screening. Methods currently used for screening and profiling in the pharmaceutical industry are limited to the monitoring of binding events. But even extraction of binding affinity requires multiple experiments at a range of concentrations. Direct measurement of function in response to (drug) stimuli will enable quicker and better selection of leads and toxic hits. The same kind of platforms could be used for advanced biomimetic sensing [64]. Different strategies have been developed to improve the stability of the freestanding membranes. They go from the preparation of tethered membranes [81, 91]—in which a spacer, a polymer [92], or a protein matrix [93] is placed between the lipid and the solid substrate to provide an aqueous environment to lodge the cytoplasmic side of the transmembrane proteins—to the development of nano black lipid membranes, in which the small size of the pore increases the mechanical stability of the membranes [94–96] and allows the formation of membrane arrays of potential use in biosensing [64, 83]. Figure 20.8 schematically illustrates some of the ways that membranes can be supported on a surface and Figure 20.9 illustrates several ways in which they can be used as biosensor platforms.

Supported lipid bilayers on a hydrophilic semiconductor or oxide substrate can be used for direct assembly of a supported lipid bilayer (SLB), and the support will act as a working electrode (WE) for electrochemical measurements (Figure 20.9a). Alternatively, the bilayers can be formed on covalently attached hydrophobic molecules with a hydrophilic linker—often derived from lipids—used to tether and support the lipid membrane to a gold WE [97, 98] (Figure 20.9b). Free-spanning membranes or black lipid

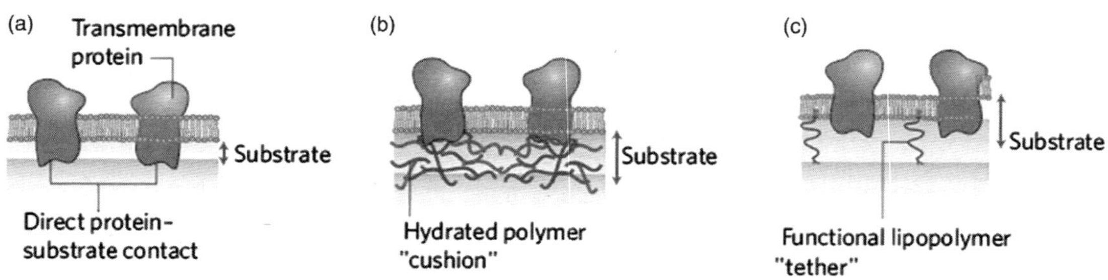

FIGURE 20.8. Supported membranes. (**a**) Solid-supported membrane. (**b**) Membranes on a polymer cushion. (**c**) Lipopolymer tethers. (Figure adapted from reference 152).

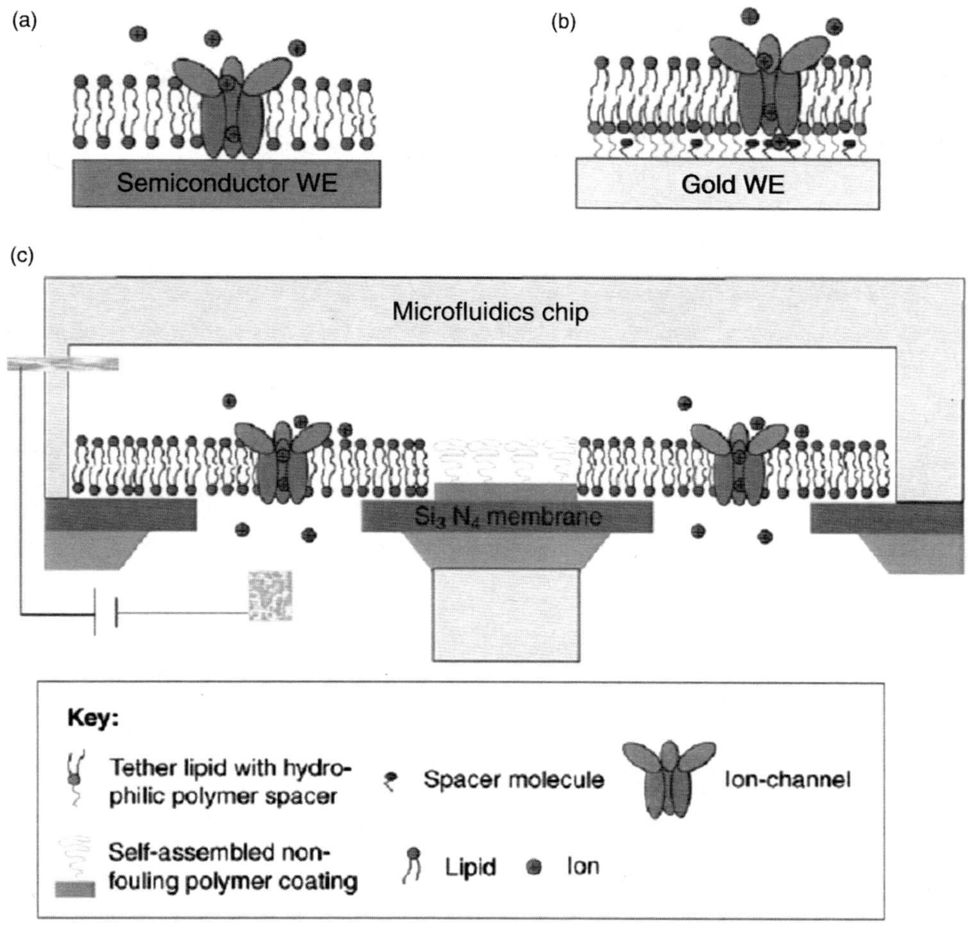

FIGURE 20.9. A selection of membrane sensor platforms. (**a**) Supported lipid bilayer on hydrophilic support. (**b**) Tethered supported lipid bilayer. (**c**) Free-spanning membrane or black lipid membrane [64].

membranes can also be used in self-assembled membrane sensor arrays [99]. Lipid membranes containing ion channels separated into functional spots, for example by nonfouling polymer barriers, span apertures in a solid support, providing free liquid access on both sides to perform conducting electrochemical measurements. The schematics in Figure 20.9c show how such a platform combined with microfluidics could be used for parallel voltage clamp measurements on different single transmembrane proteins.

Using a polymer cushion or a protein monolayer for spacing the membranes away from the substrate has the additional advantage of increasing the stability of the supported membrane. Figures 20.10 and 20.11 illustrate the use of nanostructured two dimensional arrays of proteins, the naturally occurring bacterial S-layer, to improve supported membrane technology. Figure 20.10 shows the structure of the S-layer, and Figure 20.11 illustrates various ways in which it can be used to support lipid membranes.

Progress has also been made in developing air-stable lipid membranes. Unprotected solid supported lipid bilayers are known to delaminate from the supporting substrate upon passage through an air–water interface [100]. This is problematic when developing practical biosensors based upon supported lipid bilayers because the membrane must be constantly hydrated. It is therefore highly advantageous if the system can be dried after fabrication and rehydrated just prior to use. Systems affording air stability include hybrid bilayers [101], protein-stabilized lipid bilayers [100], and polymerized membranes formed using synthetic diacetylene-containing phospholipids [102, 103]. Some of these strategies compromise the mobility of the lipids on the surface, but recently an air-stable system has been developed that maintains high lipid mobility and is still capable of binding analyte proteins to ligands presented at the lipid bilayer surface [104].

20.2.2.4 Polysaccharides. Carbohydrates are also important components of biological materials. Although their tertiary structures are less defined than those of proteins and DNA, several polymeric carbohydrate structures as chitin,

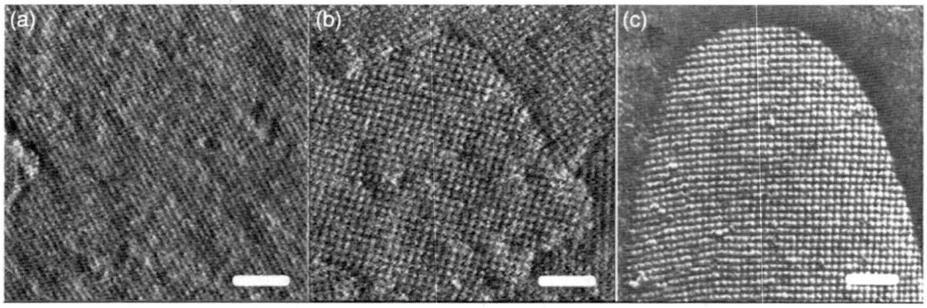

FIGURE 20.10. Atomic force microscopical images the S-layer protein from *Geobacillus stearothermophilus* PV72/p2 exhibiting an oblique S-layer lattice (**a**) and the S-layer protein from *Lysinibacillus sphaericus* CCM 2177 exhibiting a square S-layer lattice (**b**). In the latter image, crystalline patches forming the closed S-layer lattice are visible. The bars correspond to 50 nm. (c) Electron micrograph of a freeze-etched and Pt/C-shadowed preparation of a Gram-positive organism exhibiting a square S-layer lattice. The bar corresponds to 100 nm [153].

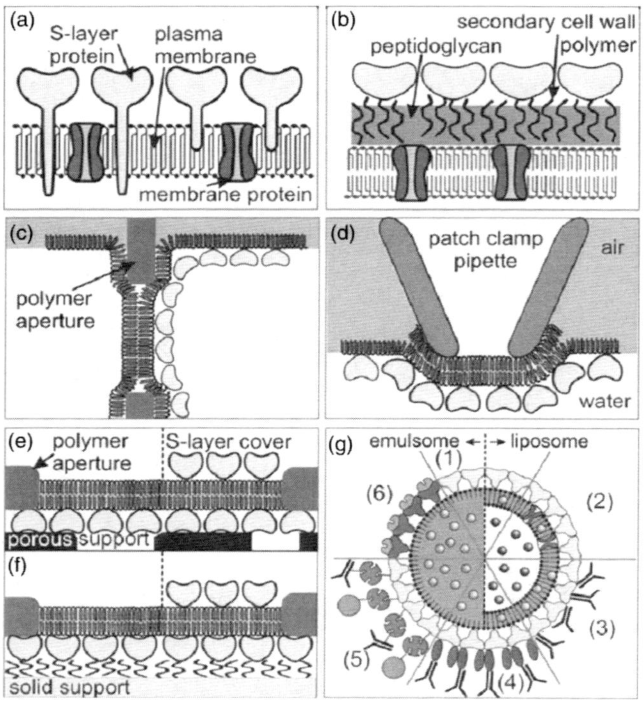

FIGURE 20.11. Supramolecular structure of an archaeal (**a**) and Gram-positive bacterial cell envelope (**b**). Schematic illustrations of various S-layer-supported lipid membranes: (**c**) A painted membrane. (**d**) A bilayer lipid membrane generated across an orifice of a patch clamp pipette. (**e**) A lipid membrane generated on an S-layer ultrafiltration membrane (SUM). (**f**) A solid support covered by a layer of modified secondary cell wall polymer (SCWP) with a closed S-layer lattice assembled. (**g**) Schematic drawing of (1) an S-layer-coated emulsome (*left part*) and S-liposome (*right part*) with entrapped water-soluble or lipid-soluble functional molecules and (2) functionalized by reconstituted integral membrane proteins. Immobilization of functional molecules (e.g., IgG) by direct binding (3), via the Fc-specific ligand protein A (4), or via biotinylation (5). Emulsomes and liposomes coated with S-layer fusion proteins incorporating functional domains (6) [153].

cellulose, or hyaluronic acid play an important role in maintaining cellular integrity and have important, although not yet well understood, roles in cell signaling and communication. Synthetic carbohydrate-based polymers are being increasingly explored as biodegradable, biocompatible, and biorenewable materials for use as water absorbents, chromatographic supports, and medical devices [105]. Moreover, synthetic polymers bearing sugar residues can also offer a good surface for cell attachment and, thus, might be applied to study cell recognition events in antimicrobial/viral and tissue engineering. Different carbohydrate polymers, namely, dextran [106], cellulose [107], chitosan [108], polyelectrolytes [109, 110], and lipopolymer tethers [111], have been explored as polymer cushions for supported membranes. Alternatively, supported lipid bilayer (SLB) are being used to create well-defined films with good control of their molecular attachment to the substrate [112]. This confinement to a solid support makes these model coats accessible to characterization with a range of surface-sensitive techniques that are not easily applicable on living cells. These model systems contribute to the basic understanding of the biological role played by these materials and their interaction with different proteins.

20.3 SURFACE PATTERNING

The protocols described above refer to modifications that can be either homogeneously covering the surface or localized in defined regions. The ability to pattern surfaces with monolayers and multilayers is central to numerous chemical studies in fields ranging from sensor design to microelectronics. Protocols developed to pattern monolayers can also be applied to localize the surface modifications used to incorporate biological materials onto surfaces. Depending on the interest, patterning can order single molecules at the subnanometer scale or form vast assemblies over macroscopic areas. Moreover, this chemistry can range from simple homogeneous systems to mixed monolayers with complex patterns [113]. Soft lithographic techniques were one of the first approaches to micropatterning surfaces with organic materials [114]. Microcontact printing was originally developed for the patterning of alkane thiols onto gold substrates [115] and involves the use of a stamp such as poly(dimethylsiloxane) (PDMS), which has been molded against a lithographically patterned surface [114]. The stamp then transfers chemically or biologically relevant materials to a solid substrate. More recently, other more sophisticated methods have also become available. Patterns can be created by direct scratching the surface at the nanoscale using the tip of a scanning probe microscope [116], or using electrochemical patterning techniques [117]. Photopatterning and click chemistry are also highly attractive methods [118]. Patterning SAMs on glass, gold, carbon, or plastic, as well as in additional substrates, continues to become easier to do and better understood. Increasing efforts have been put into the development of novel interfaces to mimic protein–protein interaction at the cell membrane [119] and allow spatial control of the density and activity of biomolecules on two-dimensional (2D) solid supports. Although serial processes such as dip-pen nanolithography (DPN) have paved the way for nanopatterning proteins in the sub-100-nm range with high quality [120], cost- and time-efficient production of areas adequate to the demands of cell-biology experiments or high-throughput protein and DNA screening is still challenging. Parallel processes such as soft lithography or nanoimprint lithography [121] are capable of producing large-area patterns but lack the resolution to reach the single-protein or macromolecular length scale. A way to determine the spacing of proteins on the surface is to biofunctionalize gold nanoparticle arrays whose spacing is controlled by means of block-copolymer micelle nanolithography (BCMN) [122].

Surfaces modified with lipids are amenable to additional patterning strategies. They retain the same two-dimensional fluidity at the liquid–solid interface that lipid membranes possess in vivo [123] and that makes them especially attractive for material patterning. The first method for patterning surfaces with solid-supported phospholipid bilayers was developed in 1997 [124]. A typical formation procedure involved the patterning of photoresist on fused quartz wafers by means of standard photolithographic techniques. Small unilamelar lipid vesicles (SUVs) were then fused onto the substrate between the barriers, creating a lithographically patterned array of essentially identical planar supported phospholipid membranes. Each membrane was confined within its own two-dimensional corral. The bilayers retained two-dimensional fluidity within a given corral, but the barriers did not allow mixing between neighboring patches as was demonstrated by fluorescence microscopy [125] (Figure 20.12).

An alternative strategy applicable to solid-supported as well as polymer-supported systems consists in microcontact printing grid-like diffusion barriers using hydrophobic species that attach to the support surface [62, 126, 127] (Figure 20.13). These barriers then effectively separate the subsequently deposited membrane into isolated compartments.

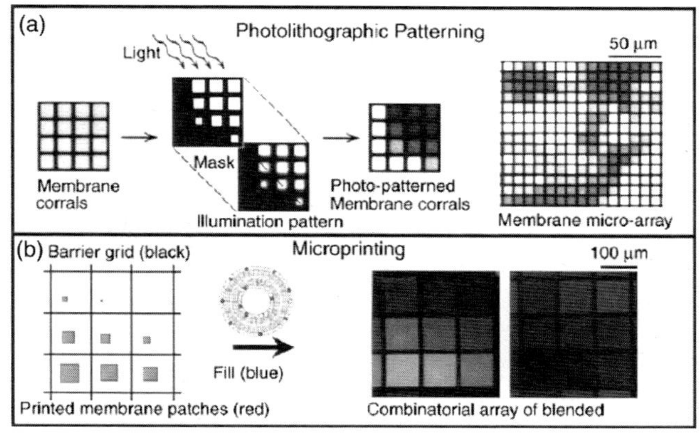

FIGURE 20.12. Composition arrays generated by photopatterning. (**a**) A mask is used to selectively bleach different sized areas of a membrane array. After diffusive mixing within each corral, a concentration array is observed. (**b**) Bilayer patches of different size are miconcontact printed, and the empty space in each corral is backfilled with SUVs to form a continuous bilayer of different composition in each corral. Shown here is an epifluorescence image of printed Texas Red-labeled membranes backfilled with Cascade Blue-labeled lipids [127].

In addition to simple membrane patterning, spatially addressed arrays of solid-supported phospholipid bilayers have also been produced. Spatial addressing enables complete control over the chemical composition of each address in a supported bilayer array. This was first achieved by pipetting from pulled capillaries. Additional methods include microcontact printing [128], laminar flow deposition [129] (Figure 20.14), and robotic pin printing [130].

The same surface modification protocols that have been described to modify flat inorganic surfaces with biological materials can be extended to the modification of surfaces with other geometries such as nanotubes or nanoparticles when required for their use in medical applications [131–133]. The self-assembling capacity of the lipid membranes and their flexibility make them particularly versatile and adaptable to use in the modification of surfaces of different geometries [134]. They can be used to align molecules in an ordered array on many surfaces, including carbon nanotubes [135, 136] or nanoparticles of different composition such as metals [137], quantum dots [138, 139], gold nanoparticles [140], or silica beads [141, 142].

20.4 SOME EXAMPLES

All the above-mentioned strategies for surface biofunctionalization find wide applications in fields such as biosensor

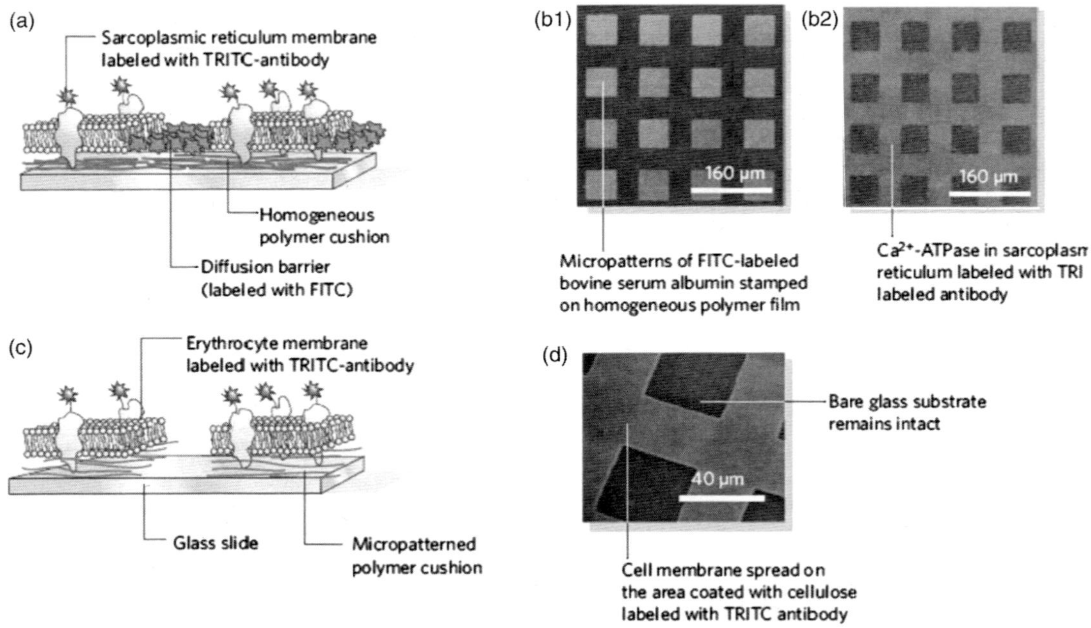

FIGURE 20.13. Membrane patterning. (**a**) Sarcoplasmic reticulum membranes confined by diffusion barriers, established by the microcontact printing of water-soluble protein (bovine serum albumin labeled with FITC. (**b1**) On a homogeneous cellulose film. (**b2**) The cytoplasmic domain of Ca^{2+-}ATPase is visualized with TRITC-labeled antibody. (**c**) Incubation of human erythrocyte ghosts with lithographically micropatterned cellulose films results in selective spreading of cell membranes on the area coated with cellulose. (**d**) Cytoplasmic domain of the proteins in the erythrocyte membrane (band III) is visualized with antibodies conjugated with a fluorescent dye (TRITC) [152].

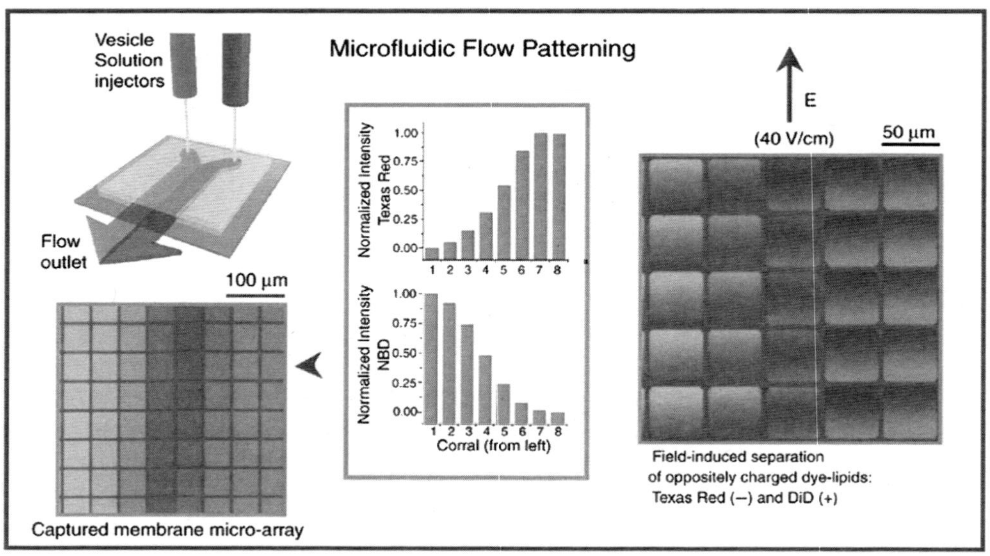

FIGURE 20.14. Addressing by laminar flow in a microfluidic channel. Diffusive mixing in a microchannel under laminar flow conditions provides a concentration gradient of different dye-labeled vesicles. The concentration of vesicles in the gradient is reflected in the surface concentration of each membrane in the resultant array. The array shown is a mixture of Texas Red-labeled lipids and DiD-labeled lipids. Since the dyes have opposite charge, they can be separated in an electric field. See color insert [127].

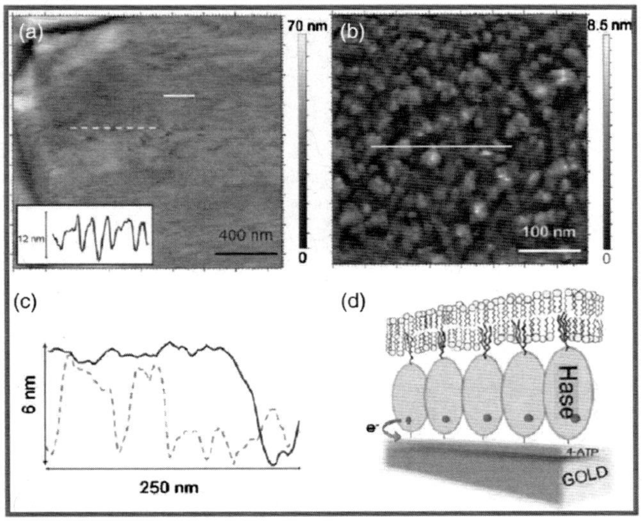

FIGURE 20.15. (**a**) Tapping-mode AFM topography of a 4-ATP-modified gold plate to which membrane hydrogenase has been covalently immobilized in the presence of phospholipids and CAL-BIOSORB adsorbent. The inset represents the z-axis profile across the dashed white line. (**b**) AFM topography of the same system after 2 min of incubation with 1 μM Triton X100. (**c**) z-axis profiles across the solid lines in panel (a) (black line) and panel (b) (cyan line). (**d**) Scheme depicting the covalent and oriented immobilization of hydrogenase molecules with the subsequent formation of a membrane on top of the proteins. The direct electron transfer between the active center of the enzyme and the electrode is represented [145].

development, single-molecule or single-cell biophysical studies, and medical research.

Proper binding of proteins to surfaces is important for biosensor development. It is quite common to use redox enzymes, proteins that oxidize or reduce different ligands, to detect their presence by reading the electrochemical current generated by the enzymatic oxidation or reduction of the molecule. If the enzyme is coupled to the electrode, the current generated by the enzymatic activity can be detected. It is of interest to attach the proteins with the correct orientation to favor direct electron transfer between the active site and the electrode, avoiding this way the need for soluble redox mediators to shuttle the electrons from the active enzyme to the electrode surface. Several works illustrate how self-assembled monoloayers on gold or on carbon can be properly used to orient redox proteins (soluble or membrane proteins) on electrode surfaces achieving direct electron transfer [133,143–147] (see Figure 20.15).

The controlled oriented attachment of proteins on surfaces is also of great use for biophysical studies. For example, adequate attachment of kinesins (a motor protein used in cells for transport of nanoscale material) on a solid surface allows harnessing their activity to move and assemble synthetic components in a "molecular shuttle" (Figure 20.16).

Controlled lipid and protein assemblies on surfaces have contributed to the development of biosensors for high-throughput screening. Besides the use of supported membrane arrays on flat, polymer covered or porous membranes [64, 65], liposome arrays on surfaces are also being explored [68, 148, 149].

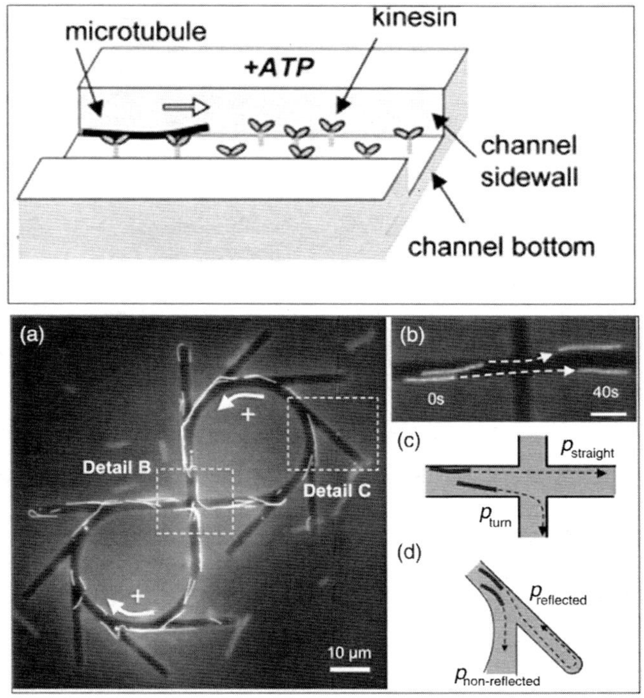

FIGURE 20.16. (**Top**) The motor protein kinesin adsorbed at the bottom of microfabricated channels translates microtubules. Detail shown in (B) illustrates the sorting "figure 8"-shaped track shown in (A). This Fluorescent microscope image shows two microtubules before (0 s) and after (40 s) crossing the junction. Dashed lines denote the paths of the microtubules. (C) In crossing junctions, microtubules generally continue straight through the intersection while only a few are redirected along one of the perpendicular paths. (D) At unidirectional reflector junctions, microtubules were trapped in the arm more often than turning. Fluorescent microscope composite image shows two microtubules before (0 s) and after (40 s) crossing the junction. Dashed lines denote the paths of the microtubules. (**C**) In crossing junctions, microtubules generally continue straight through the intersection while only a few are redirected along one of the perpendicular paths. (**D**) At unidirectional reflector junctions, microtubules were trapped in the arm more often than turning. [154].

More sophisticated surface modifications that combine polymer chemistry with gold-particle nanopatterning and peptide modifications allow us to perform biophysical studies of cell behavior that address the effect of substrate elasticity and intermolecular spacing on cell adhesion properties [150, 151] (see Figure 20.17).

These are only very few examples of the very large number of applications in which the biofunctionalization of a surfaces is an essential initial step. Most of the work described refers to passive surface modifications that orient and spatially distribute the biological molecules of interest. It is expected that

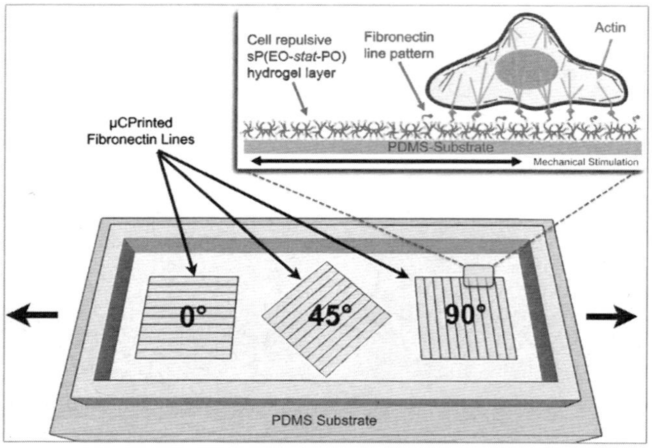

FIGURE 20.17. Schematic diagram of the PDMS stretchable substrate (not to scale). The entire surface of the culture medium reservoir (70 × 50 × 5 mm) was passivated with an NCO-sP(EOstat-PO) coating. Insert in upper-right-hand corner shows the layer structure of bovine fibronectin (5 mg/mL) onto the NCO-sP(EO-stat-PO) and PDMS. Adhesive lines have orientations of 0°, 45°, and 90° relative to the strain direction (*pointing arrows*) [155].

this field will continue to grow and develop toward increasing the complexity and functionality of these hybrid surfaces to provide active and responsive surfaces that will find multiple medical and biosensing applications.

REFERENCES

1. Lejeune, M., Valsesia, A., Kormunda, M., Colpo, P., Rossi, F. (2005). Structural characterization of nanopatterned surfaces. *Surface Science*, **583**, L142–L146.
2. Jonkheijm, P., Weinrich, D., Schroder, H., Niemeyer, C. M., Waldmann, H. (2008). Chemical strategies for generating protein biochips. *Angewandte Chemie International Edition*, **47**, 9618–9647.
3. Wilson, D. L., Martin, R., Hong, S., Cronin-Golomb, M., Mirkin, C. A., Kaplan, D. L. (2001). Surface organization and nanopatterning of collagen by dip-pen nanolithography. *Proceedings of the National Academy of Sciences*, **98**, 13660–13664.
4. Lee, K.-B., Lim, J.-H., Mirkin, C. A. (2003). Protein nanostructures formed via direct-write dip-pen nanolithography. *Journal of the American Chemical Society*, **125**, 5588–5589.
5. Silberzan, P., Leger, L., Ausserre, D., Benattar, J. J. (1991). Silanation of silica surfaces. A new method of constructing pure or mixed monolayers. *Langmuir*, **7**, 1647–1651.
6. Howarter, J. A., Youngblood, J. P. (2006). Optimization of silica silanization by 3-aminopropyltriethoxysilane. *Langmuir*, **22**, 11142–11147.
7. Sagiv, J. (1980). Organized monolayers by adsorption. 1. Formation and structure of oleophobic mixed monolayers on solid surfaces. *Journal of the American Chemical Society*, **102**, 92–98.
8. Pathak, S., Singh, A. K., McElhanon, J. R., Dentinger, P. M. (2004). Dendrimer-activated surfaces for high density and high activity protein chip applications. *Langmuir*, **20**, 6075–6079.
9. Strother, T., Cai, W., Zhao, X., Hamers, R. J., Smith, L. M. (2000). Synthesis and characterization of DNA-modified silicon (111) surfaces. *Journal of the American Chemical Society*, **122**, 1205–1209.
10. Pallu, S., Bourget, C., Bareille, R., et al. (2005). The effect of cyclo-DfKRG peptide immobilization on titanium on the adhesion and differentiation of human osteoprogenitor cells. *Biomaterials*, **26**, 6932–6940.
11. Yang, W., Butler, J. E., Russell, J. N., Hamers, R. J. (2004). Interfacial electrical properties of DNA-modified diamond thin films: Intrinsic response and hybridization-induced field effects. *Langmuir*, **20**, 6778–6787.
12. Liberelle, B., Banquy, X., Giasson, S. (2008). Stability of silanols and grafted alkylsilane monolayers on plasma-activated mica surfaces. *Langmuir*, **24**, 3280–3288.
13. Kim, K., Yang, H., Jon, S., Kim, E., Kwak, J. (2004). Protein patterning based on electrochemical activation of bioinactive surfaces with hydroquinone-caged biotin. *Journal of the American Chemical Society*, **126**, 15368–15369.
14. Huang, N.-P., Michel, R., Voros, J., et al. (2000). poly(L-lysine)-g-poly(ethylene glycol) layers on metal oxide surfaces: Surface-analytical characterization and resistance to serum and fibrinogen adsorption. *Langmuir*, **17**, 489–498.
15. Kessel, C. R., Granick, S. (1991). Formation and characterization of a highly ordered and well-anchored alkylsilane monolayer on mica by self-assembly. *Langmuir*, **7**, 532–538.
16. Love, J. C., Estroff, L. A., Kriebel, J. K., Nuzzo, R. G., Whitesides, G. M. (2005). Self-assembled monolayers of thiolates on metals as a form of nanotechnology. *Chemical Reviews*, **105**, 1103–1170.
17. Wagner, P., Hegner, M., Kernen, P., Zaugg, F., Semenza, G. (1996). Covalent immobilization of native biomolecules onto Au(111) via *N*-hydroxysuccinimide ester functionalized self-assembled monolayers for scanning probe microscopy. *Biophysical Journal*, **70**, 2052–2066.
18. Hodneland, C. D., Lee, Y.-S., Min, D.-H., Mrksich, M. (2002). Selective immobilization of proteins to self-assembled monolayers presenting active site-directed capture ligands. *Proceedings of the National Academy of Sciences*, **99**, 5048–5052.
19. Ulman, A. (1996). Formation and structure of self-assembled monolayers. *Chemical Reviews*, **96**, 1533–1554.
20. Brockman, J. M., Frutos, A. G., Corn, R. M. (1999). A multistep chemical modification procedure to create DNA arrays on gold surfaces for the study of protein–DNA interactions with surface plasmon resonance imaging. *Journal of the American Chemical Society*, **121**, 8044–8051.
21. Nishikawa, M., Yamamoto, T., Kojima, N., Kikuo, K., Fujii, T., Sakai, Y. (2008). Stable immobilization of rat hepatocytes as hemispheroids onto collagen-conjugated polydimethylsiloxane (PDMS) surfaces: Importance of direct oxygenation through PDMS for both formation and function. *Biotechnology and Bioengineering*, **99**, 1472–1481.

22. Soper, S. A., Henry, A. C., Vaidya, B., Galloway, M., Wabuyele, M., McCarley, R. L. (2002). Surface modification of polymer-based microfluidic devices. *Analytica Chimica Acta*, **470**, 87–99.
23. Fixe, F., Dufva, M., Telleman, P., Christensen, C. B. V. (2004). Functionalization of poly(methyl methacrylate) (PMMA) as a substrate for DNA microarrays. *Nucleic Acids Research*, **32**, e9.
24. Gray, J. J. (2004). The interaction of proteins with solid surfaces. *Current Opinion in Structural Biology*, **14**, 110–115.
25. Merkel, J. S., Michaud, G. A., Salcius, M., Schweitzer, B., Predki, P. F. (2005). Functional protein microarrays: Just how functional are they? *Current Opinion in Biotechnology*, **16**, 447–452.
26. Patel, N., Davies, M. C., Hartshorne, M., et al. (1997). Immobilization of protein molecules onto homogeneous and mixed carboxylate-terminated self-assembled monolayers. *Langmuir*, **13**, 6485–6490.
27. Choi, H. J., Kim, N. H., Chung, B. H., Seong, G. H. (2005). Micropatterning of biomolecules on glass surfaces modified with various functional groups using photoactivatable biotin. *Analytical Biochemistry*, **347**, 60–66.
28. MacBeath, G., Schreiber, S. L. (2000). Printing proteins as microarrays for high-throughput function determination. *Science*, **289**, 1760–1763.
29. Groll, J., Amirgoulova, E. V., Ameringer, T., et al. (2004). Biofunctionalized, ultrathin coatings of cross-linked star-shaped poly(ethylene oxide) allow reversible folding of immobilized proteins. *Journal of the American Chemical Society*, **126**, 4234–4239.
30. Mateo, C., Torres, R., Fernández-Lorente, G., et al. (2003). Epoxy-amino groups: A new tool for improved immobilization of proteins by the epoxy method. *Biomacromolecules*, **4**, 772–777.
31. Fernandez-Lafuente, R., Rosell, C. M., Rodriguez, V., et al. (1993). Preparation of activated supports containing low pK amino groups. A new tool for protein immobilization via the carboxyl coupling method. *Enzyme and Microbial Technology*, **15**, 546–550.
32. Martzen, M. R., McCraith, S. M., Spinelli, S. L., et al. (1999). A biochemical genomics approach for identifying genes by the activity of their products. *Science*, **286**, 1153–1155.
33. Riggs, P. (2000). Expression and purification of recombinant proteins by fusion to maltose-binding protein. *Molecular Biotechnology*, **15**, 51–63.
34. Bouchard, B. A., Furie, B., Furie, B. C. (1999). Glutamyl substrate-induced exposure of a free cysteine residue in the vitamin K-dependent g-glutamyl carboxylase is critical for vitamin k epoxidation. *Biochemistry*, **38**, 9517–9523.
35. Schmitt, J., Hess, H., Stunnenberg, H. G. (1993). Affinity purification of histidine-tagged proteins. *Molecular Biology Reports*, **18**, 223–230.
36. Zhang, Y., So, M.-K., Loening, A. M., Yao, H., Gambhir, S. S., Rao, J. (2006). HaloTag protein-mediated site-specific conjugation of bioluminescent proteins to quantum dots. *Angewandte Chemie International Edition*, **45**, 4936–4940.
37. Schmid, E. L., Keller, T. A., Dienes, Z., Vogel, H. (1997). Reversible oriented surface immobilization of functional proteins on oxide surfaces. *Analytical Chemistry*, **69**, 1979–1985.
38. Gizeli, E., Glad, J. (2004). Single-step formation of a biorecognition layer for assaying histidine-tagged proteins. *Analytical Chemistry*, **76**, 3995–4001.
39. Zhen, G., Falconnet, D., Kuennemann, E., et al. (2006). Nitrilotriacetic acid functionalized graft copolymers: A polymeric interface for selective and reversible binding of histidine-tagged proteins. *Advanced Functional Materials*, **16**, 243–251.
40. Gershon, P. D., Khilko, S. (1995). Stable chelating linkage for reversible immobilization of oligohistidine tagged proteins in the BIAcore surface plasmon resonance detector. *Journal of immunological Methods*, **183**, 65–76.
41. Gamsjaeger, R., Wimmer, B., Kahr, H., et al. (2004). Oriented binding of the his_6-tagged carboxyl-tail of the L-type Ca^{2+} channel a_1-subunit to a new NTA-functionalized self-assembled monolayer. *Langmuir*, **20**, 5885–5890.
42. Tinazli, A., Tang, J., Valiokas, R., et al. (2005). High-affinity chelator thiols for switchable and oriented immobilization of histidine-tagged proteins: A generic platform for protein chip technologies. *Chemistry—A European Journal*, **11**, 5249–5259.
43. Mammen, M., Choi, S.-K., Whitesides, G. M. (1998). Polyvalent interactions in biological systems: Implications for design and use of multivalent ligands and inhibitors. *Angewandte Chemie International Edition*, **37**, 2754–2794.
44. Green, N. M., Anfinsen, C. B., Edsall, J. T., Richards, F. M. (1975). *Avidin Advances in Protein Chemistry*. Academic Press, San Diego, pp. 85–133.
45. Wittmann, C., Smith, C., Milea, J., Nguyen, G. (2005). *Immobilization of Nucleic Acids Using Biotin-Strept(avidin) Systems Immobilisation of DNA on Chips II*. Springer, Berlin, pp. 63–90.
46. Chattopadhaya, S., Tan, L. P., Yao, S. Q. (2006). Strategies for site-specific protein biotinylation using in vitro, in vivo and cell-free systems: Toward functional protein arrays. *Nature Protocols*, **1**, 2386–2398.
47. Yin, J., Liu, F., Li, X., Walsh, C. T. (2004). Labeling proteins with small molecules by site-specific posttranslational modification. *Journal of the American Chemical Society*, **126**, 7754–7755.
48. Lesaicherre, M.-L., Lue, R. Y. P., Chen, G. Y. J., Zhu, Q., Yao, S. Q. (2002). Intein-mediated biotinylation of proteins and its application in a protein microarray. *Journal of the American Chemical Society*, **124**, 8768–8769.
49. Lueking, A., Cahill, D. J., Müllner, S. (2005). Protein biochips: A new and versatile platform technology for molecular medicine. *Drug Discovery Today*, **10**, 789–794.
50. Jung, Y., Lee, J. M., Jung, H., Chung, B. H. (2007). Self-directed and self-oriented immobilization of antibody by protein G-DNA conjugate. *Analytical Chemistry*, **79**, 6534–6541.
51. Duburcq, X., Olivier, C., Malingue, F., et al. (2004). Peptide–protein microarrays for the simultaneous detection of pathogen infections. *Bioconjugate Chemistry*, **15**, 307–316.

52. Niemeyer, C. M., Sano, T., Smith, C. L., Cantor, C. R. (1994). Oligonucleotide-directed self-assembly of proteins: Semisynthetic DNA–streptavidin hybrid molecules as connectors for the generation of macroscopic arrays and the construction of supramolecular bioconjugates. *Nucleic Acids Research*, **22**, 5530–5539.
53. Sarikaya, M., Tamerler, C., Jen, A. K. Y., Schulten, K., Baneyx, F. (2003). Molecular biomimetics: Nanotechnology through biology. *Nature Materials*, **2**, 577–585.
54. Tamerler, C., Sarikaya, M. (2009). Molecular biomimetics: Nanotechnology and bionanotechnology using genetically engineered peptides. *Philosophical Transactions of the Royal Society A: Mathematical, Physical and Engineering Sciences*, **367**, 1705–1726.
55. Niemeyer, C. M. (2007). Functional devices from DNA and proteins. *Nano Today*, **2**, 42–52.
56. Pirrung, M. C. (2002). How to make a DNA chip. *Angewandte Chemie International Edition*, **41**, 1276–1289.
57. Kozlov, I. A., Melnyk, P. C., Stromsborg, K. E., Chee, M. S., Barker, D. L., Zhao, C. (2004). Efficient strategies for the conjugation of oligonucleotides to antibodies enabling highly sensitive protein detection. *Biopolymers*, **73**, 621–630.
58. Boozer, C., Ladd, J., Chen, S., Jiang, S. (2006). DNA-directed protein immobilization for simultaneous detection of multiple analytes by surface plasmon resonance biosensor. *Analytical Chemistry*, **78**, 1515–1519.
59. Wacker, R., Niemeyer, C. M. (2005). *Current Protocols in Nucleic acid Chemistry*, Vol. 21. John Wiley & Sons, Hoboken, NJ.
60. Rothemund, P. W. K. (2006). Folding DNA to create nanoscale shapes and patterns. *Nature*, **440**, 297–302.
61. Tanaka, M, (2006). Polymer-supported membranes: Physical models of cell surfaces. *MRS Bulletin*, **31**, 513–520.
62. Tamm, L. K., Groves, J. T. (2009). Supported membranes in structural biology. *Journal of Structural Biology*, **168**, 1–2.
63. Chan, Y.-H. M., Boxer, S. G. (2007). Model membrane systems and their applications. *Current Opinion in Chemical Biology*, **11**, 581–587.
64. Reimhult, E., Kumar, K. (2008). Membrane biosensor platforms using nano- and microporous supports. *Trends in Biotechnology*, **26**, 82–89.
65. Bally, M., Bailey, K., Sugihara, K., Grieshaber, D., Voros, J., Stadler, B. Liposome and lipid bilayer arrays towards biosensing applications. *Small*, **6**, 2481–2497.
66. Torres, A. J., Wu, M., Holowka, D., Baird, B. (2008). Nanobiotechnology and cell biology: Micro- and nanofabricated surfaces to investigate receptor-mediated signaling. *Annual Review of Biophysics*, **37**, 265–288.
67. Nielsen, C. H. (2009). Biomimetic membranes for sensor and separation applications. *Analytical and Bioanalytical Chemistry*, **395**, 697–718.
68. Höök, F., Stengel, G, Dahlin, AB, et al. (2008). Supported lipid bilayers, tethered lipid vesicles, and vesicle fusion investigated using gravimetric, plasmonic, and microscopy techniques. *Biointerphases*, **3**, FA108–FA116.
69. Tien, H. T., Ottova, A. L. (2000). *Membrane Biophysics: As Viewed from Experimental Bilayer Lipid Membranes (Planar Lipid Bilayers and Spherical Liposomes)*. Elsevier, Amsterdam.
70. Handrea, M., Sahre, M., Neubauer, A., Sleytr, U. B., Kautek, W. (2003). Electrochemistry of nano-scale bacterial surface protein layers on gold. *Bioelectrochemistry*, **61**, 1–8.
71. Hennig, M., Neumann, J., Wixforth, A., Radler, J. O., Schneider, M. F. (2009). Dynamic patterns in a supported lipid bilayer driven by standing surface acoustic waves. *Lab on a Chip*, **9**, 3050–3053.
72. Ottova, A., Tien, H. T. (2005). *The Lipid Bilayer Principle: A Historic Perspective and Some Highlights Advances in Planar Lipid Bilayers and Liposomes*. Academic Press, San Diego, pp. 1–76.
73. Zasadzinski, J. A., Viswanathan, R., Madsen, L., Garnaes, J., Schwartz, D. K. (1994). Langmuir–blodgett films. *Science*, **263**, 1726–1733.
74. White, R. J., Zhang, B., Daniel, S., et al. (2006). Ionic conductivity of the aqueous layer separating a lipid bilayer membrane and a glass support. *Langmuir*, **22**, 10777–10783.
75. Brian, A. A., McConnell, H. M. (1984). Allogeneic stimulation of cytotoxic T cells by supported planar membranes. *Proceedings of the National Academy of Sciences*, **81**, 6159–6163.
76. Cho, N. J., Frank, C. W., Kasemo, B., Höök, F. Quartz crystal microbalance with dissipation monitoring of supported lipid bilayers on various substrates. *Nature Protocols*, **5**, 1096–1106.
77. Richter, R. P., Bérat, R., Brisson, A. R. (2006). Formation of solid-supported lipid bilayers: An integrated view. *Langmuir*, **22**, 3497–3505.
78. Reimhult, E., Zäch, M., Höök, F., Kasemo, B. (2006). A multitechnique study of liposome adsorption on au and lipid bilayer formation on SiO_2. *Langmuir*, **22**, 3313–3319.
79. Sackmann, E. (1996). Supported membranes: Scientific and practical applications. *Science*, **271**, 43–48.
80. Devadoss, A., Palencsár, M. S., Jiang, D., Honkonen, M. L., Burgess, J. D. (2005). Enzyme modification of platinum microelectrodes for detection of cholesterol in vesicle lipid bilayer membranes. *Analytical Chemistry*, **77**, 7393–7398.
81. Naumann, R., Schiller, S. M., Giess, F., et al. (2003). Tethered lipid bilayers on ultraflat gold surfaces. *Langmuir*, **19**, 5435–5443.
82. Giess, F., Friedrich, M. G., Heberle, J., Naumann, R. L., Knoll, W. (2004). The protein-tethered lipid bilayer: A novel mimic of the biological membrane. *Biophysical Journal*, **87**, 3213–3220.
83. Tiefenauer, L. X., Studer, A. (2008). Nano for bio: Nanopore arrays for stable and functional lipid bilayer membranes (Mini Review). *Biointerphases*, **3**, FA74–FA79.
84. Dostalek, J., Knoll, W. (2008). Biosensors based on surface plasmon-enhanced fluorescence spectroscopy (Review). *Biointerphases*, **3**, FD12–FD22.
85. Khan, T. R., Grandin, H. M., Mashaghi, A., Textor, M., Reimhult, E., Reviakine, I. (2008). Lipid redistribution in

phosphatidylserine-containing vesicles adsorbing on titania. *Biointerphases*, **3**, FA90–FA95.

86. Howard, A. D., McAllister, G., Feighner, S. D., et al. (2001). Orphan G-protein-coupled receptors and natural ligand discovery. *Trends in Pharmacological Sciences*, **22**, 132–140.
87. Mueller, P., Rudin, D. O., Tien, H. T., Wescott, W. C. (1962). Reconstitution of cell membrane structure in vitro and its transformation into an excitable system. *Nature*, **194**, 979.
88. Bamberg, E., Alpes, H., Apell, H. J., Bradley, R., Harter, B., Quelle, M. J., Urry, D. W. (1979). Formation of ionic channels in black lipid membranes by succinic derivatives of gramicidin A. *Journal of Membrane Biology*, **50**, 257.
89. Gomezlagunas, F., Pena, A., Lievano, A., Darszon, A. (1989). Incorporation of ionic channels from yeast plasma membranes into black lipid membranes. *Biophysicals Journal*, **56**, 115.
90. Bezrukov, S. M. VI (1993). Probing alamethicin channels with water-soluble polymers. Effect on conductance of channel states. *Biophysicals Journals*, **64**, 16–25.
91. Naumann, R. L. C., Knoll, W. (2008). Protein tethered lipid bilayer: An alternative mimic of the biological membrane (Mini Review). *Biointerphases*, **3**, FA101–FA107.
92. Jeon, T.-J., Malmstadt, N., Poulos, J. L., Schmidt, J. J. (2008). Black lipid membranes stabilized through substrate conjugation to a hydrogel. *Biointerphases*, **3**, FA96–FA100.
93. Schuster, B., Pum, D., Sleytr, U. B. (2008). S-layer stabilized lipid membranes (Review). *Biointerphases*, **3**: FA3-FA11.
94. Steltenkamp, S., Muller, M. M., Deserno, M., Hennesthal, C., Steinem, C., Janshoff, A.(2006). Mechanical properties of pore-spanning lipid bilayers probed by atomic force microscopy. *Biophysical Journal*, **91**, 217–226.
95. Mey, I., Stephan, M., Schmitt, E. K., et al. (2009). Local membrane mechanics of pore-spanning bilayers. *Journal of the American Chemical Society*, **131**, 7031–7039.
96. Steinem, C., Janshoff, A. (2010). Multicomponent membranes on solid substrates Interfaces for protein binding. *Current Opinion in Colloid & Interface Science*, **15**, 479–488.
97. Atanasov, V., Knorr, N., Duran, R. S., et al. (2005). Membrane on a chip: A functional tethered lipid bilayer membrane on silicon oxide surfaces. *Biophysical Journal*, **89**, 1780–1788.
98. Terrettaz, S., Ulrich, W.-P., Guerrini, R., Verdini, A., Vogel, H. (2001). Immunosensing by a synthetic ligand-gated ion channel. *Angewandte Chemie International Edition*, **40**, 1740–1743.
99. Schmitt, E. K., Vrouenraets, M., Steinem, C. (2006). Channel activity of OmpF monitored in nano-BLMs. *Biophysical Journal*, **91**, 2163–2171.
100. Holden, M. A., Jung, S.-Y., Yang, T., Castellana, E. T., Cremer, P. S (2004). Creating fluid and air-stable solid supported lipid bilayers. *Journal of the American Chemical Society*, **126**, 6512–6513.
101. Plant, A. L. (1999). Supported hybrid bilayer membranes as rugged cell membrane mimics. *Langmuir*, **15**, 5128–5135.
102. Morigaki, K., Kiyosue, K., Taguchi, T. (2004). Micropatterned composite membranes of polymerized and fluid lipid bilayers. *Langmuir*, **20**, 7729–7735.
103. Ross, E. E., Rozanski, LJ, Spratt, T., Liu, S., O'Brie, D. F., Saavedra, S. S. (2003). Planar supported lipid bilayer polymers formed by vesicle fusion. 1. influence of diene monomer structure and polymerization method on film properties. *Langmuir*, **19**, 1752–1765.
104. Albertorio, F., Diaz, A. J., Yang, T., et al. (2005). Fluid and air-stable lipopolymer membranes for biosensor applications. *Langmuir*, **21**, 7476–7482.
105. Wang, Q., Dordick, J. S., Linhardt, R. J. (2002). Synthesis and application of carbohydrate-containing polymers. *Chemistry of Materials* **14**: 3232–3244.
106. Elender, G., Kühner, M., Sackmann, E. (1996). Functionalisation of Si/SiO_2 and glass surfaces with ultrathin dextran films and deposition of lipid bilayers. *Biosensors and Bioelectronics*, **11**, 565–577.
107. Hillebrandt, H., Wiegand, G., Tanaka, M., Sackmann, E. (1999). High electric resistance polymer/lipid composite films on indium–tin–oxide electrodes. *Langmuir*, **15**, 8451–8459.
108. Baumgart, T., Offenhäusser, A. (2002). Polysaccharide-supported planar bilayer lipid model membranes. *Langmuir*, **19**, 1730–1737.
109. Zhang, L., Longo, M. L., Stroeve, P. (2000). Mobile phospholipid bilayers supported on a polyion/alkylthiol layer pair. *Langmuir*, **16**, 5093–5099.
110. Kügler, R., Knoll, W. (2002). Polyelectrolyte-supported lipid membranes. *Bioelectrochemistry*, **56**, 175–178.
111. Shen, W. W., Boxer, S. G., Knoll, W., Frank, C. W. (2000). Polymer-supported lipid bilayers on benzophenone-modified substrates. *Biomacromolecules*, **2**, 70–79.
112. Richter, R. P., Hock, K. K., Burkhartsmeyer, J., et al. (2007). Membrane-grafted hyaluronan films: A well-defined model system of glycoconjugate cell coats. *Journal of the American Chemical Society*, **129**, 5306–5307.
113. Cremer, P. S. (2010). Surface patterning: More than just scratching the surface. *Journal of the American Chemical Society*, **133**, 167–169.
114. Xia, Y. N., Whitesides, G. M. (1998). Soft lithography. *Angewandte Chemie International Edition*, **37**, 551–575.
115. Kumar, A., Biebuyck, H. A., Whitesides, G. M. (1994). Patterning self-assembled monolayers: Applications in materials science. *Langmuir*, **10**, 1498–1511.
116. Liu, G.-Y., Xu, S., Qian, Y. (2000). Nanofabrication of self-assembled monolayers using scanning probe lithography. *Accounts of Chemical Research*, **33**, 457–466.
117. Luo, W., Chan, E. W. L., Yousaf, M. N. Tailored electroactive and quantitative ligand density microarrays applied to stem cell differentiation. *Journal of the American Chemical Society*, **132**, 2614–2621.
118. Orski, S. V., Poloukhtine, A. A., Arumugam, S., Mao, L., Popik, V. V., Locklin, J. (2010). High density orthogonal surface immobilization via photoactivated copper-free click chemistry. *Journal of the American Chemical Society*, **132**, 11024–11026.
119. Mossman, K. D., Campi, G., Groves, J. T., Dustin, M. L. (2005). Altered TCR signaling from geometrically repatterned immunological synapses. *Science*, **310**, 1191–1193.

120. Salaita, K., Wang, Y., Mirkin, C. A. (2007). Applications of dip-pen nanolithography. *Nature Nanotechnology*, **2**, 145–155.
121. Maury, P., Escalante, M., Péter, M., Reinhoudt, D. N., Subramaniam, V., Huskens, J. (2007). Creating nanopatterns of his-tagged proteins on surfaces by nanoimprint lithography using specific NiNTA–histidine interactions. *Small*, **3**, 1584–1592.
122. Glass, R., Moeller, M., Spatz, J. P. (2003). Block copolymer micelle nanolithography. *Nanotechnology*, **14**, 1153–1160.
123. Castellana, E. T., Cremer, P. S. (2006). Solid supported lipid bilayers: From biophysical studies to sensor design. *Surface Science Reports*, **61**, 429–444.
124. Groves, J. T., Ulman, N., Boxer, S. G. (1997). Micropatterning fluid lipid bilayers on solid supports. *Science*, **275**, 651–653.
125. Groves, J. T., Ulman, N., Cremer, P. S., Boxer, S. G. (1998). Substrate–membrane interactions: Mechanisms for imposing patterns on a fluid bilayer membrane. *Langmuir*, **14**, 3347–3350.
126. Tanaka, M., Tutus, M., Kaufmann, S., Rossetti, F. F., Schneck, E., Weiss, I. M.(2009). Native supported membranes on planar polymer supports and micro-particle supports. *Journal of Structural Biology*, **168**, 137–142.
127. Groves, J. T., Boxer, S. G. (2002). Micropattern formation in supported lipid membranes. *Accounts of Chemical Research*, **35**, 149–157.
128. Majd, S., Mayer, M. (2005). Hydrogel stamping of arrays of supported lipid bilayers with various lipid compositions for the screening of drug–membrane and protein–membrane interactions. *Angewandte Chemie International Edition*, **44**, 6697–6700.
129. Yoshina-Ishii, C., Boxer, S. G. (2003). Arrays of mobile tethered vesicles on supported lipid bilayers. *Journal of the American Chemical Society* **125**, 3696–3697.
130. Yamazaki, V., Sirenko, O., Schafer, R., et al. (2005). Cell membrane array fabrication and assay technology. *BMC Biotechnology*, **5**, 18.
131. Chen, X., Lee, G. S., Zettl, A., Bertozzi, C. R. (2004). Biomimetic engineering of carbon nanotubes by using cell surface mucin mimics. *Angewandte Chemie International Edition*, **43**, 6111–6116.
132. Niemeyer, C. M. (2001). Nanoparticles, proteins, and nucleic acids: biotechnology meets materials science. *Angewandte Chemie International Edition*, **40**, 4128–4158.
133. Katz, E., Willner, I. (2004). Integrated nanoparticle–biomolecule hybrid systems: Synthesis, properties, and applications. *Angewandte Chemie International Edition*, **43**, 6042–6108.
134. Ye, J.-S., Cui, H.-F., Ottova, A., Tien, H. T., Liu, A. L. (2006). Chapter 8 Interfaces, bifaces, and nanotechnology. *Advances in planar lipid bilayers and liposomes.* Academic Press, San Diego, pp. 251–267.
135. Richard, C., Balavoine, F., Schultz, P., Ebbesen, T. W., Mioskowski, C. (2003). Supramolecular self-assembly of lipid derivatives on carbon nanotubes. *Science*, **300**, 775–778.
136. Ye, J.-S., Cui, H.-F., Wen, Y., et al. (2005). Self-assembly of bilayer lipid membrane at multiwalled carbon nanotubes towards the development of photo-switched functional device. *Electrochemistry Communications*, **7**, 81–86.
137. Ye, J.-S., Ottova, A., Tien, H. T., Sheu, F.-S. (2003). Nanostructured platinum–lipid bilayer composite as biosensor. *Bioelectrochemistry*, **59**, 65–72.
138. Dubertret, B., Skourides, P., Norris, D. J., Noireaux, V., Brivanlou, A. H., Libchaber, A. (2002). In vivo imaging of quantum dots encapsulated in phospholipid micelles. *Science*, **298**, 1759–1762.
139. Geissbuehler, I., Hovius, R., Martinez, K. L., Adrian, M., Thampi, K. R., Vogel, H. (2005). Lipid-coated nanocrystals as multifunctionalized luminescent scaffolds for supramolecular biological assemblies. *Angewandte Chemie International Edition*, **44**, 1388–1392.
140. Bhattacharya, S., Srivastava, A. (2003). Synthesis and characterization of novel cationic lipid and cholesterol-coated gold nanoparticles and their interactions with dipalmitoylphosphatidylcholine membranes. *Langmuir*, **19**, 4439–4447.
141. Mornet, S., Lambert, O., Duguet, E., Brisson, A. (2004). The formation of supported lipid bilayers on silica nanoparticles revealed by cryoelectron microscopy. *Nano Letters*, **5**, 281–285.
142. Thanh, N. T. K., Green, L. A. W. (2010). Functionalisation of nanoparticles for biomedical applications. *Nano Today*, **5**, 213–230.
143. Abad, J. M., Vélez, M., Santamaría, C., Guisán, J. M., Matheus, P. R., Vázquez, L., Gazaryan, I., Gorton, L., Gibson, T., Fernández, V. M. (2002). Immobilization of peroxidase glycoprotein on gold electrodes modified with mixed epoxy–boronic acid monolayers. *Journal of the American Chemical Society*, **124**, 12845–12853.
144. Alonso-Lomillo, M. A., Rüdiger, O., Maroto-Valiente, A., Vélez, M., Rodríguez-Ramos, I., Muñoz, F. J., Fernández, V. M., De Lacey, A. L. (2007). Hydrogenase-coated carbon nanotubes for efficient H_2 oxidation. *Nano Letters*, **7**, 1603–1608.
145. Gutiérrez-Sánchez, C., Olea, D., Marques, M., Fernández, V. M., Pereira, I., Vélez, M., De Lacey, A. L. (2011). Oriented immobilization of a membrane-bound hydrogenase onto an electrode for direct electron transfer. *Langmuir*, **27**, 6449–6457.
146. Madoz-Gúrpide, J., Abad, J. M., Fernández-Recio, J., Vélez, M., Vázquez, L., Gómez-Moreno, C., Fernández, V. M. (2000). Modulation of electroenzymatic NADPH oxidation through oriented immobilization of ferredoxin: $NADP^+$ reductase onto modified gold electrodes. *Journal of the American Chemical Society*, **122**, 9808–9817.
147. Jeuken, L. J. C., Connell, S. D., Henderson, P. J. F., Gennis, R. B., Evans, S. D., Bushby, R. J. (2006). Redox enzymes in tethered membranes. *Journal of the American Chemical Society*, **128**, 1711–1716.
148. Ye, Q., Konradi, R., Textor, M., Reimhult, E. (2009). Liposomes tethered to omega-functional PEG brushes and induced

formation of PEG brush supported planar lipid bilayers. *Langmuir*, **25**, 13534–13539.

149. Stadler, B., Falconnet, D., Pfeiffer, I., Höök, F., Voros, J. (2004). Micropatterning of DNA-tagged vesicles. *Langmuir*, **20**, 11348–11354.
150. Aydin, D., Louban, I., Perschmann, N., et al. (2010). Polymeric substrates with tunable elasticity and nanoscopically controlled biomolecule presentation. *Langmuir*, **26**, 15472–15480.
151. Selhuber-Unkel, C., Erdmann, T., López-García, M., Kessler, H., Schwarz, U. S., Spatz, J. P. (2010). Cell adhesion strength is controlled by intermolecular spacing of adhesion receptors. *Biophysical Journal*, **98**, 543–551.
152. Tanaka, M., Sackmann, E. (2005). Polymer-supported membranes as models of the cell surface. *Nature*, **437**, 656–663.
153. Schuster, B., Sleytr, U. B. (2009). Composite S-layer lipid structures. *Journal of Structural Biology*, **168**, 207–216.
154. Clemmens, J., Hess, H., Doot, R., Matzke, C. M., Bachand, G. D., Vogel, V. (2004). Motor-protein "roundabouts": Microtubules moving on kinesin-coated tracks through engineered networks. *Lab on a Chip*, **4**, 83–86.
155. Ahmed, W. W., Wolfram, T., Goldyn, A. M., Bruellhoff, K., Rioja, B. A., Mller, M., Spatz, J. P., Saif, T. A., Groll, J., Kemkemer, R. (2010). Myoblast morphology and organization on biochemically micro-patterned hydrogel coatings under cyclic mechanical strain. *Biomaterials*, **31**, 250–258.

21

CARBON NANOTUBE DERIVATIVES AS ANTICANCER DRUG DELIVERY SYSTEMS

Chiara Fabbro, Tatiana Da Ros, and Maurizio Prato

21.1 INTRODUCTION

The rapid development of carbon nanotube (CNT)-based technology in many fields has made this novel material very popular in the scientific community. CNTs are tubular structures made of only carbon atoms arranged in a benzenoid network similar to graphite, but rolled up in a cylindrical shape and therefore subjected to a curvature, with important consequences in their chemical reactivity. The number of cylinders that constitute a CNT can vary from only one to several ones, giving rise to different kinds of CNTs, namely single-walled CNTs (SWCNTs), double-walled CNTs (DWCNTs), up to generic multiwalled CNTs (MWCNTs), where the distance between two walls is ~0.35 nm.

The coiling-up direction of the graphene sheet determines different kinds of CNT. In fact, the angle between C–C bonds and the axis of the tube can vary, and the so-called helicity of a CNT will depend on this angle. Thus, different kinds of CNTs are described according to a chiral vector, as explained in Figure 21.1.

The two limits of the huge variety of possibilities are zigzag and armchair CNTs, with the former having $\theta = 0°$ and $m = 0$ and the latter with $\theta = 30°$ and $n = m \neq 0$. In between, all the CNTs with $0° < \theta < 30°$ are defined as chiral. Obviously, with increasing n and/or m, the diameter of the tube increases. The description of this feature is important not only to define structurally different CNTs using a common vocabulary, but mainly because the electronic properties vary a lot, depending on the chiral vector. In fact, infinite-length armchair SWCNTs are metallic, whereas infinite-length zigzag or chiral CNTs are semiconducting, even though these are not strict rules. In the case, for example, of small-diameter tubes, where the curvature plays an important role, some exceptions exist [1].

Additionally, as a result of the 1D nature of CNTs, electrons can be conducted without being scattered. The absence of scattering of the electrons during conduction is known as ballistic transport, and it allows the nanotube to conduct without dissipating energy as heat.

Besides these unique electronic properties, ideal CNTs exhibit chemical and thermal stability, together with extremely high tensile strength and elasticity.

Nevertheless, they still present to date some important drawbacks that need to be addressed by the research in the field. One of the major problems is related to batch-to-batch lack of reproducibility. In fact, commercial samples of CNTs, produced by the same supplier, do not necessarily contain identical materials. This lack of reliability is probably due to the big heterogeneity of a CNT sample, which makes it subject to too many unpredictable variables during the synthetic process. This variability is associated with CNT length, diameter, helicity, impurity content (i.e., metal and carbonaceous impurities), and presence of defect. The possible defective structures are vacancies (i.e., incomplete bonding defects), topological changes, such as pentagons/heptagons interrupting the hexagonal network (Figure 21.2), and doping with elements other than carbon [2].

The other big limitation associated with CNT research is their poor dispersibility. In fact pristine, as-produced CNTs are insoluble in water and in any organic solvent, and a correct

Organic Nanomaterials: Synthesis, Characterization, and Device Applications, First Edition. Edited by Tomás Torres and Giovanni Bottari.

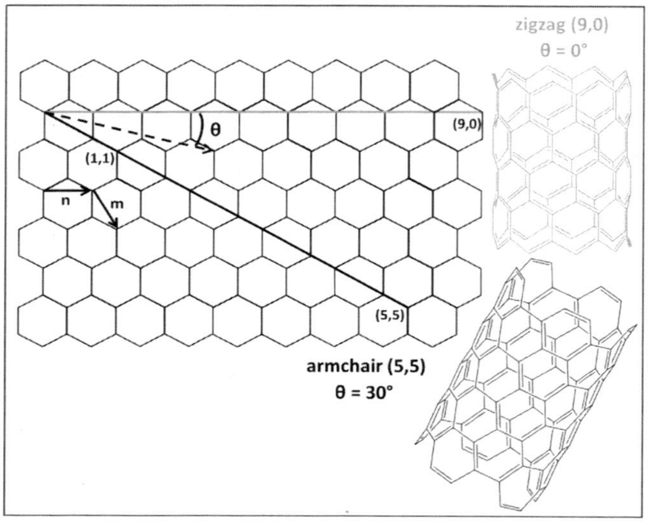

FIGURE 21.1. Graphical description of the CNT chiral vector θ. The vector refers to the direction of rolling up of a graphene sheet. Thus, each CNT can be described by the couple of values (*n*, *m*), as in the two examples depicted.

functionalization is required in order to improve dispersibility for a better manipulation of the material.

Several synthetic protocols exist for CNTs. The main ones are arc discharge, originally developed for fullerene synthesis, laser ablation, and catalytic carbon vapor deposition (CCVD) [3, 4]. The first two methodologies involve high-power energy to vaporize solid-state carbon (graphite) and form CNTs at high temperature (>1000°C), while CCVD uses carbon precursors in the gas phase and takes place at relatively low temperatures (500–1000°C). In general the use of a proper metallic catalyst can switch from the production of MWCNTs to SWCNTs. Usual impurities deriving from the synthetic process are metallic nanoparticles and carbonaceous material other than the desired CNTs, represented mainly by amorphous carbon. Among the three different synthetic methods, CCVD provides a better control in the growth of CNTs and can be scaled up to industry level because of the lower costs. Different CCVD processes have been developed, among which one of the most popular is the high-pressure carbon monoxide disproportionation process (HiPCO), used for producing SWCNTs [5]. This technique exploits carbon monoxide (CO) as the carbon source, and $Fe(CO)_5$ as the catalyst precursor. The process takes place at 800–1000°C through the disproportionation of CO to CO_2 and C for the CNT growth on the iron nanoparticles. Optimized conditions led to the production of SWCNTs with diameters ranging from 0.7 to 1.4 nm, with an acceptable degree of purity.

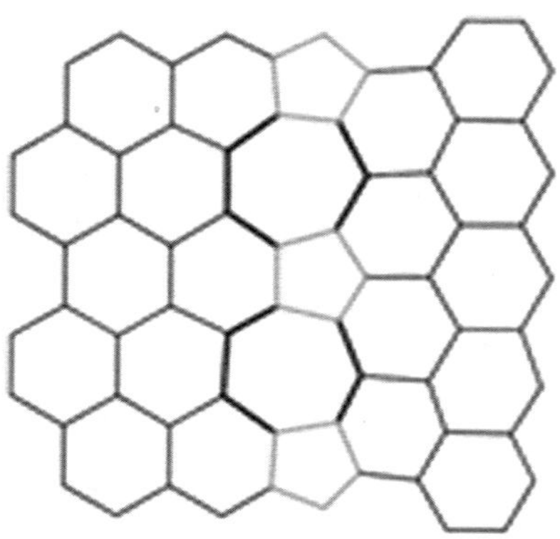

FIGURE 21.2. Example of pentagon/heptagon defects in the hexagonal CNT network, inducing local strain. Note the change in tube helicity between left and right side of the defect.

21.2 CNT FUNCTIONALIZATION

As already mentioned, CNT usefulness is strongly hindered by the difficulty in their handling. In fact, as produced, CNTs are not soluble in water or in all common organic solvents. For this reason, in most cases, a chemical modification is needed prior to application. By means of functionalization, the strong inter-tube interactions are partly removed. Moreover, specific groups can be introduced to increase solubility in certain solvents, as well as to provide an anchor point for further modifications.

The main distinction to be underlined when speaking of CNT functionalization is between noncovalent and covalent approaches (Figure 21.3). They both present pros and cons, which should be carefully taken into account when deciding which is the best route to follow, according to the desired application.

21.3 NONCOVALENT FUNCTIONALIZATION

Among noncovalent CNT functionalization strategies, one possibility is based on small aromatic molecules, bound to the tubes by means of π–π stacking [6,7], while another one is based on the wrapping of polymeric molecules around the tubes. The latter method can involve both biological macromolecules, such as nucleic acids [8], lipids [9, 10], or peptides [11, 12] and synthetic polymers [13], and it occurs via π–π stacking and/or van der Waals interactions. Another, quite particular, approach for the noncovalent modification of CNTs is the filling of their inner cavity [14–16]. The main advantage of the noncovalent modification of CNTs is that the electronic and mechanical properties of the tubes are preserved. Therefore this should be the first choice when CNTs are exploited in fields such as molecular electronics. On the other hand, the drawback of a noncovalent approach is the possible reversibility of the bonds involved, which can

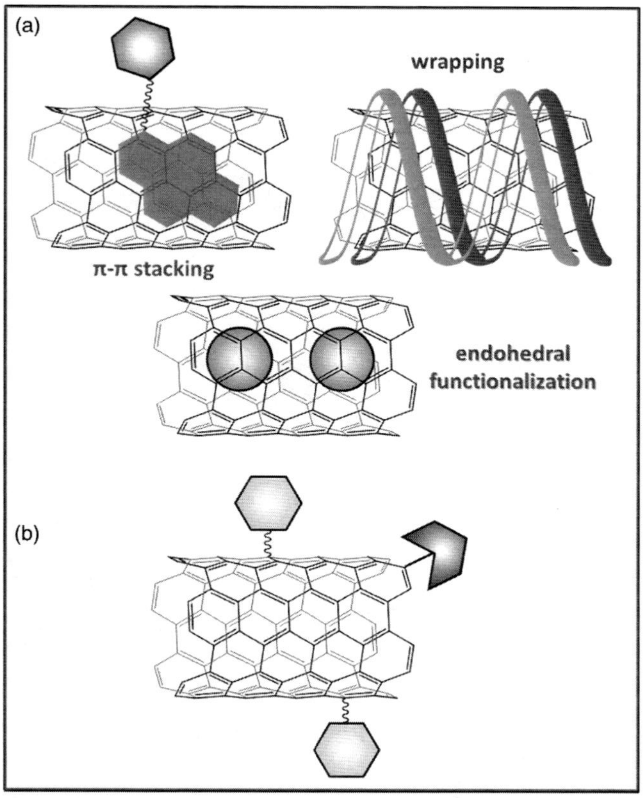

FIGURE 21.3. Noncovalent (**a**) and covalent (**b**) approaches for CNT functionalization.

result in the loss of the functionalization. This risk needs to be considered, for example, in the case of most biomedical applications, since it is difficult to exactly foresee the fate of a noncovalently bound molecule when the functionalized CNTs are administered in vivo. For this reason, a covalent approach should be preferred if CNTs will serve as drug delivery systems.

21.4 COVALENT FUNCTIONALIZATION

The reactivity of CNTs in terms of covalent chemistry on the C backbone is due to local strain, which is caused by two main reasons. The first one is the curvature-induced pyramidalization of the conjugated carbon atoms, and the second is the π-orbital misalignment between adjacent pairs of conjugated carbon atoms [17]. Since the pyramidalization angle gives a good measure of the local weakening of π-conjugation and of the strain energy of pyramidalization, it is clear why this effect is important for fullerenes and for CNTs caps, which are half-fullerene, while it is less relevant in the case of the flatter CNT sidewalls (Figure 21.4, left part).

On the other hand, π-orbital misalignment accounts for strain in CNT sidewalls much more than in fullerenes, where it is very little (in fullerene C_{60} π orbital alignment is perfect, Figure 21.4, right part). Therefore C_{60} and CNT reactivity toward addition reactions are both driven by geometry-dependent strain, but for different reasons. Furthermore, since the pyramidalization angles and the π-orbital misalignment angles of CNTs scale inversely with the diameter of the tubes, a higher reactivity is expected for smaller tubes than for larger ones, if considering just strain-induced reactivity. Nevertheless, there is another parameter that could deeply influence the reactivity of CNT sidewalls: the presence of defects, such as pentagon–heptagon pairs in the hexagons network, which results in a locally enhanced chemical reactivity of the graphitic nanostructures and can cause an unpredictable chemical behavior of CNTs with different diameters. As already anticipated, covalent sidewall functionalization may cause a partial loss of the high conductivity and of the remarkable mechanical properties of CNTs, since it generates sp^3 carbon sites on CNTs, which interrupt the conjugation of π electrons. The possibilities for a covalent modification of the unsaturated carbon network of CNTs, though being limited and requiring usually harsh conditions, have had a rapid development in the last years, starting from fluorination, one of the first covalent reactions ever performed on SWCNTs in 1998 [18]. In this case the degree of functionalization could reach such high levels that in the end the tubes can become insulators. In a second step, CNTs can be further modified using Grignard reagents or with organolithium compounds [19, 20]. Other existing possibilities include addition of both alkyl [21] and aryl [22, 23] radicals, as well as nucleophilic [24–26] or electrophilic [27] additions and cycloadditions [28–31].

Moreover, CNTs could be oxidized, using concentrated acid mixtures, where nitric acid or hydrogen peroxide plays the role of the oxidant [32, 33]. Also other oxidative agents have been used, such as phosphomolybdic acid [34], potassium permanganate [35], or molecular oxygen, by heat treatment of the CNTs in O_2 atmospheres [36, 37]. This treatment generates new defects on the tubular structure, introducing hydroxyl groups, which are then further oxidized to carboxyl groups that allow a further derivatization of CNTs. Esterification and amidation reactions are in fact widely exploited possibilities for the covalent modification of oxidized CNTs.

21.5 CNT TOXICITY

One of the big concerns regarding CNTs is, of course, their toxicity. First of all, the big diversity of this material should be taken into account, both for the as-produced CNTs and for the modified ones. In fact not only commercial pristine samples can differ according to the synthetic methodologies and the supplier, but, more importantly, CNTs always undergo some chemical modification (covalent or noncovalent) when intended to be used for biomedical applications, thus leading

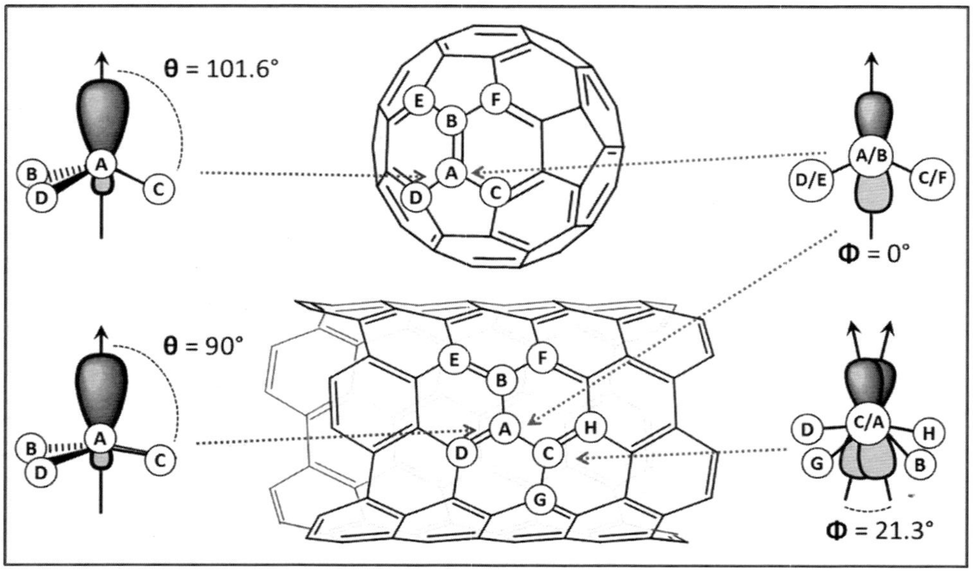

FIGURE 21.4. Representation of pyramidalization angle ($\theta - 90°$) and of π-orbital misalignment (Φ) for fullerene C_{60} and for a (5,5) SWCNT.

to changes in their water dispersibility and in toxicological behavior. Therefore results should always be considered as referred to the particular kind of CNTs they have been obtained with, and could not be generalized.

When considering CNT toxicity, it is important to distinguish among (i) biocompatibility related to possible biomedical application of CNT-based systems and (ii) safety related to CNT manipulation by all the workers in the field. In the first case it will be referred to processed CNTs, likely administered in solution via injection (this topic will be reviewed later), while in the second case the main risk is related to inhalation, and it refers not only to processed CNTs but, primarily, to pristine ones.

Pristine MWCNTs have been compared to asbestos, since they tend to form fibrous aggregates whose dimensions (in the micrometre range) and shape are similar to those reported to be carcinogenic for asbestos. Once the latter nondegradable and bio-persistent particles are inside the body, they deposit. Thus, unsuccessful inflammation and scavenging occur toward them (frustrated phagocytosis), leading to continuous oxidative stress. Furthermore, there is a contribution, given by iron impurities in the samples, that accelerates the generation of reactive oxygen species (ROS). This being true for asbestos, the possibility for CNTs to share the same carcinogenic mechanism was investigated [38]. Experiments on p53 heterozygous mice (reported to be sensitive to asbestos and to fast develop mesothelioma), intraperitoneally administered with MWCNTs, showed the induction of mesothelioma, leading to death of the animals within 25 weeks. Nevertheless, it is necessary to underline that these experiments were carried out with micrometer-sized MWCNTs particles (10–20 μm) and there is no evidence, as the authors pointed out, that the same effect would occur using pure nanometer-sized-CNTs. In a similar study CNTs, introduced into the abdominal cavity of mice, were proved to cause asbestos-like pathogenicity (inflammation and granulomas formation) in a length-dependent manner; that is, only fibers long enough (>20 μm) were able to cause pathogenic response during the 7-day experiment [39].

Nonetheless, being based on intraperitoneal injection, none of these studies addresses the question if and how much CNTs accumulate in the mesothelium after inhalation. To answer this question, other studies on inhaled CNT effects should be considered. Mitchell et al. [40] exposed male mice to MWCNTs (5–15 μm long) aerosol in a whole-body inhalation chamber. The aerosol was generated by a jet mill and particles bigger than 3 μm were removed in order to work within a range respirable by a rodent. It is necessary to underline the difference between "inhalable," which refers to the fraction of airborne material which enters nose and mouth during breathing and is therefore available for deposition in the respiratory tract, and "respirable," referring to the fraction able to penetrate to the gas exchange region of the lung. The authors observed the presence of black particles inside alveolar macrophages, but no lung inflammation or tissue damage were evidenced with exposure to up to 5 mg/m^3 for a maximum of 14 days. Nevertheless, the treatment caused systemic immunosuppression, with a response already reported after inhalation of environmental pollutants. Other authors also reported accumulation of MWCNTs in subpleural macrophages [41]. In this case CNTs were longer (0.5–50 μm compared to 5–15 μm) and the treatment consisted in a single inhalation exposure at a concentration of 30 mg/m^3 for 6 h with a nose-only inhalation chamber. The authors also observed focal subpleural fibrosis, which could be consistent with the induction of mesothelioma, even if asbestos normally causes pleural inflammation and diffuse pleural fibrosis.

It is important to analyze the real risk for people working with CNTs to inhale them. In 2003, Baron et al. [42] studied the tendency of SWCNTs to form aerosols, in order to evaluate a typical workplace exposure to this material. Vigorously agitating SWCNTs (produced by HiPCO process) led to the generation of particles below 10-μm aerodynamic diameter, and also particles smaller than 100 nm, even if it was unclear whether these smaller particles consisted predominantly of nanotubes, catalyst particles, or compact carbonaceous particles. However, generation rates were approximately two orders of magnitude lower than those from a similar volume of fumed alumina, another low-density material comprised of nanometer-sized primary particles. A field evaluation of the airborne contamination, while manipulating the bulk SWCNTs, was also conducted. The bulk material was removed from the reactors and manipulated (poured from one container into another one) in a closed area with an inlet of filtered air, with a continuous monitoring of aerosol number and mass concentration.

There was no clear evidence of increased aerosol mass concentrations during the handling of unrefined nanotube material in the field. The aerosol nanotube concentrations during material handling were estimated to be lower than 53 μg/m^3, a value that is below those associated with (a) re-suspension of ambient dust due to personal movement and (b) cleaning operations following handling of material. Importantly, these experiments could therefore lead to the conclusion that a typical aerosol concentration of CNTs during handling of the material is far below the ones used in the in vivo inhalation studies conducted so far, thus making those results not relevant. Nevertheless, the CNTs are not the same, so it is necessary to be careful in making comparison.

It is still not possible to reach a clear, unambiguous conclusion on CNTs inhalation toxicity; therefore, precautions should be taken to prevent any unnecessary release of respirable particles when handling CNTs, especially the pristine ones, but without getting into a panic. Also, to avoid exposure and guarantee respiratory protection, an adequate personal protective equipment should be used.

21.6 CNT AS DELIVERY SYSTEM FOR CANCER

Anticancer therapy, though often being very effective, still suffers from some severe drawbacks that render vital and urgent the need for further research. One of the main limitations of almost every cancer treatment so far is the lack of selectivity for tumor tissues. In fact, cytotoxic drugs (or other therapies) exert their action not solely on cancer cells, but also on healthy organs, resulting in severe side effects for patients, thereby limiting their compliance and the maximum administrable dose. Another major problem related to antineoplastic chemotherapy is multi-drug resistance (MDR) [43]. This phenomenon can be often ascribed to the activity of an efflux pump, namely P-glycoprotein (P-gp), able to recognize the drug and transport it out of the cell once it has been internalized, thus preventing it from exerting its cytotoxic action [44]. Importantly, the P-gp-mediated MDR often characterizes residual tumor cells after chemotherapy and tumor stem cells. Once this capability is acquired, it can be directed toward many different drugs to which the tumor had never been exposed before, thus heavily hampering chemotherapy's efficacy.

Nanotechnology can help in overcoming these limitations; in fact a nanovector could be conceived to be targeted toward cancer, either only in a nonspecific way or with a specific targeting agent attached to the carrier itself. The nonspecific targeting is based on the enhanced permeability and retention (EPR) effect (Figure 21.5) [45, 46]. Cancer tissue is usually characterized by a rapid and defective angiogenesis, resulting in leaky blood vessels with large fenestrations, due to endothelial cell disorganization. Moreover, the smooth-muscle layer in the vascular wall is frequently absent or abnormal, leading to passive dilatation of vessels. The consequence of these characteristics is an enhanced extravasation of macromolecules in the tumor tissue, whereas low-molecular-weight molecules can rapidly diffuse in the circulation and undergo renal clearance. Moreover, slow venous return and poor lymphatic drainage lead to retention of accumulated macromolecules in the tumor. For these reasons, drug delivery based on nano-sized systems (macromolecules) would accumulate more in the tumor than the free drug (small molecule). Moreover, this effect can be further enhanced in a specific way—that is, binding to the nanocarrier targeting agents as antibodies or other molecules able to recognize specific tumor markers. At the same time, the delivery of a carrier-driven drug could involve different metabolic and cellular pathways, thus giving the possibility to elude MDR [47,48].

Carbon nanotubes (CNTs) are among the most promising possible drug nanovectors currently under study. They are made almost entirely of carbon, a part from residual metallic particles coming from the synthetic procedure, which are usually removed by a purification step when a biological application is intended. They combine therefore the biocompatibility of liposomes or polymeric nanoparticles, with the stability of inorganic nanoparticles, such as gold or silica ones.

When a material is proposed for biomedical applications, it is fundamental to establish its toxicity profile and its in vivo fate. Even though the question is still under debate, due to the plethora of variables such as the different CNT

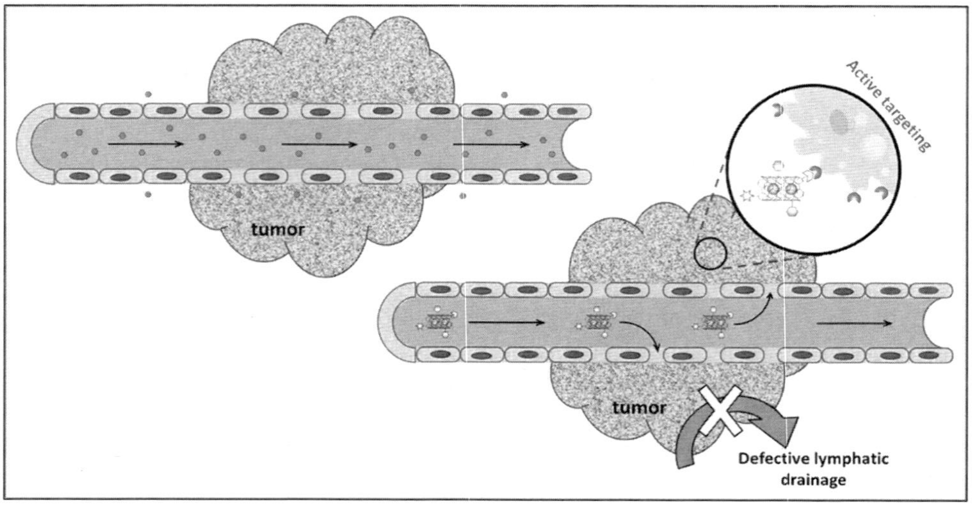

FIGURE 21.5. Enhanced permeability and retention effect for a model CNT-based drug delivery system. See color insert.

functionalizations, it appears that adequately functionalized CNTs are noncytotoxic. In fact, it has been demonstrated that highly water-soluble functionalized CNTs are uptaken by immune system cells without being toxic and preserving their activity [49], and it has been found that CNTs are more toxic as agglomerates than as good dispersions [50]. Coherently with these observations, other authors reported correlations between a higher degree of functionalization for SWCNTs and a lower toxicity toward human dermal fibroblasts. Interestingly, covalently modified SWCNTs appeared to be less cytotoxic than the ones stabilized trough surfactants [51]. Importantly, numerous studies by different groups have proven so far how surface-modified CNTs are well tolerated in vivo [52–58]. In general they have a blood clearance half-life on the order of hours. Moreover, tissue biodistribution studies showed that they are eliminated into urine, via glomerular filtration, or into feces with low residual amounts in the body. All these results indicate how the in vivo behavior of these materials could be modulated by the degree and kind of functionalization, two critical aspects that need to be accurately controlled.

Furthermore, some recent works showed in vitro enzymatic degradation of SWCNTs, thus presenting another possibility for the elimination of this carrier from the body once the therapeutic function has been exerted [59–61]. Importantly, it has been shown that MWCNTs take more time to be degraded than SWCNTs, and the process seems to occur gradually from the external walls toward the inner ones. Moreover, oxidized MWCNTs were more easily degraded than pristine tubes, indicating that defects introduced through oxidation probably facilitate the attack from the enzyme [62]. Anyway further studies, especially in vivo, are necessary to assess if CNT metabolic digestion is actually possible.

21.7 UPTAKE MECHANISM

Carbon nanotubes are able to enter cells, as recognized by the scientific community, but the way it happens is still controversial. In fact, although many studies have been carried out so far to understand which is the uptake mechanism, still there is not a unique answer to the question. The type of cell should be taken into account when comparing different results; and, more importantly, the type of CNTs could play a crucial role. In fact, the great heterogeneity of this material implies the possibility of heterogeneous behavior in a biological environment. CNTs can differ a lot in size and functionalization, thus giving a plethora of derivatives, from almost naked to differently coated ones.

Among the existing hypotheses, one is phagocytosis, a process employed by specialized cells; in fact, different authors observed engulfment of CNTs by phagocytes. Cherukuri et al. [63] treated mouse peritoneal macrophage-like cells with SWCNTs and visualized the nanotubes inside the cells exploiting their spontaneous NIR fluorescence. They observed the tubes confined in small vesicles, probably phagosomes, derived from an active ingestion process, which is consistent with their observation of a temperature-dependence of the uptake. Porter et al. [64] studied the uptake of SWCNTs by human-monocyte-derived macrophage cells and found them inside lysosomes and phagosomes, while the treatment of the same cell line with MWCNTs led to frustrated phagocytosis, an incomplete phagocytosis, probably due to the big size of the nanotubes used (diameter of about 70 nm and length ranging from few to several micrometers) [65]. Four weeks after the subcutaneous implantation of 200 nm long MWCNTs in rats, they were found inside lysosomes of macrophages [66]. On the other hand, frustrated phagocytosis with consequent cellular stress was observed

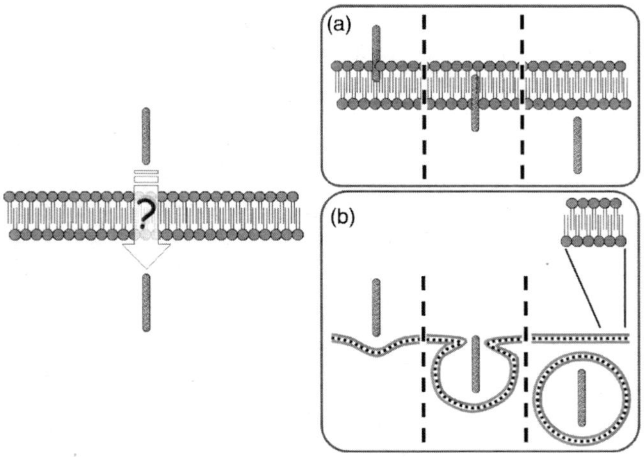

FIGURE 21.6. Nano-needle (**a**) and endocytosis (**b**) penetration mechanism. See color insert.

for human monocytes treated with long MWCNTs (also in this case from few to several micrometers) [67]. Other authors also used long MWCNTs (200–400 μm) and observed their internalization by mouse microglia cells and murine glioma cells [68]. Even if they did not specifically investigate the uptake mechanism, their results suggest two main possibilities: phagocytosis, supported by the higher internalization observed for microglia cells, which are phagocytic cells, and a needle-like penetration, supported by TEM images clearly showing single nanotubes piercing the cell membrane.

The needle-like penetration, consisting in a direct insertion of CNTs across cell membranes, and the endocytosis represent the two more important mechanisms for the uptake of CNTs by mammalian cells (Figure 21.6).

Kam et al. [69] proposed for the first time the endocytosis internalization pathway for CNTs in 2004, treating human leukaemia cells and human T cells with SWCNTs functionalized with biotin and fluoresceinated streptavidin. The intracellular fluorescence distribution was found to overlap the signal of an endosome marker. Moreover, the uptake was blocked at 4°C, consistently with energy-dependent endocytosis. Similarly, in other works by the same authors, the lack of intracellular fluorescence at 4°C and the co-staining with endosome markers suggested an endocytotic process for SWCNTs functionalized with fluorescently labeled DNA or proteins (MW <80 kDa) [70, 71]. The authors speculated about a hydrophobic interaction between the naked SWCNTs surface and the cell membrane as the driving force for the internalization. In a later work, they expanded the studies to precisely understand the uptake mechanism [72]. The internalization inhibition was observed with NaN_3 preincubation of the cells, confirming endocytosis, since ATP-production and therefore energy-dependent activities are strongly reduced by this chemical agent. Moreover, with a well-designed experimental set, endocytosis was found to happen through clathrin-coated pits, rather than through caveolae or lipid–raft pathway. Endocytosis has been suggested as the uptake pathway also by other authors. DNA-wrapped SWCNTs were tracked during in vitro trafficking exploiting their intrinsic NIR fluorescence [73]. Each trajectory was then associated with a specific kind of motion, recognizing that SWCNTs were adsorbed onto the membrane and internalized through endocytosis. Once inside the cell, they diffused consistently with confinement inside vesicles, and they finally underwent exocytosis. Moreover, colocalization experiments showed an overlap between SWCNTs and lysosomes inside the cells, confirming again the endocytotic uptake. The authors proposed this mechanism as a result of the SWCNTs coverage by the proteins present in the medium. This coating should lead to a subsequent clustering in aggregates, big enough to reach the radius cut-off thermodynamically required for being endocytosed [74]. Wang et al. [75] functionalized MWCNTs with the enzymatic toxic domain of ricin (RTA) in a noncovalent way and also studied its uptake mechanism by HeLa cells, observing clathrin-mediated endocytosis. The internalization of SWCNTs-QD-streptavidin by CD3 positive-leukemia cells was also studied, monitoring the intracellular QDs fluorescence [76]. In this case the system was rather complex; in fact the uptake was mediated by the recognition of three components. First, a biotinylated anti-CD3-antibody bound CD3 receptor on cell membrane and subsequently the streptavidin-bearing SWCNTs attached the biotinylated antibody, showing high internalization. On the contrary, poor internalization was observed when the antibody was absent, with the use of non-biotinylated antibody, when experiments were carried out at 4°C and when the cells did not express CD3-receptor on the membrane (CD3 negative cell line). All these data suggest a receptor-mediated endocytosis for this construct.

To target cancer cells, Bhirde et al. [77] described the use of a drug delivery system based on a cisplatin–SWCNT conjugate, exploiting the interaction of the epidermal growth factor (EGF) with its receptor (EGFR), overexpressed on the surface of many cancer cells. The authors demonstrated the efficient internalization of the SWCNT-EGF conjugate in vitro. Moreover, they proved that it is actually mediated by the ligand–receptor interaction, since the uptake was drastically reduced knocking down the receptor expression. These findings are consistent with receptor-mediated endocytosis. Nevertheless, a certain degree of internalization observed even without a receptor makes also other possibilities plausible, as stated by the authors themselves. Another paper conceiving two options for the uptake mechanism was published in 2008 [78]. The authors prepared SWCNTs bearing taxoids and biotin molecules, as a drug delivery system targeted toward cancer cells overexpressing the biotin receptor, demonstrating the actual internalization of the whole conjugate, hindered by both incubation at 4°C and presence of NaN_3 and strongly reduced by saturation of accessible receptors via biotin pretreatment, in accordance with receptor-mediated endocytosis. On the other hand, they observed a strong decrease in the uptake of SWCNT derivatives without biotin at low temperatures (4°C), but no changes when NaN_3 was added. This behavior is consistent with a non-endocytotic process, such as the needle-like diffusion through membranes, which is temperature-dependent but energy-independent.

It is necessary to consider that temperature can affect not only endocytosis but also membrane piercing; in fact, as simple diffusion of a molecule in a liquid is influenced by temperature, the same could happen to CNTs in biological media. More importantly, the fluidity of cellular membranes, which are made of phospholipidic bilayers, is strongly dependent on temperature. As a consequence, the mobility of an object in the bilayer will increase with temperature. Thus, it is not surprising that a decrease in temperature has sometimes been correlated with a decrease in cellular uptake of CNTs, and this experimental observation alone should not necessarily mean endocytosis, unless it is corroborated by other evidences. Therefore some cases of CNTs' internalization ascribed to endocytosis by the authors could be also differently interpreted [79, 80]. The same observation can be applied to the work published by Cheng et al. [81]. In this well-conceived study of cellular internalization of fluorescently labeled SWCNTs derivative by several mammalian cell lines, the authors compared the results obtained in normal conditions with the ones obtained at 4°C and in the presence of sodium azide. While for the former the uptake was almost completely blocked, the latter showed an intracellular fluorescent signal weaker but comparable with the normal conditions, differently from what stated by the authors (for a better understanding see Figure 2B of reference 81). Therefore it is possible to conclude that both an energy-dependent and an energy-independent mechanism are contributing to the CNTs uptake by cells in this case. Moreover, the author co-incubated cells with labeled Dextran, known to be uptaken by endocytosis, and found it in the endosomes, while the fluorescence from the SWCNTs was in the nucleus, indicating that either the retention time inside the endosomes or the entire uptake pathway was not the same for the two compounds.

The nano-needle penetration mechanism for CNTs was proposed in 2004. Internalization of functionalized CNTs was found even in energy-depleting conditions, and ultrathin transverse sections of HeLa cells incubated with the CNTs were analyzed by TEM, observing MWCNTs inside cells and in the act of crossing the membrane [82, 83]. Therefore the authors hypothesized that the cylindrical shape of CNTs let them directly insert across biological membranes, like a nano-needle. This description is in accordance with what is observed for some peptides and proteins exhibiting nonclassical transport activity [84] and also with what has been predicted for nanotube-shaped objects [85]. Another theoretical study, instead, reported that the energy cost for a SWCNT to insert into a model phospholipid bilayer is too high to occur only because of thermal motion energy [86]. Nevertheless, further studies on the penetration of a patterned CNT were performed by the same authors. Since probably CNTs in a biological environment will not have a homogeneous surface, but will be covered by biomolecules, they studied a CNT covered by alternating bands of hydrophilic and hydrophobic regions. This model can be also representative of a CNT functionalized with hydrophilic groups, since usually functionalization does not cover the whole CNT surface. In this case the energy required by CNTs to traverse the bilayer was much lower, thus resulting in enhanced penetration [87]. The same internalization pathway was proposed for a huge selection of cell types, since a comprehensive study revealed effective CNTs internalization even by cells unable to carry out endocytosis or under conditions that prevent energy-dependent processes [88]. The cells were treated with differently functionalized SWCNTs and MWCNTs, which were not coated by macromolecules. As speculated by the authors, the discrepancy between these results and the endocytotic hypothesis can lie in the different kind of CNTs derivatives used.

Zeineldin et al. [89] wrapped PL-PEG2000 onto SWCNTs and found that when the PEG integrity was preserved, the coating prevented the nonspecific uptake by cells (different ovarian cancer cell lines), while the PEG fragmentation (by sonication for a prolonged period, i.e., 1 hour) allowed CNTs internalization by cells. A further functionalization of the PEGylated, not sonicated, SWCNTs with a specific targeting molecule led to selective uptake by cancer cells overexpressing the receptor for that molecule. Although the authors did not speculate on possible uptake mechanisms, these findings underline the importance of CNTs coatings in determining their biological fate.

CNT size seems to have a role as well. A recent work compares the internalization by human hepatocellular carcinoma cells of different CNTs: namely MWCNTs with a diameter of 10–30 nm (long and short, i.e., 1–2 μm or 0.5–1 μm long) and SWCNTs with a diameter of 1–3 nm (long and short, i.e., 100–200 nm or 50–100 nm long) [90]. All the CNTs were noncovalently functionalized with chitosan. The authors reported internalization only for SWCNTs, and with different mechanism according to size. For long SWCNTs the mechanism observed was mainly clathrin-mediated endocytosis, while for short SWCNTs they describe also endocytosis via caveolae and direct insertion across cell membrane.

Evidences supporting the piercing hypothesis were reported elsewhere. SWCNTs were observed translocating across the lipid bilayer and entering the cytoplasm [64, 91]. SWCNTs and MWCNTs were found across the plasma, lysosomal, and nuclear membranes [92], and MWCNTs were observed by TEM both inside cytoplasmic vacuoles of epidermal keratinocytes (consistently with endocytosis) and piercing the nuclear membrane [93]. Similarly, as aforementioned, Kateb et al. [68] presented TEM visual confirmation of membrane insertion of MWCNTs.

A very interesting study, published in 2008, confirmed the ability of CNTs to cross biological membranes, considering the question from a rather different point of view. MWCNTs were visualized while traversing the fenestrated endothelium in the glomerular filtration system of mice and crossing the renal filtration membrane. In order to do so, the tubes had to adopt an adequate spatial conformation, facing the filtration barrier perpendicularly, since the fenestration size is compatible only with the traverse dimension of these MWCNTs, to let them pass and pierce the underlying basal membrane [61].

A further hint for understanding CNT ability to cross cell membranes could be given by another paper [94]. During an in vitro experiment, MWCNTs were guided through a magnetic field toward mammalian cells and they cross the membrane, spearing it and thus translocating inside the cells their cargo—that is, a vector containing the sequence for a fluorescent protein, which allowed the recognition of cells internalizing it. SEM pictures of cell surface confirmed the presence of the CNTs that actually look like spearing it. This work aimed to prove that when actively guiding CNTs toward cells, it is possible to induce a temporary permeabilization of the membrane, an effect of its nanomechanical penetration. The system was thought to be exploited for delivery purposes, since, despite the great efficiency of transfection, it leads to minor perturbation of cells. A recently published work confirmed this property, suggesting the possible use of CNTs as cellular endoscopes [95]. The authors of this paper prepared an endoscope ending with a MWCNT (with a diameter in the range of 50–200 nm or more and a length of 50–60 μm) and found that, thanks to its narrow cylindrical shape, it produced no cell damage, and it resulted in less mechanical stress for the cell compared to common conical glass pipettes, even for prolonged periods.

In conclusion, many different studies aimed at understanding CNT uptake mechanism by cells, considering the question from different points of view. The discrepancy between the results present in the literature can lie in the different kind of CNTs derivatives used. Considering all the works published so far, we could conclude that both endocytosis and needle-like piercing of the membrane exist and can contribute to internalization of CNTs, with a predominance of needle penetration when internalization is driven by the CNTs—that is, when CNTs are functionalized with nonbulky molecules—whereas, when CNTs are coated with big molecules or with molecules that trigger a specific receptor-mediated endocytosis, this mechanism prevails.

21.8 DELIVERY OF ANTINEOPLASTIC CHEMOTHERAPEUTIC DRUGS

The most studied possibility in CNT-based cancer therapy is drug delivery, in which CNTs are used as vehicles for a chemotherapeutic agent, in order to improve its delivery to the target.

Doxorubicin is a drug widely used in clinical cancer treatment, and it is very interesting to study its delivery through nanocarriers. So far, a number of works in which carbon nanotubes were used for this purpose has been published. In one of the first papers, doxorubicin was bound to copolymer-coated MWCNTs forming a supramolecular complex based on π–π stacking (Figure 21.7a) [96]. The cytotoxicity of

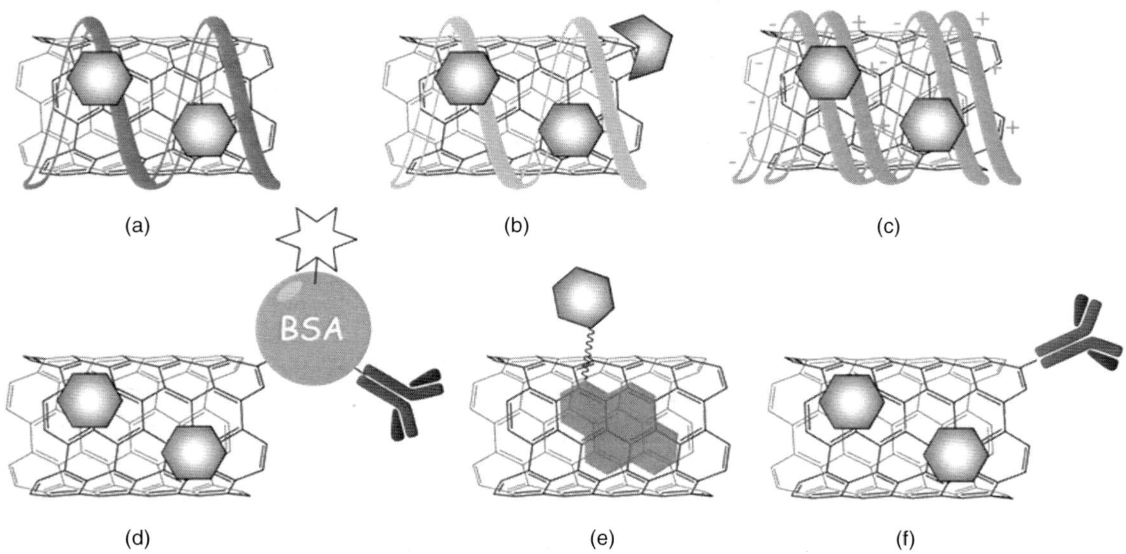

FIGURE 21.7. Doxorubicin-bearing conjugates [96–103]. For a detailed description of the different constructs a–f, see paragraph 21.8.

this derivative was tested in vitro on human breast cancer cells showing an increased mortality rate in comparison with doxorubicin alone.

This drug was also bound, through supramolecular chemistry, to two different water-soluble SWCNTs: noncovalently functionalized with a phospholipidic-PEG surfactant and oxidized SWCNTs covalently PEGylated through amidation of the carboxylic moieties [97]. The authors suggested π–π stacking and hydrophobic interactions as the binding forces between doxorubicin and nanotubes. Moreover, they report a pH-dependent interaction: The loading was made at pH 9, a condition at which doxorubicin is deprotonated and thus has low water solubility, reaching a doxorubicin/nanotubes weight ratio of ~4:1; a decrease in the pH led to the release of the drug from the SWCNT carrier due to the its consequent higher hydrophilicity. This mechanism could be very useful in cancer therapy due to the acidic environment of extracellular tissues in tumor, potentially leading to selective drug release in vivo. The cytotoxicity of the phospholipidic-PEG-wrapped SWCNT-doxorubicin construct was tested on human glioblastoma cancer cells. The derivative induced cell death similarly to free doxorubicin at a concentration of 10 μM, although the observed IC_{50} value was higher (~8 μM for the carbon nanotube conjugate compared to ~2 μM for free doxorubicin). Furthermore, a targeted doxorubicin delivery was tested using a cyclic arginine–glycine–aspartic acid (RGD) peptide, which acts as recognition moiety for integrin $\alpha_V\beta_3$ receptors, overexpressed in a wide range of solid tumors. The targeting agent was bound on the PEG chain of the noncovalent SWCNT derivative, and then doxorubicin was loaded (Figure 21.7b). This conjugate, tested on integrin $\alpha_V\beta_3$-positive cells, showed enhanced drug delivery compared to the derivative without RGD, according to the degree of brightness in confocal fluorescence experiments and to the IC_{50} value, which was smaller (~3 μM) for the RGD-bearing derivative than for the one void of any targeting agent. The behavior of the phospholipidic-PEG-wrapped SWCNT–doxorubicin complex was studied also in vivo into SCID mice bearing Raji lymphoma xenografts [98]. The authors observed higher tumor uptake, probably due to the prolonged circulation half-life of the construct, and greater inhibition of tumor growth with respect to the free drug but the therapeutic efficacy was not better than DOXIL (liposomal doxorubicin). On the other hand, differently from DOXIL, SWCNT-doxorubicin caused neither significant toxicity nor mortality, allowing the use of higher doses. Treatments with 10 mg/kg (instead of 5 mg/kg) of SWCNT-doxorubicin (dose normalized on drug) led to improved efficacy, still without causing any severe toxicity, whereas 5 mg/kg of free doxorubicin or DOXIL resulted in a strong decrease in the weight of the animals and mortality rates of 20% and 40%, respectively. The possible reasons for this behavior involve different aspects. The bigger size of the SWCNT construct with respect to free doxorubicin probably slowed down glomerular filtration, and, moreover, the PEG coating could have hidden doxorubicin from macrophages. Both these effects could determine prolonged blood circulation, thus enhancing tumor accumulation because of enhanced permeability and retention (EPR) effect. Moreover, the slightly acidic microenvironment of tumors should facilitate doxorubicin detachment from the SWCNTs, as previously reported [99], therefore leading to selective drug release. Nevertheless, the same authors observed a slow in vivo dissociation of doxorubicin from the SWCNTs after administration, and this fact can suggest that noncovalent constructs cannot guarantee proper in vivo stability. An in vivo targeting study was also performed with PEGylated SWCNTs functionalized with RGD peptide [99]. After a tail vein injection into U87MG tumor-bearing mice, the nanotubes accumulated in tumor (13% ID/g in 24 h), with no obvious toxicity or negative health effects for the animals. Zhang et al. [100] prepared a different kind of SWCNT-based vector for doxorubicin. First CNTs were oxidized, thus introducing carboxylic groups, and then they were derivatized via wrapping with two different polysaccharides, namely sodium alginate, which is anionic, and chitosan, which is cationic. Doxorubicin was then loaded on the modified SWCNTs, and the interaction was mediated probably by electrostatic interactions and π–π stacking (Figure 21.7c). The release of doxorubicin in acidic condition (pH 5.5) was assessed, in order to mimic the typical environment of lysosomes or cancerous tissues, finding a release of almost 40% in 72 h, while the construct was stable at pH 7.4. Finally, an identical conjugate bearing folic acid (linked to chitosan) was prepared, to target tumors overexpressing the folate receptor. In vitro tests on HeLa cells were performed to evaluate cellular uptake, by fluorescence microscopy, and cell viability. The compound led to stronger reduction in cell viability than the drug itself, proving its efficacy as targeted delivery system.

A triple functionalized SWCNT derivative was prepared, bearing doxorubicin, fluorescein, and a targeting antibody that recognizes carcinoembryonic antigen (CEA), a tumor marker relevant for a variety of adenocarcinomas (Figure 21.7d) [101]. Doxorubicin was noncovalently loaded on oxidized SWCNTs by π–π stacking. Afterwards a fluorescein-labeled BSA was covalently attached through its amines to the carboxylic groups present on the oxidized SWCNTs, via amidation reaction. Finally the antibody was tethered through its amines to the BSA carboxylic groups again with an amidation step. In this way the authors designed a construct that should target CEA-overexpressing cancer cells, be internalized by endocytosis, and release doxorubicin at lysosomal acidic pH. The fluorescein was introduced to localize the SWCNT construct by the fluorescent signal after doxorubicin release. The internalization by WiDr human colon cancer cells was studied by confocal microscopy. After a 4 h treatment, the fluorescently labeled SWCNTs were found in the cytoplasm, while doxorubicin was localized in the nucleus, where the drug exerts its activity. As a control, the fluorescein-labeled BSA did not show internalization to the same high extent, indicating the good ability of CNTs to act as vectors. Nevertheless, no information on the cytotoxicity of the complex is reported, nor is any stability study to evaluate the release of doxorubicin from CNTs. With a different approach, doxorubicin was linked to pyrene, and this system was noncovalently bound to SWCNTs via π–π bonding driven by the polyaromatic unit (Figure 21.7e) [102]. The linker between doxorubicin and pyrene was a short chain with a carbamate group, designed to be enzymatically cleaved inside lysosomes. In vitro studies on a mouse melanoma cell line were performed to assess the internalization of the construct and its cellular toxicity. The results showed accumulation in the lysosomes and cell growth inhibition (via apoptosis) in a time- and dose-dependent manner, approaching the doxorubicin profile for the 72-h treatment. Furthermore, the therapeutic efficacy was evaluated in vivo on melanoma-bearing mice. Free doxorubicin and SWCNT-doxorubicin induced similar reduction of tumor volume, with a significant decrease of systemic toxicity for the CNT construct, according to animal body weight, organs weight, and histopathological analyses. Also in this case the release of the drug from the CNTs was studied, comparing the effect of incubation with cell lysate to the use of normal buffer. Though the former system was significantly more efficient, proving that intracellular enzymes play a fundamental role, a release was observed also in the latter case, demonstrating how a noncovalent construct is not completely stable in a biological environment. With the double aim of obtaining a targeted drug delivery system and of overcoming MDR, SWCNTs were functionalized via amidation with an antibody directed toward P-glycoprotein, one of the main actors of MDR [103]. Subsequently the Ab-SWCNTs construct was noncovalently loaded with doxorubicin, via π–π stacking (Figure 21.7f). This construct was tested in vitro on human leukemia cells, overexpressing glycoprotein P (P-gp) on their surface and, therefore, resistant to doxorubicin. Cells were exposed to NIR radiation to trigger doxorubicin release, as a consequence of NIR absorption by SWCNTs. This system was indeed able to triplicate the amount of drug released in 24 h, but it should be underlined that a partial release was observed without NIR radiation as well. The Ab-CNT construct became bound to the cell membrane and doxorubicin was localized inside cells, in much a higher amount with respect to treatments with free compound. Doxorubicin alone or together with anti P-gp antibody (at a concentration of 5 μg/mL) could partially inhibit cell growth; but, after increasing the incubation time after treatment (from 24 h to 48 and 72 h), cells gradually recovered and started growing again. Instead, very interestingly, the SWCNT construct, used at an equivalent concentration of drug, inhibited cell growth more efficiently than the free drug and in a time-dependent way, probably due to a slow drug release from the vector surface.

Methotrexate is another well-known anticancer drug, which suffers, as many others, from low bioavailability and toxic side effects. Preliminary studies of CNT-based drug delivery with this drug were performed by Pastorin et al. [104]. MWCNTs were covalently functionalized with methotrexate and fluorescein at the same time (Figure 21.8a), and results obtained on human Jurkat T lymphocytes showed a rapid internalization of the compound, even though the efficacy of the construct was lower than the free drug. In a more recent work the same drug was bound to MWCNTs through cleavable linkers (Figure 21.8b). The use of a peptide recognized by intracellular proteases led to a significant decrease in MCF-7 breast cancer cell viability if compared to free methotrexate [105]. Therefore it is possible to conclude that the lower efficacy in the first approach was probably due to the lack of drug release once the construct was internalized.

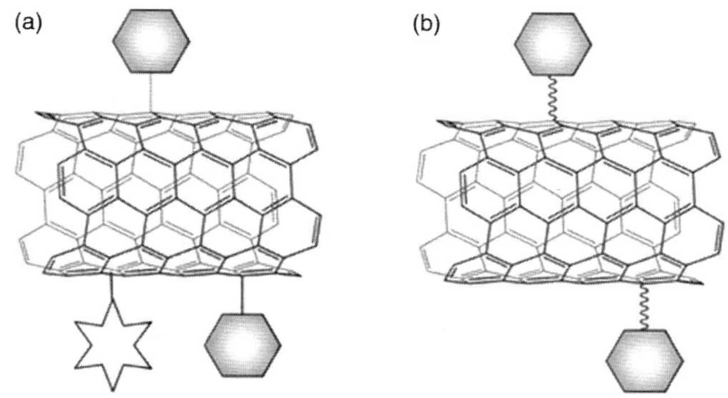

FIGURE 21.8. Methotrexate-bearing conjugates [104, 105]. For a detailed description of the two constructs a and b see paragraph 21.8.

Taxanes represent another class of antineoplastic drugs that, to date, have already been conjugated to CNTs for new drug delivery approaches. Paclitaxel (PTX), which belongs to this class, was covalently bound through an ester bond to the PEG of noncovalently PEGylated SWCNTs and i.v. injected into xenograft tumor mice (PTX-resistant 4T1 murine breast cancer mice model) to test the in vivo efficacy [106]. Results were very promising, showing a tumor growth inhibition value of almost 60% for the SWCNT-PTX derivative, much higher than the one reported for Taxol and PEG-PTX (28% and 21%, respectively). A higher apoptosis level and a lower proliferation active cells level were observed after treatment with the CNT derivative, when compared to the drug alone. It is important to underline the lack of toxicity in mice for these PEGylated SWCNTs themselves, demonstrated by the authors in another work [57]. The evaluation of the pharmacokinetics showed a longer plasmatic half-life for SWCNT-PTX, coherently with the enhanced hydrophilicity of the conjugate with respect to the drug itself, and much higher PTX presence in RES organs (liver/spleen) and intestine 2 h after injection. This is a predictable behavior for nanomaterials in general and could give concern for the toxicity toward these organs. Nevertheless, the authors reported differences between the biodistribution of SWCNT and PTX, indicating a rapid release of the drug from the conjugate probably due

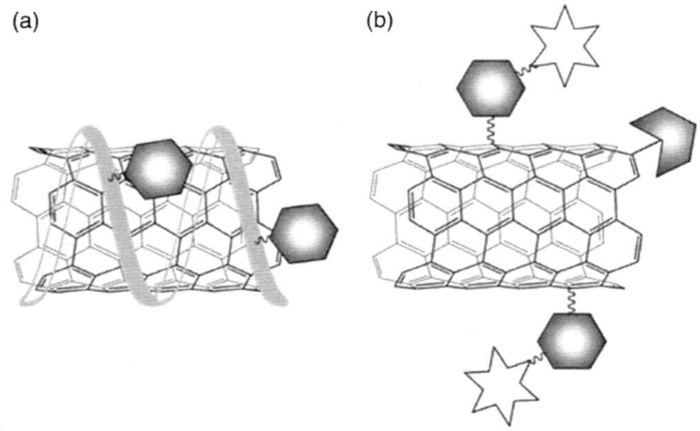

FIGURE 21.9. Taxane-bearing conjugates [78,106]. For a detailed description of the two constructs a and b, see paragraph 21.8.

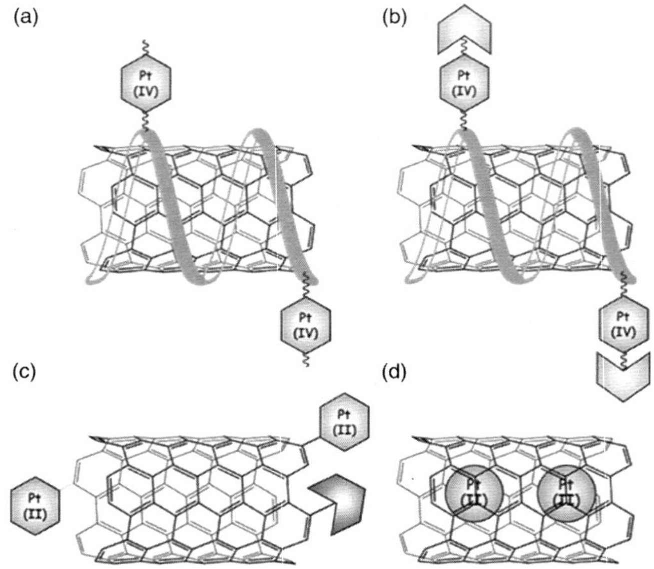

FIGURE 21.10. Platinum analogues-bearing conjugates [77, 107, 108, 110, 111]. For a detailed description of the different constructs a–d, see paragraph 21.8.

to ester cleavage by carboxylesterases. As a consequence the drug seemed to be rapidly excreted, lowering its toxicity. Importantly, tumor PTX levels were 10 times higher for SWCNT-PTX derivative than for free taxol 2 h after injection and tumor-to-normal organ/tissue PTX uptake ratio was bigger, thus indicating a better selectivity of the CNTs-delivered drug.

Chen et al. [78] performed another study with the same class of drugs. They prepared a covalent derivative of SWCNTs bearing a taxoid molecule and biotin as targeting unit (Figure 21.9b) toward cancer cells overexpressing its receptor, thus leading to receptor-mediated endocytosis. This construct was intended to be a prodrug, which frees its cargo, once inside the cell, upon reduction of the disulfide bond by endogenous thiols such as glutathione (GSH), whose concentrations are typically more than 10^3 times higher in tumor tissues than in blood plasma. Therefore, the system should act specifically on cancer cells, leading the taxoid to carry out its mitosis inhibition action only in the desired sites. The authors demonstrated the actual internalization of the whole conjugate in leukemia L1210FR cells, by means of confocal fluorescence microscopy, thanks to the presence of a fluorescent tag covalently bound to the taxoid molecule. Moreover, in the case of a control SWCNT derivative without biotin, they observed a temperature-dependent but energy-independent internalization. This is coherent with a nonendocytotic mechanism, in accordance with the hypothesis of a needle-like diffusion of carbon nanotubes through cell membranes. On the other hand, for the biotin-bearing conjugate, the mechanism observed was endocytosis, and the internalization degree was far higher if compared to the one for cells not expressing the biotin receptor. Furthermore, the authors proved the efficient release of the drug by GSH and the binding of the conjugate to the microtubule network, where it exerts its cytotoxic action, with an IC_{50} value smaller than the one for the drug alone, probably due to the increase in the drug uptake.

Also platinum analogues have been investigated in CNT-based drug delivery for cancer treatment. In a first work, Lippard and co-workers [107] reported the preparation of a Pt(IV)-based SWCNTs prodrug (Figure 21.10a) able to deliver and release the drug inside cells, leading to intracellular concentrations six times higher than the ones reached treating the cells with the free drug.

More recently the same group prepared another Pt-based targeted prodrug using SWCNTs as a longboat [108]. The derivative was prepared by binding folic acid, whose receptors are overexpressed in many cancer cells, to a Pt(IV) compound and also tethering this conjugate to CNTs through an amidic coupling with terminal amines of wrapped phospholipid-PEG chains (Figure 21.10b). The Pt(IV), once inside cells, can be reduced thanks to acidic endosomal pH, therefore losing the two axial ligands and leading to the active Pt(II) compound. The authors demonstrated for this derivative internalization via folate receptor-mediated endocytosis (FRME) and the ability to kill cancer cells (human choriocarcinoma and human nasopharyngeal carcinoma) with IC_{50} values more than eight times lower than the ones for cisplatin alone. Moreover, the formation of the major reaction product of cisplatin with DNA was detected, using a specific monoclonal antibody, proving the actual ability of the system to act as a prodrug, generating the cytotoxic derivative once internalized and thus killing in a selective way folate receptor overexpressing cells. Also cisplatin-delivery systems based on drug–SWCNT covalent bioconjugates were investigated (Figure 21.10c) [77]. The interaction of the epidermal growth factor (EGF) with its receptor (EGFR), overexpressed on the cell surface of a big variety of cancers, was exploited to target CNTs both in vitro and in vivo. The efficient in vitro internalization in head and neck squamous carcinoma cells of the SWCNTs–EGF conjugate was proved and resulted mediated by the EGF–EGFR ligand–receptor interaction. Furthermore, quantum dot (Qdot)-functionalized SWCNTs were administrated to a tumor-bearing athymic mice to study the short-term biodistribution. The results showed a much higher accumulation within the tumor mass of the EGF conjugate compared to the control without EGF. Small amounts of CNTs were found within spleen, lungs, liver, kidneys, and heart, regardless of the presence of EGF. Finally, the animals were treated with SWCNT–cisplatin–EGF showing a decrease in the tumor growth in comparison with the untargeted SWCNTs–cisplatin conjugate, demonstrating that SWCNTs can selectively deliver cisplatin in vivo toward EGFR overexpressing cancer.

Another interesting approach for the functionalization of CNTs, exploited also in oncology studies, is the filling of the nanotubes. This method was first investigated by Green's group in 1994 and involves the opening of MWCNTs' caps by a nitric acid treatment and the filling of the inner cavity through a wet chemistry approach [109]. It was applied to fill the tubes with different materials, among which was carboplatin, a widely used chemotherapeutic agent for cancer treatment (Figure 21.10d) [110]. MWCNTs were opened with nitric acid and subsequently filled, thereby sonicating them in a carboplatin solution at different temperatures. Finally, they were cleaned from particles deposited on the outer surface, washing them in water and in ethanol. The derivatives were characterized through different techniques to confirm the presence of the drug inside the tubes, and it was shown that the best temperature for the filling protocol was 90°C, leading to a drug loading of 30% in weight. This derivative was then tested on human bladder cancer cells to evaluate its cytotoxicity, and the test showed reduced cell viability compared to the drug alone. More recently, also cisplatin was encapsulated into CNTs, in this case single-walled ones, the diameter of which is big enough to host the drug [111]. The SWCNTs were opened by concentrated acid treatment and were subsequently annealed at 850°C under high vacuum in order to remove oxygen functional groups introduced in the first step. The filling with cisplatin was achieved by stirring at 40°C the SWCNTs in a DMF solution of the drug for 48 h. The release of the drug in physiological solution was studied, finding that it started after 24 h and continued up to 72 h, and the inhibition of cell growth on two prostate cell lines (DU145 and PC3) obtained with the construct was comparable to the one caused by the free drug.

Another antineoplastic drug, namely 10-hydroxycamptothecin (HCPT), was covalently attached to MWCNTs to study the influence of CNTs on drug efficacy [112]. MWCNTs were first oxidized with a 36-h acid treatment, and the carboxylic groups thus introduced underwent an amidation

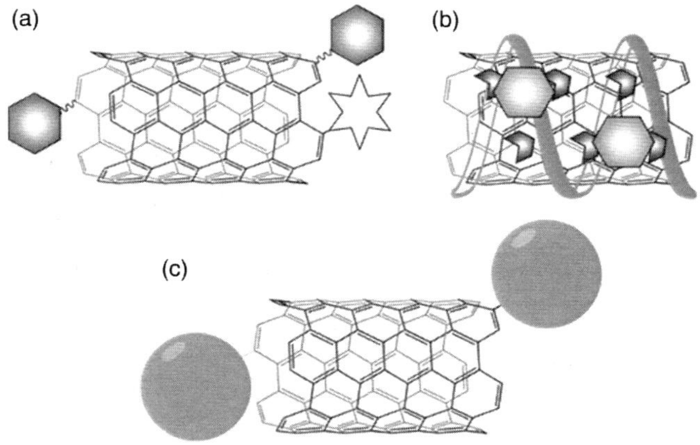

FIGURE 21.11. Anticancer conjugates bearing (**a**) HCPT, (**b**) gemcitabine, and (**c**) TCV [112–114]. For a detailed description of the different constructs a–c, see paragraph 21.8.

step to introduce a spacer bearing a terminal protected amine. After deprotection of the amino group, an ester derivative of HCPT was bound to the construct via an amide bond (Figure 21.11a). The ester linkage, hydrolytically unstable, was introduced to subsequently trigger the release of the drug. The authors observed a small drug release (<15% after 128 h) under both acidic or basic conditions, and a much higher release (80% after 128 h), when incubating the construct with fetal bovine serum, which contains esterases that could catalyze the hydrolysis. Furthermore, a fluorescein molecule was bound to the unreacted amine after HCPT conjugation, for in vitro tests on human gastric carcinoma cells. Fluorescence confocal microscopy allowed the intracellular localization of the constructs, showing a uniform distribution in the cytoplasm. WST-1 assay revealed a significantly improved cytotoxicity with respect to equal concentrations of HCPT, whereas CNTs alone did not cause any decrease in cell viability. Finally, a different construct bearing DTPA was prepared, in order to chelate technetium, a radioactive nuclide, for in vivo biodistribution studies in tumor-bearing mice. High uptake was found in the liver, spleen, lung, kidney, stomach, femur, and tumor. In the latter the maximum uptake level (3.6% ID/g) was reached within 4 h post injection. The observed blood circulation half-life was of 3.6 h, versus the 30 min reported for HCPT. Since the in vitro drug release tests showed that 4 h after incubation with serum only 12% of the HCPT was released, the authors considered that the majority of the conjugate could reach the different organs (and the tumor) without releasing a big amount of drug in the bloodstream. The in vivo antitumor performance of the HCPT-bearing MWCNTs was also studied, finding a tumor growth inhibition much more efficient than HCPT alone, without causing any severe toxicity to the animals.

Yang et al. [113] delivered a gemcitabine-MWCNTs construct to lymphatic vessels, exploiting the EPR effect and the specific guidance of an external magnetic field. SWCNTs were functionalized with poly(acrylic acid) and then decorated with magnetite nanoparticles ($FeO{\cdot}Fe_2O_3$) by a co-precipitation step with Fe^{2+} and Fe^{3+}. Three hours after subcutaneous injection in SD rats, the construct was able to reach popliteal lymph nodes without accumulation in the major organs such as liver, spleen, kidney, heart, and lung, simply by EPR effect. The same CNTs were loaded with the antineoplastic drug gemcitabine by physical adsorption (Figure 21.11b). The system was guided in vivo by applying a permanent magnet on the projection surface of one popliteal lymph node and very high accumulation of gemcitabine was detected in the lymph node, with a maximum after 24 h. At the same time, the blood plasma concentration was lower if compared to gemcitabine alone, to the treatment without the magnetic field, or to a control represented by nanosized activated carbon decorated with magnetic nanoparticles.

CNTs-based antitumor immunotherapy was explored by Meng et al. [114]. This technique employs tumor cell vaccines (TCV), which are made of inactivated cancer cells or dendritic cells presenting tumor antigens, in order to trigger the immune response of the patient against the tumor itself. With the aim of improving the efficacy of TCV, tumor lysate proteins were covalently coupled to oxidized MWCNTs via an amidic bond (Figure 21.11c). The conjugate was injected subcutaneously to H22 hepatoma-bearing mice treated with TCV (controls were performed with TCV + CNTs only and TCV + tumor lysate proteins only). The cure rate was significantly increased with respect to animals treated only with TCV or with TCV + tumor lysate proteins, even if a partial effect was exerted by CNTs alone as well. To assess whether or not the immunity was tumor-specific, the animals survived after treatment with tumor lysate proteins-bearing MWCNTs were challenged again with subcutaneous injection of tumor cells. In the case of H22 cells, they rejected the tumor, while they did not reject mouse breast cancer, proving how the therapeutic system made them develop a specific immunity.

21.9 CONCLUSIONS

Nanomedicine is increasing its importance in the scientific stage, offering new solutions to unsolved clinical problems. In this context, the use of carbon nanotubes, in particular in anticancer drug delivery, is really promising despite the fact that some actual limitations are slowing down the development of this technology. In fact, problems such as batch-to-batch lack of reproducibility, low homogeneity, and poor dispersibility of the pristine material still represent a serious drawback in the preparation of clinical applicable derivatives. Nevertheless, the continuous development of the carbon cage functionalization, improving CNT dispersibility and biocompatibility, paves the way to their use in clinics. However, two important topics still are under debate among scientists in the field, namely this material's toxicity, and its cellular uptake mechanism. In this chapter these questions are carefully overviewed, and a full description of the existing literature on the application of CNTs in anticancer drug delivery is given.

REFERENCES

1. Lu, X., Chen, Z. (2005). Curved pi-conjugation, aromaticity, and the related chemistry of small fullerenes ($< C_{60}$) and single-walled carbon nanotubes. *Chemical Reviews*, **105**, 3643–3696.
2. Charlier, J. C. (2002). Defects in carbon nanotubes. *Accounts of Chemical Research*, **35**, 1063–1069.
3. Govindaraj, A., Rao, C. N. R. (2006). Synthesis, growth mechanism and processing of carbon nanotubes. In: Dai, L., ed. *Carbon Nanotechnology*. Elsevier B.V. Amsterdam, pp. 15–51.
4. Fu, Q., Liu, J. (2006). Catalytic chemical vapor deposition of single-walled carbon nanotubes. In: Dai, L., ed. *Carbon Nanotechnology*. Elsevier B.V.. Amsterdam, pp. 81–107.
5. Nikolaev, P., Bronikowski, M. J., Bradley, R. K., Rohmund, F., Colbert, D. T., Smith, K. A., Smalley, R. E. (1999). Gas-phase catalytic growth of single-walled carbon nanotubes from carbon monoxide. *Chemical Physics Letters*, **313**, 91–97.
6. Ehli, C., Rahman, G. M. A., Jux, N., Balbinot, D., Guldi, D. M., Paolucci, F., Marcaccio, M., Paolucci, D., Melle-Franco, M., Zerbetto, F., Campidelli, S. P., Prato, M. (2006). Interactions in single wall carbon nanotubes/pyrene/porphyrin nanohybrids. *Journal of the American Chemical Society*, **128**, 11222–11231.
7. Chen, R. J., Zhang, Y., Wang, D., Dai, H. (2001). Noncovalent sidewall functionalization of single-walled carbon nanotubes for protein immobilization. *Journal of the American Chemical Society*, **123**, 3838–3839.
8. Zheng, M., Jagota, A., Semke, E. D., Diner, B. A., McLean, R. S., Lustig, S. R., Richardson, R. E., Tassi, N. G. (2003). DNA-assisted dispersion and separation of carbon nanotubes. *Nature Materials*, **2**, 338–342.
9. Qiao, R., Ke, P. C. (2006). Lipid–carbon nanotube self-assembly in aqueous solution. *Journal of the American Chemical Society*, **128**, 13656–13657.
10. Richard, C., Balavoine, F., Schultz, P., Ebbesen, T. W., Mioskowski, C. (2003). Supramolecular self-assembly of lipid derivatives on carbon nanotubes. *Science*, **300**, 775–778.
11. Zorbas, V., Smith, A. L., Xie, H., Ortiz-Acevedo, A., Dalton, A. B., Dieckmann, G. R., Draper, R. K., Baughman, R. H., Musselman, I. H. (2005). Importance of aromatic content for peptide/single-walled carbon nanotube interactions. *Journal of the American Chemical Society*, **127**, 12323–12328.
12. Li, X., Chen, W., Zhan, Q., Dai, L., Sowards, L., Pender, M., Naik, R. R. (2006). Direct measurements of interactions between polypeptides and carbon nanotubes. *Journal of Physical Chemistry B*, **110**, 12621–12625.

13. Star, A., Stoddart, J. F., Steuerman, D., Diehl, M., Boukai, A., Wong, E. W., Yang, X., Chung, S.-W., Choi, H., Heath, J. R. (2001). Preparation and properties of polymer-wrapped single-walled carbon nanotubes. *Angewandte Chemie International Edition*, **40**, 1721–1725.

14. Dujardin, E., Ebbesen, T. W., Krishnan, A., Treacy, M. M. J. (1998). Wetting of Single Shell Carbon Nanotubes. *Advanced Materials*, **10**, 1472–1475.

15. Wilson, M., Madden, P. A. (2001). Growth of ionic crystals in carbon nanotubes. *Journal of the American Chemical Society*, **123**, 2101–2102.

16. Smith, B. W., Monthioux, M., Luzzi, D. E. (1999). Carbon nanotube encapsulated fullerenes: A unique class of hybrid materials. *Chemical Physics Letters*, **315**, 31–36.

17. Niyogi, S., Hamon, M. A., Hu, H., Zhao, B., Bhowmik, P., Sen, R., Itkis, M. E., Haddon, R. C. (2002). Chemistry of single-walled carbon nanotubes. *Accounts of Chemical Research*, **35**, 1105–1113.

18. Mickelson, E. T., Huffman, C. B., Rinzler, A. G., Smalley, R. E., Hauge, R. H., Margrave, J. L. (1998). Fluorination of single-wall carbon nanotubes. *Chemical Physics Letters*, **296**, 188–194.

19. Saini, R. K., Chiang, I. W., Peng, H., Smalley, R. E., Billups, W. E., Hauge, R. H., Margrave, J. L. (2003). Covalent sidewall functionalization of single wall carbon nanotubes. *Journal of the American Chemical Society*, **125**, 3617–3621.

20. Boul, P. (1999). Reversible sidewall functionalization of buckytubes. *Chemical Physics Letters*, **310**, 367–372.

21. Ying, Y., Saini, R. K., Liang, F., Sadana, A. K., Billups, W. E. (2003). Functionalization of carbon nanotubes by free radicals. *Organic Letters*, **5**, 1471–1473.

22. Bahr, J. L., Yang, J., Kosynkin, D. V., Bronikowski, M. J., Smalley, R. E., Tour, J. M. (2001). Functionalization of carbon nanotubes by electrochemical reduction of aryl diazonium salts: A bucky paper electrode. *Journal of American Chemical Society*, **123**, 6536–6542.

23. Hudson, J. L., Casavant, M. J., Tour, J. M. (2004). Water-soluble, exfoliated, nonroping single-wall carbon nanotubes. *Journal of American Chemical Society*, **126**, 11158–11159.

24. Holzinger, M., Vostrowsky, O., Hirsch, A., Hennrich, F., Kappes, M., Weiss, R., Jellen, F. (2001). Sidewall functionalization of carbon nanotubes. *Angewandte Chemie International Edition*, **40**, 4002–4005.

25. Graupner, R., Abraham, J. R., Wunderlich, D., Vencelová, A., Lauffer, P., Röhrl, J., Hundhausen, M., Ley, L., Hirsch, A. (2006). Nucleophilic-alkylation-reoxidation: A functionalization sequence for single-wall carbon nanotubes. *Journal of American Chemical Society*, **128**, 6683–6689.

26. Syrgiannis, Z., Hauke, F., Röhrl, J., Hundhausen, M., Graupner, R., Elemes, Y., Hirsch, A. (2008). Covalent sidewall functionalization of SWNTs by nucleophilic addition of lithium amides. *European Journal of Organic Chemistry, 2008*, 2544–2550.

27. Tessonnier, J.-P., Villa, A., Majoulet, O., Su, D. S., Schlögl, R. (2009). Defect-mediated functionalization of carbon nanotubes as a route to design single-site basic heterogeneous catalysts for biomass conversion. *Angewandte Chemie International Edition*, **48**, 6543–6546.

28. Yinghuai, Z., Peng, A. T., Carpenter, K., Maguire, J. A., Hosmane, N. S., Takagaki, M. (2005). Substituted carborane-appended water-soluble single-wall carbon nanotubes: New approach to boron neutron capture therapy drug delivery. *Journal of the American Chemical Society*, **127**, 9875–9880.

29. Menard-Moyon, C., Dumas, F., Doris, E., Mioskowski, C. (2006). Functionalization of single-wall carbon nanotubes by tandem high-pressure/$Cr(CO)_6$ activation of Diels–Alder cycloaddition. *Journal of the American Chemical Society*, **128**, 14764–14765.

30. Ogrin, D., Chattopadhyay, J., Sadana, A. K., Billups, W. E., Barron, A. R. (2006). Epoxidation and deoxygenation of single-walled carbon nanotubes: Quantification of epoxide defects. *Journal of the American Chemical Society*, **128**, 11322–11323.

31. Georgakilas, V., Kordatos, K., Prato, M., Guldi, D. M., Holzinger, M., Hirsch, A. (2002). Organic functionalization of carbon nanotubes. *Journal of the American Chemical Society*, **124**, 760–761.

32. Liu, J., Rinzler, A. G., Dai H. J., Hafner, J. H., Bradley, R. K., Boul, P. J., Lu, A., Iverson, T., Shelimov, K., Huffman, C. B., Rodriguez-Macias, F., Shon, Y. S., Lee, T. R., Colbert, D. T., Smalley, R. E. (1998). Fullerene pipes. *Science*, **280**, 1253–1256.

33. Ziegler, K. J., Gu, Z., Peng, H., Flor, E. L., Hauge, R. H., Smalley, R. E. (2005). Controlled oxidative cutting of single-walled carbon nanotubes. *Journal of the American Chemical Society*, **127**, 1541–1547.

34. Warakulwit, C., Majimel, J. R., Delville, M.-H. L. N., Garrigue, P., Limtrakul, J., Kuhn, A. (2008). Controlled purification, solubilisation and cutting of carbon nanotubes using phosphomolybdic acid. *Journal of Materials Chemistry*, **18**, 4056–4061.

35. Aitchison, T. J., Ginic-Markovic, M., Matisons, J. G., Simon, G. P., Fredericks, P. M. (2007). Purification, cutting, and sidewall functionalization of multiwalled carbon nanotubes using potassium permanganate solutions. *Journal of Physical Chemistry C*, **111**, 2440–2446.

36. Chiang, I. W., Brinson, B. E., Huang, A. Y., Willis, P. A., Bronikowski, M. J., Margrave, J. L., Smalley, R. E., Hauge, R. H. (2001). Purification and characterization of single-wall carbon nanotubes (SWNTs) obtained from the gas-phase decomposition of CO (HiPco Process). *Journal of Physical Chemistry B*, **105**, 8297–8301.

37. Strong, K. L., Anderson, D. P., Lafdi, K., Kuhn, J. N. (2003). Purification process for single-wall carbon nanotubes. *Carbon*, **41**, 1477–1488.

38. Takagi, A., Hirose, A., Nishimura, T., Fukumori, N., Ogata, A., Ohashi, N., Kitajima, S., Kanno, J. (2008). Induction of mesothelioma in p53+/− mouse by intraperitoneal application of multi-wall carbon nanotube. *Journal of Toxicological Sciences*, **33**, 105–116.

39. Poland, C. A., Duffin, R., Kinloch, I., Maynard, A., Wallace, W. A. H., Seaton, A., Stone, V., Brown, S., MacNee, W.,

Donaldson, K. (2008). Carbon nanotubes introduced into the abdominal cavity of mice show asbestos-like pathogenicity in a pilot study. *Nature Nanotechnology*, **3**, 423–428.

40. Mitchell, L. A., Gao, J., Wal, R. V., Gigliotti, A., Burchiel, S. W., McDonald, J. D. (2007). Pulmonary and systemic immune response to inhaled multiwalled carbon nanotubes. *Toxicological Sciences*, **100**, 203–214.
41. Ryman-Rasmussen, J. P., Cesta, M. F., Brody, A. R., Shipley-Phillips, J. K., Everitt, J. I., Tewksbury, E. W., Moss, O. R., Wong, B. A., Dodd, D. E., Andersen, M. E., Bonner, J. C. (2009). Inhaled carbon nanotubes reach the subpleural tissue in mice. *Nature Nanotechnology*, **4**, 747–751.
42. Baron, P. A., Maynard, A. D., Foley, M. (2003). Evaluation of aerosol release during the handling of unrefined single walled carbon nanotube material. *NIOSH DART-02-191 Rev. 1.1.*
43. Tan, D. S. W., Gerlinger, M., Teh, B.-T., Swanton, C. (2010). Anti-cancer drug resistance: Understanding the mechanisms through the use of integrative genomics and functional RNA interference. *European Journal of Cancer*, **46**, 2166–2177.
44. Ambudkar, S. V., Dey, S., Hrycyna, C. A., Ramachandra, M., Pastan, I., Gottesman, M. M. (1999). Biochemical, cellular, and pharmacological aspects of the multidrug transporter. *Annual Review of Pharmacology and Toxicology*, **39**, 361–398.
45. Matsumura, Y., Maeda, H. (1986). A new concept for macromolecular therapeutics in cancer chemotherapy: Mechanism of tumoritropic accumulation of proteins and the antitumor agent smancs. *Cancer Research*, **46**, 6387–6392.
46. Iyer, A. K., Khaled, G., Fang, J., Maeda, H. (2006). Exploiting the enhanced permeability and retention effect for tumor targeting. *Drug Discovery Today*, **11**, 812–818.
47. Oh, K. T., Baik, H. J., Lee, A. H., Oh, Y. T., Youn, Y. S., Lee, E. S. (2009). The reversal of drug-resistance in tumors using a drug-carrying nanoparticular system. *International Journal of Molecular Sciences*, **10**, 3776–3792.
48. Jabr-Milane, L. S., van Vlerken, L. E., Yadav, S., Amiji, M. M. (2008). Multi-functional nanocarriers to overcome tumor drug resistance. *Cancer Treatment Reviews*, **34**, 592–602.
49. Dumortier, H. L. N., Lacotte, S. P., Pastorin, G., Marega, R., Wu, W., Bonifazi, D., Briand, J.-P., Prato, M., Muller, S., Bianco, A. (2006). Functionalized carbon nanotubes are non-cytotoxic and preserve the functionality of primary immune cells. *Nano Letters*, **6**, 1522–1528.
50. Wick, P., Manser, P., Limbach, L. K., Dettlaff-Weglikowska, U., Krumeich, F., Roth, S., Stark, W. J., Bruinink, A. (2007). The degree and kind of agglomeration affect carbon nanotube cytotoxicity. *Toxicology Letters*, **168**, 121–131.
51. Sayes, C. M., Liang, F., Hudson, J. L., Mendez, J., Guo, W., Beach, J. M., Moore, V. C., Doyle, C. D., West, J. L., Billups, W. E., Ausman, K. D., Colvin, V. L. (2006). Functionalization density dependence of single-walled carbon nanotubes cytotoxicity in vitro. *Toxicology Letters*, **161**, 135–142.
52. Singh, R., Pantarotto, D., Lacerda, L., Pastorin, G., Klumpp, C. D., Prato, M., Bianco, A., Kostarelos, K. (2006). Tissue biodistribution and blood clearance rates of intravenously administered carbon nanotube radiotracers. *Proceedings of the National Academy of Sciences USA*, **103**, 3357–3362.
53. Lacerda, L., Ali-Boucetta, H., Herrero, M. A., Pastorin, G., Bianco, A., Prato, M., Kostarelos, K. (2008). Tissue histology and physiology following intravenous administration of different types of functionalized multiwalled carbon nanotubes. *Nanomedicine*, **3**, 149–161.
54. Lacerda, L., Herrero, M. A., Venner, K., Bianco, A., Prato, M., Kostarelos, K. (2008). Carbon-nanotube shape and individualization critical for renal excretion. *Small*, **4**, 1130–1132.
55. McDevitt, M. R., Chattopadhyay, D., Jaggi, J. S., Finn, R. D., Zanzonico, P. B., Villa, C., Rey, D., Mendenhall, J., Batt, C. A., Njardarson, J. T., Scheinberg, D. A. (2007). PET imaging of soluble yttrium-86-labeled carbon nanotubes in mice. *PloS One*, **2**, e907–e907.
56. Kang, B., Yu, D., Dai, Y., Chang, S., Chen, D., Ding, Y. (2009). Biodistribution and accumulation of intravenously administered carbon nanotubes in mice probed by Raman spectroscopy and fluorescent labeling. *Carbon*, **47**, 1189–1192.
57. Liu, Z., Davis, C., Cai, W., He, L., Chen, X., Dai, H. (2008). Circulation and long-term fate of functionalized, biocompatible single-walled carbon nanotubes in mice probed by Raman spectroscopy. *Proceedings of the National Academy of Sciences USA*, **105**, 1410–1415.
58. Ruggiero, A., Villa, C. H., Bander, E., Rey, D. A., Bergkvist, M., Batt, C. A., Manova-Todorova, K., Deen, W. M., Scheinberg, D. A., McDevitt, M. R. (2010). Paradoxical glomerular filtration of carbon nanotubes. *Proceedings of the National Academy of Sciences USA*, **107**, 12369–12374.
59. Allen, B. L., Kichambare, P. D., Gou, P., Vlasova, I. I., Kapralov, A. A., Konduru, N., Kagan, V. E., Star, A. (2008). Biodegradation of single-walled carbon nanotubes through enzymatic catalysis. *Nano Letters*, **8**, 3899–3903.
60. Allen, B. L., Kotchey, G. P., Chen, Y., Yanamala, N. V. K., Klein-Seetharaman, J., Kagan, V. E., Star, A. (2009). Mechanistic investigations of horseradish peroxidase-catalyzed degradation of single-walled carbon nanotubes. *Journal of the American Chemical Society*, **131**, 17194–17205.
61. Kagan, V. E., Konduru, N. V., Feng, W., Allen, B. L., Conroy, J., Volkov, Y., Vlasova, I. I., Belikova, N. A., Yanamala, N., Kapralov, A., Tyurina, Y. Y., Shi, J., Kisin, E. R., Murray, A. R., Franks, J., Stolz, D., Gou, P., Klein-Seetharaman, J., Fadeel, B., Star, A., Shvedova, A. A. (2010). Carbon nanotubes degraded by neutrophil myeloperoxidase induce less pulmonary inflammation. *Nature Nanotechnology*, **5**, 354–359.
62. Zhao, Y., Allen, B. L., Star, A. (2011). Enzymatic degradation of multiwalled carbon nanotubes. *Journal of Physical Chemistry A*, **115**, 9536–9544.
63. Cherukuri, P., Bachilo, S. M., Litovsky, S. H., Weisman, R. B. (2004). Near-Infrared fluorescence microscopy of single-walled carbon nanotubes in phagocytic cells. *Journal of the American Chemical Society*, **126**, 15638–15639.
64. Porter, A. E., Gass, M., Muller, K., Skepper, J. N., Midgley, P. A., Welland, M. (2007). Direct imaging of single-walled

carbon nanotubes in cells. *Nature Nanotechnology*, **2**, 713–717.

65. Cheng, C., Müller, K. H., Koziol, K. K. K., Skepper, J. N., Midgley, P. A., Welland, M. E., Porter, A. E. (2009). Toxicity and imaging of multi-walled carbon nanotubes in human macrophage cells. *Biomaterials*, **30**, 4152–4160.
66. Sato, Y., Yokoyama, A., Shibata, K.-I., Akimoto, Y., Ogino, S.-i., Nodasaka, Y., Kohgo, T., Tamura, K., Akasaka, T., Uo, M., Motomiya, K., Jeyadevan, B., Ishiguro, M., Hatakeyama, R., Watari, F., Tohji, K. (2005). Influence of length on cytotoxicity of multi-walled carbon nanotubes against human acute monocytic leukemia cell line THP-1 in vitro and subcutaneous tissue of rats in vivo. *Molecular BioSystems*, **1**, 176–182.
67. Brown, D. M., Kinloch, I. A., Bangert, U., Windle, A. H., Walter, D. M., Walker, G. S., Scotchford, C. A., Donaldson, K., Stone, V. (2007). An in vitro study of the potential of carbon nanotubes and nanofibres to induce inflammatory mediators and frustrated phagocytosis. *Carbon*, **45**, 1743–1756.
68. Kateb, B., Van Handel, M., Zhang, L., Bronikowski, M. J., Manohara, H., Badie, B. (2007). Internalization of MWCNTs by microglia: Possible application in immunotherapy of brain tumors. *NeuroImage*, **37**, S9–S17.
69. Kam, N. W. S., Jessop, T. C., Wender, P. A., Dai, H. (2004). Nanotube molecular transporters: internalization of carbon nanotube-protein conjugates into mammalian cells. *Journal of the American Chemical Society*, **126**, 6850–6851.
70. Kam, N. W. S., Connell, M. O., Wisdom, J. A., Dai, H. (2005). Carbon nanotubes as multifunctional biological transporters and near-infrared agents for selective cancer cell destruction. *Proceedings of the National Academy of Sciences USA*, **102**, 11600–11605.
71. Kam, N. W. S., Dai, H. (2005). Carbon nanotubes as intracellular protein transporters: Generality and biological functionality. *Journal of the American Chemical Society*, **127**, 6021–6026.
72. Kam, N. W. S., Liu, Z., Dai, H. (2006). Carbon nanotubes as intracellular transporters for proteins and DNA: An investigation of the uptake mechanism and pathway. *Angewandte Chemie International Edition*, **45**, 577–581.
73. Jin, H., Heller, D. A., Strano, M. S. (2008). Single-particle tracking of endocytosis and exocytosis of single-walled carbon nanotubes in NIH-3T3 cells. *Nano Letters*, **8**, 1577–1585.
74. Jin, H., Heller, D. A., Sharma, R., Strano, M. S. (2009). Size-dependent cellular uptake and expulsion of single-walled carbon nanotubes: Single particle tracking and a generic uptake model for nanoparticles. *ACS Nano*, **3**, 149–158.
75. Wang, M., Yu, S., Wang, C., Kong, J. (2010). Tracking the endocytic pathway of recombinant protein toxin delivered by multiwalled carbon nanotubes. *ACS Nano*, **4**, 6483–6490.
76. Bottini, M., Cerignoli, F., Dawson, M. I., Magrini, A., Rosato, N., Mustelin, T. (2006). Full-length single-walled carbon nanotubes decorated with streptavidin-conjugated quantum dots as multivalent intracellular fluorescent nanoprobes. *Biomacromolecules*, **7**, 2259–2263.
77. Bhirde, A. A., Patel, V., Gavard, J., Zhang, G., Sousa, A. A., Masedunskas, A., Leapman, R. D., Weigert, R., Gutkind, J. S., Rusling, J. F. (2009). Targeted killing of cancer cells in vivo and in vitro with EGF-directed carbon nanotube-based drug delivery. *ACS Nano*, **3**, 307–316.
78. Chen, J., Chen, S., Zhao, X., Kuznetsova, L. V., Wong, S. S., Ojima, I. (2008). Functionalized single-walled carbon nanotubes as rationally designed vehicles for tumor-targeted drug delivery. *Journal of the American Chemical Society*, **130**, 16778–16785.
79. Jia, N., Lian, Q., Shen, H., Wang, C., Li, X., Yang, Z. (2007). Intracellular delivery of quantum dots tagged antisense oligodeoxynucleotides by functionalized multiwalled carbon nanotubes. *Nano Letters*, **7**, 2976–2980.
80. Liu, Y., De-Cheng, W., Zhang, W.-D., Jiang, X., He, C.-B., Chung, T. S., Goh, S. H., Leong, K. W. (2005). Polyethylenimine-grafted multiwalled carbon nanotubes for secure noncovalent immobilization and efficient delivery of DNA. *Angewandte Chemie International Edition*, **44**, 4782–4785.
81. Cheng, J., Fernando, K. A. S., Veca, L. M., Sun, Y.-P., Lamond, A. I., Lam, Y. W., Cheng, S. H. (2008). Reversible accumulation of PEGylated single-walled carbon nanotubes in the mammalian nucleus. *ACS Nano*, **2**, 2085–2094.
82. Pantarotto, D., Briand, J.-P., Prato, M., Bianco, A. (2004). Translocation of bioactive peptides across cell membranes by carbon nanotubes. *Chemical Communications*, 16–17.
83. Pantarotto, D., Singh, R., McCarthy, D., Erhardt, M., Briand, J.-P., Prato, M., Kostarelos, K., Bianco, A. (2004). Functionalized carbon nanotubes for plasmid DNA gene gelivery. *Angewandte Chemie International Edition*, **43**, 5242–5246.
84. Kueltzo, L. A., Middaugh, C. R. (2003). Nonclassical transport proteins and peptides: An alternative to classical macromolecule delivery systems. *Journal of Pharmaceutical Sciences*, **92**, 1754–1772.
85. Lopez, C. F., Nielsen, S. O., Moore, P. B., Klein, M. L. (2004). Understanding nature's design for a nanosyringe. *Proceedings of the National Academy of Sciences USA*, **101**, 4431–4434.
86. Pogodin, S., Baulin, V. A. (2010). Can a carbon nanotube pierce through a phospholipid bilayer? *ACS Nano*, **4**, 5293–5300.
87. Pogodin, S., Slater, N. K. H., Baulin, V. A. (2011). Surface patterning of carbon nanotubes can enhance their penetration through a phospholipid bilayer. *ACS Nano*, **5**, 1141–1146.
88. Kostarelos, K., Lacerda, L., Pastorin, G., Wu, W., Wieckowski, S. B., Luangsivilay, J., Godefroy, S., Pantarotto, D., Briand, J.-P., Muller, S., Prato, M., Bianco, A. (2007). Cellular uptake of functionalized carbon nanotubes is independent of functional group and cell type. *Nature Nanotechnology*, **2**, 108–113.
89. Zeineldin, R., Al-Haik, M., Hudson, L. G. (2009). Role of polyethylene glycol integrity in specific receptor targeting of carbon nanotubes to cancer cells. *Nano Letters*, **9**, 751–757.
90. Kang, B., Chang, S., Dai, Y., Yu, D., Chen, D. (2010). Cell response to carbon nanotubes: Size-dependent intracellular uptake mechanism and subcellular fate. *Small*, **6**, 2362–2366.
91. Porter, A. E., Gass, M., Bendall, J. S., Muller, K., Goode, A., Skepper, J. N., Midgley, P. A., Welland, M. (2009). Uptake

of noncytotoxic acid-treated single-walled carbon nanotubes into the cytoplasm of human macrophage cells. *ACS Nano*, **3**, 1485–1492.

92. Cheng, C., Porter, A. E., Muller, K., Koziol, K., Skepper, J. N., Midgley, P., Welland, M. (2009). Imaging carbon nanoparticles and related cytotoxicity. *Journal of Physics Conference Series*, **151**, 1–15.
93. Monteiro-Riviere, N. A., Nemanich, R. J., Inman, A. O., Wang, Y. Y., Riviere, J. E. (2005). Multi-walled carbon nanotube interactions with human epidermal keratinocytes. *Toxicology Letters*, **155**, 377–384.
94. Cai, D., Mataraza, J. M., Qin, Z.-H., Huang, Z., Huang, J., Chiles, T. C., Carnahan, D., Kempa, K., Ren, Z. (2005). Highly efficient molecular delivery into mammalian cells using carbon nanotube spearing. *Nature Methods*, **2**, 449–454.
95. Singhal, R., Orynbayeva, Z., Sundaram, R. V. K., Niu, J. J., Bhattacharyya, S., Vitol, E. A., Schrlau, M. G., Papazoglou, E. S., Friedman, G., Gogotsi, Y. (2011). Multifunctional carbon-nanotube cellular endoscopes. *Nature Nanotechnology*, **6**, 57–64.
96. Ali-Boucetta, H., Al-Jamal, K. T., McCarthy, D., Prato, M., Bianco, A., Kostarelos, K. (2008). Multiwalled carbon nanotube–doxorubicin supramolecular complexes for cancer therapeutics. *Chemical Communications*, 459–461.
97. Liu, Z., Sun, X., Nakayama-Ratchford, N., Dai, H. (2007). Supramolecular chemistry on water-soluble carbon nanotubes for drug loading and delivery. *ACS Nano*, **1**, 50–56.
98. Liu, Z., Fan, A. C., Rakhra, K., Sherlock, S., Goodwin, A., Chen, X., Yang, Q., Felsher, D. W., Dai, H. (2009). Supramolecular stacking of doxorubicin on carbon nanotubes for in vivo cancer therapy. *Angewandte Chemie International Edition*, **48**, 7668–7672.
99. Liu, Z., Cai, W., He, L., Nakayama, N., Chen, K., Sun, X., Chen, X., Dai, H. (2007). In vivo biodistribution and highly efficient tumour targeting of carbon nanotubes in mice. *Nature Nanotechnology*, **2**, 47–52.
100. Zhang, X., Meng, L., Lu, Q., Fei, Z., Dyson, P. J. (2009). Targeted delivery and controlled release of doxorubicin to cancer cells using modified single wall carbon nanotubes. *Biomaterials*, **30**, 6041–6047.
101. Heister, E., Neves, V., Tîlmaciu, C., Lipert, K., Beltran, V. S., Coley, H. M., Silva, S. R. P., McFadden, J. (2009). Triple functionalisation of single-walled carbon nanotubes with doxorubicin, a monoclonal antibody, and a fluorescent marker for targeted cancer therapy. *Carbon*, **47**, 2152–2160.
102. Chaudhuri, P., Soni, S., Sengupta, S. (2010). Single-walled carbon nanotube-conjugated chemotherapy exhibits increased therapeutic index in melanoma. *Nanotechnology*, **21**, 025102 (025111pp).
103. Li, R., Wu, R. A., Zhao, L., Wu, M., Yang, L., Zou, H. (2010). P-glycoprotein antibody functionalized carbon nanotube overcomes the multidrug resistance of human leukemia cells. *ACS Nano*, **4**, 1399–1408.
104. Pastorin, G., Wu, W., Wieckowski, S. B., Briand, J.-P., Kostarelos, K., Prato, M., Bianco, A. (2006). Double functionalisation of carbon nanotubes for multimodal drug delivery. *Chemical Communications*, 1182–1184.
105. Samorì, C., Ali-Boucetta, H., Sainz, R., Guo, C., Toma, M. F., Fabbro, C., Da Ros, T., Prato, M., Kostarelos, K., Bianco, A. Enhanced anticancer activity of multi-walled carbon nanotube-methotrexate conjugates using cleavable linkers. *Chemical Communications*, **46**, 1494–1496.
106. Liu, Z., Chen, K., Davis, C., Sherlock, S., Cao, Q., Chen, X., Dai, H. (2008). Drug delivery with carbon nanotubes for in vivo cancer treatment. *Cancer Research*, **68**, 6652–6660.
107. Feazell, R. P., Nakayama-Ratchford, N., Dai, H., Lippard, S. J. (2007). Soluble single-walled carbon nanotubes as longboat delivery systems for platinum (IV) anticancer drug design. *Journal of the American Chemical Society*, **129**, 8438–8439.
108. Dhar, S., Liu, Z., Thomale, J. R., Dai, H., Lippard, S. J. (2008). Targeted single-wall carbon nanotube-mediated Pt(IV) prodrug delivery using folate as a homing device. *Journal of the American Chemical Society*, **130**, 11467–11476.
109. Tsang, S. C., Chen, Y. K., Harris, P. J. F., Green, M. L. H. (1994). A simple chemical method of opening and filling carbon nanotubes. *Nature*, **372**, 159–162.
110. Hampel, S., Kunze, D., Haase, D., Krämer, K., Rauschenbach, M., Ritschel, M., Leonhardt, A., Thomas, J. R., Oswald, S., Hoffmann, V., Büchner, B. (2008). Carbon nanotubes filled with a chemotherapeutic agent: a nanocarrier mediates inhibition of tumor cell growth. *Nanomedicine*, **3**, 175–182.
111. Tripisciano, C., Kraemer, K., Taylor, A., Borowiak-Palen, E. (2009). Single-wall carbon nanotubes based anticancer drug delivery system. *Chemical Physics Letters*, **478**, 200–205.
112. Wu, W., Li, R., Bian, X., Zhu, Z., Ding, D., Li, X., Jia, Z., Jiang, X., Hu, Y. (2009). Covalently combining carbon nanotubes with anticancer agent: preparation and antitumor activity. *ACS Nano*, **3**, 2740–2750.
113. Yang, D., Yang, F., Hu, J., Long, J., Wang, C., Fu, D., Ni, Q. (2009). Hydrophilic multi-walled carbon nanotubes decorated with magnetite nanoparticles as lymphatic targeted drug delivery vehicles. *Chemical Communications*, 4447–4449.
114. Meng, J., Duan, J., Kong, H., Li, L., Wang, C., Xie, S., Chen, S., Gu, N., Xu, H., Yang, X.-D. (2008). Carbon nanotubes conjugated to tumor lysate protein enhance the efficacy of an antitumor immunotherapy. *Small*, **4**, 1364–1370.

22

POROUS NANOMATERIALS FOR BIOMEDICAL APPLICATIONS

HENNING LÜLF, ANDRÉ DEVAUX, EKO ADI PRASETYANTO, AND LUISA DE COLA

22.1 INTRODUCTION

Nanotechnology is a booming field of research that focuses on the development of nanometer-scaled (1–100 nm) functional materials with unique electronic, optical, or magnetic properties [1–3]. It has been shown that for nanomaterials the size, the shape, and of course the chemical composition as well as the organization at the nanoscale is critical to be able to control their properties and to manipulate them, or even to control their response to external stimuli. These nano systems can be built up from a variety of components such as semiconductors, metals, and metal oxides, as well as from polymers, lipids, or small organic or organometallic molecules through self-assembly. The general design principle revolves around synthesizing or assembling the components into the desired composition, shape, and size while further properties can be added by surface modifications. A general scheme of a functionalized nanomaterial for biomedical applications is given in Figure 22.1. Furthermore, some of these materials can be realized with biocompatible components and therefore have spawned a huge interest in application areas as diverse as cell biology, pharmaceutics, and biotechnology [4–6], molecular or cellular labeling, high contrast medical imaging [7, 8], detection and tracking, or targeted drug delivery [9]. The most interesting aspect versus a purely molecular approach is that several functions can be combined in a single type of nanoparticle and a high density of desired components can be organized in a rather small space. As an example, one can think of a system that combines specific cellular recognition with an imaging agent and drug payload that is released through an external trigger.

However, nano-sized particles are presenting quite a few challenges not only with regard to their design and synthesis, but also in how they interact with living entities. This behavior is dependent not solely on the particle composition but also on its shape and size. It has been recently observed that the in vivo biodistribution and clearance behavior of silica nanoparticles depends strongly on their aspect ratio [10]. Short rod-like particles accumulate mostly in the liver and are excreted faster than longer ones, which are mostly found in the spleen. The cytotoxicity of nanoparticles is also an important topic currently under investigation. One often finds conflicting reports on whether a material with a specific composition is toxic or not. In the case of silica nanoparticles, for example, this difference in assertions could be traced to the morphology and dimensions of the material [11–13]. Therefore, a systematic investigation of a nanomaterials' cytotoxicity has to be performed for every variation in size, shape, or surface composition.

Another challenge is the uptake of nanoparticles into cells or bacteria. Depending on the size, shape, and surface composition, the particles will follow different internalization pathways like endocytosis, macropinocytosis, or phagocytosis. The ultimate intracellular fate of the particle and associated biological responses depend strongly on the uptake mechanism. Using the "right" size and surface modification agents allow controlling the internalization process to some degree. A recent review by Chan and co-workers [14] summarizes the strategies employed so far to control the intracellular delivery of nanoparticles.

When looking at the many types of multifunctional nanomaterials available, the so-called nanocontainers shows

Organic Nanomaterials: Synthesis, Characterization, and Device Applications, First Edition. Edited by Tomás Torres and Giovanni Bottari.

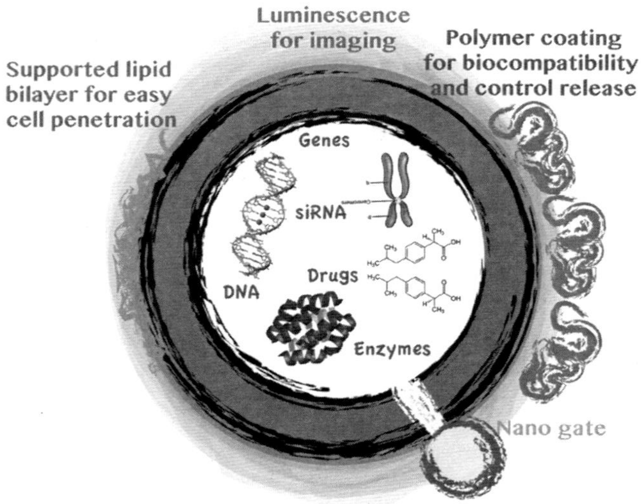

FIGURE 22.1. Schematic representation of a porous nanomaterial suitable for biomedical applications. The cavities are filled with a payload (e.g., drugs, enzymes, RNA, or DNA), while the surface is modified to improve the biocompatibility and to provide other functions such as targeting or easier cell penetration. The pores can be engineered in such a way that the payload release occurs in a controlled manner.

particular promise for nanomedical applications (see Figure 22.2). A major reason for this interest is their ability to entrap a large variety of guest species such as proteins, enzymes, peptides, hormones, metabolites, DNA, complexes, metal clusters, or other types of molecules [15, 16]. The encapsulation protects the guest species from chemical or biological degradation as well as from other unwanted interactions with the in vitro or in vivo environment. Furthermore, since RNA and DNA, as well as many drugs, are not taken up by the cells or are not water-soluble, solubilizing groups can be covalently bound to the outer surface of the nanocontainers. It is also possible to add receptor moieties to these surfaces to target specific cell types or, after cellular uptake, organelles or compartments of the cell [17, 18]. Once the container is in position, the entrapped molecules could be released in a controlled manner.

Many types of nanocontainers have been described in the literature such as polymer-based micelles or vesicles [19,20], liposomes [21], metal–organic frameworks (MOFs) [22,23], carbon nanotubes [24], porous crystals composed of complex salts [25], classical zeolites or zeolite-like materials [15], and mesoporous silica or organosilicates [26–28]. This list should by no means be taken as exhaustive. This chapter will focus only on mesoporous silicates and classical zeolites featuring micropores as host materials. Despite the obvious differences between the two classes of porous materials, one can find many analogies and similarities [29].

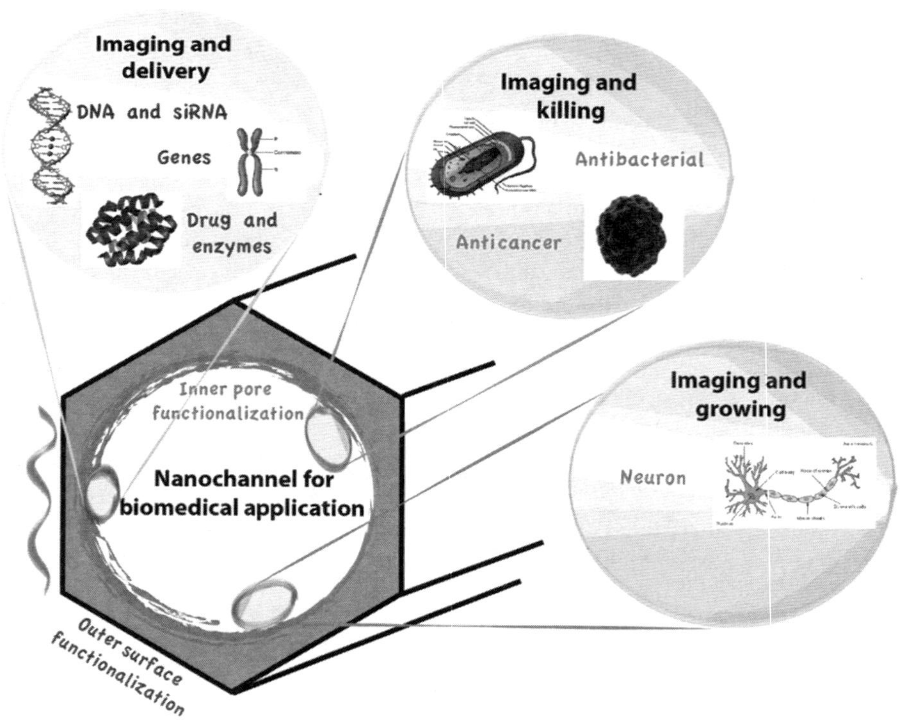

FIGURE 22.2. Schematic illustration of multifunctional nanocontainers and their potential bio applications. See color insert.

Zeolite L is a representative of the classical alumosilicate zeolites with micropores (pore diameter of 0.71 nm). It is a synthetic zeolite built up from cancrinite cages stacked into columns that are interconnected to form a channel system with hexagonal symmetry. The crystals have the shape of a straight hexagonal prism, which usually looks like a cylinder. The channels are running parallel to the long axis of the crystal. The size and aspect ratio (length/diameter) of the crystals can be tuned by variations of the synthetic conditions. It is thus possible to prepare crystals with lengths varying from 30 to 10,000 nm and with shapes that are cylindrical, cubic, or disc-like [30]. While classical zeolites are a fully crystalline material, mesoporous silica is amorphous with easily tunable pore sizes between 2 and 30 nm [31, 32]. The synthetic conditions for mesoporous silica are more flexible than for zeolites, resulting in a large variety of particles, ranging from 20 nm to mm in size, with different shapes and pore sizes. In mesoporous silica, functional groups can also be introduced during synthesis by the so-called in situ or direct synthesis method [33, 34]. This flexibility and relative ease of preparation make mesoporous silica a good candidate for nanomedical applications, and many reports can be found in the literature [35–38]. On the other hand, very few publications exist where classical zeolites have been used in such a fashion [39–41]. This might be due to the fact that synthesis of zeolite particles in the required nano-sized regime is quite challenging. More general overviews of zeolite L and mesoporous silica as host material have been recently published [15, 42].

All porous materials mentioned above exhibit a wide variety of intrinsic properties, morphologies, and cavities or channel systems. They can, however, be approximated by considering them as being built up from three main components: the internal cavities, the pore openings, and the external surfaces (see also Figure 22.2). Each component offers unique functionalities, and we will shortly discuss diverse strategies that allow for their exploitation.

The cavities or channel systems of porous materials can host a variety of guest species such as molecules, complexes, quantum dots, metal clusters, DNA, or smaller proteins (see Figure 22.3A). The inserted guests are shielded from the environment through the host and usually exhibit an increased chemical or thermal stability. The entrapment, which can be performed by stopper molecules (Figure 22.3B), also helps in avoiding adverse reactions by preventing interactions with external reactive species. The reopening of the channels can be triggered by a chemical reaction or by an external input (e.g., light; see Figure 22.3C).

The dimensions of the internal cavities can lead to geometrical constraints that can, in turn, lead to a preferred arrangement of the guests. For example, a one-dimensional channel system with a small diameter, as found in zeolite L, allows for the realization of high concentrations of oriented guest molecules that usually behave as monomers [43]. Such

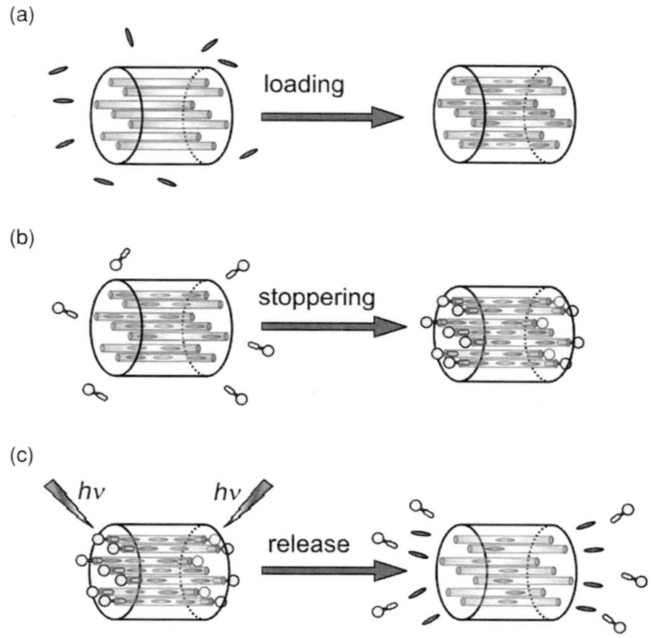

FIGURE 22.3. Loading, stoppering, and triggered release from a nanocontainer. (**A**) The empty channels can be loaded either by ion exchange or from gas phase adsorption, depending on the nature of the guest molecule (cationic or neutral). (**B**) The channel entrances can be plugged by means of a so-called stopper molecule, where only the tail part can enter the pores while the head group is too bulky. (**C**) If the host–guest material pores are closed by stoppers that are sensitive to a specific impulse such as light irradiation, pH changes, or magnetic fields, a triggered release can occur.

materials are ideal candidates for observing energy or electron transfer in confined spaces. On the other hand, for some applications the overlap of molecules, and in particular their electronic interactions, could be desirable and appropriate molecules, and loading degree can result in the formation of aggregates [43]. Hosts with a two- or three-dimensional pore system allow for an easier guest transport and minimize the risks of channel blocking. The faster material transport offered by higher-dimensionality materials is not always most desirable for drug delivery systems. In fact a control of the loading and of the release kinetics is necessary. One interesting characteristic of linear channels is the possibility to realize "sandwich-like" loadings where well-defined zones of different guest species are present (see Figure 22.4).

If the molecules cannot pass each other in the channels, guests located in the middle of the channel will be able to leave only after their neighbors have been released. By playing around with the release kinetics, one could create a system where different active components exit the channels one after another with specific time delays, as illustrated in Figure 22.4. In a real case, for example, multiple drugs could be released in a certain amount and delay, leading to a simple dispenser that could be implanted or in contact with the skin.

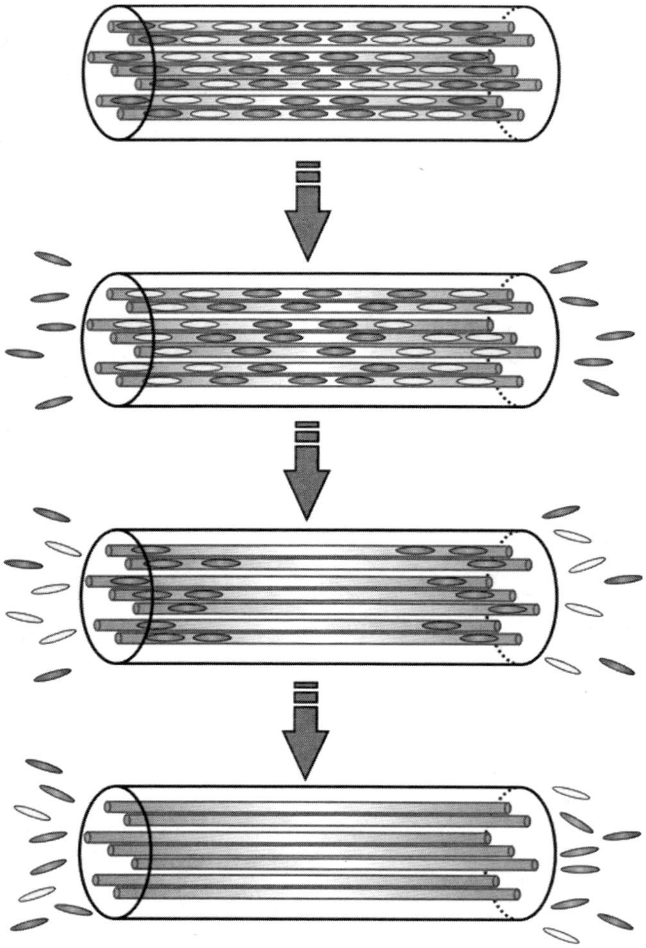

FIGURE 22.4. Sequential or delayed release of guest species from a nanocontainer. If a nanocontainer with a one-dimensional channel system is sequentially loaded with guest species that are too large to pass each other inside the cavities, one obtains well-defined regions where only one type of guest is present. The release can now only occur in a specific sequence: First the guests located closest to the pore openings will leave the nanocontainer (**second from top**), followed by the next species (**second from bottom**), while the guests will exit the channels last (**bottom**).

Furthermore, the spacing between the drugs could be a small layer of proteins, sugars, or other natural substances that are slowly or quickly metabolized in order to control the release of the next drug.

Another parameter that can be modulated is the chemical composition of the inner cavity walls. These can play a role not only in tuning the physicochemical characteristics of the guests, but also in controlling the host–guest interactions. A large variety of chemical properties can be influenced and include the inner pH value, the presence of coadsorbed solvent or gas molecules, the ionic composition and strength, or the hydrophobicity [15,44]. Mesoporous materials are best suited for such a fine-tuning of their internal characteristics by appropriate chemistries due to their larger cavity system, as illustrated in Figure 22.5. A good example was recently reported by Bein [45a] where the polarity inside mesoporous silica was tuned by grafting selected hydrophobic molecules.

In a similar way it is possible to selectively functionalize the coat of zeolites and the channel entrances (see Figure 22.5) with different chemical groups and asymmetrize the system [45b]. The second key component playing an important role in the properties of a porous material is the pore opening. Because these are located between the interior and the exterior of the host, they are responsible for the loading or release of guest species. Controlling their structure and being able to functionalize them selectively is of fundamental importance for applications such as drug delivery [46,47]. An interesting concept is the so-called stopper molecules that were first developed for zeolite L [48]. These stoppers consist of a head group that is too large to pass the pore opening and a tail with high affinity for the channel system (see also Figure 22.3B). Thus, the molecule cannot fully enter the zeolite and efficiently blocks the pore opening. Depending on their nature, these stoppers can be attached either by physisorption, electrostatic interactions, or hydrogen or covalent bonds. The blocking induced by the stoppers not only prevents leakage of inserted guest species, but also can prevent small molecules such as water or oxygen from entering the channels and interfering with the guests' properties. Completely blocking the pore openings of a mesoporous material with a single molecular species, however, is quite difficult, and other strategies have to be employed. Functionalization of the pore openings is not limited to sealing the interior, but can also be envisaged to serve as molecular valves that open or close under selected conditions (as shown schematically in Figure 22.3C) [49–51], or even to chemically modify the guests as they leave the host system. An important point to keep in mind is that any chemistry to be employed for the modification of pore openings should be very specific and spatially well-controlled. Should this not be the case, one can easily end up with functional groups grafted in unwanted places.

The last component to be discussed is the external surface, where further functionalities can be added by grafting molecules onto it. The external surface plays an important role in how the host–guest material can interact with the environment. Many reported modifications are aimed at improving the dispersability or biocompatibility of the material [52, 53], to enable targeting of biomolecules, bacteria, or cells [54], to add chemical, luminescent, magnetic, or radioactive labels [7, 8], or even to promote self-assembly on various supports [55–57]. Using the outer surface of the nanocontainer allows tailoring the material for the desired application and represents a huge advantage as this process is independent of the filling. Thus, once a model nanocontainer has been designed, most changes to the "payload" can be integrated without having to make changes to the delivery platform. It is obvious that all mentioned modifications can

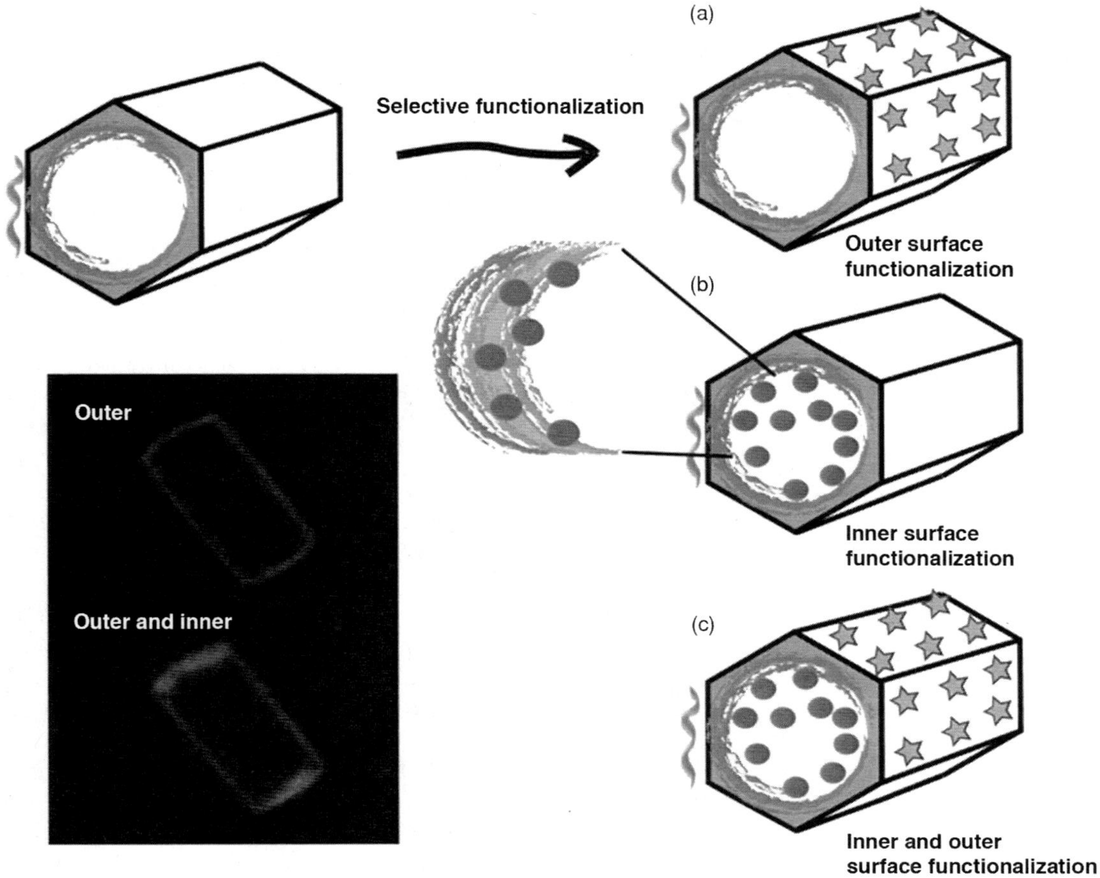

FIGURE 22.5. Selective modification of porous nanomaterials. (**a**) Grafting molecules to the outer surface only, (**b**) selective functionalization of the inner cavity walls, and (**c**) modification of both inner and outer surface in a single material. The epifluorescence microscopy image in the left part of panel c shows a zeolite L crystal where the coat surface (green emission) and the pore openings (red emission) bear different functionalizations.

be implemented on the same nanocontainer so that multiple functionalities can be obtained in a single object, as shown in Figure 22.5C.

As can be seen from the properties outlined above, multifunctional nanomaterials are fascinating and helpful tools for different research fields. They have the potential to become a beautiful tool for biomedical applications. There is, however, still plenty of work to be done to elucidate their interactions with living organisms and to discover ways to control their behavior. Our aim in this chapter is to give the reader an overview of the achievements obtained with mesoporous silicates and microporous zeolites. We will mostly discuss our work in connection with other reports, and this chapter is divided following the medical applications of these materials. In particular, we will start with the functionalization of porous systems with biocompatible molecules. We will then step to their interactions with bacteria and cells, along with their in vivo applications. Finally, we will conclude this chapter with describing their use for cell growth.

22.2 MICRO- AND MESOPOROUS MATERIALS MEETS BIOLOGY

Since the most common hard meso- and micoroporous materials are inorganic in nature, their biocompatibility must be implemented on their surfaces. Therefore a strong research effort devoted to the adsorption or covalent binding of proteins, enzymes, or other biomolecules is necessary to translate this abiotic system for in vitro and in vivo applications. Depending on the final aim, adsorption of biological molecules can have desired (for applications in biosensors, immunoassays, or protein purification systems) or undesired (deposition of biomaterial on contact lenses or dialysis membranes) effects, and a method to predict the adsorption properties on different substrates is a big step toward new materials in bioengineering. Since most of the mesoporous and microporous nanomaterials discussed in this chapter are silica-based, their surface is strongly polar and the overall charge is negative. This important observation leads to unspecific binding of polar molecules but also to a masking effect on the real surface functionalization in solutions containing serum or proteins that compensate the overall charge [58, 59]. The creation of biocompatible nanocontainers leads to important uses of them such as imaging [8, 60–63], drug delivery [64–67], and combination of diagnostic and therapy (theranostics). Also, reports exist where nanocontainers have shown antibacterial effects [68–71], were used in sensing [72], and have applications in biotechnology [73–77] where nanocontainers have been used as supports to immobilize active molecules (like proteins or enzymes).

The immobilization of antibodies on zeolites has been reported by us in order to create a more specific receptor on the surface of the nanomaterials. This specificity is important if we wish to target a certain type of cell, bacteria, or virus. However, the interactions between biomolecules or even more complex systems such as cells depend on many factors that have been rationalized only in part. Intuitively, factors like hydrophilicity and hydrophobicity, electrostatic and structural interactions, size and shape of the nanomaterials, and of course their chemical compositions play an important role. The change in protein conformation may also complicate the prediction of their adsorption [78]. Mitchell and Perez-Ramirez [79] studied the uptake of different lipase enzymes to mesoporous zeolites and mesoporous silica particles for catalytic applications. They have shown that the enzyme uptake strongly depends on the mesoporous surface. Furthermore, Hu et al. [80] studied the influence of the crystal plane and size of nanoparticles (in this case zeolite L) on the protein absorption behavior. Due to their larger surface areas, smaller crystals show larger protein adsorption capacities. In addition, a larger protein adsorption was observed at the (001) crystals plane for zeolite L, which can be explained by abundant exposed pore openings.

Tavolaro et al. [81] investigated the influence of a series of different factors on the immobilization of cytochrome c on zeolites and have shown that the electrostatic interaction seems to be the most important factor governing the immobilization. The adsorption could also be used to enrich a solution by sequestering a desired protein. An elegant example was published by Cao et al. [82], who showed that it is possible to use nanozeolites for carrying out such a challenging task. In contrast to mesoporous silica materials [83, 84], zeolites exhibit a multilayer protein adsorption mechanism on the their surface [85], which endows them with larger protein adsorption capacity. In addition, high chemical and environmental stability of zeolite materials favors their applications under harsher conditions compared to other nanomaterials [80].

When going from protein or enzyme adsorption to cell adsorption on zeolite surfaces, the mechanisms of adsorption become much more complicated due to the complex constitution of cellular surfaces, where a variety of different active sites, such as proteins, polysaccharides, or lipids, is present. Kubota et al. [86] have tried to simplify the complex interactions between cells and zeolite surfaces by approximating them to colloidal particles interactions, where van der Waals and electric double-layer interactions are taken into account. Modification of the zeolite surface with surfactants can also be used to enhance cellular adhesion, as has been shown by Joshi et al. [87].

With this knowledge a better design of hybrid systems can be achieved in order to facilitate the uptake or adhesion of cells on the nanocontainers. However, we must remember that size matters and that shape is also relevant for cell uptake and the toxicity of the material [10, 88, 89].

22.3 IN VITRO STUDIES: INTERACTION OF ZEOLITES AND MESOPOROUS MATERIALS WITH CELLS

In addition to just adsorbing proteins, enzymes, or cells on zeolite surfaces, it has been shown that these nanocrystals can also be used as external tool to self-assemble living systems such as bacteria (see Figure 22.6) [90]. The self-assembly was achieved by using nonpathogenic *E. coli* (strain JM 109) and zeolite L crystals with a mean length and mean diameter of 1.0×1.0 μm, which were filled with fluorescent dyes. The outer membrane of *E. coli* is studded with lipopolysaccharides that are phosphorylated at multiple locations, which leads to a negatively charged outer membrane. The zeolite L crystals were amino-functionalized only at the channel entrances, and these groups were protonated under the experimental conditions. This led to a positively charged system, which can bind electrostatically to the negatively charged bacteria membrane. Furthermore, it has been shown that by using this setup and the right amounts of bacteria and modified crystals, it is possible to self-assemble bacteria and in particular two bacteria with a container as a junction (see Figure 22.6).

This experimental background was then exploited for theranostic applications using smaller zeolites featuring different functionalities that are able to target, label, and photoinactivate pathogenic and antiobiotic-resistant bacteria (*Neisseria gonorrhoeae* and *E. coli*) [91]. A highly luminescent green dye was inserted into the channels of zeolite L nanocrystals to label the cells. The outer surface was functionalized with a photosensitizer, a silicon phthalocyanine, that produces toxic singlet oxygen upon irradiation with red light, and protonated amino groups were employed to bind the nanomaterial to the cells. A schematic view of this trifunctional nanomaterial and its components is given in Figure 22.7.

Another example of using nanomaterials for photodynamic therapy (PDT) applications was given by Hocine et al. [92], where crystalline silicalite, MCM-41, and nanoparticles with radial porosity were used. This time the porous

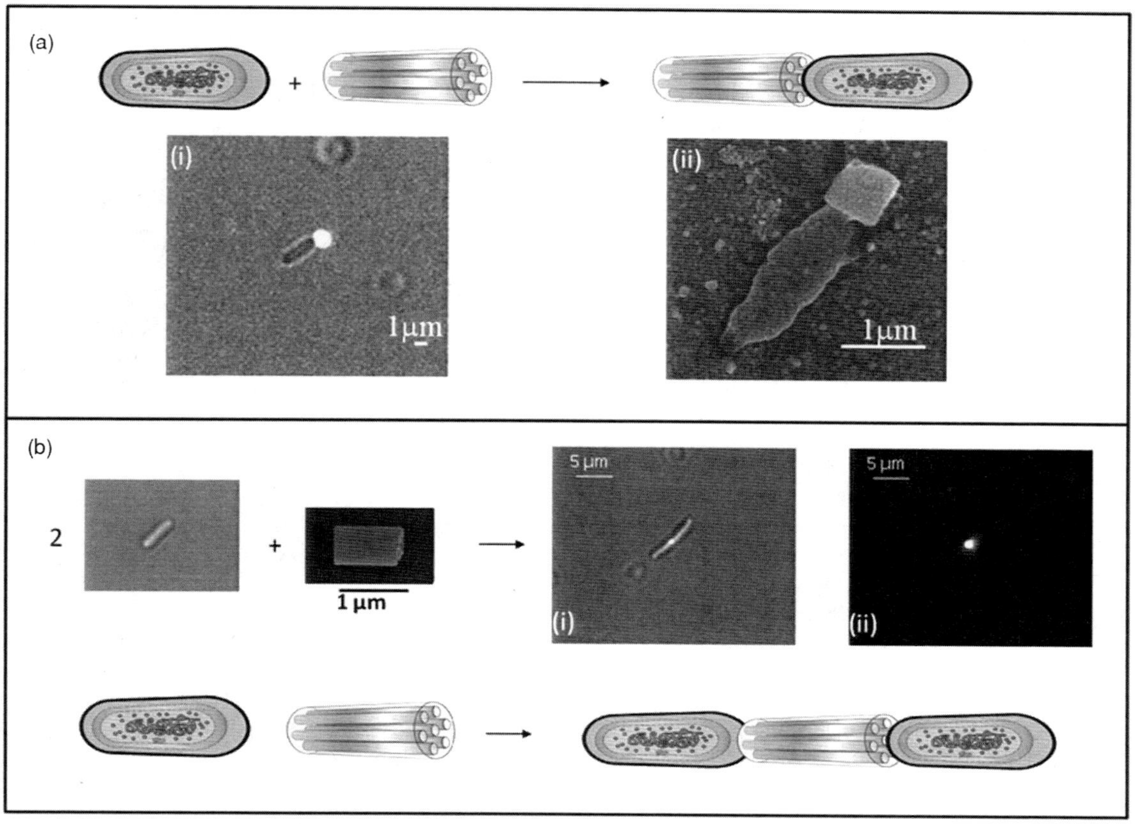

FIGURE 22.6. (**A**) Illustration of a 1:1 zeolite L/bacteria assembly. (**i**) Optical microscope image taken upon irradiation with white and blue light. (**ii**) SEM image of the assembly taken upon evaporation of the solvent and coating with silver (5 nm). (**B**) Self-assembly of two bacteria by using a zeolite L crystal as junction. (**i**) Optical image taken upon simultaneous irradiation with white and blue light. (**ii**) Irradiation of the same sample with blue light. The fluorescence of the dye-filled junction is observed.

materials have been internalized by cancer cells. Therefore water-soluble porphyrins were encapsulated covalently inside the nanoparticles and PDT experiments were performed on MDA-MB-231 breast cancer cells. It has been shown that the reported nanoparticles show significant cell death after irradiation, which was, however, not correlated with the quantum yield of singlet oxygen production. Therefore, other parameters like endocytosis or membrane damage due to the size of the particles must play an important role in cell death.

Even though many studies have recently focused on the uptake of particles by cells, not much is known on the mechanism of internalization and on the cytotoxicity of such materials. The shape, surface properties, and functionalization play an important role as shown by Huang et al. [89]. It has been reported that nanoparticles without further functionalization are mediated by a clathrin-pitted mechanism [93], while amino-functionalized nanoparticles are uptaken via a caveolae-dependent mechanism [94] that is clathrin-independent, and finally the uptake of mannose functionalized nanoparticles is mediated by mannose receptors [95]. However, such behavior has been tested only for cell line and in specific conditions.

Due to their porous nature, an obvious application for nanocontainers lies in the frame of drug release. The zeolites have been very little investigated with regard to this application. To the best of our knowledge, only a few reports deal with drug release from zeolite pores.

Rivera and Farís [96] studied the potential application of clinoptilite–surfactant composites as supports for bioactive molecules. Therefore the antibiotic sulfamethoxazole (see Figure 22.8) was incorporated in a clinoptilite–benzalkonium chloride composite, and it has been shown that 80% of the drug can be released from the system in aqueous media within 24 hours.

Arruebo et al. [97] reported a magnetic triggered drug delivery system based on magnetite and FAU zeolites containing doxorubicin, an antitumorial drug. In vitro experiments have shown that this system is able to store and release remarkable amounts of the drug.

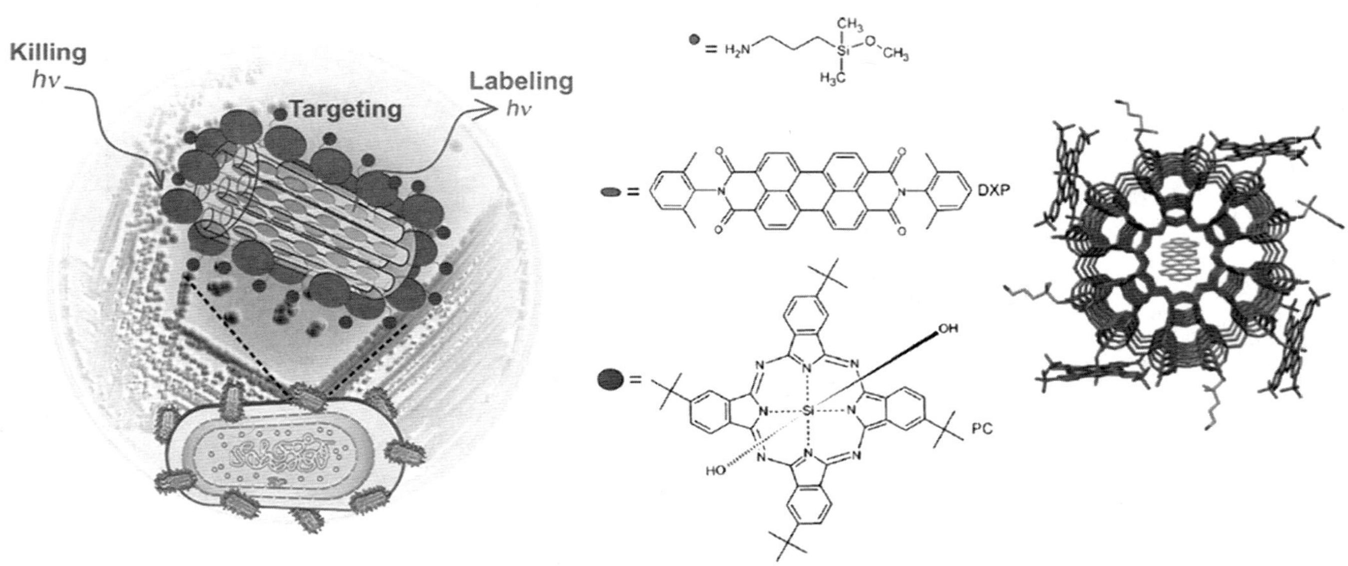

FIGURE 22.7. **Left**: Schematic representation of a trifunctional nanomaterial based on zeolite L that can target, label, and kill bacteria. **Middle**: Components used for realizing this nanomaterial. **Right**: Scheme showing a top view one channel of zeolite L loaded with the label (DXP) and with targeting groups (3-aminopropyl-dimethyl-methoxysilane) and killing agent (singlet oxygen producing silicon pthalocyanine) bound to the outer surfaces.

A slightly different approach to drug release from loaded zeolites was reported by Fatouros et al. [40] in 2011. They have applied a combination of experiments and theoretical modeling in order to assess the potential use of zeolite BEA as drug delivery agent. Molecular dynamics have been used to examine the diffusion of the two drugs salbutamol and theophylline (see Figure 22.9).

Experimental studies regarding the loading and release of these drug molecules are in agreement with the molecular dynamics model. These results suggest that modeling can have an important role in screening zeolite–drug combinations for drug-release applications.

Contrary to zeolites, in which internal functionalization is difficult due to narrow cavity sizes, the interior walls of mesoporous silica can be functionalized in order to covalently bind drugs or to stabilize noncovalently the entrapment of desired molecules (see Figure 22.5).

Indeed many studies have been reported for drug release from nano-sized mesoporous systems. An early study on mesoporous silica as a drug carrier was reported by Vallet-Regi et al. [98] in 2001, using ibuprofen as a model drug; here the material was studied in simulated body fluid. The attempt to control the kinetics of the drug release by functionalization of the surface of mesoporous silica was later reported by the same groups [99]. They prepared MCM-41-type mesoporous materials modified with different functional groups (chloro, phenyl, benzyl, mercaptopropyl, cyanopropyl, and butyl groups) (Figure 22.10) and studied the adsorption and desorption kinetics of ibuprofen. Since then, many studies on ex vivo drug delivery were reported and the parameters controlling the release profile were examined and reviewed [100–104].

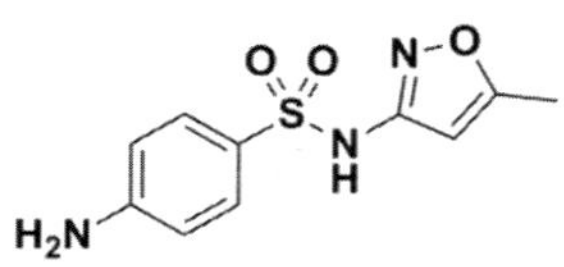

FIGURE 22.8. Structure of sulfamethoxazole.

From this simple behavior it is clear that mesoporous silica nanoparticles are promising materials for release and uptake of molecules. However, investigations on the stability of certain materials show a degradation of the silica particles in physiological condition [105, 106]. In order to stabilize the silica and to introduce a system capable of triggered release from the porous materials, several interesting groups have been grafted at the pore entrance [107–110] or on the surface [111–113]. Those types of functionalization enables drug delivery via porous nanomaterial responsive to external triggers such as temperature, light [102], magnetic [5, 114, 115], pH [116–118], enzyme [108], or other specific chemical property of the target cell and release of biomolecules at the desired site has been achieved [110]. A particularly elegant example was reported by Lin and coworkers [110], demonstrating that the release of insulin and cyclic adenosine monophosphate (cAMP) from mesoporous particles can be regulated in response to saccharide triggers.

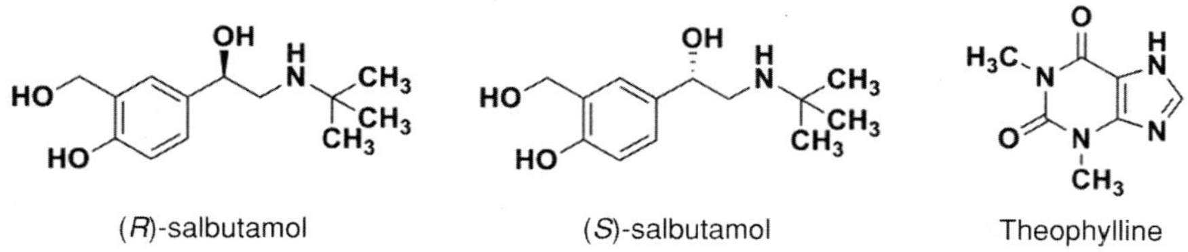

FIGURE 22.9. Structures of salbutamol and theophylline. The active form of salbutamol is (*R*)-salbutamol.

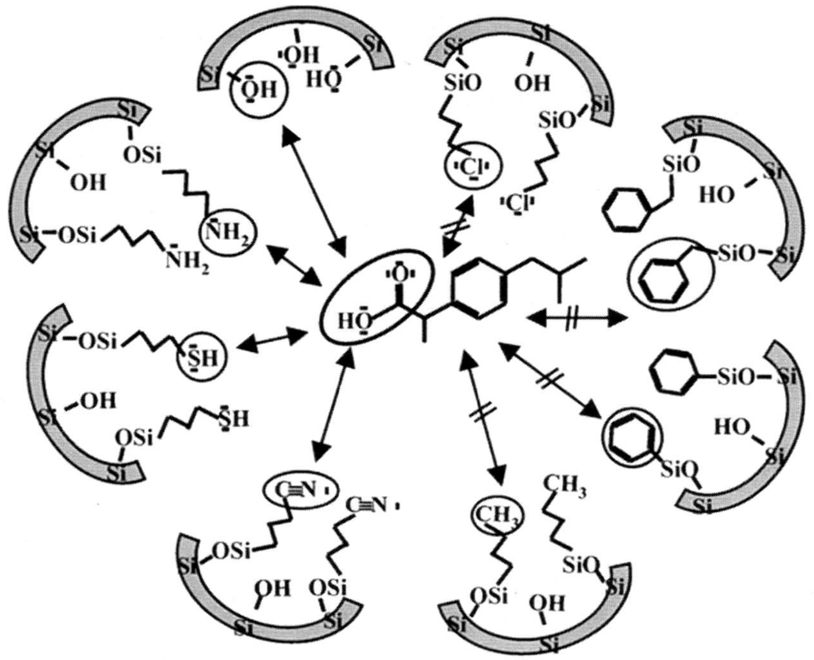

FIGURE 22.10. Schematic representation of surface-modified mesoporous silica with various functional groups and their interaction with ibuprofen [99].

In this report, the authors use covalently bonded gluconic acid-modified insulin (FITC-G-Ins) proteins on the exterior surface of MSN which also served as caps while cAMP molecules were encapsulated inside the mesopores. Upon a change in glucose level, the insulin was cleaved and cAMP can leave the nano channel (see Figure 22.11).

An increase in complexity is achieved by functionalizing both the inside walls and the surface of a nanomaterial with different biocompatible groups. We have chosen the work described by Ashley et al. [113] as an example in which mesoporous silica nanoparticles have been loaded with siRNA or chemo-therapeutic drugs. The drug-loaded

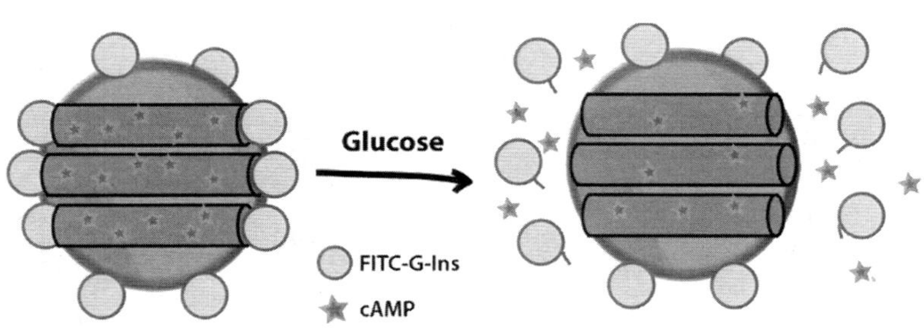

FIGURE 22.11. Schematic representation of the glucose-responsive MSN-based delivery system for controlled release of bioactive G-Ins and cAMP.

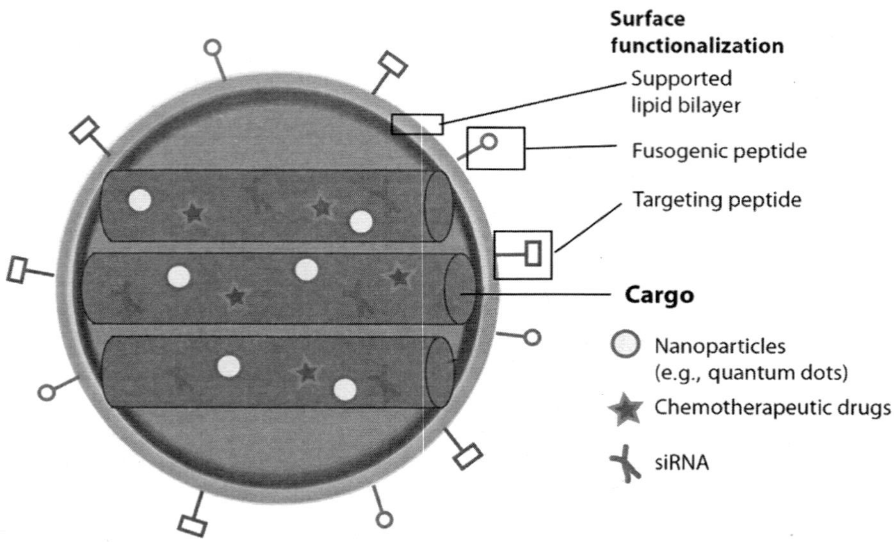

FIGURE 22.12. Pictorial representation of a nanoporous particle-supported lipid bilayer, depicting the diverse types of therapeutic and diagnostic agents that can be loaded within the nanoporous silica particle.

core is then enveloped by a single lipid bilayer, which is further functionalized with poly(ethylene glycol) (to reduce nonspecific interactions with its environment), peptides (to direct binding to distinct target cells), and pH-responsive peptides, causing disruption of endosomes and bilayer coating upon particle internalization into acidic intracellular compartments, allowing drug delivery into the cytosol of the target cell (see Figure 22.12).

An important class of drugs is those related to antibacterial activity. Also in this case, mesoporous materials have played an important role [66,119–122]. An interesting example has been reported by Santamaria and co-workers [66]. They designed a porous stainless steel reservoir that mimics the pins used in external frame fracture fixation. Its hollow interior was filled with mesoporous silica particles with a linezolid antibiotic loaded into them. A schematic of the system is shown in Figure 22.13. The mechanical properties of the porous drug-eluting pin and release behavior were tested and found to be satisfactory. The bactericidal studies showed an excellent reduction in bacterial growth by a factor of up to 1000 after 48 hours in comparison with an unloaded pin.

22.4 IN VIVO STUDIES: IMAGING AND DRUG DELIVERY

As already mentioned in the introduction, the use of mesoporous materials for imaging and theranostic is potentially very interesting due to the high density of labels and/or drugs as well as the multimodality that can be implemented on these systems. An early example, reported by Bresinka and Balkus [62] in 1994, describes the use of zeolite Y in combination with the well-known paramagnetism of Gd(III). They have shown that Gd(III) can be exchanged with sodium ions present in NaY–zeolites. The GdNaY zeolites are reasonably stable at low pH, and in vivo studies were done by Young et al. [61] in order to evaluate efficacy and safety of the system. They found that the Gd-zeolites significantly improved the efficacy scores and that they are not absorbed from the gastrointestinal tract and thus are nontoxic. A company was started on this discovery (Pharmacyclics), and the product is marketed under the trade name Gadolite®.

In 2008 we reported the first example of using zeolite crystals as a dual-probe image system (optical and magnetic). The channels of nanometer sized (about 30 nm) zeolite L crystals were loaded with a fluorescent dye (pyronine), which can be

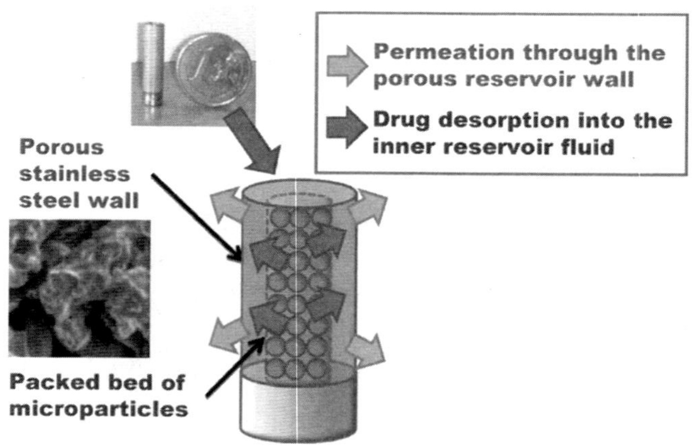

FIGURE 22.13. Hollow porous implants filled with mesoporous silica particles as a two-stage antibiotic-eluting device [66].

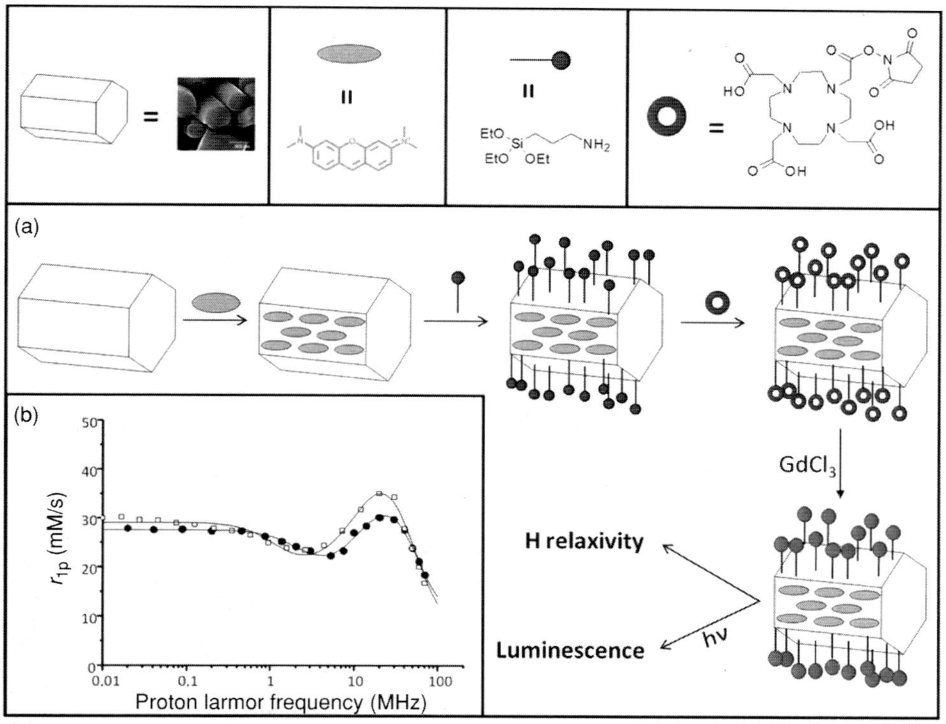

FIGURE 22.14. (**i**) Synthesis of Gd-DOTA functionalized zeolite L crystals. In the first step, pyronine molecules are introduced into the channels via ion exchange. In a second step, a linker (APTES) is bound to the entire surface of the zeolite L crystals. The chelating ligand (DOTA) is then covalently attached to the linker. Finally, Gd(III) is complexed to the surface-bound DOTA. The Gd(III)-DOTA complexes are represented. (**ii**) 1/T NMRD profile of a 1 mM solution of Gd-DOTA-zeolite L measured at 25°C (•) and 37°C (□).

used as an optical imaging unit. In a second step the outer surface was modified with a highly stable gadolinium(III) complex acting as probe for magnetic imaging. A pictorial view of the system and its 1/T1 NMRD profile is shown in Figure 22.14.

Imaging can also be done by scintillation methods using radioisotopes in PET (positron emission tomography). We pioneered this field by filling zeolite L crystals with the positively charged γ-emitter $^{111}In^{3+}$. In order to avoid leaking of the indium, the channels were closed with tailor made stopcock molecules. The zeolite crystals have been coated with PEG chains to render them biocompatible. The designed system and obtained 2D static scintigraphic images are shown in Figure 22.15.

The zeolites were injected into rats, and two-dimensional static scintigraphic images were recorded (see Figure 22.15B). The biodistribution of the particles after 24 hours was monitored, but the ultimate fate of the zeolites is still unknown. Mesoporous materials have also been studied for imaging [115, 123–125], and an elegant example combining imaging and drug delivery in mesoporous silica nanoparticles (MSNs) has been described by Zink and co-workers [123]. They studied the biodistribution of mesoporous silica nanoparticles using human cancer xenografts with in vivo imaging and fluorescent microscopy imaging, as well as with inductively coupled plasma mass spectroscopy. The experiments showed that MSNs preferentially accumulate in tumors. Therefore, the drug-delivery capability of MSNs was demonstrated by following tumor growth in mice treated with camptothecin-loaded MSNs.

In vivo studies related to the toxicity of inorganic nanomaterials are a big and controversial field. Regarding mesoporous materials, the early studies of Langer and co-workers [126] showed that the "naked" mesoporous systems were very toxic even at rather low concentrations. However, we should mention that the size of the materials employed was larger than 150 nm and that the toxicity was observed only when the materials were injected directly into the bloodstream. Other studies indicated [123, 127–130] that the toxicity of the materials is very much related to their dose. Tamanoi and co-workers [123] showed that mesoporous systems with a diameter of 100–120 nm and covered with folic acid are toxic in mice after several days of treatment with a dose of 100 mg per kilogram by intravenous injection. On the other hand, decreasing the dose to 1 mg per mouse did not show any significant toxicity.

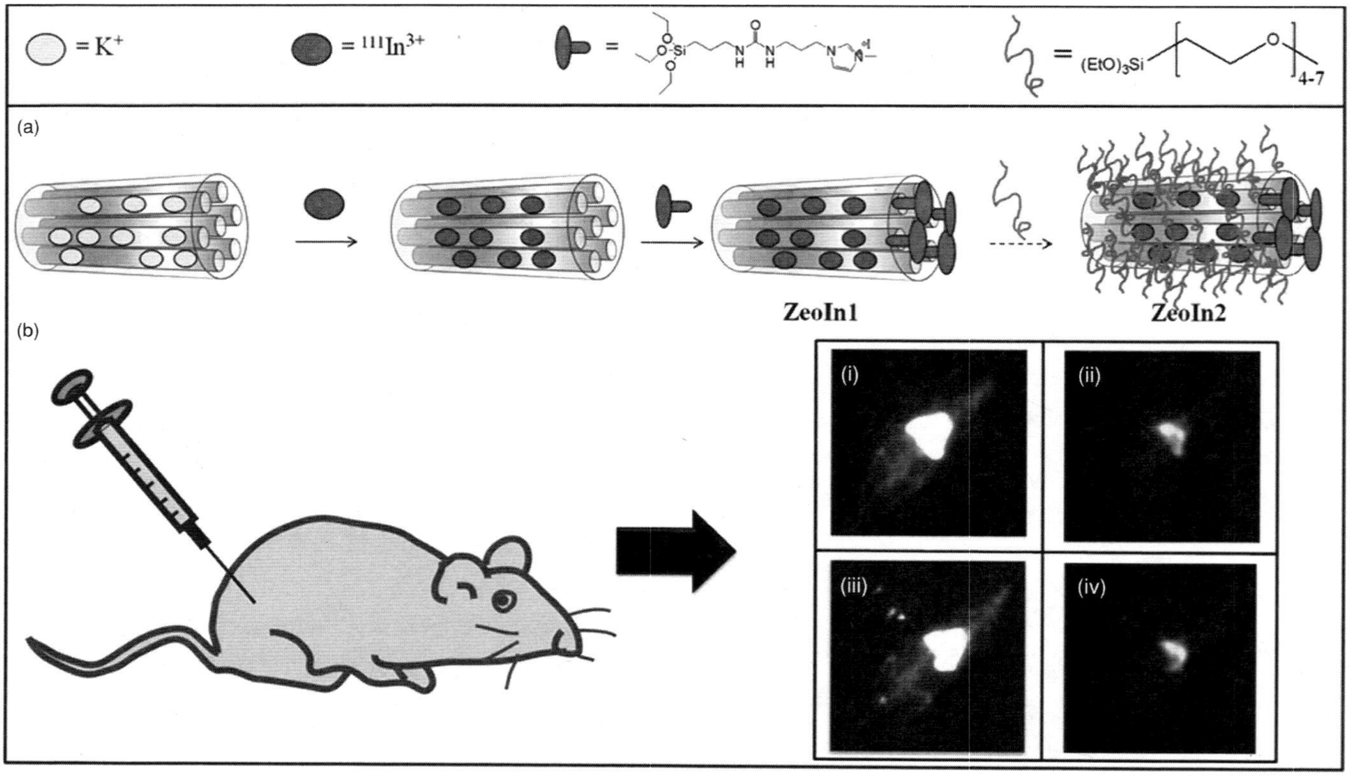

FIGURE 22.15. (**A**) Schematic functionalization of the zeolite L crystals. (**B**) Two-dimensional static scintigraphic images are obtained 70 min after injection of the zeolites. ZeoIn1 injected into rat 1 in (i) high contrast and (ii) low contrast. ZeoIn2 injected into rat 2 in (iii) high contrast and (iv) low contrast.

The cytotoxicity of nanozeolites has only been studied in vitro. It has been shown that their toxicity depends on the size, composition, and shape. The pure-silica zeolite silicalite-1 shows no cytotoxicity, whereas aluminum-containing zeolites like ZSM-5, zeolite L, and zeolite A show dose-dependent toxic effects [131].

However, many more studies are necessary since the fate of the porous silica materials in the body is still not known. On the other hand, it has been demonstrated that for mesoporous silicon, degradation is completed within 20 hours [132, 133].

22.5 SELF-ASSEMBLY OF ZEOLITES INTO ORDERED MONOLAYERS

Self-assembly is a process in which a disordered system assembles itself into an ordered structure [134–136]. The first examples of self-assembly were reported at the molecular level [137–139], and numerous examples are known in the literature about the self-assembly of molecules [136, 140–142]. In recent years, more and more attention was devoted to the self-assembly of all kinds of nano-objects [143–145], like nanoparticles [145], rods [145, 146], and plates [147]. Analogue to the self-assembly of molecules, nano-objects were assembled into 1D and 2D arrays using the well-established concepts of supramolecular chemistry. Although several examples are known in the literature, self-assembly of nanostructures is much more difficult than self-assembly of molecules, due to the larger size, shape, and often lack of specificity and recognition sites. The situation is different when an assembly is made on a surface since the nature or functionalization of the substrate can favor the deposition of the nanoparticles and can lead to high coverage and the formation of homogeneous monolayers [143, 148–152]. Most self-assembled monolayers (SAM) based on nano-objects are made up from spherically shaped objects. A few examples were reported in which the assembly involves porous systems of nonspherical shape. Amongst them, and very relevant to this chapter, are the reports in 1995 by Calzaferri and co-workers [153] and Lee et al. [151], who described a clever strategy to form zeolite monolayers.

In order to prepare a zeolite monolayer, covalent bonds between the zeolite crystals and the substrate, usually glass, need to be formed. Therefore the substrate is functionalized with molecules acting as covalent linkers to the zeolite crystals (see Figure 22.16). A variety of linker molecules have

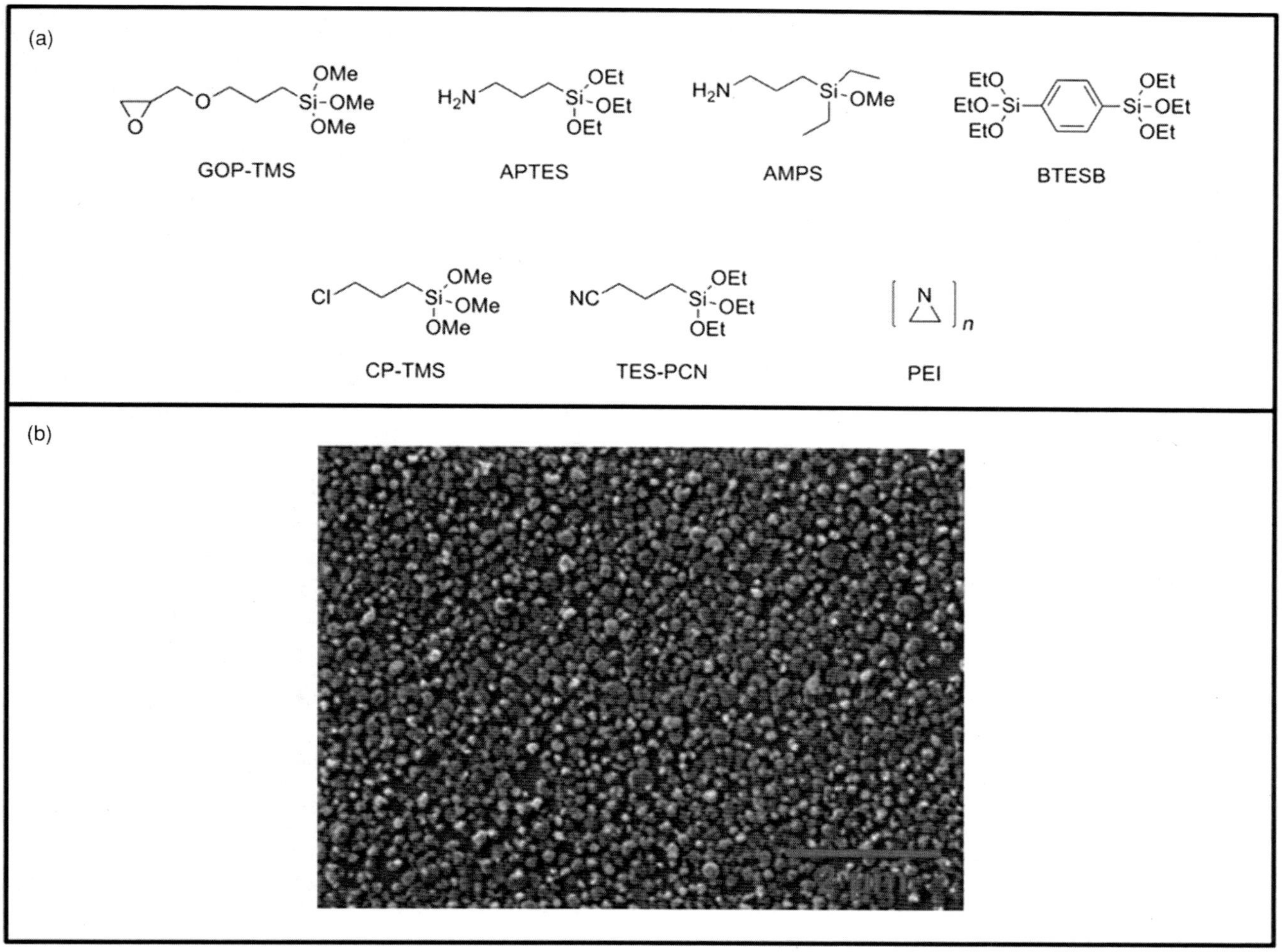

FIGURE 22.16. (**1**) A selection of commonly used linker molecules for zeolite monolayer preparation. (**2**) SEM image of a zeolite L monolayer prepared in our laboratory from small disc shaped crystals (150–230 nm) applying the sonication method.

been reported, and several of them were studied in detail by Yoon and co-workers [154–156]. Depending on the target substrate, different reagents have to be employed to functionalize the zeolites or the substrate itself. The most common substrates used for SAMs of zeolites are glass or quartz, which are typically functionalized with alkoxysilanes bearing reactive head groups to form covalent bonds with the zeolites [48, 148–150, 152]. In order to obtain high-quality monolayers, it is important that the interaction between the substrate and the bases of the zeolites must be stronger than with the coat or any other interactions between the crystals.

Highest packing densities for zeolite L monolayers are achieved by using disc-shaped zeolite L crystals and the sonication method [151]. The formation of such high packing densities can be explained by a phenomenon called surface migration. During the formation of a densely packed monolayer, the zeolites undergo dynamic cycles of bond forming/bond breaking reactions until the crystals are packed very close to each other, allowing the formation of inter-crystal hydrogen bonds.

In recent years many authors have reported several methods to prepare monolayers from a variety of different kinds of zeolites (e.g., zeolite A, ZSM-5, zeolite L) [143, 148–152] and even of mesoporous materials [157, 158].

The motivations to build such particular type of substrate range from the intention to have high-density functional groups, which can be attached to the zeolites, to the desire to have the channel perpendicular to the surfaces and therefore prone to the release of chemicals. Indeed the presence of thousands of molecules inside each nano-container along with the fact that each of them could be addressed as a single object is very appealing and has so far been almost unexplored. The orientation of the channels is of critical importance since the surface will block one side if the channels are oriented perpendicularly to it. The release could in this case be directional. In this respect, materials with one-dimensional channels are the best choice, and the shape of the container should favor a specific arrangement on the surface. Amongst the applications that can be envisaged with this type of organized layer, biomedical uses—such as cell growth, differentiation, and eventually communication—are the most relevant for this chapter.

22.6 FUNCTIONALIZATION OF ZEOLITE MONOLAYERS

In order to use zeolite monolayers as a substrate for cellular growth, ensuring the biocompatibility of the surface is crucial. This can be provided by coating the zeolite L surfaces with all kinds of bio-friendly molecules. If one thinks about attaching these "bio carpets" to the zeolite L surface, there are two major strategies. One is simply via physisorption or noncovalent interactions such as hydrogen or electrostatic bonds; and the other, which requires more advanced techniques, is by covalent functionalization of the zeolite surface or, even more controlled, of the channel entrances with the desired biomolecules.

An attachment via physisorption can be achieved in relatively simple ways, like dropcasting, spin coating, and layer-by-layer deposition. These methods rely on the fact that biomolecules are well-soluble in water while our nanocontainers do not decompose or dissolve in these conditions. While the versatility of these methods is an advantage, the stability of the coating could be an issue. Other problems, related to the noncovalent functionalization, are the control of the thickness and the density. Also, the control of the orientation of the biomolecules versus the surface is impossible. More interesting, but technically more demanding, is the covalent binding of biomolecules to the monolayer surface, which could be done following two different methodologies. The most common and straightforward involves the functionalization of nanocrystals before assembling them into monolayers. Numerous examples of such nanoparticle functionalization are known. However, in such a case, spatial control is a challenge and the nanocontainers are generally functionalized over all surfaces. More interesting is the functionalization of only the top part of the zeolites or the channel entrances, which can be achieved on an already prepared monolayer. Such functionalization is obtained by using soft lithography [159] and, in particular, microcontact printing (MCP). This simple technique allows for spatial control of multiple chemical reactions in a simple and rapid process at room temperature. MCP on flat substrates is used with increasing frequency in cell biology due to its versatility, low cost, and compatibility with cells [160, 161]. A material that is most commonly used for soft lithography in combination with biology is polydimethylsiloxane (PDMS). This choice is justified by the properties of PDMS such as softness, flexibility, biocompatibility, insulating, unreactiveness, transparency to UV–vis light, permeability to gases, and moderate permeabiliy to water [160, 161].

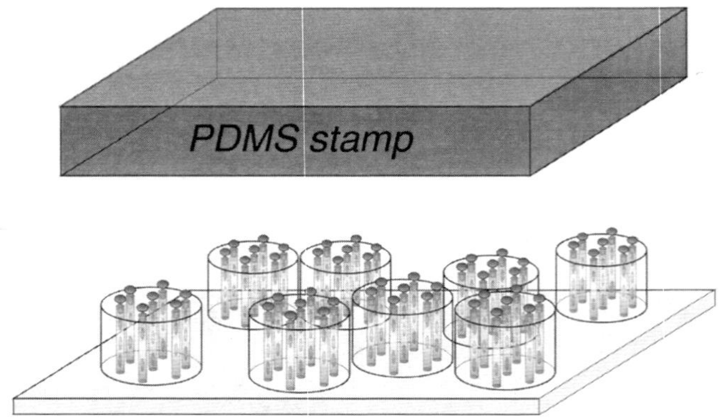

FIGURE 22.17. Schematic illustration of microcontact printing. The PDMS stamp is inked with the molecules we wish to print and with a moderate pressure pushed on the monolayer, resulting in the functionalization of the upper side.

The working principle of MCP on a monolayer of zeolites is depicted schematically in Figure 22.17. A zeolite monolayer on a glass surface with the channel entrances modified by reactive groups can be further functionalized with desired biochemical molecules following a few simple steps. A PDMS stamp is "inked" with the desired biomolecules by immersion in a concentrated solution. After evaporation of the solvent, the PDMS stamp is pressed with moderate strength onto a surface for a certain time in order to promote by simple contact the covalent reaction between the groups attached to the zeolites (e.g., amino) and the biomolecules possessing a complementary group (e.g., carboxylate). The covalent bonds formed during the contact printing allow the transfer of the desired molecule from the PDMS stamp to the channel entrances of the zeolites. By using patterned PDMS stamps, it is possible to print different chemical functionalities with a resolution of a few micrometers. This strategy can also be applied to transferring a monolayer from one substrate to another one [56]. The quality of the transfer depends strongly on the elastomer used as well as on the receiving substrate, which should be as flat as possible. We have demonstrated that it is possible to print dye molecules on top of zeolites monolayers [162].

We have also reported cell growth on zeolite L monolayers functionalized with Arg–Gly–Asp (RGD) tripeptide [163]. It has been shown that cells such as HeLa, 3T3 cells, and human macrophages adhere better on the RGD-functionalized than on the amino-functionalized control surfaces. Also, a stripe-patterned monolayer has been prepared and it has been shown that cells grow and proliferate only along the RGD functionalized stripes. The principle is schematically shown in Figure 22.18 along with fluorescence microscope images of samples prepared in this way. The pattern can be shaped in different size and geometry; and, as already mentioned,

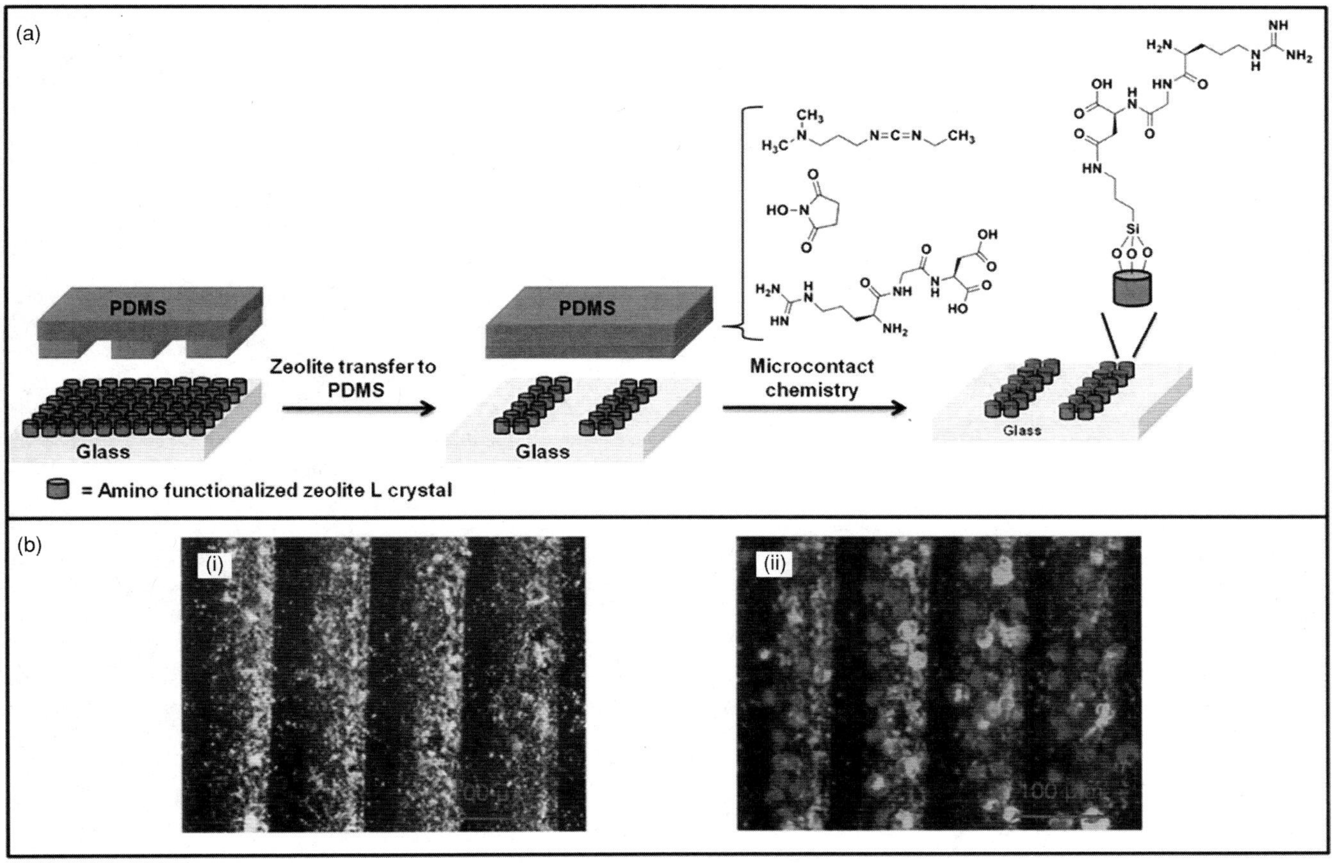

FIGURE 22.18. (**1**) Schematic illustration of creating a stripe patterned zeolite L monolayer functionalized with RGD. (**2**) HeLa cell adhesion on a patterned RGD functionalized zeolite L monolayer. Fluorescence microscopy images: (**a**) excitation at 460–490 nm; (**b**) excitation at 400–440 nm.

different functionalities and fillings can be achieved with the zeolites. The functionalization of the channel entrances could also lead to the growth of soft materials combining the hard porous scaffold of the microporous systems with more cell compatible networks. Finally, the release of the molecules entrapped into the channels could modify the cell adhesion or even cell composition and cause a cellular response.

In concluding this short overview on some of the fascinating materials and applications that the nanoworld can offer, we wish to stress that a constant improvement of the size, monodispersity, and chemical composition is ongoing. Also, hybrid systems in which inorganic and organic building blocks can be amalgamate together are surely of great interest and could open unexpected perspectives and could represent the ultimate choices.

REFERENCES

1. Eustis, S., El-Sayed, M. A. (2006). Why gold nanoparticles are more precious than pretty gold: Noble metal surface plasmon resonance and its enhancement of the radiative and nonradiative properties of nanocrystals of different shapes. *Chemical Society Reviews*, **35**, 209–217.
2. Lu, A. H., Salabas, E. L., Schuth, F. (2007). Magnetic nanoparticles: Synthesis, protection, functionalization, and application. *Angewandte Chemie International Edition*, **46**, 1222–1244.
3. Smith, A. M., Nie, S. M. (2010). Semiconductor nanocrystals: structure, properties, and band gap engineering. *Accounts of Chemical Research*, **43**, 190–200.
4. Chopra, N., Gavalas, V. G., Hinds, B. J., Bachas, L. G. (2007). Functional one-dimensional nanomaterials: Applications in nanoscale biosensors. *Anal. Letters*, **40**, 2067–2096.
5. Corchero, J. L., Villaverde, A. (2009). Biomedical applications of distally controlled magnetic nanoparticles. *Trends in Biotechnology*, **27**, 468–476.
6. Wang, Z., Lu, Y. (2009). Functional DNA directed assembly of nanomaterials for biosensing. *Journal of Materials Chemistry*, **19**, 1788–1798.
7. Tsotsalas, M. M., Kopka, K., Luppi, G., Wagner, S., Law, M. P., Schaefers, M., De Cola, L. (2010). Encapsulating (111)In in nanocontainers for scintigraphic imaging:

Synthesis, characterization, and in vivo biodistribution. *ACS Nano*, **4**, 342–348.

8. Tsotsalas, M., Busby, M., Gianolio, E., Aime, S., De Cola, L. (2008). Functionalized nanocontainers as dual magnetic and optical probes for molecular imaging applications. *Chemical Materials*, **20**, 5888–5893.
9. Cai, W., Chen, X. (2007). Nanoplatforms for targeted molecular imaging in living subjects. *Small*, **11**, 1840–1854.
10. Huang, X., Li, L., Liu, T., Hao, N., Liu, H., Chen, D., Tang, F. (2011). The shape effect of mesoporous silica nanoparticles on biodistribution, clearance, and biocompatibility in vivo. *ACS Nano*, **5**, 5390–5399.
11. Cauda, V., Argyo, C., Bein, T. (2010). Impact of different PEGylation patterns on the long-term bio-stability of colloidal mesoporous silica nanoparticles. *Journal of Materials Chemistry*, **20**, 8693–8699.
12. Sauer, A. M., Schlossbauer, A., Ruthardt, N., Cauda, V., Bein, T., Brauchle, C. (2010). Role of endosomal escape for disulfide-based drug delivery from colloidal mesoporous silica evaluated by live-cell imaging. *Nano Letters*, **10**, 3684–3691.
13. Suteewong, T., Sai, H., Cohen, R., Wang, S.T., Bradbury, M., Baird, B., Gruner, S.M., Wiesner, U. (2011). Highly aminated mesoporous silica nanoparticles with cubic pore structure. *Journal of the American Chemical Society*, **133**, 172–175.
14. Chou, L. Y. T., Ming, K., Chan, W. C. W. (2011). Strategies for the intracellular delivery of nanoparticles. *Chemical Society Reviews*, **40**, 233–245.
15. Calzaferri, G., Devaux, A. (2011). Manipulation of energy transfer processes within the channels of L-zeolite. In: Ramamurthy, V., Inoue, Y., eds. *Supramolecular Effects in Photochemical and Photophysical Processes*, John Wiley & Sons, Hoboken, NJ, pp. 285–387.
16. Moller, K., Bein, T. (2011). Pores within pores—How to craft ordered hierarchical zeolites. *Science*, **333**, 297–298.
17. Buck, S. M., Koo, Y. E. L., Park, E., Xu, H., Philbert, M. A., Brasuel, M. A., Kopelman, R. (2004). Optochemical nanosensor PEBBLEs: Photonic explorers for bioanalysis with biologically localized embedding. *Current Opinion in Chemical Biology*, **8**, 540–546.
18. Sukhorukov, G. B., Rogach, A. L., Zebli, B., Skirtach, A. G., Kohler, K., Antipov, A. A., Gaponik, N., Susha, A. S., Winterhalter, M., Parak, W. J. (2005). Nanoengineered polymer capsules: tools for detection, controlled delivery, and site-specific manipulation. *Small*, **1**, 194–200.
19. Blanazs, A., Armes, S. P., Ryan, A. J. (2009). Self-assembled block copolymer aggregates: from micelles to vesicles and their biological applications. *Macromolecular Rapid Communications*, **30**, 267–277.
20. Rijcken, C. J. F., Soga, O., Hennink, W. E., van Nostrum, C. F. (2007). Triggered destabilization of polymeric micelles and vesicles by changing polymers polarity: An attractive tool for drug delivery. *Journal of Controlled Release*, **120**, 131–148.
21. Andresen, T. L., Jensen, S. S., Jorgensen, K. (2005). Advanced strategies in liposomal cancer therapy: Problems and prospects of active and tumor specific drug release. *Progess in Lipid Research*, **44**, 69–97.
22. Muller, M., Devaux, A., Yang, C., De Cola, L., Fischer, R. A. (2010). Highly fluorescent metal–organic frameworks by means of host–guest chemistry. *Abstracts of Papers, 239th ACS National Meeting, San Francisco, CA, United States, March 21–25, 2010*, INOR-370.
23. Yaghi, O. M., O'Keeffe, M., Ockwig, N. W., Chae, H. K., Eddaoudi, M., Kim, J. (2003). Reticular synthesis and the design of new materials. *Nature*, **423**, 705–714.
24. Lu, F., Gu, L., Meziani, M. J., Wang, X., Luo, P. G., Veca, L. M., Cao, L., Sun, Y. P. (2009). Advances in bioapplications of carbon nanotubes. *Advanced Materials*, **21**, 139–152.
25. Mauro, M., Schuermann, K. C., Pretot, R., Hafner, A., Mercandelli, P., Sironi, A., De Cola, L. (2010). Complex Ir (III) salts. A new class of luminescent porous crystalline materials. *Angewandte Chemie International Edition*, **49**, 1222.
26. Corma, A., Garcia, H. (2004). Supramolecular host–guest systems in zeolites prepared by ship-in-a-bottle synthesis. *European Journal of Inorganic Chemistry*, **6**, 1143–1164.
27. Inagaki, S., Ohtani, O., Goto, Y., Okamoto, K., Ikai, M., Yanabaka, K. I., Tani, T., Okada, T. (2009). Light harvesting by a periodic mesoporous organosilica chromophore. *Angewandte Chemie International Edition*, **48**, 4042–4046.
28. Schulz-Ekloff, G., Wohrle, D., van Duffel, B., Schoonheydt, R. A. (2002). Chromophores in porous silicas and minerals: Preparation and optical properties. *Microporous and Mesoporous Materials*, **51**, 91–138.
29. Bruehwiler, D., Calzaferri, G., Torres, T., Ramm, J. H., Gartmann, N., Dieu, L. Q., Lopez-Duarte, I., Martinez-Dias, M. (2009). Nanochannels for supramolecular organization of luminescent guests. *Journal of Materials Chemistry*, **19**, 8040–8067.
30. Zabala Ruiz, A., Bruehwiler, D., Ban, T., Calzaferri, G. (2005). Synthesis of zeolite L: Tuning size and morphology. *Chemical Monthly*, **136**, 77–89.
31. Zhao, D., Feng, J., Huo, Q., Melosh, N., Fredrickson, G. H., Chmelka, B. F., Stucky, G. D. (1998). Triblock copolymer syntheses of mesoporous silica with periodic 50 to 300 angstrom pores. *Science*, **279**, 548–552.
32. Kresge, C. T., Leonowicz, M. E., Roth, W. J., Vartuli, J. C., Beck, J. S. (1992). Ordered mesoporous molecular sieves synthesized by a liquid-crystal template mechanism. *Nature*, **359**, 710–712.
33. Park, S. E., Prasetyanto, E. A. (2008). Organocatalytic application of direct organo-functionalized mesoporous catalysts prepared by microwave. *Topics in Catalysis*, **52**, 91–100.
34. Park, S.-E., Prasetyanto, E. A. (2011). Morphosynthesis and Catalysis of the Organofunctionalized Mesoporous Materials. In: Wyman, E. B., Skief, M. C., eds. *Organosilanes: Properties, Performance and Applications*, Nova Science Publishers Inc, New York, pp. 101–129.
35. Allen, T. M., Cullis, P. R. (2004). Drug delivery systems: Entering the mainstream. *Science*, **303**, 1818–1822.

36. Manzano, M., Vallet-Regi, M. (2010). New developments in ordered mesoporous materials for drug delivery. *Journal of Materials Chemistry*, **20**, 5593–5604.
37. Petros, R. A., DeSimone, J. M. (2010). Strategies in the design of nanoparticles for therapeutic applications. *Nature Reviews Drug Discovery*, **9**, 615–627.
38. Slowing, I. I., Trewyn, B. G., Giri, S., Lin, V. S. Y. (2007). Mesoporous silica nanoparticles for drug delivery and biosensing applications. *Advanced Functional Materials*, **17**, 1225–1236.
39. Arruebo, M., Fernandez-Pacheco, R., Irusta, S., Arbiol, J., Ibarra, M. R., Santamaria, J. (2006). Sustained release of doxorubicin from zeolite-magnetite nanocomposites prepared by mechanical activation. *Nanotechnology*, **17**, 4057–4064.
40. Fatouros, D. G., Douroumis, D., Nikolakis, V., Ntais, S., Moschovi, A. M., Trivedi, V., Khima, B., Roldo, M., Nazar, H., Cox, P. A. (2011). In vitro and in silico investigations of drug delivery via zeolite BEA. *Journal of Materials Chemistry*, **21**, 7789–7794.
41. Rivera, A., Farias, T. (2005). Clinoptilolite-surfactant composites as drug support: a new potential apllication. *Microporous and Mesoporous Materials*, **80**, 337–346.
42. Devaux, A., Cucinotta, F., Kehr, S., De Cola, L. (2010). Functionalization and assembling of inorganic nanocontainers for optical and biomedical applications in functional supramolecular architectures. In: Samori, P., Cacialli, F., eds. *Functional Supramolecular Architectures for Organic Electronics and Nanotechnology*. Wiley-VCH, Weinheim, pp. 283–342.
43. Busby, M., Devaux, A., Blum, C., Subramaniam, V., Calzaferri, G., De Cola, L. (2011). Interactions of perylene bisimide in the one dimensional channels of zeotlite L. *Journal of Physical Chemistry C*, **115**, 5974.
44. Albuquerque, R. Q., Calzaferri, G. (2007). Proton activity inside the the zeolite L. *Chemistry—A European Journal*, **13**, 8939–8952.
45. (a) Bein, T. (2007). Tuning functionality and morphology of periodic mesoporous materials. *Studies in Surface Science and Catalysis*, **170A**, 41–53. (b) Busby, M., Kerschbaumer, H., Calzaferri, G., De Cola, L. (2008). Orthogonally bi-functional fluorescent zeolite-L micro-crystals. *Advanced Materials*, **20**, 1614–1618.
46. Fisher, K. A., Huddersman, K. D., Taylor, M. J. (2003). Comparison of micro- and mesoporous inorganic materials in the uptake and release of the drug model fluorescein and its analogues. *Chemistry—A European Journal*, **9**, 5873–5878.
47. Pearce, M. E., Mai, H. Q., Lee, N., Larsen, S. C., Salem, A. K. (2008). Silicalite nanoparticles that promote transgene expression. *Nanotechnology*, **19**, 175103–175109.
48. Calzaferri, G., Maas, H., Pauchard, M., Pfenniger, M., Megelski, S., Devaux, A. (2002). Supramolecularly organized luminescent dye molecules in the channels of zeolite L. In: Neckers, D. C., von Bunau, G., Jenks, W. S., eds. *Advances in Photochemistry* vol. 27, John Wiley & Sons, New York, NY, pp. 1–150.
49. Meng, H., Xue, M., Xia, T., Zhao, Y. L., Tamanoi, F., Stoddart, J. F., Zink, J. I., Nel, A. E. (2010). Autonomous in vitro anticancer drug release from mesoporous silica nanoparticles by pH-sensitive nanovalves. *Journal of the American Chemical Society*, **132**, 12690–12697.
50. Thomas, C. R., Ferris, D. P., Lee, J. H., Choi, E., Cho, M. C., Kim, E. S., Stoddart, J. F., Shin, J. S., Cheon, J., Zink, J. I. (2010). Noninvasive remote-controlled release of drug molecules in vitro using magnetic actuation of mechanized nanoparticles. *Journal of the American Chemical Society*, **132**, 10623–10625.
51. Vallet-Regi, M., Balas, F., Arcos, D. (2007). Mesoporous materials for drug delivery. *Angewandte Chemie International Edition*, **46**, 7548–7558.
52. Bruns, A. A., Vider, J., Ow, H., Herz, E., Penate-Medina, O., Baumgart, M., Larson, S. M., Wiesner, U., Bradbury, M. (2009). Fluorescent silica nanoparticles with efficient urinary excretion for nanomedicine. *Nano Letters*, **18**, 442–448.
53. Suarez, S., Devaux, A., Banuelos, J., Bossart, O., Kunzmann, A., Calzaferri, G. (2006). Transparent zeolite–polymer hybrid materials with adaptable properties. *Advanced Functional Materials*, **17**, 2298–2306.
54. Rosenholm, J. M., Meinander, A., Peuhu, E., Niemi, R., Eriksson, J. E., Sahlgren, C., Linden, M. (2009). Targeting of porous hybrid silica nanoparticles to cancer cells. *ACS Nano*, **3**, 197–206.
55. Choi, J., Ghosh, S., Lai, Z., Tsapatsis, M. (2006). MFI zeolite membranes from a- and randomly oriented monolayers. *Angewandte Chemie International Edition*, **45**, 1154–1158.
56. Cucinotta, F., Popovic, Z., Weiss, E. A., Withesides, G. M., De Cola, L. (2009). Microcontact transfer printing of zeolite monolayers. *Advanced Materials*, **21**, 1142–1145.
57. Hashimoto, S., Samata, K., Shoji, T., Taira, N., Tomita, T., Matsuo, S. (2009). Preparation of large-scale 2D zeolite crystal array structures to achieve optical functionality. *Microporous and Mesoporous Materials*, **117**, 220–227.
58. Mahmoudi, M., Lynch, I., Ejtehadi, M. R., Monopoli, M. P., Bombelli, F. B., Laurent, S. (2011). Protein–Nanoparticle interactions: Opportunities and challenges. *Chemical Reviews*, **111**, 5610–5637.
59. Becker, M., De Cola, L., Studer, A. (2011). Site-specific immobilization of proteins at zeolite L crystals by nitroxide exchange reactions. *Chemical Communications*, **47**, 3392–3394.
60. Norek, M., Neves, I. C., Peters, J. A. (2007). H-1 relaxivity of water in aqueous suspensions of Gd^{3+}-loaded NaY nanozeolites and AITUD-1 mesoporous material: The influence of Si/Al ratio and pore size. *Inorganic Chemistry*, **46**, 6190–6196.
61. Young, S. W., Qing, F., Rubin, D., Balkus, K. J., Engel, J. S., Lang, J., Dow, W. C., Mutch, J. D., Miller, R. A. (1995). Gadolinium zeolite as an oral contrast agent for magnetic-resonance-imaging. *Journal of Magnetic Resonance Imaging*, **5**, 499–508.
62. Bresinska, I., Balkus, K. J. (1994). Studies of Gd(III)-exchanged Y-type zeolites relevant to magnetic-resonance-imaging. *Journal of Physical Chemistry*, **98**, 12989–12994.

63. Csajbok, E., Banyai, I., Vander Elst, L., Muller, R. N., Zhou, W. Z., Peters, J. A. (2005). Gadolinium(III)-loaded nanoparticulate zeolites as potential high-field MRI contrast agents: Relationship between structure and relaxivity. *Chemistry—A European Journal*, **11**, 4799–4807.

64. Dahm, A., Eriksson, H. (2004). Ultra-stable zeolites—A tool for in-cell chemistry. *Journal of Biotechnology*, **111**, 279–290.

65. Andersson, L. I. M., Eriksson, H. (2007). De-aluminated zeolite Y as a tool to study endocytosis, a delivery system revealing differences between human peripheral dendritic cells. *Scandinavian Journal of Immunology*, **66**, 52–61.

66. Perez, L. M., Lalueza, P., Monzon, M., Puertolas, J. A., Arruebo, M., Santamaria, J. (2011). Hollow porous implants filled with mesoporous silica particles as a two-stage antibiotic-eluting device. *International Journal of Pharmaceutics*, **409**, 1–8.

67. Horcajada, P., Ramila, A., Perez-Pariente, J., Vallet-Regi, M. (2004). Influence of pore size of MCM-41 matrices on drug delivery rate. *Microporous and Mesoporous Materials*, **68**, 105–109.

68. Chen, G., Beving, D. E., Bedi, R. S., Yan, Y. S., Walker, S. L. (2009). Initial bacterial deposition on bare and zeolite-coated aluminum alloy and stainless steel. *Langmuir*, **25**, 1620–1626.

69. Marambio-Jones, C., Hoek, E. (2010). A review of the antibacterial effects of silver nanomaterials and potential implications for human health and the environment. *Journal of Nanoparticle Research*, **12**, 1531–1551.

70. Lalueza, P., Monzoacuten, M., Arruebo, M., Santamariacutea, J. (2011). Bactericidal effects of different silver-containing materials. *Materials Research Bulletin*, **46**.

71. Lalueza, P., Monzon, M., Arruebo, M., Santamaria, J. (2010). Antibacterial action of Ag-containing MFI zeolite at low Ag loadings. *Chemical Communications*, **47**, 680–682.

72. Mintova, S., Mo, S. Y., Bein, T. (2001). Humidity sensing with ultrathin LTA-type molecular sieve films grown on piezoelectric devices. *Chemistry of Materials*, **13**, 901–905.

73. Zhang, Y. H., Wang, X. Y., Shan, W., Wu, B. Y., Fan, H. Z., Yu, X. J., Tang, Y., Yang, P. Y. (2005). Enrichment of low-abundance peptides and proteins on zeolite nanocrystals for direct MALDI-TOF MS analysis. *Angewandte Chemie International Edition*, **44**, 615–617.

74. Zhang, D., Chen, Y., Chen, H. Y., Xia, X. H. (2004). Silica-nanoparticle-based interface for the enhanced immobilization and sequence-specific detection of DNA. *Analytical and Bioanalytical Chemistry*, **379**, 1025–1030.

75. Corma, A., Fornes, V., Rey, F. (2002). Delaminated zeolites: An efficient support for enzymes. *Advanced Materials*, **14**, 71–74.

76. Platas-Iglesias, C., Vander Elst, L., Zhou, W. Z., Muller, R. N., Geraldes, C., Maschmeyer, T., Peters, J. A. (2002). Zeolite GdNaY nanoparticles with very high relaxivity for application as contrast agents in magnetic resonance imaging. *Chemistry—A European Journal*, **8**, 5121–5131.

77. Xu, F., Wang, Y. J., Wang, X. D., Zhang, Y. H., Tang, Y., Yang, P. Y. (2003). A novel hierarchical nanozeolite composite as sorbent for protein separation in immobilized metal-ion affinity chromatography. *Advanced Materials*, **15**, 1751–1753.

78. Walczyk, D., Bombelli, F. B., Monopoli, M. P., Lynch, I., Dawson, K. A. (2010). What the cell sees in bionanoscience. *Journal of the American Chemical Society*, **132**, 5761–5768.

79. Mitchell, S., Perez-Ramirez, J. (2011). Mesoporous zeolites as enzyme carriers: Synthesis, characterization, and application in biocatalysis. *Catalysis Today*, **168**, 28–37.

80. Hu, Y.-Y., Zhang, Y.-H., Ren, N., Tang, Y. (2009). Crystal plane- and size-dependent protein adsorption on nanozeolite. *Journal of Physical Chemistry C*, **113**, 18040–18046.

81. Tavolaro, P., Tavolaro, A., Martino, G. (2009). Influence of zeolite PZC and pH on the immobilization of cytochrome c: A preliminary study regarding the preparation of new biomaterials. *Colloids and Surfaces B: Biointerfaces*, **70**, 98–107.

82. Cao, J., Hu, Y., Shen, C., Yao, J., Wei, L., Yang, F., Nie, A., Wang, H., Shen, H., Liu, Y., Zhang, Y., Tang, Y., Yang, P. (2009). Nanozeolite-driven approach for enrichment of secretory proteins in human hepatocellular carcinoma cells. *Proteomics*, **9**, 4881–4888.

83. Vinu, A., Murugesan, V., Hartmann, M. (2004). Adsorption of lysozyme over mesoporous molecular sieves MCM-41 and SBA-15: Influence of pH and aluminum incorporation. *Journal of Physical Chemistry B*, **108**, 7323–7330.

84. Hudson, S., Magner, E., Cooney, J., Hodnett, B. K. (2005). Methodology for the immobilization of enzymes onto mesoporous materials. *Journal of Physical Chemistry B*, **109**, 19496–19506.

85. Zhang, Y., Liu, Y., Kong, J., Yang, P., Tang, Y., Liu, B. (2006). Efficient proteolysis system: A nanozeolite-derived microreactor. *Small*, **2**, 1170–1173.

86. Kubota, M., Nakabayashi, T., Matsumoto, Y., Shiomi, T., Yamada, Y., Ino, K., Yamanokuchi, H., Matsui, M., Tsunoda, T., Mizukami, F., Sakaguchi, K. (2008). Selective adsorption of bacterial cells onto zeolites. *Colloids and Surfaces B: Biointerfaces*, **64**, 88–97.

87. Joshi, P., Rayalu, S., Bansiwal, A., Juwarkar, A. A. (2007). Surface modified zeolite, a novel carrier material for Azotobacter chroococcum. *Plant and Soil*, **296**, 151–158.

88. Meng, H., Yang, S., Li, Z., Xia, T., Chen, J., Ji, Z., Zhang, H., Wang, X., Lin, S., Huang, C., Zhou, Z. H., Zink, J. I., Nel, A. E. (2011). Aspect ratio determines the quantity of mesoporous silica nanoparticle uptake by a small GTPase-dependent macropinocytosis mechanism. *ACS Nano*, **5**, 4434–4447.

89. Huang, X., Teng, X., Chen, D., Tang, F., He, J. (2010). The effect of the shape of mesoporous silica nanoparticles on cellular uptake and cell function. *Biomaterials*, **31**, 438–448.

90. Popovic, Z., Busby, M., Huber, S., Calzaferri, G., De Cola, L. (2006). Assembling micro crystals through cooperative coordinative interactions. *Angewandte Chemie International Edition*, **46**, 8898–8902.

91. Strassert, C. A., Otter, M., Albuquerque, R. Q., Hoene, A., Vida, Y., Maier, B., De Cola, L. (2009). Photoactive hybrid nanomaterial for targeting, labeling, and killing

antibiotic-resistant bacteria. *Angewandte Chemie International Edition*, **48**, 7928–7931.

92. Hocine, O., Gary-Bobo, M., Brevet, D., Maynadier, M., Fontanel, S., Raehm, L., Richeter, S., Loock, B., Couleaud, P., Frochot, C. L., Charnay, C., Derrien, G., Smaihi, M., Sahmoune, A., Morere, A., Maillard, P., Garcia, M., Durand, J.-O. (2010). Silicalites and mesoporous silica nanoparticles for photodynamic therapy. *International Journal of Pharmaceutics*, **402**, 221–230.
93. Huang, D. M., Hung, Y., Ko, B. S., Hsu, S. C., Chen, W. H., Chien, C. L., Tsai, C. P., Kuo, C. T., Kang, J. C., Yang, C. S., Mou, C. Y., Chen, Y. C. (2005). Highly efficient cellular labeling of mesoporous nanoparticles in human mesenchymal stem cells: implication for stem cell tracking. *FASEB Journal*, **19**, 2014–2016.
94. Slowing, I., Trewyn, B. G., Lin, V. S. Y. (2006). Effect of surface functionalization of MCM-41-type mesoporous silica nanoparticleson the endocytosis by human cancer cells. *Journal of the American Chemical Society*, **128**, 14792–14793.
95. Brevet, D., Gary-Bobo, M., Raehm, L., Richeter, S., Hocine, O., Amro, K., Loock, B., Couleaud, P., Frochot, C., Morere, A., Maillard, P., Garcia, M., Durand, J.-O. (2009). Mannose-targeted mesoporous silica nanoparticles for photodynamic therapy. *Chemical Communications*, 1475–1477.
96. Rivera, A., Farias, T. (2005). Clinoptilolite-surfactant composites as drug support: A new potential application. *Microporous and Mesoporous Materials*, **80**, 337–346.
97. Arruebo, M., Fernandez-Pacheco, R., Irusta, S., Arbiol, J., Ibarra, M. R., Santamaria, J. (2006). Sustained release of doxorubicin from zeolite–magnetite nanocomposites prepared by mechanical activation. *Nanotechnology*, **17**, 4057–4064.
98. Vallet-Regi, M., Ramila, A., del Real, R. P., Perez-Pariente, J. (2000). A new property of MCM-41: Drug delivery system. *Chemical Materials*, **13**, 308–311.
99. Horcajada, P., Ramila, A., Ferey, G., Vallet-Regi, M. (2006). Influence of superficial organic modification of MCM-41 matrices on drug delivery rate. *Solid State Sciences*, **8**, 1243–1249.
100. Vallet-Regi, M., Balas, F., Arcos, D. (2007). Mesoporous materials for drug delivery. *Angewandte Chemie International Edition*, **46**, 7548–7558.
101. Bernardos, A., Aznar, E., Coll, C., Martinez Manez, R., Barat, J. M., Marcos, M. D., Sancenon, F., Benito, A., Soto, J. (2008). Controlled release of vitamin B2 using mesoporous materials functionalized with amine-bearing gate-like scaffoldings. *Journal of Controlled Release*, **131**, 181–189.
102. Lin, Q., Huang, Q., Li, C., Bao, C., Liu, Z., Li, F., Zhu, L. (2010). Anticancer drug release from a mesoporous silica based nanophotocage regulated by either a one- or two-photon process. *Journal of the American Chemical Society*, **132**, 10645–10647.
103. Petkar, K. C., Chavhan, S. S., Agatonovik-Kustrin, S., Sawant, K. K. (2011). Nanostructured materials in drug and gene delivery: A review of the state of the art. *Critical Reviews in Therapeutic Drug Carrier Systems*, **28**, 101–164.
104. Wang, S. (2009). Ordered mesoporous materials for drug delivery. *Microporous and Mesoporous Materials*, **117**, 1–9.
105. Cauda, V., Argyo, C., Bein, T. (2010). Impact of different PEGylation patterns on the long-term bio-stability of colloidal mesoporous silica nanoparticles. *Journal of Materials Chemistry*, **20**, 8693–8699.
106. Lin, Y.-S., Abadeer, N., Haynes, C. L. (2011). Stability of small mesoporous silica nanoparticles in biological media. *Chemical Communications*, **47**, 532–534.
107. Climent, E., Bernardos, A., Martinez-Manez, R., Maquieira, A., Dolores Marcos, M., Pastor-Navarro, N., Puchades, R., Sancenon, F., Soto, J., Amoros, P. (2009). Controlled delivery systems using antibody-capped mesoporous nanocontainers. *Journal of the American Chemical Society*, **131**, 14075–14080.
108. Bernardos, A., Mondragon, L., Aznar, E., Marcos, M. D., Martinez-Manez, R., Sancenonn, F., Soto, J., Barat, J. M., Perez-Paya, E., Guillem, C., Amoros, P. (2010). Enzyme-responsive intracellular controlled release using nanometric silica mesoporous supports capped with "saccharides". *ACS Nano*, **4**, 6353–6368.
109. Vivero-Escoto, J. L., Slowing, I. I., Wu, C.-W., Lin, V. S. Y. (2009). Photoinduced intracellular controlled release drug delivery in human cells by gold-capped mesoporous silica nanosphere. *Journal of the American Chemical Society*, **131**, 3462–3463.
110. Zhao, Y., Trewyn, B. G., Slowing, I. I., Lin, V. S. Y. (2009). Mesoporous silica nanoparticle-based double drug delivery system for glucose-responsive controlled release of insulin and cyclic AMP. *Journal of the American Chemical Society*, **131**, 8398–8400.
111. Gao, Q., Xu, Y., Wu, D., Sun, Y., Li, X. (2009). pH-Responsive drug release from polymer-coated mesoporous silica spheres. *Journal of Physical Chemistry C*, **113**, 12753–12758.
112. Shi, X., Wang, Y., Varshney, R. R., Ren, L., Zhang, F., Wang, D.-A. (2009). In-vitro osteogenesis of synovium stem cells induced by controlled release of bisphosphate additives from microspherical mesoporous silica composite. *Biomaterials*, **30**, 3996–4005.
113. Ashley, C. E., Carnes, E. C., Phillips, G. K., Padilla, D., Durfee, P. N., Brown, P. A., Hanna, T. N., Liu, J., Phillips, B., Carter, M. B., Carroll, N. J., Jiang, X., Dunphy, D. R., Willman, C. L., Petsev, D. N., Evans, D. G., Parikh, A. N., Chackerian, B., Wharton, W., Peabody, D. S., Brinker, C. J. (2011). The targeted delivery of multicomponent cargos to cancer cells by nanoporous particle-supported lipid bilayers. *Nature Materials*, **10**, 389–397.
114. Rosenholm, J. M., Sahlgren, C., Linden, M. (2010). Towards multifunctional, targeted drug delivery systems using mesoporous silica nanoparticles—Opportunities & challenges. *Nanoscale*, **2**, 1870–1883.
115. Hsiao, J.-K., Tsai, C.-P., Chung, T.-H., Hung, Y., Yao, M., Liu, H.-M., Mou, C.-Y., Yang, C.-S., Chen, Y.-C., Huang, D.-M. (2008). Mesoporous silica nanoparticles as a delivery system of gadolinium for effective human stem cell tracking. *Small*, **4**, 1445–1452.

116. Xia, W., Chang, J., Lin, J., Zhu, J. (2008). The pH-controlled dual-drug release from mesoporous bioactive glass/polypeptide graft copolymer nanomicelle composites. *European Journal of Pharmaceutics Biopharmaceutics*, **69**, 546–552.
117. Gao, W., Chan, J. M., Farokhzad, O. C. (2010). pH-Responsive nanoparticles for drug delivery. *Molecular Pharmaceutics*, **7**, 1913–1920.
118. Schlossbauer, A., Dohmen, C., Schaffert, D., Wagner, E., Bein, T. (2011). pH-Responsive release of acetal-linked melittin from SBA-15 mesoporous silica. *Angewandte Chemie International Edition*, **50**, 6828–6830.
119. Izquierdo-Barba, I., Vallet-Regi, M., Kupferschmidt, N., Terasaki, O., Schmidtchen, A., Malmsten, M. (2009). Incorporation of antimicrobial compounds in mesoporous silica film monolith. *Biomaterials*, **30**, 5729–5736.
120. Dai, C., Yuan, Y., Liu, C., Wei, J., Hong, H., Li, X., Pan, X. (2009). Degradable, antibacterial silver exchanged mesoporous silica spheres for hemorrhage control. *Biomaterials*, **30**, 5364–5375.
121. Ehlert, N., Badar, M., Christel, A., Lohmeier, S. J., Luessenhop, T., Stieve, M., Lenarz, T., Mueller, P. P., Behrens, P. (2011). Mesoporous silica coatings for controlled release of the antibiotic ciprofloxacin from implants. *Journal of Materials Chemistry*, **21**, 752–760.
122. Trewyn, B. G., Whitman, C. M., Lin, V. S. Y. (2004). Morphological control of room-temperature ionic liquid templated mesoporous silica nanoparticles for controlled release of antibacterial agents. *Nano Letters*, **4**, 2139–2143.
123. Lu, J., Liong, M., Li, Z., Zink, J. I., Tamanoi, F. (2010). Biocompatibility, biodistribution, and drug-delivery efficiency of mesoporous silica nanoparticles for cancer therapy in animals. *Small*, **6**, 1794–1805.
124. Bharali, D. J., Klejbor, I., Stachowiak, E. K., Dutta, P., Roy, I., Kaur, N., Bergey, E. J., Prasad, P. N., Stachowiak, M. K. (2005). Organically modified silica nanoparticles: A nonviral vector for in vivo gene delivery and expression in the brain. *Proceedings of the National Academy of Sciences USA*, **102**, 11539–11544.
125. Al-Salam, S., Balhaj, G., Al-Hammadi, S., Sudhadevi, M., Tariq, S., Biradar, A. V., Asefa, T., Souid, A.-K. (2011). In vitro study and biocompatibility of calcined mesoporous silica microparticles in mouse lung. *Toxicology Science*, **122**, 86–99.
126. Hudson, S. P., Padera, R. F., Langer, R., Kohane, D. S. (2008). The biocompatibility of mesoporous silicates. *Biomaterials*, **29**, 4045–4055.
127. Radu, D. R., Lai, C.-Y., Jeftinija, K., Rowe, E. W., Jeftinija, S., Lin, V. S. Y. (2004). A polyamidoamine dendrimer-capped mesoporous silica nanosphere-based gene transfection reagent. *Journal of the American Chemical Society*, **126**, 13216–13217.
128. Trewyn, B. G., Nieweg, J. A., Zhao, Y., Lin, V. S. Y. (2008). Biocompatible mesoporous silica nanoparticles with different morphologies for animal cell membrane penetration. *Chemical Engineering Journal*, **137**, 23–29.
129. Lu, J., Liong, M., Zink, J. I., Tamanoi, F. (2007). Mesoporous silica nanoparticles as a delivery system for hydrophobic anticancer drugs. *Small*, **3**, 1341–1346.
130. Roy, I., Ohulchanskyy, T. Y., Pudavar, H. E., Bergey, E. J., Oseroff, A. R., Morgan, J., Dougherty, T. J., Prasad, P. N. (2003). Ceramic-based nanoparticles entrapping water-insoluble photosensitizing anticancer drugs: A novel drug carrier system for photodynamic therapy. *Journal of the American Chemical Society*, **125**, 7860–7865.
131. Kihara, T., Zhang, Y., Hu, Y., Iviao, Q., Tang, Y., Miyake, J. (2011). Effect of composition, morphology and size of nanozeolite on its in vitro cytotoxicity. *Journal of Bioscience and Bioengineering*, **111**, 725–730.
132. Tasciotti, E., Liu, X., Bhavane, R., Plant, K., Leonard, A. D., Price, B. K., Cheng, M. M.-C., Decuzzi, P., Tour, J. M., Robertson, F., Ferrari, M. (2008). Mesoporous silicon particles as a multistage delivery system for imaging and therapeutic applications. *Nature Nanotechnology*, **3**, 151–157.
133. Godin, B., Gu, J., Serda, R. E., Bhavane, R., Tasciotti, E., Chiappini, C., Liu, X., Tanaka, T., Decuzzi, P., Ferrari, M. (2010). Tailoring the degradation kinetics of mesoporous silicon structures through PEGylation. *Journal of Biomedical Materials Research*, **94A**, 1236–1243.
134. Huck, W. T. S. e. (2005). *Nanoscale assembly techniques*. Springer, New York, p. 217.
135. Whitesides, G. M., Boncheva, M. (2002). Beyond molecules: Self-assembly of mesoscopic and macroscopic components. *Proceedings of the National Academy of Sciences USA*, **99**, 4769–4774.
136. Whitesides, G. M., Grzybowski, B. (2002). Self-assembly at all scales. *Science*, **295**, 2418–2421.
137. Whitesides, G. M., Kriebel, J. K., Love, J. C. (2005). Molecular engineering of surfaces using self-assembled monolayers. *Science Progress*, **88**, 17–48.
138. Mrksich, M., Whitesides, G. M. (1995). Patterning self-assembled monolayers using microcontact printing—A new technology for biosensors. *Trends in Biotechnology*, **13**, 228–235.
139. Whitesides, G. M., Mathias, J. P., Seto, C. T. (1991). Molecular self-assembly and nanochemistry—A chemical strategy for the synthesis of nanostructures. *Science*, **254**, 1312–1319.
140. Badjic, J. D., Nelson, A., Cantrill, S. J., Turnbull, W. B., Stoddart, J. F. (2005). Multivalency and cooperativity in supramolecular chemistry. *Accounts of Chemical Research*, **38**, 723–732.
141. Cram, D. J. (1988). The design of molecular hosts, guests, and their complexes (Nobel lecture). *Angewandte Chemie International Edition English*, **27**, 1009–1020.
142. Lehn, J.-M. (1988). Supramolecular chemistry—Scope and perspectives molecules, supermolecules, and molecular devices (Nobel lecture). *Angewandte Chemie International Edition in English*, **27**, 89–112.
143. Yoon, K. B. (2007). Organization of zeolite microcrystals for production of functional materials. *Accounts of Chemical Research*, **40**, 29–40.

144. Daniel, M. C., Astruc, D. (2004). Gold nanoparticles: Assembly, supramolecular chemistry, quantum-size-related properties, and applications toward biology, catalysis, and nanotechnology. *Chemical Reviews*, **104**, 293–346.

145. Hurst, S. J., Payne, E. K., Qin, L. D., Mirkin, C. A. (2006). Multisegmented one-dimensional nanorods prepared by hard-template synthetic methods. *Angewandte Chemie International Edition*, **45**, 2672–2692.

146. Liu, K., Zhao, N., Kumacheva, E. (2011). Self-assembly of inorganic nanorods. *Chemical Society Reviews*, **40**, 656–671.

147. Clark, T. D., Tien, J., Duffy, D. C., Paul, K. E., Whitesides, G. M. (2001). Self-assembly of 10-mu *m*-sized objects into ordered three-dimensional arrays. *Journal of the American Chemical Society*, **123**, 7677–7682.

148. Zabala Ruiz, A., Li, H., Calzaferri, G. (2006). Organizing supramolecular functional dye-zeolite crystals. *Angewandte Chemie International Edition in English*, **45**, 5282–5287.

149. Ha, K., Park, J. S., Oh, K. S., Zhou, Y. S., Chun, Y. S., Lee, Y. J., Yoon, K. B. (2004). Aligned monolayer assembly of zeolite crystals on platinum, gold, and indium–tin oxide surfaces with molecular linkages. *Microporous and Mesoporous Materials*, **72**, 91–98.

150. Lee, J. S., Lim, H., Ha, K., Cheong, H., Yoon, K. B. (2006). Facile monolayer assembly of fluorophore-containing zeolite rods in uniform orientations for anisotropic photoluminescence. *Angewandte Chemie International Edition*, **45**, 5288–5292.

151. Lee, J. S., Ha, K., Lee, Y. J., Yoon, K. B. (2005). Ultrasound-aided remarkably fast assembly of monolayers of zeolite crystals on glass with a very high degree of lateral close packing. *Advanced Materials*, **17**, 837–841.

152. Leiggener, C., Calzaferri, G. (2004). Monolayers of zeolite A containing luminescent silver sulfide clusters. *ChemPhysChem*, **5**, 1593–1596.

153. Li, H., Devaux, A., Popovic, Z., De Cola, L., Calzaferri, G. (2005). Carboxyester functionalised dye–zeolite L host–guest materials. *Microporous and Mesoporous Materials*, **95**, 112–117.

154. Choi, S. Y., Lee, Y. J., Park, Y. S., Ha, K., Yoon, K. B. (2000). Monolayer assembly of zeolite crystals on glass with fullerene as the covalent linker. *Journal of the American Chemical Society*, **122**, 5201–5209.

155. Kulak, A., Lee, Y. J., Park, Y. S., Yoon, K. B. (2000). Orientation-controlled monolayer assembly of zeolite crystals on glass and mica by covalent linkage of surface-bound epoxide and amine groups. *Angewandte Chemie International Edition*, **39**, 950–953.

156. Chun, Y. S., Ha, K., Lee, Y. J., Lee, J. S., Kim, H. S., Park, Y. S., Yoon, K. B. (2002). Diisocyanates as novel molecular binders for monolayer assembly of zeolite crystals on glass. *Chemical Communications*, 1846–1847.

157. Kim, E., Seo, S., Seo, M. L., Jung, J. H. (2010). Functionalized monolayers on mesoporous silica and on titania nanoparticles for mercuric sensing. *Analyst*, **135**, 149–156.

158. Feng, X., Fryxell, G. E., Wang, L.-Q., Kim, A. Y., Liu, J., Kemner, K. M. (1997). Functionalized monolayers on ordered mesoporous supports. *Science*, **276**, 923–926.

159. Xia, Y. N., Whitesides, G. M. (1998). Soft lithography. *Angewandte Chemie International Edition*, **37**, 551–575.

160. Whitesides, G. M., Ostuni, E., Takayama, S., Jiang, X. Y., Ingber, D. E. (2001). Soft lithography in biology and biochemistry. *Annual Review of Biomedical Engineering*, **3**, 335–373.

161. Weibel, D. B., DiLuzio, W. R., Whitesides, G. M. (2007). Microfabrication meets microbiology. *Nature Reviews Microbiology*, **5**, 209–218.

162. Kehr, N. S., Schäfer, A., Ravoo, B. J., De Cola, L. (2010). Asymmetric printing of molecules and zeolites on self assembled monolayers. *Nanoscale*, **2**, 601–605.

163. Kehr, N. S., Riehemann, K., El-Gindi, J., Schäfer, A., Fuchs, H., Galla, H.-J., De Cola L. (2010). Cell adhesion and cellular patterning on self-assembled monolayer of zeolite L. *Advanced Functional Materials*, **20**, 2248–2254.

23

DICATIONIC GEMINI NANOPARTICLE DESIGN FOR GENE THERAPY

MAHMOUD ELSABAHY, ILDIKO BADEA, RONALD VERRALL, MCDONALD DONKURU, AND MARIANNA FOLDVARI

23.1 GENE THERAPY AND CHALLENGES

Gene therapy is a medical technique whereby genetic defects can be repaired by inserting a normal or functional gene to replace or correct a faulty gene or repress an overexpressing gene. Hence, gene therapy can be used for the treatment of a wide range of diseases, such as cancer, viral infections, and dermatological diseases [1–4]. However, the therapeutic benefits from nucleic acid-based therapy are still unsatisfactory. Introducing nucleic acids into blood or topical application on the skin or mucosal surfaces encounters many barriers that alter their cellular biodistribution and intracellular bioavailability. Following administration, they rapidly degrade in biological fluids by extracellular nucleases before they can reach the surface of the target cells. This definitely influences their activity and interaction with the cells, compromising the therapeutic outcome of nucleic acids. In addition, nucleic acids have very limited cellular uptake because of their hydrophilic nature and high molecular weight. A small fraction that can be taken up by the cells is usually internalized into vesicles (i.e., endosomes), which convert later into lysosomes. Accumulation and subsequent digestion of the nucleic acids inside the lysosomes preclude them from reaching their cytoplasmic or nuclear targets and is an important barrier to their efficacy. In addition, the nuclear membrane is an additional barrier for the efficient translocation of the plasmid DNA (pDNA) into the nucleus (Figure 23.1). The successful transfection with pDNA or other small nucleic acids (e.g., antisense oligonucleotides, short interfering RNA) must include packaging the DNA/RNA in a delivery system that can compact the nucleic acids and facilitate their transfer through the cell membrane. Many strategies have been investigated to increase the intracellular bioavailability of nucleic acids [5,6]. These delivery systems consist of two types: viral and nonviral. The former have been used in a majority of gene therapy clinical trials, but there is concern and evidence that their use can elicit a strong immune response [7–9]. Due to these events, emphasis over the past decade has begun to shift to the development of nonviral delivery systems. These delivery systems have significant advantages over viral systems, because they are safer to use and easier to prepare and can transfer larger-sized plasmids into cells. However, in order for nonviral delivery systems to be useful in gene therapy, they must have higher transfection efficiencies than have been realized to date. The focus of our research group over the last decade has been to design cationic nonviral systems that improve the efficiency of the gene delivery process.

23.2 GEMINI SURFACTANTS AS NOVEL BIOMATERIALS

23.2.1 Cationic Gemini Surfactants: General Structure and Possible Modifications

A variety of cationic lipids have been described as nucleic acid delivery agents. One of the first studies in which a gene was delivered by a nonviral cationic lipid system was reported by Felgner et al. [10], wherein a cationic system comprising a mixture of DOTMA/DOPE (*N*-[1-(2, 3-dioleyloxy)propyl]-*N*,*N*,*N*-trimethylammonium chloride/1,2-dioleoyl-*sn*-glycerol-3-phosphoethanolamine)

Organic Nanomaterials: Synthesis, Characterization, and Device Applications, First Edition. Edited by Tomás Torres and Giovanni Bottari.

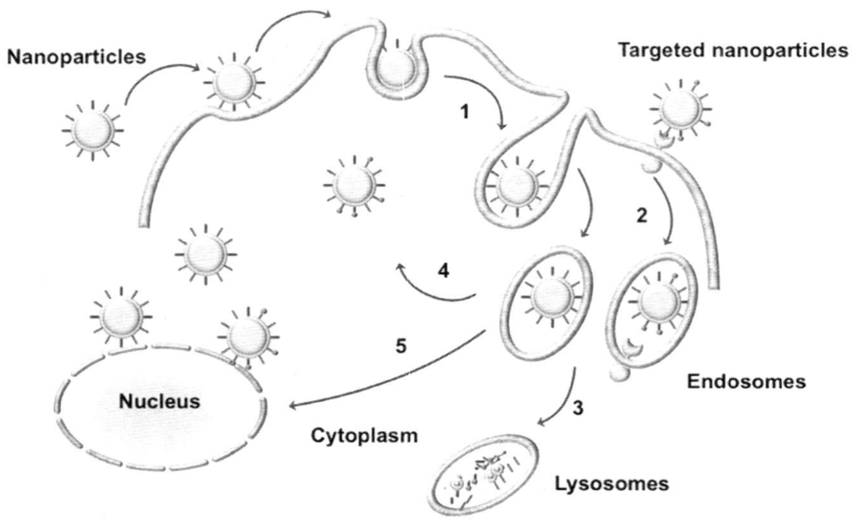

FIGURE 23.1. Intracellular fate of nanoparticles for nucleic acids delivery. The process is initiated either by nonspecific binding to the cell surface or by specific ligand–receptor internalization, which results in endocytosis (1) or receptor-mediated endocytosis (2), respectively. These systems should possess either excess positive charge or targeting moieties to facilitate cell binding and subsequent uptake. The internalized vesicles (i.e., endosomes) either convert into lysosomes (3), where extensive degradation can take place, or crumble and release their contents into the cytoplasm (4). Cytoplasmic release is essential to allow the encapsulated nucleic acids to reach their cytoplasmic targets or to further translocate to the nucleus (5). Incorporation of fusogenic lipids, peptides, or membrane-destabilizing polymers facilitates the escape from the endosomes, whereas tagging with a nuclear homing sequence enhances nuclear translocation.

was shown to deliver DNA in vitro. This system is commercially available as Lipofectin®. It contains the fundamental features of the inaugural type of cationic systems: a monocationic head group and two lipid tails. Such lipid systems can form vesicles, aggregates, or other specific types of supramolecular structures, which condense and encapsulate DNA into particles that can be readily internalized by cells.

One family of building blocks for the design of cationic-lipid-based delivery systems is the "gemini" surfactants, a term used by Menger to describe a family of surfactants comprised of two head groups (anionic, cationic, or neutral) that are linked covalently at or near the head groups by a spacer group [11, 12]. They have become of widespread interest to develop nonviral carriers for gene therapy because of their rather unique solution properties imparted by their molecular structure. In a number of ways they are superior to the conventional monomeric surfactants. They have remarkably low critical micelle concentrations (CMC), higher surface activity, low Krafft temperature, and unusual rheological properties, and they self-assemble into a variety of aggregate structures ranging from worm-like micelles to vesicles. Their order-of-magnitude-lower CMC compared to conventional surfactants means that they potentially impart lower cellular toxicity. More importantly, compared to monomeric surfactants, their self-assembly into various aggregate structures can be effectively adjusted through tailoring their molecular structures by independent or collective variation in head group size, valence, spacer length, carbon tail length, and degree of unsaturation. For these reasons, our interest in developing nonviral carriers has focused on the use of cationic gemini compounds because they can interact with biomolecules, nucleic acids in particular, through electrostatic and hydrophobic interactions, van der Waals forces, and hydrogen bonding. The most-studied series of gemini surfactants are the cationic *N*, *N*-bis(dimethylalkyl)-α,ω-alkane-diammonium surfactants, having a general structure $[C_mH_{2m+1}(CH_3)_2N^+(CH_2)_sN^+(CH_3)_2 \quad C_mH_{2m+1}.2X^-]$, abbreviated as the *m-s-m* series, where *m* and *s* refer to the number of carbon atoms in the alkyl tails and in the polymethylene spacer group, respectively, and X is the counterion (Figure 23.2 and Table 23.1).

For example, the 12–3–12 surfactant has two dodecyl tails and a trimethylene spacer group. On top of the basic structural template, certain functional groups, such as amino groups or amino acid groups, when incorporated into the template, are expected to greatly augment the molecular properties leading to higher potency. Modifications of the chemistry and architecture of the gemini surfactants allow a "fine-tuning" of the physicochemical behavior, which in turn modulates the binding and compaction of DNA/RNA and ultimately plays a major role in the nucleic acids release in cells.

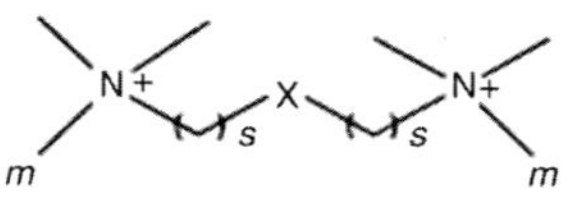

Gemini Surfactant	*m*	*s*	X
12–3–12	12	1	CH_2
12–7–12	12	3	CH_2
12–7N–12	12	3	N-CH_3
12–7NH–12	12	3	NH
16–7NH–16	16	3	NH
18–7NH–18	18	3	NH
12–7NG–12	12	3	*N*-glycine
12–7NK–12	12	3	*N*-lysine
12–7NGK–12	12	3	*N*-glycine–lysine
12–7NKK–12	12	3	*N*-lysine–lysine
12–5N–12	12	2	*N*-CH_3
Py–3–12	Pyrene–12	1	CH_2

FIGURE 23.2. General structure of gemini surfactants with some examples of the common gemini surfactants.

23.2.2 Synthesis of Gemini Surfactants

Synthetic schemes that have been used to prepare a wide range of anionic and cationic gemini surfactants have been summarized in a comprehensive review [13]. Briefly, gemini surfactants of the *m–s–m* family with $12 \geq s \geq 3$ can be synthesized by refluxing (24–48 h) the appropriate α,ω-dibromoalkane with 2 molar equivalents (plus a 10% excess to ensure as much as possible a complete diquaternization) of the appropriate *N,N*-dimethylalkyl amine [14, 15]. The 12–2–12 surfactant is best obtained by synthesizing 1 equivalent of *N,N,N′,N′*-tetramethylethylenediamine with 2 equivalents (plus a slight excess) of 1-bromododecane. Apparently, the reaction of one amine with dibromoalkane having $s = 2$ results in a compound where the unreacted CH_2Br groups are very reactive and readily convert into CH_3 [15]. Upon cooling, target compounds can be recovered by filtration and purified by recrystallizing from either acetonitrile (12–2–12) or a mixture of acetonitrile/ethyl acetate. Gemini surfactants with $s > 12$ can be also synthesized. For instance, the 12–16–12 compound was synthesized by reaction of 1 equivalent of 1,16-dibromohexadecane with 2 equivalents (plus a slight excess of *N,N*-dimethyldodecylamine) in 2-propanol [14].

Synthesis of gemini surfactants with *N*-substituted moieties in the spacer can be carried out by refluxing one equivalent of *N,N,N′,N′,N″*-pentamethyldiethylenetriamine, 3,3′-iminobis(*N,N*-dimethylpropylamine), 2,6,10-trimethyl-2,6,10-triazaundecane, or 1,1,4,7,10,10-hexamethyltriethylenetetramine with 2.2 equivalents of 1-bromododecane in acetonitrile for 24 h. The resulting surfactants, abbreviated 12-5N-12, 12-7NH-12, 12-7N-12, 12-8N-12, respectively, can be purified by recrystallization from acetonitrile or a mixture of acetonitrile/ethyl acetate [16]. Modification of the 12-7NH-12 compound by chemically coupling amino acid moieties to the N–H group in the spacer has been also carried out to test the effect of the amino acids coupling on the toxicity and transfection efficiency of gemini surfactants (Figure 23.3).

Hydroxylated gemini surfactants have been synthesized to examine the effect of the spacer hydrophilicity on the transfection properties of the formed nanoparticles. For instance, 1,4-bis(dodecyl-*N,N*-dimethylammonium bromide)-2-butanol and 1,4-bis(dodecyl-*N,N*-dimethylammonium bromide)-2,3-butanediol, abbreviated 12-4(OH)-12 and 12-4$(OH)_2$-12, respectively, have been synthesized by reacting appropriate amounts of 1,4-dibromo-2-butanol, or 1,4-dibromo-2,3-butanediol with 2 molar equivalents (plus a slight excess) of *N,N*-dimethyldodecylamine in high-performance liquid chromatography (HPLC) grade 2-propanol for 24 h. Several recrystallizations from a mixture of ethyl acetate/methanol (10:1 v/v) provided pure materials confirmed by CH&N analysis, nuclear magnetic resonance (NMR) spectroscopy and surface tension measurements [17]. Other modified gemini surfactants have also been successfully synthesized. For example, pyrene (py)-3-12 (N^1-dodecyl, N^1,N^1,N^3,N^3-tetramethyl-N^3-(6-(pyren-6-yl) hexyl)propane-1,3-diammonium dibromide and py-6-12 ((N^1-dodecyl,N^1,N^1,N^6,N^6-tetramethyl-N^6-(6-(pyren-6-yl) hexyl) hexane-1,6-diammonium dibromide) have been synthesized as shown in Figure 23.4 [18]. Purity of the gemini compounds is usually determined by using several of the following techniques: integrated ^{1}H NMR, electrospray mass spectrometry (ES-MS), and surface tension and CH&N analyses.

TABLE 23.1. Structural Characteristics of Gemini Surfactants with Varied Spacers and Alkyl Tails

	Variations in Structure		
Surfactant Series	Spacer(s)	R Group	Symbol
12–*s*–12 series			
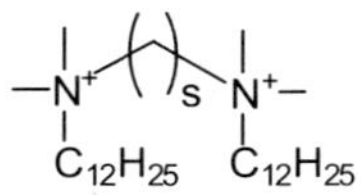	3	Unchanged	12–3–12
	4	Unchanged	12–4–12
	6	Unchanged	12–6–12
	8	Unchanged	12–8–12
	10	Unchanged	12–10–12
	12	Unchanged	12–12–12
	16	Unchanged	12–16–12
12–spacer–12 series			
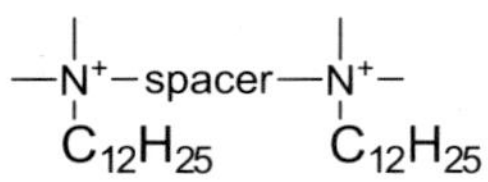	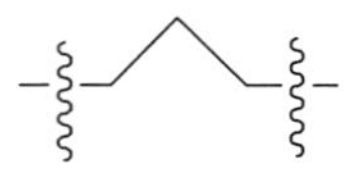	Unchanged	12–3–12
	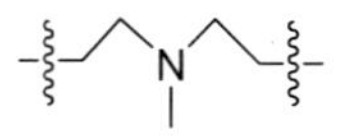	Unchanged	12–5N–12
	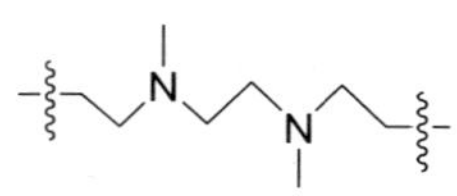	Unchanged	12–8N–12
	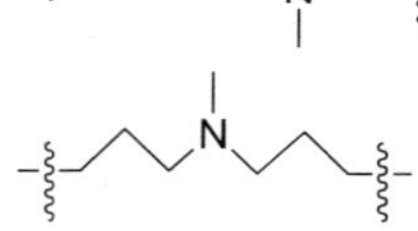	Unchanged	12–7N–12
	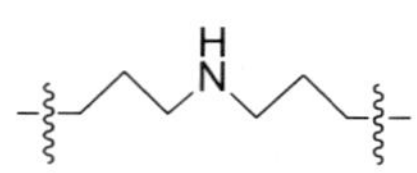	Unchanged	12–7NH–12
***m*–7NH–*m* series**			
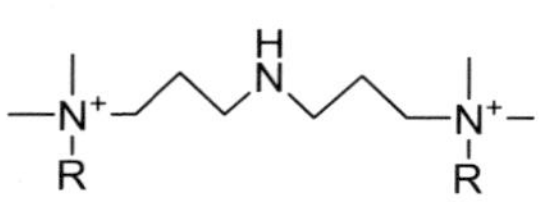	Unchanged	*n*-$C_{12}H_{25}$	12–7NH–12
	Unchanged	*n*-$C_{16}H_{33}$	16–7NH–16
	Unchanged	*n*-$C_{18}H_{37}$	18–7NH–18
	Unchanged	*n*-$C_{18}H_{35}$ (oleyl)	18:1–7NH–18:1
***m*–3–*m* series**			
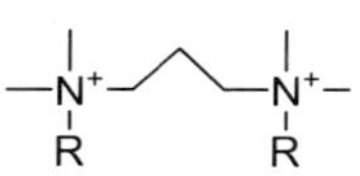	Unchanged	*n*-$C_{12}H_{25}$	12–3–12
	Unchanged	*n*-$C_{16}H_{33}$	16–3–16
	Unchanged	*n*-$C_{18}H_{37}$	18–3–18
	Unchanged	*n*-$C_{18}H_{35}$ (oleyl)	18:1–3–18:1
***m*–7–*m* series**			
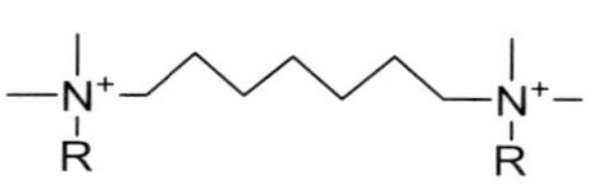	Unchanged	*n*-$C_{12}H_{25}$	12–7–12
	Unchanged	*n*-$C_{16}H_{33}$	16–7–16
	Unchanged	*n*-$C_{18}H_{37}$	18–7–18
12–7(AA)*n*–12 series			
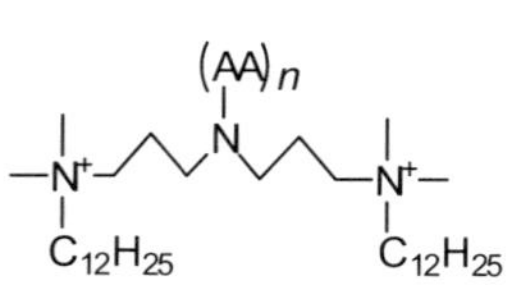	(AA)*n* = None	Unchanged	12–7NH–12
	G	Unchanged	12–7NG–12
	K	Unchanged	12–7NK–12
	GK	Unchanged	12–7NGK–12
	KK	Unchanged	12–7NKK–12

(AA)*n*: amino acid sequence—G (glycine), K (lysine), GK (glycine–lysine), KK (lysine–lysine)

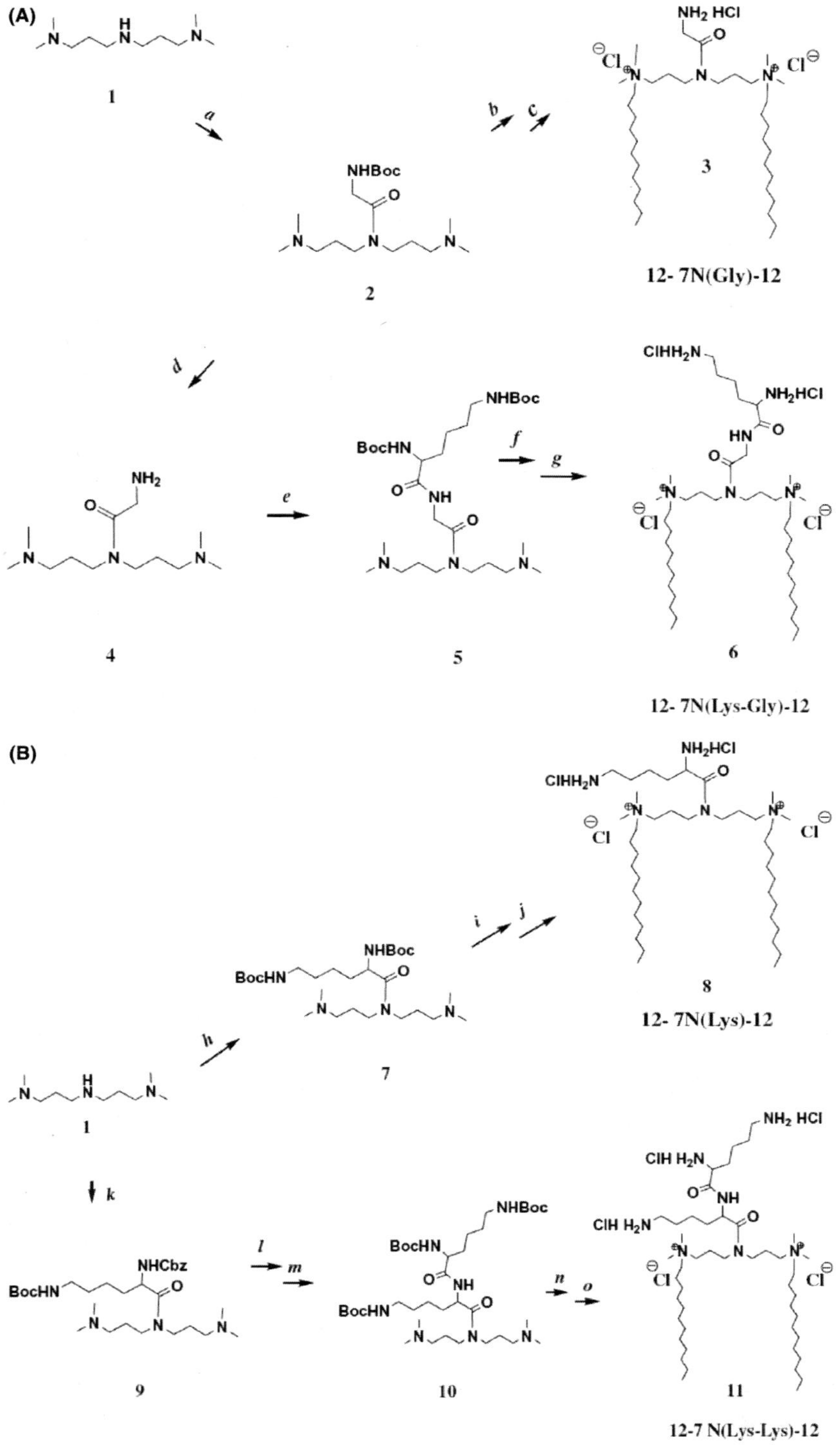

FIGURE 23.3. Synthetic scheme for the preparation of amino acid functionalized gemini surfactants: (**A**) 12-7N(Lys-Gly)-12. (*a*) Boc-glycine, EEDQ, $CHCl_3$; (*b*) 1-bromododecane, MeCN; (*c*) (i) TFA, $CHCl_3$; (ii) Cl^- ion exchange; (*d*) (i) TFA, $CHCl_3$; (ii) NaOH, EtOH; (*e*) Boc-Lys(Boc)-OH,EEDQ,$CHCl_3$; (*f*) 1-bromododecane, MeCN; (*g*) (i) TFA, $CHCl_3$; (ii) Cl^- ion exchange, reproduced from reference 40, with permission. (**B**) 12–7N(Lys-Lys)–12. (*h*) Boc-Lys(Boc)-OH, EEDQ, $CHCl_3$; (*i*) 1-bromododecane, MeCN; (*j*) (i) TFA, $CHCl_3$; (ii) Cl^- ion exchange; (*k*) Cbz-Lys(Boc)-OH, EEDQ,$CHCl_3$; (*l*) $HCOONH_4$, 10% Pd/C; (*m*) Boc-Lys(Boc)-OH, EEDQ,$CHCl_3$; (*n*) 1-bromododecane, MeCN; (*o*) (i) TFA, $CHCl_3$; (ii) Cl^- ion exchange. Reproduced from reference 40, with permission.

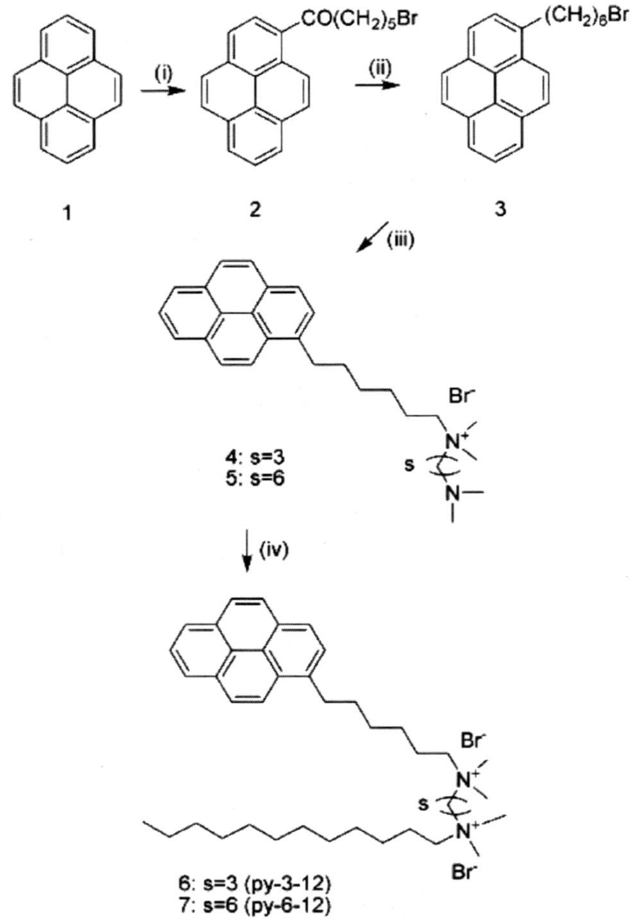

FIGURE 23.4. Synthesis of py–3–12 and py–6–12: (i) $Br(CH_2)_5COCl$, $AlCl_3$, CH_2Cl_2, dry ice; (ii) Et_3SiH, TFA, ice bath; (iii) *N,N,N′,N′*-tetramethylpropanediamine, CH_3CN, 45°C, 3 days; (iv) 1-$Br(CH_2)_{11}CH_3$, CH_3CN, reflux, 2 days. Reproduced from reference 18, with permission.

23.3 RATIONAL DESIGN OF GEMINI NANOPARTICLES

Gemini nanoparticles are composed of gemini surfactants and nucleic acids (DNA or RNA) with or without neutral lipid vesicles (Figure 23.5). DOPE lipids are probably the most employed lipids to prepare the gemini nanoparticles. Cationic gemini surfactants spontaneously neutralize, condense, and encapsulate DNA into small particles (whose size can range from <100 nm to ~1000 nm) by electrostatic, hydrophobic, and van der Waals' interactions [19,20]. The interactions, in the classical sense, give rise to particles with internal DNA content and external vesicle membrane formed by the gemini surfactant hydrophobic tails [21,22]. These nanoparticles can be prepared at different charge ratios (usually $\rho_{+/-}$ from 1 to 10), defined as the molar ratio of the quaternary amine groups of the gemini surfactant to the negatively charged phosphate groups of the nucleic acids. Although the composition of the nanoparticles, the type and molar ratio of the lipids, and charge ratio have great effect on their morphology and transfection properties, the challenging part in the rational design of these nanoparticles relates to the structure of the gemini surfactants. We have taken a bottom-up approach to design nonviral gemini delivery systems that allows component-by-component testing, development, and optimization of safer and more effective nonviral gene delivery systems.

23.3.1 Effect of Alkyl Chain Length and Structure

Alkyl Chain Length. Longer alkyl tails (longer than C_{12}) are expected to encapsulate DNA more efficiently at lower concentrations for both improved gene delivery and lowered cytotoxicity. For instance, transfection was observed for 16–3–16 and 18:1–3–18:1, both in the absence and presence of DOPE, whereas for the 12–*s*–12 series virtually no transfection was observed in the absence of DOPE [23].

Asymmetric Gemini Surfactants. In the Past decade there has been an increased interest in asymmetric compounds, *m*–*s*–*n* where $m \neq n$, and in alkyl tails having different chemical composition. The objective was to determine to what extent the asymmetry in the chains would influence the diversity of thermodynamic parameters and aggregation behavior. Series of *m*–2–*n*, where $m + n = 24$ and 32 [24–27] and *m*–6–*n* where $m + n = 32$ [28, 29], were studied and the thermodynamic properties showed no unusual trends as compared to the symmetric series with similar total carbon tail number. However, compounds where *s* and *m* were fixed and *n* varied so that $m + n$ increased the asymmetry, the micellization process was affected cooperatively by tail length growth and increasing asymmetry [30]. In an attempt to provide insight as to how the degree of asymmetry in the gemini compounds affected interaction with DNA/gemini complexes, a series of gemini surfactants (*m*–*s*–*n*) with different degrees of asymmetry (*m*/*n*) were studied [31]. The results indicated that the interaction tends to be weaker as *m*/*n* is increased, which is consistent with an expected disruption of tail–tail hydrophobic interaction in the self-assembled aggregates at higher *m*/*n*.

Pyrene-Substituted Gemini Surfactants. These modified surfactants are synthesized by inserting a pyrene group at the terminal end of one of the alkyl chains of the gemini surfactants. Pyrene was chosen as fluorescent moiety because of its relatively high quantum yield, although other fluorescent moieties can be used if different emissive wavelength regions are required. The objective was to provide a qualitative view of the fate of the complexes through real-time fluorescence microscopy. By maintaining the hydrophobic tails approximately the same length, it would ensure that the probe gemini

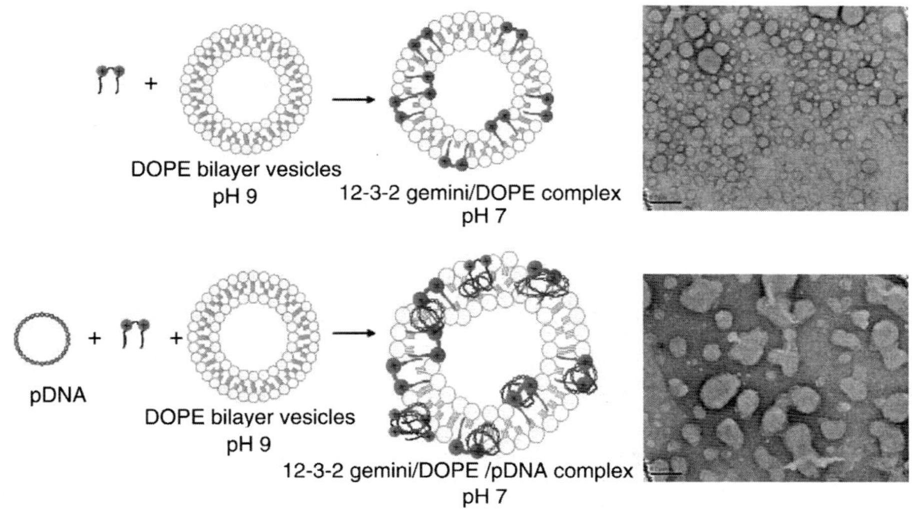

FIGURE 23.5. Schematic illustration of gemini nanoparticle formation and comparative electron micrographs of the structural morphology of gemini nanoparticles and plasmid–gemini nanoparticles prepared from 12–3–12/DOPE (**A**) and 12–3–12/DOPE/plasmid (**B**) by negative staining. Complexation of the plasmid DNA in the gemini/DOPE nanoparticles results in slightly larger particle size distribution and "fuzzy" particles, which is indicative of the presence of an additional polymorphic phase [92] and confirms our previous results on the structural features of these nanoparticles by small-angle x-ray scattering [93]. See color insert.

would insert itself within the regular gemini to form mixed aggregates. The physicochemical and photophysical properties of these gemini were examined under dilute and self-assembled conditions, and characteristics of their interaction with salmon-sperm DNA were monitored by using fluorescence and UV–vis techniques. The broadening and red shift of the absorbance bands of pyrene and the hyperchromism in the presence of DNA are evidence of associated pyrene compounds. Evidence of pyrene association (i.e., excimer formation) is expected to occur at concentrations of pyrene compounds greater than 1×10^{-5} M and, since the CMC for the pyrenyl gemini are ~0.22–0.30 mM [18], would be observed in the post-micelle region. Indeed, the fluorescence spectrum of py-3-12 at 3.5 μM (<CMC) is characteristic of pyrene in monomer form, and at a concentration of 0.35 mM (>CMC) a peak characteristic of pyrene excimers is also observed at 480 nm [18]. Fluorescence emission spectra of DNA/py-*s*-12 systems were obtained by titrating DNA into 4.8 μM py-*s*-12. A marked decrease in monomer surfactant fluorescence emission occurs as DNA concentration increased. This can result from two sources: quenching of the pyrene fluorescence by DNA bases (exciplex formation) [32–34] or pyrene excimer formation arising from proximity of pyrene rings as DNA-surfactant lipoplexes form. There is also a concomitant emission peak that appears at 480 nm, and its intensity increases as more DNA is added to the point where the DNA is mostly in the complexed form. An emission peak at 480 nm is more typically associated with excimer formation between two monomers, but conclusive evidence of excimer versus exciplex formation in these systems was lacking from these results. In a follow-up thermodynamic study comparing the binding of asymmetric pyrenyl-gemini surfactants to DNA with members of the *m–s–m* series to DNA, there was evidence that the pyrenyl–gemini intercalated with the dsDNA and disrupted the typical interactions observed between DNA and the *m–s–m* series [35]. A comprehensive fluorescence lifetime study is currently in progress to further elucidate whether exciplex or excimer formation dominates over the course of the complex formation or whether there may be a transition between the two as the ratio of pyrenyl–gemini to DNA increases. The unique nature of the interaction may make these compounds useful as a means of modulating DNA release during transfection. An investigation of the morphology of the pyrenyl-gemini nanoparticles as transfection agents, their cytotoxicity, and intracellular fate is currently being carried out.

23.3.2 Effect of Spacer Length and Structure

Spacer Hydrophilicity. A less hydrophobic spacer would reduce the unfavorable hydrocarbon–water contacts and make it easier for the spacer to locate at the micelle–water interface and improve the formation of lipoplexes of surfactant/nucleic acid/lipid. For instance, in vitro transfection of COS-7 cells with plasmid–gemini–DOPE nanoparticles showed that 12–4(OH)–12 enhanced transfection efficiencies beyond 12–3–12 by approximately 4*x*, while the incorporation of a second OH group, 12–4(OH)$_2$–12 had about half the effect compared to 12–4(OH)–12 [36].

Spacer Length. From studies of the 12–3–12 series, it is apparent that transfection efficiency is greatest when $s \leq 4$ which can be interpreted in terms of an optimized electrostatic interaction between the N^+–N^+ centers of the gemini and negatively charged phosphate groups of DNA. Studies of polyamine–DNA complexes have shown that a spacing of 4.9 Å, equivalent to the trimethylene spacing between amine groups in spermine, is suitable to interact with adjacent phosphate groups of DNA [37, 38]. The N^+–N^+ separations for 12–3–12 and 16–3–16 have been calculated to be 5.25 Å using Chem Office Chem 3D Pro software (Table 23.2) [36]. Hence, it was hypothesized that a gemini series in which the spacer was substituted with N–CH_3 and N–H moieties separated by three methylene segments could enhance transfection efficiency beyond that observed with gemini *m–s–m* compounds, such as 12–3–12 and 16–3–16, by increasing the binding of the surfactant to DNA and increasing the compaction of DNA.

pH-Sensitive Gemini Surfactants. The 12–7NH–12 and 12–7N–12 gemini have a single N–H or N–CH_3 substituent, respectively, with a three-methylene spacer between adjacent N centers, whereas the 12–5N–12 and 12–8N–12 gemini have one or two N–CH_3 centers, with a two-methylene spacer between adjacent N centers. The results of transfection studies with COS-7 cells [39] showed that 12–7N–12 and 12–7NH–12 compounds have superior transfection efficiencies compared to 12–5N–12 and 12–8N–12. Small-angle x-ray scattering (SAXS) measurements show the formation of multiple phases for the 12–7N–12 and 12–7NH–12 formulations that can readily allow for membrane fusion which may facilitate DNA release within the cell [39]. The significant increase in transfection of 12–7NH–12 over 12–7N–12 could result from a pH-dependent change in morphology arising from the protonated N–H group [39]. In addition, N–H presents possibilities to chemically coupling amino acids, polypeptides, or appropriate tethers to which other molecules can be attached to introduce cell-specificity. While substitution within the groups on the quaternary centers could achieve this end, some separation of the cell specific group from the electrostatic point of interaction with DNA/RNA favors location of the former outside any aggregate or lipoplex surface and is easier to signal with the cell surface.

Amino Acids-Substituted Gemini Surfactants. Modification of the 12–7NH–12 compound by chemically coupling amino acid moieties to the N–H group in the spacer was found to prolong and increase gene expression relative to 12–7NH–12 [40]. Amino acid (glycine and lysine)- and dipeptide (glycyl–lysine and lysyl–lysine)-substituted spacers of 12–7NH–12 were synthesized (Figure 4), and their efficiency of gene delivery in the presence of DOPE as a helper lipid was assessed in epithelial cells (COS-7, PAM 212, Sf 1 Ep). The surfactants having glycine and glycyl–lysine moieties showed consistently high gene expression, greater than the parent 12–7NH–12 in all cell lines and greater than Lipofectamine Plus^{TM} in COS-7 and Sf 1 Ep cell lines. Amino acid substitution on the gemini spacer introduces additional terminal amino acid groups that can positively affect the transfection efficiency of the delivery system [41], either by enhancing hydration of the molecules, leading to improved DNA interactions with the cell surface [42], or because the terminal amino groups, having theoretical pK_a values > 9.5, are highly protonated at physiological pH, they contribute to more efficient gemini–DNA electrostatic interactions and DNA compaction. The conformational orientation of amino acid molecules can drive liposomal fusion in a "flip-flop" manner by exposing hydrophobic regions after strong association with the negatively charged surface molecules of the cell [43–45].

23.4 CHARACTERIZATION OF GEMINI NANOPARTICLES

Physicochemical Characterizations. Optimization of the desired physicochemical characteristics of the gemini nanoparticles begins with a comprehensive characterization of the physical chemistry properties of the constituent components—for example, the gemini compounds themselves. Table 23.3 shows the values of important surfactant properties, such as CMC and area of head group (a_0) obtained from surface tension measurements at the air–water interface and pK_a values of surfactants for the various gemini series studied to date in our laboratory. The series having the substituted amino groups in the spacer (Table 23.3) [16] have pK_a values in the range 4.1–5.2. Their head group areas are slightly greater relative to analogues with no amino substitution in the spacer due to steric/conformational constraints imposed by the amino substituents, especially in the case of the 12–8N–12 and 12–7N–12 series. In the 12–s–12 series, the CMC has a maximum value at $s = 4$ and decreases thereafter as the length of the alkyl spacer increases. This is a result of the progressively lengthened and more flexible spacer being able to fold into the micelle core to minimize water contact. This behavior is consistent with the trend in a_0 values as a function of spacer length. Initially, a_0 increases with spacer length reaching a maximum value at approximately *s* equals 10 and as the spacer begins to more easily fold into the micelle core [14, 46, 47] a_0 decreases as s increases. The CMC values for the *m–3–m*, *m–7–m* and *m*–7NH–*m* series (Table 23.3) show two general trends. First, the CMC values for each series decrease with increasing alkyl tail length, consistent with trends reported in the literature [48]. Second,

TABLE 23.2. Gemini Surfactant Structures and Illustration of Surfactant–DNA Close-Match Interaction Models for the Different Inter-Nitrogen Distances

Surfactants	Atoms	$d_{\text{N-to-N}}$ (Å)	Model
***m*–3–*m* series** 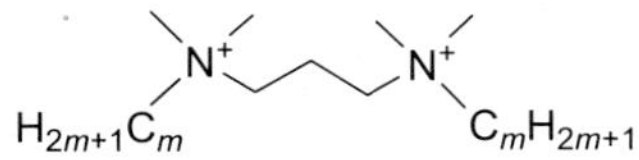 m = 12, 16, 18, 18:1	N1–N2	5.25	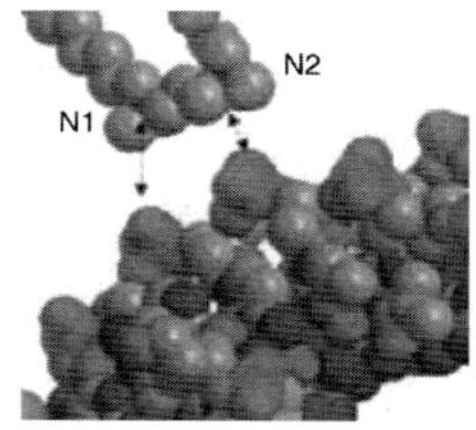
12–5N–12 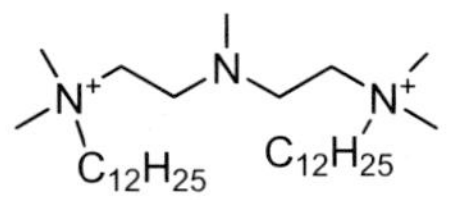	N1–N2	3.90	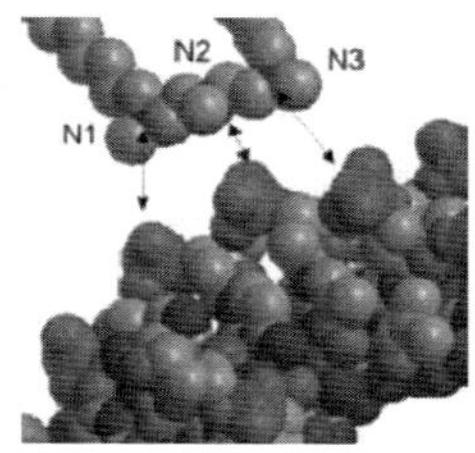
	N1–N3	7.52	
	N2–N3	3.88	
12–8N–12 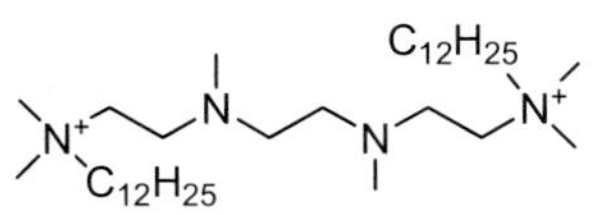	N1–N2	3.86	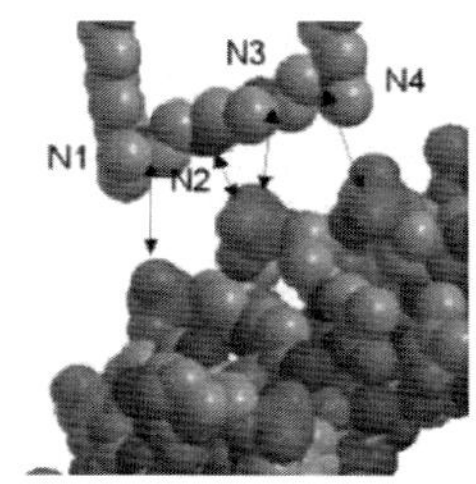
	N1–N3	7.40	
	N1–N4	11.17	
	N2–N3	3.86	
	N2–N4	7.42	
	N3–N4	3.86	
***m*–7NH–*m* series** 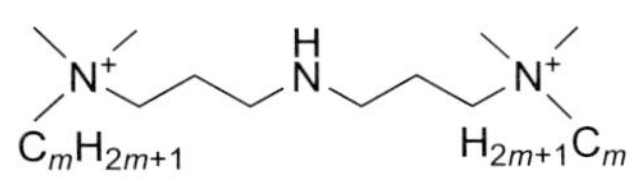 m = 12, 16, 18, 18:1	N1–N2	5.08	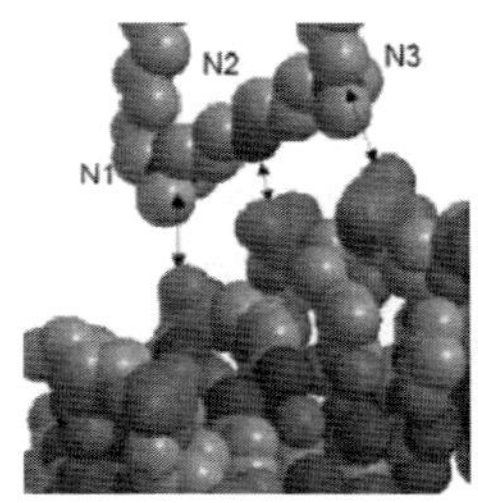
	N1–N3	10.15	
	N2–N3	5.08	
12–7N–12 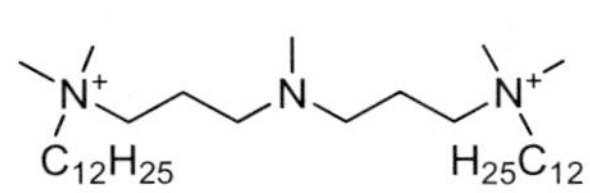	N1–N2	5.14	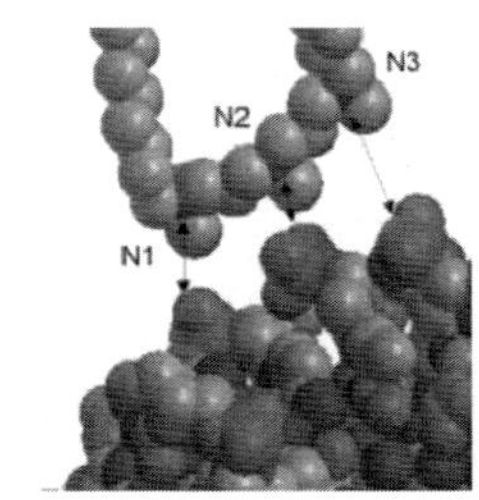
	N1–N3	10.18	
	N2–N3	5.14	
***m*–7–*m* series** 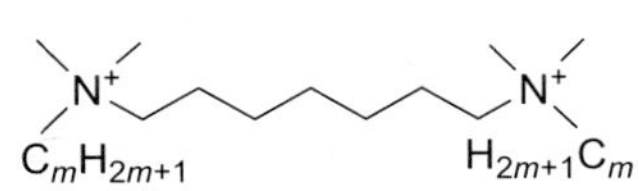 m = 12, 16, 18	N1–N2	10.08	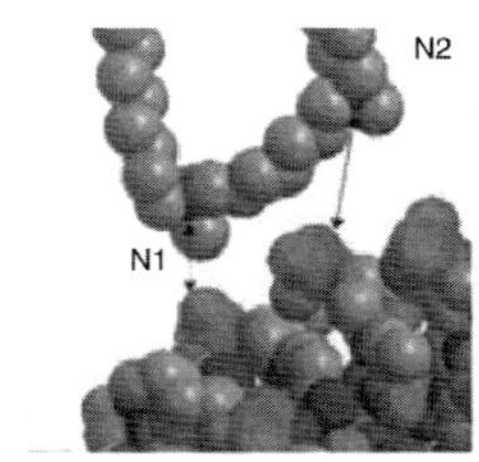

Calculated from Energy Minimized Model of Molecules using ChemOffice Chem 3D Pro Software. Molecules were energy minimized (MM2 force field) in the gas phase to a minimum root-mean-square gradient of 0.05 [39].

TABLE 23.3. Physicochemical Parameters of Gemini Surfactants Having Structural Variety

Gemini Surfactant	CMC [μM]	a_0 [nm^2 molecule]	pK_a	Reference
12–s–12 series				
12–3–12	980 ± 40	0.98	NA	14
12–4–12	1170 ± 40	1.10	NA	14
12–6–12	1090 ± 40	1.40	NA	14
12–8–12	840 ± 30	1.78	NA	14
12–10–12	620 ± 30	2.16	NA	14
12–12–12	360 ± 30	2.14	NA	14
12–16–12	120 ± 10	1.44	NA	14
12–spacer–12 series				
12–3–12	980 ± 40	0.98	NA	14
12–5N–12	1140 ± 40	1.30	5.2 ± 0.4	16
12–8N–12	1100 ± 100	1.95	4.1 ± 0.4	16
12–7N–12	1170 ± 40	2.06	4.5 ± 0.4	16
12–7NH–12	1290 ± 70	1.56	5.0 ± 0.4	16
***m*–7NH–*m* series**				
12–7NH–12	1290 ± 70	1.56	5.00 ± 0.4	16–49
16–7NH–16	160 ± 50	1.44	5.04 ± 0.3	49
18–7NH–18	13 ± 1	1.36	5.06 ± 0.3	49
18:1–7NH–18:1	44 ± 7	1.22	4.99 ± 0.2	
***m*–3–*m* series**				
12–3–12	980 ± 40	0.98	NA	14
16–3–16	30 ± 2	1.21	NA	46
18–3–18	13 ± 1	0.82	NA	49
18:1–3–18:1	15 ± 0.4	0.72	NA	94
***m*–7–*m* series**				
12–7–12	850 ± 70	1.24	NA	49
16–7–16	16 ± 9	1.19	NA	49
18–7–18	11 ± 1	1.26	NA	49
12–7N(AA)$_n$–12 series				
12–7NH–12	1290 ± 70	1.56	5.0 ± 0.4	16
12–7NG–12	2940 ± 30	ND	ND	40
12–7NK–12	3810 ± 30	ND	ND	40
12–7NGK–12	3720 ± 30	ND	ND	40
12–7NKK–12	7600 ± 200	ND	ND	40

CMC, critical micelle concentration.
a_0, head group area of one molecule at air–water interface.
NA, not applicable; the parameter is not expected in the compound.
ND, not determined; data for the parameter not available for the compound.

for gemini surfactants having the same alkyl tail length, the CMC decreases across the three series in the order: *m*–7NH–*m* > *m*–3–*m* >*m*–7–*m*. The reasons for this include: The *m*–7NH–*m* series has a greater aqueous solubility because of the amino moiety; the *m*–7–*m* series has a 7-methylene spacer that contributes to moderate hydrophobic character whereas the *m*–3–*m* series, because of the shorter methylene spacer, has a weaker hydrophobic character [49]. The CMC values of the 12–7N(AA)$_n$–12 series are significantly greater than the other series because the hydrophilic amino acid/dipeptide substituents on the spacer increase their aqueous solubility. The order of the CMC values follows 12–7NH–12 < 12–7NG–12 < 12–7NK–12 < 12–7NGK–12 < 12–7NKK–12 (Table 23.3).

Cytotoxicity. In the absence of plasmid or helper lipid (DOPE), gemini compounds generally exhibit high cytotoxicity with cell viability of 10–20%. Incorporating the gemini surfactants into the nanoparticles significantly increases cell viability [50]. The high cytotoxicity of the gemini compounds alone may be due to the fact that in the absence of DNA, the positively charged gemini derivatives can bind strongly to the negatively charged cell surface proteins impairing important membrane and other cellular functions [51]. The inclusion of amino acid substituents at the N–H center of the spacer in 12–7NH–12 improves transfection efficiency without increasing the cytotoxicity of the nanoparticles; this may occur because the amino acids conjugated to the gemini are inherently biocompatible.

23.5 TRANSFECTION PROPERTIES—STRUCTURE–ACTIVITY *IN VITRO* STUDIES

Structure of Gemini Surfactants. The greatest transfection efficiencies for the *m–s–m* series were recorded for spacer groups having $s \leq 4$ or $s > 12$ (Figure 23.6). The distance between nitrogen centers in gemini surfactants affects the interaction of the surfactant molecules with DNA and consequently affects transfection efficiency. Previous reports showed that an inter-nitrogen spacing of 4.9 Å allows suitable, close-matching electrostatic interaction between gemini surfactants and adjacent phosphate groups on DNA, which are 6.5–7.1 Å apart (Table 23.2) [52–54]. The short spacer $s = 3$, having a spacing of 5.23 Å, thus enables the best interaction with DNA phosphate groups, followed by the spacer $s = 4$, having a spacing of 6.21Å [55]. It is interesting also to note that very long spacers such as $s = 12$ and 16 are flexible and thus can fold so as to associate with the alkyl tails, thereby generating a shorter inter-nitrogen spacing for potentially enhancing DNA transfection. However, the improvement in transfection efficiency that is attributable to folding of long spacers ($s = 12$, 16) still falls below the transfection efficiencies obtained for short spacers, $s = 3$, 4.

The 12–7NH–12 surfactant presents a clear contrast to 12–7N–12 and the other N–CH_3 spacer-substituted surfactants due to the higher pH-sensitivity and lower steric hindrance stemming from having an –N(H)–(imino) group instead of an –N(CH_3)–substituent within the spacer. Thus, as expected, the *m*–7NH–*m* surfactants have yielded the most efficient transfections in both COS-7 and PAM 212 cells (Figures 23.7 and 23.8).

Variation of the alkyl tails alone produced minor enhancement in transfection efficiency compared to the stronger enhancement seen from modification of the head-group–spacer region. Statistically significant increases, as a result of alkyl tail variation, occurred only in the *m*–7NH–*m* group (Figure 23.8). The tails $m = 16$ or 18 yielded better results than $m = 12$, 18:1 ($p < 0.01$). Alkyl tail lengths in the gemini series are generally known to yield more efficient transfection when they fall within the range of 12–18 C atoms but not within the range of 10 C atoms or fewer [56]. However, there is an apparent debate in the literature as to what constitutes an optimal alkyl tail length (saturated or unsaturated, asymmetric or not) within the 12–18 C atoms range [20, 57]. For instance, a study of phospholipids has indicated that shorter alkyl tails allowed for better transfection in vitro, whereas longer tails were needed for better transfection in vivo [58]. Due to the generally moderate differences in transfection

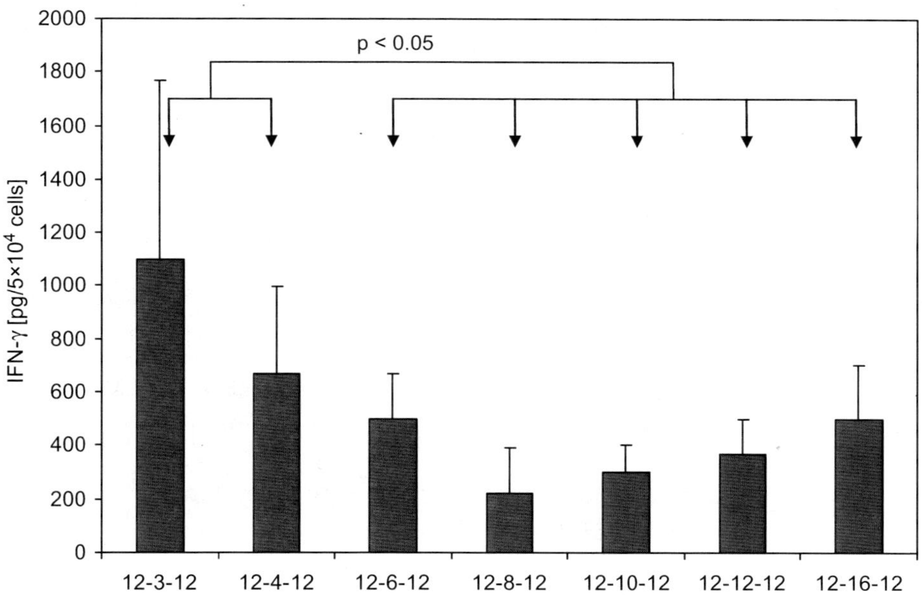

FIGURE 23.6. Gene expression in PAM 212 keratinocyte cells transfected with gemini nanoparticles containing a proportion of DOPE. Results are expressed as mean of triplicates ± SD ($n = 3$). Adapted from reference 23.

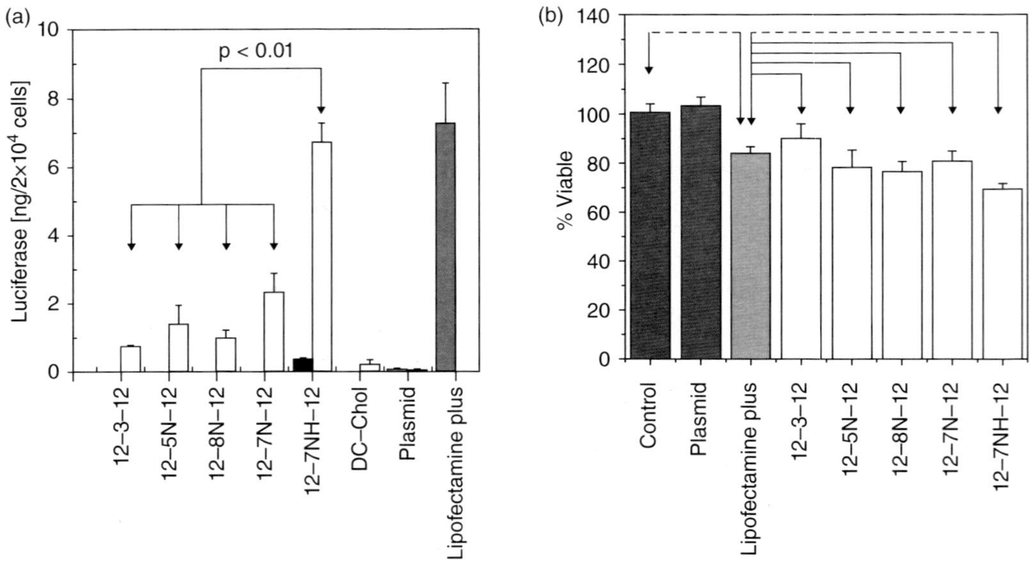

FIGURE 23.7. Gene expression and cytotoxicity results of COS-7 cells transfected with gemini nanoparticles. (**A**) Cell transfection using gemini nanoparticles with a component of DOPE (white bars) and without a component of DOPE (black bar). For both types of nanoparticles, $\rho_{+/-} = 10{:}1$. Results for controls: DC-Chol with (white) and without DOPE (black); plasmid with DOPE (white), plasmid without DOPE (black), and Lipofectamine Plus™ (shaded) are also shown. Results are expressed as mean measurements ± SD ($n = 6$). (**B**) The viability of cells determined after transfection as in panel A. Data are also presented for controls: untransfected cells (black bars), cells transfected using only plasmid (i.e., naked DNA; black bars), cells transfected with Lipofectamine Plus™ (gray bar). Columns connected by solid lines show no significant difference, and those connected with dashed lines show a significant difference ($p < 0.01$). In all cases, results are expressed as a mean of measurements ± SD ($n = 4$). Reproduced from reference 39, with permission.

efficiencies between alkyl tail lengths within the range C_{12}–C_{18}, advanced rational design efforts are more concentrated on implementing changes around the head-group–spacer region of the gemini surfactants targeted for gene delivery.

With regards to unsaturated alkyl tails, the 18:1–3–18:1 surfactant, made by incorporating mono-unsaturated oleyl tails in the *m*–3–*m* template, yielded the highest transfection for that series, whereas 18:1–7NH–18:1 created from the *m*–7NH–*m* template yielded the lowest transfection efficiency in its group (Figure 23.8). This apparent paradox interestingly fits within the context of recent literature because there have been reports that C_{18} tails are better than $C_{18:1}$ tails [56, 59] and other reports that find oleyl tails to be better than their saturated analogues [60–62]. Oleyl tails, when they result in higher transfection, are believed to do so through their enhancement of the membrane fluidity of transfection complexes, an effect correlated to endosomal escape [63]. On the other hand, a loss of the double bond via oxidation, either during transfection complex preparation or during storage, may reduce transfection efficiency [20].

The amino acid (glycine and lysine)- and dipeptide (glycyl–lysine and lysyl–lysine)-substituted spacers of the gemini surfactant showed a significant improvement in transfection in PAM 212 cells ($p < 0.001$) over the amino acid–free 12–7NH–12 surfactants (Figure 23.9). This result importantly demonstrates the 12–7N(AA)*n*–12 surfactants to be yet another potent class of surfactants for application to gene delivery.

Composition of the Nanoparticles. The addition of DOPE to the binary 12–7NH–12/plasmid system causes a significant increase in transfection due to the ability of DOPE to facilitate the release of the complexed DNA through membrane fusion and/or destabilization of the endosomal membrane [64, 65]. The charge ratio corresponding to $\rho_{+/-} = 10{:}1$; yielded the best transfection for all surfactants. Transfections involving lower $\rho_{+/-}$ values were at least 50% lower in luciferase expression and with similar cytotoxicity to complexes prepared at $\rho_{+/-} = 10{:}1$, thus, $\rho_{+/-} = 10{:}1$ may be regarded as an optimal ratio.

23.6 *IN VIVO* STUDIES

In this section, we focus on the application of cationic gemini nanoparticles as a gene delivery system for the treatment of localized scleroderma. Localized scleroderma differs from systemic sclerosis in that only skin and occasionally subcutaneous tissues are involved in the disease. Although

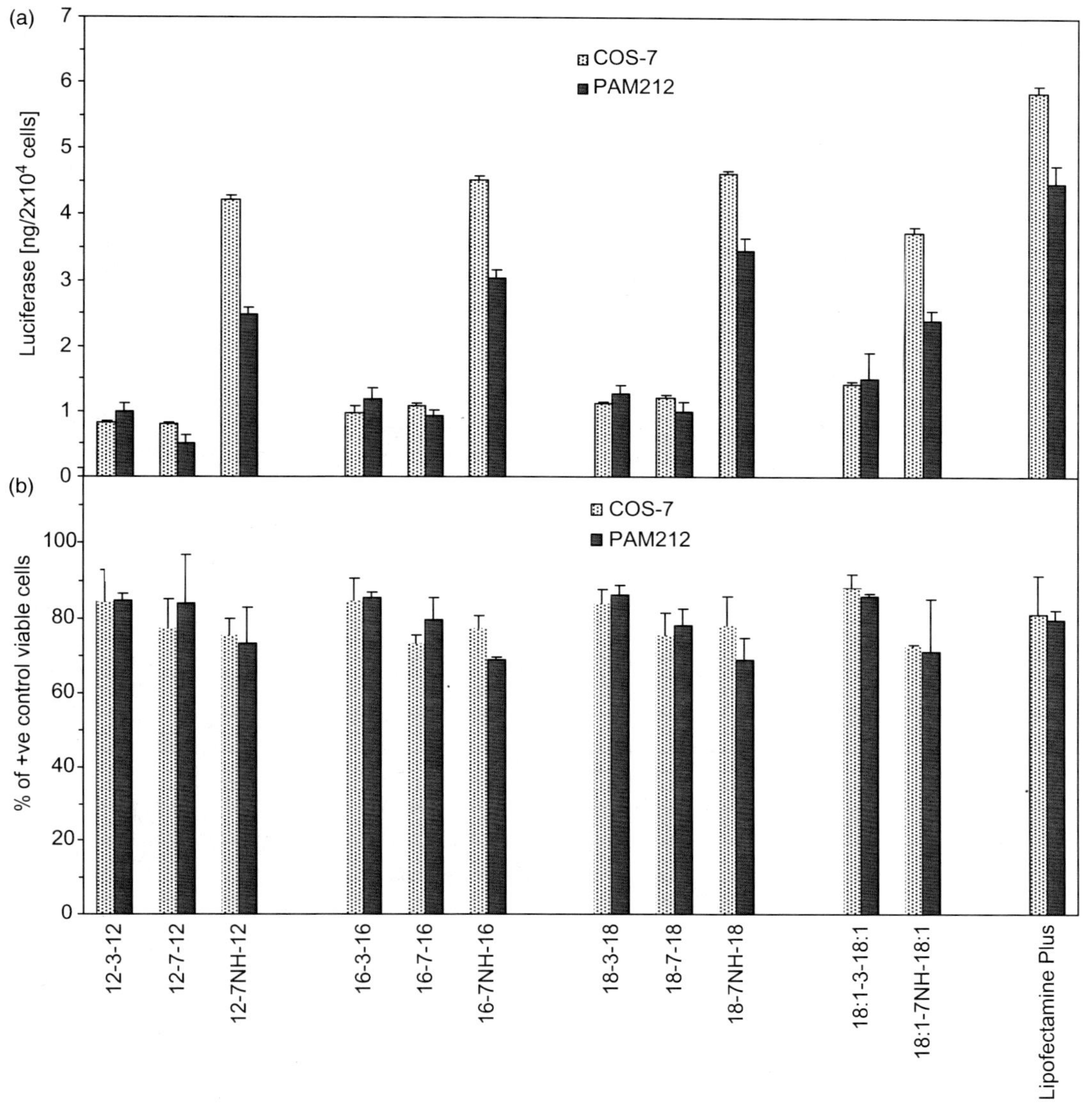

FIGURE 23.8. Gene expression and cytotoxicity in two epithelial cell lines transfected with gemini nanoparticles derived from *m*–3–*m*, *m*–7–*m*, *m*–7NH–*m* surfactants. (**A**) Gene expression in transfected PAM 212 and COS-7 cells. (**B**) Evaluation of cytotoxicity in transfected PAM 212 and COS-7 cells. The labels on the horizontal axis, from left to right, represent the bars from left to right as shown in parts A and B. Results are expressed as mean measurements ± SD ($n = 4$). Reproduced from reference [49], with permission.

rarely life-threatening, it can be disfiguring and disabling and, consequently, can adversely affect quality of life. Clinical signs of fibrosis are caused by sclerotic fibroblasts (myofibroblasts) in the dermis, which are capable of multiple passages, thus their accumulation might be responsible for the fibrosis characteristic of the disorder [66]. These fibroblasts produce excessive collagen even without an immune stimulus, suggesting the dysfunction of some regulatory genes associated with phenotypic selection. Overall, there is an indication of the benefit of using collagen synthesis inhibitors [67–69], even though the long-term effects are not always clear. IFNγ decreases collagen production in vitro in overproliferative fibroblasts from hypertrophic scars, and in vivo it improves the clinical signs of skin conditions such as scleroderma, morphea, and keloids [70, 71]. The limitation of the treatment by IFNγ in clinical trials is related to the nontargeted administration method. Subcutaneous or intramuscular injection of IFNγ does not provide sufficient levels of this cytokine within the specific target areas of the skin; therefore, the main challenge is the delivery and targeting of IFNγ to the epidermal and dermal layers of the skin.

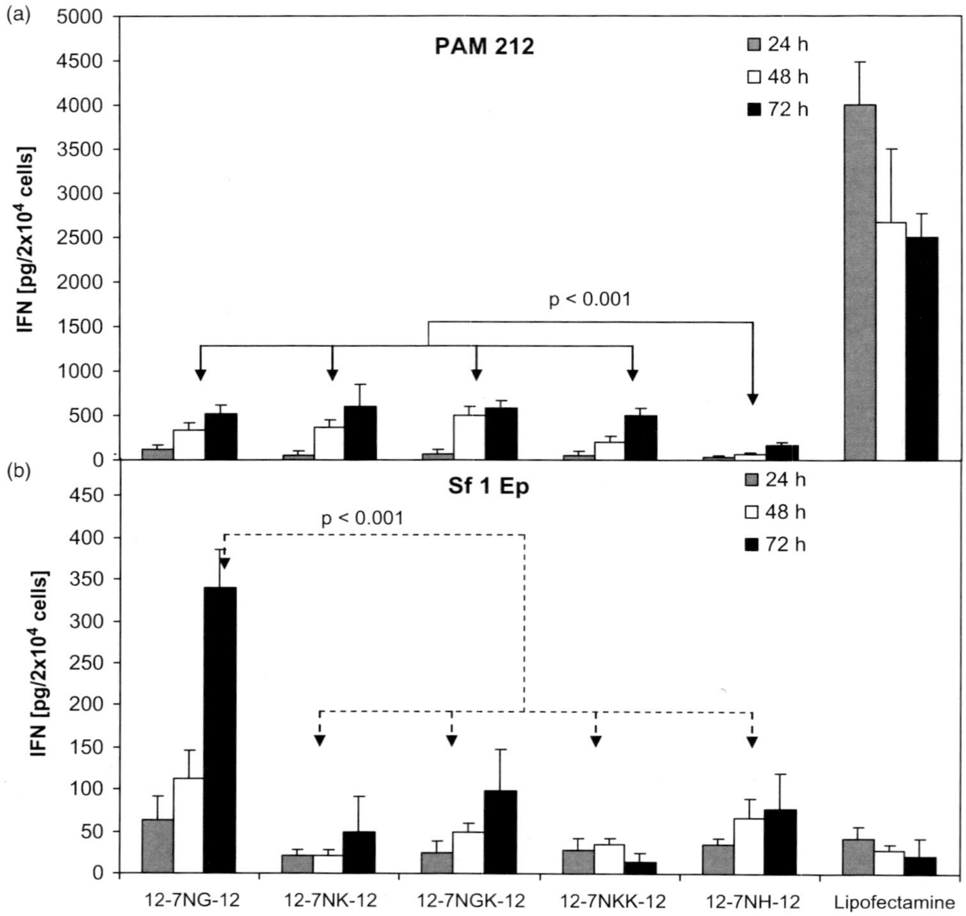

FIGURE 23.9. Gene expression in two epithelial cell types transfected with gemini nanoparticles derived from 12-7N(AA)n-12 surfactants. (**A**) PAM 212 cells; (**B**) Sf 1 Ep cells. The labels on the horizontal axis, from left to right, represent the bars from left to right as shown in parts A and B. Results are expressed as mean triplicates ± SD. Reproduced from reference 40, with permission.

Delivery of IFNγ to the skin could enhance the local concentration at the target site while minimizing the systemic exposure [72, 73]. Administration of plasmids coding for IFNγ could have several advantages over protein administration: (1) generation of the biologically active protein in the skin, using the bioreactor function of the keratinocytes, (2) increase of the half-life by sustained gene expression, (3) possibility of insertion of two or more genes in the same vector (i.e., IFNγ and Relaxin), and (4) reduction of systemic side effects, due to the dermal localization of the generated protein [74, 75]. In addition, topical treatment could avoid aggravating the lesions by invasive procedures, it could be self-administered by the patient, and the treatment could be easily terminated at the point where clinical assessment confirmed remission of the lesions. However, these advantages are contingent upon successful delivery of the DNA into the skin. The delivery of plasmids through intact skin has been a challenging task. Topical delivery of the DNA molecules through intact skin of mice was demonstrated using novel lipid-based delivery systems such as liposomes [4, 76] biphasic vesicles [77], cationic nanoparticles [78], ethanol-in-fluorocarbon microemulsion [79], or water-in-oil nanoemulsions [80].

Topical application of gemini cationic nanoparticles led to significantly higher IFNγ expression in the skin of CD1 mice compared to topical naked DNA (DNA-s-t) and blank nanoparticles, NP16-t (359.4 versus 135.69 and 105.87 pg IFN/cm^2) (Figure 23.10) [23]. Application of the gemini nanoparticles induced three fold higher levels of IFNγ than the 3ß-[*N*-(*N*′,*N*′-dimethylaminoethane)-carbamoyl]cholesterol hydrochloride (Dc-chol)-based formulation (NPDc-DNA-t). Dc-chol was selected as a control for the in vivo studies since laboratory [81, 82] and clinical trials [83] showed its ability to deliver pDNA in vivo in various tissues.

Following the preliminary studies in CD1 mice [23], where background levels of IFNγ can contribute to the overestimation of our therapeutic IFNγ gene, we evaluated

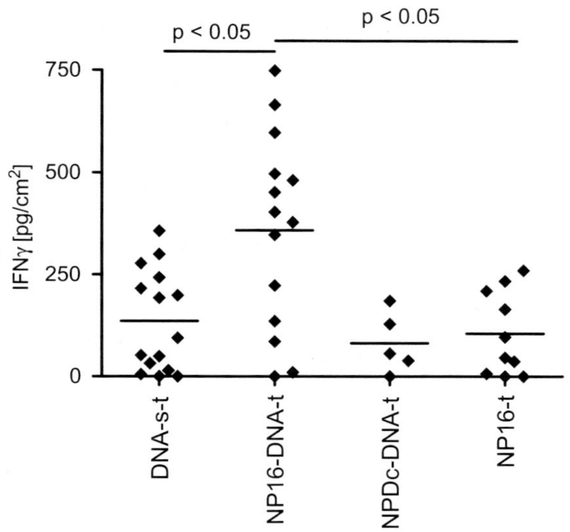

FIGURE 23.10. IFNγ expression in the skin of CD-1 mice. Adapted from reference 23.

gene delivery in a more suitable model, the IFNγ-deficient mouse [84]. Topical treatment with the gemini nanoparticles in the IFNγ-deficient mice resulted in significantly higher levels of IFN-γ expression in the skin (278 pg/cm^2) compared to the untreated animals ($p < 0.001$) (Figure 23.11). The NP16-DNA formulations prepared with the micelle-forming gemini surfactant showed higher transgene expression, compared to the NPDc nanoparticles prepared with the bilayer stabilizing cholesterol derivative. The SAXS profile of the NP16-DNA formulation show a complex polymorphic structure: a dominant lamellar structure and *Fd3m* cubic phase [85]. Interestingly, the NPDc-DNA formulation does not show the same structure, but rather

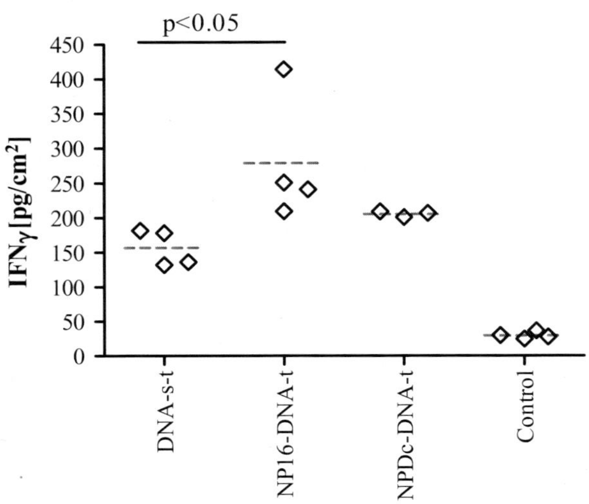

FIGURE 23.11. IFNγ expression in the skin of IFNγ-deficient mice, reproduced from reference 84, with permission.

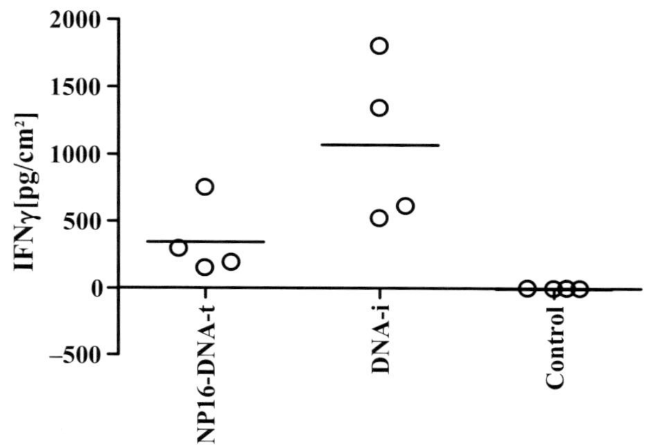

FIGURE 23.12. IFNγ expression in the skin of Tsk/+ mice. Adapted from reference 86.

appears to consist of two lamellar phases (L_a and L_b). We hypothesized, based on in vitro transfection data, that the efficiency of the gemini surfactants depended on the ability of the system to assume mixed polymorphic (lamellar–cubic–hexagonal) phases. The greater in vivo efficiency of the structurally complex NP16 nanoparticles (with lamellar-cubic phases) compared to the only lamellar-phase forming Dc-chol suggests that the same criterion is valid in vivo as well.

To study the cutaneous modifications in scleroderma, the tight skin-1 (Tsk/+) mouse model has been proposed due to increased collagen synthesis in the skin and accumulation in the dermis. IFNγ-coding plasmid was administered either by intradermal injection of the plasmid solution or topically on the skin using the gemini nanoparticle formulation [86]. The 20-day treatment consisted of either (a) topical treatment with the nanoparticles (NP16-DNA-t) once a day (plasmid dose 100 μg) or (b) two intradermal injections (DNA-i), weekly, for four weeks (plasmid dose 2.5 μg per site). Both topical nanoparticle and intradermal treatments increased IFNγ expression in the skin of Tsk/+ mice (Figure 23.12). Intradermal injections with plasmid solutions appeared to produce higher transgene expression (1069 ± 612 pg/cm^2), compared with topical application of the nanoparticles (345 ± 275 pg/cm^2), with no statistically significant difference. No IFNγ was detected in the serum of any treatment groups.

We have demonstrated that in-situ-generated IFNγ from pDNA injected or in the topically applied formulations can significantly reduce collagen levels in a Tsk/+ mouse model. The absence of IFNγ in the serum might suggest minimal systemic exposure to the protein; hence, side effects previously observed in the clinical trials could be avoided by topical administration of pDNA via the nanoparticulate delivery system. The present data suggest that topical application of IFNγ-coding pDNA using cationic gemini delivery vehicles significantly enhances cutaneous

IFNγ levels after application to intact skin and appears to be comparable or more efficient than other non-viral delivery systems tested in vivo [87–89]. Additionally, its effect on collagen synthesis indicates the feasibility of this application to treat cutaneous involvement of scleroderma in humans.

23.7 NANOPARTICLE DESIGN FOR SUBCELLULAR INTELLIGENCE

The rapid degradation of nucleic acids by intra- and extracellular nucleases and the limited cellular and nuclear uptake limit the effectiveness of gene therapy products (Figure 23.1). Foldvari and co-workers have taken a bottom-up approach to design gemini nanoparticles for subcellular intelligence, which means to engineer nanoparticles capable of overcoming the different cellular (e.g., specificity, toxicity, and uptake) and subcellular (lysosomal and cytoplasmic degradation and inefficient nuclear translocation) barriers. To achieve these goals, different constituents need to be incorporated into the nanoparticles (i.e., multifunctional nanoparticles). Gemini nanoparticles are composed of gemini surfactants and nucleic acids with or without neutral lipid vesicles. The nucleic acids are selected based on the target gene or the desired protein to be expressed, depending on the aimed therapeutic benefits. The incorporation of neutral lipids (e.g., DOPE) decreases the toxicity of the nanoparticles probably by decreasing the charge density and, importantly, facilitates the endosomal escape of the nanoparticles through membrane fusion and/or destabilization of the endosomal membrane. The selection of cationic gemini surfactant was essential to compact the nucleic acids, which is a prerequisite for cellular binding and endocytosis, and to protect the nucleic acids against intra- and extracellular enzymatic degradation. For the optimal stability of nucleic acids, highest cellular uptake and least cytotoxicity, $\rho_{+/-}$ around 10 was selected. Alkyl chain length of $\geq C_{12}$ and spacer $\leq C_4$ resulted in optimized electrostatic interaction with the nucleic acids at low concentrations resulting in both high transfection and low cytotoxicity. Spacer substitution with N–CH_3 or N–H moieties separated by three methylene segments further enhanced the transfection efficiency by increasing the binding of the surfactant to DNA/RNA (i.e., better compaction of the nucleic acids). The significant increase in transfection of the NH– over N–CH_3-based gemini surfactants is due to the lower steric hindrance and the pH-dependent change in morphology, which can play a role in the endosomal destabilization. In addition, N–H could be used to introduce amino acids or polypeptides. These amino acid-modified surfactants showed higher transfection efficiency and lower cytotoxicity than the parent unmodified ones. The prospective research will involve using these active centers (i.e., NH group) to chemically couple targeting moieties to permit active targeting and nuclear localization. Active targeting can be achieved via conjugation of targeting ligand to the carrier so as to favor its recognition by cell membrane-bound receptor proteins, which enhance the uptake by the targeted (e.g., diseased) cells while decreasing the toxicity to other cells (e.g., healthy tissues). These targeting moieties can be also used as nuclear localization sequences to allow for the active transport of nucleic acids through the nuclear pore complex [90]. These sequences are composed of short amino acid sequences that behave as specific targeting signals to enable the active transport to the nucleus, which enhances the protein-expression efficiency, in case of using pDNA [91]. Functionalization of gemini surfactants with fluorescent probes (e.g., pyrene) can also provide useful information on the intracellular fate of the nanoparticles, which aids in designing nanoparticles for subcellular intelligence. Furthermore, the speculated mechanism of intercalation between the pyrene-substituted gemini surfactants and nucleic acids may modulate the complexation fashion and affect the intracellular release of the nucleic acids cargos [35].

23.8 SUMMARY

Cationic gemini surfactants have become of widespread interest to develop nonviral delivery systems for gene therapy due to their ability to compact nucleic acids, in the presence or absence of helper lipids, into nanometer-sized particles and facilitate their intracellular delivery. These surfactants have rather unique solution properties imparted by their molecular structure as compared to their monomeric counterparts. The self-assembly into various aggregate structures and transfection efficiency of these surfactants can be effectively adjusted through tailoring their molecular structures. In addition, certain functional groups, such as amino groups or amino acid groups, can be incorporated into the template to greatly augment the characteristic molecular properties or trigger site-specific targeted delivery. Modifications of the chemistry and architecture of the gemini surfactants allow a "fine-tuning" of the physicochemical behavior, which in turn modulates the binding and compaction of DNA and ultimately plays a major role in the DNA release in cells. Other factors, such as the type of the lipids and charge ratio, dictate the transfection properties of these nanoparticles. We have taken a bottom-up approach to design nonviral gemini delivery systems that allows component-by-component testing, development, and optimization of safer and more effective nonviral gene delivery systems. Additionally, in-situ-generated IFNγ from topically applied gemini nanoparticles could significantly reduce collagen levels in a scleroderma mouse model. The gemini nanoparticles reported in this chapter can provide potential nonviral delivery carriers for topical gene therapy applications.

ACKNOWLEDGMENTS

Grant support received from the Natural Sciences and Engineering Research Council of Canada (NSERC) and Canadian Institutes of Health Research (CIHR) is gratefully acknowledged.

REFERENCES

1. Elsabahy, M., Nazarali, A., Foldvari, M. (2011). Non-viral nucleic acid delivery: Key challenges and future directions. *Current Drug Delivery*, **8**, 235–244.
2. Wilson, D. S., Dalmasso, G., Wang, L., Sitaraman, S. V., Merlin, D., Murthy, N. (2010). Orally delivered thioketal nanoparticles loaded with TNF-alpha-siRNA target inflammation and inhibit gene expression in the intestines. *Nature Materials*, **9**, 923–928.
3. Rodrigo-Garzon, M., Berraondo, P., Ochoa, L., Zulueta, J. J., Gonzalez-Aseguinolaza, G. (2010). Antitumoral efficacy of DNA nanoparticles in murine models of lung cancer and pulmonary metastasis. *Cancer Gene Therapy*, **17**, 20–27.
4. Li, J., Li, X., Zhang, Y., Zhou, X. K., Yang, H. S., Chen, X. C., Wang, Y. S., Wei, Y. Q., Chen, L. J., Hu, H. Z., Liu, C. Y. (2010). Gene therapy for psoriasis in the K14-VEGF transgenic mouse model by topical transdermal delivery of interleukin-4 using ultradeformable cationic liposome. *Journal of Gene Medicine*, **12**, 481–490.
5. Semple, S. C., Akinc, A., Chen, J., Sandhu, A. P., Mui, B. L., Cho, C. K., Sah, D. W., Stebbing, D., Crosley, E. J., Yaworski, E., Hafez, I. M., Dorkin, J. R., Qin, J., Lam, K., Rajeev, K. G., Wong, K. F., Jeffs, L. B., Nechev, L., Eisenhardt, M. L., Jayaraman, M., Kazem, M., Maier, M. A., Srinivasulu, M., Weinstein, M. J., Chen, Q., Alvarez, R., Barros, S. A., De, S., Klimuk, S. K., Borland, T., Kosovrasti, V., Cantley, W. L., Tam, Y. K., Manoharan, M., Ciufolini, M. A., Tracy, M. A., de Fougerolles, A., MacLachlan, I., Cullis, P. R., Madden, T. D., Hope, M. J. (2010). Rational design of cationic lipids for siRNA delivery. *Nature Biotechnology*, **28**, 172–176.
6. Donkuru, M., Badea, I., Wettig, S., Verrall, R., Elsabahy, M., Foldvari, M. (2010). Advancing nonviral gene delivery: Lipid- and surfactant-based nanoparticle design strategies. *Nanomedicine (London)*, **5**, 1103–1127.
7. Lehrman, S. (1999). Virus treatment questioned after gene therapy death. *Nature*, **401**, 517–518.
8. Young, L. S., Searle, P. F., Onion, D., Mautner, V. (2006). Viral gene therapy strategies: From basic science to clinical application. *Journal of Pathology*, **208**, 299–318.
9. Waehler, R., Russell, S. J., Curiel, D. T. (2007). Engineering targeted viral vectors for gene therapy. *Nature Reviews Genetics*, **8**, 573–587.
10. Felgner, P. L., Gadek, T. R., Holm, M., Roman, R., Chan, H. W., Wenz, M., Northrop, J. P., Ringold, G. M., Danielsen, M. (1987). Lipofection: A highly efficient, lipid-mediated DNA-transfection procedure. *Proceedings of the National Academy of Science USA*, **84**, 7413–7417.
11. Menger, F. M., Keiper, J. S. (2000). Gemini surfactants. *Angewandte Chemie International Edition in English*, **39**, 1906–1920.
12. Menger, F. M., Mbadugha, B. N. (2001). Gemini surfactants with a disaccharide spacer. *Journal of the American Chemical Society*, **123**, 875–885.
13. Ikeda, J. (2004). *Synthesis of Gemini (Dimeric) and Related Surfactants*. Marcel Dekker, New York, pp. 9–33.
14. Wettig, S. D., Verrall, R. E. (2001). Thermodynamic studies of aqueous *m–s–m* gemini surfactant systems. *Journal of Colloid and Interface Science*, **235**, 310–316.
15. Zana, R., Benrraou, M., Rueff, R. (1991). Alkanediyl-α,ω-bis(dimethylalkylammonium bromide surfactants 1. Effect of the spacer chain length on the critical micelle concentration and micelle ionization degree. *Langmuir*, **7**, 1072–1075.
16. Wettig, S. D., Wang, C., Verrall, R. E., Foldvari, M. (2007). Thermodynamic and aggregation properties of aza- and imino-substituted gemini surfactants designed for gene delivery. *Physical Chemistry Chemical Physics*, **9**, 871–877.
17. Wettig, S. D., Nowak, P., Verrall, R. E. (2002). Thermodynamic and aggregation properties of gemini surfactants with hydroxyl substituted spacers in aqueous solution. *Langmuir*, **18**, 5354–5359.
18. Wang, C., Wettig, S. D., Foldvari, M., Verrall, R. E. (2007). Synthesis, characterization, and use of asymmetric pyrenyl-gemini surfactants as emissive components in DNA–lipoplex systems. *Langmuir*, **23**, 8995–9001.
19. Templeton, N. S. (2001). Developments in liposomal gene delivery systems. *Expert Opinion on Biological Therapy*, **1**, 567–570.
20. Niculescu-Duvaz, D., Heyes, J., Springer, C. J. (2003). Structure–activity relationship in cationic lipid mediated gene transfection. *Current Medicinal Chemistry*, **10**, 1233–1261.
21. Kirby, A. J., Camilleri, P., Engberts, J. B., Feiters, M. C., Nolte, R. J., Soderman, O., Bergsma, M., Bell, P. C., Fielden, M. L., Garcia Rodriguez, C. L., Guedat, P., Kremer, A., McGregor, C., Perrin, C., Ronsin, G., van Eijk, M. C. (2003). Gemini surfactants: New synthetic vectors for gene transfection. *Angewandte Chemie International Edition in English*, **42**, 1448–1457.
22. Tranchant, I., Thompson, B., Nicolazzi, C., Mignet, N., Scherman, D. (2004). Physicochemical optimisation of plasmid delivery by cationic lipids. *Journal of Gene Medicine*, **6**, S24–35.
23. Badea, I., Verrall, R., Baca-Estrada, M., Tikoo, S., Rosenberg, A., Kumar, P., Foldvari, M. (2005). In vivo cutaneous interferon-gamma gene delivery using novel dicationic (gemini) surfactant-plasmid complexes. *Journal of Gene Medicine*, **7**, 1200–1214.
24. Oda, R., Huc, I., Schmutz, M., Candau, S. J., MacKintosh, F. C. (1999). Tuning bilayer twist using chiral counterions. *Nature*, **399**, 566–569.
25. Oda, R., Huc, I., Candau, S. J. (1997). Gemini surfactants, the effect of hydrophobic chain length and dissymmetry. *Chemical Communications*, 2105–2106.
26. Oda, R., Huc, I., Homo, J. C., Heinrich, B., Schmutz, M., Candau, S. (1999). Elongated aggregates formed

by cationic gemini surfactants. *Langmuir*, **15**, 2384–2390.

27. Huc, I., Oda, R. (1999). Gemini surfactants: Studying micellisation by 1H and 19F NMR spectroscopy. *Chemical Communications*, 2025–2026.
28. Bai, G., Wang, J., Wang, Y., Ye, J., Yan, H., Thomas, R. K. (2002). Thermodynamics of hydrophobic interaction of dissymmetric gemini surfactants in aqueous solutions. *Journal of Physical Chemistry B*, **106**, 6614–6616.
29. Wang, X., Wang, J., Wang, Y., Ye, J., Yan, H., Thomas, R. K. (2003). Micellization of a series of dissymmetric gemini surfactants in aqueous solution. *Journal of Physical Chemistry B*, **107**, 11428–11432.
30. Fan, Y., Li, Y., Cao, M., Wang, J., Wang, Y., Thomas, R. K. (2007). Micellization of dissymmetric cationic gemini surfactants and their interaction with dimyristoylphosphatidylcholine vesicles. *Langmuir*, **23**, 11458–11464.
31. Jiang, N., Wang, J., Wang, Y., Yan, H., Thomas, R. K. (2005). Microcalorimetric study on the interaction of dissymmetric gemini surfactants with DNA. *Journal of Colloid and Interface Science*, **284**, 759–764.
32. Yamana, K., Iwase, R., Furutani, S., Tsuchida, H., Zako, H., Yamaoka, T., Murakami, A. (1999). 2′-Pyrene modified oligonucleotide provides a highly sensitive fluorescent probe of RNA. *Nucleic Acids Research*, **27**, 2387–2392.
33. Nakamura, M., Fukunaga, Y., Sasa, K., Ohtoshi, Y., Kanaori, K., Hayashi, H., Nakano, H., Yamana, K. (2005). Pyrene is highly emissive when attached to the RNA duplex but not to the DNA duplex: The structural basis of this difference. *Nucleic Acids Research*, **33**, 5887–5895.
34. Kawai, T., Ikegami, M., Arai, T. (2004). Exciplex formation between pyrene and guanine in highly polar solvents. *Chemical Communications*, 824–825.
35. Wettig, S. D., Deubry, R., Akbar, J., Kaur, T., Wang, H., Sheinin, T., Joseph, J. W., Slavcev, R. A. (2010). Thermodynamic investigation of the binding of dissymmetric pyrenyl-gemini surfactants to DNA. *Physical Chemistry Chemical Physics*, **12**, 4821–4826.
36. Wettig, S. D., Verrall, R. E., Foldvari, M. (2008). Gemini surfactants: A new family of building blocks for non-viral gene delivery systems. *Current Gene Therapy*, **8**, 9–23.
37. Vijayanathan, V., Thomas, T., Shirahata, A., Thomas, T. J. (2001). DNA condensation by polyamines: A laser light scattering study of structural effects. *Biochemistry*, **40**, 13644–13651.
38. Ruiz-Chica, J., Medina, M. A., Sanchez-Jimenez, F., Ramirez, F. J. (2001). Fourier transform Raman study of the structural specificities on the interaction between DNA and biogenic polyamines. *Biophysical Journal*, **80**, 443–454.
39. Wettig, S. D., Badea, I., Donkuru, M., Verrall, R. E., Foldvari, M. (2007). Structural and transfection properties of amine-substituted gemini surfactant-based nanoparticles. *Journal of Gene Medicine*, **9**, 649–658.
40. Yang, P., Singh, J., Wettig, S., Foldvari, M., Verrall, R. E., Badea, I. (2010). Enhanced gene expression in epithelial cells transfected with amino acid-substituted gemini nanoparticles. *European Journal of Pharmaceutics and Biopharmaceutics*, **75**, 311–320.
41. Wheeler, C. J., Sukhu, L., Yang, G., Tsai, Y., Bustamente, C., Felgner, P., Norman, J., Manthorpe, M. (1996). Converting an alcohol to an amine in a cationic lipid dramatically alters the co-lipid requirement, cellular transfection activity and the ultrastructure of DNA–cytofectin complexes. *Biochimica et Biophysica Acta*, **1280**, 1–11.
42. Bennett, M. J., Aberle, A. M., Balasubramaniam, R. P., Malone, J. G., Nantz, M. H., Malone, R. W. (1996). Considerations for the design of improved cationic amphiphile-based transfection reagents. *Journal of Liposome Research*, **6**, 545–565.
43. Hong, S., Leroueil, P. R., Janus, E. K., Peters, J. L., Kober, M. M., Islam, M. T., Orr, B. G., Baker, J. R., Jr., Banaszak Holl, M. M. (2006). Interaction of polycationic polymers with supported lipid bilayers and cells: Nanoscale hole formation and enhanced membrane permeability. *Bioconjugate Chemistry*, **17**, 728–734.
44. Mecke, A., Majoros, I. J., Patri, A. K., Baker, J. R., Jr., Holl, M. M., Orr, B. G. (2005). Lipid bilayer disruption by polycationic polymers: The roles of size and chemical functional group. *Langmuir*, **21**, 10348–10354.
45. Tribet, C., Vial, F. (2008). Flexible macromolecules attached to lipid bilayers: Impact on fluidity, curvature, permeability and stability of the membranes. *Soft Matter*, **4**, 68–81.
46. Alami, E., Beinert, G., Marie, P., Zana, R. (1993). Alkanediyl-.alpha.,.omega.-bis(dimethylalkylammonium bromide) surfactants. 3. Behavior at the air–water interface. *Langmuir*, **9**, 1465–1467.
47. Diamant, H., Andelman, D. (1994). Dimeric surfactants: Spacer chain conformation and specific area at the air/water interface. *Langmuir*, **10**, 2910–2916.
48. Zana, R., Benrraou, M., Rueff, R. (1991). Alkanediyl-a,w-bis(dimethylalkylammonium bromide) surfactants. 1. Effect of the spacer chain length on the critical micelle concentration and micelle ionization degree. *Langmuir*, **7**, 1072–1075.
49. Donkuru, M., Wettig, S. D., Verrall, R. E., Badea, I., Foldvari, M. (2012). Designing pH-sensitive gemini nanoparticles for non-viral gene delivery into keratinocytes. *Journal of Material Chemistry*, **22**, 6232–6244.
50. Singh, J., Yang, P., Michel, D., Verrall, R., Foldvari, M., Badea, I. (2011). Amino-acid substituted gemini surfactant-based nanoparticles as safe and versatile gene delivery agents. *Current Drug Delivery*, **8**, 299–306.
51. Schuber, F. (1989). Influence of polyamines on membrane functions. *Biochemical Journal*, **260**, 1–10.
52. Saenger, W. (1984). *Principles of Nucleic Acid Structure*. Springer-Verlag, New York.
53. Morgan, J. E., Blankenship, J. W., Matthews, H. R. (1987). Polyamines and acetylpolyamines increase the stability and alter the conformation of nucleosome core particles. *Biochemistry*, **26**, 3643–3649.
54. Ruiz-Chica, J., Medina, M. A., Sanchez-Jimenez, F., Ramirez, F. J. (2001). Fourier transform Raman study of the structural specificities on the interaction between DNA and biogenic polyamines. *Biophysical Journal*, **80**, 443–454.

55. Menger, F. M., Keiper, J. S., Azov, V. (2000). Gemini surfactants with acetylenic spacers. *Langmuir*, **16**, 2062–2067.

56. Castro, M., Griffiths, D., Patel, A., Pattrick, N., Kitson, C., Ladlow, M. (2004). Effect of chain length on transfection properties of spermine-based gemini surfactants. *Organic and Biomolecular Chemistry*, **2**, 2814–2820.

57. Balasubramaniam, R. P., Bennett, M. J., Aberle, A. M., Malone, J. G., Nantz, M. H., Malone, R. W. (1996). Structural and functional analysis of cationic transfection lipids: The hydrophobic domain. *Gene Therapy*, **3**, 163–172.

58. Floch, V., Loisel, S., Guenin, E., Herve, A. C., Clement, J. C., Yaouanc, J. J., des Abbayes, H., Ferec, C. (2000). Cation substitution in cationic phosphonolipids: A new concept to improve transfection activity and decrease cellular toxicity. *Journal of Medicinal Chemistry*, **43**, 4617–4628.

59. Roosjen, A., Smisterova, J., Driessen, C., Anders, J. T., Wagenaar, A., Hoekstra, D., Hulst, R., Engberts, J.B.F.N. (2002). Synthesis and characteristics of biodegradable pyridinium amphiphiles used for in vitro DNA delivery. *European Journal of Organic Chemistry*, 1271–1277.

60. McGregor, C., Perrin, C., Monck, M., Camilleri, P., Kirby, A. J. (2001). Rational approaches to the design of cationic gemini surfactants for gene delivery. *Journal of the American Chemical Society*, **123**, 6215–6220.

61. Fielden, M. L., Perrin, C., Kremer, A., Bergsma, M., Stuart, M. C., Camilleri, P., Engberts, J. B. F. N. (2001). Sugar-based tertiary amino gemini surfactants with a vesicle-to-micelle transition in the endosomal pH range mediate efficient transfection in vitro. *European Journal of Biochemistry*, **268**, 1269–1279.

62. Delepine, P., Guillaume, C., Floch, V., Loisel, S., Yaouanc, J. J., Clement, J. C., Des Abbayes, H., Ferec, C. (2000). Cationic phosphonolipids as nonviral vectors: In vitro and in vivo applications. *Journal of Pharmaceutical Sciences*, **89**, 629–638.

63. Hyvonen, Z., Ruponen, M., Ronkko, S., Suhonen, P., Urtti, A. (2002). Extracellular and intracellular factors influencing gene transfection mediated by 1,4-dihydropyridine amphiphiles. *European Journal of Pharmaceutical Sciences*, **15**, 449–460.

64. Dass, C. R. (2002). Biochemical and biophysical characteristics of lipoplexes pertinent to solid tumor gene therapy. *International Journal of Pharmaceutics*, **241**, 1–25.

65. Chesnoy, S., Huang, L. (2000). Structure and function of lipid–DNA complexes for gene delivery. *Annual Review of Biophysics and Biomolecular Structure*, **29**, 27–47.

66. Roumm, A. D., Whiteside, T. L., Medsger, T. A., Jr., Rodnan, G. P. (1984). Lymphocytes in the skin of patients with progressive systemic sclerosis. Quantification, subtyping, and clinical correlations. *Arthritis and Rheumatism*, **27**, 645–653.

67. Seibold, J. R., Korn, J. H., Simms, R., Clements, P. J., Moreland, L. W., Mayes, M. D., Furst, D. E., Rothfield, N., Steen, V., Weisman, M., Collier, D., Wigley, F. M., Merkel, P. A., Csuka, M. E., Hsu, V., Rocco, S., Erikson, M., Hannigan, J., Harkonen, W. S., Sanders, M. E. (2000). Recombinant human relaxin in the treatment of scleroderma. A randomized, double-blind, placebo-controlled trial. *Annals of Internal Medicine*, **132**, 871–879.

68. Nagler, A., Pines, M. (1999). Topical treatment of cutaneous chronic graft versus host disease with halofuginone: A novel inhibitor of collagen type I synthesis. *Transplantation*, **68**, 1806–1809.

69. Yamamoto, T., Takagawa, S., Kuroda, M., Nishioka, K. (2000). Effect of interferon-gamma on experimental scleroderma induced by bleomycin. *Archives of Dermatological Research*, **292**, 362–365.

70. Harrop, A. R., Ghahary, A., Scott, P. G., Forsyth, N., Uji-Friedland, A., Tredget, E. E. (1995). Regulation of collagen synthesis and mRNA expression in normal and hypertrophic scar fibroblasts in vitro by interferon-gamma. *Journal of Surgical Research*, **58**, 471–477.

71. Widom, R. L. (2000). Regulation of matrix biosynthesis and degradation in systemic sclerosis. *Current Opinion in Rheumatology*, **12**, 534–539.

72. du Plessis, J., Egbaria, K., Ramachandran, C., Weiner, N. (1992). Topical delivery of liposomally encapsulated gamma-interferon. *Antiviral Research*, **18**, 259–265.

73. Short, S. M., Paasch, B. D., Turner, J. H., Weiner, N., Daugherty, A. L., Mrsny, R. J. (1996). Percutaneous absorption of biologically-active interferon-gamma in a human skin graft-nude mouse model. *Pharmaceutical Research*, **13**, 1020–1027.

74. Meng, X., Sawamura, D., Ina, S., Tamai, K., Hanada, K., Hashimoto, I. (2002). Keratinocyte gene therapy: Cytokine gene expression in local keratinocytes and in circulation by introducing cytokine genes into skin. *Experimental Dermatology*, **11**, 456–461.

75. Badea, I., Taylor, M., Rosenberg, A., Foldvari, M. (2009). Pathogenesis and therapeutic approaches for improved topical treatment in localized scleroderma and systemic sclerosis. *Rheumatology (Oxford)*, **48**, 213–221.

76. Alexander, M. Y., Akhurst, R. J. (1995). Liposome-medicated gene transfer and expression via the skin. *Human Mol Genetics*, **4**, 2279–2285.

77. Babiuk, S., Baca-Estrada, M. E., Pontarollo, R., Foldvari, M. (2002). Topical delivery of plasmid DNA using biphasic lipid vesicles (Biphasix). *Journal of Pharmacy and Pharmacology*, **54**, 1609–1614.

78. Cui, Z., Mumper, R. J. (2002). Topical immunization using nanoengineered genetic vaccines. *Journal of Control Release*, **81**, 173–184.

79. Cui, Z., Fountain, W., Clark, M., Jay, M., Mumper, R. J. (2003). Novel ethanol-in-fluorocarbon microemulsions for topical genetic immunization. *Pharmaceutical Research*, **20**, 16–23.

80. Wu, H., Ramachandran, C., Bielinska, A. U., Kingzett, K., Sun, R., Weiner, N. D., Roessler, B. J. (2001). Topical transfection using plasmid DNA in a water-in-oil nanoemulsion. *International Journal of Pharmaceutics*, **221**, 23–34.

81. Caplen, N. J., Kinrade, E., Sorgi, F., Gao, X., Gruenert, D., Geddes, D., Coutelle, C., Huang, L., Alton, E. W., Williamson, R. (1995). In vitro liposome-mediated DNA transfection of epithelial cell lines using the cationic liposome DC-Chol/DOPE. *Gene Therapy*, **2**, 603–613.

82. Nomura, T., Yasuda, K., Yamada, T., Okamoto, S., Mahato, R. I., Watanabe, Y., Takakura, Y., Hashida, M. (1999). Gene expression and antitumor effects following direct interferon (IFN)-gamma gene transfer with naked plasmid DNA and DC-chol liposome complexes in mice. *Gene Therapy*, **6**, 121–129.

83. Gill, D. R., Southern, K. W., Mofford, K. A., Seddon, T., Huang, L., Sorgi, F., Thomson, A., MacVinish, L. J., Ratcliff, R., Bilton, D., Lane, D. J., Littlewood, J. M., Webb, A. K., Middleton, P. G., Colledge, W. H., Cuthbert, A. W., Evans, M. J., Higgins, C. F., Hyde, S. C. (1997). A placebo-controlled study of liposome-mediated gene transfer to the nasal epithelium of patients with cystic fibrosis. *Gene Therapy*, **4**, 199–209.

84. Badea, I., Wettig, S., Verrall, R., Foldvari, M. (2007). Topical non-invasive gene delivery using gemini nanoparticles in interferon-gamma-deficient mice. *European Journal of Pharmaceutics and Biopharmaceutics*, **65**, 414–422.

85. Winter, R., Kohling, R. (2004). Static and time-resolved synchrotron small-angle x-ray scattering studies of lyotropic lipid mesophases, model biomembranes and proteins in solution. *Journal of Physics Condensed Matter*, **16**, S327–S352.

86. Badea, I., Virtanen, C., Verrall, R., Rosenberg, A., Foldvari, M. (2011). Effect of topical interferon-γ gene therapy using gemini nanoparticles on pathophysiological markers of cutaneous scleroderma in Tsk/ +mice. Unpublished data.

87. Ciotti, S. N., Weiner, N. (2002). Follicular liposomal delivery systems. *Journal of Liposome Research*, **12**, 143–148.

88. Rakhmilevich, A. L., Turner, J., Ford, M. J., McCabe, D., Sun, W. H., Sondel, P. M., Grota, K., Yang, N. S. (1996). Gene gun-mediated skin transfection with interleukin 12 gene results in regression of established primary and metastatic murine tumors. *Proceedings of the National Academy of Sciences USA*, **93**, 6291–6296.

89. Wu, M. H., Yokozeki, H., Takagawa, S., Yamamoto, T., Satoh, T., Kaneda, Y., Katayama, I., Nishioka, K. (2004). Hepatocyte growth factor both prevents and ameliorates the symptoms of dermal sclerosis in a mouse model of scleroderma. *Gene Therapy*, **11**, 170–180.

90. Liu, G., Li, D., Pasumarthy, M. K., Kowalczyk, T. H., Gedeon, C. R., Hyatt, S. L., Payne, J. M., Miller, T. J., Brunovskis, P., Fink, T. L., Muhammad, O., Moen, R. C., Hanson, R. W., Cooper, M. J. (2003). Nanoparticles of compacted DNA transfect postmitotic cells. *Journal of Biological Chemistry*, **278**, 32578–32586.

91. Kalderon, D., Richardson, W. D., Markham, A. F., Smith, A. E. (1984). Sequence requirements for nuclear location of simian virus 40 large-T antigen. *Nature*, **311**, 33–38.

92. Foldvari, M., Elsabahy, M., Taheri-Araghi, S., Kohandel, M., Sivaloganathan, S., Badea, I., Verrall, R. (2011). Ultrastructural and molecular modeling study of plasmid–gemini surfactant-phospholipid nanoparticles and analysis of particle structure–transfection efficiency correlation. Unpublished data.

93. Foldvari, M., Badea, I., Wettig, S., Verrall, R., Bagonluri, M. (2006). Structural characterization of novel gemini non-viral DNA delivery systems for cutaneous gene therapy. *Journal of Experimental Nanoscience*, **1**, 165–176.

94. Li, X., Wettig, S. D., Wang, C., Foldvari, M., Verrall, R. E. (2005). Synthesis and solution properties of gemini surfactants containing oleyl chains. *Physical Chemistry Chemical Physics*, **7**, 3172–3178.

24

SENSING Hg(II) IONS IN WATER: FROM MOLECULES TO NANOSTRUCTURED MOLECULAR MATERIALS

Imma Ratera, Alberto Tárraga, Pedro Molina, and Jaume Veciana

24.1 INTRODUCTION

Mercury pollution is a global problem, and a major source of human exposure stems from contaminated natural water, although mercury bioaccumulation also occurs in plants, fishes, and other animals. Because of the increasing concern about health and environmental protection, the EU has approved a regulation where 1 ppb is the maximum allowed concentration for Hg(II) ions in drinkable water; this has stimulated the research of new mercury detection methods applicable to the environmental and biological milieus.

A wide variety of mercury determination techniques have been developed; the majority of these techniques are based on analytical instrumentation methods but are not well-suited for quick in-field detection of Hg(II) or for in vivo studies of Hg(II) in biology and toxicology. The use of mercury-responsive small-molecule ligands or hybrid molecularly nanostructured materials that provide immediate optical feedback can overcome such limitations because their use does not require sophisticated instrumentation or sample preparation [1].

Thus, this chapter focuses on present mercury sensors with an emphasis on newly developed photoactive molecules and nanomaterials which represent an exciting and truly revolutionary approach to mercury detection that benefit from the versatility of chemical synthesis.

The design of artificial chemosensors that have the ability to selectively bind and sense pollutant species, such as Hg(II) metal cations, in aqueous or biological media still remains a challenging task with many potential applications, especially in environmental sciences. Such chemosensors are generally formed by two moieties, namely a binding site and a signaling subunit, connected through a covalent bond. They are designed to exhibit a reversible easy-to-observe output by the signaling subunit when the interaction between the binding site and the metal cation takes place. Another approach developed for the selective determination of Hg(II) metal cations is based on the use of synthetic chemodosimeters. They can perform an irreversible chemical reaction with the Hg(II) metal cation leading to the formation of a fluorescent or colored species displaying a unique spectroscopic change. While a quite large number of coordinating groups acting as suitable binding sites for this metal cation have been described, three main groups of reporter subunits have extensively been employed, including organic dyes, fluorescent groups, and redox moieties. Thus, incorporating these signaling probes into various acyclic, heterocyclic, and macrocyclic ligand frameworks leads to receptor systems of different nature capable of responding to Hg(II) recognition through different channels: chromogenic, fluorescent, and redox. Taking into account that the most popular binding motifs currently being used to prepare Hg(II) detectors are those containing N-functionalized groups as well as different types of N-heterocyclic systems and aza- and polyaza-macrocyclic frameworks, highlights of these molecules will be reviewed in this chapter in the following order: nitrogen acyclic receptors, nitrogen heterocyclic receptors, and aza-macrocyclic-based receptors.

Organic Nanomaterials: Synthesis, Characterization, and Device Applications, First Edition. Edited by Tomás Torres and Giovanni Bottari.

Moving one step forward toward real applications, Section 24.3 addresses approaches to Hg(II) detection that relay on nanostructured materials. This section is divided into different parts according to the support used to anchor the receptor, nanoparticles, surfaces, or polymeric matrices.

24.2 Hg(II) RESPONSIVE SMALL-MOLECULE RECEPTORS

24.2.1 Nitrogen Acyclic Receptors

To date, the detection of Hg(II) ions with a single N heteroatom has rarely been reported. Recently, two terfluorene derivatives **1** bearing two 4-[bis(2-hydroxyethyl) amino]phenyl (X = OH) or two 4-(diethylamino)-phenyl (X = H) pendent groups have been described to exhibit high sensitivity and selectivity for the detection of Hg(II) with ratiometric fluorescence response. In these molecules, the detection of Hg(II) ions is realized by only a single amino nitrogen atom as receptor unit, and the sensory mechanism is attributed to a new metal-coordination-inhibited spiroconjugation-like charge-transfer emission [2].

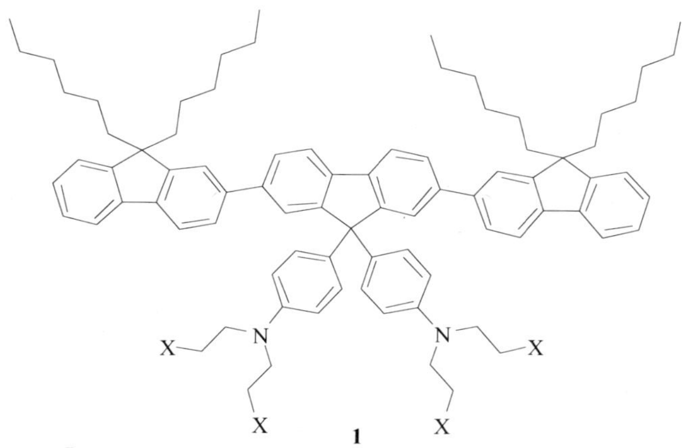

1

On the other hand, a new molecular sensor **2** in which two pyrenyl moieties are connected by a 2-aza-1,3-butadiene bridge as the metal-binding group has also been described. However, in terms of selectivity and sensitivity, this ligand showed a good sensing ability not only toward Hg(II) but also toward Cu(II) and functions both as chromogenic (naked-eye detection) and fluorogenic probe for these metal cations. Interestingly, the different emission wavelengths of the pyrene excimer showed by the two different complexes formed with these metal cations have been used to distinguish both heavy pollutant metals [3].

Cao and co-workers [4] designed and prepared a fluorescent sensor **3** for Hg(II) in which a pyrene unit is linked to a indolium moiety through an amide bridge. This sensor displays significant fluorescence quenching upon addition of this metal cation in pH 7.4 HEPES buffer without excimer formation. The ^{1}H NMR spectra collected upon titration with Hg(II) and Job's plot analysis indicates the formation of a complex with a 2:1 (ligand–metal cation) stoichiometry in which the nitrogen atom of the amide is involved in the coordination of the metal cation by using its lone electron pair.

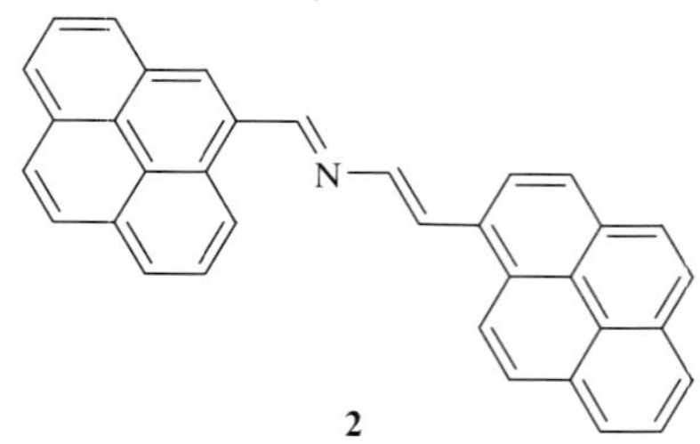

2

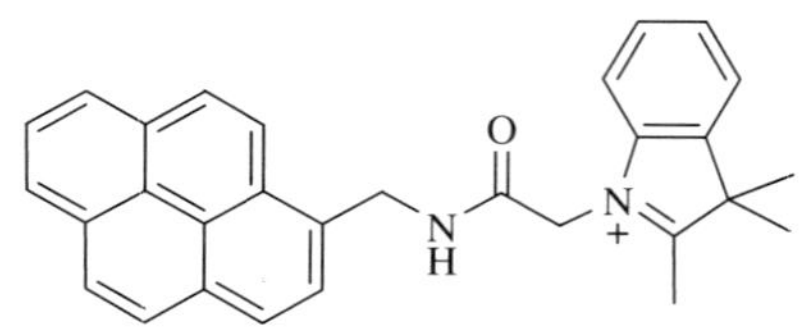

3

N-Unsubstituted cyclic imides **4**, as well as nucleobase thymine derivatives, react specifically with Hg(II) and form imide–Hg–imide complexes through an imido proton–metal exchange process like **5**. The reaction is reversible and occurs rapidly at moderate to high pH. In this context, several chemosensors containing thymine moieties have been reported [5].

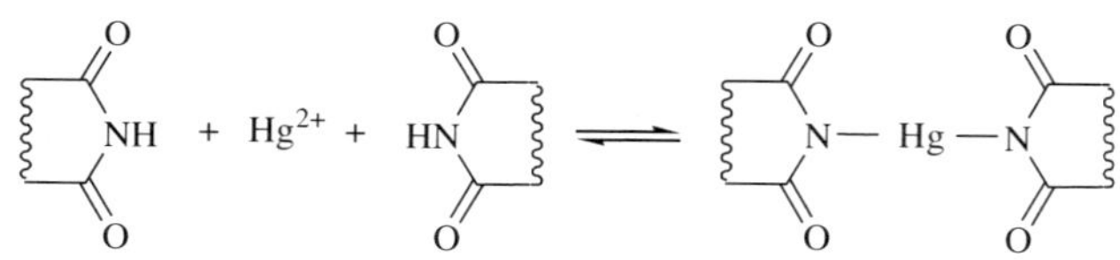

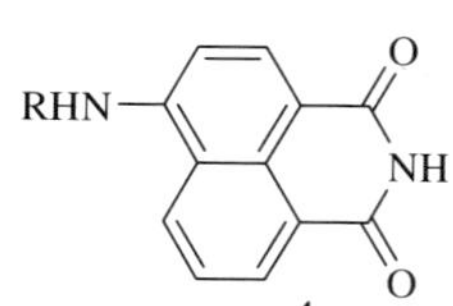

4

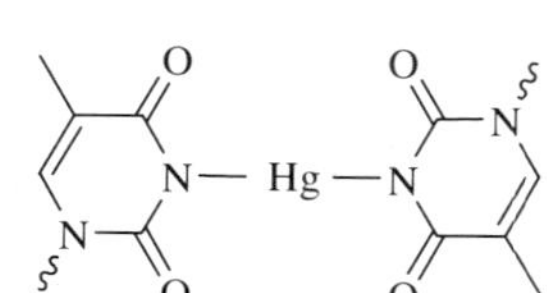

5

The sulfonamide moiety present in the dansyl fluorophore has also been used to detect metal ions by chelation enhanced fluorescence. Thus, a chemosensor **6** containing dimerized L-cysteine residues connecting to two dansyl fluorophores has been demonstrated to act as a highly selective ratiometric fluorescent sensor for detection of Hg(II) in 100% aqueous solution [6].

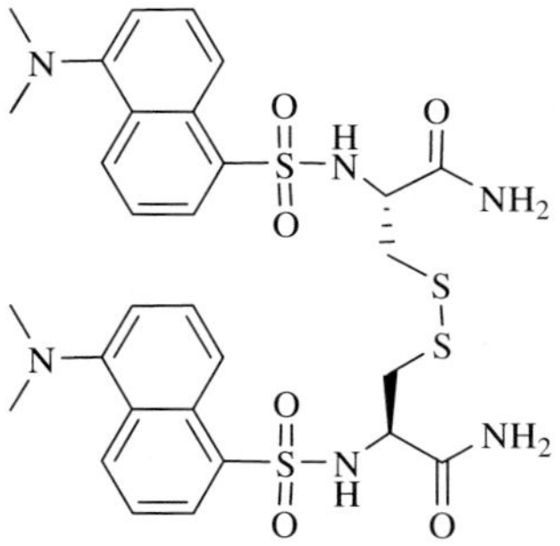

6

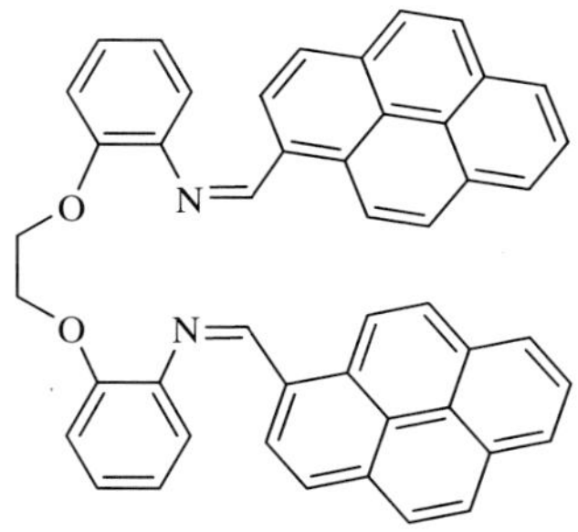

7

A 2,2′-(ethylenedioxy)-diimino-based molecular framework **7** has also been designed as a platform for the construction of an efficient Hg(II)-selective ratiometric and “off–on” fluorescent chemosensor [7]. The “off–on”–type signaling behavior of this fluoroionophore is due to the metal-ion-induced conformational changes, which might be partly attributed to the participation of the imino group of the Shiff base and O atoms of ether–methylene in the complex formation. The change in the conformation forces the two pyrene moieties to become relatively folded with face-to-face orientation resulting in the switch of the pyrene monomer emission to an excimer emission.

Selective chemosensors for Hg(II) ions using a diazo moiety as the reporter functionality, such as **8**, have also been described [8]. Interestingly, azo-phenol receptors, having an azo core incorporated with naphthol and substituted benzothiazole units, such as **9**, constitute an efficient and cost-effective chemosensor for the selective detection of Hg(II) in aqueous medium and on a test paper strip, acting as a real-time, sensitive, and naked-eye visible colorimetric sensor for this metal ion. The formation of the complex takes place in a 2:1 stoichiometry (ligand–metal cation) through a binding mode that involves the coordination of the chemosensor with the metal cation by the phenolic oxygen atom, the azo nitrogen atom adjacent to the naphthol ring, and the thiazole nitrogen of benzothiazole ring [9].

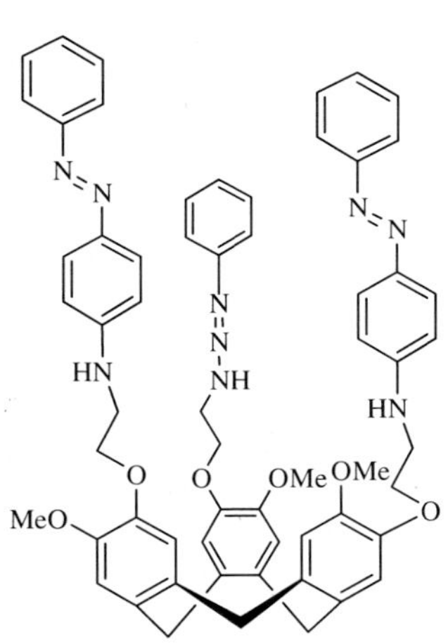

8

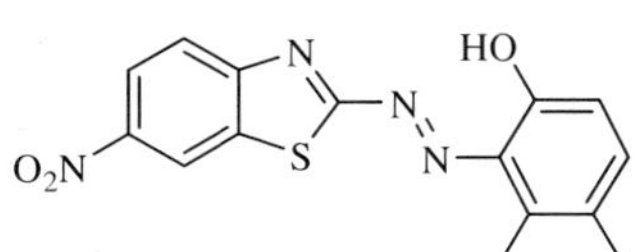

9

Indole–ferrocene–indole triad **10**, in which the redox-active organometallic fragment is linked to the indole rings by 2-azadiene moieties, acting as recognition subunits, also behaves as highly selective dual (redox and chromogenic) chemosensor for Hg(II) cations. The binding mode and the observed selectivity of this ligand toward Hg(II) in comparison to other metal cations was established by means of quantum chemical calculations [10].

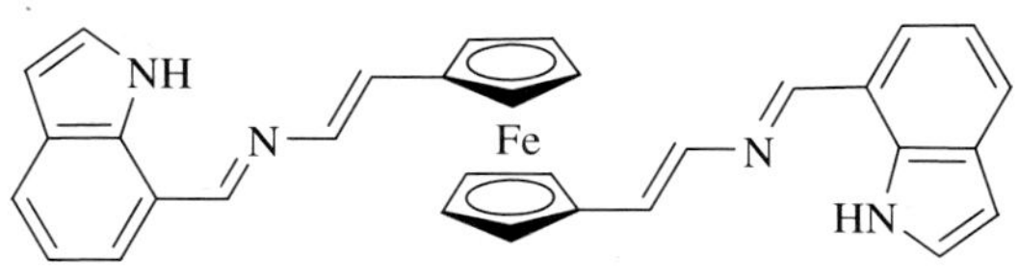

10

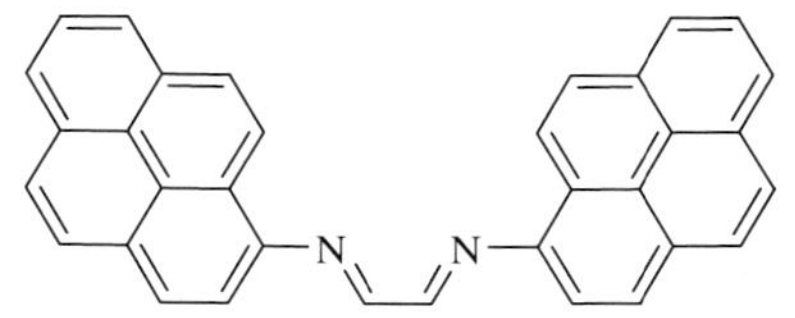

11

An azine-based receptor **11** has recently been published [11]. Confocal scanning microscope experiments revealed that this chemosensor can be used to detect Hg(II) present in Epithelial cell line KB 31 cells through an increase in intracellular fluorescence.

A new type of mononuclear ferrocenophane-based thiourea chemosensor **12** acting as a dual binding site for anions and Hg(II) metal cation has also been described [12]. The rigidity of this new structural motif imposes two different binding sites: (i) a sulfur-rich face as that selectively senses the thiophilic Hg(II) metal cations and (ii) a face with four directional N–H bonds able to recognize anions. Remarkably, the selective recognition of the metal cation, involving the formation of a 1:1 (ligand:metal) complex, is achieved either by electrochemical or by spectral measurements.

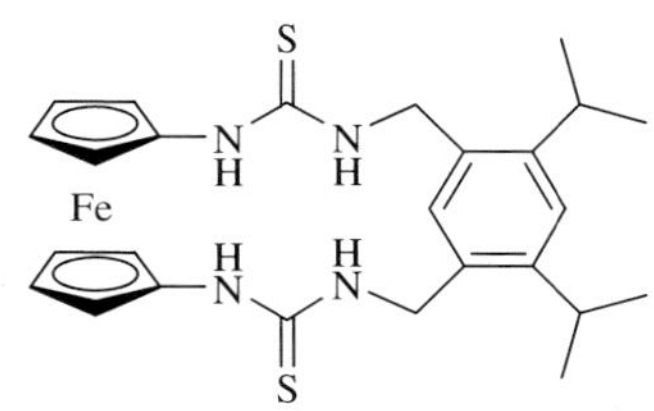

12

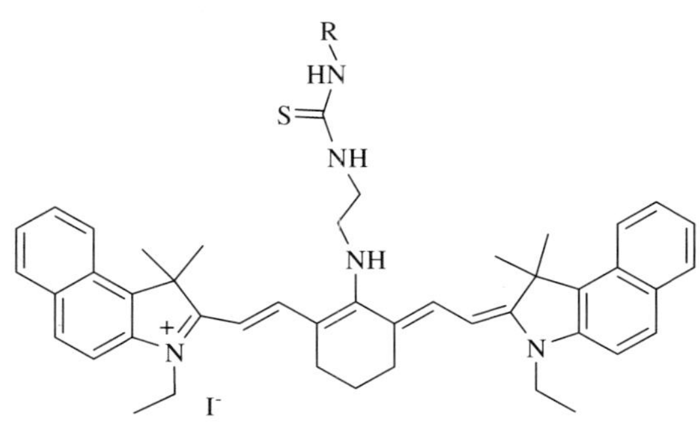

13

On the other hand, three selective chemodosimeters for Hg^{2+} **13** (R = phenyl, *n*-butyl or benzoyl) have been successfully designed by connecting a tricarbocyanine dye, as NIR (near-infrared region) fluorescence labels, to three different thiourea substituents [13]. Upon treatment with Hg(II), they exhibit large redshifts, which result in a clear color change from deep blue to pea green, which can easily be monitored by the naked eye for a convenient indicator paper. In emission spectra, they display a characteristic turn-off mode at 780 nm and turn-on mode at 830 nm with titration of Hg(II) ions. Additionally, experiments with living SW1116 cells show that these three tricarbocyanine dyes with low toxicity can exhibit special characteristics that are favorable for visualizing intracellular Hg(II) and $MeHg^+$ ions in biological systems, including excellent membrane permeability, minimal interfering absorption and fluorescence from biological samples, low scattering, and deep penetration into tissues.

24.2.2 Nitrogen Heterocyclic Receptors

The ability of terpyridine to selectively and sensitively detect Hg(II) in aqueous environments has been demonstrated by using the new colorimetric mercury sensors **14**. The response time is instantaneous, and a detection limit of 2 ppb was achieved even in the presence of excess metal ion competitors of concern as drinking water pollutants. The development of paper strips coated with this chemosensor demonstrated the suitability of this method for detecting Hg(II) in a very simple way [14].

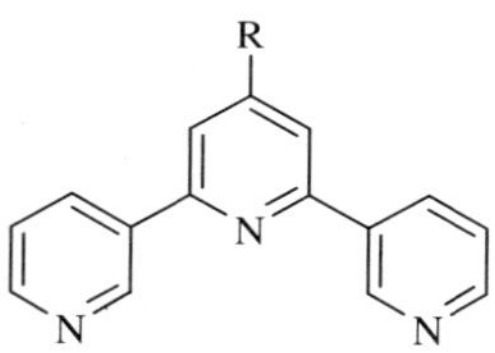

14

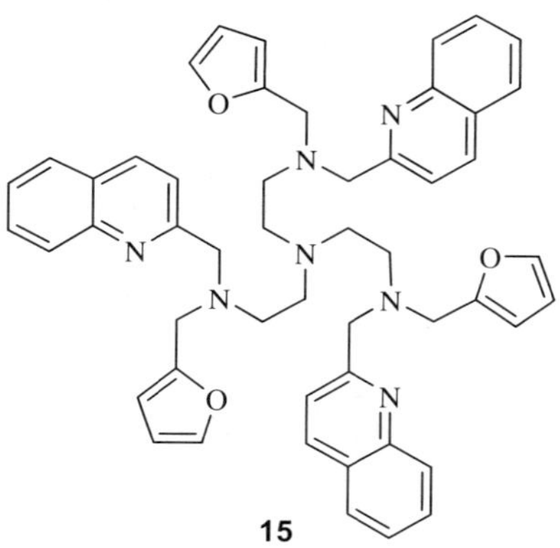

15

Tripodal fluoroionophore **15**, bearing three quinoline moieties as coordinating signaling units, also shows selectivity toward Hg(II) and senses this metal cation even in acidic conditions [15].

In this context, two new small-molecule Hg(II)-responsive probes bearing an 8-benzyloxiquinoline **16** [16] or a 2-amino isoquinoline **17** [17] moiety have also been described, although they function in a different manner: through a fluorescence quenching or enhancement mechanism, respectively.

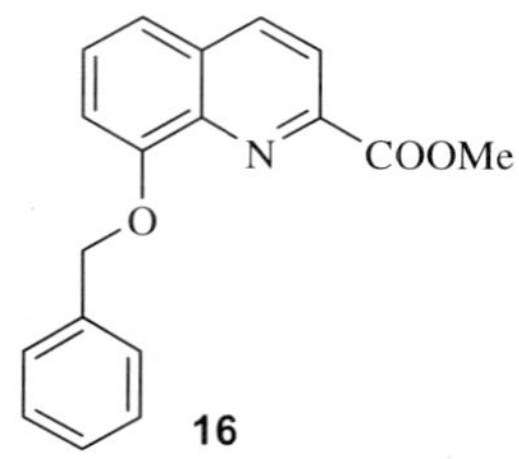

16

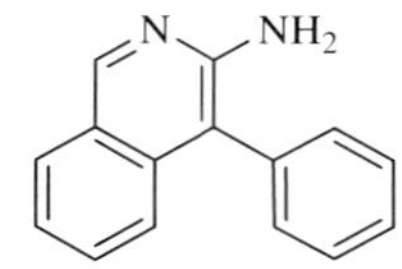

17

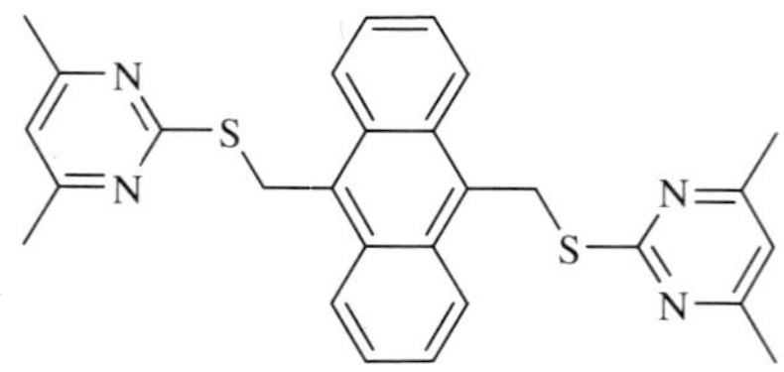

18

The pyrimidine–anthracene–pyrimidine triad **18** shows strongly both electronic and fluorescent responses toward Hg(II) and Cu(II) in acetonitrile solutions. DFT calculations suggest that the two "arms" of the anthracene skeleton "bend" over toward the same side of the anthracene moiety giving rise to a mononuclear complex in a distorted planar geometry, in which Hg(II) binds with the two pairs of donors (S and one of the N of the pyrimidine moiety) present in the subtituents linked to the 9 and 10 position of the anthracene unit [18].

Click chemistry has been used to synthesize a fluorescent sensor that responds to both Cu(II) and Hg(II) (**19**). This sensor has properties for practical applications, including selectivity for detecting these metal cations in the presence of excess competing metal ions, as well as optimum functionality in neutral aqueous solution and even seawater. Furthermore, the sensor can be used to distinguish Cu(II) and Hg(II) effectively, even when both are present as a mixture, through the addition of specific anions. Thus, this sensor-and-anion system represents a new method for detecting these metals simultaneously and independently [19].

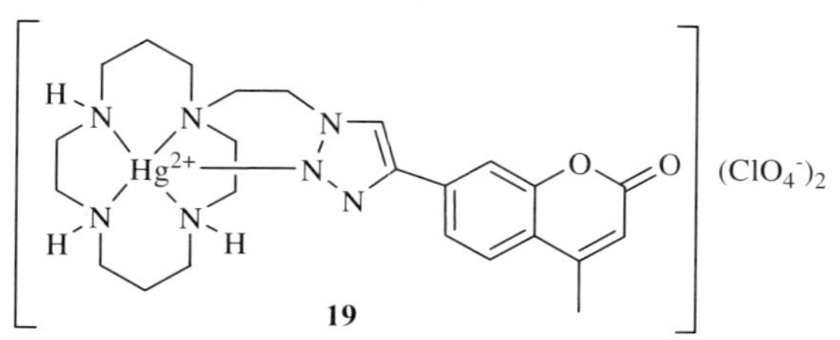

19

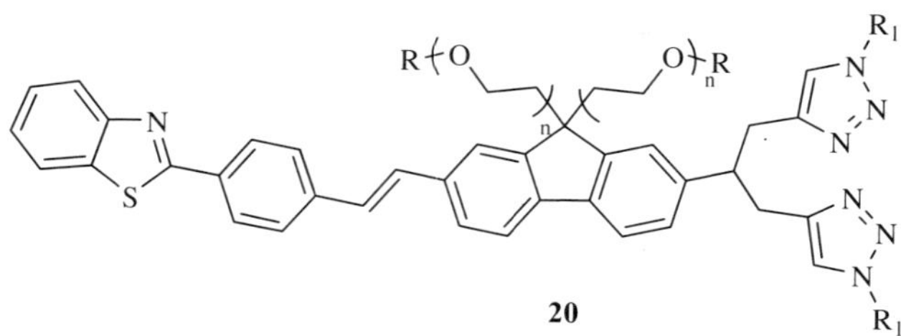

20

Similarly, a series of hydrophilic bis(1,2,3-triazolyl) fluorine-based derivatives **20** containing a 1,2,3-triazole-based recognition moiety have also been synthesized through click chemistry and behave as fluorescence sensing probes but not only for Hg(II) metal ions but also for Zn(II) metal cations [20].

Owing to their remarkable photophysical properties, such as high molar extinction coefficients, high fluorescence quantum yields, excellent photostability, easy structural modification, and appropriate redox potential, boron–dipyrromethene (BODIPY) derivatives have been widely employed for detecting different types of analytes. In this context, the derivative **21**, containing a tridentate diaza-oxa ligand of Hpyramol as a high selective fluorescent turn-on chemosensor for Hg(II), has been described [21]. Such significantly enhanced fluorescene upon addition of Hg(II) is probably attributed to the formation of a 2:1 (ligand/metal) complex in which the PET quenching process is efficiently suppressed by coordination with the Hg(II) ion. Moreover, confocal fluorescence microscopy experiments have established the utility of this chemosensor in monitoring Hg(II) within living cells.

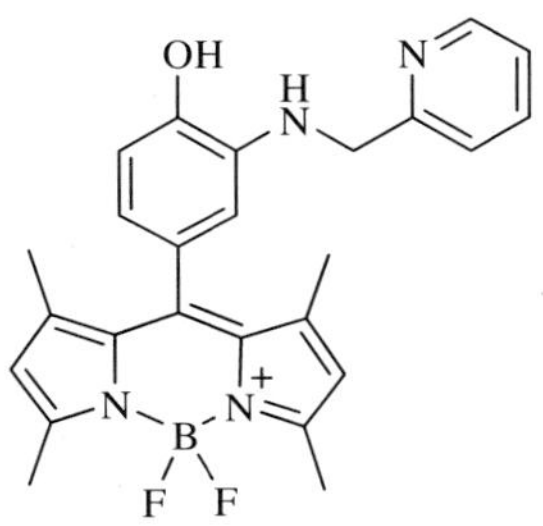

21

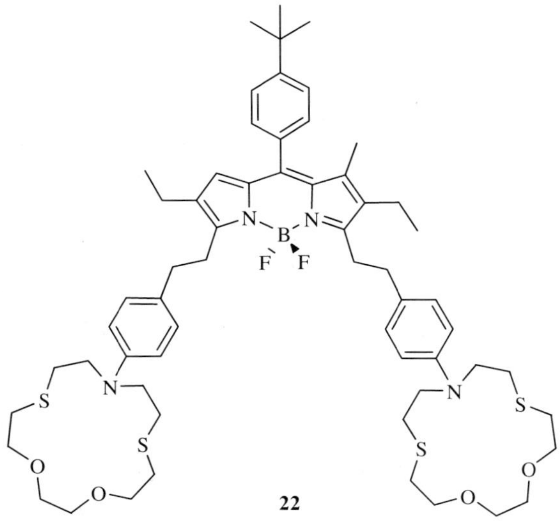

22

BODIPY derivatives in which one or two dithiodioxomonoaza-based crown-containing phenyl units are conjugated to the BODIPY core as a chelating unit, such as **22,** have also been used to detect Hg(II) cations by using both the absorption and emission spectra [22].

On the other hand, several kinds of ferrocene-azaheterocycle derivatives have also been reported to behave as multichannel molecular chemosensors for Hg(II) metal cations in both aqueous and organic media. Among them, several types of dyads should be highlighted: ferrocene-quinoxaline [23] or ferrocene-azaquinoxaline derivatives **23** [24], ferrocene-imidazoquinoxalines **24** [25], ferrocene-imidazophenazine **25** [26], and ferrocene-benzimidazole derivatives, **26** [27].

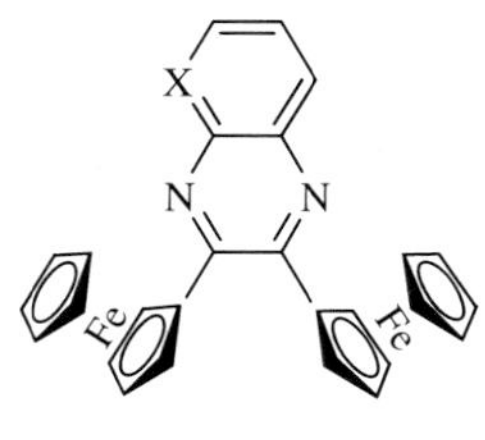

23: X = CH or N

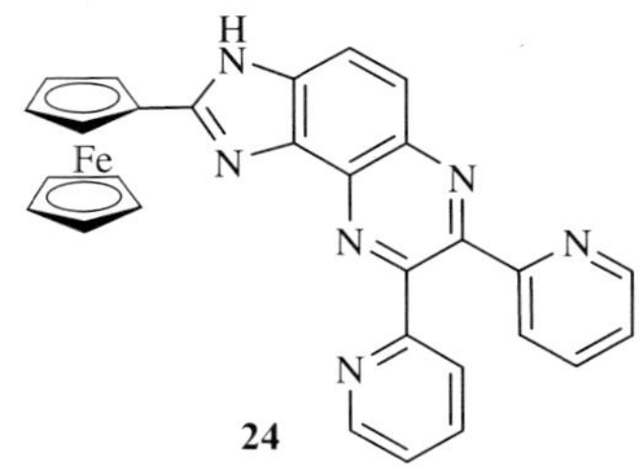

24

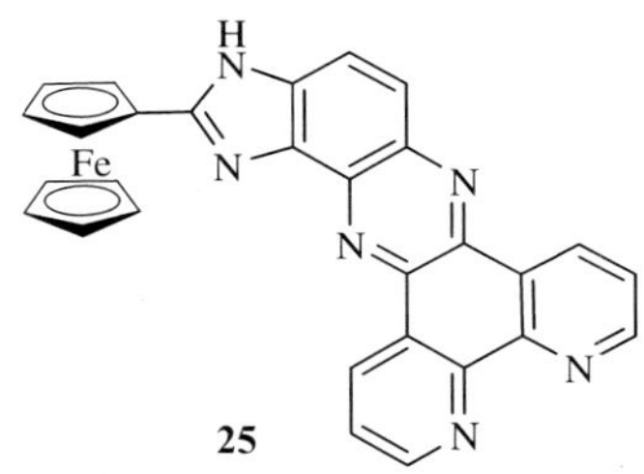

25

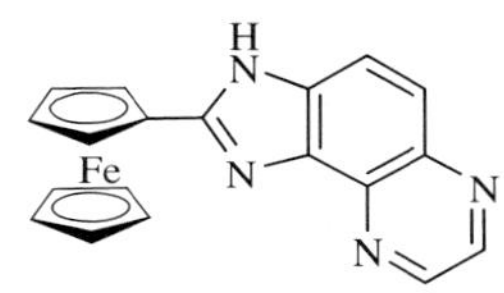

26

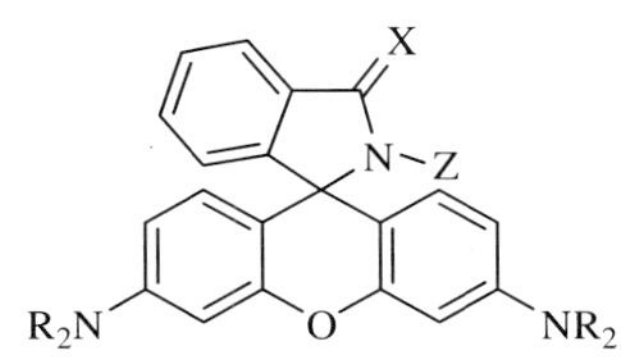

X = O or S

27

Rhodamines are dyes extensively employed in biolabeling and material sciences because of their photophysical properties and the apparent easiness with which such spirocyclic systems undergo ring-opening induced by the complexation of metal cations. Thus, based on this irreversible ring-opening reaction, several types of rhodamine-based chemodosimeters **27** for sensing Hg(II) have been developed [28].

24.2.3 Aza-Macrocyclic Based Receptors

Sensing of Hg(II) metal cations has also been achieved by using small-molecule fluorescent sensors that consist of an ionophore–chromophore system, with monoaza-crown ethers being the ionophore moiety.

Within this category, a water-soluble fluorescent ion probe, with high selectivity for Hg(II) cations, belonging to the turn-on class of sensors has been reported [29]. Thus, the probe **28** that consists of a monoaza-15-crown-5-moiety, acting as recognition unit, connected via a thiourea linker to a 7-amino-4-methylcoumarin, which functions as the chormophore of the system, is a remarkable example. It operates via a photoinduced electron-transfer process, providing a dye that is a potential tool for the detection and quantification of Hg(II) in environmental and biological samples.

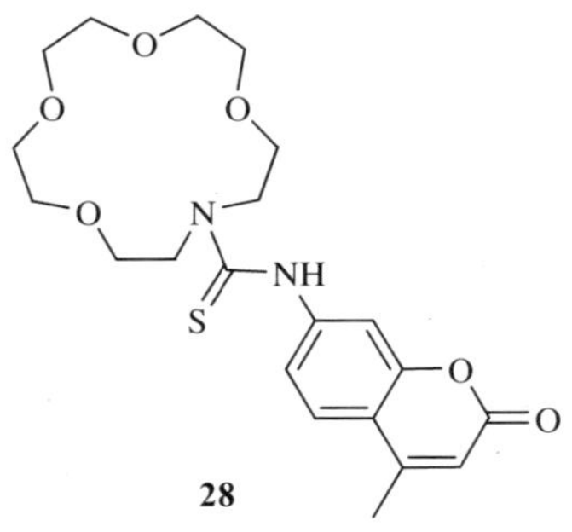

28

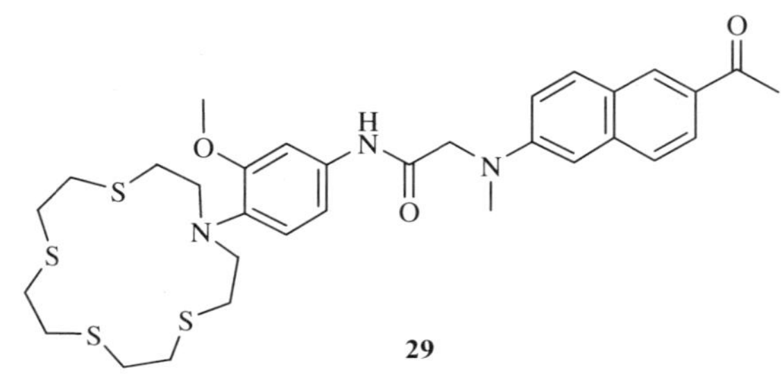

29

Similarly, the probe **29** that can selectively detect Hg(II) in live cells and fish organs at 80- to 150-μm level, under physiological conditions and without interference from other metal ions, is derived from a 6-dimethylamino-2-acetylnaphthalene (acedan) reporter and 4,7,10,13-tetrathia-1-aza-15-crown-5 binding unit [30].

Platinum(II) terpyridyl complexes connected to thioether-containing macrocycles, such as **30**, have also been proved to exhibit a highly sensitive and selective colorimetric responses to Hg(II) metal cations, through modulation of the relative strength of internal charge-transfer (ICT) and metal-to-ligand charge (MLCT) transitions [31].

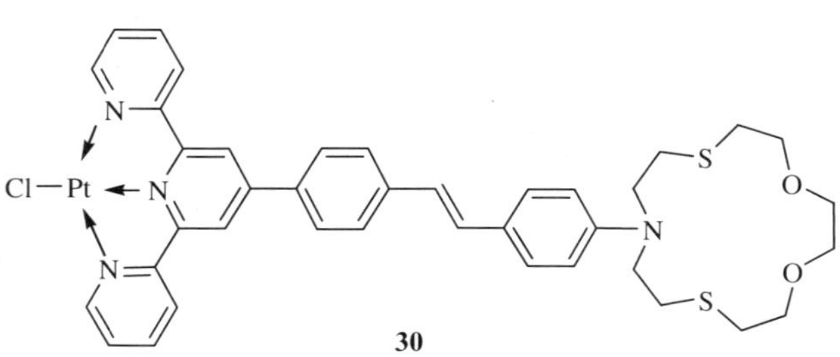

30

A newly 1,4,8,11-tetraazacyclotetradecane functionalized with a diazo moiety as the reporter functionality, such as **31**, is found to bind specifically to Hg(II) with an associated change in color visually detected. With biologically benign β-cyclodextrin, it forms an inclusion complex that shows a much higher solubility in water, and this helps in developing a more intense color on binding to Hg(II) in acetonitrile. The nontoxic nature of this probe was checked with the living cells of a Gram-negative bacterium, *Pseusomonas putida*. Furthermore, experiments revealed that these two reagents could be used as staining agents for the detection of Hg(II) present in this microorganism [32].

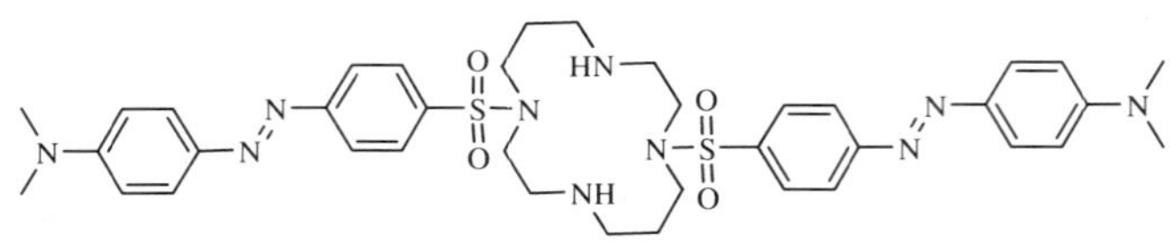

31

On the other hand, laterally nonsymmetric aza-oxa cryptands, derivatized with a electron-withdrawing moiety, such as fluorofore 7-nitrobenz-2-oxa-1,3-diazole, give rise to receptors that bind Hg(II) metal cations preferentially, showing a large enhancement of fluorescence [33]. Responses of these probes to Hg(II) were determined in polyvinyl chloride (PVC)-based membrane electrodes.

All the molecular probes described in this section offer a great possibility for the obtaining of Hg(II) sensors for real applications if they are supported or nanostructured on a solid support.

24.3 NANOSTRUCTURED MOLECULAR MATERIALS AS Hg(II) SENSORS

In this section we describe the work performed on the design of Hg(II) sensors based on molecular nanostructured materials. There are several examples of probes that employ nanostructured materials as mercury (II) responsive elements. These approaches are based on functionalized nanoparticles, nanoporous materials, and nanostructured membranes or films acting themselves as the Hg(II) probes or the media that support the mercury-responsive units. The use of nanostructured materials have the advantage to increase the surface area and thus, in a lot of cases, increase the level of sensitivity of the devices obtained, as well as to enable operating in aqueous media using molecular probes that in solution only work in the presence of organic solvents.

24.3.1 Hybrid Nanoparticles as Hg(II) Sensors

Several kinds of nanoparticles have been used as responsive elements for sensing Hg(II) ions in aqueous media. Among them we will mention the most common and efficient ones:

those composed of gold, carbon, metal oxide core–shells, or quantum dots. In the following subsections we will describe them separately.

24.3.1.1 Hybrid Gold Nanoparticles. The possibility to functionalize gold nanoparticles (AuNPs) with both solubilizing and binding groups to afford water compatibility allows also an increase of specificity, thereby increasing the interest in the use of such nanoparticles for sensing purposes [34].

Gold nanorods have been employed to detect dissolved Hg(II) by optical absorption spectroscopy [35].

The nanorods have an average aspect ratio of 1.6 and, in water, exhibit absorption bands at 520 and 612 nm attributed to the transversal and longitudinal modes, respectively. In the presence of sodium borohydride ($NaBH_4$) in the analyzed solution, which reduces all dissolved Hg(II) to Hg(0), an amalgamation process between the formed Hg(0) and gold occurs. This amalgamation induces changes in the optical absorption spectrum of the nanorods since their shape changes (Figure 24.1). The absorption peak resulting from the longitudinal mode, which is centered at 599 nm in the presence of $NaBH_4$, undergoes a blue shift that directly correlates to Hg(II) concentration at low ($\sim 10^{-11}$ M) levels. At higher concentrations ($\sim 10^{-4}$ M), only one absorption band centered at 530 nm is observed, consistent with the conversion of the nanorods to nanospheres. Such hybrid nanorods respond in less than 10 min, and ppt levels of Hg(II) ion can be monitored.

More recently, 13-nm-diameter spherical AuNPs functionalized at their surface with rhodamine B (RB) as a chromophores have been utilized for turn-on Hg(II) detection in aqueous media. These functionalized NPs operate on the basis of a modulation of the fluorescence resonance energy transfer (FRET) between the RB and the AuNPs [36]. Emission from the RB–AuNPs is weak ($\lambda_{ex} = 510$ nm, $\lambda_{em} = 575$ nm) because the AuNPs quench the organic chromophores effectively by both energy- and electron-transfer processes. However, the addition of Hg(II) cation to the RB-AuNPs causes a ~400-fold fluorescence enhancement, which results from the Hg(II)-induced release of the rhodamine chromophores from the AuNP surfaces. Thus, when the AuNPs are only functionalized with RB, a fluorescence turn-on also occurs for many other metal ions including Mn(II), Fe(II), Cr(III), Cd(II), and Pb(II). However, additional functionalization of the AuNPs with thiophilic chelates, including mercaptopropionic acid (MPA), helps to promote a Hg(II) selectivity. When the AuNP surfaces are ~20% covered by a thiophilic chelate and 80% with the RB chromophore, they exhibit a reduced sensitivity to Mn(II), Fe(II) and Cr(III) but remain responsive to Hg(II), Pb(II), and Cd(II). Introduction of pyridine-2,6-dicarboxylic acid (PDCA) to solutions of RB–AuNP–MPA provides a better Hg(II)-selective response. Thus, the RB–AuNP–MPA detectors along with PDCA provide a linear response to Hg(II) ions from 15 to 250 nM and have a lower detection limit of 10 nM (2 ppb).

Another strategy to detect Hg(II) ions is with AuNPs decorated with DNA molecules since thymine units (T) in DNA strands were demonstrated to specifically bind Hg(II) ions forming stable T–Hg–T complexes. Thus, a variety of colorimetric sensors based on DNA-functionalized AuNPs (DNA–AuNPs) have been developed for the selective detection of this ion. Mirkin and co-workers [37] reported a DNA-functionalized AuNPs (DNA–AuNPs) for colorimetric Hg(II) detection in aqueous media. Individual DNA–AuNPs with complementary sequences form DNA-linked aggregates when combined in aqueous solution [38]. These aggregates have sharp melting temperatures (T_m) and they

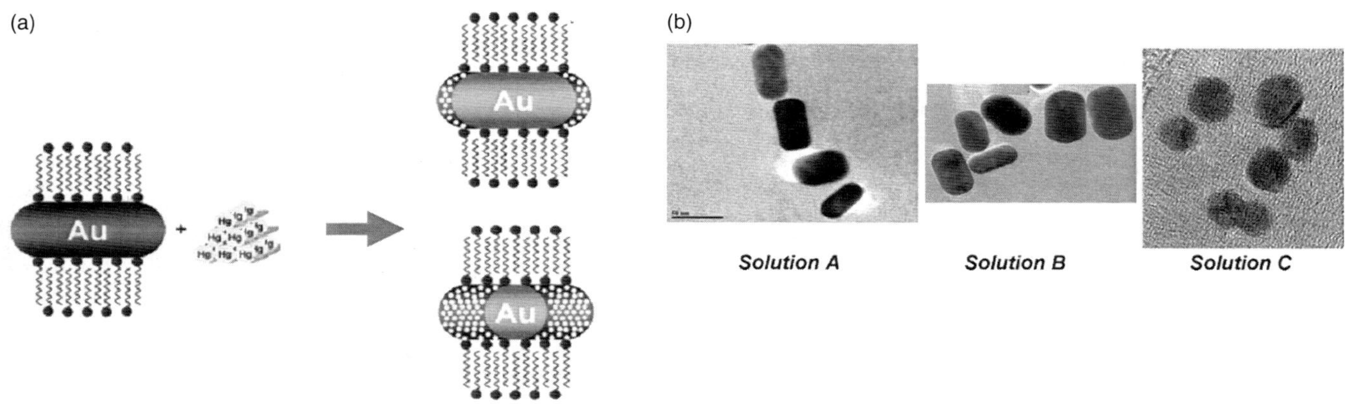

FIGURE 24.1. (**a**) Schematic diagram showing the amalgamation of Hg with Au nanorods. (**b**) TEM analysis of multiple Au nanorods in the absence and the presence of Hg(II). Solution A, no Hg(II); Solution B, 1.25×10^{-5} M Hg(II); Solution C, 1.57×10^{-4} M Hg(II). All solutions were prepared in 1.67×10^{-3} M $NaBH_4$. Reproduced from reference 35 with permission from the Amercian Chemical Society.

provide a color change from purple to red upon dissociation. These features were utilized in the design of a Hg(II) detection method that employs two kinds of DNA–AuNPs, with the thiolated DNA sequences 5′-HS–C_{10}–A_{10}–T–A_{10}-3′ and 5′-HSC_{10}–T_{10}–T–T_{10}-3′, that exhibit a thymine–thymine (T–T) mismatch.

The grounds for this method are based on the following points: (i) Hg(II) ions have high affinity for T–T mispairs, (ii) addition of Hg(II) to a DNA duplex containing T–T mispairs results in stabilization of the duplex, and (iii) T_m values are very sensitive to base-pair mismatches and their values decrease relative to those of perfectly complementary stands. For such reasons, the addition of Hg(II) to the DNA–AuNP aggregates harboring a T–T mismatch causes an increase in the T_m that is linearly dependent on the concentration of Hg(II) increasing ~5°C per 1 μM Hg(II). As a result, a purple-to-red color change occurs, resulting from desaggregation of the hybrid DNA–AuNPs, at a particular Hg(II)-dependent temperature. The response is highly selective for Hg(II) ions since K(I), Ca(II), Mg(II), Ba(II), Mn(II), Fe(II), Fe(III), Co(II), Ni(II), Cu(II), Zn(II), and Cd(II) have no effect on the T_m value. However, the addition of Pb(II) to the DNA–AuNP aggregates causes a negligible increase in the T_m of ~0.8°C. The detection limit for Hg(II) with this method is 100 nM (Figure 24.2).

Although developed sensors based on DNA–AuNPs possess good sensitivity and selectivity, the chemical synthesis and modification of DNA are relatively expensive and complicated, thereby decreasing their use in real applications. Additionally, the low stability against nuclease and the nonspecific interactions resulting from the negatively charged backbone of DNA also limit the application of such sensors in real complex samples. Besides T-rich DNA strands, small thymine derivatives have also been demonstrated to bind Hg(II) ions with high affinity and good selectivity. Compared with DNA strands, small organic molecules are more stable and cheaper. Thus, recently a new thymine derivative, the thymine acetamidoethanethiol (T-SH), was designed as a recognition unit to conjugate with AuNPs for the detection of Hg(II) ions. The resulting hybrid T–S–AuNPs were induced to aggregate via the formation of a stable T–Hg–T complex, resulting in a color change of the solution from red to blue-gray. The sensitivity and selectivity of this sensor were investigated. As low as 0.5 μM of Hg(II) ions can be easily monitored by the naked eye using this sensor that also shows a high selectivity. The high stability and selectivity as well as an easy operation enable this sensor for the rapid on-site detection of Hg(II) pollution [39].

24.3.1.2 Hybrid Carbon-Based Nanoparticles. More recently, carbon-based nanostructures (CNSs) including single-walled carbon nanotubes (SWCNTs) [40], graphene oxide (GO) [41], and nano-C_{60} [42] have been also used to quench the fluorescence of a dye contained on a single-labeled fluorescent oligonucleotide probe. This kind of CNSs have been used as an effective fluorescent-sensing platform for sensitive and selective Hg(II) ions detection. Recently, researchers have also demonstrated the first use of relatively inexpensive carbon nanoparticles (CNPs) obtained from a candle soot, as a fluorescent-sensing platform for highly sensitive and selective detection of Hg(II) ions [43]. The general concept used in this approach is based on the fact that the fluorescently labeled single-stranded DNA are adsorbed by CNPs via π–π stacking interactions between the DNA bases and the

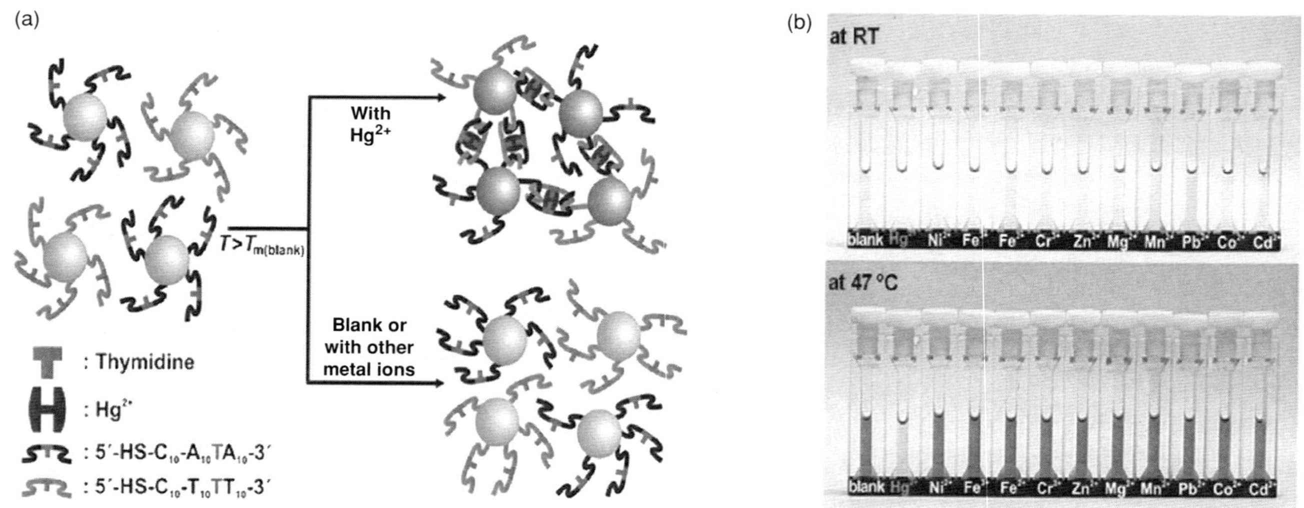

FIGURE 24.2. (**a**) Scheme of the colorimetric detection of Hg(II) ions using hybrid DNA–Au NPs. (**b**) Color change of the aggregates in the presence of various representative metal ions upon heating from room temperature to 47°C. Reproduced from reference 37 with permission from the Wiley-VCH. See color insert.

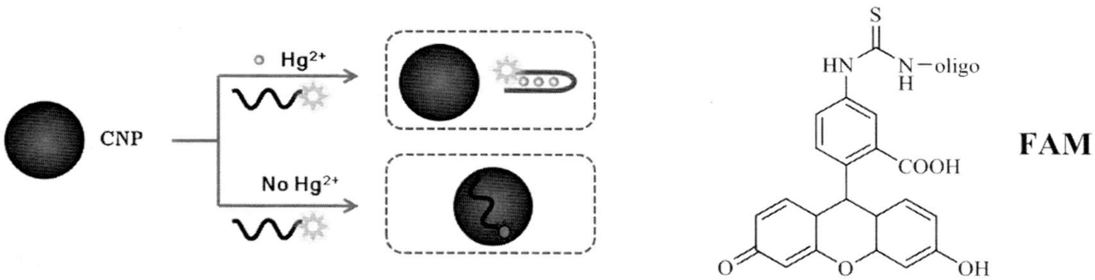

FIGURE 24.3. Scheme (not to scale) illustrating the CNP-based fluorescent Hg(II) detection based on conformational change of a Hg(II)-specific T-rich and fluorescent labeled ssDNA: 5′-TTC TTT CTT CCC CTT GTT TGT T-FAM-3′. Reproduced from reference 43 with permission from Elsevier.

aromatic subunits of the CNPs, which leads to a substantial fluorescence quenching. However, in the presence of Hg(II) ions the T–Hg(II)–T-induced hairpin structures do not adsorb on the CNPs and thus retains the dye fluorescence. Figure 24.3 presents a schematic illustration of the CNP-based fluorescent detection of Hg(II) ions. A detection limit as low as 10 nM was achieved. This CNP-based sensor exhibits a remarkable selectivity against other possible metal ions.

24.3.1.3 Hybrid Quantum Dots. Semiconductor quantum dots (QDs) have also attracted an extensive research interest as optical labels for sensing events. The unique photophysical properties of QDs, such as high fluorescence quantum yields, narrow emission bands, high Stokes shifts, and high stability against photobleaching, make them superior sensing materials. Specifically, the size-controlled luminescence features of QDs enable the multiplexed detection of different analytes both in solution and in on-chip formats. Accordingly with such unique properties, several studies with hybrid NPs using CdS, CdSe, and CdTe as QDs have been performed to develop Hg(II) sensors. The specific electron-transfer quenching of the thymine- or cytosine-modified QDs by the ions enabled us to use Hg(II) and Ag(I) ions as inputs that activate logic gates and to implement QDs as optical readout signals for logic gate operations [44]. Figure 24.4 outlines the principle for the selective analysis of Hg(II) or Ag(I) ions by two QDs, showing a rigid hairpin structure formed in the presence of Hg(II) or Ag(I) ions, in which the T or C residues of the spatially separated nucleotides are linked by the ions.

Carbon-based dots show some common properties with QDs and through their surface functionalization acquire a strong photoluminescence in both solution and solid state. In general, the photoluminescence has been attributed to the presence of surface energy traps, likely related to the abundant surface defect sites that become emissive upon functionalization. In addition, the surface emissive sites of the carbon dots are likely quantum-confined in the sense that a large surface-to-volume ratio is required for a strong photoluminiescence. Since they can be functionalized with several molecules in a number of layers accordingly with the desirable application, these nanoparticles show a great potential for in vivo fluorescence sensing applications. As a representative example, we may mention carbon dots obtained by laser ablation (UV-pulsed laser irradiation) of carbon targets immersed in water in the presence of NH_2–polyethyleneglycol (NH_2-PEG200) or *N*-acetyl-l-cysteine (NAC) [45]. This methodology yields the functionalization of C-NPs with NH_2-PEG200 and NAC which becomes fluorescent with excitation and emission wavelengths at 340 and 450 nm, respectively. The fluorescence intensity of both nanoparticles is quenched by the presence of Hg(II) and Cu(II) ions with a Stern–Volmer constant at a pH = 6.8 of 1.3×10^5 and $5.6 \times 10^4 M^{-1}$, respectively.

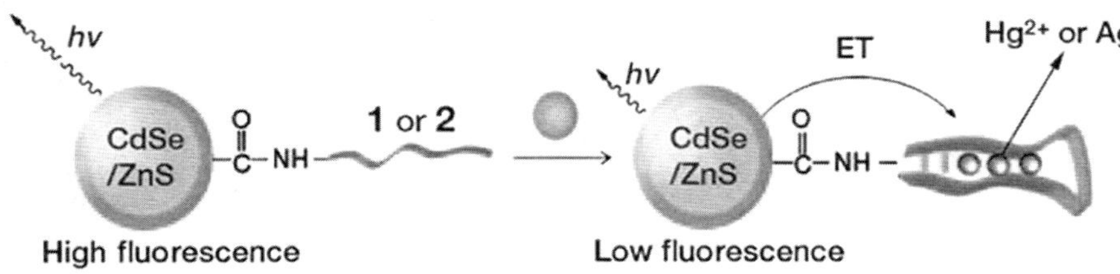

FIGURE 24.4. Schematic representation of the optical analysis of Hg(II)/Ag(I) ions hybrid QDs modified with DNA containing thymine units:

1: 5′-TTCTTTCTTCCCTTGTTTGTT-3′

2: 5′- CTCTCTTCTCTTCAAAAAACAACACAACACAC-3′.

Reproduced from reference 44 with permission from Wiley-VCH.

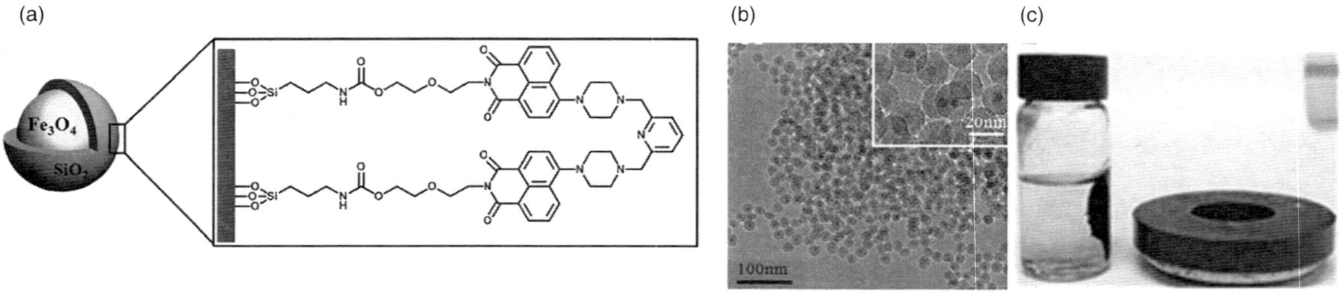

FIGURE 24.5. (**a**) Scheme of the aminonaphthalimide-functionalized Fe_3O_4@SiO_2 core–shell magnetic nanoparticles. (**b**) TEM image of this chemosensor. (**c**) Photograph of a magnet attracting the NPs in water. Reproduced from reference 47 with permission from the Royal Society of Chemistry.

24.3.1.4 Hybrid Core-Shell Nanoparticles. Functionalized magnetic nanoparticles, composed of both inorganic and organic components, have recently been examined as promising platforms for detection and separation applications. This unique class of nanomaterials not only can retain the beneficial features of both the inorganic and organic components, but also can provide the ability to systematically tune the properties of the hybrid materials through the combination of appropriate functional components [46]. Thus, Jung and co-workers [47] recently reported the preparation of the aminonaphthalimide-functionalized Fe_3O_4@SiO_2 core–shell magnetic nanoparticles and their applications for the detection and separation of Hg(II) and methylmercury ions in drinking water (Figure 24.5). Upon the addition of increasing Hg(II) concentrations, such NPs showed a large quenching effect with a photoinduced electron transfer (PET) effect in the emission spectrum, which is exclusive of these ions. The fluorescence quenching that occurred upon exposure to Hg(II) was fully reversible, as the addition of EDTA (0.01 N, 1 mL) restored the emission band. The fluorescence change was also reproducible over several cycles of detection–separation. Moreover, these NPs showed a detection limit of 1.2 ppb, more than sufficient to sense the Hg(II) concentration in drinking water established by the U.S. EPA.

Very recently, Alarcón and co-workers [48] reported a synthetic procedure for the preparation of structured inorganic–organic hybrid materials consisting of boehmite- SiO_2 core–shell nanoparticles functionalized with anthracene-containing amines covalently attached to the nanoparticles surface. Such functionalized NPs show a very high sensing performance for Hg(II) detection in pure water reaching a detection limit of 0.2 ppb. The fluorescence intensities of these NPs showed a clear diminution when the pH of the colloidal solution was varied from 2.0 to 11.0, as a consequence of a PET effect from the amine lone-pair electrons to the excited fluorophore due to deprotonation of the ammonium groups. Interestingly, these NPs show a good selectivity upon addition of equimolar amounts of Cu(II), Zn(II), Cd(II), Pb(II), and Hg(II) in aqueous solution at pH 5, giving a highly selective response with only Hg(II). This remarkable efficiency and selectivity for Hg(II) at pH 5 can be ascribed to the ability of Hg(II) to form stable complexes with low coordination numbers. The detection limits of such NPs were between 0.2 ppb and 0.4 ppb. Two additional advantages of these systems are their stability over a wide pH window and the feasibility to be recovered by a simple procedure.

24.3.2 Solid-Supported Hg(II) Sensors

The immobilization of indicator dyes on solid supports is also an alternative for the preparation of optical chemical sensors because they provide several advantages with respect methods using NPs. The indicator dyes can be either physically or chemically immobilized on the support matrixes. Both of these methodologies have their advantages and disadvantages. Physical entrapment is the simplest method, but the resulting sensors have relatively short lifetimes because of the leaching of dye molecules into the analyte solution. Among all the available matrices for physical entrapment, the polymeric materials have proved to be the most suitable ones. By contrast, chemical immobilization by covalent binding of the indicator dyes on support matrixes is the most efficient technique for obtaining optical chemosensors, since they have well-reproducible responses and long lifetimes. However, the immobilization process involved in the reaction between the indicator dyes and support matrixes might cause some problems. Recent advances in surface chemistry and the appearance of new characterization and processing techniques such as atomic force microscopy (AFM), self-assembled monolayers (SAMs), and the microcontact printing (μCP) technique have made the covalently modified substrates one of the most interesting subjects for the development of sensors. In contrast to the use of physisorption as an anchoring strategy, the specificity and intrinsic strength of the covalent bond proportionate to the heterogeneous sensors not only the stability and reliability required but also higher density of active sites. Accordingly, an increase of the potential recognition properties of the sensors is usually obtained.

In this section, different materials, such as modified glass or metal surfaces with a self-assembled monolayer (SAM) and polymeric and mesoporous materials grafted with the indicator dyes will be described as supporting media for heterogeneous Hg(II) chemosensors.

24.3.2.1 Hybrid Gold and Glass Surfaces. Among the available optical sensing techniques to be used with hybrid gold surfaces, the surface plasmon resonance (SPR) is one of the most sensitive. Highly specific SPR sensors are usually based on the proper modification of a metal surface like gold with a SAM containing ligand molecules as recognizing elements. Most of the SPR-based sensors available in the market or reported in the literature are focused on the recognition and detection of large biomolecules, and only a very few have been shown to work with analytes of low molecular weights. Recently, Díez-Gil et al. [49] developed the first reusable Hg(II) SPR sensor chip able to perform "on flow" detection of concentrations of this metal ion in water down to the picomolar range, reaching detection levels of atomic adsorption techniques without losing the receptor core intrinsic selectivity versus the most common mono-, di-, and trivalent cations (Figure 24.6).

The SPR sensor was obtained by applying rational chemical design principles of metal–organic and surface chemistry where the 1-(4′-oxyphenyl)-4-(1′-pyrenyl)-2,3-diaza-1,3-butadiene receptor [49] unit was extensively modified with the purpose of ameliorating its self-assembling properties without modifying its sensing characteristics. In order to increase the final wettability and flexibility of the sensing unit, a layer of hydrophilic triethyleneglycol spacer was attached to the receptor core followed by a hydrophobic alkyl linker, used as packing driving force for the assembling. Moreover, a thioacetyl unit was used to anchor the molecule to the gold surface of the sensor chip. This design also prevents unspecific absorptions, increases its selectivity, and optimizes the sensitivity and reliability of the sensor by formation of a laterally well-packed SAM (Figure 24.6). Upon injection of different Hg(II) ion solutions in the concentration range from 10^{-11} to 10^{-7} M, in ionic strength controlled media (aqueous solution of 0.1 M NaCl), an increase of the SPR signal was observed which is linearly related to the Hg(II) concentration (Figure 24.6). The lack of any sample pretreatment and the low sample volume ($\sim$1 mL) required by this method makes it suitable for in-field analysis.

The development of a carboxylic substituted polydiacetylene (PCDA) liposome-based microarray for selective and sensitive mercury detection has also been performed [50]. In this development, researchers have identified the terminal epoxy groups of PCDA-epoxy molecules as a universal functional group for an efficient immobilization of liposomes onto an amine-modified solid glass substrate (Figure 24.7). These groups have also been used for the convenient post-tethering of the amine-modified ssDNA aptamer 50-TTCTTTCTTCCCCTTGTTTGTT-30, on the liposome surface as a selective receptor for the recognition of mercury ions (Figure 24.7). The PCDA mercury sensors were designed in such a way that when the ssDNA aptamers recognize and wrap around mercury ions, the steric repulsion between the resulting bulky T–Hg–T complexes perturbs the conjugated ene-yne backbones of the PCDA liposomes, producing a color change from blue to red together with an intense red fluorescence emission. The detection limit of the PCDA microarray is 5 mM, and the specificity of the ssDNA aptamer toward Hg(II) provides an excellent selectivity of

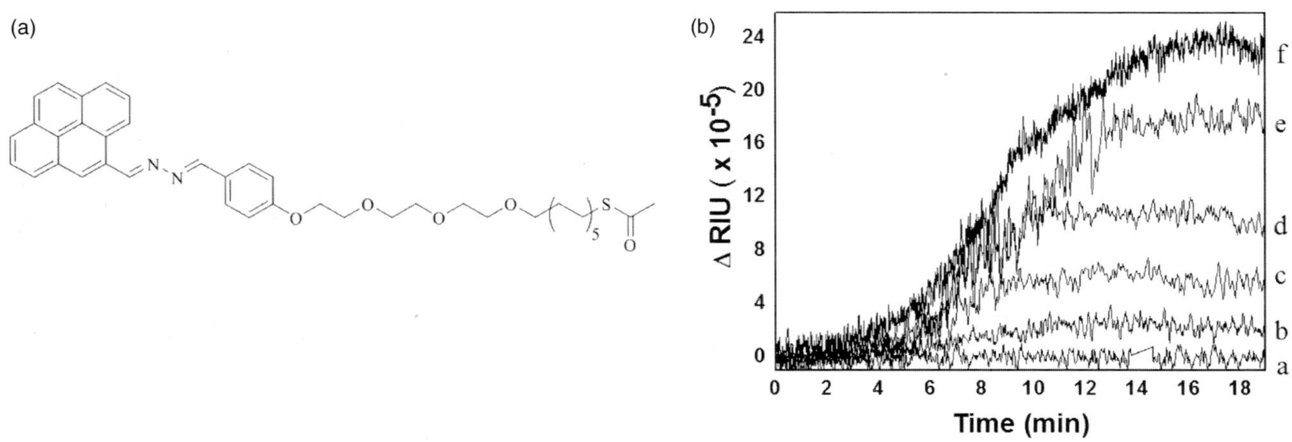

FIGURE 24.6. (**a**) Structure of the building block for the Hg(II) recognition using SAMs on Au and the SPR technique. (**b**) Time-dependent SPR signal variation after the injection of different concentrations of Hg(II) ions in water (a, base-line; b, 10^{-11} M; c, 10^{-10} M; d, 10^{-9} M; e, 10^{-8} M; f, 10^{-7} M). Reproduced from reference 49 with permission from the Royal Society of Chemistry.

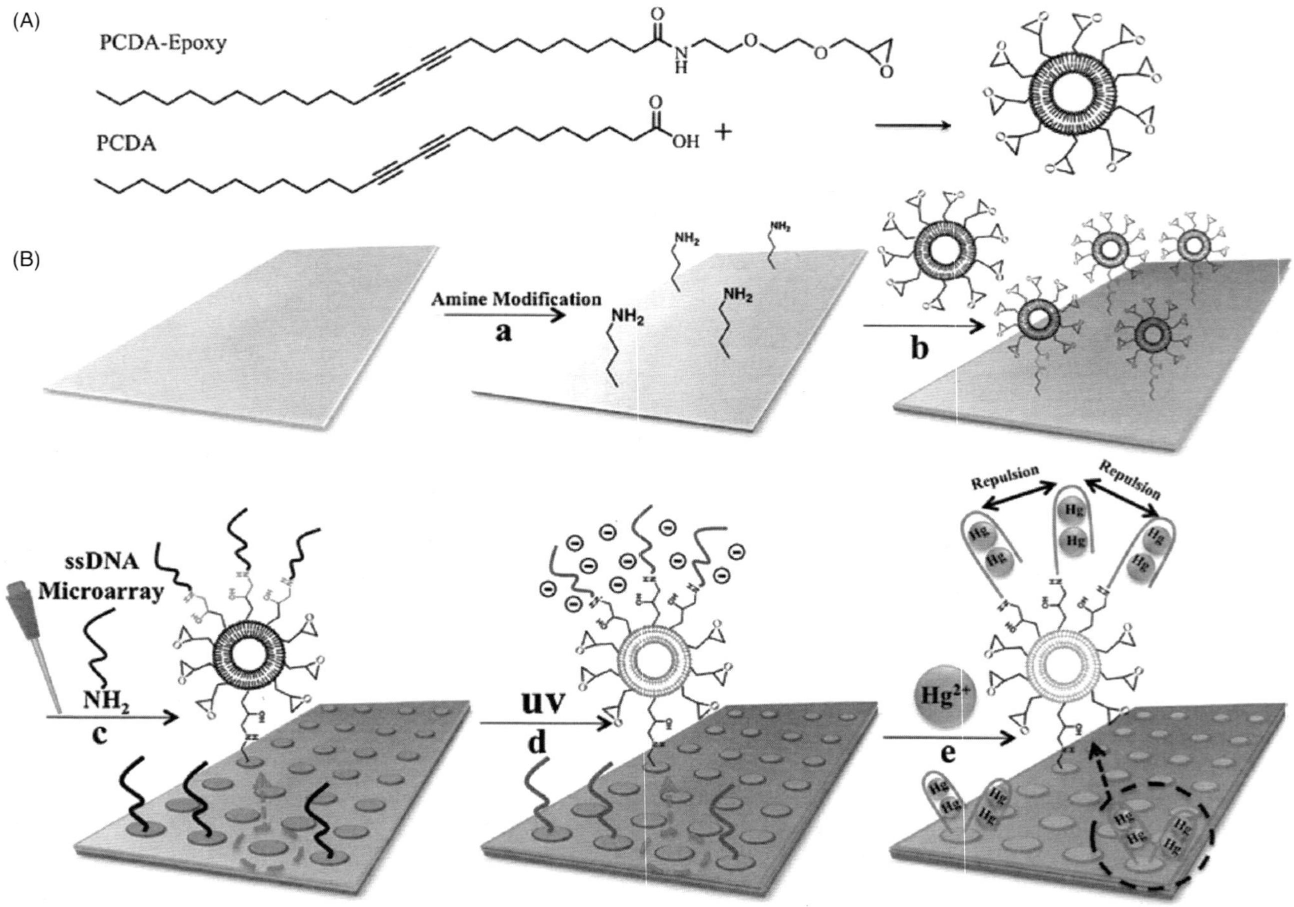

FIGURE 24.7. (**A**) Chemical structure of the diacetylene monomers, PCDA and PCDA-Epoxy. (**B**) Schematic illustration of the PCDA liposome-based microarray for mercury detection. (**a**) Surface modification of the glass substrate with an amine functionality. (**b**) Immobilization of the epoxy terminated liposomes onto the amine glass slide through the epoxy-amine coupling. (**c**) Post-tethering of the ssDNA aptamer by means of a microarrayer. (**d**) Photopolymerization of the PCDA liposomes using a 254-nm UV lamp. (**e**) Recognition of the target mercury ions resulting in a red fluorescent emission. Reproduced from reference 50 with permission from the Wiley–VCH.

this microarray. A point worth noting is that the developed epoxy-based PCDA liposome design is an excellent platform that can be readily applicable to the design of other sensors.

24.3.2.2 Hybrid Mesoporous Materials. The use of mesoporous silica as a support for grafting fluoroionophores has also opened new opportunities to achieve efficient optical sensors for heavy metals in water. Mesoporous materials offer the possibility of immobilizing small molecules showing large surface-area-to-volume ratios, uniform porosities, and a photo-insensitive property in the visible region.

Recently, a novel method has been developed to prepare optical disc-like sensors by the immobilization of two different organic groups on mesoporous silica monoliths. The first organic moiety is introduced to enhance the polarity of the inorganic mesoporous disc-like monoliths, and the second one acts as the probe center. In this regard, the authors described a Hg(II) ion optical sensor based on a disc-like monolith by a two-step inclusion procedure via a grafted controlled surface modification of the monoliths with *N*-trimethoxysilylpropyl-*N*,*N*,*N*-trimethylammonium chloride (TMAC) and then by immobilization of the tetraphenylporphyrine tetrasulfonic acid (TPPS) probe. This approach leads to a facile handling of signal readout with visual detection of ultratrace (10^{-9} M) concentrations of mercury ions (Figure 24.8) [51].

Results revealed that the sensing disc-like monoliths could be used as simple preconcentrators to yield high adsorption capacity and preconcentration efficiency, leading to simultaneously visual inspection and complete removal of Hg(II) ions over a wide, adjustable range concentration. The new

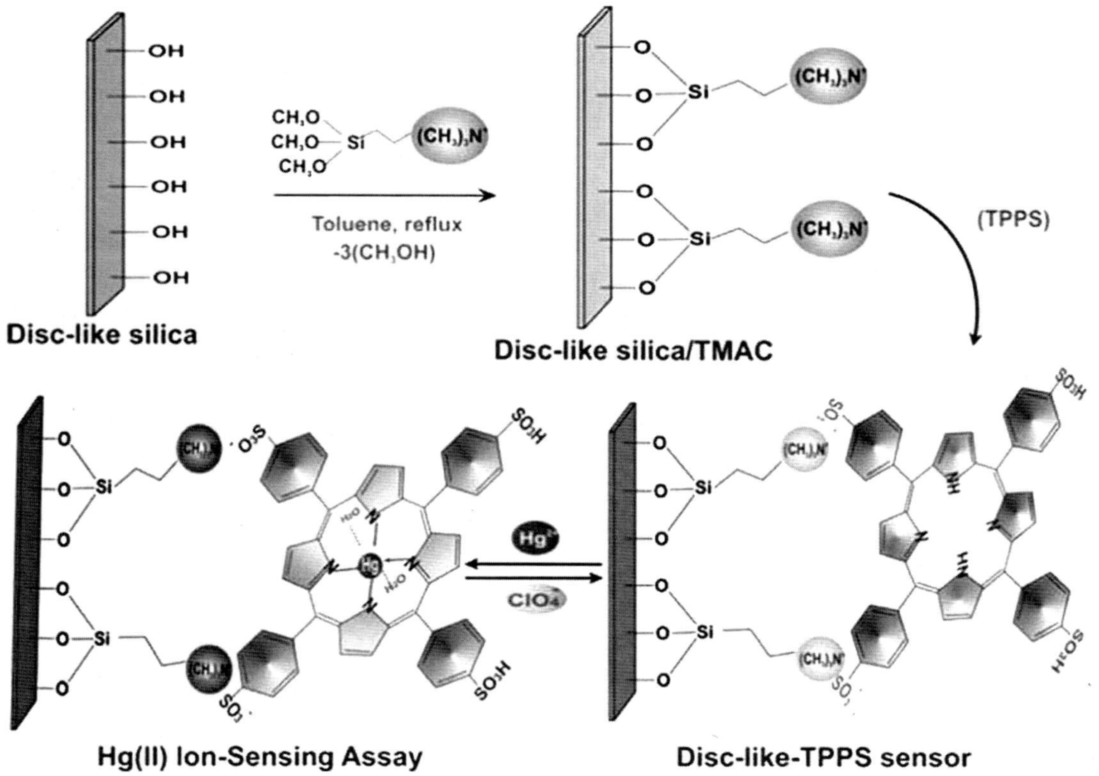

FIGURE 24.8. Mesoporous silica sensor for Hg(II) ion sensing. The grafting technique controlled the immobilization of two organic moieties, TMAC and TPPS, into the mesopore disc-like monoliths. The reversibility of the [Hg–TPPS] complex was occurred by using of 0.01 M ClO_4^- as stripping agent for several times without leaching of the TPPS probe from the pore surface. Reproduced from reference 51 with permission from Springer.

class of optical cage sensors exhibited a long-term stability of recognition and signaling functionalities of Hg(II) ions with an extraordinary sensitivity, selectivity, reusability, and fast kinetic detection and quantification of Hg(II) ions in aqueous media.

Based on a mesoporous material, Martínez-Máñez, Rurack, and co-workers have also designed dual-function hybrid materials for simultaneous chromofluorogenic detection and removal of Hg(II) ions in aqueous environments [52, 53]. In this case, mesoporous UVM-7 with homogeneously distributed pores of about 2–3 nm in size, a large specific surface area, and nanoscale particles was used as an inorganic support. The mesoporous solid is decorated with thiol groups that were treated with squaraine dye III to give a 2,4-bis(4-dialkylaminophenyl)-3-hydroxy-4-alkylsulfanylcyclobut-2-enone (APC) derivative that is covalently anchored to the inorganic silica matrix. Reaction of Hg(II) with the APC fragment of this hybrid material released the squaraine dye into the solution, which turned deep blue and strongly fluoresced. The response was noteworthy because both absorption and emission were found at the far end of the visible spectral window (642 and 670 nm, respectively) [54]. This material not only acts as a chemodosimeter that signals the presence of Hg(II) ions down to ppb concentrations, but also, at the same time, is an excellent adsorbent for the removal of mercury cations from aqueous solutions. The amount of adsorbed mercury ranges from 0.7 to 1.7 mmol g^{-1}, depending on the degree of functionalization. In addition, the hybrid material can be regenerated for both sensing and removal purposes (Figure 24.9).

Regarding the development of chromo- and fluorogenic indication systems for mercury derivatives, a large number of examples have been reported for the inorganic Hg(II) form, but only a very few studies have targeted to detect the extremely toxic methylmercury (CH_3Hg^+) ions. Following a similar strategy, Martínez-Mañez et al. [55] developed an organically capped mesoporous inorganic material for the selective CH_3Hg^+ determination through signal amplification. Inspired by gated ion channels and pumps, the proposed sensing mechanism relies on the opening of a pore of a mesoporous material that is controlled by the interaction of a certain molecular stimulus (the target species, CH_3Hg^+) at the receptors that close the gate. Although this reaction itself can already induce an optical response, a second process is

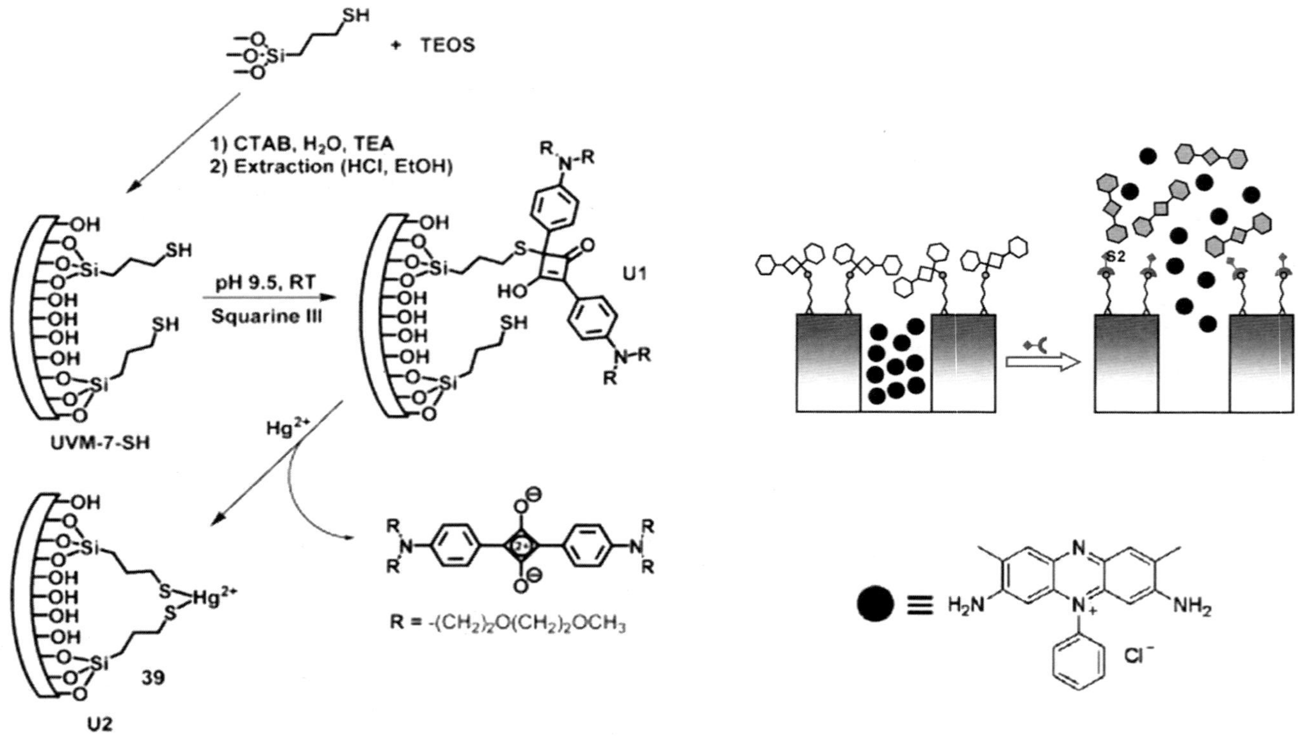

FIGURE 24.9. (**Left**) Scheme of the mechanism of formation and Hg(II) detection by UVM-7 mesoporous grafted with squaraine dye. (**Right**) Chromogenic detection of methylmercury ions using a mesoporous solid with safranine molecules (black circles) entrapped by APC groups (gray). Reproduced from reference 55 with permission from Wiley–VCH.

implemented in the system that leads to a strong signal amplification: The pores of the hybrid material are loaded with a large amount of dye molecules, which are only liberated upon the analyte-induced opening of the pores.

The sensing procedure is shown in Figure 24.9. The inorganic support is a calcined MCM-41 mesoporous solid that features homogeneous porosity, a facile surface functionalization, and a high loading capacity. The solid is first loaded with a dye, like safranine O, and is then capped with 2,4-bis(4-dialkylaminophenyl)-3-hydroxy-4-alkylsulfanylcyclobut-enone (APC) groups. Once reacted, the bulky APC groups significantly inhibit the release of the included dye by closing the pores. As the same authors have previously shown, the APC groups react with mercury ions, resulting in the coordination of the cation to the thiol group and the liberation of the restored squaraine chromophore. This concept suggests that the system should work equally well with thiophilic methylmercury ions as a "key" to unlock the "gate" (Figure 24.9). The use of lipophilicity partitioning guaranteed the required selectivity for CH_3Hg^+, which has not yet been accomplished by other low-cost and time-effective sensory methods, while reaching comparable sensitivity in the ppb range. This sensitivity is related to the signal amplification features of this method. The approach is very versatile and simple considering that the nanostructured material can be embedded into a hydrophobic sensing matrix, which can, for example, be dipped into a suspended and digested fish sample, thus extracting and indicating the lipophilic target species.

Lee et al. [56] have designed a novel example for chromogenic sensing of Hg(II) ions by using the hybrid nanotubes enabling a heterogeneous "naked-eye" colorimetric and spectrophotometric detection. This sensor was obtained by immobilization of the azo-coupled macrocyclic receptor shown in Figure 24.10 on a silica nanotube via a sol–gel reaction. The system recognized and separated Hg(II) ions with a high degree of selectivity among other metal ions. Furthermore, its extraction ability was also estimated by measuring the amount of Hg(II) ions adsorbed by ion chromatography showing that 95% of the ions were extracted. This suggests that the hybrid material is potentially useful as a stationary phase for the separation of Hg(II) ions in liquid chromatography. In order to extend the above performance to a portable chemosensor kit, the hybrid material was coated as a thin film onto a glass substrate. The supported sensor also changed from yellow to violet when dipped into Hg(II) solution, indicating the potential usefulness of the coated glass plates as portable chemosensors for Hg(II) ions. On the other hand, no

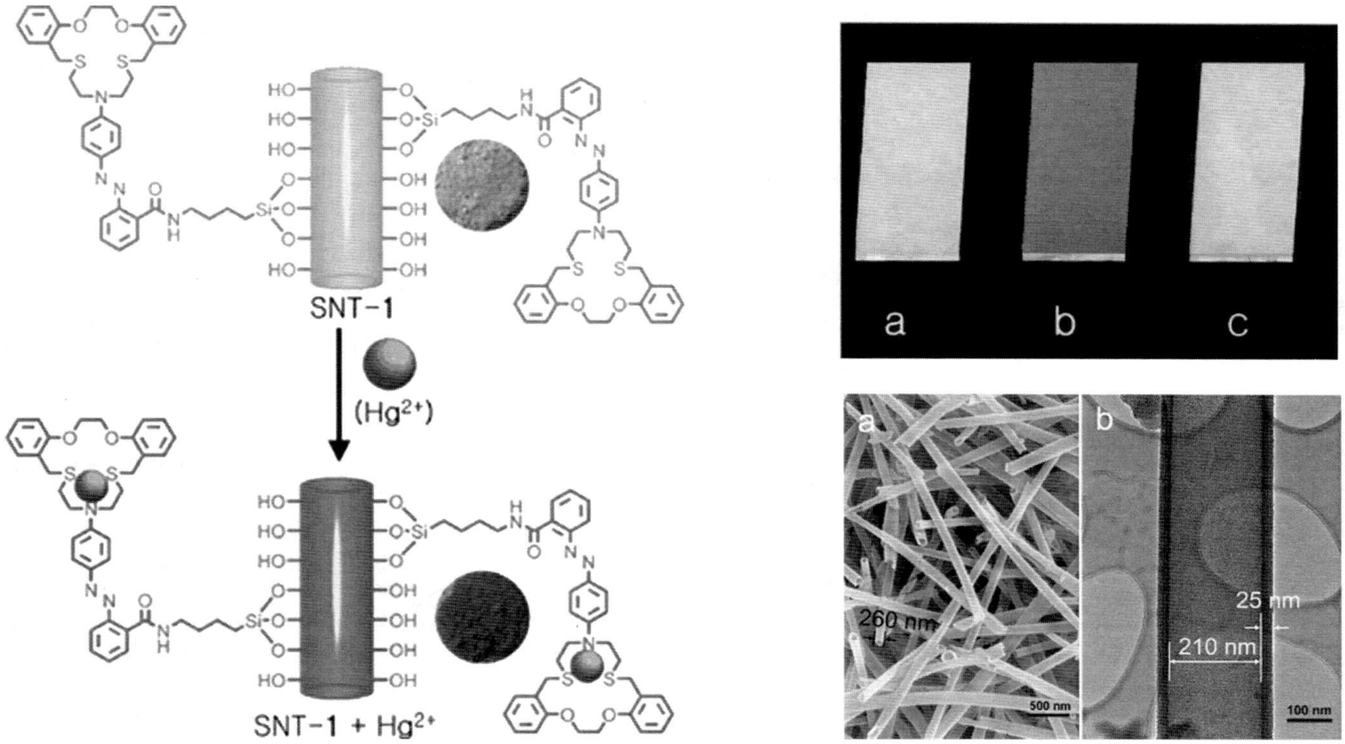

FIGURE 24.10. (**Left**) Scheme of mechanism of action of functionalized silica nanotubes for the mercury detection. (**Top, right**) Color changes of glass plates coated the functionalized silica nanotubes: (**a**) before immersion and after immersion with (**b**) Hg(II) (0.01 mM) and (**c**) Cu(II) mM) solution in water. (**Bottom, right**) (**a**) SEM and (**b**) TEM images of the functionalized silica nanotubes. Reproduced from reference 56 with permission from Wiley–VCH.

significant change in color was observed in other metal–ion solutions. Furthermore, these findings may lead to the development of a new type of tailor-made selective-ion sensing systems built by finely controlling the properties of both the molecular receptor and the solid support.

24.3.2.3 Polymeric Materials. Interaction of Hg(II) with the fluoroionophore chlorophenyl imino propenyl aniline (CPIPA) embedded in a plasticized polyvinyl chloride (PVC) membrane has been shown to be a very sensitive and highly selective irreversible optical chemical sensor for mercury ions [57]. The CPIPA in such polymeric matrix exhibited a high fluorescence quantum yield, long excitation and emission wavelength, and large Stokes' shift values that make it compatible with solid state optics. The sensor membranes were tested for the determination of mercury ions in aqueous solutions by batch and flow-through methods and both methods allow determination of Hg (II) in the working range of 10^{-9}–10^{-5} M with a detection limit of 4 ppb. In addition, the sensor exhibited excellent selectivity for Hg (II) with respect to several common alkali, alkaline earth and transition metal ions.

Conjugated polymers (CPs) have also been employed for the optical sensing of several analytes including metal ions such as Hg(II) [58]. For example, a Hg(II) sensor based on a CP that relies on a polythiophene containing thymine moieties and works in the turn-off mode has been developed [59]. There is also a Hg(II) sensor that works on the turn-on mode based on the conjugated polymer, poly(3-(3′-*N*,*N*,*N*-triethylamino-1′-propyloxy)-4-methyl-2,5-thiophene hydrochloride) (PMNT), and a label-free, mercury-specific oligonucleotide (MSO) probe that contains thymine groups [60]. In this case, the PMNT binds to the Hg(II)-free MSO and the Hg(II)–MSO complex exhibits distinguishable optical properties in response to the target-induced conformational change. Even with the naked eye, micromolar Hg(II) ion concentrations could be identified; by using the fluorometric method, the detection limit was improved to the nanomolar range (42 nM).

Cellulose is an abundant, inexpensive, biodegradable, and renewable biopolymer exhibiting very good mechanical properties and a large surface area; it is also water-absorbing, providing a good alternative to synthetic polymers. Cellulose can be modified to alter and tailor its chemical and physical properties, thus being a candidate for a sensor support where the sensing agent can be attached covalently to the surface, or just deposited by physicochemical treatments. Poplin et al. [61] took advantage of the fact that either functional probe molecules or cellulose can be dissolved in an ionic liquid to describe a simple approach for the development of a cellulose-based sensor. The sensor was based on the encapsulation of 1-(2-pyridylazo)-2-naphthol (PAN) in a cellulose support by the co-dissolution in 1-butyl-3-methylimidazolium chloride followed by regeneration with water to form strips which exhibit a proportionate (1:1) response to Hg(II) in aqueous solution. These cellulose-based sensors have several advantageous features, including low cost, high flexibility, a huge diversity of physical forms that can be realized, and easy preparation and handling.

Following the strategy of the covalent immobilization of the molecular probe on the cellulose matrix, a novel optical sensor for Hg(II) ions was described. The sensor is based on the immobilization of 4-hydroxy salophen on a transparent triacetylcellulose membrane [62]. When the sensing membrane is in contact with Hg(II) ions at pH 6.0, there is a change of color from white-yellow to orange. The sensing membrane is capable of determining Hg(II) ions with high selectivity over a dynamic range of 10^{-6} and 10^{-2} M with a limit of detection of 1.3×10^{-7} M. The membrane revealed very good selectivity with respect to many cations including alkali, alkaline earth, and transition metal ions.

Díez-Gil et al. [63] have achieved the preparation of test paper strips for high-selectivity detecting Hg(II) ions in mixed acetonitrile–water solutions using the simple physisorption of the excellent chromogenic sensor bis(ferrocenyl) azine onto a solid cellulose fiber by the dipping technique. Depending on the amount of mercury ions in contact with the detecting molecule, a spectacular color change in the cellulose indicator is produced, being possible to determine the concentration of Hg(II) ions either by naked eye or spectroscopically, although the detection limit was still far from the acceptable by the EPA in potable water. The same authors, in order to increase the limit of detection and the performance of the sensor material, described a novel method for homogeneously supporting a mercury molecular indicator on commercial nanopore-size membranes which consists of the coating of a nanoporous cellulose membrane surface by means of the filtration of an aqueous suspension of nanoparticles of the molecular probe [64]. Remarkably, this procedure promotes the coalescence of the suspended nanoparticles of the fluorescent and highly Hg(II) selective 1-(4′-methoxyphenyl)-4-(1′-pyrenyl)-2,3-diaza-1,3-butadiene **32** on the membrane surface yielding a nanocomposite membrane with large surface area that can be used as a cheap analytical test filter for the highly selective and reliable quantitative detection of Hg(II) ions in water at concentrations down to the ppb level (Figure 24.11). Thus, the sensor material was achieved by a simple process that uses the "reprecipitation method" for preparing aqueous nanodispersions of the solid azine and their deposition on the surface of a commercial membrane with ~100-nm pore size by filtering the suspensions through the membrane.

As has been described in previous sections, DNA-based sensors with gold nanoparticles have emerged as highly selective, simple, and stable, taking advantage of the strong coordination of Hg(II) to the thymine–thymine (T–T) mismatched pair. To further reduce the operational cost and avoid the involvement of Au-NPs in the assays, a G-quadruplex DNAzyme, which is a DNA mimic of horseradish peroxidise (HRP) when bound to hemin, can be used as the signaling unit because it catalyzes H_2O_2-mediated oxidation and

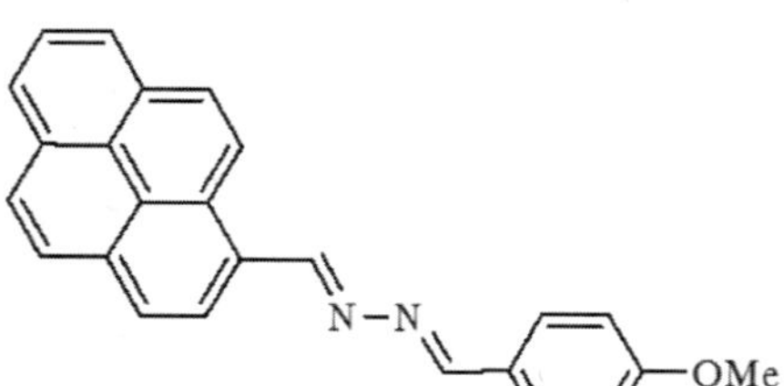

32

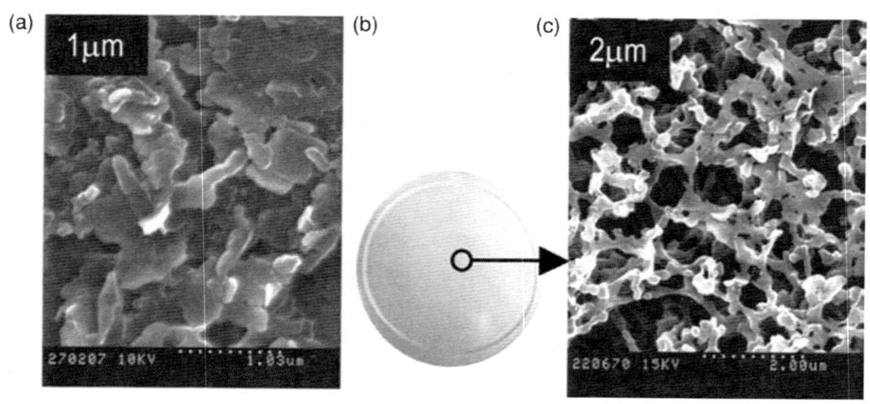

FIGURE 24.11. (**a**) SEM image of a polycarbonate membrane surface after filtering a suspension of nanoparticles. (**b**) Photograph of a mixed cellulose ester membrane loaded by filtering a suspension of nanoparticles of the receptor, showing a homogeneous light-green colored surface due to the uniform distribution of the mercury indicator. (**c**) SEM image of a mixed cellulose ester membrane loaded with the receptor showing the high homogeneity of the coating of the fibers of cellulose. Reproduced from reference 64 with permission from the Royal Society of Chemistry.

generates color from reducing molecules such as 2,2′-azino-bis(3-ethylbenzthiazoline-6-sulfonic acid) (ABTS).

Following this strategy, researchers have reported a straightforward method for the visual detection of Hg(II) on a piece of a nylon membrane with improved sensitivity and selectivity. This method can be adapted into a portable device for in-field monitoring of Hg(II) ions contamination in water. Thus, a DNA probe was designed to have two G-rich sequences on the 5′ and 3′ ends with a T-rich portion in between. Once deposited on the membrane and provided with the hemin cofactor, the G-quadruplex constructed from the G-rich sequences could catalyze the oxidation of 3,3′,5,5′-tetramethylbenzidine (TMB) and produce an insoluble oxidation product on the membrane, showing blue color. When Hg(II) ions exist in the sample, the T-rich sequence becomes rigid due to the formation of T–T mismatches mediated by Hg(II) ions and prevents the folding of the G-quadruplex and thus the TMB oxidation, leading to a fainter blue color observable on the nylon membrane [65]. Thus, this is a simple and sensitive approach to detect Hg(II) ions colorimetrically using blotting membrane with detection limit as low as 0.1 nM using a scanner.

24.4 SUMMARY

In recent years, remarkable progress in the design, synthesis, and characterization of Hg(II) probes has been done. New small-molecule fluorescent and colorimetric Hg(II) detectors that exhibit a range of sensing characteristics have been published. Through both rational design but also serendipity, a number of these probes have been obtained for practical applications. This chapter addresses the recent literature on Hg(II) probes based on small-molecule receptors with a special focus on its nanostructuration to obtain sensitive solid-supported sensors. The chapter is divided into two main parts: (i) the use of Hg(II) probes in solution either in the form of small-molecular receptors or as hybrid nanoparticle dispersed in the aqueous media and (ii) the use of Hg(II) probes confined on the surface of a solid material or inside a nanostructured one. In the first part we report the coordination of Hg(II) ions to S or N atom-based receptors and mercury-mediated reactions describing a number of turn-on Hg(II) molecular detectors based on such strategies that operate in aqueous/organic solvent mixtures. These probes employ a variety of chromophores, metal binding-ligands, and detection strategies, and they are generally used in solution, a fact that limits its real use for sensing purposes. This drawback is solved when such molecular receptors are nanostructured on the surface of nanoparticles or organized/grafted on surfaces or incorporated into a solid matrix. Any of these two approaches enable the obtaining of highly sensitive solid-supported probes that work in aqueous media showing promising applications. Thus, this chapter makes a special emphasis on the newly developed molecular nanomaterials that represent an exciting approach to mercury detection that benefits from the versatility of chemical synthesis and nanoscience. According to the different solid support used to anchor the receptor, there are functionalized nanoparticles (gold, carbon, core–shell, or quantum dots), glass or gold surfaces, mesoporous materials, or different polymeric matrices. The use of nanostructrued molecular materials have the great advantage to increase the surface area of the sensor and, thus, increase the level of sensitivity of the materials obtained. Another advantage is the possibility to operate in aqueous media using molecular probes that in solution only work in the presence of organic solvents. In fact, some of the mercury sensor materials described in this chapter, in addition to sensors, provide the scientific community with a versatile toolkit for monitoring this metal ion in a wide variety of contexts. The fundamental chemistry and nanoscience described may also provide new frameworks and translation into devices for addressing related problems.

REFERENCES

1. (a) Nolan, E. M., Lippard, S. J. (2008). Tools and tactics for the optical detection of mercuric ion. *Chemical Reviews,* **108**, 3443–3480. (b) Quang, D. T., Kim, J. S. (2010). Fluoro- and chromogenic chemodosimeters for heavy metal ion detection in solution and biospecimens. *Chemical Reviews*, **110**, 6280-6301.
2. Zhu, L., Qin, J., Yang, C. (2010). New metal-coordination-inhibited charge transfer emission for terfluorenes: Highly sensitive and selective detection for Hg^{2+} with ratiometric "turn-on" fluorescence response. *Chemical Communications*, **46**, 8755–8757.
3. Martínez, R., Espinosa, A., Tárraga, A., Molina, P. (2010). A new bis(pyrenyl)azadiene-based probe for the colorimetric and fluorescent sensing of Cu(II) and Hg(II). *Tetrahedron*, **66**, 3662–3667.
4. Yang, Y., Gou, X., Blecha, J., Cao, H. (2010). A high selective pyrene based fluorescent sensor toward Hg^{2+} detection. *Tetrahedron Letters*, **51**, 3422–3425.
5. (a) Fang, C., Zhou, J., Liu, X., Cao, Z., Shangguan, D. (2011). Mercury(II)-mediated formation of imide–Hg–imide complexes. *Dalton Transactions*, **40**, 899–903. (b) Xiangjun, L., Qi, C., Bing, T., Cheng, X., Shangguan, D. (2009). Highly selective phthalocyanine-thymine conjugate sensor for Hg^{2+} based on target induced aggregation. *Analytical Chemistry*, **81**, 3699–3704. (c) Guo, L., Hu, H., Sun, R., Chen, G. (2009). Highly sensitive fluorescent sensor for mercury ion based on photoinduced charge transfer between fluorophore and pi-stacked T-Hg(II)-T base pairs. *Talanta*, **79**, 775–779. (d) Chen, Y., Bai, H., Hong, W., Shi, G. (2009). Fluorescent detection of mercury ions in aqueous media with the complex of a cationic oligopyrene derivative and oligothymine. *Analyst*, **134**, 2081–2086.

6. Joshi, N. P., Lohani, C. R., Lee, K-H. (2010). A highly sensitive and selective detection of Hg(II) in 100% aqueous solution with fluorescent labeled dimerized Cys residues. *Organic & Biomolecular Chemistry*, **8**, 3220–3226.
7. Zhou, Y., Zhu, C.-Y., Gao, X.-S., You, X.-Y., Yao, C. (2010). Hg^{2+}-selective ratiometric and "off–on" chemosensor based on the azadiene-pyrene derivative. *Organic Letters*, **12**, 2566–2569.
8. (**a**) Bambang Kuswandi, N., Verboom, W. (2009). Selective chemosensor for Hg(II) ions based on tris[2-4-phenyldiazenyl]phenylaminoethoxy]cyclotriveratrylene in aqueous samples. *Analytica Chimica Acta*, **655**, 75–79. (b) Min, G., Wei, X., Mingyun, G., Jianhua, S., Gui, Y. (2009). New azaobenzene dye colorimetric and ratiometric chemosensors for mercury(II) ion. *Chinese Journal of Chemistry*, **27**, 1773–1776.
9. Misra, A., Shahid, M. (2010). Chromo and fluorogenic properties of some azo-phenol derivatives and recognition of Hg^{2+} ion in aqueous medium by enhanced fluorescence. *Journal Physical Chemistry*,**114**, 16726–16739.
10. Romero, T., Caballero, A., Espinosa, A., Tárraga, A., Molina, P. (2009). A multiresponsive two-arm ferrocene-based chemosensor molecule for selective detection of mercury. *Dalton Transactions*, 2121–2129.
11. Moorthy, S., Mandal, A. K., Saha, S., Suresh, E., Mandoli, A., Di Liddo, R., Parnigotto, P. P., Das, A. (2010). Azine-based receptor for recognition of Hg^{2+} ion: Crystallographic evidence and imaging application in live cells. *Organic Letters*, **12**, 5406–5409.
12. Otón, F., Espinosa, A., Tárraga, A., Ratera, I., Wurst, K., Veciana, J., Molina, P. (2009). Mononuclear ferrocenophane structural motifs with two thiourea arms acting as a dual binding site for anions and cations. *Inorganic Chemistry*, **48**, 1566–1576.
13. Guo, Z., Zhu, W., Zhu, M., Wu, X., Tian, H. (2010). Near-infrared cell-permeable Hg^{2+}-selective ratiometric fluorescent chemodosimeters and fast indicator paper for $MeHg^+$ based on tricarbocyanines. *Chemistry: A European Journal*, **16**, 14424–14432.
14. Shunmugam, R., Gabriel, G. J., Smith, C. E., Aamer, K. A., Tew, G. N. (2008). A highly selective colorimetric aqueous sensor for mercury. *Chemistry: A European Journal*, **14**, 3904–3907.
15. Ahamed, B. N., Ravikumar, I., Ghosh, P. (2009). A new chemosensor that signals Hg(II), Cu(II) and Zn(II) at different emission wavelengths: Selectivity toward Hg(II) in acetonitrile. *New Journal of Chemistry*, **33**, 1825–1828.
16. Vaswani, K. G., Keränen, M. D. (2009). Detection of aqueous mercuric ion with a structurally simple 8-hidroxiquinoline derived on–off fluorosensor. *Inorganic Chemistry*, **48**, 5797–5800.
17. Wan, Y., Niu, W., Behof, W. J., Wang, Y., Boyle, P., Gorman, C. B. (2009). Aminoisoquinolines as colorimetric Hg^{2+} sensors: The importance of molecular structure and sacrificial base. *Tetrahedron*, **65**, 4293–4297.
18. Wang, Y., Qian, G., Xiao, Z., Wang, H., Long, L., Wang, H., Li, Z., Liu, X. (2010). Synthesis and characterization of anthracene-based fluorophore and its interactions with selected metal ions. *Inorganica Chimica Acta*, **363**, 2325–2332.
19. Lau, Y. H., Price, J. R., Todd, M. H., Rutledge, P. J. (2011). A click fluorophore sensor that can distinguish Cu(II) and Hg(II) via selective anion-induced demetallation. *Chemistry: A European Journal*, **17**, 2850–2858.
20. Nguyen, D. M., Frazer, A., Rodriguez, L., Belfield, K. D. (2010). Selective fluorescente sensing of zinc and mercury ions with hydrophilic 1,2,3-triazolyl fluorene probes. *Chemistry of Materials*, **22**, 3472–3481.
21. Lu, H., Xiong, L., Liu, H., Yu, M., Shen, Z., Li, F., You, X. (2009). A highly selective and sensitive fluorescent turn-on sensor for Hg^{2+} and its application in live cell imaging. *Organic & Biomolecular Chemistry*, **7**, 2554–2558.
22. (a) Atilgan, S., Kutuk, I., Ozdemir. (2010). A near IR distiryl BODIPY-based ratiometric fluorescent chemosensor for Hg(II). *Tetrahedron Letters*, **51**, 892–894. (b) Atilgan, S., Ozdemir, T., Akkaya, E. U. (2010). Selective Hg(II) sensing with improved Stokes shift by coupling the internal charge transfer process to excitation energy transfer. *Organic Letters*, **12**, 4792–4795.
23. Zapata, F., Caballero, A., Molina, P., Tárraga, A. (2010). A ferrocene-quinoxaline derivative as a highly selective probe for colorimetric and redox sensing of toxic mercury(II) cations. *Sensors*, **10**, 11311–11321.
24. Zapata, F., Caballero, A., Espinosa, A., Tárraga, A, Molina, P. (2009). A selective redox and chromogenic probe for Hg(II) in aqueous environment based on a ferrocene–aza-quinoxaline dyad. *Inorganic Chemistry*, **48**, 11566–11575.
25. Alfonso, M., Tárraga, A., Molina, P. (2010). Ferrocene-based multichannel molecular chemosensors with high selectivity and sensitivity for Pb(II) and Hg(II) metal. *Dalton Transactions*, **39**, 8637–8645.
26. Alfonso, M., Tárraga, A., Molina, P. (2011). Ferrocene-based heteroditopic receptors displaying high selectivity toward lead and mercury metal cations through different channels. *Journal of Organic Chemistry*, **76**, 939–947.
27. Alfonso, M., Sola, A., Caballero, A., Tárraga, A., Molina, P. (2009). Heteroditopic ligands based on ferrocenyl benzimidazoles fused to an additional diaza heterocyclic ring system. *Dalton Transactions*, 9653–9658.
28. (a) Huang, J., Xu, Y., Qian, X. (2009). A rhodamine-based Hg^{2+} sensor with high selectivity and sensitivity in aqueous solution: A NS2-containing receptor. *Journal of Organic Chemistry*, **74**, 2167–2170. (b) Lin, W., Cao, X., Ding, Y., Yuan, L., Yu, Q. (2010). A reversible fluorescent Hg^{2+} chemosensor based on a receptor composed of a thiol atom and an alkene moiety for linving cell fluorescence imaging. *Organic & Biomolecular Chemistry*, **8**, 3618–3620. (c) Lin, W., Cao, W., Ding, Y., Yuan, L., Long, L. (2010). A highly selective and sensitive fluorescent probe for Hg^{2+} imaging in live cells based on a rhodamine–thioamide–alkyne scaffold. *Chemical Communications*, **46**, 3529–3531. (d) Lee, Y. H., Lee, M. H., Zhang, J. F., Kim, J. S. (2010). *Journal for Organic Chemistry*, **75**, 7159–7165. (e) Jiang, L., Wang, L., Zhang, B., Yin, G., Wang, R.-Y. (2010). Cell compatible fluorescent chemosensor for

Hg^{2+} with high sensitivity and selectivity based on the rhodamine fluorophore. *European Journal of Inorganic Chemistry,* 4438–4443. (f) Kumar, M., Kumar, N., Bhalla, V., Singh, H., Sharma, P. R., Kaur, T. (2011). Naphthalimide appended rhodamine derivative: Through bond energy transfer for sensing Hg^{2+} ions. *Organic Letters*, **13**, 1422–1425. (g) Wang, H., Li, Y., Xu, S., Li, Y., Zhou, C., Fei, X., Sun, L., Zhang, C., Li, Y., Yang, Q., Xu, X. (2011). Rhodamine-based highly sensitive colorimetric off-on fluorescent chemosensor for Hg^{2+} in aqueous solution and for live cell imaging. *Organic & Biomolecular Chemistry*, **9**, 2850–2855.

29. Voutsadaki, S., Tsikalas, G. K., Klontzas, E., Froudakis, G. E., Katerinopoulos H. E. (2010). A "turn-on" coumarin-based fluorescent sensor with high selectivity for mercury ions in aqueous media. *Chemical Communications*, **46**, 3292–3294.
30. Lim, C. S., Kang, D. W., Tian, Y. S., Han, J. H., Hwang, H. L., Cho, B. R. (2010). Detection of mercury in fish organs with a two-photon fluorescent probe. *Chemical Communications*, **46**, 2388–2390.
31. Chung, S.-K., Tseng, Y.-R., Chen, C.-Y., Sun, S.-S. (2011). A selective colorimetric Hg^{2+} probe featuring a styryl dithiaazacrown containing platinum(II) terpyridine complex through modulation of the relative strength of ICT and MLCT transitions. *Inorganic Chemistry*, **50**, 2711–2713.
32. Mahato, P., Ghosh, A., Saha, S., Mishra, S., Mishra, S. K., Das, A. (2010). Recognition of Hg^{2+} using diametrically disubstituted cyclam unit. *Inorganic Chemistry*, **49**, 11485–11492.
33. Sadhu, K. K., Sen, S., Bharadwaj, P. K. (2011). Cryptand derived fluorescence signaling systems for sensing Hg(II) ion: A comparative study. *Dalton Transactions*, **40**, 726–734.
34. (a) Prodi, L. (2005). Luminescent chemosensors: From molecules to nanoparticles. *New Journal of Chemistry*, **29**, 20–31. (b) Murphy, C. J. (2002). Optical sensing with quantum dots. *Analytical Chemistry*, **74**, 520A–526A. (c) Walt, D. R. (2002). Nanomaterials: Top-to-bottom functional design. *Nature Materials*, **1**, 17–18.
35. Rex, M., Hernandez, F. E., Campiglia, A. D. (2006). Pushing the limits of mercury sensors with gold nanorods. *Analytical Chemistry*, **78** (2), 445–451.
36. Huang, C.-C., Chang, H.-T. (2006). Selective gold-nanoparticle-based "turn-on" fluorescent sensors for detection of mercury(II) in aqueous solution. *Analytical Chemistry*, **78**, 8332–8338.
37. Lee, J.-S., Han, M. S., Mirkin, C. A. Colorimetric detection of mercuric ion (Hg^{2+}) in aqueous media using DNA-functionalized gold nanoparticles. (2007). *Angewandte Chemie International Edition*, **46**, 4093–4096.
38. Han, M. S., Lytton-Jean, A. K. R., Oh, B.-K., Heo, J., Mirkin, C. A. (2006). Colorimetric screening of DNA-binding molecules with gold nanoparticle probes. *Angewandte Chemie International Edition*, **45**, 1807–1810.
39. Liu, X., Cheng X., Bing, T., Fang, C., Shangguan, D. (2010). Visual detection of Hg^{2+} with high selectivity using thymine modified gold nanoparticles. *Analytical Science*, **26**, 1169–1172.
40. Zhang, L., Li, T., Li, B., Li, J., Wang, E., (2010). Carbon nanotube–DNA hybrid fluorescent sensor for sensitive and selective detection of mercury(II) ion. *Chemical Communicantions*, **46**, 1476–1478.
41. He, S., Song, B., Li, D., Zhu, C., Qi, W., Wen, Y., Wang, L., Song, S., Fang, H., Fan, C. (2010). A graphene nanoprobe for rapid, sensitive, and multicolor fluorescent DNA analysis. *Advanced Functional Materials*, **20**, 453–459.
42. Li, H., Zhai, J., Sun, X. (2011). Nano-C_{60} as a novel, effective fluorescent sensing platform for mercury(II) ion detection at critical sensitivity and selectivity. *Nanoscale*, **3**, 2155–2157.
43. Li, H., Zhai, J., Tiana, J., Luoa, Y., Sun, X. (2011). Carbon nanoparticle for highly sensitive and selective fluorescent detection of mercury(II) ion in aqueous solution. *Biosensors and Bioelectronics*, **26**, 4656– 4660.
44. Freeman, R., Finder, T., Willner, I. (2009). Multiplexed analysis of Hg^{2+} and Ag^+ ions by nucleic acid-functionalized CdSe/ZnS quantum dots and their use for logic gate operations. *Angewandte Chemie International Edition*, **48**, 7818–7821.
45. Gonçalves, H., Jorge, P. A. S., Fernandes, J. R. A., Esteves da Silva, J. C. G. (2010). Hg(II) sensing based on functionalized carbon dots obtained by direct laser ablation. *Sensors and Actuators B*, **145**, 702–707.
46. Jung, J. H., Lee J. H., Shinkai, S. (2011). Functionalized magnetic nanoparticles as chemosensors and adsorbents for toxic metal ions in environmental and biological fields. *Chemistry Society Reviews*, **40**, 4464–4474.
47. Park, M., Seo, S., Lee, I. S., Jung, J. H. (2011). Ultraefficient separation and sensing of mercury and methylmercury ions in drinking water by using aminonaphthalimide-functionalized $Fe3O_4@SiO_2$ core/shell magnetic nanoparticles. *Chemical Communications*, **46**, 4478–4480.
48. Delgado-Pinar, E., Montoya, N., Galiana, M., Albelda, M. T., Frías, J. C., Jiménez, H. R., García-España, E., Alarcón, J. (2010). Preparation of Hg^{2+} selective fluorescent chemosensors based on surface modified core–shell aluminosilicate nanoparticles. *New Journal of Chemistry*, **34**, 567–570.
49. Díez-Gil, C., Martínez, R., Ratera, I., Hirsh, T., Espinosa, A., Tárraga, A., Molina, P., Wolfbeis O. S., Veciana J. (2011). Selective picomolar detection of mercury(II) using optical sensors. *Chemical Communications*, **47**, 1842–1844.
50. Lee, J., Jun, H., Kim, J. (2009). Polydiacetylene–liposome microarrays for selective and sensitive mercury(II) detection. *Advanced Materials*, **21**, 3674–3677.
51. El-Safty, S. A. (2009). Organic–inorganic hybrid mesoporous monoliths for selective discrimination and sensitive removal of toxic mercury ions. *Journal of Materials Science, 44*, 6764–6774.
52. Quang, D. T., Kim, J. S. (2010). Fluoro- and chromogenic chemodosimeters for heavy metal ion detection in solution and biospecimens. *Chemical Reviews*, **110**, 6280–6301.
53. Ros-Lis, J. V., Casasús, R., Comes, M., Coll, C., Marcos, M. D., Martínez-Máñez, R. Sancenón, F., Soto, J., Amor, P., El Haskouri, J., Garr. N., Rurack, K. (2008). A mesoporous 3D

hybrid material with dual functionality for Hg^{2+} detection and adsorption. *Chemistry an European. Journal*, **14**, 8267–8278.

54. Ros-Lis, J. V., Marcos, M. D., Martinez-Mañez, R., Rurack, K., Soto, J. (2005). A Regenerative chemodosimeter based on metal-induced dye formation for the highly selective and sensitive optical determination of Hg^{2+} ions. *Angewandte Chemie International Edition*, **44**, 4405–4407.
55. Climent, E., Marcos, M. D., Martínez-Máñez, R., Sancenón, F., Soto, J., Rurack, K., Amorós, P. (2009). The determination of methylmercury in real samples using organically capped mesoporous inorganic materials capable of signal amplification. *Angewandte Chemie International Edition*, **48**, 8519–8522.
56. Lee, S. J., Lee, J.-E., Seo, J., Jeong, I. Y., Lee, S. S., Jung, J. H. (2007). Optical sensor based on nanomaterial for the selective detection of toxic metal ions. *Advanced Functional Materials*, **17**, 3441–3446.
57. Ertekin, K., Oter, O., Ture, M., Denizalti, S., Cetinkaya, E. (2010). A long wavelength excitable fluorophore; chloro phenyl imino propenyl aniline (CPIPA) for selective sensing of Hg (II). *Journal of Fluorescence*, **20**, 533–540.
58. Thomas, S. W., Joly, G. D., Swager, T. M. (2007). Chemical sensors based on amplifying fluorescent conjugated polymers. *Chemical Reviews*, **107**, 1339–1386.
59. Tang, Y. L., He, F., Yu, M. H., Feng, F. D., An, L. L., Sun, H., Wang, S.,Li, Y. L., Zhu, D. B. (2006). A Reversible and Highly Selective Fluorescent Sensor for Mercury(II) Using Poly(thiophene)s that Contain Thymine Moieties *Macromolecular Rapid Communication*, **27**, 389–392.
60. Liu, X., Tang, Y., Wang, L., Zhang, J., Song, S., Fan, C., Wang, S. (2007). Optical detection of mercury(II) in aqueous solutions by using conjugated polymers and label-free oligonucleotides. *Advanced Materials*, **19**, 1471–1474.
61. Poplin, J. H., Swatloski, R. P., Holbrey, J. D., Spear, S. K., Metlen, A., Grätzel, M., Nazeeruddin, M. K., Rogers, R. D. (2007). Sensor technologies based on a cellulose supported platform. *Chemical Communications*, 2025–2027.
62. Ensafi, A. A., Far, A. K., Meghdadi, S. (2008). Highly selective optical sensor for mercury assay based on covalent immobilization of 4-hydroxy salophen on a triacetylcellulose membrane. *Sensors and Actuators B*, **133**, 84–90.
63. Díez-Gil, C., Caballero, A., Ratera, I., Tárraga, A., Molina, P., Veciana, J. (2007). Naked-eye and selective detection of mercury (II) ions in mixed aqueous media Using a Cellulose-based Support. *Sensors*, **7**, 3481–3488.
64. Díez-Gil, C., Martínez, R., Ratera, I., Tárraga, A., Molina, P., Veciana, J. (2008). Nanocomposite membranes as highly selective and sensitive mercury(II) detector. *Journal of Materials Chemistry*, **18**, 1997–2002.
65. Li, J., Yao, J., Zhong, W. (2009). Membrane blotting for rapid detection of mercury(II) in water. *Chemical Communications*, 4962–4964.

25

ORGANIC NANOMATERIALS FOR EFFICIENT BULK HETEROJUNCTION SOLAR CELLS

PAVEL A. TROSHIN AND NIYAZI SERDAR SARICIFTCI

25.1 INTRODUCTION

25.1.1 Photovoltaic Devices for Solar Energy Conversion

Solar energy has primary importance among all renewable energy resources. It is a remarkable fact that sun gives to the earth surface in one hour so much energy that it would be enough to cover the annual need of the present population of our planet. Photovoltaic devices allow for the most simple and efficient conversion of the solar energy to electricity. The power conversion efficiencies of the best laboratory devices based on silicon reach 24.4% [1]. The first generation of solar cells based on crystalline silicon is known since the middle of the last century (1954 by AT&T Bell Labs). However, for a long time, wide-scale distribution of such devices has been limited to aerospace applications by their extremely high cost. Today, the crystalline silicon modules can be installed at the cost of 2–3 USD per every watt of nominal power generated at maximal (peak) (denoted as watt-peak, W_p) [2].

The second generation of solar cells is based mainly on the amorphous and multicrystalline silicon and metal chalcogenides such as CdTe, $CuIn_xGa_{(1-x)}Se_2$ (CIGS) and other similar thin-film absorbers. The best devices of the CIGS type yield power conversion efficiencies of 15–20% [3]. The power conversion efficiencies of such cells in general stay below the level achieved for the first-generation solar cells. However, second-generation solar cells are less expensive than the devices of the first generation. Therefore, the resulting economical effect is quite significant. The price of the generated energy becomes much lower, which stimulated growth of the thin-film photovoltaic technologies market [4]. Unfortunately, toxicity of cadmium telluride or gallium arsenide and expected shortage of indium and gallium world production will have to be considered for these technologies as an energy solution for the future [5]. At the same time, the efficiency of the cells based on amorphous silicon remained stuck at around 10.1%, which resulted in decrease of their market share since 2009–2010 [6].

The tandem and triple-junction solar cells are expected be able to operate at potentially ultrahigh light power conversion efficiencies (>50%) for concentrated sun light. Power conversion efficiencies of the best tandem solar cells are near 43% while their price remains much higher than the target 100–150 USD/m^2 in spite of using inexpensive solar light concentrators [7].

Organic solar cells offer another alternative that yields moderate power conversion efficiencies of 8–10% at very low module costs (40–60 USD/m^2) and lifetimes of 5–10 years. Indeed, laboratory prototypes of organic solar cells with the active areas as large as 1–8 cm^2 demonstrated certified power conversion efficiencies of 8–10% (see Table 25.1 below). Lifetimes of 7 years were projected for conjugated polymer-based devices using accelerated tests [8]. At the same time, Heliatek confirmed continuous operation lifetimes of 16,000 hr (100 mW/cm^2, AM1.5) for their small molecular-based double-junction devices [9]. Further improvements of organic solar cells in terms of performance, lifetime, module design and production technologies might lead to a breakthrough in the renewable energies. Ultimately, the energy generated by solar light conversion should become less expensive than the energy produced by combustion of fossil fuels.

Organic Nanomaterials: Synthesis, Characterization, and Device Applications, First Edition. Edited by Tomás Torres and Giovanni Bottari.

In this chapter, we will focus on the organic semiconductor nanomaterials used for construction of efficient bulk heterojunction solar cells. The content of this chapter and provided citations are organized for didactical purposes only and do not reflect the chronology of the research in the field and/or have no claim of completeness. The further-interested reader is referred to monographs and reviews addressing different aspects of organic photovoltaics [10–15].

25.1.2 Construction of Bulk Heterojunction Solar Cell

The bulk heterojunction concept was invented in 1992 and was reported in a patent application [16]. A fast photoinduced charge separation between the [60]fullerene (or its functionalized derivatives) and conjugated polymer was revealed and the first *p–n* bulk heterojunction plastic solar cell was realized [17, 18]. According to the bulk heterojunction concept, *p*-type and *n*-type materials are mixed together and self-organize on nanoscale to form three-dimensional interpenetrating networks capable of efficient charge generation and transport. The strongest advantage of bulk heterojunction solar cells is significantly increased interface between *p*-type and *n*-type materials compared to the planar heterojunction devices known before [19, 20]. Typically, a size of the interconnected grains formed by the two phases should stay in the range of the exciton diffusion length (5–20 nm). Such nanomorphology allows more or less all excitons generated in the active layer to reach the interface and contribute to the generation of free charge carriers. That is why internal quantum efficiency of some bulk heterojunction solar cells approaches 100% [21].

A schematic layout of the architecture of an organic bulk heterojunction solar cell is shown in Figure 25.1. The active layer of this device comprises interpenetrated phases of electron donor material and electron acceptor material capable of efficient hole and electron transport in opposite directions toward respective electrodes. In order to avoid charge recombination at the electrodes, some buffer layers of pristine materials (or some other charge selective materials) should be introduced under the electrodes. For instance, the positive electrode, which extracts holes from the active layer, should form a direct contact only with a *p*-type material. At the same time, just a pure phase of the *n*-type material should be adjacent to the negative electron-collecting electrode.

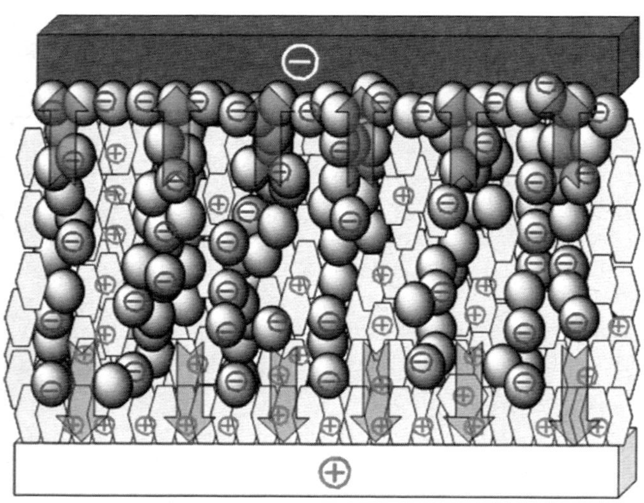

FIGURE 25.1. Schematic layout of bulk heterojunction solar cell.

Bulk heterojunction solar cells were explored intensively during the last decade. Many different materials were evaluated as *p*-type and *n*-type semiconductor components for construction of the devices. Blending *p*-type and *n*-type materials together produced polymer/polymer, polymer–small molecule, polymer–carbon nanotube, and polymer–inorganic nanoparticles, as well as small-molecule–small-molecule composites. Several types of these composite systems showed comparably high performances in organic solar cells (Table 25.1).

25.1.3 Operation Principle of Organic Bulk Heterojunction Solar Cells

Here we describe the operation principle of organic bulk heterojunction solar cells just briefly. The interested reader might be referred to specialized reviews and monographs [43, 44]. Any organic bulk heterojunction solar cell comprises a blend of *p*-type and *n*-type semiconductor materials which harvests photons and generates free charge carriers. Organic *p*-type material typically serves as electron donor, while an *n*-type component works as an electron acceptor. Photoinduced charge separation between donor and acceptor is a fundamental principle of operation of organic photovoltaic devices as well as the natural photosynthetic systems.

At the first step (step I in Figure 25.2), absorption of photons in the donor–acceptor blend leads to the generation of excitons D* and A*. In an ideal case, donor and acceptor components have complementary absorption spectra, and both contribute significantly to the photon harvesting. However, typical electron acceptor components such as [60]fullerene and its derivatives are quite poor visible light absorbers because they have symmetry-forbidden transitions above 450 nm. On the contrary, [70]fullerene and its derivatives possess relatively strong absorptions in the visible range (up to ~700 nm) that make them very promising *n*-type materials for organic photovoltaics. Unfortunately, high costs of [70]fullerene and its functional derivatives limit industrial application of this type of materials. The excitons D* and A* have to diffuse in order to reach the interface between the donor and acceptor materials where charge separation might take place (Step II in Figure 25.2). Therefore it is crucially important to adjust the morphology of

TABLE 25.1. Some Advanced Material Composite Systems Used in Bulk Heterojunction Solar Systems

Description of the Materials	Power Conversion Efficiency (%)	Reference
Fullerene/Polymer		
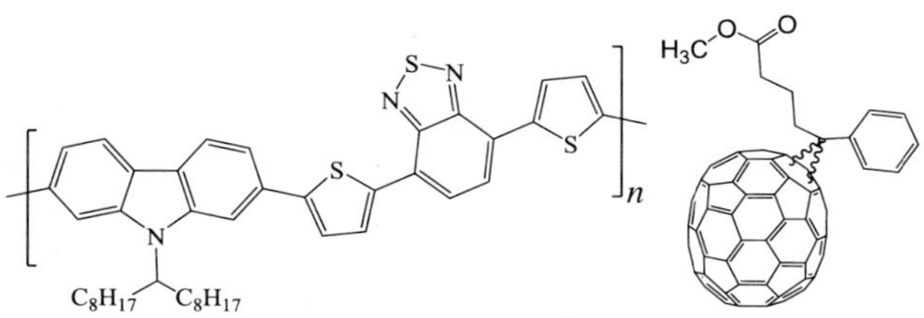	7.1	22
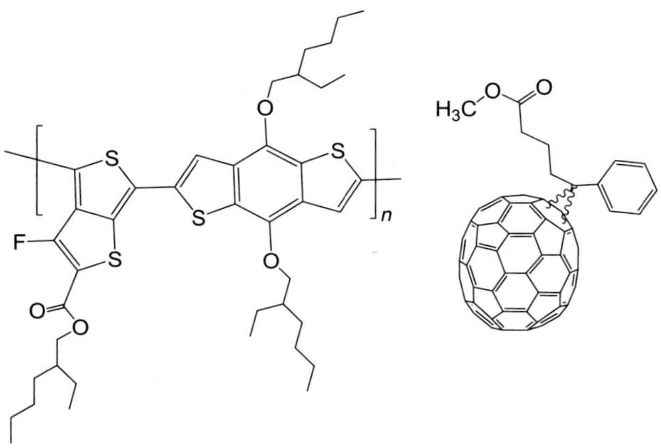	7.4	23
	8.37	24
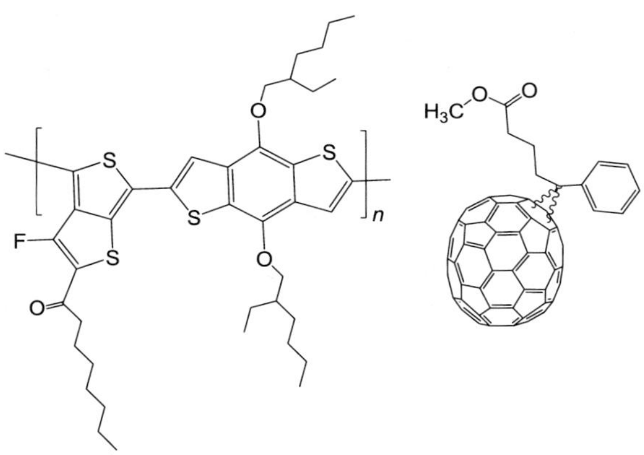	7.7	25
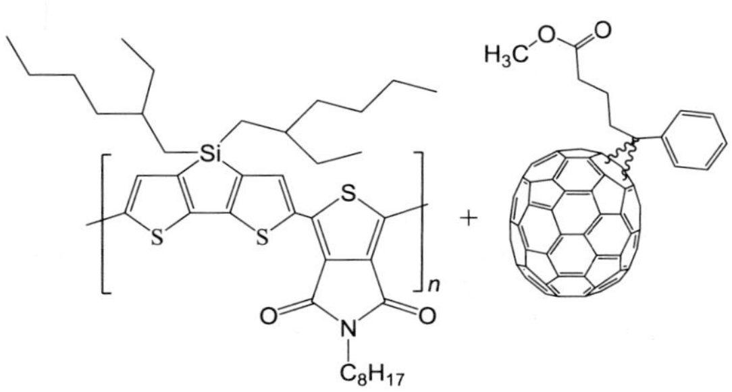	7.3	26
Undisclosed material combination from Konarka Technologies	8.3	27

(*Continued*)

TABLE 25.1. (*Continued*)

Description of the Materials	Power Conversion Efficiency (%)	Reference
Solution Processible Small Molecule/Small Molecule		
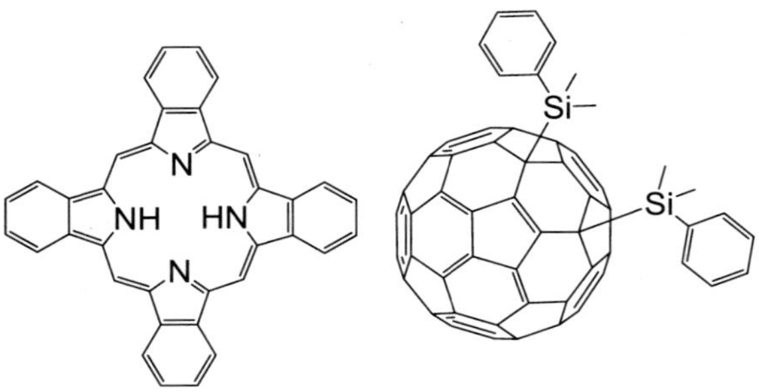	5.2	28
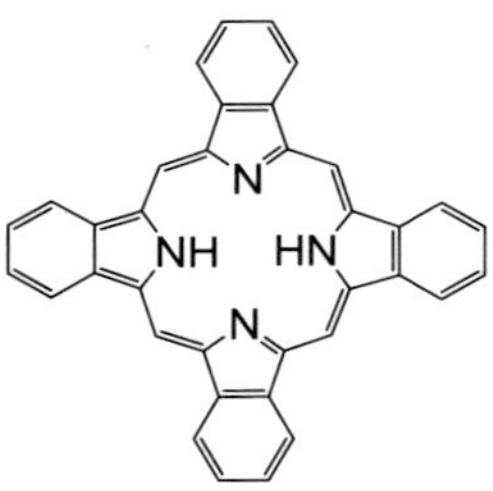 + **Undisclosed fullerene derivative (proprietary of Mitsubishi Chemical)**	9.2–10.2	29, 30
Evaporation Processible Small Molecule/Small Molecule		
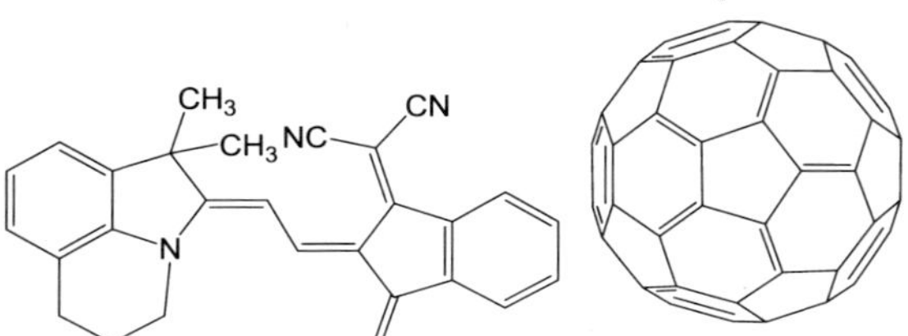 **C_{60}**	4.9	31
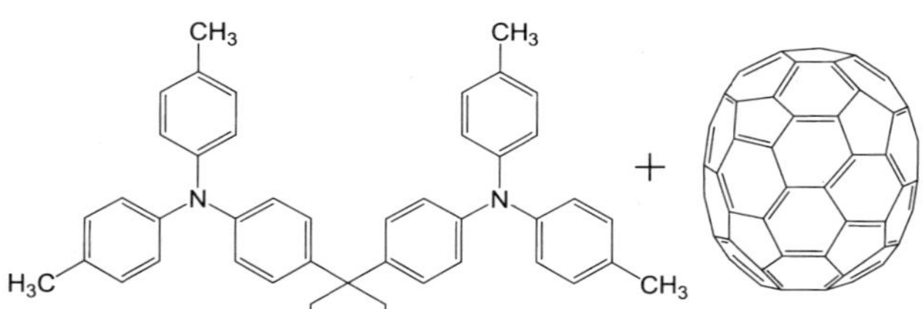 **C_{70}**	5.2	32
Undisclosed material combination from Heliatek, double-junction device	8.3	27
Polymer/Nanoparticle		
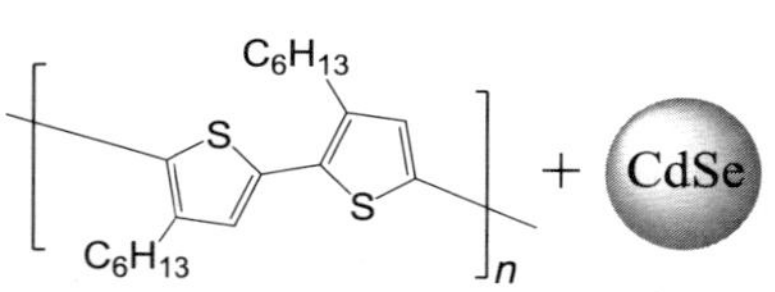 **P3HT** **QDs**	2.9	33
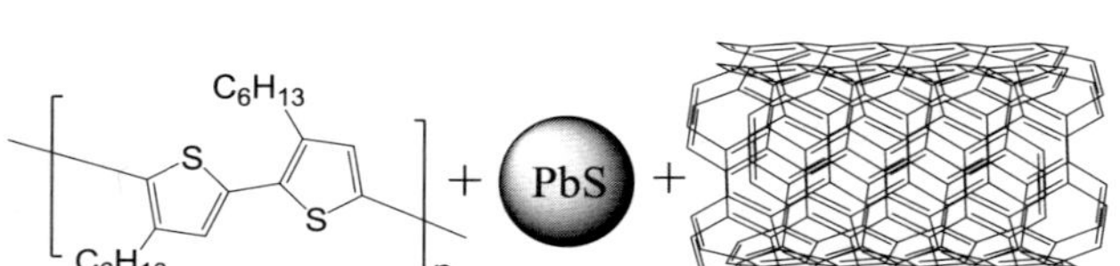 **P3HT** **QDs** **MWCNT**	3.03	34

TABLE 25.1. (*Continued*)

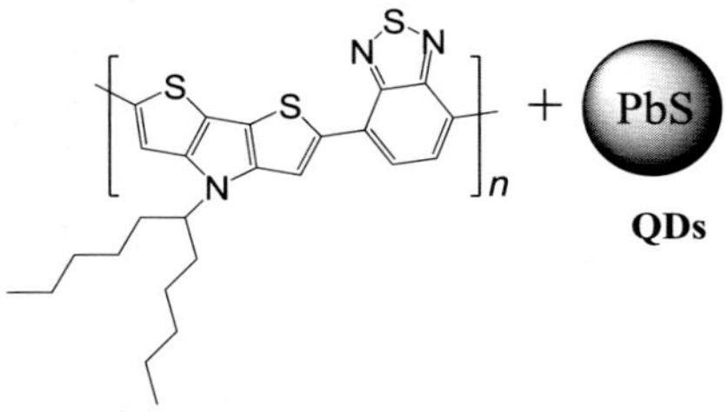	3.8	35
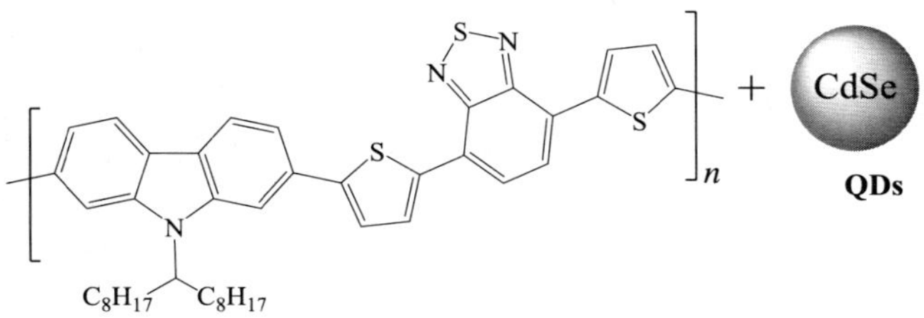	3.13	36
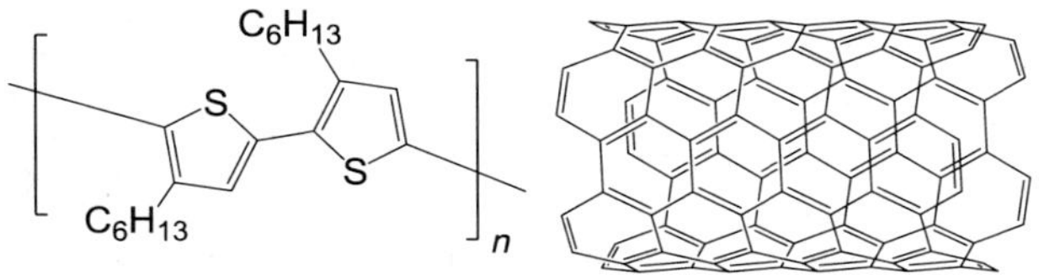 **P3HT** **SWCNT**	0.22	37, 38
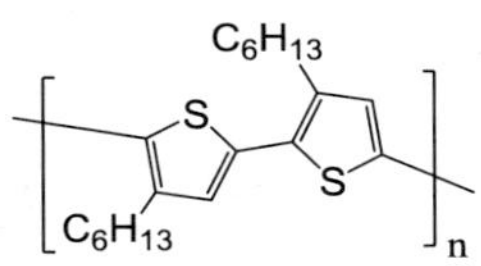 **P3HT** 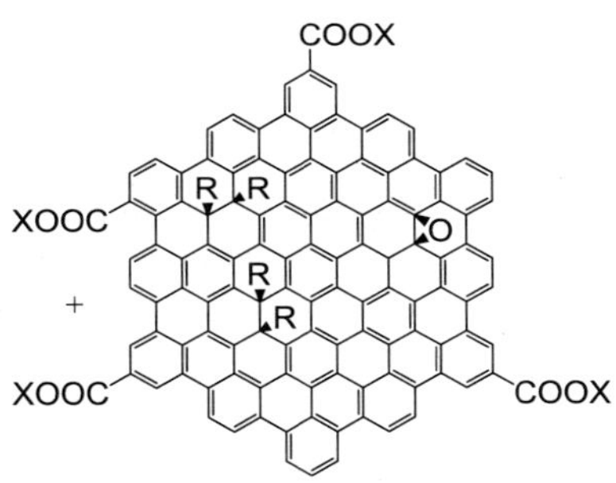 **SPFGraphene**	0.6–0.8	39, 40
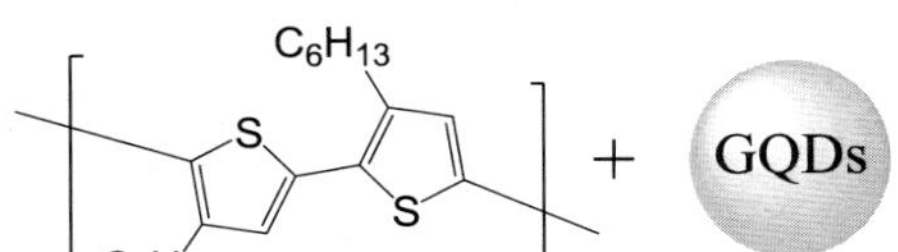 **Graphene QDs** **P3HT**	1.28	41
Polymer/Polymer		
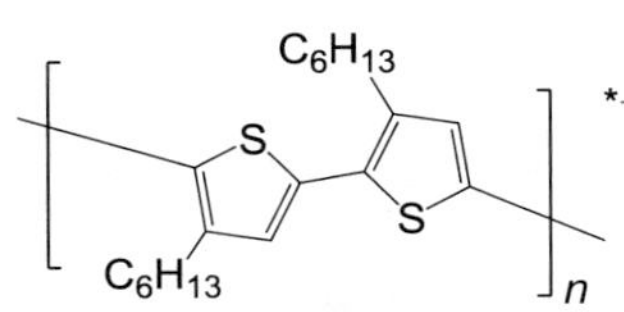 **P3HT** 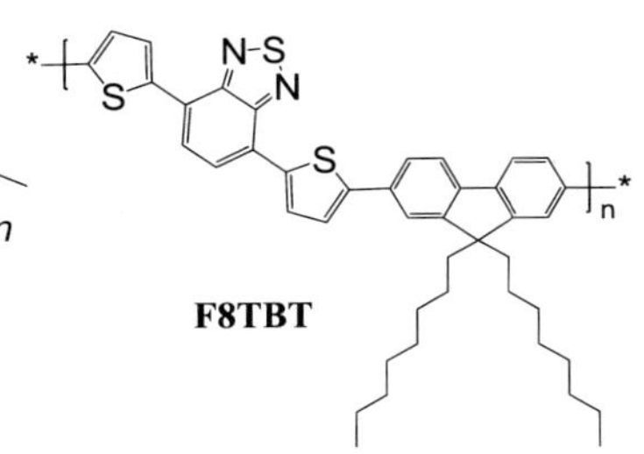	1.8	42

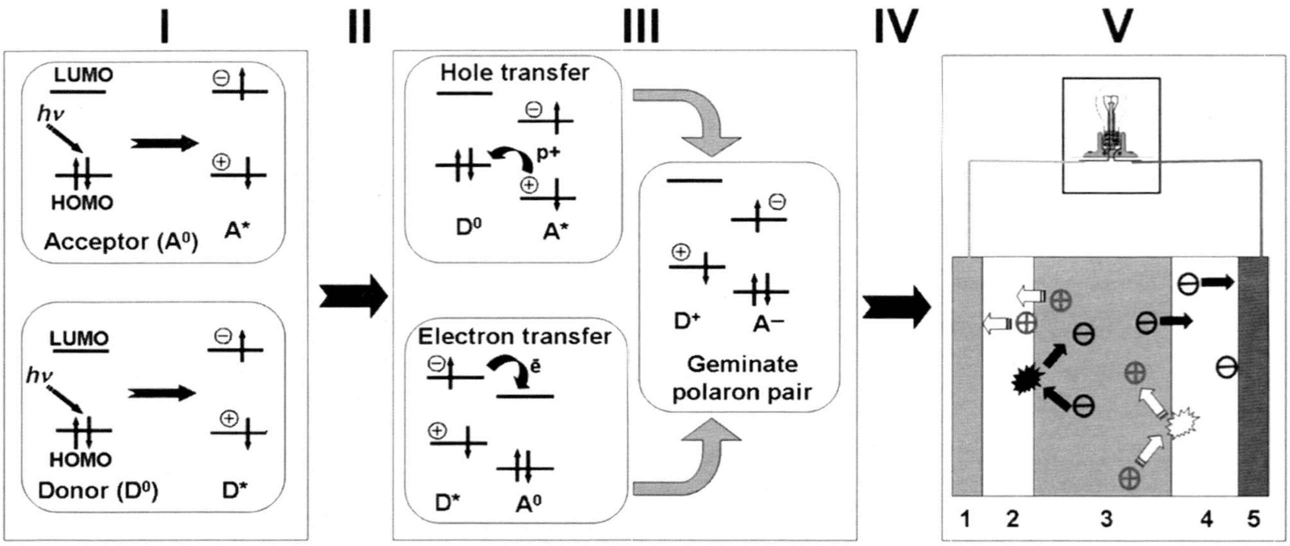

FIGURE 25.2. Operation mechanism of organic heterojunction solar cells. **Step I:** Absorption of photons and generation of excitons. **Step II:** Migration of the excitons toward the donor–acceptor interface. **Step III:** Charge separation at the donor–acceptor interface and formation of geminate ion pairs. **Step IV:** Dissociation of the ion pairs and moving of the charges in the opposite directions in the internal electric field in the device. **Step V:** Collection of the charges at the electrodes in the device. 1, positive electrode (ITO); 2, electron-blocking hole-transporting layer (PEDOT:PSS, MoO_3, WO_3, V_2O_5); 3, active layer of the device; 4, hole-blocking electron-transporting layer (TiO_x, ZnO); 5, negative electrode (Al, Ag, Ca, Mg, Ba). Reproduced from reference 14, with permission of Wiley–VCH.

these devices so that more or less all excitons can reach the donor–acceptor interface and contribute to the charge generation [45]. This can be achieved by keeping the size of the *p*-type and *n*-type material domains comparable to the exciton diffusion length which stays near 5–20 nm for organic materials [46–48].

The exciton D* is quenched via electron transfer to the LUMO level of the acceptor molecule (A^0) at the donor–acceptor interface (step III in Figure 25.2). On the contrary, exciton A* is quenched via hole transfer to the HOMO level of the donor molecule (D^0). Both pathways result in the formation of the same charge-separated state $D^+ \cdots A^-$. Positive and negative charges in this ion pair are bound by Coulomb attraction forces and are also denoted as a "geminate polaron pair" [49–51]. This pair can dissociate in the electric field induced by the potential jump at the heterojunction and/or by the difference in the electrode work functions. At the same time, the energy difference between the donor and acceptor LUMO levels (in the case when electron transfer takes place) or HOMO levels (in the case of hole transfer) released as a heat also helps to some extent with dissociation of geminate polaron pairs [52].

The charges generated in the blend should be transported to the respective electrodes (step IV in Figure 25.2). Holes should move in the phase of *p*-type material while electrons are transported in the *n*-type counterpart. Therefore it is essential to form percolated pathways for charge transport in both phases (see Figure 25.1).

At the final stage (step V in Figure 25.2) the generated charges have to be collected at the electrodes. In bulk heterojunction devices the lower work function metal (e.g., calcium, lithium, magnesium, barium, etc.) should match well the LUMO energy level of the acceptor component to collect negative charges (electrons) easily. At the same time, the high work function electrode (e.g., gold, platinum, nickel) should be adjusted to the HOMO energy level of the donor component to collect positive charge carriers (holes) without large barriers [43, 44]. However, the choice of electrode materials is somewhat complicated in the case of real devices. For instance, an indium–tin oxide electrode (ITO) can extract both positive and negative charges from the active layer of the device. At the same time, top electrodes composed of aluminum or silver can also extract both holes and electrons. Such poor selectivity of the charge collection results in a low photovoltaic performance of the device because of the massive charge recombination at the electrodes. To avoid this loss, some buffer layers should be introduced at the interfaces between the electrodes and the active layer. Electron blocking layers composed of evaporated or solution-processed vanadium (V) oxide, molybdenum (VI) oxide, tungsten (VI) oxide, and nickel (II) oxide were extensively utilized [53–55]. Titanium dioxide, cesium carbonate, zinc oxide, or fullerene derivatives behave as electron-transporting and hole-blocking materials [55–60]. There are recent reviews summarizing information on the available buffer layer materials used in organic bulk heterojunction devices [61–64]. Optimized bulk heterojunction cells with appropriate buffer layers can give high fill factors of 65–75% typically providing also high-power conversion efficiencies.

25.1.4 Characterization of Organic Photovoltaic Devices

The main tool for characterization of organic photovoltaic cells is measuring their current–voltage characteristics (I–V curves) in the dark and under illumination. Under standard conditions the light-on curves of the photovoltaic devices should be measured using standardized light sources (solar simulators) with a known irradiation spectrum which has to be close as much as possible to the true *AM*1.5 (Air Mass 1.5) spectrum. The light intensity should be set to 100 mW/cm^2, and the photovoltaic cell should be kept at 25°C. From the experimental I–V curve (Figure 25.3a) it is possible to extract all main parameters of the device: short-circuit current density I_{SC}, which is determined as current at the zero applied voltage; open-circuit voltage V_{OC}, which is measured at the point where current is equalized to zero, fill factor, determined using the equation shown as inset in Figure 25.3a; and, finally, the power conversion efficiency of the device (η). The power conversion efficiency of photovoltaic cell is calculated as electrical power produced by the device, divided by the power of the light irradiating the device area. The electrical power is calculated as a maximal product of the current and voltage in the fourth quadrant. The current and voltage at the maximal power point on the *I–V* curve are defined as I_{max} and V_{max}, respectively.

The characterization of photovoltaic devices remains incomplete without measuring their *external quantum efficiency (EQE) spectra*, also called *incident photon to collected electron efficiency (IPCE) spectra*. To obtain such spectra, the devices are irradiated with monochromatic light of known intensity, which generates a photocurrent that has to be measured with high accuracy. The calculation of an average fraction of incident photons that produced free electrons collected at the device electrodes gives us EQE (IPCE) values at each wavelength (see example in Figure 25.3b). If we recalculate the fraction of the produced in the device charge carriers per every absorbed (but not incident) photon, we can obtain internal quantum efficiency (IQE) spectra. IQE is always higher than EQE because it does not account for the optical losses (e.g., scattering, reflection). Integration of an EQE spectrum of a photovoltaic device over a known AM1.5 solar irradiation spectrum provides a very precise way for calculation of the device short-circuit current density. This method should always be used for checking the I_{SC} values

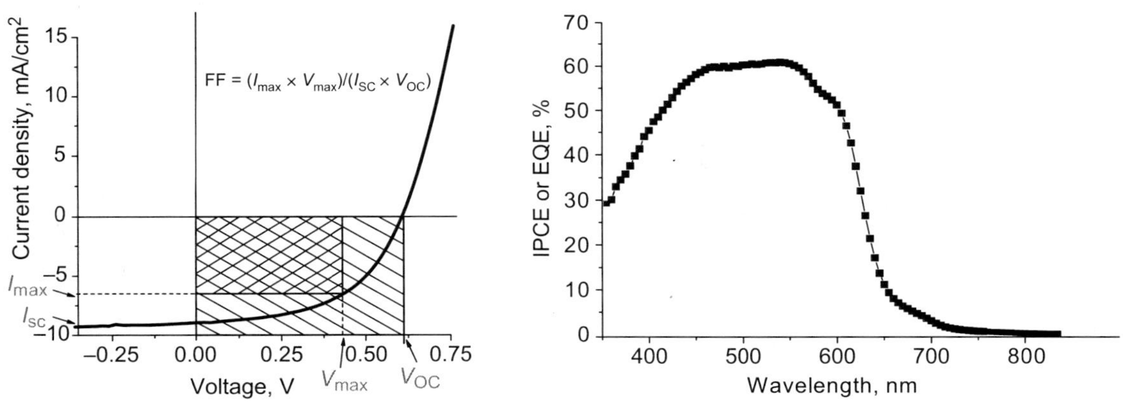

Light power conversion efficiency η

$$\eta = \frac{P_{electrical}}{P_{light}} = \frac{I_{max} \cdot V_{max}}{P_{light}} \cdot 100\% = \frac{FF \cdot I_{SC} \cdot V_{OC}}{P_{light}} \cdot 100\%$$

External quantum efficiency (EQE) or incident photon to collected electron efficiency (IPCE)

$$EQE_\lambda = IPCE_\lambda = \frac{N(\text{col. electr.})_\lambda}{N(\text{photons})_\lambda} \cdot 100\%$$

FIGURE 25.3. (**a**) Typical current–voltage characteristic of organic P3HT–PCBM solar cell, along with equation used for calculation of the maximal light power conversion efficiency η. (**b**) The external quantum efficiency (EQE), also called incident photon to collected electron efficiency (IPCE) spectrum, of a typical P3HT–PCBM device, along with a simple equation used for calculation of EQE at each wavelength. Reproduced from reference 14, with permission of Wiley–VCH.

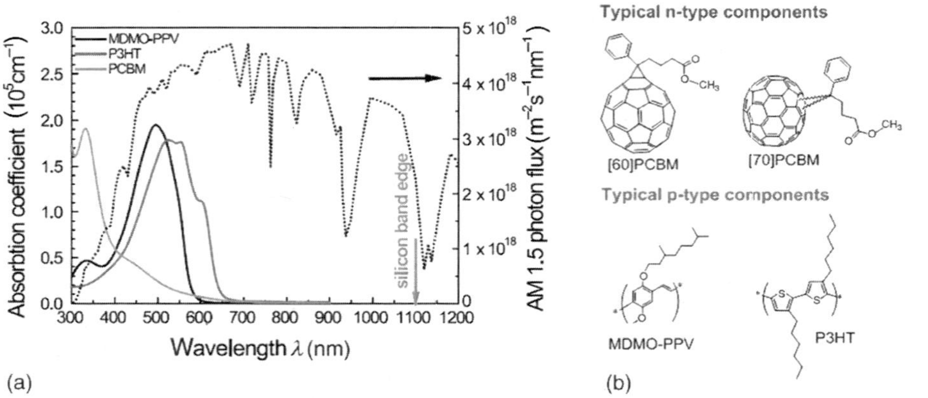

FIGURE 25.4. (**a**) The absorption spectra of some common photoactive materials used for fabrication of organic solar cells compared to the solar AM1.5 emission spectrum. (b) The molecular structures of some *p*-type and *n*-type materials used for solution-processible fullerene–polymer bulk heterojunction solar cells. Reproduced from reference 88, with permission of Springer.

obtained from the *I–V* measurements, which appear to be less accurate because of the non-accounted spectral mismatches between the irradiation of the used light sources (solar simulators) and true AM1.5 spectrum.

25.2 MAJOR TRENDS IN THE DESIGN OF NOVEL PHOTOACTIVE MATERIALS FOR BULK HETEROJUNCTION SOLAR CELLS

Efforts of the research are focused mainly on achieving higher power conversion efficiencies in the devices by using novel materials and their combinations while the stability issue remains non-addressed in the vast majority of publications. Since the power conversion efficiency depends linearly on short-circuit current, open-circuit voltage and fill factor, each of these parameters also should be increased.

The short-circuit current density in organic photovoltaic cells is limited by a number of photons absorbed in the active layer of the device. Therefore, the active layer materials should exhibit wide absorption spectra to harvest the solar light efficiently and to produce high current densities. Absorption profiles of such well-known material combinations as MDMO-PPV–[60]PCBM and P3HT–[60]PCBM cover only a small part of the solar irradiation spectrum as illustrated in Figure 25.4. Design of low-bandgap electron donor materials capable of the efficient light harvesting has been one of the most intensively developing research directions during the last decade [65].

Considering the overlap between the absorption spectrum of the device active layer of a certain thickness and the solar AM1.5 emission spectrum, it is possible to estimate numerically the maximal short-circuit current density of any photovoltaic device [66–68].

The maximal open-circuit voltage in bulk heterojunction organic solar cells is defined by the energy offset between the HOMO level of the donor material (*p*-type component) and the LUMO level of the acceptor (*n*-type component) (Figure 25.5) [69]. However, real V_{OC} values are lower than theoretically predicted ones by approximately 0.3 eV [70]. This dependence was well-illustrated using sets of fullerene derivatives with different LUMO energy levels and electron donor polymers with different HOMO energy levels [70, 71].

The fill factor of photovoltaic devices depends strongly on (a) the charge transport characteristics of the photoactive blend and (b) charge transfer through the interfaces between the active layer and the electrodes. In general, reasonably high fill factors can be obtained only for the systems where electron and hole mobilities are balanced in the photoactive blend (means close to each other as much as possible) [32, 72, 73]. At the same time, the energy levels of buffer layers and work functions of the electrode materials have to be well-aligned to facilitate collection of each type of charge carrier [74]. The fill factors of 65–75% are reached for organic photovoltaic devices [12].

For a certain combination of electron donor and electron acceptor materials with known electronic properties, it is possible to estimate maximal short-circuit current and

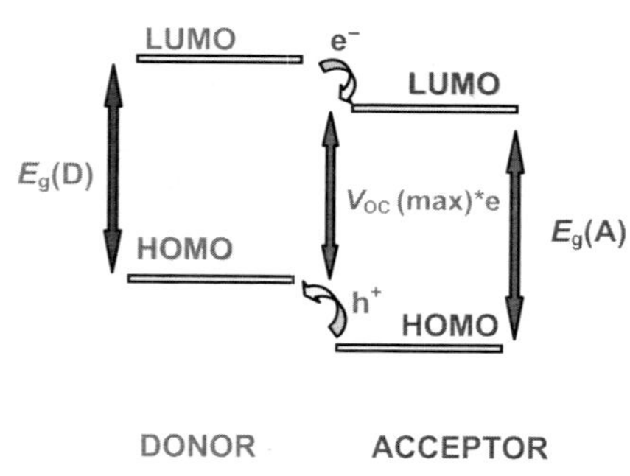

FIGURE 25.5. Open-circuit voltage in organic bulk heterojunction solar cells.

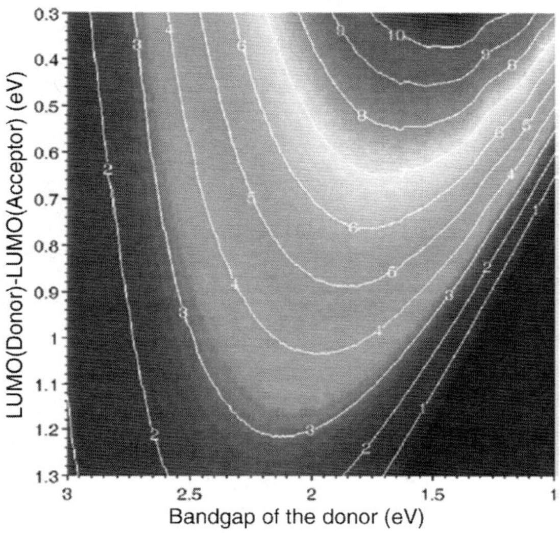

FIGURE 25.6. Theoretically feasible power conversion efficiencies of single-junction organic bulk heterojunction solar cells based on the materials with different electronic properties. Reproduced from reference 96, with permission of the Royal Society of Chemistry.

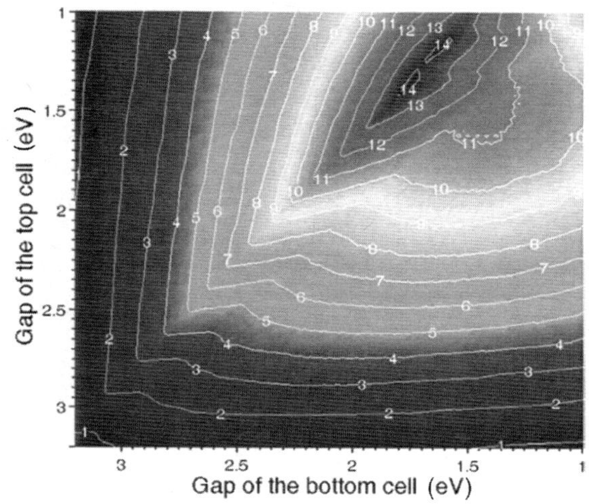

FIGURE 25.7. Theoretically feasible power conversion efficiencies of double junction organic solar cells. Reproduced from reference 96, with permission of the Royal Society of Chemistry.

open-circuit voltage. Taking these values into account and assuming some reasonable fill factor (FF = 65%) and external quantum efficiency (EQE = 65%) values, it is possible to calculate theoretically achievable power conversion efficiency. Extensive modeling was performed by many groups, which allowed estimating ultimate efficiencies of single-junction and double-junction organic bulk heterojunction cells [70, 75, 76].

For a single-junction cell, the most illustrative diagram shows a correlation between the bandgap of the donor component and offset between the LUMO level energies of the donor and acceptor materials (Figure 25.6). This diagram shows that optimal electron donor material should have a bandgap of ~1.5 eV, which corresponds to the absorption band edge of ~820 nm. At the same time, the offset between the LUMO level energies of the donor and acceptor components should be minimized to 0.2–0.3 eV to enable maximal V_{OC} achievable for the system. The modeling performed recently by Kotlarski and Blom [76] suggests that the optimal donor bandgap is ~1.7 eV and the optimal active layer thickness should be around 100 nm. The optimized composite system (with donor–acceptor LUMO offset of 0.3 eV) is expected to yield $V_{OC} = 1.0$ V, $I_{SC} = 15.6$ mA/cm^2, FF = 74%, and $\eta = 11.5\%$. The second efficiency maximum of 9.9% was found for the systems with the donor bandgap of 1.9 V (similar to the bandgap of P3HT) and the active layer thinkness of 200 nm.

It is very likely that on-grid applications of organic solar cells will become feasible only when the module efficiences will reach 10% in combination with the lifetime of 10 years and the cost below 100 USD/m^2 [77]. The power conversion efficiencies of single cells have to be boosted up to 13–15% to meet this severe module efficiency requirement. It is clear that classical organic single junction devices will fail to produce 13–15% efficiencies. Therefore, the research community pays more and more attention to tandem organic solar cells, especially the double-junction devices. Theoretical modeling suggests that indeed double-junction organic solar cells can produce power conversion efficiencies of 14–15% if optical and electronic properties of the materials in both subcells are optimized.

The diagram shown in Figure 25.7 suggests that 14% efficiency can be obtained for double-junction organic solar cells whose subcells are composed of the materials with the bandgaps of 1.60–1.65 eV and 1.20–1.25 eV (the first maximum) or 1.75–1.80 eV and 1.30–1.35 eV (the second maximum) [75].

Somewhat different modeling results for double-junction organic solar cells were reported by Kotlarski and Blom [76]. First of all, they confirmed that in the optimal situation the front cell (one that is closer to transparent electrode) should have a higher bandgap compared to the back cell (one that is closer to the reflecting metal electrode). The optimal bandgap for the front-cell donor component was estimated to be around 1.9 eV, and the optimal active layer thinkness was around 150 nm. At the same time, the back cell should have the bandgap of 1.5 eV and the thickness of the light absorbing layer has to be near 90 nm. These results are comparable to the ones reported by Ameri et al. [75] and presented in Figure 25.7, though the absolute values of the bandgaps are somewhat different.

The second efficiency maximum was obtained for the system where the front cell materials have a smaller bandgap (E_g = 1.4 eV) compared to the back cell components

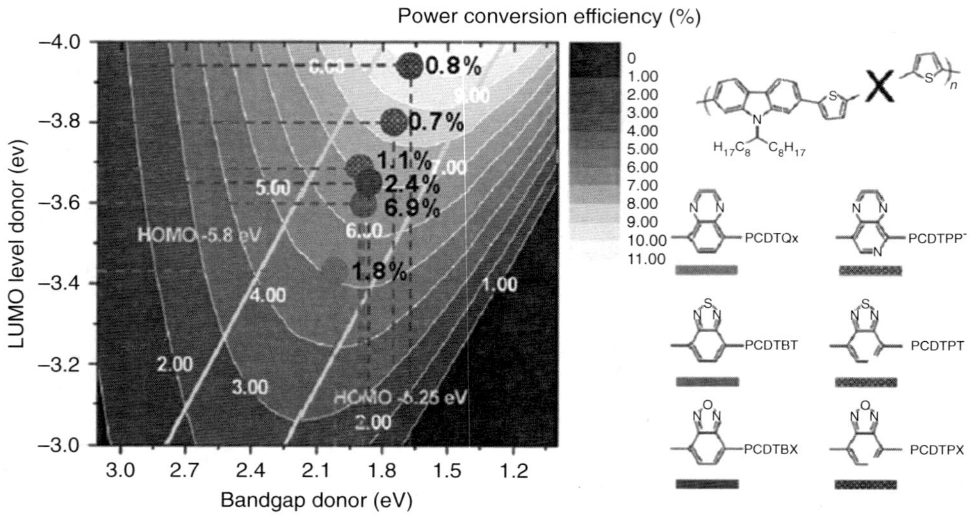

FIGURE 25.8. The power conversion efficiencies of solar cells based on different polymer–PCBM combinations: experiment versus theory. Reproduced from reference 101, with permission of the American Chemical Society.

(E_g = 1.9 eV). However, the thinckness of the front cell active layer (40 nm) should be much smaller compared to the back cell active layer (200 nm) to provide the maximal efficiency of 13.1%.

The presented modeling results are very useful guidelines for material chemists who design novel photoactive electron donor and electron acceptor materials for bulk heterojunction organic solar cells. It becomes clear what kind of electronic and optical properties one should attain by synthesizing, for instance, a new donor polymer that is supposed to be used in organic solar cells in combination with the fullerene-based material [60]PCBM. Assuming that the LUMO level energy of PCBM is equal to −3.9 eV as reported [78,79], the optimal LUMO and HOMO energies of the polymer should be around −3.6 eV and −5.1 eV, respectively (taking optimal polymer bandgap of 1.5 eV according to references 70 and 75). If we assume the lower LUMO energy for PCBM (−4.3 eV according to the report of Scharber et al. [70]), the optimal LUMO and HOMO energies of the donor component will transform to −4.0 and −5.5 eV.

Unfortunately, the correlations between the theoretical predictions and the experimental data are not always perfect. A good illustration of this disagreement between the theory and the experiment is reported by Blouin et al. [80]. A range of carbazole-based electron donor polymers was synthesized and investigated in bulk heterojunction solar cells in combination with [60]PCBM. Some polymers possessing optimal frontier energy levels were good candidates for 8–10% efficient organic solar cells. However, the experimental power conversion efficiencies were 5–10 times lower than the theoretically estimated ones. This situation is illustrated by Figure 25.8, where the black numbers on the diagram correspond to the solar cells efficiencies experimentally obtained for every polymer. It is seen from the Figure 25.8 that only one polymer (PCDTBT) shows the performance that corresponds well to the theoretical predictions.

The theoretical modeling results of Scharber et al. [70] and Ameri et al. [75] were based on the assumption of high external quantum efficiency (EQE = 65%) and high fill factors (FF = 65%) that experimentally are not achievable for many systems. Insufficiently high fill factors and EQEs typically account for the disagreement between the theoretically predicted efficiencies and experimental results. Nanomorphology of the active layer in each case determines the charge carrier mobility in the bulk and therefore has to be optimized.

25.3 ACTIVE LAYER NANOMORPHOLOGY AS A MAJOR FACTOR LIMITING PHOTOVOLTAIC PERFORMANCE OF BULK HETEROJUNCTION SOLAR CELLS

Importance of the active layer nanomorphology of organic solar cells was first recognized in 2001 for the solar cells based on the MDMO-PPV/[60]PCBM composite. The power conversion efficiency of the devices was increased from 0.9% to 2.5% simply by replacing the solvent used for active layer casting from toluene to chlorobenzene [81]. In the subsequent study by Hoppe et al. [82] it was demonstrated that MDMO-PPV/[60]PCBM films cast from toluene are rather inhomogeneous and comprise round-shaped distinct features approaching in size 500–600 nm which were later shown to be the PCBM crystallites [83]. On the contrary, films cast from chlorobenzene were much more homogeneous, and the cluster size in that case did not exceed 50 nm. Illustrative AFM images recorded for the toluene-cast

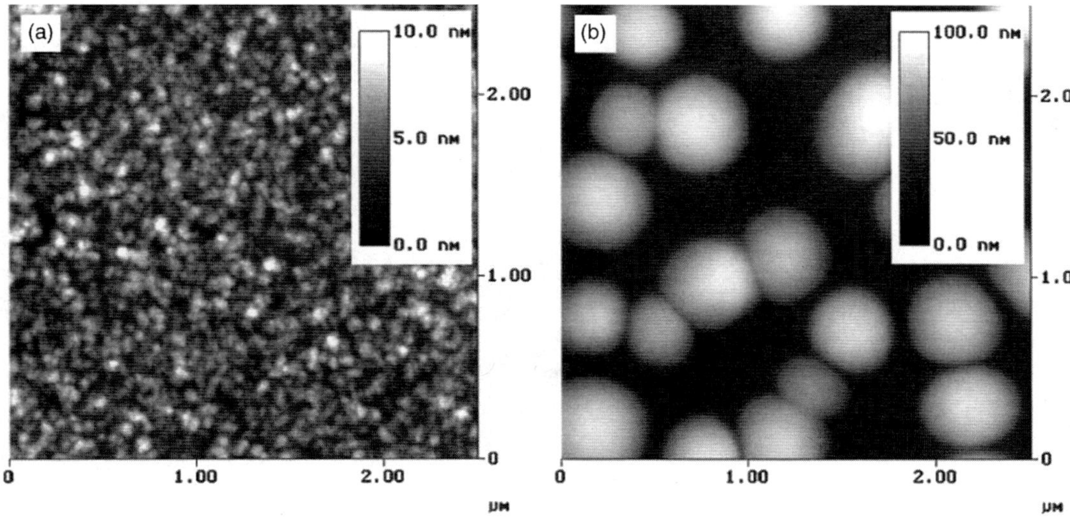

FIGURE 25.9. AFM images of MDMO-PPV–[60]PCBM (1:4 w/w) composite films cast from (**a**) chlorobenzene and (**b**) toluene. Reproduced from reference 103, with permission of Wiley–VCH.

and cholorobenzene-cast MDMO-PPV–[60]PCBM films are shown in Figure 25.9.

It was mentioned above that the charge generation in bulk heterojunction solar cells occurs at the interface between the donor (in this case MDMO-PPV) and acceptor ([60]PCBM) components of the blend. The excitons generated in the active layer have to diffuse to the interface where charge separation takes place. It is known that characteristic exciton diffusion lengths in organic semiconductors typically do not exceed 20 nm [46–48]. Therefore, the domains of individual materials formed in the blend as a result of the phase separation should not be larger than the exciton diffusion lengths (L_{ex}) in these materials. The round-shaped clusters in the case of the toluene-cast MDMO-PPV/[60]PCBM films are at least 25–30 times larger than the typical L_{ex} values. Therefore a vast majority of excitons generated inside these clusters recombine since they cannot reach the donor–acceptor interface where charge separation takes place. The situation is much more positive in the case of chlorobenzene-cast films where the average cluster size matches quite well the L_{ex} value. The smaller degree of phase separation in this case allows the majority of excitons to reach the fullerene–polymer interface and contribute to the charge carrier generation. Improved morphology of the chlorobenzene-cast MDMO-PPV–[60]PCBM blends results in the superior device performances: EQEs approach 50% and light power conversion efficiencies come close to 2.5% [81].

It has been shown recently that the morphology of the MDMO-PPV-based composites can be controlled by changing the molecular structure of the fullerene derivatives [68]. Indeed, small variations in the length of alkyl chains attached to the carboxylic group in the structures of investigated [70]fullerene derivatives induce significant changes in the composite morphology as concluded from the AFM images of the blends (Figure 25.10). It was shown experimentally that both short-circuit current density, and power conversion efficiency of the devices decrease rapidly with increase in the lateral size of the clusters (round-shaped features) revealed on the surface of the films by AFM.

A clear correlation between the solar cell parameters (I_{SC} and η) and the size of the clusters in the fullerene–polymer composites observed in reference 68 is a good experimental confirmation of the theoretical considerations presented above. Indeed, recombination of excitons inside the large clusters leads to the expectable drop in the device current density and power conversion efficiency.

Next generation of organic solar cells based on the P3HT–[60]PCBM blends is also governed by the active layer morphology. Padinger et al. [84] have shown that thermal annealing of P3HT–[60]PCBM solar cells results in a dramatic improvement of their performance providing power conversion efficiency of 3.5% for the first time. This work attracted tremendous attention in the field and stimulated subsequent investigation of the P3HT–[60]PCBM composite solar cells. It was shown that morphology of P3HT–[60]PCBM solar cells can be tuned by thermal annealing [85], solvent vapor annealing [86, 87], and use of chemical additives [88].

Electron tomography has revealed a three-dimensional structure of the P3HT–[60]PCBM blends with nanometer resolution [89]. It was demonstrated that thermal and solvent vapor annealing results in the formation of genuine 3D nanoscale interpenetrating networks of donor and acceptor materials with high crystalline order (Figure 25.11). These favorable morphological changes account for a considerable increase in the power conversion efficiency of the devices after thermal or solvent-assisted annealing. Electron tomography allows for precise determination of concentration gradients of both P3HT and [60]PCBM through the thickness

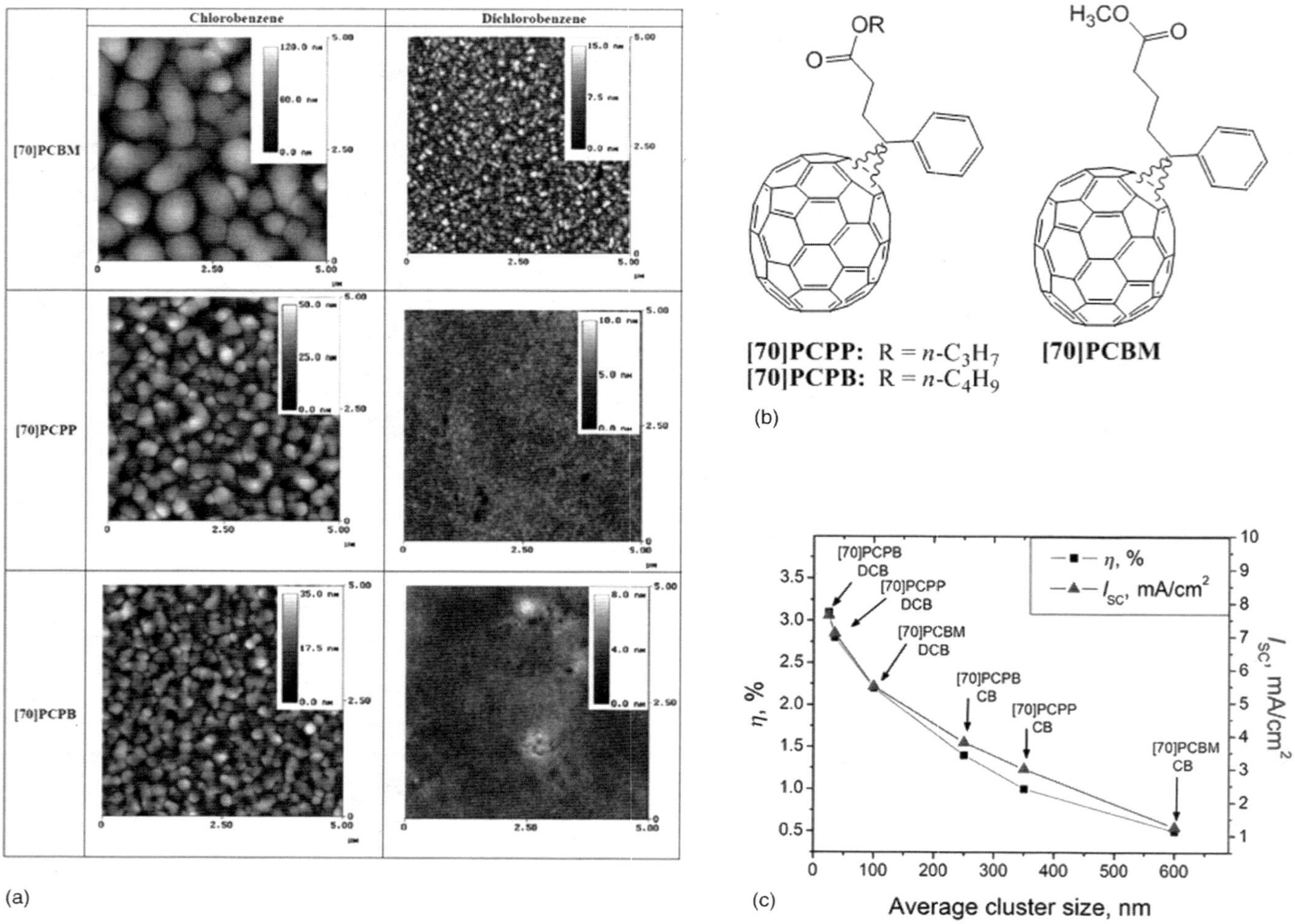

FIGURE 25.10. (**a**) AFM images of the blends of MDMO-PPV with different [70]fullerene derivatives. (**b**) Molecular structures of the investigated fullerene derivatives. (**c**) The correlation between the lateral size of the clusters derived from the AFM images (film surface profiles), short-circuit current density (I_{sc}), and light power conversion efficiency (η) of the devices. CB corresponds to the blends processed from chlorobenzene; DCB corresponds to the systems processed from 1,2-dichlorobenzene. Reproduced from reference 89, with permission of Wiley–VCH.

of the photoactive layer. It has proved in particular that the concentration of P3HT is higher at the ITO–PEDOT:PSS electrode for the annealed blend films.

A thorough understanding of the morphology effects in organic bulk heterojunction solar cells based on the P3HT–[60]PCBM composite resulted in gradual improving of their light power conversion efficiency up to the level of 4.0–4.5% [66, 90–91]. Unusual results came out from a systematic study of a library of fullerene derivatives in bulk heterojunction solar cells in combination with P3HT [92]. It was found that all fullerene derivatives presented in Figure 25.12 have virtually the same frontier energy levels (in particular, LUMO energies affecting V_{OC} of solar cells) regardless the structure of organic addend attached to the cages of fullerenes C_{60} and C_{70}. However, even very slight modifications of the molecular structures of fullerene derivatives induce significant changes in their physical properties such as solubility in organic solvents. This is well illustrated by the solubility values measured for all fullerene derivatives in chlorobenzene which are presented in Figure 25.12 near numbers of the compounds in brackets.

The solubility of the fullerene derivatives affected strongly the morphology of their blends with P3HT (Figure 25.13). For instance, the films composed of the least-soluble fullerene derivative **7** (solubility $S = 5$ mg/mL) and P3HT showed large aggregates approaching 30–100 μm in size. An increase in the solubility of the fullerene component by a factor of two for the compound **6** ($S = 10$ mg/mL) resulted in the remarkable decrease in the cluster size in the blend films by a factor of 10. Use of the fullerene-based materials with the solubility of 22 mg/mL (compound **9**) and 30 mg/mL (compound **10**) led to further improvement of the blend

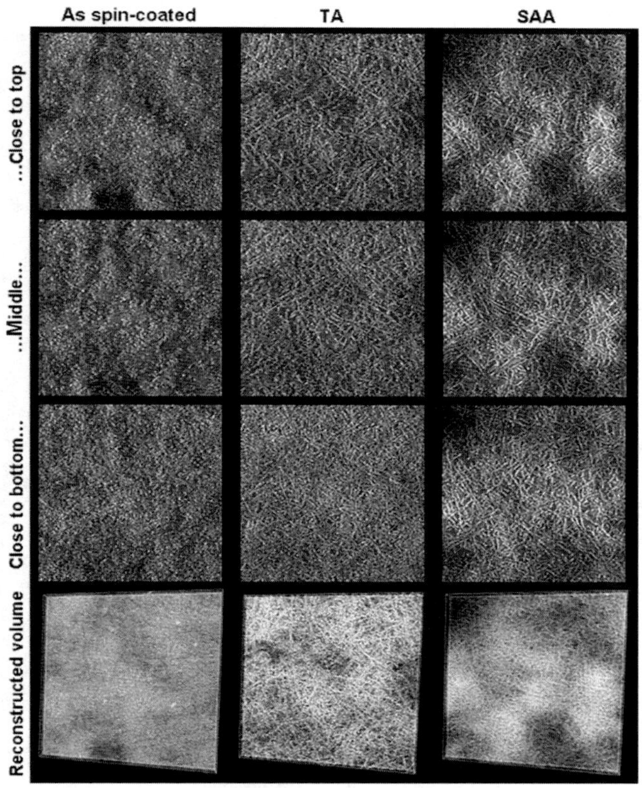

FIGURE 25.11. Results of electron tomography applied to P3HT–[60]PCBM photoactive layers: as spin-coated, thermally annealed at 130°C for 20 min (TA), and solvent-assisted annealing for 3 h (SAA). The first three rows contain slices taken out of a reconstructed volume of the corresponding film. All slices are lying in the horizontal (*X*, *Y*) plane of the film at a different depth (*Z* location): one slice close to the top of the film (i.e., to the electron collecting electrode), another one in the middle of a film, and the third one close to the bottom of the film (the hole collecting PEDOT:PSS/ITO electrode). The dimensions of the slices are around 1700 nm × 1700 nm. Images in the fourth row are snapshots of the corresponding film's whole reconstructed volume—that is, a stack of all of the slices through the whole thickness of a film, with dimensions of around 1700 nm × 1700 nm × 100 nm. Reproduced from reference 112, with permission of the American Chemical Society.

morphology erasing any signs of the phase segregation at least on the micrometer scale.

Strong variation of the film morphology induced by different solubility of the fullerene derivatives (molecular structures are shown in Figure 25.12) affected significantly photovoltaic performance of the P3HT–fullerene composites. Rather clear correlations of the short-circuit current density, open-circuit voltage, fill factor, and light power conversion efficiency with the solubility of the fullerene-based materials were revealed (Figure 25.14).

The presented examples prove that supramolecular assembling of the fullerene derivative and the polymer in the blends is governed by relative solubility of the materials. In the case of P3HT, the best solar cell performances were obtained with the use of fullerene derivatives that exhibited solubility values close to the solubility of P3HT itself (70–90 mg/mL). Therefore it was suggested that balanced active layer morphology in fullerene–polymer composites can be achieved through a combination of electron donor and electron acceptor materials with similar solubilities. As a main consequence of this conclusion, any novel electron donor material might require a specific fullerene counterpart with fitting solubility to be combined in order to achieve the highest photovoltaic performance.

Further investigation of the library of fullerene derivatives in combination with poly(3-alkylthiophenes) with different side chains revealed unexpected results [93]. It was found that the dependence of the solar cell parameters upon the fullerene component solubility might have an unexpected double-branched shape as shown for poly(3-pentylthiophene) in Figure 25.15. The fullerene derivatives with the solubility fitting the range of 20–60 mg/mL form two separate groups. One group of the fullerene-based compounds clearly outperforms another one in solar cells as follows from the Figure 25.15. Moreover, very similar dependences were revealed for poly(3-alkylthiophenes) with heptyl, octyl, decyl, and dodecyl side chains as well.

The observed double-branched behavior was related to variations in the active layer morphology induced by peculiarities of the molecular structures of fullerene derivatives. For instance, fullerene derivatives **3** and **8** have very similar solubility in chlorobenzene. However, the films of their blends with poly(3-decylthiophene) (P3DT) identically annealed at 140°C for 5 minutes have very different morphologies (Figure 25.16). Fullerene derivative **8** seems to be well compatible with the P3DT polymer, which results in homogeneous film structure and relatively high photovoltaic performance ($\eta = 2.0\%$). On the contrary, compound **3** seems to be weakly miscible with the polymer, which results in a large-scale phase separation in the blend (Figure 25.16) and poor photovoltaic performance ($\eta = 0.3\%$).

The obtained results suggest strongly that the morphology and photovoltaic performance of the fullerene/polymer composites is governed to a large extent by supramolecular interactions between the components. On the one hand, the best-performing systems (upper "branch" in Figure 25.15) comprise fullerene derivatives that are well compatible (like **8**–P3DT composite above) with the polymer due to some attractive intermolecular interactions. On the other hand, the worse performing systems (lower "branch" in Figure 25.15) are based on badly compatible fullerene and polymer components (like **3**–P3DT composite above) that might be a consequence of some missing intermolecular interactions between the components.

It would not be a big exaggeration to say that there are several thousands of known conjugated polymers that were synthesized and applied or were supposed to be applied as

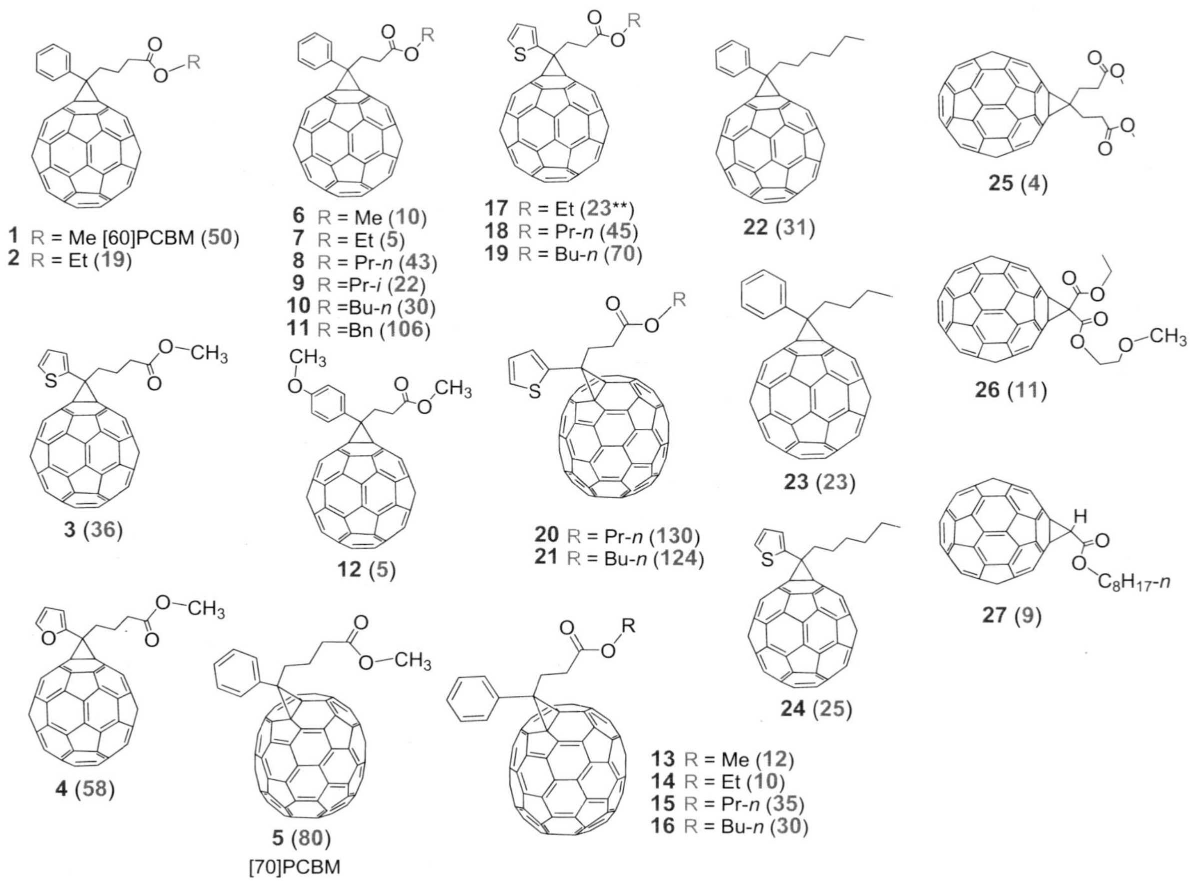

FIGURE 25.12. Molecular structures of fullerene derivatives forming a library of acceptor materials investigated in solar cells in combination with P3HT. Solubility values determined in chlorobenzene are given in parentheses near the numbers of the compounds. Reproduced from reference 115, with permission of Wiley–VCH.

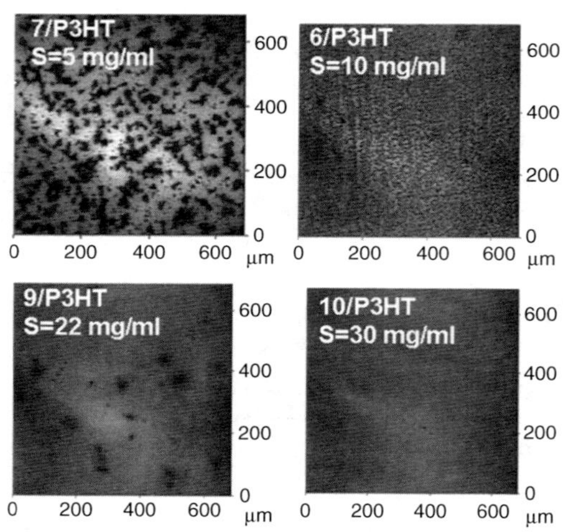

FIGURE 25.13. Optical microscopy images for the blends of four different fullerene derivatives with P3HT. Obvious improvement of the active layer morphology with increase in the solubility of the fullerene derivatives can be observed. Reproduced from reference 115, with permission of Wiley–VCH.

electron donor components in organic bulk heterojunction solar cells. However, nearly all these polymers were investigated in solar cells in combination with conventional commercially available fullerene derivatives [60]PCBM and [70]PCBM (molecular structures are shown in Figure 25.4). In the view of the experimental results reported in references 92 and 93, it is very unlikely that two conventional fullerene derivatives ([60]PCBM and [70]PCBM) have appropriate solubility and good compatibility[1] making them suitable electron acceptor components for a very broad range of electron donor polymers. It has already been shown that replacement of PCBM with better suiting fullerene counterpart might improve photovoltaic performance of some conjugated polymers by a factor of 2–3 [93,94]. One more example

[1] We relate the term "good compatibility" to the existence of attractive intermolecular interactions between the fullerene derivative and the polymer, preventing large-scale phase separation like it was illustrated above for 3/P3DT composite. On the contrary, the term "bad compatibility" refers to the systems with insufficient attractive interactions between the components that finally result in their large-scale segregation in the blends.

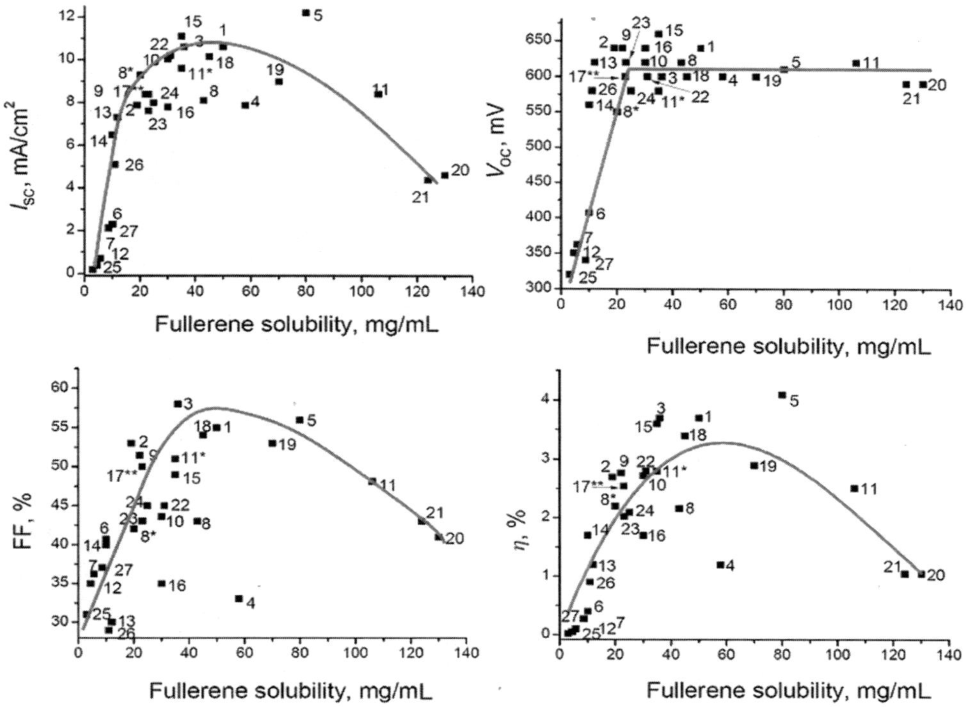

FIGURE 25.14. The short-circuit current (I_{SC}), open-circuit voltage (V_{OC}), fill factor (FF), and light power conversion efficiency (η) as functions of the solubility of the fullerene-based materials. Reproduced from reference 115, with permission of Wiley–VCH.

is provided by the polymer AnE-PV*stat*, which yields power conversion efficiencies of 2.4–3.5% in solar cells in combination with [60]PCBM and [70]PCBM. However, the composites of this polymer with fullerene derivatives **8** and **18** showed power conversion efficiencies of 5.0–5.1% in optimized bulk heterojunction solar cells (Figure 25.17). The superior device performance using fullerene derivatives **8** and **18** was correlated with the improved active layer morphology [95].

It seems that every conjugated polymer requires its own fullerene-based counterpart with optimized molecular structure (to provide necessary compatibility) and appropriate solubility. This is an important message suggesting the need for revisiting many theoretically promising electron donor copolymers that showed power conversion efficiencies of 1–3% with PCBM and consequently were discarded. It is very probable that many previously abandoned polymers will show state-of-the-art solar cell performances if appropriate fullerene-based counterparts are provided.

25.4 ADVANCED ELECTRON ACCEPTOR MATERIALS FOR BULK HETEROJUNCTION SOLAR CELLS

Fullerene derivatives [60]PCBM and [70]PCBM were running horses for many years in the field of the research related to the polymer–fullerene bulk heterojunction solar cells. However, many promising electron donor conjugated polymers were developed that are characterized by insufficiently low LUMO energies that result in considerably large $\Delta E_{ET} = E_{LUMO}(D) - E_{LUMO}(A)$ energy band offset (D states for donor polymer and A for acceptor fullerene derivative). As shown in Figure 25.6, the optimal ΔE_{ET} should be ~0.3 eV to ensure efficient electron transfer. The systems with the larger

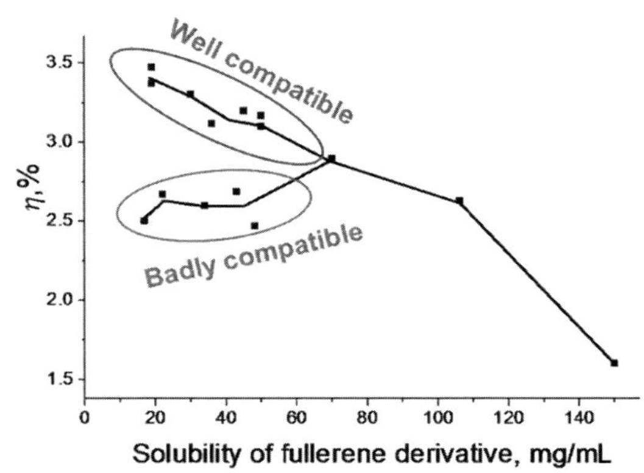

FIGURE 25.15. Unusual double-branched dependence of the solar cell power conversion efficiency upon the solubility of the fullerene derivatives combined with poly(3-pentylthiophene) as a donor polymer. Reproduced from reference 116, with permission of the authors.

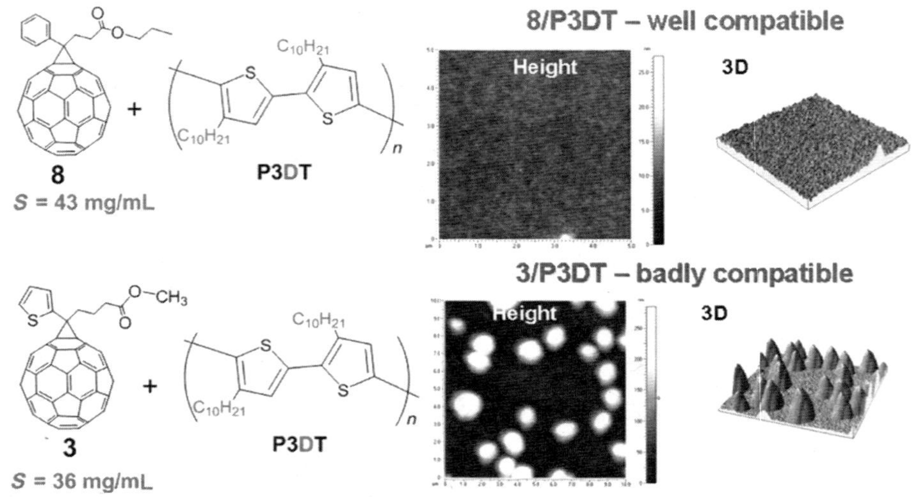

FIGURE 25.16. The blends of fullerene derivatives **3** and **8** with P3DT reveal very different morphologies in spite of the similar solubility of these compounds in chlorobenzene. Molecular compositions of the blends are shown (on the left side) together with the height AFM images and their 3D profiles (on the right side). Reproduced from reference 116, with permission of the authors.

ΔE_{ET} values suffer from the losses in the open-circuit voltage. The P3HT–PCBM is one of the most illustrative systems where high LUMO energy of P3HT limits severely the V_{OC} of the P3HT–PCBM solar cells. If we take HOMO–LUMO energy levels for [60]PCBM and P3HT given in reference 70, the maximal obtainable V_{OC} in the solar cells based on this system is ~0.8 V (Figure 25.18). At the same time, the ΔE_{ET} is unnecessary large and approach ~1.1 eV.

The experimental V_{OC} values (V_{OC}(exp)) for the reasonably efficient P3HT–PCBM cells vary in the range of 0.58–0.68 V [91]. We believe that the experimental V_{OC} values are described with a reasonably good accuracy by the model of Vandewal et al. [96], who proposed the following empiric equation for estimating V_{OC}(exp):

$$V_{OC}(\text{exp}) \approx E_g/e - 0.43\text{V} \quad (25.1)$$

where E_g is a bandgap of the charge-transfer state formed in the polymer–fullerene system, which is also ~0.1–0.2 eV smaller than the LUMO(A)–HOMO(D) offset defining $V_{OC}(\text{max})^*e$.

Therefore, we can modify Eq. (25.1), leading to a more convenient equation:

$$V_{OC}(\text{exp}) \approx V_{OC}(\text{max}) - (0.53 \div 0.63)\text{V} \quad (25.2)$$

If we take the LUMO energy value of −3.9 eV for PCBM (following references 78 and 79), we can obtain more optimistic $V_{OC}(\text{max}) = 1.2$ V and the $\Delta E_{ET} = 0.7$ eV. Following the Eq. (25.2), we can obtain $V_{OC}(\text{exp}) = 0.57$–$0.67$ V, which shows excellent agreement with the values published in the literature [91].

To make efficient P3HT-based organic solar cells, one needs to design fullerene derivatives with higher LUMO level energies to bring ΔE_{ET} value as close to the desired 0.3 eV as possible. This idea was first illustrated in 2001 by Brabec et al. [69]. A set of fullerene derivatives demonstrating higher open circuit voltage in solar cells compared to [60]PCBM was designed by Kooistra et al. [97] in 2007. The best V_{OC} of 925 mV was provided in MDMO-PPV based solar cells by the compound **F1** (Figure 25.19) bearing

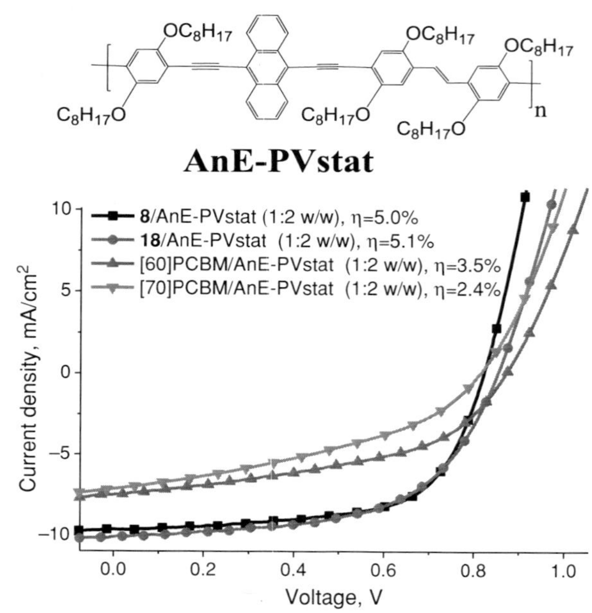

FIGURE 25.17. (**Top**) Molecular structure of the AnE-PV*stat* polymer. (**Bottom**) *I–V* curves for solar cells based on the composites of AnE-PV stat with different fullerene derivatives.

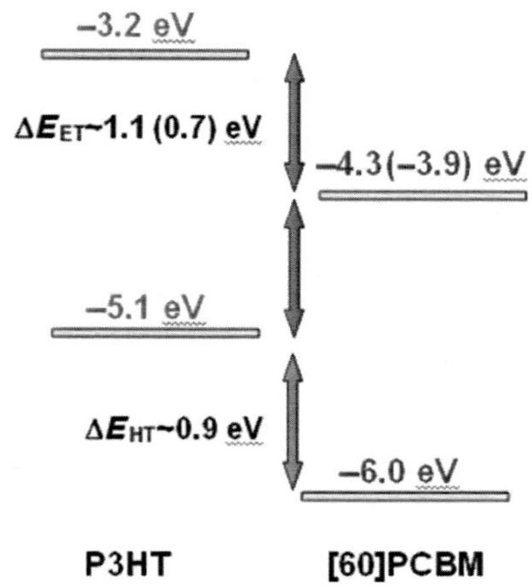

FIGURE 25.18. Frontier energy level diagram for a P3HT–PCBM system.

three methoxy groups attached to the PCBM phenyl group. The reference [60]PCBM provided a V_{OC} of 870 mV. Unfortunately, no solar cells efficiencies are given in reference 97, which strongly suggests that the observed improvement in V_{OC} was counterbalanced by decreased FF and/or I_{SC} values.

A much more successful attempt to design superior fullerene-based acceptor materials was reported by Riedel et al. [98] in 2005. The developed material **F2** (Figure 25.19) showed in combination with P3HT the V_{OC} value of 0.65 V, which is ~100 mV higher than the V_{OC} obtained for the [60]PCBM–P3HT blends under identical conditions. However, the power conversion efficiency of the **F2**/P3HT devices did not exceed 2.3% due to the lowered current densities. A similar compound **F3** (Figure 25.19) reported recently produced V_{OC} of 690 mV and power conversion efficiency of 2.6% in solar cells with P3HT used as an electron donor component [99].

A dramatic progress was made by application of bis-functionalized derivatives of C_{60} and C_{70} fullerenes as electron acceptor components in organic solar cells. The first success in this field was reported in a patent application filled in 2007 by a joint team of inventors from Plextronics and Nano-C companies [100]. They reported, in particular, bisindene fullerene adduct **F4** (Figure 25.20) providing for $V_{OC} = 0.84$ V, $I_{SC} = 9.43$ mA/cm^2, FF = 64%, and $\eta = 5.1\%$ in bulk heterojunction solar cells based on its composite with P3HT. Virtually the same results were reported by independent group in 2010 [78]. Subsequent efforts allowed the same group to bring the efficiency of the **F4**/P3HT solar cells up to the level of 6.5% [101], which, up to our best knowledge, was never certified or independently reproduced. The Yang group investigated the **F4**/P3HT system [102] and reported the efficiency of 4.5%, which is close to the values of 5.1–5.4%, reported in references 78 and 100.

There is some uncertainty regarding the other types of bis-functionalized fullerene derivatives—in particular, bis-[60]PCBM **F6** (Figure 25.20). Bis-PCBM was reported in 2008 by Lennes et al. [103] as a fullerene-based material yielding improved photovoltaic performance in solar cells with P3HT.

The theoretical modeling performed by Nelson and coworkers [104] suggests an existence of a considerable energetic disorder in the mixtures of bis-PCBM isomers possessing somewhat different LUMO level energies. The energetic disorder is expected to lead to the charge trapping in the system reducing FF and I_{SC} values of the devices. This conclusion was supported by experimental data reported by different groups [105, 106]. A number of other explored products bearing two to four cyclopropane addends on the fullerene cage (**F7–F12**) produced poor performances in organic bulk heterojunction solar cells [107, 108].

A family of different cyclopentadiene-type fullerene derivatives was synthesized and investigated by Niinomi et al. [111]. It was shown that LUMO energies of the compounds **F13–F20** (Figure 25.21) range from −3.4 eV to −3.3 eV, thus approaching very closely the LUMO energy of P3HT (−3.2, −3.3 eV). The resulting driving force $\Delta E_{ET} = 0.1 \div 0.2$ eV might still be sufficient for the efficient electron transfer in the view of the results reported

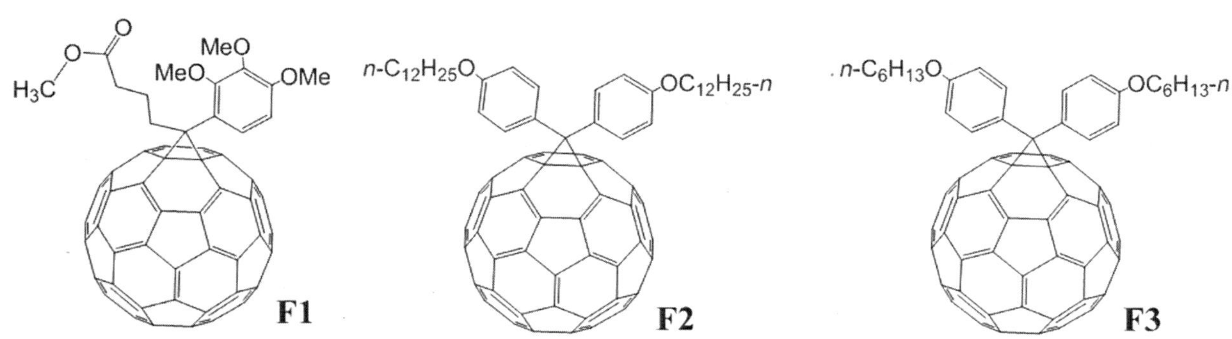

F1 F2 F3

FIGURE 25.19. Molecular structure of diphenylmethanofullerenes that showed improved open-circuit voltages in organic bulk heterojunction solar cells compared to the reference [60]PCBM material.

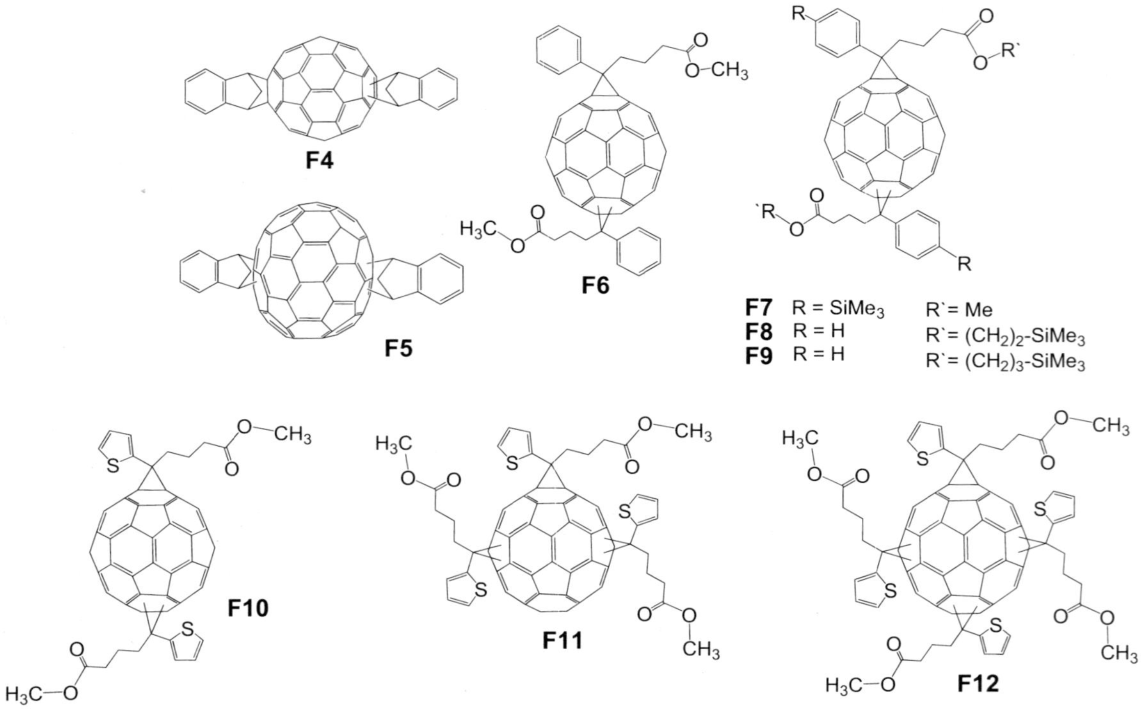

FIGURE 25.20. Molecular structures of some fullerene bis- and polycycloadducts applied as electron acceptors in bulk heterojunction solar cells.

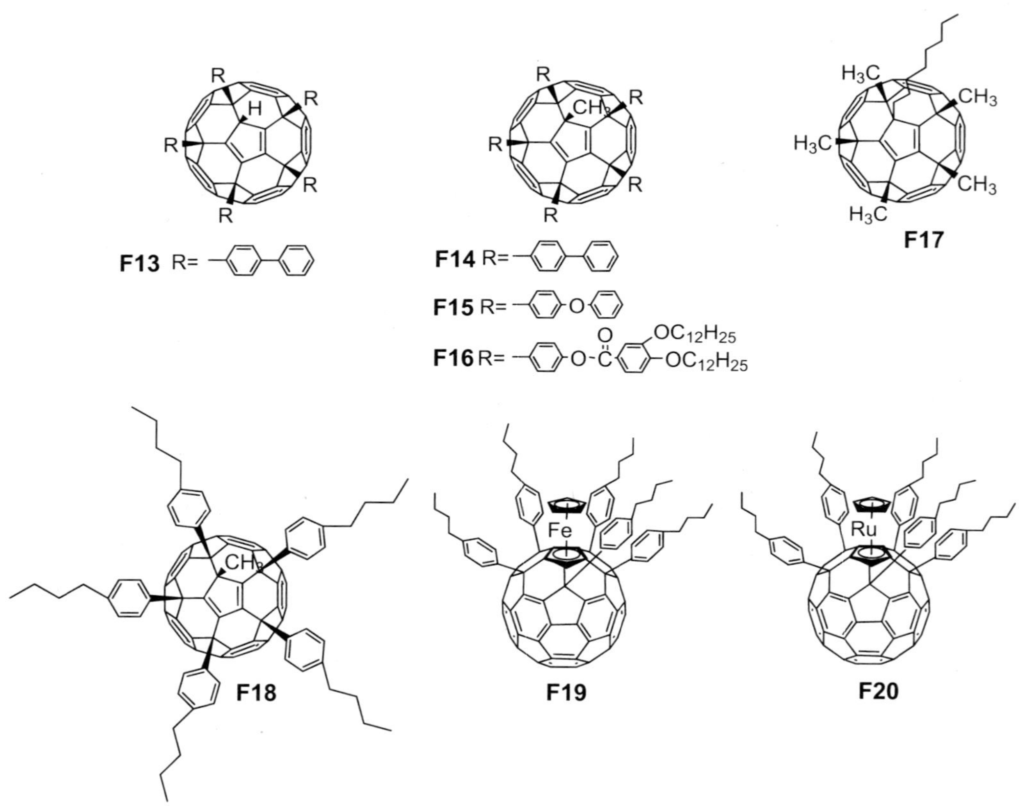

FIGURE 25.21. Molecular structures of some cyclopentadiene-type fullerene derivatives with high LUMO level energies investigated as electron acceptor materials in bulk heterojunction solar cells.

TABLE 25.2. The Parameters of Photovoltaic Cells Based on P3HT and Different Fullerene Derivatives

Fullerene Derivative	I_{SC} (mA/cm^2)	V_{oc} (V)	FF (%)	η (%)	Reference
[60]PCBM	10.6	0.64	55	3.7	92
F4	9.43	0.84	64	5.1	100
	10.61	0.84	72	6.5	101
	9.7	0.84	67	5.4	78
	8.5	0.82	65	4.5	102
F5	11.34	0.81	63	5.79	109
	10.6	0.85	74	6.7	110
F6	9.14	0.724	68	4.5	103
F7	2.56	0.75	29	0.57	107
F8	2.27	0.78	35	0.58	107
F9	5.86	0.77	34	1.53	107
F10	5.91	0.72	41	1.72	108
F11	1.88	0.64	28	0.34	108
F12	0.48	0.57	32	0.09	108
F13	0.2	0.25	43	0.02	111
F14	1.9	0.72	34	0.45	111
F15	2.8	0.76	51	1.08	111
F16	0.2	0.30	34	0.02	111
F17	0.8	0.42	37	0.12	111
F18	3.8	0.63	38	0.91	111
F19	2.5	0.73	38	0.68	111
F20	2.1	0.61	32	0.40	111

recently by Gong et al. [112]. However, the composite systems comprising P3HT and fullerene derivatives **F13–F20** showed rather poor photovoltaic performance. The origins for poor performance of these compounds are still poorly understood.

Table 25.2 summarizes the data on photovoltaic performance of different fullerene derivatives possessing higher LUMO level energies compared to [60]PCBM. It is seen from this table that only the fullerene derivatives **F4** and **F5** bearing two indene fragments attached to the C_{60} and C_{70} cages, respectively, outperform clearly [60]PCBM in bulk heterojunction solar cells using P3HT as an electron donor component. All other adducts show either comparable (like **F6**) or lower performances compared to the reference [60]PCBM. Even though the efficiency of ~5.0% seems to be the most realistic value for the **F4**/P3HT and **F5**/P3HT solar cells, there is no doubt that **F4** and **F5** are superior fullerene-based materials compared to [60]PCBM and [70]PCBM.

25.5 ADVANCED ELECTRON DONOR MATERIALS FOR BULK HETEROJUNCTION SOLAR CELLS

It was shown above that the optimal LUMO and HOMO energies of the electron donor counterpart for [60]PCBM (or [70]PCBM) should be around −3.6 eV and −5.1 eV, respectively (taking optimal polymer bandgap of 1.5 eV according to references [70 and 75] as illustrated in Figure 25.6). The design of such material with well-matching energy levels is a challenge for material chemists. The evolution of conjugated polymers during the last few years resulted in the development of many different promising structures. It is not possible to discuss all of them in the present chapter; therefore we will focus below mainly on the materials that provided power conversion efficiencies above 6.0% in bulk heterojunction solar cells.

The carbazole–thiophene–benzothiadiazole copolymer **P1** (abbreviated as PCDTBT) designed by the Leclerc group [80] was the first disclosed polymer providing certified power conversion efficiencies of 6.0% in bulk heterojunction solar cells [21] (Figure 25.22). This material has low HOMO energy resulting in high open-circuit voltages in bulk heterojunction solar cells using PCBM as the electron acceptor component. At the same time, it has approximately the same bandgap as P3HT, thus enabling appreciably high short-circuit current densities of 8–10 mA/cm^2. A thorough investigation of **P1**–[70]PCBM composites showed recently that their morphology can be turned by introducing small amounts of polar solvents to the blend solutions used for film casting. Recently, the additives of DMSO (methylsulfoxide) and DMF (dimethylformamide) boosted the efficiency of the solar cells based on the **P1**–[70]PCBM composites up to the level of 7.0%. Very similar efficiency has been achieved for the **P1**/[70]PCBM system by Cao group using additional charge selective buffer layer and Al negative electrode [24]. An excellent environmental stability of **P1** is a strongest

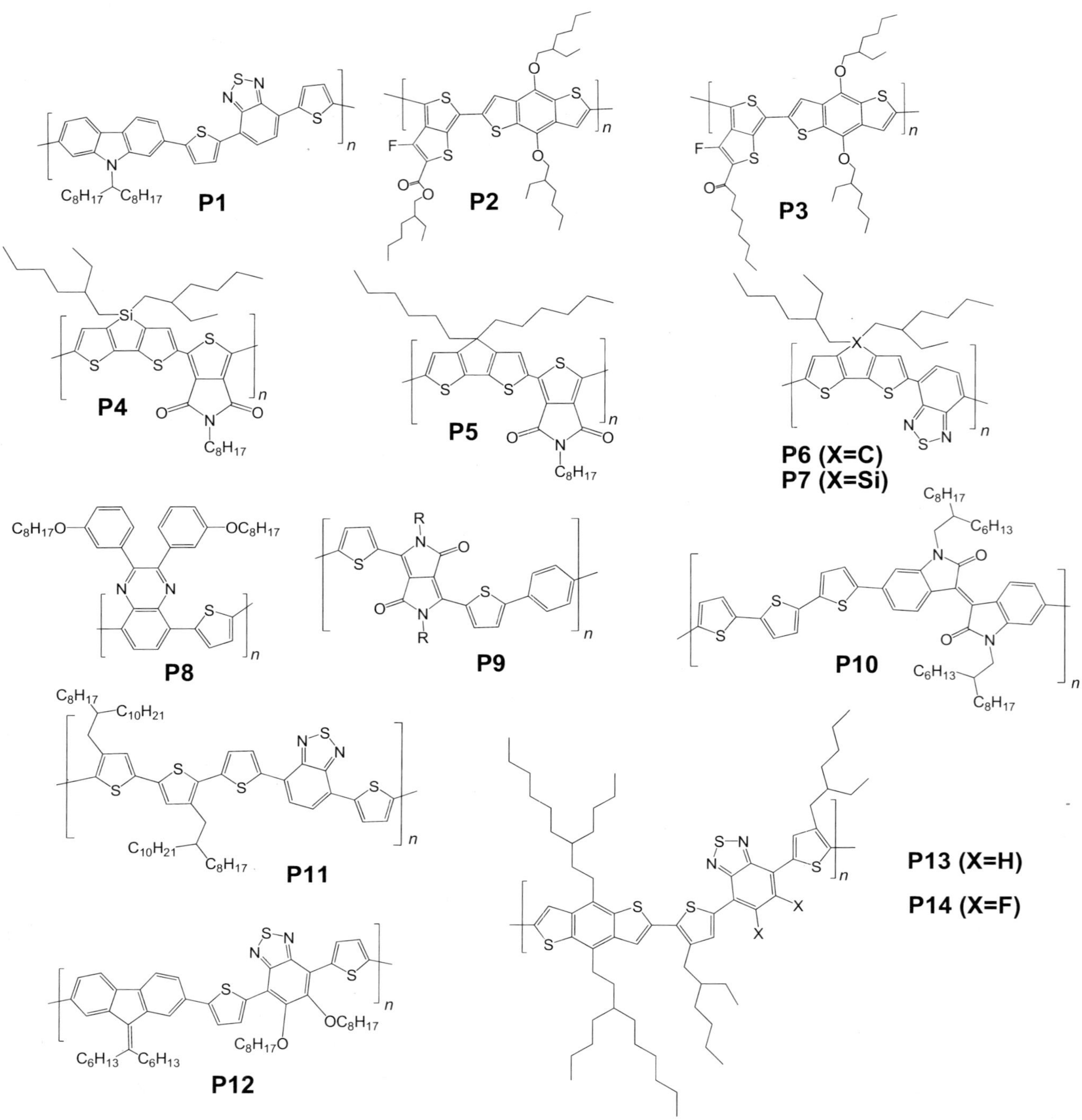

FIGURE 25.22. Molecular structures of some promising conjugated polymers investigated as electron donor materials in organic bulk heterojunction solar cells.

advantage of this polymer. The solar cells comprising **P1** as an electron donor polymer showed estimated operation lifetime of 7 years in accelerated tests [8].

A group of very promising conjugated polymers comprising alternating thieno[3, 4-*b*]thiophene and benzodithiophene units was introduced by Yu et al. [113, 114]. Starting with appreciably high initial efficiencies of ~5.0%, subsequent structural evolution of these materials resulted in the development of conjugated polymers **P2** and **P3** providing power conversion efficiencies well above 7.0%. A record certified efficiency of 8.37% has been reported recently for the solar cells based on **P2**–[70]PCBM composite modified with the charge-selective buffer layer at the interface between the photoactive blend and electron-collecting Al electrode [24]. It is obvious that polymers **P2** and **P3** are capable of giving high

TABLE 25.3. Performance of Organic Bulk Heterojunction Solar Cells Based on the Selected Combinations of Electron Donor Polymers and Fullerene Derivatives

Polymer–Fullerene	I_{SC} (mA/cm^2)	V_{oc} (V)	FF (%)	η (%)	Reference
P1–[70]PCBM	11.8	0.91	66	7.1	22
	12.1	0.90	62	6.55	24
P2–[70]PCBM	15.75	0.756	70.15	8.37	24
	14.5	0.74	69	7.4	23
P3–[70]PCBM	15.2	0.76	67	7.7	25
P4–[70]PCBM	12.2	0.88	68	7.3	115
P5–[70]PCBM	14.1	0.75	61	6.4	120
P6–[70]PCBM+ODT	16.2	0.62	55	5.5	119
P7–[70]PCBM	12.7	0.680	55	5.1	118
	14.9	0.576	61	5.2	117
P8–[70]PCBM	10.5	0.89	64	6.0	121
P9–[70]PCBM	10.3	0.8	65	5.5	122
P10–[70]PCBM	13.3	0.7	69	6.3	123
P11–[70]PCBM	12.3	0.722	70.5	6.26	124
P12–[70]PCBM	9.9	0.89	70	6.2	125
P13–[70]PCBM	10.03	0.87	57.3	5.0	126
P14–[70]PCBM	12.91	0.91	61.2	7.2	126

efficiencies of organic bulk heterojunction solar cells. However, the operation stability of these materials and devices remain poorly investigated.

Another promising electron donor material **P4** is a copolymer of Si-modified cyclopentadithiophene and 5-octyl-4*H*-thieno[3,4-*c*]pyrrole-4,6(5*H*)-dione (DOPT) [115]. A family of cyclopentadithiophene-based copolymers **P6** and **P7** has been intensively studied since 2006 [116]. However, the maximal power conversion efficiencies of bulk heterojunction solar cells based on these polymers were in the range of 5–6% [117–119]. The replacement of the benzothiadiazole unit in **P7** with DOPT block resulted in **P4** polymer which showed power conversion efficiency of 7.3% after substantial device and polymer optimization [26]. At the same time, a similar polymer **P5** comprising alternating cyclopentadithiophene and DOPT units showed appreciably high power conversion efficiency of 6.4% [120].

Somewhat lower power conversion efficiencies (5.0–6.5%) were obtained for the copolymers **P8** [121], **P9** [122], **P10** [123], **P11** [124], and **P12** [125]. Very elegant chemical structures and relatively simple synthetic routes developed to produce these conjugated polymers make them promising electron donor materials for organic photovoltaic cells potentially available in bulk quantities.

Directed chemical structure design plays an important role in the development of novel conjugated polymers for photovoltaic applications. One of the illustrative examples is provided by the polymers **P13** and **P14** [126]. The dithienobenzene–benzothiadiazole copolymer **P13** showed moderate performance in organic bulk heterojunction solar cells defined by the solar light power conversion efficiency of 5.0%. The introduction of two fluorine atoms in the benzothiadiazole ring in the polymer **P13** resulted in a new polymer **P14**. Such modification has changed the electronic structure of the polymer: The HOMO level of **P14** is lowered by 0.14 eV compared to **P13**, while the LUMO level of **P14** is 0.2 eV lower than that of **P13**. Such modification of the electronic structure of the polymer brought an expected increase in the V_{OC} from 0.87 V to 0.91 V accompanied also by the considerable improvements in the short-circuit current density and the fill factor (Table 25.3) resulting in the power conversion efficiency of 7.2%.

The data presented in Table 25.3 give an overview of the current status of the research in the field of organic fullerene–polymer solar cells. It is clearly seen that the intensive development of novel conjugated polymers during the last 5 years resulted in tremendous progress in the field and brought the solar cell efficiencies from 4.0–4.5% level (the best P3HT-based devices) to the current level of 7–8%.

It would be unfair not to mention low molecular weight materials that boosted the performance of organic bulk heterojunction solar cells well beyond 8–9% recently [29, 30]. Many low molecular weight materials are not soluble in organic solvents, however their inherent stability allows one to purify and process them by vacuum sublimation. Some merocyanine dyes showed promising performances when used as electron donor materials in organic solar cells. In particular, the **D1**/C_{60} combination (Figure 25.23) provided power conversion efficiencies close to 5.0% [31]. Similar performances were obtained using oligothiophene donor molecule **D2** and diphenylamine–thiophene–benzothiadiazole hybrid **D3** also combined with C_{60} [127].

A tandem bulk heterojunction solar cell structure based on coevaporated C_{60} fullerene, fluorinated phthalocyanine **D4**, and oligothiophene **D5** bearing dicyanovinyl groups was reported recently [128]. A power conversion efficiency of 6.0% was achieved in solar cells based on these simple

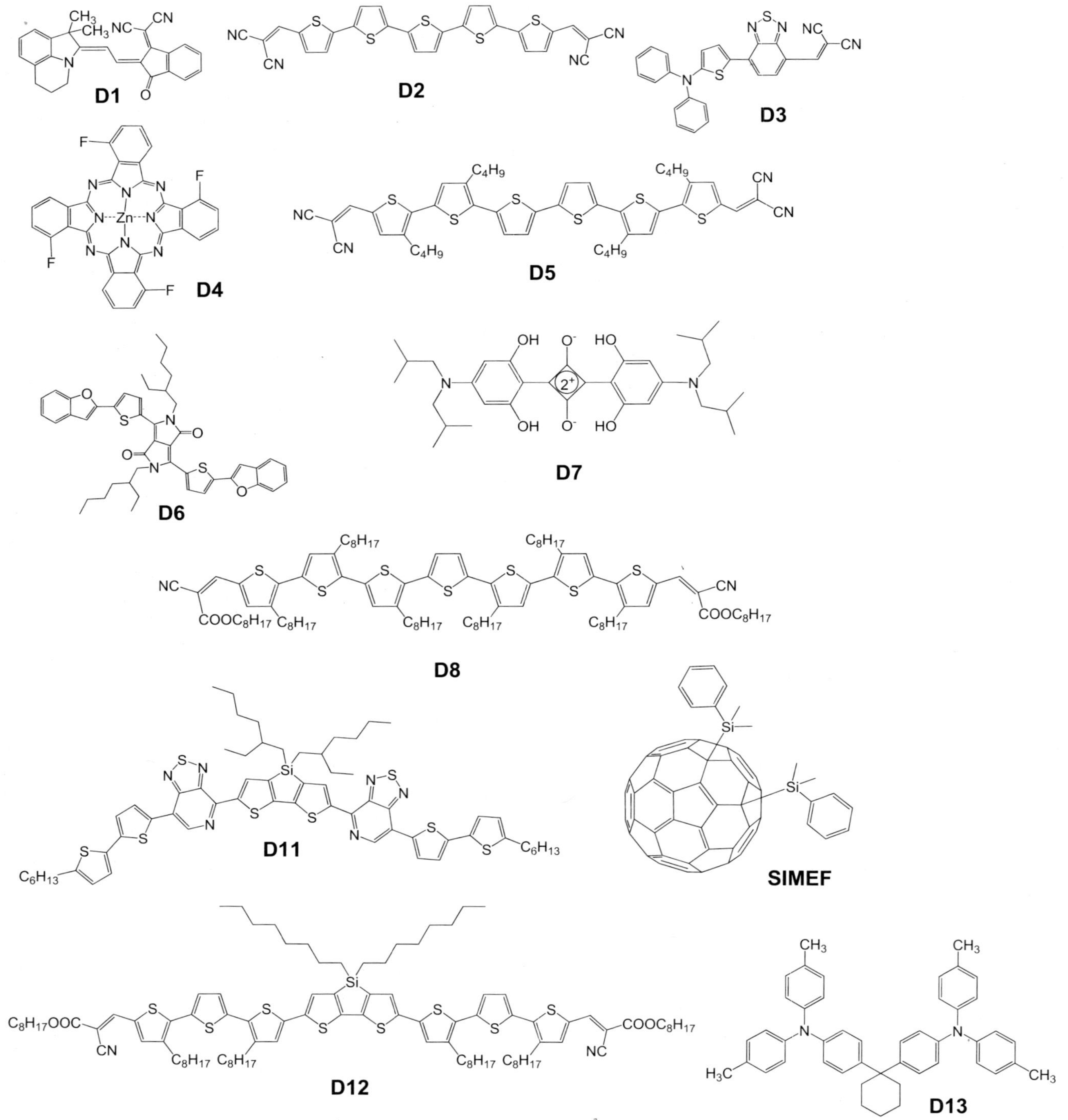

FIGURE 25.23. Molecular structures of some low molecular weight compounds applied as electron donor materials in organic bulk heterojunction solar cells.

material combinations by using very sophisticated device structure (Figure 25.24).

Heliatek has certified 8.3% organic double-junction solar cells[27]. An improved efficiency of 9.8% has been claimed by Heliatek for small-molecular photovoltaic cells very recently [129]. Unfortunately, the material combinations used to produce these high-efficiency devices are not disclosed by now.

Small electron donor molecules might possess reasonably high solubility in organic solvents, enabling wet

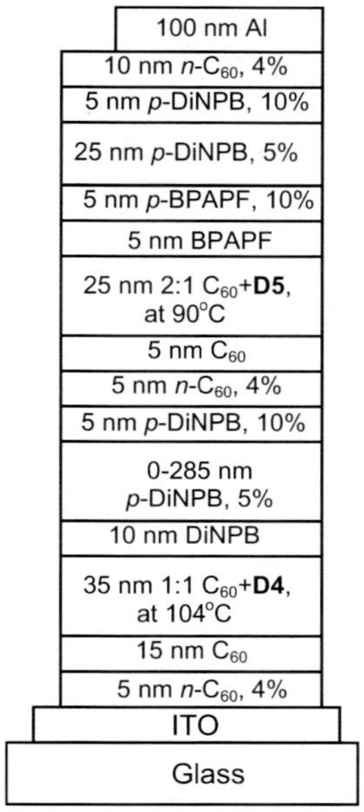

FIGURE 25.24. A schematic structure of a double-junction tandem device based on **D4**/C_{60} and **D5**/C_{60} composites. For definitions of the used transport layers reader is referred to the original publication [128].

processing of organic photovoltaic devices. A combination of diketopyrrolopyrrole compound **D6** with [70]PCBM was initially one of the most efficient low-molecular-weight material combinations developed for this purpose [130, 131]. However, recently reported squaraine dye **D7** outperformed **D6** significantly in bulk heterojunction solar cells with [70]PCBM [132]. Similarly high performances were reached also using oligothiophene donor molecule **D8** modified with cyanovinyl units on both sides [133].

A very elegant and highly promising approach to design solution-processible organic solar cells was demonstrated by Matsuo et al. [28]. The key material is a soluble porphyrin precursor **D9** that eliminates ethylene upon heating, producing insoluble tetrabenzoporphyrin **D10** (Figure 25.25). When the composite of **D9** with the fullerene derivative SIMEF was subjected to annealing, a highly ordered columnar structure was formed (Figure 25.26). Such composite morphology is very favorable for photovoltaic applications. To build a

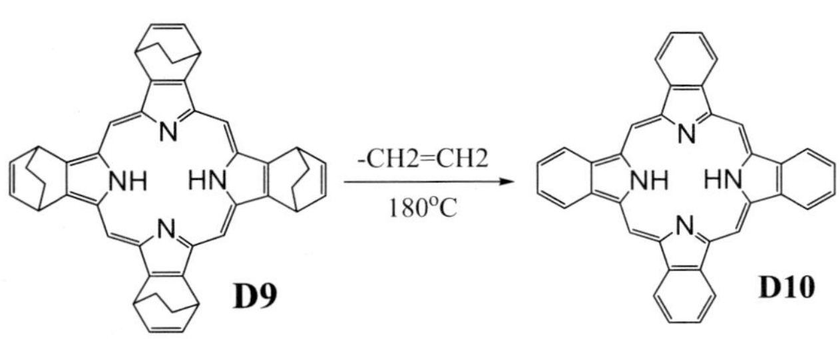

FIGURE 25.25. Conversion of soluble **D9** precursor to insoluble tetrabenzoporphyrin **D10**.

photovoltaic device, a soluble precursor **D9** was spin-coated on a PEDOT:PSS-covered ITO slide and then converted to insoluble **D10** by thermal annealing at 180°C thus producing a continuous bottom donor **D10** layer. A 3:7 w/w blend of **D9** and SIMEF was spin-coated above the bottom **D10** layer. The resulting film was also sintered at 180°C thus forming intermediate bulk heterojunction layer with permanently fixed morphology. The morphology of the composite can be easily revealed by washing the fullerene component SIMEF away and exposing a well-ordered columnar network of the **D10** material (Figure 25.26). The device was completed by the deposition of the top SIMEF layer followed by the hole-blocking layer of bathocuproine or similar material and aluminum top electrode. The fabricated devices yielded an impressive power conversion efficiency of 5.2% with an EQE of 35–45% in the whole visible range. Subsequent research of Mitsubishi Chemical in that direction brought record power conversion efficiencies of 9–10% in 2011 [29, 30]. The devices produced using this approach are expected to enter a commercialization stage rather soon.

New soluble small-molecule **D11** featured by the Heeger group provided a power conversion efficiency of 6.7% when combined with [70]PCBM in organic solar cells [136]. Very similar material **D12** yielding a solar cell power conversion efficiency of 5.84% was reported recently by Zhou et al. [135].

An overview of small-molecule-based solar cells presented in Table 25.4 proves that low-molecular-weight compounds have a large and still not fully explored potential to be used as advanced photoactive materials for construction of efficient bulk heterojunction devices.

Finally, we would like to mention a recent work of C. Tang and co-workers [32] which featured Schottky barrier-type photovoltaic devices based on organic semiconductors. The efficiency of the devices based on a C_{70}/MoO_3 Schottky junction exceeded 5% when a small amount (5%) of organic dopant **D13** was added to the layer to improve its electrical properties. The emerging Schottky barrier approach promises to bring the efficiency of organic solar cells closer to the performance of conventional photovoltaic cells based on inorganic semiconductors.

25.6 CONCLUSION AND OUTLOOK

The last 5–10 years of intensive research resulted in the development of many novel photoactive *p*-type and *n*-type organic semiconductors, which are key materials for organic solar cells. At the same time, the OPV community learned how to manage the nanoscale ordering of the materials to create the appropriate nanoscale *p*–*n* heterojunctions in the entire volume of the active layer of the device. New materials and new knowledge brought up commercially interesting certified solar cell efficiencies of 8–10% reported recently by several independent groups.

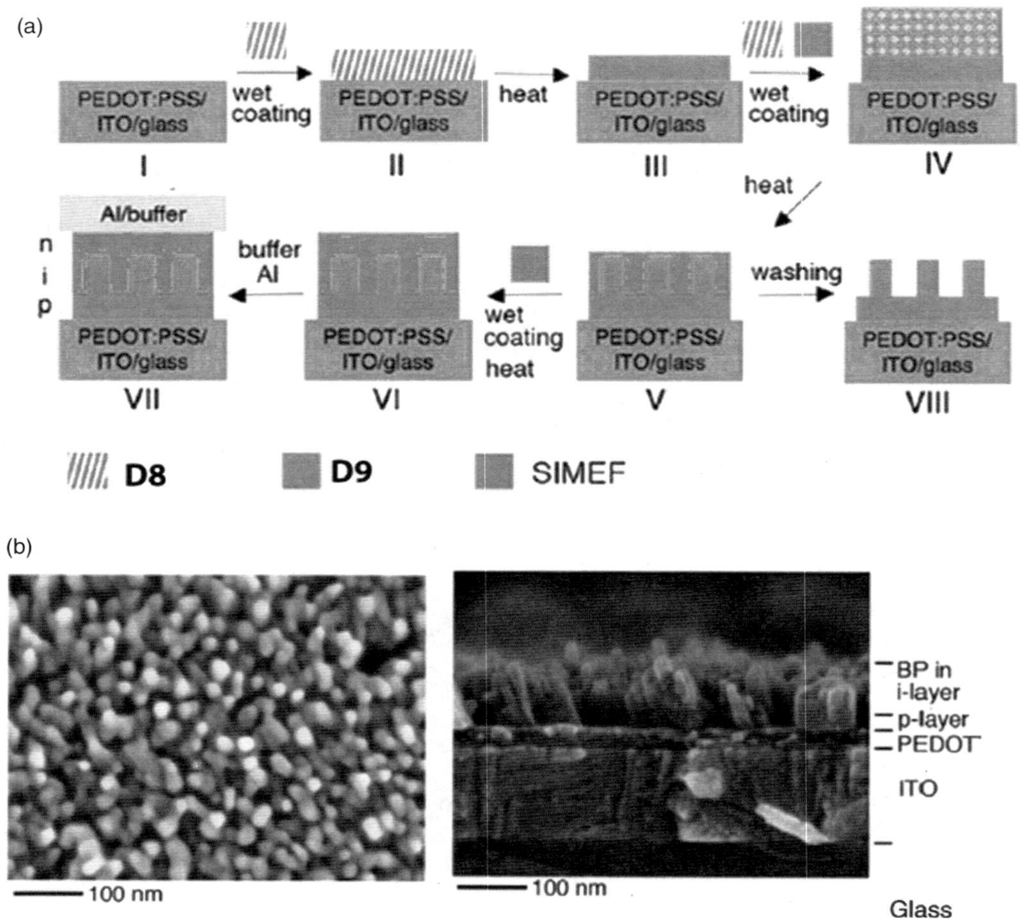

FIGURE 25.26. (**a**) Technology for the solar cell construction based on SIMEF and **D9/D10**. (**b**) Top view (*left*) and side view (*right*) SEM images of the exposed porphynin **D10** network after removing SIMEF material. Reproduced from reference 49, with permission of the American Chemical Society. See color insert.

TABLE 25.4. Selected Examples of Organic Bulk Heterojunction Solar Cells Based on Low-Molecular-Weight Materials

Material Combination	Processing Route	I_{SC} (mA/cm^2)	V_{oc} (V)	FF (%)	η (%)	Reference
D1–C_{60}	Evaporation	—	—	—	4.9	31
D2–C_{60}	Evaporation	11.1	0.97	49	5.2	127
D3–C_{60}	Evaporation	14.68	0.79	50	5.81	134
D4–C_{60}+D5–C_{60}	Evaporation	6.18	1.589	61	6.07	128
D6–[70]PCBM	Solution	10.0	0.92	48	4.4	130
D7–[70]PCBM	Solution	12.0	0.92	50	5.5	132
D8–[70]PCBM	Solution	10.74	0.86	55	5.1	133
D10–[70]PCBM	Solution	10.3	0.75	65	5.0	28
D11–[70]PCBM	Solution	14.4	0.78	59.3	6.7	136
D12–[70]PCBM	Solution	11.5	0.80	64	5.84	135
D13–C_{70}–MoO_3	Evaporation	11.43	0.91	50.23	5.2	32

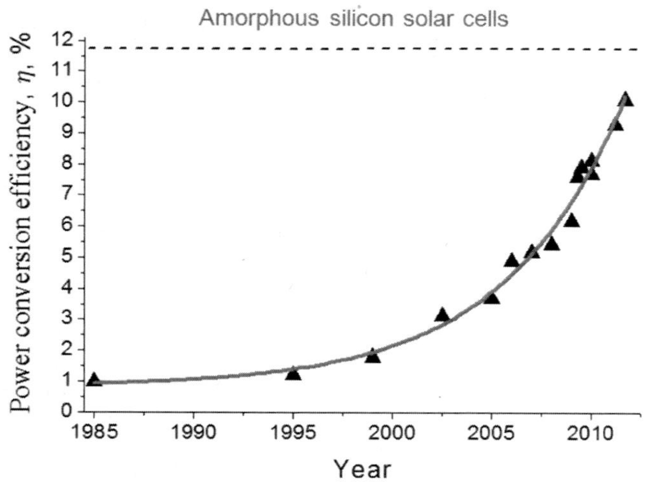

FIGURE 25.27. Progress of organic bulk heterojunction solar cells.

The progress in organic photovoltaics is reflected on the graph shown in Figure 25.27. If this trend continues in future, the organic solar cells will reach the level of amorphous silicon and go beyond it within the next couple of years, presenting serious competition to the classic solar energy conversion technologies.

ACKNOWLEDGMENTS

This work was supported by the Russian Foundation for Basic Research (RFBR grant 13-03-01170) and Russian Ministry for Science and Education (contract No. 8709).

REFERENCES

1. Wang, A., Green, M.A., Ferrazza, F. (1998). Novel 19.8% efficient "honeycomb" textured multicrystalline and 24.4% monocrystalline silicon solar cells. *Applied Physics Letters*, **73**, 1991–1993.
2. Shaheen, S. E., Ginley, D. S., Jabbour, G. E. (2005). Organic-based photovoltaics: Toward low-cost power generation. *MRS Bulletin*, **30**, 10–19.
3. Green, M. A., Emery, K., Hishikawa, Y., Warta, W. (2010). Solar cell efficiency tables (version 35). *Progress in Photovoltaics: Research and Applications*, **18**, 144–150.
4. Global market outlook for photovoltaics until 2014. European photovoltaic industry association. Available at http://www.epia.org/fileadmin/EPIA_docs/public/Global_Market_Outlook_for_Photovoltaics_until_2014.pdf. Accessed May 2010.
5. Tao, C. S., Jiang, J., Tao, M. (2011). Natural resource limitations to terawatt-scale solar cells. *Solar Energy Materials & Solar Cells*, **95**, 3176–3180.
6. Amorphous Silicon: Losing the Shakeout? Available at http://www.greentechmedia.com/articles/read/amorphous-silicon-solar-losing-the-shakeout/ Accessed on 20 November 2011.
7. Green, M. A., Emery, K., Hishikawa, Y., Warta, W. (2010). Solar cell efficiency tables (version 36). *Progress in Photovoltaics: Research and Applications*, **18**, 346–352.
8. Service, R. F. (2011). Outlook brightens for plastic solar cells. *Science*, **332**, 293.
9. http://www.novaled.com/oledcompetence/opv.html. Accessed 22 October 2011.
10. Sun, S., Sariciftci, N. S. (2005). *Organic photovoltaics. Mechanism, Materials, and Devices*. CRC Press, Boca Raton, FL.
11. Wurfel, P. (2005). *Physics of Solar Cells*. Wiley–VCH Verlag GmbH&Co, KgaA, Weinheim.
12. Brabec, C., Scherf, U., Dyakonov, V. (2008). *Organic Photovoltaics: Materials, Device Physics, and Manufacturing Technologies*. Wiley–VCH Verlag GmbH&Co, KGaA, Weinheim.
13. Heeger, A., Sariciftci, N. S., Namdas, E. B. (2010). *Semiconducting and Metallic Polymers*. Oxford Graduate Texts, Oxford University Press, New York.
14. Troshin, P. A., Sariciftci, N. S. (2012). *Supramolecular chemistry for organic photovoltaics*. In: Gale, P. A., Steed, J. W., eds. *Supramolecular Chemistry: From Molecules to Nanomaterials*. John Wiley & Sons, Hoboken, NJ.
15. Brabec, C., Dyakonov, V., Parisi, J., Sariciftci, N. S. (2003). *Organic Photovoltaics: Concepts and Realization* (Springer Series in Materials Science). Springer, Berlin.
16. Sariciftci, N. S., Heeger, A. J., (inventors). The Regents of the University of California (Oakland, CA), assignee. Conjugated polymer–acceptor heterojunctions; diodes, photodiodes, and photovoltaic cells. US patent 5,331,183. 1994 July 19.
17. Sariciftci, N. S., Smilowitz, L., Heeger, A. J., Wudl, F. (1992). Photoinduced electron transfer from a conducting polymer to buckminsterfullerene. *Science*, **258**, 1474–1476.
18. Yu, G., Gao, J., Hummelen, J. C., Wudl, F., Heeger, A. J. (1995). Polymer photovoltaic cells: Enhanced efficiencies via a network of internal donor–acceptor heterojunctions. *Science*, **270**, 1789–1791.
19. Tang, C. W. (1986). Two-layer organic photovoltaic cell. *Applied Physics Letters*, **48**, 183–185.
20. Tang, C. W. (1978). Multilayer organic photovoltaic elements. US patent 4164431, filled 18.05.1978.
21. Park, S. H., Roy, A., Beaupre, S., Cho, S., Coates, N., Moon, J. S., Moses, D., Leclerc, M., Lee, K., Heeger, A. J. (2009). Bulk heterojunction solar cells with internal quantum efficiency approaching 100%. *Nature Photonics*, **3**, 297–303.
22. Chu, T.-Y., Alem, S., Tsang, S.-W., Tse, S.-C., Wakim, S., Lu, J., Dennler, G., Waller, D., Gaudiana, R., Tao, Y. (2011). Morphology control in polycarbazole based bulk heterojunction solar cells and its impact on device performance. *Applied Physics Letters*, **98**, 253301.
23. Liang, Y., Xu, Z., Xia, J., Tsai, S.-T., Wu, Y., Li, G., Ray, C., Yu, L. (2010). For the bright future—Bulk heterojunction polymer solar cells with power conversion efficiency of 7.4%. *Advanced Materials*, **22**, E135–E138.

24. He, Z., Zhong, C., Huang, X. , Wong, W.-Y., Wu, H., Chen, L., Su, S., Cao, Y. (2011). Simultaneous enhancement of open-circuit voltage, short-circuit current density, and fill factor in polymer solar cells. *Advanced Materials*, **23**, 4636–4643.

25. Chen, H.-Y., Hou, J., Zhang, S., Liang, Y., Yang, G., Yang, Y., Yu, L., Wu, Y., Li, G. (2009). Polymer solar cells with enhanced open-circuit voltage and efficiency. *Nature Photonics*, **3**, 649–653.

26. Chu, T.-Y., Lu, J., Beaupre, S., Zhang, Y., Pouliot, J.-R., Wakim, S., Zhou, J., Leclerc, M., Li, Z., Ding, J., Tao, Y. (2011). Bulk heterojunction solar cells using thieno[3,4-c]pyrrole-4,6-dione and dithieno[3,2-b:2 ',3 '-*d*]silole copolymer with a power conversion efficiency of 7.3%. *Journal of the American Chemical Society*, **133**, 4250–4253.

27. Green1, M. A., Emery, K., Hishikawa, Y., Warta, W. (2011). Solar cell efficiency tables (version 37). *Progress in Photovoltaics: Research and Applications*, **19**, 84–92.

28. Matsuo, Y., Sato, Y., Niinomi, T., Soga, I., Tanaka, H., Nakamura, E. (2009). Columnar structure in bulk heterojunction in solution-processable three-layered p-i-n organic photovoltaic devices using tetrabenzoporphyrin precursor and silylmethyl[60]fullerene. *Journal of the American Chemical Society*, **131**, 16048–16050.

29. Mitsubishi chemical to commercialize printable solar cells next year. http://techcrunch.com/2011/04/06/mitsubishi-chemical-to-commercialize-printable-solar-cells-next-year last accessed. Accessed 2011 November 20.

30. Mitsubishi Chemical Corp develops spray-on solar cells. Available at http://www.whatsonxiamen.com/tech998.html. Accessed 20 Novvember 2011.

31. Kronenberg, N. M., Steinmann, V., Bürckstümmer, H., Hwang, J., Hertel, D., Würthner, F., Meerholz, K. (2010). Direct comparison of highly efficient solution- and vacuum-processed organic solar cells based on merocyanine dyes. *Advanced Materials*, **22**, 4193–4197.

32. Zhang, M., Wang, H., Tian, H., Geng, Y., Tang, C. W. (2011). Bulk heterojunction photovoltaic cells with low donor concentration. *Advanced Materials*, **23**, 4960–4964.

33. Sun, B., Greenham, N. C. (2006). Improved efficiency of photovoltaics based on CdSe nanorods and poly(3-hexylthiophene) nanofibers. *Physical Chemistry Chemical Physics*, **8**, 3557–3560.

34. Wang, D., Baral, J. K., Zhao, H., Gonfa, B. A., Truong, V.-V., Khakani, M. A. E., Izquierdo, R., Ma, D. (2011). Controlled fabrication of pbs quantum-dot/carbon-nanotube nanoarchitecture and its significant contribution to near-infrared photon-to-current conversion. *Advanced Functional Materials*, **21**, 4010–4018.

35. Seo, J., Cho, M. J., Lee, D., Cartwright, A. N., Prasad, P. N. (2011). Efficient heterojunction photovoltaic cell utilizingm nanocomposites of lead sulfide nanocrystals and a low-bandgap polymer. *Advanced Materials*, **23**, 3984–3988.

36. Dayal, S., Kopidakis, N., Olson, D. C., Ginley, D. S., Rumbles, G. (2010). Photovoltaic devices with a low band gap polymer and CdSe nanostructures exceeding 3% efficiency. *Nano Letters*, **10**, 239–242.

37. Kymakis, E., Koudoumas, E., Franghiadakis, I., Amaratunga, G. A. J. (2006). Post-fabrication annealing effects in polymer-nanotube photovoltaic cells. *Journal of Physics D: Applied Physics*, **39**, 1058–1063.

38. Kazaoui, S., Minami, N., Nalini, B., Kim, Y., Hara, K. (2005). Near-infrared photoconductive and photovoltaic devices using single-wall carbon nanotubes in conductive polymer films. *Journal of Applied Physics*, **98**, 084314.

39. Yu, D., Yang, Y., Durstock, M., Baek, B., Dai, L. (2010). Soluble P3HT-grafted graphene for efficient bilayer-heterojunction photovoltaic devices. *ACSNano*, **4**, 5633–5640.

40. Liu, Z., He, D., Wang, Y., Wu, H., Wang, J. (2010). Solution-processable functionalized graphene in donor–acceptor-type organic photovoltaic cells. *Solar Energy Materials and Solar Cells*, **94**, 1196–1200.

41. Li, Y., Hu, Y., Zhao, Y., Shi, G., Deng, L., Hou, Y., Qu, L. (2011). An electrochemical avenue to green-luminescent graphene quantum dots as potential electron-acceptors for photovoltaics. *Advanced Materials*, **23**, 776–780.

42. McNeill, C. R., Abrusci, A., Zaumseil, J., Wilson, R., McKiernan, M. J., Burroughes, J. H., Halls, J. J. M., Greenham, N. C., Friend, R. H. (2007). Dual electron donor/electron acceptor character of a conjugated polymer in efficient photovoltaic diodes. *Applied Physics Letters*, **90**, 193506.

43. Blom, P. W. M., Mihailetchi, V. D., Koster, L. J. A., Markov, D. E. (2007). Device physics of polymer: Fullerene bulk heterojunction solar cells. *Advanced Materials*, **19**, 1551–1566.

44. Nelson, J. (2003). *The Physics of Solar Cells*. Imperial College Press, London.

45. Peet, J., Heeger, A. J., Bazan, G. C. (2009). "Plastic" solar cells: Self-assembly of bulk heterojunction nanomaterials by spontaneous phase separation. *Accounts of Chemical Research*, **42**, 1700–1708.

46. Theander, M., Yartsev, A., Zigmantas, D., Sundstrom, V., Mammo, W., Andersson, M. R., Inganas, O. (2000). Photoluminescence quenching at a polythiophene/C-60 heterojunction. *Physical Review B*, **61**, 12957–12963.

47. Markov, D. E., Tanase, C., Blom, P. W. M., Wildeman, J. (2005). Simultaneous enhancement of charge transport and exciton diffusion in poly(*p*-phenylene vinylene) derivatives. *Physical Review B*, **72**, 045217.

48. Markov, D. E., Amsterdam, E., Blom, P. W. M., Sieval, A. B., Hummelen, J. C. (2005). Accurate measurement of the exciton diffusion length in a conjugated polymer using a heterostructure with a side-chain cross-linked fullerene layer. *Journal of Physical Chemistry A*, **109**, 5266–5274.

49. Mihailetchi, V. D., Koster, L. J. A., Hummelen, J. C., Blom, P. W. M. (2004). Photocurrent generation in polymer–fullerene bulk heterojunctions. *Physical Review Letters*, **93**, 216601.

50. Morteani, A. C., Sreearunothai, P., Herz, L. M., Friend, R. H., Silva, C. (2004). Exciton regeneration at polymeric semiconductor heterojunctions. *Physical Review Letters*, **92**, 247240.

51. Koster, L. J. A., Smiths, E. C. P., Mihailetchi, V. D., Blom, P. W. M. (2005). Device model for the operation of polymer/fullerene bulk heterojunction solar cells. *Physical Review B*, **72**, 085205.

52. Nelson, J. (2003). Diffusion-limited recombination in polymer–fullerene blends and its influence on photocurrent collection. *Physical Review B*, **67**, 155209.

53. Tao, C., Ruan, S., Xie, G., Kong, X., Shen, L., Meng, F., Liu, C., Zhang, X., Dong, W., Chen, W. (2009). Role of tungsten oxide in inverted polymer solar cells. *Applied Physics Letters*, **94**, 043311.

54. Zhao, D. W., Liu, P., Sun, X. W., Tan, S. T., Ke, L., Kyaw, A. K. K. (2009). An inverted organic solar cell with an ultrathin Ca electron-transporting layer and MoO_3 hole-transporting layer. *Applied Physics Letters*, **95**, 153304.

55. Liao, H.-H., Chen, L.-M., Xu, Z., Li, G., Yang, Y. (2008). Highly efficient inverted polymer solar cell by low temperature annealing of Cs_2CO_3 interlayer. *Applied Physics Letters*, **92**, 173303.

56. Kim, J. Y., Kim, S. H., Lee, H. H., Lee, K. H., Ma, W. L., Gong, X., Heeger, A. J. (2006). New architecture for high-efficiency polymer photovoltaic cells using solution-based titanium oxide as an optical spacer. *Advanced Materials*, **18**, 572–576.

57. Hau, S. K., Yip, H.-L., Baek, N. S., Zou, J., O'Malley, K., Jen, AK.-Y. (2008). Air-stable inverted flexible polymer solar cells using zinc oxide nanoparticles as an electron selective layer. *Applied Physics Letters*, **92**, 253301.

58. Mor, G. K., Shankar, K., Paulose, M., Varghese, O. K., Grimes, C. A. (2007). High efficiency double heterojunction polymer photovoltaic cells using highly ordered TiO_2 nanotube arrays. *Applied Physics Letters*, **91**, 152111.

59. Wei, Q., Nishizawa, T., Tajima, K., Hashimoto, K. (2008). Self-organized buffer layers in organic solar cells. *Advanced Materials*, **20**, 2211–2216.

60. Hsieh, C.-H., Cheng, Y.-J., Li, P.-J., Chen, C.-H., Dubosc, M., Liang, R.-M., Hsu, C.-S. (2010). Highly efficient and stable inverted polymer solar cells integrated with a cross-linked fullerene material as an interlayer. *Journal of the American Chemical Society*, **132**, 4887–4893.

61. Chen, L.-M., Hong, Z., Li, G., Yang, Y. (2009). Recent progress in polymer solar cells: Manipulation of polymer:fullerene morphology and the formation of efficient inverted polymer solar cells. *Advanced Materials*, **21**, 1434–1449.

62. Huang, F., Wu, H., Cao, Y. (2010). Water/alcohol soluble conjugated polymers as highly efficient electron transporting/injection layer in optoelectronic devices. *Chemical Society Reviews*, **39**, 2500–2521.

63. Steim, R., Kogler, F. R., Brabec, C. J. (2010). Interface materials for organic solar cells. *Journal of Materials Chemistry*, **20**, 2499–2512.

64. Chen, L.-M., Xu, Z., Hong, Z., Yang, Y. (2010). Interface investigation and engineering – achieving high performance polymer photovoltaic devices. *Journal of Materials Chemistry*, **20**, 2575–2598.

65. Gendron, D., Leclerc, M. (2011). *Energy & Environmental Science*, **4**, 1225–1237.

66. Hoppe, H., Shokhovets, S., Gobsch, G. (2007). Inverse relation between photocurrent and absorption layer thickness in polymer solar cells. *Physica Status Solidi (RRL)*, **1**, R40–R42.

67. Hoppe, H., Sariciftci, N. S. (2008). Polymer Solar Cells. In: Marder, S. R., Lee, K.-S., eds. *Photoresponsive Polymers II. Advances in Polymer Science*. Springer, Berlin, pp. 1–86.

68. Troshin, P. A., Hoppe, H., Peregudov, A. S., Egginger, M., Shokhovets, S., Gobsch, G., Sariciftci, N. S., Razumov, V. F. (2011). [70]Fullerene-based materials for organic solar cells. *ChemSusChem*, **4**, 119–124.

69. Brabec, C. J., Cravino, A., Meissner, D., Sariciftci, N. S., Fromherz, T., Rispens, M. T., Sanchez, L., Hummelen, J. C. (2001). Origin of the open circuit voltage of plastic solar cells. *Advanced Functional Materials*, **11**, 374–380.

70. Scharber, M. C., Muhlbacher, D., Koppe, M., Denk, P., Waldauf, C., Heeger, A. J., Brabec, C. J. (2006). Design rules for donors in bulk-heterojunction solar cells—towards 10% energy-conversion efficiency. *Advanced Materials*, **18**, 789–794.

71. Kooistra, F. B., Knol, J., Kastenberg, F., Popescu, L. M., Verhees, W. J. H., Kroon, J. M., Hummelen, J. C. (2007). Increasing the open circuit voltage of bulk-heterojunction solar cells by raising the LUMO level of the acceptor. *Organic Letters*, **9**, 551–554.

72. Mihailetchi, V. D., Wildeman, J., Blom, P. W. M. (2005). Space-charge limited photocurrent. *Physical Review Letters*, **94**, 126602.

73. Tress, W., Petrich, A., Hummert, M., Hein, M., Leo, K., Riede, M. (2011). Imbalanced mobilities causing S-shaped IV curves in planar heterojunction organic solar cells. Imbalanced mobilities causing S-shaped IV curves in planar heterojunction organic solar cells. *Applied Physics Letters*, **98**, 063301.

74. Tress, W., Leo, K., Riede, M. (2011). Influence of hole-transport layers and donor materials on open-circuit voltage and shape of *I–V* curves of organic solar cells. *Advanced Functional Materials*, **21**, 2140–2149.

75. Ameri, T., Dennler, G., Lungenschmied, C., Brabec, C. J. (2009). Organic tandem solar cells: A review. *Energy & Environmental Science*, **2**, 347–363.

76. Kotlarski, J. D., Blom, P. W. M. (2011). Ultimate performance of polymer: Fullerene bulk heterojunction tandem solar cells. *Applied Physics Letters*, **98**, 053301.

77. LuxResearch Report on OPV, https://portal.luxresearchinc.com/research/document_excerpt/7827; http://www.luxresearchinc.com/blog/2011/07/flexible-opv-modules-sharpen-the-market%E2%80%99s-growth-curve-through-2020/, Accessed 20 Novmber 2011.

78. He, Y., Chen, H.-Y., Hou, J., Li, Y. (2010). Indene-C60 bisadduct: A new acceptor for high-performance polymer solar cells. *Journal of the American Chemical Society*, **132**, 1377–1382.

79. Kanai, K., Akaike, K., Koyasu, K., Sakai, K., Nishi, T., Kamizuru, Y., Nishi, T., Ouchi, Y., Seki, K. (2009).

Determination of electron affinity of electron accepting molecules. *Applied Physics A*, **95**, 309–313.

80. Blouin, N., Michaud, A., Gendron, D., Wakim, S., Blair, E., Neagu-Plesu, R., Belletete, M., Durocher, G., Tao, Y., Leclerc, M. (2008). Toward a rational design of poly(2,7-carbazole) derivatives for solar cells. *Journal of the American Chemical Society*, **130**, 732–742.
81. Shaheen, S. E., Brabec, C. J., Sariciftci, N. S., Padinger, F., Fromherz, T., Hummelen, J. (2001). 2.5% efficient organic plastic solar cells. *Applied Physics Letters*, **78**, 841–843.
82. Hoppe, H., Niggemann, M., Winder, C., Kraut, J., Hiesgen, R., Hinsch, A., Meissner, D., Sariciftci, N. S. (2004). Nanoscale morphology of conjugated polymer/fullerene-based bulk- heterojunction solar cells. *Advanced Functional Materials*, **14**, 1005–1011.
83. Hoppe, H., Glatzel, T., Niggemann, M., Hinsch, A., Lux-Steiner, M. C., Sariciftci, N. S. (2005). Kelvin probe force microscopy study on conjugated polymer/fullerene bulk heterojunction organic solar cells. *Nano Letters*, **5**, 269–274.
84. Padinger, F., Rittberger, R. S., Sariciftci, N. S. (2003). Effects of postproduction treatment on plastic solar cells. *Advanced Functional Materials*, **13**, 85–88.
85. Ma, W., Yang, C., Gong, X., Lee, K., Heeger, A. J. (2005). Thermally stable, efficient polymer solar cells with nanoscale control of the interpenetrating network morphology. *Advanced Functional Materials*, **15**, 1617–1622.
86. Li, G., Shrotriya, V., Huang, J., Yao, Y., Moriarty, T., Emery, K., Yang, Y. (2005). High-efficiency solution processable polymer photovoltaic cells by self-organization of polymer blends. *Nature Materials*, **4**, 864–868.
87. Li, G., Yao, Y., Yang, H., Shrotriya, V., Yang, G., Yang, Y. (2007). "Solvent annealing" effect in polymer solar cells based on poly(3-hexylthiophene) and methanofullerenes. *Advanced Functional Materials*, **17**, 1636–1644.
88. Moulé, A. J. Meerholz, K. (2008). Controlling morphology in polymer–fullerene mixtures. *Advanced Materials*, **20**, 240–245.
89. van Bavel, S. S., Sourty, E., de With, G., Loos, J. (2009). Three-dimensional nanoscale organization of bulk heterojunction polymer solar cells. *Nano Letters*, **9**, 507–513.
90. Dennler, G., Scharber, M. C., Brabec, C. J. (2009). Polymer–fullerene bulk-heterojunction solar cells. *Advanced Materials*, **21**, 1323–1338.
91. Dang, M. T., Hirsch, L., Wantz, G. (2011). P3HT:PCBM, best seller in polymer photovoltaic research. *Advanced Materials*, **23**, 3597–3602.
92. Troshin, P. A., Hoppe, H., Renz, J., Egginger, M., Mayorova, J. Y., Goryachev, A. E., Peregudov, A. S., Lyubovskaya, R. N., Gobsch, G., Sariciftci, N. S., Razumov, V. F. (2009). Material solubility–photovoltaic performance relationship in the design of novel fullerene derivatives for bulk heterojunction solar cells. *Advanced Functional Materials*, **19**, 779–788.
93. Troshin, P. A., Susarova, D. K., Khakina, E. A., Goryachev, A. E., Borshchev, O.V., Ponomarenko, S. A., Razumov, V. F., Sariciftci, N. S. (2012). Material solubility and molecular compatibility effects in the design of the fullerene/polymer composites for organic bulk heterojunction solar cells. *Journal of Materials Chemistry*, **22**, 18433–18441.
94. Susarova, D. K., Khakina, E. A., Troshin, P. A., Goryachev, A. E., Sariciftci, N. S., Razumov, V. F., Egbe, D. A. M. (2011). Photovoltaic performance of PPE–PPV copolymers: Effect of the fullerene component. *Journal of Materials Chemistry*, **21**, 2356–2361.
95. Troshin, P. A., Mukhacheva, O. A., Usluer, O., Rathgeber, S., Goryachev, A. E., Akkuratov, A. V., Susarova, D. K., Sariciftci, N. S., Razumov, V. F., Egbe, D. A. M. (2013). Improved photovoltaic performance of the PPV-PPE-type copolymer using optimized fullerene-based counterparts. *Advanced Energy Materials*, **3**, 161–166.
96. Vandewal, K., Gadisa, A., Oosterbaan, W. D., Bertho, S., Banishoeib, F., van Severen, I., Lutsen, L., Cleij, T. J., Vanderzande, D., Manca, J. V. (2008). The relation between open-circuit voltage and the onset of photocurrent generation by charge-transfer absorption in polymer: Fullerene bulk heterojunction solar cells. *Advanced Functional Materials*, **18**, 2064–2070.
97. Kooistra, F. B., Knol, J., Kastenberg, F., Popescu, L. M., Verhees W. J. H., Kroon, J. M., Hummelen, J. C. Increasing the open circuit voltage of bulk-heterojunction solar cells by raising the LUMO level of the acceptor. *Organic Letters*, **9**, 551–554.
98. Riedel, I., von Hauff, E., Parisi, J., Martin, N., Giacalone, F., Dyakonov, V. (2005). Diphenylmethanofullerenes: New and efficient acceptors in bulk heterojunction solar cells. *Advanced Functional Materials*, **15**, 1979–1987.
99. Bolink, H. J., Coronado, E., Aliaga, A. F., Lenes, M., Rosa, A. L., Filippone, S., Martin, N. (2010). Polymer solar cells based on diphenylmethanofullerenes with reduced sidechain length. *Journal of Materials Chemistry*, **21**, 1382–1386.
100. Laird, D., Richter, H., Vejins, V., Scott, L., Lada, T. (2009) Fullerenes and derivatives thereof and improved methods of making fullerenes suitable for organic photovoltaic devices. WO 2009/086210, 09 July 2009.
101. Zhao, G., He, Y., Li, Y. (2010). 6.5% Efficiency of polymer solar cells based on poly(3-hexylthiophene) and indene-C_{60} bisadduct by device optimization. *Advanced Materials*, **22**, 4355–4358.
102. Yang, J., Zhu, R., Hong, Z., He, Y., Kumar, A., Li, Y., Yang, Y. (2011). A robust inter-connecting layer for achieving high performance tandem polymer solar cells. *Advanced Materials*, **23**, 3465–3470.
103. Lenes, M., Wetzelaer, G. J. A. H., Kooistra, F. B., Veenstra, S. C., Hummelen, J. C., Blom, P. W. M. (2008). Fullerene bisadducts for enhanced open-circuit voltages and efficiencies in polymer solar cells. *Advanced Materials*, **20**, 2116–2119.
104. Frost, J. M., Faist, M. A., Nelson, J. (2010). Energetic disorder in higher fullerene adducts: A quantum chemical and voltammetric study. *Advanced Materials*, **22**, 4881–4884.
105. Faist, M. A., Keivanisis, P. E., Foster, S., Wobkenbergvg, P. H., Anthopoulos, T. D., Bradley, D. C., Durrant, J. R., Nelson, J. (2010). Effect of multiple adduct fullerenes on charge generation and transport in photovoltaic blends with

poly(3-hexylthiophene-2,5-diyl). *Journal of Polymer Science: Part B: Polymer Physics*, **49**, 45–51.

106. Susarova, D. K., Goryachev, A. E., Troshin, P. A., Razumov, V. F. (2011). Synthesis and photovoltaic performance of various bis-adducts of [60]fullerene. E-MRS 2011 Spring Meeting; 2011 May 9-13; Nice, France.
107. Jeon, M., Jung, J., Lee, J., Lim, J., Shin, W. S., Moon, S. J., Lee, C., Yoon, S. C. (2010). Silyl substituted methanofullerenes as electron acceptors in organic photovoltaic cells. *Molecular Crystals and Liquid Crystals*, **519**, 266–275.
108. Choi, J. H., Son, K.-I., Kim, T., Kim, K., Ohkuboa, K., Fukuzumi, S. (2010). Thienyl-substituted methanofullerene derivatives for organic photovoltaic cells. *Journal of Materials Chemistry*, **20**, 475–482.
109. Zhao, B. G., He, Y., Li, Y., Peng, B. (2010). High-Yield Synthesis and Electrochemical and Photovoltaic Properties of Indene-C_{70} Bisadduct. *Advanced Materials*, **20**, 3383–3389.
110. Fan, X., Cui, C., Fang, G., Wang, J., Li, S., Cheng, F., Long, H., Li, Y. (2011). Efficient Polymer Solar Cells Based on Poly(3-hexylthiophene):Indene-C_{70} Bisadduct with a MoO_3 Buffer Layer. *Adv. Funct. Mater*, **21**, DOI: 10.1002/adfm.201102054
111. Niinomi, T., Matsuo, Y., Hashiguchi, M., Satoand, Y., Nakamura, E. (2009). Penta(organo)[60]fullerenes as acceptors for organic photovoltaic cells. *Journal of Materials Chemistry*, **19**, 5804–5811.
112. Gong, X., Tong, M., Brunetti, F. G., Seo, J., Sun, Y., Moses, D., Wudl, F., Heeger, A. J. (2011). Bulk heterojunction solar cells with large open-circuit voltage: Electron transfer with small donor-acceptor energy offset. *Advanced Materials*, **23**, 2272–2277.
113. Liang, Y., Wu, Y., Feng, D., Tsai, S.-T., Son, H.-J., Li, G., Yu, L. (2009). Development of new semiconducting polymers for high performance solar cells. *Journal of the American Chemical Society*, **131**, 56–57.
114. Liang, Y., Feng, D., Wu, Y., Tsai, S.-T., Li, G., Ray, C., Yu, L. (2009). Highly efficient solar cell polymers developed via fine-tuning of structural and electronic properties. *Journal of the American Chemical Society*, **131**, 7792–7799.
115. Zou, Y., Najari, A., Berrouard, P., Beaupre, S., Aıch, B.-R., Tao, Y., Leclerc, M. (2010). A thieno[3,4-*c*]pyrrole-4,6-dione-based copolymer for efficient solar cells. *Journal of the American Chemical Society*, **132**, 5330–5331.
116. Muhlbacher, D., Scharber, M., Morana, M., Zhu, Z., Waller, D., Gaudiana, R., Brabec, C. (2006). High photovoltaic performance of a low-bandgap polymer. *Advanced Materials*, **18**, 2884–2889.
117. Scharber, M. C., Koppe, M., Gao, J., Cordella, F., Loi, M. A., Denk, P., Morana, M., Egelhaaf, H.-J., Forberich, K., Dennler, G., Gaudiana, R., Waller, D., Zhu, Z., Shi, X., Brabec, C. J. (2010). Influence of the bridging atom on the performance of a low-bandgap bulk heterojunction solar cell. *Advanced Materials*, **22**, 367–370.
118. Hou, J., Chen, H.-Y., Zhang, S., Li, G., Yang, Y. (2008). Synthesis, characterization, and photovoltaic properties of a low band gap polymer based on silole-containing polythiophenes and 2,1,3-benzothiadiazole. *Journal of the American Chemical Society*, **130**, 16144–16145.
119. Peet, J., Kim, J. Y., Coates, N. E., Ma, W. L., Moses, D., Heeger, A. J., Bazan, G. C. (2007). Efficiency enhancement in low-bandgap polymer solar cells by processing with alkane dithiols. *Nature Materials*, **6**, 497–500.
120. Li, Z., Tsang, S.-W., Du, X., Scoles, L., Robertson, G., Zhang, Y., Toll, F., Tao, Y., Lu, J., Ding, J. (2011). Alternating copolymers of cyclopenta[2,1-*b*;3,4-*b*] dithiophene and thieno[3,4-c]pyrrole-4,6-dione for high-performance polymer solar cells. *Advanced Functional Materials*, **21**, 3331–3336.
121. Wang, E., Hou, L., Wang, Z., Hellström, S., Zhang, F., Inganäs, O., Andersson, M. R. (2010). An easily synthesized blue polymer for high-performance polymer solar cells. *Advanced Materials*, **22**, 5240–5244.
122. Bijleveld, J. C., Gevaerts, V. S., Nuzzo, D. D., Turbiez, M., Mathijssen, S. G. J., de Leeuw, D. M., Wienk, M. M., Janssen, R. A. J. (2010). Efficient solar cells based on an easily accessible diketopyrrolopyrrole polymer. *Advanced Materials*, **22**, E242–E246.
123. Wang, E., Ma, Z., Zhang, Z., Vandewal, K., Henriksson, P., Inganas, O., Zhang, F., Andersson, M. R. (2011). An easily accessible isoindigo-based polymer for high-performance polymer solar cells. *Journal of the American Chemical Society*, **133**, 14244–14247.
124. Ong, K.-H., Lim, S.-L., Tan, H.-S., Wong, H.-K., Li, J., Ma, Z., Moh, L. C. H., Lim, S.-H., de Mello, J. C., Chen, Z.-K. (2011). A versatile low bandgap polymer for air-stable, high-mobility field-effect transistors and efficient polymer solar cells. *Advanced Materials*, **23**, 1409–1413.
125. Du, C., Li, C., Li, W., Chen, X., Bo, Z., Veit, C., Ma, Z., Wuerfel, U., Zhu, H., Hu, W., Zhang, F. (2011). 9-alkylidene-9*H*-fluorene-containing polymer for high-efficiency polymer solar cells. *Macromolecules*, **44**, 7617–7624.
126. Zhou, H., Yang, L., Stuart, A. C., Price, S. C., Liu, S., You, W. (2011). Development of fluorinated benzothiadiazole as a structural unit for a polymer solar cell of 7% efficiency. *Angewandte Chemie*, **50**, 2995–2998.
127. Fitzner, R., Reinold, E., Mishra, A., Mena-Osteritz, E., Ziehlke, H., Körner, C., Leo, K., Riede, M., Weil, M., Tsaryova, O., Weiß, A., Uhrich, C., Pfeiffer, M., Bäuerle, P. (2011). Dicyanovinyl-substituted oligothiophenes: Structure–property relationships and application in vacuum-processed small-molecule organic solar cells. *Advanced Functional Materials*, **21**, 897–910.
128. Riede, M., Uhrich, C., Widmer, J., Timmreck, R., Wynands, D., Schwartz, G., Gnehr, W.-M., Hildebrandt, D., Weiss, A., Hwang, J., Sundarraj, S., Erk, P., Pfeiffer, M., Leo, K. (2011). Efficient organic tandem solar cells based on small molecules. *Advanced Functional Materials*, **21**, 3019–3028.
129. http://www.energyharvestingjournal.com/articles/heliatek-achieves-new-world-record-for-organic-solar-cells-00003979.asp?sessionid=1. Accessed on 15 December 2011.

130. Walker, B., Tamayo, A. B., Dang, X.-D., Zalar, P., Hwa, S. J., Garcia, A., Tantiwiwat, M., Nguyen, T.-Q. (2009). Nanoscale phase separation and high photovoltaic efficiency in solution-processed, small-molecule bulk heterojunction solar cells. *Advanced Functional Materials*, **19**, 3063–3069.

131. Dang, X.-D., Tamayo, A. B., Seo, J., Hoven, C. V., Walker, B., Nguyen, T.-Q. (2010). Nanostructure and optoelectronic characterization of small molecule bulk heterojunction solar cells by photoconductive atomic force microscopy. *Advanced Functional Materials*, **20**, 3314–3321.

132. Wei, G., Wang, S., Sun, K., Thompson, M. E., Forrest, S. R. (2011). Solvent-annealed crystalline squaraine: $PC_{70}BM$ (1:6) solar cells. *Advanced Energy Materials*, **1**, 184–187.

133. Liu, Y., Wan, X., Wang, F., Zhou, J., Long, G., Tian, J., You, Jingbi., Yang, Y., Chen, Y. (2011). Spin-coated small molecules for high performance solar cells. *Advanced Energy Materials*, **1**, 771–775.

134. Lin, L.-Y., Chen, Y.-H., Huang, Z.-Y., Lin, H.-W., Chou, S.-H., Lin, F., Chen, C.-W., Liu, Y.-H., Wong, K.-T. (2011). A low-energy-gap organic dye for high-performance small-molecule organic solar cells. *Journal of the American Chemical Society*, **133**, 15822–15825.

135. Zhou, J., Wan, X., Liu, Y., Long, G., Wang, F., Li, Z., Zuo, Y., Li, C., Chen, Y. (2011). A planar small molecule with dithienosilole core for high efficiency solution-processed organic photovoltaic cells. *Chemistry of Materials*, **23**, 4666–4668.

136. Sun, Y., Welch, G. C., Leong, W. L., Takacs, C. J., Bazan, G. C., Heeger, A. J. (2011). Solution-processed small-molecule solar cells with 6.7% efficiency. *Nature Materials*, DOI: doi:10.1038/nmat3160.

26

MESOSCOPIC DYE-SENSITIZED SOLAR CELLS

Mohammad Khaja Nazeeruddin, Jaejung Ko, and Michael Grätzel

26.1 INTRODUCTION

The most pressing problems of our planet are rapid decline of natural energy resources, increasing population, and global environment causing the greenhouse effect. Nuclear power is not our planets' preferred energy, and therefore we have no option but to develop new technologies to power our planet. In this respect, solar energy is one of the best available options. The sun is recognized as a major natural resource with which we are blessed in abundance (2200 thermal kilowatt hours (kWh) per square meter) and that should be fully exploited for the benefit of mankind. Consequently the development of nanomaterials-based technologies to convert solar energy into electricity is paramount. In this chapter we discuss the new technology based on mesoscopic dye sensitized solar cells (DSCs).

Dye-sensitized solar cells are unique because they are the only photovoltaic device that separates light absorption and the charge transportation during the photoelectric conversion process mimicking photosynthesis in green plants. Hence, both materials can be individually tailored to optimize light absorption and charge transportation properties. These are unlike silicon cells, in which both light absorption and the charge transportation carried by the same material require highest purity. Dye-sensitized solar cells are a top runner among the third-generation solar cell owing due to its low cost and easy fabrication using environmentally benign materials [1]. Recently, it has also been endorsed by the millennium technology prize committee as having an excellent price-to-performance ratio with short payback time and is now regarded as an easy alternative for conventional silicon-based photovoltaic devices [2].

In a dye-sensitized solar cell, sunlight is absorbed by a dye monolayer located at the junction between the electron and hole transporting phases where the former is a wide bandgap oxide semiconductor, typically TiO_2 anatase, and the latter is iodide/triiodide redox (I^-/I_3^-) system [3, 4]. Upon photoexcitation, the dye injects an electron and a hole into the *n*- and *p*-type materials, respectively, generating free charge carriers, which then travel through the nanostructures to be collected as current at the external contacts. A schematic representation of the dye-sensitized solar cell is shown Figure 26.1, which broadly consists of five components: (1) a mechanical support coated with transparent conductive oxides; (2) a mesoscopic semiconductor film, usually TiO_2; (3) a sensitizer adsorbed on the surface of the semiconductor; (4) an electrolyte containing a redox mediator; and (5) a counterelectrode capable of regenerating the redox mediator. Titanium dioxide became the semiconductor of choice for mesoscopic film because it is a low-cost, widely available, and nontoxic material. The most commonly used redox mediator is the iodide/triiodide redox couple. Ruthenium complexes such as [bis-dithiocyanate di(4,4′-dicarboxylic acid 2,2′-bipyridine)$_2$ruthenium(II)] were employed as a sensitizer very early on and are still now the most commonly used sensitizer, yielding 11.2% efficiency. The success of the DSC is the high surface area obtained by the semiconductor film made of nanoparticles, which leads to increased dye loading when compared to single crystals, thereby resulting in increased optical density and, thus efficient light harvesting. While the total efficiency of the dye-sensitized solar cell depends on optimization and compatibility of each of its constituents, the initial requirement is for the device to be able to gather as much photons from sunlight as possible.

Organic Nanomaterials: Synthesis, Characterization, and Device Applications, First Edition. Edited by Tomás Torres and Giovanni Bottari.

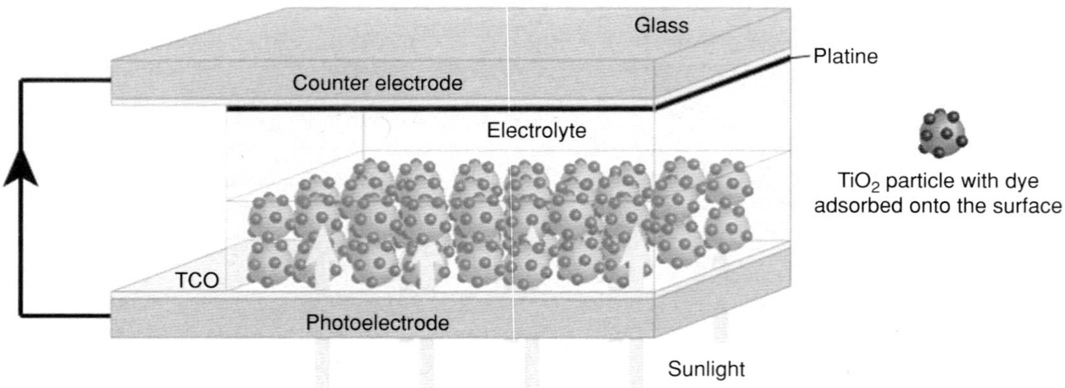

FIGURE 26.1. Schematic representation of a dye-sensitized solar cell.

The operating principles of the dye-sensitized solar cell are shown in Figure 26.2. The first step is the absorption of a photon by the sensitizer S [Eq. (26.1)], leading to the excited sensitizer S^* which injects an electron into the conduction band of the semiconductor, leaving the sensitizer in the oxidized state S^+ [Eq. (26.2)]. The injected electron flows through the semiconductor network to arrive at the back contact and then through the external load to the counterelectrode to reduce the redox mediator [Eq. (26.3)], which, in turn, regenerates the sensitizer [Eq. (26.4)]. This completes the circuit. Under illumination, the device constitutes a regenerative and stable photovoltaic energy conversion system.

$$S_{(\text{adsorbed})} + h\nu \rightarrow S^*_{(\text{adsorbed})} \quad (26.1)$$

$$S^*_{(\text{adsorbed})} \rightarrow S^+_{(\text{adsorbed})} + e^-_{(\text{injected})} \quad (26.2)$$

$$I_3^- + 2 \cdot e^-_{(\text{cathode})} \rightarrow 3I^-_{(\text{cathode})} \quad (26.3)$$

$$S^+_{(\text{adsorbed})} + \tfrac{3}{2}I^- \rightarrow S_{(\text{adsorbed})} + \tfrac{1}{2}I_3^- \quad (26.4)$$

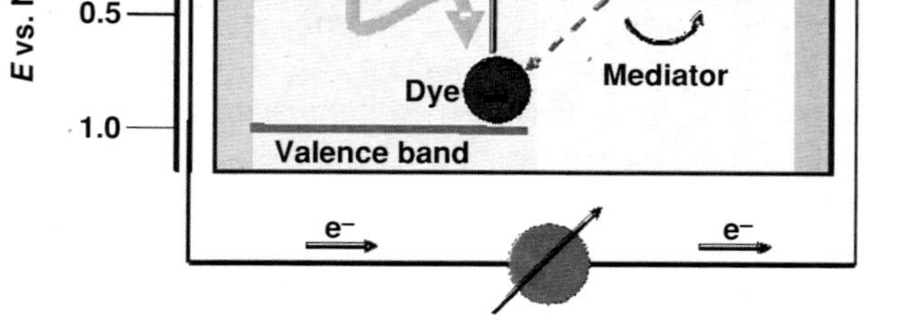

FIGURE 26.2. Operating principles and energy level diagram of a dye-sensitized solar cell.

Some undesirable reactions resulting in losses in the cell efficiency occur. They are the recombination of the injected electrons either with oxidized sensitizer [Eq. (26.5)] or with the oxidized redox couple at the TiO_2 surface ([Eq. (26.6)].

$$S^+_{(\text{adsorbed})} + e^-_{(\text{TiO}_2)} \rightarrow S_{(\text{adsorbed})} \quad (26.5)$$

$$I_3^- + 2 \cdot e^-_{(\text{TiO}_2)} \rightarrow 3I^-_{(\text{anode})} \quad (26.6)$$

The total efficiency of the dye-sensitized solar cell depends on optimization and compatibility of each of these constituents, in particular on the semiconductor film along with the dye spectral responses [5]. A very important factor is the high surface area and the thickness of the semiconductor film, which leads to increased dye loading and thus optical density, resulting in efficient light harvesting [6].

The incident monochromatic photon-to-current conversion efficiency (IPCE), sometimes referred to also as the external quantum efficiency (EQE), is an important characteristic of a device. In particular, using devices with same architecture, it is possible to compare the light-harvesting performance of sensitizers. It is defined as the number of electrons generated by light in the external circuit divided by the number of incident photons as a function of excitation wavelength as in [Eq. (26.7)] [7]:

$$\begin{aligned} \text{IPCE}(\lambda) &= \frac{\text{Photocurrent density}}{\text{Wavelength} \times \text{Photon flux}} \\ &= \text{LHE}(\lambda) \times \varphi_{\text{inj}} \times \eta_{\text{coll}} \end{aligned} \quad (26.7)$$

where LHE(λ) is the light-harvesting efficiency at wavelength λ, φ_{inj} is the quantum yield for electron injection from the excited sensitizer in the conduction band of the TiO_2, and η_{coll} is the efficiency for the collection of electrons.

The overall conversion efficiency (η) of the dye-sensitized solar cell is determined by the photocurrent density (J_{ph}), the open-circuit potential (V_{OC}), the fill factor (*ff*) of the cell, and the intensity of the incident light (I_S) [Eq. (26.8)] [8].

$$\eta_{\text{global}} = \frac{J_{ph} \cdot V_{OC} \cdot ff}{I_S} \tag{26.8}$$

The open-circuit photovoltage is determined by the energy difference between the Fermi level of the solid under illumination and the Nernst potential of the redox couple in the electrolyte (Figure 26.2). However, the experimentally observed open-circuit potential (V_{OC}) for various sensitizers is smaller than the difference between the conduction-band edge and the redox couple. This is generally due to the competition between electron transfer and charge recombination pathways. Knowledge of the rates and mechanisms of these competing reactions are vital for the design of efficient sensitizers and thereby improvement of the devices [9]. The fill factor *ff* is defined as the ratio of the maximum power P_{max} obtained with the device and the theoretical maximum power, that is, $P_{th} = I_{SC} \cdot V_{OC}$ (I_{SC} is the short-circuit current and V_{OC} is the open-circuit voltage). The fill factor *ff* can then take values between 0 and 1. It reflects electrical and electrochemical losses occurring during operation of the DSSC.

The performance of the DSSC device depends mainly on the optical and electrochemical properties of the sensitizer: first through the absorption spectra and then through the ground and excited redox potentials, which set the maximum injection efficiency, resulting in a short-circuit current. The dye in the ground state is singlet in nature, where, upon absorbing light, an electron is photoexcited from the highest occupied molecular orbital (HOMO) of the dye to the lowest unoccupied molecular orbital (LUMO). During the excitation from HOMO to LUMO, the singlet spin state remains unchanged for less than 50 fs. In the excited state the dye can inject an electron from singlet state (i.e., in less than 50 fs) or can undergo intersystem crossing resulting in a triplet state [10]. The excited state singlet and triplet lifetimes have been measured, which are in the femtoseconds and nanoseconds domains. Most of the ruthenium sensitizers excited singlet and triplet states are energetically favorable for injecting electrons on the TiO_2 conduction band [11]. However, in the DSC, the excited dye either relaxes back to the ground state or injects an electron onto the conduction band or reacts with the electrolyte system. The loss of the excited-state electrons to the redox system is energetically more favorable; nevertheless, the kinetics of electron injection onto the conduction band is extremely fast compared to the back reaction to the electrolyte by 10^{13} versus 10^2 s^{-1}. The optimized single-junction dye-sensitized solar cell yields a power conversion efficiency of 13% [12]. However, the AIST (Japan) certified cell efficiency is 11.4% [13]. Also, stable DSC cells with efficiency 9% [14], as well as the production of W-contact submodules with more than 8% efficiency, have been reported [15]. In spite of this progress, a substantial amount of research work is still paramount to fill the gap between today's benchmark conversion efficiency and the Shockley–Queiser limit of η 32% predicted for a single junction cell [16]. In this respect, design and development of novel nanomaterials, sensitizers, redox mediators, and counterelectrodes are essential, which are discussed in the following sections.

26.2 MESOSCOPIC NANOMATERIALS

The best dye-sensitized solar cell efficiency was obtained by preparing TiO_2 nanoparticles using hydrothermal autoclaving and electrodes by screen-printing deposition as reported by Barbé et al. [5]. The synthesis of TiO_2 nanoparticles were done by hydrolysis of $Ti(OCH(CH_3)_2)_4$ in 0.1 M nitric acid solution under vigorous stirring. During this process a white precipitate forms instantaneously, which was subjected to 80°C for 8 h under vigorous stirring. This process is called peptization, after which the solution was filtered using a glass frit. Water was added to the filtrate in such a way that the final solid concentration was around 5% by weight. The particles were grown to 10–25 nm under hydrothermal conditions using a titanium autoclave by heating at 200–250°C for 12 h. Then the particles were re-dispersed using a titanium ultrasonic horn two times. The resulting colloidal suspension was evaporated to a final TiO_2 concentration of 11%. The authors have investigated three types of colloids, which were prepared by controlling the pH during precipitation while undergoing hydrolysis by adding acidic (0.1 M nitric acid) and basic (0.1 M ammonia) water solutions. The influence of pH on the morphology of the particles, during hydrothermal growth was studied in samples that were hydrolyzed in acid and then autoclaved at a temperature of 250°C for 12 h in (i) 0.1 M nitric acid solution, (ii) ammonia at pH 11, and (iii) 0.1 M triethylamine at pH 13. SEM micrographs and Brunauer–Emmett–Teller (BET) results both show that the particles were larger under basic conditions than under acidic conditions, suggesting that Ostwald ripening is more important in a basic environment. This observation implies that the transient species that are formed in a basic environment under hydrothermal conditions are more stable than those that are formed at low pH (acidic conditions) [5].

To shorten the procedure, Ito et al. [17] have developed an elegant procedure using commercially available TiO_2 powders P25. The powder was mixed with ethylcellulose, which was dissolved beforehand in an ethanol solution. The fabrication scheme for TiO_2 pastes was described in Figure 26.3. In each step in the mortar, liquids were added drop by drop and the TiO_2 dispersions in the mortar were transferred and washed with excess of ethanol (100 mL) to a big beaker and stirred with a magnet tip at 300 rpm. The homogenization of the colloids was performed with using a Ti-horn-equipped

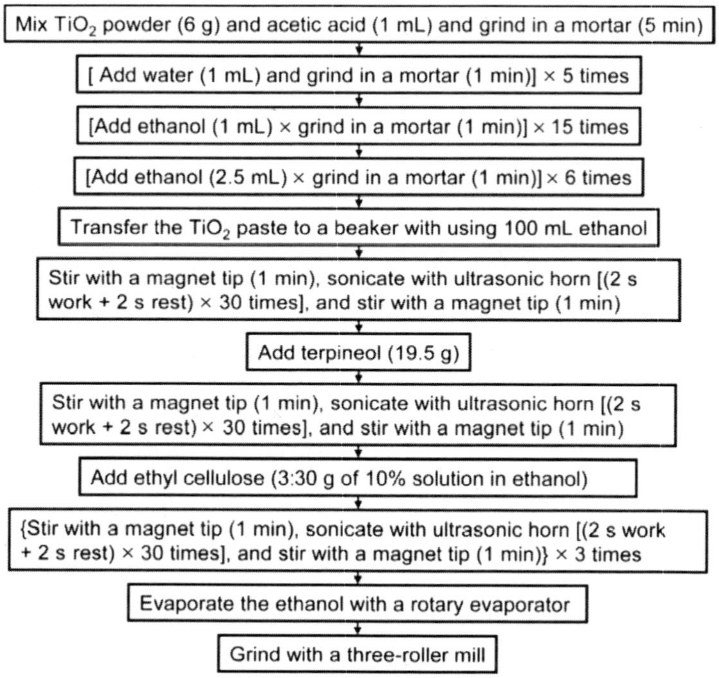

FIGURE 26.3. Scheme of screen-printing paste fabrication from a nanocrystalline TiO_2 powder.

ultrasonic sonicator (Vibra cell 72408, Bioblock scientific). To this suspension, anhydrous terpineol (Fluka) and the ethyl cellulose solution in ethanol were added, followed by stirring and sonication. The contents in dispersion were concentrated by using an evaporator at 35°C, and the resulting paste was charged onto a three-roller mill grinder (EXAKT). The TiO_2 paste is coated onto a conducting FTO/glass substrate (Solar-4mm, Nippon Sheet Glass, Japan) by screen printing using 90T 75 screen-printing mesh. The sintered TiO_2 electrodes were used to assemble DSC. A linear relationship between the coating times and the thicknesses was confirmed with a surface profiler (Alfastep550). Each paste gave a thick layer over 17 μm without cracking and peering-up on the exterior. The photovoltaic characteristics of these films were slightly lower ($J_{SC} = 16.25$ mA cm^{-2}, $V_{OC} = 779$ mV, $FF = 0.730$, and $\eta = 9.24\%$) than the TiO_2 films screen printed using a hydrothermal procedure.

Recently, Caruso and co-workers [18] developed mesoporous beads with a surface area of 108.0 m^2 g^{-1} and tunable pore sizes (pore diameters varying from 14.0 to 22.6 nm) through a facile combination of sol–gel and solvothermal processes. The mesoporous TiO_2 beads have a diameter of 830 nm and are composed of anatase TiO_2 nanocrystals. The scanning electron microscopy (SEM) images of the calcined mesoporous TiO_2 beads screen-printed film prepared after the solvothermal process with different ammonia concentrations is shown in Figure 26.4. The advantages of these beads is large size that allows scattering of light within the film, at the same time providing

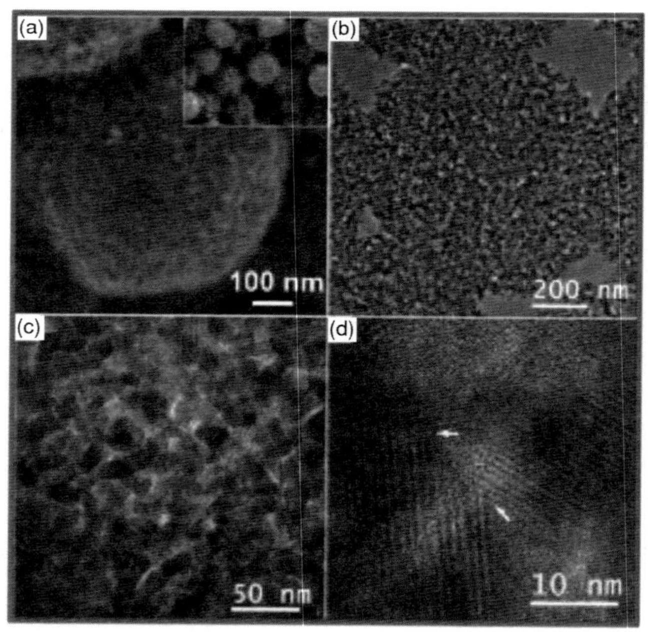

FIGURE 26.4. (**a**) Scanning electron micrograph of the screen-printed film composed of TiO_2 porous beads. (**b, c**) TEM images of the ultramicrotomed titania bead. (**d**) HRTEM image of the intergrowth of the anatase crystals, as indicated by the white arrows, within the titania bead.

TABLE 26.1. Cell Characteristics Recorded at 1-equiv Sunlight Intensity for 12-μm-Thick Films Based on Degussa P25 Titania or Mesoporous Titania Beads Before and After $TiCl_4$ Post-Treatment

	P25				Beads			
Treatment	J_{sc} (mA/cm^2)	V_{oc} (mV)	*ff*	η(%)	J_{sc} (mA/cm^2)	V_{oc} (mV)	*ff*	η(%)
No post-$TiCl_4$	12.27	756	0.75	7.1	16.50	735	0.74	9.1
Post-$TiCl_4$	15.00	761	0.75	8.5	18.44	745	0.77	10.6

enough surface area to cover sensitizer. Using these beads, dye-sensitized solar cell efficiency reached over 10% using only single-layer TiO_2 films. Usually to reach such efficiencies, a double layer consisting of a transparent 20- to 30-nm nanocrystalline TiO_2 layer followed by 400-nm scattering layers were used. The double-layer configuration makes the fabrication process complicated as well as more costly. The current–voltage characteristics of both Degussa P25 titania or mesoporous titania beads were measured at 1-equiv sunlight intensity for 12-μm-thick films based on before and after $TiCl_4$ post-treatment [19]. The data shown in Table 26.1 are obtained using a heteroleptic ruthenium sensitizer consisting of Na-*cis*-Ru(4,4-(5-hexylthiophen-2-yl)-2,2-bipyridine)(4-carboxylic-acid-4-carboxylate-2,2-bipyridine) (thiocyanate)2 (coded C101) [20], and an electrolyte composed of 1 M DMII, 50 mM LiI, 30 mM I2, 0.5 M *tert*-butylpyridine, and 0.1 M GuNCS in a solvent mixture of 85% acetonitrile with 15% valeronitrile [19].

The superior optical characteristic of the beads is also seen in the incident photon-to-current conversion efficiency (IPCE) shown in Figure 26.5. When normalized the IPCE to 100% the difference between the P25 and beads films are clearly visible. Particularly the impact of beads in the red part of the spectrum is sticking, where there is an important enhancement in light conversion efficiency from 600 nm until far in the absorption tail (to 800 nm). The improved light-harvesting characteristic originates from the combination of an increased dye loading on the beads and on their higher diffusion reflectance properties, which attest to their superior light-scattering properties [19]. The dye-sensitized solar cells with a single TiO_2 bead layer achieves an efficiency of 10.5%, which is either comparable or superior to the double-layer configuration consisting of 9-μm-thick transparent nanocrystalline TiO_2 and 5-μm 400-nm CCIC TiO_2 films.

26.3 MOLECULAR ABSORBERS

Sensitization of large bandgap semiconductors to the visible and the near-infrared solar spectrum can be obtained by anchoring molecular absorbers. In dye-sensitized solar cells, mesoporous titania semiconductor film anchored with sensitizer is one of the main components. The sensitizer should not only absorb visible and near IR light but also exhibit thermal and photochemical stability. The anchoring groups of the sensitizer should have strong binding properties, and the LUMO of the sensitizer and the conduction band of semiconductor oxide overlap efficiently, thereby facilitating electron transfer from the excited-state dye to the conduction band. The sensitizer in the excited state should have directionality where the excited electron should migrate toward an anchoring group and toward the positive charge on the opposite side. Also, the sensitizer should have the LUMO just above the conduction band and the HOMO below the redox level of the hole transporting material or redox couple. Moreover, the sensitizer

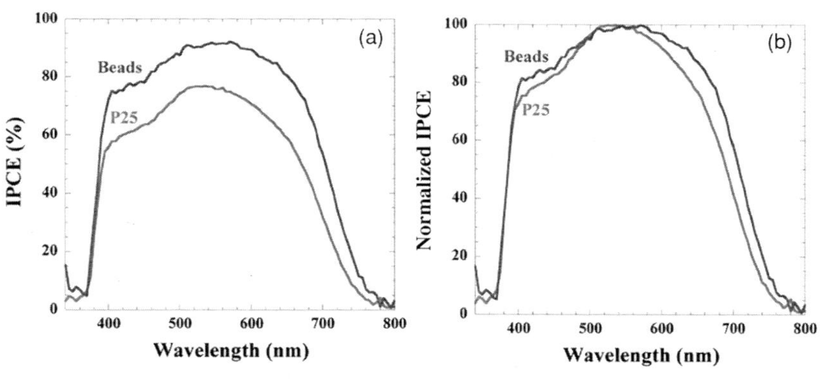

FIGURE 26.5. (**a**) IPCE spectrum of 12-μm-thick films composed of P25 particles or mesoporous beads with $TiCl_4$ post-treatment sensitized using the C101 dye. (**b**) Normalized IPCE spectrum of titania beads and P25 titania particles.

should have electrochemical robustness to undergo millions of cycles. To incorporate all the requirements, engineering of sensitizers at the molecular level is required, which is the topic of this section.

The photophysical and photochemical properties of group VIII metal complexes using terpyridine and bipyridine ligands have been thoroughly investigated during the last three decades [21]. The main thrust behind these studies is to understand the energy and electron-transfer processes in the excited state and to apply this knowledge to potential practical applications such as dye-sensitized solar cells and light-driven information processing [22]. Ruthenium(II) complexes have been used extensively as charge-transfer sensitizers on nanocrystalline TiO_2 films [23]. The choice of ruthenium metal is of special interest for a number of reasons: (a) Because of its octahedral geometry, one can introduce specific ligands in a controlled manner; (b) the photophysical, photochemical, and the electrochemical properties of these complexes can be tuned in a predictable way; (c) the ruthenium metal possesses stable and accessible oxidation states from I to IV [24]. However, iron is an inexpensive and abundant metal, but the photophysical and electrochemical properties of its coordination complexes are difficult to tune in a predictable fashion [25]. The other notable disadvantage of this metal is weakest ligand field splitting compared to ruthenium and osmium. On the other hand, osmium is in the third row of the transition metal ions and therefore has a stronger ligand field splitting compared to ruthenium. Moreover, the spin–orbit coupling in osmium complexes leads to an enhanced response in the red region [26]. But, the low abundance of this metal restricts its use for large-scale applications.

26.3.1 Ruthenium Sensitizers

While several transition-metal complexes as well as metal-free organic dyes have been tested [27], the best photovoltaic performances both in terms of conversion yield and long-term stability has so far been achieved with polypyridyl complexes of ruthenium. The ruthenium complex *cis*-$RuL_2(NCS)_2$ (**1**) (where L = 2,2′-bipyridyl-4,4′-dicarboxylic acid), known as N3 dye, has become the paradigm of heterogeneous charge-transfer sensitizer for dye-sensitized solar cells [8]. The role of the carboxylate groups is to immobilize the sensitizer to the TiO_2 film surface via the formation of bidendate coordination, while the (NCS) groups enhance visible light absorption by destabilizing HOMO of the metal t_{2g} orbitals. The N3 dye exhibits metal-to-ligand charge-transfer transitions (MLCT) at 400 and 535 nm with molar extinction coefficient of 1.45×10^4 $M^{-1}cm^{-1}$ and 1.41×10^4 $M^{-1}cm^{-1}$, respectively. The MLCT excitation of the dye involves the transfer of an electron from the metal t_{2g} orbital to the π^* orbital of the ligand, which is anchored onto the surface through carboxylic aid groups. The N3 dye, upon anchoring onto the TiO_2 surface, releases its protons, which intercalate within the TiO_2 matrix. Since the conduction band of the TiO_2 is known to have a Nernstian dependence on pH [28], it is expected that the dye, which contains four protons, releases onto the TiO_2 surface (depending on the number of groups anchored), thereby influencing the energy level of the TiO_2 conduction band and, hence, the efficiency of the device. The fully protonated N3 sensitizer **1** charges the TiO_2 surface positively by transferring its protons upon adsorption. The electric field associated with the surface dipole generated in this fashion enhances the adsorption of the anionic ruthenium complex and assists electron injection from the excited state of the sensitizer into the titania conduction band, favoring high photocurrents (18–19 mA/cm^2). However, the open-circuit potential (0.65 V) is lowered due to positive shift of the conduction band edge induced by the surface protonation. The open-circuit photovoltage is determined by the energy difference between the Fermi level of the solid under illumination and the Nernst potential of the redox couple in the electrolyte. Since, power conversion efficiency (η) of the dye-sensitized solar cell is the product of photocurrent density (J_{ph}), open-circuit potential (V_{OC}), and fill factor (*ff*) of the cell, divided by the intensity of the incident light (I_S) [Eq. (26.8)]. Therefore, to obtain high light-to-electric-power conversion efficiencies, the short-circuit photocurrent (i_{sc}) and open-circuit potential (V_{oc}) of the solar cell have to be optimized. The open-circuit potential can be tuned by either (a) controlling the pH or number of protons carried by the sensitizer or (b) tuning the redox couples oxidation potential more positively, which is discussed in the electrolyte section.

To see the influence of protons, the performance of the three sensitizers **1**, **2**, and **3** that contain different degrees of protonation were studied on nanocrystalline TiO_2 electrodes [29]. Figure 26.6 shows the photocurrent action spectra obtained with a monolayer of these complexes coated on TiO_2 films. The sensitizer **3** that carries no proton-sensitized

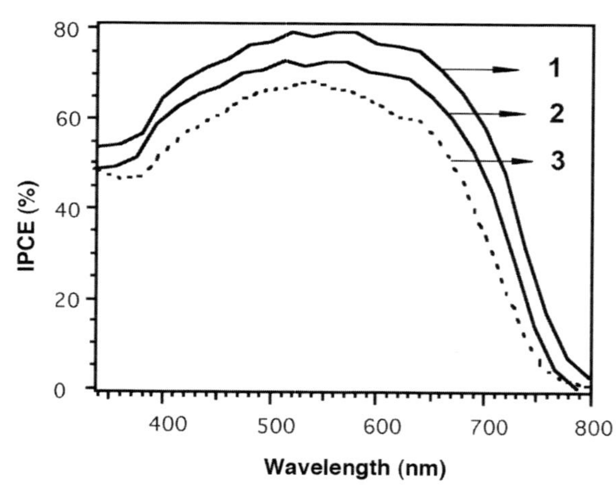

FIGURE 26.6. Photocurrent action spectra of nanocrystalline TiO_2 films sensitized by complexes **1**, **2**, and **3**. The incident photon-to-current conversion efficiency is plotted as a function of wavelength.

films exhibited lower short-circuit current than did **1**; however, the open-circuit potential is significantly higher than that of **1**, due to the relative negative shift of the conduction band edge induced by the adsorption of the anionic complex, but as a consequence the short-circuit photocurrent is lower. Between those two extremes, there should be an optimal degree of protonation of the sensitizer for which the product of short-circuit photocurrent and open-circuit potential is maximized.

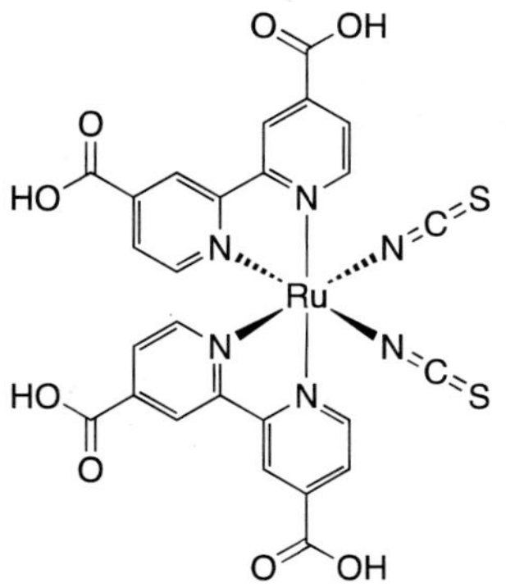

1
N3

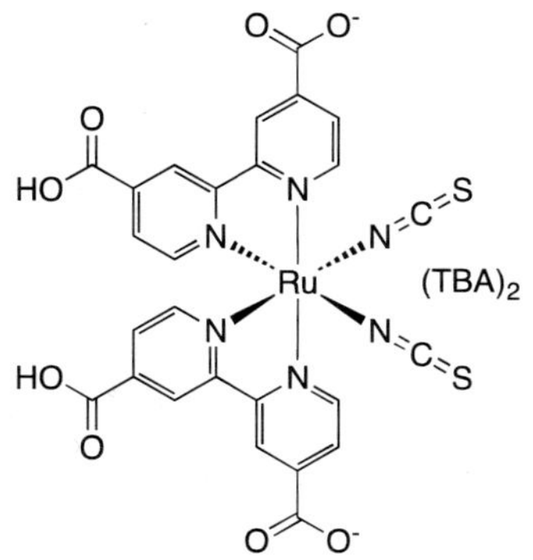

2
N719

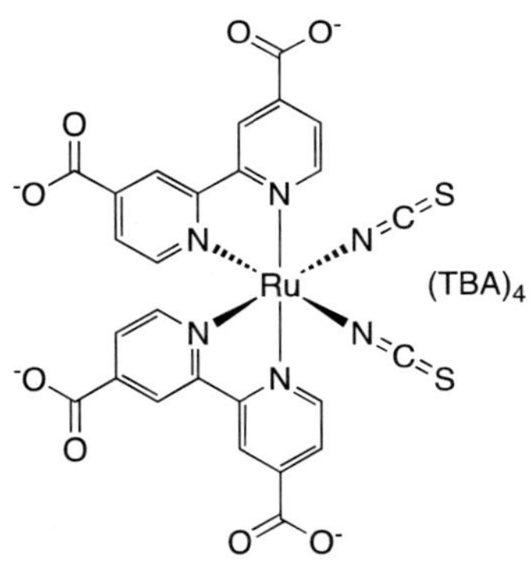

3

The incident monochromatic photon-to-current conversion efficiency (IPCE) is plotted as a function of excitation wavelength. The IPCE value in the plateau region is 80% for complex **1**, while for complex **3** it is only about 66%. This difference is even more pronounced in the red region. Thus, at 700 nm the IPCE value is twice as high for the fully protonated complex **1** as compared to the deprotonated complex **3**. Consequently the short-circuit photocurrent drops from 18–19 mA/cm^2 for complex **1** to only about 12–13 mA/cm^2 for complex **3**. However, there is a tradeoff in photovoltage, which is 0.9 V for complex **3**, as compared to 0.65 V for complex **1**. Nevertheless, this is insufficient to compensate for the current loss.

Photovoltaic performance data obtained with a sandwich cell under illumination by simulated AM 1.5 solar light using

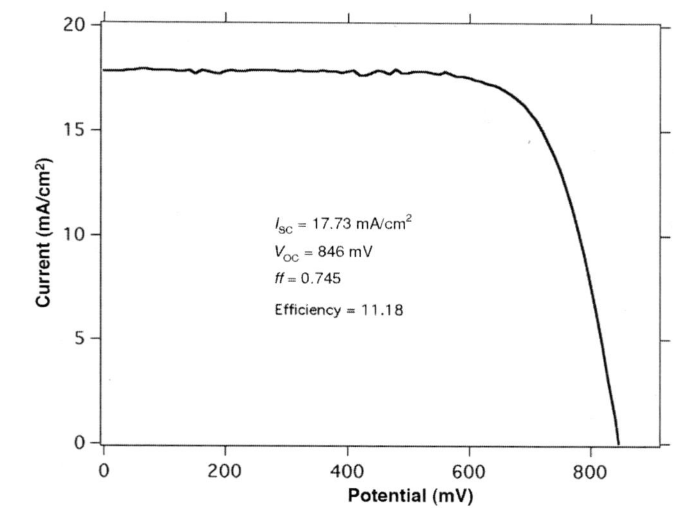

FIGURE 26.7. Photocurrent–voltage curve of a solar cell based on complex **2**. The cell was equipped with an antireflective coating. The conversion efficiency in full AM 1.5 sunlight illumination (100 mW cm^{-2}) is 11.18%. The cell is masked with black plastic to avoid diffusive light, leaving an active cell area of 0.158 cm^2.

complex **2** with additives are shown in Figure 26.7. At 1 sunlight the sensitized solar cell exhibited 17.73 ± 0.5 mA current, 846 mV potential, and fill factor 0.75, yielding an overall conversion efficiency of 11.18%. Hence, the photovoltaic performance of complex **2** carrying two protons is superior to that of compounds **1** and **3**, which contain four or no protons, respectively. The doubly protonated form of the complex is therefore preferred over the other two sensitizers for sensitization of nanocrystalline TiO_2 films.

26.3.1.1 Hydrophobic Sensitizers. Due to the chemical nature of the anchoring groups, the water-induced desorption of the sensitizer from the TiO_2 surface is a crucial aspect in dye-sensitized solar cells as it impacts the long-term stability of the device. To overcome this problem, alkyl chains are grafted onto 2,2′-bipyridine conferring hydrophobic properties to the complexes (**4**–**8**). The absorption spectra of these complexes show broad features in the visible region and display maxima around 530 nm. The performance of these hydrophobic complexes as charge-transfer photosensitizers in a nanocrystalline TiO_2-based solar cell shows excellent stability toward water-induced desorption [30].

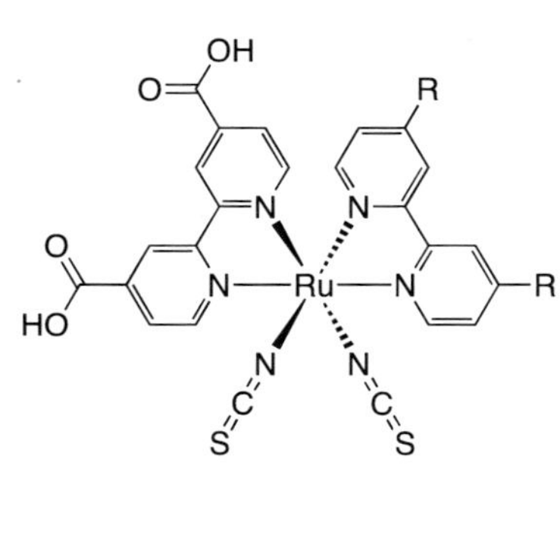

4: R = CH_3
5: R = C_6H_{13}
6: R = C_9H_{19}
7: R = $C_{13}H_{27}$
8: R = $C_{18}H_{37}$

In addition, these sensitizers that have C6–C12 alkyl chains suppress considerably the recombination reactions as described in Eqs. (26.5) and (26.6). The rate of electron transport in dye-sensitized solar cells is a major element of the overall efficiency of the cells. The injected electrons into the conduction band from optically excited dye can traverse the TiO_2 network and can be collected at the transparent conducting glass or can react either with oxidized sensitizer molecule or with the oxidized redox couple (recombination). The reaction of injected electrons into the conduction band with the oxidized redox mediator gives undesirable dark currents, reducing significantly the charge-collection efficiency and thereby decreasing the total efficiency of the cell. Several groups have tried to reduce the recombination reaction by using sophisticated device architecture such as composite metal oxides as the semiconductor with different bandgaps [31]. Gregg et al. [32] have examined surface passivation by deposition of insulating polymers. We have studied the influence of spacer units between the dye and the TiO_2 surface, with little success [33]. Nevertheless, by using TiO_2 films containing hydrophobic sensitizers that contain long aliphatic chains (**4–8**), the recombination reaction was suppressed considerably [34]. The most likely explanation for the reduced dark current is that the long chains of the sensitizer interacts laterally to form an aliphatic network, like a shield, thereby preventing triiodide from reaching the TiO_2 surface.

26.3.1.2 Sensitizers with Extended π-System Showing High Molar Extinction Coefficient. Due to its ease of synthesis, along with availability of chemicals coupled with excellent performances, complex **1** has become a paradigm in the area of dye-sensitized nanocrystalline TiO_2 films [35]. Therefore the vast majority of sensitizers are based on its design. In spite of this, the main drawback of this sensitizer is the lack of absorption in the red region of the visible spectrum and also relatively low molar extinction coefficient, which is 14,500 $M^{-1}cm^{-1}$. Therefore, sensitizers with high molar extinction coefficient have been particularly sought after. A new series of high molar extinction coefficient sensitizers (**9–11**) featuring alkyloxy groups has been synthesized and utilized in dye-sensitized solar cells. The purpose of 4,4′-di-(2-(3,6-dimethoxyphenyl)ethenyl)-2,2′-bipyridine ligand that contains extended π-conjugation with substituted methoxy groups is to enhance molar extinction coefficient of the sensitizers and to provide directionality in the excited state by fine-tuning the LUMO level of the ligand with the electron-donating alkoxy groups.

The absorption spectra of complexes (**9–11**) are dominated by the metal-to-ligand charge-transfer transitions in the visible region, as well as by the lowest allowed MLCT bands appearing at 400 and 545 nm. The molar extinction coefficients of these bands being close to 35,000 and

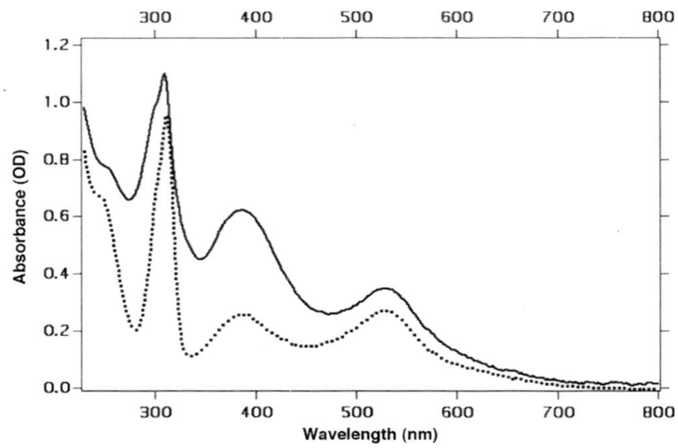

FIGURE 26.8. Comparison of absorption spectra of complexes **1** and **11** in ethanol.

19,000 $M^{-1}cm^{-1}$, respectively, are significantly higher than N3 (Figure 26.8).

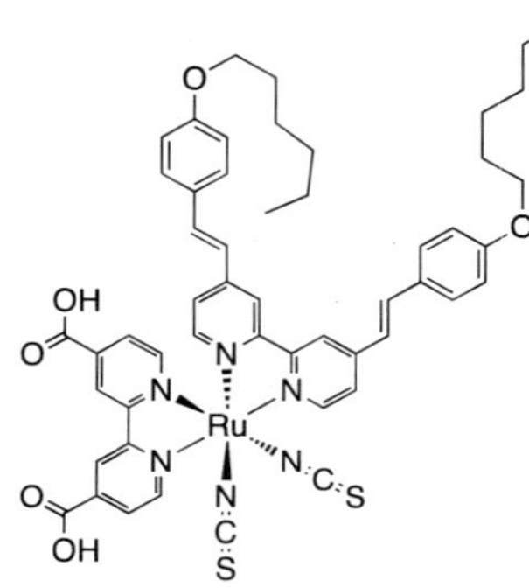

9

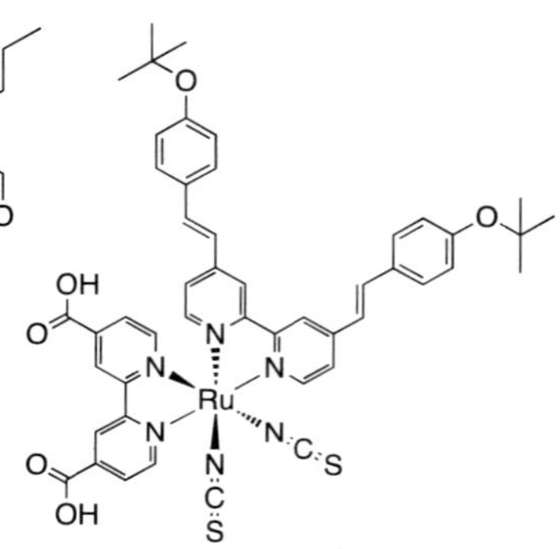

10

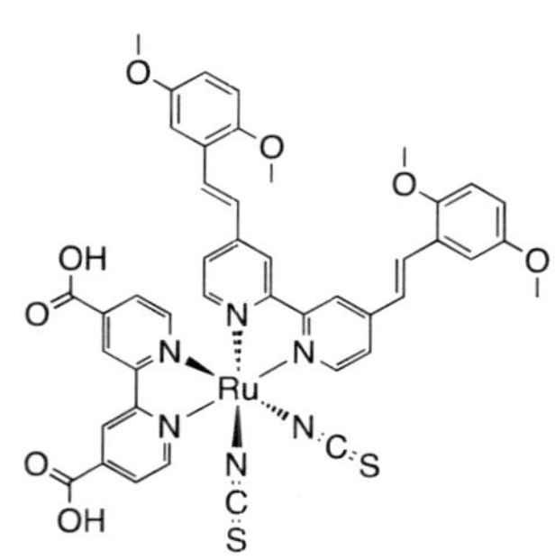

11

The photovoltaic data of these sensitizers using an electrolyte containing 0.60 M butylmethylimidazolium iodide (BMII), 0.03 M I_2, 0.10 M guanidinium thiocyanate, and 0.50 M *tert*-butylpyridine in a mixture of acetonitrile and valeronitrile (volume ratio: 85:15) exhibited a short-circuit photocurrent density of 16.50 ± 0.2 mA/cm^2, with an open-circuit voltage 790 ± 30 mV and a fill factor of 0.72 ± 0.03, corresponding to an overall conversion efficiency of 9.6%

under standard AM 1.5 sunlight, and demonstrated stable performance under light and heat soaking at 80°C [36].

Recently, a new design consisting of a ligand incorporating thiophene moieties to shift the spectral response into near-IR regions and to enhance the molar extinction coefficient has been developed [37]. Because the extinction coefficients of those sensitizers are much higher than **N719**, it is possible to decrease the thickness of the semiconductor film. This results in an enhanced open-circuit voltage as well as the fill factor, translating into high efficiencies of 11.4% [38]. To enhance further the molar extinction coefficient of ruthenium sensitizers, two thiophene units were incorporated (CYC B1 and CYC B11), resulting in slightly red shifted absorption maxima and high molar extinction coefficient values of over 24,000 M^{-1} cm^{-1} [39].

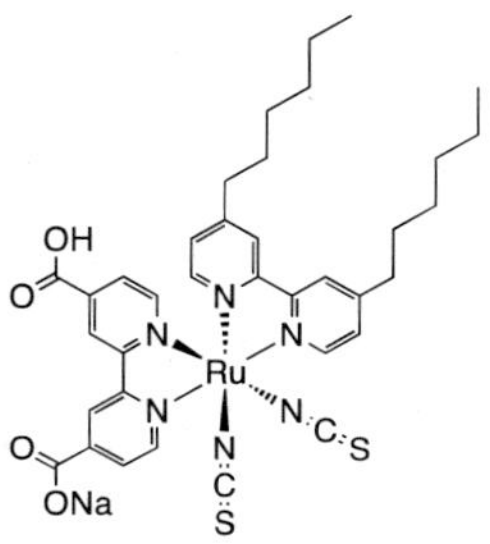

Z907

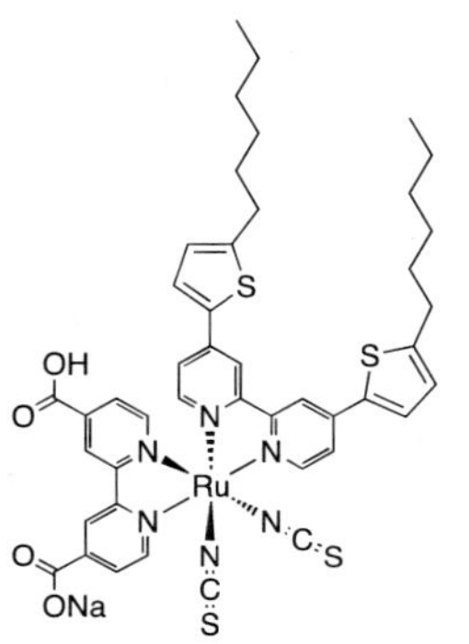

C101

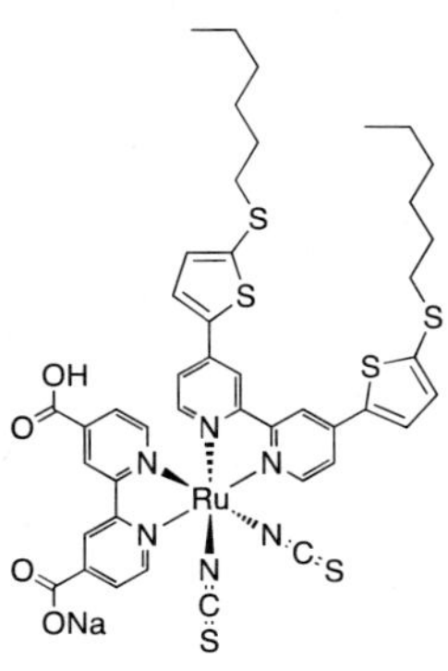

C106

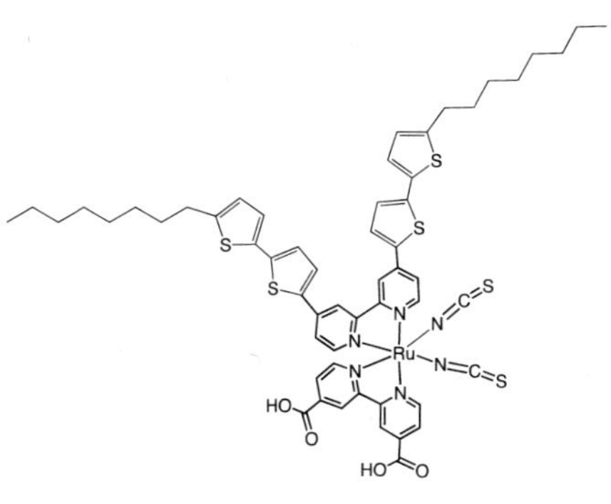

CYCB1

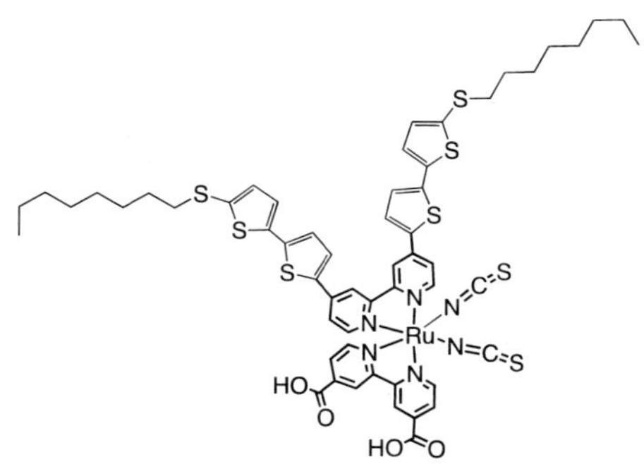

CYCB11

The absorption spectral properties of ruthenium sensitizers can be tuned toward the red part of the visible spectrum by introducing a ligand having a low-lying π^* molecular orbital (LUMO) and/or shifting the HOMO level by introducing strong donor ligands. The LUMO of a ligand depends on π-conjugation; therefore 4,4′,4″-tricarboxylic acid-2,2′;6,2″-terpyridine has a lower LUMO compared to the 4,4′-dicarboxylic acid-2,2′-bipyridine. The metal t_{2g} orbitals can be destabilized through the introduction of a donor thiocyanate type of ligand. The former lowers the energy of the lowest unoccupied molecular orbital (LUMO), while the latter destabilizes the highest occupied molecular orbital (HOMO) of the sensitizer, ultimately reducing the HOMO–LUMO gap. However, the extension of the spectral response into the near-infrared region by lowering the LUMO energy is limited to energy levels below which charge injection into the TiO_2 conduction band can no longer occur [40]. On the other hand, a near-infrared response by destabilization of Ru t_{2g} (HOMO) levels close to the redox potential of the redox mediator also proves not useful because of problems associated with regeneration of the oxidized dye following the electron injection into the TiO_2. Therefore, the optimum ruthenium sensitizers should exhibit excited-state oxidation potential of at least −0.9 V versus SCE, in order to inject electrons efficiently into the TiO_2 conduction band [41], while the ground-state oxidation potential should be about 0.5 V versus SCE, in order to be regenerated rapidly via electron donation from the electrolyte (iodide/triiodide redox system or a hole conductor). The panchromatic ruthenium complex N749 (so-called "black dye"), in which the ruthenium center is coordinated to a tricarboxylic acid terpyridine ligand and three thiocyanate ligands, has been synthesized [42].

Figure 26.9 shows the photocurrent action spectrum of a cell containing N719 and N749 sensitizers, where the incident photon-to-current conversion efficiency is plotted as a function of wavelength. It is evident that the response of the N749 extends 100 nm further into the infrared region than that of N719. The photocurrent onset is close to 920 nm—that is, near the optimal threshold for single-junction converters. The IPCE rises gradually from 920 until at 700 nm

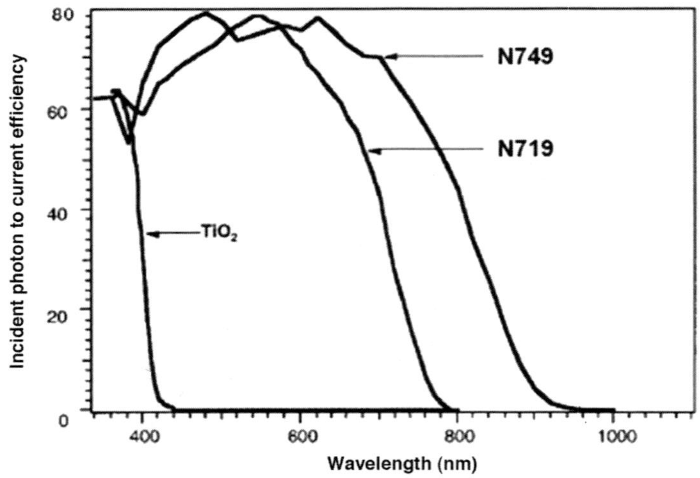

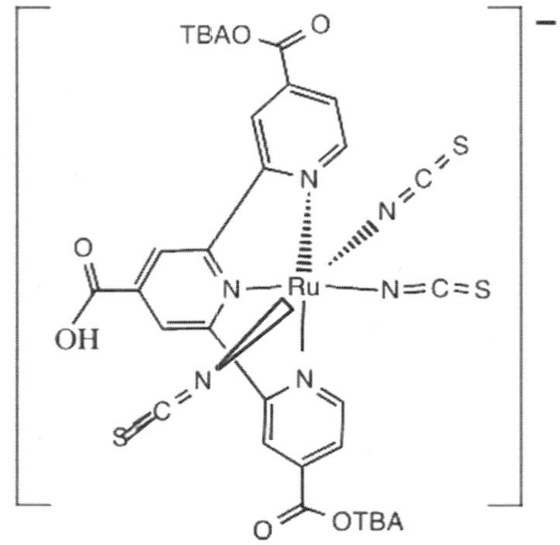

FIGURE 26.9. IPCE obtained with the N749 attached to nanocrystalline TiO_2 films. The incident photon-to-current conversion efficiency is plotted as a function of the wavelength of the exciting light. IPCE for bare TiO_2 and TiO_2 sensitized with N719 has been included for comparison.

it reaches a plateau of over 80%. From the overlap integral of the curves in IPCE with the AM 1.5 solar emission, one predicts the short-circuit photocurrents (J_{sc}) of N719- and N749-sensitized cells to be 16.5 and 20.5 mA/cm^2, respectively [43]. Routinely, experimental photocurrents obtained with N749 are in the range of 18–21 mA/cm^2 [42b]. The open-circuit potential (V_{oc}) is 720 mV, and the fill factor (*ff*) is 0.7, yielding for the overall solar (global AM 1.5 solar irradiance 1000 W m^{-2})-to-electricity conversion efficiency (η) a value of 10.4% [42b]. With the N749 dye, conversion efficiency of 11.4% has been achieved using high-haze TiO_2 electrodes by Han and colleagues [13, 44].

Ruthenium metal complexes other than 2,2′-bipyridine and 2,2′;6,2″-terpyridine have been much more rarely investigated [21]. 2,2′:6′,2″:6″,2‴-Quaterpyridine (qpy) ligands for DSC sensitizers are largely unexplored, in contrast to the extensive studies of 4,4′-dicarboxylic acid 2,2′-bipyridine-based sensitizers. This is likely due to the synthetic challenges associated to the qpy ligands, and very few reports have been published on DSC Ru(II)-sensitizers based on such tetradentate ligands [45]. Although their overall solar-to-electric-power efficiencies were not among the highest-ranked, these studies have pointed out the possible panchromatic response of the corresponding complexes extending from the NIR to the UV region, rendering them as alternative promising sensitizers with enhanced solar harvesting capability over the conventional bpy-based sensitizers.

To address the common issues related to the rather weak absorption of Ru(II) dyes in the red and NIR region, a number of heteroleptic Ru(II) sensitizers with extended bpy ancillary ligands π-conjugated with electron-rich benzenoid cores carrying donor primary organic functionalities such as alkoxy have been introduced [38, 39]. Ru(II) complexes based on such ligands are generally endowed with enhanced optical properties with respect to the prototypical N3 or N719 dyes, which also translate into higher photocurrents when employed in DSC devices. In particular, encouraging performances have recently been reached by using thiophene-based derivatives as electron-rich donor end-groups [39]. In order to address some of the major issues so far reported in the literature, one approach would be to synergetically combine the superior optical properties of π-donor conjugated bpy with the panchromatic response of quaterpyridine ligands. However, π-donor conjugated 2,2′:6′,2″:6″,2‴-qpys have so far been scantily investigated and even more rarely used as DSC sensitizers. Recently, researchers reported a first example of heteroarylvinylene π-conjugated 4,4‴-bis[(*E*)-2-(3,4-ethyl enedioxythien-2-yl)vinyl]-4′,4″-bis(carboxy)-2,2′:6′,2″:6″,2‴-quaterpyridine ligand and its *trans*-dithiocyanato Ru(II) sensitizer (N1044, Figure 26.10), which exhibits a panchromatic spectral response extending from the UV throughout the entire visible spectral region [46].

The photovoltaic performance of N1044 has been explored by using a double-layer (20-nm particle layer +

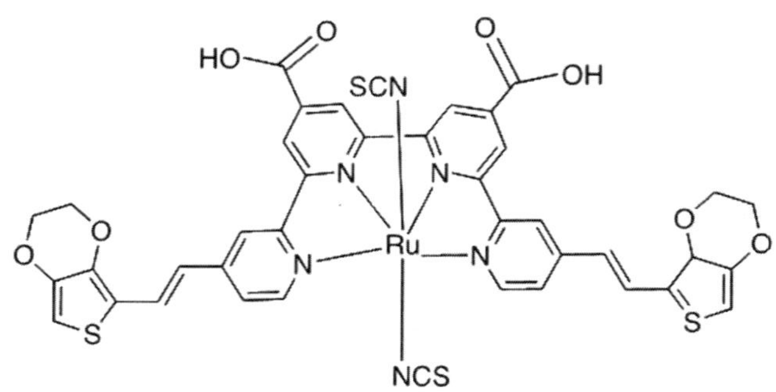

FIGURE 26.10. Molecular structure of N1044 sensitizer.

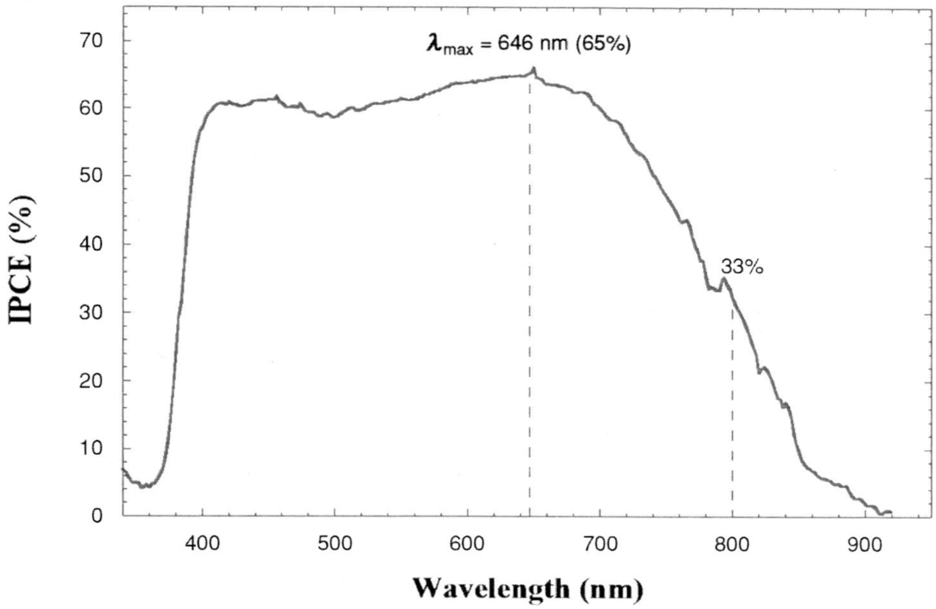

FIGURE 26.11. IPCE spectrum of the N1044-sensitized cell using volatile electrolyte.

diffusive/reflective layer) photo-anode configuration of anatase TiO_2. Chenodeoxycholic acid was added to the DMF solution of the dye as a co-adsorbing agent to prevent an excessive formation of dye aggregates both in solution and on the nanostructured film. Figure 26.11 presents the incident photon-to-electron conversion efficiency (IPCE) spectrum, and the Figure 26.12 the photovoltaic characteristics of current–voltage curves of N1044 at different sun intensity are shown. The N1044 complex delivers a maximum IPCE of 65% at 646 nm. Despite the fact that this value is lower than what was reported for other Ru(II) sensitizers exceeding 90% [35], at one equivalent sunlight illumination (100 mW cm^{-2}, AM 1.5 G), N1044 delivers an excellent short-circuit current density as high as $J_{sc} = 19.15$ mA cm^{-2} and a good fill factor of 0.67. Nevertheless, the low V_{oc}, likely due to the presence of Li^+ ions in the electrolyte and to the dye adsorption

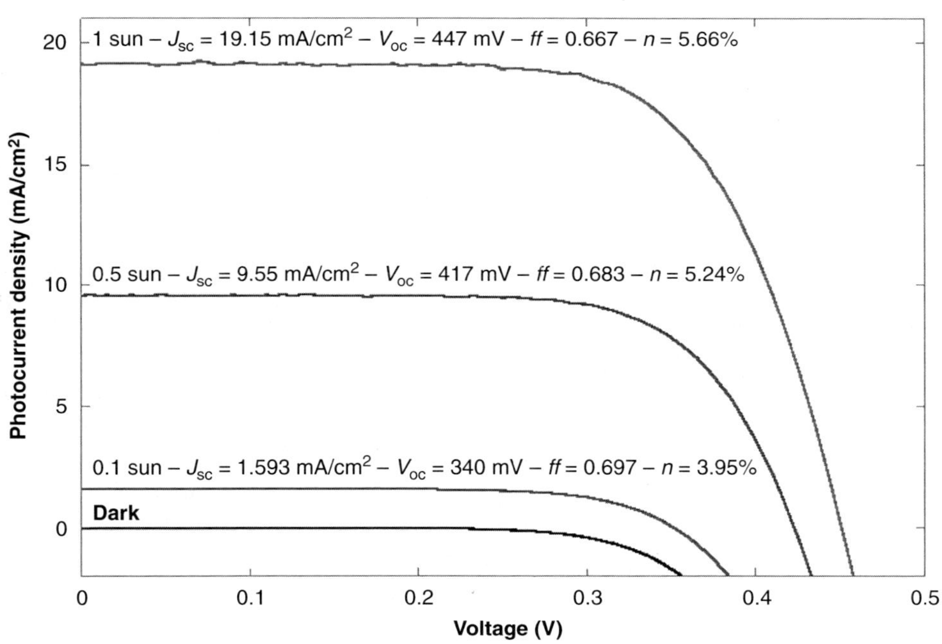

FIGURE 26.12. Current–voltage plots of N1044 at 0.1, 0.5, and 1 sunlight.

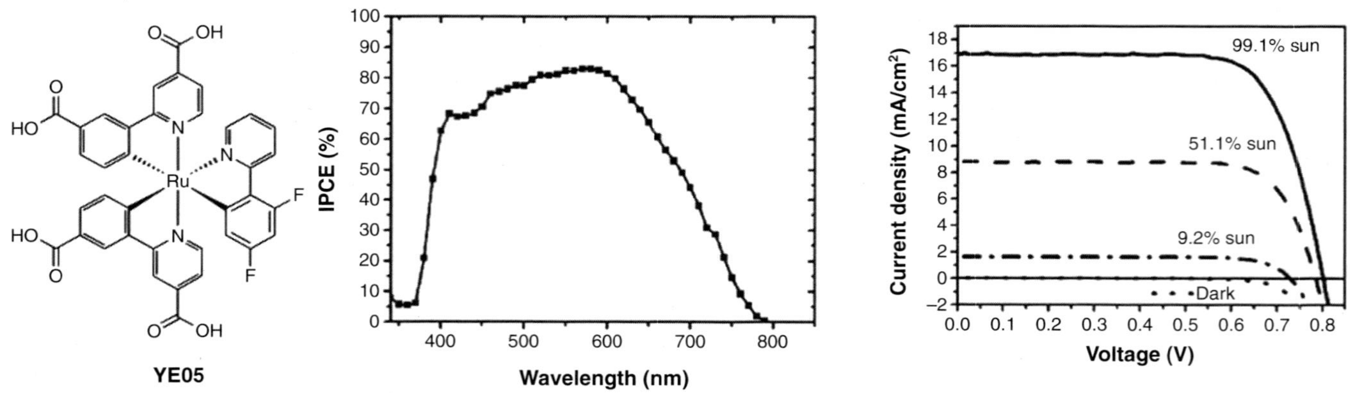

FIGURE 26.13. YE05 chemical structure and IPCE spectrum (**left**) and photocurrent voltage curves (**right**) under various light intensities of AM 1.5 sunlight.

characteristics discussed above, reduces the overall cell performance leading to a power conversion efficiency of 5.7%. These values can be compared with $J_{sc} = 17.6$ mA cm^{-2}, $V_{oc} = 849$ mV, and $ff = 0.73$, which deliver a record efficiency at 1 sunlight of 11.2% for the prototype dye N719 [35]. In agreement with the measured high current density and broad absorption in the visible range, the IPCE curve shows a panchromatic response with a maximum of conversion obtained at 646 nm and still 33% efficiency at 800 nm, where the vast majority of dye sensitizers have zero or negligible response. It should be noted how the absorption tail is recorded up to 910 nm, which is deeply inside in the NIR region. A bathochromic shift of about 30 nm is found as a consequence of the dye absorption onto the TiO_2 surface, which entails the formation of a carboxylate group as well as the difference of solvation shell strength when using acetonitrile/valeronitrile-based electrolytes.

As mentioned above, the poor V_{oc} can possibly be related to the dye adsorption onto TiO_2, whereby the loose packing between dye aggregates of different belts would lead empty space for oxidized species in the electrolyte (e.g., I_2) to access the TiO_2 surface, thus leading to increased recombination between injected electrons and the electrolyte. We can also speculate that this class of dyes might generally suffer from sizable recombination between injected electrons and the oxidized dye cation, since the dye HOMO, localized across the metal-NCS groups, lies very close to the surface with little screening from the dye aromatic ligands.

26.3.2 Cyclometallated Ruthenium Complexes

Thiocyanate ligands are monodentate, and in coordination chemistry they are usually considered weaker than the bidentate ligand. Besides monodentate property they are ambidentate, thus producing linkage isomers. Efforts have been made in the past to replace the thiocyanate ligands without great success as the efficiencies obtained for the devices remain well below 10%. However, a promising result was obtained recently by replacing thiocyanate by a cyclometallated 2,4-difluorophenyl-pyridine, yielding the complex **YE05** (Figure 26.13) [47]. This type of ligand is widely used in iridium complexes for organic light-emitting devices (OLEDs) [48]. The donor strength of the carbon anion ligand of 2,4-difluorophenyl-pyridine is slightly higher than that of the combined two thiocyanate ligands, which was measured by a cyclic voltammogram. Therefore, to reduce the donor strength of the carbanion, electron acceptor groups such as fluoro groups were substituted at the 2 and 4 positions. The UV–vis absorption spectra of YE05 exhibit three absorption bands, which are red-shifted when compared to **N719** as can be seen in the IPCE spectrum, which reaches a maximum over 80% at 600 nm extending to 800 nm. The lowest-energy MLCT band in YE05 is red-shifted by 25 nm when compared to N719, with overall remarkable high molar extinction coefficient exceeding substantially that of N719 over the whole visible domain. This is due to the cyclometallated ligand, which is a stronger donor than the two thiocyanate groups. It results in the stronger destabilization of the highest occupied molecular orbital (HOMO) when compared to the lowest unoccupied molecular orbital (LUMO). The presence of the two fluorine atoms allows the fine-tuning of the redox potential of the sensitizer. Overall, YE05 produces a short-circuit photocurrent of 17 mA/cm^2, a V_{oc} of 800 mV, and a fill factor of 0.74, corresponding to a conversion efficiency of 10.1% under AM 1.5 standard sunlight. Thus, YE05 emerges as a paradigm prototype for thiocyanate-free cyclo-metalated ruthenium complexes, exhibiting remarkable spectral and stability properties.

26.4 REDOX MEDIATORS

In dye-sensitized solar cells, light is absorbed by a dye monolayer located at the junction between a nanostructured

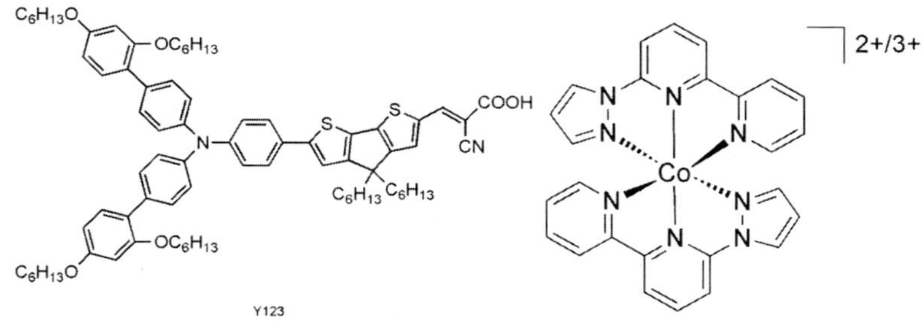

FIGURE 26.14. Chemical structures of Y123 dye and tridentate cobalt complex.

electron-transporting (n-type) and hole-transporting (p-type) phase. The former is a wide bandgap semiconductor oxide (typically TiO_2, anatase), and the latter is typically the triiodide/iodide (I_3^-/I^-) redox electrolyte [49]. Upon photoexcitation, the dye injects an electron into the n-type material and the hole is captured by the electrolyte. The electrons then travel through the nanostructure to be collected as current at the external contact, while the holes are transported to the cathode by the redox shuttle. The main drawback of the commonly used triiodide/iodide redox (I_3^-/I^-) system is a large mismatch between its oxidation potential ($E^0(I_3^-/I^-) = 0.35$ V versus the normal hydrogen electrode (NHE)) and the oxidation potential (E^0 $(S^+/S) = \sim 1.0$ V versus NHE), of the sensitizer, which limits the open-circuit voltage (V_{oc}) to 0.7–0.8 V [49]. The large expenditure needed for efficient dye regeneration is due to the complex regeneration kinetics with the I_3^-/I^- redox couple involving formation of intermediates, that is, the I_2^- radical [50]. Additionally, this redox couple is very corrosive toward metals such as Ag, Au, and Cu, thereby excluding the use of such materials as current collectors in DSC modules requiring long-term stability. Therefore the development of noncorrosive redox mediators, with reduced mismatch between the oxidation potential of the dye and the redox couple, is paramount to enhancing open-circuit potential.

Previous studies to replace the I_3^-/I^- redox system by cobalt polypyridine complexes have drawn attention because of their low visible light absorption, higher redox potential, and reduced corrosiveness toward metallic conductors offered by these redox couples [51]. However, the overall efficiency obtained with these cobalt electrolytes were inferior compared to the I_3^-/I^- couple, especially under full sunlight. This has been attributed to slow mass transport [52] and faster back reaction of photoinjected electrons with the oxidized redox species [53] coupled with the slow regeneration of the Co(II) species at the cathode. Recently, cobalt redox shuttles have attracted renewed attention after Feldt et al. [54] increased the power conversion efficiency to 6.7% by employing a newly designed D–π–A organic sensitizer, coded D35, in conjunction with the cobalt (III/II) tris-bipyridyl complex, $[Co(III)(bpy)_3](PF_6)_3/[Co(II)(bpy)_3](PF_6)_2$ couple. Most recently, Feldt et al. [55] have achieved over 1000 mV with cells based on cobalt phenanthroline redox shuttles, but the power conversion efficiencies were below 3.6%.

Because cobalt is a first-row transition metal, dissociation and exchange of ligands may occur rapidly in the case of Co(II). Tridentate ligands are expected to improve significantly the stability of the cobalt complex compared to bidentate ligands. Another advantage of tridentate ligands over bidentate ligands is the absence of isomers. Thus in the case of dissymmetric bidentate ligands, useful for fine-tuning of the redox properties, *facial* and *meridional* isomers coexist. One particular attractive feature of cobalt complexes is the facile tuning of their redox potential, which can be adjusted to match the oxidation potential of the sensitizer minimizing energy loss in the dye regeneration step. Yum et al. [56] have reported a redox relay system $[Co(III/II)(bpy\text{-}pz)_2](PF_6)_{3/2}$ [bpy-pz = 6-(1*H*-pyrazol-1-yl)-2,2′-bipyridine]), which they used as a redox mediator in dye-sensitized solar cells. They achieved an unprecedented output voltage exceeding 1000 mV due to its high oxidation potential of 0.86 V versus NHE of the cobalt redox mediator. The designed complex $[Co(bpy\text{-}pz)_2]^{3+/2+}$ (Figure 26.14), whose redox potential is offset by only 230 mV from that of the *3-{6-{4-[bis(2′,4′-dihexyloxybiphenyl-4-yl)amino-]phenyl}-4,4-dihexyl-cyclopenta-[2,1-b:3,4-b]dithiphene-2-yl}-2-cyanoacrylic acid* (coded **Y123**) dye, gave a power conversion efficiency (PCE) of over 10% under different solar intensities from 10 to 100 mW cm^{-2}. The cobalt (III) and (II) complexes based on 6-(1H-pyrazol-1-yl)-2,2′-bipyridine tridentate ligand, used as redox mediators in combination with the high molar extinction coefficient sensitizer Y123 in mesoscopic dye-sensitized solar cells yielding the <1-sunlight condition, a very high open-circuit voltage, ~1000 mV along with a short-circuit current of 13.06 mA/cm^2, and fill factor 0.77 were obtained, resulting in a power conversion efficiency of above 10% (Figures 26.15 and 26.16). The high open-circuit voltage was obtained by stabilizing the HOMO level of the cobalt (II) complex in order to minimize the mismatch between the redox couple and the sensitizer HOMO level to just 230 mV. This resulted in higher-power conversion efficiency compared to the iodide/triiodide redox couple. These promising results

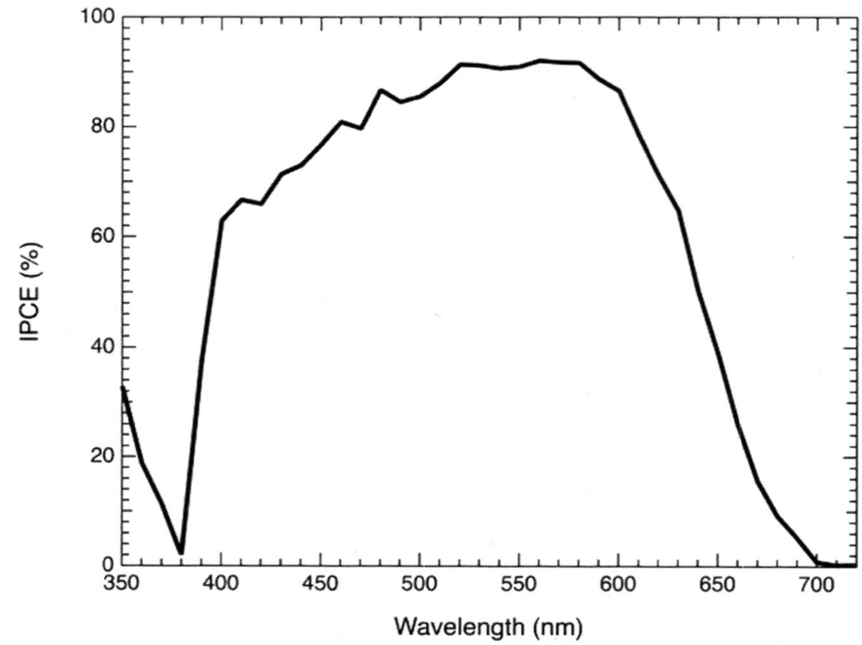

FIGURE 26.15. Photovoltaic characteristics of DSC based on a $[Co(bpy\text{-}pz)_2]^{3+/2+}$ system. IPCE of the DSC employing the double-layered TiO_2 (5.6 + 5 μm) and Pt counterelectrode.

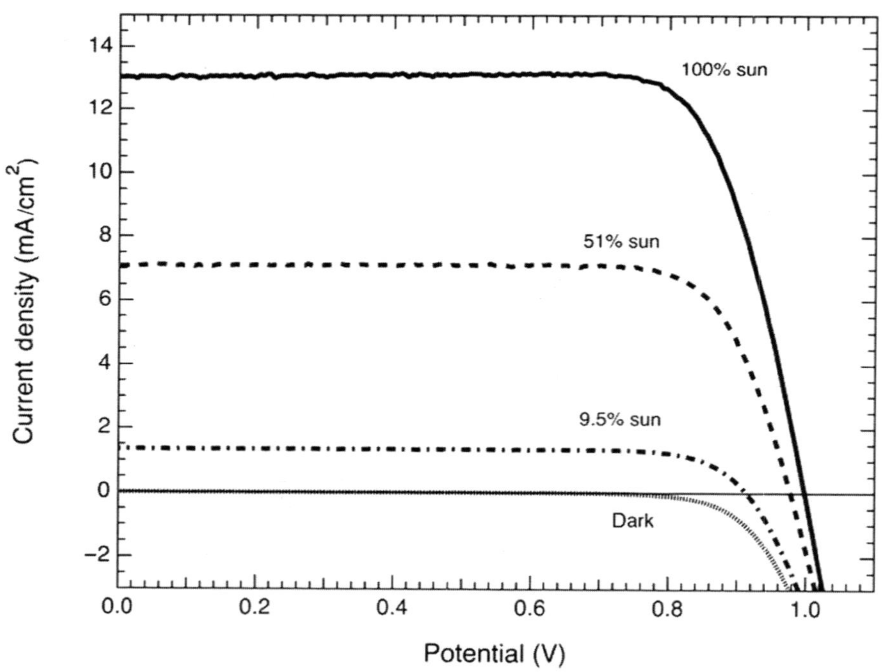

FIGURE 26.16. Photovoltaic characteristics of DSC based on a $[Co(bpy\text{-}pz)_2]^{3+/2+}$ system. *J–V* characterization of the DSC employing the double-layered TiO_2 (5.6 + 5 μm) and Pt counterelectrode.

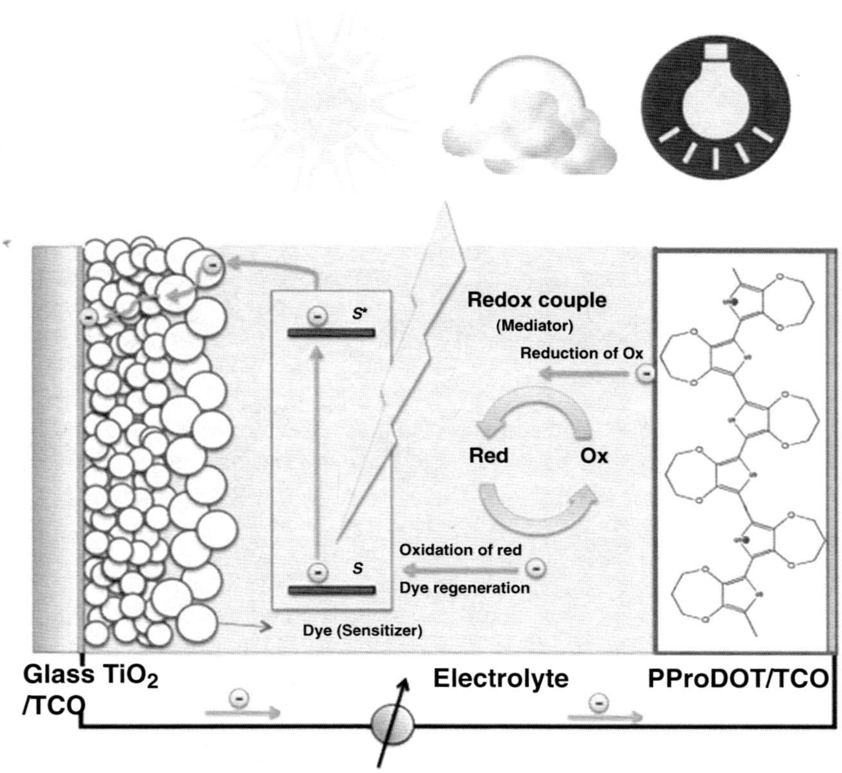

FIGURE 26.17. Schematic diagram of DSSCs illustrating PProDOT as a counterelectrode. The diagram is for illustration and is not to scale. See color insert.

confirm that the molecularly engineered cobalt redox shuttle is a legitimate alternative to the commonly used I_3^-/I^- redox shuttle.

26.5 COUNTERELECTRODE

The counterelectrode (CE) is one of the major components that reduce the oxidized species (i.e., tri-iodide (I_3^-) ions) generated after the dye regeneration. Since transparent conductive oxide (TCO) substrates exhibit insufficient electron-transfer kinetics for I_3^- reduction, a thin layer of platinum (Pt) is coated onto the TCO substrate (platinized counterelectrode) and used to catalyze the cathodic reduction of I_3^- ions that is redox couple regeneration [57]. However, the one-electron redox couple cobalt complexes are sluggish on the platinum-sputtered counterelectrode. Therefore, efforts have been made to replace the platinized counterelectrode by various carbon nanotubes and graphite-based and other carbon-based materials and polymers. Most recently, silver thin films or nanoparticles were coated on the rear side of TCO glass to increase the efficiency of CEs to 6.05% [58]. Microporous polyaniline (PANI) on TCO has shown an efficiency of 7.1% in DSSCs [59], but it is well known that PANI is not a material of choice because of its instability and self-oxidation properties. Due to the high conductivity, electrochemical stability, and high catalytic activity, polythiophenes can replace platinized counterelectrodes in DSSCs [60]. The commercial product PEDOT:PSS (Baytron P) is an aqueous dispersion, so its long-term stability is ruled out and additionally it can trap electrons. An electrode coated with poly(3,4-ethylenedioxythiophene) (PEDOT) on a porous material has been shown as an excellent oxygen reduction catalyst [61], and recently we have shown $\geq$8% efficiency using PEDOT as a counterelectrode in DSSCs [62]. Since poly(3,4-alkylenedioxythiophene) is an insoluble material, it has limited processability. With electropolymerization, thin layers of π-conjugated polymers can be easily coated on any conducting and flexible substrate in a cost-effective way at ambient condition. For electropolymerization, ionic liquids are ideal solvents for a long life of electrochemical processes; thus stable π-conjugated polymers can now be realized. Figure 26.17 shows schematic representation of dye-sensitized solar cell using Poly (3,4-propylenedioxythiophenes) (PProDOT) as counter electrode. Poly(3,4-propylenedioxythiophenes) (PProDOT) has especially useful redox switching properties due to their electron-rich character that yields especially low switching potentials for electrochromic devices, better stability, easy processability; additionally, PProDOT switch

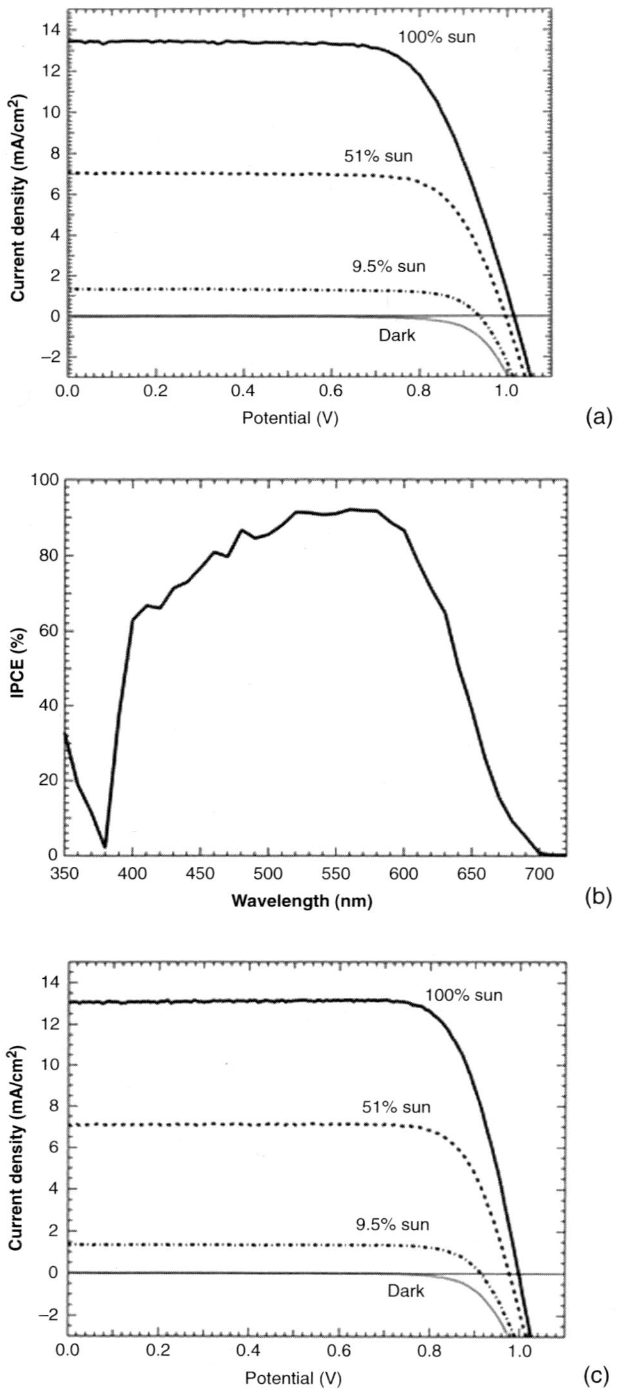

FIGURE 26.18. Photovoltaic characteristics of DSC based on a $[Co(bpy\text{-}pz)_2]^{2+/3+}$ system. (**a**) *J–V* characterization and (**b**) IPCE of the DSC employing the double-layered TiO_2 (5.6 + 5 μm) and Pt counterelectrode. (**c**) *J–V* characterization of the DSC employing the double-layered TiO_2 (4.0 + 4.5 μm) and the PProDOT cathode instead of Pt.

rapidly and efficiently between their neutral and *p*-doped forms with a minimum of side reactions [63]. DSCs used in conjunction with iodide/triiodide and PProDOT gave the record performance of these PProDOT-based counterelectrodes in DSSCs and resulted in the highest power conversion efficiency (>9%), which is significantly higher than the data reported using various types of counterelectrodes. Also, the data are remarkably comparable to the platinized counterelectrode.

Using one electron redox couple and platinum-sputtered counterelectrode, comparatively high charge-transfer resistance at the counterelectrode was observed even though promising PCE was obtained. Figure 26.18a shows photocurrent-voltage characteristics of the DSCs employing a 5.6 um thin transparent nanoporous TiO2 (anatase) film, the Y123 sensitizer, and the redox electrolytes under simulated sunlight at various intensities, 9.5, 51, and 100 mW cm-2. The [Co(bpy-pz)2]3+/2+ redox system gave a photocurrent density of (Jsc) 12.54 mA cm-2, an open-circuit potentials (Voc) of 1020 mV and a fill factor (FF) of 0.69, yielding a PCE (h) of 8.87%. The Figure 26.18b exhibit the incident photon-to-current conversion efficiency (IPCE) spectra of both redox systems showing high values, i.e., 70–90% in the 440 and 620 nm wavelength range. However, using a high-surface area cathode material (i.e., nanoporous poly(3,4-propylenedioxythiophene) (PProDOT) [64] layers), the RCT was reduced significantly. With the incorporation of the PProDOT as cathode material, the FF also was drastically improved due to a lower RCT, and thus the power conversion efficiency improved at 100-sunlight illumination (Figure 26.18c).

26.6 CONCLUSIONS

Dye-sensitized solar cells have become a credible alternative to solid-state *p–n* junction devices. Conversion efficiencies of >12% have already been obtained with a single junction cell of laboratory scale, but there is ample room for further improvement of the power conversion efficiency. Future research will focus on combining ruthenium sensitizers with one-electron redox mediators to obtain a J_{SC} of 22 mA/cm^2 and an open-circuit potential of 1 V, with fill factor of 0.75, which would yield close to 16.5% efficiency under 1 sunlight. The dye-sensitized solar cells with 16.5% efficiency armed with various beautiful colors and flexibility are well-suited for a whole realm of applications ranging from the low power market to large-scale applications. Their excellent performance in diffuse light gives them a competitive edge over silicon in providing electric power for stand-alone electronic equipment for both indoor and outdoor applications. Integration of DSC in building architecture has already started and will become a fertile field of future commercial development.

REFERENCES

1. Grätzel, M. (2009). Recent advances in sensitized mesoscopic solar cell. *Accounts of Chemical Research*, **42**, 1788–1798.
2. http://www.millenniumprize.fi.
3. O'Regan, B., Grätzel, M. (1991). A low-cost, high-efficiency solar cell based on dye-sensitized colloidal TiO_2 films. *Nature*, **353**, 737–740.
4. Grätzel, M. (2001). Photoelectrochemical cells. *Nature*, **414**, 338–344.
5. Barbé, C. J., Arendse, F., Comte, P., Jirousek, M., Lenzmann, F., Shklover, V., Grätzel, M. (1997). Nanocrystalline titanium oxide electrodes for photovoltaic applications. *Journal of the American Ceramic Society*, **80**, 3157.
6. Rothenberger, G., Comte, P., Grätzel, M. (1999). A contribution to the optical design of dye-sensitized solar cells. *Solar Energy Materials and Solar Cells*, **58**, 321–336.
7. Hagfeldt, A., Grätzel, M. (1995). Light-induced redox reactions in nanocrystalline systems. *Chemical Reviews*, **95**, 49.
8. Nazeeruddin, M. K., Kay, A., Rodicio, I., Humphry-Baker, R., Muller, E., Liska, P., Vlachopoulos, N., Grätzel, M. (1993). Conversion of light to electricity by *cis*-X_2Bis(2,2′-bipyridyl-4,4′-dicarboxylate)ruthenium(II) charge-transfer sensitizers (X = Cl-,Br-,I-,CN- and SCN-) on nanocrystalline TiO_2 electrodes. *Journal of the American Chemical Society*, **115**, 6382.
9. (a) Cahen, D., Hodes, G., Grätzel, M., Guilemoles, J. F., Riess, I. (2000). Nature of photovoltaic action in dye-sensitized solar cells. *Journal of Physical Chemistry B*, **104**, 2053. (b) Ferber, J., Luther, J. (2001). Modeling of photovoltage and photocurrent in dye-sensitized titanium dioxide solar cells, *Journal of Physical Chemistry B*, **105**, 4895.
10. Willig, F., Burnfeindt, B., Schwarzburg, K., Hannappel, T., Stock, W. (1997). Experimental results and basic considerations concerning injection and transport of electrons in the dye sensitized colloidal sponge type anatase electrode. *Proceedings of the Indian Academy of Sciences (Chemical Sciences)*, **109**, 415–428.
11. Monat, J. E., Rodriguez, J. H., McCusker, J. K. (2002). Ground- and excited-state electronic structures of the solar cell sensitizer bis(4,4′-dicarboxylato-2,2′-bipyridine)bis (isothiocyanato)ruthenium(II). *Journal of Physical Chemistry A*, **106**, 7399–7406.
12. Yella, A., Lee, H.-W., Tsao, H. N., Yi, C., Chandiran, A. K., Nazeeruddin, M. K., Diau, E. W.-G., Yeh, C.-Y., Zakeeruddin, S. M., Grätzel, M. (2011). Porphyrin-sensitized solar cells with cobalt (II/III)–based redox electrolyte. *Science*, **334**, 629–634.
13. Han, L., Islam, A., Chen, H., Malapaka, C., Chiranjeevi, B., Zhang, S., Yang, X., Yanagida, M. (2012). High-efficiency dye-sensitized solar cell with a novel co-adsorbent. *Energy Environ. Sci.*, **5**, 6057–6060.
14. Kuang, D., Klein, C., Ito, S., Moser, J.-E., Humphry-Baker, R., Evans, N., Duriaux, F., Grätzel, C., Zakeeruddin, S. M., Grätzel, M. (2007). Stable, high-efficiency ionic-liquid-based mesoscopic dye-sensitized solar cells. *Advanced Materials*, **19**, 1133–1137.
15. Han, L., Fukui, A., Chiba, Y., Islam, A., Komiya, R., Fuke, N., Koide, N., Yamanaka, R., Shimizu, M. (2009). Integrated dye-sensitized solar cell module with conversion efficiency of 8.2%. *Applied Physics Letters*, **94**, 013305.
16. Frank, A. J., Kopidakis, N., Lagemaat, V. D. (2004). Electrons in nanostructured TiO_2 solar cells: Transport, recombination and photovoltaic properties. *Journal of Coordination Chemistry Reviews*, **248**, 1165–1179.
17. Ito, S., Chen, P., Comte, P., Nazeeruddin, M. K., Liska, P., Péchy, P., Grätzel, M. (2007). Fabrication of screen-printing pastes from TiO_2 powders for dye-sensitised solar cells. *Progress in Photovoltaics*, **15**, 603–612.
18. Chen, D. H., Huang, F. Z., Cheng, Y. B., Caruso, R. A. (2009). Mesoporous anatase TiO_2 beads with high surface areas and controllable pore sizes: A superior candidate for high-performance dye-sensitized solar cells. *Advanced Materials*, **21**, 2206–2210.
19. Sauvage, S., Chen, D., Comte, P., Huang, F., Heiniger, L. P., Cheng, Y. B., Caruso, R. A., Grätzel, M. (2010). Dye-sensitized solar cells employing a single film of mesoporous TiO_2 beads achieve power conversion efficiencies over 10%. *ACSNANO*, **4**, 4420.
20. Gao, F., Wang, Y., Shi, D., Zhang, J., Wang, M. K., Jing, X. Y., Humphry-Baker, R., Wang, P., Zakeeruddin, S. M., Grätzel, M. (2008). Enhance the optical absorptivity of nanocrystalline TiO_2 film with high molar extinction coefficient ruthenium sensitizers for high performance dye-sensitized solar cells. *Journal of the American Chemical Society*, **130**, 10720–10728.
21. Juris, A., Balzani, V., Barigelletti, F., Campagna, S., Belser, P., von Zelewsky, A. (1988). Ruthenium(II) polypyridine complexes: Photophysics, photochemistry, electrochemistry, and chemiluminescence. *Coordination Chemistry Reviews*, **84**, 85.
22. Nazeeruddin, M. K., Gra1tzel, M. (2007). *Photofunctional Transition Metals Complexes (Book Series): Structure and Bonding* **123**, 113–175.
23. Moser, J. E., Noukakis, D., Bach, U., Tacchibana, Y., Klug, D. R., Durrant, J. R., Humphry-Baker, R., Grätzel, M. (1998). Comment on measurement of ultrafast photoinduced electron transfer from chemically anchored Ru-dye molecules into empty electronic states in a colloidal Anantase TiO_2 film. *Journal of Physical Chemistry B*, **102**, 3649–3650.
24. Meyer, T. J. (1986). Ru polypyridyl complexes, photophysical properties. *Pure and Applied Chemistry*, **50**, 1293.
25. Ferrere, S. (2000). New photosensitizers based upon $Fe(L)_2(CN)_2]$ and $Fe(L)_3]$ (L = substituted 2,2′-bipyridine): Yields for the photosensitization of TiO_2 and effects on the band selectivity. *Chemical Materials*, **12**, 1083.
26. Fleming, C. N., Maxwell, K. A., DeSimone, J. M., Meyer, T. J., Papanikolas, J. M. (2001). Ultrafast excited-state energy migration dynamics in an efficient light-harvesting antenna polymer based on Ru(II) and Os(II) polypyridyl complexes. *Journal of the American Chemical Society*, **123**, 10336–10347.
27. Hagfeldt, A., Boschloo, G., Sun, L., Kloo, L., Pettersson, H. (2010). Dye-sensitized solar cells. *Chemical Reviews*, **110**, 6595–6663.

28. (a) Yan, S. G., Hupp, J. T. (1996). Ultrafast excited-state energy migration dynamics in an efficient light-harvesting antenna polymer based on Ru(II) and Os(II) polypyridyl complexes, semiconductor-based interfacial electron-transfer reactivity: Decoupling kinetics from pH-dependent band energetics in a dye-sensitized titanium dioxide/aqueous solution system. *Journal of Physical Chemistry*, **100**, 6867. (b) Tachibana, Y., Haque, S. A., Mercer, I. P., Moser, J. E., Klug, D. R., Durrant, J. R. (2001). Modulation of the rate of electron injection in dye-sensitized nanocrystalline TiO_2 films. *Journal of Physical Chemistry B*, **105**, 7424–7431.
29. Nazeeruddin, M. K., Zakeeruddin, S. M., Humphry-Baker, R., Jirousek, M., Liska, P., Vlachopoulos, N., Shklover, V., Fischer, C. H., Grätzel, M. (1999). Acid-base equilibria of (2,22-bipyridyl-4,42-dicarboxylic acid)ruthenium(II) complexes and the effect of protonation on charge-transfer sensitization of nanocrystalline titania. *Inorganic Chemistry*, **38**, 6298.
30. Wang, P., Zakeeruddin, S. M., Moser, J. E., Nazeeruddin, M. K., Sekiguchi, T., Grätzel, M. (2003). A stable quasi-solid-state dye-sensitized solar cell with an amphiphilic ruthenium sensitizer and polymer gel electrolyte. *Nature Materials*, **2**, 402.
31. (a) Tennakone, K., Kumara, G. R. R. A., Kottegoda, I. R. M., Perera, V. P. S. (1999). *Chemical Communications*, 15. (b) Chandrasekharan, N., Kamat, P. V. (2000). Improving the photoelectrochemical performance of nanostructured TiO_2 films by adsorption of gold nanoparticles. *Journal of Physical Chemistry B*, **104**, 10851.
32. Gregg, B. A., Pichot, F., Ferrere, S., Fields, C. L. (2001). Interfacial recombination processes in dye-sensitized solar cells and methods to passivate the interfaces. *Journal of Physical Chemistry B*, **105**, 1422–1429.
33. Nazeeruddin, M. K., Grätzel, M. (2002). In: Licht, S., ed. *Encyclopedia of Electrochemistry: Semiconductor Electrodes and Photoelectrochemistry*, Vol. 6. Wiley–VCH, Wienheim, pp. 407–431.
34. Schmidt-Mende, L., Durrant, J. R., Nazeeruddin, M. K., Grätzel, M. (2005). Effect of hydrocarbon chain length of amphiphilic ruthenium dyes on solid-state dye-sensitized photovoltaics. *Nano Letters*, **5**, 1315–1320.
35. Nazeeruddin, M. K., De Angelis, F., Fantacci, S., Selloni, A., Viscardi, G., Liska, P., Ito, S., Bessho, T., Grätzel, M. (2005). Combined experimental and DFT–TDDFT computational study of photoelectrochemical cell ruthenium sensitizer. *Journal of the American Chemical Society*, **127**, 16835–16847.
36. Wang, P., Klein, C., Humphry-Baker, R., Zakeeruddin, S. M., Grätzel, M. (2005). Stable 8% efficient nanocrystalline dye-sensitized solar cell based on an electrolyte of low volatility. *Applied Physics Letters*, **86**, 123508-123501–123508-123503.
37. Cao, Y., Bai, Y., Yu, Q., Cheng, Y., Liu, S., Shi, D., Gao, F. W. P. (2009). Dye-sensitized solar cells with a high absorptivity ruthenium sensitizer featuring a 2-(hexylthio)thiophene conjugated bipyridine. *Journal of Physical Chemistry C*, **113**, 6290.
38. Cao, Y., Bai, Y., Yu, Q., Cheng, Y., Liu, S., Shi, D., Gao, F., Wang, P. (2009). Dye-sensitized solar cells with a high absorptivity ruthenium sensitizer featuring a 2-(hexylthio)thiophene conjugated bipyridine. *Journal of Physical Chemistry*, **113**, 6290–6297.
39. Chen, C. Y., Wang, M. K., Li, J. Y., Pootrakulchote, N., Alibabaei, L., Ngoc-le, C. H., Decoppet, J. D., Tsai, J. H., Gratzel, C., Wu, C. G., Zakeeruddin, S. M., Gratzel, M. (2009). Highly efficient light-harvesting ruthenium sensitizer for thin-film dye-sensitized solar cells. *ACS Nano*, **3**, 3103–3109.
40. (a) Hara, K., Sugihara, H., Tachibana, Y., Islam, A., Yanagida, M., Sayama, K., Arakawa, H., Fujihashi, G., Horiguchi, T., Kinoshita, T. (2001). Dye-sensitized nanocrystalline TiO_2 solar cells based on ruthenium(II) phenanthroline complex photosensitizers. *Langmuir*, **17**, 5992–5999. (b) Yanagida, M., Yamaguchi, T., Kurashige, M., Hara, K., Katoh, R., Sugihara, H., Arakawa, H. (2003). Panchromatic sensitization of nanocrystalline TiO_2 with *cis*-bis(4-carboxy-2-2′-(4-carboxypyridyl)]quinoline)bis(thiocyanato-N)ruthenium(II). *Inorganic Chemistry*, **42**, 7921–7931. (c) Islam, A., Sugihara, H., Arakawa, H. (2003). Molecular design of ruthenium(II) polypyridyl photosensitizers for efficient nanocrystalline TiO_2 solar cells. *Journal of Photochemistry and Photobiology A—Chemistry*, **158**, 131–138.
41. Nazeeruddin, M. K., Grätzel, M. (2002). *Encyclopedia of Electrochemistry: Semiconductor Electrodes and Photoelectrochemistry*, Vol. 6. Wiley-VCH, Weinheim.
42. (a) Nazeeruddin, M. K., Péchy, P., Grätzel, M. (1997). Efficient panchromatic sensitization of nanocrystalline TiO_2 films by a black dye based on a trithiocyanato-ruthenium complex. *Chemical Communications*, 1705–1706. (b) Nazeeruddin, M. K., Péchy, P., Renouard, T., Zakeeruddin, S. M., Humphry-Baker, R., Comte, P., Liska, P., Cevey, L., Costa, E., Shklover, V., Spiccia, L., Deacon, G. B., Bignozzi, C. A., Grätzel, M. (2001). Engineering of efficient panchromatic sensitizers for nanocrystalline TiO_2-based solar cells. *Journal of the American Chemical Society*, **123**, 1613-1624.
43. Grätzel, M. (2004). Conversion of sunlight to electric power by nanocrystalline dye-sensitized solar cells. *Journal of Photochemistry and Photobiology A—Chemistry*, **168**, 235–235.
44. (a) Chiba, Y., Islam, A., Watanabe, Y., Komiya, R., Koide, N., Han, L. Y. (2006). Dye-sensitized solar cells with conversion efficiency of 11.1%. *Japanese Journal of Applied Physics 2*, **45**, L638–L640. (b) Islam, A., Chen, H., Malapaka, C., Chiranjeevi, B., Zhang, S., Yang, X., Yanagida, M. (2012). High-efficiency dye-sensitized solar cell with a novel co-adsorbent. *Energy & Environmental Science*, **5**, 6057.
45. (a) Barolo, C., Nazeeruddin, M. K., Fantacci, S., Di Censo, D., Ito, S., Comte, P., Liska, P., Viscardi, G., Quagliotto, P., De Angelis, P., Grätzel, M. (2006). Synthesis, characterization and DFT-TDDFT computational study of a ruthenium complex containing a functionalized tetradentate ligand. *Inorganic Chemistry*, **45**, 4642–4653. (b) Renouard, T., Fallahpour, R.-A., Nazeeruddin, M. K., Humphry-Baker, R., Gorelsky, S. I., Lever, A. B. P., Grätzel, M. (2002). Novel ruthenium sensitizers containing functionalized hybrid tetradentate ligands: Synthesis, characterization, and INDO/S analysis. *Inorganic Chemistry*, **41**, 367.

46. Abbotto, A., Sauvage, F., Barolo, C., De Angelis, F., Fantacci, S., Grätzel, M., Manfredi, N., Marinzi, C., Nazeeruddin, M. (2011). Panchromatic ruthenium sensitizer based on electron-rich heteroarylvinylene pi-conjugated quaterpyridine for dye-sensitized solar cells. *Dalton Transactions*, **40**, 234–242.

47. Bessho, T., Yoneda, E., Yum, J.-H., Guglielmi, M., Tavernelli, I., Imai, H., Rothlisberger, U., Nazeeruddin, M. K., Grätzel, M. (2009). New paradigm in molecular engineering of sensitizers for solar cell applications. *Journal of the American Chemical Society*, **131**, 5930.

48. Baranoff, E., Yum, J.-H., Grätzel, M., Nazeeruddin, M. K. (2009). Cyclometallated iridium complexes for conversion of light into electricity and electricity into light. *Journal of Organometallic Chemistry*, **694**, 2661.

49. Boschloo, G., Hagfeldt, A. (2009). Characteristics of the iodide/triiodide redox mediator in dye-sensitized solar cells. *Accounts of Chemical Research*, **42**, 1819–1826.

50. Clifford, J. N., Palomares, E., Nazeeruddin, M. K., Grätzel, M., Durrant, J. R. (2007). Dye dependent regeneration dynamics in dye sensitized nanocrystalline solar cells: Evidence for the formation of a ruthenium bipyridyl cation/iodide intermediate. *Journal of Physical Chemistry C*, **111**, 6561–6567.

51. Sapp, S. A., Elliott, C. M., Contado, C., Caramori, S., Bignozzi, C. A. (2002). Substituted polypyridine complexes of cobalt(II/III) as efficient electron-transfer mediators in dye-sensitized solar cells. *Journal of the American Chemical Society*, **124**, 11215–11222.

52. Nusbaumer, H., Moser, J. E., Zakeeruddin, S. M., Nazeeruddin, M. K., Grätzel, M. (2001). CoII(dbbip)$_2^{2+}$ complex rivals tri-iodide/iodide redox mediator in dye-sensitized photovoltaic cells. *Journal of Physical Chemistry B*, **105**, 10461–10464.

53. Wang, H. X., Nicholson, P. G., Peter, L., Zakeeruddin, S. M., Grätzel, M. (2010). Transport and interfacial transfer of electrons in dye-sensitized solar cells utilizing a Co(dbbip)$_2$ redox shuttle. *Journal of Physical Chemistry C*, **114**, 14300–14306.

54. Feldt, S. M., Gibson, E. A., Gabrielsson, E., Sun, L., Boschloo, G., Hagfeldt, A. (2010). Design of organic dyes and cobalt polypyridine redox mediators for high-efficiency dye-sensitized solar cells. *Journal of the American Chemical Society*, **132**, 16714–16724.

55. Feldt, S. M., Wang, G., Boschloo, G., Hagfeldt, A. (2011). Effects of driving forces for recombination and regeneration on the photovoltaic performance of dye-sensitized solar cells using cobalt polypyridine redox couples. *Journal of Physical Chemistry C*, **115**, 21500–21507.

56. Yum, J. H., Baranoff, E., Kessler, F., Moehl, T., Ahmad, S., Bessho, T., Marchioro, A., Ghadiri, E., Moser, J.-E., Yi, C., Nazeeruddin, M. K., Gratzel, M. (2012). A cobalt complex redox shuttle for dye-sensitized solar cells with high open-circuit potentials. *Nature Communications*, **3**, DOI: 10.1038/ncomms1655.

57. Papageorgiou, N. (2004). Counter-electrode function in nanocrystalline photoelectrochemical cell configurations. *Coordination Chemistry Reviews*, **248**, 1421–1446.

58. Ji, W. W., Zhang, X. D., Zhao, Y., Yang, Z. H., Cai, N., Wei, C. C., Sun, J., Xiong, S. Z. (2010). A new structure of counter electrode used for dye-sensitized solar cells. *Physica Status Solidi (c)*, **7**, 1124–1127.

59. Li, Q. H., Wu, J. H., Tang, Q. W., Lan, Z., Li, P. J., Lin, J. M., Fan, L. Q. (2008). Application of microporous polyaniline counter electrode for dye-sensitized solar cells. *Electrochemical Communication*, **10**, 1299–1302.

60. Ahmad, S., Deepa, M., Singh, S. (2007). Electrochemical synthesis and surface characterization of poly(3,4-ethylenedioxy thiophene) films grown in an ionic liquid. *Langmuir*, **23**, 11430–11433.

61. Winther-Jensen, B., Winther-Jensen, O., Forsyth, M., MacFarlane, D. R. (2008). High rates of oxygen reduction over a vapor phase-polymerized PEDOT electrode. *Science*, **321**, 671–674.

62. Ahmad, S., Berger, R., Khan, H. U., Butt, H. J. (2010). Electrical field assisted growth of poly(3-hexylthiophene) layers employing ionic liquids: Microstructure elucidated by scanning force and electron microscopy. *Journal of Materials Chemistry*, **20**, 5325–5334.

63. Sinha, J., Sahoo, R., Kumar, A. (2009). Processable, regioregular, and "click"able Monomer and Polymers Based on 3,4-Propylenedioxythiophene with Tunable Solubility. *Macromolecules*, **42**, 2015–2022.

64. Ahmad, S., Yum, J. H., Butt, H. J., Nazeeruddin, M. K., Grätzel, M. (2010). Electrical field assisted growth of poly(3-hexylthiophene) layers employing ionic liquids: Microstructure elucidated by scanning force and electron microscopy. *Chemphyschem*, **11**, 2814–2819.

INDEX

Organic Nanomaterials: Synthesis, Characterization, and Device Applications, First Edition. Edited by Tomás Torres and Giovanni Bottari.